International PostGraduate Conference on Applied Science and Physics (GRASP2017)

AA006728

Journal of Physics: Conference Series
Volume 1049

Johor, Malaysia
7 December 2017

ISBN: 978-1-5108-6790-1
ISSN: 1742-6588

Printed from e-media with permission by:

Curran Associates, Inc.
57 Morehouse Lane
Red Hook, NY 12571

Some format issues inherent in the e-media version may also appear in this print version.

Copyright© (2017) by the Institute of Physics
All rights reserved. The material featured in this book is subject to
IOP copyright protection, unless otherwise indicated.

Printed by Curran Associates, Inc. (2018)

For permission requests, please contact the Institute of Physics
at the address below.

Institute of Physics
Dirac House, Temple Back
Bristol BS1 6BE UK

Phone: 44 1 17 929 7481
Fax: 44 1 17 920 0979

techtracking@iop.org

Additional copies of this publication are available from:

Curran Associates, Inc.
57 Morehouse Lane
Red Hook, NY 12571 USA
Phone: 845-758-0400
Fax: 845-758-2633
Email: curran@proceedings.com
Web: www.proceedings.com

Table of contents

Volume 1049

International PostGraduate Conference on Applied Science & Physics 2017

7 December 2017, Johor, Malaysia

Accepted papers received: 22 June 2018
Published online: 20 July 2018

Preface

Papers

IOP Conf. Series: Journal of Physics: Conf. Series **1049** (2018) 011001 doi:10.1088/1742-6596/1049/1/011001

Preface

Nik Hisyamudin Muhd Nor, Al Emran Ismail, Amir Khalid, Mohd Azlis Md Jalil, Mohd Helmy Abd Wahab, Nabihah Ahmad, Fariza Mohamad, Nabilah Ibrahim, Saifulnizan Jamian, Saiful Azhar Ahmad Tajudin, Alias Masek.

The International PostGraduate Conference on Applied Science & Physics 2017 (GRASP2017) was held in Universiti Tun Hussein Onn Malaysia, Johor Malaysia. It was aimed to provide a better platform for communication and discussion among academics and post-graduate students from various research fields. It was also aimed to provide a place for sharing knowledge and ideas, networking, presenting research findings and exploring other potential benefits that might be contributed for a better future.

We would like to express fullest gratitude and appreciation all participants, respective authors and technical reviewers for sharing their knowledge and experience for this conference.

Content from this work may be used under the terms of the Creative Commons Attribution 3.0 licence. Any further distribution of this work must maintain attribution to the author(s) and the title of the work, journal citation and DOI.
Published under licence by IOP Publishing Ltd

Peer review statement

All papers published in this volume of *Journal of Physics: Conference Series* have been peer reviewed through processes administered by the proceedings Editors. Reviews were conducted by expert referees to the professional and scientific standards expected of a proceedings journal published by IOP Publishing.

Content from this work may be used under the terms of the Creative Commons Attribution 3.0 licence. Any further distribution of this work must maintain attribution to the author(s) and the title of the work, journal citation and DOI.
Published under licence by IOP Publishing Ltd

IOP Conf. Series: Journal of Physics: Conf. Series **1049** (2018) 012001 doi:10.1088/1742-6596/1049/1/012001

A Technical Review of MANET Testbed Using Mobile Robot Technology

Farkhana Muchtar*[1,2], Abdul Hanan Abdullah[1], Muhammad Shafie Abd Latiff[1], Suhaidi Hassan[3], Mohd Helmy Abd Wahab[4] Gaddafi Abdul-Salaam[5]

[1] Faculty of Computing, Universiti Teknologi Malaysia, Skudai, 81310, Johor, Malaysia
[2] School of Computer Sciences, Universiti Sains Malaysia, 11800, Gelugor, Pulau Pinang, Malaysia
[3] School of Computing, Universiti Utara Malaysia, 06010, Sintok, Kedah, Malaysia
[4] Faculty of Electronic and Electrical Engineering, Universiti Tun Hussein Onn, 86400 Batu Pahat, Johor, Malaysia
[5] Department of Computer Science, Kwame Nkrumah University of Science and Technology, Ghana

E-mail: farkhana@gmail.com, hanan@utm.my, shafie@utm.my, suhaidi@uum.edu.my, helmy@uthm.edu.my, gaddafi.ict@knust.edu.gh,

Abstract. Recently, MANET (Mobile Ad Hoc Network) researchers have shown increased interest in using mobile robot technology for their testbed platforms. Despite the existence of review papers that discuss the usage of mobile robot technology pertaining to a MANET testbed from the perspective of a MANET researcher, said findings are rather lacklustre as it is not the sole purpose of said reviews. Hence, this review aims to comprehensively discuss the technical aspects that were facilitated with mobile robot technology for previous undertaken research. Only selected MANET testbed relevant to this review are compiled, namely Mobile Emulab, MiNT-m, MiNT-2, Proteus, w-ilab.t, ARUM, Sensei-UU, Kansei, CONET-IT, CONE, Roomba MADNet, Explorebots, SCORPION, MOTEL, iRobotSense, IoT-Lab, and NITOS. The discussions in this article are divided into three parts, namely hardware, software as well as mobile robot positioning and localization. The technical content of this review is expected to be a source of reference for other MANET researchers who are interested to use mobile robots for real mobility in their MANET testbeds.

1. Introduction

MANET (Mobile Ad Hoc Network) researchers have shown increased interest in using mobile robot technology for their testbed platforms. This paper therefore extends the discussions of our previous paper titled *"Mobile Ad hoc Network Testbed Using Mobile Robot Technology"* [1] by scrutinizing it as a technical review.

This paper is organized as follows. In section 2, the relevancy of selected MANET testbed facilities in this article is detail out. Section 3 describes three aspect of this paper, namely software, hardware and mobile robot positioning as well as localisation of robot technology used in MANET testbed. Finally, section IV provide some concluding remark.

Content from this work may be used under the terms of the Creative Commons Attribution 3.0 licence. Any further distribution of this work must maintain attribution to the author(s) and the title of the work, journal citation and DOI.
Published under licence by IOP Publishing Ltd

2. Related Work

It is to be noted that most previous discussions on MANET testbed facilities did not reveal technical aspects of mobile node facilities used in MANET testbed laboratories, especially those that are conducted in public testbed laboratories. Testbed facilities that do provide technical information are mostly from private and community testbed facilities and this is most likely due to the need to publish articles and technical reports as it is part of their requirement. However, the remaining testbed facilities merely states some technical specifications of mobile nodes in the testbed facilities without elaborating much details. Thus, the aim of this article is to comprehensively discuss these technical aspects of mobile robot usage in MANET testbeds which includes exploring the trends and advancements of robot technology usage chosen by robot-based MANET testbeds. These conditions are very much needed to deduce suitable construction of a good robot-based MANET testbed facilities that can be applied in the future.

The selection of robot-based MANET testbeds that were chosen in this review was purely based from the perception of MANET researchers and not from a robotic expertise perspective as was reported in most review articles. The focus and perspectives of the selected surveys were determined based on the discussion of the types of suitable, non-suitable and irrelevant robot-based MANET testbeds that were specific to this review.

In this article it was decided that MSNs, opportunistic networks and DTNs were a subset of MANETs because all of the mentioned wireless ad-hoc networks had mobility criteria and most importantly, wireless multi-hop ad-hoc communication. Therefore, 17 identified types of MANET platform mobile robot will be discussed in this article, namely, Mobile Emulab, MiNT-m, MiNT-2, Proteus, w-ilab.t, ARUM, Sensei-UU, Kansei, CONET-IT, CONE, Roomba MADNet, Explorebots, SCORPION, MOTEL, iRobotSense, IoT-Lab, NITOS.

On the other hand, testbed facilities devoted to research of multiple mobile robots using MANET-based communication were dismissed in this article. Examples of research in MANET-based multiple robot communication are CENTIBOTS [2], Robomote [3] and Mobile Multirobot Systems [4]. Based on a review of the aforementioned research articles, it was deduced that the testbed facilities that they developed were more focused towards research on multiple robot communication through MANETs with very limited discussion on the MANET research itself. Hence, the testbed facilities devoted to research of multiple mobile robots using MANET-based communication were dismissed.

3. Technical Review

In this section, the discussion is divided into three parts, namely (i) hardware platform (ii) software platform and (iii) mobile robot positioning and localization.

3.1. Hardware Platform

3.1.1. Robot Platforms and Chassis. Much of our review is centred on testbed platforms using readymade robot platforms such as iRobot Roomba/Create and LEGO Mindstorms. iRobot Roomba and Create are one of the most popular robot platforms due to several factors including the ability to carry loads up to 2 kg, presence of basic sensors for mobile robots such as bump sensors, obstacle avoidance sensors and wheel encoders, ease of controllability via iRobot Roomba Open Interface (ROI), large rechargeable li-ion battery (44000 I) capacity, and a self-recharging docking station as well as their competitive prices as compared to other platforms [5].

iRobot Roomba was introduced in 2002 and became among the earliest household robots made available to the public. It functioned as an autonomous robotic vacuum cleaner while at the same time it became a favourite robot for robotics researchers and hobbyists that used it to be the main platform for mobile robots. iRobot Create was later introduced in 2007 and was based on the iRobot platform due to the popularity of iRobot Roomba among robotics researchers and hobbyists. iRobot Create was sold at a lower prices because it did not have some of the components attached to iRobot Roomba such as vacuum cleaning (although the component could be added separately) [5].

iRobot Roomba and Create are usually controlled with a robot controller (usually a combination of an embedded PC and microcontroller) through the iRobot Roomba Open Interface (ROI) protocol whereby communication is enabled via a serial port on iRobot Roomba/Create are MINT-m [6], MINT-2 [7], Proteus Roomba [7], w-ilab.t [8], Roomba MADNet [9], SCORPION [10] and iRobotSense [11].

Apart from the iRobot Roomba and Create, LEGO Mindstorms is another preferred choice as robot platforms for mobile robots in the MANET testbeds. Examples of MANET testbeds that used LEGO Mindstorms robot platform are Sensei-UU [12] and Cone Testbed [13].

The Acroname Garcia robot platform also was a popular choice in several MANET testbeds such as the Emulab Mobile (TrueMobile) [14] and Kansei Testbed [15].

There are also other readymade robot platforms that were utilised, such as e-puck that was used in the MOTEL testbed [16], the Pioneer 3AT robot platform used in CONET-IT [17], the Lynxmotion 4WD rover robot platform used in ARUM [18] and the Rogue ATV robot platform used in Explorebot [19].

3.1.2. Robot Controllers. The selection of robot controllers used for mobile robots in previous MANET testbeds varies. Most have used a combination of an embedded PC board, microcontroller and motor controller. Examples of testbeds that combined an embedded PC board and microcontroller are MINT-m, MiNT2, Proteus, CONET (custom version), Kansei as well as NITOS.

There were testbeds that used smartphones as their robot controller such as Sensei-UU [12], custom made circuit boards such as w-ilab.t [8] while others utilised ready-made robot controllers like the LEGO NXT robot controller, a robot controller for LEGO Mindstorms that was used in the CONE testbed [13] and also Atom Bot Board Lynxmotion [18], a robot controller for the Lynxmotion based robot platform used in the ARUM testbed.

Some isolated cases like Roomba MADNet used an openwrt-based wireless router as a robot controller [7], a solution that was quite rare in mobile robot design.

Some mobile robots have used laptops/netbooks as their robot controller, e.g. ARUM [18], CONET-IT (Pioneer 3AT platform) [17] and IoT-lab (turtlebot2 platform) [20] where they were capable of carrying laptop loads weighing between 1.5 kg to 2 kg.

3.1.3. Supporting Components and Sensors. Apart from main robot platform/chassis and robot controllers, mobile robot s for MANET testbed also required other components such as for mobile robot localisations, obstacles and collision avoidance, testbed miniaturization and mobile devices such as sensor nodes to carry out MANET experiments.

i. Sensors and Other Components for Robot Localisation

There are many methods of robot localisation used in robot-based MANET testbeds that have been reviewed in this work and is the main factor for the selection of sensors and components used. Robot-based MANET testbeds that chose centralized visual localisation methods used ceiling-mounted video cameras with some opting for IR cameras such as ARUM [18] or fish eye cameras such as in the MOTEL testbed [21] to track the location and movement of mobile robots. Most testbeds used colour pattern recognition to identify different mobile robots identities such as Mobile Emulab [22], MINT-m [23] and MOTEL [21], where each mobile robot was equipped with a board with different colour patterns from one another. ARUM on the other hand, selected a different method that used motion capture-based localisation that used reflective balls arranged in different pattern for each mobile robot. Aside from centralized visual localisation, there have been mobile robots in MANET testbeds that used local visual localisation methods such as NITOS [24] that utilised webcams fixed to the mobile robot and at predetermined locations, black and white image pattern boards were placed that acted as location marks.

There have been some robot-based MANET testbed facilities such as Sensei-UU [12] and ARUM [18] that used fixed circuit-based mobility and they were usually equipped with infrared sensors to detect black linex placed on the floor to guide their movements.

For testbeds that used the iRobot Roomba and iRobot Create, wheel encoders that were available on Roomba or Create were fully utilised as an aid in determining the distance, speed and direction of their movements using the odometry method to support the localisation mechanism as used in the MINT-2 mobile robot [7], w-iLab.t [25] and Explorebot [19].

Robot-based MANET testbeds that used the Acroname Garcia robot platform for mobile robots such as Mobile Emulab (TrueMobile) [14] and Kansei Testbed [26], were equipped with ultrasonic sensors and IR sensors that could be used in obstacle and collision avoidance mechanisms. The Proteus mobile robot in Pharos testbed facilities had a variety of different settings to suit the experiments that were to be conducted. For experiments conducted outdoors, a combination of GPS and a magnetometer sensor were used, and for indoor experiments, a combination of visual localisation and IMU were chosen to be the components for robot localisation [27].

Mobile robots in the CONET-IT laboratory were equipped with laser range finders (LIDAR), Microsoft Kinect 3D camera sensors, GPS and IMU sensors to enable accurate robot localisation to be done independently on each mobile robot unit [28].

ii. Radio Signal Attenuators
Other than sensors and components used for robot localisation, obstacles and collisions, there were other supporting components such as radio signal attenuators that were used for testbed miniaturization such as the ones used in Mobile Emulab [22], MiNT-m [23], MINT-2 [7], ARUM [18], Kansei [26] and Sensei-UU [12]. The original purpose of the radio signal attenuators were to reduce the noise and interference of the radio signals. However, for testbed miniaturization, radio signal attenuators could also be used to reduce the wireless signal range that enabled the multiple hop ad hoc network to occur in a small testbed area. All the testbeds mentioned used fixed radio signal attenuators.

iii. Mobile Devices
The final component to be viewed in the MANET testbed is the type of components that are used as mobile devices. The selection of a mobile device carried piggyback by the mobile robot in the testbed is dependent on the purpose and scope of the testbed facilities. For example, the MANET testbed dedicated to WSN research or IoT with mobility used sensor nodes (or mote), the Mobile Emulab used Mica2 sensor nodes for mobile device, Sensei-UU used TelosB sensor node, CONET-IT used several different sensor nodes like TelosB, Iris, MicaZ, Mica2, CM5000 and CM5000-SMA, Explorebots used Mica2 sensor nodes, MOTEL testbeds used TelosB, MicaZ and Scatterweb sensor nodes while IoT- Lab used WSN430, M3 and A8 sensor nodes.

Some parts of the MANET testbeds used embedded computer boards such as MINT-m that used the Routerboard RN-230, MINT-2 that used the Soekris net5501 board, Pharos/Proteus and SCORPION testbeds that used a mini-ITX x86 computer, ARUM that used a laptop or Macbook, Roomba MADNet that used a Liniksys WRTSL54GS wireless router and NITOS that used an Alix motherboard. All of the above used low power consumption computer boards to represent the mobile devices. The CONE testbed was a bit unique as it used a Nokia mobile phone to represent the mobile device, specifically, the Nokia mobile phone models: Nokia N97 mini, Nokia N97 and Nokia 5800 XpressMusic that had the same firmware and capabilities.

Additionally, some robot-based MANET testbeds combined sensor nodes and embedded computer board as the mobile device such as the w-iLab.t lab testbed that combined eZ430 sensor nodes and an embedded board computer powered with an Intel Atom processor chip and the Kansei testbed that used Extreme Scale Mote (XSM) sensor nodes and a Stargate board as the mobile device in the MANET testbed.

3.2. Software Platforms

3.2.1. Robotic Software. Much of the earlier generations of the MANET testbeds with real mobility facilities like MINT and Mobile Emulab used in-house robotic software platforms as open source robot frameworks did not exist at that time. Later generation testbeds such as Pharos/Proteus, ARUM and Kansei combined in-house software and an open source software robot framework Player/Stage or ROS (Robot Operating System).

The latest trend in testbed platforms such as the w-ilab.t, NITOS and CONET robots combined an open source framework with ROS with the network testbed platform such as the GENI interface and the OMF-OML framework with some additional customization that were suitable to their needs.

3.2.2. Testbed Management and Monitoring Platforms. Testbed platforms from earlier generations of MANET testbeds that used mobile robots had their own custom made testbed management and monitoring platforms. This was because, at their time of development, a standard network testbed framework had not existed as the development of wireless network testbeds was still considered new.

Normally, the components to control the experiment process and the components that control the mobile robots are isolated but are integrated through the testbed core management component. Modular architecture enables the implementation of two different components separately using different technologies and tools. Some testbed platforms used different programming languages for the experiment control process components and mobile robot control components.

Some MANET testbed platforms integrated simulator and testbed platforms to compare data obtained from the simulator with data obtained from the testbed, such as MINT that used a hybrid simulator that was modified from the ns-2 simulator [23], libAra that used a DES-testbed platform [29] and Pharos testbed that used a combination of an OMNeT ++ simulator and a click modular router [30]

The Emulab testbed in its early phases of development, used its own version of testbed practice management that included mobile emulab facilities. On a similar note, the MINT testbed used MOVIE, an in-house testbed GUI based management and monitoring platform that was developed based on the NAM toolkit (GUI component of ns-2 network simulator).

3.3. Mobile Robot Positioning and Localisation

It has been observed that a majority of MANET testbed facilities that used mobile robots utilised mobile robot localisation technology to monitor the current location of the mobile robots and guide the mobile robots' positioning. Earlier generations of MANET testbed however, used visual based localisation methods where several cameras were mounted on the overhead ceiling within the testbed area. The cameras served to recognize different mobile nodes and identified their positions and movements. Among the testbeds that used visual based localisation methods were MINT and Mobile Emulab.

Proteus on the other hand, used several different methods as mobile robots in Proteus were used in many different fields including robotics research. The combination of GPS and compass sensors were an option for Proteus outdoor testbed localisation. For indoor testbeds however, various localisation methods were chosen such as ultrasound scan localisations, IR bacon RSSI based localisations as well as visual localisations.

It was found that only a few testbeds did not utilise the robot localisation method to control mobile robot positioning and navigation. In fact, there were some that used only random movement approaches that excluded the monitoring of the location and movement of each mobile robot used in the testbed. The CONE testbed is one of the examples of a testbed that did not utilise a localisation method to control the mobility of the mobile robots in the testbed. In the CONE testbed, the movement of mobile robots were limited to green lines that acted as boundaries in the testbed area and the mobile robots were equipped with ultrasound sensors that provides avoidance abilities during their run.

To facilitate the readers understanding and flow of our discussion for the topic abovementioned, we presents it in tabular form as shown in Table 1. Table 2 on the other hand provide pictures of the type of robot used in MANET-Testbed.

Table 1. Technical Summary of Robot-Based MANET Testbed

Testbed	Robot Platform	Robot Controller	Sensor and Other Components	Robot Software	Testbed Management System	Mobility and Localisation User Defined Mobility
Mobile Emulab	Acroname Garcia	Intel Stargates Board (X-Scale 400MHz CPU)	• Ceiling Mounted Cameras • Colour Pattern B oard • IR Sensors • Ultrasonic Sensors	• Inhouse	• Emulab Framework	• Centralised visual localisation
MiNT-m	iRobot Roomba	Routerboard RN-230	• Ceiling Mounted Cameras • Colour Pattern board	• Inhouse	• MoVIE	• Centralised visual localisation
MiNT-2	iRobot Create	Net5501 Board	• RFID Reader • Wheel Encoder • IMU	• Inhouse	• Desktop GUI interface	• RFID based localisation
Proteus	Traxxas Stampede RC Car Chassis	Mini-ITX x86 Board and Arduino	• GPS • Compass Sensor	• Player • ROS	• Command Line Interface	• Random waypoint mobility • User defined mobility
w-ilab.t	iRobot Roomba	In-House Board	• Wheel encoder • Bump sensors • IR Sensors • Wi-Fi interface	• ROS	• Emulab Framework • GENI Framework • OMF/ OML Framework	• Random mobility with position logging • User controlled mobility • Odometry • RSSI based localisation
ARUM	Lynxmotion 4WD Rover	Lynxmotion Atom Bot Board	• IR Cameras • Reflective Ball as Marker	• Inhouse	• Inhouse	• Fixed circuit-based mobility • Motion capture based monitoring
Sensei-UU	LEGO Mindstorms	Smartphone	• IR Sensors	• Inhouse	• Inhouse	• Fixed circuit-based mobility • RSSI based monitoring

Kansei	Acroname Garcia	Stargate Board	• Ceiling Mounted Cameras • IR sensors • Ultrasonic Sensors	• Inhouse	• GENI Framework	• Centralised visual localisation
CONET-IT	Pioneer 3AT	Netbook	• LIDAR • Microsoft Kinect • GPS • IMU Sensors	• ROS	• CONET Framework • GENI Framework	• Random waypoint • User control mobility
	RC Car Chassis	PC-104 Embedded PC	• GPS • IMU Sensors • IR Sensors			
CONE	LEGO Mindstorms	LEGO NXT	• Ultrasonic Sensors • Colour Sensors	• Inhouse	• Inhouse	• No localisation • Random move in fixed area
Roomba MADNet	iRobot Roomba/ Create	Linksys WRTSL54GS Wireless Router		• Inhouse	• Inhouse	• No localisation • Random move based on wireless signal RSSI
Explorebots	Rogue ATV	Rabbit 3000 Microprocessor		• Inhouse	• Inhouse	• No localisation • Movement controlled by user
SCORPION	iRobot Create	Unknown		• Inhouse	• Inhouse	• No localisation • Random move in fixed area
MOTEL	e-puck	iRobot Create Controller		• Inhouse	• Inhouse	• Centralised Visual Localisation
iRobotSense	iRobot Create	Asus X200CA	• Compass Sensor (HMC 6352) • Wheel Encoder	• Inhouse	• Inhouse	• No localisation • Random move
IoT-Lab	Turtlebot2	Netbook	• Microsoft Kinect • IR Sensors • 1-axis Gyrometer • Bumper Sensors • Cliff Sensors • Wheel Drop Sensors	• ROS	• GENI Framework	• Fixed circuit-based mobility
	Wifibot	Intel Atom Based SBC	• Microsoft Kinect • IR Sensors			
NITOS	iRobot Create	Alix Motherboard	• Wheel Odometry • Webcam • Black and White Pattern Boards	• Inhouse	• GENI Framework	• Fixed waypoint mobility

Table 2. Robot Used in MANET Testbed

Picture of Robot Used in MANET Testbed					
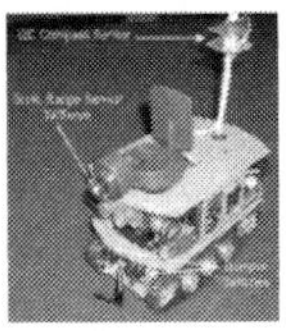	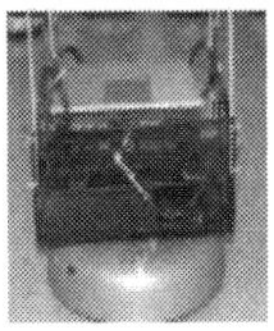		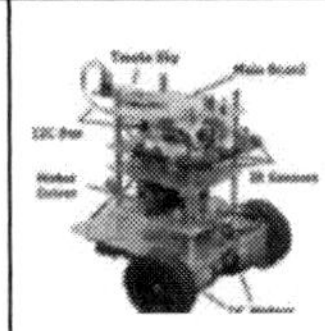		
Explorebots	MiNT-m	CONET-IT	CONET-IT	ARUM	Kansei
		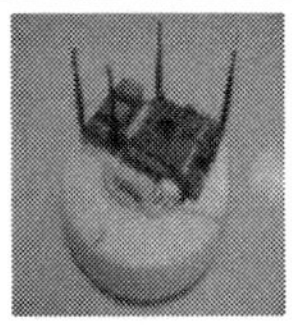	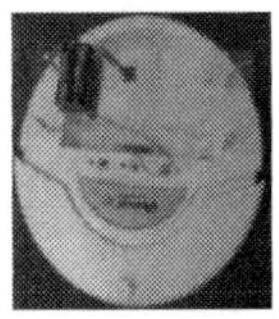		
NITOS	Roomba MADNet	MiNT-2	iRobotSense	w-ilab.t	Motel

CONE	Sensei-UU	Mobile Emulab	IoT-lab	IoT-lab	Proteus

4. Conclusion

The targeted objective of this work was achieved, that is to comprehensively discuss on the technical aspects of mobile robot usage in MANET testbeds. Further extension of this paper can be referred to in our paper entitled *"A Critical Review of MANET Testbed Using Mobile Robot Technology"* [30].

Armed with the wealth of relevant information provided in this paper, the content of this paper is expected to be a source of reference for MANET researchers who are at a crossroad when selecting the preferred mobile robot technology and approach to suit their own specific needs.

5. Acknowledgements

The authors would like to thank the Universiti Teknologi Malaysia (UTM) for supporting this research under VOT NUMBER: Q.J130000.2528.14H81.

References

[1] Farkhana Muchtar, Abdul Hanan Abdullah, Mohd Helmy Abd Wahab, Radzi Ambar, Hafizul Fahri Hanafi, Siti Nor Zawani Ahmmad 2017 Mobile Ad hoc Network Testbed Using Mobile Robot Technology *J. Phys.: Conf. Ser*

[2] Stewart B Ko J Fox D and Konolige K The Revisiting Problem in Mobile Robot Map Building: A Hierarchical Bayesian Approach, *arXiv preprint*: 1212.2515 [cs]

[3] Dantu K and Sukhatme G S 2007 *IEE International Conf. on Robotics and Automation, 2007* pp 3665–3672

[4] Antonelli G Arrichiello F Caccavale F and Marino A 2014 *International Journal of Robotics Research* **33**(7) pp 1029–1043

[5] Nattharith P and Guzel M 2014 *Electrical Engineering Congress (iEECON' 2014) International,* pp 1–4

[6] De P, Raniwala A Krishnan R Tatavarthi K Modi J Syed N A, Sharma S and Chiueh T-C 2006 *Proc. of the 4th International Conference on Mobile Systems, Applications and Services, MobiSys '06* (New York, NY, USA: ACM), pp 124–137

[7] Munishwar V Singh S Mitchell C Wang X Gopalan K and Abu-Ghazaleh N 2009 *IEEE International Conference on Pervasive Computing and Communications (PerCom 2009)* pp 1–3

[8] Abdelhadi A Rechia F Narayanan A Teixeira T Lent R Benhaddou D Lee H and Clancy T-C Position Estimation of Robotic Mobile Nodes in Wireless Testbed using GENI arXiv *preprint arXiv:* 1511.08936

[9] Reich J Misra V Rubenstein D and Zussman G 2012 *IEEE Journal on Selected Areas in Communications* **30**(5) pp 935–950

[10] Bromage S Engstrom C Koshimoto J Bromage M Dabideen S Hu M Menchaca-Mendez R Nguyen D Nunes B Petkov V Sampath D Taylor H Veyseh M Garcia-Luna-Aceves J J Obraczka K Sadjadpour H and Smith B 2009 *Communications Review* **13**(1) pp 65–68

[11] Janansefat S Senturk I Akkaya K and Gloff M 2012 *37th Annual IEEE Conference on Local Computer Networks (LCN)*

[12 Rensfelt O,Hermans F Gunningberg P Larzon L-A and Björnemo E 2011 *The Computer Journal* **54**(12) pp 1973–1986

[13] Vingelmann P Heide J Pedersen M V Zhang Q and Fitzek F H P 2014 *Computer Networks* **74** Part B pp 34–47

[14] Johnson D Stack T Fish R Flickinger D Stoller L Ricci R and Lepreau J 2006 INFOCOM 2006. *25th IEEE International Conference on Computer Communications* pp 1–12

[15] Arora A Ertin E Ramnath R Nesterenko M and Leal W 2006 *IEEE Internet Computing* **10**(2) pp 35–47

[16] Foerster A Foerster A Garg K Giordano S and Gambardella L M *Adhoc & Sensor Wireless Networks* **24**(3)

[17] Dios J R M-d Jimenez-Gonzalez A Bernabe A d S and Ollero A 2014 *Electrical and Computer Engineering* (Switzerland: Springer International Publishing) pp 23–39

[18] Killijian M O Roy M and Severac G 2012 *Advances in Autonomous Mini Robots,* (ed) U RÃŒckert, S Joaquin, W Felix *et al* (Heidelberg: Springer Berlin) pp 221–234

[19] Dahlberg T A Nasipuri A and Taylor C 2005 *Proc. of the 2005 ACM SIGCOMM Workshop on Experimental Approaches to Wireless Network Design and Analysis, E-WIND '05,* (New York, NY, USA: ACM) pp 76–81

[20] Quilez R Zeeman A Mitton N and Vandaele J 2015 *Self-Adaptive Systems* 1(2): e2

[21] Förster A Förster A Garg K Puccinelli D Giordano S and Gambardella L M 2012 *Wireless Integration of Sensor networks in Hybrid architectures* **13**

[22] Johnson D 2010 Design and Implementation of a Mobile Wireless Sensor Network Testbed Masters thesis, University of Utah

[23] De P 2007 *Mint: A Reconfigurable Mobile Multi-hop Wireless Network Testbed* Ph.D. Thesis State University of New York at Stony Brook, Stony Brook, NY, USA

[24] Niavis H Kazdaridis G Korakis T and Tassiulas L Social Informatics and Telecommunications Engineering 2012 (eds) T Korakis M Zink and M Ott *et al*, LNIC, vol 44 (Heidelberg: Springer Berlin) pp 421–424

[25] Becue P Jooris B Sercu V Bouckaert S Moerman I and Demeester P 2012 T (eds.) T Korakis M Zink and M Ott *et al, Social Informatics and Telecommunications Engineering, LNICST,* vol 44 (Heidelberg: Springer Berlin) pp 425–426

[26] Ertin E Arora A Ramnath R Nesterenko M Naik V Bapat S Kulathumani V Sridharan M Zhang H and Cao H 2006 *The 5th International Conf on Information Processing in Sensor Networks, IPSN'06* pp 399–406

[27] N. A. Paine 2010 *Design and development of a modular robot for research use Master Thesis* University of Texas at Austin

[28] Jimenez-Gonzalez A Martinez-De Dios J R and Ollero A 2011 *Sensor (Basel, Switzerland)* **11**(12) pp 11516–11543

[29] Frey M Grosse F and Günes M 2013 *The 7th International Conf. on Performance Evaluation Methodologies and Tools, ValueTools'13* (Turin, Italy: ACM)

[30] Petz A Fok C-L and Julien C 2012 *Proc. of the 9th ACM International Workshop on Vehicular Inter-networking, Systems, and Applications, VANET '12* (New York, NY, USA: ACM) pp 21–26

[31] Farkhana Muchtar, Abdul Hanan Abdullah, Marina Md Arshad, Mohd Helmy Abd Wahab, Siti Nor Zawani Ahmmad and Gaddafi Abdul-Salaam 2017 A Critical Review of MANET Testbed Using Mobile Robot Technology *J. Phys.: Conf. Ser*

IOP Conf. Series: Journal of Physics: Conf. Series **1049** (2018) 012002 doi:10.1088/1742-6596/1049/1/012002

Performance of Kaolin Clay on Hot-mix Asphalt Properties

M K I Mohd Satar[1], R P Jaya[1,2], M H Rafsanjani[1], N Che' Mat[1], M R Hainin[1], Md. M A Aziz[1], M E Abdullah[3], D S Jayanti[4]

[1]Faculty of Civil Engineering, Department of Geotechnics and Transportation, Universiti Teknologi Malaysia, 81310 Johor Bahru, Malaysia
[2]Faculty of Civil Engineering and Earth Resources, Universiti Malaysia Pahang, 26600 Pekan, Pahang, Malaysia
[3]Faculty of Civil and Environmental Engineering, Universiti Tun Hussein Onn Malaysia, 86400 Batu Pahat, Johor Bahru, Malaysia
[4]Faculty of Agricultural, Syiah Kuala of University, 23111 Banda Aceh, Indonesia

E-mail: ramadhansyah@utm.my

Abstract. Kaolin clay is a waste product with numerous applications in construction. This study explored the addition effect of kaolin clay on the properties of hot-mix asphalt. Four replacement levels of kaolin clay were considered by weight of binder, i.e., 2%, 4%, 6%, and 8%. The performance of kaolin clay on the hot-mix asphalt was evaluated through a Marshall stability and flow test, including stiffness, density, voids in total mix, and voids in filled with asphalt. Test results showed that kaolin clay can be satisfactorily used as filler replacement material to increase the asphalt mixture properties. Generally, asphaltic concrete with 2% kaolin clay replacement level exhibits excellent performance with good stability and stiffness.

1. Introduction

Road damage is one of the main problems faced by road users and can be the cause of an accident. An increase in traffic capacity can damage a new road after just a few years. Road damage is due to the deformation of the road. The deformation can either be permanent or temporary. Rutting, a permanent deformation, comprises the majority of deformations that occur on roads nowadays [1]. Various studies have been conducted to ensure asphalt concrete can be used for a long time [2]. As an alternative material, asphalt concrete mixture has been modified using a variety of natural resources to strengthen asphaltic concrete pavement [3]. For instance, the presence of a filler in the asphaltic concrete mixture is important because of the possible interactions with asphalt [4]. During the process of mixing the asphalt and the aggregate together, the mastic asphalt filler forms because of the fineness of the filler [5]. The interaction between the asphalt and the fillers result in certain mastic properties [6], thereby affecting the performance of the mixture [7]. Owing to the increased surface area, the filler can absorb more asphalt, and asphalt interaction can lead to a different performance of the asphaltic concrete mixtures [8]. One common filler is kaolin clay [9], which is an important raw material in various industrial sectors. Kaolin is clay that is composed of kaolin stone. Kaolin clay is one of the main minerals in the world and one of the most widely used minerals [10]. It is mainly

[1] To whom any correspondence should be addressed.

Content from this work may be used under the terms of the Creative Commons Attribution 3.0 licence. Any further distribution of this work must maintain attribution to the author(s) and the title of the work, journal citation and DOI.
Published under licence by IOP Publishing Ltd

formed after kaolin stone is completely weathered and rain washed [11]. Pure kaolin is soft, easy to disperse suspended in water, has high whiteness, good plasticity and cohesiveness, excellent electrical insulation, good acid soluble cationic, low resistance, physical properties, and improved exchange with other chemicals [12]. In the current study, the potential of kaolin clay as a binder replacement in asphalt mixture was investigated.

2. Materials and Method

2.1. Aggregate

The local natural aggregates from granite used in this investigation are nearly always massive and hard. The coarse and fine aggregates indicated a specific gravity of 2.66 g/cm^3, and water absorption rates were 0.48% and 0.86%, respectively. A gradation test (Figure 1) was conducted following the procedure outlined in BS EN 933-1:2012 [13] and JKR/SPJ/2008-S4 [14].

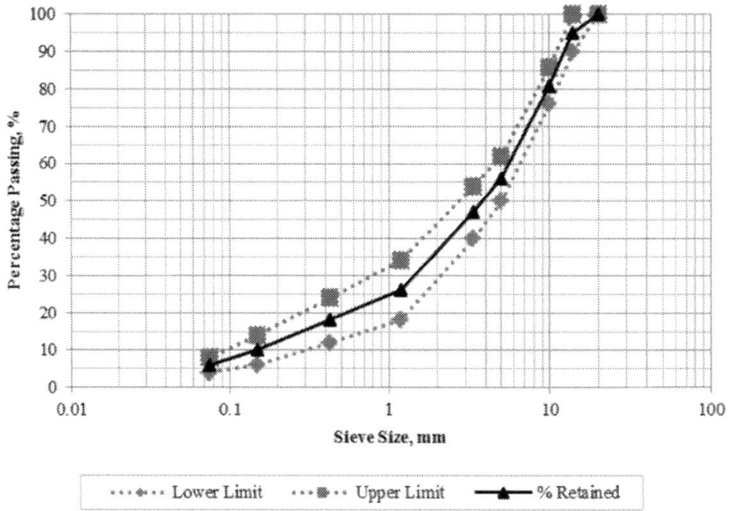

Figure 1. Aggregate gradation used in this study [14]

2.2. Bitumen

Bitumen 60/70 penetration grade was used in this study. The bitumen properties according to a laboratory test were 65 dmm penetration, 51 °C softening point, 1.03 relative density, and 600 cp viscosity at 135 °C.

2.3. Kaolin clay

This study used kaolin clay primarily composed of SiO_2 and Al_2O_3 (over 90%). In general, kaolin clay has a low amount of Fe_2O_3 (0.63%) and about 0.9% of MgO. K_2O (2.5%) and Na_2O (0.02%), which are referred to as alkalis, are also found in kaolin clay. In this investigation, five replacement levels of kaolin clay by weight of bitumen were considered namely 0%, 2%, 4%, 6%, and 8%. The control sample (0%) was designated as KC0, while the 2%, 4%, 6%, and 8% levels were designated respectively as KC1, KC2, KC3, and KC4.

2.4. Sample preparation

Bitumen was heated in the oven until fluid condition was reached, and then prepared kaolin clay was added gradually to the bitumen and blended for three to five minutes. Mixing was continued for a specific duration to produce a homogenous mixture. Three specimens were prepared for each percentage of kaolin clay. After the specimens were compacted using a Marshall hammer, they were tested for stability, flow, density, stiffness, voids in total mix (VTM), and voids in filled with asphalt (VFA).

2.5. Marshall Stability test
In the laboratory, the aggregate, binder, and plastic were individually mixed and compacted at 180±0.5 °C. The mixes were compacted with 75 blows on each side with the standard Marshall hammer to avoid disintegration of materials. After compaction, the specimens were removed from the molds and allowed to cool down. The stability value of each test specimen was then determined in accordance with BS EN 12697-34:2012 [15]. In the stability test, the specimens were prepared with the specified temperature by immersing in a water bath at 60±1 °C for 45 minutes. Next, they were placed in the Marshall Stability testing machine and loaded at a constant deformation rate of 50.8 mm/minute until the maximum load was reached.

3. Results and Discussion

3.1. Stability and Flow
Figure 2 graphically presents the results of the stability and flow test on asphaltic concrete containing kaolin clay. Stability is shown to increase with increasing kaolin clay content up to 2% and then decrease with further additions. The increase in stability may be attributed to the reaction of carbon and hydrogen [6], as well as that of one or more of the elements sulfur, oxygen, and nitrogen [16] with silicon dioxide and aluminium oxide [9]. The results also indicate that stability is sensitive toward variances in kaolin clay content. High percentages of kaolin clay have a decreasing net effect on mix stability. The stability drops from 11,884 N to 10,753 N as the kaolin clay replacement increases from 2% to 8%. The control specimen produced low stability compared with KC1, KC2, KC3, and KC4. However, when the kaolin clay replacement levels are compared, the 2% KC asphalt mixture exhibited the highest stability. High stability values indicate stiff and resistant mixtures. Thus, 2% replacement of KC is considered the optimal limit. The differential movement between initial and maximum loading is also recorded as the flow. The results shown in Fig. 2 indicate that flow for the corresponding kaolin clay content initially increased between 0% to 2% and then started to decrease until 6% KC was reached. Generally, the flow value of KC0 was 3.05 mm, while the 2%, 4%, 6%, and 8% KC replacement levels had corresponding flow values of 3.22, 3.13, 2.65, and 3.20 mm, respectively. Hence, the flows of all the specimens are in the range recommended by the local specification JKR [14] and the Asphalt Institute [17].

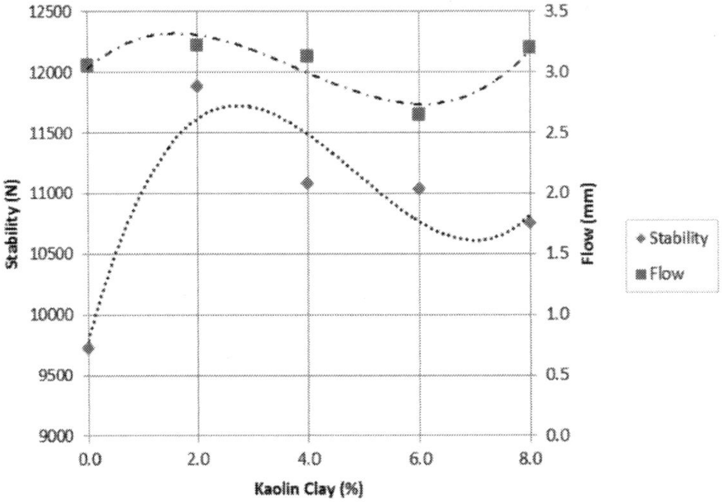

Figure 2. Stability and flow at varying kaolin clay content

3.2. Stability and Density

Figure 3 graphically illustrates the relationship between stability and density of asphaltic concrete at different levels of kaolin clay content. The plotted readings show that the stability and density values increase up to a peak level and then decrease with further additions. However, the rate of increase in stability and density with respect to the kaolin clay content of all the mixes is higher than the control mix because the fine kaolin clay particles help form a denser asphalt mix and increase stability and density. For instance, the maximum density of KC0 was 2.283 Mg/cu.m, which was lower than the maximum density values of KC1, KC2, KC3, and KC4. Hence, a 2% increase in kaolin clay content can cause a 1.2% increase in density. Among all kaolin clay replacements investigated, the KC2 mixes recorded the highest mix density, followed by KC1, KC3, and KC4. For stability, the maximum value recorded was for KC1, followed by KC2, KC3, and KC4.

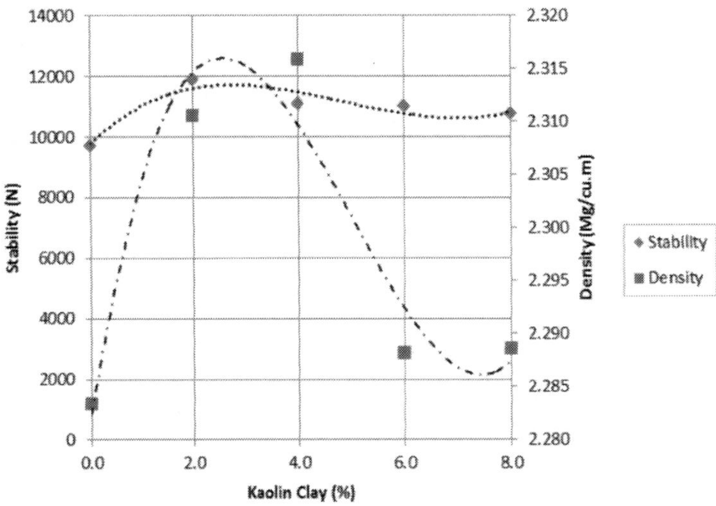

Figure 3. Stability and density at varying kaolin clay content

3.3. Stiffness and Density

Figure 4 graphically illustrates the relationship between the stiffness and density of asphaltic concrete incorporating kaolin clay. The results are the average of three specimens prepared in optimum bitumen content (OBC). They show that the stiffness and density of the asphalt mixture increases with increasing kaolin clay content up to a peak level and then decreases with further additions. The rate of increase in stiffness with respect to the kaolin clay content of the asphalt concrete mix is higher than the control mix. This outcome may be due to kaolin clay being an active filler that improves the stiffness properties of asphalt mixture. For instance, the maximum stiffness for KC1 is 3690 N/mm, while KC0 exhibits a maximum stiffness of 3183 N/mm, which means the addition of 2% kaolin clay to the asphalt mix can increase stiffness by up to 13.7%. Another example is when kaolin clay content increases from 4% to 6%, the maximum stiffness increases from 3540 N/mm to 4169 N/mm and then starts to decrease at 8% replacement.

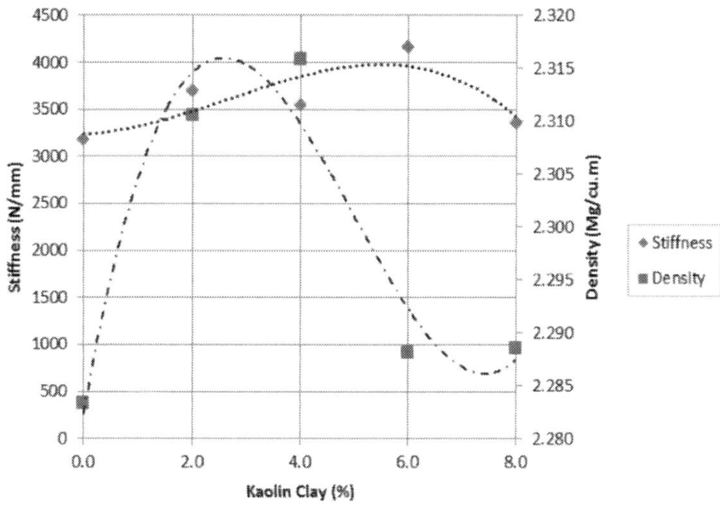

Figure 4. Stiffness and density at varying kaolin clay content

3.4. Voids in total mix

Figure 5 presents the effects of kaolin clay on asphaltic concrete. Generally, VTM decreases gradually with the decrease of kaolin clay up to 4% and then increases with further additions. The results clearly show that KC3 exhibits higher VTM than KC0, KC1, KC2, and KC4. Mixes with low VTM inhibits high susceptibility toward rutting [18,19]. Hence, asphaltic concrete incorporating kaolin clay significantly affects the performance of a mixture because of low void content. For instance, at 0% KC, the VTM is 4.5%, while at 2%, 4%, 6%, and 8% KC, the corresponding VTM are 3.5%, 3.1%, 4.7%, and 4.5%, respectively. However, at all VTM investigated, KC2 mixes record the lowest VTM. VTM is the total volume of small pockets of air between the coated aggregate particles throughout a compacted paving mixture. According to the JKR specification [14], VTM should range from 3% to 5%. Therefore, the VTM values of all specimens are in the range recommended by the JKR specification [14].

IOP Conf. Series: Journal of Physics: Conf. Series **1049** (2018) 012002 doi:10.1088/1742-6596/1049/1/012002

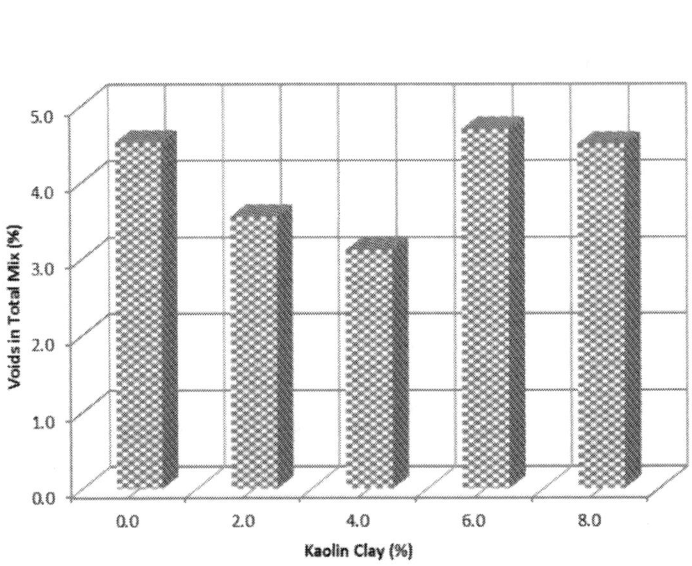

Figure 5. VTM at different kaolin clay content

3.5. Voids in filled with asphalt

From the investigation results shown in Figure 6, the VFA increases as the percentage of kaolin clay increases up to 4% and then decreases with further additions. This phenomenon is due to the constant optimum bitumen content used in mixes for all the percentages of kaolin clay. The results also revealed that the control specimen has the lowest VFA value. VFA is the percentage of the volume of VMA that is filled with asphalt. The control specimen (0% KC mix) exhibits the lowest VFB, followed by 2%, 4%, 6%, and 8%. Samat [20] and Jeffry et al. [21] reported that when the VFA exceeds approximately 80% to 85%, the dense mix becomes unstable and rutting is likely to occur. Hence, the VFB values are within 76% to 84%. According to the JKR specification [14], the VFB of the dense mix should range from 75% to 85%, which can be achieved by incorporating 3.0% to 5.0% bitumen content. However, the local specification does not state the limiting values for replacing or adding the main binder with other materials.

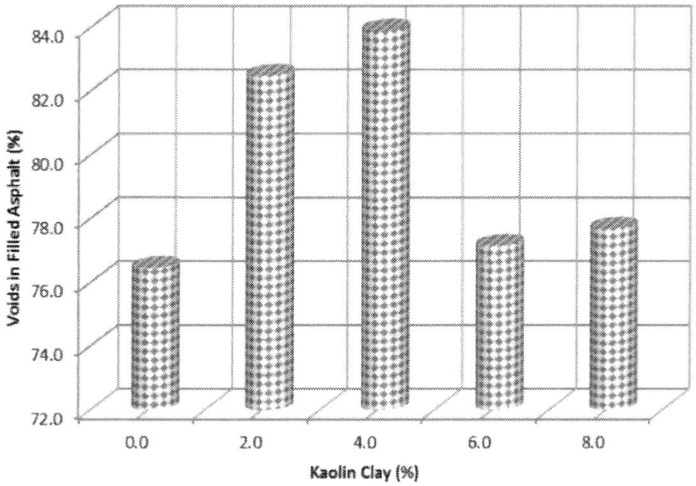

Figure 6. VFA at different kaolin clay content

4. Conclusions

The obtained data suggest that kaolin clay as filler replacement effectively improves the properties of asphaltic concrete. The low replacement level of kaolin clay has an advantageous effect in improving stability and stiffness. Results also indicate that the addition of kaolin clay in asphalt mixtures provides additional improvements in VTM and VFA. Finally, the use of 2% kaolin clay results in the good performance of asphalt mixtures.

5. References

[1] Ramadhansyah P J, Nurfatin Aqeela M, Siti Nur Amiera J, Norhafizah M, Norhidayah A H, and Dewi S J 2016 *ARPN J. Eng. Appl. Sci.* **11** 7457-7462.
[2] Hainin M R, Jaya R P, Ali Akbar N A, Jayanti D S and Yusoff N I M 2014 *J. Eng. Res.* **2** 34-46.
[3] Jayanti D S, Mirza J, Jaya R P, Abu Bakar B H, Hassan N A and Hainin M R 2016 *P. I. Civil Eng-Mar. Eng.* **169** 76-85.
[4] Abdullah M E, Zamhari K A, Hainin M R, Oluwasola E A, Yusoff N I M and Hassan N A 2016 *J. Clean. Prod.* **122** 326-334.
[5] Ting T L, Jaya R P, Hassan N A, Yaacob H, Jayanti D S and Ariffin M AM 2016 *J. Teknologi.* **78** 85-89.
[6] Abdullah M E, Abd Kader S A, Putra Jaya R, Yaacob H, Abdul Hassan N and Che Wan C N 2017 *MATEC Web of Conferences* **103** 1-7.
[7] Abd Kader S A, Putra Jaya R, Yaacob H, Hainin M R, Abdul Hassan N, Wan Ibrahim M H, Ali Mohamed A and Ichwana 2017 *MATEC Web of Conferences* **103** 1-7.
[8] Lotfy, O. Karahan, E. Ozbay, K.M. Hossain and M. Lachemi, *Constr. Build. Mater.* 83, 102-107 (2015).
[9] Hainin M R, Ramadhansyah P J, Chan T H, Norhidayah A H, Nazri F M and Ichwana 2017 *Mater. Sci. Forum* **889** 265-269.
[10] El-Shafie M, Ibrahim I and El Rahman A A 2012 *Egypt. J. Pet.* **21** 149-154.
[11] Mohd Ezree A, Nurul Asyiqin A, Ramadhansyah P J, Norhidayah A H, Haryati Y and Mohd Rosli H 2017 *IOP Conf. Ser.: Mater. Sci. Eng.* **204** 1-7.
[12] You Z, Mills-Beale J, Foley J M, Roy S, Odegard G M, Dai Q and Goh S W 2011 *Constr. Build. Mater.* **25** 1072-1078.

[13] BS EN 933-1 2012 Tests for geometrical properties of aggregates. Determination of particle size distribution. Sieving method, British European Standard, UK.
[14] JKR/SPJ/2008-S4 2008 Standard Specification for Road Works Section 4: Flexible Pavement.
[15] BS EN 12697-34 2012 Bituminous mixtures. Test methods for hot mix asphalt. Marshall Test, British European Standard, UK.
[16] Iskender E 2016 *Meas.* **93** 359-371.
[17] Asphalt Institute 2007 The Asphalt Handbook, 7th edition, Asphalt Institute, Lexington.
[18] Abdullah M E, Madzaili A H, Jaya R P, Yaacob H, Hassan N A and Nazri F M 2017 *IOP Conf. Ser.: Mater. Sci. Eng.* **222** 1-5.
[19] Jeffry S N A, Jaya R P, Hassan N A, Yaacob H, Mirza J and Drahman S H 2018 *Constr. Build. Mater.* **158** 1-10.
[20] Samat M M 2006 The Effects of Drain Asphalt Modified Additive (DAMA) on the Engineering Properties and Performance of Porous Asphalt, Master thesis, Universiti Sains Malaysia.
[21] Jeffry S N A, Jaya R P, Manap N, Miron N A and Hassan N A 2016 *Key. Eng. Mater.* **700** 227-237.

Acknowledgments
The support provided by Malaysian Ministry of Higher Education and Universiti Teknologi Malaysia in the form of a research grant vote number Q.J130000.2622.14J09 and Q.J130000.2522.18H05 for this study is highly appreciated.

IOP Conf. Series: Journal of Physics: Conf. Series **1049** (2018) 012003 doi:10.1088/1742-6596/1049/1/012003

Analysis of Non-invasive Video Based Heart Rate Monitoring System obtained from Various Distances and Different Facial Spot

Norwahidah Ibrahim[1], Razali Tomari[1], Wan Nurshazwani Wan Zakaria[1], Nurmiza Othman[2]

[1]Department of Robotic and Mechatronics, Faculty of Electric & Electronics, University Tun Hussein Onn Malaysia
[2]Department of Electronics, Faculty of Electric & Electronics, University Tun Hussein Onn Malaysia

E-mail: wahidahibrahim93@gmail.com, mdrazali@uthm.edu.my, shazwani@uthm.edu.my, nurmiza@uthm.edu.my

Abstract. Heart rate (HR) is one of the crucial indicators for human psychological. In recent works, it has been shown that a standard camera is able to detect illumination changes in the face skin due to the human cardiac pulse and this can be used to estimate the human HR. However most of previous systems work on near distance mode with a single face patch, thus the expediency of the camera based remote heart rate estimation for long range distances remains ambiguous. This paper has proposed a solution by analyzing an optimal framework that able to works properly under the mentioned issues. Initially, presumable facial landmarks are estimated by applying cascaded of regression mechanism. Then, the region of interest (ROI) was selected based on the facial landmarks in the location where non rigid motion is minimal. Temporal photoplethysmograph (PPG) signal is extracted from the ROI and the unwanted signal such as environment illumination signal or motion artifact signal is eliminated by using Independent Component Analysis (ICA) filter. Then, PPG signal is further processed using series of temporal filter to exclude frequencies outside the range of interest prior to estimate the HR. Since, the HR is estimated independently from multiple local regions, a histogram analysis is constructed to calculate the average HR estimation accurately. From the experiments, it can be concluded that the HR can be detected up to 5 meters range with 94% accuracy using full face region.

1. Introduction

Nowadays, there are many non-contact heart rate monitoring methods that had been developed. Generally, there are three types of existing non-contact heart rate monitoring methods. The first method of non-contact heart rate monitoring system is electromagnetic based heart rate monitoring system. In general electromagnetic signal would be transmitted to the targeted person and the back-scattered signal would be measured and thus the estimation of heart rate would be made from the back-scattered signal obtained. Continuous-wave (CW) and wide band pulsed radar (UWB) are two

[1] To whom any correspondence should be addressed.
[2] To whom any correspondence should be addressed.

Content from this work may be used under the terms of the Creative Commons Attribution 3.0 licence. Any further distribution of this work must maintain attribution to the author(s) and the title of the work, journal citation and DOI.
Published under licence by IOP Publishing Ltd

types of electromagnetic based monitoring systems [1-3]. For laser based heart rate monitoring system, typically a device called laser Doppler vibrometer (LVDi) was used [4]. LVDi detects the signal from skin movement that related with blood pressure pulse under carotid artery by measuring a small skin surface displacement cause by expansion and contraction of artery [5].

Recently, video based remote HR monitoring framework becomes subject of interest due to vast amount of data available from a cost effective device. Generally, in video-based HR system, PPG signal will be extracted from the face video recorded and multiple signal processing filters will be applied in order to obtain HR from PPG signal. PPG signal is an optically volumetric signal measurement of human body that usually cause by fluctuation of air or blood in the body.

One of the early works was started back in 2008 by Verkruysse et al [6]. They stated that the signal emitted through a face video recording under a normal or ambient lighting was rich enough to measure HR of a human being. They started by recording subjects' face video by using a standard recording camera with 640*480 resolution, a frame rate of 30 frame per second (30 FPS) and a standard graphical mode of Video Graphic Array (VGA). For this experiment, the ROI is selected manually and the raw signal is computed frame by frame based on the mean value of Red, Green and Blue (RGB) colour channel. They concluded that the green channel contain the strongest PPG signal compare to red and blue channel.

The research work then evolved with the work of Poh et al [7], instead of using a video recorder, Poh and his colleagues try a different approach by using a standard laptop webcam to record the face video of subjects. They used a face detector to detect the subjects' face frame by frame rather than defining the ROI manually. Another different approach that they made was, they used three colours channel RGB to obtain the raw PPG signal information. Then, ICA is used to separate the raw PPG signal from unwanted noise that may affect the HR reading. Lastly, they estimated the signal obtained as frequency and the highest frequency is selected as the HR estimation. Balakrishnan et al [8] proposed a HR estimation system by extracting raw PPG signal from subtle head motion of subject. They claimed that raw PPG signal can be extracted using this method due to blood incursion towards head. However, the system is not robust to motion artifact and subjects had to remain stationary during the recording.

Even though all of the mentioned works had very promising results and accuracies, all of the experiments conducted by the researchers mention above are made in controlled environments such as the distance of the subject between camera was limited, the subjects of the experiment were not allowed to exhibit any movements and lastly, all of experiments only use one ROI and no comparison between ROIs had been made. Thus, researchers are focusing on creating algorithms that can tackle more realistic condition. This leads to another different approach by Philips Research Group that tried to overcome the problem of moving subject with respect to illumination source. They stated that, an optimal fixed combination of band passed RGB signal can be found based on ratio of normalized colour signal when assuming standardized skin, eliminating noise due to specular reflection [9].

Recently, ROI selection had become increasing popular research topic. This is because it had been reported that ROI has a significant influence for HR estimation accuracy [7]. So intelligent *ROI* selecting and tracking algorithm had been developed, in order to achieve robust algorithm to motion artifact [10-14]. Facial landmarks detection is used, so that more detailed face region could be selected. Feng et al [15] discovered a set of points on the forehead that can subsequently be used to update ROI. However, they improved the algorithm and used the cheeks area instead. M.Kumar et al [13], tried a different method by dividing the face region into smaller ROIs. Raw PPG signal is extracted from all the region selected and is combined using a weighted average.

Although, there were many works and researches that was done during these past few years regarding the HR estimation system, there are still a lot of improvement that can be made in order to obtain the optimal algorithm for the system. In this paper, we investigate the capability of remote HR for acquiring data from various distances and provide an optimal framework that capable to satisfy the constraint.

1.1 PPG General Algorithm
There are general algorithms of existing PPG approach prior to our project. In this section we are going discuss and classify the chosen approach accordingly. We divide this section into three important steps :
i) ROI Detection.
ii) ROI Selection.
iii) Signal Extraction, Filtering and HR estimation.

1.1.1 ROI Detection
ROI detection is important to the HR system algorithm because PPG algorithm is based on human face. Hence it is crucial to detect the face of subject in the video frame. Most of commonly used ROI detection is Viola and Jones (VJ) algorithm [16]. VJ algorithm is a machine learning based algorithm that used Haar cascade features to detect face. This algorithm is popular due to its high rate detection, fast processing time and it could easily be used as it is available in most of software programming such as OpenCV and Matlab.

Another different method is to use skin detection algorithm. The advantage of using skin detection is that additional area such as neck or hands of the subject will be included as the ROI. However, due to the same feature of the skin detection a problem could occur. The drawback of using skin detection is the presence of unwanted noise interference in the PPG signal extracted due to object that have similar colour to the subjects' skin. Facial landmark tracking algorithm is used to tackle the motion artifact and illumination issues. Furthermore, facial landmark tracking method will result in more detailed ROI selection due to basic points provided by the tracking system. Two popular facial landmark algorithm that used by many researchers in this field is Active Appearance Models (AAM) [17] and Discriminative Response Map Fitting (DRMF) [18].

There are many works evolution or improvement that were made for the past few years in order to obtain the most optimal method for facial detection. In this paper, the ROI detection used was VJ algorithm with Haar cascade feature and facial landmarks tracking. The reason that we chose VJ algorithm because it is a very promising algorithm due to its high detection rate and it is practical for real time application, in case if this project is going through an improvised version of real time application. We combine the face detection with facial landmark tracking system to obtain more detailed ROI selection for the project. ROI selection give a massive influence to the HR estimated result, thus selecting more detailed ROI would leads to more accurate results.

1.1.2 ROI Selection
Poh et al [7] stated that the ROI selection will influence the estimated HR reading because the raw PPG signal strength varies according to area of face region. In most publication, the common face areas selected are forehead and cheeks [19-21]. These areas are chosen because there would be less non-rigid motion involve compared to another face part. Lam et al [22] stated that the most optimal area that would yield a good PPG signal is located at the center of the face which include nose and mouth are but excluding the eyes area. Since our project involve with variation of distances, we are going to use all of the mentioned ROIs in order to determine the most optimal patch for our system.

1.1.3 Signal Extraction, Filtering and HR Estimation

Basically, there are many algorithms that can be used to extract the raw PPG signal from the ROIs selection. However, the most used method is BSS using ICA algorithms. Many researchers [7-8] used this method due to its ability to separate mixed signal into independent signals effectively. Another known method to extract raw PPG signal was to use Fast Fourier Transform straightly without BSS. Due to its effectiveness to separate mixed signal into independent signals and elimination most of unwanted signal from PPG signal, for this project we proposed to use the BSS with ICA as our algorithm for PPG signal extraction.

Filtering is an important step in the HR estimation system because despites using BSS as signal extraction tools and noise removal algorithm, the signal may contain a few unwanted noises that could not be eliminated using BSS. Hence further filtering process is required in order to obtain genuine PPG signal. There are three types of filtering algorithm that commonly used for the HR estimation system. The first one is bandpass filtering. Bandpass filter can eliminate both high and low unwanted frequencies. However, this method requires a definite frequency range that is feasible for human HR. A common frequency range for human HR is 0.7Hz – 4Hz (42bpm – 240bpm).

Another way is to use both detrending filter and moving average filter. Similar to high pass filter, detrending filter remove the unwanted long-running trend from PPG signal, which results in smoother signal. Moving average filter on the other hands, represent a low pass filter. Moving average filter is commonly used to smooth out short-term fluctuations and highlight longer-term trends or cycles. Both filters need to be applied to the PPG signal in order remove both high and low unwanted frequency.

A different approach was to use adaptive bandpass filter. Adaptive bandpass filter is known for its ability to progressively change the cutoff frequencies according to previous HR estimation. Hence, will produce a consistent HR results. However, the drawback of this filter is that it is depending on the previous HR estimated which means that it could not eliminate unwanted noise independently. With that being said, for this project, we used the detrending and moving average filtering due to is effectiveness and its promising result for our HR algorithm.

Lastly, frequency analysis can be used to estimate the HR readings. PPG signal is converted to power spectrum density (PSD) because PSD displayed a strong frequency that corresponds to human HR [22]. There are many frequency analyses that can be used to convert raw PPG signals to PSD such as Discrete Fourier Transform (DFT), Fast Fourier Transform (FFT), Welch's method and Short-time Fourier Transform (STFT). Once the signal had been converted, the frequency that exhibits the highest spectral power is chosen as HR estimation reading.

As a conclusion, since previous works only used certain part of face at a time in their experiments, thus there were no proper comparison between ROI were made. This paper aims to propose optimal ROI by investigating the relationship of face region for PPG signal extraction and the distance of subjects from the camera.

2. SYSTEM OVERVIEW

The proposed framework for the system consist of five important steps that are facial tracker, signal extraction, BSS using ICA, signal filtering and lastly a histogram analysis. Figure 1 shows the block diagram of the system flow.

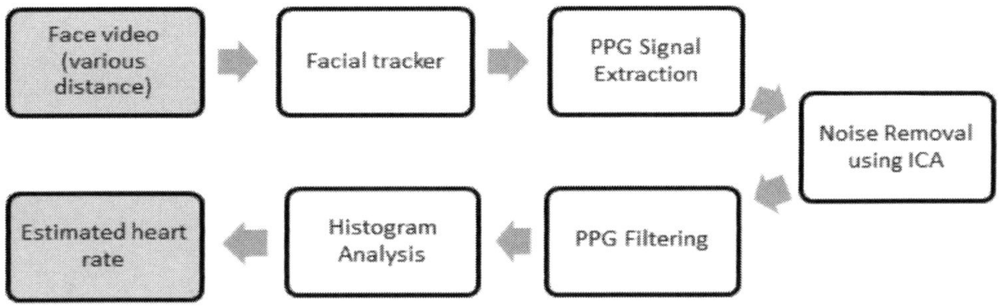

Figure 1. Overall system flow for the proposed framework

2.1. Facial Tracker

Facial tracker is one of the important steps in this system to detect the face and extract facial landmarks information. For the facial tracker in this system, AdaBoost-based cascade with Haar like feature [16] was used. This classifier works by constructing a strong classifier (positive images) as linear combination weak classifiers (negative images). From the detected face region, facial landmarks are detected by using cascaded of pose regression model [23]. This facial tracker would track rigid and non-rigid facial landmark, and 49 points of important facial point would be displayed. The current shape estimate of the facials location can be estimated by using following formulation let $S = (x1, x2...xp)$ denotes the coordinate of all p facial landmarks in a bounding box I and $rt(.,.)$ be the regressor cascade.

$$S(t+1) = S(t) + rt(I, S(t)) \qquad (1)$$

Based on the facial landmarks location obtained, a ROI of face region was constructed. In this paper, four ROIs were investigated to determine the most suitable patch that can be used in various distances requirement as shown in Figure 2(b). The first patch was on the right cheek of the face. Right cheek was chosen because according to [24] cheek is the most practical for ROI selection since they are rarely covered by clothing of facial hair and less non-rigid motion is generated in this area compared to other regions.

Next, the second patch was selected on the center of the face that includes eyes and nose region but excluding the forehead area since the forehead tends to be covered by hair. This area was chosen because previous study [25] reported that the center of face to be the most suitable ROI for video based HR. The third patch was the whole face since basically the larger the region, the possibility to extract the PPG information from a far distance is high. Means that the information loss could be reduce even if the distance of the subject to the camera was farther apart. Finally, the fourth patch was selected at the lower part of the face that includes nose and mouth but excluding eyes and chin area. This region was chosen because of there are less non-rigid motion in it and wider ROI dimension compared to the patch placed on the right cheek.

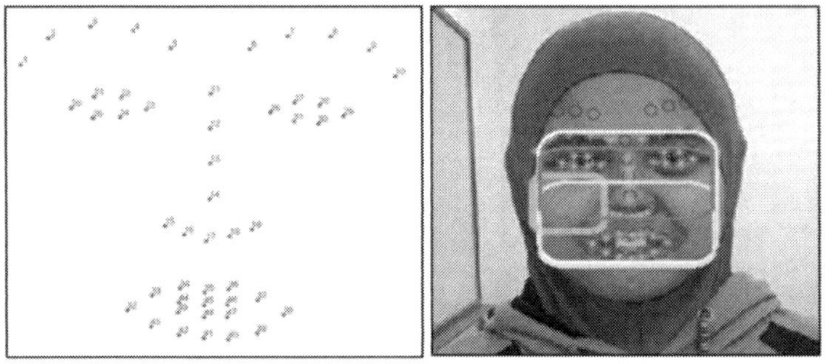

(a) (b)

(Right Cheek Lower Face Part Center of Face Whole face)

Figure 2. Sample of facial tracker result (a) 49 facial landmarks (b) ROI for extracting the PPG signal

2.2. PPG Signal Extraction and Noise Removal

Raw PPG signal was extracted from ROI constructed based on the 49 points facial landmarks location. The raw PPG signal was extracted from the green channel since the green channel provide strongest reading for the PPG signal due to the sensitivity of light absorption of hemoglobin. The raw PPG signal that was extracted was a mixed signal that contained unwanted signal such as background illumination and motion artifact. So, in order to eliminate the unwanted signal, ICA based BSS method was used to separate the mixed signal.

ICA is one of the BSS techniques that capable to recover unobserved signal from a set of observed mixture in which the mixing process is unknown and assume to be linearly correlated. The general mathematical model of ICA can be represented according to equation (2) below, where x is the observed signal (mixed signal), A is the unknown mixing matrix of two independent signals and s is the independent component.

$$x = As \tag{2}$$

Thus, based on ICA mathematical model, we can say that blood volume variation cause by cardiac pulse and the illumination changes are two important factor that affect the reading of raw PPG signal. The mathematical model of the factor influences the PPG signal can be represented by a linear correlation define by equation (3) in which s represents the green channel signal and y represents the variation of illumination. Ideally if the y parameter can be estimated, then the pure cardiac pulse signal can be obtained. However, in practice the signal of y cannot be measured directly.

$$PPG_{raw} = s + y \tag{3}$$

The raw PPG traces obtained is fed to the ICA by assuming that the rectified PPG signal and the illumination components are independent. After solving for the components, the ICA will produce two separated signals. Average Euclidean distance measure is performed to find the similarity score between the signals, and the one with the lowest value will be labeled as the rectified PPG signal while the other one will be denoted as illumination variation.

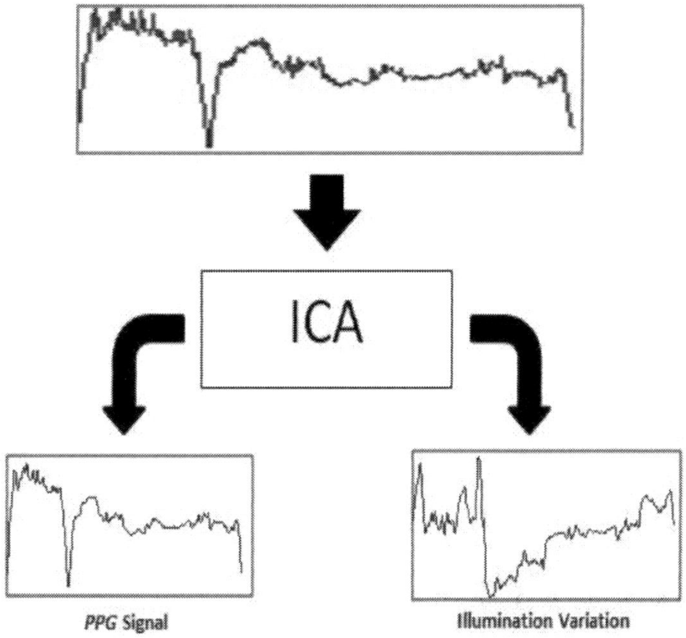

Figure 3. Sample of ICA implementation of BSS. Original signal in the left side and two ICA components on the right side

2.3. Signal Processing and Filtering

The PPG signal obtained after the BSS was considered as a refined but not a pure signal. The signal obtained still contained some unwanted noise that could affect the HR reading. Thus, further signal processing and filtering was applied to the refined signal. In this paper, there were two step of the PPG signal filtering process. Firstly, a detrending filter was applied to the PPG signal in order to reduce slow and non-stationary trends of the raw signal. Then a moving average filter was applied to decrease the noise that presence in the signal. Next, the filtered PPG signal was change to frequency domain and the power spectrum density of the signal with the frequency within the range of 0.7Hz to 4Hz that represent the HR value range from 42bpm to 240bpm was computed using the Welch's method. The highest frequency of the signal amplitude was multiplied by 60 in order to get the estimated heart rate value. After that, a confidence ratio computational was made because, a high ratio that show PG signal had dominant frequency and therefore providing more confidence that the heart rate estimated was correct. Illustration of the generated signal as a result of the elaborated process can be seen in Figure 4.

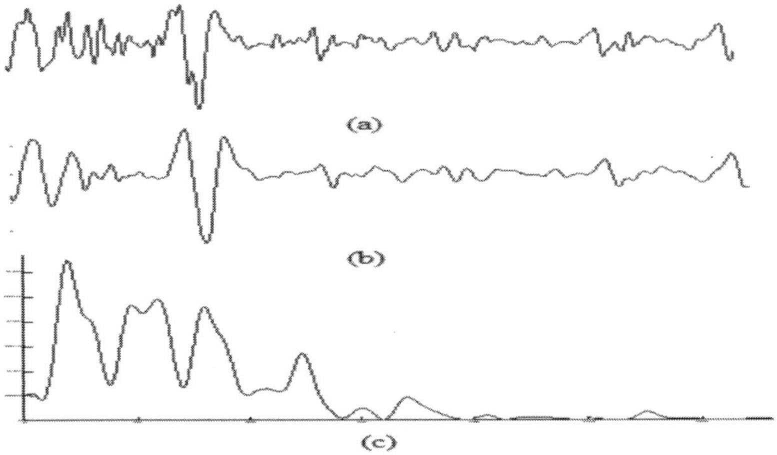

Figure 4. Filtering step. (a) Detrending. (b) Moving Average (c) Pwelch with frequency range from 0.7Hz – 4Hz

2.4. Histogram Analysis

The HR value obtained from a single sequence calculation is still subject to the variation and hence a histogram based analysis is performed. The histogram analysis was made in order to increase the accuracy of the heart rate value estimated. Histogram analysis could be done by repeatedly selecting another pair of random points from ROI, extract the PPG signal from the green channel, apply filter to the PPG signal, compute the confidence ratio from the PPG signal, compute the heart rate value and was added to a histogram. Lastly, the final result of the heart rate estimation value will be obtained. The purpose of using histogram analysis was to obtain more accurate heart rate estimation value from the proposed system.

3. RESULTS AND ANALYSIS

This section, will discuss on the performance of the proposed system of this paper. In the experiment, we conduct the face videos recording in an indoor environment under controlled light condition, the videos were recorded with 1440*1080 pixels resolution and 60 FPS is taken from participant with different distance setting of 1 meter, 3 meters and 5 meters respectively. During the experiment, the subjects were allowed to exhibit normal static motion along the process. There were eight subjects that were involved in this experiment and the subjects were all in different races and skin colours. Figure 5 and 6, shows the experimental set up for the system that was used for the proposed framework.

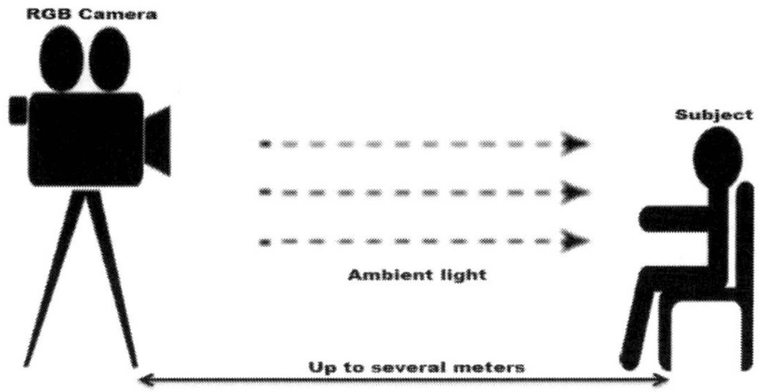

Figure 5. General experiment set up for this project

There were three parts of analyses that conducted to test the performance of the proposed system. The first analysis target was to determine the heart rate value using the system proposed for various distances, the second analysis was to find patches region that will give more accurate heart rate readings and last analysis was to observe the effect of different skin tones to HR estimated. Since the analyses were focusing on heart rate for various distances, the distance for both experiments ranges up until five meters. Quantitative assessment such as accuracy and percent of error are calculated using equation below, where measured value is heart rate obtained from camera and actual value is heart rate obtained by using pulse oximeter (ground truth).

$$Percentage\ of\ Error = \left(\frac{|measured\ value - actual\ value|}{actual\ value}\right) * 100\% \quad (3)$$

$$\text{Accuracy (\%)} = 100\% - \text{Percent of Error} \quad (4)$$

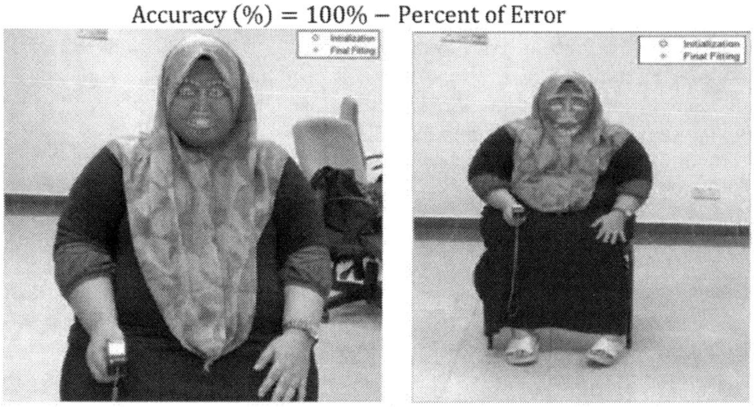

Figure 6. . Experiment setup to determine the HR value with detected facial landmarks for different distances

The analysis is conducted to determine an optimal dependency between distances and face ROI that gives the highest accuracy of the estimated HR value. Previously, like mentioned in the system overview there were four ROI selected for this experiment which are right cheek, center face, whole face and lower face. The entire patch will be used fully in each of the video taken at three different distances as explained above and the result is showcase in Table 1. From the table, for 1 meter reading the most accurate HR was found to be generated from the whole face region with 94.11% accuracy. Clearly this happen because of the region selected was the biggest region compared among the other region.

Larger ROI means that more information about the raw PPG signals could be extracted, hence resulting in more accurate HR reading. Moving to a 2 meter distance, it is found that the most accurate HR was obtained from the center of face area with 90.09% accuracy. Apparently, same as the whole face region, the center part is considered larger part of region and it can generate an accurate HR reading even from a far distance. For a 5 meter case which consider as the farthest distance from the camera, the most accurate HR was also obtained from the whole of the face area with 91.53% accuracy. Again, this was due to the larger region for PPG signal extraction and bigger dimension as mentioned previously that make the obtained PPG to be more accurate.

From the obtained data, for a single person HR reading over a 5 meters range, it can be deduce that the ROI that produce the most inaccurate reading was the patch that placed at the right cheek of the face with an average of 84.11% of accuracy while the optimal ROI is found to be the whole region of the face with an average of 91.86% of accurate HR estimation. One of the reasons that contribute to the inaccuracy of the HR reading is the region size of the ROI which practically contribute to information loss during the signal extraction process. However, generally the entire selected path is capable to produce around 80% of accurate HR estimation in each of the distance variation.

Table 1. Accuracy table for each patch corresponding to the distances

Patch (*ROI*) / Range (m)	Right Cheek	Center of Face	Lower Part of Face	Whole face
	Accuracy (%)	Accuracy (%)	Accuracy (%)	Accuracy (%)
(*HR* accuracy results for near distance)				
1	90.41	92.91	93.34	**94.11**
3	84.87	**90.09**	89.78	89.95
(*HR* accuracy results for farther distance)				
5	77.05	84.69	87.36	**91.53**
Average	**84.11**	**89.23**	**90.16**	**91.86**

Last but not least, the third analysis that was done was regarding the effect of HR estimated corresponding to the skin tones of the subjects. As mentioned previously, the volunteered subjects for this experiment came from different races such as Malay, Chinese and Indian. For the HR result corresponding to the skin tones, was shown in table 2.

Table 2. HR Estimation according to skin tones

Subject	Distance	Ground Truth (*bpm*)	Estimated *HR* with respect to *ROI* (*bpm*)			
			Lower Face	Whole Face	Right Cheek	Face Center
1 (Chinese)	1	91	91	91	94	91
	3	94	64	82	76	86
	5	94	95	95	95	71
2 (Malay)	1	65	67	67	67	67
	3	75	74	74	74	74
	5	75	71	71	71	72
3 (Malay)	1	70	71	71	62	85
	3	65	79	82	76	64
	5	69	79	77	73	79
4 (Chinese)	1	69	69	61	88	70
	3	68	68	73	68	70
	5	70	95	75	79	75
5 (Malay)	1	61	61	61	61	61
	3	56	64	64	67	64
	5	67	79	79	87	79
6 (Malay)	1	67	79	73	71	73
	3	68	72	73	92	91
	5	66	76	77	79	77
7 (Indian)	1	80	71	71	74	74
	3	68	64	71	76	76
	5	75	74	74	97	97
8 (Indian)	1	59	67	62	66	48
	3	59	55	55	69	56
	5	59	65	63	**106**	64

Based on the table 2 above, we can see that the estimated HR value did not stray further away from the original HR reading regardless of whatever skin tones involve. The only estimated HR that gives the most inaccurate result was for the 8th subject for 5m distance. This happened due to the distance of the subject from the camera was to farther apart that probably much information loss occurred during the PPG signal extraction. In addition, the ROI for the reading was at the right cheek, which is the smallest ROI compared to other selected area. This explains a lot regarding the inaccuracy that occurred for the specific HR readings.

Based on table 1, it can be concluded that the HR can be detected up to 5 meters range with 94% accuracy using the whole face region. Based on this analysis, we can conclude that the most optimal ROI with respect to distance for this experiment was the whole face region. This is because, for every distance the whole face area yield the most accurate HR results compared to another face regions selected. Finally, we can conclude that the proposed framework for this system is independent towards the subject skin tones, this system also can work properly despite of it distances variation and lastly the most optimal patch with respect to distance for this project is the whole face area.

4. CONCLUSION AND FUTURE WORKS

This paper investigate about the non-contact HR reading for various distances that correspond to the ROI selected and the effect of skin tones to the estimated HR reading. This paper tackles this problem by analyzing an optimal framework that able to works properly under the mentioned issues. For the first step, presumable set of facial landmark points are estimated by applying cascaded of regression mechanism. Then, the region of interest (ROI) was selected based on the facial landmarks in order to obtain more detailed ROI for better HR accuracy. The location of ROI selected is where non rigid motion is minimal. Temporal photoplethysmograph (PPG) signal extracted from the ROI and is calculated based on the average green pixels intensity. The unwanted signal such as environment illumination signal or motion artifact signal is eliminated by using Independent Component Analysis (ICA) filter. Lastly, the PPG signal is further processed using series of temporal filter to exclude frequencies outside the range of interest prior to estimate the HR.

We can conclude that the system able to works properly, since the accuracy values of the overall performance was above 90% which was an acceptable value. Based on the experiment conducted, we able to determine the most practical ROI were the whole face region, since it gives the most accurate reading with 91.86%. Lastly, we can also conclude that this HR algorithm is not affected by skin colour or illumination variation, based on the analysis that we made. For future work, the system is expected to be able to increase the accuracy of the heart rate reading. Although the current accuracy seems promising with more than 90%, it would really better if we could increase the percentage of the accuracy to be more than 95% for a better performance. Next, the system is also expected to be able to reduce the execution time of the system since the execution time for system proposed was quite slow. Lastly, we would also interested to study the effect of using different types of signal analysis method to obtain more accurate the HR reading. Apart from that, the system is also can be improved by reducing the processing time to make it applicable for real time data HR measurement scenario.

5. References

[1] Scalise L. Non Contact Heart Monitoring. Advances in Electrocardiograms - Methods and Analysis. 2012
[2] Das V, Boothby A, Hwang R, Nguyen T, Lopez J, Lie D. Antenna evaluation of a non-contact vital signs sensor for continuous heart and respiration rate monitoring. 2012 IEEE Topical Conference on Biomedical Wireless Technologies, Networks, and Sensing Systems (BioWireleSS). 2012
[3] Tan H, Qiao D, Li Y. Non-contact heart rate tracking using Doppler radar. 2012 International Conference on Systems and Informatics (ICSAI2012). 2012
[4] Sakamoto T, Imasaka R, Taki H, Sato T, Yoshioka M, Inoue K et al. Feature-based Correlation and Topological Similarity for Interbeat Interval Estimation using Ultra-Wideband Radar. IEEE Transactions on Biomedical Engineering. 2015
[5] Pancera E, Wiesbeck W. Correlation properties of the pulse transmitted by UWB antennas. 2009 International Conference on Electromagnetics in Advanced Applications. 2009
[6] Verkruysse W, Svaasand L, Nelson J. Remote plethysmographic imaging using ambient light. Optics Express. 2008;16(26):21434.
[7] Poh M, McDuff D, Picard R. Advancements in Noncontact, Multiparameter Physiological Measurements Using a Webcam. IEEE Transactions on Biomedical Engineering. 2011;58(1):7-11.
[8] Balakrishnan G, Durand F, Guttag J. Detecting Pulse from Head Motions in Video. 2013 IEEE Conference on Computer Vision and Pattern Recognition. 2013
[9] de Haan G, Jeanne V. Robust Pulse Rate From Chrominance-Based rPPG. IEEE Transactions on Biomedical Engineering. 2013;60(10):2878-2886.

[10] Datcu D, Cidota M, Lukosch S, Rothkrantz L. Noncontact automatic heart rate analysis in visible spectrum by specific face regions. Proceedings of the 14th International Conference on Computer Systems and Technologies - CompSysTech '13. 2013
[11] Li X, Chen J, Zhao G, Pietikainen M. Remote Heart Rate Measurement from Face Videos under Realistic Situations. 2014 IEEE Conference on Computer Vision and Pattern Recognition. 2014
[12] McDuff D, Gontarek S, Picard R. Remote Detection of Photoplethysmographic Systolic and Diastolic Peaks Using a Digital Camera. IEEE Transactions on Biomedical Engineering. 2014;61(12):2948-2954.
[13] Kumar M, Veeraraghavan A, Sabharwal A. DistancePPG: Robust non-contact vital signs monitoring using a camera. Biomedical Optics Express. 2015;6(5):1565.
[14] Tasli H, Gudi A, den Uyl M. Remote PPG based vital sign measurement using adaptive facial regions. 2014 IEEE International Conference on Image Processing (ICIP). 2014
[15] Feng L, Po L, Xu X, Li Y. Motion artifacts suppression for remote imaging photoplethysmography. 2014 19th International Conference on Digital Signal Processing. 2014
[16] Viola P, Jones M. Rapid object detection using a boosted cascade of simple features. Proceedings of the 2001 IEEE Computer Society Conference on Computer Vision and Pattern Recognition CVPR 2001.
[17] Cootes T, Edwards G, Taylor C. Active appearance models. Lecture Notes in Computer Science. 1998;:484-498.
[18] Asthana A, Zafeiriou S, Cheng S, Pantic M. Robust Discriminative Response Map Fitting with Constrained Local Models. 2013 IEEE Conference on Computer Vision and Pattern Recognition. 2013
[19] Sun Y. Motion-compensated noncontact imaging photoplethysmography to monitor cardiorespiratory status during exercise. Journal of Biomedical Optics. 2011;16(7):077010.
[20] Lewandowska M, Rumiński J, Kocejko T. Measuring pulse rate with a webcam — A non-contact method for evaluating cardiac activity. Computer Science and Information Systems (FedCSIS). Szczecin, Poland: IEEE; 2011.
[21] Takano C, Ohta Y. Heart rate measurement based on a time-lapse image. Medical Engineering & Physics. 2007;29(8):853-857.
[22] Lam A, Kuno Y. Robust Heart Rate Measurement from Video Using Select Random Patches. 2015 IEEE International Conference on Computer Vision (ICCV). 2015
[23] Dollar P, Welinder P, Perona P. Cascaded pose regression. 2010 IEEE Computer Society Conference on Computer Vision and Pattern Recognition. 2010
[24] Lempe G, Zaunseder S, Wirthgen T, Zipser S, Malberg H. ROI Selection for Remote Photoplethysmography. Bildverarbeitung für die Medizin 2013. 2013;:99-103.
[25] Pursche T, Krajewski J, Moeller R. Video-based heart rate measurement from human faces. 2012 IEEE International Conference on Consumer Electronics (ICCE). 2012

Acknowledgments

The authors would like to thank to Ministry of Education (MOE) and Universiti Tun Hussein onn Malaysia (UTHM) for supporting this research under Fundamental Research Grant Scheme (Vot. no. 1582). This paper was also partly sponsored by the Centre for Graduate Studies UTHM.

Effect of grinding period on physical properties of modified bitumen using palm oil fuel ash (POFA)

Amira Maryam Md Sukaimi, Raja Nurul Afiqah Raja Zulkefli, Haryati Yaacob, Ramadhansyah Putra Jaya, Norhidayah Abdul Hassan, Mohd Naqiudin Mohd Warid, Mohd Khairul Idham Mohd Satar

Faculty of Civil Engineering, Department of Geotechnics and Transportation, UniversitiTeknologi Malaysia, 81310 UTM Johor Bahru, Malaysia.

E-mail: amira.maryam9@gmail.com, haryatiyaacob@utm.my

Keywords: POFA, microstructural, modified bitumen, penetration test, softening point

Abstract. Malaysia is having rapid economic development in recent years, and this has placed high performance demand on its highway infrastructure. The increased in number of traffic has led to severe pavement deterioration. Due to this, better and stronger pavement materials are needed in order to cater the increasing number of axle loading. Palm oil fuel ash (POFA) which comes from the ash of burning mesocarp of the palm oil fruitlets has been used as a modifier to improve the properties of bitumen. POFA is a biomass that had pozzolanic properties (siliceous material). By using POFA in bitumen, it will promote sustainability and lesser landfill to dump this waste. The objective of this research is to evaluate the effect of grinding period of POFA and different percentage of POFA towards physical properties of modified bitumen. A few sets of samples of POFA modified bitumen at different grinding time and POFA content have been prepared and tested with penetration test and softening point test. The microstructural characteristics of POFA were also investigated by using Field Emission Scanning Electron Microscope (FESEM). From the result, it shows that grinding time and POFA content does not influence the physical performance of the bitumen.

1. Introduction

Deterioration of road structure happened more these days due to the increased of traffic volume and low efficiency on the road maintenance process. To increase on the long term durability of the pavement, the bituminous layers should be improved with regard to performance properties, such as resistance to permanent deformation, fatigue, and aging. Study on the application and usage of bitumen modifier using waste product in order to increase the quality of the bitumen has become current focus in bitumen modification [1-4].

The American Society for Testing and Materials (ASTM) defines bitumen as a class of black color (solid, semisolid or viscous) cementitious substance which occurs in nature or obtains in petroleum processing, composed principally of high molecular weight hydrocarbons of which bitumen. Bitumen has certain weaknesses whereby it can 'flow' at high temperature, leading to pavement rutting, and it can fracture at low temperature, leading to pavement cracking, the adhesion between bitumen and aggregate breaks down under a combination of ageing and water attack. The inherent weaknesses of

Content from this work may be used under the terms of the Creative Commons Attribution 3.0 licence. Any further distribution of this work must maintain attribution to the author(s) and the title of the work, journal citation and DOI.
Published under licence by IOP Publishing Ltd

bitumen cause highway maintenance works quite difficult and expensive. Bitumen is shear-sensitive and hence making it prone to shove and rut at higher temperatures. On the other hand, at low temperatures it becomes brittle and tends to undergo fracture cracking and potholing. Cracking is generally reduced by making the asphalt binder either less viscous or less temperature susceptible. Reducing low temperature viscosity leads to soft bitumen and the mechanical properties of mixture deteriorate. According to McGennis [5], modified bitumen is produced for the purpose of altering and improving the properties of the bitumen to enhance the long term performance of pavements. While the modifier may affect many properties, the majority of modifiers are used in an attempt to reduce temperature dependency, oxidative hardening of bitumen, and the moisture susceptibility of asphalt mixtures [5, 6]

Malaysia is the world's leading producer and exporter of palm oil and palm oil products with a reputation for quality and reliability. In a report published by the Department of Environmental, Ministry of Science, Technology and the Environment Malaysia [7], Malaysia Palm Oil production in the year 2009 was 17.56 million tons. From 1kg of oil palm fruitlets can result in 0.34kg of crude palm oil (CPO) and 0.1kg Kernel Oil or in total results in 0.44kg of oil. This means that 1kg of oil palm fruitlets results in 0.44kg of the mesocarp waste. If the product of palm oil in year 2009 was 17.56 million tons (44% of the oil palm fruitlets), it is means that there are 7.73 million tons of the waste of mesocarp in year 2009. Salihuddin [8] is the pioneer researcher to mark this as Palm Oil Fuel Ash or POFA. Palm Oil Fuel Ash (POFA) is a by-product produced after burning of both palm oil husk and palm oil shell as fuel in the boiler at palm oil mill. Generally after combustion about 5% POFA by weight of solid wastes is produced. In other words, the physical characteristics of POFA are very much influenced by the operating system in palm oil factory. In practice, POFA produced in Malaysia Palm Oil Mill is dumped as a waste without any profitable return [9].

Previous research had also proved that bitumen modification can soften binders and mixtures at high temperature to minimize rutting and reduce the detrimental effects of load induced moisture damages, improve fatigue resistance, particularly in environments where higher strains are imposed on the asphalt concrete mixture [10-12]. By using modification bitumen will also improve bitumen-aggregate bonding to reduce stripping, improve pavement durability with an accompanying net reduction in life cycle costs, and improve overall performance as viewed by the highway user. POFA content is most feasible to be used as bitumen modifier and based on the penetration test result of 5% OPFA-MB (Oil Palm Fuel Ash bitumen modifier) can be graded as binder penetration 60/70 PEN [13] because it is resistant to high temperature rutting and resistant to low temperature thermal cracking. The addition of POFA was also found to increase the Penetration Index (PI) of the bitumen even when subjected to aging [14]. POFA was also found to improve the performance of modified asphalt concrete in terms of elasticity modulus, stiffness, indirect tensile strength and creep resistance [15, 16]

2. Material and method

The purpose of this study was to observe the effect of adding palm oil fuel ash in bitumen and act as bitumen modifier by using different percentages of POFA. The testing procedures were based on The American Society for Testing and Material (ASTM) standard guide for laboratory works. POFA had been grinded by using Los Angeles Abrasion machine at different grinding hours (1, 2, 3, and 4 hours). POFA modified bitumen were prepared by mixing different percentage of POFA (at different grinding hour) into 80/100 penetration bitumen. Physical tests, namely penetration tests (ASTM D5) and softening point test (ASTM D36) had been conducted on each samples.

2.1 Sieve Analysis

Sieve analysis had been carried out in order to get the POFA that passing 75μm sieving size. These POFA were then grinded at different grinding hours (1, 2, 3, and 4 hours).

2.2 Modification process of bitumen

Four different percentages of POFA has been chosen based on previous findings by Rusbintardjo G. (2011), namely 0% (control), 5%, 6%, and 7% by the total weight of the bitumen content. POFA has been mixed with bitumen 80/100 penetration at a mixing temperature of 160°C, for about of 60 minutes and stirring speed of 800rpm. POFA was poured into 400g of bitumen at room temperature before mixing. For every 100g of bitumen, an equal amount of POFA was added then the mixtures of bitumen-POFA were heated in the oven just to make it as liquid later then place it on a hot plate under the drive shaft rotor propeller of the mixer at the speed of 800rpm. This procedure was conducted so that POFA was thoroughly mixed with bitumen.

2.3 Field Emission Scanning Electron Microscope (FESEM)
After grinding process of POFA, samples were examined using Field Emission Scanning Electron Microscope (FESEM) to check its morphological structure. FESEM which is available at University Lab Management Unit, Universiti Teknologi Malaysia are able to visualize a very small object and give topography information to its user.

3. Result and discussion

3.1 Morphological structure of POFA
The morphology of POFA with grinding hour of 0, 2 and 4 has been observed under FESEM and shown in Figure 1. Under the magnification of 1000, it can be seen that POFA is generally has irregular shape with porous and spongy texture for all samples. It was expected that, further grinding of POFA for 4 hours crushed the particle into smaller size, reduce porosity texture and it has more irregular shape compared to ungrounded POFA. The finding obtained is similar to previous study by other researchers [17, 18].

Figure 1. Microstructural images of POFA (a) Microstructure of POFA passing 75μm (0hour) (b) Microstructure of POFA after grinding for 2 hours (c) Microstructure of POFA after grinding for 4 hours

IOP Conf. Series: Journal of Physics: Conf. Series **1049** (2018) 012004 doi:10.1088/1742-6596/1049/1/012004

3.2 Penetration test

Figure 2 shows the results of penetration test for different percentages of POFA which are 5%, 6% and 7%. On each percentage, the effect of grinding time has been further looked. Generally from this Figure 2, it can be seen that by grinding the POFA from 1-4 hour, the penetration values has decreased slightly showing that the specimens has become harder. However for 5% and 6 % of POFA, with the increased of grinding period from 1 to 3 hour the penetration has shown an increment from 7.24 to 8.20 and 7.01 to 8.37 respectively. When the grinding time increased to 4 hours, the penetration has reduced back to 7.21 and 7.43 respectively. In addition, analysis of variance (ANOVA) test has been run in order to determine the significant of effect of grinding time towards penetration value.

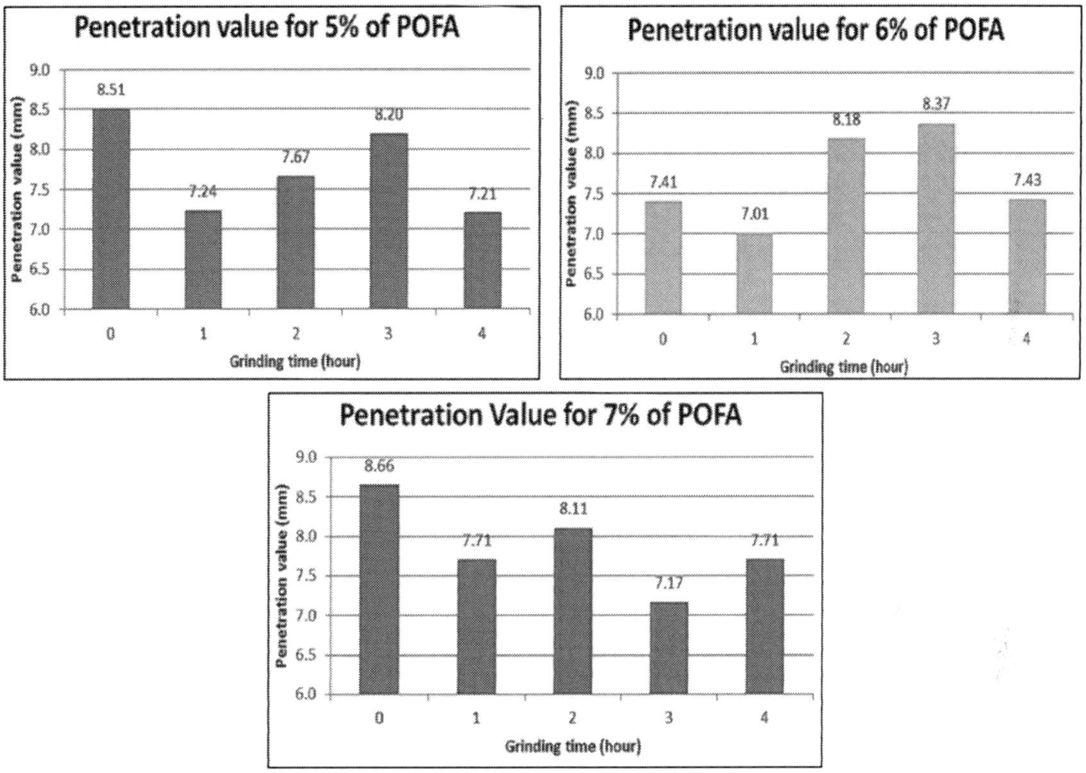

Figure 2. Graph of penetration value based on different percentage of POFA

Table 1 shows ANOVA of Penetration Value by different percentage of POFA to grinding hour, the P value are higher than the P critical which is 0.05. The hypothesis on the different hour of grinding will give smaller penetration value was rejected because based on this ANOVA, it shows it has no significant difference between each of the grinding time.

Table 1. ANOVA of Penetration Value by different percentage of POFA to grinding hour

Variables	P_{value}	P_{crit}	Condition	Significant	Remarks
0 hour of grinding	0.78	0.05	$P > P_{crit}$	No	Reject the hypothesis
1 hour of grinding	0.40	0.05	$P > P_{crit}$	No	Reject the hypothesis
2 hour of grinding	0.76	0.05	$P > P_{crit}$	No	Reject the hypothesis
3 hour of grinding	0.74	0.05	$P > P_{crit}$	No	Reject the hypothesis
4 hour of grinding	0.40	0.05	$P > P_{crit}$	No	Reject the hypothesis

Figure 3 shows the results of the penetration test for different grinding times which are 0 hour, 1 hour, 2 hours, 3 hours and 4 hours. At each grinding time, the effect of different percentage of POFA has been further looked and compared with control sample (without POFA). In general, the graphs show inconsistent result with some samples of POFA modified bitumen having higher penetration value compared to the control sample.

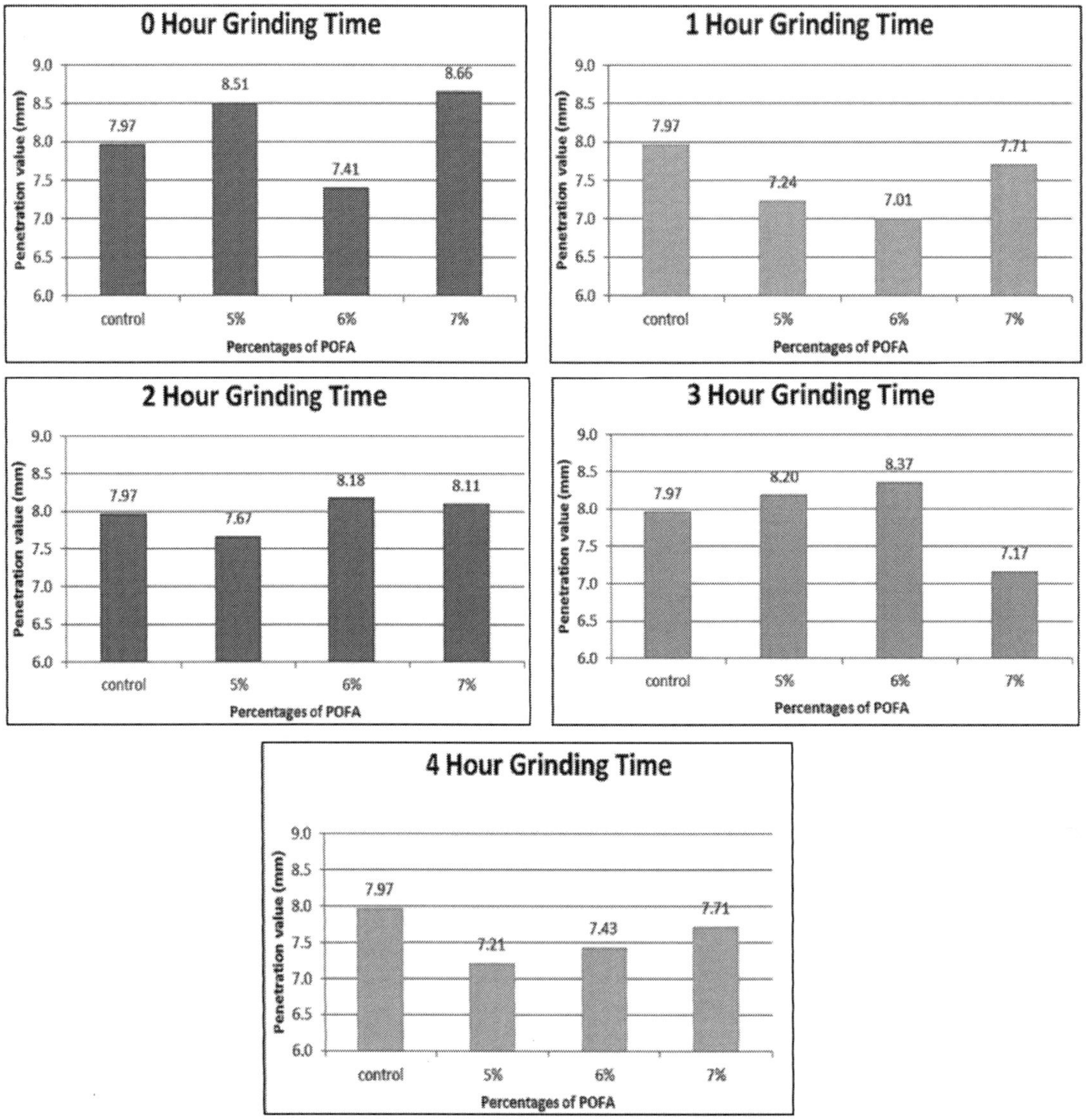

Figure 3. Graph of penetration value based on different grinding time on POFA

Further analysis has been done by using POFA, where from Table 2 shows that ANOVA results of each percentages of POFA at certain grinding hour are not having the significant change with each other. The hypothesis is rejected because the samples with different percentages of POFA modified bitumen are not having significant effect with the controlled sample.

Table 2. ANOVA of Penetration Value by grinding period to different percentage of POFA

Variables	P_{value}	P_{crit}	Condition	Significant	Remarks
5% of POFA	0.36	0.05	$P > P_{crit}$	No	Reject the hypothesis
6% of POFA	0.68	0.05	$P > P_{crit}$	No	Reject the hypothesis
7% of POFA	0.19	0.05	$P > P_{crit}$	No	Reject the hypothesis

3.3 Softening Point Test

Figure 4 shows the results of softening point for different percentages of POFA which are 5%, 6% and 7%. For each percentage of POFA the effect of grinding time (0, 1, 2, 3 and 4 hours) towards softening point value has been further looked. At 5% POFA, it was recorded that softening point value increased from 39°C to 41.5°C with an increase of grinding time from 0 to 3 hours. When the grinding time increased to 4 hours, the softening point has reduced back to 40°C. However, for 6% of POFA, the temperature maintain at 40°C except at 3 hour grinding time, the value increased to 42°C and reduced back to 40°C at 4 hours grinding time. For 7% POFA, inconsistent value has been recorded. Further investigation has been done by conducting the ANOVA test analysis.

Based on Table 3 ANOVA of softening point value by different percentage of POFA to grinding hour it shows all of the grinding times are not have the significant change. The hypothesis on the different grinding time will give higher temperature on softening point test is rejected because it has no significance value at every temperature.

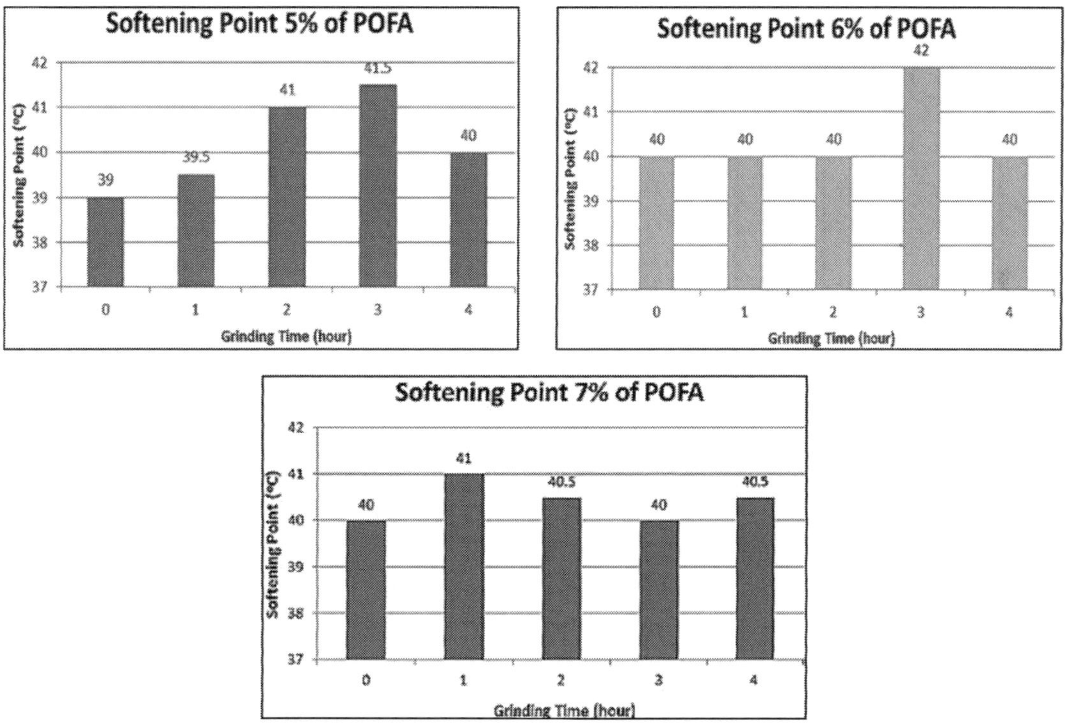

Figure 4. Graph of softening point value based on different percentage of POFA

Figure 3 shows the results of the penetration test for different grinding times which are 0 hour, 1 hour, 2 hours, 3 hours and 4 hours. At each grinding time, the effect of different percentage of POFA has been further looked and compared with control sample (without POFA). In general, the graphs show inconsistent result with some samples of POFA modified bitumen having higher penetration value compared to the control sample.

Figure 5 shows the results of the softening point test at different percentage of POFA. For 1, 3 and 4 hours grinding time, there is a slight increase of softening point values with an increase of POFA content, compared to control sample. However, similar to penetration test, inconsistent results were found in softening point values in certain cases. Therefore ANOVA test has been done in order to summarize the significant of POFA towards softening point value.

Table 3. ANOVA of Softening Point Value by different percentage of POFA to grinding hour

Variables	P_{value}	P_{crit}	Condition	Significant	Remarks
0 hour of grinding	0.89	0.05	$P > P_{crit}$	No	Reject the hypothesis
1 hour of grinding	0.62	0.05	$P > P_{crit}$	No	Reject the hypothesis
2 hour of grinding	0.61	0.05	$P > P_{crit}$	No	Reject the hypothesis
3 hour of grinding	0.61	0.05	$P > P_{crit}$	No	Reject the hypothesis
4 hour of grinding	0.46	0.05	$P > P_{crit}$	No	Reject the hypothesis

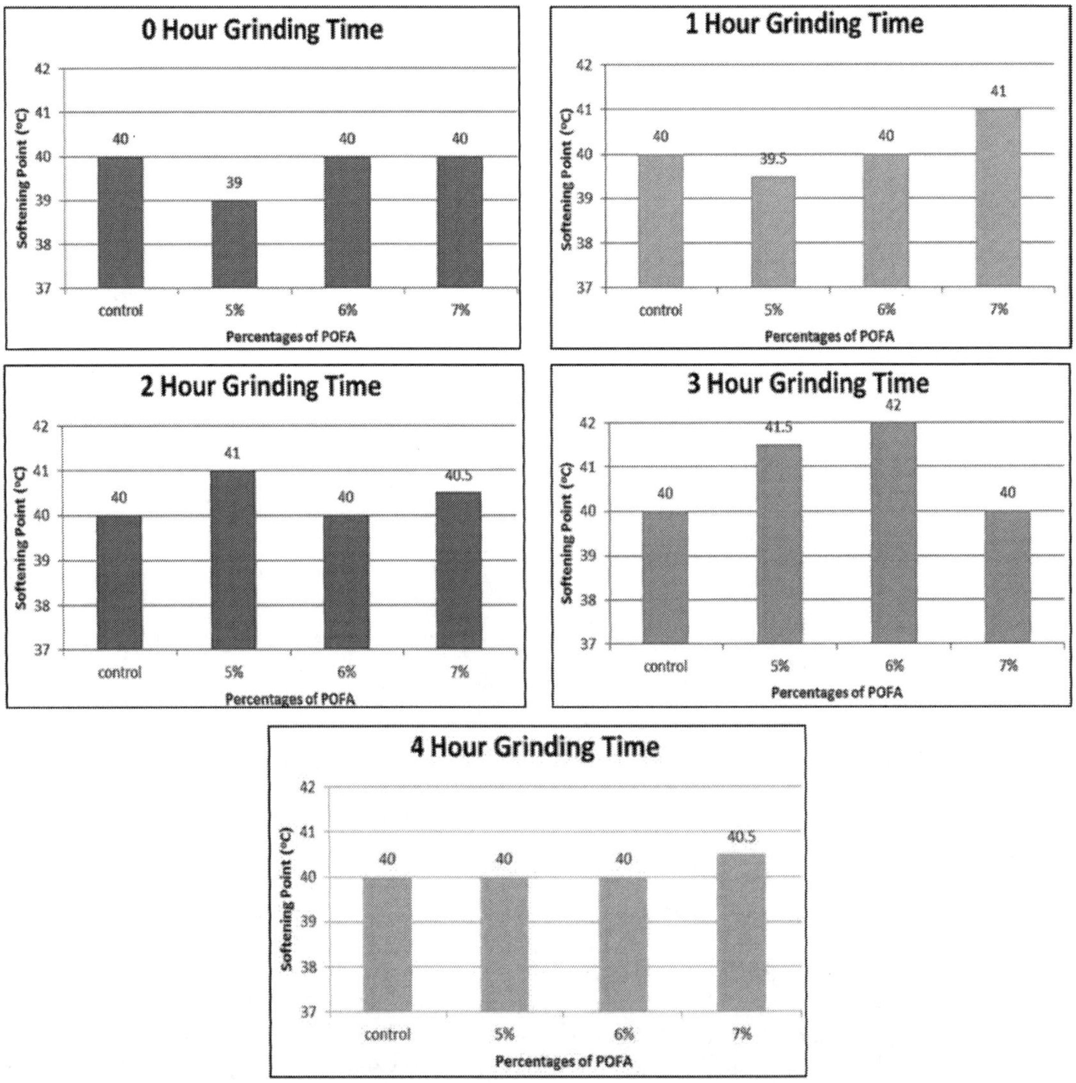

Figure 5. Graph of softening point value based on different grinding time of POFA

From Table 4 ANOVA of Softening Point Value by grinding period to different percentage of it can be seen that all of the percentages have P value that are bigger than the P critical. Therefore, the hypothesis was rejected because each of the percentages of POFA was not significance to controlled sample.

Table 4. ANOVA of Softening Point Value by grinding period to different percentage of POFA

Variables	P_{value}	P_{crit}	Condition	Significant	Remarks
5% of POFA	0.24	0.05	$P > P_{crit}$	No	Reject the hypothesis
6% of POFA	0.43	0.05	$P > P_{crit}$	No	Reject the hypothesis
7% of POFA	0.70	0.05	$P > P_{crit}$	No	Reject the hypothesis

3.4 Penetration Index
Based on Table 5, the penetration index value was found to be within the range of -3.1 to -2. Penetration index less than -2, means that the bitumen has high temperature susceptibility. Asphalt mixture with this high temperature susceptibility bitumen will face with permanent deformation at high temperature, and it will easily crack at low temperature.

Table 5. Penetration Index according to the percentages of POFA and the grinding time.

Penetration Index	Grinding time (hour)					
	Control sample	0	1	2	3	4
5%		-3.1	-3	-2.5	-2.20	-3
6%	-2.6	-3.00	-2.97	-2.50	-2.00	-3.05
7%		-2.50	-3.00	-2.8	-3.00	-3.00

4. Conclusion
From the results obtained, the following conclusions can be drawn:

(i) From softening point and penetration test results, addition of POFA into bitumen have either increased or decreased the value respectively for certain samples.
(ii) However, ANOVA analysis shows that for all conditions, the p value was found to be more than 0.05. It can be conclude here that the grinding time and percentage of POFA does not contribute to any significant effect towards penetration and softening point result.

5. Acknowledgements
The authors would like to extend their gratitude to Malaysia Ministry of Higher Education (MOHE) and the Universiti Teknologi Malaysia Research grants (Q.J130000.2522.17H68) for the financial support in this research project.

References

[1] Ting T L, Jaya R P, Hassan N A, Yaacob H, Jayanti D S, Ariffin M A M 2016 A review of chemical and physical properties of coconut shell in asphalt mixture *J. Teknol.* **78** 85-89
[2] Romastarika R, Jaya R P, Yaacob H, Nazri F M, Agussabti, Ichwana and Jayanti D S 2017 Effect of black rice husk ash on the physical and rheological properties of bitumen *Int. Conf.*

on Applied Physics and Engineering (Pulau Pinang, Malaysia, 2-3 November 2016) (*AIP Conference Proceedings* vol 1875) ed Jumidali *et al* (AIP Publishing) pp 030012-1-8

[3] Yaacob H, Ali Mughal M, Jaya R P, Hainin M R, Jayanti D S, Che Wan C N 2016 Rheological properties of styrene butadiene rubber modified bitumen binder *J. Teknol.* **78** 121–26

[4] Farouk A I B, Hassan N A, Mahmud M Z H, Mirza J, Jaya RP, Hainin M R, Yaacob H and Yusoff N I M 2017 Effects of mixture design variables on rubber–bitumen interaction: properties of dry mixed rubberized asphalt mixture *Mater. Struct.* **50** 12

[5] McGennis R B, Shuler S and Bahia H U 1994 *Background Of Superpave Asphalt Binder Test Methods FHWA, Report No. FHWA-SA-94-069*

[6] Read J and Whiteoak D 2003 *The Shell Bitumen Handbook* (London: Thomas Telford) p 64

[7] Department of Environmental, Ministry of Science, Technology and the Environment Malaysia 1999 Industrial process and the environment of crude palm oil industry (Kuala Lumpur: Aslita Sdn. Bhd)

[8] Salihuddin Radin Sumadi 1993 *Relationship between Engineering Properties and Microstructural Characteristics of Mortar containing Agricultural Ash* Doctor Philosophy, Universiti Teknologi Malaysia, Skudai.

[9] Salihuddin Radin Sumadi and Mohd Warid Hussin 1995 Palm Oil Fuel Ash (POFA) as a future partial cement replacement material in housing construction *J. Ferro.* **25** 25-34

[10] King G, King H, Pavlovich R D, Epps A L and Kandhal P S 1999 Additives in asphalt *J. Assoc. Asph. Pav. Tech.* **68** 32-69

[11] Abdullah M E, Kader S A A, Jaya R P, Yaacob H, Hassan, N A and Wan C N C 2017 Effect of Waste Plastic as Bitumen Modified in Asphalt Mixture *Int. Symp. on Civil and Environmental Engineering (Melaka, Malaysia, 5-6 December 2016) (MATEC Web Conference* vol 103*)* ed Zainorizuan *et al* (EDP Sciences) pp 09018-1-7

[12] Rusbintardjo G, Hainin M R and Yusoff N I M 2013 Fundamental and rheological properties of oil palm fruit ash modified bitumen *J. Constr. Build. Mater.* **49** 702-11

[13] Rusbintardjo G 2011 *Bitumen Modification Using Oil Palm Fruit Ash for Stone Mastic Asphalt* Doctor Philosophy, Universiti Teknologi Malaysia, Skudai.

[14] Hainin M R, Jaya R P, Ali Akbar N A, Jayanti D S and Yusoff N I M 2014 Influence of palm oil fuel ash as a modifier on bitumen to improve aging resistance *J. Eng. Res.* **2(1)** 34-46

[15] Borhan M N, Ismail A and Rahmat R A 2010 Evaluation of palm oil fuel ash (POFA) on asphalt mixtures *Aust. J. Basic Appl. Sci.* **4** 5456-63

[16] Maleka A M, Alkali I A and Jaya R P 2014 The indirect tensile strength of palm oil fuel ash (POFA) modified asphaltic concrete *Appl. Mech. Mater.* **587-589** 1270-75

[17] Tangchirapat W, Jaturapitakkul C and Chindaprasirt P 2009 Use of palm oil fuel ash as a supplementary cementitious material for producing high-strength concrete *Const. Build. Mater.* **23** 2641-46

[18] Zarina Y, Al Bakri A M M, Kamarudin H, Khairul Nizar I and Rafiza A R 2013 Review on the various ash from palm oil waste as geopolymer material *Rev. Adv. Mater. Sci.* **34** 37-43

IOP Conf. Series: Journal of Physics: Conf. Series **1049** (2018) 012005 doi:10.1088/1742-6596/1049/1/012005

Performance Analysis between Quantum Computers and Silicon Computers: A Preliminary Investigation

M. Aisar M.M.I[1], S.S.M.Fauzi*[,1], H. Baharin[2], W.A.W.M. Sobri[1], A.J. Suali[1], R.A.JM. Gining[1], M.N.F.Jamaluddin[1]

[1]Faculty of Computer and Mathematical Sciences, Universiti Teknologi MARA, Perlis Campus, 02600 Arau, Perlis, Malaysia.
[2]Institute of Visual Informatics, Universiti Kebangsaan Malaysia, Malaysia.

[*,1]E-mail: shukorsanim@uitm.edu.my

Abstract. The purpose of this study was to perform a preliminary investigation on the performance between a high-performance silicon computer and a quantum computer in terms of programming control statements. The study hypothesizes that the quantum computer would outperform the silicon computer in terms of processing performance. There were three types of experimental performance analysis scenarios conducted under the context of basic programming statements which are the '*for loop*', '*while*', and, '*if else*' control statements. To obtain the average of the completion times, the experiments were repeated five times. The obtained results suggest that the silicon computers outperformed the quantum processing capabilities.

1. Introduction

Computers have progressed at an increasingly fast rate over the recent years. To think that back in the early 90s, the ENIAC machine, developed by John W. Mauchly and J. Presper Eckert at the University of Pennsylvania was the first major computer that operated on vacuum tubes. These days however, human-kind is employing silicon computers with transistors for a multitude of tasks. Furthermore, in certain things, an even more state-of-the-art computer which is known as the quantum computer is beginning to be utilized.

Silicon computers have and still is a part of our lives from the late 90s. We use them to simplify so many things such as calculations, scheduling, and planning. There are several classes such as personal computers, workstations, minicomputers, mainframes, and supercomputers [1]. Their processing capabilities differ based on their classes and that brings about a point of concern. Moore's Law states that the processing power of computers will double every two years and carry on indefinitely [2]. There will be a breaking point for this and that is why we need preparations to move on to better and improved technology that can compensate for this growth which is the '*Quantum Computer*'.

[*,1] Corresponding author: shukorsanim@uitm.edu.my.

Content from this work may be used under the terms of the Creative Commons Attribution 3.0 licence. Any further distribution of this work must maintain attribution to the author(s) and the title of the work, journal citation and DOI.
Published under licence by IOP Publishing Ltd

Quantum computers are the next step in computer evolution [3]. The technology that will one day step-up to replace our silicon computers in terms of performance. Unfortunately, they are still in development. However, D-Wave Systems has released a flagship model known as D-Wave Two or Vesuvius [4]. Regrettably, the price tag is far beyond the capabilities of most people and even if they are purchased, there is no definitive use for them albeit from solving relatively complex algorithms. This study aimed to do a performance analysis on how a quantum computer and silicon computer initiate simple algorithms involving control statements. There were several tests conducted with the various control statements of programming to differentiate the completion times between the computer generations.

This paper is organized to several sections. Our introductory sections discussed in brief about silicon and quantum computing. It then follows by motivation of the study. Next, method used in this study is discussed. It then follows by the results and discussion. Finally, we conclude our paper.

2. Motivation

Quantum computers are such powerhouses of the computer world [3]. However, they are not getting the attention they deserve. What this means is that development on quantum computers are not a priority compared to the furthering of the current technology of the silicon computer.

While it is relatively true that if it is not broke, do not fix it. There comes a time when change is necessary. Just as we have stepped away from vacuum tube computers. Now comes the time where we as humans must begin the journey away from silicon computers. This however is not a journey to be undertaken lightly as quantum computers are not to be associated with how we understand computers today in their structural entirety.

At the present moment, quantum computers have a major downside which is mainly focused around instability because even minor interference can interfere with its processes as calculations are taking place at the quantum level. To counter this, most experimental system have to rely on absolute zero cooling as well as heavy shielding. This is both monumentally expensive and inconvenient to work with.

The study conducted was for the purpose of exposing the processing capabilities of quantum computers within the context of control statements algorithm when compared to a silicon computer. Thus, the main objective of this study was to analyze the effects of executing various control statement algorithms experiments between a quantum computer and silicon computer.

3. Method

Two phases involved in this study includes experimental design and, testing and evaluation.

3.1. Experimental Design Phase

Experimental design was the first phase in this study. Within this phase, the programming algorithm guidelines for the experiment were made by standardizing the algorithms which were executed on both the quantum and silicon computers. In addition, the experiment environment was then prepared by utilizing similarly stable control statement (*for, while and if-else*) algorithm which would then be executed through QScript and C.

3.2. Testing and Evaluation Phase

Testing and evaluation was the second phase of this study. During this phase, the prepared algorithms from the experimental design phase were executed on both the quantum and silicon computer programming environments. For the quantum computer, the program algorithms were executed online via *Qscript* on the "*Google Quantum Playground*", while on the silicon computer, it was executed via *C-Free 5.0* including an optional online *C* compiler. The experiments were repeated several times for all three experiments on both machines. During the execution of the experiments, the results, which were the completion times were recorded. Every algorithm was executed a minimum of five to any amount of times as deemed necessary to account for random inconsistencies which may occur.

4. Results

This section details the results throughout the experimentation phase of the study. Data from all three experiment mediums which include the physical silicon computer and the quantum computer were illustrated.

4.1. For loop experiment

Figure 1 and Figure 2 are both results from the '*for loop*' experiments conducted.
Results from Figure 1 suggest the silicon computer complete each cycle with a difference of about three seconds due to internet latency. The quantum computer on the other hand provided erratic results from the experiment which ranged from the fastest completion of 14.83 minutes, to the slowest time of 21.25 minutes. The quantum machine completed the experiment in terms of minutes.

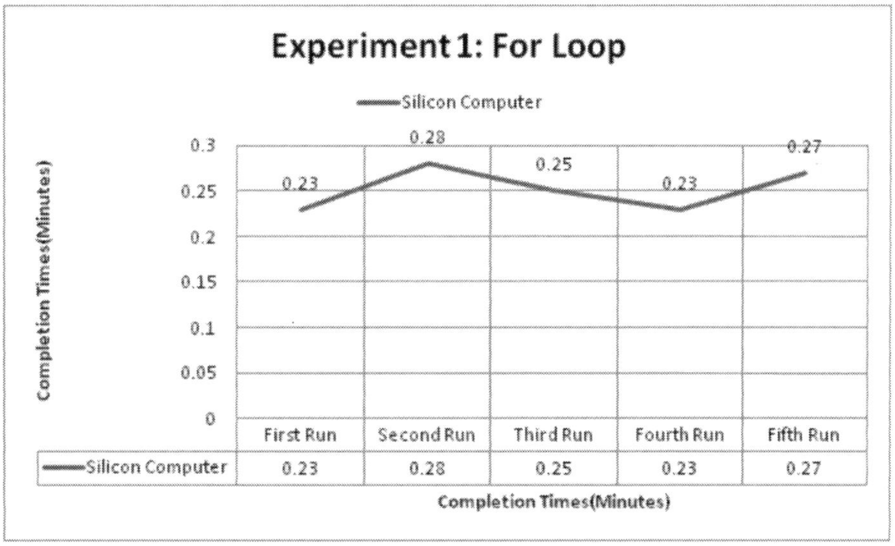

Figure 1: Physical Computer '*For Loop*' Experiment

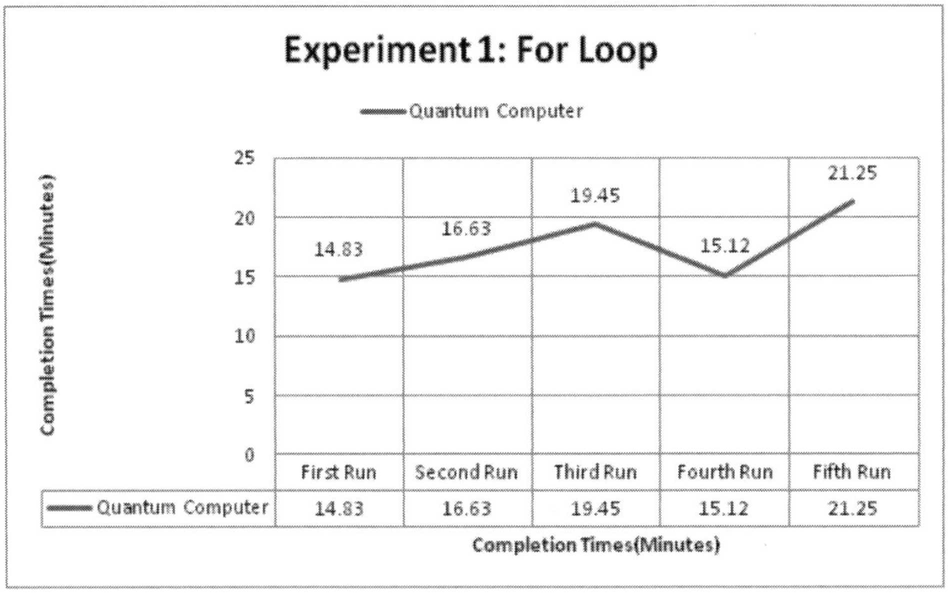

Figure 2: Quantum Computer 'For Loop' Experiment

4.2. While Experiment

Figure 3 and Figure 4 are results from the '*while*' experiment, the second that was conducted. The results from the silicon computers were relatively similar to the first '*for loop*' experiment in completion times with a difference of about three seconds as well between them most likely due to internet latency. Unlike the first experiment, there are no anomalies present in the completion times of the silicon machines. The silicon machines completed the experiment in terms of seconds.

The quantum computer on the second experiment also produced sluggish completions times of 14.1 minutes, to the slowest time of 17.85 minutes. The completion times are faster on average when comparing to the quantum computer results of the first experiment. This is most likely due to the '*while*' experiment not executing the extra processing requirements of the layered approach of the '*for loop*' whereby there are loops within loops. The quantum machine completed the experiment in terms of minutes.

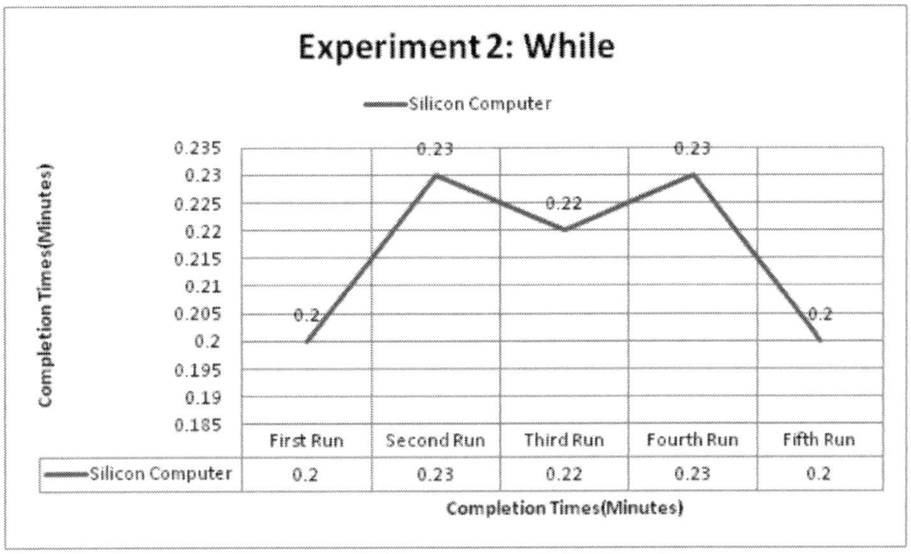

Figure 3: Physical Computer '*While*' Experiment

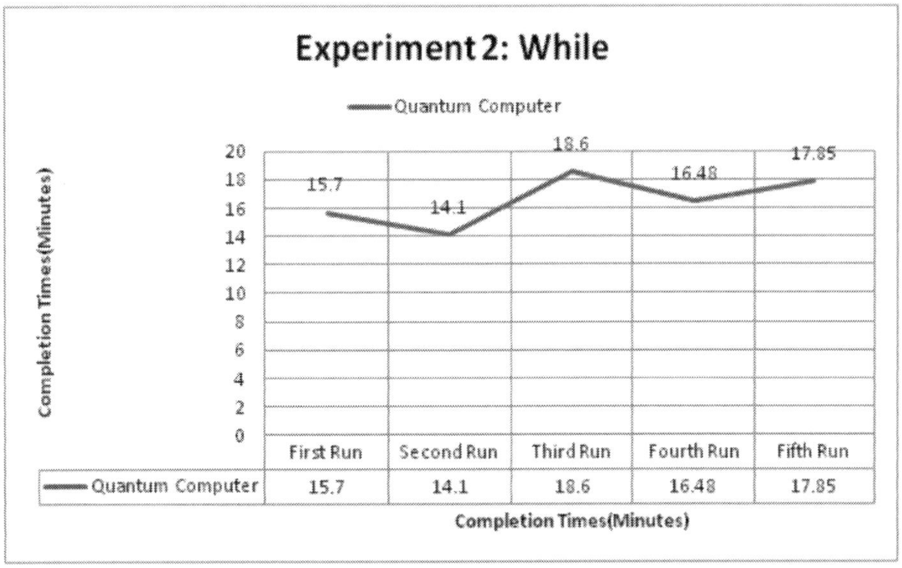

Figure 4: Quantum Computer 'While' Experiment

4.3. If-else Experiment

Figure 5 and Figure 6 are results from the '*if else*' experiment. The completion times of the physical silicon computer is relatively similar to the second experiment in average.

The quantum computer had the slowest average completion time out of all three experiments with the fastest completion time of this experiment being 16.35 minutes and the slowest being 20.45 minutes. The results are unexpected partly due to the physical silicon machine displaying an average completion time similar to the second experiment albeit having the slowest silicon machine completion times. These could be due to unexpected interference with the Internet connection during the course of this experiment. The quantum machine completed the experiment in terms of minutes.

IOP Conf. Series: Journal of Physics: Conf. Series **1049** (2018) 012005 doi:10.1088/1742-6596/1049/1/012005

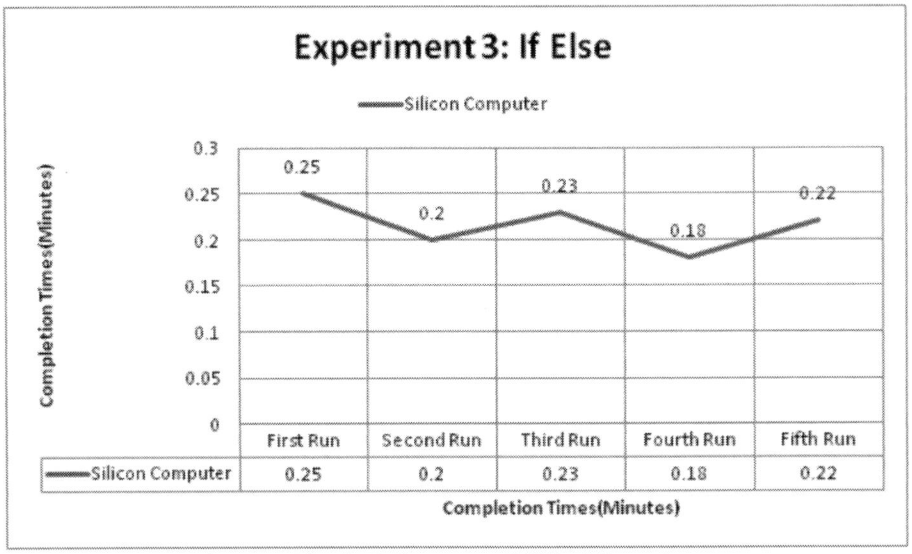

Figure 5: Physical Computer 'If Else' Experiment

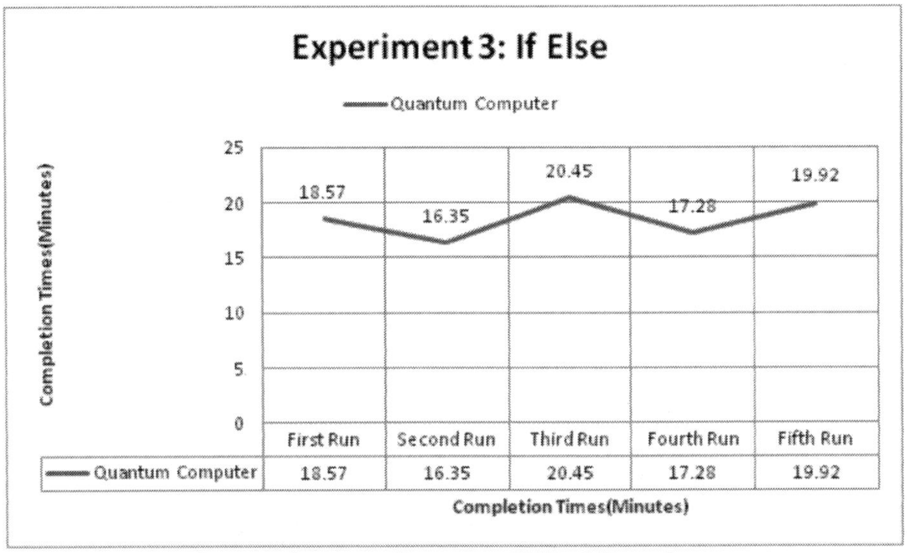

Figure 6: Quantum Computer 'If Else' Experiment

5. Discussion

All three of these experiments were conducted with both a physical silicon computer environment as well as a virtual quantum computer environment. The virtual capability of the quantum machine is stated at 22 *qubits* of processing power with about 2.2 *qubits* best effort basis per user while under heavy usage. The physical capability of the silicon machine on the other hand, is running at a processor with the following statistic: *Intel® Core™ i7-4720HQ CPU @ 2.60GHz*. Based on the bare minimum processing capability afforded by the quantum machine when compared to physical silicon machine, it was assumed during the course of the study that the quantum machine would outperform the silicon machines in completion times for the experiments but the results of this study proved otherwise.

IOP Conf. Series: Journal of Physics: Conf. Series **1049** (2018) 012005 doi:10.1088/1742-6596/1049/1/012005

Physical silicon machines were able to compile, execute, and complete the simple control statement algorithms within terms of seconds with a minor difference in completion times between three to four seconds between the two machines most likely due to latency between executing the program algorithms and displaying the results. The results from the silicon machines were in terms of seconds. The average completion times between the three experiments were between 11 seconds to 20 seconds from physical environments of the silicon machines.

The quantum machine on the other hand was only available in a virtual-state and was also able to compile, execute and complete the simple control statement algorithms. The results from the quantum machine were in terms of minutes. The average completion times between the three experiments ranged between 14 minutes to 21 minutes overall.

Quantum computing itself is a relatively new technology which has recently picked up the pace within the recent years. On paper, the hardware specifications claim to outperform silicon computing significantly. Research previously conducted on quantum computing such as those discussed in chapter two of this study clearly portray the paper specifications are not merely just for show. However, this study has not been able to remotely achieve any significant evidence of quantum computing superiority in terms of executing simple control statement algorithms.

There may be a plethora of variables which had influenced the end-results of the study but it comes down to the availability of the quantum computer itself. This study was conducted without a physical quantum computer and substituting the variable with a virtual quantum computer hosted publically by Google which is known as, *Google's Quantum Playground*. The latency of the Internet most definitely affected the results and the availability of only a single compiler executing *QScript* was also likely to influence the experiments. Experiments previously conducted to analyse the performance of quantum computers may have utilized entirely difference compilers built to specifically execute and complete algorithms of specific natures. *QScript* on the other hand was just a general compiler offered to the public to experiment with the limitations of *Google's Quantum Playground*.

The additional inclusion of a stop watch was also needed as to properly record the results of executing the *QScript* algorithms since there was no time comparison functions during the course of this study. Nevertheless, *Google's Quantum Playground* improves as days goes by and it may most likely portray significantly better results in the near future as it receives updates to both the architecture of *QScript* and the minimum *qubit* limitations per user on best effort basis. However, as of now, it may be better to utilize a physical quantum machine but the steep pricing on physical units may be deterrence to most people.

6. Limitation of the Study

Throughout the entirety of the study, there were several limitations imposed.

The main limitations were the lack of a physical quantum computer, the reliance on a publically shared virtual quantum computer hosted by Google with functioned under best-effort-basis, and the utilization of only a single experimental compiler known as *QScript*.

The next limitation is the utilization of *Google's Quantum Playground* as an alternative quantum computer for use in the study. It was fortunate there even was a publically available quantum computer for public use globally. However, this was the issue as the quantum computer only offered a maximum performance capability of 22 qubits.

The final limitation was the quantum computer's compiler offered by *Google's Quantum Playground*, the *QScript* compiler. It is estimated that no matter how powerful a computer hardware may be, it is always limited by its software utilization of said hardware.

7. Conclusion

The study was undertaken to explore the performance capabilities of the new quantum computers when compared to our currently utilized silicon computers. Initially, the study predicted that the quantum computers would surpass the silicon computers without much effort.

Remarkably, that was entirely the opposite of the findings this study. After completion of all experiments, the results were used to determine which computer is more capable performance-wise in executing control statements when comparing the completion times between a quantum computer and silicon computer. The quantum computer had sluggish results with a massive gap in completion times when compared to the silicon computer.

Acknowledgment

This research work was supported by Ministry of Education (MOE), Malaysia and Universiti Teknologi MARA, Malaysia under Research Acculturation Grant Scheme (RAGS), Project code: "600-RMI/RAGS 5/3 (8/2014)".

References

[1] Bell, G. "Bell's Law for the birth and death of computer classes: A theory of the computer's evolution". *IEEE Solid-State Circuits Society Newsletter*, *13*(4), 8-19, 2008.
[2] Schaller, R. R. "Moore's law: past, present and future". *IEEE spectrum*, *34*(6), 52-59, 1997.
[3] Ladd, T. D., Jelezko, F., Laflamme, R., Nakamura, Y., Monroe, C., & O'Brien, J. L. "Quantum computers". *Nature*, *464*(7285), 45-53, 2010.
[4] N.A. "D-Wave Systems Breaks the 1000 Qubit Quantum Computing Barrier". Retrieved December 31, 2016, from http://www.dwavesys.com/press-releases/d-wave-systems-breaks-1000-qubit-quantum-computing-barrier.

IOP Conf. Series: Journal of Physics: Conf. Series **1049** (2018) 012006 doi:10.1088/1742-6596/1049/1/012006

A State of the Art: Software Configuration Management Tools for Global Software Development

S.S.M.Fauzi*,1, A.J.Suali1, W.A.W.M.Sobri1
Faculty of Computer and Mathematical Sciences
Universiti Teknologi MARA, Perlis Campus, 02600 Arau, Perlis, Malaysia

E-mail: shukorsanim@uitm.edu.my

Abstract. Global Software Development (GSD) has emerged widely in today's software development environment. The emergence of GSD has create difficulty to several technical areas such as Requirement Engineering, Validation and Verification; not to mention Software Configuration Management (SCM) practices. To date, many SCM tools has been proposed by the researchers to reduce the difficulties faced by GSD project developers when dealing with SCM practices. Therefore, there is a need to update on the current trend of SCM tools in GSD context in order to assist practitioners in selecting the best tools for their project. To perform the study, Systematic Mapping Studies (SMS) has been used to identify and provide an overview of the SCM tools which has been developed specifically for GSD environment. This paper contributes the following: 1) This paper synthesized the SCM tools in GSD and 2) Current trend of SCM tools in today's software development environment also is identified.

1. Introduction

Global Software Development (GSD) has become an enabler in today's world IT industry. Software and Information Industry Association (SIIA) in their survey has identified seven important factors that lead the decision makers to move towards GSD environment. Cost Advantage remains one of the most essential aspects, whereas increase speed to market, increase productivity, growth strategy, improve operational effectiveness, competitive pressure and improve product quality also has drive IT companies to shift to this trend. Numerous Information Technology (IT) mainstream companies lead by Lucent [1] , IBM, Motorola [2] , and Philips [3] have make a move to this new era of software development environment.

Several researchers have identified a bundle of issues and challenges when dealing with the GSD environment. Prior study has identified three main threats to GSD environment 1) Communication – the exchange of complete and unambiguous information that is the sender and receiver can reach a common understanding 2) Coordination – the act of integrating each task with each organizational unit, so the unit contributes to the overall objective 3) control issues – the process of adhering to goals, policies, standards or quality levels [4].

*,1 Corresponding author: shukorsanim@uitm.edu.my.

Content from this work may be used under the terms of the Creative Commons Attribution 3.0 licence. Any further distribution of this work must maintain attribution to the author(s) and the title of the work, journal citation and DOI.
Published under licence by IOP Publishing Ltd

Software Configuration Management (SCM) also suffered with the emergence of GSD environment [5]. That has lead various researchers to introduce a range of strategies, approaches, techniques and tools that can be used to improve SCM in GSD environment [6].

By looking at the GSD development trend which is incredibly complex, we believe that there is an evolutionary of SCM tools which has been developed specifically for GSD environment. Therefore, the objective of the study is to identify and synthesize the SCM tools which have been developed for GSD environment. Next, we also would like to identify the current trend of SCM tools development.

Next section of this paper will explain about the methodology used to conduct this research. It then follows by the discussion of the results achieved in this study. Finally, we will conclude our paper.

2. Methodology

We have carried out a systematic mapping study; an approach that provides a structure of research results that has been published by categorizing them in a visual summary or the map [7].

2.1. Developing Research Questions

While the objective of this paper is to identify the SCM tools for GSD environment, the research questions which drive this study are:

RQ1: *What is the evolution of SCM tools in GSD environment?*
RQ2: *What is the current trend of SCM tools?*

2.2. Search Strings and Source Selection

We have performed two separate stage of extensive search to answer our research questions. In stage 1, we have identified a keyword which has been alienated into two categories as shown below. Categories A comprise a keyword which is related to "Software Configuration Management", while category B consists of the keyword which has interrelated to "Global Software Development". Both categories were combined using the Boolean "AND" expression: ***(A1 OR A2 OR A3 OR A4 OR A5 OR A6 OR A7 OR A8 OR A9) AND (B1 OR B2 OR B3 OR B4 OR B5 OR B6 OR B7 OR B8 OR B9 OR B10 OR B11)***.

A1- Configuration management tool
A2- Software configuration management tool
A3 - Software change management tool
A4 - Software change request tool
A5 - Change management tool
A6 - Change request tool
A7 - Change control tool
A8 - Version control tool
A8 - Versioning tool
A9 - Modification request tool

B1- Global software development
B2 - Global software engineering
B3 - Globally distributed development
B4 - Geographically distributed development
B5 - Distributed software development
B6 - Collaborative software development
B7 - Collaborative software engineering
B8 - Offshore software development
B9 - Off shoring
B10 - Dispersed locations
B11 - Dispersed teams

IOP Conf. Series: Journal of Physics: Conf. Series **1049** (2018) 012006 doi:10.1088/1742-6596/1049/1/012006

We then performed a search to several electronic databases; for instance:

1) IEEEXplore (http://ieeexplore.ieee.org)
2) ACM Digital Library (http://www.portal.acm.org/dl.cfm)
3) Compendex EI (http://www.engineeringvillage2.org)
4) INSPEC (http://www.engineeringvillage2.org)
5) Google Scholar (http://scholar.google.com)
6) Elsevier ScienceDirect (http://www.sciencedirect.com)

2.3. Screening of Papers

As a first step, we have decided to analyze the title, abstract and keywords for each paper. In some condition, we also have to read through the entire of paper to determine its relevancy to the research question. However, both situations are lie on the inclusion and exclusion criteria that we have governed before the screening process. It will helps us to ensure only paper which has relevancy to the research questions is being considered. Thus, the inclusion criteria that we have applied are:

I1.Papers published should directly relate to SCM tools and the studies should have focus on GSD environment context.

As such, the exclusion criteria that we applied for this research consist of:

E1.The paper which lies on difference aspect of SCM. For instance cultural, time zone, language issues; which is extremely differ from our focus;
E2.We excluded any posters, panels, abstract, power point presentation or even the summaries of articles.
E3.Papers which consist of opinions, claims, personal evaluation and reported recommendations will not be considered since there is a lack of empirical work that focus unswervingly to SCM in GSD environment context.
E4. Some authors may publish more than one paper related to their tools. Therefore, we have decided to exclude any redundancy paper which mainly discuss about the same subject.

3. Results and Discussion

3.1. Overview of studies

Details of the papers found have been entered to online bibliographic information; RefWorks. Our early record indicates there are 125 papers related to our research. However, only 7 papers are selected and it has resulted 8 SCM tools. Table 1 depicts most of the SCM tools have been developed in 2007 (33%). We can consider that, the attentiveness to develop SCM tools for GSD project developers just started in 4 years back.

Most of the SCM tools has been published in a leading conference; for instance International Workshop on Cooperative and Human Aspects of Software Engineering - CHASE and International Conference on Software Engineering – ICSE (25%) , whereas International Conference on Global Software Engineering – ICGSE, SIGCHI Conference on Human Factors in Computing Systems – CHI, ACM SIGSOFT International Symposium on Foundations of Software Engineering – ESEC/FSE and European Conference on Computer Supported Cooperative Work – ECSCW (13%).

Table 1. Number of papers published

Year	Percentage	Frequency
1999	13	1
2000	0	0
2001	0	0
2002	0	0
2003	13	1
2004	0	0
2005	0	0
2006	13	1
2007	38	3
2008	13	1
2009	0	0
2010	13	1

Based on our systematic study, we found that most of the researchers employed several methods to develop their tools; for instance – survey, observation, interview and also video recording. We also notice that, researchers have tested their tools in two difference environment 1) academic environment - engage a certain number of students which then been alienated to several groups or pairs 2) iterative development experiment; actual environment experiment - to gather the actual practices of GSD project developers, lab experiment – to develop and simulate the tools.

3.2. SCM tools in GSD environment (RQ1)

Tools are becoming one of the essential parts in GSD environment. It can assist GSD project developers to collaborate, facilitate the works and even to get communicate among other project developers [8]. As such, there are numerous tools which have been developed. To name a few; collaboration tools [8][9][10] and communication tools [11].

In SCM context, adequate tools to support GSD project developers performing their task are vital. This is for reason that, GSD environment has limit the group awareness [6]. This therefore can lead to a several difficulties such as problem to make a change [3][5][12][13] and even to control the versioning of the artifacts [5].

Our study reveal numerous of SCM tools which have been developed to cater the need of today's development environment. Table 2 shows the list of tools which has been developed specifically to accommodate the GSD environment needs. Subsequently, we extend our discussion on the tools listed in Table 2 in the next paragraph.

Table 2. List of SCM tools in GSD environment context

Title of Paper	Author	Tools	Year	Source	Focus of the tools
Supporting distributed software development with fine-grained artefact management	[14]	ADAMS	2006	ICGSE	Focus on high level documentation versioning
Palantir: Raising awareness among configuration management workspaces	[15]	Palantir	2003	ICSE	To raise the awareness among the project developers in dispersed location

FASTDash: A visual dashboard for fostering awareness in software teams	[16]	FASTDash	2007	CHI	Determine which team members have source files checked out, which files being viewed and what are the task currently being done by other project developers
CHIME: A metadata-based distributed software development environment	[17]	CHIME	1999	ESEC/FSE	Intend to collaborate large software development project which are geographically dispersed via the use of virtual environment technology
Syde: A tool for collaborative software development	[18]	Syde	2010	ICSE	To establish group awareness by sharing change and conflict information across project developer's workspace
Continuous coordination within the context of cooperative and human aspects of software engineering	[11]	WAV	2007	CHASE	Visualizes the developers and artifacts in a project using 3D metaphor
Continuous coordination within the context of cooperative and human aspects of software engineering	[11]	Ariadne	2007	CHASE	Highlight on the socio-technical relationships between source code and the developers implementing the artifacts
Semi-synchronous conflict detection and resolution in asynchronous software development	[19]	CollabVS	2007	ECSCW	Focus on semi-synchronous distributed model that allows project developers to detect early conflict

Advanced Artifact Management System (ADAMS) - ADAMS is developed to accommodate software artifacts; not only source code, but all the documentations in software development. It can be associated with a template which has been defined in each project. Next, ADAMS also has a facility to create and store traceability links between each artifacts. Similar like other version control system, this tool also enable group of people working on the same artifacts. The developer will be notified once there is any changes happened in the artifacts.

Columbia Hypermedia Immersion Environment (CHIME) - CHIME is dedicated to support large software development project which are geographically dispersed. It comprises three main components; Groupspace which contain all the associated software artifacts and tools for the project. Groupviews is used to describe multiuser. As is ADAMS, Groupviews also will keep other team members informed about the work done by others. While Software Immersion is a third component in CHIME and is used to create a sub task from Groupviews. This will allow other team members to learn any difficulties faced.

Workspace Activity Viewer (WAV) - WAV has been developed using the data gathered from the actual GSD environment. It can visualize the developers and dependencies of the artifacts by using the stack of cylinders. Each cylinders will represent each modified artifacts and which developers has made a changes to the artifacts.

Ariadne - Ariadne is dedicates to deal with socio-technical relationship between source code and developer who own and implement the code. The tool can help the team members to get an idea on what has been done by other team members at other location, as well.

FastDash - FastDash is designed specifically for project of 3 to 8 team members. The specific purpose of this tool is to create a real time awareness and access to key information such as which code is changing, who is changing the code and how the code is being used.

Palantir - Palantir is another tool that provides insight to other team members' workspaces. This tool concentrates more on the collection, distribution, organization and presentation of the relevant information. It means that Palantir can react as a platform to inform the developers about the changes made by other developers in a certain artifacts, able to measure the severity of the changes and organize the report in a graphical way.

Syde – Syde has the ability to detect merge conflicts by showing the differences between two versions of the artifacts and provide information about the conflict and changes to the project developers in a real time.

CollabVS – CollabVS used a semi-synchronous model which allows a group of project developers to create a code asynchronously. Next, this tool also has the ability to detect early conflict, and resolve the conflict before project developers completed the task.

3.3. Trends in SCM Tools (RQ2)

Prior study has discussed in detail about the evolution of SCM tools from the early generation starting from 1970s to 2000 [20]. First, in 1970s until 1980s, the SCM tools have been used by a single person to manage a mainframe. Thus it requires applying versioning and building technique. It then continues (1980s-1990s), where at that time the software development involve only local team which develops large scale software and it runs using UNIX platform. Hence, workspace management has been introduced in SCM tools. From 1990s-2000s, the development environment has shift to distributed environment. The software is developed by project developers in difference locations (perhaps in difference building). Thus, it requires explicit process support in SCM tools.

Now, the SCM tools have emerged more widely to GSD environment. It involves a large scale software development but in difference locations (perhaps difference continents with a difference time zone). The GSD environment has created a complexity to the SCM practices [21]. Thus, based on our systematic study, we found eight SCM tools have introduced additional features in SCM tools which therefore help to improve the SCM practices in GSD environment. This features lead to another generation of SCM tools. Hence, it also creates a new trend in SCM tools.

As depicts in Table 3, two additional features introduced in today's SCM tools are; dependency and group awareness. Group awareness is referred as understanding the activities of other project developers [22]. In SCM context, these activities includes the information on who creates the artifacts, who make changes, which artifacts affected with the changes, what has been changed and what other project developers do. Dependency refers to the situation where project developers are not able to proceed their work since waiting other project developers to complete their task. Both conditions thus, resulting a delay and time interval to make a changes in GSD environment.

Thus, several tools have been proposed to resolve the problems. [19] has introduced their tools Palantir and CollabVS. These tools provide an insight for project developers on other project developer's workspace. To a certain extent, other group of researchers also has made an improvement for the tool by introducing 3D metaphor which allows virtual communication between dispersed team. Such studies include the Columbia Hypermedia Immersion Environment (CHIME); an internet and intranet-based distributed software development which allow users interact with project artifacts by "walking around" the virtual world [17]. [11] introduce Workspace Activity Viewer (WAV) and Ariadne; a similar tools which can visualizes the developers and artifacts in a project using 3D metaphor and gives managers an overview of ongoing activities in a project. The same goes to FastDash by [16] which has a capability to help a developer to quickly determine which team members have source files checked out, which files are being viewed and what methods and classes are currently being changed.

Table 3. Features incorporated in SCM tools

Tools	Features incorporated	Team size
ADAMS	versioning control system	8
Palantir	group awareness	3
FASTDash	group awareness	3 to 8
CHIME	versioning control system	Not mentioned
Syde	group awareness	Not mentioned
WAV	dependency	Not mentioned
Ariadne	group awareness	Not mentioned
CollabVS	group awareness	8

Open source tools that focus on GSD also emerged rapidly. For instance; Git (*Git.*), Darcs (*Darcs.*) , Bazaar (*Bazaar.*) , Mercurial (*Mercurial.*). Whereas, other popular open source tools for instance Subversion (*Subversion.*) and CVS (*CVS.*) has been used widely in collocated software development environment. All of the mentioned tools are purposely used as version control system. Such tools for trackers are Jira (*Jira.*) and Bugzilla (*Bugzilla.*).

4. Conclusion

We consider following to be our initial contributions. First, we have identified the SCM tools which have been proposed by various researchers which then can be applied to GSD context. Second, we have recognized the current trend of SCM tools in today's software development environment; GSD. Our future work will focus on comparing the feature of the tools identified.

References

[1] Herbsleb, J. D., & Grinter, R. E. (1999). Splitting the organization and integrating the code: Conway's law revisited. *ICSE '99: Proceedings of the 21st International Conference on Software Engineering,* Los Angeles, California, United States. 85-95.

[2] Battin, R. D., Crocker, R., Kreidler, J., & Subramanian, K. (2001). Leveraging resources in global software development. *IEEE Software, 18*(2), 70-77.

[3] Kommeren, R., & Parviainen, P. (2007). Philips experiences in global distributed software development. *Empirical Software Engineering, 12*(6), 647-660.

[4] Ågerfalk, P. J., Fitzgerald, B., Holmström, H., Lings, B., Lundell, B., & Conchúir, E. Ó. (2005). A framework for considering opportunities and threats in distributed software development. *International Workshop on Distributed Software Development,* Paris. 47–61.

[5] Prikladnicki, R., Jorge, A., & Evaristo, R. (2006). A reference model for global software development: Findings from a case study. *Global Software Engineering, 2006. ICGSE '06. International Conference on,* 18-28.

[6] Sangwan, R., Mullick, N., Bass, M., Paulish, D. J., & Kazmeier, J. (2006). *Global software development handbook* Auerbach Publications Taylor & Francis Group.

[7] Petersen, K., Feldt, R., Mujtaba, S., & Mattsson, M. (2008). Systematic mapping studies in software engineering. *12th International Conference on Evaluation and Assessment in Software Engineering (EASE),* Department of Informatics, University of Bari, Italy.

[8] Lanubile, F., Ebert, C., Prikladnicki, R., & Vizcaíno, A. (2010). Collaboration tools for global software engineering. *IEEE Software, 27* (2), 52-55.

[9] Aranda, G. N., Vizcaíno, A., & Piattini, M. (2009, July). Which Groupware Tool is the most suitable for this group?. In *Fourth IEEE International Conference on Global Software Engineering, 2009.* pp. 400-405.

[10] Lamersdorf, A., & Munch, J. (2009, July). TAMRI: a tool for supporting task distribution in global software development projects. In F*ourth IEEE International Conference on Global Software Engineering, 2009.* 322-327.

[11] Al-Ani, B., Trainer, E., Ripley, R., Sarma, A., Van Der Hoek, A., & Redmiles, D. (2008). Continuous coordination within the context of cooperative and human aspects of software engineering. *Proceedings of the 2008 International Workshop on Cooperative and Human Aspects of Software Engineering,* 1-4.

[12] Herbsleb, J. D., & Mockus, A. (2003). An empirical study of speed and communication in globally distributed software development. *IEEE Transactions on Software Engineering, 29; 29*(6), 481-494.

[13] Cataldo, M., Bass, M., Herbsleb, J. D., & Bass, L. (2007). On coordination mechanisms in global software development. *Second IEEE International Conference on Global Software Engineering, 2007,* 71-80.

[14] Bernd Bruegge, Andrea De Lucia, Fausto Fasano, & Genoveffa Tortora. (2006). Supporting distributed software development with fine-grained artefact management. *International Conference on Global Software Engineering, 2006.* 213-222.

[15] Sarma, A., Noroozi, Z., & van der Hoek, A. (2003). Palantir: Raising awareness among configuration management workspaces. *ICSE '03: Proceedings of the 25th International Conference on Software Engineering*. 444-454.

[16] Biehl, J. T., Czerwinski, M., Smith, G., & Robertson, G. G. (2007). FASTDash: A visual dashboard for fostering awareness in software teams. *CHI '07: Proceedings of the SIGCHI Conference on Human Factors in Computing Systems*. 1313-1322.

[17] Dossick, S. E., & Kaiser, G. E. (1999). CHIME: A metadata-based distributed software development environment. *ESEC/FSE-7: Proceedings of the 7th European Software Engineering Conference Held Jointly with the 7th ACM SIGSOFT International Symposium on Foundations of Software Engineering*. 464-475.

[18] Hattori, L., & Lanza, M. (2010, May). Syde: A tool for collaborative software development. In *Proceedings of the 32nd ACM/IEEE International Conference on Software Engineering-Volume 2* (pp. 235-238). ACM.

[19] Dewan, P., & Hegde, R. (2007). Semi-synchronous conflict detection and resolution in asynchronous software development. *ECSCW 2007*, 159-178.

[20] Estublier, J., Leblang, D., Hoek, A. v. d., Conradi, R., Clemm, G., Tichy, W., et al. (2005). Impact of software engineering research on the practice of software configuration management. *ACM Transaction on Software Engineering Methodology, 14*(4), 383-430.

[21] Prikladnicki, R., Audy, J. L. N., & Evaristo, R. (2003). Global software development in practice lessons learned. *Software Process: Improvement and Practice, 8*(4), 267-281.

[22] Dourish, P., & Bellotti, V. (1992). Awareness and coordination in shared workspaces. *Proceedings of the 1992 ACM Conference on Computer-Supported Cooperative Work,* 114.

The Impact of Authentic Leadership on Knowledge Sharing Behavior with the Moderating role of Islamic Work Ethics

M.Javaid[1]., N.H.Abdullah[2] A.Zeb[3] and K.Hussain[4]
Facility of Technology Management and Business, Universiti Tun Hussein Onn Malaysia, 86400 Parit Raja, Johor, Malaysia.

E-mail: hp160047@siswa.uthm.edu.my

Abstract: This paper aims to explore the effects of authentic leadership on knowledge-sharing behavior and moderating role of Islamic work ethics. There is limited research devoted to the authentic leadership and knowledge-sharing behaviour within the Telecom sectors of Pakistan. The objective of this study is to analyze the impact of authentic leadership on knowledge sharing behavior with moderating role of Islamic work ethics. This is a quantitative research study. Convenience sample method was used as a sampling technique for this research. The data was collected from different two organizations (Mobilink and Ufone) and questionnaire was distributed to 220 employees, working in various multinational organizations of Islamabad (Pakistan). The data was analyzed using SPSS. The result showed that authentic leadership at Telecom organizations was developed positive behavior through knowledge-sharing. Therefore, the association between authentic leadership, Islamic work ethics and knowledge-sharing behavior is positive. The contribution of this study is Islamic work ethics using first time between authentic leadership and knowledge sharing behavior which not used so far.

1. Introduction

Nowadays, knowledge-sharing has been considered as an important organizational asset. Knowledge-sharing is involves the exchange of knowledge among employees, exchange the experiences and skills through the whole organization [1]. In knowledge based economy knowledge-sharing plays a significant role for survival and better performance [2]. An organization may have loyal customers, latest technology, financial resources and fixed assets but behavior of employees is the core competency in an organization [3]. Knowledge-sharing is a key enabler of knowledge management. Knowledge-sharing behaviors is not only depends to achieve the objective of the groups or department, but it also lead to superior performance [4]. Therefore, organizations need talented leaders who motivate to employees to displays positive behavior. The importance of knowledge-sharing for innovation and initiatives formulated to promote knowledge-sharing realized by the leaders [5].

Organizations need to pay attention to facilitate knowledge-sharing behavior because there is a lack of research between authentic leadership and knowledge-sharing behavior [6]. In Telecom organization, knowledge-sharing has been under discussion of many researchers [7]. In fact, there are some researcher who explores the relationship of authentic leadership and knowledge-sharing

Content from this work may be used under the terms of the Creative Commons Attribution 3.0 licence. Any further distribution of this work must maintain attribution to the author(s) and the title of the work, journal citation and DOI.
Published under licence by IOP Publishing Ltd

behavior [8]. In those studies, suggested initial support on the positive relationship between authentic leadership and knowledge-sharing behaviors.

Further, several studies have examined the enablers of knowledge-sharing in universities, organizations and public sectors [10]. There is no empirical research to test the effect of authentic leadership on knowledge-sharing behavior using Islamic work ethics as moderator in Telecom organization, Pakistan. Authentic Leadership is a relatively new theory that still needs to be validated [11]. Therefore, this study examines the effect of authentic leadership on knowledge-sharing behavior and tests the outcomes of Islamic work ethics at organizational level.

2. Related Works

2.1 Authentic leadership and Knowledge sharing behavior:

Authentic leadership refers to the leader personally treating each employee and affecting each employee with authenticity to engage in the organization. Authentic leadership understood as a pattern of behavior role to reflect the effects on working employees [12]. The authentic leadership maintains relation between their followers based on sincerity so as to be consistent with positive behaviors [13]. Authentic leader is to be passionately smart to the development of behavior [14]. Authentic leadership has positive influence to their employees when they share knowledge between them for the productivity of organization. Leader behavior involves communicative actions that are observable and words that are understandable [16]. Authentic leader's mission is to serve collective interests [15], they will foster and shape in their collaborators the behaviors such as knowledge-sharing behaviors that help the organization for productivity. According to [17] indicates that willingness of leaders to share knowledge is a critical factor in building positive behavior among employees. An authentic leader behavior can influence on others employees to create such positive behavior as a standard that is accepted by the people [18]. Authentic leader could also influence on employees behavior such as knowledge sharing behavior.

Knowledge-share (KS) is defined as the method of establishing and distribution of the shared knowledge understanding for the correct knowledge to the individuals at the correct time in an organization [19]. Knowledge-sharing indicates interest, expertise and abilities by which you can come back with your friends, family, people, team and organization. Knowledge-sharing has become the main and most appreciated source in this rapidly moving situation. The aim of knowledge-sharing is to control the features of the acquisition process, ask for help and protection of knowledge, without these procedures, knowledge has no worth. Knowledge must be encouraged, nurtured, established and grouped somehow in command to produce the worth [20]. According to [21], knowledge is information that is organized and analyzed and can be applied to the resolution of problems or decision making process. It plays a big role in increasing the competitive advantage of the organization and is essential to the improvement of productivity of organization [22].

Several studies in past research have shown that, authentic leadership is related positive with other variables like job satisfaction, performance, and collaborators' organizational citizenship behaviors [23]. Authentic leaders is positively related to a series of desirable results such as, knowledge-sharing indicates interest, expertise and abilities by which you can come back with your friends, family, people, team and organization. [24] studied the relationship between knowledge-sharing, leadership roles, and collaborative culture. In his study, he looked into eight leadership roles and name of roles are facilitator, innovator, producer, coordinator, mentor, broker, director, and monitor. The finding of this research showed that facilitator, mentor, and innovator roles were positively correlated with knowledge-sharing, and there was a negative relationship between the leadership role of monitor and knowledge sharing. In another study, [25] also found that empowering leadership was positively related to knowledge-sharing being positively related to performance. There has been less research devoted to the authentic leadership and knowledge-sharing behaviour within the Telecom sectors of Pakistan. Despite this fact, authentic leadership is hypothesized to create knowledge-sharing behaviour among employees [26]. It seems logical to suppose that authentic leadership have positive impact on knowledge-sharing behaviour among employees.

On the basis of above discussion, it is hypothesized that:

H1: There is a positive relationship between the authentic leadership and knowledge sharing behavior.

2.2 Islamic work ethics and Knowledge sharing behaviour:

Islamic Work Ethics is according to the lessons of the Qur'an and the Sunnah of the Prophet Mohammed (peace be upon him) [27]. Muslim religious people follow the Holy Quran and they beliefs that every aspects of human life is given in this book. Human being is the topic of Holy Quran and it mostly covers the matters of human life [28]. Muslims in the world are increasingly becoming inspired by the principles of Islam. In general, the principles of Islamic work ethics (IWE) focus on the relations between human beings and their creator. Islamic work ethics are based on the principles and teachings of Islam. [29] Indicates that Islam is supposed to be a complete way of life. Most of the research on the ethics of religions was conducted in the West, with emphasis on the Protestant work ethic (PWE) [30]. Both PWE and IWE have a large emphasis on effort, devotion, commitment, imagination and prevention of prosperity accumulates by using illegal resources and collaboration in the organization. Therefore, Islamic work ethics has made a positive role on organizational behavior, and it is expected that it will positively impact on employee's knowledge-sharing behavior.

Knowledge-sharing is realistic in major religions, cultures and ideologies. The Quran and Sunnah have laid and encouraged Muslim people to share knowledge in order to contribute to the development and welfare of society, including the workplace [31]. Knowledge-sharing behavior is considered to be very critical to an organization success [32], and scholars have suggested that organizations should encourage and involve their employees in knowledge-sharing. Knowledge-sharing as a procedure in which a person exchanges knowledge with others and then jointly creates new knowledge [33]. Prophet Muhammad (PBUH) emphasized the sharing of knowledge and said, "Do you know who is most generous? God is the most Generous. After then in human kind, I am most generous, and after me the most generous people will be those who will acquire knowledge and then share knowledge to others [35].

In past studies, management and organization have devoted minimal attention to Islamic work ethic (IWE) and Islamic work ethics plays a role to influence a variety of desirable job outcomes and organizational innovations [36]. [37], suggested that managers working scored are high related to IWE. Similarly, [38] reported that employees working in government organizations showed stronger support of the IWE. [39], found that IWE within a university environment is significantly related to individuals' attitudes towards computer use ethics, job satisfaction and organizational commitment. [40] examined the influence of IWE on innovation reported positive relationship between the constructs. [41] found a strong and highly significant correlation between the IWE and individualism scales. However, very few studies have been described on the relationship between the IWE and knowledge-sharing behavior but no studies conducted in Telecom sectors, Pakistan.
On the basis of above discussion, it is hypothesized that:

H2: There is a positive relationship between Islamic work ethics and knowledge sharing behavior (KSB).

2.3 The moderating role of Islamic work ethics:

Pakistan in terms of population is the second largest Muslim country in the world [42]. It is important to evaluate the impacts of Islamic work ethics between leadership and knowledge-sharing behaviour in Telecom sectors, Pakistan. Because there are majority of Muslims working in telecom sectors in Pakistan and this relation not used so far in the context of Telecom, Pakistan. The Islamic work ethic may be defined in the Islamic context as the set of moral principles that distinguish right from wrong [43]. Islamic work ethic keeps positively in nature [44]. Therefore, Islamic work ethics has not been used as moderator between authentic leadership and knowledge-sharing behaviour so far, but on the basis of previous literature, this effort is done to develop this relationship.

Previous research has established that Islamic work ethics plays positive role as a moderator. For example, a study conducted by [45] in Pakistan revealed that IWE moderate the relationship between ethical leadership and job satisfaction. They reported significant combined effect of ethical leadership and Islamic work ethics (IWE) on organizational commitment of employees. [46], in one of the earliest studies on the moderating role of IWE, reported that when IWE is taken as a moderator on the positive relationship of Job satisfaction and organizational commitment. [47] studied the moderating role of IWE on the positive relationship of knowledge-sharing capability and innovation capability and found out that when the IWE and knowledge-sharing capability interact with each other their impact reduce relationship due to moderation. [48], studied the combined effect of job stress and IWE on job satisfaction and turnover intention. Results of the study revealed that IWE moderates the relationship of job stress and job satisfaction. Furthermore, the Islamic work ethics (IWE) focuses on fairness and kindness at work [49] it may moderate relationship between authentic leadership and knowledge-sharing.

This research would help organizational managers to understand the role of IWE in strengthening knowledge-sharing behaviour among the Muslim employees of Telecom, Pakistan. This study intended to investigate the how Islamic work ethics moderator between authentic leadership and knowledge-sharing behaviour in Telecom sectors, Pakistan. The importance of Islamic work ethics (IWE) as a core human process in organizations and assessment of knowledge-sharing behaviour in Telecom sectors Pakistan is necessary.
Based on the above discussion, following hypotheses is developed:

H3: *Islamic work ethic will lead the positive relationship of authentic leadership and KSB such that the relationship will be stronger when Islamic work ethic is high.*

3. Methodology

3.1 Research Design
This was a descriptive study where the impacts of authentic leadership on knowledge-sharing behavior were measured and Islamic work ethics used as a moderator. This is survey study because the questionnaires were distributed to employees of telecom two companies in Pakistan.

3.2 Sample
Convenience non probability sampling technique was used for data collection. At the beginning of the questionnaires, the purpose of data collection was described. Furthermore, it also assured to the employees that responses were remain anonymous. Data were collected voluntarily. 220 questionaries were distributed and 193 were returned back.

3.3 Measures
Knowledge-sharing behavior was based on 5-items scales by [50]. Authentic leadership questionnaires were developed by [51]. Islamic work ethics was based on the 17-items scale adopted by [52]. 5-point Likert scale (Strongly Agree = 5, Agree = 4, Neutral = 3, Disagree = 2, Strongly Disagree = 1) was used to measure the all constructs.

3.4 Data Analysis
Data were analyzed using SPSS version 20. To determine the reliability of instruments was used to Cronbach's alpha. To determine the reliability of measurement were used to Cronbach alpha, descriptive, correlation and regression analysis.

3.5 Finding & Discussion

Table 1. Alpha Reliabilities of scale Used in the Study

Variable/ scale	**Cronbach's Alpha**	***Number of Items***
Authentic leader	.724	16
Knowledge sharing behaviour	.771	05
Islamic work ethics	.794	17

The alpha reliability of 16 items of AL items was found to be 0.724, which falls in the accepted range. The alpha reliability of 5 items Knowledge sharing behavior was found to be 0.771, which falls into the acceptable range. The alpha reliability of 17 items of Islamic work ethics was found to be 0.794, which falls in the acceptable range.

3.6 Descriptive statistic

Table 2. Mean and Standard deviation of the study

Variable name	Mean	Std.Deviation
Authentic leader	2.52	.576
Islamic work ethics	3.77	.391
Knowledge sharing behaviour	3.76	.490

The mean score of responders on knowledge-sharing scale was 3.76, standard deviation in the score was 0.49, mean score of authentic leadership scale was 2.52, standard deviation was 0.57, and mean of Islamic work ethics was 3.77, standard deviation was 0.39 (table 2).

3.7 Correlation analysis:

Table 3. Correlation

	Variable name	1	2	3
1	AL	1		
2	IWE	.384**	1	
3	KSB	.307**	.521**	1

Results reveal that authentic leadership was significant positively correlated to knowledge sharing behavior ($r = 0.307$, $p < 0.01$). This provided initial support to hypothesis 1 and Islamic work ethics was significantly positive related to knowledge sharing behavior ($r = 0.521$ $p < 0.01$). This provided initial support to hypothesis 2.

3.8 Regression Analysis:

Table 4. Regressions analysis of all variables

Knowledge sharing behaviour			
Predictors	**B**	**R^2**	**ΔR^2**
Step 1			
Control Variables		.025	
Step 2			
AL	.162		
IWE	.584	.299	.274**
Step 3			
AL x IWE	.162	.344	.045**

AL x IWE = Interaction Term (authentic leadership and Islamic work ethics).

A hierarchical regression analysis was used to test the hypotheses of the study. The basic purpose of the research was examining of hypotheses, hence a technique for hierarchical regression research designed by [53]. Table 4 shows a series of regression. Results revealed that authentic leadership had a significant positive impact on knowledge-sharing behavior where the beta value was found to be β =.162. Then IWE was testing with the knowledge-sharing and analysis show the positive result between these two constructs. IWE was found to have a significant positive impact on knowledge-sharing behavior, where the beta value was found to be $\beta = 0.584$. Furthermore, IWE was analyzed as a moderator between authentic leadership and knowledge-sharing behavior. Results revealed that the interactive effect of authentic leadership and IWE on knowledge sharing behavior was found to be significant $\beta = 0.162$ (Table 4).

4. Conclusion

4.1 Implication to Managers:
The current study is the first study conducted in Pakistan, to investigate knowledge-sharing behavior of employees in Telecom sectors. It is suggested that knowledge-sharing behavior leads to improved performance in Telecom sectors. The result indicates that when the leaders share their knowledge to employees then they will work more effective and productive. It was discovered that the IWE and behaviors can make positive relation between each other in organization. The result also suggested that knowledge-sharing behavior can also facilitate to employees in maintaining a healthy work in organization. Moreover, the evidence of the significant moderating effect of Islamic work ethics also points that it is significance if Islamic work ethics succeed high then employees of telecom sectors do respond good and act excellent.

The findings of this study have many implications for Telecom sectors. Therefore the company administrator and experts are suggested to keep this in mind about the authentic leader must be supportable and helpful with the employees. It makes more training sections between manager and employees to learn them how they can get success and convey the knowledge to each other during work. Management should create a climate of tolerance for error because errors in authentic leadership and knowledge-sharing can be powerful sources of learning [54]. The manager suggest to keep in mind that they can hire those people in telecom sectors who are more responsible and sharing their knowledge with others because positive sharing behavior make the positive result from the data. Leaders should ensure the implementation of supportive management system. This perception is of great importance with regard to their knowledge sharing behavior.

Further, managers should take steps to ensure the maintenance of psychological safety by supporting individuals' job autonomy and by promoting innovative values. It also necessary that build in their importance on the basis of Islamic values. Islamic work ethics more reinforced other than Ethics. Managers may also provide more opportunities for employees to have direct contact with the beneficiaries of their work to better understand the impact of their work on others [55].

4.2 Future Recommendation:

This study is providing directions for future research. We employed a convenience sample in this research; future researchers expend the data collection sample. We have adapted a scale previously used in the literature but in future may be develop a more refined scale. We focus in this study on IWE as a moderator, but in future we will use another variable as a moderator between these two constructs. Future research should also effort to achieve a larger population sample size. However, other organizational design and psychological elements may play a role and should be considered in future studies.

4.3 *Conclusion:*

The current study enhances the knowledge management literature through incorporating the moderating effect of Islamic work ethics between authentic leadership and knowledge-sharing behavior relationship. The evidence of this study, influence indirectly IWE on the relationship between authentic leadership and knowledge-sharing prove that IWE has major contribution toward the promotion of knowledge-sharing in Telecom sectors. Assurance of good behavior between authentic leadership and employees performance is improved the productivity of the organization. This study present research aids leadership in planning policies and strategies to extend knowledge-sharing and following the Islamic rules among Telecom sectors employees and to create a supportive knowledge-sharing culture.

5. Acknowledgment

"This paper was partly sponsored by the center for Graduate Studies UTHM"
"Kertas kerja ini ditaga sebahagiannya oleh Pusat Pengajian Siswazah"

Reference

[1] Lin, H. F. (2007). Knowledge sharing and firm innovation capability: an empirical study. *International Journal of manpower*, *28*(3/4), 315-332.

[2] Wu, G. C., Cheng, Y. H., & Huang, S. Y. (2010). The study of knowledge transfer and green management performance in green supply chain management. *African Journal of Business Management*, *4*(1), 44.

[3] Muqadas, F., Ilyas, M., & Aslam, U. (2016). ANTECEDENTS OF KNOWLEDGE SHARING AND ITS IMPACT ON EMPLOYEES'CREATIVITY AND WORK PERFORMANCE. *Pakistan Business Review*, *18*(3), 655-674.

[4] Valsania, S. E., Moriano, J. A., & Molero, F. (2016). Authentic leadership and intrapreneurial behavior: cross-level analysis of the mediator effect of organizational identification and empowerment. *International Entrepreneurship and Management Journal*, *12*(1), 131-152.

[5] Mura, M., Lettieri, E., Radaelli, G., ... & Spiller, N. (2016). Behavioural operations in healthcare: a knowledge sharing perspective. *International Journal of Operations & Production Management*, *36*(10), 1222-1246.

[6] González, L. M., Giachetti, R. E., & Ramirez, G. (2005). Knowledge management-centric help desk: specification and performance evaluation. *Decision support systems*, *40*(2), 389-405.

[7] Shah, S. R., Shah, S. R., Mahmood, K., & Mahmood, K. (2016). Contributing factors in knowledge sharing for performance of university students in teachers' training programs. *Library Management*, *37*(8/9), 496-506.
[8] Edú-Valsania, S., Moriano, J. A., & Molero, F. (2016). Authentic leadership and employee knowledge sharing behavior: Mediation of the innovation climate and workgroup identification. *Leadership & Organization Development Journal*, *37*(4), 487-506.
[9] Jiang, X., Flores, H. R., Leelawong, R., & Manz, C. C. (2016). The effect of team empowerment on team performance: A cross-cultural perspective on the mediating roles of knowledge sharing and intra-group conflict. *International Journal of Conflict Management*, *27*(1), 62-87.
[10] Abu-Shanab, E., Haddad, M., & Knight, M. B. (2014). Knowledge sharing practices and the learning organization: A study. *IUP Journal of Knowledge Management*, *12*(2), 38.
[11] Wang, H., Sui, Y., Luthans, F., Wang, D., & Wu, Y. (2014). Impact of authentic leadership on performance: Role of followers' positive psychological capital and relational processes. *Journal of Organizational Behavior*, *35*(1), 5-21.
[12] Fu, W., & Deshpande, S. P. (2012). Factors impacting ethical behavior in a Chinese state-owned steel company. *Journal of Business Ethics*, *105*(2), 231-237.
[13] Norman, S. M., Avolio, B. J., & Luthans, F. (2010). The impact of positivity and transparency on trust in leaders and their perceived effectiveness. *The Leadership Quarterly*, *21*(3), 350-364.
[14] Avolio, B. J., & Gardner, W. L. (2005). Authentic leadership development: Getting to the root of positive forms of leadership. *The leadership quarterly*, *16*(3), 315-338.
[15] Lord, R. G., & Brown, D. J. (2001). Leadership, values, and subordinate self-concepts. *The Leadership Quarterly*, *12*(2), 133-152.
[16] Walumbwa, F. O., Avolio, B. J., Gardner, W. L., Wernsing, T. S., & Peterson, S. J. (2008). Authentic leadership: Development and validation of a theory-based measure. *Journal of management*, *34*(1), 89-126.
[17] Caldwell, C., & Hayes, L. A. (2010). Leadership, trustworthiness, and ethical stewardship. *Journal of Business Ethics*, *96*(4), 497-512.
[18] De Vries, R. E., Bakker-Pieper, A., & Oostenveld, W. (2010). Leadership= communication? The relations of leaders' communication styles with leadership styles, knowledge sharing and leadership outcomes. *Journal of business and psychology*, *25*(3), 367-380.
[19] Van den Hooff, B., & de Leeuw van Weenen, F. (2004). Committed to share: commitment and CMC use as antecedents of knowledge sharing. *Knowledge and process management*, *11*(1), 13-24.
[20] Yang, S. C., & Farn, C. K. (2009). Social capital, behavioural control, and tacit knowledge sharing—A multi-informant design. *International Journal of Information Management*, *29*(3), 210-218.
[21] Bolloju, N., Khalifa, M., & Turban, E. (2002). Integrating knowledge management into enterprise environments for the next generation decision support. *Decision Support Systems*, *33*(2), 163-176.
[22] Hendriks, P. (1999). Why share knowledge? The influence of ICT on the motivation for knowledge sharing. *Knowledge and process management*, *6*(2), 91.
[23] Hannah, S. T., Avolio, B. J., & Walumbwa, F. O. (2011). Relationships between authentic leadership, moral courage, and ethical and pro-social behaviors. *Business Ethics Quarterly*, *21*(4), 555-578.
[24] Yang, J. T. (2007). Knowledge sharing: Investigating appropriate leadership roles and collaborative culture. *Tourism management*, *28*(2), 530-543.
[25] Srivastava, A., Bartol, K. M., & Locke, E. A. (2006). Empowering leadership in management teams: Effects on knowledge sharing, efficacy, and performance. *Academy of management journal*, *49*(6), 1239-1251.

[26] Avolio, B. J., & Walumbwa, F. O. (2014). 16 Authentic Leadership Theory, Research and Practice: Steps Taken and Steps that Remain. *The Oxford handbook of leadership and organizations*, *331*.
[27] Salin, A. P., Saiful, A., Manan, A., Khadijah, S., Kamaluddin, N., & Nawawi, A. (2017). THE ROLE OF ISLAMIC ETHICS TO PREVENT CORPORATE FRAUD. *International Journal of Business & Society*, *18*.
[28] Razimi, M. S. B. A., Noor, M. M., & Daud, N. M. (2014). The concept of dimension in human resource management from Islamic management perspective. *Middle-East Journal of Scientific Research*, *20*(9), 1175-1182.
[29] Ahmad, M. S. (2011). Work ethics: an Islamic prospective. *Journal of Human Sciences*, *8*(1), 850-859.
[30] Weber, M. (1958). Science as a Vocation. *Daedalus*, *87*(1), 111-134.
[31] Rahimi, H., ArbabiSarjou, A., Allameh, S. M., & Aghababaei, R. (2011). Relationship between knowledge management process and creativity among faculty members in the university. *Interdisciplinary Journal of Information, Knowledge, and Management*, *6*, 17-33.
[32] Faizuniah Pangil, Aizzat Mohd Nasurddin, (2016). Knowledge and the Importance of Knowledge Sharing in Organizations. *Business Management Research*, 1983, 349-357.
[33] Van den Hooff, B., & de Leeuw van Weenen, F. (2004). Committed to share: commitment and CMC use as antecedents of knowledge sharing. *Knowledge and process management*, *11*(1), 13-24.
[34] Van Den Hooff, B., & De Ridder, J. A. (2004), Knowledge sharing in context: the influence of organizational commitment, communication climate and CMC use on knowledge sharing. *Journal of knowledge management*, *8*(6), 117-130.
[35] Murtaza, G., Abbas, M., Raja, U., Roques, O., Khalid, A., & Mushtaq, R. (2016). Impact of Islamic work ethics on organizational citizenship behaviors and knowledge-sharing behaviors. *Journal of Business Ethics*, *133*(2), 325-333.
[36] Kumar, N., & Che Rose, R. (2012). The impact of knowledge sharing and Islamic work ethic on innovation capability. *Cross Cultural Management: An International Journal*, *19*(2), 142-165.
[37] Ali, A. J., & Al-Owaihan, A. (2008). Islamic work ethic: a critical review. *Cross cultural management: An international Journal*, *15*(1), 5-19.
[38] Yousef, D. A. (2001). Islamic work ethic–A moderator between organizational commitment and job satisfaction in a cross-cultural context. *Personnel review*, *30*(2), 152-169.
[39] Lam, M. K., Lee, K. T., & Mohamed, A. R. (2010). Homogeneous, heterogeneous and enzymatic catalysis for transesterification of high free fatty acid oil (waste cooking oil) to biodiesel: a review. *Biotechnology advances*, *28*(4), 500-518.
[40] Kumar, N., & Che Rose, R. (2010). Examining the link between Islamic work ethic and innovation capability. *Journal of Management Development*, *29*(1), 79-93.
[41] Khalil, M., & Abu-Saad, I. (2009). Islamic work ethic among Arab college students in Israel. *Cross cultural management: An international Journal*, *16*(4), 333-346.
[42] Khan, S., Jaffer, N. N., Khan, M. N., Rai, M. A., Shafiq, M., Ali, A., ... & Ali, S. H. (2007). Human papillomavirus subtype 16 is common in Pakistani women with cervical carcinoma. *International journal of infectious diseases*, *11*(4), 313-317.
[43] Beekun, R. I. (1997). *Islamic business ethics* (No. 2). International Institute of Islamic Thought (IIIT).
[44] Ali, A. J., & Al-Owaihan, A. (2008). Islamic work ethic: a critical review. *Cross cultural management: An international Journal*, *15*(1), 5-19.
[45] Batool, S. M. K., Gul, A., & Shahzad, K. (2013). Combined effects of ethical leadership and Islamic work ethics (IWE) on organizational commitment and job satisfaction. *Journal of Islamic Business and Management*, *3*(1), 35-50.
[46] Yousef, D. A. (2001). Islamic work ethic–A moderator between organizational commitment and job satisfaction in a cross-cultural context. *Personnel review*, *30*(2), 152-169.

[47] Kumar, N., & Che Rose, R. (2012). The impact of knowledge sharing and Islamic work ethic on innovation capability. *Cross Cultural Management: An International Journal*, *19*(2), 142-165.
[48] Ajmal, M. U., & Irfan, S. (2014). Understanding the moderating role of Islamic work ethics between job stress and work outcomes. *IOSR Journal of Business and Management (IOSR-JBM)*, *16*(1), 62-67.
[49] Ahmad, M. S. (2011). Work ethics: an Islamic prospective. *Journal of Human Sciences*, *8*(1), 850-859.
[50] Bock, G. W., Zmud, R. W., Kim, Y. G., & Lee, J. N. (2005). Behavioral intention formation in knowledge sharing: Examining the roles of extrinsic motivators, social-psychological forces, and organizational climate. *MIS quarterly*, 87-111.
[51] Voon, M. L., Lo, M. C., Ngui, K. S., & Ayob, N. B. (2011). The influence of leadership styles on employees' job satisfaction in public sector organizations in Malaysia. *International Journal of Business, Management and Social Sciences*, *2*(1), 24-32.
[52] Ali, A. J. (1992). The Islamic work ethic in Arabia. *The Journal of psychology*, *126*(5), 507-519.
[53] Cohen, L., Manion, L., & Morrison, K. (2013). *Research methods in education*. Routledge.
[54] Husted, K., & Michailova, S. (2002). Knowledge sharing in Russian companies with Western participation. *Management International*, *6*(2), 17.
[55] Piccolo, R. F., Greenbaum, R., Hartog, D. N. D., & Folger, R. (2010). The relationship between ethical leadership and core job characteristics. *Journal of Organizational Behavior*, *31*(2-3), 259-278.

Investigating the Capacitive Inter-Winding Response of Power Transformer

S. Al-Ameri[1], M. F. M. Yousof, M. R. Ramly, and H. Ahmad

Universiti Tun Hussein Onn Malaysia, 86400 Parit Raja, Malaysia

E-mail: mgammal10@gmail.com

Abstract. This paper presents an investigation on the occurrence of oscillation in the frequency response of transformer due to aging process. The frequency response was obtained using frequency response analysis (FRA) test and capacitive inter-winding test configuration. It has been reported that, there is abnormal occurrence to the frequency response after some period of operating time. The response shows a small oscillation at the mid-frequency region. Although, this oscillation was not there at first measured response. Therefore, three case studies are presented to investigate this occurrence, the measurement, the transfer function mathematical equation and the developed circuit. The case studies of the transfer function and the developed circuit are simulated using MATLAB and being compared with the actual measured response. The parameters that control the mid–frequency oscillation are also studied. The oscillation occurred at mid-frequency region is related with the actual condition of the transformer experiencing aging process. The new finding provides an understanding on the factors that affecting transformer aging. It is believed that, thermal, moisture and oil degradation are the main contributors of transformer aging process. Therefore, the significance of this study is to present the relationship between occurrence of mid-frequency oscillation and transformer aging condition.

1. Introduction

Power transformer is an important equipment in the power system network. Operating at very high loading condition to meet the demands can accelerate the deterioration of power transformer. This is due to the increase of mechanical, electric, and thermal stresses during the operation. All these stresses are called transformer aging process. In this study, the main concern is regarding the aging of power transformer. It is important to understand the factors that cause acceleration of transformer aging. Transformer may still be working as usual despite the aging process. However, its health condition is deteriorating thus need to be continuously monitored. By observing the transformer's condition and performing appropriate maintenance, its lifespan can be maximized and the power system reliability can be achieved.

In general, there are several tests available to diagnose the transformer's condition such as Dissolved Gas Analysis (DGA), Frequency Domain Spectroscopy (FDS) and Partial Discharge test. However, it has been proposed in the literature that for detecting mechanical changes, frequency response analysis (FRA) is the best method to be used. Beside mechanical or physical changes, it was

[1] mgammal10@gmail.com

Content from this work may be used under the terms of the Creative Commons Attribution 3.0 licence. Any further distribution of this work must maintain attribution to the author(s) and the title of the work, journal citation and DOI.
Published under licence by IOP Publishing Ltd

also mentioned that FRA is sensitive towards moisture content, temperature and oil condition [1][2][3][4].
There are four different test configurations to conduct FRA measurement. The first two are the end to end open circuit test and end to end short circuit test which were demonstrated in [5] and [6]. In this paper, we propose an investigation of using different FRA test configuration which is the capacitive inter-winding test. Although this test was mentioned in the CIGRE, IEC, and IEEE standards, it has not been further investigated for its application and potential in examining the transformer condition. This is probably due to the fact that main reason of FRA test is to assess winding damage which can be determined from the end to end test. For this reason, the capacitive inter-winding test is rarely discussed in the literature.

In this paper demonstrates actual FRA measurement on a transformer which is subjected to an accelerated aging process in the laboratory. These measurements are taken before the aging, after 14 days and after 28 days of aging. From the measurement, an oscillation was occurred in the response due to the aging process. Beside the measurement, frequency responses are also simulated using transfer function and RLC circuit. Both transfer function and RLC circuit are for the first time being proposed in this paper. Comparison between three case studies (measurement, transfer function and circuit) are conducted in order to understand the occurrence of the oscillation in mid-frequency region of the capacitive inter-winding response. Although the equation and the circuit have been modelled, it is important to compare the simulated FRA traces with the actual measured response. This methodology is conducted to investigate the factors which affecting the mid-frequency oscillation in the response.

2. Frequency response analysis of power transformer

2.1. Frequency response analysis

There is a general agreement that, frequency response analysis (FRA) is a well technique to diagnosis transformer winding damage. The main function of FRA is to determine the damage on the active parts (winding, leads, and core). Another application of FRA is to verify the mechanical integrity of the transformer after relocation [2][7]. Also, it has been reported that, FRA is sensitive to non-mechanical parameters such as temperature and moisture content of the winding insulation [3]. FRA is a comparative test which requires two measurements to be compared (before and after faulty conditions). Any differences between two measured responses suggests mechanical change has been occurred.

2.2. Capacitive inter-winding measurement

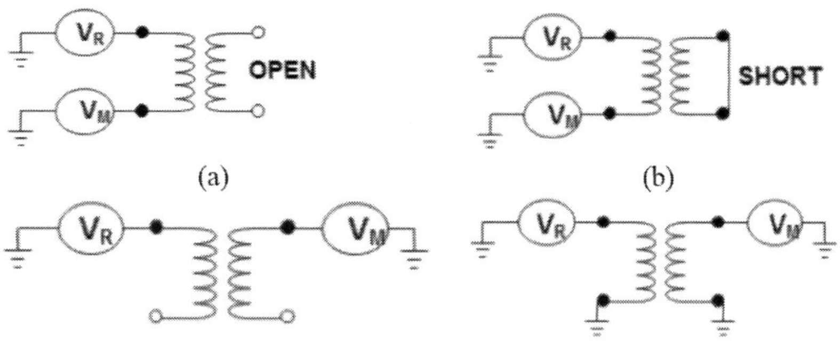

Figure 1. FRA measurement configuration for (a) end to end open circuit (b) end to end short circuit (c) capacitive inter-winding (d) inductive inter-winding [8].

Measurement configuration refers to the connection between winding terminals and FRA equipment. There are four test configurations as shown in Figure 1. V_R and V_M in the figure are the reference voltage and measured voltage respectively. Many previous literatures only focus on end to end open and short circuit tests. In this study, capacitive inter-winding test is presented. As presented in Figure 1(c), the signal is applied to one end of a winding and the response is measured at one end of another winding of the same phase [2]. This test configuration is applied to measure the response due to the influence of the insulation between two windings.

3. Methodology

In this paper, three case studies are presented to investigate the occurrence of oscillation in mid-frequency region of transformer response. These studies are from the actual FRA measurement, the transfer function equation, and the developed circuit. The simulated result from the transfer function and the developed circuit are compared with the result from the actual FRA measurement. Then, the parameters and their values that have effect on the mid-oscillation of the frequency response are determined. By using MATLAB, the proposed mathematical equation of the transfer function is constructed. The resonant frequency and damping effect of the transfer function can influence the frequency response. On the other hand, a RLC circuit is also proposed and has been developed using Simulink to represent the frequency response. The three cases are discussed in details in relation with the aging process of the transformer.

3.1. Actual measurement of response from the transformer

The first case is based on the actual measurement of the response from 5 kVA oil filled transformer as shown in Figure 2. The measurement was taken in three different occasions to determine the transformer aging process. The first measurement was taken before the aging process started (in new condition), second was taken after 14 days and the third was taken after 28 days of operating under accelerated aging condition. This work is a continuation of the previous study which has been presented in [3]. The study investigated transformer aging process using FRA and FDS. However, it did not include further investigation on circuit simulation to transformer aging process. Therefore in this paper, the measured results are compared with the simulated responses from transfer function and developed circuit.

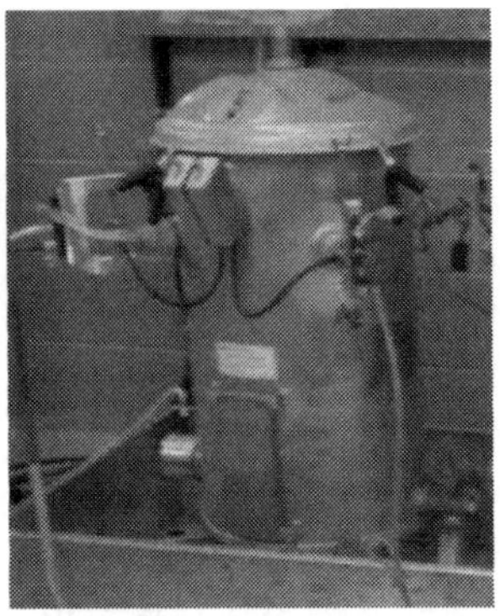

Figure 2. Single phase transformer 5 kVA used in this study.

3.2. Simulation of response from transfer function equation

In the second case study, a transfer function which is a mathematical equation representing the response is constructed. This is required to understand the behaviour of the mid-frequency oscillation which occurred in the actual transformer measurement. The value of parameters of the transfer function are defined in the results section.

3.3. Simulation of response from lumped circuit

In the third case study, it shows the methodology to obtain frequency response using the proposed RLC lumped circuit. The circuit is constructed in MATLAB Simulink to simulate the response. After that, the capacitance, resistance and inductance are adjusted to produce the response which is similar to the actual measured response. Then, examination on the capacitance and inductance values which effect the mid-frequency oscillation is performed.

4. Results and discussions

The measurement of FRA using capacitive inter-winding test configuration was conducted on a single phase transformer. After 14 days of aging, there was an unusual variation occurred on the frequency response. It was absorbed that an oscillation has appeared in the middle region of the frequency response. The oscillation was not seen on the initial response before the transformer experiences the aging process (in the first measurement). It is believed in this occurrence might be due to the aging of insulation.

4.1. Actual FRA measurement

The actual FRA results of the tested transformer are shown in Figure 3 and Figure 4. The measurements were taken before the aging, after 14 days and 28 days of aging process. Based on the results, it shows an oscillation has occurred at about 2 kHz and -80 dB. It can be seen in Figure 3 that the oscillation amplitude has increased from 14 days to 28 days of aging. The phase plot in Figure 4 also presented an oscillation at frequency of about 2 kHz. Due to aging, it causes the drop in phase plot from about 80° to 20° after 28 days.

The main concern now is to investigate how the mid-frequency oscillation occur and what aspect influence it. It is believed that temperature and moisture content increasing the insulation permittivity as mentioned in [3] and [9]. Additionally, the oscillation could possibly also be due to core magnetization as the core influence is usually dominant in low frequency response. Besides that, a change of inter-winding capacitance and insulation resistance may also have trigged the response oscillation. According to [1] and [10], FRA can be influenced by paper insulation of the winding which is subjected to the transformer operating temperatures. The higher the operating temperature is, the more degradation it can cause on the paper. FRA measurement is also sensitive to several other factors as stated in [1][11][12] such as temperature, moisture content, core magnetizing and measurement leads. Further study in [4] investigate on the influence of temperature and moisture on FRA trace. It is concluded that this significant movement is due to changes on the transformer shunt capacitance.

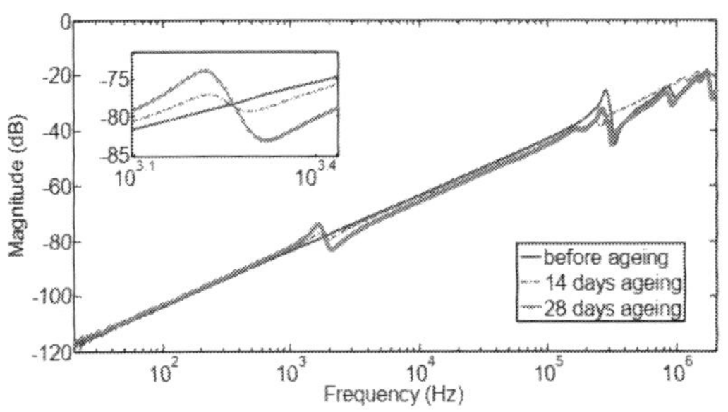

Figure 3. Magnitude plot of frequency response measurement on a single phase transformer.

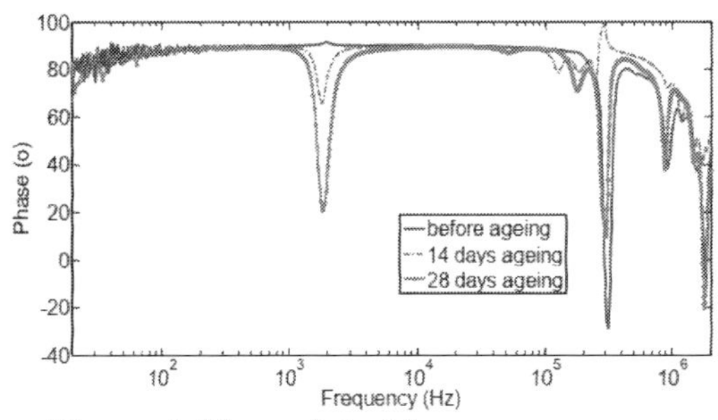

Figure 4. Phase plot of frequency response measurement on a single phase transformer.

4.2. Transfer function equation

To simulate the capacitive inter-winding response, we are proposing the transfer function as in (1). It consist of the combination of three terms. The first term controls the overall positive slope of the response. The other two terms control the peak and valley of the oscillation in the mid-frequency region.

$$H(s) = \left(\frac{s+35}{100000000}\right).\left(\frac{\omega_{na}}{s^2+2\zeta_a\omega_{na}.s+\omega_{na}}\right).\left(\frac{s^2+2\zeta_b\omega_{nb}.s+\omega_{nb}}{\omega_{nb}}\right) \quad (1)$$

The simulated responses from (1) are shown in Figure 5 and Figure 6. These are relatively similar with the measured responses especially on the oscillation in the magnitude plot and the phase drop in the phase plot. The parameter values for obtaining the frequency response from the transfer function is shown in Table 1.

Based on the transfer function, the damping factor, ζ controls the mid-frequency oscillation. To explain the relationship of the aging process and the damping factor, the damping is adjusted to a certain value that produces oscillation which similar in terms of its magnitude and frequency compared with the actual frequency measurement. Data in Table 1 shows that the decrease of damping value, ζ cause an increase of oscillation's magnitude. This indicates that ζ_a and ζ_b control the resonance and anti-resonance of the mid-frequency oscillation. In general, this study suggests that the damping factor might have relationship with the moisture content, temperature and oil quality which are the contributors to the aging of transformer as has been discussed in the previous section.

Table 1. Value of parameters used for the transfer function

	Natural frequency(ω_{na})	Natural frequency(ω_{nb})	Damping(ζ_a)	Damping(ζ_b)
Before ageing	100×10^6	110×10^6	11,200	11,200
14 days ageing	100×10^6	110×10^6	2,600	2,600
28 days ageing	100×10^6	110×10^6	800	800

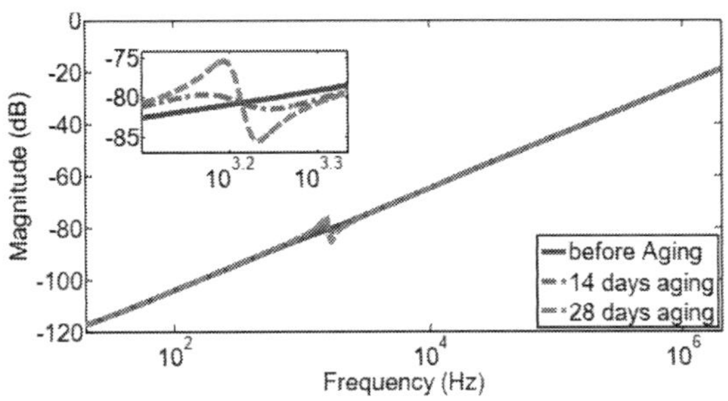

Figure 5. Transfer function magnitude plot.

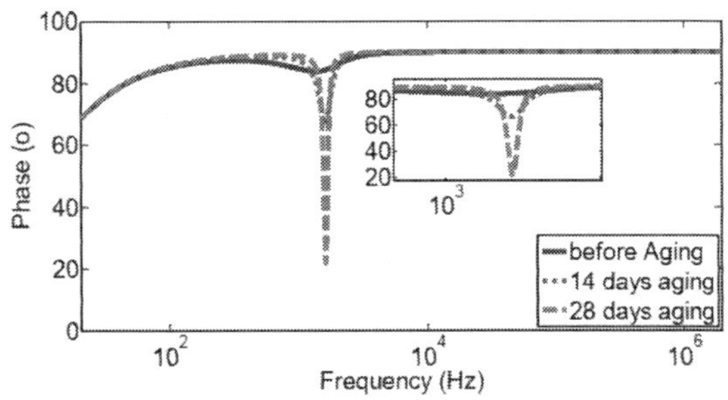

Figure 6. Transfer function phase plot.

4.3. Developed circuit

A circuit model describing the electrical behaviour of the transformer's capacitive inter-winding frequency response is proposed. It is a combination of RLC parameter connected in parallel and series. It is developed to simulate the response which is similar to the actual measured frequency response. This was achieved by adjusting the value of circuit components. With this approach, the effect of capacitance on the mid-frequency oscillation can be investigated. The circuit parameters may have real and imagery values. Figure 7 shows the proposed combination of RLC circuit that can produce similar response which contains the mid-frequency oscillation. In the circuit, reference voltage, Vref is measured across a 50Ω resistor and the output voltage, Vout is measured across another 50Ω resistor. Besides the RLC circuit, other components such as the 50Ω resistor are the equivalent circuit of FRA test equipment.

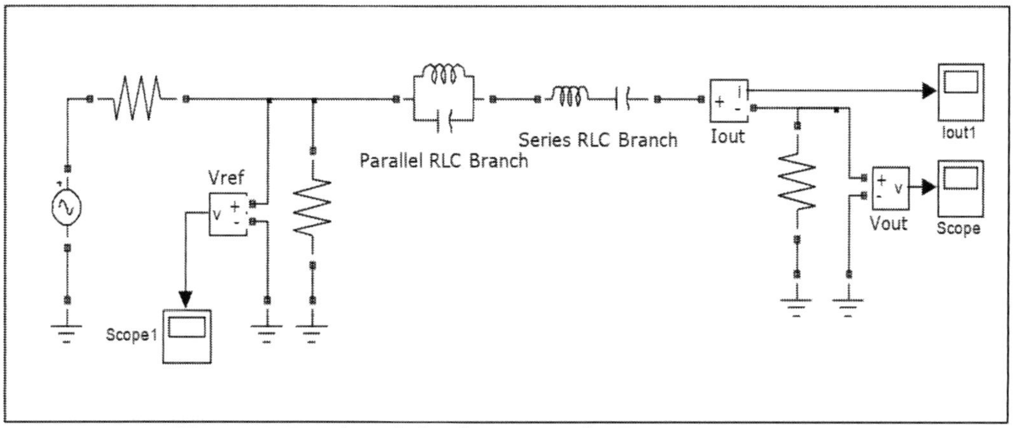

Figure 7. Developed circuit presents the Frequency Response

The simulated responses obtained from the circuit are shown in Figure 8 and Figure 9. The best values used in the circuit to obtain these responses are given in Table 2. The parameters are series capacitance, Cs, series inductance, Ls, parallel capacitance, Cp and parallel inductance, Lp. Each of the parameters controls different factor in the simulated frequency response. In this case, both capacitances (Cp and Cs) are a complex value which contains real and imaginary part. This is mainly due to the complex permittivity (ε = ε' + iε") of the insulation system as capacitance is dependent of the permittivity (C= εA/d). The use of complex capacitance is actually important to achieve the mid-frequency oscillation. The values of complex capacitance which can provide response similar to the actual frequency measurement are shown in Table 3.

Table 2. Value of parameters used for the circuit

Component	Cs (F)	Ls (H)	Cp (F)	Lp (H)
value	$4.444e^{-10} - i\,9.0909e^{-18}$	$2e^{-7}$	$4.65e^{-9} - i\,5.5e^{-9}$	0.1

Table 3. Complex capacitance for all three aging conditions

Cp	Before aging	14 days aging	28 days aging
Imaginary Capacitance (F)	$i\,x\,5.5e^{-9}$	$i\,x\,0.77e^{-9}$	$i\,x\,0.333e^{-9}$
Real Capacitance (F)	$4.65e^{-9}$	$5.71e^{-9}$	$5.71e^{-9}$

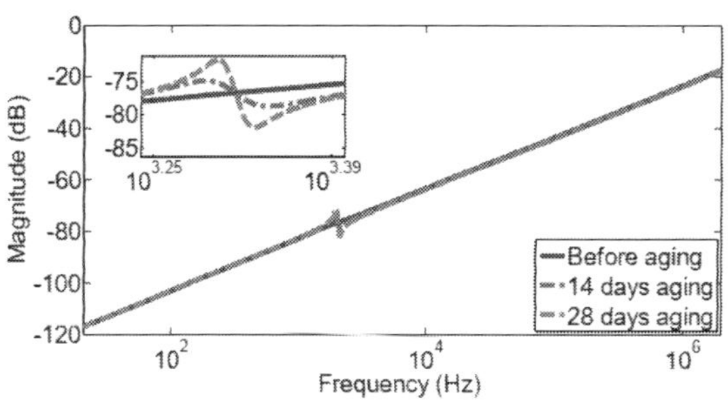

Figure 8. Magnitude response from the developed circuit

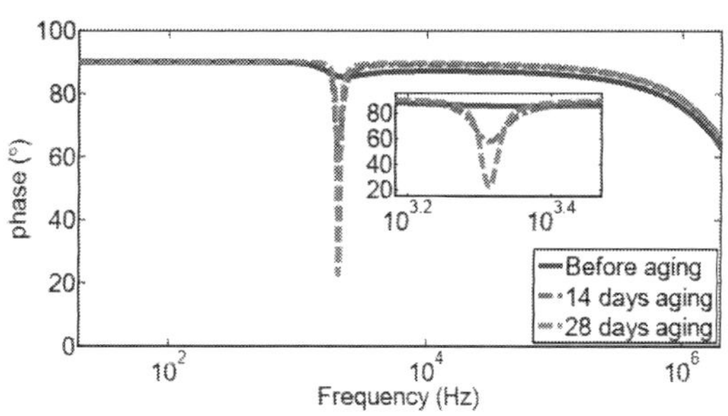

Figure 9. Phase plot from the developed circuit

Figure 8 and Figure 9 present the magnitude and phase frequency responses which contain the same oscillation as in the measured response. As in typical FRA measurement of the capacitive inter-winding test, a continuous positive slope can be seen in the simulated magnitude response. Regarding the oscillation in the mid-frequency region, the simulated responses showed almost similar trend as in the measured response. This is mainly possible by changing the imaginary part of the complex capacitance. This is actually can be supported by studies presented in [3] and [9] where both mentioned that the imaginary part of complex permittivity basically corresponds to the aging of insulation material. Therefore in general, this finding suggest that the aging process in transformer can be modelled by this proposed circuit.

5. Conclusion

This paper investigate the transformer aging process using the capacitive inter-winding configuration of FRA test. The FRA measurement was taken from an actual 5 kVA single phase transformer. This study demonstrated two simulations using transfer function and RLC lumped circuit for further understanding the effect of transformer aging on its frequency response. In order to better interpret the changes on the FRA response, it is useful to understand which elements are responsible for the frequency region and the shapes of the response curve. This can be realized by examining the character's response of the various elements and their combination. The proposed techniques presented are used to evaluate the mid-frequency oscillation by using MATLAB simulation. The frequency response of actual measured was being compared with two simulated techniques from the transfer function and the developed RLC circuit. The obtain results from frequency response of both simulated techniques show a relatively similar with the frequency response from actual measurement. Findings show damping, ζ in transfer function and complex capacitance in the developed circuit are controlling the oscillation which occurred in the mid-frequency region. It is believed that, these parameters are representing the changes in winding capacitance mainly the insulation. According to the theory, the change of winding capacitance is due to increasing of moisture content and temperature which are the main contributors of transformer aging. Further investigation is recommended on the transformer aging process using FRA and capacitive inter-winding.

6. Acknowledgments

This work is supported financially by the University Tun Hussein Onn Malaysia under Special Contract Grant vot U425 and GPPS PhD student grant project vot U708.

7. References

[1] A. A. Reykherdt and V. Davydov, 2011, "Case studies of factors influencing frequency response analysis measurements and power transformer diagnostics," *Electr. Insul. Mag. IEEE*, vol. **27**, no. 1, pp. 22–30.

[2] P. Picher, April 2008, "Mechanical Condition Assessment of Transformer Windings Using Frequency Response Analysis (Fra)," *Tutorial of Cigre WGA2.26 Evaluation*, pp. 30–34.

[3] M. F. M. Yousof, C. Ekanayake, and T. K. Saha, 2014, "Examining the Ageing of Transformer Insulation Using FRA and FDS Techniques," *The University of Queensland Brisbane, Australia,* pp. 1258–1265.

[4] M. Bagheri and B. Phung, 2013, "Influence of temperature and moisture on frequency response analysis of transformer winding," *IEEE Transactions on Dielectrics and Electrical Insulation (EIC)* , vol. **21**, no. 3, pp. 40–44.

[5] J. Bak-Jensen, B. Bak-Jensen, and S. D. Mikkelsen, 1995, "Detection of faults and ageing phenomena in Transformers by transfer functions," *IEEE Trans. Power Deliv.*, vol. **10**, no. 1, pp. 308–314.

[6] M. Prameela, G. R. Murthy, P. Nirgude, and B. Gunasekaran, 2010, "Experimental Investigations to Identify SFRA Measurement Sensitivity for Detecting Faults in Transformers," *16th national power system conference*, pp. 716–721.

[7] T. Committee, I. Power, and E. Society, March. 2013, *IEEE Guide for the Application and Interpretation of Frequency Response Analysis for Oil-Immersed Transformers IEEE Power and Energy Society.*

[8] P. Dinesh Chhajer and V. Naranjo, 2013, "Frequency Response Analysis : Failure Mode Analysis," *Megger 46th AnnualTsdos.*

[9] C. Ekanayake, S. M. Gubanski, A. Graczkowski, and K. Walczak, 2006, "Frequency response of oil impregnated pressboard and paper samples for estimating moisture in transformer insulation," *IEEE Trans. Power Deliv.*, vol. **21**, no. 3, pp. 1309–1317.

[10] M. Kohtoh, S. Kaneko, S. Okabe, and T. Amimoto, 2009, "Aging effect on electrical characteristics of insulating oil in field transformer," *IEEE Trans. Dielectr. Electr. Insul.*, vol. **16**, no. 6, pp. 1698–1706.

[11] N. Abeywickrama, S. Member, Y. V Serdyuk, and S. M. Gubanski, 2008, "Effect of Core Magnetization on Frequency Response Analysis (FRA) of Power Transformers," vol. **23**, no. 3, pp. 1432–1438.

[12] Z. Moravej and S. Bagheri, 2015, "Condition Monitoring Techniques of Power Transformers : A Review," *Journal of operation and Automation in power Engineering*, vol. **3**, no. 1, pp. 71–82.

IOP Conf. Series: Journal of Physics: Conf. Series **1049** (2018) 012009 doi:10.1088/1742-6596/1049/1/012009

A review on recent wheel/rail interface friction management

C T Foo, B Omar[1] and A S Jalil[2]
Faculty of Mechanical and Manufacturing Engineering (FKMP),
Universiti Tun Hussein Onn Malaysia, 86400, Parit Raja, Batu Pahat, Jojor, Malaysia

E-mail: davfoo.ct@gmail.com

Abstract. Traction is influenced greatly by the climatic conditions, especially under wet conditions induced by rain or snowfall which could cause the loss of adhesion and could reduce the coefficient of traction (COT) to a dangerous level. This is why good adhesion is needed to ensure safe and good traction of train. Poor adhesion between wheel and rail can cause various problems and it could be fatal. Friction modifiers are applied to the contact interface to control adhesion, whether to enhance friction or lubricates, to ensure safe running of the rolling stock. In handling poor adhesion issues, sand is the most popular and the main adhesion enhancer used globally on railway networks. Sand particles enhance adhesion dramatically in contaminated wheel/rail contacts such as rail covered by autumn leaf layers or oil stain, and in wet rails. In this review paper, a compilation of some of the research of over the last decade is presented. The topics that will be presented in this paper are the adhesion characteristics of wheel/rail contacts under wet and dry condition, and the influence of sanding on rails covered with leaf layers.

1. Introduction

The wheel/rail contact is a complex and imperfect link due to its small area and loaded with highly concentrated stresses. The conical wheel shape makes the wheelset a mechanical amplifier, limited by the transverse play, with partially sliding surfaces [1]. Even pure rolling bodies involves sliding though it is minimal. In sliding bodies, the friction force is contributed by the coefficient of friction which depends on the roughness, adhesion of the sliding surfaces, and ploughing caused by asperities [2].Error! Reference source not found.

Figure 1 shows the wheel/rail contact area, is divided into two regions; stick and micro-slip regions. Tangential force, also known as tractive force, and longitudinal creep exist because of the micro-slip that occurs in the rear region of the contact region as traction is transferred to rail due to sliding friction in relation to micro-slip. This causes the initiation of yield under the surface. The higher the tractive force, the more the slip increases and the smaller the stick region becomes, resulting in a rolling-sliding contact and also the yielding point to approach the surface. When first yield occurs at the surface, the critical point lies at the boundary between the stick and micro-slip zones. Maximum tractive force is attained at the critical point and is the maximum value of sliding friction which can be

[1] davfoo.ct@gmail.com

Content from this work may be used under the terms of the Creative Commons Attribution 3.0 licence. Any further distribution of this work must maintain attribution to the author(s) and the title of the work, journal citation and DOI.
Published under licence by IOP Publishing Ltd

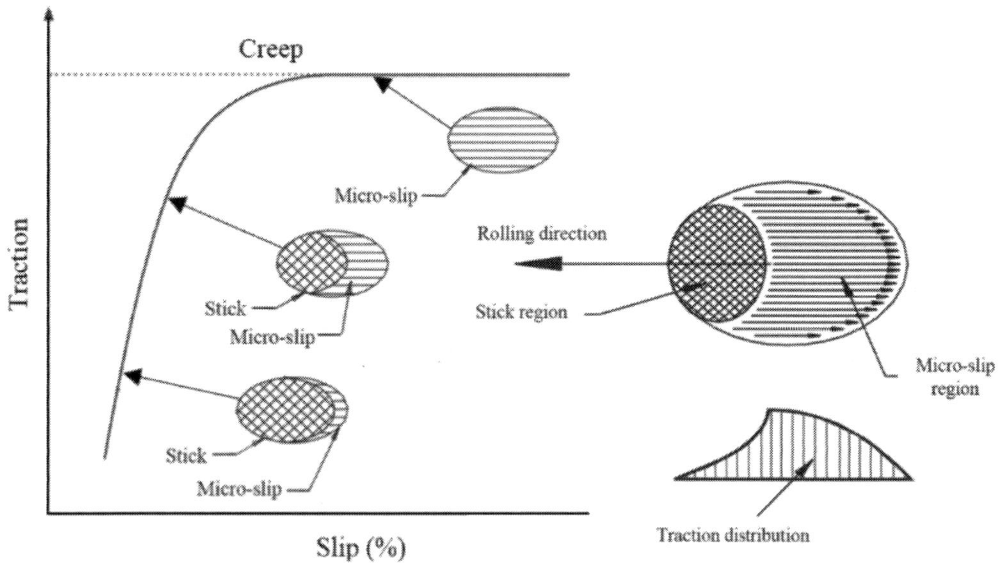

Figure 1. Creep curve in the wheel/rail contact adopted from Olofsson & Lewis [4].

converted into adhesive weight, which is responsible for the transfer of tractive force. After the critical point, capability of transferring tractive force reduces with the increase in micro-slip. When tractive force is greater than adhesive weight, the saturation value of the tractive force is reached and, resulting in the contact patch dominated by slip. The contact is then considered as pure sliding. [2], [3]

The wheel/rail traction characteristics are significantly related to the upper limit of the running speed and running stability. Traction is influenced greatly by the climatic conditions, especially under wet conditions induced by rain or snowfall which causes the loss of adhesion and it may cause wheel slippage and damage to both the wheel and rail [5]. In tropical countries, wetness and humidity is capable of reducing the coefficient of traction (COT) to a value that is even lower than the allowable minimum COT for stable and safe running of the vehicle, which is 0.20 [6]. The same issue of poor adhesion is also faced by countries with four seasons, where autumn leaves and snowfall could reduce the COT to a dangerous level. The presence of a third-body layer in the wheel/rail contact will dramatically influence the traction creep curve. Good adhesion is needed to ensure safe and good traction of train. While, poor adhesion between wheel and rail can cause various problems such as producing high frequency noises [7], severe rolling contact fatigue (RCF) [8], exacerbate the development of rail corrugation [9], high energy dissipation [10], unstable traction before accelerating [11], and on the other hand causes poor braking performance resulting in extended stopping distances [12]. Furthermore, excessive tangential friction force at sharp curves could cause wheel climb derailment [5].

In handling poor adhesion issues, sand is the most popular and the main adhesion enhancer used globally on railway networks. Sand particles enhance adhesion dramatically in contaminated wheel/rail contacts such as rail covered by autumn leaf layers, rainfall, or oil stain, because of the intrinsic interlocking action of crushed sand particles that embedded into the softer part of the wheel and rail surfaces, and therefore, enhancing traction [13]. Sanding does not only restores adhesion but is also able to remove dried leaf layers covering the rail [14], [15]. Even though sanding had been widely used to improve traction in poor adhesion condition, a satisfactory method to enhance adhesion and avoid severe wear of wheel/rail interface at the same time is still uncertain [16], [17]. Sand tends to increase the wear rate of the rail by a factor between two to ten due to the damage caused by asperities in the wheel/rail interface [18]. This is because sand could cause indents and scratches on the wheel and rail surfaces because of its hardness [14] and gives a high value of mass loss [16]. Arias-Cuevas had conducted extensive laboratory tests to study the influence of sanding parameters on traction [15],

[19]–[21]. Although numerous studies regarding the influence of sand on adhesion and wear had been done recently, the influence of different types and different compositions of sand used for sanding is also unclear as different types of sand exhibit different hardness level which may interact differently in the contact interface. Wang et al. (2014) and Cao et al. (2016) conducted a laboratory test to investigate the influence of alumina particles as replacement for sand in sanding, which yields quite favourable results.

2. Dry and wet wheel/rail contact

Friction modifier (FM) is applied to the wheel/rail contact to manage and generate the required friction coefficients. Generally, there are three categories of FM [4]; (1) Low coefficient FMs such as lubricants which are applied at the wheel flange-gauge corner interface to give friction coefficients less than 0.2, (2) high FMs with intermediate friction coefficients of 0.2~0.4 which are used in wheel thread-rail top applications, and (3) very high FMs such as friction enhancers which are used to increase adhesion for both traction and braking, especially at where adhesion loss problems occur.

When FMs are added into the contact interface, the lubrication effect will kick-in immediately. Figure 2 shows the general trend of COT under dry condition, wet condition and lubricated contact. Applying FM is able to reduce the COT instantly. The fluctuation of COT for lubricated contact after the application of FM is due to the uneven spreading of FM in the interface. After the FM is evenly spread, the COT becomes stable and constant. This is most probably why rail vehicles travelling in the middle of rain and snowfall is hazardous as the steel wheel loses its grip on the rail and causes slippage and difficulties in braking and accelerating. This explains why the climate of a location, such as ambient temperature and humidity, can greatly affect the safety of train operation, as high temperature and humidity can also cause the reduction of COT [11].

The traction curve for contact under dry condition shown in Figure 2 can be divided into three regions; initial stage (steep positive slope), wear stage (steep negative slope), and stable stage (gradual decline). During the initial stage, the oxide film on the disc surface would be gradually reduced due to the increase in sliding distance. Therefore, the shear stress increases at the interface resulting in rapid increase of COT. At the same time, the effective hardness of the disc surface will decrease. At the end

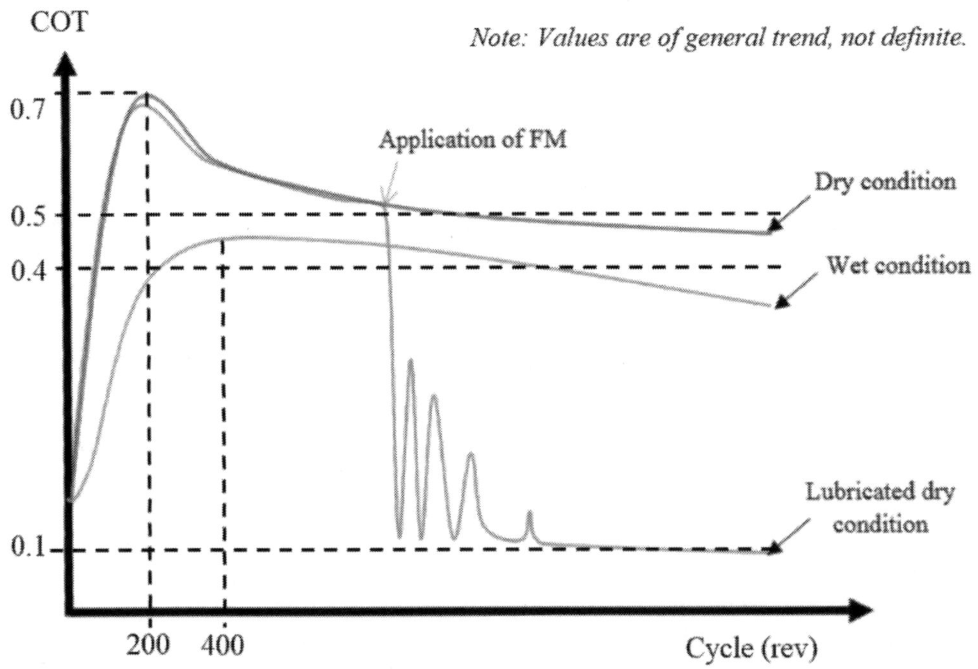

Figure 2. General pattern of COT after the application of FM (FM).

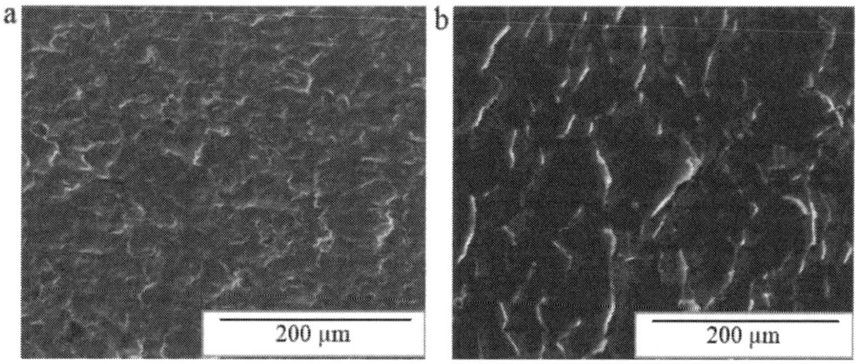

Figure 3. Micrographs of wear surface under (a) dry condition, and (b) wet condition [23].

of this first stage, the hardness and shear stress varies negligibly due to the formation of mixed layer of steel and the destroyed oxide film which had embedded in the surfaces, thus, resulting in a lower rate of increase in COT. After the COT reaches a maximum value (running-in), the contact tends to experience a decline in COT, which is the wear stage. It is common that the COT decreases after running-in for dry condition. This is because a significant wear on the surface had begun to occur. Subsurface plastic deformation causes the hardness to increase by strain hardening. Although the COT is almost constant with a slight decline in the stable stage, the hardness increases gradually due to the continuous gradual increase of shear stress on the surface. [5]

On the other hand, the COT under wet condition, unlike dry condition, increases gradually and then becomes steady. After that, it decreases marginally. Furthermore, rail wear is mild under wet condition compared to that of dry condition because of the lubricating effect of the fluid which reduces abrasion. The traction curve under wet condition is divided into two stages; initial stage and stable stage. In this initial stage, the COT will increase rapidly (but with lower gradient compared to dry condition) with a slight increase in surface roughness. The effective hardness is nearly constant as the wear of the oxide layer on the steel surfaces is minimal due to the tribochemical reaction film with water and the lubricity of the fluid. The oxide film would then be destroyed more moderately, forming a tribochemical reaction film of oxide and hydroxide mixture, which would cause the decrease of the effective hardness. At the end of the initial stage, the rate of increase in COT becomes slower because of the constant surface roughness and slight decrease in effective hardness. In the stable stage, the surface roughness tends to decrease slightly, while the hardness increases slightly. This explains the marginal decrease in COT. [5]

Mass loss of wheel and rail are caused by both abrasion and material removal. Under dry condition, delamination and wear scars can be seen on the rolling surface while there is some peeling on the wear surface under water-lubricated condition as shown in Figure 3 [23]. The mass loss of rail disc after the application of FM tends to be higher than that in dry tests due to flaking and cracks growth assisted by fluid. The wear rates of wheel and rail under dry conditions remains almost constant, as it stabilized after a certain number of fatigue cycles. However, the wear rate of rails with FM is only constant in the beginning and becomes almost double as the number of fatigue cycle increases. The crack angle and crack depth under wet condition is also doubled of that under dry condition. The wear mechanism responsible for the greater wear rates is delamination that leads to spalling [24]. It is the hydro-pressure within the crack that is responsible for the crack elongation which eventually leads to flaking. As fluid enters the cracks on rail initiated by shear and pressed by the rolling contact of the wheel, the contact pressure is transmitted hydraulically via the crack to the subsurface of the rail as shown in Figure 4. The fluid is then entrapped and pressurized inside the crack which then results in the generation of squeeze film pressure. [25]. When two discs slide on each other, a tangential force that varies with the slip will appear. These tangential forces cause compressive and tensile circumferential stresses in and around the contact zone, and the stresses are oppositely located in the wheel and rail.

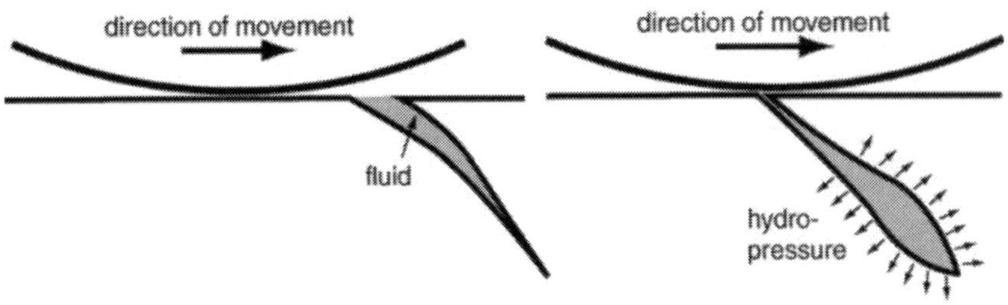

Figure 4. Mechanism of crack propagation by the pressure of a trapped fluid [26].

Prior to the contact zone the compressive stresses close the tips of the cracks in the wheel avoiding the fluids to enter in the cracks. On the other hand, the tensile stresses open the cracks tips in the rail and facilitate the entrance of fluids; right after the contact zone the stresses are reversed and the crack in the rail closes pressurizing the fluid inside [24].

Abrasion usually causes more removal of wheel material than rail. This is why mass loss of wheel disc tended to be higher than that of rail disc. Another reason why crack elongation assisted by fluid pressure does not occur on the wheel is because fluid on the wheel surface is being removed by the centrifugal force of the rotating wheel. Furthermore, rain and snowfall only falls on the rail and not on the wheel. FMs are also applied on the rail instead of the wheel. There are three FM application methods, which are mobile lubricators, wayside lubricators, and on-board lubricators [4]. Mobile lubricators are railway vehicles designed to apply lubricant to the gauge corner of the track. Wayside lubricators are mounted next to the track and spray lubricant to the rail gauge corner. Only on-board lubricators apply lubricant on to the wheel flange, which is then transferred to the gauge corner of the rail.

Rough surface has high asperity which could provide a better grip than smooth surface. Therefore, greater surface roughness gives higher maximum COT to a wheel/rail contact. This is confirmed by the recent work of Chen et al. [17], both theoretical and experimental analysis, where under wet condition the COT reduces as the surface roughness decreases regardless of the contact pressure. However, based on the theoretical analysis, which employed a numerical model that adopted Elasto-Hydrodynamic Lubrication (EHL) theory and Greenwood–Williamson's (G–W) stochastic model, different trends of COT can be obtained by varying the surface roughness, as shown in Figure 5.

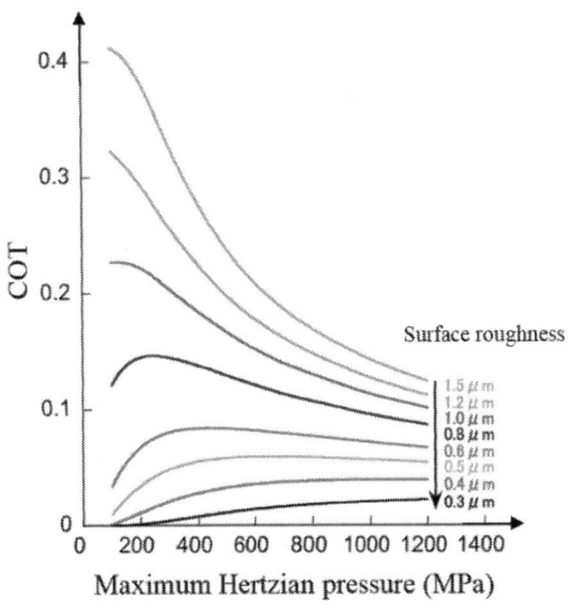

Figure 5. Theoretical simulation result of influences of Hertzian pressure and surface roughness on COT [17].

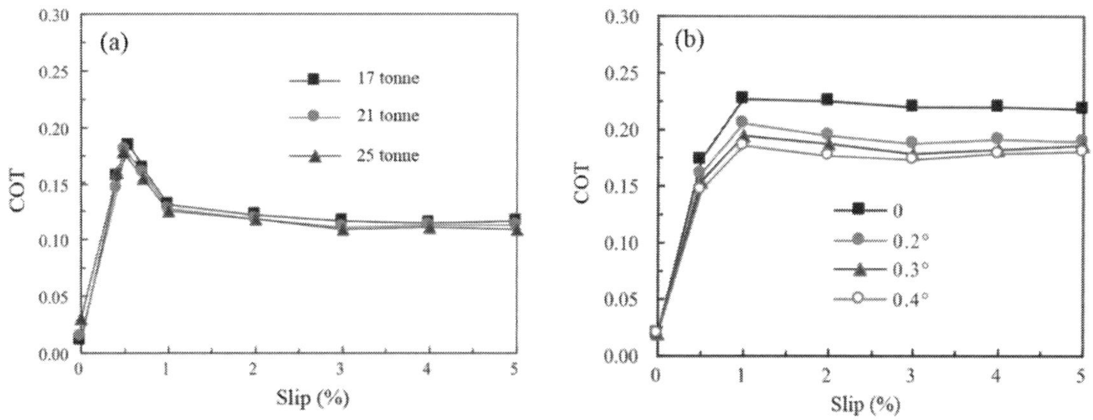

Figure 6. The influence of slip on COT under wet condition with (a) various axle and (b) various angle of attack [23].

The experimental analysis of Chen et al. [17] showed that the COT seems to decrease when axle load increases in slow rolling speed condition, regardless of surface roughness. However, at faster rolling speed the COT for high surface roughness decreases drastically while for low surface roughness decreases slightly before increasing as contact pressure increases. The COT will not be affected much in the case of the high-speed region. However, under low-speed condition such as metro and commuter, the maximum COT will decrease with an increase in rolling speed [11]. From both theoretical and experimental analysis, though the results are slightly different, the influence of axle load on the COT was clearly found to be dependent on angular velocity and surface roughness. The increase in load causes the reduction of COT because the increase in axle load results in the expansion of the contact area between the wheel and rail and can transfer more tractive force. However, the ratio between tractive force and normal force would decrease, which will reduce the COT. But when water is added into a dry contact interface, the COT dropped dramatically to a similar value, regardless of the axle load [23]. Though the COT varies as slip increases, the effect of axle load on traction is minimal and almost constant, as shown in Figure 6(a). If water of higher temperature is applied into the contact interface, the maximum COT is higher compared to that of lower water temperature, but with the same declining trend [27].

The wheel/rail contact on a curved rail is simulated by adjusting the angle of attack of rail disc in a twin disc roller rig machine. Sharp curves would result in a greater reduction in COT than that of a curve with larger curvature radius, in both dry and water-lubricated condition as shown in Figure 6(b). This is because the longitudinal creep force of the contact area would reduce by decreasing the curve radius, which leads to the loss of adhesion [23]. According to the experiment conducted by Baek et al. [11] with setting of 800 MPa of contact pressure, there are three types of transient traction characteristics dependent on slip, as shown in Figure 7. In cases where slip is less than 0.3%, the COT increased linearly and then became steady. In cases where slip is above 0.3%, the peak appeared at a certain sliding distance. In cases of slip greater than 2.8%, the COT increased gradually after the peak.

The dependency of COT on the presence of iron oxide layer on the wheel is unclear. However, the COT under dry condition is undeniably high regardless of the thickness of iron oxides and surface roughness. A rough rail surface with thick oxide layer on the wheel can help to gain extra adhesion during taking off, but at the cost of severe wear. The surface of the rail tends to suffer severe surface delamination if the oxide layer is too thick. On the other hand, the presence of the iron oxide layer is very critical under wet condition. Thin oxides are able to protect the interface from wear and ploughing for rails with smooth surface roughness even though the adhesion is high. While, thick oxides break off becoming debris, enhances the ploughing effect and increases wear. Surprisingly, the COT of thick oxide layer under wet condition is low because wear debris tends to destroy more of the

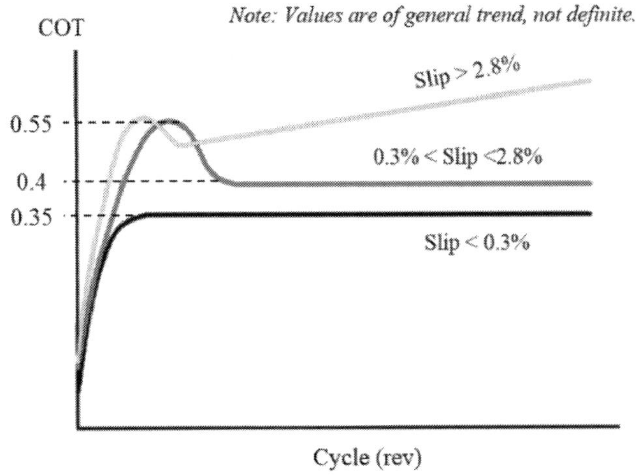

Figure 7. Schematic patterns of the COT curve [11].

surface to produce more debris. This process continues on until the quantity of wear debris is enough to form a lubricating paste with water which eventually reduces COT. [28]

3. Sanding

Figure 8 shows the mechanism of a sanding apparatus. In practice, the sanders are fitted on only certain axles of a train so that the amount of sand laid on the rails is first used by the wheels of sanding axles, while the following wheels roll over the remaining crushed sand [19]. An effective sanding at the leading wheels could condition the railhead so that the following wheels will not experience poor adhesion problems. Sand does not only enhance adhesion but also acts as a solid lubricant under dry wheel/rail contacts at different slips, feed rates, and particle sizes. Therefore, a proper adjustment on the most influencing parameters, which are the feed rate, particle size, and wheel slip, are needed in order to provide good friction management on the railhead.

Leaf layer does not only cause the wheel to lose traction but also insulates the rail from conducting electricity [14], [19]. Furthermore, the leaf stalks will cause long indentations and scratches to the wheel and rail surfaces. The adhesion loss caused by leaf is even lower than that of oil stain. The optimum value of COT is 0.35 under dry condition [7], [29] while the minimum allowable COT is 0.2 [24]. Any wheel/rail contact with COT below 0.2 will be considered as unsafe. Sand is able to restore

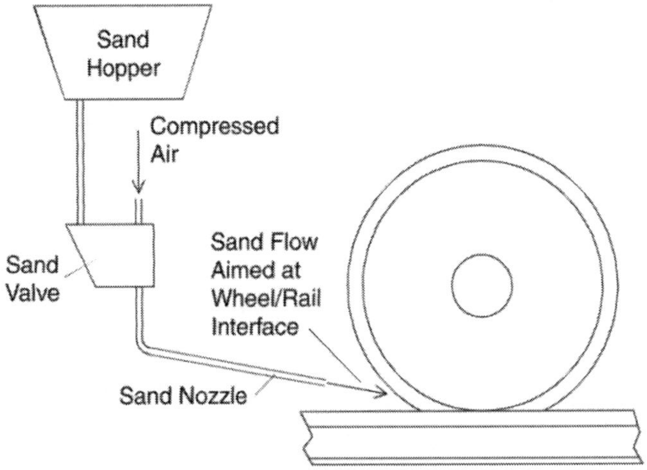

Figure 8. Sanding apparatus [4].

both adhesion and electrical conductivity by effectively removing the leaf layers from the rail surface. There are two types of sanding method, which are initial sanding and continuous sanding. Initial sanding is used to provide the initial adhesion of the contact interface by ejecting sand particles into the interface for only once, while continuous sanding is the ejection of sand into the interface over a period of time in order to maintain desired lubricity along the travel.

Figure 9 shows the typical influence of sanding on the COT for rail discs with leaf layers contamination in a typical twin disc roller rig test. Typical values of slip found in actual operations of traction and braking are 1%, 5%, and 10% and was found that the baseline of COT under dry uncontaminated contact of twin disc roller rig test is approximately 0.6 for most slip conditions in actual operations and 0.3 for 0.5% slip, while the baseline of COT in leaf-contaminated rail disc in most slips, depending on the number of cycle, ranged from 0.15~0.25 which is considered unsafe [15], [21], [30]–[32]. In initial sanding the COT increases due to the removal of leave layers as shown in Figure 10 (left). Increasing the number of sanding axles is able to remove the leaf layers even faster and also provides a larger initial adhesion improvement after initial sanding. On the other hand, the influence of slip becomes weaker when fewer number of sanding axle is used. However if more sanding axles are used at high slips for initial sanding, a black powder of broken leaf layers which can be easily rubbed away with fingers, as shown in Figure 10 (right), will form on the disc which will cause the reduction of traction gradually and then be removed eventually due to the abrasive action of the entrapped sand, restoring the adhesion to clean condition [21]. Furthermore, the adhesion recovery is faster at higher slips.

Since sand are ejected continuously into the interface in continuous sanding, the adhesive behaviour of the sand are also different from that of initial sanding even though the sand may have the same particle size ejected into the interface at the same feed rate and slip. An investigation on particle size was conducted by O. Arias-Cuevas, Li, & Lewis [19] on improving adhesion of rail with leaf layers by initial sanding, shows that particle size that ranged between 0.3~0.6 mm (medium-size sand) gives the best result in restoring traction, better than particle size of range 0.06~0.3 mm (small-sized sand), while sand with particle size of range 0.85~1.6 mm (large-sized sand) restores the COT the least among all the other tested sands. Having more particles travelling through the interface is expected to ease the removal of contamination. But medium-sized sand performed better than small-sized sand, which could probably due to the balance in size to promote good particle entrapment and to remove the leaf layers is satisfied. However, another investigation on particle size which was also conducted

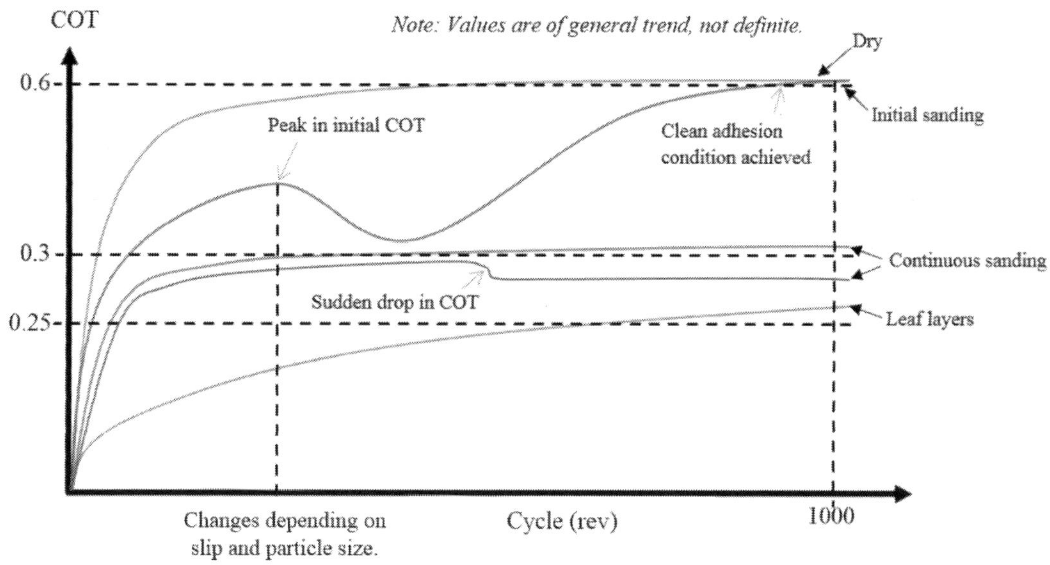

Figure 9. General pattern of the influence of different sanding conditions on COT.

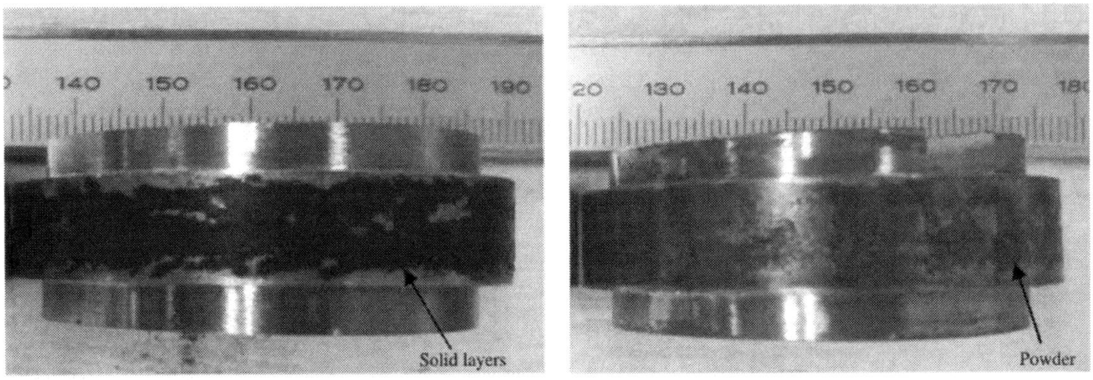

Figure 10. Photographs of wheel disc with remaining solid leaf layer (left) and powder of broken layer (right) [21].

by Oscar Arias-Cuevas et al. [15], but with continuous sanding, shows that larger sand particle sizes promotes greater COT. Instead of enhancing adhesion, smaller particle size sand, including medium-sized sand, promotes lubrication with COT lower than that of leaf contaminated baseline, which also confirms the previous study of O. Arias-Cuevas, Li, & Lewis [19].

Furthermore, The investigation of O. Arias-Cuevas, Li, & Lewis [19], which is by initial sanding, also discovered that medium size sand is the best in removing leaf layers and at the same time poor at electrical insulating. But the investigation of Oscar Arias-Cuevas et al. [15] with continuous sanding shows that smaller sand particle sizes, ranged between 0.06~0.6 mm, caused more electrical insulation than larger sands, leading to open circuit conditions. This is due to the breaking of larger particles into smaller particles to withstand the contact load before forming a compacted crushed sand coating on the rolling surfaces with the broken particles that enter the disc interface. When using smaller sized sand, more particles can enter the contact without being broken up which will form the sand coating faster. Electrical insulation is not only dependent on the particle size of the sand but also the feed rate. As the feed rate increases, the disc interface will experience a transition from partial lubrication to full lubrication regime due to the increase in sand particle in the interface [19]. The additional sand particles that enters the interface at higher feed rate will speed up the formation of sand coating on the disc surface.

Large sand experiences breaking of sand particles from the contact load which will cause the COT to be slightly unstable. The larger sand particles will first bear the load and break up until the number of particles to withstand the load is sufficient. Small sand provides steady COT partly due to the formation of compacted crushed sand coating on the disc [15]. The coating of compacted crushed sand on the disc surfaces is formed by entrapment of crushed sand particles in the softer steel surfaces and are repeatedly compressed under contact load. Furthermore, larger particle size at the same slip forms lesser sand coating. This may be caused by two reasons: (1) Larger sand particles has better abrasive action and could remove more coating, and (2) less sand particles will be entrapped in the interface at larger particle sizes, hindering the coating formation.

Higher slips will also increases the effect of adhesion because slip increases with tractive force [14], as shown in the traction creep curve, until the saturation point reached where the traction is maximum before the contact is dominated by slip contact [2]. At higher slips condition, the COT is almost constant with only slight deviations as shown in Figure 11 [14]. Theoretically, a pure rolling contact, where there is 0% slip, could provide zero COT. However, the experiment of Wang et al. [23] and Baek et al. [11] yielded a slightly positive result, which could possibly due to the error in setting the datum for zero traction force, small vibration, and asperity differences between the two contact surfaces.

In terms of sanding mechanism, the increase of slip enhances the interlocking action of sand due to higher abrasion as a result of sliding of crushed sand particles for a longer distances when travelling

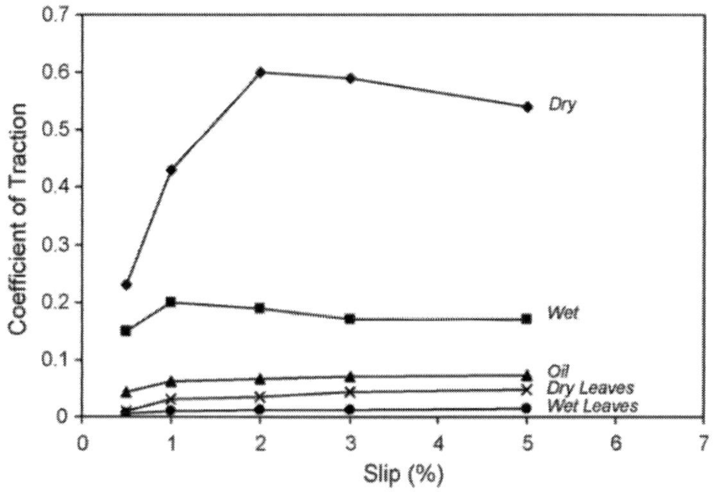

Figure 11. Effect of slip on COT in different rail conditions [14].

through the contact interface at higher slips. This interlocking of sand yields higher COTs. Such increase in tractive force is able to remove leaf layers, coatings of FMs, thin film coatings, and compacted crushed sand coating. Therefore, there seemed to be a threshold dependent on both sand particle size and slip above which no sand coating would be formed. This explains the sudden drop of COT, as shown in Figure 9, for continuous sanding [15]. Increasing slip could reduce the probability of sand coating formation as embedded crushed sand particles abrade the surface of the opposite disc due to the relative motion, but lead to surface wear. This abrasive behaviour also promotes the removal of rail and wheel material on the rolling surfaces. In addition, high strains on the disc surfaces, as a result of the high tangential stresses occurring at higher slips, causes surface strain hardening and could have led to greater grain deformation.

Besides slip and particle size, feed rate of sand also influences the COT. Feed rate of 7.5 g/m is representative of the ones used in practice [18]. As solid lubricants, sand can reduce the COT of 0.6 for dry condition to a range of 0.1~0.4 according to the grain size. There will be an instant reduction in COT when sand is fed into the contact. Higher feed rates of sand during sanding promote the continuous reduction of COT at all slips as shown in Figure 12, due to the flooding of particles which reduces abrasion and promotes coating formation [19]. However, the rate of reduction is only severe for low slips while the rate of reduction for higher slips is lesser due to the increased abrasion which prevents the formation of sand coating.

Although larger particle sizes can enhance adhesion, it also causes indentations and cracks on the disc surfaces while smaller particle sizes only cause abrasion and formation of coating. In addition, larger particle sizes give higher strains near the disc surfaces. These strains, if high enough, could form cracks on the rolling surfaces. The subsurface plastic deformation layer is deeper and greater for large sand particle size while smaller sand shows only little plastic deformation. This is because the elastic limit is exceeded and introduces residual stresses when the contact stress is first applied. Continuous application of load will subject the materials to the combined action of contact stresses and residual stresses which causes a process known as shakedown, where initial plastic deformation introduces residual stresses which make the steady cyclic state purely elastic which are protective to the rail. One of the effects of shakedown is strain hardening of material due to the exceeding of shakedown limit which causes orthogonal plastic shear in the subsurface elements. In repeated rolling cycles the plastic deformation will accumulates and causes the surface layers to be displaced 'forward' as shown in the result of Figure 13, which shows that the grain deformation in the near-surface region. The deformation increases in larger particle sizes and caused more strain hardening of the disc surface for both wheel and rail, which means that larger particle size can cause a greater effect of shakedown.

However, large sand seems to cause less rough surfaces than small sand. This could be related to the balance existing between the number of particles entering the disc interface and their effective abrasive size. [3], [15]

It is undeniable that slip had a clear influence on the subsurface deformation as the sliding distance of sand particle in the interface is elongated. Figure 14 shows the plastic deformation of wheel and rail discs tested at 1%, 5%, and 10% slip. The deformation in 1% slip is almost unnoticeable, while 5% and 10% slip caused severe deformation [15]. With the increase in slip ratio, the tangential force grows and the plastic deformation results in an increase of surface hardness. The maximum shear

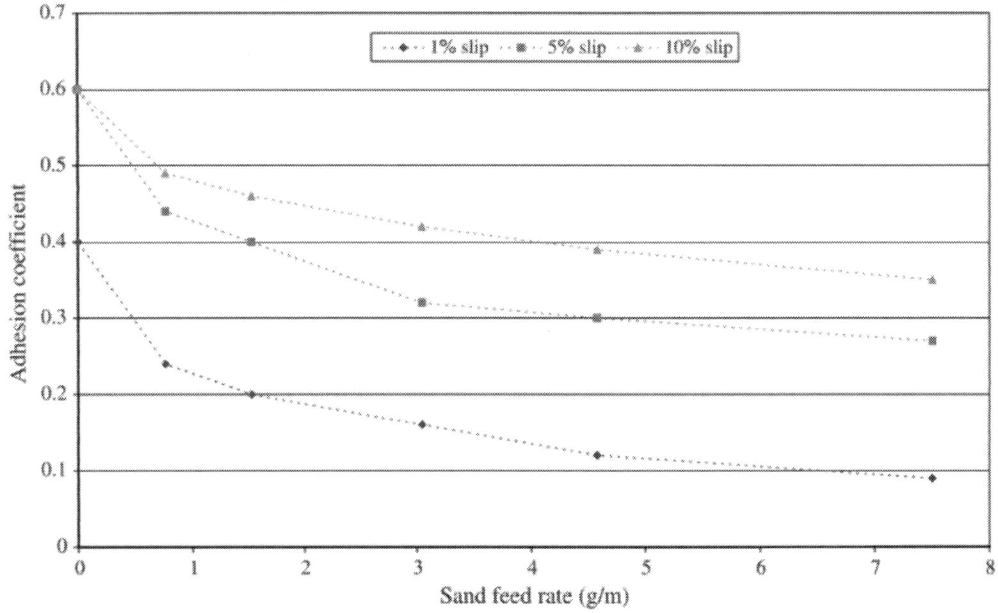

Figure 12. Average COT of S sand (0.06~0.3 mm) during feed at 1%, 5%, and 10% slip and different feed rates under dry conditions [19].

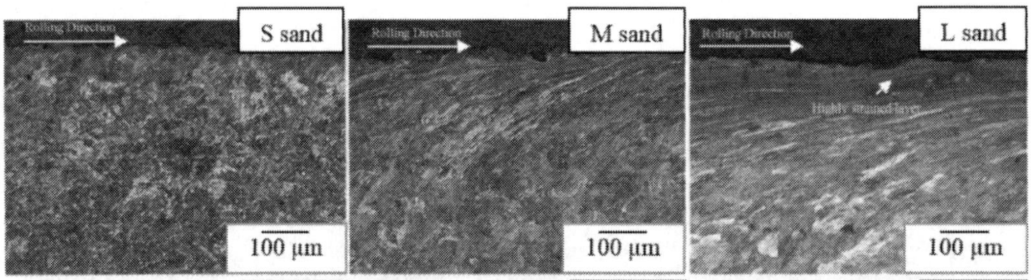

Figure 13. Subsurface micrographs of rail (top) and wheel (bottom) after tests with S (0.06~0.3 mm), M (0.3~0.6 mm), and L sand (0.85~1.6 mm) at 5% slip, 7.5 g/m feed rate [15].

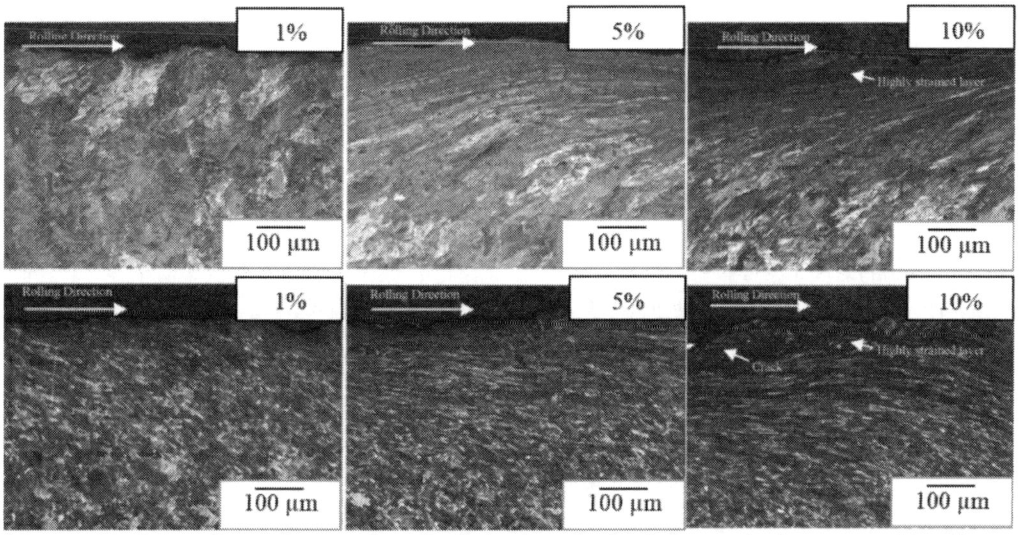

Figure 14. Subsurface micrographs of rail (top) and wheel (bottom) after tests with railway sand (0.25~1.4 mm) at 1%, 5%, and 10% slip [15].

stress is located on the subsurface, while the point closer to the surface at higher slip will have a harder surface. The hardness of the disc surfaces for both wheel and rail increases dramatically, up to double its original hardness, after tests are done for all slips [33]. Cracks, deep indentations, and spalling defects can be found on the disc which may cause break-off as a flake. Such micro-fracture contributes to the mass loss and surface roughness as a result of wear due to fatigue. The wear rate of rail increases linearly with the increase of slip, but the wear rate of wheel will have a great increase at a certain point of slip [33].

4. Summary

Friction management in wheel/rail contact is extremely important as poor friction management could cause various problems and issues as mentioned earlier. There are numerous studies and journals on friction management of wheel/rail contact over the past ten years while this review paper only covers some of the journals on contacts under dry/wet condition and sanding in rail contaminated with leaf layers.

Even though FM and sand had been used for many years in friction management of wheel/rail contact, applying liquid FMs and sand causes more wear on the rail surfaces when used to improve adhesion. A satisfactory method to enhance adhesion and avoid severe wear at the same time is still uncertain. The influence of different types and different compositions of sand used for sanding is also unclear as different types of sand exhibit different hardness level which may interact differently in the contact interface. Furthermore, Leaves are actually good lubricating agent due to its high lubricity effect on wheel/rail contacts. Perhaps leaves can be used as dry solid FMs, instead of water-based FM which promotes crack growth, to control friction at high friction rail sections for smoother traction.

Besides applying FM, there are also other approaches used to improve traction in low adhesion condition. A new method of friction enhancer, which is mixing sand into traction gels was proposed by S. R. Lewis, Lewis, Cotter, Lu, & Eadie [30]. Different chemical compositions of thin film rail coating such as polytetrafluoroethylene (PTFE) [34], molybdenum disulphide (MoS_2) and tungsten disulphide (WS_2) [35] had been used to reduce friction of the rail. Thin films bonded with graded coating [36], and laser cladding layer [37] were also tested and are still testing on their tribological characteristics in order to find a better solution to improve traction in low adhesion condition.

The dependency of COT on the presence of iron oxide layer on the wheel is also uncertain for both contacts under dry and wet condition. Thin oxide layer seems to be able to lubricate the interface under wet condition, but the inconsistent COT shows independency to the thickness of the oxide layer, especially under dry condition.

The reason why the increase in ambient temperature could cause the reduction in COT is also unknown. Perhaps this phenomena occurs because of the conversion of heat energy from the fluid into kinetic energy stored in the steel molecule which results in metal expansion, thus, reducing the surface hardness. The contact patch is then enlarged and reducing the ratio of tangential force to normal force, resulting in reduction of COT.

5. Acknowledgment

This research was funded by Postgraduate Research Grant (GPPS/U799).

6. References

[1] Ayasse J and Chollet H 2006 *Handbook of Railway Vehicle Dynamics* (New York: Taylor& Francis Group) pp 85

[2] Lewis R and Olofsson U 2009 *Wheel–Rail Interface Handbook* (Sweden: Woodhead Publishing Limited) pp 34

[3] Johnson K L 1985 *Contact Mechanics* vol 37 9th ed (UK: The Press Syndicate of the University of Cambridge pp 202

[4] Olofsson and Lewis R 2006 *Handbook of Railway Vehicle Dynamics* (New York: Taylor& Francis Group) pp 121

[5] Baek K S, Kyogoku K and Nakahara T 2007 An experimental investigation of transient traction characteristics in rolling-sliding wheel/rail contacts under dry-wet conditions *Wear.* **263** 169

[6] Fulford C R 2004 *Review of Low Adhesion Research* (UK: The Rail Safety and Standards Board)

[7] Eadie D T, Santoro M and Kalousek J 2017 Railway noise and the effect of top of rail liquid friction modifiers: Changes in sound and vibration spectral distributions in curves *Wear* **258** 1148

[8] Steenbergen M 2017 Rolling contact fatigue: Spalling versus transverse fracture of rails *Wear* **380** 96

[9] Kalousek J 2002 Keynote address: Light to heavy, snail to rocket *Wear* **253** 1

[10] Alarcón G I, Burgelman N, Meza J M, Toro A and Li Z 2016 Power dissipation modeling in wheel/rail contact: Effect of friction coefficient and profile quality *Wear* **366** 217

[11] Baek K S, Kyogoku K and Nakahara T 2008 An experimental study of transient traction characteristics between rail and wheel under low slip and low speed conditions *Wear* **265** 1417

[12] Arias-Cuevas O and Li Z 2011 Field investigations into the performance of magnetic track brakes of an electrical multiple unit against slippery tracks. Part 2: Breaking force and side effects *Proc. Inst. Mech. Eng. (Part F J. Rail Rapid Transit* vol. 226) pp 72

[13] Astle-Fletcher M W 1963 Mechanical methods of improving rail adhesion *Proceeding Inst. Mech. Eng. Conf. Proc. (electronic material* vol 178*)* pp 185

[14] Gallardo-Hernandez E A and Lewis R 2008 Twin disc assessment of wheel/rail adhesion *Wear* **265** 1309

[15] Arias-Cuevas O, Li Z and Lewis R 2010 A laboratory investigation on the influence of the particle size and slip during sanding on the adhesion and wear in the wheel-rail contact *Wear* **271** 14

[16] Wang W J, Liu T F, Wang H Y, Liu Q Y, Zhu M H and Jin X S 2014 Influence of friction modifiers on improving adhesion and surface damage of wheel/rail under low adhesion conditions *Tribol. Int.* **75** 16

[17] Chen H, Namura A, Ishida M and Nakahara N 2016 Influence of axle load on wheel/rail adhesion under wet conditions in consideration of running speed and surface roughness *Wear* **366** 303

[18] Lewis R and Dwyer-Joyce R S 2006 Wear at the wheel/rail interface when sanding is used to increase adhesion *Proc. Inst. Mech. Eng. (Part F-Journal Rail Rapid Transit* vol 220*)* pp 29
[19] Arias-Cuevas O, Li Z and Lewis R 2010 Investigating the lubricity and electrical insulation caused by sanding in dry wheel-rail contacts *Tribol. Lett.* **37** 623
[20] Arias-Cuevas O and Li Z 2011 Field investigations into the performance of magnetic track brakes of an electrical multiple unit against slippery tracks. Part 1: Adhesion improvement *Proc. Inst. Mech. Eng. (Part F J. Rail Rapid Transit* 225*)* pp 613
[21] Arias-Cuevas O, Li Z, Lewis R and Gallardo-Hernández E A 2009 Laboratory investigation of some sanding parameters to improve the adhesion in leaf-contaminated wheel–rail contacts *F J. Rail* **224** 139
[22] Cao X, Huang W L, He C G, Peng J F, Guo J, Wang W J, Liu Q Y and Zhu M H 2016 The effect of alumina particle on improving adhesion and wear damage of wheel/rail under wet conditions *Wear* vol **348** pp 98
[23] Wang W J, Shen P, Song J H, Guo J, Liu Q Y and Jin X S 2011 Experimental study on adhesion behavior of wheel/rail under dry and water conditions *Wear* **271** 2699
[24] Maya-Johnson S, Santa J F and Toro A 2017 Dry and lubricated wear of rail steel under rolling contact fatigue-Wear mechanisms and crack growth *Wear* **380** 240
[25] Fletcher D I, Hyde P and Kapoor A 2008 Modelling and full-scale trials to investigate fluid pressurisation of rolling contact fatigue cracks *Wear* **265** 1317
[26] Ekberg A and Kabo E 2005 Fatigue of railway wheels and rails under rolling contact and thermal loading-an overview *Wear* **258** 1288
[27] Chen H, Ishida M, Manura A, Baek K S, Nakamura T, Leban B and Pau M 2011 Estimation of wheel/rail adhesion coefficient under wet condition with measured boundary friction coefficient and real contact area *Wear* **271** 32
[28] Zhu Y, Chen X, Wang W J and Yang H 2015 A study on iron oxides and surface roughness in dry and wet wheel-rail contacts *Wear* **328** 241
[29] Polach O 2005 Creep forces in simulations of traction vehicles running on adhesion limit *Wear* **258** 992
[30] Lewis S R, Lewis R, Cotter J, Lu X and Eadie D T 2016 A new method for the assessment of traction enhancers and the generation of organic layers in a twin-disc machine *Wear* **366** 258
[31] Arias-Cuevas O, Li Z, Lewis R and Gallardo-Hernández E A 2010 Rolling-sliding laboratory tests of friction modifiers in dry and wet wheel-rail contacts *Wear* **268** 543
[32] Li Z, Arias-Cuevas O, Lewis R and Gallardo-Hernández E A 2009 Rolling–Sliding Laboratory Tests of Friction Modifiers in Leaf Contaminated Wheel–Rail Contacts *Tribol. Lett.* **33** 97
[33] Ma L, He C G, Zhao X J, Guo J, Zhu Y, Wang W J, Liu Q Y and Jin X S 2016 Study on wear and rolling contact fatigue behaviors of wheel/rail materials under different slip ratio conditions *Wear* **366** 13
[34] Lu X, Wong k C, Wong P C, Mitchell K A R, Cotter J and Eadie D T 2006 Surface characterization of polytetrafluoroethylene (PTFE) transfer films during rolling-sliding tribology tests using X-ray photoelectron spectroscopy *Wear* **261** 1155
[35] Wong K C, Lu X, Cotter J, Eadie D T, Wong P C and Mitchell K A R 2008 Surface and friction characterization of MoS2 and WS2 third body thin films under simulated wheel/rail rolling-sliding contact *Wear* **264** 526
[36] Guler M A, Alinia Y and Adibnazari S 2012 On the rolling contact problem of two elastic solids with graded coatings *Int. J. Mech. Sci.* **64** 62
[37] Lewis S R, Fretwell-Smith S, Goodwin P S, Smith L, Lewis R, Aslam M, Fletcher D I, Murray K and Lambert R 2016 Improving rail wear and RCF performance using laser cladding *Wear* **366** 268

IOP Conf. Series: Journal of Physics: Conf. Series **1049** (2018) 012010 doi:10.1088/1742-6596/1049/1/012010

Lighting system design according to different standards in office building: A technical and economic evaluations

K R Wagiman[1,2,a] and M N Abdullah[2]

[1] Institute of Industrial Training, Jalan Batang Melaka, 77500 Selandar, Melaka, Malaysia
[2] Faculty of Electrical and Electronic Engineering, Universiti Tun Hussein Onn Malaysia, 86400 Batu Pahat, Johor, Malaysia

E-mail: khairulrijalw@gmail.com

Abstract. Lighting system is an important sub-system in office buildings. A good design can contribute energy saving and visual comfort to occupants. This paper presents a comparison of lighting system design according to different standards: European Standard EN12464-1 and Malaysian Standard MS1525 in terms of average illuminance level and illuminance uniformity. In the design, the illuminance measurement is according to the Illuminating Engineering Society of North America (IESNA) and the European Standard EN12464-1. Three types of luminaires are considered which are: T8, T5 and light emitting diode (LED) lamps. The technical and economic evaluations are taken into account in this paper. The technical evaluations are considered the number of luminaires, average illuminance level and illuminance uniformity. The energy consumption (EC) and lighting energy numeric indicator (LENI) are considered according to the European Standard EN15193. Meanwhile, the economic evaluation involves billings and luminaires costs and billing savings. The case study is an office room and it is modelled and simulated using DIALux software. The results show that the design according to the European Standard EN12464-1 with uniformity based method that used T5 lamps had contributed to the highest number of luminaires and costs. Whereas, the design according to the Malaysian Standard MS1525 with illuminance level based method that used LED lamps had contributed to the lowest number of luminaires and costs. Thus it can provide the higher billing savings. It can be concluded the LED has great potential to replace T8 and T5 lamps.

Content from this work may be used under the terms of the Creative Commons Attribution 3.0 licence. Any further distribution of this work must maintain attribution to the author(s) and the title of the work, journal citation and DOI.
Published under licence by IOP Publishing Ltd

1. Introduction

Lighting system is one of important sub-systems in buildings beside heating, ventilation and air condition (HVAC) and plug loads. Lighting system accounts 20 - 45% of their energy consumption [1]. In U.S, lighting consumed 10% of total electricity energy in commercial buildings [2]. Numerous lighting system design guides were published to help engineers to design in order to fulfil the standards. The standards and green building rating systems such as Leadership in Energy and Environment Design (LEED) are incorporated the guides. Some examples of the guides are The Installer's Guide to Lighting Design [3], Lighting Design Guide [4] and The Lighting Handbook [5].

Dubois et al. [6] reviewed the electrical lighting retrofit to minimize energy used in buildings. They outlined lighting retrofit strategies which are: lamp, ballast and luminaires replacement, retrofits with light emitting diode (LED) lamps, retrofits with T8 and T5 lamps and reduction of maintained illuminance levels. Ganandran et al. [7] studied the potential energy saving of the lighting system by retrofit LED bulb in several buildings of Universiti Tenaga Nasional, Malaysia. In the study, they calculated the electricity savings, payback period and potential benefit to environment. Vahl et al. [8] carried out techno-economic analysis of lighting retrofit CFLs and LED lamps. In the study, they analysed the equivalent annual cost (EAC) and net present value (NPV). Mahlia et al. [9] studied the economic perspective in term of retrofit T8 and T5 lamps was conducted in Universiti Malaya, Malaysia. The economic perspective includes payback period and life cycle cost. Gan et al. [10] examined the feasible potential cost-benefit analysis for three types of lamps: T8 lamps, T5 lamps and LED tube on several blocks of Universiti Teknikal Malaysia Melaka, Malaysia. Halim et al. [11] evaluated lighting retrofit LED in term of economic evaluation. In the study, they calculated simple payback period and NPV.

Aforementioned studies are only considered in term of economic evaluations. In this paper, the lighting system design will be based on different lighting system standards which are the European Standard EN12464-1 and the Malaysian Standard MS1525. The results will be presented in terms of technical and economic evaluations. In technical evaluation, the illuminance levels and illuminance uniformity will be compared based on different standards. Meanwhile, for economic evaluation the billings and luminaires costs will be calculated and compared for both standards. Moreover, the type of luminaires which are T8, T5 and LED lamps will be taken into analysis in order to compare their performances.

1.1. Standards related to lighting system design

Standard is documents of guidelines, specifications and requirement about technical system. In terms of energy and environmental management standards, the standards help to minimize energy consumption, reduce environmental impact, reduces waste and more sustainable. Various standards are established by standard organizations or institutions such as American National Standards Institute (ANSI), Illuminating Engineering Society of North America (IESNA), European Standard, Malaysian Standard and so on according to their subject matter experts. For lighting, energy efficiency and renewable energy subject, some standards were established such as the IES Lighting Handbook by IESNA, the European Standard EN12464-1 by European Committee for Standardization and the Malaysian Standard MS1525 by SIRIM Berhad.

1.1.1. European Standard EN12464-1

The European Standard EN12464-1: Light and lighting – Lighting of work places – Part 1: Indoor work places [12] was published on November 2011and it was produced by European Committee for Standardization. The standard members are the nation of Austria, Belgium, Czech Republic, Denmark, Finland, France, Germany, Greece, Iceland, Ireland, Italy, Luxembourg, Malta, Netherlands, Norway, Portugal, Spain, Sweden, Switzerland and United Kingdom. The main objective of the standard is to establish technical guidelines for design, maintenance and verification that focus to indoor lighting system in work places.

1.1.2. Malaysian Standard MS1525

The Malaysian Standard MS1525: Energy efficiency and use of renewable energy for non-residential buildings – Code of practice was revised in 2014 [13] and it is one of the Malaysian Standard which was developed by SIRIM. The standard helps architects and engineers to design the building, lighting and electrical systems, air conditioning and mechanical ventilation systems and energy management systems. The standard provides technical guidelines for both interior and exterior of lighting systems. The summary of maintained illuminance level ($\overline{E}_m$) and illuminance uniformity (U_o) for different standards in general office is presented in Table 1.

Table 1. Summary of maintained illuminance level ($\overline{E}_m$) and illuminance uniformity (U_o) for different standards in general office

Ref.[a]	Standard	$\overline{E}_m$ (lx)	U_o
[12]	EN12464-1	500	0.6
[13]	MS1525	300-400	Not specified

[a] Reference

2. Interior lighting system design

In building design process, after an architect has completed the building design, the mechanical and electrical (M&E) engineers take place for design processes. The design includes HVAC and electrical circuit and lighting systems. The lighting system design involves several considerations such as luminaires, design methods (illuminance and uniformity based), illuminance level measurement, daylighting and sensor placement.

2.1. Luminaire specifications

Luminaire consists of lamp that emits amount of visible light. Scientifically, it is also known as luminous flux (Φ) which is measured in lumen (lm). In luminaire data sheet, the specifications provides commonly such as lumen output/luminous flux, input power/wattage, luminous efficacy, colour temperature, Colour Rendering Index (CRI) and polar intensity diagram.

2.2. Illuminance level based

Illuminance level based method also known as lumen method is the most widely used method to calculate illumination for interior lighting design. Moreover, most of the lighting design softwares utilized this method for calculating the number of luminaire such as DIALux. This method is recommended by [14] in order to determine the illuminance level by considering the luminous flux, room area and

IOP Conf. Series: Journal of Physics: Conf. Series **1049** (2018) 012010 doi:10.1088/1742-6596/1049/1/012010

others parameters related to lamp. Illuminance (E) describes the amount of luminous flux on the surface of the room. The SI unit of illuminance is lux (lx). The total number of luminaires (Nl) in the room can be expressed in equation (1) as follows:

$$Nl = \frac{E_{avg} \cdot A}{n_l \cdot \Phi_n \cdot UF \cdot MF \cdot BF} \tag{1}$$

where, E_{avg} is the average illuminance according to the standard measured in lux (lx), A is the area of the room in meter square (m^2), n_l is the number of lamp for each luminaire, Φ_n is the nominal lumen output for each lamp in lumen (lm), UF is utilization factor provided by luminaire manufacturer based on room index, MF is the maintenance factor and BF is the ballast factor. The typical value of UF and BF are 0.9 and 1, respectively. The most common value of MF is 0.8 that used in lighting system design and is also stated in OSRAM [15].

2.3. *Illuminance uniformity based*

Illuminance uniformity which also called as uniformity represents the quality of the illuminance distribution on surface or area either interior or exterior. The most widely used of uniformity is global uniformity (U_o) and express in equation (2).

$$U_o = \frac{E_{min}}{E_{avg}} \tag{2}$$

where, E_{min} is the minimum value calculated of illuminance and E_{avg} is the average illuminance value calculated.

2.4. *Illuminance level measurement*

To determine the value of illuminance level, two ways are used that are real field measurement and lighting software package. The real measurement is carried out using lux meter based on measurement point of the area. Meanwhile, the method by using lighting software package takes lesser time to calculate the value of illuminance for each measurement point compared to real measurement method. Beside time consuming, the other advantage of using software package is faster generated results such as average, minimum and maximum illuminance values, global uniformity (U_o) and graphical contour illuminance level.

The measurement points of illuminance is forms of grid either square or rectangular grids. According to European Standard EN12464-1 [12], the maximum grid size (p) can be determined by using equation (3) as follows:

$$p = 0.2 \times 5^{\log_{10}(d)} \tag{3}$$

where, d is longer dimension of the area (m)

On the other hand, according to the IESNA [14], the illuminance measurement grid size is square (0.66 x 0.66 m^2).

3. Methodology

This section describes and discusses the methods that are used in this study. The sub-section includes building and lighting system design, illuminance level measurement, analyse the energy and the economic performances.

3.1. *Building and lighting system design*

The case study is an actual office room, department of Mechanical Technology, Institute of Industrial Training, Selandar, Melaka, Malaysia. The office dimensions are (20 x 8 x 2.7 m^3) and it is illuminated by 24 luminaires (8 by 3 grids). Each luminaire consists of 2x36W T8 fluorescent lamps.

The building and lighting system designs were carried out using DIALux software. The DIALux is the most widely used lighting software which was developed by DIAL GmbH, Denmark with over 600,000 users worldwide [16]. The software is free and it can perform both interior and exterior lighting designs.

In this study, the lighting system design is according to two standards which are European Standard EN12464-1 and Malaysian Standard MS1525. Two parameters were considered that are average illuminance level (E_{avg}) and illuminance uniformity (U_o). Different types of luminaires will be considered in this study which are: T8 fluorescent, T5 fluorescent and LED lamps and the luminaire grid size will be determined based on lighting design parameters and standards. The luminaires specifications are provided in Table 2.

3.2. *Illuminance level measurement*

The illuminance level measurement was performed on horizontal plane with 0.75 m high from the floor using DIALux software. The measurement is according to two standards that are: the IESNA and the European Standard EN12464-1. According to the IESNA, the illuminance level measurement grid is (0.66 x 0.66 m^2). Whereas, according to European Standard EN12464-1, the illuminance level measurement grid will be calculated using equation (3).

3.3. *Analyse the energy performance*

In this study, the energy performance that was analysed are annual energy consumption (EC) and lighting energy numeric indicator (LENI) which are according to the EN 15193 [17]. The EC and LENI are measured in kWh/year and kWh/year/m^2, respectively. The LENI is the sum of energy consumed include daytime, night-time and parasitic energy used divided by the area. The calculation of EC and LENI were performed using DIALux software.

Table 2. Luminaires specifications

Type of luminaire	Total luminous flux (lm)	Total power (W)
T8	6700	72
T5	6600	70
LED	7450	50

3.4. *Analyse the economic perspective*

In economic analysis evaluation, three parameters were considered in this study are: energy billings and luminaires costs and billing savings. In this case, the latest electric tariff from Tenaga Nasional Berhad which is the national electricity provider is RM 0.365/kWh [18]. Based on market survey, the prices of luminaires are as follows: T5 fluorescent lamp (RM 35) and LED (RM 52). Moreover, the simple payback period (SPP) was evaluated in this paper. SPP can be described as estimated project cost divided by estimated billing savings.

4. Results and discussion

This section presents the simulation results from DIALux. The simulation results include 3D model of office with luminaires, technical and economic analyses.

4.1. Building and lighting design using DIALux

The 3D model of office with luminaires on the ceiling was performed in DIALux is illustrated in Figure 1. The model includes 4 workstations, 1 set coffee table with 2 chairs, copier, a table with printer, 5 file cabinets and 24 luminaires.

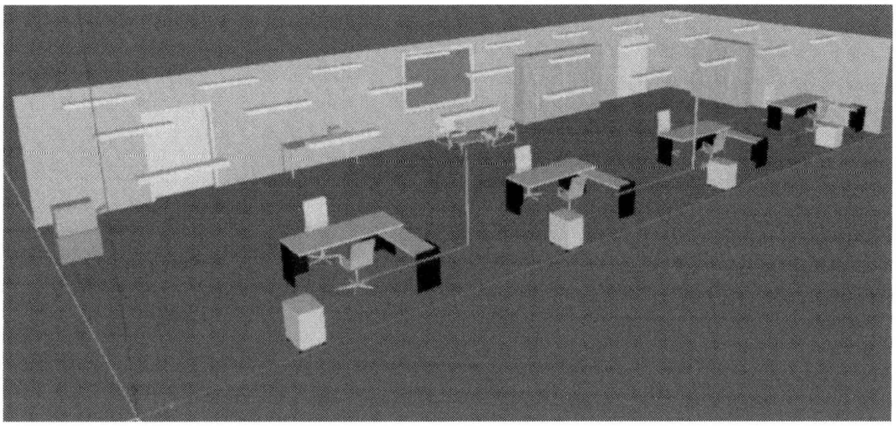

Figure 1. 3D model of office room with luminaires on the ceiling in DIALux

4.2. Technical analysis results

The results of illuminance level, uniformity and luminaire grid for different standards, design methods and type of luminaires are shown in Table 3. Based on Table 3, the MS1525 with T8 lamp is existing luminaires in the office room and the rest are feasible analysis in this paper. The existing design clearly showed that it is over design in term of number of luminaires due to the E_{avg} is significantly higher than 300-400 lx according to Malaysian Standard for general office. Because of this, the lighting system consumed higher electricity energy and waste it consequently. According to Malaysian Standard, they consider the E_{avg} value only, thus, the value of E_{avg} always lesser than E_{avg} value that design according to European Standard for all type of luminaires. The illuminance uniformity based method is significantly contributed higher number of luminaires for both standards and the whole type of luminaires compared to illuminance level based method. Consequently, the illuminance level and uniformity explicitly showed that higher values. The higher luminaires grid which contributed from design according to the EN12464-1 with uniformity based method that used T5 lamps is (8 by 5) grids equal to 40 luminaires. Meanwhile, the lowest luminaires grid is design according to the MS1525 with illuminance level based method that used LED of (3 by 3) grids equal to 9 luminaires.

For illuminance measurement grid size according to the IESNA, it is clearly showed that the E_{avg} and U_o values always less than which used EN12464-1. The smaller size of illuminance measurement grid the more number of measuring point are obtained thus, the points almost covered the overall horizontal workspace plane.

Simulation results of energy consumption and LENI for different standards, design methods and type of luminaires are presented in Table 4. Based on Table 4, the highest energy consumption and LENI which contributed from design according to the EN12464-1 with uniformity based method used T5 lamps are 7950 kWh/year and 50 kWh/year/m^2, respectively. Meanwhile, the lowest energy consumption and

LENI are design according to the MS1525 with illuminance level based method used LED of 1150kWh/year and 7 kWh/year/m^2, respectively.

Table 3. Simulation results from DIALux of illuminance level, uniformity and luminaire grid for different standards, design methods and type of luminaires (ToL)

Standard/design method/ToL	IESNA (M)		EN 12464-1 (M)		Luminaires grid
	E_{avg}	U_o	E_{avg}	U_o	
MS1525- T8	502	0.58	514	0.79	8 x 3
MS1525(E) - T5	342	0.50	349	0.62	5 x 3
EN 12464-1 (E) - T5	536	0.57	546	0.77	6 x 4
EN 12464-1 (U) - T5	879	0.61	897	0.78	8 x 5
MS1525(E) - LED	320	0.33	328	0.46	3 x 3
EN 12464-1 (E) - LED	528	0.54	540	0.67	4 x 3
EN 12464-1 (U) - LED	966	0.61	985	0.78	7 x 4

E is illuminance level based design
U is uniformity based design
M is illuminance measurement grid according to the standard

4.3. Economic analysis results

Luminaires and billings costs and billing savings for different standards, design method and type of luminaires are shown in Figure 2. Based on Figure 2, the highest luminaires and billings cost which contributed from design according to EN12464-1 with uniformity based method used T5 lamps are RM 1400 and RM 2902 per year, respectively. Meanwhile, the lowest luminaires and billings costs are design according to MS1525 with illuminance level based method used LED of RM 468 and RM 2412 per year, respectively. Moreover, it is contributed to the highest billing savings up to RM 1194 per year. It showed that the LED is leading the lowest cost in term of billings and luminaires costs and the highest billing savings.

Table 5 shows simple payback period (SPP) for different standards, design methods and type of luminaires (ToL). As can be seen from Table 5, the lowest SPP is design according to MS1525 with illuminance level based method used LED of 0.39 year about 7 months. Meanwhile, the highest SPP is design according to EN12464-1 with uniformity based method used T5 lamps of 4.7 years about 57 months. Lighting design based on illuminance level using LED lamps showed that SPP less than 1 year for both standards.

The lighting system design according to EN12464-1 with uniformity based on method that used T5 lamps explicitly showed that it leads the number of luminaires grid for both technical and economic evaluation. The luminaires grid is propotional to E_{avg}, U_o, EC and LENI. LED lamps are highly recommended to retrofit T8 lamp due to superior performance in technical evaluation and very economic. This finding supports previous studies [7-10] and has conluded that LED has great potential to replace flourescent lamps and reduce energy consumption consequently.

Table 4. Simulation results of energy consumption and LENI for different standards, design methods and type of luminaires (ToL)

Standard/design method/ToL	EC (kWh/year)	LENI (kWh/year/m^2)
MS1525- T8	4400	27
MS1525(E) - T5	3000	19
EN12464-1 (E) - T5	4750	30
EN12464-1 (U) - T5	7950	50
MS1525(E) - LED	1150	7
EN12464-1 (E) - LED	1900	12
EN12464-1 (U) - LED	3550	22

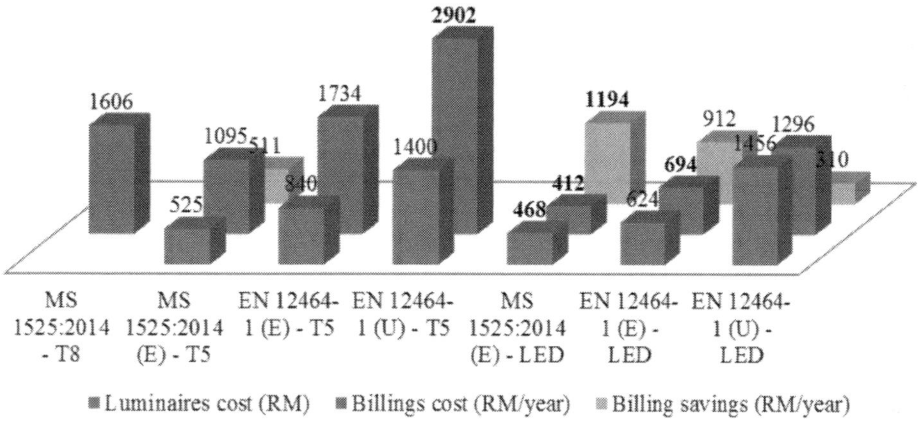

Figure 2. Luminaires and billings costs and billing savings for different standards, design methods and type of luminaires

Table 5. Simple payback period (SPP) for different standards, design methods and type of luminaires (ToL)

Standard/design method/ToL	SPP (year)
MS1525:2014 (E) - T5	1.03
EN16434-1 (E) - T5	-
EN16434-1 (U) - T5	-
MS1525:2014 (E) - LED	0.39
EN16434-1 (E) - LED	0.68
EN16434-1 (U) - LED	4.70

5. Conclusion

The aim of this paper is to compare the lighting system design according to different standards in terms of technical and economic evaluations. The findings showed that the lighting system design according to the EN12464-1 with uniformity based method that used T5 lamps contributed higher luminaires grid and higher values consequently of E_{avg} and U_o which contributed higher billings and luminaires costs. The second finding had stated that the illuminance measurement grid according to the IESNA contributed lower values of E_{avg} and U_o due to smaller size of grid compared to the EN12464-1. It was also showed that LED was the vast potential to replace fluorescent lamps due to superior performance in term of technical and economic.

6. References

[1] M.-C. Dubois and Å. Blomsterberg, "Energy saving potential and strategies for electric lighting in future North European, low energy office buildings: A literature review," *Energy Build.*, vol. 43, no. 10, pp. 2572–2582, 2011.

[2] U.S. Energy Information Administration, "Commercial Buildings Energy Consumption Survey (CBECS)," 2012. [Online]. Available: https://www.eia.gov/consumption/commercial/reports/2012/energyusage/index.php. [Accessed: 05-Aug-2017].

[3] Society of Light and Lighting, *The installer's guide to lighting design*. 2014.

[4] EATON, *Lighting Design Guide*. Eaton's Cooper Lighting and Safety Business, 2013.

[5] ZUMTOBEL, *The Lighting Handbook*. ZUMTOBEL Lighting GmbH, 2013.

[6] M.-C. Dubois, F. Bisegna, N. Gentile, M. Knoop, B. Matusiak, and W. O. and E. Tetri, "Retrofitting the Electric Lighting and Daylighting Systems to Reduce Energy Use in Buildings: A Literature Review," *Energy Res. J.*, vol. 6, no. 1, pp. 25–41, 2015.

[7] G. S. B. Ganandran, T. M. I. Mahlia, H. C. Ong, B. Rismanchi, and W. T. Chong, "Cost-Benefit Analysis and Emission Reduction of Energy Efficient Lighting at the Universiti Tenaga Nasional," *Sci. World J.*, vol. 2014, pp. 1–12, 2014.

[8] F. P. Vahl, L. M. S. Campos, and N. Casarotto Filho, "Sustainability constraints in techno-economic analysis of general lighting retrofits," *Energy Build.*, vol. 67, pp. 500–507, 2013.

[9] T. M. I. Mahlia, H. A. Razak, and M. A. Nursahida, "Life cycle cost analysis and payback period of lighting retrofit at the University of Malaya," *Renew. Sustain. Energy Rev.*, vol. 15, no. 2, pp. 1125–1132, 2011.

[10] C. K. Gan, A. F. Sapar, Y. C. Mun, and K. E. Chong, "Techno-economic analysis of LED lighting: A case study in UTeM's faculty building," *Procedia Eng.*, vol. 53, pp. 208–216, 2013.

[11] Mohd Firdaus Mohd Ab Halim *et al.*, "Lighting Retrofit Scheme Economic Evaluation," *Indones. J. Electr. Eng. Comput. Sci.*, vol. 5, no. 3, pp. 496–501, 2017.

[12] European Committee for Standardization, *EN12464-1. Light and lighting - Lighting of work places - Part 1: Indoor work places*. 2002.

[13] Department of Standards Malaysia, *MS 1525:2014. Energy efficiency and use of renewable energy for non-residential buildings - Code of practice*. 2014.

[14] IESNA, *IES Lighting Handbook Reference and Application*. New York: Illuminating Engineering Society of North America (IESNA), 2000.

[15] OSRAM, *Lighting Solutions for Office Worlds*. OSRAM GmbH.

[16] DIAL GmbH, "DIALux Lighting Design Software," 2017. [Online]. Available: https://www.dial.de/en/dialux/. [Accessed: 14-Jul-2017].

[17] European Committee for Standardization, *EN 15193:2007. Energy Performance of Buildings - Energy Requirements for Lighting*. 2007.

[18] Tenaga Nasional Berhad, "TNB Commercial Tariffs," 2017. [Online]. Available: https://www.tnb.com.my/commercial-industrial/pricing-tariffs1/. [Accessed: 01-Aug-2017].

Acknowledgments

This work was supported by Short Term Research Grant (U639) funded by the Universiti Tun Hussein Onn Malaysia (UTHM), Malaysia.

The Influence of Information Technology and Operational Competencies toward Supply Chain Agility: Findings from Textile Manufacturer

A. H. Nor Aziati [1], Yong Man Ling, Md Fauzi Ahmad[2], Nor Hazana Abdullah[3]

Faculty of Technology Management and Business, Universiti Tun Hussein Onn Malaysia

E-mail: [1]aziati@uthm.edu.my; [2]fauzi@uthm.edu.my; [3]hazana@uthm.eduu.my

Abstract. Supply chain agility become a critical strategic for business communities to gain the superior competitive advantage in the volatile and turbulent market. Previous studies on supply chain agility focus to the industries such as electrical and electronics industries. The limited study particularly in the context of supply chain agility in Malaysia motivates the study to further investigate the practices. The purpose of this study is to determine the relationship between information technology and operational competencies towards supply chain agility in firm. Data collection instrument used was a questionnaire which administrated to a total sample size of 60 executive officers, directors, CEO, managers and senior staff from 15 textiles and apparel companies from southern Malaysia. The respond rate is 63.3% which is only 38 respondents were responded to the survey. Sample selection was based on purposively sampling. The data were analysed by using mean, standard deviation and correlation between independent and dependent variables. The analyses involved statistical methods such as reliability, normality and nonparametric correlation. The finding showed that information technology and operational competencies have a positive relationship with supply chain agility respectively.

1. Introduction

In today's business environment, the competitions among the firms are very intense. The fast-changing on global market demands transform the supply chain level into an agile mode from the organization to its partners in supply chain [1]. Firms have realized supply chain agility (SCA) is essential in their business for survival and gain the competitive advantage in the uncertain market [2]. The firms are able to respond instantly and easily in the volatile market by achieving SCA. Firms must form an alliance with their suppliers and customers to achieve a competitive edge. The cooperation among these parties in the supply chain is able to form a streamline operation, as well as achieve a level of agility beyond individual companies.

As stated by [3], the firm's SCA directly impacts its ability to product, and delivers innovative products to their customers in a timely and cost effective manner to a constantly changing global competitive environment. An agile supply chain model consists of information-driven virtual integration, process integration and performance management, centralized and collaborative planning,

[1] aziati@uthm.edu.my

Content from this work may be used under the terms of the Creative Commons Attribution 3.0 licence. Any further distribution of this work must maintain attribution to the author(s) and the title of the work, journal citation and DOI.
Published under licence by IOP Publishing Ltd

and market sensitivity and responsiveness [4]. There are few empirical studies showed that successful SCA concept applied will give positive impacts to business performance [5]. However, SCA is a philosophically simple concept but difficult extensively to imply in a whole supply chain network due to the disparity in ability of the parties in a supply chain. Hence, the study aims to examine the two most influential competencies; information technology competence and operational competence will show a positive relationship towards supply chain agility of the firm.

1.1. Research background

Based on Ministry of International Trade and Industry report in 2014, textiles and textile products industry in Malaysia comprises into four sub-sectors, namely primary textiles which cover activities such as polymerization, spinning, weaving, knitting and wet processing; made-up garments; made-up textiles; and textile accessories. This industry are involved the upstream supply chain which the supplier to manufacturer and downstream supply chain which counted the end consumer.

Maslow's hierarchy of need states that textile and apparel are significant components of physiological needs. This implied that the textile and apparel are vital to human as there were the basic consumption which gives physical protection to individuals. Based on statistics [6] reported the textiles, clothing and footwear manufacturing have increased by 13% to RM12.12 billion to the growth in export for 2014. Textile and apparel industry in Malaysia has growth to be one of the export earners. It has been recognized in Malaysia's Third Industrial Master Plan (IMP3) for further development because it is potential to growth. In 2013, 1.4 per cent of Malaysia's total exports of manufactured goods are contributed by textile and textile product industry and this industry was ranked 11th top export earner to Malaysia. According to Malaysian Investment Development Authority (MIDA), the export of textile to USA, Japan, Indonesia, Turkey and China brought along RM10.3 billion income to Malaysia. The textile and clothing industry in Malaysia is planning to focus on high value fashion, dyeing and finishing, and technical textiles to sustain growth [7].

Textile and apparel industries has the market characteristics of short product life cycle, volatile demand patterns, low predictability, high impulsive purchases and complex supply chain [8]. The supply chain agility (SCA) is the integration of these four features which are responsiveness, competence, flexibility and speed for firms to react effectively and efficiently in the turbulent and volatile market. In other words, textile and apparel industries can indicates the effectiveness of competencies of information technology and operational toward SCA on business performance. Therefore this study tries to address this gap and test the level of SCA in firms.

1.2. Problem statement

A firm that wants to achieve supply chain agility is not simply as implementing rules or procedures [2]. According to [9] there are eleven barriers faced by firm instead of to achieve superior supply chain agility. Supply chain agility is about the collaboration among the parties involved in a supply chain. Every decision made such as location or allocation decisions, inventory control, production planning, transportation mode selection, and supplier selection in manufacturing will affect the agility in a supply chain network. Besides, the issues like organizational resistance to change, inter-functional conflicts, joint production planning, profit sharing, team oriented performance measures, shifts in channel power, information sharing, real time communication, and technical compatibility may also influence the SCA.

According to the directory of Federal of Malaysian Manufacturer, the textile and apparel industry in Malaysia is made up of 25 % large enterprises, 44% of medium enterprise, 30% small enterprises and 1% micro enterprises [10]. As a corollary, these small and medium size textile and apparel company might not aware on the information technology integration among the supply chain units due to their limited IT resources. Besides, they might face difficulties to coordinate with external supply chain partners due to the incompatible technology platforms and standards [11]. It is because they have limited capital to invest in IT to develop a complete information system with their supply chain partners. The information about the internal supply chain system for instance purchasing, materials management, manufacturing and distribution flows inconsistently towards the external supply chain coordination.

Additionally, it is costly for small and medium size textile and apparel companies to adopt supply chain technology since they have resource limitation [8]. Thus, the poor IT infrastructures and facilities caused the textile and apparel companies failed to integrate either the departments in their own company or with their supply chain partners. The information sharing and spares information among units in firm or within firms has increase the difficulty to exchange data. Hence, the firms are enable to retrieve accurate customer's demand patterns of customer. This phenomenon will affect the cost of production indirectly, communication and the list of factors which related to the SCA. Yet, many organizations unrealized the importance of information systems in their supply chain structure. Thus it caused insufficient applications of IT in virtual enterprise.

Meanwhile, many company unwilling to share critical information with their suppliers and caused a poor correlation among the firm with their supply chain partners. According to [12], "over reliance on technology without willingness to share critical information pertaining to supply chain will not make the firms meaning fully connected." In other word, it caused failed on supply chain integration. Insufficient resources are available or dedicated to information technology, lack of trust among partners in supply chain network, and history of opportunistic behaviour limits the willingness to share more information than is necessary [13]. Besides, the firm armed with the trade secret is a tool for them to compete with others when there is a change in market.

Most of the textile and apparel companies in Malaysia are small and medium enterprises, trained their workers internally to enhance their skills and knowledge about supply chain technologies. Thus, these issues have direct impact toward the whole supply chain flexibility. Hence, the textile and apparel companies might face retardation on respond to the changes in market and making decision to transform the uncertainties into opportunities. It is critical for these firms to adhere with SCA concept to be competitive in future. The present study had four principal objectives, namely to:

- To determine the influence of information technology and operational competencies on supply chain agility.
- To identify degree of importance of information technology and operational competencies toward supply chain agility.

1.3. Scope of research

This study is focused on the effects of information technology and operational competence on enhancing the firm's SCA. The study is concentrated on the textile and apparel industry at southern Malaysia. A statistical analysis is performed and the questionnaire survey question is distributed amongst top level management and executive staff from various units such as operation, procurement, sourcing, logistics, warehousing or transporter at textile and apparel company. These groups of staff became target respondents because they are involved in the decision making on supply chain and understand the operations of the company with their supply chain partners. This study is organizational level of analysis.

1.4. Significance of study

The result of this study will become the stepping stone for future researcher on the field on supply chain agility toward firm's performance. This study contributes the body of knowledge and SCA awareness among textile and apparel industry in Malaysia. Besides, this study can be extended the other industries with different market characteristics for future study. On the other hand, this study provides a significant endeavour in modelling the relationship of the information technology and operational competencies on a firm's SCA and help them enhance their firm performance. This study is also beneficial to the parties in a supply chain's corporate strategies when they integrate their competencies in the cooperation will let them experience the changes on the firm's responses to the market and also the chain's performance. The firms may able to get quick responses when facing a sudden change in the market by obtaining the right information within a short period and able to fully use their available resources to transform the uncertainties in market into business opportunities.

Moreover, this study may help and educate firms to understand the importance and criteria of each competency in the area of supply chain management which tools help their company sustain in the market and the key to grab the chance in the volatile market. It will also serve as a future reference for researchers on the subject of SCA.

2. Literature review

2.1 Supply Chain Agility

The word "agility" is defined by [14] as the ability respond to the anticipated or unexpected changes in a proper ways and due time, survive unprecedented treats and able to take advantage of the changes by transform them into business opportunity. Supply chain agility is a new paradigm in supply chain management. There was no single accepted definition of supply chain agility (SCA). Most of the definitions of SCA shared the common elements with concept of agility for instance [5] defined SCA as the organization's ability to respond to unexpected market changes and convert these changes to business opportunities with the supply chain's capability to use resources in responding proactively or reactively to such changes, all in timely and flexible manner [15]. SCA is measured through four dimensions as presented in Table 1.

Table 1. The items of measurement for supply chain agility.

Dimension	Item of measurements	Adapted Sources
Supply Chain Agility	1. Market sensitivity includes the items of - market sensitiveness - new product introduction - minimizing resistance to change - minimizing uncertainty	[8] [9] [16]
	2. Virtual includes the - data accuracy - cost minimization - lead time reduction - trust development	
	3. Process integration - process integration - service level improvement - centralized and collaborative planning	
	4. Network based includes - delivery speed - use of IT tools - quality improvement - customer satisfaction	

2.2 IT competence

Information technology (IT) increases the capabilities of firms to sense and respond to the changes in market by allowing better inter-firms coordination [2]. Moreover, IT allow firms to communicate with the partners in a supply chain network and increase the volume and complexity of information about the accurate product demand, inventory, financial. IT also provides real-time supply chain information, inventory level, status of delivery, and scheduling plan and production planning to firms for controlling their supply chain activities [12]. Thus, IT competence is the fundamental support to supply chain agility. IT competence is the extent to which a firm is knowledgeable about the effectively utilizes IT to manage information within the firm [17]. IT competence is measured by two elements which are IT integration and IT flexibility.

IT integration is defined as the extent to which IT is used to coordinate and integrate information within a firm's function and with the firm's supply chain partners. [5]. IT integration enables firm to obtain more adequate and accurate data and information to making decision in the changing business environment. Firm apply IT integration intra and inter- organizational is aimed to achieve real-time transmission and processing of information required for supply chain decision making [12]. Hence, IT integration can enhance the supply chain agility capability of firm.

Firms only manage to change their business strategic according to the change of business environment when confronted with time and other pressures with accommodation adequate IT infrastructures [18]. Flexibility due to IT infrastructure is about the connectivity, compatibility and modularity of information technology components. IT flexibility can be defined as the ability of IT infrastructure to adapt to both incremental and revolutionary changes in the business or business process with minimal penalty to current time, effort, cost or performance [2]. It is mean the firm able to enhance their IT infrastructure based on market condition and changes according to the uncertainty happen in market in appropriate time. Table 2 lists the items of measurement of information technology competence.

Table 2. The items of measurement for information technology competence

Dimension	Item of measurements	Adapted Sources
Information Technology Competence	1. IT integration: To achieve real-time transmission and processing of information required for supply chain decision making the item of measurement include - efficient resource flow - inventory status - service capabilities - real time information exchange 2. IT flexibility The ability of IT infrastructure to adapt to both incremental and revolutionary changes in the business or business process with minimal penalty to current time, effort, cost or performance. The items include - connectivity - compatibility - modularity	[2] [5] [12]

2.3 Definition of Operational Competence

Operational competence is defined as the preparedness, skill, or capability that enables manufacturers to pursue a product-market specific business strategy [19]. Firms with operational competence can help to maintain an efficient flow of information and materials among different parties and functions along the supply chain, maintain continuous search of knowledge and reserve a wide variety of operational resources that can be activated quickly [14]. There are three dimensions contain in operational competence which are supply chain integration, supply chain flexibility and supply chain learning orientation.

The definition of supply chain integration is the formation of a network in which separate supply chain partners collaboratively manage intra and inter-organizational processes to arrive at mutually acceptable outcomes [20]. According to [21], firm not only can achieve efficient flows of information but also provide low cost and high speed of products and services, resources, and cash to provide maximum value to the customers.

Supply chain flexibility can be defined as the supply chain's ability to provide products and services in a timely and cost effective manner to an uncertain and rapid changing global environment [5]. According to the authors too, the supply chain flexibility in procurement, manufacturing, distribution and product development give impacts on supply chain agility. Supply chain flexibility can be categorized into strategic flexibility and manufacturing flexibility. Strategic flexibility is the competence of firm to recognize the changes in business environment, facilitate resources quickly to adapt the changes and able to identify and act promptly when it is time suspend or reverse such resource commitment [22]. Manufacturing flexibility which meant by [23] is the competence of firms to manage their manufacturing resources and meet different customer requirement. It is summarized from the definition, strategic flexibility is related to the firm's ability to respond whilst manufacturing flexibility is related to firm's operational ability and capability to make strategic decisions.

On the other hand, supply chain learning orientation is the ability of firm for learning continuously and has well prepared for any challenges in their business and environment. Commitment to learning which referred to the value placed on learning and open-mindedness which pertains to the willingness to evaluate critically the supply chain operations and accept new ideas. Besides, a shared vision which relates to the consensus on direction and focus of learning among organizational members [2].

Table 3. The items of measurement for operational competence

Dimension	Item of measurements	Adapted Sources
Operational Competence	1. Supply chain integration includes - effective flows of information, products and services, resources and cash - efficient flows of information, products and services, resources and cash - provide maximum value to customer with low cost and high speed. 2. Supply Chain Flexibility includes - the identify changes in the environment - the commit resources quickly to new courses of action in respond to change - recognition and act promptly when time to halt. 3. Supply Chain Learning Orientation is the - commitment to learning - open-mindedness - shared vision	[2] [14] [19] [21] [22] [24]

2.4 Conceptual Framework and Hypotheses

Based on the literature review discussed in previous section, a conceptual framework in figure 1 and several hypotheses have been developed as shown below.

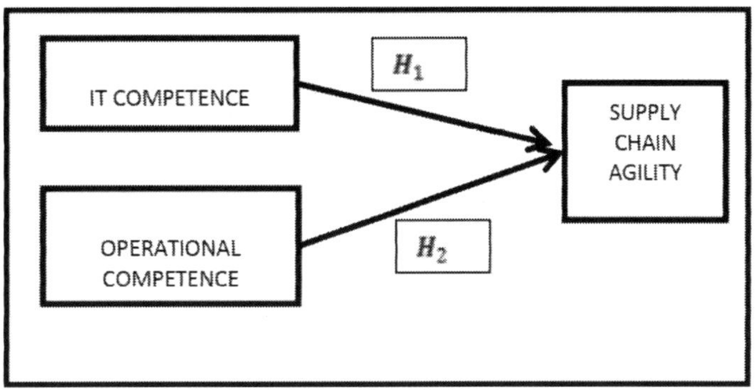

Figure 1: Conceptual Framework

Table 4. Research hypotheses

H_1 :	IT competence is positively related to supply chain agility.
H_2 :	Supply chain flexibility is positively related with supply chain agility.

3. Methodology

The study employed quantitative research using survey method. The collected data were analysed by using Statistical Package for Social Science (SPSS) version 22.0. The researchers choose quantitative compared into this qualitative because survey method rarely used on research of supply chain agility. The purpose of using survey questionnaire was to see the perception of the respondents towards supply chain agility in the selected company.

A flow diagram is established to ensure this research can be carried out smoothly. There are several stages found in the flow diagram. First, the research problem is identified and this research as the basis for solving those problems. Secondly, the literature reviews on the research topic are reviewed for better understanding of the terms to be investigated in this research. Then, the conceptual framework and several hypotheses are developed. After that, the questionnaire of this research is designed and undergoes the stage of pre-testing for further refinement. Next, data is analysed after the distributed questionnaires have been collected. Finally, the interpreted result is combined and reported.

3.1 Sampling design

Sampling is essential in the study because it is almost always impossible to study a large population. Purposive sampling technique is used in this research in which the researcher decides the tasks need to be identified and find the people who can provide the information by virtue of knowledge voluntary [25]. The target respondents need to be top level management who are from the unit of operation, procurement, sourcing, logistics, warehousing, transporter or executive staff in textile and apparel companies in southern Malaysia.

3.2 Population and sample

The population of this study is the 300 textile and apparel companies located at southern Malaysia listed in the Federation of Malaysian Manufacturers directory (FMM) and Malaysia External Trade Development Corporation (MATRADE) directory. Specially, the targeted respondents are the top level management who are the director, CEO, managers, senior executive who are direct involved in supply chain to give a response on the present study.

The unit of analysis in this study is the individual from the firms. The respondents were assumed to be knowledgeable and familiar with the operations related to the issues under investigation. The questionnaires were sent to the office of the respondent who willing to answered. Total of 38 respondents completed the survey questionnaire out of 60 respondents from the sampling frame.

3.3 Pre-testing and Data collection instrument

A pre-testing was undertaken to assess the reliability of the attributes, and to ensure that the wordings of the questionnaire were clear. As suggested by [26] pre-testing in this research was done by six experts in the related fields of this research topic with accompaniment of researcher. Some problems were identified with the wordings and implications of some questions, thus some minor revisions were made to avoid confusion. There are three sections consisted in the questionnaire which are demographic information of the respondents, IT competence with 9 items, operational competence with 13 items and SCA with 10 items. The measurement of these items is using a five-point Likert Scales which rank from 5=strongly agree until 1=strongly disagree. Total 60 samples of survey questionnaire has been distributed to 15 textile and apparel companies with each company received 4 surveys to avoid biases.

4 Data Analysis and Findings

The Statistical Package for the Social Sciences version 22.0 was used to analyze the data. Descriptive statistics analysis was used to measure respondent's perception. The means, standard deviations, and the difference scores were computed for each attribute. The means were computed by adding up the scores allocated by respondents for each attribute and dividing the total value by the number of respondents. Correlation analysis was employed for hypotheses testing. From the results, some conclusion on the competence toward SCA would be drawn.

4.1 Demographic analysis

Male respondents showed the highest frequency which is 26 person (68.4%) and only 12 person of female answered the questionnaire. Besides, most of the respondents are Malay which incurred 55.3% whilst Chinese respondent only 39.5% and the least is India respondents which is only 5.3%. Meanwhile, the age of majority of the respondents were in the range of 40-49 years old which are 28.9% with 12 respondents. 28.9% or 11 respondents are aged in the range of 30-19 years old, 21.1% or 8 respondents are below 30 years old and only 18.4% or 7 respondents are more than 50 years old.

There are 26.3%, 23.7%, 21.1%,18.4% and 10.5% respondents are work in units of others, procurement/sourcing, operation, warehouse/distributor and logistic respectively. Next, from the data collected, it is noticed that 57.9% (22 respondents) are work as manager, 7 people of them (18.4%) are directors and senior manager respectively, and only 5.3% are CEO. Due to this research is focused on top level management; most of the respondents (34.2 %) have 6-10 years of experience. Next is 8 respondents (21.1%) with 11-15 years of experience, 5 respondents (13.2%) with 16-20 and 26 above years experiences separately, 4 of them only below than 5 years of experience, and lastly only 4 (10.5%) are worked for 21-25 years.

About the company's position on supply chain, there are 60.5% of respondents are main manufacturer and main supplier to main manufacturer respectively. However, there are 65.8% of them are secondary supplier to main manufacturer and only 11 （28.9%） of them from the sample size are distributor or dealer. In term of the engagement of company with suppliers, 11 respondents (28.9%) have 16-20 suppliers. The other are 9 respondents (23.7%) engage with the range of 21 to 25 respondents, 18.4% with 26 and above suppliers, 6 respondents (15.8%) with 6 to10 suppliers and lastly 13.2% have 11 to 15 suppliers.

From the aspect of the frequency of the company provides IT training for staff, it is discovered that there is none of them are always provide the training. 15 of respondents are sometime provided the training for staff. The rest of 9 respondents (23.7%), 8 (21.1%) and 6 (15.8%) are never, rarely and often provide training for their staff. In the perspective of frequency to upgrade company's IT infrastructures, the highest is 2 years per time which are 9 respondents (23.7%). In contrast, 21.1% are upgraded their IT infrastructure more than 5 years once. 7 respondents upgraded 3 years once, 4 of them are never upgrade or upgrade their IT infrastructure every 5 years. The other of 6 respondents are equally separated in the frequency of once a year and every 4 years for upgrading.

4.2 Reliability analysis

Validity and reliability of the adapted or modified scale were established. Validity tests how well an instrument that is developed measures the particular concept it is supposed to measure. Reliability of a scale on the other hand indicates the stability and consistency with which the instrument measures the concept and helps to assess the goodness of a measure [27]. Reliability analysis was conducted by computing the Cronbach's alpha. The result showing that variables in the study had acceptable reliability with Cronbach's alpha ranged from 0.835 to 0.924. Nunnally and Bernstain [28] stated that values of Cronbach's alpha that are above 0.7 (ranging from 0.6-0.9) for all performance measures is acceptable.

Table 5. Reliability Analysis

Variables	Cronbach's Alpha (Original)	Cronbach's Alpha after Item Delete	Number of Item Delete	Number of Item Left
IT Competence	0.835	0.835	None	9
Operational Competence	0.820	0.820	None	13
Supply Chain Agility	0.924	0.924	None	10

Table 5 indicates the reliability of all variables which Cronbach's Alpha values are higher than 0.7. IT competence is 0.835, operational competence is 0.820 and supply chain agility showed the Cronbach's Alpha value of 0.924. These results indicate that all measured items have achieved internal consistency and measuring the right construct. Thus, none of the items being deleted due to the high Cronbach's Alpha value shown.

4.3 Descriptive analysis

Respondents are required to answer each question in the survey by scoring a point of 1 to 5 according to their opinion. Then, the data is computed into mean and standard deviation. Morgan et al. [29] stated that the mean value obtained gives the average some of their response for each of the items that the respondents answered. While standard deviation is used to measure the dispersion of the data in which how close the entire set of data is to the average value. The lower the value of standard deviation, the closer is the data to the average value. Table 6 indicates the level of mean measurement which is ranked by the central tendency level according to [30].

Table 6. Level of Mean Measurement

Mean Range	Central Tendency Level
High	3.68 – 5.00
Moderate	2.34 – 3.67
Low	1.00 – 2.33

(Source, Wiersma, 1995)

4.3.1 Descriptive analysis of IT Competence items.

Table 7 shows that mean and standard deviation value of IT competence items. The mean values are in the range of 2.32 to 3.71 which indicated moderate mean score. Meanwhile, the standard deviation values are in the range of 0.98-1.32. The results show the data point are far grouping the mean value.

Table 7. Descriptive analysis of IT competence

Item of measurement	Mean	Std. Deviation	Interpretation
	Statistic	Statistic	
ITI_analyze	3.50	1.25	Medium
ITI_changeinfo	3.55	1.08	Medium
ITI_UGinventory	3.24	1.05	Medium
ITI_measure	2.66	1.21	Medium
ITI_monitor	3.47	1.22	Medium
ITF_staff_support	3.71	0.98	High
ITF_bintegrete	3.39	1.10	Medium
ITF_MIS	2.32	1.04	Low
ITF_Itmanager	2.34	1.32	Medium

Average mean score	3.13		Medium

4.3.2 Descriptive analysis of Operational Competence items

Table 8 shows that mean and standard deviation value of operational competence items. The mean values are in the range of 2.71 to 3.92. It is meant that the respondents are agreed on the items of measurement. However, the standard deviation of operational competence items is between 0.81 and 1.17 which showed the items are grouping far from to the mean.

Table 8. Descriptive analysis of operational competence

Item of measurement	Mean	Std. Deviation	Interpretation
	Statistic	Statistic	
OCI_communication	3.13	1.04	Medium
OCI_extendSC	2.71	1.18	Medium
OCI_marketing	3.39	1.13	Medium
OCLO_proTraining	2.84	1.20	Medium
OCLO_SClearning	3.76	0.88	High
OCLO_NewSCtech	3.29	0.98	Medium
OCLO_NewSCknow	3.45	1.16	Medium
OCF_seasonality	3.92	0.97	High
OCF_breakdown	3.87	0.81	High
OCF_delivery	3.42	1.06	Medium
OCF_newmarket	3.37	1.17	Medium
Average mean score	3.38		Medium

4.3.3 Descriptive analysis of Supply Chain Agility items.

Table 9 illustrates the mean and standard deviation values on the items of supply chain agility. The range of mean scores is 2.68 to 3.55 which classified as a moderate range of mean score. The range value of standard deviation under SCA items are from 0.92 to 1.25 which mean the respondents' answer is far grouping from the mean value.

Table 9. Descriptive analysis of supply chain agility

Item of measurement	Mean	Std. Deviation	Interpretation
	Statistic	Statistic	
SCA_leadtime	3.55	0.98	Medium
SCA_uncertainty	2.87	0.99	Medium
SCA_dataaccuracy	3.45	1.18	Medium
SCA_newPintro	2.68	1.23	Medium
SCA_delivery	3.42	0.92	Medium
SCA_Pcost	3.34	1.1	Medium
SCA_cusServ	3.45	1.13	Medium
SCA_marketneeds	3.39	1.13	Medium
SCA_quality	3.29	1.25	Medium
SCA_collaborative	3.24	1.2	Medium
Average mean score	3.268		Medium

4.4 Correlation analysis

Correlation is the term used to measure the strength of relationships between variables. Moreover, it is a statistical technique to determine the relationship between the independent variables and dependent variables. Correlation with the significance value below 0.05 is considered significant. Thus, the null hypotheses is rejected. Table 10 highlighted the final results.

Table 10. Descriptive analysis of supply chain agility

			IT competence	Operational competence	SCA
Spearman's rho	IT competence	Correlation Coefficient	1.000	.632**	.491**
		Sig. (1-tailed)		.000	.001
		N	38	38	38
	Operational competence	Correlation Coefficient	.632**	1.000	.529**
		Sig. (1-tailed)	.000		.000
		N	38	38	38
	SCA	Correlation Coefficient	.491**	.529**	1.000
		Sig. (1-tailed)	.001	.000	
		N	38	38	38

The significant value of IT competence towards supply chain agility is 0.001 which is less than 0.005. It indicates that the correlation is significant. The Spearman's rho correlation coefficient is 0.491 or 49.1% which stated that there is a positive relationship between IT competence and SCA. Thus, H0 is rejected. In addition, the degree of importance of IT competence toward SCA is 49.1%.

The significance value of operational competence towards SCA is 0.000 <0.005. It is meant that the correlation between variables is significance. Meanwhile, the Spearman's rho correlation coefficient is 0.529 or 52.9%. It indicates that there is a positive relationship between operational competence and SCA. Thus, H0 is rejected. Moreover, the degree of importance of operational competence toward SCA is 52.9%.

5 Discussion

This study is conducted to determine the influences between the independent variables (IT and operational competence) towards the dependent variable (supply chain agility). Moreover, it is objectively to prove there is a significance correlation between those two competencies towards supply chain agility.

5.1 Relationship between IT competence towards SCA

H_1 shows the relationship between IT competence and SCA. The values of Spearman's rho correlation coefficient is +0.491**. It is meant that the relationship has moderate correlation among two variables. This result indicates that the competence of IT of a firm will influence the achievement of SCA.

This result congruent with the study of [2] which stated that IT integration and IT flexibility are enablers of supply chain agility. Moreover, the study of [11] found that information systems had a positive effect on supply chain agility based on network management to enhance the firm's ability to transform the uncertainty into business opportunities.

5.2 Relationship between Operational competence towards SCA

H_2 shows the relationship between operational competence and SCA. The values of Spearman's rho correlation coefficient is +0.529**. It is meant that the relationship between two variables was positively correlated. The present study supports the finding of [2] who did prove that the operational competence had positively correlated with SCA. Besides, [5] also recognize the supply chain flexibility had significant direct impacts on supply chain agility. The author mentioned that the higher level of supply chain flexibility might leads to higher SCA. It is also suggested that the alignment of IT infrastructure and flexibility can behave as complementary capabilities that facilitate agility [30].

6 Recommendation and Conclusion

This study was conducted to identify the influence of IT and operational competencies of textile and apparel industry in Malaysia toward its company's SCA. The findings from literature reviews revealed that IT and operational competencies plays an important role for textile and apparel companies to reached supply chain agility. Similarly, this study was expected to show a positive correlation between the two competencies with supply chain agility respectively in Malaysian textile and apparel industry. However, researcher discovers that some of the practice of textile and apparel companies in local is divergence with overseas. Most of the companies in Malaysia are not develop their owned information system with their supply chain partners and caused a failed business integration. Beside the importance of providing staff for IT training and upgrading IT infrastructure is not notified by local textile and apparel companies. However, many top level management are willing to provide their staff new supply chain knowledge and technology if within their ability.

This study has certain limitations in its findings and conclusions. Data are collected from average of two respondents in each related company. It might contained bias although the selected respondents with substantial knowledge on the issue discussed. The results obtained from 15 companies may not be generally applied considering the sample size. However, researcher convinced that this study able to provide a concept of how the IT and operational competencies enhance the firm's SCA.

The textile and apparel company in Malaysia should improve on the integration of information system with their supply chain partners. It is because the information might update on time and can be used to analyse the solutions for uncertainties. Besides, the companies should no longer only focus on solely on internal competence development but need to expand their competence align with their partner in supply chain. It is an insight for the top level management strategically plan for the advancement of supply chain agility to achieve a satisfactory supply chain with a good inter-organization collaborations.

References

[1] Zhelyazkov, G. (2010). Agile Supply Chain: Zara's case study Zara ' s Case Study, (2009).
[2] Ngai, E. W. T., Chau, D. C. K., & Chan, T. L. a. (2011). Information technology, operational, and management competencies for supply chain agility: Findings from case studies. Journal of Strategic Information Systems, 20(3), 232–249. doi:10.1016/j.jsis.2010.11.002
[3] Swafford, P. M., Ghosh, S., & Murthy, N. (2006). The antecedents of supply chain agility of a firm: Scale development and model testing. Journal of Operations Management, 24, 170–188. doi:10.1016/j.jom.2005.05.002
[4] Yang, J. (2014). Supply chain agility: Securing performance for Chinese manufacturers. International Journal of Production Economics, 150, 104–113. doi:10.1016/j.ijpe.2013.12.018
[5] Swafford, P. M., Ghosh, S., & Murthy, N. (2008). Achieving supply chain agility through IT integration and flexibility. International Journal of Production Economics, 116, 288–297. doi:10.1016/j.ijpe.2008.09.002
[6] Malaysia External Trade Statistic (2014), International Trade Statistic Reports 2014. Retrieved from https://www.wto.org/english/res_e/statis_e/its2014_e/its14_merch_trade_product_e.pdf

[7] Malaysian Knitting Manufacturers Association (2009), International Trade and Industry Report 2008. Malaysian Textiles and Apparel Industry. Retrieved from http://www.mkma.org/Notice%20Board/2009/MsiaT&AIndReport2008.htm
[8] Lee, K. L., Udin, Z. M., & Hassan, M. G. (2014a). A review of relational capabilities on supply chain performance in textile and apparel industry. International Conference on Management & Business Sustainability 2014: 4th International Conference on Technology and Operations Management, 169–178. Retrieved from http://stmlportal.net/ictom04/papers/P17.pdf
[9] Agarwal, A., & Shankar, R. (2007). Modelling the Barriers of Integration in an Agile Modelling the Barriers of Integration in an Agile. Telecommunication Systems, 016, 1–42
[10] Lee, K. L., Udin, Z. M., & Hassan, M. G. (2014b). Global supply chain capabilities in Malaysian textile and apparel industry. International Journal of Supply Chain Management, 3(2), 31–40. doi:10.11648/j.ijber.s.2014030601.13
[11] Judith, N. M., & David, K. (2014). Role Of Information Systems Competence In Supply Chain Agility In Service Industry, 2 (11), 1–22
[12] Prajogo, D., & Olhager, J. (2012). Supply chain integration and performance: The effects of long-term relationships, information technology and sharing, and logistics integration. International Journal of Production Economics, 135(1), 514–522. doi:10.1016/j.ijpe.2011.09.001
[13] Fawcett, S. E., Osterhaus, P., Magnan, G. M., Brau, J. C., & McCarter, M. W. (2007). Information sharing and supply chain performance: the role of connectivity and willingness. Supply Chain Management: An International Journal, 12(5), 358–368. doi:10.1108/13598540710776935
[14] Zhang, Z., & Sharifi, H. (2000). A methodology for achieving agility in manufacturing organizations. International Journal of Operations & Production Management, 20(4), 496–512. doi:10.1108/01443570010314818
[15] Li, X., Chung, C., Goldsby, T. J., & Holsapple, C. W. (2008). A unified model of supply chain agility: the work-design perspective. The International Journal of Logistics Management. doi:10.1108/09574090810919224
[16] Christopher, M., Lowson, R., & Peck, H. (2004). Creating agile supply chains in the fashion industry. International Journal of Retail and Distribution Management, 32(8), 367–376.
[17] Tippins, Michael J. & Sohi, Ravipreet S. (2003). IT Competency and Firm Performance: Is Organizational Learning a Missing Link? Strategic Management Journal 24 (8), pp. 745–761
[18] Fink, L. & Neumann, S. (2009). Exploring the perceived business value of the flexibility enabled by information technology infrastructure. Information & Management 46, 90–99
[19] Halley, A., & Beaulieu, M. (2009). Mastery of operational competencies in the context of supply chain management. Supply Chain Management: An International Journal, 14(1), 49–63. doi:10.1108/13598540910927304
[20] Kim, D., & Cavusgil, E. (2009). The impact of supply chain integration on brand equity. Journal of Business & Industrial Marketing. doi:10.1108/08858620910986730
[21] Flynn, B. B., Huo, B., & Zhao, X. (2010). The impact of supply chain integration on performance: A contingency and configuration approach. Journal of Operations Management, 28(1), 58–71. doi:10.1016/j.jom.2009.06.001
[22] Shimizu, K., & Hitt, M. a. (2004). Strategic flexibility: Organizational preparedness to reverse ineffective strategic decisions. Academy of Management Executive, 18(4), 44–59. doi:10.5465/AME.2004.15268683
[23] Duclos, L. K., Vokurka, R. J., & Lummus, R. R. (2003). A conceptual model of supply chain flexibility. Industrial Management & Data Systems, 103, 446–456. doi:10.1108/02635570310480015

[24] Braunscheidel, M. J., & Suresh, N. C. (2009). The organizational antecedents of a firm's supply chain agility for risk mitigation and response. Journal of Operations Management, 27(2), 119–140. doi:10.1016/j.jom.2008.09.006
[25] Lewis, J.L. & S.R.J. Sheppard. (2006). Culture and communication: can landscape visualization improve forest management consultation with indigenous communities? Landscape and Urban Planning 77, 291–313.
[26] Olson, K. (2010). An Examination of Questionnaire Evaluation by Expert Reviewers Field Methods 22 (4), pp. 295–318
[27] Sekaran, U., 2000. Research Methods for Business: A Skill-building Approach. Wiley, New York, NY.
[28] Nunnally, J. C., & Bernstein, I. H. (1994). Psychometric Theory (3rd ed.), New York: McGraw-Hill
[29] Morgan, G. A., Leech, N. L., Gloeckner, G. W., & Barrett, K. C. (2012). IBM SPSS for Introductory Statistics: Use and Interpretation, Fifth Edition. Routledge.
[30] Tallon, P. P., & Pinsonneault, A. (2011). Competing Perspectives on the Link between Strategic Information Technology Alignment and Organizational Agility: Insights from a Mediation Model. MIS Quarterly, 35(2), 463–486. Retrieved from http://content.ebscohost.com/ContentServer.asp?T=P&P=AN&K=60461965&S=R&D=buh&EbscoContent=dGJyMNLr40SeqLM40dvuOLCmr0qeprVSsKu4Sa6WxWXS&ContentCustomer=dGJyMPGtsVGyqa5IuePfgeyx44Dt6fIA

IOP Conf. Series: Journal of Physics: Conf. Series **1049** (2018) 012012 doi:10.1088/1742-6596/1049/1/012012

Dual Frequencies Band and Enhanced Wideband Effect of Dual Layer Microstrip Patch Antenna with Parasitic

H. Nornikman[#1], **B. H. Ahmad**[1], **M. Z. A. Abd Aziz**[1], **H. A. Bakar**[1]
Centre for Telecommunication Research and Innovation (CeTRI),
Faculty of Electronics and Computer Engineering (FKEKK),
Universiti Teknikal Malaysia Melaka (UTeM), Durian Tunggal, Melaka,

E-mail: nornikman@utem.edu.my[#]

Abstract. In current telecommunication situation, the antenna with wideband or multi-band effect are important for cater many users' demands. In this research, the works explores the effect of parasitic element on the above of the dual layer microstrip patch antenna. Firstly, a basic microstrip patch antenna with a single layer of substrate (Design A) had been simulated using the CST Microwave Studio software. After that, the addition of another layer of substrate with parasitic element is shown as Design B. This proposed microstrip patch antenna resonates at two different frequencies of 2.396 GHz and at 2.543 GHz of frequency with – 36.8 dB and – 35.644 dB of return loss, respectively. The gain performance for both resonant frequencies at 2.396 GHz and at 2.543 is 5.035 dB and 4.814 dB, respectively. The bandwidth of the Design B antenna is 0.097 GHz (2.354 GHz – 2.451 GHz), had been increase compare with Design A performance of 0.039 GHz (2.380 GHz– 2.419 GHz). The second resonant frequency for Design B is only 0.028 GHz. (2.526 GHz– 2.554 GHz).

1.Introduction

Future telecommunication systems would be accomplished of accommodating higher data rates than the recent systems because of the various user demands from different segments. There is several of antenna design that had been investigating and fabricated earlier. The categories of antenna basically depend on the frequency range or band, the location and the application that the user desires to apply. One of the favorite types is the microstrip patch. Microstrip patch antenna is used in many applications such as handphone, Wi-Fi connection, and satellite communication due to their light weight & low profile.

In the previous design, there are many techniques to create the dual-band or multi-band frequency effect to the microstrip patch antenna. For examples, in [1] using fractal-like geometrical structure and in [2], the researcher using modified fractal slot antenna that fed by coplanar waveguide (CPW) technique. In [3], the author used the U-shaped slot and also two mitered corners to achieve dual frequency bands. In [4], dual band L-shaped had been apply in microstrip patch antenna. Lastly, T-shaped parasitic elements effect a dual-band frequency for the microstrip antenna [5], shorted parasitic element for dual-band monopole antenna [6], novel parasitic-element at printed antenna [7], and parasitic element at WLAN Yagi-Uda antennas [8]. There are other antenna designs that apply of parasitic element at their design such as in these several papers [9-17].

[1] To whom any correspondence should be addressed.

Content from this work may be used under the terms of the Creative Commons Attribution 3.0 licence. Any further distribution of this work must maintain attribution to the author(s) and the title of the work, journal citation and DOI.
Published under licence by IOP Publishing Ltd

The multi-band range antenna is highly in demand by users because of multi frequency or multiband application in one device. For example, the common dual applications that using a single antenna is the combination of the Wireless Local Area Network (WLAN) at 2.4/5.2 GHz and Worldwide Interoperability for Microwave Access (WiMAX) at 3.6/5.8 GHz [18-24]. The WLAN application is the common communication band that used currently, especially in the mobile phone, smart television and personal computer, iPad and laptop. This technology is easy to use and fabricated in the printed microstrip patch antenna design.

Multilayer or dual-layer substrate in microstrip patch antenna design is another technique to improve the performance of the antenna. For example, in [25], it stated that the dual layer stacked is effect to exist the ultra-wideband to the rectangular microstrip patch antenna. This technique gives a wider bandwidth of 56.8% in the between 3.06 GHz and 5.49 GHz of the resonant frequencies to the antenna performance. In [26], the two layers substrate and loop antenna with air gap had been effect two different frequencies of the antenna at 1.5 GHz and 2.0 GHz

There are many examples of microstrip patch antenna design with parasitic element had been research before. For example, in [27], the simple single shaped of parasitic element with two orthogonal bowties had been increase the bandwidth of the circular polarized cross dipole antenna in between 1.9 GHz and 3.9 GHz. It also effects to enhance a 3-dB axial ratio of the antenna. In the other paper [28], the parasitic element had been increase the gain of the antenna. The design is using parasitic element for dual feed microstrip array antenna. In this case, the parasitic element had been effected to increase 2 dB (from 12.9 dB to 14.9 dB) of gain to the antenna design.

In [29], the parasitic element effects the triband frequencies to the patch antenna. The band effect is for Universal Mobile Telecommunications System (UMTS) band (1.9 GHz to 1.979 GHz), ISM band (2.4 GHz to 2.4835GHz) and the Long-Term Evolution band (LTE) (2.5 GHz to 2.57 GHz). This parasitic element also effect to improve the bandwidth of the patch antenna. In [30], the author successfully shows the improvement of the bandwidth of the compact transverse bilateral helical antenna using parasitic element. It stated that the antenna achieves 105% wider bandwidth in a 50-Ω reference system. Beside multiband effect, the parasitic element also can effect to miniaturize the size of the printed microstrip patch antenna. In [31], the microstrip patch antenna with compact size designed by using two rectangular U-shaped parasitic elements. This antenna size is 40.0 mm x 30.0 mm and resonates at the three different frequencies of 2.6 GHz, 6.0 GHz, and 8.5 GHz with bandwidth of 50 MHz, 22.8 MHz and 30 MHz, respectively.

In this paper, dual frequencies band microstrip patch antenna had been design using parasitic elements and dual layer of FR-4 substrate. The performance of the resonant frequencies return loss, bandwidth, gain and radiation pattern are the parameters that consider in this paper.

2.Antenna Design

This section illustrates the dimension and configuration structure of the proposed microstrip patch antenna. Figure 1 represents the design of the dual layer microstrip patch with parasitic element. This design contains two main layers of substrates with dimension of 56.0 mm width x 50.2 mm length. The first layer consists a basic patch antenna (38 mm width x 27.8 mm length) and feedline (3.1 mm width x 17.7 mm length).

For the second layer (above layer), the substrate is consisted with a parasitic element (33 mm width x 21.6 mm length) and a cut-off substrate (5.1 mm width x 25.1 mm length). This proposed antenna was designed using FR-4 substrates with permittivity, $\varepsilon r = 4.3$.

The design started with the basic rectangular patch antenna, shown by Design A. This basic rectangular patch is in the antenna with single layer of FR-4 substrate with 0.035 mm thickness of copper at the ground plane, shown as Figure 1(a).

The Design B is combination of the Figure 1(a) and Figure 1(b) with the patch antenna and parasitic elements. The parasitic element is located at the upper layer and did not connect with the patch antenna in the bottom layer. This parasitic element is effect to improve the bandwidth of the antenna and to create another resonant frequency.

IOP Conf. Series: Journal of Physics: Conf. Series **1049** (2018) 012012 doi:10.1088/1742-6596/1049/1/012012

Figure 1(c) shows the perspective view of the proposed antenna with dual layer of FR-4 substrate that consist a patch antenna and parasitic element. The full dimension copper ground is located at the back of the first layer of the substrate, same in the Design A.

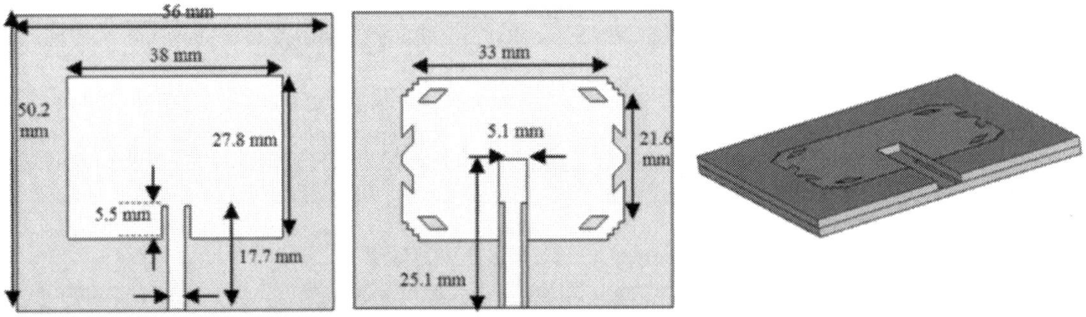

Figure 1. Dual layer microstrip patch antenna with parasitic element, (a) first layer of patch antenna and feedline – represent Design A, (b) Above layer with parasitic element and cut-off substrate – the combination of two layer represent Design B, (c) Perspective view of the antenna with dual layer of FR-4 substrate

3.Result

This section represents the result of the proposed microstrip patch antenna. The significant parameters that are considered in this research are resonant frequency (in GHz), return loss (in dB), bandwidth (in GHz), and antenna gain (in dB). Figure 2 shows return loss performance of the proposed antenna. It shows that, the resonant frequency of Design A antenna is at 2.4 GHz with a return loss of – 27.051 dB. With the addition of the second layer with parasitic elements (Design B), it creates a new resonant frequency at 2.543 GHz with – 35.644 dB. The second resonant frequency is narrower compared with the first resonant frequency.

It also remains the first resonant frequency at 2.396 GHz (shifted 0.004 GHz from 2.4 GHz) with return loss of - 35.644 dB. So, the addition of the parasitic element had been effected to increase the bandwidth from 0.039 GHz to 0.097 GHz while effects to enhance the return losses at the resonance frequencies while create new resonant frequencies for the antenna. Table 1 shows the resonant frequency, return loss performance of the proposed antenna. It shows that the Design B had been enhanced its bandwidth at the first resonant frequency compare with Design A, from 39 MHz to 97 MHz. The bandwidth of the second resonant frequency at 2.543GHz for Design B is only 28 MHz.

This section represents the result of the proposed microstrip patch antenna. The significant parameters that are considered in this research are resonant frequency (in GHz), return loss (in dB), bandwidth (in GHz), and antenna gain (in dB). Figure 2 shows return loss performance of the proposed antenna. It shows that, the resonant frequency of Design A antenna is at 2.4 GHz with a return loss of – 27.051 dB. With the addition of the second layer with parasitic elements (Design B), it creates a new resonant frequency at 2.543 GHz with – 35.644 dB. The second resonant frequency is narrower compared with the first resonant frequency.

It also remains the first resonant frequency at 2.396 GHz (shifted 0.004 GHz from 2.4 GHz) with return loss of - 35.644 dB. So, the addition of the parasitic element had been effected to increase the bandwidth from 0.039 GHz to 0.097 GHz while effects to enhance the return losses at the resonance frequencies while create new resonant frequencies for the antenna. Table 1 shows the resonant frequency, return loss performance of the proposed antenna. It shows that the Design B had been enhanced its bandwidth at the first resonant frequency compare with Design A, from 39 MHz to 97 MHz. The bandwidth of the second resonant frequency at 2.543GHz for Design B is only 28 MHz.

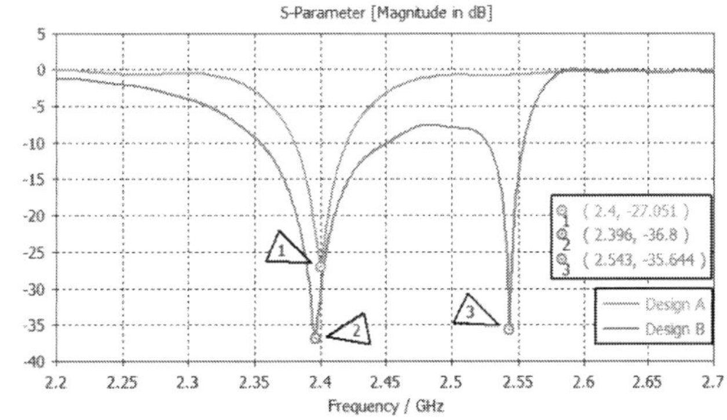

Figure 2. Return loss performance of the basic microstrip patch antenna for Design A and proposed microstrip patch antenna with parasitic element, Design B

Table 1. Resonant frequency, return loss and bandwidth of the microstrip patch antenna for Design *A* and Design *B*

Design	Resonant Frequency (GHz)	Return loss (dB)	Bandwidth (GHz), Frequency range (GHz)
A	2.400	- 27.051	0.039 (2.380 – 2.419)
B	2.396	- 36.800	0.097 (2.354 – 2.451)
	2.543	- 35.644	0.028 (2.526 – 2.554)

Figure 3 shows the radiation pattern effect for the resonant frequencies of 2.396 GHz and 2.543 GHz of Design *B* antenna at phi = 90^0. It shows that both resonant frequencies show the circular shape of the radiation pattern towards to the above part of the graph.

Figure 4 shows the 3D radiation pattern and gain performance effect of 2.396 GHz and 2.543 GHz of Design *B* antenna with 5.035 dB and 4.814 dB of antenna gain, respectively. Compare with the gain performance of Design *A* at 2.4 GHz, it shows the result of 5.014 dB, nearly the gain performance of Design *B* at 2.396 GHz.

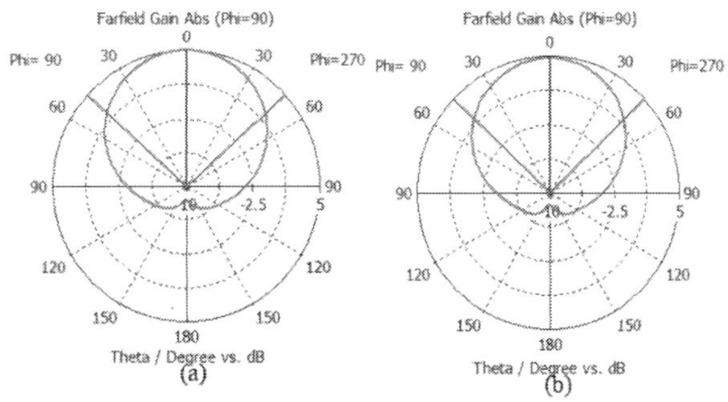

Figure 3. Radiation pattern effect for the resonant frequencies at phi = 90^0 of (a) first resonant frequency of 2.396 GHz and (b) second resonant frequency of 2.543 GHz of Design *B* antenna.

Table 1 shows the resonant frequency, return loss performance of the proposed antenna. It shows that the Design *B* had been enhanced its bandwidth at the first resonant frequency compare with Design *A*, from 39 MHz to 97 MHz. The bandwidth of the second resonant frequency at 2.543GHz for Design *B* is only 28 MHz.

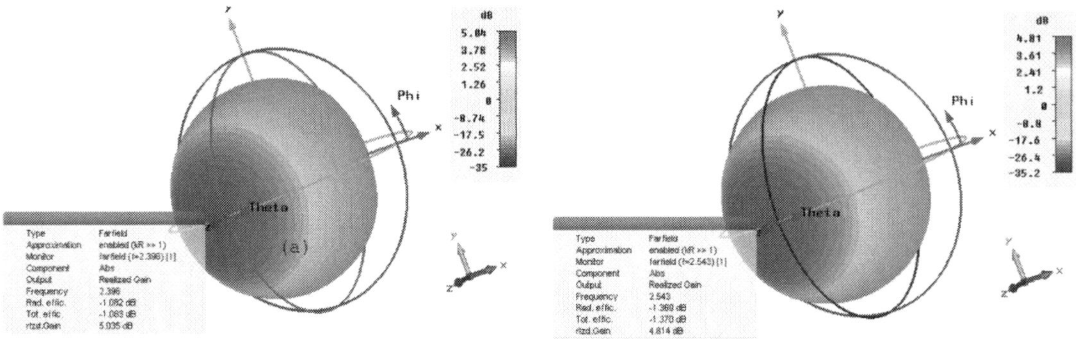

Figure 4. 3D radiation pattern and gain performance effect at the resonant frequencies of (a) 2.396 GHz and (b) 2.543 GHz of Design B dual-band microstrip patch antenna with parasitic elements.

Figure 5 shows the surface current distribution (for 0^0 and 90^0) of the dual-band antenna with parasitic element two different resonant frequencies at 2.396 GHz and 2.543 GHz. The figure represent that the surface current is computed a large number at patch antenna and at the parasitic element, while a little number of amount the feedline. This antenna is successfully to radiate at these two different frequencies, based on the surface current distribution. The current is pump in from the power source via SMA connector to the feedline at the bottom of the patch antenna. Then from feedline, the current is distributed to the other place of the patch antenna.

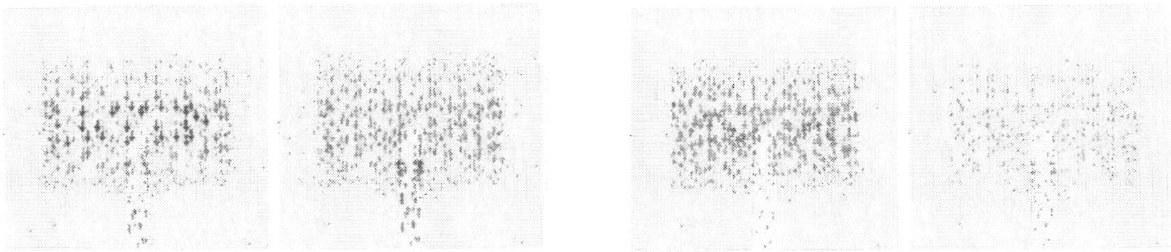

Figure 5. The surface current distribution of the dual-band antenna with the parasitic element. Two different resonant frequencies at (a) 2.396 GHz (for 0^0 and 90^0) and (b) 2.543 GHz (for 0^0 and 90^0).

4. Conclusion

After simulation work of dual-band antenna with the parasitic element done in this research, the author founded that the addition of the parasitic element structure successfully to enhanced the performance of the return loss of the first resonant frequency of 2.4 GHz (Design B). The addition of parasitic elements to the patch antenna also successfully creates a new resonant frequency at 2.543 GHz. It also shows the increasing the bandwidth from 39 MHz for Design A to 97 MHz to Design B. For the gain antenna, this addition structure did not give the significant effect, with only the increment of 21 MHz at the first resonant frequency.

Acknowledgments
The authors would like to thank Centre for Telecommunication Research and Innovation (CeTRI), Faculty of Electronics and Computer Engineering (FKEKK), Universiti Teknikal Malaysia Melaka (UTeM), and Ministry of Higher Education (MOHE) and Government of Malaysia. The authors would also like to thanks Centre for Research and Innovation Management Universiti Teknikal Malaysia Melaka (CRIM-UTeM).

References

[1] R. S. Aziz, M. A. S. Alkanhal, A. F. A. Sheta 2011 *Progress In Electromagnetics Research B* **29** 339

[2] C. Mahatthanajatuphat, P. Akkaraekthalin, S. Saleekaw, M. Krairiksh 2009 *Progress In Electromagnetics Research* **95** 59

[3] Y. Zhuo, L. Yan, X. Zhao, K.-M. Huang 2011 Progress In Electromagnetics Research Letters **26** 153

[4] B. Kelothu, K. R. Subhashini, G. Lalitha Manohar 2012 *Students Conference on Engineering and Systems (SCES)* 1

[5] J.-W. Kim, H.-J. Ham, J.-M. Woo, D.-K. Lee 2014 *20th International Conference on Microwaves, Radar, and Wireless Communication (MIKON)* 1

[6] C. -Y. Huang, P. -Y. Chiu 2005 *Electronics Letters*, **41**(21) 1154

[7] K.-C. Lin, C.-H. Lin, Y.-C. Lin 2012 *IEEE Transactions on Antennas and Propagation* 61(1), 488

[8] F. Fezai, C. Menudier, M. Thevenot, T. Monediere 2013 *IEEE Antennas and Wireless Propagation Letters*, **12** 413

[9] K.-L. Wong, L.-C. Chou, C.-M. Su 2005 *IEEE Transactions on Antennas and Propagation* **53**(1), 539

[10] C. Wood 1980 Optics and Antennas, **127**(4), 231

[11] J. Bao, Q. Huang, X. Wang, X. Shi 2014 *International Journal of Antennas and Propagation* 2014 1

[12] C.-Y. Huang, P.-Y. Chiu 2005 *Elctronics Letters* **41** 21 1154

[13] C. Menudier, M. Thevenot, E. Arnaud, A. Oueslati, F. Fezai, 2016 *46th European Microwave Conference (EuMC)* 16603645

[14] Y. Luo, Q.-X. Chu 2015 *IEEE Antennas and Wireless Propagation Letters* **15** 564

[15] Buckley J. L., K. G. Mccarthy, L. Loizou, B. O'flynn, C. O'mathuna *IEEE Antennas and Wireless Propagation Letters*, 2015 15, 630

[16] L. Peng, Y.-J. Qiu, L.-Y. Luo, X. Jiang, 2016 Wireless Personal Communications, **91** 3, 1163

[17] V. Iyer, S. Kulkarni, G. Zucchelli, S. N. Makarov, *2016 10th European Conference on Antennas and Propagation (EuCAP)*, 16037787, 2016

[18] H. Zhai, Z. Ma, Y. Han, C. Liang, 2013 *IEEE Antennas and Wireless Propagation Letters*, **12**, 65

[19] R. Rajkumar, K. U. Kiran 2016 *International Journal of Electronics and Communications* **70** 5, 559

[20] M. Jusoh, T. Aboufoul, T. Sabapathy, A. Alomainy, M. R Kamarudin, 2014 *IEEE Antennas and Wireless Propagation Letters* **13** 860

[21] O. Hazila, S. A. Aljunid, F. Malek, A. Sahadah, *2010 IEEE Student Conference on Research and Development (SCOReD)* 47

[22] M. S. Zulkefli, F. Malek, M. F. Jamlos, M. H. Mat, S. H. Ronald, 2011 *Loughborough Antennas & Propagation Conference* 1

[23] K. Paramayudha, Y. Wahyu, Y. Taryana, A. B. Adipurnama, H. Wijanto, 2016 *International Seminar on Intelligent Technology and Its Applications (ISITIA)* 363

[24] K. Rojanasaroch, T. Laohapensaeng, 2016 *International Symposium on Intelligent Signal Processing and Communication Systems (ISPACS)* 1-4

[25] M.A. Matin, B.S. Sharif and C.C. Tsimenidis, 2007 *IET Microwaves, Antennas & Propagation*, **1** 6 1192

[26] Y. Rikuta, H. Arai, Y. Ebine, Proceedings, 2000 Asia-Pacific Microwave Conference

[27] H. H. Tran, I. Park, T.-K. Nguyen 2017 *IEEE Antennas and Wireless Propagation Letters* 2016 1

[28] H. Satow, E. Nishiyama, I. Toyoda 2015 *International Symposium on Antennas and Propagation (ISAP)* 1

[29] T. Faradi, A. Diallo, P. L. Thuc, P. Daragon, R. Staraj, 2015 *9th European Conference on Antennas and Propagation (EuCAP)*, **63**(3) 937

[30] A. Celebi, M. Kenkel, T. T. Y. Wong, 2015 IEEE Transactions on Antennas and Propagation, 937

[31] S. S. Asif, A. Iftikhar, M. N. Rafiq, B. D. Braaten, M. S. Khan, D. E. Anagnostou, T. S. Teeslink *2015 IEEE International Symposium on Antennas and Propagation & USNC/URSI National Radio Science Meeting*, 617

IOP Conf. Series: Journal of Physics: Conf. Series **1049** (2018) 012013 doi:10.1088/1742-6596/1049/1/012013

Design a Three-Layer Patch Antenna with Defected Ground Structure (DGS) and 3-Shaped Slot Array

M. M. Saad*[1], M. N. Husain[1], M. Z. A. Abd Aziz[1], K. A. Abd. Rashid[1], H. Nornikman[1]#
[1]Centre for Telecommunication Research and Innovation (CeTRI),
Faculty of Electronics and Computer Engineering (FKEKK),
Universiti Teknikal Malaysia Melaka (UTeM), Durian Tunggal, Melaka, Malaysia

E-mail: majdi@psp.edu.my*, nornikman@utem.edu.my#

Abstract. A three-layer patch antenna with 3-shaped slot array and defected ground structure is offered in this paper. The work started with the patch antenna with single structure of 3-shaped slot. Then, it followed by addition of the four set of the 3-shaped slot array twin structures at each layer of patch antenna. The propose antenna had three layers of FR-4 substrate – a upper layer with 3-shaped slot at front and back side, a middle layer with 3-shaped slot at front and back side, while bottom layer contain a rectangular patch with a single 3-shaped slot structure. The last part is to add the defected ground structure (DGS) at the ground plane to optimize the performance of the antenna. The Design *D* antenna operates at tri-band at 2.452 GHz, 4.996 GHz, and 3.052 GHz with return loss of - 12.111 dB, - 16.138 dB and - 10.175 dB with 4.035 dB, 3.587 dB, and 3.841 dB of gain, respectively.

1.Introduction

Conservative basic microstrip patch antennas had some limits and limitations with single resonant frequency effect, narrow impedance, antenna with low gain and not miniaturized size for many applications. There are several of improvement techniques for conservative basic microstrip patch antennas stated. The example improvement technique is multi-layers' structure with air gap to increase the antenna gain while slot structure can effect to reduce the size of the antenna.
Defected ground structure (DGS) effect the reduction dimension size of the structure, to control the resonant frequencies location of the antenna while also can enhancing the bandwidth performance [1]. Er-rebyiy [2] in his design used DGS to shift the resonant frequency from 10 GHz to 3.5 GHz without change the antenna size. This DGS structure also effect to reduce the mutual coupling reduction in his design. There are several works that apply the DGS structures on the antenna such as in this paper [3-6].

The example of air gap technique used between two substrates in antenna designs is shown in Kasabegoudar work [7]. He compares the performance of an antenna design with and without air gap. This technique allows the antenna to be functioning at lower frequencies due to reactive loading. Ding [8] in his paper stated to apply the multi-layer of different substrate to improve the bandwidth and the gain of the antenna. There are several works that apply the DGS structures on the antenna such as in this paper [3-6].

In this paper, a three-layer patch antenna with 3-shaped slot array and defected ground structure had been design in four steps works of Design *A*, Design *B*, Design *C* and Design *D*.

Content from this work may be used under the terms of the Creative Commons Attribution 3.0 licence. Any further distribution of this work must maintain attribution to the author(s) and the title of the work, journal citation and DOI.
Published under licence by IOP Publishing Ltd

2.Antenna Design

Figure 1 illustrates the proposed multi-layer patch antenna with 3-shaped slot array structure and DGS structure in CST Microwave Studio simulation software. These antennas contain three RF-4 substrate layers with minimal height of air gap between all substrate. The initial design (Design *A*) of this proposed antenna is shown in Figure 1(b) that contains a rectangular patch with the 3-shaped slot at the centre. The Design *B* follows by the addition of two more layers – middle layer and upper layer. At this stage, also four sets of twin 3-shaped slots were adding to create more resonant frequencies to the antenna design. The dimension of the both FR-4 substrate of middle layer and upper layer is smaller compare with the bottom layer. For the last stage of Design *C* in Figure 1(c), the DGS structure were add at the ground plane part of the bottom layer to optimize and control the resonant frequency of the antenna.

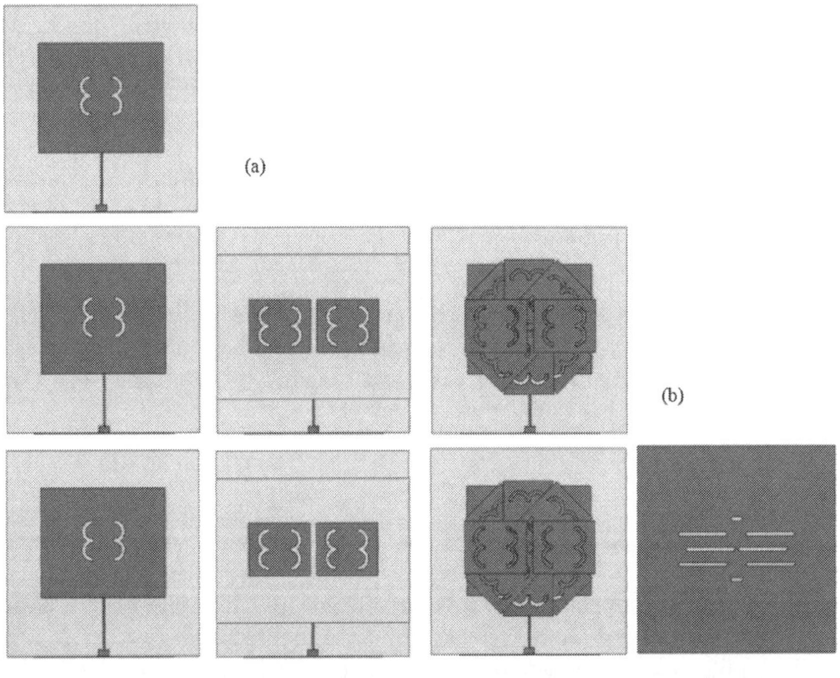

Figure 1. Proposed antenna, (a) Design *A*, (b) Design *B*, (c) Design *C*

3.Result

The main results in analysis section of the proposed antenna were done in term of the return loss with the resonant frequency and the antenna gain. Figure 2 shows the return loss and resonant frequencies for three stages of proposed antenna – Design *A*, Design *B* and Design *C*. The first stage of Design A effect to creates two resonant frequency, effect by the patch structure at 2.4 GHz region with – 26.918 dB of resonant frequency while the 3-shaped structure at the patch effect to resonate at 5.668 GHz with – 20.494 dB.

For Design B, it shows that there four narrowband frequencies, but in case, only two are accepted because other two not achieve better than – 10 dB performance. At 2.392 GHz and 5.668 GHz, it shows the performance of – 26.918 dB and – 20.494 dB, with gain of 3.537 dB and 3.339 dB respectively. For the last design of Design C, it shows the improvement of the gain performance for all resonant frequency with 4.035 dB, 3.587 dB and 3.841 dB at 2.452 GHz, 4.996 GHz and 6.052 GHz.

Table 1 shows the resonant frequency; return loss and gain of proposed antenna of Design *A*, Design *B* and Design *C*.

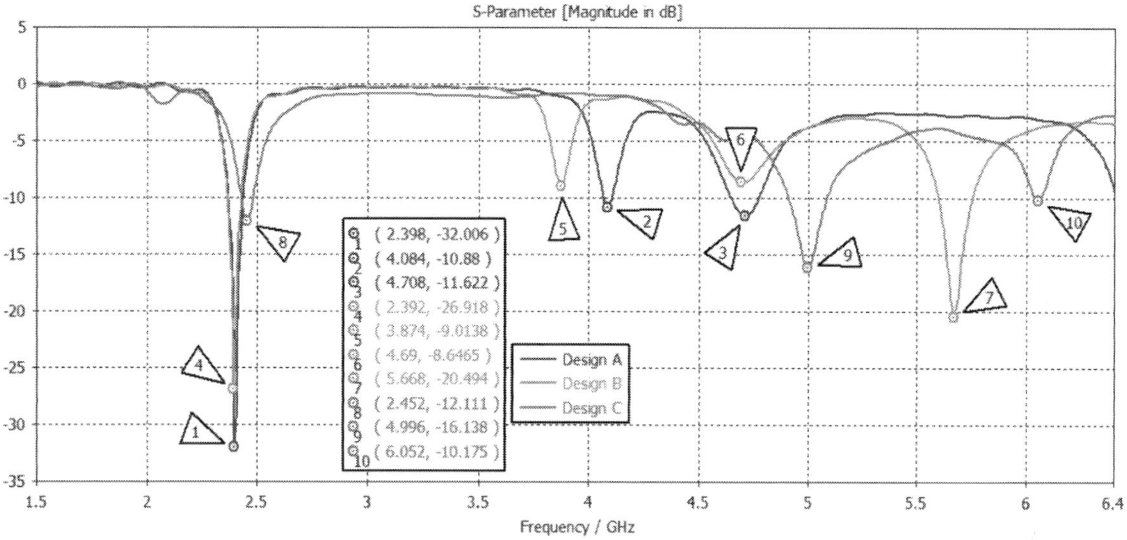

Figure 2. (a) Comparison of the return loss of fractal patch antenna of Design *A*, Design *B*, Design *C*, and Design *D*

Table 1. Resonant frequency, return loss and gain of proposed antenna of Design *A*, Design *B* and Design *C*

Design	Resonant frequency (GHz)	Return loss (dB)	Gain (dB)
A	2.398	- 32.006	3.348
	4.084	- 10.880	2.152
	4.708	- 11.622	2.531
B	2.392	- 26.918	3.537
	5.668	- 20.494	3.339
C	2.452	- 12.111	4.035
	4.996	- 16.138	3.587
	6.052	- 10.175	3.841

Figure 3 shows surface current of proposed antenna of Design *C* at 2.452 GHz, 4.996 GHz, and 6.052 GHz. It shows that, for the resonant frequency at 2.452 GHz, the current is focus at the defected ground structure at ground plane of the antenna. Figure 4 shows the radiation pattern at phi = 0 for several resonant frequencies.

IOP Conf. Series: Journal of Physics: Conf. Series **1049** (2018) 012013 doi:10.1088/1742-6596/1049/1/012013

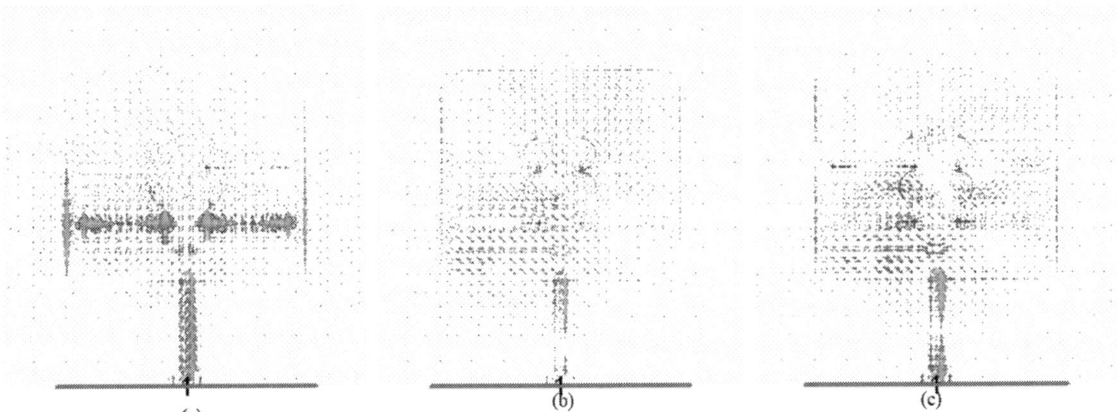

Figure 3. Surface current of fractal patch antenna of Design *C* at (a) 2.452 GHz, (b) 4.996 GHz and 6.052 GHz.

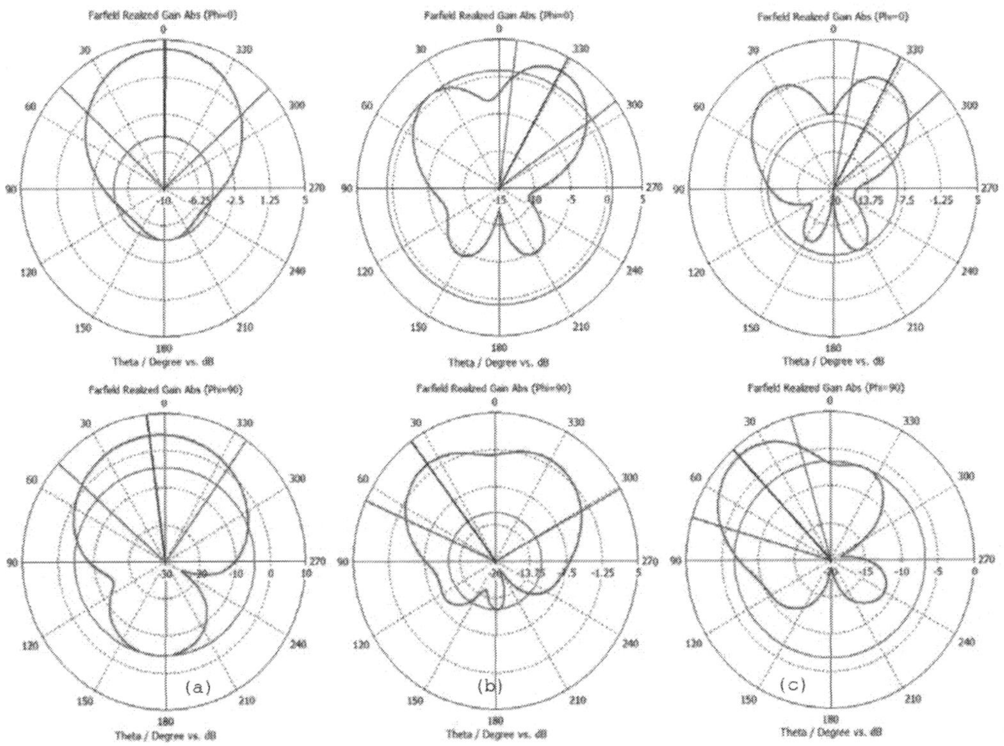

Figure 4. Radiation pattern of fractal patch antenna of Design *C* at different resonant frequency of 2.452 GHz, 4.996 GHz, and 6.052 GHz

4. Conclusion

After simulation work done, it shows that the proposed antenna operates at a multi-band frequency at 2.452 GHz, 4.996 GHz, and 6.052 GHz with higher gain compare with the basic design. It shows also the improvement of the gain of the antenna effect of the air gap and three-layer substrate.

Acknowledgments
The authors would like to thank Centre for Telecommunication Research and Innovation (CeTRI), Faculty of Electronics and Computer Engineering (FKEKK), Universiti Teknikal Malaysia Melaka (UTeM), and Ministry of Higher Education (MOHE) and Government of Malaysia. The authors would also like to thank Centre for Research and Innovation Management Universiti Teknikal Malaysia Melaka (CRIM-UTeM).

References

[1] M. K. Khandelwal, B. K. Kanaujia, and S. Kumar 2017 *International Journal of Antennas and Propagation* **2017** 1

[2] R. Er-rebyiy, J. Zbitou, A. Tajmouati, M. Latrach, A. Errkik, L. El Abdellaoui, 2017 *International Conference on Wireless Technologies, Embedded and Intelligent Systems (WITS)* 1

[3] A. K. Dubey 2017 *International Conference on Computer, Communications and Electronics* (Comptelix), 13

[4] R. Panwar and D. Bhatia, 2015 *International Conference on Computational Intelligence and Communication Networks (CICN)*, 20

[5] M. A. Belen, İ. Ö. Evranos, F. Güneş and P. Mahouti, 2017 *8th International Conference on Recent Advances in Space Technologies (RAST)*, 199

[6] R. Kiruthika, T. Shanmuganantham and R. K. Gupta 2016 *International Conference on Control, Instrumentation, Communication and Computational Technologies (ICCICCT)* 292

[7] E. S. Rameswarudu, M. Srinubabu and S. R. Rao, *2016 International Conference on Wireless Communications, Signal Processing and Networking (WiSPNET)* 2094

[8] V. G. Kasabegoudar and A. Kumar 2013 *Progress In Electromagnetics Research C* **36** 105

[9] Z. Ding, S. Xiao, M.-C. Tang, C. Liu, B.-Z. Wang, Q. Zeng, 2015 *IEEE MTT-S International Conference on Numerical Electromagnetic and Multiphysics Modeling and Optimization (NEMO)* 1

IOP Conf. Series: Journal of Physics: Conf. Series **1049** (2018) 012014 doi:10.1088/1742-6596/1049/1/012014

Co-planar Truncated Corner Patch Antenna with Fishbone-Shaped Slot Structure

K. A. Abd. Rashid*[1], M. N. Husain[1], M. Z. A. Abd Aziz[1], M. M. Saad[1], H. Nornikman[1#]
Centre for Telecommunication Research and Innovation (CeTRI),
Faculty of Electronics and Computer Engineering (FKEKK),
Universiti Teknikal Malaysia Melaka (UTeM), Durian Tunggal, Melaka,

E-mail: qamarz78@yahoo.com*, nornikman@utem.edu.my[#]

Abstract. This work proposed the co-planar truncated rectangular antenna with fishbone-shaped slot structure at the patch for GSM 1.8 GHz and WLAN 5.2 GHz applications. This work provides in six different develop, including a stage from Design *A* to Design *F*. Firstly, a basic rectangular patch antenna is done for Design A while the second stage is rectangular patch with cut-out truncated shaped at the below part for Design *B*. Then, it followed by the added co-planar effect for Design *C*, added fishbone slot structure for Design *D*, mirror effect of truncated patch for Design *E* and lastly cut-off triangular truncated corner at the patch for Design *F*. The performance of proposed antenna of Design *F* at resonant frequencies of 1.8 GHz and 5.2 GHz is – 25.207 dB and – 16.625 dB, respectively.

1.Introduction

Microstrip patch antennas are the right nominees for numerous wireless applications in the market. Nevertheless, these devices facing several restrictions such as it has low impedance bandwidth and single resonant frequency effect in their initial form. To cater their problem, previous researchers stated several methods to recover these restrictions.

Co-planar waveguide (CPW) is one of the techniques to improve the bandwidth. Besides that, it also potentially has a capability to reduce antenna efficiency and gain, increase end fire radiation, increase cross-polarization levels, and limit the applicable frequency range [1]. Khan [2] in this paper stated that his CPW structure effected the ultra-wideband region between 3 GHz and 9 GHz to his antenna design. Fishbone slot is another technique to create dual or tri band resonant frequency, such as referred at paper by Kang [3]. The other works on the CPW antenna shown in these several papers [4-9].

That antenna operates at two different application of S-DMB range (2.630 GHz –2.655 GHz) and WLAN application (5.125 GHz –5.325 GHz). Another works by Liu [8] apply the fishbone shaped slot array on his antenna. Truncated corner is another technique to reduce the size, increase the return loss of the antenna while control the location of the wanted resonant frequency such as in the paper of Mishra [10].

In this work, a co-planar truncated corner patch antenna with fishbone-shaped slot structure had been design with several step developments from basic antenna to the complete fabricated antenna.

Content from this work may be used under the terms of the Creative Commons Attribution 3.0 licence. Any further distribution of this work must maintain attribution to the author(s) and the title of the work, journal citation and DOI.
Published under licence by IOP Publishing Ltd

2. Antenna Design

Figure 1 represent the co-planar truncated corner patch antenna with fishbone-shaped slot structure development stage that contain six different design from basic antenna of Design *A* to the complete proposed antenna of Design *F*. The design starts with the basic antenna design of Design *A*. This antenna had a ground plane at the back part. Then, it followed by the Design *B* that had been cut-out truncated shaped at the bottom part that effect the resonant frequency. For the Design *C*, the co-planar structure had been effect to give a wideband range to the antenna. Then, it followed by the Design *D* and Design *E* that included the fishbone-shaped slot structure into the antenna. The last part is larger cut-out of triangular truncated corner to control the location of the resonant frequency of the antenna.

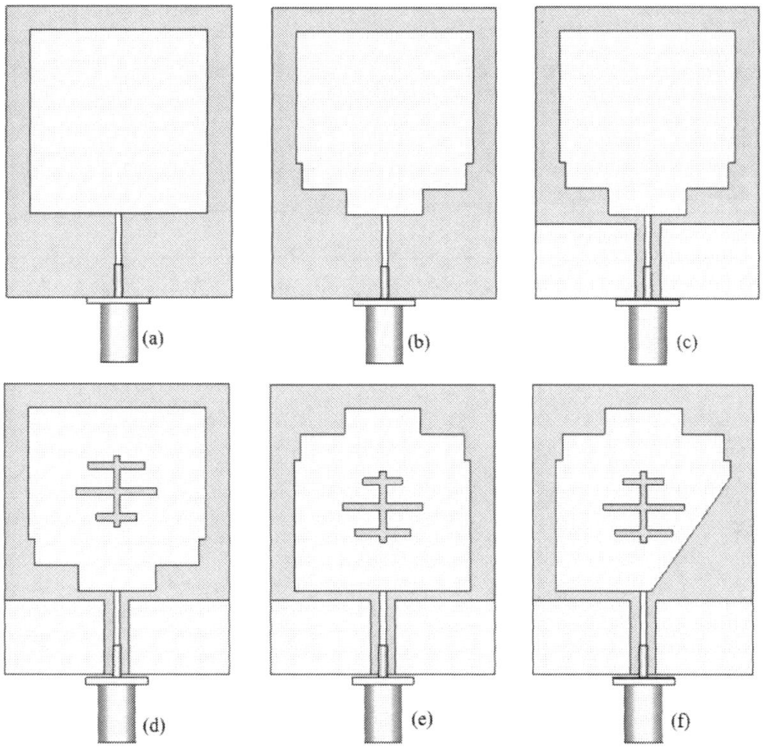

Figure 1. Co-planar truncated corner patch antenna with fishbone-shaped slot structure development stage, (a) Design *A*, (b) Design *B*, (c) Design *C*, (d) Design *D*, (e) Design *E* and (f) Design *F*

3. Result

This segment displays the return loss, radiation pattern, surface current, and gain of co-planar truncated corner patch antenna with fishbone-shaped slot structure. Figure 2 and Table 1 shows the performance comparison of the proposed antenna. Design *A* generates three different location of resonant frequency at 2.4 GHz, 4.7 GHz and 5.29 GHz. The first resonant frequencies are effect from the patch antenna itself with – 39.451 dB of return loss and 1.020 dB of gain while other two did not give the significant effect performance to the antenna. This is because the first and second resonant had a negative gain.

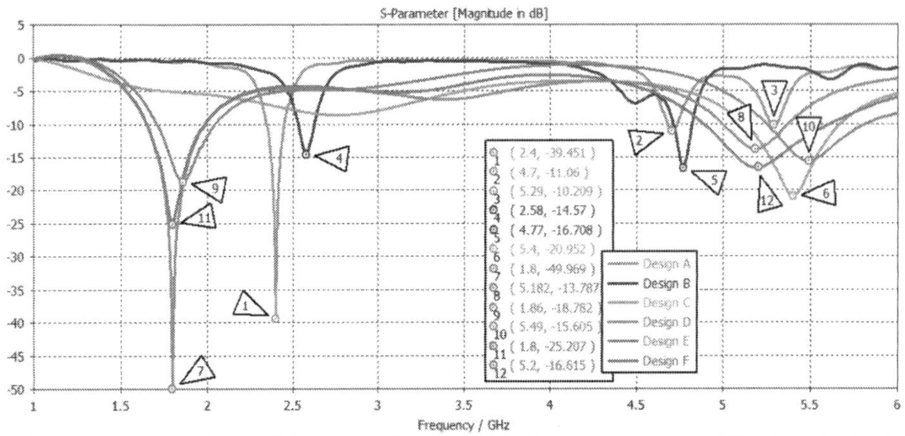

Figure 2. Comparison of the return loss of fractal patch antenna of Design *A*, Design *B*, Design *C*, and Design *D*

Table 1. Comparison of the return loss of fractal patch antenna of Design A, Design B, Design C, and Design D

Design	Resonant frequency (GHz)	Return loss (dB)	Gain (dB)
A	2.4	- 39.451	1.020
	4.7	- 11.060	- 2.243
	5.29	- 10.209	- 1.539
B	2.58	- 14.570	1.082
	4.77	- 16.708	0.771
C	5.4	- 20.952	5.310
D	1.8	- 49.969	1.537
	5.182	- 13.787	5.224
E	1.86	-18.782	1.494
	5.49	- 15.605	5.231
F	1.8	- 25.207	1.441
	5.2	- 16.625	5.983

For Design *B*, its shows that the truncated shaped at the patch effect to shifted the first resonant frequency to 2.58 GHz with resonant frequency of – 14.57 dB. For the second resonant frequency, it shows that the improvement of return loss from – 11.06 dB to – 16.708 dB. In design C that consist of co-planar technique, the antenna had been achieved the wideband range shape performance but not achieved wanted – 10 dB at the first resonant frequency. It only shows the good performance at the second resonant frequency at 5.4 GHz with return loss of – 20.952 dB. It shows that also the co-planar technique extremely increases the gain of the antenna to 5.310 dB.

The fishbone-shaped slot structure in Design *D* is shown to give back the first resonant frequency at 1.8 GHz. The second resonant frequency at this stage is shifted from 5.4 GHz to 5.182 dB. Then, mirror effect of the truncated patch in Design *E* illustrate to shifted the first resonance frequency to 1.86 GHz while the second resonant frequency at 5.49 GHz with – 18.782 dB and – 15.606 dB, respectively. For the last part of the Design F, it shows that the controlled mechanism structure of the cut-off triangular truncated effect to shifted back the first resonant frequency from 1.86 GHz to 1.8 GHz again with resonant frequency of – 25.207 dB. It also shifted the second frequency from 5.49 GHz to 5.186 GHz with resonant frequency of – 16.625 dB with the highest gain for all design with 5.983 dB.

Figure 3 shows the surface current of fractal patch antenna of Design *F* at 1.8 GHz and 5.186 GHz. It shows that the main contribution part of the current flow is at the feedline of the antenna and radiate to the patch part. Figure 4 shows the radiation pattern of fractal patch antenna of Design *F* at different resonant frequency of 1.8 GHz and 5.186 GHz

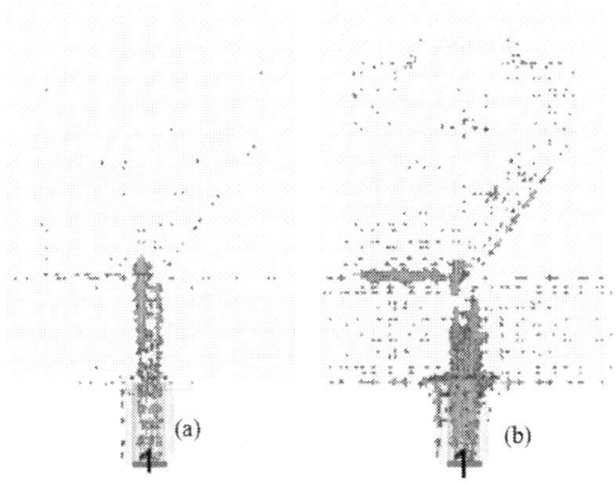

Figure 3. Surface current of fractal patch antenna of Design *F* at (a) 1.8 GHz and (b) 5.186 GHz

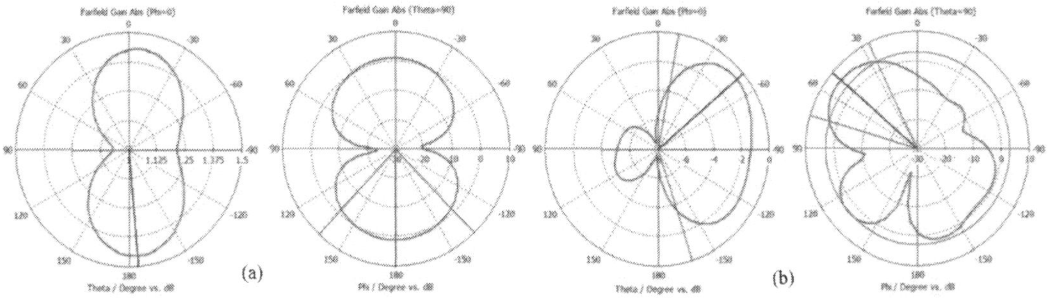

Figure 4. Radiation pattern of fractal patch antenna of Design *F* at different resonant frequency of (a) 1.8 GHz and (b) 5.186 GHz

4. Conclusion

The simulation work proves that the combination for all technique such as the fishbone-shaped slot, truncated shaped and co-planar give effect to several performances of the proposed antenna. This antenna successfully to use for GSM 1.8 GHz and WLAN 5.2 GHz applications with higher gain at second resonant frequency with 5.983 dB. The best and wanted design are shown by Design *F*.

Acknowledgments
The authors would like to thank Centre for Telecommunication Research and Innovation (CeTRI), Faculty of Electronics and Computer Engineering (FKEKK), Universiti Teknikal Malaysia Melaka (UTeM), and Ministry of Higher Education (MOHE) and Government of Malaysia. The authors would also like to thank Centre for Research and Innovation Management Universiti Teknikal Malaysia Melaka (CRIM-UTeM).

References

[1] Y.-C. Chen, S.-Y. Chen, and P. Hsu 2011 *IEEE Antennas Wireless Propagation Letters*, **10** 1124

[2] R. Khan, A. A. Al-Hadi, M. Elahi 2016 *IEEE Asia-Pacific Conference on Applied Electromagnetics (APACE)* 282

[3] S.-Y. Kang, K.-Sik Min, C. Yoon, H.-D. Park 2008 Microwave and Optical Technology Letters, **50**(1) 2509

[4] M. N. Osman, M. K. A. Rahim, M. F. M. Yusoff, M. R. Hamid and M. Jusoh, *2015 IEEE Conference on Antenna Measurements & Applications (CAMA)* 1

[5] N. M. Jizat, S. K. A. Rahim, Y. C. Lo and M. M. Mansor, 2014 *IEEE Asia-Pacific Conference on Applied Electromagnetics (APACE)*, 20

[6] M. Roshayati, P. J. Soh, A. Ismahayati, G. A. E. Vandenbosch and H. Mirza, *2010 Fourth Asia International Conference on Mathematical/Analytical Modelling and Computer Simulation*, 571

[7] S. Ben Mbarek, F. Choubani and B. Cretin 2017 *11th European Conference on Antennas and Propagation (EUCAP)* 2812

[8] Y. Liu, K. Li, Y. Jia, Y.Hao, S. Gong, Y. J. Guo 2016 *IEEE Transactions on Antennas and Propagation*, **64**(1) 326

[9] M. Ben Hassen, S. B. Mbarek and F. Choubani, 2017 *International Conference on Information and Digital Technologies (IDT)*, 62

[10] S. Mishra, P. Wankhade, A. Sahu, *2016 Symposium on Colossal Data Analysis and Networking (CDAN)* 1

A mixed-method analysis on students' critical thinking and problem solving skill development in Malaysian public universities

S A Rodzalan[1] and M M Saat[2]

[1] Department of Technology Management, Universiti Tun Hussein Onn Malaysia, Parit Raja Batu Pahat, Johor, Malaysia
[2] Department of Accounting and Finance, Universiti Teknologi Malaysia, Skudai Johor, Malaysia

E-mail: shazaitul@uthm.edu.my

Abstract. Employers nowadays are looking for potential fresh graduates who are able to think outside the box, so that they can produce with quality works. However, they are failed to do so and this situation result in high unemployment rate among graduates, mainly in Malaysia. With this concern, this paper aims to examine students' critical thinking and problem solving skill development after undergoing an industrial training. Further investigation also examines relationship of the skill development with respect to supervisors' leadership styles. Both quantitative and qualitative methods were employed, with three hypotheses were tested. A total of 2,000 undergraduate students from six public universities in Malaysia involved in pre- and post-training surveys. In assessing students' critical thinking and problem solving skill, eleven items in Belbin inventory on Thinking role was used. Meanwhile, nine employers and twelve students were participated in the interviews and they were asked regarding their perception of industrial training in assisting the development of critical thinking and problem solving skill. Descriptive analysis (such as mean), paired samples t-test, correlation and content analysis were utilized for the data analysis. Overall, results reveal that students developed in their critical thinking and problem solving skill upon completion of the industrial training. The results also demonstrate that there is a relationship between transformational leadership style the skill development. The findings of this study provide the necessity for industrial training stakeholders to improve industrial training program, mainly by implementing appropriate leadership styles.

1. Introduction

One of challenges faced by higher learning institutions is to produce graduates who possess with both hard and soft skills (and known as generic skills), as required by employers nowadays. Specifically, the employers seek potential employees who are able to think outside the box as it provides an advantage for the organization to stay compete with others. Therefore, university graduates as one of the sources for human capital should meet with the employers' requirement. Unfortunately, most of them were still lacking to think critically in solving the problems as reported by past studies [1, 2]. An option to develop students' generic skills is through sending them to industrial training. The industrial training exposes students with real working environment, provides practical work, enhances students'

Content from this work may be used under the terms of the Creative Commons Attribution 3.0 licence. Any further distribution of this work must maintain attribution to the author(s) and the title of the work, journal citation and DOI.
Published under licence by IOP Publishing Ltd

knowledge and skills as well as serves with employment opportunities. Specifically, industrial training can be defined as students' placement in an organisation by taking part in supervised practical work in selected industries, either outside or inside the country, within a certain period of time before they are awarded certificate, diploma or degree [3].

It is undeniable that industrial training could develop students' generic skills [4], however, little research has looked into the development of specific skills such as critical thinking and problem solving skill. The limited research of this area has motivated this study to be conducted. Thus, this pre- and post-training study aims to examine the development of students' critical thinking and problem solving skill after attending the industrial training program. In addition, it also attempts to examine the relationship of this skill development with regards to the factor of supervisors' leadership styles refer to transformational, transactional and laissez-faire.

The next sub-section discusses on model of critical thinking and problem solving, and leadership model. It also includes past studies related to this skill and its relationship with the factor supervisors' leadership styles. The following section then explained methods used and continues by reporting the findings. Subsequently, it concludes with a brief discussion and outlines both practical and empirical implications.

1.1. Model of critical thinking and problem solving skill

In describing students' thinking process, this study implements model by [5] which is the main focus in learning and teaching context. The model suggests that students engaged with learning process through repetitive activities, memorizing, understanding and reflecting. All of these processes require them to think in order to achieve effective learning outcomes and thereby enhance problem solving skill. However, the order of thinking differs at each stage. For example, reflecting stage requires higher order thinking (critical thinking) as compared to repetitive or memorizing stage. There are factors which may influence students' thinking process. These factors are teacher-student relationship, collective or collaborative studying, deep approach and transformational learning. In the context of this study, lecturer plays main role in providing clear instruction and conduct interesting activities in the class. The lecturer should provide students with challenging tasks that require them to think critically, instead on focusing in rote learning [6].

Moving from repetitive activities or memorizing into understanding stage, students may influence with the second factor, which is collective or collaborative studying. By working collaboratively, students will experience the process of analyzing the problems and express their ideas to other team members [7] which later enhance their understanding of knowledge learned. At this understanding stage, students will attempt to make sense of the knowledge learned through deep approach to learn. Students who integrate with this approach have commitment to understand the knowledge, and thereby reflected in using variety of methods.

As mentioned by [8], reflection process is one of essential elements in critical thinking. Reflection requires one's thought in making inferences, analogies, evaluations and explore deep understanding regarding the specific knowledge, and these relate to problem solving [9]. The outcome of reflecting process then results in the transformational learning as it influences the whole learning process. In short, transformative learning requires students to have broader view based on their past experience through critical reflection process.

1.2. Full Range Leadership Model (FRLM)

The FRLM represents three leadership styles that start with the lowest leader interaction of laissez-faire behaviour to transactional leader behaviour and reaching the highest transformational leader behaviour. These three leadership styles are represented by eight distinct factors of leadership behaviours. In particular, [10] proposed that the leader who avoids involvement in leading process is called laissez-faire leader. The laissez-faire is the most ineffective leadership style as the leader avoids taking decision, delay to deal with problems and ignore with the responsibilities and authority given

[10]. In this situation, the laissez-faire leaders can be said to have no leadership behaviour as they hands over all the work to their followers. In the context of this study, trainees who being supervised by laissez-faire supervisor will become more independent and being given freedom in accomplishing the tasks. However, lack of supervision may undermine their generic skills development as less interaction between trainee and supervisor and worse, it prevents them to explore new knowledge [11].

Secondly, the transactional leadership style is described as leaders who either actively or passively monitor the work by providing followers a contingent reward for attaining targeted goals. As developed by [10], transactional leadership style has been characterized as an exchange relationship between leader and followers. Here, "exchange" means that the followers will receive rewards (such as financial benefits and promotion) for achieving targeted goals. This type of leader is more focusing on meeting specific goals [12], without taking any consideration to change followers' personal values and commitment to work. In addition, [13] suggested that leader monitors followers' work and takes corrective action in timely manner by exhibiting management by-exception (active) behaviours. In contrast, passive leader will only intervene when the problems are occurred. Thus, in relation to this study, trainees may have low generic skills development and vice versa when they were being supervised by passive supervisor. Moreover, the exchange of contingent rewards may extrinsically motivate the trainees' commitment to work in short term.

Unlike transactional leader, transformational leaders being characterized as leaders who are respected, charismatic and able to motivate employees by inspiring them, stimulate their thinking and treat them individually. Specifically, there are four main behaviours of transformational leadership style which are idealized influence, inspirational motivation, intellectual stimulation and individualized consideration [10]. Firstly, the idealized influence is the degree where the followers view their leaders as a role model to be inspired with. The leaders who maintains with high moral standards may influence their followers to emulate them [10]. Secondly, the leaders' ability to drive commitment of their followers in attaining the goals is known as inspirational motivation behaviour. Apart from that, leaders used intellectual stimulation in order to encourage followers to think critically and solve the problems in new ways [10]. Most of the time, the leaders will provide their followers with interesting and challenging tasks and ask them to solve in their own way [14]. Thus, if supervisor assists the trainees to think critically and find other alternatives to solve the problems, it may result in improving their critical thinking and problem solving skill. Lastly, the individual consideration refers to the extent where leaders recognize followers need by treating them individually, but all are treated equitably. By having this behaviour, the leaders delegate the work to followers with the aims to provide learning opportunities and coach them if they need it. But in the scope of this study, supervisor should play a role to closely supervise and monitor trainees' work as they are newcomers to work in real workplace. Such closely supervision through regular interaction may enhance trainees commitment in their work, and thereby improve their critical thinking and problem solving skill [13].

1.3. Critical thinking and problem solving skill development

As people become adults and matured, they started to encounter with more complicated problems, especially when it comes to problems at the workplace. The notion also being supported by [15] where employees should able to think outside the box in facing those complicated problems. Thus, it is not surprisingly when educational system in worldwide concern to emphasise on students' critical thinking and problem solving skill [16]. With the concern, higher education in Malaysia has also emphasised to inculcate concept of higher-order thinking or HOT into teaching and learning process, mainly in tertiary education [17].

For that reason, a number of studies were conducted to investigate the effects of university experience (in terms of coursework, tests and assignments, classroom learning and others) regarding critical thinking and problem solving skill [18, 19]. Furthermore, there were studies that highlighted the importance of this skill, in particular to university students [21-25]. A longitudinal study by [18] indicated that students developed their critical thinking skill after undergoing learning process in the classroom. However, the study used a small number of sample size (n=34 students) which may affect towards the validity of the study. In avoiding this issue, the present study intends to have a large number of sample sizes in order to ensure the findings are more valid and accurate. [20] found similar finding where students developed their critical thinking skill after answering high-ordered questions. In their study, they had designed three different methods with undergraduate students became the sample size of study. In contrast, a longitudinal study by [19] showed almost half of total 2,322 American College students made no significant development during first two years of college, while more than quarter of the sample size also made no significant development after an entire four year college degree.

In relation to the importance of critical thinking and problem solving skill, [21] studied on the importance of critical thinking skill according to the disciplines. In their findings, engineering discipline was still vague with the concept of critical thinking skill as compared to humanities discipline. A study in Ghana found that both employers and graduates agreed with the importance of critical thinking skill for accounting education [23]. This study used survey method of 46 employers and 164 accounting students. There are studies which also investigated on critical thinking and problem solving skill development within the context of industrial training. Industrial training is a platform where students are able to transfer their knowledge to contexts beyond education, which involve practical tasks that require some reflection on knowledge learned in university, add up with assistance from the experts [26]. Such opportunities permit students to improve knowledge, skills and attitude as an early preparation into the future career. In assessing the development of critical thinking and problem solving skill towards working experience, [32] conducted interviews with 12 students who had completed their training. Based on the interview sessions, students were able to experience how to solve problems effectively.

Other past study showed the occurrence of reflection process throughout work-based experienced (WBE) module acquired by students [26]. Thus, it aligns with the critical thinking model which also stressed out the reflection process where students are able to think critically, and look the problem in various perspectives [5]. As they are able to think critically, it may motivate them to work harder and simultaneously increase their job satisfaction. In contradict with the encouraging result of both skill, [25] found that graduates were still shortage with problem solving skill although it is important. This perception was based on the interview sessions with 30 employers. Furthermore, a cross sectional study on 118 students also produced with negative results, where students were unable to think critically and solve problems during industrial training [4].

1.4. Relationship between supervisors' leadership styles and critical thinking and problem solving skill

With regards to supervisors' leadership styles, a number of past studies did confirm that a leader with transformational leadership was able to enhance follower's thinking skill [13, 21, 27, 28]. It is closely linked to one of the factors in transformational leadership, which is intellectual stimulation [10]. Such leader always encourages and helps followers to view problems in other perspectives by stimulating them to be innovative and creative in solving the problems [27, 28]. On the other hand, transactional and laissez-faire leadership styles seem to have no influence on students' critical thinking and problem solving skill. This is due to the transactional leader only intervenes when needed and provide rewards when the followers attaining goals, while the laissez-faire leader gives total freedom to the followers to complete the tasks given [10].

2. Research method

A pre- and post-training survey (before and after industrial training) was conducted to examine the development of students' critical thinking and problem solving skill. Based on the population of undergraduate students from Malaysian public universities, students from three fields of study (social science, science and engineering) were selected as the sample size. A total of 2,000 questionnaires were distributed in the pre-training survey, but only 1,227 questionnaires were completed and usable for data analysis. The researcher has self-administered the pre-training survey during industrial training briefing sessions. For the post-training survey, a total of 485 students completed the similar questionnaire through online medium. In addition to the pre- and post-training survey, the researcher also conducted interview sessions with nine employers and twelve students (similar students who have participated in the survey).

Both pre- and post-training survey comprises of eleven items related to critical thinking and problem solving skill. Meanwhile, nineteen items of supervisors' leadership styles are also included in the post-training survey. In measuring the development of critical thinking and problem solving skill, students had to rate their acceptance based on seven point scale ranging from "strongly disagree" (1) to "strongly agree" (7). Students were asked to rate the items of supervisors' leadership styles based on five point Likert scale; Not at all' (0); 'Once in a while' (1), 'Sometimes' (2), 'Fairly often' (3) and 'Frequently' (4). All items used in obtaining the development of critical thinking and problem solving skill data were adapted from Thinking roles of Belbin Team Role Self-Perception Inventory (BTRSPI) [29]. In regards to supervisors' leadership styles, a Multifactor Leadership Questionnaire (MLQ) by [30] was employed. They categorized nineteen items according to three leadership styles refer to transformational, transactional and laissez-faire. With respect to interview sessions, the employers and students were asked regarding their perception on the effectiveness of industrial training in developing critical thinking and problem solving skill. The data from interview sessions provide support for the quantitative findings.

In analyzing the data, paired samples t-test was used to measure critical thinking and problem solving skill development. Besides, correlation analysis was conducted in identifying the relationship between the skill development and supervisors' leadership styles. Meanwhile, content analysis was used in analyzing the qualitative data.

3. Findings

The following analysis was to report students' demographic profile, related to gender, age, races and academic discipline. A total of 2,000 students from six public universities in Malaysia answered the pre-training questionnaire with 1,227 usable questionnaires in this pre-training. However, the number of respondents in the post-training decreased approximately 40% from the pre-training. In pre- and post-training, female respondents constitute more than half of the sample compared to males. This is consistent with statistics provided by the Ministry of Higher Education Malaysia (Statistics of Malaysian Higher Education, 2009). The proportion of ethnicity in pre- and post-training was almost similar, as Malay respondents dominate the sample followed by Chinese respondent and the remaining were Indians and other races. The distribution of respondents in regards to field of study appears that the ratio is almost the same for pre- and post-training. Briefly, the majority of respondents were engineering students, the next largest consists of science student, followed by social science students.

With regards to demographic profile of employers who have being interviewed, there are three males (SS1, SS2, E1; *SS: Social Science, E: Engineering*) and six females (SS3, S1, S2, S3, E2, E3; *S: Science*) employers were participated in semi structured interviews. Out of nine interviewees, there are four employers with experienced of eight to twelve years in supervising the trainees, while the rest are six years and below. Meanwhile, there are five males (SS4, S3, E1, E2, E5) and seven females (SS1, SS2, SS3, S1, S2, E3, E4) students were participated in the semi structured interviews. Out of four social science students, two of them were assigned in accounting department (SS1, SS2), while the other two in audit and finance (SS3) and human resource department (SS4). Meanwhile, there are two science students who undergone their industrial training under research and development department (S2, S3), but only one Science student has been placed in administrative department (S1). On the other hand, there are three out of five engineering students were assigned in project site (E2, E3, E4) while the other two engineering students were placed in consultation (E1) and testing and inspection department (E5).

3.1. *Critical thinking and problem solving skill development*

Paired samples t-test results show that students who undertook industrial training have developed in critical thinking and problem solving skill in all eleven items (positive mean difference). Students developed significantly ($p<0.05$) in three items related to have creative approach to solve problem (1), feelings seldom interfere with judgement if given to work with limited time and unfamiliar people (9) and approach the problem in analytical way (10), while the other two items; able to see pattern in solving problems where others would see items as unconnected (2) and able to deny propositions that contain of invalid facts (8) were reported to have highly significant development ($p<0.01$).

Using the paired samples t-test, it is found that students have developed significantly in critical thinking and problem solving skill after finishing their industrial training. This can be seen from the mean score of pre-training is lower than mean score in post-training (5.11 to 5.21).

3.2. Relationship between supervisors' leadership styles and critical thinking and problem solving skill development

This section highlights that there is a positive relationship between transformational leadership style and critical thinking and problem solving skill *($r=0.244$, $p<0.05$)*. However, correlation analysis for *Transactional* and *Laissez-faire* show different results as it appears that there is no relationship occurs between these two leadership styles and critical thinking and problem solving skill. This suggests that as supervisors implement more with this leadership style, the more students will develop in critical thinking and problem solving skill. The findings indicate that the development of students' critical thinking and problem solving skill is only associated with transformational leadership styles, but not with the other two leadership styles.

3.3. Employers' perception on critical thinking and problem solving skill development

A number of employers raised an issue of students' attitude related to their thinking and problem solving skill. For instance, SS2 reiterated '... most of the trainees always follow the guideline given to them without having any initiatives to solve the problems'. In other words, students are unable or most probably avoid thinking critically as they have being given a certain way or guideline on how to do and solve the problem by their supervisor. In fact, SS2 also asserted that 'trainees are less independent and very most situation they need close supervision, and I observed only few of them are able to solve on their own'.

Out of nine employers, only E1 highlighted that trainees were unable to develop critical thinking and problem solving skill due to lack of creative approach and unable to link theory into practice. This is evident based on his observation, '... usually, I notice they (trainees) prefer to solve the tasks in simple ways, whereas there are many ways to solve it'. Again, he suggested that the lecturers most probably should adopt learning process which is more fun and attractive, despite focusing on the traditional method of rote learning. Based on these observations, it is evident that students' attitude in relying on the guideline and less independent may undermine their development in critical thinking and problem solving skill.

On the other hand, four employers mentioned that they felt trainees had developed their critical thinking and problem solving skill throughout the industrial training period. These employers reported that this development resulted as they were able to reflect theory into practical work, and were able to understand tasks given. Employer, E2 said: 'When I ask them how to solve the problem, they are able to solve it in their own ways...I am very satisfied'. However, SS3 employer argued that this development may take time. She further explained that in the beginning, the trainees faced difficulties to understand the relationship between practical and theoretical aspect, but they were able to overcome this situation as they observed what have being done by their supervisor.

3.4. Students' perception on critical thinking and problem solving skill development

Out of twelve trainees, nine of them agreed that they have development in their critical thinking and problem solving skill. For example, SS2 asserted that, 'Once I enter this company, I realised that I need to think outside the box because most of the work is complicated. In fact, the guideline given also just tells you the basic solution'. She added that, it requires her to think creatively in solving the problems. Moreover, S2 reiterated that: '...I must find other alternatives when inspecting the quality of the equipment ...'. By having similar opinion, E1 shared his experienced:

> 'During industrial training, I learn how to speak with others, but most importantly, I noticed that I am able to become a good problem solver. For example, there is a case where I need to handle the process of filing. Worse, they handled me the failed project. So, I need to start from the scratch, which requires me to think critically. At first, it quite difficult, but I still managed to solve it. The best thing, my supervisor was impressed with my ideas.'

On the other hand, there are three trainees (SS3, E3 and E4) claimed that they have less development in critical thinking and problem solving skill. This is evidenced based on claimed made by SS3, '... I just follow the guideline, no need to think other method in solving the problem'. However, she realised that job rotation during six months of industrial training allow her to think critically. As she mentioned, 'In accounting department, I just follow the guideline in order to ensure the account is OK, but when transferred to taxation department, I need to think critically. It was two different things to do'.

All the trainees agreed that the supervisor's approach in stimulating their thinking also one of the factors that lead to develop the skill in solving problem critically. SS4 stressed out that, 'Even though my supervisor provides the trainees with close supervision, yet, she did not allow you to have similar solution as others. We need to find other approach in solving the problems.' He also highlighted that his supervisor only provide some hint to solve the problems. In other words, SS4's supervisor intends to stimulate the trainees' thinking, by teaching them not to rely only to one solution. Other than that, SS2 emphasised that passion and interest while working is also important factors to develop skill in solving problem critically. She said, 'My work is quite challenging and require a lot of critical thinking. Because I love my work, I managed to handle it.'

4. Discussion and recommendations

This study aims to examine the development of students' critical thinking and problem solving skill after undergoing industrial training. The industrial training appears to be an imperative method in developing students' critical thinking and problem solving skill. Result of paired samples t-test shows that there is an increment of overall mean score in the post-training. This supports result from a pre- and post-training study by [31], where he found that students' perceived development in their critical thinking and problem solving skill. In similar vein, data from qualitative finding based on stakeholders (employers and students) perceptions also support the quantitative findings, where students were able to think critically in solving the problem. Nevertheless, employers stressed out the issue of students' attitude who avoid thinking critically.

When examining the results of the eleven critical thinking and problem solving skill items, students highly agreed that they evaluated both advantages and disadvantages of other people's ideas in solving the problems. Most probably, they have being taught to evaluate pro and cons before making decision in solving the problem. They also have a creative approach and innovative look in solving problem. Furthermore, they agreed that they were able to make accurate judgement if they were allocated with sufficient time. This agreement supports the observation made by the employers and consistent with a study by [4]. Besides, students perceived that they were able to work under pressure and unfamiliar people which give good impression to future employers as they were able to cope with real life situations [32]. The results from this study seem to support the model by [5] where students required to understand and reflect what they have learned in classroom as these processes lead to higher order thinking in solving the problems.

Correlation analysis showed that there is a relationship between transformational leadership and critical thinking and problem solving skill. However, the other two leadership styles (transactional and laissez-faire) show no relationship with this skill development. This expected finding might be related the factor of intellectual stimulation of transformational leadership where supervisors stimulate their students to be innovative and creative by looking old situations in new perspectives [13]. On top of that, intellectual stimulation, as part of transformational leadership is identified to help fosters subordinates' motivation through engagement of inner interest and psychological needs [33]. As a result, the students will apply skills that they acquired from supervisors and thus show development upon completion of their industrial training.

5. References

[1] ManpowerGroup 2012 2012 Talent Shortage Survey Retrieved June 4, 2013 from http://www.manpowergroup.us/campaigns/talent-shortage 2012/pdf/2012_Talent_Shortage_Survey_Results_US_FINALFINAL.pdf
[2] Omar N H, Manaf A A, Mohd R H, Kassim A C and Aziz K A 2012 Graduates' employability skills based on current job demand through electronic advertisement *Asian Social Science* **8** 103–110
[3] MOHE 2006 Development of soft skills for institutions of higher learning. Serdang, Selangor: Universiti Putra Malaysia Press
[4] Sulaiman H and Sapuan N H 2010 Kemahiran generik yang diperoleh pelajar sarjana muda teknologi serta pendidikan (kemahiran hidup) pkpg semasa menjalani latihan industri. Universiti Teknologi Malaysia
[5] Haller C, Fisher R and Gapp R 2007 Reflection as a means of understanding: ways in which Confucian heritage students learn and understand organisational behaviour *Multicultural Education & Technology Journal* **1** 6–24
[6] Ibrahim M A and Noordin S 2003 Perbandingan pencapaian para pelajar dalam pentaksiran kerja amali dengan peperiksaan bertulis. Buletin Persatuan Pendidikan Sains dan Matematik Johor
[7] Forgarty R and McTighe J 1993 Educating teachers for higher ordering thinking: the three-story intellect *Theory into Practice* **32** 161–169
[8] Colley B M, Bilics A R and Lerch C M 2012 Reflection: A key component to thinking critically. *The Canadian J. for the Scholarship of Teaching and Learning* **3** 1–19
[9] Kitchenham A 2008 The evolution of John Mezirow's Transformative Learning Theory *J. of Transformative Education* **6** 104–123
[10] Bass B M 1985 *Leadership and performance beyond expectations* (New York: The Free Press)
[11] Ronald G 2004 Laissez-faire leadership. Retrieved March 28, 2013 from https://secure.sagepub.com/northouseintro2e/study/chapter/encyclopedia/encyclopedia3.2.pdf.
[12] Sosik J J and Dinger S L 2007 Relationships between leadership style and vision content: the moderating role of need for social approval, self-monitoring, and need for social power *The Leadership Quarterly* **18** 134–153
[13] Bass B M, Avolio B J, Jung D I and Berson Y 2003 Predicting unit performance by assessing transformational and transactional leadership *Journal of Applied Psychology* **88** 207-218
[14] Hinkin T R and Tracey J B 1999 The relevance of charisma for transformational leadership in stable organisations *Journal of Organisational Change Management* **12** 105–119
[15] Nasir A N M, Ali D F, Noordin M K and Nordin M S 2011 Non-technical skills for technical skilled workers in Malaysia. *Proc. of the Int. Engineering and Technology Education Conference (Kuala Lumpur, Malaysia 16–19 January)*
[16] Billing D 2007 Teaching for transfer of core/key skills in higher education: cognitive skills *Research in Higher Education* **37** 71-93
[17] Heong Y M, Yunos J M, Othman W, Hassan R, Kiong T T and Mohamad M M 2011 The needs analysis of learning higher order thinking skills for generating ideas *Procedia-Social and Behavioral Sciences* **59** 197–203
[18] Reid J R and Anderson P R 2012 Critical thinking in the business classroom *Journal of Education for Business* **87** 52–59
[19] Trounson A 2011 Students fail to improve their thinking, study finds. Retrieved November 1, 2012 from The Australian. http://www.theaustralian.com.au/higher-education/students-fail-

to-improve-their-thinking-study-finds/story-e6frgcjx-1225991311384
[20] Renaud R and Murray H 2007 The validity of higher-order questions as a process indicator of educational quality *Research in Higher Education* **48** 319–351
[21] Ahern A, O'Connor T, McRuairc G, McNamara M and O'Donnell D 2012 Critical thinking in the university curriculum – the impact on engineering education *European J. of Engineering Education* **37** 125–132
[22] Moore T 2011 Critical thinking: seven definitions in search of a concept *Studies in Higher Education* 1–17
[23] Awayiga J Y, Onumah J M and Tsamenyi M 2010 Knowledge and skills development of accounting graduates: the perceptions of graduates and employers in Ghana *Accounting Education* **19** 139–158
[24] Douglas E 2009 The practice of critical thinking among engineering students. *Proc. of the 116th ASEE Annual Conf. & Exposition (Hilton Austin Texas 14–17 June*)
[25] Zaharim A, Yusoff Y M, Omar M Z, Mohamed A and Muhamad N 2009 Employers' perceptions and expectation toward engineering graduates: a study case *Proc. of the 6th WSEAS Int. Conf. on Engineering Education (Rodos, Greece,16–19 July)* pp. 23–29
[26] Smith K, Clegg S, Lawrence E and Todd M J 2007 The challenges of reflection: students learning from work placements *Innovations in Education and Teaching International* **44** 131–141
[27] Kark R, Shamir B and Chen G 2003 The two faces of transformational leadership: empowerment and dependency *Journal of Applied Psychology* **38** 246–255
[28] Landrum N E, Howell J P and Paris L 2000 Leadership for strategic change *Leadership & Organisation Development Journal* **21** 150–156
[29] Belbin R M 1996 *Team roles at work* (Butterworth-Heinemann: Oxford)
[30] Bass B M and Avolio B J 1995 *Multifactor Leadership Questionnaire* (Redwood City, CA: Mind Garden)
[31] Nabi F 2013 Management students' self-perception towards their employability skills-a pre and post soft skills training analysis *International Research Journal of Business and Management* **IV** 1–12
[32] Shariff S M and Muhamad M 2010 Learning in an industrial practicum training program: a case study in a public university in Malaysia *World Applied Sciences J.* **11** 1361–1368
[33] Bass B M and Riggio R E 2006 *Transformational leadership* (Mahwah, NJ: Lawrence Erlbaum)

IOP Conf. Series: Journal of Physics: Conf. Series **1049** (2018) 012016 doi:10.1088/1742-6596/1049/1/012016

M-Learning: Atomic Orbitals of Elements in Periodic Table using SPATO

L S Ang[1], S S M Fauzi[2,3], M Umi Hanim[2], A Amin Zhafran[2] and M N N Najwa-Alyani[1]

[1] Faculty of Applied Sciences, Universiti Teknologi MARA (Cawangan Perlis), Perlis, Malaysia.

[2] Faculty of Computer and Mathematical Sciences, Universiti Teknologi MARA (Cawangan Perlis), Perlis, Malaysia.

E-mail: shukorsanim@perlis.uitm.edu.my

Abstract. Recently, student's enrolment in science and health studies has shown a downward trend. This may jeopardise the successful implementation of government's agenda in alleviating Malaysia's performance in the world stage. In this report, the focus is on a mobile application named SPATO, a teaching aid in the learning of atomic orbitals of elements in the Periodic Table. SPATO is developed with the aim to spur the interest of students in studying the science-related subjects in the guise as a mobile-learning tool. The Waterfall Model was used to fulfill the aims in the development of this mobile application. Students and education's staffs were involved in the testing of this mobile-learning tool. From the results, SPATO had shown to be well accepted by the users. Users found that it is easy to use, even though initially they might not familiar with the m-learning application. These results indicated that SPATO can provide an alternative and another approach for user in understanding the chemical elements.

1. Introduction

Recently, the Ministry of Education Malaysia reported that the number of student enrolment in the field of science and health studies had decreased [1], indicating the interest in the study of science in Malaysia is at its lowest ebb. Since Malaysia aims to become a developed country by the year 2020, and the agenda forwarded in the TN50 plan, the importance of the science, technology, engineering, and mathematics (STEM) studies cannot be overstated [2, 3].

Due to the rapid advancement of mobile technologies, there are new opportunities to incorporate the latest innovation in the digital front for teaching and learning. Compared to the computer-based educational software and simulation, studies using the mobile application is still new to the education world. However, some of the educational apps had shown to contribute and enhanced the effectiveness in teaching and shaping the students' characteristics [4, 5]. Mobile application can be used to attract science students by providing a flexible platform in learning a topic of interest. It is an alternative way to support the traditional methods in teaching, in which the amount of time spent on a topic could be increased [6].

One of the widely used basic information in pure science education is the atomic orbitals of the chemical element. In the current report, a mobile application SPATO (SPatial ATomic Orbitals) with certain properties will be presented. This mobile application will allow the user to view image of the

[3] To whom any correspondence should be addressed

Content from this work may be used under the terms of the Creative Commons Attribution 3.0 licence. Any further distribution of this work must maintain attribution to the author(s) and the title of the work, journal citation and DOI.
Published under licence by IOP Publishing Ltd

atomic orbital surface. The size of the orbitals can be compared between the elements of the Periodic Table (PT). Different type of orbitals can be analysed using this application.

2. Literature Review

Understanding the elements in PT is an important part in science education. PT is an organized form of sorting the chemical element according to the atomic number. It was introduced in 1869, and has undergone many changes before it exists in the form as it is known nowadays. The modern version of the PT was created by Nobel Laureate Glenn Seaborg during 5th decade of 20th century [7].

PT is widely used due to the important information that are usually incorporated in it. There are many types of printable PT that are available in the internet and text books. However, due to the space limitation, not all properties of the elements can be included. One of them is the atomic orbital. Atomic orbital is a space where the electrons have a chance to be found. It is a standard topic usually taught in the introductory chemistry classes [8, 9]. For some other countries, atomic and molecular orbitals are part of the secondary school syllabus [8]. There have been many approaches in generating the atomic orbitals, ranging from using marbles and papers [10], software [11, 12] to the 3D printer [13].

Mobile learning, also known as M-learning, is defined as the learning process by using the mobile devices, such as smartphones. It is also known as "e-learning using mobile devices" [14]. Learning the elements in the PT through M-learning can be considered as an interactive aid to the existing methods. M-Learning is supported by variety mobile communication platform such as communicate using voice, surfing to the learning portal on the Internet and learn on platform Short Message Service (SMS) [6]. This shows that M-Learning could be interactive learning platform with the combination of audio, portal web and mobile phone technologies in the one package.

Atomic orbitals can be had from performing quantum-mechanical calculations (e.g., using software Gaussian [15]) or drawn from the hydrogen-like equations of the orbitals. One of the innovations using M-learning, related to atomic orbital is developed by Bonifácio [11]. In the work, labelled QR-APTE, which is based on quick-response (QR) codes, the information of the chemical elements in the PT can be retrieved in the audio form by scanning the given QR codes. It focusses on the usability for the visually-impaired people.

3. Methods

To fulfil the aims of this study, Waterfall model was used, where the model consist of analysis, design, implementation and testing phases, as shown in Figure 1 [16]. To proceed to the next phase of the Waterfall model, the previous phase need to be completed [17].

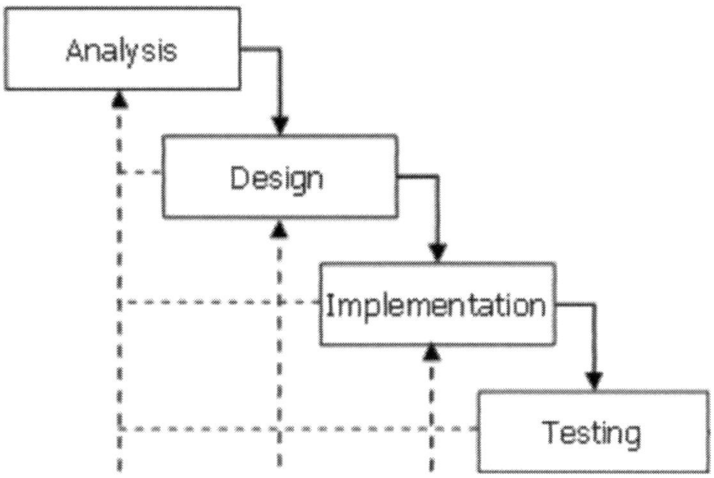

Figure 1. The waterfall model.

The explanation of each phases of the development of this project is elaborated further in the following sections:

IOP Conf. Series: Journal of Physics: Conf. Series **1049** (2018) 012016 doi:10.1088/1742-6596/1049/1/012016

3.1. Analysis

In the analysis phase, the information for the development of the mobile application were collected and recorded in requirement specification document. The information was collected from the online sources, article, journal and online databases.

3.2. Design

Based on the requirements, the designing of the interface for the android application was made. The storyboard was used as the guidance on the real interface design before implementing the interface design prototype using the Android Studio, Genymotion, Balsamiq and Adobe Photoshop CS5. The interface was designed by referring to the existing interface of another application that is being used for the PT in order to fulfil the users' requirements. It provides two different options to compare the atomic orbitals.

As shown in Figure 2(a), two different buttons were developed to indicate different options. "Explore" button would provide the detailed information about the chemical elements and the atomic orbital of the element. The "Explore" button will lead to page with 18 different buttons indicating 18 different elements in PT as shown in Figure 2(b). All of these 18 buttons would lead to another page with the information of the chemical elements (Figure 2(c)). The atomic orbital and information is placed side by side. The atomic orbitals include in this page also has a button to go to another page. Figure 2(d) shows the interface of the atomic orbitals for the chemical elements. This interface provides the zooming option of the molecular orbital. The name and coordinates was overlapping the atomic orbital for better understanding of the size of the atomic orbital of the chemical elements.

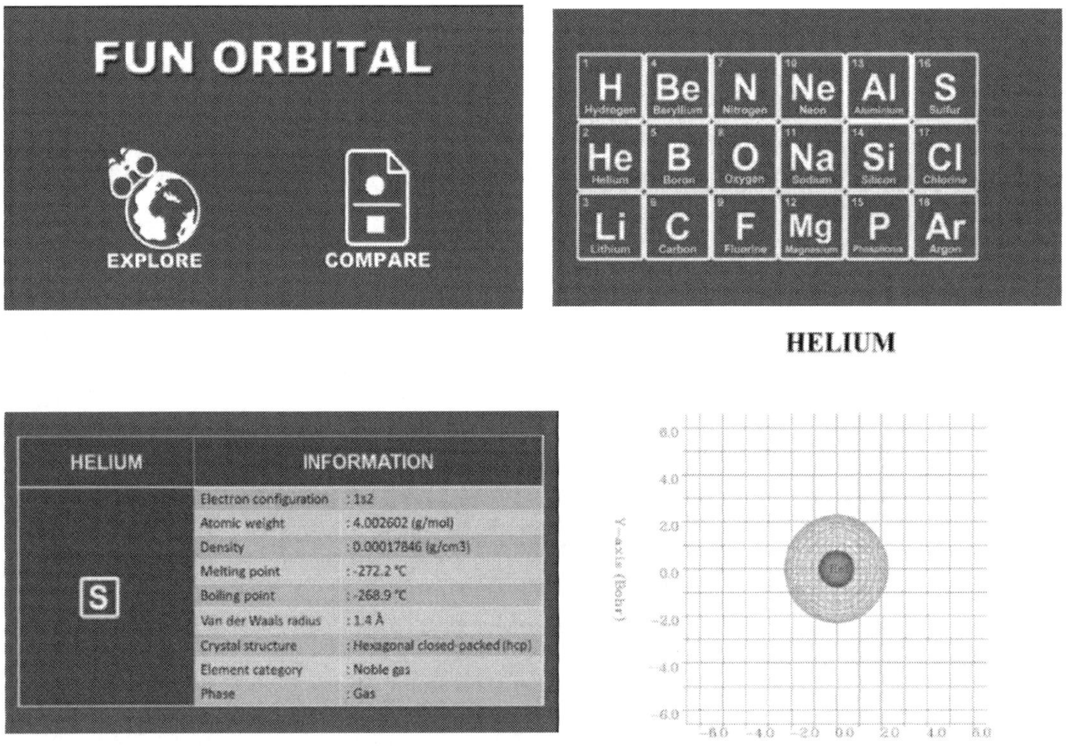

Figure 2. Interface of the application in (a) front page, (b) elements provided, (c) details of element and (d) molecular orbital of element.

The "Compare" button as shown in Figure 2(a) would lead to another interface where a list of chemical elements will be shown. To compare two different elements, Figure 2(b) would appear for second time after second element can be chosen to compare to. Last interface would show the two different elements put side by side as in Figure 3. Same zooming option was provided to the users. The coordinate scale is similar for a better comparison of the atomic orbitals.

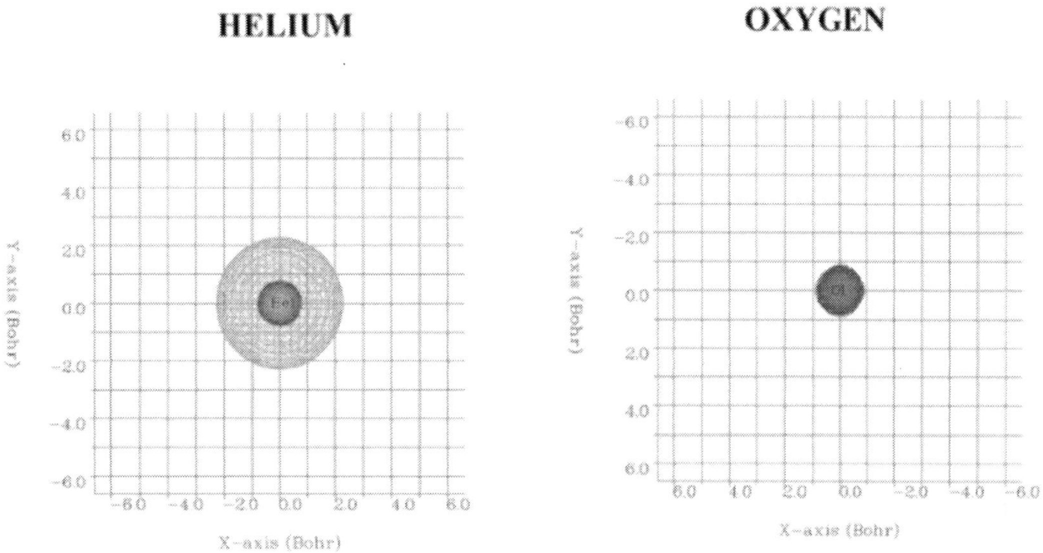

Figure 3: The comparison of molecular orbital page.

The atomic orbitals viewed in this mobile application are generated using Gaussian 09 [15] and Multiwfn [18] software. The Gaussian 09 is used to generate the atomic orbital of the chemical elements through the quantum mechanical calculations. The outputs from the calculations were fed Multiwfn, which provides the coloured atomic orbital with scales overlapping the atomic orbital.

3.3. Implementation

The project was then developed based on the input collected during the analysis and design phases. Java language was used as the writing source code to develop the mobile interface link and function. The library used is based on the Android library provided by Android Studio software.

3.4. Testing

The mobile application uses functional and acceptance testing test for any fault and failure. Functional testing is to ensure the application is working as planned while acceptance testing is to ensure the acceptance of the user to the applications. The mobile application is deployed on the user environment which involved the students as a target user, to identify the effectiveness of the mobile application.

3.4.1. Functionality testing.

Functional testing is applied to validate the functions of the application where it is aimed for a better performance to meet the requirements. There were several modules used for the functionality testing as shown in Table 1. The module for this system consists of the list of 18 elements based on the PT, the list of each kind of orbital available for viewing and zooming option of the atomic orbital image.

Table 1. Functional Testing

Test ID	Modules	Expected Results	Results
1	Show list of 18 elements based on the PT.	Users choose an element to learn the atomic orbital shape and information.	Good
2	List type of orbital.	The user chooses the type of orbital.	Good
3	Zoom image atomic orbital.	The user is able to zoom the image to view more clear the shape and coordinate of orbitals.	Good

3.4.2. Acceptance Testing.
The questionnaire was constructed to examine the m-learning interest of the research participant on this mobile application. The Technology Acceptance Model (TAM) was used to understand the implementation of the mobile application from the users' perspective [19]. TAM questionnaire was divided into four different categories, which are, perceived usefulness (PU), perceived ease of use (PEU), attitude toward usage (ATU) and behavioural intention to use (BIU).

PU section was constructed to understand the extent of the system satisfying the user needs. Meanwhile, PEU is the extent of user's understand ability and their ease when using the system. ATU discussed the attitude toward usage which is the extent of the effectiveness of the system to the users. Lastly, BIU is asking about the intention of users for adopting this system in future.

4. Results and Discussion
Two preliminary evaluations were carried out to evaluate SPATO. The first evaluation is about the functional testing of SPATO. The second one is to gauge the student's understanding on the atomic orbitals after using SPATO.

From the results of the functional test, it can be concluded that SPATO is easy to use. The user found it easy to handle this mobile application even though majority of them are not familiar with the M-learning application. From the observation, there were no users facing any problem when using the application in the testing phase.

The overall mean was classified in three scales which was negative, neutral and positive as shown in Table 2. The range of this scale is between zeros to five.

Table 2. Range of mean

Category	Range of Mean
Negative	0 – 1.66
Neutral	1.67 – 3.33
Positive	3.34 – 5

From table 3, results shows that, the overall mean for the acceptance test using TAM is 4.16. This concluded that for each category tested, positive feedbacks were obtained from the users. This also means that the users accept SPATO mobile application.

Table 3. Acceptance testing TAM result

TAM Category	Min	Max	Mean	Mean Category
Perceive Usefulness (PU)	3.9	4.2	4.05	Positive
Perceived Ease of Use (PEU)	4.1	4.5	4.3	Positive
Attitude Toward Usage (ATU)	4.1	4.4	4.25	Positive
Behavioral Intention to Use (BIU)	4	4.1	4.05	Positive

5. Conclusion
SPATO mobile application was developed to provide effective learning of atomic orbitals of the elements in the PT, with the aim to spur the interests of more students to study science courses. Generally, this mobile application has achieved the objective as to develop a convenient and suitable mobile learning of atomic orbital for elements 1 to 18 in the PT. The preliminary evaluations the effectiveness of the mobile learning among students of sciences and technology in UiTM Perlis. The limitations of the SPATO should be improved in the future.

Acknowledgment
The authors would like to thank Institute of Research Management & Innovation of Universiti Teknologi MARA for providing the Academic & Research Assimilation (ARAS) fund (project code 600-IRMI/DANA 5/3/ARAS (0194/2016)).

References

[1] Meng, C C, Idris, N and Eu, L K, *EURASIA Journal of Mathematics, Science and Technology Education*, 2014. **10**: p. 219-27.
[2] Phang, F A, *STEM boleh jayakan agenda TN50*, in *Berita Harian*. 2017: Malaysia.
[3] Sipon, S, *STEM penuhi impian jadi negara pencipta*, in *Berita Harian*. 2017: Malaysia.
[4] Falloon, G, *Computers & Education*, 2013. **68**: p. 505-21.
[5] Wu, H-K, Lee, S W-Y, Chang, H-Y and Liang, J-C, *Computers & Education*, 2013. **62**: p. 41-9.
[6] Ting, R Y-L. *Mobile learning: current trend and future challenges.* in *Fifth IEEE International Conference on Advanced Learning Technologies (ICALT'05)*. 2005.
[7] Chattopadhyay, D, *Resonance*, 2017. **22**: p. 79-87.
[8] Tsaparlis, G, *Research in Science Education*, 1997. **27**: p. 271.
[9] Autschbach, J, *J. Chem. Edu.*, 2012. **89**: p. 1032-40.
[10] Specht, Z and Raley, D, *J. Chem. Edu.*, 2014. **91**: p. 151-3.
[11] Bonifácio, V D B, *J. Chem. Edu.*, 2012. **89**: p. 552-4.
[12] Ramachandran, B and Kong, P C, *J. Chem. Edu.*, 1995. **72**: p. 406.
[13] Griffith, K M, Cataldo, R d and Fogarty, K H, *J. Chem. Edu.*, 2016. **93**: p. 1586-90.
[14] Pinkwart, N, Hoppe, H U, Milrad, M and Perez., J, *Journal of Computer Assisted Learning*, 2003. **19**: p. 383–91.
[15] Frisch, M J, Trucks, G W, Schlegel, H B, Scuseria, G E, Robb, M A, Cheeseman, J R, Scalmani, G, Barone, V, Petersson, G A, Nakatsuji, H, Li, X, Caricato, M, Marenich, A, Bloino, J, Janesko, B G, Gomperts, R, Mennucci, B, Hratchian, H P, Ortiz, J V, Izmaylov, A F, Sonnenberg, J L, Williams-Young, D, Ding, F, Lipparini, F, Egidi, F, Goings, J, Peng, B, Petrone, A, Henderson, T, Ranasinghe, D, Zakrzewski, V G, Gao, J, Rega, N, Zheng, G, Liang, W, Hada, M, Ehara, M, Toyota, K, Fukuda, R, Hasegawa, J, Ishida, M, Nakajima, T, Honda, Y, Kitao, O, Nakai, H, Vreven, T, Throssell, K, J. A. Montgomery, J, Peralta, J E, Ogliaro, F, Bearpark, M, Heyd, J J, Brothers, E, Kudin, K N, Staroverov, V N, Keith, T, Kobayashi, R, Normand, J, Raghavachari, K, Rendell, A, Burant, J C, Iyengar, S S, Tomasi, J, Cossi, M, Millam, J M, Klene, M, Adamo, C, Cammi, R, Ochterski, J W, Martin, R L, Morokuma, K, Farkas, O, Foresman, J B and Fox, D J, *Gaussian 09*. 2016: Gaussian, Inc., Wallingford CT.
[16] Bassil, Y, 2012. **2**: p. 1-3.
[17] Balaji, S and Murugaiyan, M S, *International Journal of Information Technology and Business Management*, 2012. **2**: p. 26-30.
[18] Lu, T and Chen, F, *J. Comput. Chem.*, 2012. **33**: p. 580.
[19] Dumpit, D Z and Fernandez, C J, *Int J Math Educ Sci Technol*, 2017. **14**: p. 5.

Study of factors influencing construction delays at rural area in Malaysia

M Z Ramli[1,5], M A Malek[2], M H Hanipah[3], C L Lin[1], M F Mahamad Sukri[1], M H Zawawi[1], M Z Zainal Abidin[4], and N F S Mohamad Fuad[1]

[1]Civil Engineering Department, College of Engineering, Universiti Tenaga Nasional, 43000 Kajang, Selangor.

[2]Civil Engineering Department, Institute of Sustainable Energy, Universiti Tenaga Nasional, 43000 Kajang, Selangor.

[3]Faculty of Civil and Environmental Engineering, Universiti Tun Hussein Onn Malaysia, 86400 Batu Pahat, Johor.

[4]Faculty of Architecture and Built Environment, Infrastructure University of Kuala Lumpur, 43000 Kajang, Selangor.

[1]E-mail: zakwan@uniten.edu.my

Abstract. Construction is one of an important industry which contributes to the economic growth in Malaysia. However, it has been revealed that 79.5 percent and 66.7 percent of the public and private projects were not completed within the time specified in the contracts out of 359 projects in Malaysia. Therefore, the purpose of this study is to investigate the delay factor caused project delay at rural area. A 5-points Likert scale questionnaire survey were answered by 111 respondents which having experience with rural construction project. The questionnaire data were analysed by using Relative Importance Index (RII). Five top factors were determined from this study based on their RII values which are improper construction method implemented by contractor, weather condition, difficulties in providing delivery to site, breakdown of site equipment, and poor qualification of contractor's technical staffs.

1. Introduction

Construction sector plays important role in economic development of country. In Malaysia, construction is an essential and productive sector for economic growth to improve the quality of life and living standard of Malaysian people [1]. Department of Statistics Malaysia stated that the value of construction works in 1st quarter 2017 had increase of 8.1% compared to the 4th quarter of 2015 in which major type of the activity are civil engineering works following with non-residential building, residential buildings and lastly special trade activities [2]. However, Malaysia is suffering challenges in construction industry in such of poor performance of time and budget, wastage of construction, low productivity and over-reliance on foreign labours [3]. Despite from all challenges, project delay is categorised as major component of construction management process and has become the key factor to gain success in a project [4]. Malaysian Treasury Secretary-general, Datuk Dr Wan Abdul Aziz Wan Abdullah said that construction projects of 30% or three months behind the schedule are considered as

[5] To whom any correspondence should be addressed.

Content from this work may be used under the terms of the Creative Commons Attribution 3.0 licence. Any further distribution of this work must maintain attribution to the author(s) and the title of the work, journal citation and DOI.
Published under licence by IOP Publishing Ltd

IOP Conf. Series: Journal of Physics: Conf. Series **1049** (2018) 012017 doi:10.1088/1742-6596/1049/1/012017

'Sick' project [5]. It has been revealed that 79.5 percent and 66.7 percent of the public and private projects were not completed within the time specified in the contracts out of 359 projects in Malaysia [6]. Other than that, it has been reported that 80 percent of Malaysian government projects were delayed [7]. It can be seen that public projects needs more attention in order to complete the project according time schedule given.

One of the topic that had been discuss in the Eleventh Malaysia Plan of the rural area is the Transforming Rural Areas to Uplift Wellbeing of Rural Communities which indicate issue of fewer basic infrastructure and other amenities in rural area such as road network [8]. This statement can be supported with the statistics of construction project in Malaysia 2015 by Construction Industry Development Board (CIDB). From this statistics, urban area such as Johor, Selangor and Wilayah Persekutuan shows a greater number of construction project compared to other area such as Kedah, Kelantan and Perlis for three years [9]. A project construction at rural area of East Coast Expressway Project Phase 2 (LPT2) was declared delay in completion time due to changes in the entire design and structure from a Federal highway to Toll highway [10]. Another project in rural area of Bakun Hydropower in Sarawak was reported experienced an extensive delay which cause cost overrun [11]. Due to the delay project issues, the demand of studies to overcome the problems has increase. However, lack of research has been done on project delay at rural area in Malaysia. Therefore, it has become necessity to investigate the factor caused project delay at rural area.

2. Construction Delay

Delay in construction is define as a time over-run beyond completion date of stated project timeline which can affect the owner and also contractor especially in term of budget [12]. All the construction project in the world might have completed project with time overrun by give an example of Asian Development Bank project whom awarded with the "Outstanding Award" but still facing the 9 months project delaying [13].

Every construction project should have high productivity rate. When construction project faced with the construction constraint, it will have an impact on three major element which are time, cost and quality of the project thus it will have low in a productivity rate [14]. The whole construction phase is a critical phase of any project in which many unforeseen factor could occur. Failure to identify the unforeseen factor that cause the problem in construction project will resulted in the completion time delays and cost [15].

Inappropriate planning during early stage of project will triggered delays at various stage throughout the project [14]. Therefore, an experienced contractor and labour is needed especially in large and mega project. A good planning and management is important to accelerate the project plan, to overcome time burn during delays.

2.1. Some of Factor Influence Construction Delay at Rural Area

Based on previous research, there are many factors that cause delay in construction project at rural area. The following are factors which have been studied by the researchers.

2.1.1. Accessibility

Provide an access to the site construction is one of the responsibilities for the owner as owner must make the site construction are available to perform the work for the contractor before the construction project can be started. Failure of the owner to provide the access of the worksite will cause the delay of the construction project and other organization such as contractor can claim the payment due to the delay of site access [15]. In rural construction project, limited access to the project site is one of the factor can cause the delay. A lot of matter need to be considered about the access to the site where usually rural projects are far from any headquarters thus to manage the travel time and efficient use of vehicle can be one of the challenging matter to access the site construction in rural area [16].

2.1.2. Site Condition
Site condition is one of the factor that can influence the time of the completion of the construction project which are involves underground condition and works that start at the beginning of the construction project [17]. A detailed investigation and preparation before commencement of construction project are required to avoid the unforeseen ground condition where usually the condition of natural ground cannot be predictable [18]. National Vocational Training Institute (NVTI) [19], explained that rural builders deal with some difficulties situation as not all the construction site has a flat and level ground. Some of the difficulties situation are:

a) Interfering/Obstruction object
b) Uneven ground
c) Sloping Sites

2.1.3. Space Constraint
Space constraint or the workspace congestion is regarded as one of a main cause of productivity loss on construction site [20]. Space in the construction site are limited while construction in the progress due to many circumstances or conflict such as space needed for equipment, workers to move from one place to another place, transportation of material and to store the material [21].

2.1.4. Underground Utilities
In the Rural Road Standards from the Public Works Department (PWD) of Baltimore country (2008), state that, due to the narrowness of the rural road, all the utilities should be relocated or place in order to maintain the scenic or history road characteristic [22]. In this case, all the utility are advisable to be made to underground for purposed of aesthetics and safety.

2.1.5. Fluctuation/Suitability Material
Price fluctuation of material is one of the factor that can lead to the cost overrun of the construction. In rural area of Morocco, the mobilization of the local materials to the rural site construction such as steel, cement, gravel and sand are brought with high price and it is unavoidable solution. So, they tend to use materials that taken from the soil directly to avoid the price fluctuation of local material thus can avoid the delay of construction project [23].

2.1.6. Material Delivery
Factors that cause the delay of the large construction project are comes from the consultant, contractor and owner [12]. From the study, material delivery is one of the major factor that cause the delay in construction project based from the consultant perspective. As material plays a big role in the productivity of the construction project, materials procurement in rural area are more complicated compared to the urban area. This is due to some of the procurement company have a different strategies which acquire additional cost such as for mobilization or delivery to the rural site construction [24].

2.1.7. Availability of Labour
Most of the 'ruralness' site project area has a problem to recruit and retain staff [16]. Hence, the project lack of expertise worker to handle certain work such as scheduling, project control, quality control and safety management. Besides that, a survey found out that there have a negative impact towards the employee's family life due to the employee working in a rural area of the construction site [24].

2.1.8. Productivity of Labour
Knowledge and consideration of numerous aspects that affect productivity of construction labour is required to avoid project cost overrun and delay in the completion time of the construction project [25]. In rural areas, due to the labour shortage and also pressure from the project schedule, working

overtime has become one of the factor that can decrease the labour productivity which eventually caused delay of the project [4].

2.1.9. Number of Skilled Labours
One of the factor that can cause delay of the construction project is lack of skilled or experienced labour which is out of contractor's control [13]. In order to increase the productivity of labour, skilled workers for the project construction in rural area are highly needed compared to urban area. However, due to the 'ruralness' area with limited access to the project site, it is challenging to get skilled workers [16].

2.1.10. Qualification of Contractor
Inadequate experienced contractor is one of the factor that can affect completion date of construction project. In some cases, incompetence contractor are awarded with large and complex project because of the lowest bid by the contractor [26]. Project in rural area, contractor are less interested to bid the competition compared to the urban project due to low rates of skilled labour [16].

2.1.11. Performances of Contractor
The performance of the contractor is reflected by successful to complete the construction project on time while inadequate of experienced or incompetence contractor will result of project delay [26]. In rural project, one of the challenges for the contractor is the limitation of ability to communicate and normally they do not using scheduling software to reduce the risk of delayed project [16].

2.1.12. Site Management
Poor site management is one of the source that affecting the timeline of construction project. Difficulties of material delivery to the site and challenging to handle waste or disposal from project site is related to management of project site [27].

2.1.13. Decision Making
Slow decision making during will affect the construction project in a whole. In rural construction area, delay of decision making happened due to the extensive physical distance from one to another whom involve in the construction project [28]. One of study in rural area revealed the fact that supervisor at the project site primarily used mobile phones and digital camera to monitor the project progress and also used e-mail communication to communicate to each other which make the process of decision making longer [29].

2.1.14. Feasibility Studies
One of the factor that caused delay in construction project is insufficient knowledge and feasibilities studies of the project [18]. Limited resource of historical data and information of the project area are one of the challenges facing in rural area [16].

2.1.15. Weather Condition
Climatic factor can disrupt the project's labour productivity in the construction project. From the past study for the rural and urban climate, it was found that urban areas has wetter climate compared to the rural area due to the effect of the topography and urbanization [30].

3. Methodology
A set of questionnaire was developed based on the factors that are cited in literature review in this study. The questionnaire is consists of 24 questions where 5 questions are related to the respondent's age, respondent's profession, years of experiences in the field, organisation and etc. The other 19 questions are related to the factors that causes delay in rural area. Likert Scale has been used to rank the factors that affect delay in a construction project at rural area. There are 5-point scale ranging from

"Strongly Disagree" on one end to "Strongly Agree" on the other with neither "Neither Agree nor Disagree" in the middle. Each agreement will tabulate a scale that is from the scale of 1 = Strongly Agree, 2 = Agree, 3 = Neutral, 4 = Disagree and 5 = Strongly Disagree. In this study, the method to analyse the data from the survey is Relative Importance Index (RII) and ranking factor. Relative Importance Index (RII) method is used to determine the relative importance of the various factors of causing delay in a construction project at rural area 'as in equation (1)' [31].

$$RII{=}Sum\ of\ weights\ (W1{+}W2{+}W3{+}....{+}Wn)\ /A\ x \qquad (1)$$

Where;
RII = Relative Important Index
W = weighting given to e ach factor by respondents and it ranges from 1 to 5
X = frequency of it response given for each factor
A = highest weight (i.e. 5 in this case)
N = total number of participants

4. Data Analysis and Discussion

A total of 111 respondents has been participated in this survey by professional who are currently working in construction industry as shown in **Table 1.**

Table 1. No. of Respondents according to position level.

Position Level	No. of Respondent
Non-executive	70
Executive	31
Middle Manager	7
Top Management	3
Total	**111**

4.1. Ranking of Factors

Based on the calculation of Relative Importance Index, each of the factors have obtain its RII value. The maximum value of the RII value is 1.0. **Table 2** shows the list of factors with RII values. The 5 major causes of delay at rural area are as follows;

4.1.1. Improper Construction Methods Implemented By Contractor

Zumrawi[32] stated that insufficient geotechnical investigation is the first factor that causes project delays, disputes, claims, and project cost overruns. According to report prepared after the disaster, from the outcrops exposed after the landslip, the geological structure at the site is primarily composed of sandstone and shale. The sandstone in the upper strata was weathered with well-developed joints [33].

4.1.2. Weather Condition (Heavy Rain)

Weather can have a major effect on any construction project such as highway construction due to the continuous exposure to the environment [34]. In addition, surroundings such as air temperature, rainfall and air velocity cause delays in construction project. Forecasting plan delays caused by weather circumstances such as snow, rain and cold temperature is essential as it is obviously weather plays the role to contribute delay [35]. It shows level of consequences on project activities were different depends on type of weather condition.

4.1.3. Difficulties in Providing Delivery to Site

Accessibilities is one of major challenges in rural area to manage construction project [36]. The 'ruralness' and long distance of the project complicates the process of provide delivery to the construction; for example, machinery for excavation. Machinery for excavation and bore piling machine are extremely heavy and need lorry trailer to deliver to the construction site. When these types of machinery are unable to reach to the project site, it will definitely cause delay. Bad condition of road will increase the level of difficulties in delivering material and equipment to the site [29].

Therefore, a good management to handle project is important in order to overcome problems that arise during construction phase and limit the cause of delay issue.

Table 2. Ranking of Factors

No.	Factor	Relative Important Index
1.	Accessibility of human resources.	0.654
2.	The loose control over communication and management.	0.621
3.	Availability of the required resources.	0.637
4.	Inaccurate estimations of certain work progress. (work progress timeline and inaccurate estimation slope stability)	0.674
5.	The extensive physical distance between project participants, sometimes extending over national boundaries.	0.623
6.	Lack of project pre-planning, certainty, and/or clarity concerning project process integration. (e.g. mounting details of machineries, power feed layout and connection is site, wrongly scheduling of work progress)	0.664
7.	Difficulties in providing delivery to site. (Delay in term of time arrival)	0.705
8.	'Ruralness' of the project location.	0.649
9.	Lack of management skills.	0.638
10.	Unpredictable site condition	0.640
11.	Weather condition (Heavy rain)	0.719
12.	Timely decision-making.	0.661
13.	Improper construction methods implemented by the contractor	0.732
14.	Low performance of the selected contractors. (e.g. Slow work progress by contractor)	0.681
15.	Delays and or shortages within materials and equipment.	0.668
16.	Poor qualification of contractor technical staff. (e.g. Unqualified personal doing high technical work)	0.683
17.	Poor site management.	0.676
18.	Breakdown of site equipment.	0.692
19.	Material fluctuation	0.677

4.1.4. Breakdown of Site Equipment

Poor management of plant and machinery[37] and low level of equipment of operator's skill [12] will definitely cause delay of progress construction project. Equipment or machinery is an essential tool in the construction industry which helps to proceed the heavy task job such as excavation in a construction site. If machinery such as an excavator face mechanical failure, it will affect the work progress of construction project especially if the location of the construction project is at rural area.

4.1.5. Poor Qualification of Contractor's Technical Staff

Contractor selection is a major factor for a construction project to gain success. Success of project is to complete the project within schedule and budget allocate as agreed in the contract. Great leader from contractor will manage to select and hire technical staff according qualification and experience level. Poor qualification contractor's technical staff is one of the factors that causes delay in the construction project [38]. This is because a good and qualify contractor's technical staff will able to perform task and ensure project delivery on time.

5. Conclusion

The objectives of this research study is to determine the factors that causes delay rural area and rank the factors based on perception of industrial workers. Based on the outcome of questionnaires that have done, this research study is able to find out the factors that causes delay.

These are the top 5 factors that contribute delay at rural area:

1. Improper construction method implemented by contractor.
2. Weather condition. (Heavy Rain)
3. Difficulties in providing delivery to site.
4. Breakdown of site equipment.
5. Poor qualification of contractor technical staff.

References

[1] Ali Khan R, Shahir Liew M and Ghazali Z 2014 Malaysian Construction Sector and Malaysia Vision 2020 : Developed Nation Status *Procedia - Soc. Behav. Sci.* **109** 507–13

[2] Department of Statistics Malaysia 2017 Quarterly Construction Statistics, First Quarter 2017 *Dep. Stat. Malaysia*

[3] Memon A H, Rahman I A and Azis A A A 2012 Time and cost performance in construction projects in southern and central regions of peninsular Malaysia *Int. J. Adv. Appl. Sci.* **1** 45–52

[4] Durdyev S, Omarov M and Ismail S 2017 Causes of delay in residential construction projects in Cambodia *Cogent Eng.* **134** 1–12

[5] Singh S 2007 The Star Online *Star Online* Available: http://www.thestar.com.my/news/nation/2007/06/06/treasury-puts-its-foot-down-on-project-delays/ [Accessed: 17-Aug-2017]

[6] Rohani Endut I, Akintoye A and Kelly J 2005 Cost and Time Overruns of Projects In Malaysia 243–52 *Fraunhofer IRB* Available: https://www.irbnet.de/daten/iconda/CIB10633.pdf [Accessed: 17-Aug-2017]

[7] Joshi M 2009 80 per cent of Malaysian government projects delayed, minister says *TopNews.in* Available: http://www.topnews.in/80-cent-malaysian-government-projects-delayed-minister-says-2173299 [Accessed: 17-Aug-2017]

[8] Economic Planning Unit 2016 *Strategy Paper 4: Transforming Rural Areas to Uplift Wellbeing of Rural Communities* (Putrajaya.)

[9] Construction Industry Development Board and CIDB 2015 Construction statistics quarterly bulletin 2015 *Constr. Ind. Dev. Board CIDB*

[10] Bernama Delay in LPT2 due to major amendment in design, structure *New Straits Times* Available: https://www.nst.com.my/news/2015/09/delay-lpt2-due-major-amendment-design-structure [Accessed: 17-Aug-2017]

[11] The Star 2014 Bakun project firms in deal to withdraw all claims *Star Online* Available: http://www.thestar.com.my/business/business-news/2014/08/13/bakun-project-firms-in-deal-to-withdraw-all-claims/ [Accessed: 17-Aug-2017]

[12] Assaf S A and Al-Hejji S 2006 Causes of delay in large construction projects *Int. J. Proj. Manag.* **24** 349–57

[13] Acharya N K, Yong Kim S and Dai Lee Y 2004 Factors affecting timely completion of construction projects *Proceedings of the Fifth Asia Pacific Industrial Engineering and Management Systems Conference 2004* p 22.4.1-13

[14] Sambasivan M and Soon Y W 2007 Causes and effects of delays in Malaysian construction industry *Int. J. Proj. Manag.* **25** 517–26

[15] Ahmed S M, Azhar S, Castillo M and Kappagantula P 2002 *Construction Delays in Florida: An Empirical Study. Final Report.* Department of Community Affairs, Florida, US
[16] Tran D Q, Hallowell M R and Molenaar K R 2014 Construction management challenges and best practices for rural transit projects *J. Manag. Eng.* 040140721–9
[17] Collins S A, Zack J G and Faace C 2016 Changing Trend in Risk Allocation — Differing Site Conditions *Navig. Constr. Forum* 31–50
[18] TY L, H F I W and F T K C 2006 Construction Delays in Hong Kong Civil Engineering Projects *J. Constr. Eng. Manag.* 636–49
[19] Winden J v., Keijzer M de, Hohnerlein M F and Pforte W 1986 *Rural building construction* (Nederland: National Vocational Training Institute Accra)
[20] Chavada R ., Dawood N . N . and Kassem M . 2012 Construction workspace management : The development and application of a novel nD planning approach and tool *J. Inf. Technol. Constr.* 213–36
[21] Wu I and Yen-Chang Chiu 2010 4D workspace conflict detection and analysis system *10th Int. Conf. Constr. Appl. Virtual Real.* 1–8
[22] Department of Public Work 2008 *Rural Road Standards: Design Standards and Policies for Rural Baltimore Country Roads* (Baltimore)
[23] Khaloufi B 1956 *The use of natural resources for housing in rural areas of Morocco* (Lunds Tekniska Högskola)
[24] McAnulty, S. and Baroudi B 2010 Construction challenges in remote Australian locations *Proceeding 26th Annu. ARCOM Conf.* 1247–57
[25] Dheenadhayalan M and Alean S 2015 A Study On Labour Productivity And Possibilities Of Its Improvement *Int. J. Sci. Eng. Res.* **3**
[26] Salleh R 2009 *Critical success factors of project management for Brunei constructions project : improving project performance* (Queensland University of Technology)
[27] Sidawi B and Email S A 2012 Management problems of remote construction projects and potential IT solutons : the case of Kingdom of Saudi Arabia *J. Inf. Technol. Constr.* **17** 103–20
[28] Deng Z M, Li H, Tam C M, Shen Q P and Love P E D 2001 An application of the Internet-based project management system *Autom. Constr.* **10** 239–46
[29] Sidawi B 2012 Remote Construction Projects ' Problems And Solutions : The Case Of Sec *48th ASC Annual International Conference Proceedings* (Associated Schools of Construction)
[30] Bulut Y, Toy S, Irmak M A, Yilmaz H and Yilmaz S 2008 Urban-rural climatic differences over a 2-year period in the City of Erzurum , Turkey *Atmósfera* **21** 121–33
[31] Memon A H, Rahman I A, Akram M and Ali N 2014 Significant Factors Causing Time Overrun in Construction Projects of Peninsular Malaysia *Mod. Appl. Sci.* **8** 16–28
[32] M. E Zumrawi M 2014 Effects of inadequate geotechnical investigations on civil engineering projects Effects of Inadequate Geotechnical Investigation on Civil Engineering Projects *Int. J. Sci. Res.* **3**
[33] Moh Z 2004 Site investigation and geotechnical failures *International Conference on Structural and foundational Failures*
[34] Sebt M, Rajaei H and Pakseresht M 2007 A Fuzzy Modeling Approach to Weather Delays Analysis in Construction Projects *Int. J. Civ. Eng.* **5** 169–81
[35] Thorpe D and Ebrahim K 2008 Method for calculating schedule delay considering weather conditions *Proceedings 24th Annual ARCOM Conference* pp 809–18
[36] Usman N and Ibrahim A M 2015 Challenges Associated With Remote Sites Project Management *J. Multidiscip. Eng. Sci. Technol.* **2** 3159–40
[37] Nuwan Randunupura S and Hadiwattege C 2013 Plant and equipement management to minimize delay in road construction projects *Second World Constr. Symp. 2013 Socio-Economic Sustain. Constr.* 333–42
[38] Abbas Niazai G and Gidado K 2012 Causes of project delay in the construction industry in Afghanistan 63–74

A New Design of Recycled Concrete Aggregates as an Aerated Filter for Removal of Phosphorus

[1] S.H. Adnan, [2]N.A.Roni, [3]R. Hamdan, [4]N.I.M. Yassin, [5]N.A. Kamarulzaman.

Dept. of Civil Engineering Technology, Faculty of Engineering Technology, Universiti Tun Hussein Onn Malaysia, 86400 Batu Pahat, Johor, Malaysia.

Email: norwardatun93@gmail.com

Abstract. This paper assesses a new design of recycled concrete aggregates as an aerated filter for removal of phosphorus (P). Recycled concrete aggregates (RCA) obtained from crushed concrete waste. In this study, RCA is used as an Aerated Filter for removal of P. A commercial concrete was crushed, and the RCA were sieving according to different sieve size. The highest percentage passing sieve size is size 5 mm is 99.88%. The bulk density for RCA is 1680 kg/m^3 while percentage of voids is 36%. Result for water absorption test of RCA is 1.27, the Aggregate Impact Value (AIV) test is 28.75 and the pH value for RCA is 9.30 which is alkaline. RCA with smallest size and lowest initial concentration of P which is 0mm to 5mm and 10mg/L respectively has the lowest uptake capacity which 0.996 mgL^{-1}/g but highest percentage of P removal which 99.60%.Besides cost saving, due to application of water material, the usage of RCA will ease the environmental problems that are currently perceived globally. RCA could also saves landfill space which helps reduce the needs for gravel mining and reduces pollution.

1. Introduction

Concrete is the most used construction material, with estimated annual production of 10 billion cubic meters. Since 60–80% of the concrete volume is taken by aggregates, the overall consumption of natural aggregates (sand and gravel) is very high, generating huge pressure on surrounding ecosystems. The environmental impact of aggregate extraction is particularly severe regarding sands, with distinct problems associated to different extraction or production technologies (Evangelista *et al.* 2015). Concrete is the most widely used construction material in the world due to its low cost, excellent durability, easy availability of its constituent materials, easy formability to any shape, etc.

In the following years, a large amount of experimental works have been carried out worldwide to investigate the recycling of waste concrete. Previous studies were mainly engaged in the processing of demolished concrete, mix-proportion design, mechanical properties, durability aspects and improvements. Recently, structural performances and economic aspects of using recycled concrete aggregates (RCA) are also analyzed. RCA are composite materials formed from cement, aggregates, water and eventually admixtures or partial cement replacement materials. The resulting properties are substantially different from those of natural aggregates and strongly dependent on processing and treatment.

Use of RCA fulfils the three green requirements that has been set out by the World Environmental Organization. Firstly, it can recycle and reduce natural resources and energy consumption. Secondly, it will not affect the environment and lastly, it can maintain sustainable development. Therefore, RCA is

Content from this work may be used under the terms of the Creative Commons Attribution 3.0 licence. Any further distribution of this work must maintain attribution to the author(s) and the title of the work, journal citation and DOI.
Published under licence by IOP Publishing Ltd

green concrete. Although reuse of RCA reduces the consumption of limited resources and thereby saves costs, they have some limitation for example weak interfacial behavior between aggregate and cement paste high portion of cement mortar attached that lower its quality. To facilitate the use of RCA, it is vital to identify some ways to overcome these shortcoming.

Domestic wastewater treatment is also becoming more critical issue due to diminishing water resources and resulting water shortage. Aeration is one of the method which help in the removal of various contaminants present in domestic wastewater. The characteristics of wastewater having pH, Biochemical oxygen demand (BOD), Chemical oxygen demand (COD), Oil, phosphorus and nitrogen (about 6% of the BOD level) at a range of value which is harmful to human and environment was found to be the major polluting sources. The process of aeration was found to be an effective method in reducing these contaminants. A large amount of the waste then is directly discharged into the receiving water bodies, which bear the effect of ecological damage caused by its pollutants. The objective of this study is to determine physical and chemical characterization of recycled concrete aggregates as a aerated filter for phosphorus removal.

2. Experimental Study

2.1. Materials

This study is based on recycled concrete aggregate. Aggregates were crushed at a maximum size of 20 mm. The source of these RCA is a local concrete tested cube specimens (concrete grade, 25-30 MPa). In this study, Recycled Concrete Aggregates (RCA) was obtained from thrown waste cubes produced from the Material Laboratory in Universiti Tun Hussein Onn Malaysia (UTHM). Initially, the thrown waste cubes outside laboratory were selected. Then the waste cubes were crushed by using the crushing machines in order to produce the aggregates. Figure 1 shows crushing machine which is equipment for crushing the waste cube into recycled concrete aggregate while Figure 2 shows the RCA that has been crushed.

Figure 1: Crushing Machine

Figure 2: Recycled concrete aggregates

3. Results and Discussion

3.1. Physical Characteristics of RCA
There are four types of test on physical characterization which are sieve analysis test, bulk density test, water absorption test and aggregate impact value test.

3.1.1. Sieve Analysis
This test is used to determine the grading of RCA as an aggregates. Figure 3 shows equipment of sieving the RCA which is Sieving Machine.

Figure 3: Sieving Machine

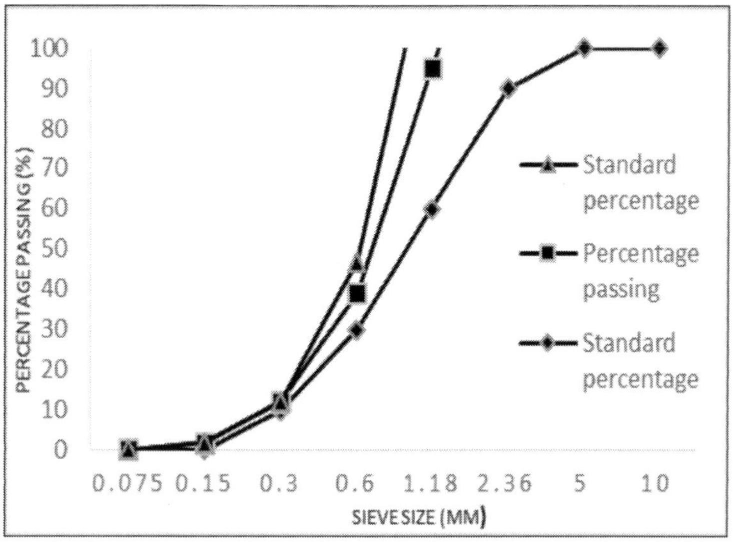

Figure 4 : Distribution curve for Percentage Passing of RCA

Based on the tests that have been carried out, the highest percentage of the size of the RCA passing sieve size is size of 5 mm. Percentage passing sieve size of 5 mm is 99.88%. Grade curve for RCA specified entering the designated area. The conclusion proof that from the aspects of grading aggregate, RCA used beyond the specifications set by BS410: 1986 "Specification for test sieve".

3.1.2. Bulk Density Test
This test method is often used to determine bulk density values that are necessary for use for many methods. The bulk density also may be used for determining mass/volume relationships for conversions in purchase agreements. Table 1 show the bulk density for RCA.

Table 1: Result for Bulk Density Test

Sample	Bulk density(kg/ m^3)
Recent Study, (2017)	1680
Faiz Uddin, (2016)	1247
Larbi *et al.* (2015)	1148
Evangelista *et al.* (2015)	1004

Value of the bulk density of the aggregate depends upon the amount of effort used to fill the container as possible, size distribution, shape and specific gravity. More graded the aggregate, greater the bulk density. Angular and flaky shape of the material reduce the bulk density. The value for this bulk density of RCA was 1680 kg/m^3.

Most study shows that the bulk density test for RCA is lower than this study. Several researchers have conducted study for determining Bulk Density of RCA and summarize in Table 1. It can be seen that the lower bulk density is 1004 kg/ m^3. It is significantly lower than those of RCA. This is because this RCA are mostly composed by mortar, and it presents a narrow size distribution. Different with Faiz Uddin (2016), the bulk density test for RCA is 1247 kg/ m^3. The RCA is generally sourced from crushed concrete structures, therefore, old mortars always adhered to the RCA. It is also known that in RCA the old adhered mortars are more porous. Meanwhile Larbi *et al* (2015) determine that bulk density for RCA is 1338 kg/ m^3. This is because the effect of the superplastizier on the density. Test clearly shows that compacted aggregate which has bigger bulk density provides better strength. Figure 5 shows the equipment for Bulk Density Test.

Figure 5: Equipment for Bulk Density

3.1.3. Water Absorption
This test helps to determine the water absorption of RCA as per IS: 2386 (Part III) – 1963.

Table 2: Result for water absorption test

Researchers	Percentage Water Absorption RCA (%)
Recent study, (2017)	1.27
Faiz Uddin, (2016)	4.9
Dina Sadek, (2012)	0.83
Nikoo *et al.* (2016)	0.59
Suraya *et al.* (2016)	5.58

According to Anchor (1998), the water absorption of aggregates should not exceed 3% of absorption. In this study, water absorption for RCA was 1.27. Hence the result shows it not exceed the maximum of 3% absorption. Water absorption gives an idea of strength of aggregates. Aggregates was composed of cement paste, which having more water absorption are more porous in nature and are generally considered suitable to be an aerated filter for removal of P. Ayaz *et al.* (2015) reported that larger the porosity, larger the specific surface area where sorption of mechanism can take place. Therefore, the RCA in this water absorption can be said as higher strength because the results shows lowers than maximum absorption.

Table 2 shows result for water absorption test for RCA from different researchers. Most study shows that the percentage water absorption for RCA is lower than this study except in Faiz Uddin (2016). Several researchers have conducted study for determining percentage water absorption of RCA and summarize in table 2. This is because the RCA from geopolymer concrete. This increased absorption due to capillary rise is expected and is due to inferior properties of the RCA. Dina Sadek (2012) identified that percentage water absorption of RCA is 0.83. This may be attributed to the highly porous of RCA and may be due to the filler effect of the fine portion of the RCA. Other researchers Nikoo *et al.* (2016) found that the percentage water absorption is 0.59. This is because the RCA is from PVC concrete. This can be attributed to the lower relative humidity in the pores of the specimens which caused the hydration reaction of cement dose not progress satisfactory. Figure 6 shows equipment for Water Absorption Test. Suraya *et al* (2016) examined that water absorption for RCA was 5.58. This is due to high absorption characteristic of RA that developed the higher osmosis pressure. This higher osmosis pressure have a tendency to absorb more water from the surrounding of the samples (Cakur and Sofyanh 2015).

Figure 6: Equipment Water Absorption Test

3.1.4. *Aggregate Impact Value (AIV) Test*

This test is conducted to determining the aggregate impact value of aggregates as per IS: 2386 (Part IV) – 1963.

Table 3: Result for aggregates impact value test

Researchers	AIV test (%)
Recent study, (2017)	28.75
Shahid *et al.* (2016)	28.00
John *et al.* (2017)	22.20
Suraya *et al.* (2016)	20.80
Shahiron *et al.* (2017)	12.70

The strength of aggregates is assessed by aggregates impact value. The aggregates impact value test provides a relative measure of resistance under a gradually applied compressive load. To achieve high quality strength, aggregate possessing low aggregate impact value should be preferred. Table 3 shows result for Aggregates Impact Value (AIV) test. In this study, the AIV test is 28.75. Based on IS: 2386 (Part IV) – 1963, it states that the aggregates impact value for wearing surface shall not exceed 45%.

Shahiron *et al.* (2017) shown that the AIV Test for RCA was 12.7 % which is the lowest percentage. This is because the recycled aggregates were treated first with 25% epoxy resin. Next, Shahid *et al.* (2017) shows that the value for AIV Test was 28% while John *et al.* (2017) state the value for AIV Test was 22.20 %. For John *et al.* (2017), the residual mortar content of the RCA particles was found using the hydrochloric acid dissolution method. Other researcher, Suraya *et al.* (2010) examined the AIV test for RCA was 20.80 %. This is due to the mortar attached to RA could most probably influence the amount of broken particles size smaller than 2.36 mm size after the blasting. All the results shows that AIV test not exceed 45% thus its proof that the RCA has achieve its strength. Figure 7 shows that the equipment used for AIV Test.

Figure 7: The equipment for AIV

3.2. *Chemical Characteristics of RCA*

There are three types of test in chemical characterization which is determine pH_{PZC} and determine pH.

3.2.1. *Determine Point Zero Charge (pH_{PZC})*

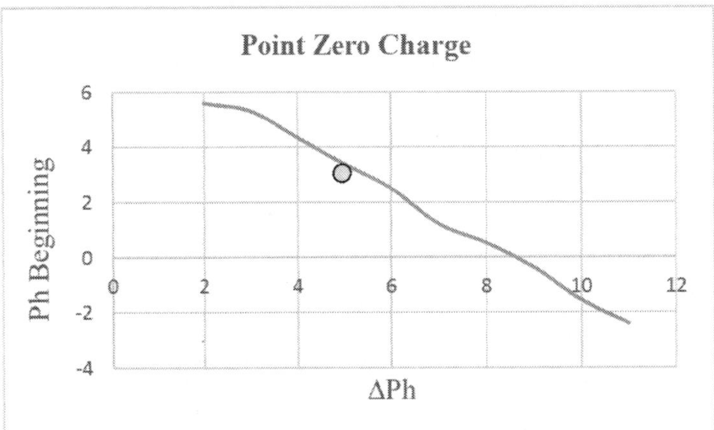

Figure 8: Graph Point Zero Charge (pH_{PZC})

Point Zero Charge (pH_{PZC}) is the pH value at which the sum of surface positive charges balances the sum of surface negative charges. In addition, the surface is positively charged at pH values below pH_{PZC} and negatively charged at pH values above pH_{PZC}. Figure 8 shows the point zero charge for RCA after shaken at 170 rpm for 24 hours. The reading point zero charge of RCA is 10.60. Its show that the pH value 9.30 lower than the point zero charge of RCA. Then, h_20 contributes H^+ is more than OH^-, therefore the surface absorption was negative charge. Figure 9 shows the equipment used in pH_{PZC} test.

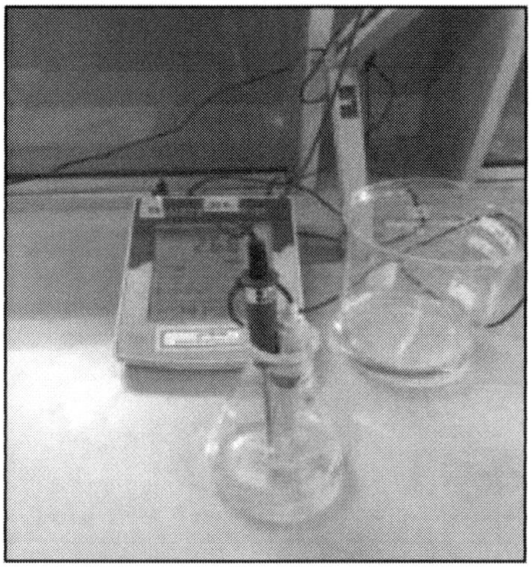

Figure 9: Equipment in pH_{PZC} test

3.2.2. Determine pH

The pH value of the filter media that is RCA was determined by the principle of extraction with boiling water. The pH of the extracts was determined and accepted as the pH value of filter media. From the boiling water extraction principle, the pH value of RCA is 9.30 which is alkaline. Regarding Izzati *et al.* 2016, RCA contains high concentration of SiO_2 which is 57.20%, while 11.60% of CaO, 3.99% of Al2O3 and 2.13% of Fe_2O_3. This shows that RCA also could be one of the potential substrates for removal of P in wastewater treatment process. Besides, RCA also gives high alkalinity content. Thus, based on the determination of pH that has been conducted, its demonstration that RCA was alkaline. The influence removal of P related to pH level and Ca content of filter material. As a calcium ions can form stable and insoluble products with phosphate, calcium-based materials are considered to be one of the potential sorbent of phosphorus removal. The most of materials with high pH also have higher Ca content which is (>15%) (Xiangling *et al.* 2016). Najwa Nasir (2016) also get pH value for steel slag as an alkali which is 10.19. Although steel slag is the product of different processes and different chemical compositions, but they share similar characteristics such as high calcium content and alkalinity, which makes them good for phosphorus removal (Zuo *et al.* 2015).

Figure 10: Equipment for determining pH

3.2.3. X-ray Fluorescent (XRF) Analysis of RCA

In this project, XRF Analysis was conducted in order to identify chemical composition in RCA. Based on the XRF analysis conducted for RCA, it showed that, RCA also contains high concentration of SiO_2 which is 57.20%, while 11.60% of CaO, 3.99% of Al_2O_3 and 2.13% of Fe_2O_3. This shows that RCA also could be one of the potential substrates for P-removal in wastewater treatment process. This is because all these chemical compositions were easily react with P (Izzati *et al* 2016). Table 4 shows the chemical composition of RCA based on the XRF analysis.

Table 4: Chemical composition of RCA

Formula	Concentration
Orig-g	8
Added-g	2
S_iO_2	57.20%
C_aO	11.60%
Al_2O_3	3.99%
Fe_2O_3	2.13%
C	1.00%
K_2O	0.98%
SO_3	0.60%
Na_2O	0.48%
MgO	0.46%
TiO_2	0.23%
Cr_2O_3	0.18%

3.2.4. Scanning Electron Microscopy (SEM) Analysis of RCA

SEM images of the surfaces of the RCA samples are shown in Figure 11.The surface of the RCA sample is covered by loosely bonded cement mortar and considerably amount of porous fine particles which were created by a crushing process. The SEM analysis has demonstrated that recycled concrete aggregate provide less porous, more dense and connected microstructure.

Figure 11: SEM analysis of RCA

3.2.5. Batch experiment process

In this project, the laboratory work has been conduct for 8 weeks. 96 samples of wastewater have been treated by using RCA. Each of the samples was treated for 24 hours. RCA that was used in this experiment were categories into two different sizes which are 0mm to 5mm and 10mm to 20mm. Besides, the initial concentration of P in synthetic wastewater were fixed into six which are 10mg/L, 20mg/L, 30 mg/L, 40 mg/L, 50 mg/L while 0mg/L as standard solution (control). The dosage of RCA are fix which is 10g.

3.2.6. Uptake Capacity of Phosphorus & Percentage of Phosphorus Removal

For this project, uptake capacity, q of Phosphorus (P) is the amount of P absorbed by a unit gram of RCA. Observation and calculation on uptake capacity of P is to show the ability of RCA as filter media in removing P from wastewater. Equation 3.0 shows the formula used in calculating the uptake capacity and equation 3.1 shows the formula used to calculate the percentage of P removal. Table 5 shows the uptake capacity of P and percentage of P removal for 0mm to 5mm and 10mm to 20mm.

$$\frac{\text{initial concentration (Co)} - \text{final concentration (Ce)}}{\text{weight of RCA}} = uptake\ capacity\ (q) \qquad \text{(Equation 3.0)}$$

$$\frac{\text{influent} - \text{effluent}}{\text{influent}} \times 100 = \%\ of\ P\ removal \qquad \text{(Equation 3.1)}$$

Table 5 shows the uptake capacity of P and percentage of P removal for 0mm to 5mm and 10mm to 20mm.

Table 5 The uptake capacity of P and percentage of P removal for 0mm to 5mm and 10mm to 20mm.

Size of RCA (mm)	Co / Influent (mg/L)	Ce / Effluent (mg/L)	Uptake Capacity (q)	Percentage of P removal (%)
0mm to 5mm	0	0	0	0
	10	0.045	0.996	99.60
	20	4.610	1.539	76.95
	30	7.526	2.247	74.91
10mm to 20mm	0	0	0	0
	10	0.555	0.945	94.45
	20	7.452	0.627	62.74
	30	14.15	0.528	52.83

Figure 12 shows the result for uptake capacity of Phosphorus (P) and percentage of P removal for RCA size 0mm to 5mm with three different initial concentration of P which is 0mg/L, 10 mg/L, 20 mg/L and 40 mg/L. For 10 mg/L, the uptake capacity of P is 0.996 and the percentage of P removal is 99.60% while the uptake capacity for 20mg/L is 1.539 and the percentage of P removal is 76.95%. The last initial concentration of 30mg/L shows the uptake capacity is 2.247 and the percentage of P removal is 74.91%. Based on the result, it shows that the uptake capacity of P increasing as the initial concentration increasing. However, the percentage of P removal decreasing as the initial concentration increasing.

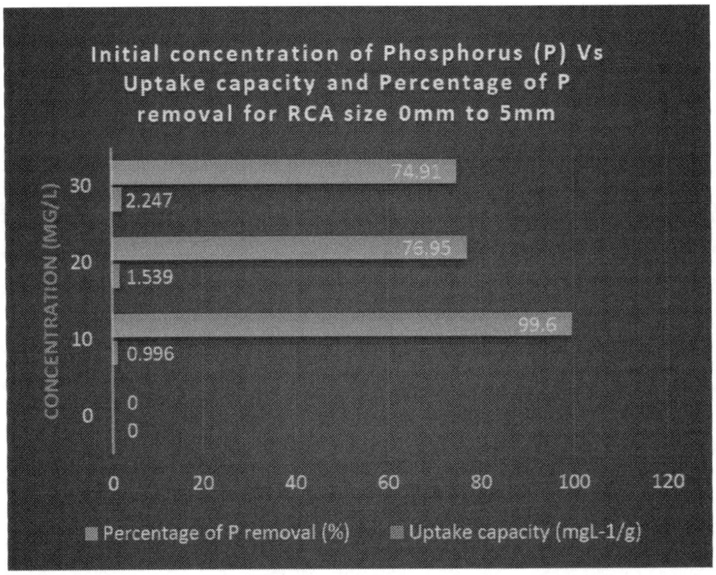

Figure 12: Initial concentration of Phosphorus (P) Vs Uptake capacity and Percentage of P removal for RCA size 0mm to 5mm

Figure 13 shows the result for uptake capacity of Phosphorus (P) and percentage of P removal for RCA size 10mm to 20mm with three different initial concentration of P which is 0mg/L, 10 mg/L, 20 mg/L and 40 mg/L. For 10 mg/L, the uptake capacity of P is 0.945 and the percentage of P removal is 94.45% while the uptake capacity for 20mg/L is 0.627 and the percentage of P removal is 62.74%. The last initial concentration of 30mg/L shows the uptake capacity is 0.528 and the percentage of P removal is 52.83%. Generally it can be seen that the uptake capacity of P increasing as the initial concentration increasing but the percentage of P removal decreasing as the initial concentration increasing. From the two graph for different size of RCA which is 0mm to 5mm and 10mm to 20mm, it is obvious shown that RCA which smallest size and lower concentration have the highest uptake capacity and percentage of P removal. This results, shows the similarities finding with Akratos and Tsihrintzis (2007) found greater removal efficiency of P removal for fine gravel is 89%, followed by medium gravel with cattail 67% and cobbles (57%). Percentage of P removal efficiency was predominantly affected by porous media size and initial concentration.

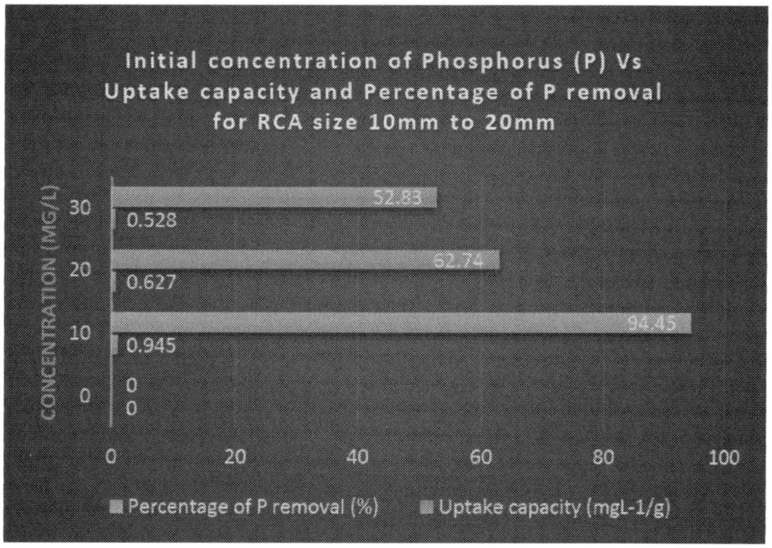

Figure 13: Initial concentration of Phosphorus (P) Vs Uptake capacity and Percentage of P removal for RCA size 10mm to 20mm

4. Conclusion

Based on the experimental result and data analysis, number of conclusions could be made.

1. Based on physical and chemical characteristic of RCA, it shows that RCA could be a new types of filter for removal of phosphorus.
2. RCA with smallest size and lowest initial concentration of P which is 0mm to 5mm and 10mg/L respectively has the lowest uptake capacity which 0.996 mgL^{-1}/g.
3. RCA with smallest size and lowest initial concentration of P which is 0mm to 5mm and 10mg/L respectively has the highest percentage of P removal which 99.60%.

References

[1] Adnan, S.H., Hamdan, R., Yassin, N.I.M., 2016. Removal of phosphorus from synthetic wastewater using recycled concrete aggregates as filter media. 2nd International Conference on Science, Technology and Social Science 2016.

[2] Adnan, S.H., Loon, L.Y., Rahman, I.A., Saman, H.M., Soejoso, M.W., 2010. *Compressive Strength of Recycled Aggregate Concrete with Various Percentage of Recycled Aggregate*.

[3] Akratos, C.S., Tsihrintzis, V.A., 2007. Effect of temperature, HRT, vegetation and porous media on removal efficiency of pilot-scale horizontal subsurface flow constructed wetlands. Ecol. Eng. 29, 173–191.

[4] Annual Book of ASTM Standard, Standard test for pH of Point Zero Charge, ASTM D 3838-05, Philadelphia PA, United State of America, 2011.

[5] ASTM (1995). D1293-95: Standard test methods for pH of water. ASTM International, West Conshohocken, Pennsylvania.

[6] ASTM C1723-16. Standard Guide for Examination of Hardened Concrete Using Scanning Electron Microscopy.

[7] ASTM Designation C 114-1. Standard Test Method of X-ray Fluorescent (XRF) Analysis.

[8] ASTM Designation C136-01. Standard Test Method for Sieve Analysis of Fine and Coarse Aggregates.

[9] ASTM Designation: C 29/C 29M-97 (2003). Standard Method of Test for Bulk Density (Unit Weight) and Voids in Aggregate.

[10] Dina, S. Physico-mechanical properties of solid cement bricks containing recycled aggregates. Journal of Advanced Research.

[11] Evangelista, L., Guedes, M., Brito, J., Ferro, A.C., Pereira, M.F., 2015. Physical, chemical and mineralogical properties of fine recycled aggregates made from concrete waste. Construction and Building Materials 86 (2015) 178–188.

[12] Evangelista, L., Guedes, M., Ferro, A.C., 2015. Physical, chemical and mineralogical properties of fine recycled aggregates made from concrete waste. Construction and Building Materials 86 (2015) 178–188.

[13] Faiz, U. 2016. Mechanical and durability properties of fly ash geopolymer concrete containing recycled coarse aggregates. International Journal of Sustainable Built Environment.

[14] IS: 2386 (Part III) – 1963. Water Absorption Test.

[15] IS: 2386 (Part IV) – 1963. Aggregate Impact Value Test.

[16] John, R.P., Gaurav, G., Singh, B., 2017. Splice strength of deformed steel bars embedded in recycled aggregate concrete. Structures 10 (2017) 130–138.

[17] Larbi, B., Abdelaziz, M., Miloud, B., 2015. Study of the physico-mechanical properties of a recycled concrete incorporating admixtures by the means of NDT methods. 7th Scientific-Technical Conference Material Problems in Civil Engineering.

[18] Nasir, N.N., 2016. Effect Of pH and Density Metal Oxide on Removal of Phosphorus Using The Filter System Vertical Steel Slag.

[19] Nikoo, H., Yassin, S., Jalal, S., 2016. Properties of recycled PVC aggregate concrete under different curing conditions. Construction and Building Materials 126 (2016) 943–950.

[20] O. Cakir, O.O.Sofyanh, Influence of Silica Fume On Mechanical and Physical Properties of Recycled Aggregate Concrete, Housing and Building National Research Center, 11, 157-166, (2015)

[21] Shahid, K., Ammar, A.S., Imran, M.K., 2016. Recycled Construction Debris as Concrete Aggregate for Sustainable Construction Materials. International Conference on Sustainable Design, Engineering and Construction.

[22] Shahiron, S., Kumanan, K., Azmi, M.,A., Zuki, S.,S., Alie, N., 2017. Utilizing Construction and Demolition (C&D) Waste as Recycled Aggregates (RA) in Concrete. 2016 Global Congress on Manufacturing and Management.

[23] Xiangling, Z., Lu, G., Hualing, H., Yinghe, J., Meng, L., Yujie, L., 2016. Removal of phosphorus

by the core-shell bio-ceramic/Zn-layered double hydroxides (LDHs) composites for municipal wastewater treatment in constructed rapid infiltration system. Water research, 280-291.

[24] Xu, D., Wang, L., Li, H., Li, Y., Howard, A., Guan, Y., ... Xu, H. (2015). The forms and bioavailability of phosphorus in integrated vertical flow constructed wetland with earthworms and different substrates.

[25] Zuo, M., Renman, G., Gustafsson, J. P., & Renman, A. (2015). Phosphorus removal performance and speciation in virgin and modified argon oxygen decarburisation slag designed for wastewater treatment.

A Review of the Dynamic Modelling of Pollution Flashover on High Voltage Outdoor Insulators

Ali A. Salem[1] **R. Abd-rahman**[2]

[1] Faculty of Electrical Engineering, UTHM, Johor, Malaysia.

Abstract. In this paper, a dynamic model that simulated and tested the flashover of polluted insulators with HV voltage is presented. Generally, developing the performance of the transmission line depends on the improvement of the insulators due to the pollutants deposited on its surface. This study focused on the classification of insulators and the factors that affect the insulators to cause flashover as well as the analysis of the insulator's electric properties. The electric field and voltage distribution along with composite insulators used in transmission systems under different conditions were also presented. The problem of nonlinearity of the electric field and voltage distribution, especially along insulators used for HV transmission systems, was investigated and a solution for improving the electric field and voltage distribution along the insulators was suggested. Findings showed that the presence of water droplets and pollution layer distributed on weather sheds enhanced the local electric field but it did not really affect the electric field distribution inside the fibre reinforced plastic core. Moreover, for the pollution case, the effect of pollution layer conductivity was significant to the electric field. For thicker pollution layers, the maximum electric field point moved from the triple junction point. Finally, the study also found that the corona ring might be necessary for the HV insulator for both ends (HV and ground) and its behaviour was affected under polluted and wet conditions.

1. Introduction

Besides the electric stress, insulators are subjected to a lot of other stresses [1]. Outdoor environmental conditions vary widely. Temperature and moisture can greatly affect the performance of insulators. For example, moisture such as rain, dew, fog and melting ice can significantly lower the surface resistance of insulators. With the presence of pollution, the insulator surface resistance is reduced even more. The reduction of surface resistance might cause increased leakage of current to flow on the surface and dry band arching to take place. In addition, large magnitudes of leakage currents flowing on the surface for an extended period might cause degradation of the insulator surface. With these factors, flashover might be initiated, which leads to the failure of a line.

In addition, temperature can affect the insulation properties of all material. For some insulators made from organic materials, radiation from sunlight results in surface degradation. Altitude can also affect insulator performance. Since air at ambient pressure and temperature is a form of primary insulation, higher altitudes reduce air density; thus, reducing the surface insulation potency. One of the major problems for high voltage (HV) transmission lines is flashover on the surface of polluted insulators. Flashovers caused by pollution are known as temporary electrical arcs, which can change into permanent electrical arcs and cause sustained interruption. Dynamic models can predict and calculate the electrical

Content from this work may be used under the terms of the Creative Commons Attribution 3.0 licence. Any further distribution of this work must maintain attribution to the author(s) and the title of the work, journal citation and DOI.
Published under licence by IOP Publishing Ltd

parameters of arcs according to the time-dependent evolution. Over the past decade, researchers and manufacturers have tried to promote mathematical models to overcome the problem of flashovers. The first model to predict flashover voltage was proposed by Obenaus [2]. Uniform polluted layers, electric field distribution along the surface of the insulator and the discharge burns in series with the polluted layer were considered under HV voltage. Then, Neumarker [3], presented the modified version of Obenaus's model by using the uniform resistance per unit length of the pollution layer instead of a fixed resistance. Some researchers have presented the arc extinction mechanism by introducing the quantitative criteria in terms of electric stress, pollution resistance and leakage current under HV voltage. A recent study showed that source strength is a major factor in the dispersion of test results from power systems and high voltage laboratories [4]. Source parameters, including inductance, capacitance and short circuit current, were investigated systematically. Results showed that flashover in service can occur at much lower levels of contamination than predicted by international standard tests.

2. Factors Influencing Flashover Discharge.

Some of the factors that affect flashover discharge, either natural or industrial factors, are shown in figure 1.

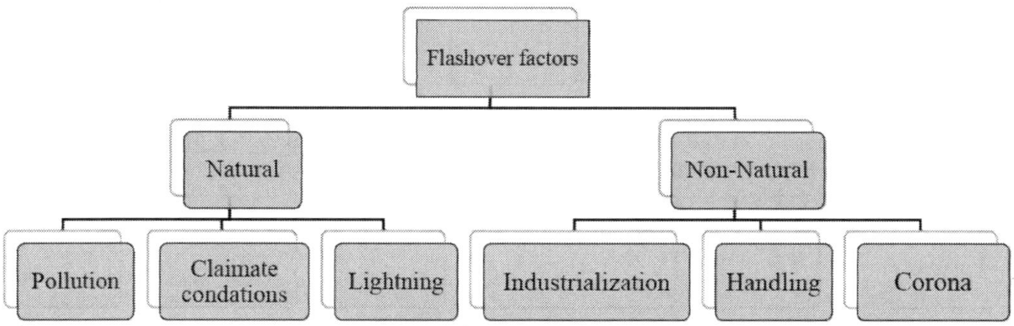

Figure 1. Flashover factors

2.1. Natural Flashover Factors

2.1.1. Pollutions

The airborne particles that deposit on the surface of insulators will form the contamination layer. The solubility of the pollution layer has different effects on insulators depending on whether the solubility grade increases or decreases; thus, there are soluble and insoluble contamination [5]. Different components of contamination have different effects on the characteristics of flashover insulators. Wetting of pollution layers leads to leakage current on the surface of an insulator, which leads to a flashover. The extent of pollution at the surface of the insulator is usually indicated by using *Equivalent Salt Deposit Density* (mg/cm2) ESDD.

2.1.2. Climate Conditions

Climate variability such as rain, fog, relative humidity, and ambient temperature can cause an increase in the contamination of the insulator's surface. Relevant experiences have shown that a flashover is very easily caused by fog, dew and drizzle, in which the biggest threat is fog. Researchers have found that weather conditions on high voltage line flashovers under heavy fog accounted for 65% of the cases and drizzling accounted for under 10%.

2.1.3. Lightning

When impulse surges are applied, regardless of whether the surge is from lightning or air gaps, the flashover process involves corona evolution, the figuration of channels and finally the stroke. The formation of corona happens in a fraction of a microsecond. This is followed by a second detonation of corona depending on the increasing rate of the applied voltage. If the voltage is high enough to evolve a leader channel that crosses the gap, a flashover will take place. The formation of the leader channel is the main criterion for a flashover of the gap [5].

2.2. Non-natural Flashover Factors

2.2.1. Industrialization
During the manufacturing stage of insulators, defects and damage do occur from inadequate quality standards. For example, when setting the end fitting to the core, care must be taken to avoid damage. Factors that could affect this are the diameter, binding stage, pre-heat temperature, moisture, and mould temperature etc. Care should be taken to maintain these variables within acceptable limits. Any neglect in these variables will lead to insulators facing more electric stresses then flashovers [7].

2.2.2. Handling
The insulators might be liable to tension or cantilever load, which could lead to cracking of the core during the installation stage. Any type of physical damage to the housing material makes the insulator rod vulnerable by exposing it directly to contamination and wetting. Cases of incisions that are ulterior in nature during the installation process might lead to flashover during operations. Insulators might be prone to damage caused by vandalism and a breakdown could occur due to tracking along the insulator or brittle fractures. Regular inspections should be carried out in regions that are prone to vandalism to prevent any disoperation due to insulators affected by gun shots or other acts of vandalism.

2.2.3. Corona
The corona phenomenon affects the electrical and mechanical performance of insulators along transmission lines. Increasing the electric field on insulator surfaces could result in a corona, which is caused by accelerating the free electrons around the insulator. The number of free fast-moving electrons increases exponentially, which then leads to the air around the insulator being ionized. This causes the corona discharge. Under dry conditions and if the electric field exceeds the threshold of 15 kV/cm, chances to a corona discharge are very high. However, in wet conditions a corona is seen at lower electric field magnitudes. The threshold electric field magnitude for the corona discharge not only depends on the condition of the surface (dry or wet) but also on the type of voltage applied (AC or dc) [8].

3. Pollution Flashover Mechanism.
Generally, the contamination often leads to flashover of insulators, especially in high voltage transmissions. The insulation property of insulators tends to decrease when the surface of the insulator is covered by a wet pollution layer. Pollution leads to an increase in the leakage current on the insulator surface accompanied by the widening of the dry band, which leads to an increase in the voltage and then to a flashover. The flashover due to insulator pollution develops in the following manner:
(1) cumulative pollutants on surface of insulator.
(2) A wet pollution.
(3) Change of leakage current and voltage.
(4) Discharge of flashover arc along the polluted area.

3.1. Ac and Dc Mathematical Flashover Model of Polluted Insulators
Many mathematical models for propagation of a flashover in polluted insulators have been proposed. Obenaus [2], proposed that the resistance of the pollution layer will be in series with the arc of flashover, as shown in the model in figure 2.

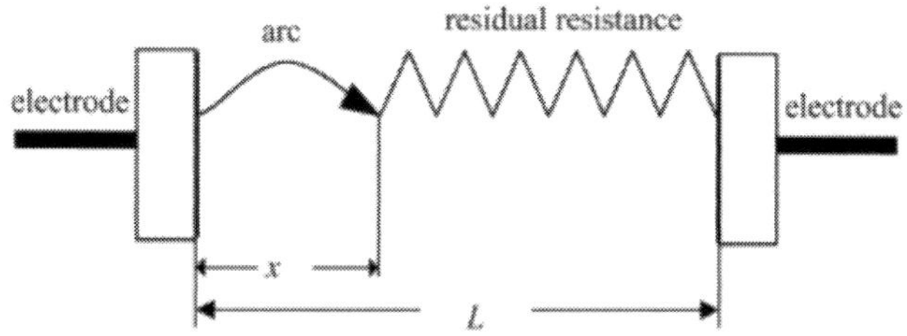

Figure 2. Obenaus's model.

The applied voltage is given as,

$$U = AXI^{-n} + R_P I \tag{1}$$

Where,

U is applied voltage, X is length of arc, I is e current, Rp is resistance of the pollution layer, A and n are characteristic constants.

When the applied voltage is more than a limited value, which is called critical voltage, the Uc arc will propagate until the flashover stage [9]. Most models used to predict the flashover voltage on covered-polluted insulators are derived by Obenaus but the Obenaus model is limited to explaining the AC arc propagation. The reignition of the arc condition was proposed by Claverie [10], and Porcheron, which can be described as.

$$U_m = \frac{800 \times X}{(I_m)^{1/2}} \tag{2}$$

Where, x is the arc length, I_m is the maximum value of the leakage current, and U_m is the maximum value of the applied voltage.

Based on Rizk,[11] the equation (2) could be written as:

$$U_m = \frac{2080 \times X}{I_m} \tag{3}$$

Recovery of the arc condition was proposed under no significant off and rekindle, as in the following:

$$U_f = \frac{531 \times L}{I_m} \tag{4}$$

The previous condition under significant off and rekindle will become:

$$U_f = \frac{1050 \times x}{I_m} \tag{5}$$

Where, x is length of arc, L is the leakage distance of the insulator, Im is the maximum leakage current and U_f is the recovery voltage.

The AC Dynamic Arc model was very well studied and applied in polluted ceramic insulators and ice-covered insulators by Farzaneh (2007) [12]. Mathematical and physical models were proposed to explain the dynamics and predict the pollution flashover voltage of composite insulators by Venkataraman (2006). To develop a model that can predict pre-flashover leakage current, the model must consider all dynamic arc parameters, such as voltage, arc length, surface resistance, etc. to better understand the flashover process. Zhang (1994) [13], has clearly explained arc initiation and propagation under AC voltage in the literature.

3.2. Arc Model

Until today, many arc models have been developed but only a few are used frequently by engineers. The fundamental one is undoubtedly the Ayerton's arc model, which is based on an approximation of experimental measurements. The voltage gradient E_{arc} can be approximated by the following equation:

$$E_{arc} = \frac{V_{arc}}{x} = \frac{AI_{arc}^{-n}x}{x} = AI_{arc}^{-n} \tag{6}$$

Where, I_{arc} is the current of the arc, and A and n are constants of the arc. As can be seen from the table, the constant A has a range from 30 to 310 and constant n from 0.1 to 1.38. Based on experiments by Canadian researchers CIGELE, the coefficient A and n were obtained and the equation (13) can be defined as:

$$E_{arc} = \begin{cases} 84.6 \times I^{-0.772} & dc \\ 208.9 \times I^{-0.449} & ac \end{cases} \tag{7}$$

The high variation was due to its dependency on several factors such as ambient atmospheric conditions, the medium in which the arc propagates, the voltage waveform, and the surface of an insulation layer. All these factors can influence the arc propagation and its behaviour.

4. Models of Pollutant Deposits on the Insulator Surface

Recently, the deposited pollutants on the surface of insulators under humidity condition were divided to a uniform distribution (fog) and a non-uniform (light rain) one.

4.1. Uniform Wetting Model (FOG)

In this case, the pollution that deposited over the insulator surface is assumed to have the same thicknesses along the surface of an insulator on all regions. The tiny fog droplets would move slowly in random motion under the uniform (fog) condition.

4.2. Non-uniform Wetting Model (Light Rain)

Three non-uniform diverse regions under light rain were divided namely low (L), medium (M), and high (H), as showed in figure 3. According to the wetting action, the regions that were categorized at the upper surface H represented a highly-wetted area. Region M, under the shed near the tip, was where the water could easily reach this region. The least wetted areas were in region L, which was down to the shank because it was protected by the sheds [14].

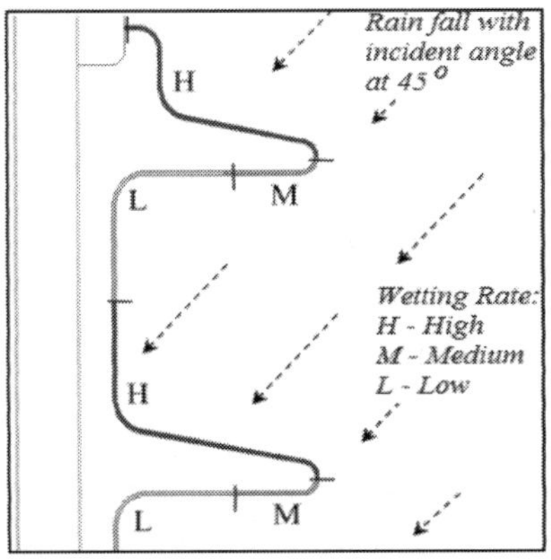

Figure 3. Divide of pollution layer under non-uniform conditions according to wetting level.

5. Electric Field Computation Techniques

The electric potential and field distribution along the insulating surface can be affected by the factors listed below.

1) Voltage magnitude. 2) Insulating material properties. 3) The leakage current. 4) Surface contamination. 5) Internal defects

The methods applied to compute the electric potential and field strength along the insulator surface can be divided into three classes, namely experimental, analytical, and numerical. In this chapter, all three types of methods are introduced, and each has its own advantage and disadvantage. Among the three methods, both the analytical and numerical methods were simulated on the PC platform, demonstrating similar procedures. Firstly, the theoretical models based on real insulator geometry were built, and then described by integral equations or differential equations [15]. The field details were achieved by solving these equations. In the free space, Maxwell's equations were used to describe all the macroscopic electromagnetic phenomena.

$$\begin{cases} \nabla \times H = J + \frac{\partial D}{\partial t} \\ \nabla \times E = -\frac{\partial B}{\partial t} \\ \nabla \times B = 0 \\ \nabla \times D = \rho \end{cases} \tag{8}$$

To show macroscopic electromagnetic properties, three formulas are given below:

$$\begin{cases} D = \varepsilon E \\ B = \mu H \\ J = \sigma E \end{cases} \tag{9}$$

Since the electrodes are conducting material, the principle of the charge conservation is shown below.

$$\nabla * J + \frac{\partial \rho}{\partial t} = 0 \tag{10}$$

Where, H is magnetic field, E is field stress, B is magnetic stress density, D is electric flux density, J is current density, ρ is electric charge density, ε *is* permittivity, μ is permeability, σ *is* conductivity

5.1. Experimental Methods

The rated voltage is energized onto the sample insulators in the experimental methods. To get detailed results, insulators were tested under different conditions to observe the corona and flashover phenomena. The experimental methods were the most accurate, and the results were closest to the performance of insulators in the field tests. However, the experimental methods were time-consuming and more costly than other methods. Therefore, the experimental methods were often used to verify simulations results, or to provide reference data.

5.2. Analytical Field Computation Methods

The analytical methods are used to calculate the field strength directly from the original differential formulas. The results of analytical methods are precise, whereas the scale of the model that the analytical method deals with is relatively small [16]. On the one hand, the formulas are difficult to generate due to complicated geometries in the practical cases, because the boundary conditions are often too uneven to be described with equations. On the other hand, as the model scale increases due to the voltage level, the analytical methods require copious amounts of computational resources to solve the differential equations. Therefore, the analytical methods were replaced with numerical methods. The field stress at the high voltage side was highest compared to that at the ground terminal. It was reported that the insulator with dead end facing experienced greater field stress at the high voltage terminal (as high as 30% for a 500-kV system) than the suspension insulator, as in figure 4.

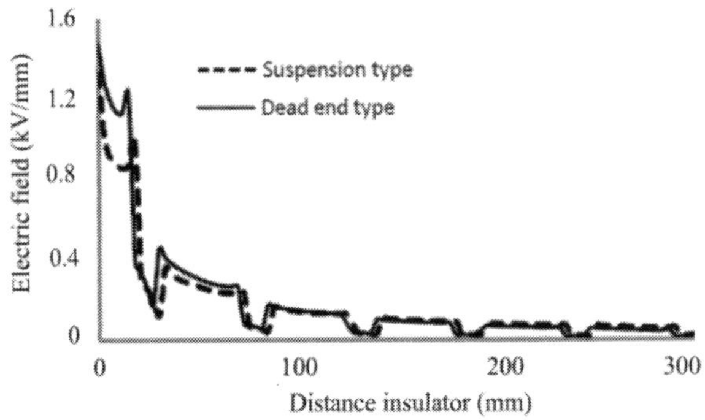

Figure 4. Field distribution for diverse types of insulators (a) Suspension (b) Dead end

5.3. Numerical Field Computation Methods

In numerical methods, the differential equations describe the electric field domain. The purpose of numerical methods is to transfer these equations into a linear matrix [17], which can be easily solved by computers. There are four steps in the process: The first step is to transfer the field functions into a set of linear independent functions with unknown variables. The second step is to cast the continuous solution domain into a discrete form, which contains several elements or fictitious charges. And the third step is to minimize the error of unknown variables from the linearization process. Finally, the solutions are taken into original equations to verify the results.

5.3.1. Finite Difference Method

The principle behind the finite difference method is to divide the field domain with regular grid, and replace the Poisson's equations with the difference equations, whose unknown variables are the potentials at the nodes of the grid. The Poisson's equations and the boundary conditions in the field D are shown below:

$$\nabla^2 u = \frac{\partial^2 u}{\partial x^2} + \frac{\partial^2 u}{\partial y^2} = F(x, y) \qquad \text{on the domain.} \tag{11}$$

$$U(x, y)|_{\Gamma} = g(\Gamma) \qquad \text{on the bound} \tag{12}$$

To build the linear matrix, the potentials at the nodes are described by the Taylor series of adjacent nodes. The accuracy requirement is set as the second order. For example, the coordinate of a two-dimension node is (x, y) and the adjacent node is (x0, y0) . The potential at (x, y) is described by the Taylor series below:

$$\varphi(x, y) = \varphi_0 + \left[(x - x_0)\left(\frac{\partial \varphi}{\partial x}\right)_0 + (y - y_0)\left(\frac{\partial \varphi}{\partial y}\right)_0\right]$$

$$+\frac{1}{2}\left[(x - x_0)^2\left(\frac{\partial^2 \varphi}{\partial x^2}\right)_0 + 2(x - x_0)(y - y_0)\left(\frac{\partial^2 \varphi}{\partial x \partial y}\right)_0 + (y - y_0)^2\left(\frac{\partial^2 \varphi}{\partial y^2}\right)_0\right] \tag{13}$$

In the equation, the subscript 0 represents (x_0, y_0) and the potential φ_0 can be calculated by the average potential values of four adjacent nodes. Therefore, the homogeneous functions with potential variables are built and the known potential values of the boundary nodes are set as V_i.

$$\begin{bmatrix} P_{11} & \cdots & P_{1j} \\ \vdots & \ddots & \vdots \\ P_{i1} & \cdots & P_{ij} \end{bmatrix} \begin{bmatrix} \varphi_1 \\ \vdots \\ \varphi_i \end{bmatrix} = \begin{bmatrix} V_1 \\ \vdots \\ V_i \end{bmatrix} \tag{14}$$

The principle behind FDM is simple, and division of the field is easy to complete. The Finite difference method has great advantages in the two-dimension field, when the shapes of boundaries are relatively simple. However, FDM can get complicated when the boundary shapes are uneven. Moreover, since the infinite field domain cannot be divided, this method cannot deal with open boundary conditions.

5.3.2. Finite Element Method

In the previous session, it was shown that FDM focuses on the nodes of the grid and derives the linear homogeneous functions directly from Poisson's equation. Different from FDM, the FEM divides the field domain according to elements. These elements are small areas in the two-dimension model and small volumes in the three-dimension model. The related equations are then derived from the coordination of each element. In addition, FEM can solve the second-class boundary condition [18], which is the derivative of potential vectors. For a two-dimension field domain:

$$\nabla^2 u = \frac{\partial^2 u}{\partial x^2} + \frac{\partial^2 u}{\partial y^2} = F(x, y) \qquad \text{in the domain.} \tag{15}$$

$$U(x, y)|_\Gamma = g(\Gamma) \qquad \text{on the boundary} \tag{16}$$

$$\frac{\partial u}{\partial n}|_\Gamma = g_2(\Gamma) \tag{17}$$

The triangle is a popular geometry of an element used in FEM. The smaller the element, the more accurate the field strength. According to the steps above, the whole field domain could be described by each point-potential of the elements. An example of a two-dimension field discretization is shown in figure 5.

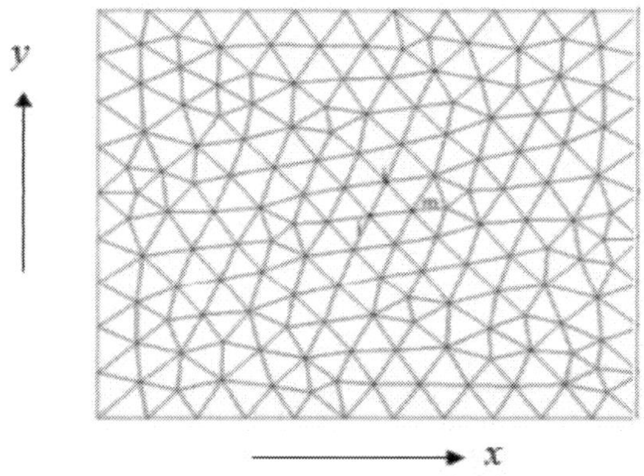

Figure 5. FEM field division and example of a triangle element ijm.

Figure 5 shows that the square field domain is divided by triangle elements. The potential in each triangle has a relationship with the coordination of the triangle vertices.

$$\varphi = a_1 + a_2 x + a_3 y \tag{18}$$

The calculated coefficients a1, a 2 and a3?

$$\varphi = \frac{1}{2\Delta}\left[(a_i + b_i x + c_i y)\varphi_i + \left(a_j + b_j x + c_j y\right)\varphi_j + (a_m + b_m x + c_m y)\varphi_m\right] \tag{19}$$

Where, Δ is the area of the triangle ijm.

The variation problem is discretized with the principle of weighted residuals. The matrix equation for the triangle elements is shown below:

$$[A]=\begin{bmatrix} A_{ii} & A_{ij} & A_{im} \\ A_{ji} & A_{jj} & A_{jm} \\ A_{mi} & A_{mj} & A_{mm} \end{bmatrix} \tag{20}$$

In the end, the discretized linear equations are represented as, $[A][\varphi]=[V]$, and the unknown potentials at the vertices of all the triangles are calculated.

Designers can divide the field domain for their own purpose with FEM. For example, small elements are set in the area where the electric field changes intensively to achieve accurate results; meanwhile relatively large elements are put in the area far from the electrode where the field strength changes slowly to reduce the requirement for calculation resources. However, the FEM calculation process is more complicated than FDM and the requirement for data storage capacity is also larger. Like FDM, the boundary for infinite distance in FEM is difficult to deal with, as the whole field domain cannot be divided.

5.3.3. Boundary Element Method

The boundary element method focuses on boundary conditions that affect the field domain. The unique characteristic of this method is the ability to decrease the dimensions of the problem. For example, a two dimension-problem can be described by the boundary line and reduced to a one-dimension problem. Three-dimension problems can be described by the boundary surface, and reduced to two-dimension problems. The procedure for the boundary element method is shown below,

- The boundary is discretized into many elements in functions with unknown potential and normal flux densities.
- The principle of weighted residuals is used to minimize the error.
- The coefficient matrix is evaluated after analysis of each element.
- The linear algebraic equations are then achieved with the proper boundary conditions to the nodes.
- In the end, the unknown potential can be calculated from the inversion of the coefficient matrix.

The major advantage of the boundary element method is to reduce the dimensions of the space, so that the orders of the homogenous function and the amount of input data are decreased. However, the coefficient matrix is an unsymmetrical full-element matrix, which consumes copious amounts of computation resources and limits the orders of the matrix. The method makes it difficult to handle the multi-media field domain, and cannot be used directly for non-linear problems.

6. Conductivity of the Pollution Layer

An enhanced electrical field combined with a high current density leads to an increase in power dissipation, which then becomes the source of energy for surface heating that initiates dry band formation [19]. The power dissipation, PE, for a conducting pollution layer with uniform volume conductivity, σ_V, can be represented as:

$$P_E = E_S J_S = \frac{J_S^2}{\sigma_V} \tag{21}$$

The evaporation rate increases with the rising of the electric field. On the other hand, the conductance decreases when subjected to a high electric field. This process is illustrated in the curve of the electric field as a function of conductance, as illustrated in figure 6 [14].

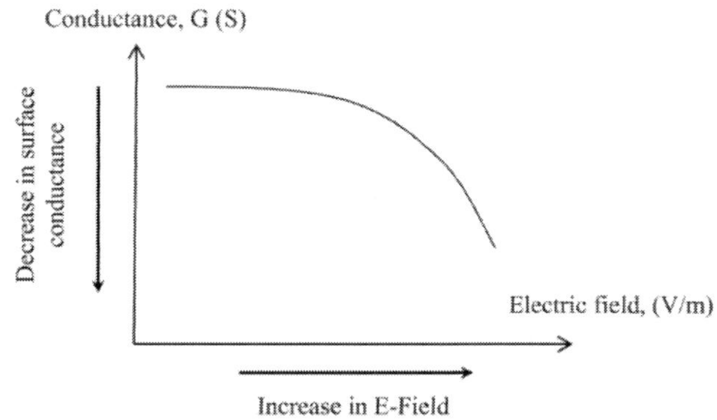

Figure 6. Conductance on insulator surface as function with electric field.

7. Conclusion

The pollution flashover is still a major argument that influences the performance of power system operations. Studies on pollution flashover considers the prediction of flashover voltage. Overall, this is used to develop the insulator design and insulation assortment. Recent studies on acceptable methods to predict the flashover could be applied to engineering practice. The characteristics of the electric field, resistive of pollution and insulators, conductivity, leakage currents during the entire pollution flashover stage under uniform and non-uniform pollution were reviewed. The spike in the electric field profile suggests a drop-in surface conductance at the high field region, and this is attributed to evaporation and heating effects that were examined in this study. This is particularly useful to locate dry band formation on the insulator surface where discharge may occur. Further studies are needed to develop a new insulator model to investigate the above characteristics and compare the experimental results with the computed results.

8. References

[1] Abimouloud and D. Korichi, "*Cap-and-Pin Polluted Insulator to Predict DC Flashover Voltage,*" pp. 2–6, 2016.
[2] Obenaus F, " *Contamination fiashover and creepage path length", ETZ 12, "Obenaus F."* pp. 135–136, 1985.
[3] F. A M- *Re''Mathematical models for poliution flashover*", Electra, 78, "F." pp. 71–103.
[4] Al-Gheilani, W. Rowe, Y. Li, and K. L. Wong, "*Stress Control Methods on a High Voltage Insulator: A Review,*" Energy Procedia, vol. 110, no. December 2016, pp. 95–100, 2017.
[5] IEEE Standard 4, *IEEE Standard for High-Voltage Testing Techniques*, IEEE Power and Energy Society, 2013.
[6] Arshad, A. Nekahi, S. McMeekin, and M. Farzaneh, "*Flashover Characteristics of Silicone Rubber Sheets under Various Environmental Conditions,*" Energies, vol. 9, no. 9, p. 683, 2016
[7] IEC 60815, *Selection and Dimensioning of High-Voltage Insulators Intended for Use in Polluted Conditions, International Electrotechnical Commission*, 2008.
[8] L. Lan, G. Yao, H. L. Wang, X. S. Wen, and Z. X. Liu, "*Characteristics of corona aged Nano-composite RTV and HTV silicone rubber,*" Annu. Rep. - Conf. Electr. Insul. Dielectr. Phenomena, CEIDP, pp. 804–808, 2013.
[9] P. S. Ghosh and N. Chatterjee, ''*Polluted Insulator Flashover Model for ac Voltage''*,IEEE Trans. Dielectr. Electr. Insul., Vol. 2, pp. 128_135, 1995.
[10] P. Claverie, *C'Predetetminationo f the behavior of the polluted ksulators*", IEEE Trans on Power Apparatus and Systems, Vol. PAS - 90, "clarve." pp. 1902–1908.
[11] RIZK, F. A. M. *Mathematical Models for Pollution Flashover. Electra.* 1981, vol. 78, pp. 71–103.

IOP Conf. Series: Journal of Physics: Conf. Series **1049** (2018) 012019 doi:10.1088/1742-6596/1049/1/012019

[12] FARZANEH, M. and W. A. CHISHOLM. *Insulators for icing and polluted environments.*, Huboken: Wiley- IEEE Press, 2009, ISBN 978-0-470-28234-2.

[13] Zhang Renyu. Insulation pollution discharge [M]. Beijing: Water Conservancy and Electric Power Press, 1994.

[14] R. A. RAHMAN, *Investigations of Zno Microvaristor for Stress Control on Polymeric Outdoor Insulators*, no. August. 2012.

[15] "*Electric Potential and Field Calculation of HVDC Composite Insulators by Charge Simulation Method*" by Jiahong He A Thesis Presented in Partial Fulfillment of the Requirements for the Degree Master of Science Approved November 2013 by the Graduate Superviso," no. November, 2013.

[16] Jiang, X., Xiang, Z., Zhang, Z., et al.: '*AC pollution flashover performance and flashover process of glass insulators at high altitude site*', IET Gener. Transm. Distrib., 2014, 8, (3), pp. 495–502.

[17] L. L. Alston, "Growth of discharges on polluted insulation". Proceedings of the Institution of Electrical Engineers, Vol.110, No. 7, pp.1260-1266, 1963.

[18] B. F. Hampton, "Flashover mechanism of polluted insulation", Proceedings of the Institution of Electrical Engineers, Vol.131, No. 4, 1964

[19] G. Karady, "*Insulator Flashover Probability Investigation Based on Numerical Electric Field Calculation and Random Walk Theory by Jiahong He A Dissertation Presented in Partial* Fulfillment of the Requirements for the Degree Doctor of Philosophy Approved June 2016 by ," no. August, 2016.

IOP Conf. Series: Journal of Physics: Conf. Series **1049** (2018) 012020 doi:10.1088/1742-6596/1049/1/012020

Impact of Human Capital Management Practices on Employees' Job Performance

Ali Zeb[1] Nor Hazana Abdullah[2] Mudaser Javaid[3] Muhammad Asad khan[4]
Faculty of Technology Management and Business, University Tun Hussein Onn Malaysia
86400 Parit Raja, Johor, Malaysia
E-mail: Alizeb_nasafi@yahoo.com

Abstract. This paper aims to analyze the relationship between the human capital management (HCM) practices which include training and development, recruitment and selection, Performance appraisal and employee's job performance in the telecom sector of Pakistan. To find the influence of HCM practices on employees job performance within telecom sector of Pakistan. This is a survey based research study. The sample of the study is employees' who are working in public and private sector telecommunication companies in Pakistan. Data are collected through questionnaires. Sample size was (n=272); therefore, descriptive statistics, correlation and regression analysis have been used. The overall results support that HCM practices have prominent and indispensable role in the performance level of employees; knowledge and skill through training and development, compatibility & competence through recruitment and selection, confidence and morale through performance appraisal. HCM practices are highly correlated with each other and also correlated with employee's job performance.

Keywords: HCM Practice, Employees' job performance, Telecom sector

1. Introduction

Nowadays, the importance of employee performance for the success of organizations cannot be overstated. Employees are the human capital of organizations and their performance is a key indicator for organization to achieve its goals. Employee performance is one of the most important constructs in both organizational psychology and human resource management [33]. According to [11] employees are most expensive asset for any organization as they can build up or destroy status of company and they can affect profitability. [2] suggest that perceived job performance is influenced by many factors like organization overall policies, training and development, working climate, relation between employer and employees. Employee's job performance in the organization can be improved by different ways. Employee's performance plays a significant role in the competitive environments, where good performance leads to success while bad performance may lead to failure. [47] claimed that

[1] Ali Zeb, *Faculty of Technology Management and Business, University Tun Hussein Onn Malaysia*

Content from this work may be used under the terms of the Creative Commons Attribution 3.0 licence. Any further distribution of this work must maintain attribution to the author(s) and the title of the work, journal citation and DOI.
Published under licence by IOP Publishing Ltd

employee's performance can be measured by various ways like, satisfaction, performance/productivity, and employee's absenteeism. It is a function that an individual can successfully perform within framework of normal constraints and available resources [23]. In past decade many studies on employees' job performance have been conducted across sectors. This is attributable to the increasing importance of human capital as the source of achieving competitive advantage. Competent human capital could transform other resources (such as money, machine, methods & methods & material) in to produce or service. Moreover, an organization requires result oriented trainings to compete in the market to be successful in the turbulent environment [28].
[27] argue that training is the most important factors to attain to attain high achievable results, and to achieve the desire results. It is important to train relevant people in the relevant area "to be adaptive to the environment". Researchers are not interested in finding only research area, but they are interesting in the importance of research areas. A lot of the researcher pleads that to understand the importance of training programs. In recent times training is used to influence employee's performance .They say that comparative achievement of organization's asset in workers development has also influenced the importance of training. Furthermore, technological advancement and change in organizations have slowly resulted in the employer's realization that success is related to the skill and competence of their employees. This phenomenon has led to increase investment in training and growing of employees [5].
[40] recruitment and selection a systematic procedure, where in first phase expectations are clarified and negotiated and information is mutually exchange. [4] emphasized that recruitment and selection as practices carried out by the association with the main determination of classifying and attracting productive workers. In the simple words that are significant part of HCM as it achieves the important purpose of drawing an important resource human investment into the organization. Research studies have found recruitment procedure is significantly and positively associated to workers and organizational [23].
Performance Appraisal is a system which can be used as a trigger to control and manage the progress of the employee's towards achieving the desired organizational goals [7]. [26] argue that employee's high moral and work targets demanded by the management will be achieved if employees are supervised and properly managed, facilitated with healthy and safety standards and compensated fairly and equally.
Expectancy theory identifies three conditions that result in high motivation to perform a task, namely, first belief or the expectancy that great effort will increases performance, second instrumentality or the belief in the systems that determine performance and offer performance-based rewards, third valence or the value assigned by individuals to specific rewards. According to above prospective that extrinsic rewards lead employee's effort in the expected direction and elicit employee's behavioural changes toward creativity when the given rewards are dependent on creative employee's performance in the organization [34].

2. Literature Review

This subsequent section reviews studies that lead to the formulation of hypotheses of this study.

2.1 Training and development and employees' job performance

Training is frequently used to fill the existing gap between present and future performance of the work force. Training is a human resource improvement function, requiring being appropriately dealt with [55]. Research authors, [45] and [16] emphasized that employees should be fully involved in training and development programs and there must be a relation between employee and employer. [16] also explained that development is process of planning and preparing employees for the future performance and upcoming competitive problems in the business field. The main aim of training and development is to train employees to work in current competitive environment; training and development programs also create creative behaviours among employees for expected job performance [11].

[11] Argues that training and other human resources management practices have been positively related with employee's job performance and this fact is opened by previous research studies. Specifications of the need for training programs for such purpose are part and parcel included in the activities of this function. Plan for implementation of such functions and also evaluating their results are part of this function [44]. [37] training is a sort of activity that is designed, loaned and systematic in nature. Training improves the level of skills, information and competency which are usual in performing work efficiently. [29] says that training ensure job performance of employees. Different organization are motivated to adopt different training methods depending upon situations and organization needs (1) it may also include the organization strategy, objectives and physical & finance availability (2) as per requirements in the firm (3) depending on the targeted groups that are needed to be trained. In general the growth and development of an organization is affected by multiple factors. However, training plays a key role in enhancing the productivity of individuals in an organization. [50] at this phenomenon has evidently supported by available literature in this field. Studies which have over viewed the progress with respect to workers progress in particular [39] the same has prolonged to a universal view of organizational progress [52], [36]. Employee's performance and the organizational performance are closely related to each other. [26] observed that employees competence and performance promote to further improving standards. This also not only further polishes the knowledge and skills of workers, effectively improve their present skills and enable them for more tough jobs. [36] each researcher on training programs and workers progress has shown ultimate results on this particular. This particular has resulted in improved performance, beneficial to both the individual and the organization as well [39].

[52] training bridges the gap and removes deficiencies in performance which results into improving performance. Furthermore, some other researchers has given examples of [52] and sorted out on job-training: the way of meeting the training deficiencies and improving standard as the method of enhancing workers progress. As per above statement when they observe that training enable organizations to identify that its employees are not motivated and also not delivering good and a thus their moral as well, information and attributes are needed to be prepared as per organization goals. This is a fact that the workers keep a specific quality of standards associated with a specific task. Much more, it is necessary to point that this is not enough and workers required consistently to change & advance need of task progress. In other sense, firms requires to have constant ideas of training programs and keeping the employees and thus not to delay for performance gaps and skills. It is observed that investigation directed at building a contingency model of transfer-oriented training intervention design would provide information important for developing training environments more conducive to positive transfer in terms of productivity effectiveness. Identification of training needs, design and implementation of training programs, transfer of training, and evaluation of programmed benefits are key activities [42] in addition to study general training variables such as types of training, selection of trainees, selection criteria, evaluation instruments etc. Good and purposeful training based upon the positive investigation of the task needs upgrades the skill of the employees and has general attitude towards the job he has to perform. At also serves many other purposes [38]. In return production costs are within limits. There is job security for the employees as their work earn recognition from the employer. In this way the employees work with a great sense of responsibility. Continuity leads to gradual skill development and in this way; an organization has a more capable and skilled staff [17], [10]. [3] an organization should adopt a comprehensive training system to train their employees in domains of skill development, decision making, team work and interpersonal relations. This will help on organization effectively achieve its objectives and goals. [36] each researcher on training programs and workers progress has shown ultimate results on this particular. This particular has resulted in improved performance, beneficial to both the individual and the organization as well [39]. [9] argued that recruitment and selection is a significance part of research study, which manage employees in small and medium business organizations with more intellectual work done directed than other functional aspect of HCM.

H1: There is significant relationship between training & development & employees' job performance.

2.2 Recruitment and selection and employees' job performance

Recruitment is also one of the factors that can affect the employee performance [48], which further suggests that the implementation of recruitment process which are not in accordance with the effective plan that has been set, will cause a lot of problems such as high level of employees absenteeism and low employees performance, often too late, and others.[50] recruitment and selection of an appropriate labour force tools therefore can be viewed as essential to the success of an initiative and a key purpose in HCM. The premeditated position of staffing and selection is habitually stated in academic literature. [32] emphasized that recruitment and selection is systematic tool of HCM practices which recruiter and applicant, use precise devices to pick from a group of applicants and individuals who are most likely to flourish in the occupation, given organization lines and permitted boundaries. Recruitment and selection are very important functions of HCM for any type of organization. These are basic terms that refer to the process of attracting and choosing outstanding employees for organization. The quality of the HCM the organization has heavily depends on the role of these two functions [12]. According to [12] the general purpose of recruitment and selection is to provide the organization with a pool of potentially qualified job candidates.

Recruitment and selection are very important functions of HCM for any type of organization. These are basic terms that refer to the process of attracting and choosing outstanding employees for organization. The quality of the HCM the organization has heavily depends on the role of these two functions [12]. According to [12]) the general purpose of recruitment and selection is to provide the organization with a pool of potentially qualified job candidates. [49]) suggest that recruitment and selection have positive relationship with employee's productivity, employee's commitment and work quality in the organization. [35] argue that when they recognized the key stages of an orderly style to staffing and selection as job analysis, job explanation, and specification of person and encourage interviewees through the several approaches of staffing. A title of book "Human Resource Management" the second edition was nearly in settlement with Foot and Hook when they labelled the staffing and selection process as containing of three phases. The three main phases affording to them are job analysis, job description and individual requirement. Good staffing and selection processes let together the organization and the candidate to appropriately evaluate the degree to which the individual hysterics the job requirement and are in keeping with all appropriate regulation. Staffing and Selection is best prospects for organization to pick applicants that they trust are most suitable for the task achievements. [31] according to him selection as a well-organized procedure of selecting applicant who have required experience and qualification to fill vacant position in an organization. This investigator explains recruitment and selection as the procedure choosing the persons form the pool of probable applicants who meets the prerequisite of the vacant position recognized in the institute.

Recruitment and selection in small and medium size organization have stated in many inspections, that to find out capable workers is one of their most imperative problems. [35] industries are frequently not the employers of choice for job searchers. Medium size business is faced with difficulties because the distinctive image those small businesses have as employers [46]. Now government's build some formulated programmed to inspire the staffing and selection of graduates by small and medium sized businesses, like [54] make a supervision organization in Turkey, to support a medium sized business which provides salaries to young graduates for a specific period of time. Small and medium size business are faced with difficulty in staffing and selection of employees with correct qualification to help attain objectives of the business [18] further he clarified that this delinquent problem is complicated by the deficiency of well-organized method for hiring and selecting employees. Well organized staffing and selection methods include situations documentation, job analysis, job explanation, individual requirement and advertisement. According to him well organized selection procedure includes the hiring process, and collecting suitable information about committed candidates, and making evaluation in the requirement of each candidate and making assessment about occupation.

IOP Conf. Series: Journal of Physics: Conf. Series **1049** (2018) 012020 doi:10.1088/1742-6596/1049/1/012020

Well organized staffing and selection made ensure by the use of selection principles on the behalf of questions queried by the selection committee and in interview. HR managers could achieve help from a better fit between job and candidate by selecting the essential applicants for vacant situations in particular department of organization [56] as cited in [53].
H2: There is significant relationship between recruitment and selection and employees' job performance.

2.3 Performance appraisal and employees' job performance
The Success of performance evaluation system depends on the idea of any firm that performance evaluation practice is associated to firm purposes and the capabilities and selective perception of concerned individuals for adoption of this tool [25]. [1] in prior research study, performance appraisal has emerged as a tool for enhancing employee's professional development and organizational growth. However, [30] also suggest that performance appraisal has positive and significant relationship with employee's performance. Performance appraisal system is used to gather information about employees of the organization and also drives organizational and individual achievements towards their common goal. The absence of proper performance appraisal results low morale and turnover of the employees [43], [1]. [41] emphasized when they doing research study in a Malaysia University sometime employees have inflated assessment of their own regarding their performance. Whenever they are appraised they perceive to be biased due to involvement of internal biasness.
[44] argued that it is significant to notice that, when there are no proper arrangements of policies to face with the results or antecedents of the performance appraisal system, the anticipated paybacks of this procedure of performance appraisal system for the business cannot be understood. Furthermore, although best performance measurement system could be better for an industry, another face of performance appraisal system may be bad if not systematically held. It basically depends upon employee assessment response; positive and negative. The impact of employee response may be harm for the firm if assessment procedures are not taken well by the employees. Performance appraisal system can dispirit the worker and for many times may top to mislaying important employees in firm just they could not yield the referral response and manager realize that they will be good in many other employees . Performance evaluation practice signifies a dignified practice of employees observing and is proposed like management instrument to expand the productivity and performance of workers. Productivity and commitment of employees can be more strengthened with progress evaluation method [8], [7]. A well-organized clarification and management of performance evaluation leads to greater professional commitment and job satisfaction among workers. When employees' in organization are more committed to work, so it is best approach toward performance assessment programs [24]. [52] Performance assessment programs will be developed in organizations and employees performance when there is availability of balance human capital management factors like compensation, training and development, and then production will be spontaneously increased.
H3: There is significant relationship between performance appraisal & employees job performance.

3. Methods
Training and development, recruitment and Selection and performance appraisal system are independent variable while employee job performance is a dependent variable. The samples of this study are employees who are working in public and private sector telecommunication companies in Pakistan. Data are collected through questionnaires. The size of the sample was 272 due to time limit and economic restraints. One hundred and ninety one 191 male and eighty one 81 female employees participated in this survey. In this survey the age of the participated respondents was between 20 years and 51 years or above. The respondents age with respect to gender as shown in Figure 1. The qualification level of respondents in this survey is from matriculation to MS/PhD with respect to grade 1 to grade 4.

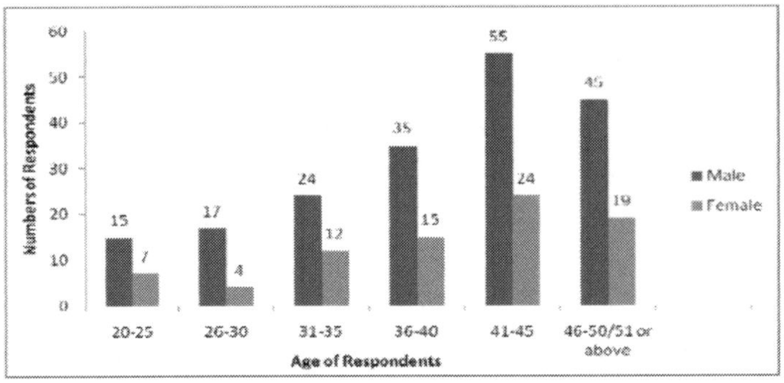

Figure 1. Respondents age with respect to gender.

3.1 Data Presentation & Statistical Analysis

In beginning data transfer to SPSS 20 versions, statistical software for social sciences for investigation is used. This research study analyzed though regression, impact of three independent variables (training & development, recruitment & selection, Performance appraisal system upon the dependent variable employee's job performance. The further relationship between dependent and independent variables was analyzed by using Pearson's Correlation. The descriptive analysis of the data is shown in table 1.

3.2 Correlation and regression analysis

Pearson correlation was conducted to determine the association between HCM Practices and employee's Job performance. In this research study Pearson correlation was used. Correlation analysis show there was statistically significant correlation between HCM practices and employees job performance. The calculated value of training and development is *(.635**)* and *(p<.005)* which shows significant relation between training and development and employee's performance which is most important factors in telecom sector of Pakistan. Correlation matrix of recruitment and selection which showing correlation coefficient of *(.659**)* and *(p<.005)* is also positively associated with employees performance in telecom sector of Pakistan. Value of correlation coefficient of performance appraisal is *(.658**)* and *(p<.005)* which ensure link between performance appraisal and employees performance as shown in table 2.

Table 3 shows regression analysis of all independent and dependent variable. When regression tools was applied on training and development and employee's performance the variance accounted, R^2 *(0.189)*, this figure shows that *(18.9%)*, of training & development can increased employees job performance in telecom sector of Pakistan. There is a statistically significant independence relationship between training & development and employees performance. When recruitment and selection regressed against employee's performance the variance accounted for, R^2 *(.432)*, this figure show that *(43.2 %)*, of recruitment and selection can increase performance. Regression coefficient verified our hypothesis that recruitment & selection and employees job performance depend upon each other in telecom sector of Pakistan. When performance appraisal regressed against employees' performance the variance accounted R^2 was *(.430)*, this figure show that *(43.0%)*, of performance appraisal increased employees job performance in telecom sector of Pakistan.

3.3 Hypothesis testing

The significance was found P<0.05 which is 0.000. This indicates that employee job performance will remain constant by 2.591 when training and development is zero. The beta coefficient for employee's job performance is 0.372. This mean when one unit increase in training and development increase, the overall performance will increases by 0.372. Hence the hypothesis $\mathbf{H_1}$ was proved to be significant as shown in Table 3. The value of significant of the model from ANOVA is 0.000. This shows that the relationship between recruitment and selection and employees job performance is statistically significant.

Table 1. Cronbach's σ coefficient, composite reliabilities and average variance extracted for each study variable.

Variable	Mean	SD	Cronbach's σ
Training & development	3.968	.415	.716
Recruitment and selection	3.914	.505	.743
Performance appraisal	3.905	.425	.751
Employees job performance	4.643	.447	.752

Table 2. Correlation analysis.

Variable	TR	*SE*	PA	EJP
Training & development	1			
Recruitment & selection	.452**	1		
Performance appraisal	.533**	.583**	1	
Employees job performance	.635**	.659**	.658**	1

** Correlation is significant at the 0.01 level (2-tailed).

Table 3. Regression analysis model summary.

Variable	β	t	p	r	r^2
Training & development	.372	7.998	.000	.438	.192
Recruitment & selection	.583	14.393	.000	.659	.434
Performance appraisal	.692	14.344	.000	.658	.432

a. Dépendent variable : EJP (Employees' job performance)

This indicates that employee job performance will remain constant by 1.781 when recruitment and selection is zero. The beta coefficient of recruitment and selection and employee job performance is 0.583. This means when 1 unit of recruitment and Selection increase, the overall employees job performance will increase by 0.583 units. Hence the $\mathbf{H_2}$ was proved to be significant as shown in Table 3. The significance was found between performance appraisal and employees job performance. This indicates that employee job performance will remain constant by 1.363 when performance appraisal is zero. The beta coefficient for employee's job performance is 0.692 this mean when one unit increase in performance appraisal increase, the overall employees job performance will increases by 0.692. Hence the $\mathbf{H_3}$ was proved to be significant as shown in Table 3.

4 Discussions

After an exclusive and extensive research study, results show that there is proper system of human capital management practices in telecom sector of Pakistan. Analysis results suggest some possible steps or remedies for employee's performance enhancement in Pakistan telecom sector. Some specific recommendations for future improvements are given below. The Training programs should conduct in order to increase the level of learning and new effective method must be introduced to evaluate the effectiveness of training programs. Training equipment's such as techniques and process need to enhance according to latest environment. For expected employees performance manager should analyze procedure and methods of recruitment & selection in organization. Performance appraisal system should be conducted in such a manner which productive and development oriented. Employee's performance will be increase positively when performance appraisal is equally conducted for booth managers and subordinates. Performance appraisal will be done in such a way which is highly result oriented.

The data which is collected through questionnaires are comprehensive, but a lot of researcher doesn't understand that what activities are basically happening in all places of the firm. Mostly, certain places of the replies may have been very true while other place is not true. In fact it is very conceivable to explain with every participant resumed assessment if this has happened. Preferably the investigation or survey should be divided into two discrete surveys, one focus on human capital management and another survey focus on financial position of the organization, to get feedback from top management in organization; this was not done due to time and financial constraints. Another additional aspect of survey which bound the availability of the financial performance department to issue profound financial information about the organization, which administration may not want to release secrete financial performance.

Acknowledgments
"This paper was partly sponsored by the Centre for Graduate Studies UTHM"
"Kertas kerja ini ditaja sebahagiannya oleh Pusat Pengajian Siswazah"

References

[1] Abdullah A, Bilau W and Enegbuma 2011 Small and medium sized construction firms job satisfaction and performance evaluation in Nigeria *Ipedr.* **15** 225–229
[2] Aktar S, Sachu M K and Ali E 2012 The Impact of rewards on employee performance in commercial banks of Bangladesh: An empirical study *IOSR J. Bus. Manag.* **6** 9–15
[3] Bartel A P 1994 Productivity gains from the implementation of employee training programs *Industrial Relations* **33** 411–425
[4] Breaugh J A and Starke M 2000 Research on employee recruitment: so many studies, so many remaining questions *J. Manage.z vol.* **26** 405–434
[5] Beardwell J and Claydo T 2007 " *Human Resource Management A Contemp. Approach, 5th ed., Pearson Educ. Essex.* 694
[6] Boxall P 2003 HR strategy and competitive advantage in the service sector *Hum. Resour. Manag. J.***13** 5–20
[7] Brown M and Benson J 3003 Rated to exhaustion ? Reactions to performance appraisal processes, 67–81
[8] Brown M and Heywood J S 2005 Performance appraisal systems: Determinants and change *Br. J. Ind. Relations.* **43** 659–679
[9] Cardon M S and Stevens C E 2004 Managing human resources in small organizations: What do we know? *Hum. Resour. Manag. Rev.* **3** 295–323
[10] Colombo E, Stanca L, January N, Bicocca M, Colombo E and Stanca L 2008 The impact of training on productivity : Evidence from a large panel of firms the Work *Pap. Dip. di Econ. Polit.Univ. di Milano Bicocca.* 134
[11] Elnaga A and Imran A 2013 The effect of training on employee performance *Eur. J. Bus. Manag.* **5** 2222–2839
[12] Gamage A S 2014 Recruitment and selection practices in manufacturing smes in japan: An analysis of the link with business performance *Ruhuna J. Manag. Financ.* **1** 37–52
[13] Ichniowski C, Shaw K and Prennushi G 1997 The effects of human resource management practices on a study of steel finishing lines productivity: A study of steel finishing lines, *Am. Econ. Rev.* 292–312
[14] Imran M and Tanveer A 2015 Impact of training and development on employees performance in banks of pakistan *Eur. J. Train. Dev. Stud.***3** 22–44
[15] Imran M, Maqbool N Shafique H 2014 Impact of technological advancement on employee performance in banking sector *Int. J. Hum. Resour. Stud.* **4** 57–70
[16] Kadiresan V, Selamat M H, Selladurai S, Ramendran C and Mohamed R M K H 2015 Performance appraisal and training and development of human resource management practices (hrm) on organizational commitment and turnover intention *Asian Soc. Sci.* **11** 162–176
[17] Konings J and Vanormelingen S 2010 The impact of training on productivity and wages

LIOCS Discuss. Pap. Ser. **4731** 1–55
[18] Priyanath H 2010 Managerial deficiencies in the small and medium enterprises (smes) in sri lanka: an empirical evidence of smes in the Ratnapura district *Sabaragamuwa Univ. J.* **6** 79–86
[19] Belias D Koustelios A, Vairaktarakis G and Sdrolias L 2015 Organizational culture and job satisfaction of greek banking institutions **175** 314–323
[20] Delmotte J, Faems D and Forrier A 2006 Unravelling the hrm – performance link : value-creating and cost-increasing effects of small business hrm **43** 319-342
[21] Sepúlveda F 2010 Training and productivity: evidence for us manufacturing industries *Oxf. Econ. Pap.***62** 504–528
[22] Singh 2004 Impact of HR practices on perceived firm performance in India *Asia Pacific Journal Human Resource.* **42** 302–315
[23] Syed Z and Jamal J 2012 Universalistic Perspective of HRM and Organisational Performance : Meta-Analytical Study **13** 47–57
[24] Shahzad K, Bashir S and Ramay M I 2008 Impact of HR practices on perceived performance of university teachers in Pakistan Int. Rev. Bus. Res. Pap. **4** 302–315
[25] Thwala W.D, Ajagbe M A, Enegbuma W I, Bilau A A and Long C S 2012 Sudanese small and medium sized construction firms : an empirical survey of job turnover J. Basic Appl. Sci. Res. **2** 7414–7420
[26] Wright P C and Geroy G D 2001 Changing the mindset: the training myth and the need for world-class performance Int. J. Hum. Resour. Manag. **12** 586–600
[27] Raffee S 2001 "Motivating security guards: a management perspective in the private security guarding industry in Singapore, scarman centre for the study of public order, Leicester University, UK," *Unpublished Dissertation*
[28] Sherman B and Snell A 1998 *Managing Human Resources, incentives and rewards,"* International Publishing, Ohio
[29] Armstrong M 1995 *A handbook of personnel Management Practices,* Kogan Page Limited London,
[30] Bekele A Z, Shigutu A D and Tensay A T 2014 The effect of employees' perception of performance appraisal on their work outcomes *Int. Jou. of Mng. and Commerc Inno.* **2** 136-173
[31] Bohlander G and Snell S 2010 *Managing Human Resources.* Mason, OH: South-Western Cengage Learning
[32] Bratton J and Gold J 2003 *Human Resource Management: Theory and Practice,* 3rd Ed., New York: Palgrave Macmillan *Brit J. of Indu Rel.* **43** 659-679
[33] Campbell J P, McHenry J J and Wise L L 1990 Modeling job performance in a population of jobs *Personnel Personel Pesology.* **43** 313-474
[34] Eisenberger R and Cameron L 1998 Reward, intrinsic interest, and creativity, New findings
[35] Foot and Hook 2010 Introducing human resource management Pearson education (UK)
[36] Guest D E 1997 Human resource management and industrial relations *Journal of Management Studies.* **24** 503–521
[37] Gordon B 1992 Are Canadian firms under investing in training? *Canadian Business Economics.*1 25–33
[38] Goldstein L L, Dunnette, and Hough L M 1991 *Training in work organizations. Handbook Of Industrial And Organizational* Psychologists *Palo Alto CA: Consulting Psychologists Press.* 507-619
[39] Harrison R 2000 *Employee Development.* Silver Lakes, Pretoria. Beekman Publishing,
[40] Herriot P 1993 Interviewing. In P. Warr (Ed.), Psychology at work, (139-159). London: Penguin *International Research Journal of Finance and Economics.* ISSN issue 39 1450-2887
[41] June M L 2004 Effects of performance appraisal politics on job satisfaction and turnover

intention *Personnel Review.* **33** 322 – 334

[42] Krishnaveni R and Sripirabaa B 2008 Capacity building as a tool for assessing training and development activity: an Indian case study International journal of training and development. **12** 121-134

[43] Laura S 1996 Is performance being managed for the long term, Quality Progress **29**141- 221

[44] McCourt W and Derek E 2003 *Global Human Resource Management: Managing People in Developing and Transitional Countries*, UK: Edward Elgar

[45] McConnell C R 2004 Managing employee performance *The health care manager.* **23** 273-283

[46] Moy W and Lee S M 2000 "The Career Choice of Business Graduates: SMEs or NMC?" *Carr. Dev. Int.* **7**339-347

[47] Osunde D C 2015 Privatization of public enterprises in nigeria: impact on employees' performance and managerial implications Int. Jou. Of Res, –Grant Haalayah A Knowledge Repository *3*

[48] Pambagio N S, Utami and Gunawan E 2013 Influence of process recruitment, selection process and competence of employee on employee performance (studies on the employees division of administration and finance pg kebon agung malang) *J. of Bus. Adm.* **4**1–7

[49] Patimah S 2015 The influence of recruitment and selection on the performance of state elementary school principal's bandar lampung. Scientific Jou. Peuradeun – *Int. Multidisp. Jou.*3165–190

[50] Boxall P F and Purcell J 2003 "Strategy and Human Resource Management," *Palgrave MacMillan,*

[51] Purcell J, Hutchinson N, Hutchinson S, Rayton B and Swart J 2003 Understanding the people and performance link: Unlocking the Black-Box. *Research Report, CIPD, London*

[52] Swart J, Mann C, Brown B and Price P 2005 "Human Resource Development: Strategy and Tactics, *Oxford Elsevier Butterworth-Heinemann Publications*

[53] Thang L.C and Quang T 2005 Human resource management practices in a transitional economy: A comparative study of enterprise ownership forms in Vietnam. *Asia Pacific business review.* **11** 25-47

[54] Tanova C 2003 Firm size and recruitment: staffing practices in small and large organizations in north Cyprus, *Career Dev. Int.*107 – 114

[55] Weil A and J Woodall 2005 HRD in France: the corporate perspective Jou. *of Eur. Indus. Training.* **29,** 529–540

[56] Zhu Y 2004 "Responding to the challenges of globalization: human resource development in Japan *Journal of World Business.* **39** 337–348

Level of Satisfaction for Occupational Safety and Health Training Activities: A Broad Spectrum Industrial Survey

Muhammad Mujtaba Asad[1], Razali Bin Hassan[1], F. Sherwani[2], Nor Hida Ibrahim[3], Qadir Mehmood Soomro[4]

[1]Faculty of Technical and Vocational Education, Universiti Tun Hussein Onn Malaysia, Parit Raja Batu Pahat, Johor.
[2]Faculty of Electrical and Electronics Engineering, Universiti Tun Hussein Onn Malaysia, Parit Raja Batu Pahat, Johor.
[3]Institut Pendidkan GuruTun Hussein Onn Malaysia, Batu Pahat, Johor.
[4]OSHTC, Hyderabad, Sindh, Pakistan

E-mail: mujtabaasad11@gmail.com

Abstract. The main objective of this research article is to identify the satisfaction level for occupational safety and health training activities among oil and gas drilling crew and staff. In this study, data has been gathered from reputed online oil and gas health and safety social and professional platforms through e-survey questionnaire. Similarly, for data analysis descriptive (Percentage %) and graphical statistical approach has been adopted for the interpretations of research findings by using SPSS 22.0 software. According to the overall findings of this paper, there is a sheer need of effective and sufficient training strategies for safety and health training activities at oil and gas extraction industries. As per respondents, there are also limited open source resources available by oil and gas companies for safety and health awareness and personal protection. Likewise, findings also indicated that, most of the participants of this online survey have reported about the communication gap between administration and drilling workforce for safety and health related issues and discussions which is one of the major causes of accidents and unexpected hazards at drilling sites.

Keywords: Safety and Health, Oil and Gas, Drilling, Training Activities

1. Introduction

Occupational safety and health awareness trainings are one of the effective preventive measures for several industrial hazardous activities [1]. According to previous researchers oil and gas drilling

[1] mujtabasad11@gmail.com

Content from this work may be used under the terms of the Creative Commons Attribution 3.0 licence. Any further distribution of this work must maintain attribution to the author(s) and the title of the work, journal citation and DOI.
Published under licence by IOP Publishing Ltd

operations is consider as eight times more hazardous than construction industries due to the rate of life threatening injuries [2]. In oil and gas extraction operation, safety, chemical and ergonomic hazard are very common at on and offshore domains. But for the protection of drilling workforce all companies use to provide appropriate trainings to drilling crew prior to performing certain task such as handling drilling fluid training, chemical safety training, drilling pipes lifting training and well control training activities according to Occupational safety and health regulations [3]. As per OSHA US from Year 2003-2010, 823 drilling workers have been died during performing their tasks at oil and gas drilling section in United Stated of America [4].According Caldwell, accidents usually occur during oil and gas extraction operation is due to lack of safety trainings, lack of communication and equipment maintenance. Correspondingly, in this paper researchers have conducted an online survey on renowned oil and gas safety and health online platforms and discussion forums for identifying the oil and gas drilling crew level of satisfaction regarding suitable safety training activities and safety resources from oil and gas

2. Problem Statement

According to US Federal statistical report in year 2006-2010 more than 1200 oil and gas drilling crew were injured and 41 crew workers lost their lives in the Gulf of Mexico [5]. Likewise, in BP Deepwater Horizon explosion 11 fatalities and several injuries were reported due to insufficient emergency and well control monitoring trainings [6]. Likewise, in past few years due to the rapid industrial need for drilling crew at on and offshore sites some oil and gas industries recruited fresh and untrained drilling crew staff at on and offshore sites [6]. Therefore, the combined fatality rate for onshore and offshore oil and gas extraction was 27.1 per 100,000 full-time workers, compared with a rate of 3.8 for all U.S [7-8]. Hence, to examine the satisfaction level of drilling crew for safe and secure drilling operation this research have been conducted through online survey to provide a broad spectrum of safety and health training activities at oil and gas drilling sites in the perspective of workers contentment.

3. Objective

The aim of this study is to achieve following research objective:

a. To identify the drilling crew level of satisfaction for occupational safety and health training activities at oil and gas industries.

4. Methodology

In this study, data has been analyzed by using SPSS version 22.0 software to determine the percentage for every question item from distribute qualitative instrument. Descriptive method percentages (%)of every item has been used for analyzing the drilling crew level of satisfaction for sufficient health and safety training activities. According to Asian marketing research, the Internet based data collection from distant respondent is determined to be the most sufficient and reliable (Asia Marketing Research, Internet Usage, Population Statistics, & Face book Information, 2012). Data for this study has been collected from 75 health and safety experts through different online oil and gas drilling (offshore and onshore domains) forums.

5. Results and Findings

In this study, demographic results have been illustrated as per respondent's industry, origin, gender and designation as oil and gas health and safety professional .Likewise identification of level of satisfaction of drilling crew for safety training activities has been reported in percentage.

5.1 Study Respondents Based Gender

Respondents of this study are 75 health and safety practitioners and crew from oil and gas drilling industries. The collected data has been analyzed by using Statistical Packages for Social Sciences (SPSS 22).In table 1 shown the number of respondents and percentage based on respondent's gender. For this study, 68 respondents were male which carrying 91%. Meanwhile, only 7 female respondents were involved in this study which is equal to 9%.

Table 1. Sprinkling of Respondents Based on Gender

Gender	Number of Respondent	Percentage (%)
Male	68	91
Female	7	9
Total	75	100

5.2 Study Respondents Based on Origin

Similarly, table 2 shows the number of respondents and percentage based on respondent's origin. For this study 12 respondents were from Europe, 17 from Asia, 03 from North America and 41 from Middle East which is carrying highest overall percentage 55%.

Table 2. Sprinkling of Respondents Based on Origin

Origin	Number of Respondent	Percentage (%)
Europe	12	16
Middle East	41	55
Asia	17	22
North America	3	4
Australia	2	3
Total	75	100

5.3 Study Respondents Based on Designation

Likewise, table 3 shows the number of respondents and percentage based on respondent's designation. In this study, 20 respondents were safety officers, 11 were manger, 10 were supervisor and 34 were drilling crew which is carrying highest percentage 45%.

Table 3. Sprinkling of Respondents Based on Designation

Designation	Number of Respondent	Percentage (%)
Drilling Crew	34	45
Safety Officer	20	27
Manager	11	15
Supervisor	10	13

Total	75	100

5.4 Study Respondents Based on Type of Industry

Lastly, table 4 shows the number of respondents and percentage based on respondent's type of industry. For this study, 34 respondents were working in offshore oil and gas industry and 41 were working in onshore oil and gas industry which is carrying the highest percentage 55%.

Table 4. Sprinkling of Respondents Based on type of Industry

Industry	**Number of Respondent**	**Percentage (%)**
Offshore	34	45
Onshore	41	55
Total	75	100

5.5 Response on Encouragement for Safety Training

Participant of this study responded on first query of questionnaire which was about "Did safety manager encourage their employees to attend outside safety training programs regarding drilling hazards?" 75% of them said that safety manager didn't encourage their employees to attend outside safety training programs regarding drilling hazards as shown in table 5.

Table 5. Table for Safety Manager Encouragement for Training

Did safety manager encourage their employees to attend outside safety training programs regarding drilling hazards?		
Response	Number of Respondent	Percentage (%)
YES	19	25
NO	56	75
Total	75	100

5.6 Response on Updated Safety Material Provided

Research participant responded on second query of questionnaire which was about "Did you company provide you updated material for safe drilling operation?" 79% of them said that the company didn't provide updated material for safe drilling operation as indicated in table 6.

Table 6. Table for Updated Safety Material Provided

Did you company provide you updated material for safe drilling operation?		
Response	Number of Respondent	Percentage (%)
YES	16	21

NO	59	79
Total	75	100

5.7 Response on Conducting Safety Meeting with Employees
Study participants responded on third query of questionnaire which was about "Did your supervisor conduct safety meetings with employees?" 88% of them said that the supervisor didn't conduct safety meetings with employees as specified in table 7.

Table 7. Table for Conducting Safety Meeting with Employees

Did your supervisor conduct safety meetings with employees?		
Response	Number of Respondent	Percentage (%)
YES	09	12
NO	66	88
Total	75	100

5.8 Response on Safety Talk
Participants responded on fourth query of questionnaire which was about "Did safety officer talk to their employees at least once a quarter regarding safety and health issues during drilling process?" 73% of them said that the safety officer didn't talk to their employees at least once a quarter regarding safety and health issues during drilling process as shown in table 8.

Table 8. Table for Response on Safety Talk

Did safety officer talk to their employees at least once a quarter regarding safety and health issues during drilling process?		
Response	Number of Respondent	Percentage (%)
YES	20	27
NO	55	73
Total	75	100

5.9 Response on Organizing Hazard Assessment Program
Participants responded on fifth query of questionnaire which was about "Has your organization developed an effective ergonomics program that includes hazard assessments and supervisory and employee training?" 53% of them said that the organization didn't developed an effective ergonomics program that includes hazard assessments and supervisory and employee training as indicated in table 9.

Table 9. Table for Organizing Hazard Assessment Program

Has your organization developed an effective ergonomics program that includes hazard assessments and supervisory and employee training?		
Response	Number of Respondent	Percentage (%)

YES	35	48
NO	40	53
Total	75	100

5.10 Response on Demonstration of Safety Regulation

Respondents of this study responded on sixth query of questionnaire which was about "Did your safety officer demonstrate international health and safety rules and regulation during training?" 83% of them said that the safety officer didn't demonstrate international health and safety rules and regulation during training as shown in table 10.

Table 10. Table for Demonstration of Safety Regulation

Did your safety officer demonstrate international health and safety rules and regulation during training?		
Response	Number of Respondent	Percentage (%)
YES	13	17
NO	62	83
Total	75	100

5.11 Response on New Worker Training

Participants responded on seventh query of questionnaire which was about "Did supervisor conduct/coordinate new employee safety orientation training?" 92% of them said that the new employees didn't get any safety orientation training specified in table 11.

Table 11. Table for New Worker Training

Did supervisor conduct/coordinate new employee safety orientation training?		
Response	Number of Respondent	Percentage (%)
YES	6	8
NO	69	92
Total	75	100

5.12 Response on Use of Multimedia Technology for Training

Participants responded on eighth query of questionnaire which was about "Does your company use any multimedia technology for safety training?" 55% of them said that that company didn't use any multimedia technology for safety training as shown in table 12.

Table 12. Table for Use of Multimedia Technology for Training

Does your company use any multimedia technology for safety training?		
Response	Number of Respondent	Percentage (%)
YES	34	45

NO	41	55
Total	75	100

5.13 Response on E-Learning Resources

Participants responded on ninth query of questionnaire which was about "Did your company provide updated E Learning resources for health and safety awareness to drilling crew?" 68% of them said that that company didn't provide updated E Learning resources for health and safety awareness to drilling crew as shown in table 13.

Table 13. Table for E-Learning Resources

Did your company provide updated E Learning resources for health and safety awareness to drilling crew?		
Response	Number of Respondent	Percentage (%)
YES	24	32
NO	51	68
Total	75	100

5.14 Response on Attending Safety Programs

Participant of this study responded on tenth query of questionnaire which was about "Did drilling crew attend all required safety training programs?" 77% of them said that the drilling crew didn't attend all required safety training programs as indicated in table 14.

Table 14. Table for Attending Safety Programs

Did drilling crew attend all required safety training programs?		
Response	Number of Respondent	Percentage (%)
YES	17	23
NO	58	77
Total	75	100

5.15 Response on Coordinate Regulatory Compliance Training

Health and safety practitioners responded on first query of questionnaire which was about "Did supervisor coordinate regulatory compliance training activities?" 72% of them said that the supervisor didn't coordinate regulatory compliance training activities as indicated in table 15.

Table 15. Table for Coordinate Regulatory Compliance Training

Did supervisor coordinate regulatory compliance training activities?		
Response	Number of Respondent	Percentage (%)
YES	21	28

NO	54	72
Total	75	100

5.16 Response on Vestibule Training Based Safety System
Participants responded on twelfth query of questionnaire which was about "Did your company provide you vestibule training based safety system for drilling process?" 75% of them said that the company didn't provide vestibule training based safety systems for drilling process. as shown in table 16.

Table 16. Table for Vestibule Training Based Safety System

Did your company provide you vestibule training based safety system for drilling process?		
Response	Number of Respondent	Percentage (%)
YES	19	25
NO	56	75
Total	75	100

5.17 Response on Training Program Include Inspections
Health and safety drilling crew responded on first query of questionnaire which was about "Does the safety training program include inspections of different area?" 87% of them said that the safety training program didn't include inspections of different area as specified in table 17.

Table 17. Table for Training Program Include Inspections

Does the safety training program include inspections of different area?		
Response	Number of Respondent	Percentage (%)
YES	10	13
NO	65	87
Total	75	100

5.18 Response on Conducting Training on Accident Investigation
Health and safety professionals responded on fourteenth query of questionnaire which was about "Is your company currently conducting training on accident investigation during drilling?" 57% of them said that the company currently conducting training on accident investigation during drilling as indicated in table 18.

Table 18. Table for Conducting Training on Accident Investigation

Is your company currently conducting training on accident investigation during drilling?		
Response	Number of Respondent	Percentage (%)
YES	32	57
NO	43	43

Total	75	100

5.19 Response on Receive Annual or Refresher Training
Health and safety practitioners responded on first query of questionnaire which was about "Do employees receive annual and/or refresher training when required by a particular OSHA standard?"64% of them said that the employees receive annual and/or refresher training when required by a particular OSHA standard as shown in table 19.

Table 19. Table for Receive Annual or Refresher Training

Do employees receive annual and/or refresher training when required by a particular OSHA standard?		
Response	Number of Respondent	Percentage (%)
YES	27	36
NO	48	64
Total	75	100

6. Conclusion

According to the overall result analysis of health and safety training activates in drilling process shows that, there is a lack of professional safety training and updated useful E-learning material for reducing work place risk and hazards in oil and gas industries. From the result shown in figure 1, it's understandable that many of oil and gas safety officials ignore the safety detail demonstration of safety precautions during well drilling, and resultantly many accidents and critical injures has been reported. Based on the response from this research, there is also lack of effective training programs and learning resources for drilling crew safety awareness. As per the response from participants, clarify that the safety providers need to motivate and encourage their employees to attend safety training programs regarding drilling hazards with vestibule training based safety system exposure.

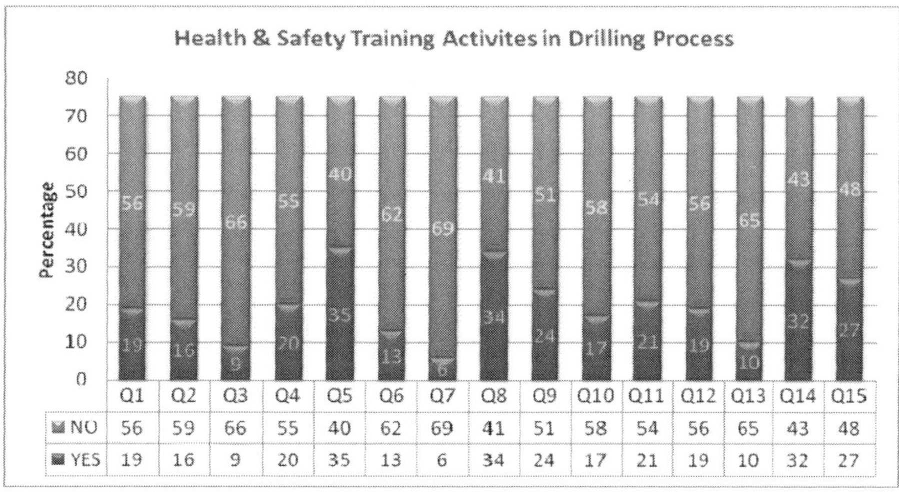

Figure 1. Overall Results

Acknowledgment

This paper was partly sponsored by the Centre of Graduate Studies (CGS) of Universiti Tun Hussein Onn Malaysia (UTHM).

References

[1] Demirkesen S, Arditi D. Construction safety personnel's perceptions of safety training practices. International Journal of Project Management. 2015 Jul 31;33(5):1160-9.

[2] Akinremi TA, Anderson R, Olomolaiye A, Adigun L. Risk Management as an Essential Tool for Successful Project Execution in the Upstream Oil Industry. InAbu Dhabi International Petroleum Exhibition and Conference 2015 Nov 9. Society of Petroleum Engineers.

[3] Witter RZ, Tenney L, Clark S, Newman LS. Occupational exposures in the oil and gas extraction industry: State of the science and research recommendations. American journal of industrial medicine. 2014 Jul 1;57(7):847-56.

[4] Caldwell B, Hinton J. Data Drilling: Changing the Way the Oil and Gas Industry Manages Safety and Risk. InSPE E&P Health, Safety, Security and Environmental Conference-Americas 2015 Mar 16. Society of Petroleum Engineers.

[5] Norazahar N, Khan F, Veitch B, MacKinnon S. Human and organizational factors assessment of the evacuation operation of BP Deepwater Horizon accident. Safety science. 2014 Dec 31;70:41-9.

[6] Asad MM, Hassan RB, Soomro QM, Sherwani F. Development of KBES with Hazard Controlling Factors and Measures for Contracting Health and Safety Risk in Oil and Gas Drilling Process: A Conceptual Action Plan. The Social Sciences. 2017;12(3):584-94.

[7] Abualfaraj N. Identifying Potential Exposure Pathways and Estimating Risk from Marcellus Shale Gas Development. Drexel University; 2016.

[8]Mendes PA, Hall J, Matos S, Silvestre B. Reforming Brazil' s offshore oil and gas safety regulatory framework: Lessons from Norway, the United Kingdom and the United States. Energy Policy. 2014 Nov 30;74:443-53. Retrieved from: https://www.cdc.gov/mmwr/preview/mmwrhtml/mm6216a2.htm

Corrigendum: Level of Satisfaction for Occupational Safety and Health Training Activities: A Broad Spectrum Industrial Survey

J. Phys. Conf. Ser **1049** (2018) 012021

Muhammad Mujtaba Asad[1], Razali Bin Hassan[1], F. Sherwani[2], Nor Hida Ibrahim[3] and Qadir Mehmood Soomro[4]

[1] Faculty of Technical and Vocational Education, Universiti Tun Hussein Onn Malaysia, Parit Raja Batu Pahat, Johor.
[2] Faculty of Electrical and Electronics Engineering, Universiti Tun Hussein Onn Malaysia, Parit Raja Batu Pahat, Johor.
[3] Institut Pendidkan GuruTun Hussein Onn Malaysia, Batu Pahat, Johor.
[4] OSHTC, Hyderabad, Sindh, Pakistan

Page 2:
In the **2. Problem Statement** the following text appears:

" According to US Federal statistical report in year 2006-2010 more than 1200 oil and gas drilling crew were injured and 41 crew workers lost their lives in the Gulf of Mexico [5]. Likewise, in BP Deepwater Horizon explosion 11 fatalities and several injuries were reported due to insufficient emergency and well control monitoring trainings [6]. Likewise, in past few years due to the rapid industrial need for drilling crew at on and offshore sites some oil and gas industries recruited fresh and untrained drilling crew staff at on and offshore sites [6]. Therefore, the combined fatality rate for onshore and offshore oil and gas extraction was 27.1 per 100,000 full-time workers, compared with a rate of 3.8 for all U.S [7-8]. Hence, to examine the satisfaction level of drilling crew for safe and secure drilling operation this research have been conducted through online survey to provide a broad spectrum of safety and health training activities at oil and gas drilling sites in the perspective of workers contentment."

This should read:

"According to US Federal statistical report in year 2006-2010 more than 1200 oil and gas drilling crew were injured and 41 crew workers lost their lives in the Gulf of Mexico **[6]**. Likewise, in BP Deepwater Horizon explosion 11 fatalities and several injuries were reported due to insufficient emergency and well control monitoring training **[5]**. Likewise, in past few years due to the rapid industrial need for drilling crew at on and offshore sites some oil and gas industries recruited fresh and untrained drilling crew staff at on and offshore sites **[6]**. Therefore, the combined fatality rate for onshore and offshore oil and gas extraction was 27.1 per 100,000 full-time workers, compared with a rate of 3.8 for all U.S [7-8]. Hence, to examine the satisfaction level of drilling crew for safe and secure drilling operation this research have been conducted through online survey to provide a broad spectrum of safety and health training activities at oil and gas drilling sites in the perspective of workers contentment."

IOP Conf. Series: Journal of Physics: Conf. Series **1049** (2018) 012022 doi:10.1088/1742-6596/1049/1/012022

Indication of Decision Making through Accident Prevention Resources among Drilling Crew at Oil and Gas Industries: A Quantitative Survey

Muhammad Mujtaba Asad[1], Razali Bin Hassan[1], Nor Hida Ibrahim[2], F. Sherwani[3], Qadir Mehmood Soomro[4]
[1]Faculty of Technical and Vocational Education, Universiti Tun Hussein Onn Malaysia, Parit Raja Batu Pahat, Johor.
[2]Institut Pendidkan GuruTun Hussein Onn Malaysia, Batu Pahat, Johor.
[3]Faculty of Electrical and Electronics Engineering, Universiti Tun Hussein Onn Malaysia, Parit Raja Batu Pahat, Johor.
[4]OSHTC, Hyderabad, Sindh, Pakistan

E-mail: mujtabaasad11@gmail.com

Abstract. The aim of the present article is to determine the decision making potential of drilling crew by utilizing the available accident prevention resources at onshore and offshore oil and gas industries. In this study descriptive quantitative approach has been used to indicate the existing accident prevention resources appropriateness for accident prevention during oil and gas drilling operation. Similarly, 75 health and safety professionals and drilling crew has been randomly selected for this online survey from around the globe. For data analysis SPSS 22.0 software has been adopted to examine the gathered data in systematic way. According to the findings, it has been indicated that there is a lack of adequate health and safety resources at onshore and offshore oil and gas industries. Similarly, there is also a sheer need for decision support system based on effective hazard controlling measures to reduce the workplace risk and hazards. Also, it has been identified that the most of drilling crew neglect the safety and health precaution and policies during operation which cause them injuries. The findings of this research can facilitate the future researchers and industries to overcome identified issues through appropriate solution of addressed problems in effective way.

1. Introduction

Oil and gas industries play a very effective role in the acceleration of world's economy from decades [1]. One of the major processes in these industries is the extraction of cured oil and natural gases from the ground and deep sea bed [2]. Likewise, oil and gas drilling process and their associated activities are considered as highly hazardous and risky [3]. Therefore oil and gas drilling operation is reported as one of the challenging and dangerous profession in the world at both oil and gas drilling domains (Onshore and Offshore) [4]. Furthermore, most of the lifethreatning events and hazardous condition

[1] mujtabasad11@gmail.com

Content from this work may be used under the terms of the Creative Commons Attribution 3.0 licence. Any further distribution of this work must maintain attribution to the author(s) and the title of the work, journal citation and DOI.
Published under licence by IOP Publishing Ltd

occur due to the negligence of safety and health precautions and regulations [4]. Also, due to the inadequate safety and health resources and efficient decision making systems based on effective hazard controlling measures [5]. Thus, hundreds of peoples died annually during oil land g as drilling operation due to improper decision making at hazardous conditions during onshore and offshore operations worldwide for accident prevention [6]. Correspondingly, this study aimed to indicate the decision making potential of oil and gas drilling crew to by using the available resources provided by their organization for safety and health awareness and hazard and risk reduction during drilling activities at onshore and offshore sites.

2. Problem statement

Petrochemical extraction process is considered as one of the most complex and risky operation in oil and gas sector [7]. Every year many onshore and offshore drilling crew reported with lifethreatning injuries and on job fatalities due to the negligence of safety standards [4]. There is numerous health and safety hazards are involved through the drilling operation such as, safety hazard, chemical hazards, ergonomic hazards and environmental hazards [8]. Therefore the drilling operation is also specified as unpredictable operation due to its hazardous nature. Similarly, the major cause which has been indicated by some previous researchers are the inadequate accident prevention materials and recourses based on suitable hazard controlling information and knowledge provided by industries to the drilling crew at both drilling domain [9]. Due to insufficient accident prevention materials drilling crew and safety professionals are unable to take appropriate decision during any hazardous condition which cause them life time disabilities and also damage of industrial equipment [10]. Therefore, this study focused on the identification of the decision making potential of drilling crew through accident prevention available resources in oil and gas extraction process.

3. Objective

This paper focused to achieve following objective:

a. To determine the decision making potential of drilling crew through accident prevention resources in oil and gas drilling operation

4. Methodology

In this research, data has been randomly collected from seventy five (75) drilling crew and safety professionals through online safety and health forums as shown in table 1. According to Asian marketing research, the Internet based data collection from distant respondent is determined to be the most sufficient and reliable [11]. Similarly, the gathered data has been analyzed by using SPSS version 22.0 to examine the percentage for each item of the distribute research questionnaire.

The questionnaire was divided in to three parts the first part was about the demographic information of the participants which included the gender, type of industry, origin and designation. Similarly, the second part contains thirteen query items for deterring the decision making potential through accident prevention resourced for oil and gas drilling operation. The question has validated from three safety and health field experts from different oil and gas industries.

Table 1. Targeted online platforms

Targeted Online Platforms
1) Gulf National Drilling Company
2) AB Drilling Company
3) Drilling Off shore Malaysia
4) Drilling Engineers
5) Drilling Fluids
6) Drilling Engineer and Drilling supervisor
7) Drilling Rigs Onshore
8) Well Drilling Institute
9) Saudi Oil and Gas Safety
10) Oil and gas Jobs and Discussion
11) HSE Oil and gas forum

5. Results and findings

Respondents for this study are 75 health and safety practitioners from oil and gas drilling industries. The collected data is analyzed by using Statistical Packages for Social Sciences (SPSS 20).In figure 1, it shows the percentage based on respondent's gender. For this study, 68 respondents are male which carrying 91%. Meanwhile, there only 7 female respondents were involved in this preliminary study which is equal to 9%.

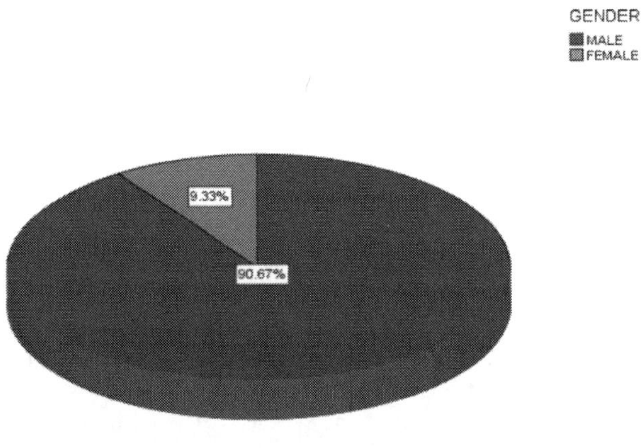

Figure 1. Pie Chart for Respondents Gender

5.1 Respondents based on origin

In figure 2, it shows the number of respondents and percentage based on respondent's origin. For this preliminary study, 12 respondents are from Europe, 17 from Asia, 03 from North America and 41 from Middle East which carrying highest percentage 55%.

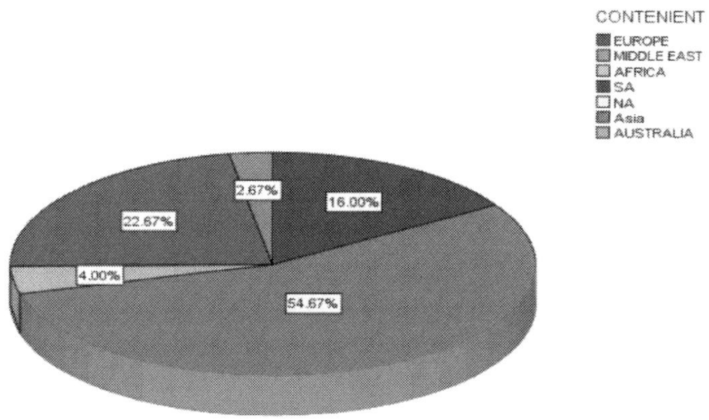

Figure 2. Pie chart for respondents origin

5.2 Respondents based on designation

In figure 3, it shows the number of respondents and percentage based on respondent's designation. For this study, 20 respondents are safety officers, 11 are manger, 10 are supervisor and 34 are drilling crew which carrying highest percentage 45%.

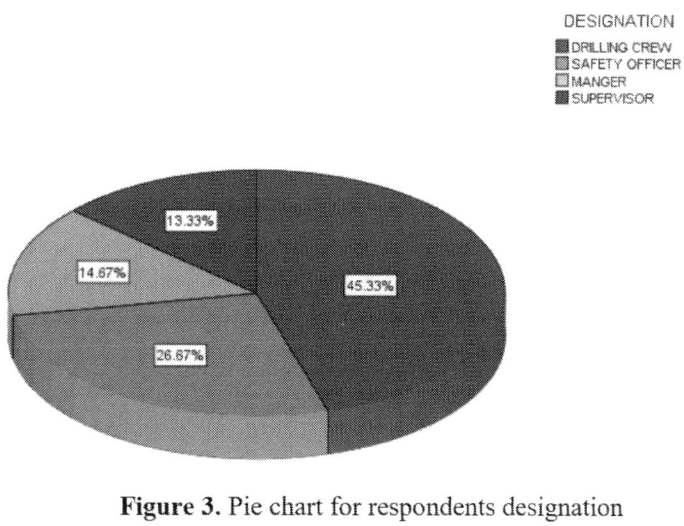

Figure 3. Pie chart for respondents designation

5.3 Respondents based on type of industry

In figure 4, it shows the number of respondents and percentage based on respondent's type of industry. For this preliminary study, 34 respondents are working in offshore oil and gas industry and 41 are working in onshore oil and gas industry which carrying the highest percentage 55%.

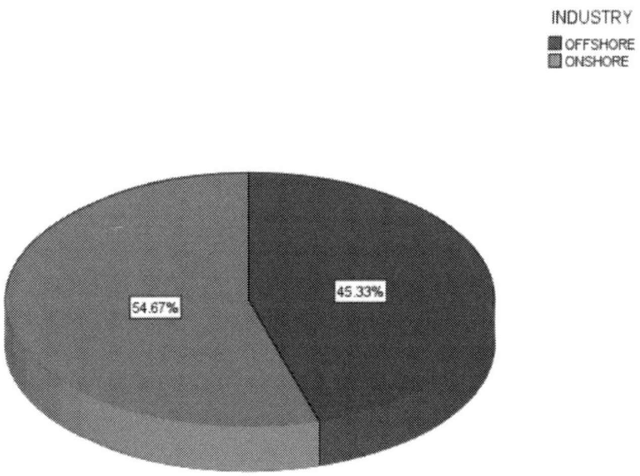

Figure 4. Pie chart for respondents type of industry

5.4 Effective hazard controls for workplace injuries

Health and safety practitioners responded on the query which was about "Does your organization use effective hazard control such as engineering controls (i.e. adjustable Workstations, mechanical lifting devises, ventilation systems, etc.) and/or administrative controls (i.e. job rotation) to prevent or control workplace injuries and illnesses?" 89% of them said that the organization didn't use engineering controls or administrative controls (i.e. job rotation) to prevent or control workplace injuries and illnesses as specified in table 2.

Table 2. First query

Does your organization use effective hazard controls such as engineering controls to prevent or control workplace injuries and illnesses?		
Response	Number of Respondent	Percentage (%)
YES	8	11
NO	67	89
Total	75	100

5.5 Safety posters, health fairs to promote health and safety

As shown in table 3, participants indicated in response of query which was about "Does your organization use safety posters, health fairs, etc. to promote health and safety in drilling?" 65% of them said that their organization didn't use safety posters, health fairs, etc. to promote health and safety in drilling already study the course of digital electronics.

Table 3. Second query

Does your organization use safety posters, health fairs, etc. to promote health and safety in drilling?		
Response	Number of Respondent	Percentage (%)
YES	26	35
NO	49	65
Total	75	100

5.6 Safety officer provide adequate resources for accident prevention

Study participants responded on the query of questionnaire which was about "Did safety officer provide adequate resources for accident prevention during drilling process?" 83% of them said that the safety officer didn't provide adequate resources for accident prevention during drilling process as indicated in table 4.

Table 4. Third query

Did safety officer provide adequate resources for accident prevention during drilling process?		
Response	Number of Respondent	Percentage (%)
YES	13	17
NO	62	83
Total	75	100

5.7 Drilling crew follow established safety rules

As revealed from table 5, health and safety professionals responded on the item which was about "Did drilling crew follow all established safety rules and procedures?" 61% of them said that the drilling crew follows all established safety rules and procedures.

Table 5. Fourth query

Did drilling crew follow all established safety rules and procedures?		
Response	Number of Respondent	Percentage (%)
YES	29	61
NO	46	39
Total	75	100

5.8 Supervisor review and respond promptly to employee safety

Participants responded on the question which was about "Did supervisor review and respond promptly to employee safety suggestions?" 83% of them said that the supervisor didn't review and respond promptly to employee safety suggestions as indicated in table 6.

Table 6. Fifth query

Did supervisor review and respond promptly to employee safety suggestions?		
Response	Number of Respondent	Percentage (%)
YES	13	17
NO	62	83
Total	75	100

5.9 Workplace injuries and illnesses during drilling reported

As shown in table 7, Health and safety practitioners responded on twenty first query of questionnaire which was about "Are all workplace injuries and illnesses during drilling reported and investigated?" 72% of the respondents said that all workplace injuries and illnesses during drilling didn't report and investigated.

Table 7. Sixth query

Are all workplace injuries and illnesses during drilling reported and investigated?		
Response	Number of Respondent	Percentage (%)
YES	21	28
NO	54	72
Total	75	100

5.10 Supervisors and managers been trained in accident investigation

Participants responded on the item which was about "Have all supervisors and managers been trained in accident investigation procedures specifically for drilling operation?" 91% of them said that supervisors and managers didn't train in accident investigation procedures specifically for drilling operation as specified in table 8.

Table 8. Seventh query

Have all supervisors and managers been trained in accident investigation procedures specifically for drilling operation?		
Response	Number of Respondent	Percentage (%)
YES	7	9
NO	68	91
Total	75	100

5.11 Supervisors required to complete a separate accident investigation report
As shown in table 9, health and safety practitioners responded on the query of questionnaire which was about "Are supervisors required to complete a separate accident investigation report?" 92% of them said that the supervisors didn't complete a separate accident investigation report

Table 9. Eighth query

Are supervisors required to complete a separate accident investigation report?		
Response	Number of Respondent	Percentage (%)
YES	6	08
NO	69	92
Total	75	100

5.12 Safety coordinator promptly notified when a serious accident occurs
On the question which was about "Did Is the safety coordinator promptly notified when a serious accident occurs?"84% of respondents said that the safety coordinator didn't promptly notify when a serious accident occurs, as shown in table 10.

Table 10. Ninth query

Did Is the safety coordinator promptly notified when a serious accident occurs?		
Response	Number of Respondent	Percentage (%)
YES	12	16
NO	63	84
Total	75	100

5.13 Organization conduct exposure monitoring
Participants responded on the query of questionnaire which was about "Does your organization conduct exposure monitoring when required by a particular OSHA Standard or when requested by employees?" 56% of them said that the organization conduct exposure monitoring when required by a particular OSHA Standard or when requested by employees as shown in table 11.

Table 11. Tenth query

Does your organization conduct exposure monitoring when required by a particular OSHA Standard or when requested by employees?		
Response	Number of Respondent	Percentage (%)
YES	42	56
NO	33	44
Total	75	100

IOP Conf. Series: Journal of Physics: Conf. Series **1049** (2018) 012022 doi:10.1088/1742-6596/1049/1/012022

5.14 Accident prevention programs

Health and safety practitioners responded on the question which was about "Are accident prevention programs and activities based upon an analysis of accident reports and injury and illness rates?" 60% of them said that the accident prevention programs and activities didn't base upon an analysis of accident reports and injury and illness rates as shown in table 12.

Table 12. Eleventh query

Are accident prevention programs and activities based upon an analysis of accident reports and injury and illness rates?		
Response	Number of Respondent	Percentage (%)
YES	30	40
NO	45	60
Total	75	100

5.15 Company use decision support systems for decision making

On the query of questionnaire which was about "Does your company use decision support system for decision making to reduce work place risk and hazards during drilling process?" 61% of participants said that their company didn't use decision support systems for decision making to reduce work place risk and hazards during drilling process, as indicated in table 13.

Table 13. Twelfth query

Does your company use decision support systems for decision making to reduce work place risk and hazards during drilling process?		
Response	Number of Respondent	Percentage (%)
YES	29	39
NO	46	61
Total	75	100

5.16 Company provide electronic application for accident prevention to employees

Health and safety professionals responded on the item which was about "Does company provide any electronic application for accident prevention to employees?" 81% of them said that they the company didn't provide any electronic application for accident prevention to employees, as shown in table 14.

Table 14. Thirteenth query

Does company provide any electronics application for accident prevention to employees?		
Response	Number of Respondent	Percentage (%)
YES	14	19
NO	61	81
Total	75	100

IOP Conf. Series: Journal of Physics: Conf. Series **1049** (2018) 012022 doi:10.1088/1742-6596/1049/1/012022

6. Conclusions

The overall findings of this study indicates that the decision making potential during hazardous condition through accident prevention resources in oil and gas drilling process indicated that there have sheer needs of electronic applications and decision support systems based on effective hazard controlling measure as per health and safety international rules and regulations for safe drilling operation. According to the quantitative findings as shown in figure 5, it has been investigated that the drilling crew didn't follow the safety instructions and precautions properly during oil and gas well drilling at both offshore and onshore oil and gas operations which cause them sever injuries and sometime death. Similarly, the findings also revealed that the most of organizations not provide adequate safety awareness materials and effective investigations training programs to the drilling crew and safety professionals.

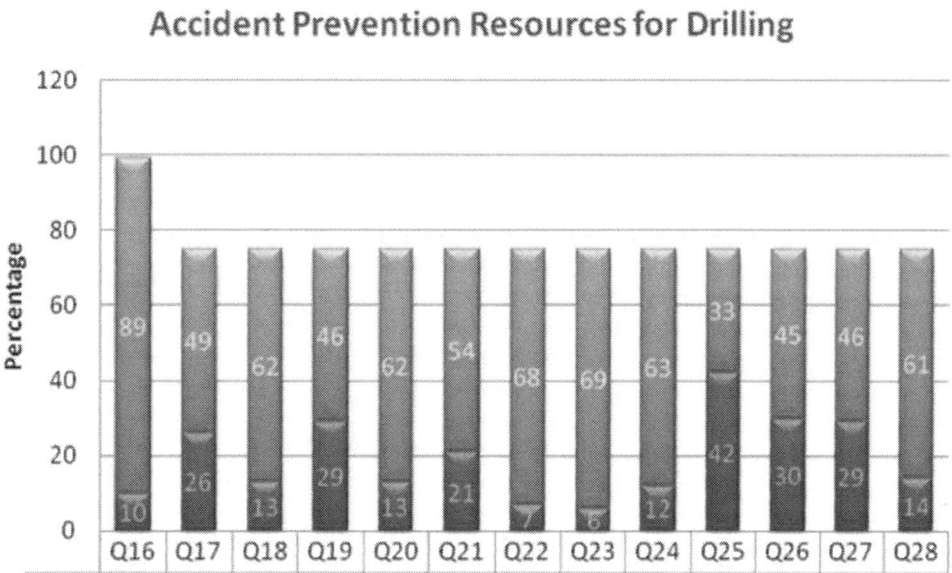

	Q16	Q17	Q18	Q19	Q20	Q21	Q22	Q23	Q24	Q25	Q26	Q27	Q28
NO	89	49	62	46	62	54	68	69	63	33	45	46	61
YES	10	26	13	29	13	21	7	6	12	42	30	29	14

Figure 5. Cumulative findings

References

[1] Wawuda S M and Mungai F. 2016 Factors Affecting Distribution of Oil Products in Kenya: A Case Study of Kenya Pipeline *International Journal of Supply Chain Management.* 22;1(1) 34 - 48

[2] Kilian L and Murphy D P 2014 The role of inventories and speculative trading in the global market for crude oil *Journal of Applied Econometrics* 1;29(3) 454-78

[3] Bamberger M and Oswald R E 2012 Impacts of gas drilling on human and animal health *New solutions: a journal of environmental and occupational health policy* 22(1) 51-77

[4] Wang Q, Chen X, Jha A N and Rogers H 2014 Natural gas from shale formation–the evolution, evidences and challenges of shale gas revolution in United States *Renewable and Sustainable Energy Reviews* 28;30 1-28

[5] Witter R Z, Tenney L, Clark S and Newman L S 2014 Occupational exposures in the oil and gas extraction industry: State of the science and research recommendations *American journal of industrial medicine* 1;57(7) 847-56

[6] Russo A 2015 The Importance of Continuous Improvement in Occupational Health and Safety Management and Regulation in the Oil and Gas Industry *Franklin Business & Law Journal* 1;2015(3)

[7] Utvik T I, Jahre-Nilsen C 2016 *The Importance of Early Identification of Safety and Sustainability Related Risks in Arctic Oil and Gas Operations* InSPE International Conference and Exhibition on Health, Safety, Security, Environment, and Social Responsibility Society of Petroleum Engineers

[8] Asad M M 2014 A Systematic Review: Development Techniques and Utilization of Expert Systems Inferences for Health and Safety Environment in Oil & Gas and Petroleum Industries In Malaysia University Conference Engineering Technology

[9] Rathnayaka S, Khan F and Amayotte P 2013 Accident modeling and risk assessment framework for safety critical decision-making: application to deepwater drilling operation Proceedings of the Institution of Mechanical Engineers, Part O: Journal of risk and reliability 227(1) 86-105.

[10] Adgate J L, Goldstein B D and McKenzie L M 2014 Potential public health hazards, exposures and health effects from unconventional natural gas development *Environmental science & technology* 24;48(15) 8307-20.

[11] Asia Marketing Research, Internet Usage, Population Statistics, & Facebook Information 2012 Internet World Stats: Usage and Population Statistics Retrieved at http://www.internetworldstats.com/asia.html

Acknowledgement
This paper was partly sponsored by the Centre of Graduate Studies (CGS) of Universiti Tun Hussein Onn Malaysia (UTHM).

IOP Conf. Series: Journal of Physics: Conf. Series **1049** (2018) 012023 doi:10.1088/1742-6596/1049/1/012023

Work-related Musculoskeletal Disorder (WRMD) among Production Operators: Studies of Differences in Age and Gender

Nor Hazana Abdullah[1], Nor Aziati Abdul Hamid, Eta Wahab, Alina Shamsuddin, Rosli Asmawi[2]
Faculty of Technology Management and Business[1]
Faculty of Mechanical and Manufacturing Engineering[2]
Universiti Tun Hussein Onn Malaysia, Johor, Malaysia

E-mail: hazana@uthm.edu.my

Abstract. Prevalence of musculoskeletal pain is a major concern among manufacturing-based companies as they are still very reliant on manual labour. A better understanding whether musculoskeletal pain differ between gender and would enable organizational intervention to improve employees' health and safe working conditions. Therefore, the objective of this study is to determine the current prevalence of musculoskeletal pain among production workers and differences of musculoskeletal pain between gender and across age groups. A cross-sectional survey involving 324 shop-floor manufacturing workers was done using Cornell Musculoskeletal Discomfort Questionnaires (CMDQ). No significance differences were observed between female and male employees in terms of right hand, left hand, lower extremity, neck and lower back. However, there is a significance difference of pain at the upper extremity with female operators experienced more pain compared to their male counterparts. No significance differences were found among age categories for lower back, right hand, left hand, upper and lower extremity. However, neck pain and left hand was significantly different. The age effects indicate non linear association.

1. Introduction

Although manufacturing sector in Malaysia experienced moderate growth over the recent years [1], its contribution to Gross Domestic Product, external trade and job creation is still significant. To sustain this sector contribution to the economy, managing employees' issues is paramount especially in terms of their health and well-being. However, work-related musculoskeletal disease (WRMD) cases are on the rise in Malaysia from only five cases in 1995 to more than 600 cases in 2014 [2][3][4]. WRMD is a costly health problems as it increases the labour cost and constitutes loss of work hours. WRMD include injuries and illnesses that affect the human body's movement such as muscles, tendons, ligaments, nerves, discs, blood vessels, and other soft tissues. It is caused by excessive force, repetition of movement or simply poor posture while working, poor work practices, poor fitness and poor health. Tan Sri Lee Lam further added that industrialization, aging workforce and lack of safety standards and regulations at workplace are major contributors of this escalating statistics of WRMD [4].

[1] hazana@uthm.edu.my

Content from this work may be used under the terms of the Creative Commons Attribution 3.0 licence. Any further distribution of this work must maintain attribution to the author(s) and the title of the work, journal citation and DOI.
Published under licence by IOP Publishing Ltd

Production operators are usually required to perform work-related tasks that are categorized as manual labour such as handling machines, assembling parts or products, quality assurance activities such as testing and repairing and prepare simple reports of their activities. This implies the importance of production ergonomics for the safety and health of the production operators. Empirical evidence has shown that proper workplace ergonomics could improve motivation, satisfaction, performance and productivity [5][6]. Despite availability of production ergonomics guidelines and best practices, majority of companies have not fully implemented them since there is no legislative guideline governing ergonomics production in Malaysia [7].

Consequently, determining the effects of individual factors such as age and gender are vital to the development of effective and efficient ergonomics programs and interventions. The health of ageing employees and those of women has significant economic implication on the nation [8]. Though there has been numerous empirical works to understand the antecedents and consequences of musculoskeletal discomforts, the impact of gender and age remains inconclusive. Furthermore, majority of these studies have been focusing on Video Display Terminal (VDU) operator [9], [10] and healthcare community [11], [12]. Thus, the purpose of this study is to compare prevalence of Work-related Musculoskeletal Disorder (WRMD) between age and gender groups among production operators.

2. Age and Gender Effects

According to Armstrong's model [13] age and gender potentially reduce an individual's ability to cope with musculoskeletal stress especially female. Furthermore, Karlqvist [14], found that female employees tend to have different working posture compared to their male counterparts that perform similar tasks. Specifically, female employees had higher shoulder elevation and rotation when working with computer mouse. Prolonged exposure to musculoskeletal stress especially among older employees would increase WRMD prevalence.

Evidence on the effects of age and gender on Work-related Musculoskeletal Disorder (WRMD) has been inconsistent. Andersson *et al.*[15] surveyed two general Swedish populations primary healthcare and found that women experienced more pain at various body parts compared to men. Prevalence of pain increased by age up to 50-59 years for both genders and then slowly decreased. Similarly, Ribeiro [16] found that gender, age, BMI , consistent exercise regime and other types of illnesses were significantly associated with musculoskeletal disorders among nurses. Dianat et al [17] to certain extent replicated this study among Iranian sewing machine operators and confirms that gender, age, BMI , and consistent exercise regime are predictive of musculoskeletal symptoms. Neupane [18] revealed similar finding among healthcare personnel where older and female healthcare personnel were more likely to report pains a various places. Weaver et al. [19] sampled university employees who worked with computers and found that young respondents experienced more wrist pain while arm and shoulder pain are different between male and female. In two population studies using secondary data, [20], [21], gender differs significantly in terms of the musculoskeletal health especially in the form of carpal tunnel syndrome.

However, Taib et al [22] surveyed 85 dentists in Malaysia and found that both age and gender are not significant predictors of musculoskeletal disorder. Rocha et al [23] study concurred as no significant differences of gender and age were observed among system analysts in Sao Paolo. This study used mixed methods of data collection which include work ergonomics assessment, individual and group semi-structured interviews and self-applied questionnaires.

Maakip et al [24] found age as significant predictor of musculoskeletal discomfort among Malaysian officer workers but not age. Likewise, in terms of mixed findings, Seppala [25] and Demure e al [26] found only gender explained variance in musculoskeletal discomforts but not age.

3. Methods

Four hundred twenty productions operators working in eleven medium-sized manufacturing companies in Johor were surveyed. Participation was voluntary and no personal data was recorded to maintain confidentiality and privacy. Those who refused certain questions were excluded from the analysis to yield a return rate of 92% (Three hundred and Twenty Four). The data were collected cross sectionally in November 2016.

3.1. Participants

Male respondents were slightly higher at 56.2% compared to female (43.8%) with the majority of respondents aged between 23 to 27 years old (31.2%) followed by those aged 28-32 years old (21.9%)., 33-37 years old (14.2%), 18-22 years old (13%), above 42 years olf (10.5%) and 38-42 years old (9.3%). It could be concluded that more than 90% of the respondents aged below 42 years old. Majority of respondents (79.6%) had worked more than 2 years. In terms of types of production tasks, 45.4% worked in packaging, 28.4% worked in quality assurance while 13% handled machines. Majority of employees were Malaysian (67.9%) while 32.1% were Non Malaysian. In terms of years of service, majority (37.6%) had worked more than four years, followed by 24.4% had worked 2 years, 17.6% had worked 3 years, 12% had worked less than a year while 8.3% reported to had worked for 1 year.

3.2. Measures

Cornell Musculoskeletal Discomfort Questionnaires [27] was used to measure the musculoskeletal pain . The questionnaires include hand assessment and total body assessment. There were a total of 54 questions were asked regarding prevalence of musculoskeletal pain, ache or discomforts in 20 body parts during the previous week. The version of female/male standing CMDQ was used in this study, where all three indicators (frequency, discomfort and interference) were averaged to yield the cumulative scores of WRMD.

3.3. Analysis

Descriptive and non-parametric tests were used to analyse the data via the use of IBM SPSS Version 22. Descriptive analysis includes the tabulation of frequency, mean and SD. Hypotheses about equality between averages of two/more quantitative variables were examined using Mann-Whitney and Kruskal-Wallis tests. Values of p less than 0.05 were considered as statistically significant.

4. Results

Figure 1 shows the mean and standard deviation (SD) of musculoskeletal pain in terms of the frequency in the last work week. As evident from the Figure 1, neck had the highest frequency and mean of discomfort, followed by lower back, lower extremity, and upper extremity. Right hand was observed to have higher frequency of discomforts compared to left hand which may imply that majority of respondent are right-handed. The prevalence areas of discomforts are consistent with previous studies such as [28]–[30].

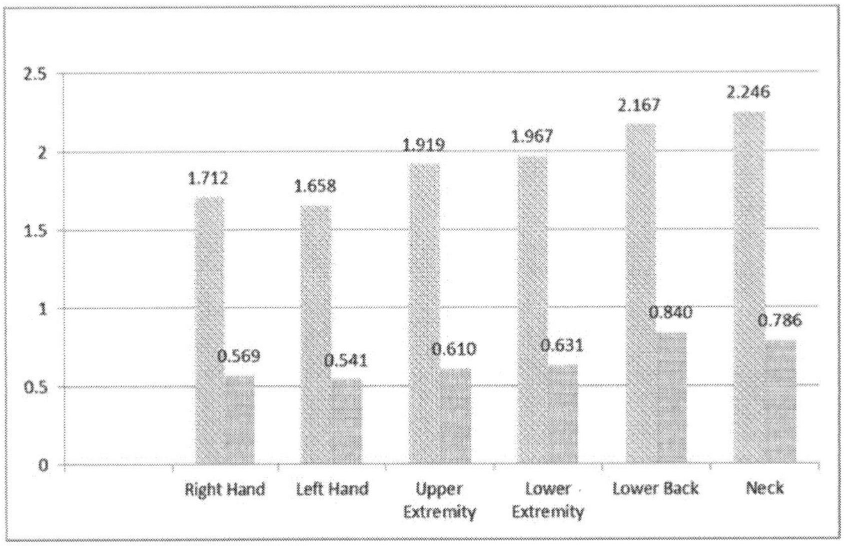

Figure 1: Mean and SD of WRDS

Table 1 shows the cross tabulation table between WRMD reported at various body parts between male and female. Male employees reported more pains in both hands (55.7% and 54.3% repectively), lower back (54.5%) and neck (53.2%). More female employees reported pains at upper extremity and and lower extremity (54.3% and 56.6%) respectively.

Table 2 shows the cross tabulation of WRMD across agre groups. It is evident that across the groups, the age category of 23-27 years old consistently had higher percentages of reported pains compared to other age categories. This indicates that pains reported is not directly related with increased age, contrary to the present understanding.

In order to test whether the descriptive differences observed from the cross tabulation tables were statistically significant, Man Whitney test were performed to find the differences between male and female while Kruskal Wallis Test was used to compare differences among age categories. Table 3 shows that only upper extremity differed between male and female operators with female reported WRMD. Table 4 shows the comparison across age categories. Only WRMD at neck and left hand were statistically different among age categories. Specifically, age groups of 38-42 years old reported higher neck pain compared to 33-37 years old; more than 42 years old; and 18-22 years old. However, 23 to 27 years old reported higher neck pain compared to 38-42 years old. This is interesting finding since the results show non-linear association between age and WRMD. In terms of left hand, age group 23-27 years old differed significantly from age group more than 42 years old.

Table 1: Cross Tabulation of WRDS and Gender

Body Parts		Gender		Total	Mean	SD
		Male	Female			
Right Hand	<= 1.44	61	46	107	1.712	0.569
		57.0%	43.0%	100.0%		
	1.45 - 1.78	62	49	111		
		55.9%	44.1%	100.0%		
	1.79+	59	47	106		
		55.7%	44.3%	100.0%		
Left Hand	<= 1.33	58	37	95	1.658	0.541
		61.1%	38.9%	100.0%		
	1.34 - 1.78	69	58	127		
		54.3%	45.7%	100.0%		
	1.79+	55	47	102		
		53.9%	46.1%	100.0%		
Upper Extremity	<= 1.59	60	41	101	1.919	0.610
		59.4%	40.6%	100.0%		
	1.60 - 2.04	76	41	117		
		65.0%	35.0%	100.0%		
	2.05+	46	60	106		
		43.4%	56.6%	100.0%		
Lower Extremity	<= 1.67	69	45	114	1.967	0.631
		60.5%	39.5%	100.0%		
	1.68 - 2.19	57	38	95		
		60.0%	40.0%	100.0%		
	2.20+	56	59	115		
		48.7%	51.3%	100.0%		
Lower Back	<= 1.67	67	45	112	2.167	0.840
		59.8%	40.2%	100.0%		
	1.68 - 2.33	36	31	67		
		53.7%	46.3%	100.0%		
	2.34+	79	66	145		
		54.5%	45.5%	100.0%		
Neck	<= 2.00	99	69	168	2.246	0.786
		58.9%	41.1%	100.0%		
	2.34+	83	73	156		
		53.2%	46.8%	100.0%		

Table 2: Cross Tabulation of WRDS and Age Categories

		Age					
		18-22 years old	23-27 years old	28-32 years old	33-37 years old	38-42 years old	Above 42 years old
Right Hand	<= 1.44	12	24	24	20	9	18
		11.2%	22.4%	22.4%	18.7%	8.4%	16.8%
	1.45 - 1.78	13	41	20	13	13	11
		11.7%	36.9%	18.0%	11.7%	11.7%	9.9%
	1.79+	17	36	27	13	8	5
		16.0%	34.0%	25.5%	12.3%	7.5%	4.7%
	Total	42	101	71	46	30	34
		13.0%	31.2%	21.9%	14.2%	9.3%	10.5%
Left Hand	<= 1.33	13	19	23	20	7	13
		13.7%	20.0%	24.2%	21.1%	7.4%	13.7%
	1.34 - 1.78	19	45	21	13	14	15
		15.0%	35.4%	16.5%	10.2%	11.0%	11.8%
	1.79+	10	37	27	13	9	6
		9.8%	36.3%	26.5%	12.7%	8.8%	5.9%
	Total	42	101	71	46	30	34
		13.0%	31.2%	21.9%	14.2%	9.3%	10.5%
Upper Extremity	<= 1.59	12	30	19	17	7	16
		11.9%	29.7%	18.8%	16.8%	6.9%	15.8%
	1.60 - 2.04	16	41	28	13	8	11
		13.7%	35.0%	23.9%	11.1%	6.8%	9.4%
	2.05+	14	30	24	16	15	7
		13.2%	28.3%	22.6%	15.1%	14.2%	6.6%
	Total	42	101	71	46	30	34
		13.0%	31.2%	21.9%	14.2%	9.3%	10.5%
Lower Extremity	<= 1.67	13	30	20	19	14	18
		11.4%	26.3%	17.5%	16.7%	12.3%	15.8%
	1.68 - 2.19	9	33	23	13	10	7
		9.5%	34.7%	24.2%	13.7%	10.5%	7.4%
	2.20+	20	38	28	14	6	9
		17.4%	33.0%	24.3%	12.2%	5.2%	7.8%
	Total	42	101	71	46	30	34
		13.0%	31.2%	21.9%	14.2%	9.3%	10.5%
Lower Back	<= 1.67	19	36	26	14	7	10
		17.0%	32.1%	23.2%	12.5%	6.3%	8.9%
	1.68 - 2.33	5	22	10	9	8	13
		7.5%	32.8%	14.9%	13.4%	11.9%	19.4%
	2.34+	18	43	35	23	15	11
		12.4%	29.7%	24.1%	15.9%	10.3%	7.6%
	Total	42	101	71	46	30	34
		13.0%	31.2%	21.9%	14.2%	9.3%	10.5%
Neck	<= 2.00	26	55	31	26	8	22
		15.5%	32.7%	18.5%	15.5%	4.8%	13.1%
	2.34+	16	46	40	20	22	12
		10.3%	29.5%	25.6%	12.8%	14.1%	7.7%
	Total	42	101	71	46	30	34
		13.0%	31.2%	21.9%	14.2%	9.3%	10.5%

Table 3: Mann Whitney Results

		Right Hand	Left Hand	Upper Extremity	Lower Extremity	Lower Back	Neck
Group	*n*						
Male	182	159.912	159.115	153.074	154.552	156.904	156.904
Female	142	165.817	166.838	174.581	172.687	169.673	169.673
U							
p-value		-.564	-.738	-2.051	-1.730	-1.490	-1.231
		.573	.461	0.04*	.084	.136	.218

*$p<0.05$

Table 4: Kruskal Wallis Results

Age Categories	*n*	Right Hand	Left Hand	Upper Exremity	Lower Extremity	Lower Back	Neck
18-22 years old	42	175.77	151.90	166.02	179.65	150.94	146.21
23-27 years old	101	175.77	180.06	160.74	170.47	159.23	158.28
28-32 years old	71	167.42	166.58	168.52	173.97	165.48	175.77
33-37 years old	46	146.27	143.10	158.50	150.79	171.76	154.93
38-42 years old	30	159.23	167.45	190.32	134.23	178.12	203.30
Above 42 years old	34	121.25	136.76	131.66	134.47	153.97	141.68
H		12.463	10.014	7.664	10.939	2.790	14.112
p-value		0.029*	.075	.176	.053	.732	0.014*

*$p<0.05$

It could be concluded that the age and gender effects differed based on localization of pains at various parts of the body.

5. Discussion and conclusions

The results clearly indicate neck had higher prevalence compared to other parts of the body. Right hand had higher prevalence than left hand. The prevalence of MSDs at these areas might indicate common work postures assumed by these operators. Production operators experienced musculoskeletal discomforts at neck, lower back and upper extremity which is consistent with the nature of the production activities which are still largely manual and repetitive. This results are consistent with findings of Raemy [31] who found that 90.8% of the industrial workers had their back being bent forward and 80.5% lifted heavy load ranging from 1 to 5 kg. Similar phenomenon was observed by Widia [32] where manual material handling were found to be significantly associated with musculoskeletal discomforts. Da Costa and Vieira [33] claimed that heavy physical work, awkward postures, lifting and whole body vibration lower legs discomforts might imply excessive standing of prolonged hours. Interestingly, neck pain is also a common discomfort in this study as being reported in studies of other types of occupations [36].

In conclusions, the effects of gender and age on WRMD were mixed based on localization of the pains. Further exploration is required to understand the nature of these linkages. However, it is worth to highlight that the effect of age is non linear which need to be scrutinized further.

References

[1] "Economic Report 2016/2017: Perform."

[2] Boon P 2013 Musculoskeletal disease increase cause for concern – BorneoPost Online | Borneo ,Malaysia, Sarawak Daily News | Largest English Daily In Borneo *Borneo Post Online* [Online] Available: http://www.theborneopost.com/2013/01/28/musculoskeletal-disease-increase-cause-for-concern/ [Accessed: 19-Jun-2017]

[3] DOSH 2017 Statistik Penyakit Muskuloskeletal Pekerjaan 1995-2009 [Online]. Available: http://www.dosh.gov.my/index.php/en/list-of-documents/statistics-socso/330-occupational-musculoskeletal-diseases-statistics-1995-2009/file [Accessed: 19-Jun-2017]

[4] Thye L L 2016 Work stress and mental health | New Straits Times | Malaysia General Business Sports and Lifestyle News *NST Online* [Online] Available https://www.nst.com.my/news/2016/04/142074/work-stress-and-mental-health. [Accessed: 20-Jun-2017]

[5] Okan D and Leblebici U 2012 Impact Of Workplace Quality On Employee's Productivity: Case Study Of A Bank In Turkey Demet Leblebici *J. Business, Econ. Financ.* **1** 1 38–49

[6] Ikonne C N and Onuoha U D 2015 Factors Influencing Job Satisfaction of Librarian in Federal and State University Libraries in Southern Nigeria *Open Access Libr. J.* **2** 2 1–9

[7] Loo H S and Richardson S 2012 Ergonomics Issues in Malaysia *J. Soc. Sci.* **8** 1 61–65

[8] World Population Ageing 2015

[9] Helland M, Horgen G Kvikstad T M, Garthus T, Richard Bruenech J and Aarås A 2008 Musculoskeletal, visual and psychosocial stress in VDU operators after moving to an ergonomically designed office landscape *Appl. Ergon.* **39** 3

[10] Helland M, Horgen G, Kvikstad T, Aarås A and Garthus T 2008 Will musculoskeletal, visual and psychosocial stress change for visual display unit (vdu) operators when moving from a single-occupancy office to an office landscape? *Int. J. Occup. Saf. Ergon.* **14** 3

[11] Palliser C R, Firth H M, Feyer A M and Paulin S M 2005 Musculoskeletal discomfort and work-related stress in New Zealand dentists *Work Stress* **19** 4

[12] Abdul Rahman H, Abdul-Mumin K and Naing L 2017 Psychosocial Work Stressors, Work Fatigue, and Musculoskeletal Disorders: Comparison between Emergency and Critical Care Nurses in Brunei Public Hospitals *Asian Nurs. Res. (Korean. Soc. Nurs. Sci).* **11** 1

[13] Armstrong T J 1993 A conceptual model for work-related neck and upper-limb musculoskeletal disorders *Scand. J. Work. Environ. Health* **19** 2 73–84

[14] Karlqvist L, Tornqvist E W, Hagberg M, Hagman M and Toomingas A 2002 Self-reported working conditions of VDU operators and associations with musculoskeletal symptoms: a cross-sectional study focussing on gender differences *Int. J. Ind. Ergon.* **30** 277–294

[15] Andersson B H B, Ingemar H M, Ejlertsson D, Göran Ph D, Leden, Ido M D, Rosenberg and Claes M D 1993 Chronic Pain in a Geographically Defined General Population:...: The

Clinical Journal of Pain *Clin. J. Pain*

[16] Ribeiro T, Serranheira F and Loureiro H 2017 Work related musculoskeletal disorders in primary health care nurses

[17] Dianat I, Kord M, Yahyazade P, Karimi M A and Stedmon A W 2015 Association of individual and work-related risk factors with musculoskeletal symptoms among Iranian sewing machine operators *Appl. Ergon.* **51** 180–188

[18] Neupane S, Nygård C H and Oakman J 2016 Work-related determinants of multi-site musculoskeletal pain among employees in the health care sector *Work* 54 3 689–697

[19] Weaver S J, Charness N, Østensvik T and Jastrzembski T 2008 Musculoskeletal pain prevalence in randomly sampled university employees: Age and gender effects *Gerontechnology* **7** 3 279–292

[20] Tanaka S, Petersen M and Cameron L 2001 Prevalence and risk factors of tendinitis and related disorders of the distal upper extremity among U.S. workers: Comparison to carpal tunnel syndrome *Am. J. Ind. Med.* **39** 3 328–335

[21] Brismée J M Differences in musculoskeletal health due to gender in a rural multiethnic cohort: a Project FRONTIER study

[22] Taib M F M, Bahn S, Yun M H and Taib M S M 2017 The effects of physical and psychosocial factors and ergonomic conditions on the prevalence of musculoskeletal disorders among dentists in Malaysia*Work* **57** 2 297–308

[23] Rocha L E and Debert-Ribeiro M 2001 Work and health: A gender study on systems analysts *Rev. Saude Publica* **35** 6

[24] Maakip I, Oakman J and Stuckey R 2016 Gender, Cultural Influences, and Coping with Musculoskeletal Pain at Work: The Experience of Malaysian Female Office Workers *J. Occup. Rehabil* 1–11

[25] Seppälä P 2001 Experience of stress, musculoskeletal discomfort, and eyestrain in computer-based office work: A study in municipal workplaces *Int. J. Hum. Comput. Interact.* **13** 3

[26] Demure B, Luippold R S, Bigelow C, Ali D, Mundt K A and Liese B 2000 Video display terminal workstation improvement program: I. Baseline associations between musculoskeletal discomfort and ergonomic features of workstations," *J. Occup. Environ. Med.* **42** 8

[27] Vahdatpour B, Bozorgi M and Taheri M R 2015 Investigating musculoskeletal discomforts and its relation to workplace ergonomic conditions among computer office workers at alzahra hospital, Isfahan, Iran," *J. Isfahan Med. Sch.* **33** 346

[28] Azma K, Hosseini A, Safarian M H and Abedi M 2015 Evaluation of the relationship between musculoskeletal discomforts and occupational stressors among nurses *N. Am. J. Med. Sci.* **7** 7

[29] Basahel A M 2015 Investigation of Work-related Musculoskeletal Disorders (MSDs) in Warehouse Workers in Saudi Arabia *Procedia Manuf.* **3**

[30] Pék E, Bánfai B, Deutsch K, Jeges S and Betlehem J 2014 The prevalence of musculoskeletal symptoms among ambulance personnel in a national subsample | Váz-, izomrendszeri megbetegedések prevalenciája mentodolgozók körében *Lege Artis Med.* **24** 10–11

[31] Zein R. M, Halim I, Azis N A, Saptari A and Kamat S R 2015 A Survey on Working Postures among Malaysian Industrial Workers *Procedia Manuf.* **2** 450–459

[32] Widia M and Zawiah Md Dawal S 2016 The Relation Of Risk Factors And Musculoskeletal Discomfort Among Manual Material Handling Workers In Malaysian Automotive Industries *Malaysian J. Public Heal. Med.* **16** 124–133

[33] Da Costa B R and Vieira E R 2010 Risk Factors for Work - Related Musculoskeletal Disorders : A Systematic Review of Recent Longitudinal Studies *Am. J. Ind. Med. J . Ind . Med.* **53** 53 85–323

IOP Conf. Series: Journal of Physics: Conf. Series **1049** (2018) 012024 doi:10.1088/1742-6596/1049/1/012024

A Systematic Review of Patient Satisfaction on Health Information Exchange in Malaysian Public Healthcare Organizations

Nurul Izzatty Ismail[1] , **Nor Hazana Abdullah**[2]
Faculty of Technology Management and Business,
University of Tun Hussein Onn, 86400 Malaysia.

E-mail: izzatty@uthm.edu.my

Abstract. Patient satisfaction is imperative in every healthcare organization. Service quality is the utmost important aspect that yields patient satisfaction in hospitals. Since Information Technology (IT) is widely used in hospitals, its applications can give rise to different issues related to patient satisfaction. This is because patient satisfaction derived from utilizing health information exchange systems (HIEs) depends on the service quality of the systems. This paper presents a systematic review of the impacts of service quality of Health Information Exchanges (HIEs) on patient satisfaction. Five databases, namely Scopus, IEEE, Proquest, Emerald and Science Direct were accessed to search for relevant studies. In total, 714 studies met the inclusion criteria. The result shows that five factors at the healthcare organizations that adopted HIEs have impacts on patient satisfaction: Tangibility, Reliability, Assistance, Safety and Empathy.

1. Introduction

Healthcare organizations are of paramount importance in providing welfare to the public. The services of healthcare organizations are expected to improve over time. In Malaysia, the majority of the public depend on the public healthcare organizations for treatments managed by the Malaysian government [1]. This is because, medical treatments and consultations at the public healthcare organizations are inexpensive compared with those of the private healthcare organizations.

The statistics show that 2,465,727 patients were admitted to the public hospitals, compared with 1,064,718 patients admitted to the private hospitals [2]. Besides that, 20,260,479 outpatients were given medical treatments and consultations at the public hospitals, compared with 3,932,361 outpatients at the private hospitals [2].

---

[1] izzatty@uthm.edu.my

Content from this work may be used under the terms of the Creative Commons Attribution 3.0 licence. Any further distribution of this work must maintain attribution to the author(s) and the title of the work, journal citation and DOI.
Published under licence by IOP Publishing Ltd

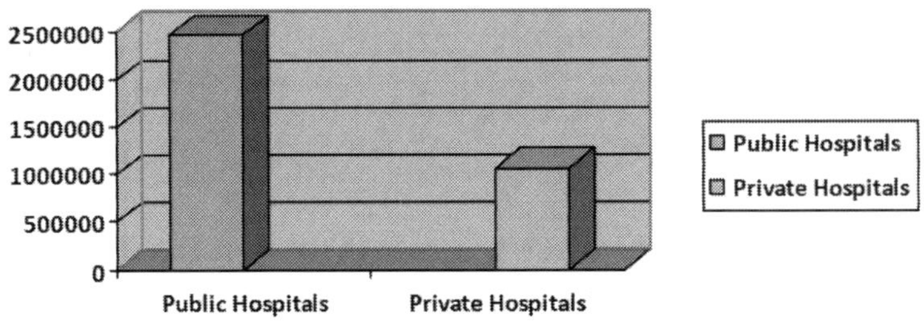

Figure 1. Numbers of public hospitals and private hospitals

Figure 1 shows the statistics of inpatients and outpatients at both the public and private hospitals. These statistics show that the public healthcare organizations play a vital role in Malaysia in providing medical treatments and consultations for the large majority of the population.

The public healthcare organizations are established by the government in every state and in almost every town of the country. In the year 2017, there were 139 public hospitals compared with 95 private hospitals located all over Malaysia [3]. Besides that, there are 2,839 public clinics compared with 1,064 private clinics in this country [4]. Figure 2 shows the statistics of both the public hospitals and public clinics in Malaysia, side by side with those operated by the private sector. It is obvious that there are a large number of public healthcare organizations in Malaysia.

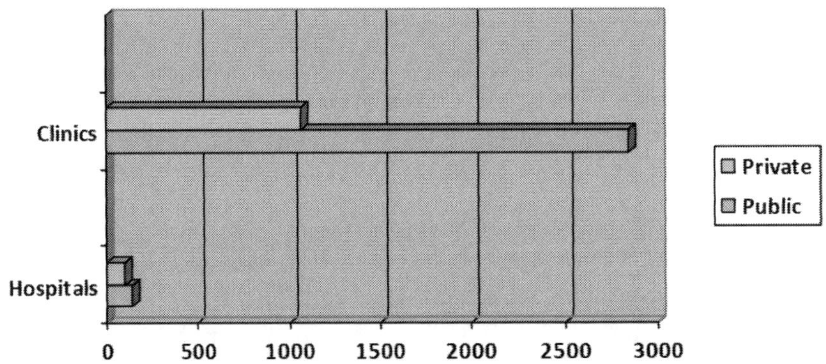

Figure 2. Numbers of public hospitals and private hospitals

Since the number of public healthcare organizations and number of patients are large, there are several issues and challenges encountered in this sector. For example, the services are inefficient in the public organizations; the patients need to wait for a longer time to obtain their medical treatments and consultations [5]. Besides that, there are several cases of professional negligence in the public healthcare system [6]. This might be caused by the manpower shortage in the public healthcare organizations [7]. These situations may affect the patient satisfaction with the public healthcare organization services.

The applications of Information Communication Technology (ICT) help to overcome the issues and challenges faced by the public healthcare organizations [8]. Thus, the Malaysian government emphasizes the importance of implementing ICT in the Malaysian healthcare organizations [9]. In Malaysia, many electronic health systems are implemented or adopted in the public healthcare organizations; for example, Hospital Information System (HIS), Malaysia Health Information

Exchange (MyHiX), Tele-Primary Care and Oral Health Clinical Information System (TPC-OHCIS), and Malaysian Health Data Warehouse (MyHDW) [10-13]. These applications may help resolve the issues and overcome the challenges faced by the public healthcare organizations, which in turn will increase the patient satisfaction level [14]. Service quality is seen as an important indicator of patient satisfaction, and hence warrants a detailed study.

2. Health Information Exchange (HIE) in Malaysia

A Health Information Exchange (HIE) may be roughly defined as a system that allows the exchange of patients' medical records between public hospitals or clinics via online network in a smooth and secure manner [15]. In Malaysia, the HIE is known as Malaysia Health Information Exchange (MyHiX). At present, there are eight public healthcare organizations utilizing the MyHiX: Hospital Putrajaya, Putrajaya, Hospital Tuanku Jaafar, Negeri Sembilan, Hospital Port Dickson, Negeri Sembilan, Hospital Bentong, Hospital Sultanah Nur Zahirah, Hospital Raja Perempuan Zainab II, Institut Kanser Negara, and Klinik Kesihatan Presint 9 Putrajaya [16]. Hence, patients' medical records can be exchanged between these public organizations.

The MyHiX toolkit, which contains the overall depiction of MyHiX was developed by MOH [17]. In addition, the characteristics and benefits of the MyHiX were studied by Hisan [18] in his research work. Furthermore, in a study of the Malaysian e-health, Shaik Allaudin [19] found that MyHiX was a new system to be implemented in Malaysia. A distributed framework for the nationwide health information exchange using the smartphone technologies was proposed by Abdulnabi et al. [20], while a security framework for HIE was proposed by Zaidan et al. [21].

3. Methods

This research employs the systematic literature review to study patient satisfaction associated with the use of MyHiX by the healthcare organizations; the findings of all related studies are identified, critically assessed and integrated. The strategic literature review covers five important activities, which are scoping, planning, identification, screening and eligibility [22].

In this paper, the systematic literature review was done to identify factors affecting MyHIX implementation. Therefore, all relevant terms related conceptually to MyHIX were also reviewed as shown in Table 1. These are the inclusion criteria:

Table 1. The inclusion criteria and description in this study

Criteria	Description
The articles were published between 2010 and 2017	This is crucial to ensure the findings are current, updated and accurate. Findings earlier than 2010 are considered outdated.
All types of studies related to the subject	This study focuses on the terms "Patient Satisfaction" and "Service Quality". In addition, the term "Health Information Exchange" is also included.
Articles published in five major databases	IEEE Explore Digital Library, Proquest, Emerald, Science Direct and Scopus.

The study gaps were scrutinized based on the above inclusion criteria. The data and information were obtained by focusing on the full text of the articles to ensure the review is comprehensive. Therefore, patient satisfaction of services provided by the hospitals using MyHIX could be identified and discussed.

4. Results and discussion

The five major databases, namely IEEE, Proquest, Emerald, Science Direct and Scopus were used in searching for articles of the relevant field of study.

Table 2. Number of studies based on databases

Database	HIE	Service Quality	Patient Satisfaction
IEEE	639	56	2
Proquest	885	1966	3821
Emerald	6789	4753	688
Science Direct	56,091	54,360	2937
Scopus	286	19	29
Total	64,690	61,154	7,477

Table 2 shows various studies focusing on the usage of HIE in healthcare organizations around the globe. The searching term "Health Information Exchange" brought in 64,690 articles. These articles discuss and explain the applications of HIE in various countries.

Besides that, the searching term "Service Quality" brought in 61,154 articles, whereas "Patient Satisfaction" brought in 7,477 articles. Only articles focusing on patient satisfaction and service quality of HIEs were selected. The various aspects covered by the articles are as follows: only five out of 714 articles focus on HIE and MyHiX, 909 studies focus on service quality, and 212 studies focus on patient satisfaction of services provided by the healthcare organizations in Malaysia. The figure indicates that there is a lack of studies in this area in Malaysia.

The full texts of these articles were reviewed as suggested by Khan et al. [22] based on the above inclusion criteria and the relevance of the articles to this study. The result shows that 714 articles are relevant and can be incorporated in this study. The findings are categorized into three areas: HIE, Service Quality and Patient Satisfaction.

4.1 Studies on Health Information Exchange

In 2014, 76% of the U.S. hospitals exchanged information among themselves; however, the number of people using the HIE was reportedly very low [23]. According to previous articles, HIE could provide connections among stakeholders, including laboratories, public health departments, hospitals, and service providers to exchange data and information [24]. The finding shows that there are various advantages of HIE; for example, Safety of Healthcare, Timeliness, Cost Reduction, Assessing Quality of Care, Research Resource, Organizational Benefits, Patient Perceptions, and Security and Privacy [21]. However, there are several issues resulting from the use of HIE; for example, usability [25-26], missing data [27-29], disruption of workflow [25; 28-29], startup costs [29; 30-31], privacy and security [29; 32], and lack of compelling business case for sustainability [29].

4.2 Studies on Service Quality in Healthcare Organizations

Service quality is defined as an assessment of how well a delivered service conforms to the customer's expectations. A high level of service quality will meet or exceed customer expectations on a consistent basis[33]. There are several studies on service quality of healthcare organizations. For example, a study by Borges et al. [34] found that the quality of services was satisfactory when the patients had high expectations of the medical and hospital service. Aliman [35] found that the strength of the perceived service quality-behavioral intentions relationship becomes weaker when the satisfaction is reflected. There are five variables to measure the quality of services in healthcare organizations: Tangibility, Reliability, Assistance, Safety and Empathy. These variables are explained as follows [36]:

Tangibility
This refers to an appearance of physical facilities, equipment, personnel and written materials.

Reliability
This refers to an ability to perform the promised service dependably and accurately.

Assistance
This refers to the willingness to help customers and provide prompt services.

Safety
This refers to the employees' knowledge and courtesy and their ability to inspire trust and confidence.

Empathy
This refers to caring, easy access, good communication, customer understanding and individualized attention given to customers.

4.3 Patient Satisfaction of Healthcare Organization Employed with HIE
Patient satisfaction has been recognized as the most important key success indicator in the healthcare industry [36-52]. It is linked to the concepts of perceptions and expectations [33]. The service will be considered excellent, if perceptions exceed expectations [53].

Table 3. Previous studies patient satisfaction in healthcare organizations

Variables	Tangibility	Reliability	Assistance	Safety	Empathy
Researchers					
[37]	✓	✓	✓	✓	✓
[38]	✓	✓	✓	✓	✓
[39]	✓	✓	✓	✓	✓
[40]	✓	✓	✓	✓	✓
[41]	✓	✓	✓	✓	✓
[42]	✓	✓	✓	✓	✓
[43]	✓	✓	✓	✓	✓
[44]	✓	✓	✓	✓	✓
[45]	✓	✓	✓	✓	✓
[46]	✓	✓	✓	✓	✓
[47]	✓	✓	✓	✓	✓
[48]	✓	✓	✓	✓	✓
[49]	✓	✓	✓	✓	✓
[50]	✓	✓	✓	✓	✓
[51]	✓	✓	✓	✓	✓
[52]	✓	✓	✓	✓	✓

According to Table 3, previous studies on patient satisfaction mostly adopted the multi-dimensional research instrument SERVQUAL [34], which is designed to capture consumer expectations and perceptions of a service along the five major variables of service quality: Tangibility, Reliability, Assistance, Safety and Empathy. This indicates that SERVQUAL could be adopted to measure the quality of services and patient satisfaction in the Malaysian healthcare organizations.

5. Conclusion
In a nutshell, based on the findings obtained from previous articles, the service quality of HIEs has an impact on patient satisfaction. An empirical study is required to test the findings, especially in

Malaysia where the electronic health systems are widely used in this country. In addition, comparative studies are recommended to be done in various developing countries. The empirical studies are significant, the findings of which can be utilized to improve the quality of services provided by the healthcare organizations that employed the HIE.

References

[1] Yap L L and Tan C L 2012 The effect of service supply chain management practices on the public healthcare organizational performance *International Journal of Business and Social Science 3*16

[2] Ministry of Health 2016 Petunjuk Kesihatan Available at: http://www.moh.gov.my/ images/gallery/publications/Petunjuk%20Kesihatan%202016pdf

[3] Ministry Of Health Senarai Hospital Kerajaan Malaysia Available at: http://www.moh.gov.my/indexphp/database_stores/store_view/3 2017

[4] Ministry Of Health Senarai Klinik Kesihatan Malaysia Available at: http://www.moh.gov.my/indexphp/database_stores/store_view/12017

[5] Pillay M S, Mohd Ghazali R J, Abd Manaf N H, Abdullah H A, Salikin A, Abu Bakar F, Umapathy M, Ali R, Bidin N and W I Wan Ismail 2011 Hospital waiting time: the forgotten premise of healthcare service delivery? International journal of health care quality assurance 247 506-522

[6] Bernama 2009, 24 July Kerajaan Bayar Pampasan RM7 Juta Untuk Kes Kecuaian Perubatan Available at: http://www.bernama.com/bernama/v3/bm/news_php?id=375694

[7] Leong Y H, Chee Yuen G A N, Tan M A F and Majid M I A 2014 Present status and future concerns of expanded newborn screening in malaysia: sustainability, challenges and perspectives The Malaysian journal of medical sciences MJMS 212 63

[8] Pai F Y and Huang K I 2011 Applying the technology acceptance model to the introduction of healthcare information systems Technological Forecasting and Social Change 784 650-660

[9] Ministry of Health Malaysia Country Plan Putrajaya, Malaysia: MOH 2011

[10] Ministry of Health Malaysian Health Data Warehouse MyHDW 2015-2016 Start up: Initiation Available at: https://myhdw.moh.gov.my/public/documents/

[11] Ministry Of Health Malaysia Malaysian Health Information Exchange Available at: http://telekesihatanmohgovmy/indexphp/20-projek-bt/7-myhix 2016

[12] Ministry Of Health Teleprimary Care Available at: http://fhmoh.gov.my/v3/indexphp/ teleprimary-care 2017

[13] Ismail N I and Abdullah N H 2016 Hospital Information System HIS Implementation in Malaysian Public Hospitals *International Information Institute Tokyo Information 19*7B 2833

[14] Ministry of Health Malaysia Annual Report 2012 Putrajaya, Malaysia: MOH 2012

[15] Washington, DC: Department of Health and Human Services 2014 What is HIE Health Information Exchange Available at: www.healthit.gov/providers-professionals/health-information-exchange/what-hie

[16] Ministry Of Health Malaysia Official Website Malaysian Health Information Exchange Available at: http://telekesihatan.moh.gov.my/indexphp/20-projek-bt/7-myhix, 2016

[17] Ministry of Health Malaysia Health Information Exchange MyHIX Toolkit Putrajaya, Malaysia 2014

[18] Hisan, A Malaysian Health Information Exchange MyHIX:Simposium Maklumat Geospatial Kebangsaan NGIS ke 5 Presentation, Putrajaya, 2012

[19] Shaik Allaudin F AeHIN Webinar 2016 National eHealth: "Malaysia Update" Presentation, Ministry of Health, Malaysia

[20] Abdulnabi M, Al-Haiqi A, Kiah M L M, Zaidan A A, Zaidan B B and Hussain M A 2017 Distributed framework for health information exchange using smartphone technologies *Journal of Biomedical Informatics* 69 230-250

[21] Zaidan B B, Haiqi A, Zaidan A A, Abdulnabi M, Kiah M M and Muzamel H A 2015 Security framework for nationwide health information exchange based on telehealth strategy *Journal of medical systems* 395 51
[22] Khan, K S, Kunz, R, Kleijnen, J, & Antes, G Five steps to conducting a systematic review Journal of the Royal Society of Medicine, 963, 118-121, 2003
[23] Hersh, W R, Totten, A M, Eden, K B, Devine, B, Gorman, P, Kassakian, S Z, & McDonagh, M S 2015 Outcomes from health information exchange: systematic review and future research needs *JMIR medical informatics*, *3*4
[24] Furukawa M F, Patel V, Charles D, Swain M and Mostashari F 2013 Hospital electronic health information exchange grew substantially in 2008–12 *Health Affairs 32*8 1346-1354
[25] Ross S E, Schilling L M, Fernald D H, Davidson A J and West D R 2010 Health information exchange in small-to-medium sized family medicine practices: motivators, barriers, and potential facilitators of adoption *International journal of medical informatics 79*2 123-129
[26] Vest J R, and Gamm L D 2010 Health information exchange: persistent challenges and new strategies *Journal of the American Medical Informatics Association 17*3 288-294
[27] Tham E, Ross S E, Mellis B K, Beaty B L, Schilling L M and Davidson A J 2010 Interest in health information exchange in ambulatory care: a statewide survey *Appl Clin Inform 1*1 1-10
[28] Rudin R, Volk L, Simon S and Bates D 2011 What affects clinicians' usage of health information exchange *Appl Clin Inform*, *2*3, 250-262
[29] Rudin R S, Motala A, Goldzweig C L and Shekelle P G 2014 Usage and Effect of Health Information Exchange A Systematic Review Usage and Effect of Health Information Exchange *Annals of internal medicine 161* 11 803-811
[30] Patel V, Abramson E L, Edwards A, Malhotra S and Kaushal R 2011 Physicians potential use and preferences related to health information exchange *International journal of medical informatics 80*3 171-180
[31] Pevnick J M, Claver M, Dobalian A, Asch S M, Stutman H R, Tomines A and Fu P 2012 Provider stakeholders perceived benefit from a nascent health information exchange: a qualitative analysis *Journal of medical systems 36*2 601-613
[32] Parker C J and Adler-Milstein J 2016 Errors Related to Health Information Exchange In *Safety of Health IT* pp 153-165
[33] Parasuraman A, Zeithaml V A and Berry L L 1985 A conceptual model of service quality and its implications for future research *the Journal of Marketing* 41-50
[34] Borges J B C, Carvalho S M R D and Silva M A D M 2010 Quality of service provided to heart surgery patients of the Unified Health System-SUS *Brazilian Journal of Cardiovascular Surgery 25*2 172-182
[35] Aliman N K and Mohamad W N 2013 Perceptions of service quality and behavioral intentions: A mediation effect of patient satisfaction in the private health care in Malaysia *International Journal of Marketing Studies 5*4 15
[36] Zeithaml V A, Parasuraman A and Berry L L 1990 *Delivering quality service: Balancing customer perceptions and expectations*
[37] Zamil A M A, Areiqat A Y and Tailakh W 2012 The impact of health service quality on patients' satisfaction over private and public hospitals in Jordan: a comparative study *International Journal of Marketing Studies* 41 123
[38] Alrubaiee L, and Alkaa'ida F 2011 The mediating effect of patient satisfaction in the patients perceptions of healthcare quality–patient trust relationship *International Journal of Marketing Studies* 31 103
[39] Zaim H, Bayyurt N and Zaim S 2013 Service quality and determinants of customer satisfaction in hospitals: Turkish experience *International Business & Economics Research Journal IBER 9*5
[40] Khamis K and Njau B 2014 Patients' level of satisfaction on quality of health care at Mwananyamala hospital in Dar es Salaam, *Tanzania BMC health services research 14*1 400

[41] Essiam J O 2013 Service quality and patients satisfaction with healthcare delivery: Empirical evidence from patients of the out patient department of a public university hospital in Ghana *European journal of business and management 5*28 52-59
[42] Wu C C 2011 The impact of hospital brand image on service quality, patient satisfaction and loyalty *African Journal of Business Management 5*12 4873
[43] Punnakitikashem P, Buavaraporn N, Maluesri P and Leelartapin K 2012 April Health care service quality: Case example of a hospital with lean implementation In *POMS 23rd Annual Conference Chicago IL USA*
[44] Rad N F, Som A P M and Zainuddin Y 2010 Service quality and patients' satisfaction in medical tourism *World Applied Sciences Journal 10*1 24-30
[45] Kitapci O, Akdogan C and Dortyol I T 2014 The impact of service quality dimensions on patient satisfaction, repurchase intentions and word-of-mouth communication in the public healthcare industry *Procedia-Social and Behavioral Sciences 148* 161-169
[46] Handayani P W, Hidayanto A N, Sandhyaduhita P I and Ayuningtyas D 2015 Strategic hospital services quality analysis in Indonesia *Expert Systems with Applications 42*6 3067-3078
[47] Li M, Lowrie D B, Huang C Y, Lu X C, Zhu Y C, Wu X H and Zhao, P 2015 Evaluating patients perception of service quality at hospitals in nine Chinese cities by use of the ServQual scale *Asian Pacific Journal of Tropical Biomedicine* 56 497-504
[48] Aman, B and Abbas, F 2016 Patient's perceptions about the service quality of public hospitals located at District Kohat *JPMA J Pak Med Assoc 66*1 72-5
[49] Meesala, A, & Paul, J 2016 Service quality, consumer satisfaction and loyalty in hospitals: Thinking for the future *Journal of Retailing and Consumer Services*
[50] Rezaei S, Matin B K, Khalil Moradi B B, Fallahi M, Shokati B and Saeidi H 2016 Measurement of quality of educational hospital services by the SERVQUAL model: the Iranian patients' perspective *Electronic physician 8*3 2101
[51] Naskar P, Naskar S and Roy S 2017 Assessment of patient's satisfaction regarding the service quality of a rural hospital of Burdwan district, West Bengal, India *International Journal Of Community Medicine And Public Health 3*8 2222-2228
[52] Neupane, R, & Devkota, M 2017 Evaluation of the Impacts of Service Quality Dimensions on Patient/Customer Satisfaction: A Study of Private Hospitals in Nepal *International Journal of Social Sciences and Management 4*3 165-176
[53] Vázquez-Casielles R, Suárez Álvarez L and Diaz Martin A M 2010 Perceived justice of service recovery strategies: Impact on customer satisfaction and quality relationship *Psychology & Marketing 27*5 487-509

Acknowledgments
This study is financially supported by Office for Research, Innovation, Commercialization and Consultancy Management (ORICC) of University of Tun Hussein Onn Malaysia.

IOP Conf. Series: Journal of Physics: Conf. Series **1049** (2018) 012025 doi:10.1088/1742-6596/1049/1/012025

Persuasive System Practices in Mobile Application Development

N A Shafin, N H Abdullah[1]
Department of Technology Management,
Faculty of Technology and Business Management,
Universiti Tun Hussein Onn Malaysia, 86400 Parit Raja, Batu Pahat, Johor

E-mail: afifahshafin@gmail.com, hazana@uthm.edu.my

Abstract. Mobile application development is at swift pace of growth across the globe. This industry also creates financial opportunities towards entrepreneurs also independent software developers. Yet, with the exponential growth of mobile applications released daily, hundreds of these software programs were being abandoned and neglected. The development of high quality mobile applications also requires the understanding of requirements needed associated with the development of the software products in order to fulfil the special needs of the products. This research was conducted with the aim to qualitative explore the persuasive technology practices followed and the processes involved which able to enhance the success level of the mobile applications. For this study, semi-structured interviews were conducted with the involvement of seven software developers and fifteen mobile applications as the case studies for the research. It has been shown that persuasive system design is appropriate to be incorporated in the process of software development as it provides a structural guideline that reflects users' requirements in the context of attitudes and behaviours. Based on this assumption, it is identified that the incorporation of persuasive context at the initial stage of software development may affect the quality of the mobile applications in a long term.

1. Introduction

Mayhew [1] and Marache-Francisco and Brangier [2] state that there are five imperative qualities for the development of high quality software products which includes software systems, interactive tools, websites and mobile applications. These qualities are usability, utility, graphic design, functional integrities and persuasive. The involvement of these qualities is the essence in creating great user experiences. As defined by Nielsen [3], usability is the attributes of satisfaction, efficiency and effectiveness of the software products, utility is the usefulness which associated to the design's functionality [4], graphic design represents the expression of personality of the software products while functional integrity is the reflections and representation of the software products' credibility. Yet, the main highlight for this research is the quality of persuasive in software development specifically in the process of mobile application development.

Persuasive as defined by scholars such as Fogg [5], Redstrom and Veludo-de-Oliveira, Pallister and Foxall [6] is the act with the intention to establish, that has the intention to establish, strengthen or change the attitude or behaviour of individuals or group of peoples without involving coercion or deception. The incorporation of persuasive in the technological domain has then establish a new

[1] To whom any correspondence should be addressed.

Content from this work may be used under the terms of the Creative Commons Attribution 3.0 licence. Any further distribution of this work must maintain attribution to the author(s) and the title of the work, journal citation and DOI.
Published under licence by IOP Publishing Ltd

branch of knowledge known as persuasive technology or captology. Persuasive technology as according to Fogg [7] and Oinas-Kukkonen, Hutchison and Michelle [8] is the involvement of technological medium to deliver the persuasive message with the intention to change the attitudes or behaviour of users. In addition, the incorporation of persuasive in software development process were thoroughly described in the framework of persuasive system design (PSD) proposed by Oinas-Kukkonen and Harjumaa [9].

The motivation behind this study is that there are low numbers of research conducted on the process of persuasive mobile application development although Oinas-Kukkonen and Harjumaa [9] has claimed that it may change and shape users' attitudes and behaviours. Furthermore, recent statistics has report that there are numbers of mobile applications released were being abandoned [10], only used once [11] or being neglected or uninstalled by the users [12]. These issues have become the new concerns among software developers as it may jeopardize the credibility of developers and affect the revenue gained by the companies from the mobile applications developed. According to studies by Feng, Hoegler and Stucky [13] and Wohl, Parush, Kim, and Warren [14], the cause of these issues is because there is declining in the quality and design of the software products. Hence, there is the needs for more in-depth studies in the process persuasive of mobile applications development. This is because, Inukollu, Keshamoni and Kang [15] and Shah [16] state that part of good mobile applications relies on well-craft product development. Besides, persuasive system framework has incorporated the five qualities of high quality software products which has been discussed earlier in this paper.

For this purpose, this study has focused on exploring the persuasive technology practices as well as the process or actions taken followed by seven mobile application developers in enhancing the quality and the success level of the software products. In doing so, this study is able to emphasize on the advantages of persuasive system design process for software development specifically mobile applications and finding the additional aspects which are required and may contribute towards high quality mobile applications. Therefore, this paper has been structured as follows. The first section (Section 1) of this paper presents the introduction for the research conducted followed by the second section (Section 2) which focuses and reviewed on the mobile application development process as well as the framework of persuasive system design. The third section (Section 3) represents the methodology of research and the fourth section (Section 4) presents and discussed the findings of the research. The final section of this paper (Section 5) will conclude and discuss on the opportunities for future research.

2. Persuasive technology in software development process

2.1. Software processes activities

As defined by Shah [16], software processes are the framework or structure that act as a guideline to develop and maintain of software development. In consonance with Sommerville [17] and Hudli, Hudli and Hudli [18], there are abundance of software processes; however, there are four fundamental sequences of activities that are involve in the production of software products which is software specification, development, validation and evolution and each of the phases feeds into the next step and are substantial for mobile applications' overall success (see Figure 1).

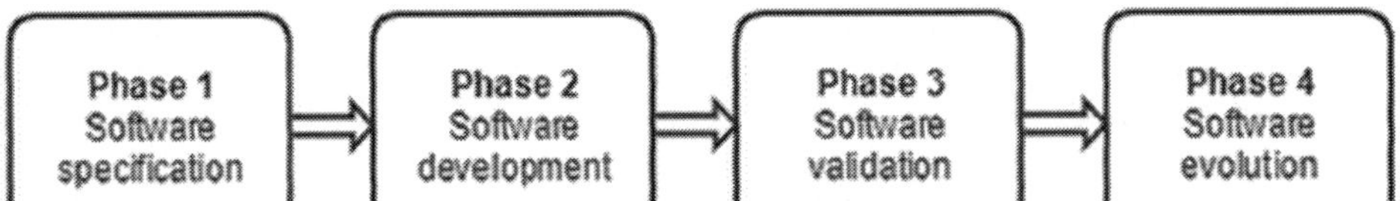

Figure 1. The generic software development process (Source: Researcher)

2.1.1. Phase 1: Software specification. The first phase of software development process is the most crucial phase as it requires the developers determine, refine in detail the specification as well as the constraints of the mobile application to be developed. During this phase, developers are required to determine, refine in detail the specification as well as the constraints of the mobile application to be developed. At this stage, ideas identification will be done and development strategy is created to ensure that the goals of the mobile application development are achievable. According to Lockton, Harrison and Stanton [19], the identification of ideas can be carried out through defining the problems aroused and the need that require the mobile application to solve. Later, the initial concept and ideas fortify by gap analysis that was proposed by Papazoglou and Van Den Heuvel [20]. Through this analysis, the proposed ideas and features will be compared to the existing software services as to devise the novelty and improvised ideas to the existent software products.

Moreover, the strategy of mobile application development is constructed to designate the objectives of the app, the constraints or scope on its operation, platform that will be targeting, justifying the target audience together with the details of the mobile application specifications. Salmre [21] mentioned that the scope of the mobile application is one of the most crucial aspects that need to be emphasized on as a successful mobile app should be prerequisite with clear and specific goals and it should be able to offer a focused experience to their users. Strategizing the fundamental purpose will also lead to determination of proper user interface (UI) design as each of mobile application require well-designed interface as it sets the best features as well as the flow of the app functionalities [22]. Salmre [21] and Sonderegger and Sauer [23] in their studies have stated that appropriate and intriguing interface will create focused and intuitive experience to the users thus it will govern the success of the mobile application. By the end of this phase, a set comprehensive mobile application requirement analysis that consists of objective abstraction and overall information technology design and architecture will be documented.

2.1.2. Phase 2: Software development. The second phase of software process is also known as the programming phase that involves the coding process for the production of the software. This phase is time consuming thus the development task will be divided into few models or units so that the software can be completed according to the schedule and interim results can be presented to the clients. The next phase in the mobile application development process is the design phase where by the ideas from the brainstorming session during the previous phase is being established into functional requirements and software architecture. According to Dix [24], functional requirement of a mobile app is a brief summarization of the services that the software product will offer to their users, the way it will respond to particular inputs and the way it will give feedbacks in particular situations and is written in general and abstract way that is understandable by the users. As for software architecture, it serves as a design plan of the requirement of the mobile application, structured discussions in between clients, developers and managers as well as part of essential tool for complexity management and is modelled using simple block diagrams that contain the details on the key system abstraction.

From these documentations, mobile application module and prototype is defined. Modules of a mobile app were determined to outline the concept of visual terms and act as visual representation for all the interfaces that will be finalized. Wasserman [25] emphasized that user experience and user interface are the most substantial facets that mobile developers need to focus on. This is because, user experience covers the interface, functions and application performance is critical as it creates the interaction architecture of design elements while user interface can create the look and the feel of the app [25]. This is supported by a research conducted by Robins and Holmes [26] which shows that, visual appearance of a mobile application could affect the way participants handled the device thus affects the users' perception on the mobile application. The end result for this phase is the completed storyboard that has incorporates all the features as well as the flow of the application that have been defined previously.

2.1.3. Phase 3: Software validation. The third phase of software process involves the software's modules or units to be integrated and tested to ensure that it is actually solving the needs addressed and gathered during the requirements phase. For the most part, the validation process will take some few testing such unit testing, integration testing, system testing, non-functional testing and acceptance testing. The user acceptance testing was done at the end of the validation phase as it is performed by the end users to ensure that software has meets their expectations. During this phase, there might be some minor or major defects aroused. Hence, the development and testing phase will be performed iteratively to correct and retested the software. Prototyping is the process where the ideas, concepts and feedbacks were organized in the way mobile app was set to be so that it can be tested later [27]. The process also allows developers to explore new ideas, validates ideas and giving room for software improvements to create better experience towards the users.

As part of the framework, functional requirements of prototypes are analysed then tested to gain new feedback. The received feedback will be used as a guideline to improve the performance of the software product and the required changes are implemented through the development phase. Thus, the development, prototyping and testing phases will be repeated for several times until the final prototype has been agreed on. This phase was emphasized as one of the crucial segment in mobile application development phase by Ohk, Park and Hong [28] when the development involves the clients. The mobile application prototype can be handy to present the overview of the software products thus allowing easier participation in voicing down their ideas.

2.1.4. Phase 4: Software evolution. The last phase of software process involves the modification of the system to reflect the software specifications and market requirements. Maintenance and upgrading the software every once in a while, will be done to adapt the changes that takes place in the user end environment as well as the technology that revolve. Furthermore, the needs for new requirements and newly aroused problems aroused once the users start using the released software. Hence, the improvements and enhancements of the software product will involve much longer time than the initial development since most maintenance required the developers to do new things which can lead to extending the time frame of the software development process. Software testing phase is one of the significant phases in any model of software development. As in this framework, the initial prototype will be tested on an emulator or simulator on a real device across multiple operating systems, versions and sizes of mobile applications' screens. The aspects that will be tested are shown in Table 2.10. Nevertheless, the types of testing required for each mobile application is variant depending on the type of the application, target users and the distribution channel [29]. This is the last stage the development phase. The application is deployed after the software product has been tested and finalized by releasing the mobile application into the appropriate mobile application market for the users.

The existing framework have discussed thoroughly on the development processes, activities involved and guidelines which are meant to instruct developers in the creation of high quality mobile applications. Yet, there is absence or rather narrow guidance for the practitioners towards persuasive system development. This is because, software processes are a part of software engineering process by which it only concerns with methods and tools for general software development based on engineering principles while persuasive approach is more concerned on the psychology for system design. Years later, Oinas-Kukkonen and Harjumaa [9] have proposed a framework known as Persuasive System Design (PSD), a model which assist researchers and practitioners to develop and evaluate persuasive systems. This framework has included the understanding of the core issues for persuasive system, comprehensive planning of persuasive context and design of system qualities. As discussed previously, the implementation of persuasive approaches is in a system are necessary as it serves as one of the determinants for an engaging and successful technology. Hence, the development of good software products should consider on the constructed persuasive system design as well as the way people should use and interact with the software.

2.2. Persuasive system design

Persuasive system design is a model which incorporates general system development lifecycle model and the persuasive design principles as the software's requirement definition. Furthermore, Harjumaa [30] identifies the objective for this model as to provide guidance for the development of a quality persuasive system through a lifecycle model. In their major study of PSD model, Oinas-Kukkonen and Harjumaa [9] have defined persuasive system as an information system or computerized software that is developed to change, shaping and reinforcing attitudes or behavior or both without involving the act of coercion or deception and it is anticipated that the development of a successful persuasive system will draw-in three potential outcomes which is; 1) voluntary reinforcement; 2) change of attitudes or behaviors or both and 3) shaping of attitudes or behaviors or both. Moreover, the persuasive design framework is also being designed for the prediction of consumer acceptance towards the products developed.

The framework of persuasive system design emphasizes on three main aspects which includes understanding the core issues in persuasive system, persuasive context and the persuasive system features. For this study, analysis of the persuasive context will be highlighted and discussed.

2.2.1. Persuasive context of persuasive system design. Persuasion context is the thorough understanding on the intent, event and strategy of persuasion which comprised of the understanding towards the role of persuader (the persuasive system), persuadee (the users of the system), the messages to be conveyed and the channels for delivering the messages [9]. The understanding of the context can be guidance for system developers to identify the inconsistencies in users' thinking and it can help to differentiate the opportune and inappropriate moments for delivering persuasive messages. The discussion on persuasive context also encompasses with the identification of the objectives or the intent of persuasion, understanding the way persuasion process occurs and determine or identify the persuasive strategies used.

2.2.1.1. The intent. According to Bondi [31] computer is an electronic device that does not have the properties or the characteristics of persuasion to affects the users' attitude or behavior. Therefore, persuasive system is designed and developed based on the aim or the intention of the developers to deliver their persuasive messages with the implementation of the human features, persuasive principles and technological characteristics in various ways.

As mentioned by Harjumaa [32], adapted from Fogg [33] seminal article, the intention of persuasive system development can be divided into three categories which are endogenous, exogenous and autogenous. Endogenous can be defined as the intention of the developers to influence the system's users while exogenous can be defined as the intentions that were caused by the external factors with the intention to influence the system's users. The external factors for an exogenous can be those who give or distribute persuasive opportunities to the other individuals. Finally, autogenous is the representation of the very person to adopt or using interactive technology to influence other system's users.

However, as discussed by Bijker et al., [34], most of technologies were not fully developed for the purpose of persuasion as most of it has carried some cues of their preferable usage practices in which also known as micro suasion. Hence, in the process of software development, intent can be understood as the goals for the information system. Software developer will determine the expected goals and create the context which will persuade the users to adopt the expected position or action.

2.2.1.2. The event. The event of a persuasive system comprises the analysis of context of use, users and the technology. Analysing the context of use require system developers to create plans and designed the system so that it can overcome the existing problems while user context involve the understanding of the target users [35]. Thorough understanding of use context will also influence the selection of persuasive design principles employed for the system as these principles were supposed to support the primary task, create a relationship in between the users and the system, make the system much more credible and create a platform in between the users to interact with each other's.

Oinas-Kukkonen and Harjumaa [9] point out that it is crucial for system developers to know the background of their potential and target users because each individual has a different level of cognitive requirement to process information which has been discussed in the third postulate of persuasive system. For example, some individuals have high cognitive requirements and there are also individuals who have a low level of cognition needs. By recognizing the characteristics of the target users, it is easier to plan for the persuasive approach to be implemented into the persuasive system.

Finally, the technological context refers to the platform that will be used for the development of persuasive system. As noted by Torning and Oinas-Kukkonen [35], the selection of the appropriate platform is important because it will affect the user context and the context of use. For example, persuasive system developed for a computer used is different compared to the mobile phones because of the differences in the characteristics as well as the features of both of the technological products. Furthermore, full utilisation of mobile application characteristics will be able to push the quality of software developed as it has multiple functions and features.

Petty and Cacioppo [36] identifies persuasion strategies are the ways persuasive message is delivered either through central or the peripheral route as described in the third core issues of persuasive system- the direct and indirect routes. The approach via central/direct route applies to a person who has the motivation to carefully evaluate the content presented on the persuasive message and the results of this approach are more permanent.

While the approach via indirect routes/peripheral route can be applied to someone who is less motivated. The strategy for the indirect routes is to use an easy or simple but strong cue to deliver the message. For an example, persuasive information is submitted to the users by using interactive yet understandable notifications in order to increase the motivation of the individual to the submission of the relevant persuasive information. Although this approach does not give a permanent result, but it is the most suitable way to be applied as the increasing and hectic of information flow, users are not able to really focus on one route of message delivered at any time. Therefore, this approach is used to help make decisions.

2.3. Persuasive system design in mobile application development

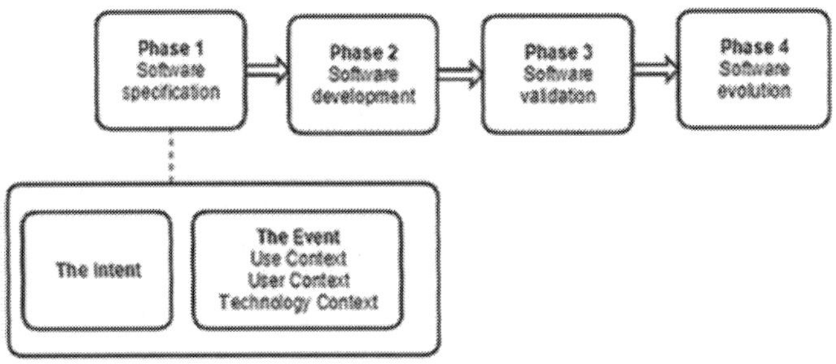

Figure 2. The incorporation of persuasive technology practices in mobile application development

The framework supports the four fundamental phases which have included the software specification, development, validation and evolution phases. The fundamental phases of software processes are adopted rather than the mobile application development life cycle (MADLC) as the software processes are much more common and accepted in every software development which includes systems and mobile applications. Moreover, as previously mentioned by Kumar and Kumar [37], the additional sub-activities were done to complement the requirement of the software itself.

This framework only emphasized on the first phase of software processes which is the software specification phase because it is the most important phase for the development of software products as mentioned by Jain and Suman [38]; Sommerville [39] and Vithani and Kumar [40]. Furthermore, it is clearly supported by Harjumaa [30] that the implementation of persuasive design practices should appear in the early stage of development as it makes easier integration of persuasiveness into the final system. This is because, the discussion of persuasive system design framework as proposed by Oinas-Kukkonen and Harjumaa [9] for the requirement definition of software qualities were require thorough understanding on the intent as well as the event of the software which is the same as activities described in the software specification phase in the software processes. However, the need to addressed the persuasive systems postulates were eliminated from this framework because it is expected to be relevant towards researchers instead of mobile developers. Other than that, several persuasive systems postulates such as the third and seventh postulates were already mentioned and applied in the event of persuasive context. Persuasive strategies were excluded from this framework because it is applied in the user and use context. This is because, the understanding for user requirements will assist the developers on providing the most convenient way to approach the users from the platform of mobile applications which were achieved in the use context.

3. Methodology

The approach employed for this research is qualitative approach as it allows the researcher to perform in-depth studies towards the research topic [41] *and* [42]. For the context of this research, qualitive method allows the researcher to explore the best practices of mobile applications development from the developers' experiences and perspectives.

Furthermore, the research design of this research follows multiple case studies as it supports the nature of the research by allowing the researcher to gain in-depth understanding on the context of the persuasive system design or any other process of development followed from the developers' perspectives and experiences hence producing detailed descriptions which will then be compared to identify the differences and similarities in between the cases. For this study, three categories of mobile applications which are successful (SMA), partially successful (PSMA) and less successful (LSMA) were identified and distinguished based on the numbers of application downloaded for the past three years. Seven participants were selected among project leader, project managers and programmers as these individuals have the adequate experiences and capabilities in ensuring the data collection process was able to meet the objective of this research.

Furthermore, qualitative technique with in-depth interviews were applied for the data collection process. As mentioned [41],[42] in-depth interviews allow the researcher to probe the participants in gaining thorough and in-depth interpretation of information based on the participants point of view. During the interview session, a set of interview guideline was used and were changed or more questions were added depending on the flow of the conversation. This was done to support the statement given hence enabling the participants to provide detailed information of features and functions incorporated in the mobile applications developed. In addition, the interviews were conducted in Malay and English language and took approximately 90 to 110 minutes. Each of the interview sessions were tape-recorded for the process of transcribing. Moreover, content and thematic analysis were conducted. NVivo software was used to analysed the content of the interview before the data were compared within the cases and in-between the cases.

Table 1. Case studies of research

Case study	**Successful mobile applications (SMA)**	**Partially successful mobile applications (PSMA)**	**Less successful mobile applications (LSMA)**
Characteristics			
Numbers of mobile applications downloaded within (2013-2015)	10,000 – 100,000	1,000 – 10,000	<1,000
Type of mobile applications	Native apps	Native apps	Native apps
Categories of mobile applications	Services and utilities	Services and utilities	Services and utilities

4. Result

Table 2. Persuasive technology practices of design process

Persuasive technology practices of design process	Case Study 1 | SMA	Case Study 2 | PSMA	Case Study 3 | LSMA
Phase 1: Software Specification			
The Intent	●●●●●●	●●●	●●●●●●
The Event			
Use Context	●●●●●●	●●●	●●●●●●
User Context	●		
Technology Context	●●●●●●		
Phase 3: Software Validation	●●●●●●	●●	●
Phase 4: Software Evolution	●●●●●●	●●	●
(+) Product Introduction Strategic partnership Initiate cross-promote strategies	●●●●●		

The content analysis revealed distinctive differences in the persuasion practices followed by the three case studies of mobile applications and there are several considerations which have been found. As shown in Table x. mobile applications of the successful category will go through a thorough software specification process, which has included the consideration towards the intent *also* the event which comprise with the user context, use context and technology context. Furthermore, the mobile applications in this category also have gone through a repetitive process for the purpose of improvement towards the overall performance of the software products.

4.1. Software specification phase

4.1.1. The Intent. The findings showed that the procedures for all the three cases were almost familiar. However, based on Table 4.7, the activities conducted for developing the mobile applications of the successful category were more in-depth and comprehensive compared to the other two categories. According to all of the mobile developers, the first stage for the development of the mobile applications started with defining the objectives, drawing out the scope and also the limitation of the software products.

4.1.2. The event. Despite the similarities, the justifications of use context for each of the categories were different. This is because; the objectives of development for the mobile applications in the successful category were decided after in-depth market research has been conducted. As mentioned by these developers, the significance for conducting the market research was to identify the gaps in between the available mobile applications in the mobile app stores. Furthermore, through the research conducted by these developers have helped them in generating more diverse functions and features as to complete and fill the gaps of the existing mobile applications. As added by developers of mobile application 4, 9, 13 and 14, the act of in-depth market research also has increased the quality and the uniqueness of their software products. Yet, as summarized in Table 4.7, most of the mobile applications were developed based on the existing or the problems faced by the users while several mobile applications such as 6, 7 and 5 were only deliberately developed by the developers. The justification of user context by conducting the market research and set forth the problem-centred mobile application have then produced new and innovative mobile application as explained in mobile application 4, 9, 3, 14 and 1. New mobile applications were the result of the innovation made on the new proposed ideas to fulfil the users' demand.

As mentioned by Oinas-kukkonen and Harjumaa [9] and Sommerville [17], the development of system with persuasive features need to consider the use context which associate the selection of pertinent persuasive design principles. Based on the findings across the three categories, the development for each of the mobile applications have involved the selection of relevant functions and features whereby each of these were the reflection of the primary functions delivered by the mobile applications. Although all of these software products did consider the context of use, yet the incorporation of the persuasive design principles were different for each of the categories. The implementations of these persuasive principles have been thoroughly discussed in the previous sub-section and the patterns for the number of persuasive design principles implemented were identified.

Oinas-kukkonen and Harjumaa [9] in their persuasive design framework also have discussed the consideration of technology context for the development of persuasive system. In this context of research, it is clearly understood that the selection of the software platform is mobile application. However, in-depth deliberations for technology context have been done in both categories of successful and partially successful mobile applications. Yet, the only different identified in both of these categories were made during the validation phase.

4.2. Software Validation Phase

In software validation phase, mobile applications will have to go through few stages of testing whereby each of all of the software's modules or units to be integrated and tested to ensure that the mobile application is able to deliver the functions and features as expected. Based on the analysis conducted, the validation process for the mobile applications of the successful category involves the mobile application to be released in the mobile application stores for several weeks or months. Over the time, the performance of the mobile applications will be assessed based on users' feedbacks and responses over the time. Furthermore, the mobile applications were also will have to go through real time performance assessments over several months until the software products have achieved the stagnant phase before another improvement or any additional modules were made. Yet, the same processes were followed in the mobile application 2 and 10 of the partially successful mobile application category. In these mobile applications, the software products were released into the mobile application market for several weeks to reach out for much more users. However, it is found that the responses given by the users were only validated once or twice before abandoning the mobile application.

4.3. Software Evolution Phase
Software evolution phase is the phase whereby the mobile applications will go through several software improvement processes and reflecting the software specifications and market requirements. During this phase, maintenance as well as upgrading the mobile applications will be made. Based on the findings, all of the mobile applications of the successful category have gone through this process as well as the mobile application 2 and 10 in the partially successful category and mobile application 6 of the less successful category. However, based on the description given by the developers of these three mobile applications, the evolution phase only happened once or twice before the mobile applications were abandoned in the mobile app stores.

Yet, in this phase, product introduction of users' acceptance was obtained from the qualitative findings. According to the mobile developers of the mobile application 4, 8, 9, 3 and 14, the product introduction was done to reach out for more users. However, it is agreed by the three mobile developers that the product introduction through the strategic partnership and cross-promote strategies were only the add factor for the mobile applications as the core functionalities were more prominent. As added by the developer of mobile application 4, 9 and 14, the product introduction was only as act to gain more users which will benefit the developers in terms of mobile application assessment, yet the primary functions and features of the mobile application that will help the continuous use of the software products.

5. Conclusion
It is obviously noticed that the incorporation of persuasive technology practices in mobile application development has shown to be associated with higher achievement as these mobile applications have implemented much more comprehensive and detailed process of development. Furthermore, product introduction by creating strategic partnership and cross-promote strategies have allow these software products to attract a lot more users hence allowing these mobile applications to went through continuous transition phase which then has lead towards quality software products. This can be proved as these six mobile applications are still being used by the users after being developed after three years of the apps being released in the mobile apps stores.

This paper aims to explore the persuasive technology practices for the development process of mobile applications and the other contribution aspects that have been employed by developers in order to have good quality mobile applications. To achieve these objectives, 15 native Android mobile applications were selected of the utilities and services category and seven semi-structured interviews were conducted based on the proposed theoretical framework as shown in Figure 2. In effort to gain in-depth and thorough description of mobile applications' development process, probing technique was used.

The content analysis from the interviews revealed that all of the mobile applications have followed the persuasive system design for software development. Yet, only one mobile application that has been conducting a study as to understand their target users' personality and the way the mobile application will be used. The analysis also revealed that, all of the software products in the successful mobile application category have the additional process known as product introduction as strategic partnership and cross-promote strategies were being done to gain and attract more users for the mobile applications. According to the developers, the advantages of having a lot more users will them to have much more insights and feedbacks for the products which will then be taken into consideration before validation process will be conducted.

However, there are several limitations of this study. This is because, this study only focusses on the software aspects of mobile applications development while a high quality mobile application should be considering other issues such as wireless communications, mobility, portability and much more which need to be cope with. Yet, the implementation of the persuasive system design and creating product introduction are imperative as it also play the important roles in software development process for better high quality mobile applications.

As conclusion, persuasive system design has provided the adequate application guideline which comprised with the consideration on the purpose of the mobile applications, users' personality and characteristics, non-functional requirements which are required to support the functions and features of the software products and the technological capabilities of the mobile applications. This study also has found that product introduction is a crucial phase as it will attract a lot more users which will attract a lot more comments and feedbacks which are beneficial for the improvement of the mobile applications. Nevertheless, these approaches should be highlighted and emphasized by the research community also for the developers to improve and provide the knowledge of the guideline should be applied in the mobile application development.

References

[1] Mayhew D J 2012 *Usability + Persuasiveness + Graphic Design = eCommerce User Experience. In: Jacko, J. A.* (3rd Ed) The Human-Computer Interaction Handbook CRC Press 1181-1194

[2] Marache-Francisco, C, and Brangier E 2015 Gamification and Human-Machine Interaction: A Synthesis. Le Travail Humain **78(2)** 165-189

[3] Nielsen J 2012 Usability 101: Introduction to Usability retrieved on 1st October 2012 from: http://www.useit.com/alertbox/20030825.html

[4] Schumacher R M and Lowry S Z 2010 NIST Guide to the Processes Approach for Improving the Usability of Electronic Health Records, National Institute of Standards and Technology

[5] Fogg B J 2003 *Persuasive Technology: Using Computers to Change What We Think and Do* Elsevier Science and Technology Books

[6] Veludo-de-Oliveira T, Pallister J and Foxall G 2013 Persuading Others to Volunteer: Is Attitude Change Enough? *E-European Adv. Consum. Res.* **10** 98–103

[7] Fogg B J 2003 *"Computers as Persuasive Social Actors," in Persuasive Technology: Using Computers to Change What We Think and Do* Elsevier Science and Technology Books pp 89-120

[8] Oinas-Kukkonen H, Hutchison D, Mitchell J C 2010 *Persuasive Technology* **6137**

[9] Harjumaa M and Oinas-Kukkonen H 2007 Persuasion theories and IT design *Int. Conf. Persuas. Technol.* 311–314

[10] Levenson H 2016 7 Common Reasons Users are Abandoning your App [Online]. Available: https://www.webanalyticsworld.net/2016/08/why-users-are-abandoning-your-mobile-app.html [Accessed: 03-Jan-2017]

[11] Pramis J 2013 Are you a rarity? Only 16 percent of people will try out an app more than twice [Online] Available: https://www.digitaltrends.com/mobile/16-percent-of-mobile-userstry-out-a-buggy-app-more-than-twice/ [Accessed: 01-Jan-2017]

[12] Dye J 2016 77 percent of users never use an app again 72 hours after installing [Online] Available: http://www.androidauthority.com/77-percent-users-dont-use-an-app-after-three-days-678107/ [Accessed: 19-Jan-2017]

[13] Feng H, Hoegler T and Stucky W 2006 Exploring the critical success factors for mobile commerce Mob. Business *ICMB'06 Int. Conf.*

[14] Wohl M J A, Parush A, Kim H A S and Warren K 2014 Building it better: Applying human-computer interaction and persuasive system design principles to a monetary limit tool improves responsible gambling *Comput. Human Behav.* **37** 124–132

[15] Inukollu V N, Keshamoni D D, Kang T and Inukollu M 2014 Factors Influencing Quality of Mobile Apps: Role Of Mobile App Development Life Cycle *International Journal of Software Engineering and Applications (IJSEA)* **5(5)** 15-34

[16] Shah U S 2016 An excursion to software development life cycle models: An old to ever-growing models *ACM SIGSOFT Software Engineering Notes* **41(1)** 1-6

[17] Sommerville I 2015 *Software Engineering* 10th Ed. Pearson Education, Inc

[18] Hudli A, Hudli S and Hudli R 2015 An Evaluation Framework for Selection of Mobile App Development Platform *Proceedings of the 3rd International Workshop on Mobile Development Lifecycle ACM* 13-16

[19] Lockton D, Harrison D and Stanton N A 2010 The Design With Intent Method: A Design Tool For Influencing User Behavior *Applied ergonomics* **41(3)** 382-392

[20] Papazoglou M P and Van Den Heuvel W J 2007 Business Process Development Life Cycle Methodology *Communications of the ACM* **50(10)** 79-85

[21] Salmre I 2005 *Writing mobile code: Essential software engineering for building mobile applications* Addison-Wesley Professional

[22] Pinzon O E and Iyengar M S 2012 *Persuasive Technology And Mobile Health: A Systematic Review. Persuasive Technology: Design for Health and Safety* Linköpings universitet: Linköping University Electronic Press 45-48.

[23] Sonderegger A and Sauer J J 2010 The influence of design aesthetics in usability testing: Effects on user performance and perceived usability *Applied Ergonomics* 41(3) 403–410

[24] Dix A 2010 Human-Computer Interaction: A Stable Discipline, A Nascent Science, and the Growth of The Long Tail *Interacting with Computers* 22(1) 13–27

[25] Wasserman A I 2010 Software engineering issues for mobile application development *In Proceedings of the FSE/SDP workshop on Future of software engineering research* 397-400

[26] Robins D and Holmes J 2008 Aesthetics and credibility in web site design *Information Processing and Management* **44(1)** 386–399

[27] Church K and de Oliveira R 2013 What's up with WhatsApp? Comparing Mobile Instant Messaging Behaviors with Traditional SMS. *Proceedings of the 15th International Conference on Human-Computer Interaction with Mobile Devices and Services (MobileHCI'13) ACM* 352–361

[28] Ohk K, Park S B and Hong J W 2015 The Influence Of Perceived Usefulness, Perceived Ease Of Use, Interactivity, And Ease Of Navigation On Satisfaction In Mobile Application. Advanced Science And Technology Letters 84(20) 88-92

[29] Kirubakaran B and Karthikeyani V 2013 Mobile Application Testing Challenges And Solution Approach Through Automation. In Pattern Recognition, Informatics and Mobile Engineering (PRIME) *IEEE* 79-84

[30] Harjumaa M 2014 On the Development of Persuasive Systems: A Framework for Designing and Evaluating Behavior Change Support Systems and Its Applicability For E-Health University of Oulu; Ph.D. Thesis

[31] Bondi A B 2016 Incorporating Software Performance Engineering Methods and Practices Into the Software Development Life Cycle. *Proceedings of the 7th ACM/SPEC on the International Conference on Performance Engineering ACM* 327–330

[32] Harjumaa M and Oinas-Kukkonen H 2007 Persuasion Theories and IT Design *Persuasive Technology* 311-314

[33] Fogg B J A 2009 Behavior Model for Persuasive Design *Proceedings of the 4th International Conference on Persuasive Technology*

[34] Bijker W E, Hughes T P, Pinch T and Douglas D G 2012 *The Social Construction of Technological System: New Directions in the Sociology and History of Technology Cambridge, MA* MIT Press

[35] Torning K and Oinas-Kukkonen H 2009 Persuasive system design: state of the art and future directions. *In Proceedings of the 4th international conference on persuasive technology ACM* 30

[36] Petty R E and Cacioppo J T 1986 *Communication and Persuasion: Central and Peripheral Routes to Attitude Change* Springer-Verlag New York Inc. New York

[37] Kumar K and Kumar S A 2013 Rule-Based Recommendation System For Selection Of Software Development Life Cycle Models *ACM SIGSOFT Software Engineering Notes* **38(4)** 1-6

[38] Jain R and Suman U 2011 A Systematic Literature Review On Global Software Development Life Cycle *ACM SIGSOFT Software Engineering Notes* 40(2) 1-14

[39] Sommerville I 2010 *Software Engineering* 9th Ed. Pearson Education, Inc. 2010.

[40] Vithani T and Kumar A 2014 Modeling the mobile application development lifecycle *Proceedings of the International MultiConference of Engineers and Computer Scientists* 596-600

[41] Gay L R, Mills G E and Airasian P W 2009 *Educational Research Competencies for Analysis and Applications* 9th edition Columbus, Ohio: Pearson

[42] Zikmund W G, Babin B J, Carr J C and Griffin M 2013 *Qualitative Research Tools. In: Business Research Method* (9th Ed) Canada: Nelson Education, Ltd 131–158

Acknowledgment
This paper was partly sponsored by the Centre for Graduate Studies UTHM.

Performance Enhancement of Inverted Suspended Circular Polarized Antenna with the Integration of Metasurfaces (MS) Structure Technique

H. A. Bakar, M. Z. A. Abd. Aziz, B. H. Ahmad and N. Hassan
Center for Telecommunication Research and Innovation (CeTRI), Faculty of Electronics and Computer Engineering (FKEKK), Universiti Teknikal Malaysia Melaka (UTeM), 76100 Durian Tunggal, Melaka, Malaysia

E-mail: hamizan421@yahoo.com

Abstract. In this paper, it is proposed to use metasurface (MS) structure for the miniaturization and antenna performance enhancement of inverted suspended circular patch with square slot. The antenna was designed based on inverted suspended circular patch with air gap between the patch substrate and copper ground plane is separated with distance of 10 mm (Design A). L-probe technique is used for the antenna design and by using Design A the MS structure is put at the top of the antenna with air gap separation of 1 mm (Design B). The proposed antenna had been designed and simulated by using Computer Simulation Technology (CST) Software. Target application for this antenna design is for Wireless Local Area Network (WLAN) at operating frequency of 2.4 GHz. Simulation and measurement antenna performances in term of return loss, resonant frequency, bandwidth, realized gain, directivity, axial ratio, total efficiency and radiation pattern at the design frequency are investigated and discussed. Based on the result, with the integration of MS, Design B antenna substrate size can be reduced by 23.45 %, yet maintaining the similar performance. Other than that, axial ratio bandwidth (ARBW) for Design B also increases up to 68.8 % from 58.4% (Design A) with impedance bandwidth 627 MHz which is 71 MHz wider than Design A.

1. Introduction

Nowadays, with the great requirement for compact wireless devices, a small effective antenna with common structures continues to be a subject of great attraction [1-4]. According to the characteristics of low profile and simple structure, planar antenna appears to be a great choice for numerous portable wireless devices. However, patch antenna restricted with narrow bandwidth, poor gain and antenna size is in general contrarily proportional to the frequency of operation. The size of radiator for patch antenna is generally 0.5 λ_o x 0.4 λ_o, whereas λ_o is the wavelength of the frequency operation [5]. Thus the size will be larger and not suitable to fit to a small wireless device.

In recent year, a lot of efforts have been made by researchers to compact [6,7] the structure and miniaturize [8,9] the size of planar antenna using metasurface structure for wireless communication system application. Metasurface (MS) is a 2-D planar equivalent of metamaterial structure and has receiving researchers interest in the last few years for planar antenna performance enhancement such as wide bandwidth and high gain [10-18]. Other than that, as in [19,20] it demonstrate that the performances of patch antenna can be enhanced simultaneously by the addition of MS at the top of the antenna design.

Content from this work may be used under the terms of the Creative Commons Attribution 3.0 licence. Any further distribution of this work must maintain attribution to the author(s) and the title of the work, journal citation and DOI.
Published under licence by IOP Publishing Ltd

IOP Conf. Series: Journal of Physics: Conf. Series **1049** (2018) 012026 doi:10.1088/1742-6596/1049/1/012026

In this work, it is proposed to use MS to miniaturize and enhanced antenna performance in term of return loss bandwidth (RLBW) and axial ratio bandwidth (ARBW). A simple design of inverted suspended circular patch with square slot (Design A) is proposed to be integrated with the MS structure. Design A is designed as circular polarization with operated frequency at 2.4 GHz which is for WLAN application. Design B antenna was designed to have a similar design as Design A (antenna without MS) for a comparison purposed. The Design B antenna consists of 8 x 5 pine tree look like MS which is placed on top of the inverted circular patch with 1 mm separation of air gap. With the integration of MS, the patch antenna substrate size can be reduced which is up to 23.45 % of size reduction, yet maintaining quite similar performances in term of axial ratio, realized gain, total efficiency and directivity.

2. Antenna design

The antenna and the MS structure was designed by using FR4 substrate with thickness, h = 1.6 mm, dielectric constant, ε_r of the substrate = 4.4, tangent loss and tan δ = 0.019. Thickness of copper, t = 0.035 mm is used for the ground plane and as the conductive material printed antenna. The basic circular polarized antenna (Design A) is an inverted suspended circular patch with square slot where both feedline and circular patch are printed at the bottom sided of the antenna substrate without any copper patch at other sided of the substrate. The inverted suspended circular patch antenna and its ground are separated by 10 mm air gap as illustrated in Figure 1 (a). The feedline was fed by using 50 Ohm SMA coaxial probe connector and represent as an L-probe technique for the antenna input port.

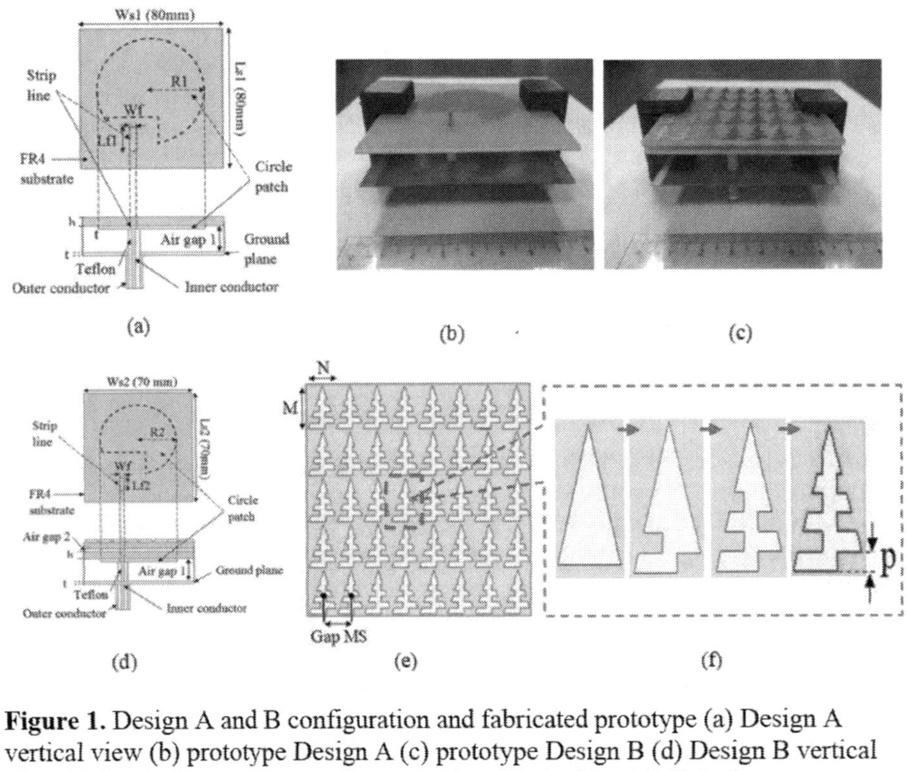

Figure 1. Design A and B configuration and fabricated prototype (a) Design A vertical view (b) prototype Design A (c) prototype Design B (d) Design B vertical view (e) 8 x 5 pine tree metasurfaces top view (f) single unit MS design process view.

The MS (Design B) is adding at the top of the antenna with separation of 1 mm air gap from the inverted suspended antenna substrate as in Figure 1 (d). The MS is designed on the front of a single-sided substrate. This MS consists of 8 x 5 homogeneous pattern of pine tree look like MS where each of the MS is separated by 8.7 mm gap between each other centre as demonstrated in Figure 1 (e). Each of the unit cells are printed periodically along the x-axis and y-axis with having a width of N and length of M. The width (W_s) and length (L_s) of substrate for the inverted suspended antenna, copper ground plane and proposed MS substrate are all equal which is 80 x 80 mm for Design A and 70 x 70 mm for Design B. The design is illustrated as in Figure 1. Optimization on the dimension of the design has been made in order to get the best performance result for both antenna designs. Both antennas are designed to operate at the frequency of 2.4 GHz. The optimize dimension for Design A and Design B antenna are tabulated as in Table 1.

Table 1. Optimize dimension of design A and B.

Design Parameter	Dimension (mm)	Description
Air gap 1	10	Air gap
Air gap 2	1	Air gap
h	1.6	Thickness of substrate
t	0.035	Thickness of copper
R1	25	Design A Circle patch radius
R2	18	Design B circle patch radius
L_{s1}	80	Length of substrate
L_{s2}	70	Length of MS substrate
L_{f1}	13	Length of feedline
L_{f2}	12	Length of MS feedline
Ws1	80	Width of substrate
Ws2	70	Width of MS substrate
W_f	3.1	Width of feed
M	13	Length of metasurface
N	8.7	Width of metasurface
P	1.75	Width of rectangular slot
GapMS	8.7	Gap between metasurface

3. Result and discussion

In this section, simulation and measurement antenna parameter which comprised of return loss (RL), resonant frequency (f_r), bandwidth (BW), return loss bandwidth (RLBW), realized gain, total efficiency, directivity, axial ratio, axial ratio bandwidth (ARBW) and radiation pattern are analysed and discussed at 2.4 GHz frequency. The antenna performance results for Design A and Design B are compared and demonstrated as in Figure 2, 3, 4, 5, 6 and tabulated as in Table 2.

3.1. Return loss (RL), Resonant frequency (f_r) and bandwidth (BW)

The comparison of simulation and measurement result for Design A and Design B are shown as in Figure 2. At 2.4 GHz frequency, the simulation return loss for Design A is -35.64 dB while its measurement result reduced to -20.28 dB. While for Design B, the simulation result is -31.94 dB and it is clearly observe that the measurement result shifted to a higher frequency resulting -11.78 dB return loss at 2.4 GHz. Simulation and measurement resonant frequency for Design A is at 2.398 GHz and

IOP Conf. Series: Journal of Physics: Conf. Series **1049** (2018) 012026 doi:10.1088/1742-6596/1049/1/012026

2.42 GHz with return loss -35.77 dB and -21.47 dB respectively. For Design B, the simulation resonant frequency is at 2.41 GHz with -32.11 dB. However the measurement result is shifted to a higher frequency at 2.56 GHz with -31.54 dB. According to Figure 2, it clearly shows that both measurement bandwidth for Design A and Design B is smaller compared to its simulation bandwidth. From observation, simulation of Design B bandwidth is the largest following by Design A. The simulation bandwidth for Design A is 556 MHz from frequency range of 2.21 GHz to 2.77 GHz. While its measurement bandwidth covering in between the frequency ranges of simulation which is from 2.28 GHz to 2.65 GHz with 362 MHz respectively. For a comparison, Design B has 71 MHz wider simulation bandwidth than Design A which ranging from 2.11 GHz to 2.74 GHz. However, the measurement bandwidth for Design B only 386 MHz covered 2.37 GHz to 2.76 GHz frequency.

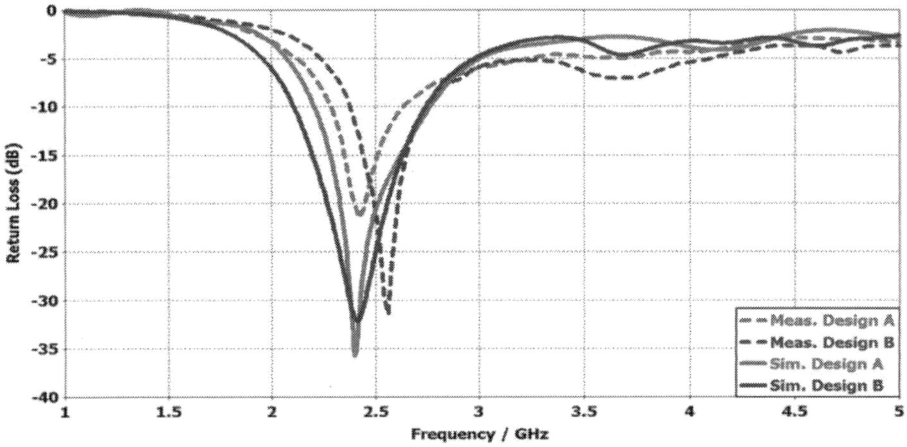

Figure 2. Comparison result of simulation and measurement return loss for Design A and B

3.2. Total efficiency and directivity

Figure 3 and Figure 4 are illustrated the simulation view of the comparison Design A and Design B antenna design for total efficiency and directivity versus frequency. The total efficiency and directivity for both design show good response at 2.4 GHz frequency. Based on Figure 3, maximum total efficiency for both designs is at 2.4 GHz frequency which achieved above -0.5 dB. Design A maximum efficiency is -0.18 dB and the total efficiency covering above 50% ranging from 2 GHz to 3.3 GHz and 3.9 GHz to 4.34 GHz. While Design B maximum efficiency is -0.23 dB with total efficiency above 50 % ranging from 1.89 GHz to 3.18 GHz and 3.54 GHz to 3.93 GHz respectively. According to Figure 4, the maximum directivity for both designs is at 2.5 GHz. The maximum directivity for Design A is 8.18 dBi while at 2.4 GHz the directivity is 7.94 dBi. Directivity for Design B at 2.4 GHz slightly decreases to 7.59 dBi with its maximum directivity is 7.73 dBi respectively.

For a comparison of Design A and Design B, directivity for Design B at 2.4 GHz slightly dropped around 0.35 dBi. However, both antennas obtained more than 7 dBi directivity at that frequency and maintaining quite similar performance result.

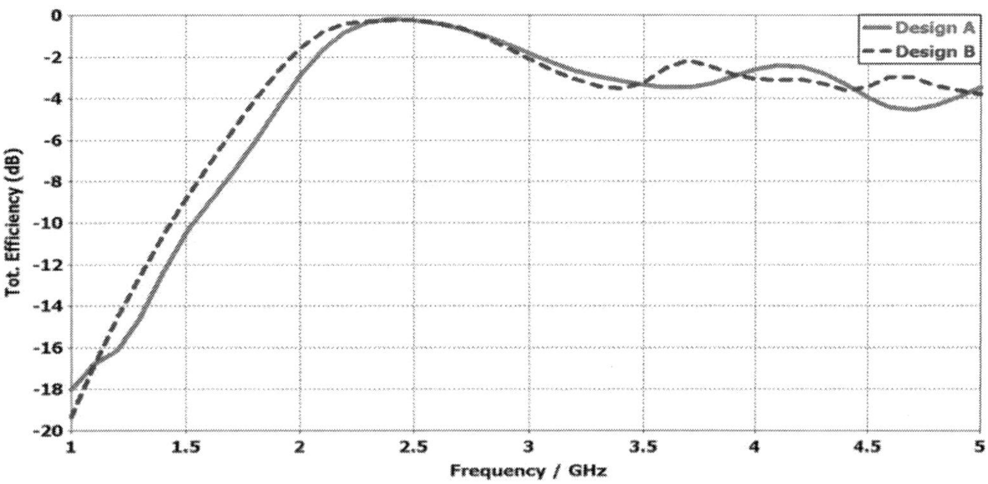

Figure 3. Comparison of simulation total efficiency for Design A and B.

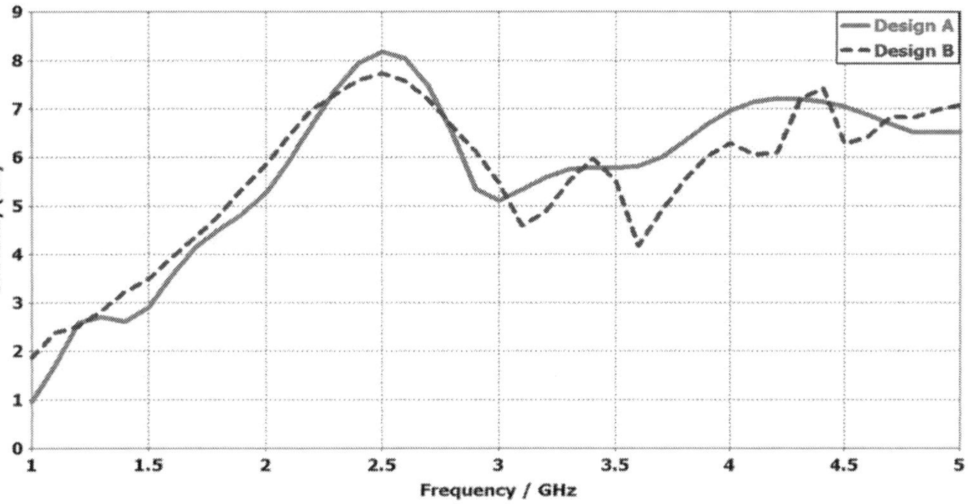

Figure 4. Comparison of simulation directivity for Design A and B.

3.3. Axial ratio

Figure 5 illustrated the simulation axial ratio for Design A and Design B. Based on Figure 5, the minimum axial ratio for Design A is at 2.1 GHz with 0.07 dB. While at 2.4 GHz, the axial ratio is 0.62 dB. Design B has a minimum axial ratio at 1.7 GHz frequency with 0.29 dB and 0.58 dB at 2.4 GHz respectively. Axial ratio bandwidth (ARBW) for Design A is 1127 MHz which covered from frequency range of 1.37 GHz to 2.50 GHz. While Design B covered from 1.44 GHz to 2.95 GHz with 1511 MHz ARBW. In comparison of ARBW for both design, Design B obtained a wider ARBW by 384 MHz. However, both design obtained axial ratio below than 1 dB at 2.4 GHz. It was indicated that both design can represent as a good circular polarization antenna at that frequency

IOP Conf. Series: Journal of Physics: Conf. Series **1049** (2018) 012026 doi:10.1088/1742-6596/1049/1/012026

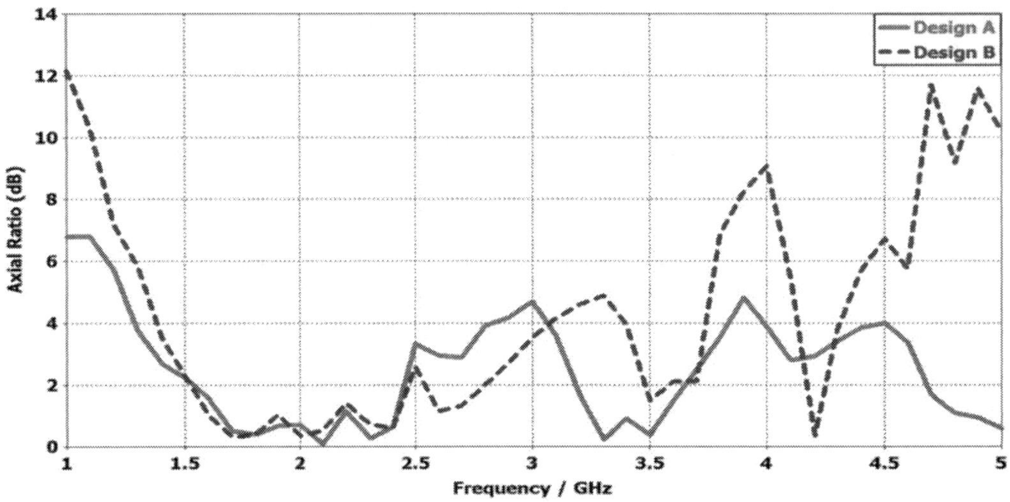

Figure 5. Comparison of simulation axial ratio for Design A and B.

3.4. Radiation pattern

The radiation pattern of the antenna is measured in far field region at 2.4 GHz frequency for phi = 0°, phi = 90° and theta = 90°. Comparison of simulated and measured radiation pattern for Design A and Design B at 2.4 GHz frequency are demonstrated as in Figure 6.

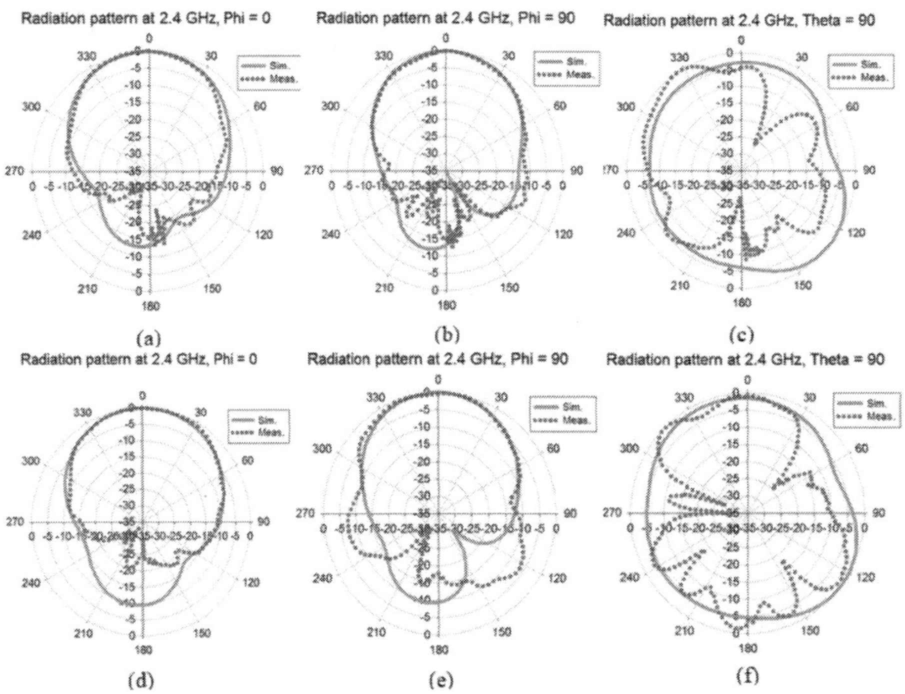

Figure 6. Comparison of simulation and measured radiation pattern for Design A (a) phi = 0° (b) phi = 90° (c) theta = 90° and Design B (c) phi = 0° (d) phi = 90° (f) theta = 90°.

The simulation and measurement result for Design A and Design B at 2.4 GHz are tabulated as in Table 2. Based on Table 2, the result of return loss (RL), realized gain, axial ratio (AR), total efficiency and directivity are demonstrated. The realized gain of Design A and Design B are quite similar which are 7.76 dB and 7.36 dB. However, the calculated realized gain based on data taken from measurement in far field result a decrease in gain performance to 5.32 dB and 3.47 dB.

Table 2. Comparison of simulation and measurement result performance for Design A and Design B at 2.4 GHz.

Antenna parameter	**Design A (80 x 80 mm)**		**Design B (70 x 70 mm)**	
	Sim.	**Meas.**	**Sim.**	**Meas.**
Return loss	-35.64	-20.28	-31.94	-11.78
Realized gain	7.76	5.32	7.36	3.47
Axial ratio	0.62	-	0.58	-
Total efficiency	-0.18	-	-0.23	-
Directivity	7.94	-	7.59	-

4. Conclusion

In this paper, using MS to miniaturize the size of inverted suspended circular patch antenna (Design A) and widen the impedance bandwidth and axial ratio bandwidth of the antenna has been proposed in this study. The MS (Design B) which consisting of 8 x 5 pine tree look like element is put atop of the Design A antenna with 1 mm air gap separation. Based on the results, it shown with the integration of MS, the inverted circular patch antenna substrate size can be reduced which is up to 23.45 % of size reduction from 80 x 80 mm to 70 x 70 mm. Other than that, Design B also have a wider antenna bandwidth up to 627 MHz compared to Design A which is 556 MHz, yet maintaining a quite similar performances in terms of realized gain, axial ratio, total efficiency and directivity at the same operating frequency of 2.4 GHz. Moreover, the axial ratio bandwidth (ARBW) for Design B also increases up to 68.8 % from 58.4% (Design A). Overall, with the integration of MS structure, the antenna (Design B) performance was enhanced simultaneously. The Design B antenna can represent as a good candidate for broadband circular polarization antenna for WLAN application at 2.4 GHz frequency.

References

[1] Ziolkowski R W, Ming-Chun T and Ning Z 2013 *In: International Workshop on Antenna Technology* 175–178
[2] Sun X L, Cheung S W and Yuk T I 2014 *Microwave Opt. Technol. Lett.* **56** 1360–1366
[3] Li L, Sing Wai C and Tung Ip Y 2014 *IET Microwaves Antennas Propag.* **8** 1021–1029
[4] Sun X L, Cheung S W and Yuk T I 2013 *Microwave Opt. Technol. Lett.* **55** 1765–1770
[5] Balanis C A 2016 *Antenna Theory; Analysis and Design 4th Edition, John wiley & Sons, Inc.*, New York
[6] Yue T, Jiang Z H and Werner D H 2016 *IEEE Transactions on Antennas and Propagation* **64** issue 5 1595-1606
[7] Majumder B, Kandasamy K, Mukherjee J and Ray K P 2015 *Electronics Letters* **51** issue 17 1310-1312
[8] Dong Y, Toyao H and Itoh T 2012 *IEEE Transactions on Antennas and Propagation* **60** issue 2 PART 2 772-785

[9] Zhu H L, Cheung S W and Yuk T I 2015 *Microwave and Optical Technology Letters* **57** issue 9 2050-2056
[10] Ta S X and Park I 2015 *IEEE Transactions on Antennas and Propagation* **63** issue 12 5929-5934
[11] Wang R-Q, Jiao Y-C, Lu L and Zhang H 2016 *Progress in Electromagnetics Research Letters* **62** 1-8
[12] Nasimuddin N, Chen Z N and Qing X 2016 *IEEE Antennas and Propagation Magazine* **58** issue 2 39-46
[13] Rajak N and Chattoraj N 2017 *Microwave and Optical Technology Letters* **59** issue 10 2575-2580
[14] Chung K L and Chaimool S 2012 *Microwave and Optical Technology Letters* **54** issue 2 529-532
[15] Islam M T, Ullah M H, Singh M J and Faruque M R I 2013 *Materials* **6** issue 8 3226-3240
[16] Hu P F and Pan Y M 2017 *Asia-Pacific Microwave Conference Proceedings* 1-3
[17] Zhao Y 2016 *AIP Advances* **6** issue 5 1-8
[18] Zhou C, Cheung S W, Li Q and Li M 2017 *Applied Physics Letters* **110** issue 21 1-5
[19] Zhu H L, Cheung S W, Liu X H, Cao Y F and Yuk T I 2014 *In: 8th European Conference on Antennas and Propagation* 2575–2577
[20] Chung K L and Chaimool S 2011 *In: Proceedings of the 5th European Conference on Antennas and Propagation* 48–52

Acknowledgments
Authors wishing to acknowledge Universiti Teknikal Malaysia Melaka (UTeM) for supporting in this research work under the grant PJP/2017/FKEKK/HI13/S01540. Other than that, special thanks to the Government of Malaysia which provide MyBrain15 program for financial support of this work under the research grant TRGS/1/2014/FKEKK/02/1/D00001 from Ministry of Higher Education (MOHE).

Influence of riparian vegetation on flow resistance in mobile bed straight compound channels

M Jumain[1*], Z Ibrahim[1], Z Ismail[1], K Entalai[1], M R Makhtar[1], M S A Rahman[1] and N Alias[1],

[1] Department of Hydraulic and Hydrology, Faculty of Civil Engineering, Universiti Teknologi Malaysia, 81310 Johor Bahru, Johor, Malaysia

E-mail: mazlinjumain@gmail.com

Abstract. Riparian vegetation is an important feature in rivers and essential in hydraulic engineering including flood management and river restoration. Deforestation activities have been pointed out as a contributing factor to the severity of flood damages. This shows that the importance of riparian vegetation in river restoration schemes in order to lessen the issues of flood risk and flood protection. The riparian vegetation creates additional resistance to flood flow which resulted in velocity reduction, increasing of flow depth and transportation of sediment in rivers. The hydraulics of flood flows in vegetated floodplain of mobile bed straight channel was experimented in the Hydraulics Laboratory, Faculty of Civil Engineering, Universiti Teknologi Malaysia. Two-lined steel rods with tandem and staggered arrays to simulate as rigid emergent vegetation were placed along the riparian zone of an asymmetrical channel. The flow resistance including Manning's n, Darcy's f and drag force, F_D were studied. The findings showed that the riparian vegetation had increased the Manning's n by 20%. Meanwhile, staggered arrangement of vegetation was found to generate the highest Darcy's f of about 16% as compared to tandem arrangement. The presence of riparian vegetation also found to induce 20% more drag force compared to the main channel roughness effect. The dramatically changes of bed morphology also led to the increase of the flow resistance in the channel.

1. Introduction

Deforestation activities on floodplain areas for agricultural, commercial or housing purposes have raised the issues of flood risk and flood protection [1]. Vegetation along river banks were pointed out to play important roles for river banks stabilisation, erosion prevention, habitat creation and riparian buffer zones development for flood protection [2]. Generally, vegetation may be regarded as kind of surface roughness which increases the flow resistance, alters velocity distribution, affects the conveyance of rivers, raises water levels and also influences a transportation rate of sediment in rivers [3]. Therefore, it is important to further understand the floodplain vegetation influence on the river hydraulics.

Several researchers have shown that vegetation properties have enormous influence on flow resistance [4, 5, 6 and 7]. Bouma et al. [8] stated that the factors affecting flow resistance including surface roughness, vegetation, channel size, shape and irregularity. Vegetation markedly reduces the flow capacity of the channel and retards the flow. In particular, drag acting on the vegetation elements

[1*] mazlinjumain@gmail.com.

Content from this work may be used under the terms of the Creative Commons Attribution 3.0 licence. Any further distribution of this work must maintain attribution to the author(s) and the title of the work, journal citation and DOI.
Published under licence by IOP Publishing Ltd

reduces the mean flow velocity and the momentum exchange between main channel and floodplain [9]. As stem density continually increases, it significantly alters the flow patterns with decreases the flow velocity and affects the turbulence properties [10, 11 and 12]. It was due to the contribution of drag forces induced by the emergent vegetation.

Vegetation characteristics such as height, density, distribution, stiffness, and type of vegetation have remarkable impact on the flow velocity as well as flow resistance. The reduced velocity will then reduce the forces necessary to cause scour and erosion processes that influenced the changes in channel morphology [13 and 14]. Simons and Richardson suggested that the flow resistance at low flows was due to the grain roughness, while at higher flows was due to bed form roughness from dune formation [15]. Knight and Brown [16] reported that the bed will deform under the action of flow, changing its roughness, and then affecting the flow itself.

Experimental investigations on the influence of riparian vegetation on flow resistance in straight compound mobile bed channels had been undertaken. The experiment in the physical model was conducted in the Hydraulic Laboratory in Universiti Teknologi Malaysia. The study was limited to an asymmetric straight mobile bed channel which involved shallow and deep flood inundation conditions.

2. Methodology

The experiments were carried out in the Hydraulics Laboratory, Faculty of Civil Engineering, Universiti Teknologi Malaysia where the experimental flume was 12 m long, with 0.5 m wide and 0.1 m deep rectangular main channel and a 0.5 m wide floodplain. The flume bed was set at a gradient of 0.1%. Figures 1 and 2 illustrate the layout of experimental set-up and the cross-sectional configuration of the flume.

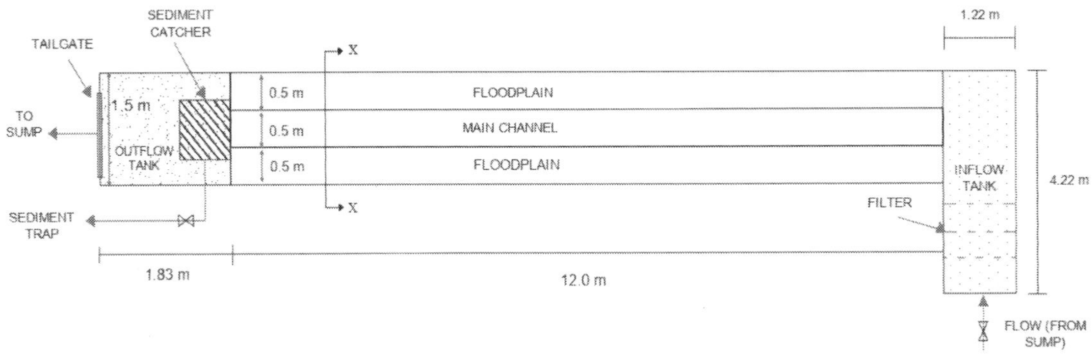

Figure 1. Plan view of the flume.

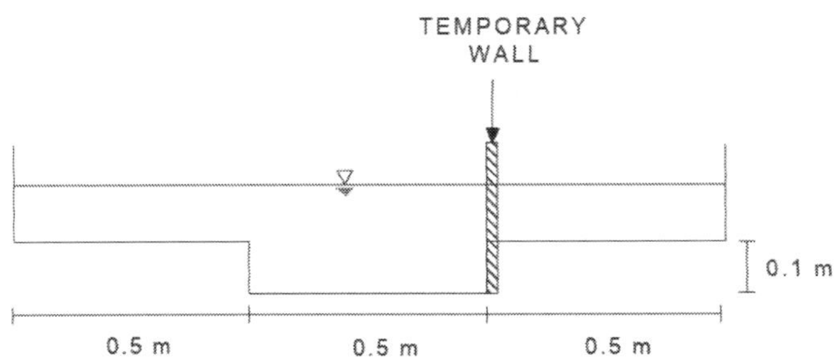

Figure 2. Cross-sectional view of the flume.

Uniform graded sand with mean diameter, d_{50} of 0.8 mm was filled in the main channel to form as bed material. A similar uniform sediment size was also used by many researches including Knight and

Brown [16], Atabay et al. [17], Tang and Knight [18] and Bousmar et al. [19]. It is hard to find a river bed with a uniform size of sediment particles. The main reason for using uniform graded sand in this research was to minimise the "sheltering" and "hiding" effects. Ismail [20] did mention that as bed forms propagate to the downstream, sediment moves from the ridge of the bed forms to the trough. In the trough, the sediment is sheltered and overlaid by the advancing grains from the upstream bed forms.

Two-line artificial rigid emergent riparian vegetation was placed on the floodplain as resistance elements. Steel rods with 5 mm diameter were used as artificial resistance elements in this study. The steel rods were arranged in two-line either with tandem and staggered arrays as illustrated in Figure 3. The rod spacing L of $2d$ was studied where d represented the diameter of steel rod. The resistance elements were placed at the edge along the floodplain at a distance between 4 m to 8 m from the channel inlet.

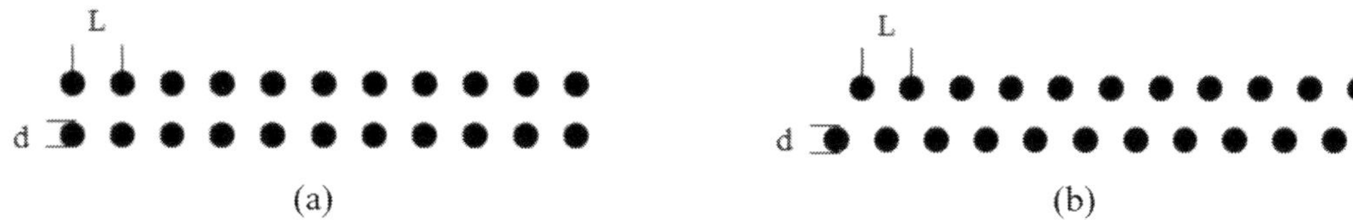

Figure 3. Arrangement of steel rods in (a) tandem and (b) staggered

A re-circulating flume system was used in the study and the discharge was measured using a Portaflo PF 330 flow meter. A flow depth in the channel was controlled by an adjustable a tailgate located at downstream of flume. Meanwhile, flow depth and bed morphology levels were measured using a point gauge or digital water surface profiler attached on a mobile carrier. The digital profiler was operated with an accuracy of $\pm$ 0.1 mm and the 3-dimensional velocities were measured using a Nortek Vectrino+ Acoustic Doppler Velocimeter (ADV) with a sampling rate of 100 *Hz*. The maximum recording time for point velocity was 2 minutes. For most turbulent statistics, sufficient record length for measurement was 60 s to 90 s [21].

The measurement sections were located at normalised longitudinal distance x/L of 0.38, 0.50 and 0.63 for non-vegetated floodplain, while at normalised longitudinal distance x/L of 0.25, 0.38, 0.50, 0.63 and 0.75 along the compound straight channel for vegetated floodplain. For both of shallow and deep relative flow depths, the calculated Reynolds number (Re) exceeded 2000 and the Froude number (Fr) was less than 1. Therefore, the regimes of flows were classified as sub-critical with turbulent condition. The relative depth (DR) is defined as $(H - H_{mc})/H$ where H represents the total flow depth in the main channel and H_{mc} is the depth of main channel.

3. Discussion of results

The experimental results on flow resistance include Manning's n, Darcy's f and drag force, F_D through rigid riparian vegetation in compound straight channels are discussed in this section. The classification of experimental cases in this study is shown in Table 1.

Table 1. Classification of cases studied in laboratory experiments.

Case	Condition
A	Non-vegetated
B	Vegetated in Tandem Arrangement
C	Vegetated in Staggered Arrangement

3.1. Manning's n

The flow resistance is often expressed in terms of a resistance coefficient such as Manning's n or Darcy's f. The Manning's n remains the most commonly used in flow resistance formula in hydraulic engineering. The standard equation for calculating Manning's n is as given in equation (1):

$$n = \frac{R^{2/3} S_o^{\ 1/2}}{U} \tag{1}$$

where R is hydraulic radius, S_o is channel bed slope and U is mean stream-wise velocity. The flow resistance in a compound channel is represented by the Manning's n value for each interval of normalised longitudinal distance (x/L). x is longitudinal distance and L is total length of the channel. n_{fp} and n_{mc} are the Manning's n for floodplain and main channel, respectively.

Figures 4 and 5 illustrate the variation of Manning's n_{fp} and n_{mc} along the channel for relative depths of 0.30 and 0.50. It can be seen that the Manning's n increased along the channel length in downstream direction due to the effect of surface roughness. The minimum Manning's n occurred in Case A and the maximum Manning's n resulted in the Case C with staggered arrangement of riparian vegetation. The n_{mc} values are higher of about 20% than the n_{fp} for both relative depths in all cases. As suggested by Einstein and Barbarossa [22], Engelund [23] and Simons and Richardson [15], flow resistance in low flow was due to the grain roughness; meanwhile, bed form roughness due to dune formation was important for increased of flow resistance at higher flow in mobile bed channels. Zhang et al. [24] and Ali et al. [25] reported that the roughness of mobile bed noticeably was higher than non-mobile bed channel due to variance of formation bed morphology.

As the floodplain was roughened with steel rods or vegetation, the Manning's n increased. The average increment of n_{fp} for Cases B and C were 20% and 30% as compared to Case A for the relative depths of 0.30 and 0.50. This indicates that staggered array of vegetation is pronounced to have larger value n due to higher flow resistance by vegetation with that arrangement. Ibrahim et al. [26] and Jumain et al. [27] also found that the roughness coefficient for staggered vegetation arrangement was higher than tandem arrangement. However, the percentages difference between vegetation arrangements were not too high since the study was only concentrated on two-lined emergent riparian vegetation placed at the middle of channel length.

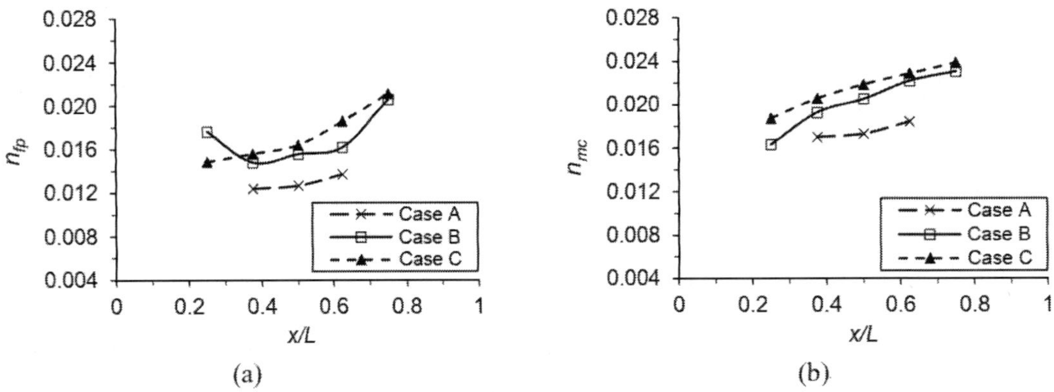

Figure 4. Manning's n profiles of (a) floodplain and (b) main channel regions for shallow relative flow depth of 0.30

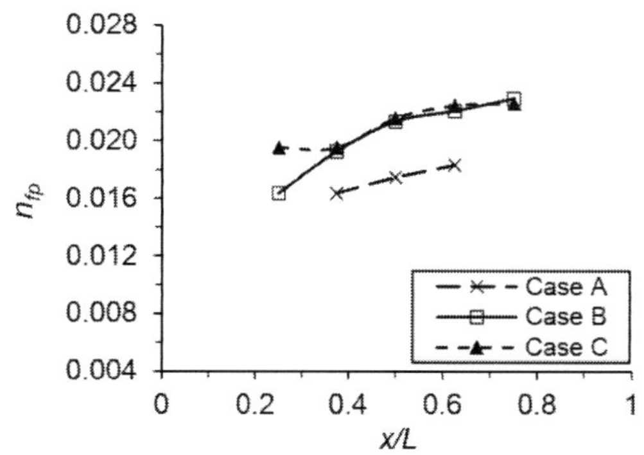

(a)

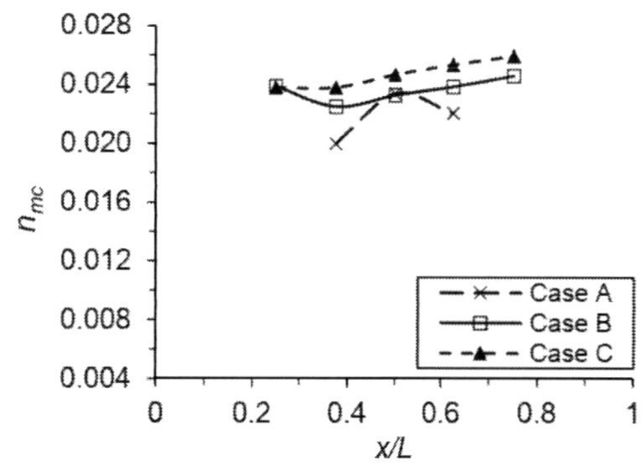

(b)

Figure 5. Manning's *n* profiles of (a) floodplain and (b) main channel regions for deep relative flow depth of 0.50

3.2. Darcy's f

The flow resistance in open channel can also be expressed in term of the Darcy's friction factor, *f* calculated using the measured stream-wise velocity data *U* as given in the equation (2):

$$f = \frac{8gRS_o}{U^2} \tag{2}$$

where g is gravitational acceleration, R is hydraulic radius, S_o is channel bed slope and U is mean streamwise velocity. The sectional mean velocities U_s for floodplain, vegetated zone and main channel were used to determine the zonal *f* values in the study.

Table 2 presents the computed *f* values in each section for all cases at relative depths of 0.30 and 0.50. As relative depth increased, the f values also increased. The highest increment of the *f* values occurred in floodplain areas. As the relative depth changed from 0.30 to 0.50, the mean increment of the *f* values for Cases B and C in floodplain zones were about 18% and 20%, while in main channel zone increased by 6% and 8% as compared to Case A. The influence of main channel bedform roughness can be seen for Cases B and C at shallow relative flow depth of 0.30, where the *f* values were larger than values in vegetated and floodplain zones. Similar finding was reported by Bousmar et al. [19].

The influence of vegetation on the flow resistance was obvious since the *f* values in the vegetated zone were larger than in main channel and floodplain. The *f* values in the vegetated zone also increased as the relative depth changed from 0.30 to 0.50. This shows that resistance induced by the vegetation to the flow increases when the flow depth rises. It can be clearly seen that the highest *f* values occur in Case C which are particularly in the vegetation zone of about 16% compared to Case B. Hence, staggered arrays of riparian vegetation generate the highest flow resistance and reduce the floodplain flow velocity.

Table 2. Zonal Darcy's f for cases studied in compound straight channels.

Case	*f* for *DR* = 0.30			*f* for *DR* = 0.50		
	Floodplain	Vegetation	Main Channel	Floodplain	Vegetation	Main Channel
A	0.039	-	0.067	0.054	-	0.077
B	0.051	0.107	0.109	0.071	0.122	0.116
C	0.064	0.118	0.119	0.076	0.136	0.128

3.3. Drag force, F_D

The presence of vegetation in main channel or floodplain produces drag force which increases flow resistance. The estimation of vegetation influence on total flow resistance must consider the drag force exerted by the vegetation on the flow. The classical approach to measure total drag force exerted by vegetation or an object is as given in equation (3):

$$F_D = \frac{1}{2}\rho\, C_D\, A_p U_d^{\,2} \tag{3}$$

where ρ is the water density, C_D is the drag coefficient, A_p is the projected area of stem in a streamwise direction, and U_d is the depth-averaged approach velocity.

The drag force can also be estimated by a force balance approach. The force balance approach considers the forces in equilibrium exerted on a control volume for quasi-uniform flow conditions. Nezu and Sanjou [28], Wu [29], Terrier [30] and Ibrahim [31] are among researchers who applied the force balance approach to evaluate the drag of the vegetation with cylindrical stems. The force balance approach is expressed in equation (4), where the drag force F_D is expressed as per unit channel length as follow:

$$F_D = \rho g H S_o - \int_P \tau_b(y)dy \tag{4}$$

where ρ is the water density, g is the gravitational acceleration, H is the water depth, S_o is channel bed slope, $\tau_b(y)$ is the boundary shear stress in y direction and P is the channel wetted perimeter.

The calculation carried out considered (i) forces across the whole channel section, $F_{D(whole)}$ and (ii) forces on the floodplain section, $F_{D(fp)}$ in non-vegetated and riparian vegetated compound straight channels are summarised in Table 3. It can be seen that the values of $F_{D(fp)}$ and $F_{D(whole)}$ were almost equal for each interval of normalised longitudinal distance (x/L) along the straight channels. It can be seen for each case at both relative flow depths. This is due to the similarity of boundary shear stress distributions along the channel length in downstream direction.

Table 3. Calculated drag force F_D using force balance approach along the channel length for shallow and deep flood inundations.

DR	*x/L*	Case A		Case B		Case C	
		$F_{D(fp)}$	$F_{D(whole)}$	$F_{D(fp)}$	$F_{D(whole)}$	$F_{D(fp)}$	$F_{D(whole)}$
	0.25	-	-	0.371	0.444	0.380	0.464
	0.38	0.234	0.418	0.288	0.448	0.314	0.467
0.30	0.50	0.238	0.418	0.295	0.448	0.320	0.468
	0.63	0.249	0.418	0.299	0.448	0.334	0.468
	0.75	-	-	0.381	0.446	0.403	0.465
	0.25	-	-	0.926	1.051	1.003	1.120
	0.38	0.770	0.991	0.858	1.056	0.932	1.124
0.50	0.50	0.781	0.992	0.875	1.056	0.944	1.124
	0.63	0.789	0.992	0.884	1.056	0.953	1.124
	0.75	-	-	0.963	1.053	1.032	1.121

As observed in Case A, the floodplain surface and main channel bed morphology also give such contribution to the drag force. In order to quantify the effect of vegetation in the compound channel, the analysis on percentage difference between non-vegetated and vegetated floodplain had been carried out. The percentage difference of drag force ΔF_D due to the emergent riparian vegetation is computed using equation (5):

$$\Delta F_D\ (\%) = \frac{\left(F_{D_{veg}} - F_{D_{NV}}\right)}{F_{D_{NV}}} x\ 100 \quad (5)$$

in which $F_{D_{veg}}$ is the drag force in vegetated case and $F_{D_{NV}}$ is the drag force in non-vegetated case.

The calculated percentage differences of ΔF_D for Cases B and C compared to Case A are as illustrated in Figures 6 and 7. The largest ΔF_D was observed in floodplain area at location $x/L = 0.50$ for Case C which was about 35%. The ΔF_D for Cases B and C in the main channel section were obtained with average difference range from 5% to 6% and 4% to 9% for relative depths of 0.30 and 0.50, respectively. Larger ΔF_D values were obtained in floodplain section. For example, ΔF_D of more than 20% for a relative depth of 0.30 in Figure 6(a) clearly indicates that the floodplain vegetation section induced more drag force compared to the main channel roughness effect in Figure 6(b). It was also found that the effect of riparian vegetation on F_D was reduced as the flood flow depth increased (Figure 7).

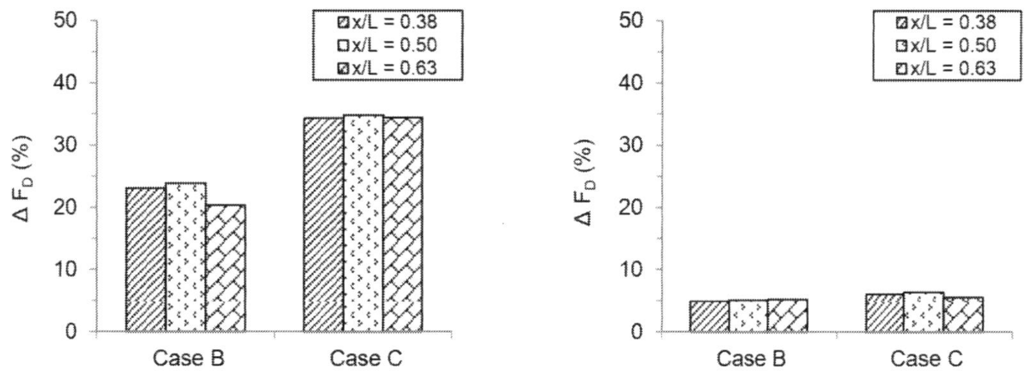

(a) (b)

Figure 6. Variation of percentage difference ΔF_D in (a) floodplain and (b) main channel sections of compound straight channel flows at relative depth of 0.30

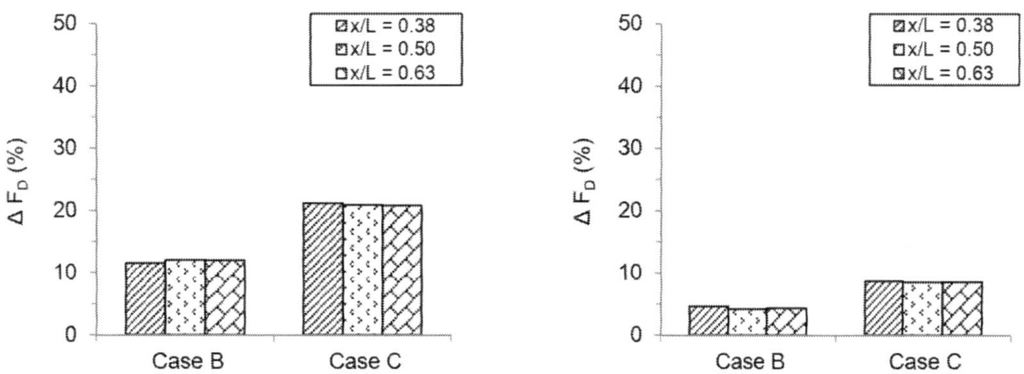

(a) (b)

Figure 7. Variation of percentage difference ΔF_D in (a) floodplain and (b) main channel sections of compound straight channel flows at relative depth of 0.50

4. Conclusion

The influence of riparian vegetation on flow resistance in mobile bed straight channels for shallow and deep overbank flows had been investigated through flume simulations. The findings of the experimental study were: (i) riparian vegetation increased the flow resistance which resulted larger

Manning's *n* in deeper flood flow, (ii) the *f* values in the vegetated compound straight channels also increased as the relative depth changed from 0.30 to 0.50. Staggered arrays of riparian vegetation was found to generate the highest *f* values of about 16% compared to tandem arrays case, (iii) a significant variation of bed morphology patterns contributed to the drag force values, (iv) the presence of riparian vegetation increased the drag force as compared to the effect of main channel bed morphology.

References

[1] Lyness J F, Myers W R C, Cassells J B C and O'Sullivan J J 2001 The influence of planform on flow resistance in mobile bed compound channels *Proceedings of the Institution of Civil Engineers Water, Maritime and Energy* **148(1)** 5-14

[2] Palmer M, Allan J D, Meyer J and Bernhardt E S 2007 River restoration in twenty-first century: century: data and experimental knowledge to inform future efforts *Restoration Ecology* **15(3)** 472-481

[3] Aberle J and Jarvela J 2013 Flow resistance of emergent rigid and flexible vegetation *J. Hydraulic Research* **51(1)** 33-45

[4] Jarvela J 2005 Effect of submerged flexible vegetation on flow structure and resistance *J. Hydrology* **307(1-4)** 233-241

[5] Nepf H M and Ghisalberti M 2008 Flow and transport in channels with submerged vegetation *Acta Geophysics* **56(3)** 753-777

[6] Wilson C, Hoyt J and Schnauder I 2008 Impact of foliage on the drag force of vegetation in aquatic flows *J. Hydraulic Engineering* **134(7)** 885-891

[7] Arroyave J A V and Crosato A 2010 Effect of river floodplain, lowering and vegetation cover *Water Management* **163(9)** 457-467

[8] Bouma T J, Temmerman S, van Duren L A and Herman P M J 2013 Organism traits determine the strength of scale-dependent bio-geomorphic feedbacks: a flume study on three intertidal plant species *Geomorphology* **180-181** 57-65

[9] Nepf H M and Vinoni E R 2000 Flow structure in depth-limited, vegetated flow *J. Hydraulic Research* **105(C12)** 28547-28557

[10] Bennett S J, Pirim T and Barkdoll B D 2002 Using simulated emergent vegetation to alter stream flow direction within straight experimental channel *Geomorphology* **44(1-2)** 115-126

[11] Zong L and Nepf H 2010 Flow and deposition in and around a finite patch of vegetation *Geomorphology* **116(3-4)** 363-372

[12] Le Bouteiller C and Venditti J G 2014 Vegetation-driven morphodynamic adjustments of sand bed. *Geophysical Research Letters* **41** 3876-3883

[13] Curran J C and Hession W C 2013 Vegetative impacts on hydraulics and sediment processes across the fluvial system *J. Hydrology* **505** 364-376

[14] Vargas-Luna A, Crosato A and Uijttewaal W S J 2015 Effects of vegetation on flow and sediment transport: comparative analyses and validation of predicting models *Earth Surface Processes and Landforms* **40(2)** 157-176.

[15] Simons D B and Richardson E V 1961 Forms of bed roughness in alluvial channels *J. Hydraulics Division, ASCE* **87(HY3)** 87-105

[16] Knight D W and Brown F A 2001 Resistance studies of overbank flow in rivers with sediment using the Flood Channel Facility *J. Hydraulic Research* **39(3)** 283-301

[17] Atabay S, Knight D W, and Seckin G 2005 Effects of overbank flow on fluvial sediment transport rates *Proc. Instn. Civ. Engrs. Water Management* **158(3)** 25-34

[18] Tang X and Knight D W 2006 Sediment transport in river models of overbank flows *J. Hydraulic Engineering* **132(1)** 77-86

[19] Bousmar D, Atabay S, Knight D W and Zech Y 2006 Stage-discharge modelling in mobile bed compound channel *Proceedings of River Flow 2006 International Conference on Fluvial*

Hydraulics Lisbon September 6-8 333-341

[20] Ismail Z 2007 *A Study of overbank flows in non-vegetated and vegetated floodplains in compound meandering channels.* Ph.D thesis, Loughborough University, **UK**

[21] Czernuszenko W, Kozial A and Rowinski P M 2007 Measurement of 3D turbulence structure in a compound channel *Archieves of Hydro-Engineering and Environmental Mechanics* **54(1)** 55-73

[22] Einstein H A and Barbarossa N L 1952 River channel roughness. *Transactions of the ASCE* **117** 1121-1132

[23] Engelund F 1966 Hydraulic resistance of alluvial streams. *J. Hydraulics Division, ASCE* **92(HY2)** 315-326

[24] Zhang G, Luo R, Cao Y, Shen R and Zhang X 2010 Impacts of sediment load on Manning coefficient in supercritical shallow flow on steep slopes *Hydrological Processes* **24** 3909-3914

[25] Ali M, Sterk G, Seeger M, Boersema M and Peters P 2012 Effect of hydraulic parameters on sediment transport capacity in overland flow over erodible beds *Hydrology and Earth System Science* **16** 591-601

[26] Ibrahim Z, Marzuki M F, Ismail Z, Jumain M, Ibrahim N H, Harun S and Katimon A 2012 Flood flow characteristics in vegetated compound straight channels: physical models observations. *Physical Modelling Application and Technologies (PHYMAT2012), National Research Institute of Malaysia, Seri Kembangan* September, 1-9

[27] Jumain M, Ibrahim Z, Ismail Z, Ibrahim N H, Makhtar M R and Nordin M R 2013 Influence of two-line emergent floodplain vegetation on a straight compound channel flow *Int. J. of Integrated Engineering* **5(4)** 58-63

[28] Nezu I and Sanjou M 2008 Turbulence structure and coherent motion in vegetated canopy open-channel flows *J. of Hydro-environment Research* **2(8)** 62-90

[29] Wu F S 2008 Characteristics of flow resistance in open channels with non-submerged rigid vegetation *J. Hydrodynamics* **20(2)** 239-245

[30] Terrier B 2010 *Flow characteristics in straight compound channels with vegetation along the main channel* Ph.D thesis Loughborough University, UK

[31] Ibrahim Z 2015 *Flow behaviour due to floodplain roughness along riparian zone in compound Channels* Ph.D thesis Universiti Teknologi Malaysia, Johor Bahru, **Malaysia**

Acknowledgments
The financial of this research were supported by Universiti Teknologi Malaysia under vote numbers 03J97 and 02K00. The authors wish to express a sincere appreciation to others who were involved either directly or indirectly during the implementation of this experimental research.

IOP Conf. Series: Journal of Physics: Conf. Series **1049** (2018) 012028 doi:10.1088/1742-6596/1049/1/012028

Permeability and Strength of Porous Concrete Paving Blocks at Different Sizes Coarse Aggregate

N H Abd Halim[1], H Md Nor[1], R P Jaya[1,3], A Mohamed[1], M H Wan Ibrahim[2], N I Ramli[3] and F M Nazri[4]
[1]Faculty of Civil Engineering, Department of Geotechnics and Transportation, Universiti Teknologi Malaysia, 81310 UTM Johor Bahru, Malaysia
[2]Faculty of Civil and Environmental Engineering, Universiti Tun Hussein Onn Malaysia, 86400 Batu Pahat, Johor Bahru, Malaysia
[3]Faculty of Civil Engineering and Earth Resources, Universiti Malaysia Pahang, 26300 Gambang, Pahang, Malaysia
[4]School of Civil Engineering, Universiti Sains Malaysia, Engineering Campus, 14300 Nibong Tebal, Penang, Malaysia

E-mail: ramadhansyah@utm.my

Abstract. Porous concrete paving block (PCPB) is a block with continuous voids which are intentionally incorporated into concrete. The permeability and strength of PCPB with different sizes of coarse aggregate was presented in this study. Three different sizes of coarse aggregate were used namely passing 10 mm retained 5 mm (as control), passing 8 mm retained 5 mm (CA 5 – 8) and passing 10 mm retained 8 mm (CA 8 – 10). Furthermore, a series of test were conducted in this study such as compressive strength, porosity and permeability. It was found that the size of coarse aggregate affects the strength and porosity of the specimens. The result also shown that PCPB with CA 8 – 10 caused in low strength, but high in porosity and permeability compared to the other blocks. Beside that PCPB using CA 8 – 10 is able to remove surface runoff efficiently.

1. Introduction

Rapid economic growth in the last two decades has turned many sub-urban areas into busy cities, which created unpredictable urbanization effects. Coupled with the global climate change and also the large ratio of road compares paved to water-permeable (porous) surface road, these in turn lead to increased volume of surface runoff and flash flood. High volume of surface runoff, low skid resistance, minimum appearances, glace and splash water is the factors contributing to the increased number of accidents during rainy weather. Recently, road pavements with a water-permeable (porous) surface have been increasingly used. Porous surface for pavements are adopted for various reasons, such as to increase the driving safety during rainy weather, to reduce the traffic noise, to recharge ground-water and to alleviate the heat island phenomena in urban city centres. It is designed to reduce surface runoff by allowing infiltration [1]. Unlike traditional installation of concrete, porous concrete usually contains a void content 15 % to 25 %, which allows water to infiltrate directly through the pavement surface to the subsurface [2].

Content from this work may be used under the terms of the Creative Commons Attribution 3.0 licence. Any further distribution of this work must maintain attribution to the author(s) and the title of the work, journal citation and DOI.
Published under licence by IOP Publishing Ltd

Porous block for Concrete Blocks Pavement nowadays enthused by many developed countries. It is known as porous concrete paving blocks, permeable or pervious interlocking concrete paving blocks. Hassani et al. [3] reported that permeable concrete block pavement is suitable for trafficking and also acts as the drainage system Other than that, porous surface in parking lots are believed to help infiltration and the cleansing of storm water [4]. Three quarters of the volume of PCPB is occupied by aggregates and its physical characteristics differ greatly from conventional concrete block [5]. The use of open-graded aggregate with little or no fine aggregates can achieve a continuous void network inside the block [6]. In addition, previous studies found that between 60 to 65 percent reduction in compressive strength of conventional concrete blocks, but increased in porosity when the fine aggregate was eliminated 100 percent of the mixed [7]. It also reported that, the compressive strengths in the range of 4 MPa to 25 MPa for 28 days (600 psi to 3,600 psi) [8,9]. In this experimental study, the effects of coarse aggregate sizes on the performance of the PCPB were extensively examined. The changes in voids content, porosity and permeability of PCPB subjected to the sizes of coarse aggregate were studied.

2. Materials and methods

2.1. Ordinary Portland cement (OPC)

The porous concrete mixtures were prepared by using Ordinary Portland cement (OPC) as a major binder. According to Hainin et al. [10], the chemical composition of the OPC was within the standard range of 70% CaO, 17.8% SiO_2, 3.9% Al_2O_3, 3.2% Fe_2O_3, 1.5% MgO, and 3.6% SO_3. The OPC similarly indicated a compound composition of 54.5% C_3S, 18.2% C_2S, 9.4% C_3A, and 10.5% C4AF.

2.2. Coarse aggregate

In this investigation, crushed granite as coarse aggregate at three different sizes was used i.e. passing 10 mm retained 5 mm (Control), passing 8 mm retained 5 mm (CA 5 – 8) and passing 10 mm retained 8 mm (CA 8 – 10). However, fine aggregate was eliminated in this investigation. Table 1 summarize the physical properties of the aggregate.

Table 1. Properties of coarse aggregate at different sizes

Size of aggregate (mm)	Type of aggregate	Specific gravity	Water Absorption (%)	Flakiness Index (%)	Elongation Index (%)
Control	Crushed granite	2.57	1.18	19.1	12.1
CA 5 – 8		2.59	1.41	27.1	3.8
CA 8 – 10		2.58	1.50	4.3	26.4

2.3. Permeability test

Permeability test was carried to determine the relative permeability of PCPB. Falling head permeability test was used by using a flexible wall permeameter (Figure 1) and the permeability was measured in term of its discharge time in seconds, which indicate the time taken for a specified volume of water permeate through the sample. The coefficient of water permeability, k, was a product of Darcy's Law equation as state in ASTM PS129 [11]. The test was conducted on a total 18 samples, with dimension of 100 mm diameter and thickness of 80 mm. The permeability was measured in term of time taken for the water level to fall between two designated points from 60 to 20 on the permeameter tube. The experiment was repeated three times for each sample and the average value was calculated.

Figure 1. Permeameter apparatus for permeability test

2.4. Compressive strength test

Compressive strength of PCPB was performed according to BS EN 1338:2003 [12]. The blocks were compressed by using a compression machine with maximum capacity of 3000 kN and a loading rate of 3.0 kN/s at 7 days and 28 days. The reported compressive strength was the average of three block measurements.

2.5. Porosity test

The porosity test was carried out in accordance with ASTM C642 [13]. The vacuum saturation method was used to determine the porosity of the PCPB. According to ASTM standard, the porosity was calculated based on equation (1).

$$P = [(W_{ssd} - W_d) / (W_{ssd} - W_{ssw})] / 100 \quad (1)$$

where P is the vacuum saturated porosity (%),Wssd is sample weight in saturated surface dry condition in air (g), Wssw the weight of saturated sample in water (g) and Wd the dry weight of the specimen after 24 h in an oven at 100±5 °C (g).

3. Results and discussion

3.1. Compressive strength and porosity

Figure 2 illustrates the relationship between porosity and compressive strength of PCPB. It is clearly shown that the porosity reduces when the compressive strength increases, also concrete porosities reduce with increasing age of PCPB and a similar result was reported by Jaya et al. [14]. At 7 days, the porosity reduces from 15.41 % to 15.01 % for an increase in strength from 16.3 MPa to 19.03 MPa. Furthermore, when tested at 28 days, porosities decreased 15.27 % to 14.69 % with the increase in strength from 23.75 MPa to 27 MPa. A statistical linear regression analysis was carried out on the data and the correlation between compressive strength and porosity is shown in Table 2. The table shows that the 'm' value at 7 days was lower than at 28 days. It can be concluded that the porosity at 7 days greater than porosity at 28 days, it is due to the PCPB become denser while the voids inside the block reducing with the increasing age [15].

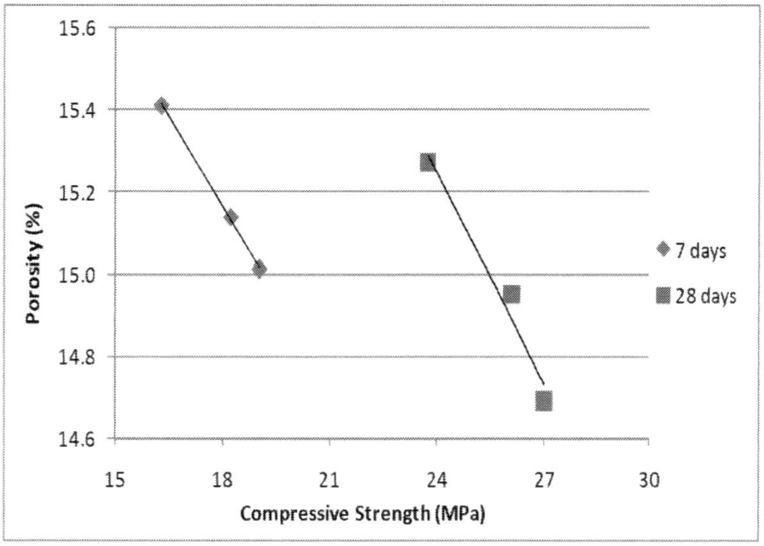

Figure 2. Compressive strength and porosity of PCPB at different curing duration

Table 2. Compressive strength and porosity correlation

Age : days	Slope, *m*	Intercept, *c*	R^2
7	0.145	17.78	0.99
28	0.170	19.33	0.97

3.2. Permeability

The result of different coarse aggregate size against the coefficient of permeability was presented in Figure 3. The results show that the coefficient of permeability ranges between 0.35 cm/s to 0.48 cm/s at 7 days and 0.32 cm/s to 0.41 cm/s at 28 days. The graphs also show the high rate of permeability particularly in early curing duration, but decreasing at a slower rate for further curing duration. There are between 8% to 17% of reduction in permeability when the curing duration increase from 7 days to 28 days. The reduction of the coefficient of permeability is due to the increasing density [16] and decreasing the porosity of the sample with the increasing curing duration. However, for PCPB with CA 8 – 10 show that at both curing duration it gives a result in high permeability compare to the other sample as in Figure 3.

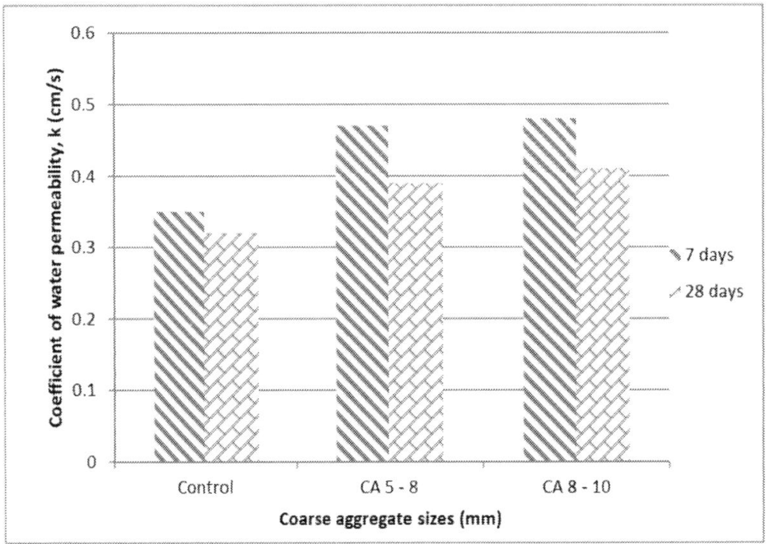

Figure 3. Coefficient of permeability at different size of coarse aggregate

4. Conclusions

a) All PCPB with different size of coarse aggregate cause a great reduction in the water volume during permeability test. However, differences in this reduction between the samples are depending on the coarse aggregate size and porosity. PCPB with CA 8 – 10 give better results in permeability and more suitable be applied at areas with high surface runoff.

b) The permeability results indicate that there are large differences in permeability for all samples when the curing duration increases. Furthermore, the permeability of sample with 28 days curing was lower than the sample with 7 days curing; indicating that age is a factor affecting permeability.

References

[1] Zachary Bean E, Frederick Hunt W and Alan Bidelspach D 2007 *J. Irrig. Drain. Eng.* **133** 583-592
[2] Wan C N C, Jaya R P and Hamzah M O 2012 *Adv. Mater. Res.* **486** 378-383
[3] Hassani A, Mohammad S and Ghoddusi P 2010 *P. I. Civil. Eng-Transp*, **163** 183-190
[4] Yang J and Jiang G 2003 *Cem. Concr.* **33** 381-386
[5] Chindaprasirt P, Hatanaka S, Chareerat T, Mishima N and Yuasa Y 2008 *Constr. Build. Mater.* **22** 894-901
[6] Jusli E, Md Nor H, Ramadhansyah P J and Zaiton H 2014 *Appl. Mech. Mater.* **554** 128-132
[7] Nur Hidayah A H, Hasanan M N and Ramadhansyah P J 2014 *Appl. Mech. Mater.* **554** 111-115
[8] Ramadhansyah P J, Mohd Ibrahim M Y, Mohd Rosli H and Mohd Haziman W I 2014 *Appl. Mech. Mater.* **554** 37-41
[9] Yusak M I M, Jaya R P, Hainin M R and Ibrahim M H W 2016 *ARPN J. Eng. Appl. Sci.* **11** 11832-11835
[10] Bakar B A, Ramadhansyah P J and Azmi M M 2011 *Mag. Concrete. Res.* **63** 313-320

[11] Hainin M R, Ramadhansyah P J, Chan T H, Norhidayah A H, Nazri F M and Ichwana 2017 *Mater. Sci. Forum* **889** 265-269
[12] ASTM PS129 2001 *Standard Provisional Test Method for Measurement of Permeability of Bituminous Paving Mixtures Using a Flexible Wall Permeameter*, American Society for Testing and Materials, West Conshohocken, PA, USA
[13] BS EN 1338 2003 *Concrete paving blocks. Requirements and test methods*, British European Standard, UK
[14] ASTM C642 2013 *Standard Test Method for Density, Absorption, and Voids in Hardened Concrete* American Society for Testing and Materials, West Conshohocken, PA, USA
[15] Jaya R P, Bakar B H A, Johari M A M and Ibrahim M H W 2011 *Cent. Eur. J. Eng.* **1** 103-112
[16] Nur Hidayah A H, Nor H M and Ramadhansyah P J 2016 *ARPN J. Eng. Appl. Sci.* **11** 11836-11839

Acknowledgments

The support provided by Malaysian Ministry of Higher Education and Universiti Teknologi Malaysia in the form of a research grant vote number Q.J130000.2522.18H05 and Q.J130000.2522.11H76 for this study is highly appreciated.

IOP Conf. Series: Journal of Physics: Conf. Series **1049** (2018) 012029 doi:10.1088/1742-6596/1049/1/012029

Waste generation and characteristization: Case study of Seberang Takir, Kuala Nerus, Terengganu, Malaysia

Izan Jaafar[1,2*], Tengku Azmina Ibrahim[3], Nurul Awanis Ahmad[3], Aeslina Abdul Kadir[4], Mohd Razali Md Tomari[5]

[1]School of Marine and Environmental Sciences, Universiti Malaysia Terengganu, 21030 Kuala Nerus, Terengganu, Malaysia
[2]Institute of Oceanography and Environment, Universiti Malaysia Terengganu, 21030 Kuala Nerus, Terengganu, Malaysia
[3]School of Ocean Engineering Universiti Malaysia Terengganu, 21030 Kuala Nerus, Terengganu, Malaysia
[4]Faculty of Civil and Environmental Engineering, Universiti Tun Hussein Onn Malaysia, 86400 Parit Raja, Batu Pahat, Johor, Malaysia
[5]Faculty of Electrical and Electronic Engineering Universiti Tun Hussein Onn Malaysia,86400 Parit Raja, Batu Pahat, Johor, Malaysia

Email: izanjaafar@umt.edu.my

Abstract. Seberang Takir is a small town in Terengganu. In 2013, the rate of population in Kuala Terengganu was 375,481 people. The population in Seberang Takir gathered in 2013 was about 32000 people. From that numbers 64.5% were in urban area while remaining 35.5% residing in rural area. The generation of solid waste is associated with number of population. The problems associated with the inefficient of waste management in Malaysia are inadequate equipment used for waste collection, lack of effective legislation for waste management and technical constraint such as solid waste management planning and operation. The purpose of this study was to determine the rate of waste generation and its characteristics, the relationship between solid waste generations and total monthly income, the relationship between solid waste generation and family size as well as to determine the willingness of community towards new approach in solid waste management. Data was collected from the distribution of questionnaire from selected residential area and analysed using Statistical Package for Social Sciences (SPSS). This study have found that the dominant composition of solid waste were food waste, followed by plastic and paper. The generation rate for dominant composition of solid waste such as food waste, plastic and paper were in the range of 24 kg/month and above respectively. The other waste such as glass, tins and other wastes were in range of 0 to 6 kg/month respectively. The result of this research also showed that there was a strong significant relationship between family size and solid waste generation but in contrast there was no significant relationship between total monthly income and solid waste generation.

1. Introduction

Solid waste is wastes arising from human and animal activities for example residential, commercial and household activities that are discarded as useless or unwanted. Solid wastes are generally classified in three general categories which are municipal waste, industrial waste and hazardous waste. Municipal waste includes waste from residential, commercial, open area such as parks, streets,

Content from this work may be used under the terms of the Creative Commons Attribution 3.0 licence. Any further distribution of this work must maintain attribution to the author(s) and the title of the work, journal citation and DOI.
Published under licence by IOP Publishing Ltd

playground and treatment plant waste while industrial wastes derived from industrial activities other than mining. Besides, hazardous waste are waste that a substantial danger immediately or over a period of time to human, plant or animal life. They are certain household waste can be classified hazardous, categorized as household hazardous waste. The examples of these wastes are solvents, paints, unused medicines, lamps, batteries and many more.

The factors influence the composition and characteristics of municipal solid waste are economic level, season, weather and culture of people living or doing business in the area the waste generation. According to [1], to date average Malaysian produced 1.64kg of solid waste a day, makes it above than the average worldwide generation of 1.2kg per person according to world bank report. Hence, this rate considered relatively high compared to other developing countries. The generation rate and composition of solid waste depends on many factors that must be taken into consideration when trying to explain waste generation, both general and site specific. Seasonal variation influences waste generation rate and waste characteristics for seasonal factors [2] while more food waste was generated in monsoon and summer seasons than in spring and winter seasons [3]. These societal changes influence the characteristic of given households, including family size, residential location and community status. Correlation between household solid-waste generation and composition and relevant socioeconomic parameters [4].

Generally MSW consists of around twenty different categories which are food waste, paper, cardboard, plastic, textile, wood waste, metals, diapers, news prints, high grade and fine paper, fruits waste, green waste, batteries, construction waste and glass these categories can grouped into organic and inorganic [5]. According to [6], the highest composition solid waste is organic waste. In Malaysia, with population of 30 million in 2014 the generation rate of 1.2 kg/per/day solid waste are relatively high compared to other developing countries. The Ministry of Housing and Local Government pointed out in 2005 that the amount of waste generation in Malaysia is projected to increase at 3.4% per year. Table 1 shows the estimated amount of waste generation until 2020. Urban area generates more paper and plastics waste compared to the rural area that generates more organic wastes.

Table 1 Estimated amount of waste generation in 1991, 1994, 2015 and 2020

Year	Population (million)	Estimated Waste (9tons/year)
1991	17 567 00	4.488 369
1994	15 917 739	5 048 804
2015	31 773 889	7 772 402
2020	35 949 239	9 092 611

Source: Ministry Of Housing and Local Government (2005)

2. Methodology

The study area was carried out at Taman Perumahan Telaga Daing (5°21'49.0"N, 103°07'26.0"E) and Taman Permint Perdana (5°20'43.8"N, 103°07'27.8"E) of Seberang Takir area.

Prior to the main survey a pilot study was carried out to test the authenticity of the questionnaires. For this study, 30 respondents took part in the pilot test. 238 respondents were participated in the actual survey selected from both study area. The total numbers of respondent was calculated based on the given equation:

IOP Conf. Series: Journal of Physics: Conf. Series **1049** (2018) 012029 doi:10.1088/1742-6596/1049/1/012029

$$n = \frac{N}{1 + Ne^2} \quad (1)$$

Where;

n the sample size

N the population size

e the level of precision (95 %; e= 0.05)

3. Results and discussion

3.1. Demographic of Respondents

238 respondents took part in this study. Respondents were consisted of 134 males (56.3%) and 104 females (43.7%). Male was more attracted to answer the survey compared to female. Majority of respondent were Muslim (100%) and Malay (100%) because they are the majority of population at Seberang Takir' district.

33.2% respondents had family monthly income which was ranged from RM600 to RM1200. This is because Seberang Takir is a small town and less developed area. 32.4% respondents had family income ranged from RM2000- RM3500. Only 3.8% respondents had monthly income which above RM3500. Hence we can say that the living standard in this area of study is considered medium.

3.2. Knowledge About Solid Waste

76.1% (181) of respondents stated that they have a knowledge about solid waste and the rest did not. This study showed that 117 of male respondents indicated having a knowledge on solid waste compare to female which was 64 respondents. The majority of respondents have heard about solid waste via television (86.1%) and the lowest percentage (13.1%) of respondents heard about solid waste when they were at school or university.

Meanwhile 45% of respondents practiced recycling while 55% of respondents did not. 67 males practiced recycling at their houses compared to female. The reasons that people did not recycle was because lack of facilities and the quantity of recycle material are varyingly uncertain [7]. Similarly in this study most respondent have responded that lack of facility was the significant reason averse them from practise recycling.

54.2% (129) households separate some of their waste while the rest (45.4%)(108) did not. By doing separation and segregation, one can earn additional income by selling the recyclables to the recyclers. The respondents also gave examples of sorted wastes into plastic bags, glasses, peelings (banana and potatoes). The relationship between income and waste separation is negative but significant at 5% confident level. This shows that households with high incomes are less likely to involve in separating waste. Thus, the majority of those who separated their waste were intended to earn some income. We can say that this activity is not important for high income household.

3.3. Characteristic of Solid Waste

Table 2 shows the percentages of waste characterisation generated as estimated by respondent. If we look at the highest generation which is above 24 kg per week, food waste (12.2 %) was the main type generated. The second largest component of solid waste was plastic with a value of about 2 %. This value is similar as reported by [8]. Generation of paper was 0.8% and none of the respondent estimated they generated more than 24kg for glass, tin and others.

Table 2 Generation of solid waste according to types

Solid waste generation per month (kg)	0	0-6	6-12	12-18	18-24	above 24
Food waste	0%	39.1%	22.7%	11.8%	10.9%	12.2%
Plastic	0%	58.0%	25.0%	9.2%	3.4%	2.0%
Glass	0%	67.2%	7.1%	2.0%	0.8%	0.0%
Paper	0.4%	58.0%	16.4%	8.4%	2.5%	0.8%
Tins	0%	58.4%	3.4%	0.8%	0.8%	0.0%
others	0%	50.8%	1.3%	0.8%	0.4%	0.0%

3.4. Generation of Solid Waste

3.4.1 Food Waste. As mentioned earlier food waste generation indicated the highest proportion among other types of solid waste. The production of organic waste was the tremendous one compared to other components because food is the essential needs for the people [9]. Respondents tend to throw organic waste rather than recycled it as compost, or used it for other usage. 10.9% (26) respondents composted food waste which is between 18 until 24 kg/month while 11.8% (28) respondents generate food waste which is between 12 to 18 kg/ month. 39.1% (93) respondents estimated that their food waste generation was in a range of 0 to 6 kg/month.

3.4.2 Plastic Waste. 2% (5) respondents estimated their plastic generation was 24kg/month and above while 3.4% (8) respondent generated plastic in a range of 18 to 24 kg/month. Plastic and paper composition was among the highest; this might be due to changes of the lifestyle where they prefer to buy variety of foods in packages [10]. A high number of respondents (53%) stated that they generated plastic in a range of 0 – 6kg.

3.4.3 Paper Waste. Similar trend with plastic waste, 58% (138) respondents stated that they generated paper in a range of 0 to 6 kg/month followed by second highest percentage (16.4%) for 6 – 12 kg. 8.4% (20) respondents generated paper between 12 to 18 kg/month while 0.8% (2) respondent estimated their generation rate of paper was 24kg/month and above. Paper found in this residential area in a form of paper scrapes, packing papers, discarded papers from people's activity.

3.5. Factors That Affect Solid Waste Generation

Rainy season and festive seasons were included as whether they were significant factors that affect waste generation. 35 % of respondents were not sure whether rainy season affect solid waste generation on the other hand the same amount of percentage 35% respondents agreed. Most of respondents were uncertain about rainy season affect solid waste generation. According to [11], the total waste generated in the dry season is similar to the waste generated during the rainy season. Organic waste, plastics, paper and metals generated in the dry season were also similar to amounts generated during the rainy season. Therefore, we can say that weather condition did not imply the difference in terms of waste generation.

A significant amount of 66 % (157) of respondents agreed that festival season affect solid waste generation while 31% (74) of respondents were not sure. This finding was also supported by study done by [12] where composition of waste can be influenced by seasonal factors. The increase in the amount of plastics at the end of the festive month was observed [13].

3.6. Relationship between Solid Waste Generation, Total Monthly Income and Family Size

3.6.1 Correlation between Solid Waste Generation and Total Monthly Income. A negative correlation was found between the income and the generation of household solid waste ($r = -0.06$, $p<0.01$). Even though it is typically expected that a family with higher income tend to generate a higher quantity of solid waste each day, here in this research, the finding was totally the opposite. The low income was also generating high amount of MSW as this group tends to process and home cooked their meal rather than eating out, hence the generation of food waste is high.

3.6.2 Correlation between Solid Waste Generation and Family Size. A significant strong positive correlation ($r = 0.45$, $p<0.01$) was found between household waste generation and family size. This means that family size gave a significant relative with solid waste generation. The larger household size was presumed to generate larger quantity of waste since more individuals are included in the household unit [14].

3.7. Willingness to Participate

The level of awareness about recycling of respondents in Seberang Takir increased but the attitude did not resembled on the importance of recycling where the level was moderate. 30.3% respondents agreed on the governmental or NGO programs reduced food waste in their residential area while 59.7% disagree. This can be expected where citizens did not perceived the changes in recycling practise eventhough much awareness campaign were done to promote. However, 84.9% of respondents stated that they willing to take the initiative to separate the solid waste if the government or NGOs to provide the appropriate bins while a small percentage of 5.0% did not. A high percentage of 73.5% of respondents willing to join the campaign that organized by government or NGO while 22.3% of respondent were not sure. Therefore, the government and NGOs should continuously take the initiative to attract public in campaign to attract and strengthen recycling practice among citizens.

4. Conclusion

The dominant composition of solid waste in this study was food waste followed by plastic and paper. There were a strong significant between family size and solid waste generation in Seberang Takir community but in contrast there are no significant between income level and solid waste generation. Lack of facilities remained a strong factor of low recycling practice. Special management should be considered in tackling amounting solid waste generation during festive seasons that can promote vectors disease and negative aesthetic values due to uncollected waste.

References

[1] Khor S. 2014 Why is it necessary to implement a compulsory waste separation system for Malaysians? Retrieved from http://says.com/my/news/

[2] Boldrin A and Christensen T H 2009 Seasonal generation and composition of Danish garden waste *Waste Manage (Oxford)* **30** 551–557

[3] Jadoon A, Batool S A and Chaudhry M N 2013 Assessment of factors affecting household solid waste generation and its composition in Gulberg Town, Lahore, Pakistan. *J Mater Cycles Waste Management* **16** 73–81

[4] Marquez M Y, Ojeda S and Hidalgo H 2008 Identification of behavior patterns in household solid waste generation in Mexicali's city: Study case. *Resources, Conservation and Recycling* **52** 1299–1306

[5] Marie L 2007 Methodology for allocating municipal solid waste to biogenic and non-biogenic energy. *Electric and Alternate Fuels U.S. Department of Energy Washington*

[6] Hasnah, Ekowati D F, Ahmad S and Lyndon N 2012 *The Management of Social-Ecological Change and Uncertainties in the Global Era* **6** 184-194
[7] Norsilan 2006 *Kajian Kitar Semula Sisa Pepejal di Dalam Kampus UKM Bangi: Permasalahan dan Keberkesanan Kitar Semula* Fakulti Kejuruteraan, Universiti Kebangsaan Malaysia
[8] Forouhar A and Hristovski K D 2012 Characterization of the municipal solid waste stream in Kabul, Afghanistan *Habitat Int* **36** 406–413
[9] Samah M A, Manaf L A, Agamuthu, Sulaiman W N A and Ahsan A 2013 Real Data Composition of Municipal Solid Waste (MSW) generated In Balakong, Selangor, Malaysia *Life Science Journal 2013* 10(4)
[10] Bolaane, B and Ali M 2004 Sampling household waste at source: lessons learnt in Gaborone. *Waste Manage Res* **22** 142–148
[11] Owamah I H, Izinyon O and Igbinewekan C P 2015 Characterization and quantification of solid waste generation in the Niger Delta Region of Nigeria: a case study of Ogbe-Ijoh community in Delta State
[12] Beukering P V and Sehker M 1999 *Analyzing Urban Solid Waste in Developing Countries. A Perspective in Bangalore* Working Paper No 24 India pp 46-67
[13] Yatim S R and Arshad M A 2015 Household Solid Waste Characteristics and Management in Low Cost Apartment in Petaling Jaya, Selangor. *Health and the Environment Journal 2010* Vol. 1, No. 2
[14] Afroz R, Hanaki K, and Tudin R 2011 Factors affecting waste generation: a study in a waste management program in Dhaka City, Bangladesh.

IOP Conf. Series: Journal of Physics: Conf. Series **1049** (2018) 012030 doi:10.1088/1742-6596/1049/1/012030

Effect of Chemical Warm Asphalt Additive on the Rutting Characteristic of Aged Binder Containing Waste Engine Oil

Nurul Hidayah Mohd Kamaruddin[1,2], Mohd Rosli Hainin[1], Norhidayah Abdul Hassan[1], Mohd Ezree Abdullah[2], Mohd Khairul Idham[1], Noor Azah Abdul Raman[1]

[1] Faculty of Civil Engineering, Universiti Teknologi Malaysia (UTM), 81310 Skudai, Johor, Malaysia.

[2] Faculty of Civil and Environmental Engineering, Universiti Tun Hussein Onn Malaysia (UTHM), 86400 Parit Raja, Batu Pahat, Johor, Malaysia.

E-mail: mrosli@utm.my

Abstract. This paper presents a laboratory study on the effect of chemical warm asphalt additive on the rutting characteristic of aged binder containing waste engine oil. The aged binder and waste engine oil were blended with 0%, 1%, 2% and 3% of chemical warm asphalt additive. The samples were tested for rutting characteristics (rutting resistance, G*/sin δ and creep recovery) using Dynamic Shear Rheometer (DSR). G*/sin δ was tested at temperature 64°C and creep recovery was tested at temperature 60°C under stress of 50 Pa. The results showed that the chemical warm asphalt additive slightly decreased the rutting resistance of the aged binder containing waste engine oil sample. Meanwhile, the elasticity of asphalt binder was improved with the addition of additive by the reduction of creep compliance. As a conclusion, results of the testing showed that chemical warm asphalt additive could be applied with aged binder and waste engine oil with the satisfactory performance.

1. Introduction

The usage of reclaimed asphalt pavement (RAP) materials were continuously applied due to the significant benefits in term of economic and environmental. In term of economic, the RAP usage derived to cost savings by reducing the new materials. Meanwhile, in term of environmental benefit, the RAP enables to reduce the disposal area required for placing milled material as well as conserves the energy associated with the production and transportation of new materials [1]. These criteria have been defined under green highway elements [2-3]. However, the major concern is the high stiffness of the aged binder in RAP. During the service life, the pavement was exposed to ultraviolet (UV), radiation and oxygen which contributed to the stiff binder. Waste engine oil (WEO) is one of the effective rejuvenator to soften the aged binder [4-5]. The soft binder give a better workability [6]. On the other hand, many researchers have studied the use of warm mix asphalt technology in pavement materials [7–11]. In addition to the role of rejuvenator, warm asphalt additive shows a great potential to improve the workability of aged binder [12-13]. However, after long term in service, the effect of warm mix asphalt was less influenced compared to aged binder to the binder properties [14]. This paper studies the effects of introducing warm asphalt additive to aged binder containing waste engine

Content from this work may be used under the terms of the Creative Commons Attribution 3.0 licence. Any further distribution of this work must maintain attribution to the author(s) and the title of the work, journal citation and DOI.
Published under licence by IOP Publishing Ltd

oil (WEO). Since the asphalt binder become softer, it was potentially prone to rutting early in service life [15]. Thus, the modified asphalt binder were observed for rutting characteristic.

2. Materials and methods

The base binder used in this study was the 80/100-penetration grade. Meanwhile, the aged binder was obtained from the extraction and recovery of reclaimed asphalt pavement (RAP). Table 1 shows the basic properties of the asphalt binder used in this study. The aged binder was blended with base binder at 50% proportion with 15% waste engine oil (WEO) based on a previous study [16]. Chemical warm asphalt additive is in solid pallet were added to the aforementioned binder at 1%, 2% and 3% respectively. The detailed label and constituents of the samples are shown in Table 2.

Table 1. Physical properties of asphalt binder.

Physical properties	Result Value		Specification
	Base Binder	Aged binder	
Penetration (25°C;0.1mm)	88.00	10.6	ASTM D5-06
Softening Point, °C	46.25	71	ASTM D36 -12
Viscosity at 135°C, Pa.s	0.493	2.865	ASTM D4402 -13
Ductility @ 25°C (cm)	>100	<100	ASTM D113-07

Table 2. Label and constituents of testing sample.

Label	Bitumen	WEO (by weight of bitumen)	Warm asphalt additive (by weight of bitumen)
A	Base binder	-	-
B	Base binder + aged binder	15%	0%
BC1	Base binder + aged binder	15%	1%
BC2	Base binder + aged binder	15%	2%
BC3	Base binder + aged binder	15%	3%

The modified binder (which refer to the sample that contained aged binder, WEO and chemical warm asphalt additive) were blended homogenously at 600 rpm at a temperature around 155 ± 5°C using Silverson-L4RT, a multipurpose high shear mixer. Subsequent the blending process, the rutting characteristics were investigated using dynamic shear rheometer (DSR) with accordance to ASTM D7175 [17]. A 25 mm diameter of circular disk-shaped with 1 mm gap setting for asphalt binder sample was used. There are two parameters used for rutting characteristics: rutting resistance, $G^*/\sin\delta$ and creep and recovery. The samples were conditioned in a Rolling Thin Film Oven (RTFO) to simulate the asphalt binder aging during construction of pavement. During this phase, rutting also possible to occur by the consolidation of the pavement layer after construction [18]. The high temperature stiffness of asphalt binder is important to minimize the rutting [18]. The samples for $G^*/\sin\delta$ was tested at temperature 64°C and creep recovery at 60°C to represent the service temperature of pavement during high temperature [19]. Meanwhile, creep recovery was run at a stress level of 50 Pa (loading for 1 seconds, recovery for 300 seconds) that represent a high stress level on pavement [20]. According to Xiao et al. [21], the compliance value is typically used to indicate the creep associated with load/stress applied. The loading is applied for a certain time and the strain induced is monitored after the loading is removed. Principally, compliance (J_{nr}) function is the ratio of the strain (γ) to the constant applied load (σ) with the unit of Pa^{-1}.

3. Results and discussions

3.1. Complex Modulus and Phase Angle

Complex modulus, G* and phase angle, δ is a basic property measured by the DSR. Complex modulus represents the stiffness and phase angle represent the elasticity of asphalt binder. Basically, a high value of G* is favourable because it indicated a higher resistance to deformation. Meanwhile, a lower δ is desirable because it reflects more elastic component to deformation [22]. Figure 1 shows the complex modulus and phase angle value for asphalt binder. The sample of base binder which abbreviated as A, has been used as a reference. This is to show the effect of the modification. As can be seen, the complex modulus (G*) of modified binder is higher than binder A. Meanwhile, the phase angle (δ) of modified binder is lower than binder A. As previously discussed, these results corresponds to greater rutting resistance by the modified binder. In the modified binder, the addition of chemical warm mix additive reduced the stiffness as asphalt binder by a slight reduction of G*. However, the additives were found enhanced the elastic behaviour of the binder B from 73.81° to 69.35°, 70.14° and 70.73° respectively.

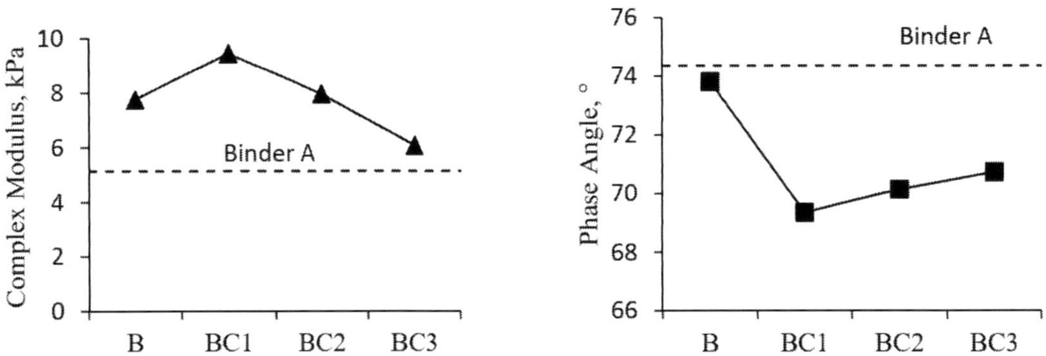

(a) (b)

Figure 1. Results of asphalt binder for (a) complex modulus (b) phase angle.

3.2. Rutting Resistance

G*/sin δ is the indicator of rutting resistance. Figure 2 shows the rutting resistance values of base binder and the modified binder. It can be seen that the modified binder showed higher G*/sin δ value compared to the base binder, A. The increased of G*/sin δ of the modified binder can be explained the addition of aged binder and WEO. The addition of chemical warm asphalt additive was observed not to compromise the aged binder and WEO. The addition of 1% warm asphalt additive to the asphalt binder displayed the increasing rutting resistance with value 10.09 kPa. In the different container, 2% of chemical warm asphalt additive was blended with binder composition of BC, the G*/sin δ decreased to 8.46 kPa and at 3% to 6.12 kPa. A similar finding has been observed in other studies [23]. A possible explanation for this reduction is the warm asphalt additive was blended at a temperature around 155°C ± 5°C which quickly dispersed the chemical warm mix additive in the asphalt binder as the melting point of chemical warm asphalt additive is 110°C [24]. The additive then uniformly blended with waste engine oil and thus, produced a slightly soft asphalt binder regardless the aged binder content.

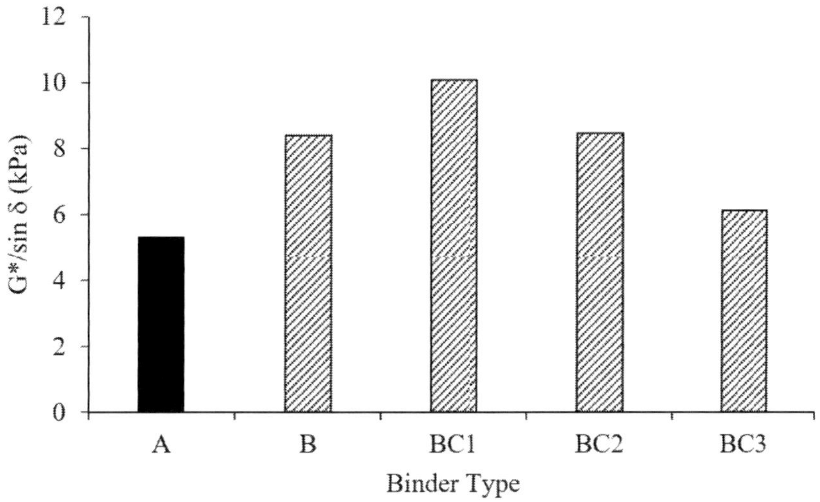

Figure 2. Rutting resistance value of asphalt binder.

3.3. Creep and Recovery

During creep phase, the stress was applied for a certain time and the resultant strain was monitored. Then, the stress was removed and the recovery strain was continuously monitored. Figure 3 shows the monitored strain during creep and recovery phase under stress of 50 Pa with total time 300 seconds at temperature 60°C. The binder A exhibit the higher strain during the creep phase. Meanwhile, the modified binder showed the overlapped strain. During the recovery phase, some of the binders showed increase strain. Since the actual resultant strain depends on the applied stress, creep compliance Jnr usually used as a deformation indicator in creep test rather than strain [25]. The lower value of creep compliance implies higher rutting resistance [20][26]. The creep compliance of the asphalt binder is presented in Table 3. From the data, as expected, the binder A showed highest J_{nr} value while the binder BC2 exhibits lowest J_{nr} value. In specific, warm asphalt additive improved the deformation resistance of the aged binder containing WEO (sample B). The Jnr value was decreased from 7.29E-04 Pa^{-1} to 6.44E-04 Pa^{-1}, 5.96E-04 Pa^{-1} and 6.65E-04 Pa^{-1} respectively with the addition 1%, 2% and 3% chemical warm mix additive. These results demonstrate that the modified binder showed the better elastic response under the applied stress and thus improved the rutting resistance. This improvement could be explained by the lower phase angle reported in section 3.1 Lower phase angle reflects to a more recoverable component of total deformation [27].

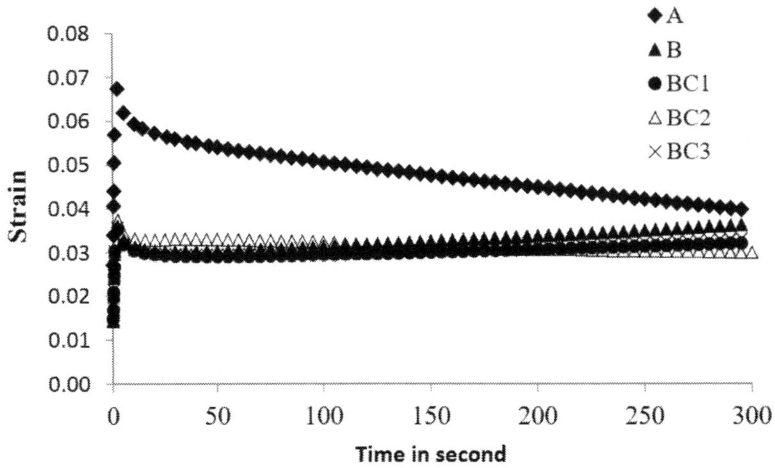

Figure 3. Strain value during creep and recovery test

Table 3. Results of creep and recovery test

Binder	Creep Compliance, J_{nr} (Pa^{-1})
A	7.88E-04
B	7.29E-04
BC1	6.44E-04
BC2	5.96E-04
BC3	6.65E-04

4. Conclusions

In this study, the effects of chemical warm asphalt additive on the rutting characteristic of aged binder containing waste engine oil were evaluated. It was found that the addition of additive, slightly decreased the rutting resistance of asphalt binder. Meanwhile, in creep recovery test, the addition of warm asphalt additive has improved the creep recovery of the aged binder containing waste engine oil. As a conclusion, results of the testing showed that warm asphalt additive could be applied with aged binder and waste engine oil. Although the rutting resistance slightly decreased with the addition of chemical warm asphalt additive, the blending binder still has sufficient stiffness and elasticity to resist the rutting.

References

[1] West R C 2015 Best Practices for RAP and RAS Management. *Quality Improvement Series 129*. Lanham, MD, United States

[2] Zakaria R, Seng F K, Majid M Z A, Zin M R, Hainin M R, Puan O C and Yaacob H 2013 Energy Efficiency Criteria for Green Highways in Malaysia *Jurnal Teknologi (Sciences and Engineering) 65*(3) 91–95

[3] Balubaid S, Bujang M, Majid M Z A, Rosli M R, Yaacob H and Ismail H H 2015 Assessment Index Tool for Green Highway in Malaysia *Jurnal Teknologi 77*(16) 99–104

[4] Oliveira J R M, Silva H M R D, Jesus C M G, Abreu L P F and Fernandes S R M 2013 Pushing the Asphalt Recycling Technology to the Limit *International Journal of Pavement Research and Technology 6*(2) 109–116

[5] DeDene C D and You Z 2014 The Performance of Aged Asphalt Materials Rejuvenated with

Waste Engine Oil *International Journal of Pavement Research and Technology* *7*(2) 145–152

[6] Dokandari P A, Kaya D, Sengoz B and Topal A 2017 Implementing Waste Oils with Reclaimed Asphalt Pavement *Proc. of the 2nd World Congress on Civil, Structural, and Environmental Engineering (CSEE'17)* Barcelona, Spain

[7] Aziz M M A, Hainin M R, Yaacob H, Feizabadi S M, Shokri M and Warid M N M 2014 Performance of Warm-Mix Asphalt in the Highway Industry *Materials Research Innovations*, *18* (Suppl 6) S6 245-S6 249.

[8] Abdullah, M. E., Zamhari, K. A., Hainin, M. R., Oluwasola, E. A., Hassan, N. A. and Yusoff, N. I. M. (2016). Engineering Properties of Asphalt Binders containing Nanoclay and Chemical Warm-Mix Asphalt Additives. *Construction and Building Materials*, *112*(June), 232–240.

[9] Abdullah, M. E., Zamhari, K. A., Hainin, M. R., Oluwasola, E. A., Nur, N. I. and Hassan, N. A. (2016). High Temperature Characteristics of Warm Mix Asphalt Mixtures with Nanoclay and Chemical Warm Mix Asphalt Modified Binders. *Journal of Cleaner Production*, *122*(February), 326–334.

[10] Abdullah, M. E., Hainin, M. R., Yusoff, N. I. M., Zamhari, K. A. and Hassan, N. (2016). Laboratory evaluation on the characteristics and pollutant emissions of nanoclay and chemical warm mix asphalt modified binders. *Construction and Building Materials*, *113*, 488–497.

[11] Kusam, A., Malladi, H., Ayebali, A. A. and Khosla, N. P. (2017). Laboratory Evaluation of Workability and Moisture Susceptibility of Warm-Mix Asphalt Mixtures containing Recycled Asphalt Pavements. *Journal of Materials in Civil Engineering*, *29*(5), 1–8.

[12] Haghshenas, H. F., Nabizadeh, H., YR, Kim. and Santosh, K. (2016). Research on High-RAP Asphalt Mixtures with Rejuvenators and WMA Additives. SPR-P1(15) M016. Lincoln, NE, United States.

[13] Sengoz, B., Topal, A., Oner, J., Yilmaz, M., Dokandari, P. A. and Kok, B. V. (2017). Performance Evaluation of Warm Mix Asphalt Mixtures with Recycled Asphalt Pavement. *Periodoca Poletechnica Civil Engineering*, *61*(1), 117–127.

[14] Buss, A., Williams, R. C., and Schram, S. (2015). The Influence of Warm Mix Asphalt on Binders in Mixes that Contain Recycled Asphalt Materials. *Construction and Building Materials,* 77, 50–58.

[15] Zaumanis, M. (2014). Warm Mix Asphalt. *Climate Change, Energy, Sustainability and Pavements, Green Energy and Technology*. (Berlin, Germany: Springer-Verlag Berlin Heidelberg) chapter 10,pp. 309–334.

[16] Kamaruddin, N. H. M., Hainin, M. R., Hassan, N. A. and Abdullah, M. E. (2014). Rutting Evaluation of Aged Binder Containing Waste Engine Oil. *Advanced Materials Research*, *911*, 405–409.

[17] ASTM. (2008). *ASTM D7175: Standard Test Method for Determining the Rheological Properties of Asphalt Binder using a Dynamic Shear Rheometer*. West Conshohocken, PA, United States.

[18] Brown, E. R., Kandhal, P. S., Roberts, F. L., Kim, Y. R., Lee, D.-Y. and Kennedy, T. W. (2009). *Hot Mix Asphalt Materials, Mixture Design and Construction* .3rd Edition. Lanham, MD, United States: NAPA Research and Education Foundation.

[19] Hassan, N. A., Kamaruddin, N. H. M., Hainin, M. R. and Abdullah, M. E. (2017). Effect of Warm Asphalt Additive on the Creep and Recovery Behaviour of Aged Binder Containing Waste Engine Oil. *IOP Conference Series: Materials Science and Engineering*, *226*, 12066.

[20] Amirkhanian, A. N., Xiao, F. and Amirkhanian, S. N. (2011). Characterization of Unaged Asphalt Binder Modified with Carbon Nano Particles. *International Journal of Pavement Research and Technology*, *4*(5), 281–286.

[21] Xiao, F., Punith, V. S. andAmirkhanian, S. N. (2012). Effects of Non-foaming WMA Additives

on Asphalt Binders at High Performance Temperatures. *Fuel, 94*, 144–155.

[22] Hunter, R. N., Self, A. and Read, J. (2015). *The Shell Bitumen Handbook* .6th Edition. London: Thomas Thelford Ltd.

[23] Abdullah, M. E., Zamhari, K. A., Shamshudin, M. K., Hainin, M. R. and Mohd Satar, M. K. I. (2013). Rheological Properties of Asphalt Binder Modified with Chemical Warm Asphalt Additive. *Advanced Materials Research*, *671–674*, 1692–1699.

[24] Norell, M., Logaraj, S. and James, A. (2009). Rediset WMX Development of New Warm Mix Technology Offering Enhanced Asphalt Properties. *Technical Paper*. Stenungsund, Sweeden.

[25] Xiao, F., Chen, M., Wu, S. and Amirkhanian, S. (2013). A Long term Ultraviolet Aging Effect on Rheology of WMA Binders. *International Journal of Pavement Research and Technology*, *6*(5), 496–504.

[26] Xiao, F., Amirkhanian, S., Wang, H. and Hao, P. (2014). Rheological Property Investigations for Polymer and Polyphosphoric Acid Modified Asphalt Binders at High Temperatures. *Construction and Building Materials*, *64*, 316–323.

[27] Mills-Beale, J., You, Z., Fini, E., Zada, B., Lee, C. H. and Yap, Y. K. (2014). Aging Influence on Rheology Properties of Petroleum-Based Asphalt Modified with Bio-Binder. *Journal of Materials in Civil Engineering*, *26*, 358–366.

Acknowledgement

The authors wish to gratefully acknowledge the support from Universiti Teknologi Malaysia (UTM) for the grant (Q.J130000.2522.11H35, Q.J130000.2522.11H76, Q.J130000.2522.17H69 and Q.J130000.2622.14J09).

Design of Circular Polarized Antenna by Using Inverted Suspended Circular Patch Design for WLAN Application at 2.4 GHz

H. A. Bakar, M. Z. A. Abd. Aziz, B. H. Ahmad and N. Hassan
Center for Telecommunication Research and Innovation (CeTRI), Faculty of Electronics and Computer Engineering (FKEKK), Universiti Teknikal Malaysia Melaka (UTeM), 76100 Durian Tunggal, Melaka, Malaysia

E-mail: hamizan421@yahoo.com

Abstract. In this paper, an inverted suspended circular patch antenna with circular polarization is proposed for wireless local area network (WLAN) operating at 2.4 GHz frequency. The antenna was designed based on inverted suspended circular patch with an air gap between substrate and ground plane layer at distance of 10 mm. L-probe technique is used in this antenna design in order to feed the signal from SMA probe to the feedline of the circular patch structure. The design and simulation process are done by using Computer Simulation Technology (CST) software. Antenna performances in term of return loss, resonant frequency, bandwidth, gain, directivity, axial ratio, total efficiency and radiation pattern at 2.4 GHz frequency are discussed. The comparison result of simulation and measurement show that the proposed circular polarized antenna obtained bandwidth more than 100 MHz with gain and return loss achieved more than 5 dB and -13 dB at 2.4 GHz.

1. Introduction

Microstrip patch antennas are popular structures in modern wireless communication system as they are compact in nature. However, in their basic form (rectangular, circular or triangular) it normally provides linear polarization, but by certain modification to the basic antenna geometry or feed it can provide circular polarization (CP). These modifications comprise modifying of the basic patch with one or more feeds, feeding the patch from its corner along the diagonal, feeding the patch at adjacent sides, trimming the corners of a square patch [1,2] and cutting a slot inside the patch [2,3].

Circular polarization is regarded as the well-known polarization schemes applied in current wireless communication system for example satellite, radar and navigation system, mainly because it can offer better mobility and weather penetration than linear polarization [4,5]. However, microstrip patch antenna is constrained with narrow bandwidth and low gain. In order to overcome this problem, a few modifications need to be done are included adding slot [6-8], fed by L-strip [9-11] and others. In this paper, L-probe technique was used for the antenna design where the antenna having a circular polarization axial ratio and high gain at 2.4 GHz frequency which is for WLAN application.

Content from this work may be used under the terms of the Creative Commons Attribution 3.0 licence. Any further distribution of this work must maintain attribution to the author(s) and the title of the work, journal citation and DOI.
Published under licence by IOP Publishing Ltd

2. Antenna Design

In this paper, the antenna was designed by using FR4 substrate with thickness, h = 1.6 mm, dielectric constant ε_r = 4.4 and tangent loss, tan δ = 0.019. The thickness of copper, t = 0.035 mm. L-probe technique is used in this antenna design and the 50 ohm probe feed SMA type was used to feed the signal from feedline to the circular patch structure. The width (W_f) and length (L_f) of feedline is 3.1 and 12 mm respectively. The width and length of substrate and copper plate layer is similar which is 80 x 80 mm. The substrate and copper plate layer is separated by an air gap with 10 mm distance. The circular patch radius is initially calculated based on equation (1) and equation (2) but later been choose as 29 mm after optimization for the best result of parametric study. Effective radius of circle patch antenna can be calculated by using equation (3).

The actual radius of the circular patch can be obtained by

$$R = \frac{F}{\sqrt{\left\{1+\frac{2h}{\pi F \varepsilon_r}\left[\ln\left(\frac{\pi F}{2h}\right)+1.7726\right]\right\}}} \tag{1}$$

Where,

$$F = \frac{8.791 \text{ x } 10^9}{f_r\sqrt{\varepsilon_r}} \tag{2}$$

The effective radius of the antenna is given by

$$R_e = R\sqrt{1+\frac{2h}{\pi R}\left[\ln\left(\frac{R}{2h}\right)+1.7726\right]} \tag{3}$$

Where,

R = radius of circle patch

R_e = effective radius of circle patch

f_r= operating frequency (GHz)

h = thickness of substrate (mm)

ε_r= dielectric permittivity of substrate

Figure 1 demonstrated the overall configuration and also the fabricated prototype for the antenna design. The front view of the inverted suspended antenna design is just an empty substrate as shown in Figure 1 (a), while at the back view having the circular patch with the feedline as shown in Figure 1 (b). Dimension of the inverted circular patch antenna is tabulated as in Table 1.

(a)

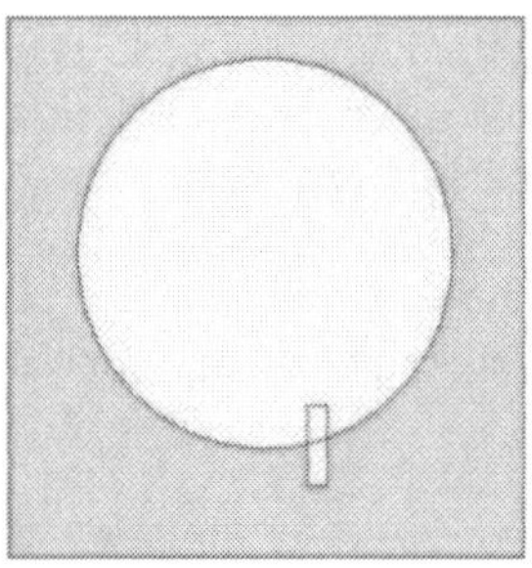

(b)

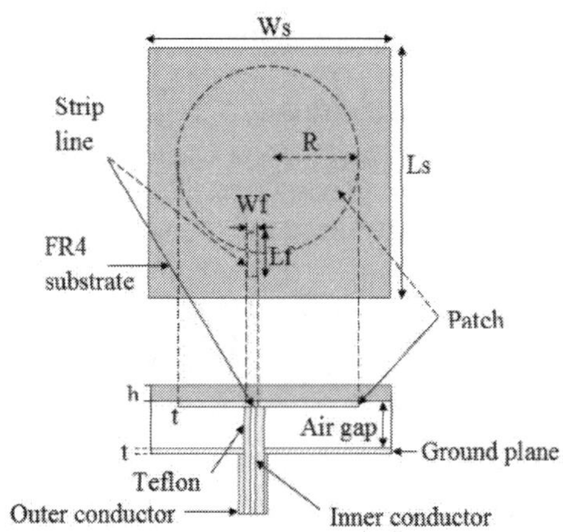

(c)

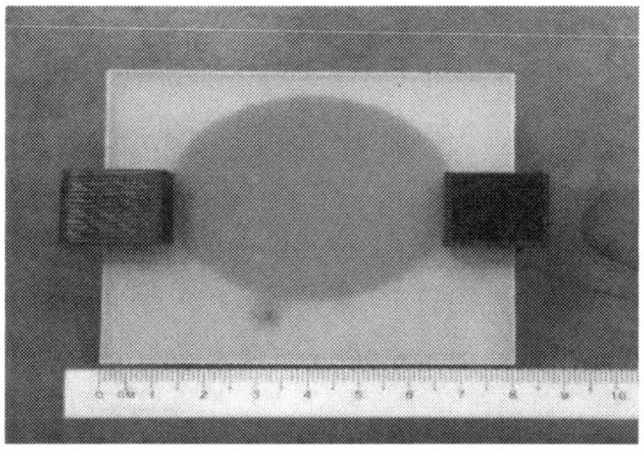

(d)

(e)

Figure 1. Fabricated structure of inverted suspended circular patch (a) front view (b) back view (c) vertical configuration view (d) prototype top view (e) prototype side view.

Table 1. Dimension of circular polarized inverted suspended circular patch antenna design.

Design Parameter	Dimension (mm)	Description
Air gap	10	Air gap
h	1.6	Thickness of FR4 substrate
t	0.035	Thickness of copper
L_s	80	Length of substrate & copper plate
L_f	12	Length of feedline
Ws	80	Width of substrate & copper plate
W_f	3.1	Width of feedline
R	29	Radius of circular patch

3. Result and discussion

In this section, the antenna parameter which comprised of resonant frequency (f_r), return loss (RL), bandwidth (BW), gain, directivity, axial ratio and radiation pattern are analysed and discussed at 2.4 GHz frequency. Other than that, simulation and measurement result of return loss, gain and radiation pattern for the proposed antenna design at 2.4 GHz also been compared and discussed.

3.1. Return loss (RL), resonant frequency (f_r) and bandwidth (BW)

Figure 2 illustrates the simulated and measured return loss performance for the antenna design which is -15.54 dB at resonant frequency 2.404 GHz and -17.80 dB at resonant frequency 2.36 GHz. While at 2.4 GHz frequency, the return loss for simulated and measured is -15.48 dB and -13.60 dB respectively. The simulated and measured bandwidth is 146 MHz from frequency range 2.33–2.48 GHz and 138 MHz which is from 2.29–2.43 GHz. From the comparison of bandwidth result, measurement frequency range is slightly shifted to a lower frequency where the bandwidth also decreases around 8 MHz compared to the simulated bandwidth.

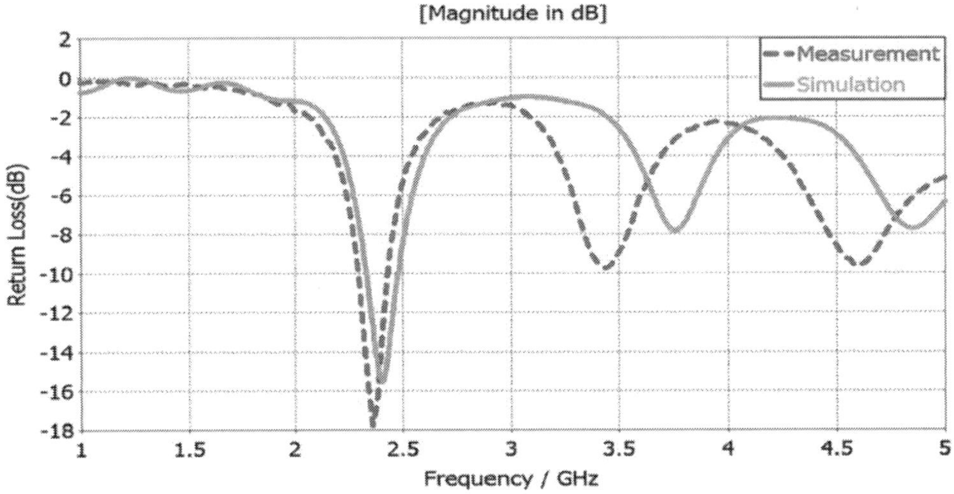

Figure 2. Comparison result of simulation and measurement return loss (S11) for inverted suspended circular patch antenna design.

3.2. Axial ratio

Figure 3 demonstrates the simulation result for axial ratio which is used to determine antenna polarization. Axial ratio for linear polarized is above 3 dB while for circular polarized it must be less than 3 dB. Based on the simulation result, the axial ratio for the antenna design at 2.4 GHz is less than 3 dB which is 0.42 dB with 1030 MHz axial ratio bandwidth (ARBW) starting from 2.15-3.18 GHz.

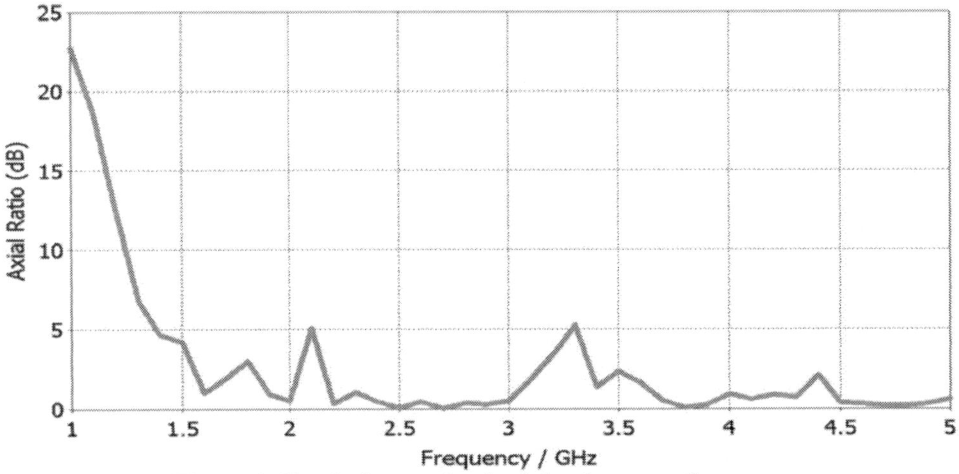

Figure 3. Simulation result of axial ratio versus frequency.

3.3. Directivity and total efficiency

Figure 4 (a) and (b) shows the simulation result of directivity and total efficiency versus frequency from frequency range of 1-5 GHz. As demonstrated in Figure 4 (a) the directivity of the antenna is more than 8 dBi starting from 1.92-2.61 GHz frequency range where the highest directivity is at 2.2 GHz with 8.54 dBi. At 2.4 GHz, the directivity of the antenna is slightly decreases to 8.46 dBi. As shown in Figure 4 (b), the highest total efficiency of the antenna is at 2.4 GHz with -0.56 dB where the efficiency is close to 0 dB. Total efficiency must be more than -3 dB to get more than 50% antenna efficiency.

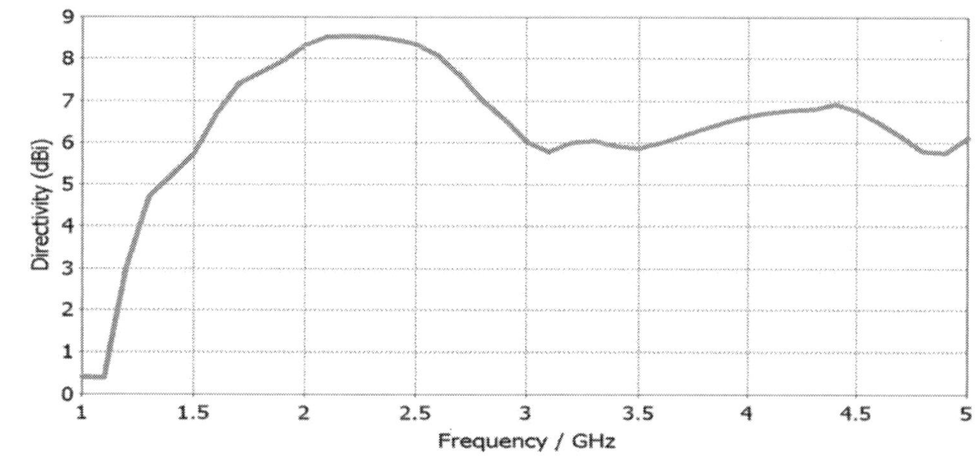

(a)

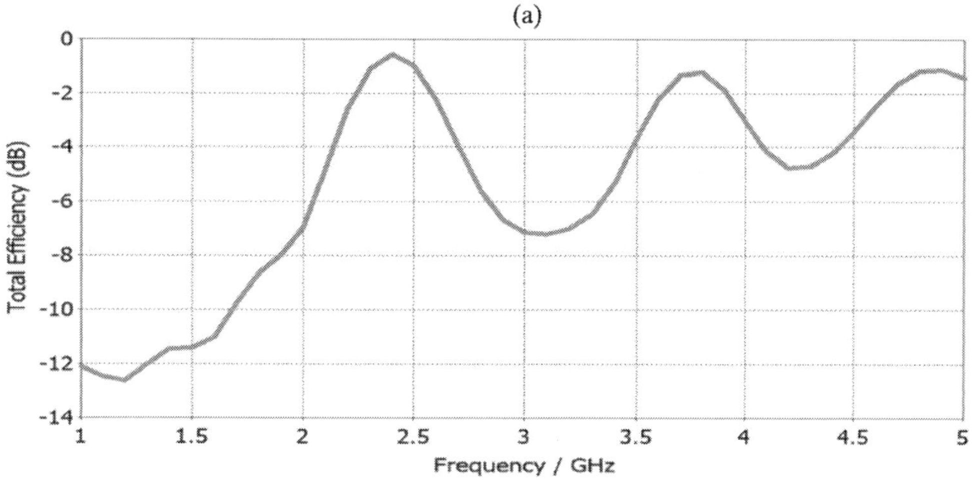

(b)

Figure 4. Simulation result of (a) directivity (b) total efficiency versus frequency.

3.4. Radiation pattern

The radiation pattern of antenna is measured in far field region. Figure 5 demonstrate the comparison of simulated and measured radiation pattern of the antenna at 2.4 GHz when phi = 0°, phi = 90° and theta = 90°. The radiation pattern obtained from measurement result is more or less similar with the simulation result.

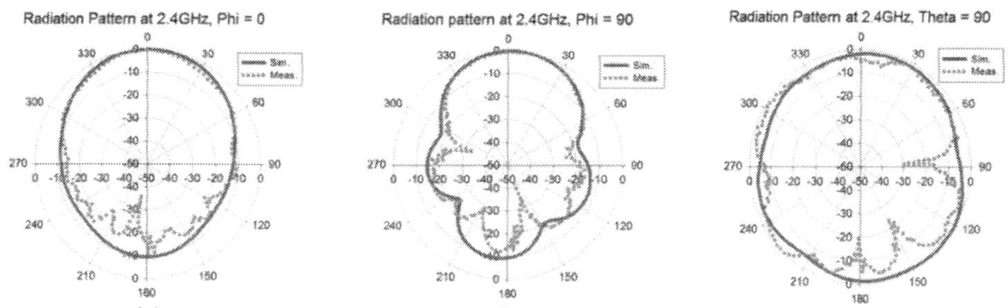

(a) (b) (c)

Figure 5. Comparison of simulation and measurement radiation pattern at 2.4 GHz (a) phi = 0° (b) phi = 90° and (c) theta = 90°.

Table 2 shows the result of axial ratio bandwidth (ARBW), axial ratio (AR), directivity and total efficiency at 2.4 GHz. While, Table 3 shows the result of return loss bandwidth (RLBW), resonant frequency (f_r), return loss and realized gain at 2.4 GHz. Based on Table 3, the simulation result of gain is 7.90 dB and calculation gain for the measurement result is 5.19 dB respectively.

Table 2. Performance result of simulation axial ratio, directivity, total efficiency and axial ratio bandwidth at 2.4 GHz

Antenna Design (simulation result)	Axial Ratio (dB)	Directivity (dBi)	Total Efficiency (dB)	Axial Ratio Bandwidth (below 3dB)	
				Frequency Range (GHz)	Bandwidth (GHz)
Sim.	0.42	8.46	-0.56	2.15-3.18	1.03

Table 3. Comparison result of simulation and measurement for return loss, realized gain, return loss bandwidth and resonant frequency at 2.4 GHz

Antenna Design (Sim. Vs. Meas.)	Return Loss (dB)	Realized Gain (dB)	Return Loss Bandwidth		Resonant Frequency (f_r)	
			Freq. Range (GHz)	BW (MHz)	Freq. (GHz)	Return Loss (dB)
Sim.	-15.48	7.90	2.33-2.48	146	2.404	-15.54
Meas.	-13.60	5.19	2.29-2.43	138	2.36	-17.80

4. Conclusion

In this paper, an inverted suspended circular patch antenna operating at frequency of 2.4 GHz for WLAN application is design based on inverted circular patch with L-probe technique of SMA connector to feed the antenna from ground to the feedline and air gap separation of 10 mm between patch and ground plane. Parameter of return loss, resonant frequency, ARBW, RLBW, realized gain, total efficiency, directivity and radiation pattern is monitor in order to analyse the performance result of the antenna. From observation on the simulation result at 2.4 GHz, axial ratio for the antenna is below 3 dB which is 0.42 dB with return loss -15.48 dB. The total efficiency is above -3dB which is -0.56 dB with directivity 8.46 dBi. The bandwidths for this antenna for simulation and measurement are more than 100 MHz which is 146 MHz and 138 MHz. Result of gain for measurement is 5.19 dB, slightly reduces compare to the simulation result which is 7.90 dB. Measurement result for resonant frequency is shifted a little bit to a lower frequency which is 2.36 GHz with -17.80 dB return loss. Overall, the simulation and measurement result for the antenna design meet a good agreement for the requirement of circular polarized antenna that can operating at 2.4 GHz for WLAN application.

References

[1] Prajakta D, Chandan B and Tazeen S 2013 *International Journal of Emerging Technology and Advanced Engineering* **3** issue 10 247-250
[2] Deshmukh A A, Singh D and Verma P 2016 *2nd International Conference on Advances in Electrical, Electronics, Information, Communication and Bio-Informatics, IEEE – AEEIC* 627-631
[3] Balanis C A, 2016 *John Wiley & Sons, Inc., New York*
[4] Cai Y M, Li K, Yin Y Z and Hu W 2014 *IEEE Antennas and Wireless Propagation Letters* **13** 674-677

[5] Wu J J, Ren X S, Li Z X and Yin Y Z 2013 *Progress In Electromagnetics Research C* **38** 1-14

[6] Abd Aziz M Z A, Mufit N A D A, Rahim M K A and Kamaruddin M R 2012 *Proceedings of International Symposium on Antennas and Propagation (ISAP) Japan* 555-558

[7] Mishra R, Mishra R G, Kuchhal P and Chaurasia R K 2016 *International Conference on Recent Advances and Innovations in Engineering (ICRAIE)* 7939468 1-5

[8] Pandey R K and Shanmuganantham T 2016 *IEEE International Conference on Circuit, Power and computing Technologies (ICCPCT)* 7530263 1-4

[9] Phoo K S, Abd Aziz M Z A, Ahmad B H, Othman M A, Suaidi M K and Abd Malek F 2014 *IEEE International Conference on Control System, Computing and Engineering* 236-240

[10] Ali M T, Nordin N, Pasya I and Tan M N M 2012 *Proceedings of 6th European Conference on Antennas and Propagation (EuCAP)* 6206166 2827-2831

[11] Wang L, Guo Y X and Sheng W 2012 *Asia-Pacific Microwave Conference Proceedings* 151-153

Acknowledgments

Authors would like to thank Universiti Teknikal Malaysia Melaka (UTeM) for supporting in this research work under the grant PJP/2017/FKEKK/HI13/S01540. Other than that, special thanks to the Government of Malaysia which provide MyBrain15 program for sponsoring this work under the research grant TRGS/1/2014/FKEKK/02/1/D00001 from Ministry of Higher Education (MOHE). Lastly, thanks to the anonymous referees whose comments led to an improved presentation of this work.

Strength Development of Marine Clay Treated by Demolished Concrete Materials with Morphological Identifications

Ayub A[1,a], Mohd Yunus NZ[2], Hezmi MA[3], Ali N[2], Abdullah RA[2], Jamaluddin N[1]
[1]Postgraduate student, Department of Geotechnics & Transportation, Faculty of Civil Engineering, Universiti Teknologi Malaysia, 81310 Skudai, Johor Bahru, Malaysia
[2]Senior Lecturer, Department of Geotechnics & Transportation, Faculty of Civil Engineering, Universiti Teknologi Malaysia, 81310 Skudai, Johor Bahru, Malaysia
[3]Lecturer, Department of Geotechnics & Transportation, Faculty of Civil Engineering, Universiti Teknologi Malaysia, 81310 Skudai, Johor Bahru, Malaysia

E-mail: azimahayub32@gmail.com

Abstract Construction and demolition (C&D) waste nowadays increasing years by years, which eventually leads to environmental and financial problems. Hence, it is suggested to recycle the concrete materials obtained from demolished building activities for geotechnical engineering applications. This study focuses on recycling the demolished concrete materials (DCM) to stabilize the Nusajaya marine clay. In this research, unconfined compressive strength (UCS) test was performed on DCM treated marine clay at 5%, 10%, and 15% and different range of curing time, 7 days, 28 days, 60 days and 90 days respectively. In additions, microstucture analysis, such as EDX, FESEM and XRD has been done in order to verify the optimum results as obtained from UCS test. The result obtained shows the strength of DCM-treated marine clay increase from 87.9kPa to 403.6kPa in 60days curing period with 15% of DCM content. EDX analysis shows the increment of Ca2+ elements meaning the cementatious agent of treated marine clay increase, hence increase the strength of soil itself. Plus, FESEM image portrays the flaky pores of the untreated marine clay, reduce because more needle-like structure of DCM contents filled up the pores particularly at 15% of DCM content. In conclusion, DCM acts as stabilizers optimally within 7-28 days curing time at 15% of DCM content.

1. Introduction

The number of construction developments increase due to the increase in demand for houses and major building, especially in Malaysia. This is attributable to the increasing of Malaysian population as well as towards of Vision 2020 achievement. While the human is too busy fulfilling the needs and pursuing visions, the effects of those rapid developments of ecosystem often being neglected [1]. As mentioned by the Board of Engineering, it has been reported that illegal dumping cases which include construction and demolition waste shown high increment from 3 cases in 2001 up to 31 cases in 2005 [2]. Not only that, statistics from Hong Kong proved that landfills need to cover 13844 tons of waste daily in 2012 and also 5.05 million ton per year.[3][4]

Marine clay could be found abundantly around the coastal and offshore areas and also beneath the ocean bed [5][6]. According to [7] most of the construction activities recently happens to be built on the clayey soil as a matter of sub grade and sub base. However, in many cases, clayey soil considers as

Content from this work may be used under the terms of the Creative Commons Attribution 3.0 licence. Any further distribution of this work must maintain attribution to the author(s) and the title of the work, journal citation and DOI.
Published under licence by IOP Publishing Ltd

problematic soil alike soft clays, musk, organic deposits and loose sand because they exhibit weak engineering properties [8] and portrays an inadequate sub grade performance which eventually may result in premature structural failure [9][10]. Other than that, marine clay is also known as weak soil due to its unstable properties such as low bearing capacity, high compressibility, soft and would swell when wetted and shrink easily when water is dissipated from it [11]. Thus, soil stabilizations are one of the best methods to improve the engineering properties of clay itself [12] and its mainly done with a purpose to modify the soil engineering properties such as to improve strength, the bearing capacity and their durability property of that particular weak soil [13]. As mentioned by [14], in some cases by a method of stabilization and soil compaction the strength of the subgrade is increased upon by mixing appropriate additives.

In conjunction with concrete and masonry constitute more than 50% of waste generated [14], reutilising this waste into fine particulate matter promotes dual benefits of saving landfill quota while at the same time reduce the usage of natural raw material for new construction activity [15]. This study, then focuses more on utilizing fine particles of demolished concrete materials to observe their stabilizing effects on Nusajaya marine clay.

2. Materials and testing programme

The main materials used for this study are natural marine clay and demolished concrete material which then will be added with different percentages; 5%,10%,15% of the soil. Each of the samples are then being cured for 7, 28 and 60 days.

2.1. Marine clay

The marine clay is obtained in the construction area of Nusajaya, Johor Bharu as shown in Figure 1. It was collected from a 2-3m depth. The clay obtained, was greenish black in colour and undesirable smell. The 2 mm marine clay used for test was air dried before any test has taken place. The fresh and original marine clay deposited on the site is taken to the Geotechnical Laboratory of Universiti Teknologi Malaysia for further experimental works. Table 1 summarizes the properties of Marine clay used in this study while Table 2 provides the chemical elements of marine clay according to [16].

Figure 1. Deposited of marine clay at Nusajaya

Table 1. Summary of marine clay properties

Marine clay properties	Value
Natural water content	50.99
Specific gravity	2.083
Free sweel index	50%
Liquid limit	47.05%
Plastic limit	23.14%
MDD	1530 kg/m^3
Optimum moisture content	24%

Table 2. Summary of chemical elements forming Marine Clay (After Mohd Yunus et al., 2014)

Chemical elements	Amount(%)
Al	13.67
Si	21.91
K	2.02
Ti	2.06
Fe	5.11
Mg	0.78

2.2 The demolished concrete materials

Grade 30 demolished concrete blocks as shown in Figure 2 are collected from the disposal site of Structural Laboratory, Faculty of Civil Engineering, Universiti Teknologi Malaysia. The demolished concrete is then being crushed and sieve up to 63μm to mix efficiently with the soil.

Figure 2. Gred 30 demolished concrete in cubes formed is used

2.3 Specimen Preparation
Soil samples which here refers to marine clay were sieved through 2 mm mesh sieve in order to prepare samples for UCS. Distilled water was used during sample preparation. All samples were prepared using 95% of the maximum dry density (MDD) from dry side and the corresponding optimum moisture (OMC) content in order to treat the samples before failure. The volume of the mold at which the samples were compacted and the MDD was used as reference during sample preparation. The predetermined values of DCM were measured to the dry mass of marine clay. Marine clay and DCM were mixed thoroughly and then distilled water was added and mixed until homogeneity was achieved. The mixing period did not exceed 5 minutes to minimise loss in water content. The mix designs of DCM were 5, 10, and 15% at which three UCS samples were prepared for each percentage to avoid errors during the test. The mixture of marine clay and tiles was compacted inside the mould (76 mm height and 38 mm diameter) using the hydraulic jack machine. A stainless steel plunger was used to extrude the samples from the mould. Samples were trimmed to remove the extra soil and wrapped with several layers of cling film in order to preserve the water content. Samples were placed inside air tight plastic containers and stored for curing inside a controlled humidity chamber. The humidity chamber temperature was 27 ± 2 C° and the humidity was 97% ± 2%. Samples were cured for 7, 14 and 28 days before testing.

Microstructural tests such as Energy Dispersive X-Ray (EDX), X-ray Diffraction (XRD), and Field-Emission Scanning Electron Microscope (FESEM) test were done to support the result as obtained from UCS test as well as to identify any possible chemical reactions happened. EDX is an analytical technique used for the elemental analysis or chemical characterisation for both treated and untreated marine clay which in this study to identify any cementatious agents presence. In addition to compound verification, X-Ray Diffraction (XRD) is then further analyses the sample in order to determine the chemical elements presence in marine clay and DCM with specific strain, preferred orientation, or even crystallographic structure by using 2theta intensity. Lastly, FESEM helps to visualise very fines topographic details and changes for treated marine clay after 60 days curing time at desire magnifications.

3.0 Results and discussion
The results show the shear strength of both untreated and treated marine clay after 7,28, and 60 days curing period with different percentage of DCM. The resulted is then supported by the microstructure analysis through SEM, EDX, and XRD.

3.1 Unconfined compressive strength
Effect of compressive strength of marine clay treated with DCM is determined based on different curing periods and various additions of DCM contents. Table 3 summarizes the value of compressive strength for both treated and untreated marine clays at 5%,10% and 15% of DCM contents after selected curing periods

Table 3. Result of shear strength value with different curing period (kPa)

	Curing Period(days)			
Percentage of DCM (%)	**0**	**7**	**28**	**60**
0	87.9	87.9	87.9	87.9
5	153.6	163.0	188.3	292.1
10	173.4	210.0	320.6	327.8
15	217.7	249.4	352.2	403.6

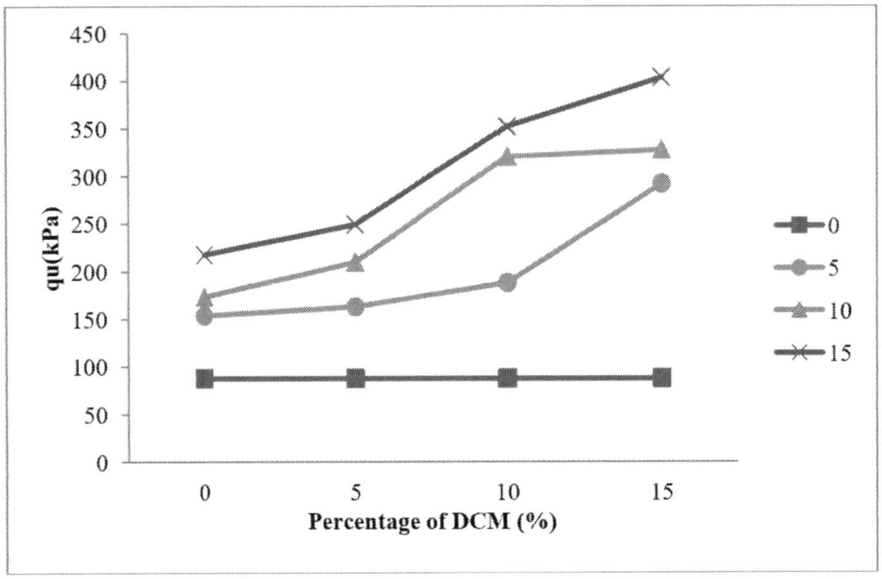

Figure 3. The effects of different DCM content on compressive strength

The results of the UCS tests carried out on DCM-treated marine clay at different curing periods are shown in Figure 3. It can be seen that strength increase with addition of demolished concrete materials with a maximum value of 403.6 kPa in 60 days of curing time. Figure 3 shows that the shear strength of all the treated specimens increased significantly compared to the strength of untreated Marine clay which in agreement with previous study [13].Basically, strength increased with higher curing periods. It was found that, DCM can provide initial strength gained, where strength increase from 87.9 kPa to 153.6 with addition of 5% DCM contents immeditely (i.e. 0 day) after mixing with marine clay. However, the slight increased in shear strength is observed at 7 curing days as compared to 0 day. At 28 days, only specimen treated with 5% DCM shows insignificant strength gained, while for 10% and 15%, the significant increased in strength is remarkable. Strength was increased from 320.6 kPa to 352.2 kPa. Investigation on strength development was continue at higher curing period of 60 days, Based on the result, it was found that, slight increased in strength of specimens treated with 10% and 15% strength a s compared to 28 curing days

The strength of treated marine clay gained due to the the frictional interaction between soil and DCM which correlation to the facts that waste materials consisting cementatious agent, particularly demolished concrete material act as an effective stabiliser for marine clay soil [17]. DCM obviously shows that it can be a great stabiliser to the marine clay even at only 5% DCM content added to the clay. Then, the cementitous agent present in the recycled concrete is well reacted with the clay itself as more curing time allocated [18].

Furthermore, it could be perceived that the strength of treated marine clay increase significantly as compared to untreated marine clay with the highest value of 403.6 kPa in 60 days. [19], quoted that the improvement of soil properties stabilized with demolished concrete could be due to it having the same reaction as lime or cement. As there is cement in concrete, this could be deemed possible. The increment of soil strength with increasing addition of cement is due to the hydration reaction of cement water which is the drying of soil-cement mix and the formation of cemementatious product [18]

Based on the graph, at 15% DCM content increase from 28 day to 60 day eventually potrays highest strength compared to other curing period. Yet, according to [20] concrete has better cementatious agent as compared to other materials like aggregate and marble tiles. However, a little difference on strength increment is due to the a slight difference of DCM content. There is no significant increase might due to the only 5% of DCM being added to each of the samples. This is because DCM less distributed into the marine clay which then resulting in poor workability as the air void in marine clay is not fully covered by the DCM content. This even common to other stabilisation process where the strength development required more cement additives to enhance more compressive strentgh as discussed by [21]. Thus, the DCM content approximately works at its best with the 10%,20%30% of DCM content.

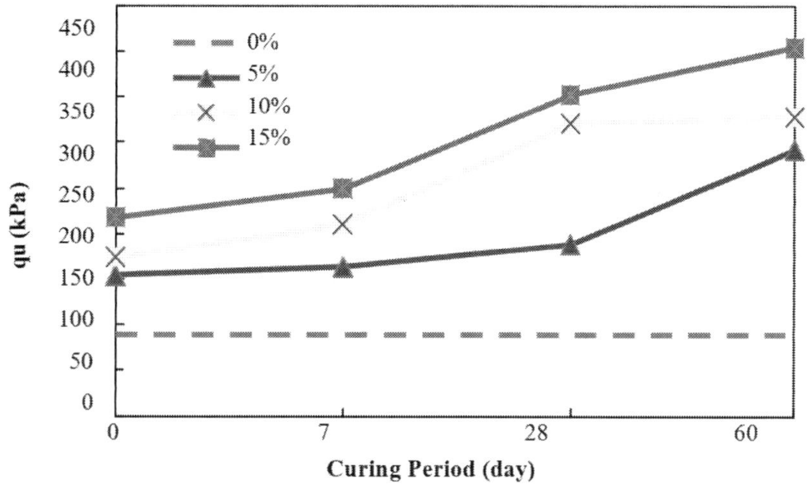

Figure 4. Effect of curing period on DCM content

Effect of DCM content on shear strength of marine clay at different curing period is shown in Figure 4. There is a huge difference between untreated marine clay to 5% treated day at 60 days curing time. It shows that, at 60 day the reaction between DCM and marine clay getting better and helps in improving the marine clay strength even at 5%, as compared to other curing time. Other than that, the graph also illustrated there is slightly diffrence between 5% DCM content and 10%DCM content for each curing period, which means, DCM acts more when more DCM contents is added. Therefore, it can be conclude here, the untreated marine clay at Nusajaya posses about 88 kPa and become higher up to 403 kPa after 60 days curing time.

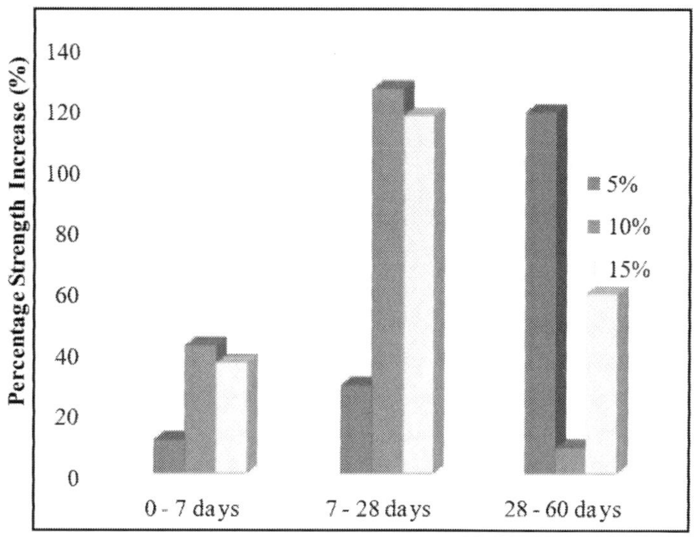

Figure 5. Strength gained at different curing period

The experimental result plotted in Figure 5 shows the strength gained from 0-7, 7-28 and 28-60 day, at different percentage of DCM. The graph obviously shows, the strength gained gradually especially at 5% from 0 to 60 day curing time. This literally defines that the DCM reacts better on marine clay from time to time. 10% of strength gained try to imposed only a DCM still not binds friendly and closely with marine clay as it appears only a bit strength gained. However, the confidence rise as more strength regained with 15% of DCM content. Hence, this result correlant as [22], mentioned, additives reacts greater when more supplementary additives (which in this case is DCM) being added to the natural soil.

3.2Microstructural Analysis

Microstructure analysis were carried out on untreated marine clay, DCM, 5%, 10% and 15% of treated marine clay at 60 days. Samples were chosen to test at 60 days as the highest strength for each addition of DCM contents was recorded at longer curing periods which is at day 60.

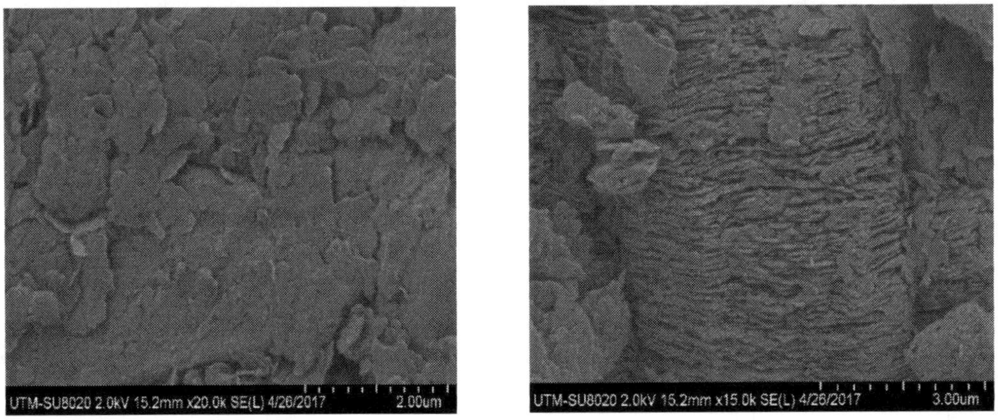

(a) (b)

Figure 6. FESEM images of Nusajaya marine clay at different magnification a) 20,000 magnfication b) 15,000 magnification

Figure 6 shows the microstructure of marine clay at different magnification. Marine clay structure appears to be flaky and have pores as could be seen in figure (a). While in figure (b) shows that the structure of the marine clay could be seen arranged as a book-like structure. The image appears as stated by [23] which soft soils like clay gives better soil structure due to presence of high organic content. Figure 7 shows the microstructure of demolished concrete material (DCM) with different magnification. The DCM structure shown in figure (a) and (b) is more of a needle-like shape with a few clumps.

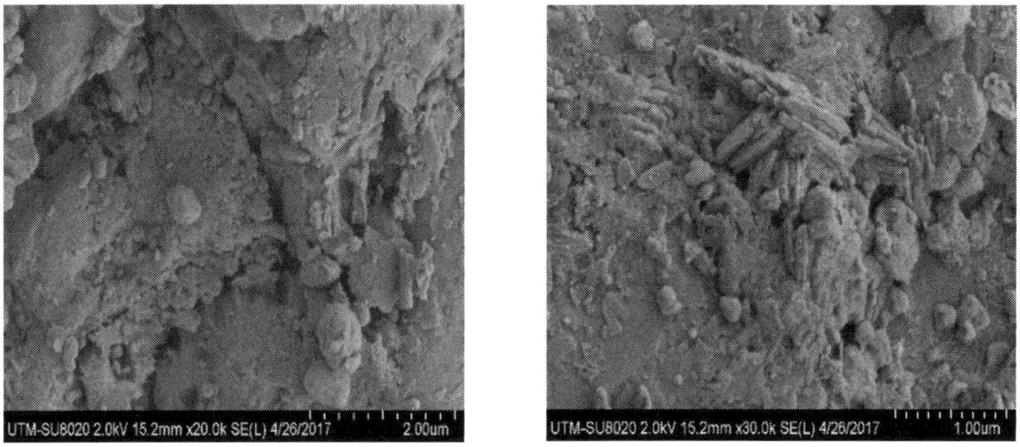

(a) (b)

Figure 7. FESEM images of Demolished Concrete Material(DCM) at different magnification (a) 20,000 maginification (b)30,000 maginification.

In Figure 8, with 5% DCM content shows that DCM is filling up the pore spaces in between the marine clay structure. But there are still pore spaces in between as there are not enough stabilizer content mixed well. Research done by [24], cement tend to fill up the pore spaces in between the soil structure thus increasing its stability. Other than that, the flakiness of marine clay structure could still be clearly seen in this figure.

Figure 8. FESEM photography of marine clay stabilized with 5% DCM at 60days.

Figure 9(a) shows that DCM is filling the pore spaces in the marine clay structure, thus making it denser and become more stable structure. At 10% DCM,it clearly appears there is a significant difference as compared to 5% addition of DCM. The pore spaces are now filled with hydration products. Moreover, figure 9(b), clearly shows an increase in DCM needle-like structure filling up the pore spaces. This is verify that significant strength increase happened at 10% DCM addition as compared to 5%. At 10%, it could be observed that the marine clay structure shows a reduction in terms of quantity. This is might due to the reaction of soil, water and cementation agent. [25] stated that stabilization of soil begins when mixing the soil with water and a stabilizer.On the other note, with the presence of water and cementation agent the soil becomes a modified soil.

(a) (b)

Figure 9. Soil structure of marine clay treated with 10% at 60 days curing time at a magnification of a)20 000x b) of 30 000x

Figure 10 shows that the pore spaces are filled with 15% DCM content. With the addition of 15% DCM, it could be observed that the flakiness of soil structure could no longer be seen. Moreover the needle-like structure of DCM could be found abundantly in figure 10(b) as compared to 10(a). With increasing percentage of DCM used, the more cementation product present in the soil structure, hence improve the strength.

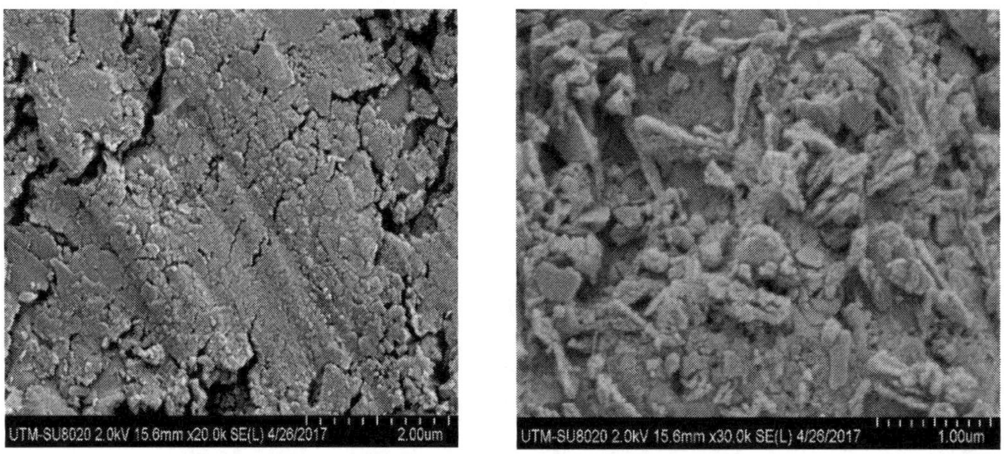

(a) (b)

Figure 10. Soil structure of marine clay treated with 15% at 60 days curing time at magnification of a)20000x b) 30000x

Table 4 shows the result from EDX analysis which the main element that consist in both marine clay and DCM itself. According to [26], the cementation compounds are Ca, Mg, Al, and Fe.The adsorption/ desorption primarily depends on valence element. The lowest cations are easily removal or replaced cation in soil particle [27]. Based on the EDX result, it could be observed that Si percentage decrease with increasing DCM percentages. While Ca percentage is increasing with the increment of stabilizer content. During ion exchange reaction of the soil with cementitious additive, cation of soil is replaced by cation of additive (Ca) and the thickness of double diffused layer is reduced. Hence, the replacement of cations results in an increase in workability and strength of soil-additive mixture [28]

Table 4. EDX analysis shows the chemical elements presence in marine clay and DCM.after 60days curing period

Element	DCM content added (%)			
	0	5	10	15
Si	19.40	17.9	15.7	8.4
Al	14.0	11.6	10.0	4.9
Ca	-	1.4	2.6	8.3
Fe	0.3	1.7	0.7	0.8
Mg	0.6	0.6	0.7	0.4

According to the XRD analysis as shown in Figure 11, the Nusajaya marine clay contain mineral quartz, illite, montmorillonite, and kaolinite. High content of kaolinite and illite may control the characteristic of Nusajaya marine clay. DCM contain minerals such as quartz, portlandite, cristobalite and calcite. Immediate reaction of treated marine clay by 5% DCM shows reduction of kaolinite and illite. Moreover, there are presents of ettringite in the mix. This is due to the result of hydration reaction. [29], mentioned that in the presents of calcium sulfate, there would be ettringite as an additional product of hydration.

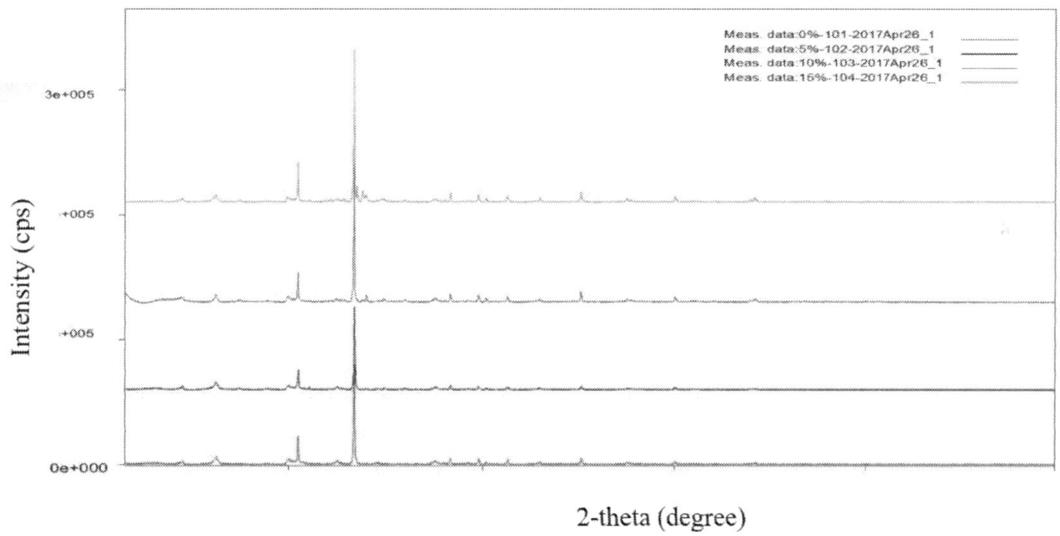

Figure 11. Comparison of XRD results for 60 days at different percentage of DCM

4.0 Conclusion

In this study, the effect of demolished concrete materials on marine clay has been determined. The comparison was made between treated and untreated marine clay at different percentages of DCM content and at different curing periods. Based on the obtained result and analysis performed the following conclusions can be drawn:

- The higher the amount of DCM content, the higher the strength of treated marine clay. A compressive strength increase from 87.9 kPa to 153.6kPa after 5% DCM content beingadded.
- The unconfined compressive strength of treated marine clay is then increased with addition of DCM.

The highest value of UCS is 403.6 kPa in 60 days with the addition of 15% DCM. But the highest strength increase for marine clay could be observed at 10% DCM addition at 28 days.

- Maximal strength gained is within the period of 7-28 days. While within the period of 28-60days, the increment of strength starts to reduce.
- Based on FESEM, the DCM filled in between the pore clay spaces. With the increasing DCM addition, it could be observed that the flakiness of the clay material has decreased, but the needle- like structure of DCM increase. This concludes that by filling the pore, DCM helps in stabilised the soil cause the strength to be increased.
- EDX result obtained portrays the presence of Ca^{2+} elements increase as the percentage of DCM increase. Ca^{2+} elements proved that presence of cementatious agent from DCM cause to increase the soil strength.

Meanwhile, XRD result shows a reduction of chemical compound for marine clay which are koalinite and illite. Calcium aluminium hydrate (CAH) exists as new chemical compound after treating with DCM. Thus, it can be concluded that DCM act as stabiliser due to the possible chemical reactions occur such as pozzolonic reaction in which cause the well deve lopment of soil strength optimally within 7-28 days.

References

[1] Souza L De, Rossi S, Calvet-mir L, Ruiz-mallén I, García-betorz S and Salvà-prat J, et al. 2017 Neglected ecosystem services : *Highlighting the socio-cultural perception of mangroves in decision-making processes.* Ecosyst Serv [Internet]. Elsevier B.V.; 2017;26:137–45. Available from: http://dx.doi.org/10.1016/j.ecoser.2017.06.013

[2] Saadi N, Ismail Z and Alias Z. 2016 A Review of Construction Waste Management and Initiatives in Malaysia *J Sustain Sci Manag* **11(2)** 101–14

[3] Alan Y. *Straight to Landfill? Why Hong Kong is Recycling Less of Your Rubbish* [Internet]. South China Morning Post. 2015 [cited 2017 Aug 3]. p. 1–7. Available from: http://www.scmp.com/lifestyle/article/1859106/why-hong-kong-recycling-less-its-rubbish-time-goes

[4] Tam V W and Lu W 2016 Construction Waste Management Profiles , Practices , and Performance : A Cross-Jurisdictional Analysis in Four Countries 1–16

[5] Saeed K A H, Kassim K A and Mohd Yunus N Z 2013 Characterization of Hydrated Lime-Stabilized Brown Kaolin Clay *Int J Eng Res Technol* **2(11)** 3722–7

[6] Al-bared M A M and Marto *A* 2017 Review On the Geotechnical and Engineering Properties of Marine Clay and the Suitable Common Stabilisation Methods. *Int Conf Seperation Technol.* 5–8

[7] Izabel K J and Sangeetha S 2017 Stabilization of Marine Clay Using Jerofix 4(3) 2014–6

[8] Ravikanth Y, Tejaram B, Pradesh A, Rahul B G and Pradesh A 2017 *Stabilisation of Debris Material by Using Soil*. Int J Civ Eng Technol [Internet]. 2017;8(4):2125–31. Available from: http://www.iaeme.com/IJCIET/issues.asp?JType=IJCIET&VType=8&IType=4

[9] Kerni V, Sonthwal V K and Jan U 2015 Review on Stabilization of Clayey Soil Using Fines Obtained From Demolished Concrete Structures *Int J Innov Res Sci Eng Technol.* **4(5)** 296–9

[10] Paul H and Cyrus S 2016 Stabilisation of Weak Subgrade Soil Using Demolished Concrete Aggregate *Indian Geotech Conf.* 1–5

[11] Ramamoorthy S. 2007 Correlation of Engineering Characteristics of Marine Clay. Universiti Teknologi Malaysia

[12] Mohd Yunus N Z, Wanatowski D, Stace R, Marto A, Kassim F and Hassan N A. 2014 Undrained & Drained Behavior of Lime- Treated Organic Clay with Calcium *EJGE* **19(1596)** 17

[13] Jain A and Chawda A 2016 Apraisal of Demolished Concrete Coarse and Fines For Stabilisation of Clayey Soil *Int J Eng Sci Res Technol.*

[14] Singh E L, Singh E S and Gill K S D 2017 Improvement in CBR Value of Soil using Waste

Concrete Fines **3(9)** 1–5

[15] Modarres A and Nosoudy Y M 2015 *Clay stabilization using coal waste and lime - Technical and environmental impacts*. *Appl Clay Sci* [Internet]. Elsevier B.V.; 2015;116–117:281–8. Available from: http://dx.doi.org/10.1016/j.clay.2015.03.026

[16] Mohd Yunus N Z, Wanatowski D and Stace L R 2012 Effectiveness of Chloride Salts on the Behaviour of Lime-Stabilised Organic. **0** 407–12.

[17] Gallen S. 1993 Appropriate Building Materials: a List of Potential Solutions. In: Building materials [Internet]. Swiss Centre for Development Cooperation in Technology and Management
p. 430. Available from: http://collections.infocollections.org/ukedu/en/d/Jsk01ae/3.3.html

[18] Xiao H W and Lee F H 2008 Curing Time Effect on Behavior of Cement Treated Marine Clay *World Acad Sci Eng Technol* **43** 71–8

[19] Abhijith B S, Murthy V S and Kavya S P 2014 Study of the Effectiveness in Improving Montmorillonite Clay Soil by Construction and Demolition Waste *J Civ Eng Environ Technol*. **1(5)** 1–4

[20] Mcdonald R, Alaa A, Carreon D and Atalay F 2006 Stabilization of Marginal Soils Using Recycled Materials. Florida Dep Transp.

[21] Cong M, Longzhu C and Bing C 2014 Analysis of strength development in soft clay stabilized with cement-based stabilizer *Comput Chem Eng* [Internet]. Elsevier Ltd; 2014;71:354–62 Available from: http://dx.doi.org/10.1016/j.conbuildmat.2014.08.087

[22] Tamadher T A, Anuar B K and Zamri B C 2007 Stabilisation of Silty Clay Soil Using *Chloride J Eng Sci Technol © Sch Eng Taylor's Univ Coll. 2(1)* 102–10

[23] Changizi F and Haddad A 2015 Strength properties of soft clay treated with a mixture of nano-SiO_2 and recycled polyester fiber

[24] Pillai R J, Bushra I and Robinson R G 2013 Undrained Triaxial Behavior of Cement Treated Marine Clay. *Geotech Geol Eng.* **31(2)** 801–8

[25] Wijeyesekera D C, Ho M H, Bai X and Bakar I 2015 Strength and Stiffness Development in Soft Soils : A FESEM aided Soil Microstructure Viewpoint

[26] Allam M M and Sridharan A. 2015 The shearing resistance of saturated clays 119–22

[27] Cherian C and Arneppali D N 2015 A Critical Appraisal of the Role of Clay Mineralogy in Lime Stabilization *Springer Int J.* 1–20

[28] Oza J B and Gundaliya P J 2013 Study of black cotton soil characteristics with cement waste dust and lime. *Procedia Eng* [Internet]. Elsevier B.V.; 2013;51(NUiCONE 2012):110–8. Available from: http://dx.doi.org/10.1016/j.proeng.2013.01.017

[29] Latifi N, Eisazadeh A and Marto A. 2014 Strength behavior and microstructural characteristics of tropical laterite soil treated with sodium silicate-based liquid stabilizer. *Environ Earth Sci.* **72(1)** 91–8.

The emerging of employment gap in the Malaysian construction industry

Abdul Rahim Abdul Hamid[1], **Nur Inayah Mohd Khazid**[2], **Riduan Yunus**[3], **Hazruwani A Halim**[4] **and Amall Raihan Abdul Razak**[5]

[1,2,4,5] Department of Structures and Materials, Faculty of Civil Engineering Universiti Teknologi Malaysia, 81310 Johor Bahru, Johor, Malaysia.

[3] Jamilus Research Centre, Faculty of Civil and Environmental, Department of Building and Construction Engineering, Universiti Tun Hussein Onn Malaysia, Batu Pahat, Johor, Malaysia.

E-mail: rahimhamid@utm.my

Abstract. Employment of young people is a big issue in the Malaysian construction industry. This is a common phenomenon not only in Malaysia but affected most developed and developing countries around the world. Sufficient workers are needed to cater to the robustness of the construction sector and to sustain the productivity that will boost the development demand. However, young people had difficulties to find decent employment in this industry due to lack of experience and skills among young people and some employer are more likely to hire foreign workers. The aim of this study is to examine the emerging of skill gaps in the Malaysian construction industry. The methodology adapted for this study comprise of primary and secondary data collection. The primary data had been collected from the questionnaire surveys on the professional within the construction industry. The data then were analyzed using the percentage score and the diagram such as bar chart had been used to visualize the results from the analysis. While, secondary data collected through literature review to gather information and important views from previous studies. The findings reveal that the construction industry prefer more than 30 years old skilled workers. Eventhough, the jobs could be filled by young people, but somehow, some barriers such as high risk, low wages, hard work and availability of foreign workers had distracted the involvement of young people in the construction industry.

1. Introduction

In year 2015 Malaysia faced a small economic downturn, however, the construction projects keep going on especially when the government announced a few big projects to be developed mostly located in Kuala Lumpur and Selangor. This indicates that the construction sector able to provide more employment opportunities for the people. In 2016, the construction sector's share of GDP was 4.5% valued at RM50.1 billion and registered 7.4% output. This sector hired 1.25 million employees, which showed a drop of 4.4% from 1.31 million in the previous year. In the second quarter of 2017, the construction sector expanded by 8.3% due to civil engineering, construction activities and non-residential buildings. By 2020, the sector is expected to contribute 5.5% to GDP at RM327 billion and employing 1.2 million employees. Despite its smaller contribution, the construction sector continues to play a key role in transforming the economy through its multiplier effect on many industries by accelerating domestic economic activities and providing comprehensive high-end quality infrastructure [1]. Thus, construction sector able to provide the path for the employment of young people since according to Malaysian Industrial Development Finance (MIDF) the rate of unemployment edge up to 3.1% in January 2015. In August 2017, the number of unemployed persons i.e. those who do not have a job but are interested to work, recorded a year-on-year (YoY) rise of 0.3%, increasing from 515,200 in August 2016 to 516,900 this year. Compared to July 2017, the number dropped by 0.4% from 519,000. The unemployment rate i.e. the percentage of unemployed persons in the labour force, fell by 0.1 percentage point from 3.5% in August last year to 3.4% this year [2].

Content from this work may be used under the terms of the Creative Commons Attribution 3.0 licence. Any further distribution of this work must maintain attribution to the author(s) and the title of the work, journal citation and DOI.
Published under licence by IOP Publishing Ltd

United Nations Educational, Scientific and Cultural Organization (UNESCO) defined young people or youth as among those who ages between 15 and 24 years. While in Malaysia, the definition of youth according to National Youth Development Policy, are those who ages between 15-40 years. However, the policy specified that youth development programs or activities should involve youth age of 18-25 years old. The youth unemployment rate in Malaysia was estimated to have reached 10.7% in 2015, more than three times higher than the national unemployment rate of 3.1%. Malaysia is among regional economies with an incidence of youth unemployment in the double-digits, despite a low overall unemployment rate. Youth represents more than half of total unemployed workers, despite only making up a third of the labour force. The youth are the most vulnerable to these trends; they are likely to be the last to be hired and the first to be made redundant, due to their lack of experience, higher information asymmetry on the labour market, and poor ability to communicate their skills effectively to employers [3]. Thus, there should be more room for the youth to participate in job market particularly in the construction sector.

Nowadays, young people had to bear with difficulties in finding decent employment. The main difficulties; chances of losing jobs during economy falls ("last in, first out"), specific barriers to entry which often cause by lack of experience and path dependence: early increase in unemployment lead to more subsequent unemployment [3-5]. In 2015, the unemployment rate expected to be higher because of slow economic growth. According to Malaysian Industrial Development Finance (MIDF) the unemployment rate increase to 3.1% in January that could be featured by high deflation in the construction sector in the fourth quarter of 2014 which having an impact on the labor market [6]. This situation will further push the young people in Malaysia difficulties in seeking for jobs since companies had a hard time during the economic downturn. Another constraint due to the foreign workers more favorable compared to local worker in construction sector. Ofori [7] stated the employers prefer foreign workers since the wages paid less compared to local workers. Moreover, according to Hamid et. al [8], foreign workers preferable because; a) foreign workers adhere to carry out any type of work they were ordered to do, b) foreign workers do not complain about their salary, c) foreign workers accepted welfare or facilities provided to them and d) foreign workers more discipline on site. Young people had lack of opportunities to enter into construction sector due to this reason, since foreign workers were less hassle to handle compared to local workers which mostly related to attitude's issues. Meanwhile, young people had less interest to join in training programs and participate in the construction sector as accordance with International Labor Organization (ILO) [9], the industry didn't have the ability to attract workers and invest in training that will cause poor impacts towards the productivity and quality of the construction products. Moreover, labor shortages and lack of skills will cause more unemployment opportunities since labor were replaced by machine through prefabrication and mechanization.

So far, there is very little published data and analysis of the current condition of employment opportunities for young people in Malaysia construction sector. Therefore, this study is important and timely in order to examine the emerging of employment gaps in the construction industry.

2. Brief Literature

2.1. Definition of employment gaps

An employment gap is a period of months or years when the job applicant was not employed at a job. Employees choose to spend time unemployed for purposes such as layoffs and downsizing, serving time in prison, or employment termination, attending school full time and having and raising children. An employment gap is either for positive or negative reasons, it will need an explanation for the potential employer [10]. Meanwhile, a website Money-zine [11] defined employment gaps as the period when someone was unemployed or absent from a work environment. This indicated the situation where the gaps occurred when the person was not having any jobs at certain period of time. On the other hand,

Willis and Cooper [12] stated that employment changes respond to the gap between the desired and actual number of workers. Employment gaps also depend on the demand and the supply of the workers required by the industry. The insufficient number of workers creates the problem of employment gaps. According to Schneer and Reitman [13], the gaps occurred in the employment histories undesirably related with the future income and satisfaction. The impact of the gaps found more serious for men rather than women since unable to follow the traditional path harmful for men more than women.

2.2. Definitions of skill gaps
Skill gaps are one of the essential factors in this study which lead to the problems in construction where the young people have big responsibilities to fill in the gaps occurred. Thus, it is important to understand the meaning of skill gaps in order to cater the problems regarding the matter. Economic Modelling Specialist International [14] defined the skill gaps as the incompatibility of the needs of employers for required skills verses the skills own by available workforce. It shows the worker unable to fulfil the skills required where this will cause the skill shortage face by the employer.

In addition, skill gap is the measurement of the differences occurred between the skills needed for specified job and skills qualified by a prospective worker as stated by Act.org [15]. Skill gap also depending on the individual workers where the skill needed for the specific job should be acquired by the related workers. Bloom et al. [16] defined skill gaps as the employee's skill insufficiencies inside a firm further occupied within the workplace. Skill gaps consist of (i) Basic skills (literacy, language, numeracy and computer skills), able to provide path for further knowledge development; (ii) Intermediate skills (related professional, apprenticeship, technician or skilled craft or trade level); (iii) Mathematics and other sciences and (iv) Leadership and management skills.

2.3. Emerging of skill gaps in construction industry
Malaysia's construction is a huge sector required a large number of people to construct a successful building and infrastructure projects. Malaysia had difficulties to employ young to work in the construction industry. According to the population survey by the Centre for Construction Research and Training, in 2010, in US the labor force is aging where the average age of the construction workers move from 36.0 to 41.5 years old from 1985 to 2010. This shows that older people in the construction industry are more than young people. While Salverda et al. [10] stated that young people actively involved in the process of earning and job mobility, where most of them were paid low which cause them to move into better paid employment and there is also declining rate to return to education. Young people with lack of experience, mostly paid low in the industry, which cause them to job hopping and some hesitate to go back into education. Education able to provide them with a higher certification in the related field and further increase their eligibility in the industry.

Aziz [17] reported that local youth in Malaysia don't have interest working in the construction industry and rather be unemployed. Hence, 80% of the entire workforce relied on foreign labor. Most of local young people perceived labor in construction industry as second class and a tough job compare to the job as skill worker. This paradigm causes a declining number of local worker labor in the construction industry. Nasir et al. [18] stated that there should be no dependant on the quantity of skilled workers, but rather the quality of the skilled workers possess in order to enhance productivity and the development of economy. While technical skills are basic knowledge to enter the industry, the lack of non-technical skills lead to slower development of skilled worker's productivity. The non-technical skills for instance communication skills, problem solving skills and decision making skills are the medium to improve the ability of skilled workers in industry.

3. Methodology of the Study

The researcher opted quantitative analysis technique [19-20] to answer the objectives of this study. The questionnaire survey was conducted among construction players in various fields through mediums such as social media, email and distributed directly to the targeted group in their companies and construction sites. The participant involves in data collection for this study include consultants, contractors and other construction players. The respondent targeted to also provide apprenticeship opportunities for young people. The data collected, grouped according to the frequency of the answer given by the respondent. The percentage calculated by dividing the number of observations with total number of observations, then the result should times with 100% in order to gain the result in percentage value. The results then, are presented in the form of charts and table.

4. Results and discussion

4.1 Respondent background

More than 100 questionnaires set were distributed by hand and email to the respondents located in the peninsular Malaysia. 51 numbers of respondent had answered the questionnaire which make up about 50% had responded to the questionnaire distributed. Most of the respondents responded through email and which 66% of the respondent are located in the state of Kuala Lumpur (37.3%) and Selangor (29.4%) while the rest are in Penang (7.8%), Pahang (7.8%), Terengganu (5.9%), Kelantan and others each (3.92%), Johor and Kedah each (2%). Construction industry involves various parties. The respondents were selected could be from various backgrounds since employability in this study not only restricted for certain professions. Based from the respondent background, most of them were from the quantity surveying from consultant and contractor organisation (33%). The second larger respondents to 21% was civil engineer followed by mechanical and electrical engineer, 12% as shown in table 1. Based from figure 1 most of the respondents involve in project management, which comprise 39% of the respondent. While 27% of the respondent work in professional service since most of the respondents came from quantity surveying background. Another respondent involves in general contracting, special trade contractor and project development.

Table 1. Respondents' Profession

Type of Profession	Count	%
M&E Engineer	6	12%
Civil Engineer	11	21%
Quantity Surveyor	17	33%
Project Manager	3	6%
Contractor	1	2%
Architect	4	8%
Landscape Architect	1	2%
Contract Executive	1	2%
Strategic Communicator	1	2%
Assistant Surveyor	1	2%
Others	5	10%

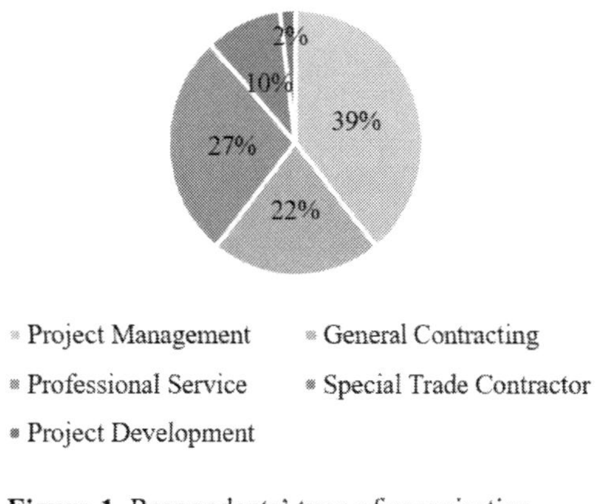

Figure 1. Respondents' type of organisation

4.2. The emerging of skill gaps in the construction industry

Based from the literature review, the demand for the skill worker was high. The phenomena didn't only occur in Malaysia but also in the UK [21-23]. Skill gaps could be caused by various factors as stated in the findings below. Based on Figure 2, the most demanding period for firm to hire new worker is throughout the year which is 64%. This reveals that the demand for worker in the construction industry is high and could be filled with young people and unemployed people. Respondents indicates that most of the workers in their organization are around 31-40 years old (61%). Thus, indicated that most organizations preferred to hire 31-40 years old and above 40 years old (6%) worker rather than 18-30 years old (33%). Figure 3 indicated the type of worker in the firm where 69% of the worker are skill worker while 31% are semi skill worker. Thus, the firm demand skill worker that the people who have an interest to participate in construction should equip themselves with related skill in order for them to be competent.

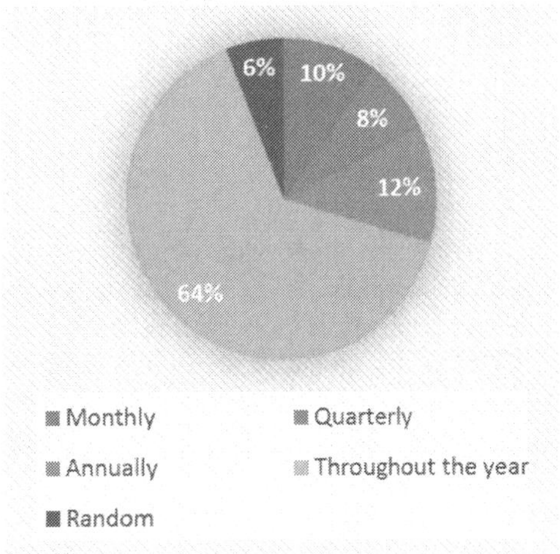

Figure 2. Most demanding period for firm to hire new worker

Figure 3. Type of worker in the firm

Figure 4 revealed that the firm had low demand (54%) for migrant in the range 0%-20%. This mean that the firm do not rely on migrant worker since the job already being filled up by the foreign workers compare to local people. Other reason might be due to high expenses required to hire migrant workers and most of the migrant workers are unskilled worker. Figure 5, 92% of the respondent agreed that skill gaps needed was catered in order for the firm to grow. Skill worker is important for the growth of a firm in order to growth and the ability to compete with other firms.

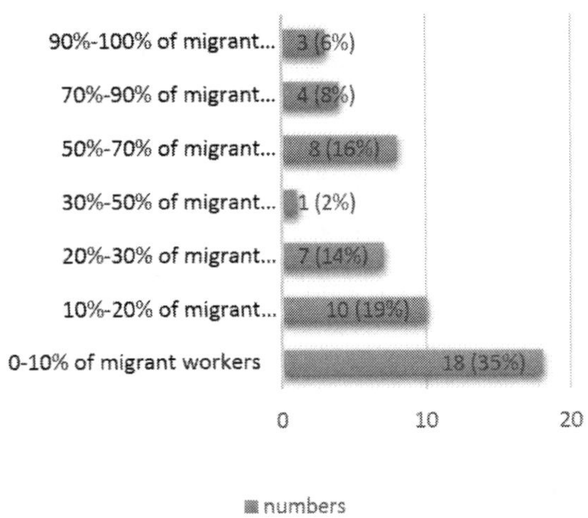

Figure 4. Reliability on migrant worker

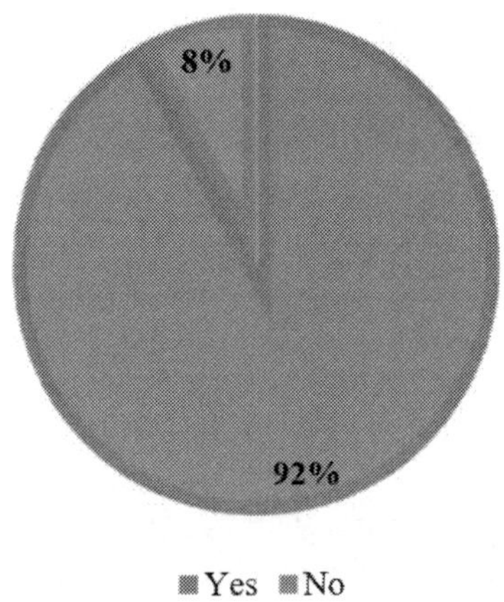

Figure 5. The effect of skill gaps to firm growth and competency

Figure 6 illustrate number of apprentices recorded low which is between 0 to 5 people. This show that the firm not really serious to recruit new workers. The reason might be due to high cost to have a big number of apprentices. The firm need to pay the apprentices while their performance might be low due to lack of experience. Figure 7 describe 84% of the respondent agree that apprenticeship is one of the way to filling up the gap existed in the construction industry where young people should be exposed and well trained in related fields in the construction industry. When the gaps could be diminished, the rate of unemployment especially among young people could be reduced and the efficiency in the construction industry could be increased.

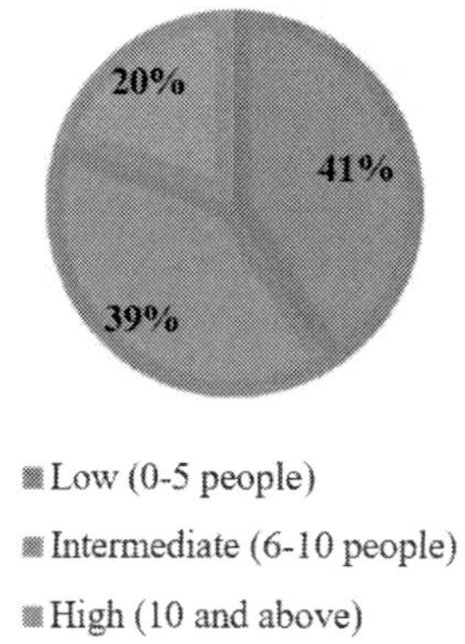

Figure 6. Number of Apprenticeship

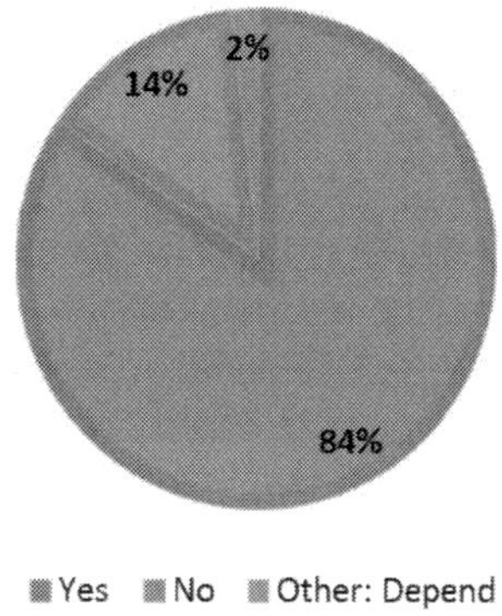

Figure 7. Apprenticeship to filling up the gap

Figure 8 states that 88% of the respondent agree it is necessary to employ skill worker to replace those who are retiring. Lack of skill worker in a firm will cause lack of efficiency and reduce the competency of the firm. If the firm unable to handle such situation, the growth of the firm will be retarded and threatened the firm's survival. Figure 9 demonstrates that most of the respondents which comprise about 57% agreed that their firm struggling to recruit the workers with the right skill. Thus, indicated that the firm demanded worker with the skill that match the industry's needs and the skill should be developed in order to make the workers possess relevant skills to adapt with the changes of the construction industry.

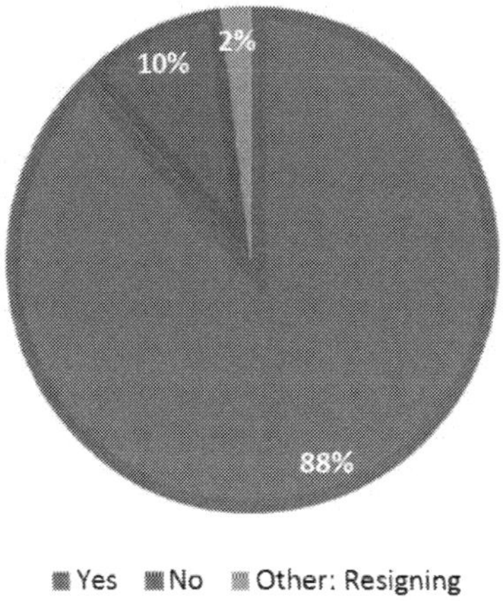

Figure 8. The necessity to employ skill worker to replace retired workers

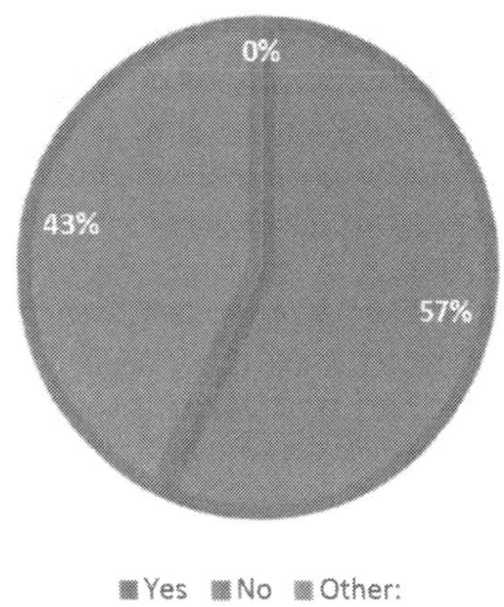

Figure 9. Firm struggling to recruit workers with right skill

Result also shows 80% of the respondent stated that their firm has incentives to employ young people while 18% of the respondent stated that their firm do not have incentives to employ young workers. Other 2% said that the seniority factor when the boss retired, only then the firm will employ new staff. Hence, there are chances for young people who are unemployed could participate in the construction industry since there are incentives by the firm. However, young people should able to possess relevant skill in order to be hired by the firm. Lastly, 92% of the respondent agreed that career advice is very important in order to inform the young people about careers in the construction industry. This is to ensure that young people know that the construction industry could offer lots of construction job for them and create the awareness of the training of the skill available. Therefore, the amount of unemployment rate for young people could be reduced and able to contribute to the growth of the construction industry in Malaysia.

5. Conclusion

There are emerging skill gaps in construction the industry where the demand for new workers in related field occur throughout the year. The reliability on migrant worker was low due to the factor that most of the migrant workers did not possess relevant skill demanded by the industry. However, the current practice is that the job could have been filled up by the migrant workers since the local youth refused to get involved in this industry. Meanwhile, the participation of young people aged from 18 to 30 years old in the construction industry was low because most firm employed those aged from 31 to 40 years old. The industry keeps demanding skill worker even though young people could fill in the gap. Only a small number of apprentices were recorded in the firm to cater for the apprenticeship demand. This indicates lack of opportunity for young people to participate in the construction industry.

6. References

[1] Malaysia Productivity Corporation 2017 *MPC 24th Productivity Report 2016/2017.* (Petaling Jaya: MPC)

[2] Theng, T.T. and Kunasekaran, T. 2017 *Malaysia Unemployment Report August 2017.* Retrieved 14 November 2017 from http://www.krinstitute.org/Indicators-@-Unemployment.aspx

[3] Ibrahim, D.H.M. and Mahyuddin, M.Z. 2017 *Bank Negara Malaysia Annual Report 2017-Youth Unemployment in Malaysia: Developments and Policy Considerations.* Retrieved November, 18, 2017 from https://www.bnm.gov.my/files/publication/ar/en/2016/cp04_003_box.pdf

[4] Clark, K.B. and Summers, L. 1982 *The dynamics of youth unemployment. The Youth Labour Market Problem: Its Nature, Causes and Consequences.* (the National Bureau of Economic Research: University of Chicago Press) 199-234.

[5] Freeman, R. and Wise, D. 1982 *The Youth Labour Market Problem: Its Nature, Causes and Consequences.* (National Bureau of Economic Research: University of Chicago Press).

[6] MIDF Amanah Investment Bank Berhad 2015 *BNM 2014 Annual Report: Moderating growth with inflation closer to trend.* (Kuala Lumpur: Malaysia Industrial Development Finance Berhad (MIDF))

[7] Ofori, G. 1996 *Foreign construction workers in Singapore.* (Geneva. Working Papers International Labour Office).

[8] Hamid, A.R.A., Singh, B., Aminah, M.Y., and Nur Ashikin, M.A. 2011 *The Employment of Foreign Worker at Construction Site.* 2nd International Conference on Construction and Project Management.

[9] International Labour Organisation 2001 *The construction industry in the twenty-first century: Its image, employment prospects and skill requirements. Tripartite Meeting.* (Geneva: International Labour Office Geneva)

[10] Heathfield, S.M. 2016) *How Do You Overcome an Employment Gap?* Retrieved October 27, 2016 from https://www.thebalance.com/how-do-you-overcome-an-employment-gap-1918103.

[11] Money-zine 2015 *Definition of Employment Gap*. Retrieved November 18, 2015, from http://www.money-zine.com/definitions/career-dictionary/employment-gaps/
[12] Willis, J.L. and Cooper, R., 2009 The cost of labor adjustment: Inferences from the gap. *Review of Economic Dynamics* **12** 632–647.
[13] Schneer, J.A. and Reitman, F., 1990. Effects of Employment Gaps on the Career of MBA's: More Damaging for Men Than for Women? *Academy of Management Journal*, **33** (2), 391-406.
[14] Economic Modelling Specialist International (EMSI), 2013. *The Skills Gap: A National Issues that Requires a Regional Approach*. Retrieved from: http://www.economicmodeling.com/wp-content/uploads/EMSI-SkillsGap-Brief.pdf.
[15] Act.org/workforce. 2011 *A Better Measure of Skills Gaps: Utilizing Act Skill Profile and Assessment Data for Strategic Skill Research. American College Test.* Retrieved from: https://www.act.org/content/dam/act/unsecured/documents/abettermeasure.pdf
[16] Bloom, N., Conway, N., Mole, K., Moslein, K., Neely, A., and Frost, C. 2004 *Solving the Skills Gap: Summary Report from the AIM/CIHE Management Research Forum*. Advanced Institute of Management Research.
[17] Aziz, A.R.A. 2001 *Site operatives in Malaysia: Examining the Foreign local asymmetry*. Unpublished report for the ILO.
[18] Nasir, A. N., Ali, D. F., Noordin, M. K., and Nordin, M. S. 2011 *Technical skills and non-technical skills-predefinition concept*. IETEC'11 Conference. Kuala Lumpur.
[19] Berry, J. 2006 *Quantitative Method in Education Research*. (Centre of Teaching Mathematics: University of Plymouth).
[20] Fink, A. 2005 *Conducting Research Literature Reviews: From Internet to Paper*. 2nd ed. (Sage Publications: Thousand Oaks).
[21] Raynsford, N., and Best, R. 2014 *No More Lost Generations: Creating Construction Jobs for Young People*. (London: House of Commons).
[22] Sung, S. 2013 *62% Malaysian businesses find hard to hire skilled workers, highest in ASEAN*. (Kuala Lumpur: Grant Thornton).
[23] Barnes, D. 2013 *Skills in the UK Construction Industry*. (Berkshire: The Chartered Institute of Building).

IOP Conf. Series: Journal of Physics: Conf. Series **1049** (2018) 012034 doi:10.1088/1742-6596/1049/1/012034

The challenges of the ageing population on the Malaysian construction industry

Abdul Rahim Abdul Hamid, Noradawiah Muhamad, Rozana Zakaria, Eeydzah Aminuddin, Amir Akmal Alwee and Adelynna Tok
Department of Structures and Materials, Faculty of Civil Engineering
Universiti Teknologi Malaysia, 81310 Johor Bahru, Johor, Malaysia.

E-mail: rahimhamid@utm.my

Abstract. Malaysia will be an ageing nation by 2035, when 15 per cent (5.6 million) of the population are classified as senior citizens. The ageing population does not only give an impact to the building design and structure, but also on the construction workforce and industry. The government has taken steps in its preparation to face the challenges of an ageing nation including ensuring adequate health care services, financials as well as infrastructure that are senior-citizen friendly. The objectives of this study are to identify the industry's awareness of the challenges as a result of the ageing population. The methodology adapted comprise of primary and secondary data collection. The primary data had been collected from the questionnaire surveys on the professional within the construction industry. The data then were analyzed using the percentage score and the diagram such as bar chart had been used to visualize the results from the analysis. While, secondary data were collected through literature review to gather information and important views from previous studies. In achieving the objective of the study, 100 questionnaire surveys have been distributed to the respondents, but only 39 questionnaires have been returned back. The discussions are done by comparing the result that is in the form of a percentage. Based on the study that has been done, the respondents are aware that the changing is needed in organization current construction projects, the additional cost is required to make an appropriate building for the elderly and there is a need of workspace to address the changing population in the industry. The challenges will be faced by the government to deal with an ageing population. The impact of ageing is the worker that close to retirement age will extend their career based on certain factors and also, loss of skill, experience and knowledge. The respondents agree that it is important to adopt the built environment to cater the needs of older people. The accessibility to the building has the highest priority to be adopted on the built environment.

1. Introduction
Under the Economic Transformation Programme (ETP), the Malaysia's construction industry is expected to be busy for this year when the project that launched for a few years ago show the committed investment and this will bring the economic benefits to the country. This ETP programme not only driven by the government project, but it is also by the private entities. In 2016, the construction sector's share of GDP was 4.5% valued at RM50.1 billion and registered 7.4% output. This sector hired 1.25 million employees, which showed a drop of 4.4% from 1.31 million in the previous year. By 2020, the sector is expected to contribute 5.5% to GDP at RM327 billion and employing 1.2 million employees. Despite its smaller contribution, the construction sector continues to play a key role in transforming the economy through its multiplier effect on many industries by accelerating domestic economic activities and providing comprehensive high-end quality infrastructure [1-2].

The research that has been done by the World Health Organization (WHO), has found that the people that aged for 60 years old and above will be increased from 605 million to 2 billion between the years 2000 to 2050 [3]. This range of age has a difference talent, experiences, skills and knowledge. Then, when the time for those people to retired from their career the industry will face the loss. When the

Content from this work may be used under the terms of the Creative Commons Attribution 3.0 licence. Any further distribution of this work must maintain attribution to the author(s) and the title of the work, journal citation and DOI.
Published under licence by IOP Publishing Ltd

IOP Conf. Series: Journal of Physics: Conf. Series **1049** (2018) 012034 doi:10.1088/1742-6596/1049/1/012034

older population increase we need to ensure to have the better quality of life and this required financing. Therefore, we need to have a good system to control the costs. In Australia, the National Ageing Research Institute (NARI) [4] was established to conduct research into ageing and improving the life and health quality of older people.

Ageing is a slow biological deterioration process [5]. It is a unique process that required one-directional relationship between the individual and time. The human organs can absorb information and process information based on the nervous system, movement and external action on the environment, energy supply and consumption. When the issues of the ageing population occur, the world are focusing more on how to maintain and securing the healthy to prevent such age-related disease. However, the needed of infrastructure and institutions as soon as possible for managing population aging also important [3].

The Chartered Institute of Building (CIOB) has conducted the study on how the ageing population impact the construction industry [6]. CIOB Britain found that the children under sixteen are less than the pensionable age people. This study has found that the ageing population does not only give an impact to the building design and structure, but also on the construction workforce and industry. The aging workforce that comes along with the globalization of business will lead to increasing of talent worldwide demands and the migration is needed for some countries [7].

In 2017, the population in Malaysia is estimated at 32.0 million with 28.7 million are citizens and 3.3 million are non-citizens [8]. Malaysia will be an ageing nation by 2035, when 15 per cent (5.6 million) of the population are classified as senior citizens. These people required the house and the government also developer will try to provide and fulfill the demand for the public. This is especially for the urban area like Kuala Lumpur, Selangor, Pulau Pinang and Johor. Rostam [9], from her studies found that there are population changes of the Extended Metropolitan Region, Klang-Langat. This is encouraging the expansion of production and services industry in that area. The housing development was increased in this area and the problem might come out whether they consider the development with the environment.

Previously, the retirement of Malaysia public work is at the range of age 56-57 years old. However, the Malaysia government expended the age of retirement to 60 years old. This is brought many feedbacks from the public, whether it is positive or negative opinion. Therefore, the private sector also needs to expend their age of retirement. From this view, all the parties that involve in development need to focus on various aspects to consider the ageing people. It is because the workers that close to retirement time might not have the capability to do the routine activities like the young people. Zaki et al. [10], state that the Malaysia has faced the skilled shortage issue in the construction industry, whether to retain the skill people and they need to train the new workers to meet the standard skill [11-14]. This has given a big impact to the contractor, especially to maintain the skilled worker and experienced worker, especially when the worker wants to return to their country or retire from the industry.

So far, there is very little published data and analysis of the current issues of ageing population on the Malaysian construction industry. Therefore, this study is important and timely in order to identify the construction industry's awareness of the challenges as a result of ageing population.

2. Brief Literature

United Nation agreed to standardize the 60 years old and above as older population or ageing population. The ageing population has a linked to the retirement age and there is a different retirement age between the man and woman based on their functional ability. Ageing population can be defined as a state when the individual is not being able to work as their normal daily activity in actively. They also need the condition that will help to maintain or gain a better healthy life. Hedge et al. [15]; Older Americans Act

(1965); the Job Training Partnership Act (1982), and the Workforce Investment Act (2000) define an older worker are the people in the categories of age 55 years and older.

The aging population will increase when the rate of birth reduced and the people living longer due to better health care [7]. In certain countries, they are not very well prepared to face the changes and the same as the organization. They might have talked about it, but there is no or little action toward ageing problem. This will make the ageing faced the unsecured situation and condition. Ilmarinen [16] define the increase in life expectancy together with lower birth as the cause of aging population and workforce in the developed or less developed countries. The public policies of the country play as an important role in managing older people. This is to make sure that the older people are secure and protected also to give a better life of the individual and stabilise the society.

The age not as a factor to adapt with the new technology or thing but the individual's personality and the working environment play the key role. From his study, Ilmarinen [17] found that the aging workforce give challenges to a few categories of group such as individual, enterprise and society. Individual might face the problems or possibilities in their functional capacity, health, competence, work motivation, work ability, work exhaustion and unemployment. The enterprise also might face the problem or possibilities from aging workforce such as productivity, competitiveness, sickness absence, tolerance for change, work organization, work environment and recruitment. The lastly is to the society, when there are attitudes toward the work and retirement, age discrimination, early retirement, work disability costs, retirement costs, health care costs and dependency ratios. There are four changes that related to aging workforce that suggests by Ilmarinen [16]:

a) Towards aging, there are societal changes in its attitudes
b) Increasing the knowledge level of managers and supervisors in age-related issues
c) Changes in the nature of working life and flexible work arrangements. Which is they do not comfortable with the routine and need the change such as working in shift
d) Changes in the health care services to meet the increasing demands of older workers.

The other factors such as discrimination against the older workers to attain job training, older employee also less willing and less able to accept to new situation, or feeling fears and less self-image during the training process. Marquie [5], state that the problem that usually faced by the ageing workers at the construction site is keeping or maintain the balance, visual information in the maintenance of equilibrium, and falls at work.

Silverstein [18] has state that the safety and performance among older workers can give a challenge. The performance of worker would not affect because of age otherwise for the works that use the capacities and the capacities faced the decrease with age. Snel and Cremer [19] have defined the challenges from the ageing population as Making companies aware of the problems of aging, Adapting jobs, Health promotion and prevention, Training, Adjustments to work careers and Creation of a forward-looking management of human resources.

3. Methodology of the Study

In order to achieve the aforementioned objective of this study, the study involved three phases of study, as shown in figure 1. The Phase 1 has three (3) steps, Phase 2 involved one (1) step and Phase 3 has two (2) steps. At the first (1) step (Determine the Problem Statement) of this study, the issues of the construction industry were defined and the problem statement was prepared. Second (2) step (Identify Objective and Scope), the objective and scope of the study were carrying out based on the problem statement that prepared in the first step. Third (3) step, the literature review is where we need to study the articles, journal papers and others collection that related to the objective and scope of the study.

Fourth (4) step, which is in Phase 2 involved the preparation of a questionnaire and data collection. The researcher opted quantitative analysis technique [20-21] to answer the objective of this study. The questionnaire survey distributed among construction players in various fields through mediums such as social media, email and distributed directly to the targeted group in their companies and construction sites. The participant involves in data collection for this study include consultants, contractors and other construction players. At this step, processes of collecting data from the respondent require to do the analysis and meet the objectives of the study.

After the fourth (4) step has been done, data (step five) that collected before can be analyzed by using SPSS and Microsoft excel software. The data collected, grouped according to the frequency of the answer given by the respondent. The percentage calculated by dividing the number of observations with total number of observations, then the result should times with 100% in order to gain the result in percentage value. The results then, are presented in the form of charts and table. Then the result will be discussed. Finally, after that analysis has been done, conclusion and suggestion of the study will be prepared for the better construction industry and next study.

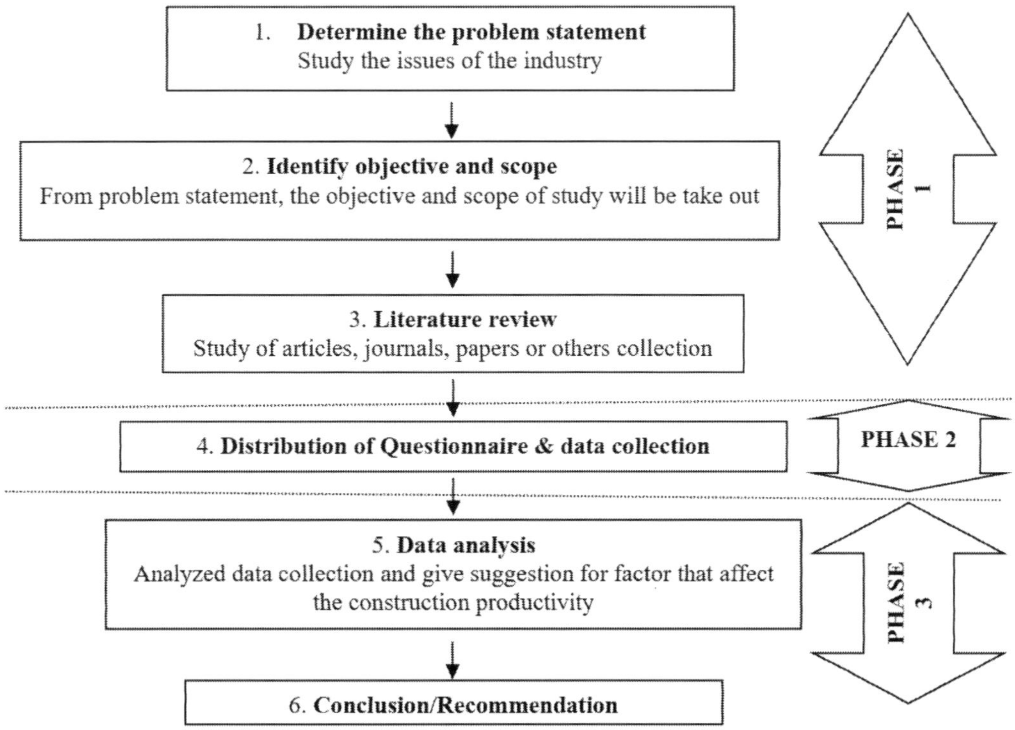

Figure 1. Brief Research Methodology

4. Results and discussion

4.1 Respondent background

More than 100 questionnaires set were distributed by hand and through the email to the respondents located in the peninsular Malaysia. 39 numbers of respondent had answered the questionnaire which make up about 39% had responded to the questionnaire distributed. Figure 2 shows the percentage of the respondent which is divided into a few ranges. The highest percentage respondent age ranges are 21 to 30 years old (71.79%). Respondents age ranges from 41 to 50 years old is the lowest (2.56%). The respondents gender is 56.41% female and 43.59% male.

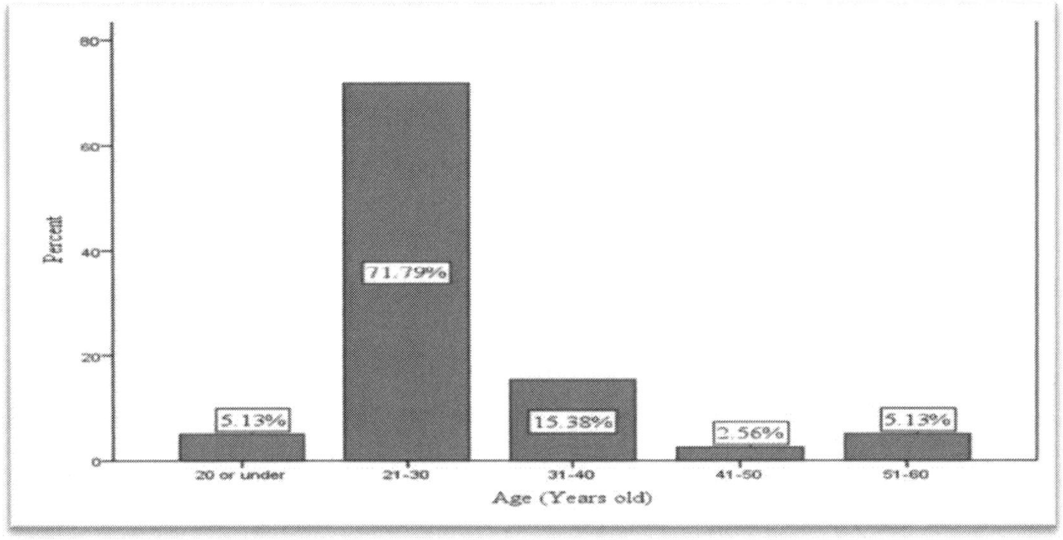

Figure 2. Respondents' age range

Figure 3 shows the percentage of job position of respondent in their working organization. The respondent positioning as an engineer shows a higher percentage of 33.33% than the quantity surveyor (25.68%), manager (12.82%) and administrative and other position.

Figure 3. Respondents' job position

Most of the respondents came from the construction company for 79.49% meanwhile consultant firm and developer only 5.13% and 15.38%. The consultant and developer firm show lower percentage than the construction company because of the difficulty to receive the feedback from this job sector. This could be because of the workload of those two-job sector. Figure 4 shows the percentage of numbers of employee in a company where the respondent is working. The numbers of employee that have more than 100 numbers of employees are the highest for more than 53.1%, where it is half of the respondent number. The organization that has employee around 21-40 numbers is the second highest with 23.08 %. Number of employees around 61-80 and 81-100 is the lowest for the 2.56%.

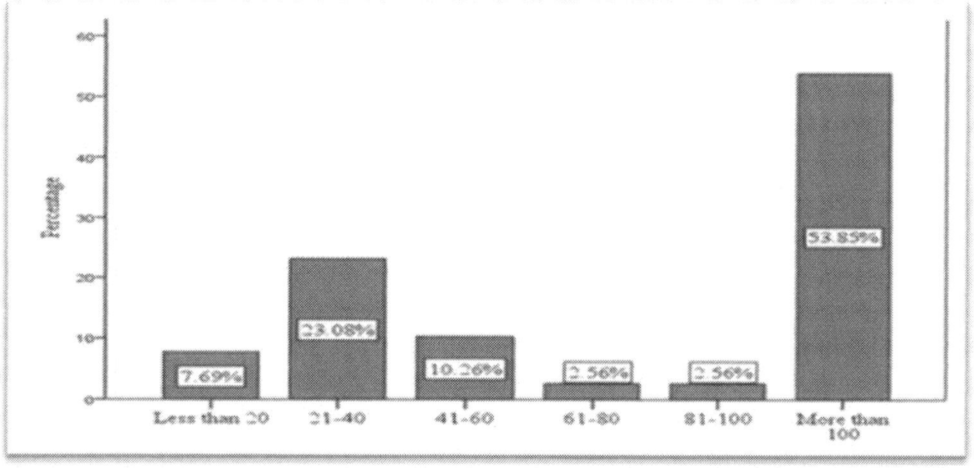

Figure 4. Respondents' employees' capacity

4.2. The awareness of the challenges as a result of ageing population in the construction industry
The highest range of age respondent that aware with the ageing population challenges is in the range of 21-30 years old (62.50%) and 31-40 (25%). Meanwhile, the lowest percentages for 4.17% are range of 20 years and below, 41-50 years and 51-60 years. From the figure, we can conclude that most of the respondent aware about the challenges facing the construction industry as a result of an ageing population. This can be factor of the surrounding when the ageing people around the respondents increase. Example, their relative, colleagues, neighbours and etc. The research and studies that has been publishing from the overseas study also encourage the awareness of the respondent. The guidelines of the design that consider the ageing people that produced by the other country also can be the factor of respondent awareness such as Lifetime Homes Standard. However, this standard not functions as mandatory in the UK [22].

The respondents also were asked whether the changing needs of an ageing population would reflect in organisation current construction projects. More than quarter of respondents, which is 38.46% of respondents agree that the organization current construction project would be reflected from the changing needs of an ageing population. In addition, the 38.46% of the respondents think that sometime the organization current project will be affected. The rest 23.08% of respondents believed no changes are required. This might happen when the person that in charge for the project are older and request for sudden retirement. The organization will need a time to replace another person to take in charge for the project. More time taken for the replacement, more profit might be loss.

Figure 5 shows the respondent opinion on the additional cost to make building appropriate for the elderly. Most of the respondents (53.85%) think that there are additional costs involve but they are not excessive. This could be due to most of the building that have more than 5 floors would need to install the lift to comply with regulations. The second highest respond is 'Yes. Excessive' for 20.51%. When the ordinary houses and design require to follow the standard to make the building more appropriate for the elderly, it causes the additional costs for the building. If the cost increases, the overall building price will be higher. This might cause the reduction of demand because the users could not afford to own it.

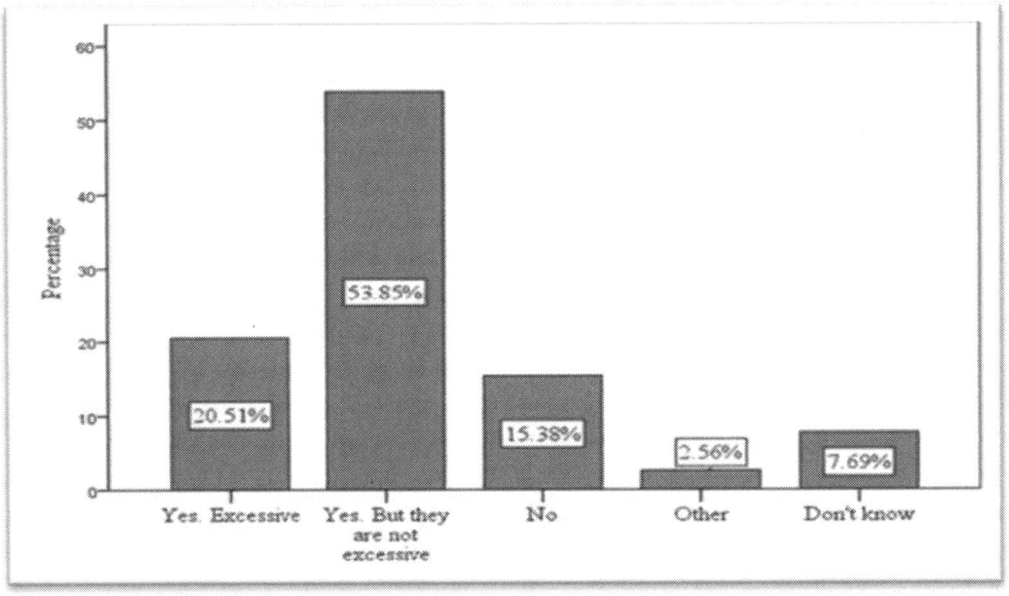

Figure 5. Additional costs to make appropriate building for the elderly

Figure 6 shows the respondent opinion about the possibility of an extension retirement age might need the workspace to be adapted to address the changing population. 59.4% respondent agrees that the work space need to be adapted. Meanwhile the 28.1% do not agree and 12.5% of respondent do not know. Most of respondent agree to have a workspace because the ageing people might face the ergonomic posture and the suitable space that limit the movement. Figure 7 shows the respondent opinion whether the Malaysia Government has dealt with the challenge of an ageing population in an effective way or not. The feedback of respondent shows that, most of the respondents (71.9%) agree that the Malaysia Government has not dealt with the challenge of an ageing population in effective ways. Even the government already mention about the loss of workforce in the industry because of the retirement and the ageing people on the industry but there is still very little action to handle the ageing issues in the industry. The 17.95% respondents agree the government has dealt with the challenges of an ageing population. This could be due to the government efforts to train the young generation such as the skill centre such as Institut Kemahiran Belia Negara (IKBN) that not qualified to go to the institution of higher learning.

From Figure 8, there is very small of percentage of respondents that were informed and comply with the housing standard. This might occur when some of the parties that involve with certain project want to gain high benefit. Currently, the Uniform Building by Law (UBBL) is the only building standard regulation but it does not cater for the ageing people. Figure 9 shows whether the housing standard should be compulsory or not to be applied in the Malaysia construction industry. From the figure, we can say that the respondents agree to apply with the standard for the housing if it is compulsory or mandatory for the industry. All buildings are required to comply with the standard that has already stated by the local authorities' codes and regulations. The certificate of completion and compliance (CCC) will be issued once the erection and completion of the building conform with the approved plans and the requirements of the provisions of Acts or by-laws, ensuring all technical conditions imposed by the local authority have been duly complied with and ensuring that the building is safe and fit for occupation.

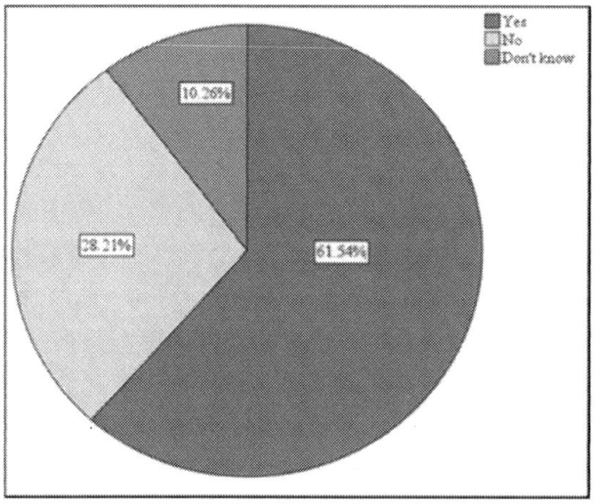

Figure 6. Needs of workspace to address the changing population

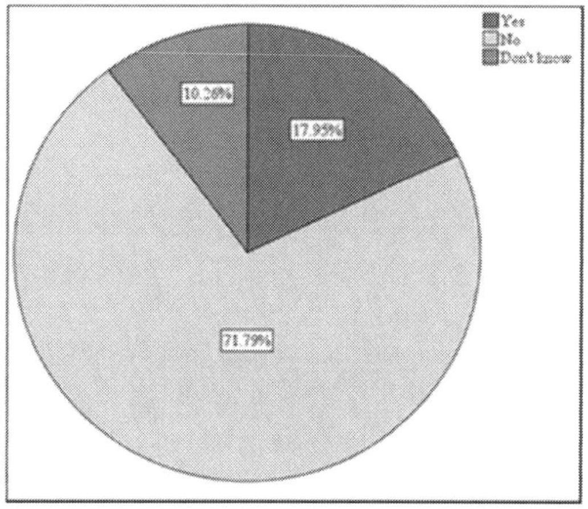

Figure 7. Opinions on Malaysia Government's dealt on the challenges of an ageing population

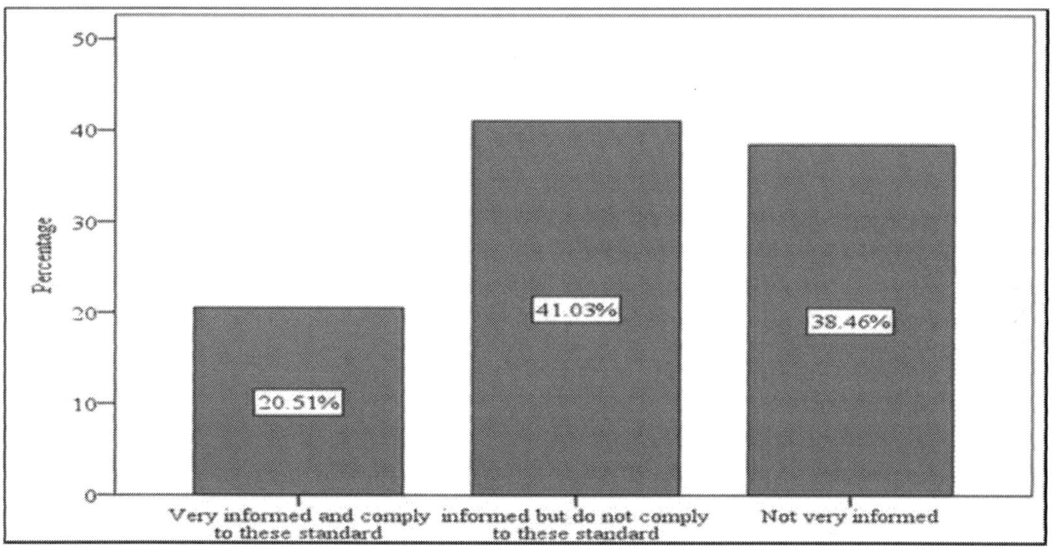

Figure 8. Awareness on the housing standard

Figure 10 show, the respondents also agree to have the similar standards as Housing standard to be adapted for commercial and public building. This is to make sure that the user feel comfortable and safe to stay, use or live in the building because the commercial and public building involve the society and the older people are also part of the society.

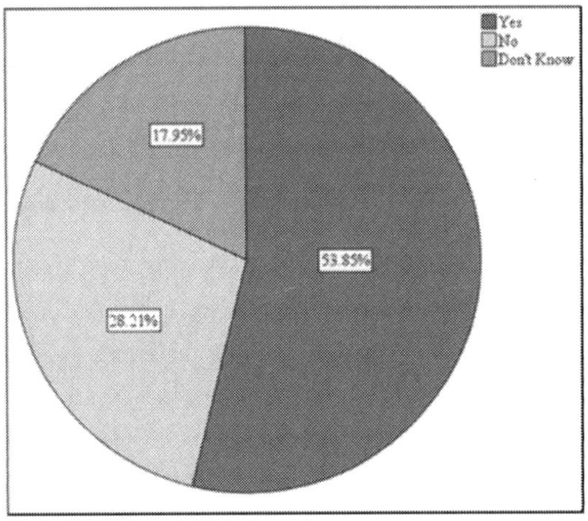

Figure 9. Opinions on the implementation of housing standard

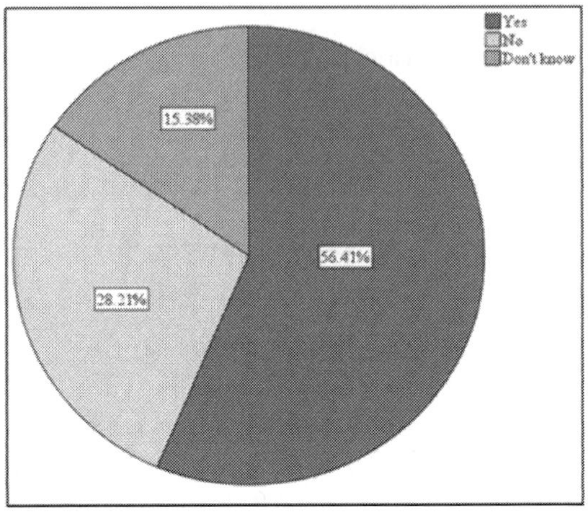

Figure 10. Opinions on the adaptation of similar standard as Housing Standard for Commercial and Public Buildings

5. Conclusion

We can conclude that most of the respondents were aware about the challenges facing the construction industry as a result of an ageing population. The respondent agreed on the changing needs of an ageing population would reflect in the organisation current construction projects. The respondent also agreed that the additional cost will incur to make building appropriate for the elderly. The respondent agreed about the possibility of an extension retirement age might need the workspace to be adapted to address the changing population. Most of the respondents agreed that the Malaysia Government has not dealt with the challenges of an ageing population in effective ways. Only small of percentage of respondent agreed that they are well inform and comply with the housing standard. The respondents agreed to apply with the standard for the housing if it is mandatory for the industry. Lastly, the respondents also agree to have the similar standards as Housing standard to be adapted for commercial and public building to cater for the elderly.

6. References

[1] Malaysia Productivity Corporation 2017 *MPC 24th Productivity Report 2016/2017* (Petaling Jaya: MPC)

[2] Hamid, A.R.A., Singh, B., Aminah, M.Y. and Nur Ashikin, M.A. 2011 *The Employment of Foreign Worker at Construction Site.* International Proceedings of Economics and Research: Construction and Project Management (IPEDR) **15** 126-130

[3] World Health Organization, 2007. *Global age-friendly cities: A guide.* World Health Organization.

[4] Australian Local Government Association, 2006. *Age Friendly Built Environments: Opportunities for Local Government.* (Deakin: Australian Local Government Association)

[5] Marquie, J.C., Cau-Bareille, D.P. and Volkoff, S., 1998. *Work with age.* (Boca Raton, FL: Taylor & Francis)

[6] CIOB. (2012). *The Impact of the Ageing Population on the Construction Industry.* Retrieved on 20 Nov 2017 form https://www.ciob.org/sites/default/files/CIOB%20research%20-

%20The%20Impact%20of%20the%20Ageing%20Population%20on%20the%20Construction%20Industry_0.pdf
[7] Field, J., Burke, R.J. and Cooper, C.L. eds., 2013. *The Sage handbook of aging, work and society*. (Thousand Oaks, CA: Sage Publications Inc)
[8] Department of Statistics Malaysia (DOSM) 2017. *Current Population Estimates, Malaysia, 2016-2017.* Available online: https://www.dosm.gov.my/v1/index.php?r=column/pdfPrev&id=a1d1UTFZazd5ajJiRWFHNDduOXFFQT09.
[9] Rostam, K., 2010. Perubahan Penduduk Wilayah Metropolitan Lanjutan Klang-Langat, Malaysia, 1957-2000. *Akademika*, **79** 2
[10] Zaki, S.B.A., Mohamed, S.F. and Yusof, Z.M., 2012. Construction skilled labour shortage–The challenges in Malaysian construction sector. *OIDA International Journal of Sustainable Development*. **4** 5 99-108
[11] Hamid, A.R.A. Hamid, Singh, B., Yusof, A.M., Ismail, A.M. and Wei, O.S. 2011 *Isu-isu berkaitan buruh asing dalam industri pembinaan: kajian literature.* Paper Proceeding of (SEPKA 2011) SEMINAR PENYELIDIKAN KEJURUTERAAN AWAM.
[12] Hamid, A.R.A, Singh, B., Yusof, W.Z.Y., Yusof, A.M. and Mustafa, N. 2011 *Problems Faced By Contractors in Managing Foreign Workers On Construction Sites.* International Proceedings of Economics and Research: Construction and Project Management (IPEDR) **15** 131-135
[13] Hamid, A.R.A., Singh, B., Yusof, A.M. and Ismail, A.M. 2012 *The Employment of Foreign Labors In The Construction Industry* Brunei International Conference on Engineering & Technology 2012 Proceedings. Rizqun International Hotel, Brunei Darussalam 92 – 99.
[14] Hamid, A.R.A. Singh, B. and Jamadi, M.A. 2013 Foreign Labour Employment in Construction Project *International Journal of Scientific and Research Publications.* **3** 11 1-7
[15] Hedge, J.W., Borman, W.C. and Lammlein, S.E., 2006. *The aging workforce: Realities, myths, and implications for organizations*. American Psychological Association.
[16] Ilmarinen, J., 2006. The ageing workforce—challenges for occupational health. *Occupational Medicine*, **56** 6 362-364.
[17] Ilmarinen, J.E., 2001. Aging workers. *Occupational and Environmental Medicine* **58** 546-552.
[18] Silverstein, M., 2008. Meeting the challenges of an aging workforce. *American journal of industrial medicine*, **51** 4 269-280.
[19] Snel, J. and Cremer, R. eds., 1995. *Work and aging: A European prospective*. (Boca Raton, Florida:CRC Press)
[20] Berry, J. 2006 *Quantitative Method in Education Research*. (Centre of Teaching Mathematics: University of Plymouth)
[21] Fink, A. 2005 *Conducting Research Literature Reviews: From Internet to Paper*. 2nd ed. (Sage Publications: Thousand Oaks)
[22] Habinteg 2010 *Lifetime Home (LTH) Revised Criteria July 2010*. Retrieved on 20th November 2017 from http://www.lifetimehomes.org.uk/data/files/For_Professionals/accessible_revisedlthstandard_final.pdf

IOP Conf. Series: Journal of Physics: Conf. Series **1049** (2018) 012035 doi:10.1088/1742-6596/1049/1/012035

A hybrid optimization technique for solving economic dispatch problem

M. H. Mansor[1], R. M. S. Raja Kechek[2], I. Musirin[3], N. A. Rahmat[4], M. S. A. Rahman[5], N. Roslan[6], M. N. Abdullah[7], S. A. Shaaya[8], N. F. Ab. Aziz[9]
[1,2,4,5,6,9]Department of Electrical Power Engineering, Universiti Tenaga Nasional, 43000 Kajang, Selangor, Malaysia
[3]Faculty of Electrical Engineering, Universiti Teknologi MARA, 40450 Shah Alam, Selangor, Malaysia
[7]Universiti Tun Hussein Onn, 86400 Parit Raja, Batu Pahat, Johor, Malaysia
[8]Department of Electronics & Communication Engineering, Universiti Tenaga Nasional, 43000 Kajang, Selangor, Malaysia

E-mail: mhelmi@uniten.edu.my

Abstract. Economic dispatch (ED) is one the main power system problems solved during operational planning phase. It is solved by power system engineers with the objective to find the best setting of dispatchable generating units output so that the total production cost is cheap and at the same time abiding the system constraints. For the past ten years, researchers and engineers have been focusing on hybrid optimization techniques to solve ED problems. The strength of each hybridized technique is combined to form an improved optimization technique. Most of the hybrid techniques are combination of two or more meta-heuristic techniques. This paper proposed a new hybrid technique termed as immune evolutionary programming (IEP) to solve economic dispatch problem. The technique is a combination of two population based optimization techniques which are artificial immune system (AIS) and evolutionary programming (EP). In order to study the effectiveness of the proposed technique, the results produced have been compared with the results produced using AIS and EP techniques. It is found that IEP succeeded to give the lowest total production cost compared to the other two techniques.

1. Introduction

Economic dispatch will always be the main topic among the power system engineers and researchers to be discussed in power system planning and operation. This happens because most of the main sources of electricity come from fossil-fuelled generating units. The fuel of the generating units can be coal, natural gas or petroleum. Therefore, the fuel price must be taken into account in dispatching the electric power. The input-output curve of each generating unit is different to one another. The difference can be caused by the type of fuel, design and efficiency of the generating units. In the process of solving ED problem, cheap generating units are forced to dispatch their power to the load first followed by expensive generating units. Basically, the generating units are ranked ascendingly based on their fuel price. ED is one of the optimization type of problems. Hence, it must be solved using optimization techniques. Optimization techniques can be divided into two categories: mathematical techniques and heuristic

[1] To whom any correspondence should be addressed.

Content from this work may be used under the terms of the Creative Commons Attribution 3.0 licence. Any further distribution of this work must maintain attribution to the author(s) and the title of the work, journal citation and DOI.
Published under licence by IOP Publishing Ltd

techniques. Several mathematical techniques have been introduced by researchers to solve ED problems such as linear programming [1]–[3], quadratic programming [4], mixed integer programming [5], etc. For the past ten years, heuristic techniques become more popular to be used for solving ED problems. This is because their capability to handle the non-smooth cost function of ED problems better than mathematical techniques. Examples of heuristic techniques that have been used for solving ED problems are evolutionary programming [5], particle swarm optimization [6], artificial bee colony algorithm [7], kinetic gas molecule [8], etc.

In recent years, the trend of solving ED problems have changed from single techniques to hybrid techniques. Normally, a hybrid technique is a combination of two or more techniques. It is proven that these hybrid techniques have capability to solve the ED problems better than single techniques. For example, Gherbi et al. [9] proposed to hybrid two meta-heuristic techniques which are firefly algorithm and bat algorithm. The technique outperformed single techniques like genetic algorithm and those two techniques that made up the hybrid technique. Instead of combining the firefly algorithm with bat algorithm, Musau et al. [10] chose to combine with Levy Flights. The hybrid technique was used to solve multi-objective economic and emission dispatch. In [11] evolutionary programming and tabu search algorithm have been hybridized to solve multi-area ED problem. Elaiw et al. [12] solved the ED problem with valve-point loading effect using a hybrid technique called DE-SQP. The technique is a combination between differential evolution and sequential quadratic programming. This is a new approach that combine between mathematical and heuristic type of techniques. Different with Elaiw et al., Sayah & Hamouda [13] combined the DE with PSO instead. The hybrid technique is termed as DEPSO which PSO acts as the main optimizer of the hybrid algorithm. Pandian & Thanushkodi [14] used a hybrid EP-EPSO technique to solve smooth and non-smooth ED problems. The technique is made up of evolutionary programming and particle swarm optimization techniques. Genetic algorithm (GA) and active power optimization (APO) have been hybridized in [15] to solve non-convex ED problem. The global optimizer for this technique is GA while an operator of APO named Newton's second order approach (NSOA) is adopted into the GA. The NSOA of APO helped the GA in searching the best optimal solution of the non-convex ED.

This paper presented a new hybrid technique termed as Immune Evolutionary Programming (IEP) for solving ED problem. The technique is made up of two heuristic techniques which are evolutionary programming (EP) and artificial immune system (AIS). EP and AIS are population type of optimization techniques. These two techniques have been popularly used to solve power system problems such as distributed generation installation [16], FACTS devices installation [17] and ED problems [18]. The proposed IEP technique have been tested on the IEEE 26-bus reliability test system (RTS) and its results have been compared with EP and AIS techniques. IEP managed to give better total production cost compared to the other two techniques.

2. Economic dispatch problem formulation

The main objective of solving economic dispatch problem is to minimize total production cost. The objective function of total production cost can be written as follows:

$$Minimize\ C_{total} = \sum_{i=1}^{n} C_i(P_i), i \in \{1,2,\dots,n\} \quad (1)$$

C_{total} is the total production cost, $C_i(P_i)$ is production cost of ith generating unit and n is number of generating units in the system. The production cost function of a single generating unit is made up of fuel cost coefficients and its real power output. Production cost function of ith generating unit can be represented as a smooth quadratic function as shown in (2).

IOP Conf. Series: Journal of Physics: Conf. Series **1049** (2018) 012035 doi:10.1088/1742-6596/1049/1/012035

$$C_i(P_i) = aP_i^2 + bP_i + c \tag{2}$$

In optimizing the objective function, some constraints must be considered. The constraints for this ED problem are:

A. Operating limits of generating units

Each dispatchable generating unit has its own cost function and it can be different from one to another. The cost function depends on the type fuel used and the capacity of the generating units. The operating limits of a generating unit can be written as follows:

$$P_{i(min)} \leq P_i \leq P_{i(max)}, i \in \{1,2, \dots, n\} \tag{3}$$

$P_{i(min)}$ and $P_{i(max)}$ are the minimum operating limit and the operating maximum limit of ith generating unit's, respectively.

B. Power balance constraint

It is a must to ensure the power generated is sufficient for the demand. Therefore, the power generated must be equal to total system loss plus demand. This power balance constraint can be written as follows:

$$P_{generate} = P_{loss} + P_{demand} \tag{4}$$

Total system loss, P_{loss} can be calculated using the equation shown in (5).

$$P_{loss} = \sum_{k=1}^{l} g_k [V_i^2 + V_j^2 - 2V_iV_j cos(\delta_i - \delta_j), k \in \{1,2, \dots, l\} \tag{5}$$

Where,

g_k is conductance of kth line,

V_i and δ_i are voltage magnitude and angle of bus i, respectively

V_j and δ_j are voltage magnitude and angle of bus j, respectively, and

l is number of lines in the system.

3. The proposed hybrid technique

The proposed hybrid technique termed as Immune evolutionary programming (IEP) is developed from two single optimization techniques which are evolutionary programming (EP) and artificial immune system (AIS). EP is the main optimizer for this algorithm and the cloning operator of AIS is adopted into EP. The structures of these two methods are shown in Figure 1 and Figure 2.

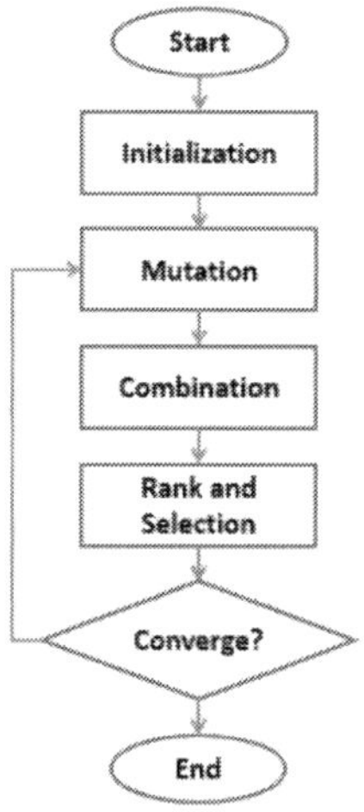

Figure 1. Standard procedure of EP algorithm.

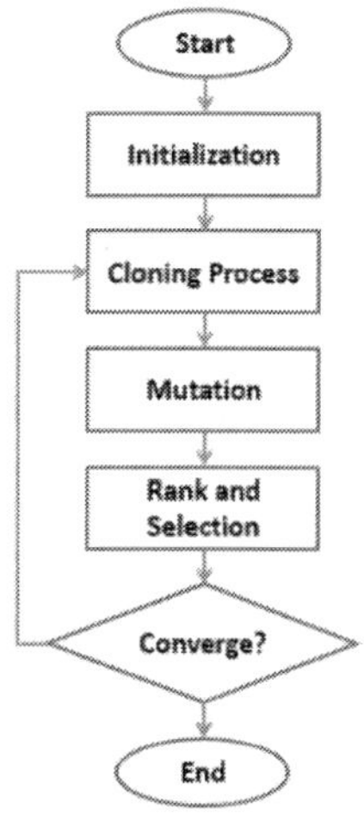

Figure 2. Standard procedure of AIS algorithm.

The advantages of their operators are gathered in the new hybrid technique in order to get better solution of ED problem. The cloning operator of AIS can be used to multiply the number of initial individuals produced from the initialization process. This will increase the chances to find the best global optima. Besides that, the combination operator of EP also plays important role in finding the fittest individuals that resulting to the best fitness value. The whole structure of IEP algorithm is shown in Figure 3.

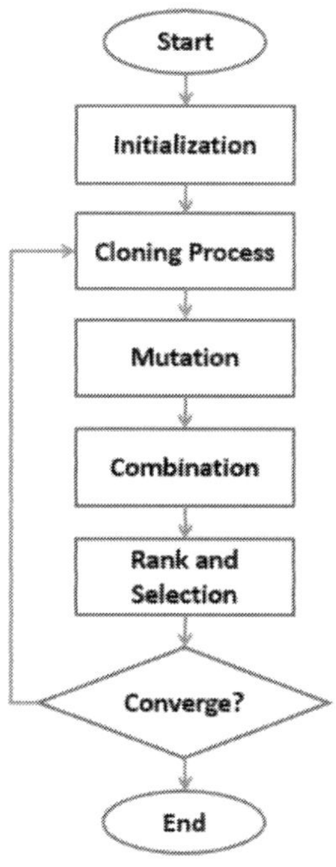

Figure 3. The IEP procedure.

The IEP algorithm starts with the initialization process. During this process, the values of decision variables are generated. For this ED problem, the decision variables are the real power output of generating units. The operating limits of the generating units are taken into account in generating the values. Then, fitness value which is total production cost is calculated based on the generated decision variables values. It is ensured that the fitness value is lower than the base value. The base value came from the non-optimal solution which is load flow solution. From this initialization process, a population of decision variables and fitness is produced. The size of the population is twenty.

Subsequently, the population produced from the initialization process will undergo the cloning process where the number of individuals are multiplied by ten to be 200 individuals. The population produced from the cloning process is called parents. After that, the parents will be mutated to produce their offsprings. Gaussian mutation technique shown in (6) is used to produce the offsprings. The size of this offspring population is also 200 individuals.

$$x_{i+m,j} = x_{i,j} + N\left(0, \beta\left(x_{jmax} - x_{jmin}\right)\left(\frac{f_i}{f_{max}}\right)\right) \tag{6}$$

Where:

$x_{i+m,j}$ is mutated parent individual (offspring)
x_{ij} is parent individual
β is mutation scale, $0 < \beta < 1$
x_{jmax} is maximum random number for every variable
x_{jmin} is minimum random number for every variable
f_i is fitness for *ith* random number
f_{max} is maximum fitness

The fourth process is combination. During this process, the parent population and offspring population are combined to be 400 individuals. From this 400 individuals, there are then ranked based on their fitness. The fittest individuals will have high chances to be selected for the next steps. Only twenty fittest individuals are selected to undergo the convergence test. The process will continue with cloning process until the IEP algorithm is converged. The IEP algorithm is said to converge when the difference between the first and the twentieth fitness values is 0.00001. This can be represented mathematically as follows:

$$C_{total(20th)} - C_{total(1st)} \leq 0.00001 \tag{7}$$

4. Results and discussion

The IEEE 26-bus reliability test system (RTS) is used for the demonstration. Prior to the IEP implementation, the ED problem has been solved using the single EP and AIS techniques. This is done for comparison purposes. Three cases have been introduced to study the performance of the IEP technique to solve ED problem. Each case has different load level. The first case has 50 % load increment which makes the load 1894.50 MW. While the second case has 70 % load increment. And, the third case has 100 % load increment.
All the optimization results produced from the proposed IEP, AIS and EP for case 1, case 2 and case 3 are presented in Table 1, Table 2 and Table 3 respectively.

Table 1. ED solution of 26-bus system with 50 % load increment.

Optimization Technique	IEP	AIS	EP
P_{G1} (MW)	101.34	113.83	113.82
P_{G2} (MW)	75.03	58.02	58.01
P_{G3} (MW)	93.96	196.53	196.53
P_{G4} (MW)	98.93	117.11	117.11
P_{G5} (MW)	112.48	51.15	51.15
P_{G26} (MW)[a]	74.33	76.84	76.84
Total Production Cost ($/h)	7.13	7.72	7.72

Table 2. ED solution of 26-bus system with 70 % load increment.

Optimization Technique	**IEP**	**AIS**	**EP**
P_{G1} (MW)	134.13	113.83	113.82
P_{G2} (MW)	57.85	58.02	58.01
P_{G3} (MW)	95.56	196.53	196.53
P_{G4} (MW)	103.31	117.11	117.11
P_{G5} (MW)	94.00	51.15	51.15
P_{G26} (MW)[a]	53.43	76.84	76.83
Total Production Cost ($/h)	6.79	7.72	7.72

Table 3. ED solution of 26-bus system with 100 % load increment.

Optimization Technique	**IEP**	**AIS**	**EP**
P_{G1} (MW)	119.28	130.56	109.71
P_{G2} (MW)	73.85	81.33	92.32
P_{G3} (MW)	107.13	91.26	186.41
P_{G4} (MW)	118.99	122.50	143.48
P_{G5} (MW)	52.36	119.24	195.15
P_{G26} (MW)[a]	106.15	94.49	91.09
Total Production Cost ($/h)	7.38	8.08	10.30

From the optimization results presented in Table 1, Table and Table 3, it can be seen that the proposed IEP technique outperformed the single AIS and EP techniques in giving the lowest production cost for the all three cases. The production costs produced in case 1 and case 2 are almost the same. This is because the load difference between case 1 and case 2 is small, unlike case 3 which has double the load compared to case 1.

5. Conclusion
In this paper, a new hybrid optimization technique termed as immune evolutionary programming has been successfully used to solve ED problem. Cloning operator of AIS has been adopted into the proposed IEP to improve its search pattern for the global optima. Meanwhile, the combination operator of EP increases its chance to find a better ED solution. It can be concluded that the objective of this ED problem to minimize the total production cost is achieved in this paper.

6. Acknowledgements
The authors would like to acknowledge the management of College of Engineering, Universiti Tenaga Nasional for their support in making this research successful. Special thanks to the head of Electrical Power Department for his continuous support and encouragement to the young researchers.

7. References
[1] B. Lazzerini and F. Pistolesi, "A linear programming-driven MCDM approach for multi-objective economic dispatch in smart grids," *IntelliSys 2015 - Proc. 2015 SAI Intell. Syst. Conf.*, pp. 475–484, 2015.
[2] A. Farag, S. Al-Baiyat, and T. C. Cheng, "Economic load dispatch multiobjective optimization procedures using linear programming techniques," *Power Syst. IEEE Trans.*, vol. 10, no. 2, pp. 731–738, 1995.

[3] R. A. Jabr, A. H. Coonick, and B. J. Cory, "A homogeneous linear programming algorithm for the security constrained economic dispatch problem," *Power Syst. IEEE Trans.*, vol. 15, no. 3, pp. 930–936, 2000.
[4] C.-H. Chen and C.-C. Chin-Chung Lin, "Simple particle swarm optimization for economic dispatch with piecewise quadratic fuel cost function," in *2009 Innovative Technologies in Intelligent Systems and Industrial Applications*, 2009, pp. 412–417.
[5] H. Daneshi, A. L. Choobbari, M. Shahidehpour, and Z. Li, "Mixed integer programming method to solve security constrained unit commitment with restricted operating zone limits," *2008 IEEE Int. Conf. Electro/Information Technol. IEEE EIT 2008 Conf.*, pp. 187–192, 2008.
[6] N. M. Jamain, I. Musirin, M. H. Mansor, M. M. Othman, and S. A. M. Saleh, "Adaptive Particle Swarm Optimization for Solving Non-Convex Economic Dispatch Problems."
[7] D. Aydin, S. Özyön, C. Yaşar, and T. Liao, "Artificial bee colony algorithm with dynamic population size to combined economic and emission dispatch problem," *Int. J. Electr. Power Energy Syst.*, vol. 54, pp. 144–153, Jan. 2014.
[8] M. Basu, "Kinetic gas molecule optimization for nonconvex economic dispatch problem," *Int. J. Electr. Power Energy Syst.*, vol. 80, pp. 325–332, 2016.
[9] Y. A. Gherbi, H. Bouzeboudja, and F. Z. Gherbi, "The combined economic environmental dispatch using new hybrid metaheuristic," *Energy*, vol. 115, pp. 468–477, 2016.
[10] M. Musau, "Multi Area Multi Objective Dynamic Economic Dispatch with Renewable Energy and Emissions," 2016.
[11] E. Engineering, S. Lecturer, and E. Engineering, "MULTI-AREA SECURITY CONSTRAINED ECONOMIC," 2009.
[12] A. M. Elaiw, X. Xia, and A. M. Shehata, "Solving dynamic economic emission dispatch problem with valve-point effects using hybrid DE-SQP," *IEEE Power Energy Soc. Conf. Expo. Africa Intell. Grid Integr. Renew. Energy Resour. PowerAfrica 2012*, 2012.
[13] S. Sayah and A. Hamouda, "A hybrid differential evolution algorithm based on particle swarm optimization for nonconvex economic dispatch problems," *Appl. Soft Comput.*, vol. 13, no. 4, pp. 1608–1619, Apr. 2013.
[14] S. M. V. Pandian and K. Thanushkodi, "Solving Economic Load Dispatch Problem Considering Transmission Losses by a Hybrid EP-EPSO Algorithm for Solving both Smooth and Non-Smooth Cost Function," *Int. J. Comput. Electr. Eng.*, vol. 2, no. 3, pp. 560–568, 2010.
[15] T. Nadeem Malik, A. ul Asar, M. F. Wyne, and S. Akhtar, "A new hybrid approach for the solution of nonconvex economic dispatch problem with valve-point effects," *Electr. Power Syst. Res.*, vol. 80, no. 9, pp. 1128–1136, 2010.
[16] M. H. Mansor, I. Musirin, M. M. Othman, S. A. Shaaya, and S. A. S. Mustaffa, "Application of Immune Log-Normal Evolutionary Programming in Distributed Generation Installation," vol. 6, no. 3, pp. 730–736, 2017.
[17] W. Ongsakul and P. Jirapong, "Optimal Allocation of FACTS Devices to Enhance Total Transfer Capability Using Evolutionary Programming," pp. 4175–4178, 2005.
[18] M. H. Mansor, "Immune Log-Normal Evolutionary Programming (ILNEP) for Solving Economic Dispatch Problem with Prohibited Operating Zones," no. 1, 2017.

Comparison of Performance between Hot and Warm Mix Asphalt as Related to Compaction Design

Siti Nur Naqibah Kamarudin[1], Mohd Rosli Hainin[1],Mohd Khairul Idham Mohd Satar[1] and Muhammad Naqiuddin bin Mohd Warid[1]
[1]Faculty of Civil Engineering, Department of geotechnics and Transportation, Universiti Teknologi Malaysia, 81310 UTM Johor Bahru, Malaysia.

E-mail: mrosli@utm.my

Abstract. Hot mix asphalt (HMA) is a conventional mixture that is commonly used in Malaysia. Production of HMA requires high temperature which produces high carbon dioxide (CO^2) emission and gives bad impact to the environment. During the construction, cool weather is one of the factor that can influence the temperature of HMA mixture. It will reduce the mixture temperature during compaction. Warm mix asphalt (WMA) is one of the methods to overcome this issue. Even though HMA mixture is produced at higher temperature, the rate of cooling also high. This study focused on evaluating the performance of HMA and WMA as related to the compaction design. The types of mixture used were HMA and WMA with 14 mm nominal maximum aggregate size. The oil based additive used in WMA was Cecabase. This study was conducted at Kg Ayer Merah, Mersing where real time temperature and density were measured during construction. Rate of cooling, degree of compaction (DOC) and rutting performance of mixtures were analysed. By comparing to HMA mixture, the results show that WMA mixture has lower rate of cooling where the temperature of WMA mixture decreases slowly but still provides higher density. WMA also has higher DOC which provides better workability of compaction compared to HMA mixture. Other than that, WMA mixture has lower resilient modulus than HMA mixture which could suggest that WMA mixture is more susceptible to rutting.

1. Introduction

Hot mix asphalt (HMA) is a conventional mixtures to construct asphalt pavements. However, nowadays all agencies are starting to acknowledge the importance of sustainability in road construction. Warm mix asphalt (WMA) is one of the strategies to achieve the sustainable construction. WMA technology could reduce carbon footprint and emissions by lowering the temperature during production and construction. WMA also has many other advantages compared to HMA such as lower cost, better performance and healthier environment [1].

In order to enhance the quality of newly laid flexible pavement, proper compaction is crucial. Appropriate temperature during compaction is very important to provide the strength of the pavement. The strength could be attained by increasing the pavement density to withstand the load of the vehicles and the weather. Reduced temperature during compaction will lessen the mixture workability thus caused an increment of voids in the pavement. The same prevails for the too high temperature where the

[1] mrosli@utm.my

Content from this work may be used under the terms of the Creative Commons Attribution 3.0 licence. Any further distribution of this work must maintain attribution to the author(s) and the title of the work, journal citation and DOI.
Published under licence by IOP Publishing Ltd

binder becomes less viscous and hindered the compaction. Therefore, measuring the cooling rate is one of the suitable methods to determine the ideal compaction of mixture. Instead of mixture temperature, there are other parameters that can influence the cooling rate of the asphalt mixture such as temperature of the existing layer, compaction process and weather. During construction, cool weather also one of the factor that will easily influence the HMA mixture where it will cause the reduction of temperature during compaction. Even though HMA have high temperature but it has high rate of cooling.

These factors affect compaction process for the pavement which is one of the most important part for road construction. Basically, low pavement density could also affects the strength and caused shorter life span and rutting. Therefore this study is very important to determine and compare the rate of cooling, the degree of compaction (DOC) and the rutting performance of HMA and WMA.

2. Literature Review

WMA is a technology that allow reduction in the temperature at which asphalt mixture is produced and placed [2]. It is produced by adding either zeolities, waxes, asphalt emulsions, or sometimes even water to the asphalt binder prior mixing. The production temperature of WMA is normally ranges from 100 to 140 ^{0}C [3]. Instead of supporting the sustainable development, WMA also improved field compaction, facilitate longer haul distances and cool weather pavement and healthier environment in term of reducing of fuel consumption and emissions. In addition, WMA has found to improve the performance in term of minimizing the thermal crack, block crack and oxidative hardening [4]. WMA also improves the aggregate coating as a result of lower asphalt binder viscosity [5-9].

Cecabase is one of the common additive used for WMA. It is a liquid chemical additive with the dosage of 0.2-0.5 % for all types of binders. Adding cecabase will not change the bitumen grade and no curing time required. Cecabase can be injected into the binder either at the mixing plant or at the asphalt terminal and relevant to be stored for a minimum one week at 160 ^{0}C. This additive improves the workability of bituminous mixtures. It can be produced and paved at lower temperature than standard HMA without any process modification or loss of mechanical properties [10].

2.1 Factors that influence performance of road construction

2.1.1 Temperature.

Previous study stated that the temperature of the asphalt mixture is important for better final quality performance [1]. Cooling rate is one of the suitable method to determine the optimum temperature of compaction by measuring the density after every roller passes. However, there are many factors influenced the cooling rate of the asphalt mixture which are the temperature of the underground, temperature of existing road, compaction process, compaction regimes, layer thickness and the weather during construction.

2.1.2 Compaction.

Compaction is the most important process during road construction with set of compactor rollers. It will affect the final quality layer because compaction is the last activity before the road is open for the public use [1]. Viscosity of the asphalt mixture is influenced by the different techniques for WMA, but on the other hand, the viscosity will increase again when the temperature of mixture is lower [11]. In order to obtain high density, optimum compaction should be reached.

2.2 Benefits of WMA

WMA is one of the potential technologies can be categorized in four groups according to the research literature overview which are environmental, production, paving and economic [12].

2.2.1 Environmental.
WMA is a technology that allows asphalt plant mixing to be produced at lower temperature [13]. It will reduce the emission of carbon dioxide by 30 %, dust emission by 50-60 % and greenhouse gasses by consuming less energy compared to HMA [1]. The workers also negatively be effected by exposure to the fumes which is produced when the reduction of emissions during asphalt paving process [6].

2.2.2 Paving.
If WMA mixes are produced at normal operating temperatures, it will allow longer haul distances and a longer construction season [14]. Lowering the paving temperature could reduce of oxidative hardening of the asphalt and improve the performance by minimizing thermal and block cracking. Low paving temperature also reduced the viscosity and improved working conditions for the workers, longer haul distances, reduced cooling rate and improve the workability [12].

2.2.3 Economic.
Less energy consumption significantly affect the cost. However, it also depends on the type of energy used during production process, the types of gaseous pollution and the maintenance cost of asphalt plant [6].

2.3 Issues in WMA
Rutting resistance performance is the major issue in WMA. Lowering the mix temperatures, in drying the aggregates and in the plant mixing, could potentially cause WMA to be more susceptible to moisture damage and rutting [14]. The in-service rutting potential of some WMA products can be increased because generally void content in WMA is lower than HMA as WMA have good workability in some products of WMA and together with a less oxidative hardening of the binder throughout the production process, despite that it can also conduct to an increased durability [15].

There are also some concern related to the production of WMA which increase the initial production cost and connected to the additional equipment needed for plants, allowing the use of specific technology or additives that bring some supplementary cost, which could be only partially compensated by lowering the operating temperature [15].

3. Methodology
This study involved several tests that have been conducted during construction and at laboratory work utilizing HMA and WMA with liquid chemical additive namely Cecabase. Asphalt binder type 80/100 was used. Table 1 shows the details of mixture design. During road construction, cooling rate method and density reading using PQI device test were conducted. After construction, coring was performed to obtain the compacted samples at one day, one month and four months.

Table 1. Mixture Design

Type of premix	section		Description
HMA (AC14)	A	A1	75 blows/ BC 5.33% / normal lay
		A2	75 blows/ BC 5.33% / double handling
	B	B1	50 blows/ BC 5.53% / normal lay
		B2	50 blows/ BC 5.53% / double handling
	C	C1	35 blows/ BC 5.83% / normal lay
		C2	35 blows/ BC 5.83% / double handling
WMA (AC14)	D		75 blows/ BC 5.35% / normal lay
	E		50 blows/ BC 5.55% / double handling
	F		35 blows/ BC 5.85% / normal lay

4. Results and Discussion

Rate of cooling was conducted to determine optimum time available for compaction which obtained when the temperature of pavement decreases to 60 ^{0} C. Based on the results, HMA has more rate of cooling compared to WMA. The ambient temperature, the temperature from old pavement, the compaction process and roller regimes and layer thickness can be influenced the cooling rate data. However, by taking an average, Table 2 shows that WMA premix achieves higher maximum density during compact compared to HMA premix. Concurrently, WMA has smaller slope graph of cooling rate than HMA. It can be summarized, temperature of WMA mixture decreases slowly but still provides higher density and keeps increasing. Different with HMA where the slope of graph is higher because of the temperature of HMA mixture decreases quickly but when the compaction is completed, the density achieved decreased. In nutshell, WMA mixture provide better cooling rate during compaction and provide higher density compared to HMA.

Table 2. Summary result for density during compact and slope graph.

section	density during compact (kg/m^3)			Slope graph of cooling rate
	initial	final	maximum	
A1	2080.5	2361.5	2419.5	4.23
A2	2033.4	2144.8	2163.4	0.59
B1	2068.2	2180.9	2189.2	0.75
B2	2097.7	2180.7	2188.4	0.40
C1	2136.3	2233.0	2233.9	0.51
D	2101.9	2254.9	2254.9	0.50
E	2238.9	2259.9	2259.9	0.36
F	2198.0	2308.0	2308.0	0.33

Table 3 shows WMA premix has higher degree of compaction (DOC) than HMA and exceed minimum requirement of 98 % [16] except section F. Concisely, WMA has higher workability of compaction compared to HMA.

Table 3. DOC of pavement after 4 months.

sample	SG coring	SG Marshall	DOC (%)
A1	2.1343	2.2102	96.57
A2	2.1600	2.2102	97.73
B1	2.0981	2.1772	96.37
B2	2.1020	2.1772	96.55
C1	2.0911	2.1763	96.09
C2	2.0883	2.1763	95.96
D	2.1551	2.176	99.04
E	2.1294	2.1632	98.44
F	2.0945	2.1595	96.99

Figure 1 plots the performance of coring samples after a month and 4 months. The performances were observed and compared. The graph demonstrate that after 4 months, the resilient modulus value is increasing from a month for both HMA and WMA. The resilient modulus was decreasing as compaction effort was decreased. High resilient modulus at 25 ^{0} C shows that the mixture is more susceptible to fatigue crack. Mixture at Section A and mixture at Section D obtain the highest resilient modulus for

HMA and WMA respectively. In overall, this test concluded that HMA is more susceptible to fatigue cracking but less susceptible to rutting compared to WMA.

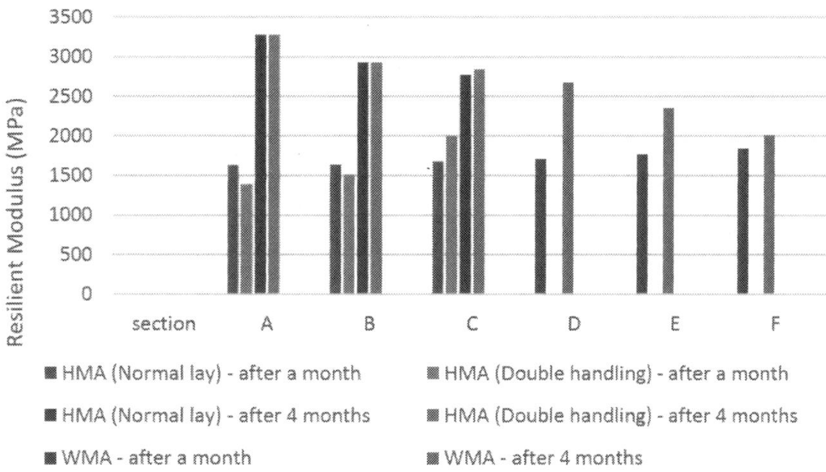

Figure 1. Performance of coring samples after a month and 4 months.

5. Conclusion

WMA technologies should be implemented because it is a good alternative and opportunity to improve road construction in Malaysia especially in its performance and environmental friendly. This is possible because WMA performs better in rate of cooling and DOC compared to HMA. WMA also has better workability in compaction and provides higher density even in low temperatures. Compaction is the most important process during road construction and it will affect the final quality layer because compaction is the last activity before the road is open for the public use. WMA also is more susceptible to rutting but less susceptible to fatigue cracking compared to HMA. These factors are the benefits of WMA. The implementation of WMA technology is uprising. Although it involves slight additional cost on the additive, the benefits on the engineering and environment are overwhelmed. More studies are suggested on mix design, long-term performance, cost benefits, and plant operations to further prove that the performance of WMA is truly as good or better than HMA [17].

6. Acknowledgement

Research work presented in this paper was supported by Universiti Teknologi Malaysia (grant number Q.J 130000.2522.15H16, Q.J 130000.2522.17H69, Q.J 130000.2522.11H35, Q.J 130000.2522.14J09). The financial assistance for this study is highly appreciated.

References

[1] Frank B, Seirgei M, and Andre D 2012. Warm Mix Asphalt - Too Cold to Handle? Learning to Deal with the Operational Consequences of Warm Mix Asphalt. *5th Eurasphalt & Eurobitume Congress, Istanbul*

[2] D'Angelo J A, Harm E E, Bartoszek J C, Baumgardner G L, Corrigan M R, Cowsert J E, and Prowell B D 2008. *Warm-mix asph.: European prac. (No. FHWA-PL-08-007)*

[3] Das A 2004. *Warm mix asphalt* 3-4

[4] Hurley G C and Prowell B D 2005. Evaluation of Sasobit for use in warm mix asphalt. *NCAT report*, 5(06)

[5] Abdullah M E, Hainin M R, Yusoff N I M, Zamhari K A and Hassan N 2016. Laboratory evaluation on the characteristics and pollutant emissions of nanoclay and chemical warm mix asphalt modified binders. *Const. and Build. Mater.* **113** 488-497

[6] Abdullah M E, Ahmad Zamhari K, Buhari R, Abu Bakar S K, Kamaruddin M N H and Nayan N, Hainin M R, Hassan A N, Hassan S A, Yusoff M N I 2014. Warm mix asphalt technology: a review. *J. Teknol.* **71** 39-52

[7] Aziz M M A, Hainin M R, Yaacob H, Feizabadi S M, Shokri M and Warid M M N 2014. Performance of warm-mix asphalt in the highway industry. *Mater. Research Innov.* **18** S6-245

[8] Abdullah M E, Zamhari K A, Hainin M R, Oluwasola E A, Yusoff N I M and Hassan N A 2016. High temperature characteristics of warm mix asphalt mixtures with nanoclay and chemical warm mix asphalt modified binders. *J. of Cleaner Product.* **122** 326-334

[9] Abdullah M E, Zamhari K A, Hainin M R, Oluwasola E A, Hassan N A and Yusoff N I M 2016. Engineering properties of asphalt binders containing nanoclay and chemical warm-mix asphalt additives. *Const. and Build. Mater.* **112** 232-240

[10] Oliviero Rossi C, Teltayev B, and Angelico R. 2017. Adhesion Promoters in Bituminous Road Materials: A Review. *Appl. Sci.* **7** 524

[11] Brian D P, Graham C H and Bob F 2012. Warm-Mix Asphalt: Best Practice. *National Asph. Pav. Assoc. (NAPA) 3rd Edition*

[12] Martins Z 2010. *Warm mix asph. investigation.* Master of science thesis, Technical University of Denmark, Kgs. Lyngby, Denmark

[13] Timothy R, Greg J, and John G n.d. *Dev. of Warm Mix Asph. Policies and Spec. in Minnesota.* Minnesota: Minnesota Department of Transportation, Office of Materials and Road Research

[14] James T and James L 2010. Evaluating constructability and properties of warm mix asphalt. *Georgia Ins. of Tech. (Atlanta), Final report*

[15] Capitão S D, Picado-Santos L G and Martinho F 2012. Pavement engineering materials: Review on the use of warm-mix asphalt. *Const. and Build. Mater.* **36** 1016-1024

[16] Jabatan Kerja Raya 2008. Standard Specification for Road Works, Section-4: Flexible Pavement. *Kuala Lumpur: Jab. Kerja Raya Malaysia*

[17] Rubio M C, Martinez G, Baena L and Moreno F 2012. Warm mix asphalt: an overview. *J. of Cleaner Product.* **24** 76-84

IOP Conf. Series: Journal of Physics: Conf. Series **1049** (2018) 012037 doi:10.1088/1742-6596/1049/1/012037

Modelling of Simple Assembly Line Balancing Problem Type 1 (SALBP-1) with Machine and Worker Constraints

N.H. Kamarudin, M.F.F. Ab. Rashid
Faculty of Mechanical Engineering, Universiti Malaysia Pahang, 26600 Pekan, Malaysia

Email : ffaisae@ump.edu.my

Abstract—This paper presents a mathematical model for Simple Assembly Line Problem Type 1 (SALBP-1) with resource constraints; machine and worker. The existing model of SALBP-1 assumes that all the workstations have similar capability, while in reality the workstation has different capability because of limitation in term of machines and worker skills. The proposed model is aimed to mathematically represent the SALBP-1 with resource constraints. Besides that, three objective functions which to minimize number of workstation, machine used and number of worker are also presented. The machine considered to have different types of machine needed to produce a product in an assembly line while worker are considered to have different abilities and skills. The model is then illustrated and validated using some examples. The proposed model for SALBP-1 with machine and worker constraints is able to minimize the resources in assembly. Therefore, the assembly cost can be reduced.

1. Introduction

Assembly line balancing is one of important activities prior to the mass production of a product. It will determine the line efficiency, productivity and as well as the assembly cost. Assembly line consists of workstations and as well as one or more dedicated machines/tools together with workers. The process of balancing the assembly task to the workstations with respect to some objective function is known as the assembly line balancing problem (ALBP)[1].

ALBP can be divided into two groups; Simple Assembly Line Balancing Problem (SALBP) and Generalised Assembly Line Balancing Problem (GALBP). SALBP involves production of single product in serial line layout on one sided workstations and GALBP consider different objectives with different assembly line such as mixed model assembly line, parallel, U-shaped and two sided lines [2]. SALBP can be categorized into different categories based on the objective function as in Table 1 [3].

The SALBP addresses the assignment of the tasks to the stations with one or more optimization criteria while satisfying the occurrence constraint, precedence constraint and cycle time constraint and other constraint which may be put into consideration such as resource constraint [4]. The issue of line balancing with the minimum or limited number of resources (machines and workers) has always been a serious problem in industry. Equipment and workers should assign to task and workstations so that maximum efficiency; maximum usage of resources and minimum number of workstations of production line can be achieved.

Content from this work may be used under the terms of the Creative Commons Attribution 3.0 licence. Any further distribution of this work must maintain attribution to the author(s) and the title of the work, journal citation and DOI.
Published under licence by IOP Publishing Ltd

Table 1. Classification of SALBP

Type	Given	Objective
SALBP-1	Cycle Time	Minimise no. of workstation
SALBP-2	Number of workstation	Minimise cycle time
SALBP-E	-	Maximise line efficiency
SALBP-F	Cycle Time & Number of workstation	Obtain feasible balance

The study on assembly line with resource constraint has been analyzed and published by several numbers of researchers. In 2005, a study on resource constrained assembly line balancing (RCALB) problem with objective function to minimize number of resources used was introduced by K. Agpak with using 0-1 integer-programming models. Resources considered in this paper is a general resources either machine or workforce [5]. Meanwhile, in 2009 I. Moon study on worker skill constraint for integrated assembly line balancing using a mixed integer linear program with a genetic algorithm in order to minimize the annual cost and cost for workers [6]. O. Mutlu in 2013, discussed on worker assignment of assembly line problem type-II and using iterative genetic algorithm (IGA) to assign worker to workstation and at the same time optimize the assembly line [7]. Machine assignment for SALBP-2 have been discussed by H. Triki (2014) where Hybrid Multi Objective Genetic Algorithm (HMOGA) is used to obtained minimum cycle time and cost [8].

In 2014, S. Jayaswal conducted a research on assign tasks to workstations, and resources (equipments and assistants) to tasks with the objectives function is to minimized total cost of workstation and resource utilization. This research is modelled to a U-shaped assembly line balancing using Simulated Annealing [9]. M. Mura and G. Dini (2016) proposes a multi-objective optimization with aim to minimize number of workstation and also number of skilled worker and equipment used in the SALBP-1 by using a Genetic Algorithm (GA) approach [10].

This paper aims to propose a model for understanding and analysing of SALBP-1 with multi-resource constraints so as to minimize number of workstation, number of machines used and also number of workers.

2. Modelling Problem of SALBP-1

Simple assembly line balancing problem is simple assembly line with a number of task are carried in designated workstation. SALBP produce a single product in a serial workstation with a fixed cycle time as illustrated in Figure 1.

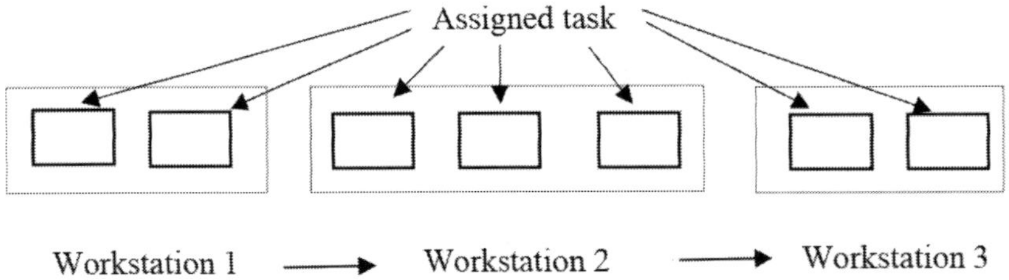

Figure 1. Simple assembly line

2.1 Precedence relation

The precedence graph of SALBP-1 with 9 tasks is presented in Figure 2. The arrows linked between task shows that a task can be performed only after its predecessor tasks are performed.

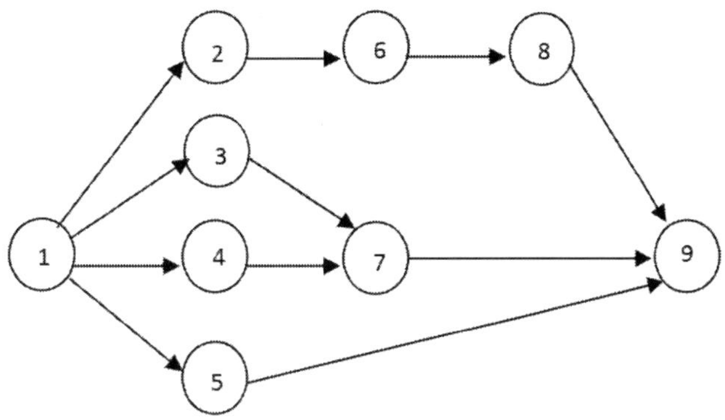

Figure 2. Precedence graph of 9 task example problem

2.2 Data presentation

Table 2 presents the precedence matrix for the above precedence diagram in a matrix form. In this matrix, when task k have precedence relation with task j, '1' will be put into the matrix, otherwise it will be '0'.

Table 2. Precedence matrix

k	*j*								
	1	2	3	4	5	6	7	8	9
1	0	1	1	1	1	0	0	0	0
2	0	0	0	0	0	1	0	0	0
3	0	0	0	0	0	0	1	0	0
4	0	0	0	0	0	0	1	0	0
5	0	0	0	0	0	0	0	0	1
6	0	0	0	0	0	0	0	1	0
7	0	0	0	0	0	0	0	0	1
8	0	0	0	0	0	0	0	0	1
9	0	0	0	0	0	0	0	0	0

Table 3 meanwhile shows the data matrix for the assembly line with resource constraints which include the task time, types of machine required and also worker skill. The worker column with tick mark shows that the worker is able to conduct the task.

2.3 Problem description

In this paper, several assumption has been made in order to describe the problem and evaluate SALBP-1 with resource constraint. The assumptions are as follows:

a) The precedence relationships are known.

b) Task can be assigned to workstation without violating the precedence relation.

c) A worker can be assigned to task depending upon skills.
d) A worker can only be assigned to one workstation.
e) Machine required to process one task may be more than one type.
f) Tasks using same type of machine can share the same machine.
g) The line is balanced for a single product.
h) The processing time is deterministic.

Table 3. Data Matrix

Task	Task Time (min)	Machine	Worker								
			1	2	3	4	5	6	7	8	9
1	6	A, B	/				/		/		/
2	2	-		/		/		/		/	
3	5	A	/		/				/		
4	7	A, B, C		/		/		/		/	
5	1	A	/				/				/
6	2	B	/	/		/	/			/	
7	3	A			/		/			/	
8	6	A, C		/		/		/			
9	5	C	/								

2.4 Mathematical Equation

The notations considered in the mathematical model are clarified as follows:

s workstation, $s = 1, 2, \ldots, n$
m machine, $m = 1, 2, \ldots, r$
w worker, $w = 1, 2, \ldots, h$
k task, $k = 1,2, \ldots, j$
C cycle time
A_s number of workstation
B_m number of machine
x_{ks} 1 if task k is assigned to workstation s
0 otherwise
y_{ms} 1 if machine m is allocated to workstation s
0 otherwise
z_{ws} 1 if worker w is applied to workstation s
0 otherwise
t_k time required by task k
E_a earliest task in precedence relation
L_a latest task in precedence relation
x_{as} 1 if earliest task i is assigned to workstation s
0 otherwise

The mathematical equation of SALBP-1 with resource constraints is presented in below equations.

$$f_1 = \sum_{k=1}^{j} t_k \tag{1}$$

The first objective function (1) is to minimize the number of workstations for a given cycle time which subject to certain constraint as in equation (2) till equation (5) as follows:

$$\sum_{s=E_k}^{L_k} x_{ks} = 1 \tag{2}$$

$$\sum_{k \in F_s} t_k \, x_{ks} \leq C \tag{3}$$

$$\sum_{s=E_a}^{L_a} s x_{as} \leq \sum_{s=E_b}^{L_b} s x_{bs} \quad \text{for } \forall \, (a,b) \in P \tag{4}$$

$$\sum_{k \in F_s} x_{ks} \leq \|F_s\| A_s \tag{5}$$

Constraint (2) is an assignment constraint, which ensures that each task is assigned only once. Constraint (3) is a cycle time constraint, which ensure that the total times in each workstation does not exceed the given cycle time. Constraint (4) is a precedence relation constraint, which guarantees that precedence relation among tasks is not violated. Constraint (5) is a workstation constraint, which guarantees that a workstation is utilized if the task(s) is/are assigned to it.

The second objective function (6) is to minimize the number of machines used which subject to certain constraint as in equation (7).

$$f_2 = \sum_{m=1}^{r} y_{ms} \tag{6}$$

$$\sum_{m=1}^{r} y_{ms} \leq B_m \tag{7}$$

Constraint (7) is a resource availability constraint, which ensure that the total number of resources in workstation is not more than the number of available machines.

The third objective function (8) is to minimize the total number of workers used in an assembly line.

$$f_3 = \sum_{w=1}^{h} z_{ws} \tag{8}$$

$$\sum_{w=1}^{h} z_{ws} \leq 1 \tag{9}$$

Constraint (9) is to restrict only one worker to be assigned to exactly one workstation depending upon his/her skills.

2.5 Fitness Function

Optimal solutions of the first, second and third objectives are obtained by solving each objective function separately using the above mathematical equation. Then, a single objective is employed to minimize the summation of normalized differences between each objective and its optimal value by determining the fitness function.

$$F = w_1 f_1 + w_2 f_2 + w_3 f_3 \tag{10}$$

Where;

$$w_1 = w_2 = w_3 = 0.33$$

w_i represent the weights of objectives and f_1, f_2 and f_3 represent respectively the normalized values derived from the equations.

$$\frac{\bar{f_i} - \bar{f}_{i\,min}}{\bar{f}_{i\,max} - \bar{f}_{i\,min}} = \frac{x - x_{min}}{x_{max} - x_{min}} \tag{11}$$

3. Numerical Example

A numerical example is used to explain the objective functions in section 2. In this example, a feasible sequence, *seq* [1, 2, 6, 8, 3, 4, 5, 7, 9] are considered in distributing task to workstation. In evaluating the optimisation objective , the assembly tasks is assigned to workstations with the ct_{max} constraint where the total processing time for tasks assigned to the workstation must not exceed the cycle time, set as C=10. The example of the assignment of assembly task for this sequence is presented in Figure 3.

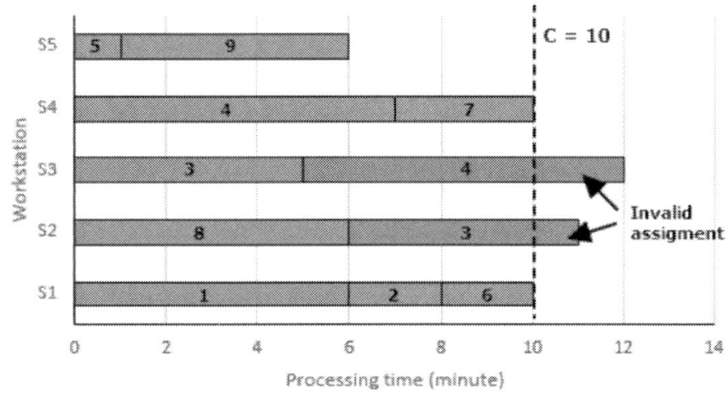

Figure 3. SALBP-1 task distribution

The task distribution is considered by distributing each task to any workstation of the assembly line, as long as the machine and worker assigned to that station can perform the task and the precedence rations are obeyed.

3.1 Problem Evaluation

Based on the precedence graph in Figure 2, a feasible assembly sequence is established to evaluate the optimization objective. As an example, a valid feasible assembly sequences (*seq*), seq_1[1, 2, 6, 8, 3, 4, 7, 5, 9] as in Figure 3 and two other sequence seq_2 [1, 5, 3, 4, 7, 2, 6, 8, 9] and seq_3 [1, 3, 4, 7, 2, 6, 8, 5, 9] are considered. For given, cycle time maximum for this assembly line is to be 10. To assign task to workstation (*w*), the processing time (p_t) which is the sum of task time (t_t) for each workstation must not exceed 10. Resources; machines (*m*) and workers (*w*) are then assigned to the task base on the compatibility by following the data given in Table 2. The results of assembly tasks assignment for seq_1, seq_2 and seq_3 are presented in Table 4, 5 and 6.

Table 4. Assembly Task Assignment for seq_1

seq_1	1	2	6	8	3	4	7	5	9
t_t	6	2	2	6	5	7	3	1	5
p_t	10			6	5	10		6	
s	1			2	3	4		5	
m	A, B			A, C	A	A, B, C		A, C	
w	5, 2			4	3	8		1	

Table 5. Assembly Task Assignment for seq_2

seq_2	1	5	3	4	7	2	6	8	9
t_t	6	1	5	7	3	2	2	6	5
p_t	7		5	10		10			5
s	1		2	3		4			5
m	A, B		A	A, B, C		A, B, C			C
w	9		3	8		2			1

IOP Conf. Series: Journal of Physics: Conf. Series **1049** (2018) 012037 doi:10.1088/1742-6596/1049/1/012037

Table 6. Assembly Task Assignment for seq_3

seq_3	1	3	4	7	2	6	8	5	9
t_t	6	5	7	3	2	2	6	1	5
p_t	6	5	10		10			6	
s	1	2	3		4			5	
m	A, B	A	A, B, C		A, B, C			A, C	
w	9	7	8		2			1	

Number of workstation: The number of workstations required when assigning task to workstation for seq_1, seq_2 and seq_3 are both 5 number of workstations.

Machines: Total 10 units of machine; 5A, 2B and 3C required for assembly task assignment seq_1 and 4A, 3B and 3C for seq_2 and 11 units of machine for seq_3; 5A, 3B and 3C.

Workers: The number of workers required to perform the assembly task assignment seq_1 is 6 while only 5 workers need to be allocate in seq_2 and seq_3.

3.2 Minimize number of workstation

The first objective function is to minimize number of workstation subjected to the equation (1).

$$f_1 = \sum_{s=1}^{n} A_s$$

$$f_{1\,min} = \frac{\sum t_i}{Ct_{max}} = \frac{37}{10} = 3.7 \sim 4$$

$$f_{1\,max} = \frac{\sum t_i}{max(t_i)} = \frac{37}{7} = 5.3 \sim 6$$

3.3 Minimize number of machine

The second objective function that subject to equation (6) are as follows.

$$f_2 = \sum_{m=1}^{r} y_{ms}$$

$$f_{2\,min} = total\ machine\ type = 3$$

$$f_{2\,max} = f_{1\,max}(total\ machine\ type) = 6(3) = 18$$

IOP Conf. Series: Journal of Physics: Conf. Series **1049** (2018) 012037 doi:10.1088/1742-6596/1049/1/012037

3.4 Minimize number of worker

$$f_3 = \sum_{w=1}^{h} z_{ws}$$

$$f_{3\,min} = f_{1\,min} = 4$$

$$f_{3\,max} = number\ of\ task = 9$$

3.5 Fitness value

Based on the above solution for sequence seq_1, seq_2 and seq_3, the fitness function for both is calculated by using equation 10 and 11 and the result obtained presented in Table 6.

Table 7. Result for test problem

Sequence	Workstation	Machine	Worker	Fitness value
seq_1	5	10	6	4.62
seq_2	5	10	5	3.96
seq_3	5	11	5	4.29

Based on the results in Table 6, all feasible sequences tested shows the same number of workstation. Feasible sequence seq_1 and seq_2 have the same number of machine used. However seq_2 have less number of worker assigned to workstation compare to seq_1. Meanwhile feasible sequence seq_3 have the highest number of machine used compare to seq_1 and seq_2 but less worker assigned to workstation compare to seq_1. seq_2 give the best fitness value followed by seq_3 and seq_1.

The model presented in this paper shows that the model is valid in presenting simple assembly line balancing type 1 (SALBP-1) with resource constraint; machine and worker. This paper considers different types of machine used in an assembly line as well as multi skilled worker which can be assigned to workstation. The number of machines and workers used in SALBP-1 can be optimize as suggested and tested in the proposed model.

4. Conclusion

In this paper, Simple Assembly Line Balancing Problem Type 1 (SALBP-1) with resource constraint was modelled and discussed. The model is design by considering three objective functions which is the minimization number of workstation, number of machines used and number of multi-skilled workers. The model is explained in details and evaluation on the model is briefly presented with a particular equation and calculation. The obtained results have described the SALBP-1 with resource constraint features. From the results of the calculation, it shows that this tested model for SALBP-1 with resource constraint could have better performance in optimization.

For future work, this proposed model of SALBP-1 with resource constraint might be extend for future research by tested using algorithm by referring to the solution procedure. The case study on real industrial problem on SALBP-1 with machine and worker constraint might be conducted to validate the propose model.

Acknowledgments

This research is supported by Ministry of Higher Education Malaysia and Universiti Malaysia Pahang under grant number RDU140103.

REFERENCES

[1] A. Scholl, N. Boysen, and M. Fliedner, "The sequence-dependent assembly line balancing problem," *OR Spectr.*, vol. 30, no. 3, pp. 579–609, 2008.

[2] S. Ozmehmet and T. Semra, "A review of the current applications of genetic algorithms in assembly line balancing," *J Intell Manuf*, vol. 19, no. 49, pp. 49–69, 2008.

[3] N. Boysen, M. Fliedner, and A. Scholl, "A classification of assembly line balancing problems," *Eur. J. Oper. Res.*, vol. 183, no. 2, pp. 674–693, 2007.

[4] H. Dwi, H. Wee, and H. Rau, "Two-sided assembly lines balancing with assignment restrictions," *Math. Comput. Model.*, vol. 57, no. 1–2, pp. 189–199, 2013.

[5] K. Ağpak and H. Gökçen, "Assembly line balancing: Two resource constrained cases," *Int. J. Prod. Econ.*, vol. 96, no. 1, pp. 129–140, 2005.

[6] I. Moon, R. Logendran, and J. Lee, "Integrated assembly line balancing with resource restrictions", *Int. J. of Prod. Research*, vol. 47, no. 19, pp. 5525-5541, 2009.

[7] Ö. Mutlu, O. Polat, and A. A. Supciller, "An iterative genetic algorithm for the assembly line worker assignment and balancing problem of type-II," *Comput. Oper. Res.*, vol. 40, no. 1, pp. 418–426, Jan. 2013.

[8] H. Triki, A. Mellouli, and F. Masmoudi, "A multi-objective genetic algorithm for assembly line resource assignment and balancing problem of type 2 (ALRABP-2)," J Intell Manuf, vol. 28, no. 371(2), pp. 371–385, 2017.

[9] S. Jayaswal and P. Agarwal, "Balancing U-shaped assembly lines with resource dependent task," *J. Manuf. Syst.*, vol. 33, no. 4, pp. 522-534, 2014.

[10] M. D. Mura and G. Dini, "Worker Skills and Equipment Optimization in Assembly Line Balancing by a Genetic Approach," *Procedia CIRP*, vol. 44, pp. 102–107, 2016.

IOP Conf. Series: Journal of Physics: Conf. Series **1049** (2018) 012038 doi:10.1088/1742-6596/1049/1/012038

Experimental and Computational Assessment of the Strength Properties of Mont Wright Tailings Matrices for Use as Road Materials

Ali A Mahmood[1,3] and Maria Elektorowicz[2]
[1] Associate, Ash System International, 930 Av. Van Dyck, Brossard, Quebec, J4W 2E6, Canada (email: alimahmood45@yahoo.ca)
[2] Professor, Dept. of Building Civil and Environmental Engineering, Concordia University, Rm: EV-6.213, 1455 De Maisonneuve Blvd. W., Montreal, Quebec, H3G 1M8, Canada (email: mariae@encs.concordia.ca)

Abstract. Mont Wright Mine tailings matrices were subjected to freezing/thawing and wetting/drying cyclic weathering tests before being tested in the unconfined compression chamber. These experimental results were subsequently compared to computational values found using the program Tailings-DEM™. Results showed that Tailings-DEM™ was capable of simulating, with reasonable accuracy, the UCS strength values of the tailings matrices tested.

1. Introduction

Mine tailings are the waste materials left after the extraction of the ore mineral. They are traditionally stored in tailing ponds adjacent to the mine opening [1]. This method of storage has created several environmental issues in the past and has prevented the use of a widely available resource.

In order to make use of mine tailings and reduce their bulky method of storage, Robinsky [2] proposed the thickened tailings disposal (TTD) system for the management of mine tailings. He showed that through the process of thickening the tailings to heavy slurry before disposal, it is possible to create a self-supporting deposit of tailings and to eliminate the traditionally used settling pond.

Robinsky [2] stated that since the conventional tailings pond has been eliminated, and without water to flow and wash out the tailings, a dam collapse becomes only a local problem and not an ecological disaster. Another advantage of the TTD system is that the system is capable of inhibiting the drainage of acid. The environmentally undesirable product is sulphuric acid. It is produced when air and water come into contact with sulphur-containing mine tailings. Once formed, it is capable of infiltrating the tailings and dissolving other undesirable metallic elements that eventually may seep through and contaminate the living environment [3].

[3]To whom any correspondence should be addressed.

Content from this work may be used under the terms of the Creative Commons Attribution 3.0 licence. Any further distribution of this work must maintain attribution to the author(s) and the title of the work, journal citation and DOI.
Published under licence by IOP Publishing Ltd

The thickened surface tailings disposal concept has been applied at various mining facilities for several decades, including the Kidd Creek Mine in the northern region of Ontario (Canada) and for managing the disposal of red muds produced by the alumina industry [3].

Success in the process of mine tailings hardening has led researchers to further think the possibility of using these hardened tailings in the construction industry. This would thereby utilize a vastly available untapped resource and hinder, albeit to a certain extent, the harmful effects of unprocessed tailings on the surrounding environment. This success coupled with historical evidence of using tailings in the mine backfilling industry further strengthened this new approach.

The earliest reported use of mine tailings as a construction material was its use as a mine backfilling material. Use of mine tailings for backfilling started in the early days of the potash industry by using dry and wet residues [4]. Different technologies were introduced in the last 80 years. One of these, the slurry backfill technology is considered an efficient backfill method, because of its combination of tailings and brine disposal. Since 1908, the slurry backfill technology has been successfully employed in the German potash industry, especially in the flat-lying deposits of the South Harz potash district [4].

Success with these early attempts led researchers to attempt a thorough investigation of mine backfilling using more advanced technologies and procedures. Garand et al. [5] discussed the effect of flocculent deposition of tailings sludge in the Bouchard-Hebert mine in northern Quebec. ASTM standard tests were performed on the tailings to determine their particle size, specific gravity and Atterberg limits. Then a flocculent (PERCOL E-10) was added to the tailings sludge before it was mechanically thickened in the paste backfill plant. These flocs had grain size distribution larger than or similar to coarse silt. It was stated that the beaches formed with these flocs could be used as competent foundation material for upstream raises [5]. This study presented a modification to Robinsky's [2] thickening principle by adding a flocculent to the tailings to enhance thickening and increase stability.

Another study by Benzaazoua et al. [6] investigated in certain detail the influence of several chemical factors on the performance of mine sulphidic paste backfill. They used four samples from three different mines located in Canada, which were mixed with four types of binders and and mixing water with six different chemical characteristics. This was done in an attempt to simulate field conditions. It was found that the paste backfill matrix texture was directly related to its strength development. Also it was found that the mixing water was an important parameter that affected the quality of the paste backfill mass and that, in contrast to the Portland cement based binder, slag-based binder hydration seemed to be inhibited by the presence of soluble sulphates [6]. The results of this study clearly demonstrated the inefficiency of choosing paste backfill mixtures without first testing the tailings and mixing water characteristics.

Theriault et al. [7] studied the surface disposal of paste tailings at the Bulyanhulu Gold mine in Tanzania. The goals of this process were, to conserve water, manage runoff, reduce risk and minimize containment dyke construction. The tailings slurry was dewatered before transportation to the paste plant where process water was added in the paste conditioner to produce a paste of desired consistency. The authors affirmed that the cycling of tailings deposition in thin layers has been successful in generating a stable paste stack. It was concluded that the paste stacks can be engineered to meet the required geotechnical and environmental objectives.

These earlier studies ascertained the applicability of using mine tailings as backfill materials and encouraged investigation of the use of mine tailings in other construction disciplines. One such discipline is exemplified by the work of Demers and Haile [8] who described the stabilization of zinc tailings (named Jarosite). Jarosite had been deposited in ponds sustained by clay dykes until 1998. Since then, Jarosite was thickened first using vacuum filters. Then lime, cement and water were added to the thickened Jarosite to make a product that was termed Jarofix. Laboratory tests and stack modeling were performed to determine the feasibility of using the cured Jarofix to build containment dykes and service roads. Field tests were performed, as well, after which the authors concluded that Jarofix was chemically inert. It was observed that cured Jarofix was an excellent fill material for raising containment dykes. This research was carried out inside the tailings storage facility and hence limited testing was done in this case.

Another example is the work of Zou and Sahito [9] who studied the applicability of using a new type of binder termed HiFa Bond for shotcreting underground support. The authors used mine tailings as aggregates in shotcrete. The tailings were mixed with sand, polymer and steel fibers. Test results showed that mine tailings had potential for shotcreting for underground support.

Meanwhile, Celik et al. [10] investigated the effects of the mineralogical composition and chemical properties of gold tailings on the compressive strength of Ordinary Portland Cement (OPC). The authors used silica fume and 2 types of fly ash, with the tailings as OPC additives. It was shown that the compressive strength values of mortars prepared by these additives were acceptable as they were within the range accepted in European standards. The authors presented the conclusion that the gold tailings could be used as an additive in OPC production. No weathering resistance or leaching testing had been done in this study.

Another study by Roy at al. [11] investigated the use of gold tailings in the making of bricks. The authors mixed the tailings with 4 types of soils and Portland cement. The quality of bricks was measured in terms of compressive strength, linear shrinkage and water absorption tests. The authors performed cost analysis, as well. They showed that the quality of these bricks was improved when mill tailings were mixed with these soils. The study showed that soil-tailings bricks passed the required criteria defined in terms of compressive strength, linear shrinkage and water absorption. The cost of producing the tailings-soil bricks was shown to be less than that of traditional clay bricks. However, there were no leaching or weathering environmental tests performed.

Swami et al. [12] investigated the use of Kimberlite diamond mine tailings in the construction of road layers. For this purpose, the authors performed a field and laboratory testing program. Physical tests including Proctor, California Bearing Ratio (CBR) and unconfined compression were executed before and after stabilization with cement and bitumen. Chemical properties of the tailings were also determined before performing laboratory and field experiments. The field experiments consisted of constructing a test road section using these tailings, which were evaluated for use in sub-base, base and wearing courses. It was shown that Kimberlite tailings could be used in cement bound sub-base layers. The study did present a new approach in the area of road construction. However, the testing regime implemented was not comprehensive and there were no weathering or leaching tests conducted. These tests are considered pivotal in cold environments such as that of Canada. Also the experimental program lacked correlation with numerical analysis methods.

Other researchers such as Fall et al. [13], Fall et al. [14], Benzaazoua et al. [15], Ercikdi et al.[16], Fall and Pokharel [17], Ercikdi et al. [18] and Helinski et al. [19] investigated the use of mine tailings in the most traditional manner by using them as cemented paste backfill. While, the research group of Yang et al. [20] investigated the use of mine tailings in the glass-ceramic/ceramic tile industry.

It appears from the above that although several research studies discussed the applicability of using mine tailings in several types of construction applications, none had comprehensively addressed the issue of using mine tailings as construction materials for temporary roads in cold climates. Simulation of the strength properties of these mine tailings using a powerful numerical technique was lacking in these studies, as well. The objective of this study is the development of a new sustainable approach to road construction using mining waste in cold climate by formulation of tailings binder matrices and investigating the simulation of their unconfined compressive strength properties. This will be done after exposing them to freezing/thawing and wetting/drying cycles to simulate environmental conditions. Their strength properties will further be analyzed using the Discrete Element Method program: Tailings-DEM™. Conclusions will be drawn from the comparison of the numerical and mechanical strength properties.

2. Binding materials

The binding materials for the mine tailings, which have been tested in this study, consist of Type I ordinary Portland cement (OPC), fly ash, slag and Calsifrit.

Unconfined compressive strength of fly ash lies in the range 1.38 to 6.895 MPa with permeability between 10^{-5} to 10^{-8} cm/s, [21]. The fly ash was obtained from St. Laurence Cement Company in Longueil, Quebec.

Ground granulated blast furnace slag used in this study was obtained from Lafarge North America Cement Company in Montreal, Quebec under the trade name NewCem®. It is a light grey odorless powder produced in accordance with ASTM C 989 [22]. It is composed of (30-50) % Calcium Oxide, (0-20) % Magnesium Oxide with less than 1% Crystalline Silica. Its specific gravity is between 2-3, water pH 8-11. Its boiling point is above 1000 °C [23].

Calsifrit™ is a totally amorphous siliceous material, a matrix of calcium and sodium fluoro-aluminosilicate. This homogeneous solid substance has a blackish grey color, possesses a high reactivity potential and shows cementitious properties when finely ground. This product was obtained from the manufacturer NovaFrit International. Table 4 shows its chemical composition. Calsifrit has a pH (1% solution in water) of 6.5-7.5, a melting point of 800 °C, a specific gravity of 1.3 with a moisture content of less than 10% [24].

3. The tailings

Bench scale and pilot tests carried out by Alcan International Ltd. on copper and gold tailings had shown the applicability of the thickening technology to a variety of tailings [25]. Also, Alumina tailings were investigated by the latter authors. Therefore in this study, another type of mine tailings was selected and obtained for evaluation; Mont Wright tailings from Quebec Cartier mineral company, from Mont-Wright (iron) mine, in Quebec.

This type of tailings was chosen because of its proximity to the northern territories of Eastern Canada; close to the area where Plan Nord will be employed. The adverse environmental impact that tailings cause to underground water resources and the living environment warrant a detailed investigation into the potential of reusing them sustainably in construction. Table 1 shows the physical properties of Mont Wright tailings.

Table 1. Physical properties for Mont Wright tailings

Type of Tailings	Mont Wright
D_{10} (mm)	0.147
D_{50} (mm)	0.255
D_{60} (mm)	0.3
D_{30} (mm)	0.2
C_u	2.04
C_z	0.91
$P_{4.75mm}$ (%)	100
$P_{0.075mm}$ (%)	2.05
Initial moisture content (%)	4.27
USCS	SP

Where: D_{10} = diameter corresponding to 10% finer, D_{50} = diameter corresponding to 50% finer, D_{60} = diameter corresponding to 60% finer, D_{30} = diameter corresponding to 30% finer, C_u = uniformity coefficient = D_{60} / D_{10}, C_z= coefficient of gradation = $D_{30}{}^2$ /(D_{10} D_{60}), $P_{4.75mm}$ (%) = percentage passing sieve no.4, $P_{0.075mm}$ (%) = percentage passing ASTM sieve # 200, SW-SM = well graded sand with silt, SM = silty sand, SP = poorly graded sand, USCS = unified soil classification system, USCS [26].

The mineralogical composition of Mont Wright tailings is: Sandstone (quartzite), Mica schist, Amphibolite, Gabroïque Granite, Gabroïque Feldsphate, Specular Hematite, Specular Magnetite, Quartz, Diopside, Tremolite, Actinolite and Gruenerite.

Atomic absorption analysis performed on the tailings revealed that the tailings contained Fe 3660 mg/kg, whilst the other heavy metals (Cu, Cr, Zn, Pb, Ni) where no detected.

4. Preparation of cylindrical specimens

Cylindrical specimens of Mont Wright tailings matrices measuring (44 diameter x 74 height) *mm* were molded for this purpose in accordance with ASTM D 4842 [27]. These cylindrical specimens were, then, cured in a moisture chamber for 28 *days* in the case of the wetting and drying samples, and for 43 *days* for the freezing and thawing samples. Unconfined compression tests were conducted on these tailings matrices after exposing them to 12 cycles of drying at 60 °*C* and wetting by being submerged in room temperature distilled water. The remaining samples were compressed after undergoing 12 cycles of freezing at -20 °*C* in a freezing cabinet and thawing in distilled water at room temperature (22 °*C*). Table 2 shows the codes implemented for the cylindrical specimens of the tailings matrices.

Table 2. Tailings matrices specimen codes.

Tailings Type	Weathering Type	Code	Specimen	Binder/ Tailings	OPC/ Binder (%)	Calsifrit/ Binder (%)	Fly ash/ Binder (%)	Slag/ Binder (%)
Mont Wright	Wetting/ Drying	MC1	1, 2, 3, 5, 6	0.5	100	0	0	0
		MC3	1, 2, 3, 4, 5, 6	0.5	90	10	0	0
		MC6	1, 2, 3, 4, 5, 6	0.5	75	25	0	0
		MCF1	1, 2, 3, 4, 5, 6	0.5	75	0	25	0
		MCF3	1, 2, 3, 4, 5, 6	0.5	75	10	15	0
		MCF5	1, 2, 3, 4, 5, 6	0.5	75	20	5	0
		MCS1	1, 2, 3, 4, 5, 6	0.5	75	0	0	25
		MCS3	1, 2, 3, 4, 5, 6	0.5	75	10	0	15
		MCS5	1, 2, 3, 4, 5, 6	0.5	75	20	0	5
Mont Wright	Freezing / Thawing	MC'1	1, 2, 3, 4, 5, 6	0.5	100	0	0	0
		MC'3	1, 3, 4, 5, 6	0.5	90	10	0	0
		MC'6	1, 2, 3, 4, 5, 6	0.5	75	25	0	0
		MCF'3	1, 2, 3, 4, 5, 6	0.5	75	10	15	0
		MCF'5	1, 2, 3, 4, 5, 6	0.5	75	20	5	0
		MCS'1	1, 2, 3, 4, 5, 6	0.5	75	0	0	25
		MCS'3	1, 2, 3, 4, 5, 6	0.5	75	10	0	15
		MCS'5	1, 2, 3, 4, 5, 6	0.5	75	20	0	5

5. Unconfined compression testing after weathering

It is known, that, solidified waste in the field is under continuous weathering cycles. Therefore, strength testing will not be complete without investigating the effect of weathering cycles on compression strength.

After undergoing the weathering tests, the specimens of the tailings matrices were tested under uniform conditions to find their strength in the UCS apparatus. In this test, speed of descent of the platen was kept constant at 70 *mm/min.* with the exception of a few samples where it had to be changed to 50 *mm/min.* in order to increase the maximum load the machine can apply, since these latter samples had higher compression strengths than the others. Experimental results of this test are discussed elsewhere [28].

6. Computational and experimental results

It is necessary to verify the accuracy and reliability of the proposed computational model (Tailings-DEM^{TM}) using laboratory-approved experimental results. Therefore the UCS experimental results [28] were used in the verification and analysis of this program.

Figures 1 to 6 show the UCS experimental and computational results comparison for Mont Wright tailings matrices for the matrix binder combinations shown in Table 2. The following shows these matrix binder combinations in groups of figures:

1) Figure 1 shows the 100% OPC,

2) Figure 2 shows the 90% OPC, 10% Calsifrit,

3) Figure 3 shows the 75% OPC with 25% Calsifrit,

4) Figure 4 shows the 75% OPC with 25% fly ash,

5) Figure 5 shows the 75% OPC, 15% fly ash and 10% Calsifrit,

6) Figure 6 shows the 75% OPC and 25% slag,

IOP Conf. Series: Journal of Physics: Conf. Series **1049** (2018) 012038 doi:10.1088/1742-6596/1049/1/012038

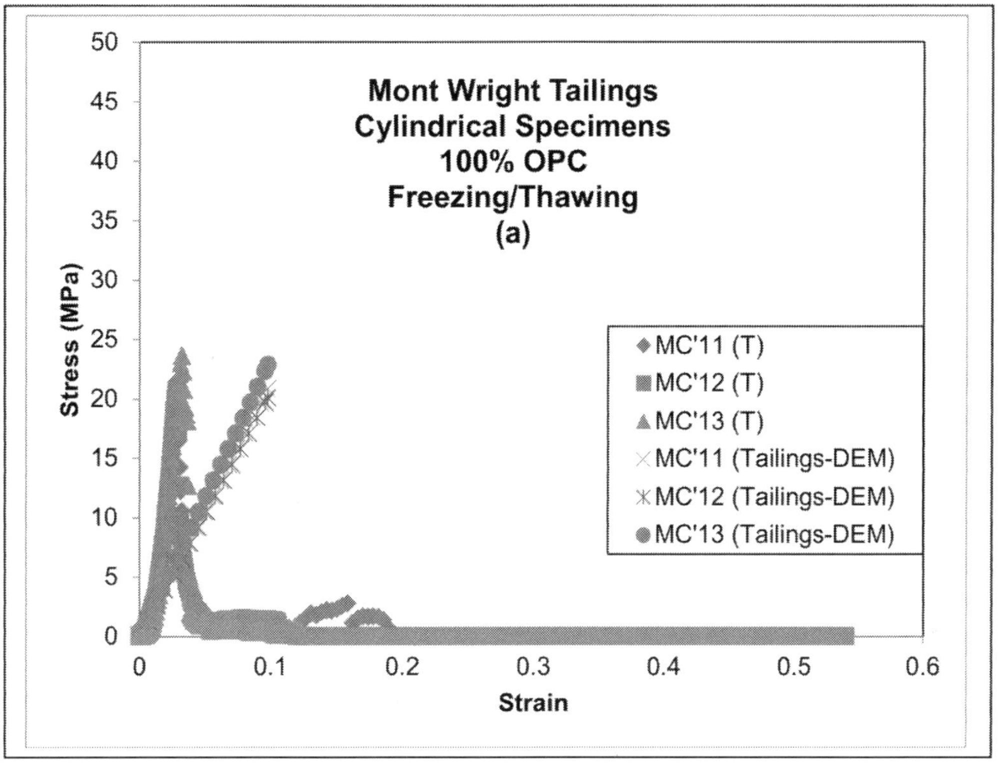

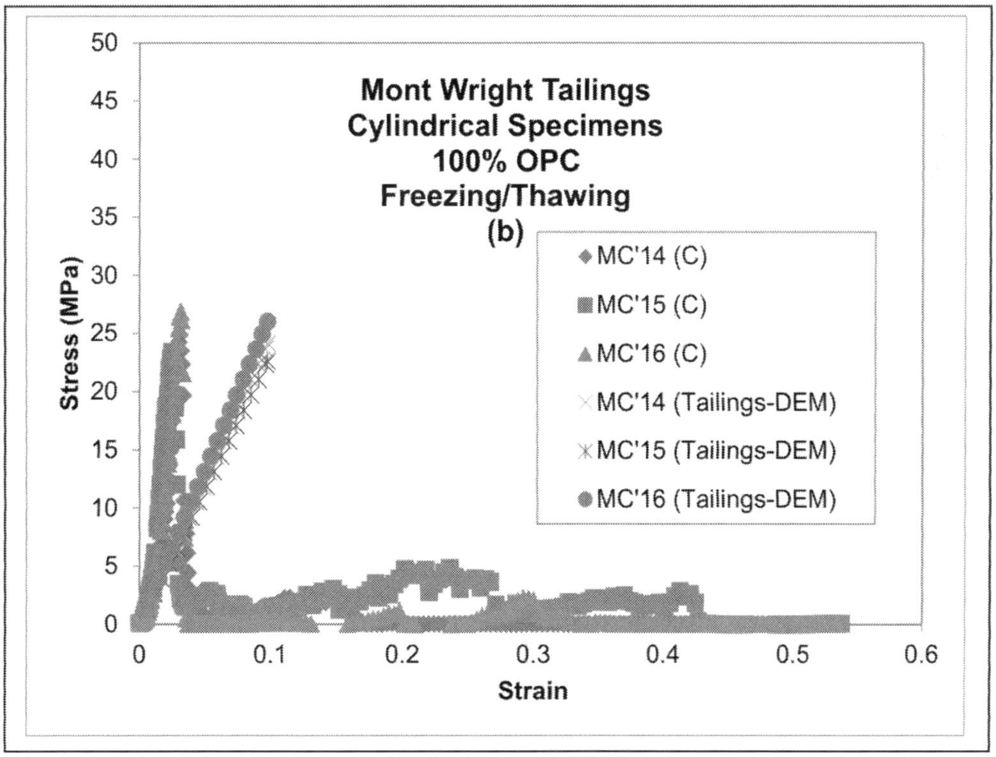

Figure 1. Computational and experimental UCS values for Mont Wright samples MC'11(a)-MC'16(b) after freezing/thawing

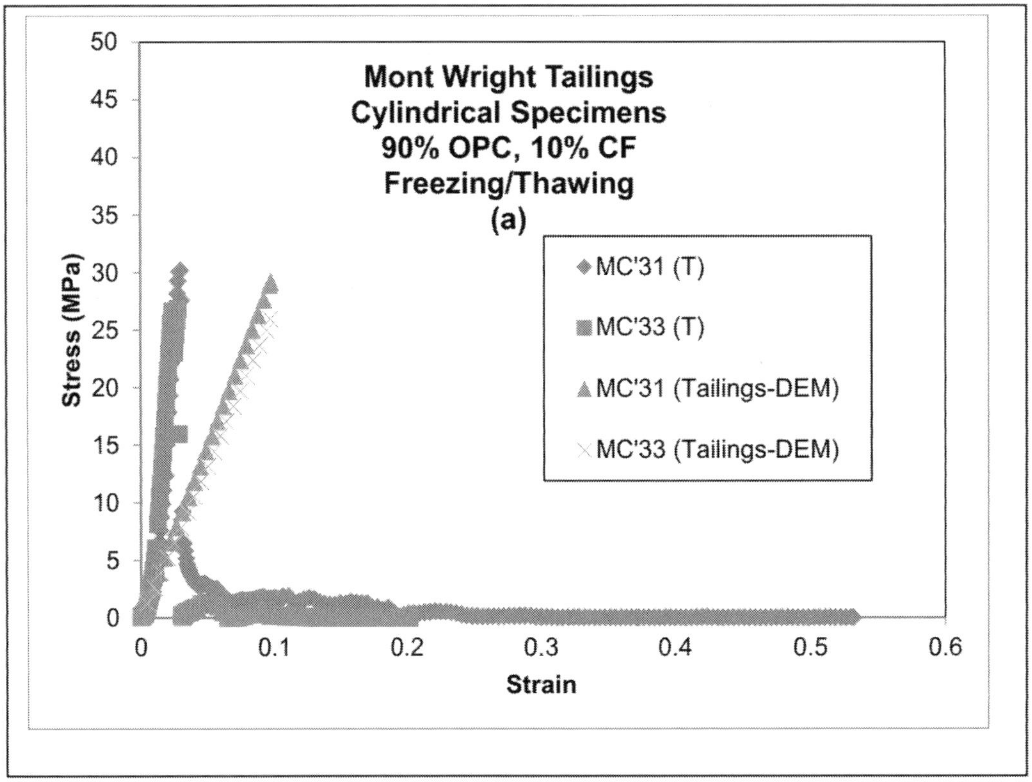

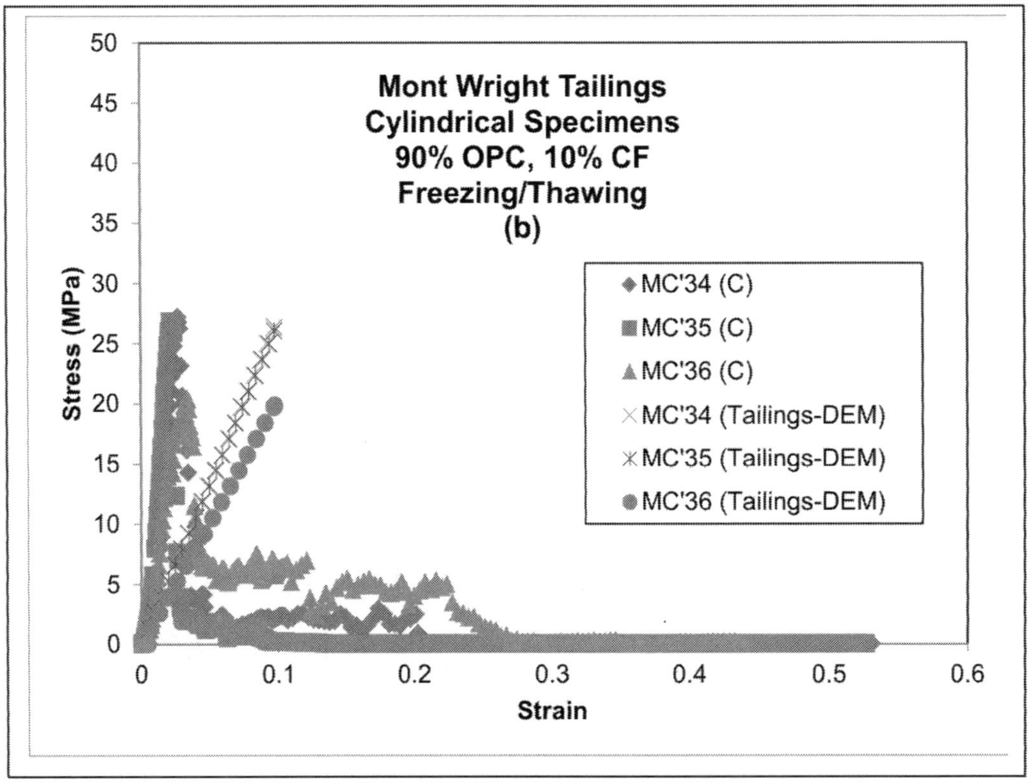

Figure 2. Computational and experimental UCS values for Mont Wright samples MC'31(a)-MC'36(b) after freezing/thawing

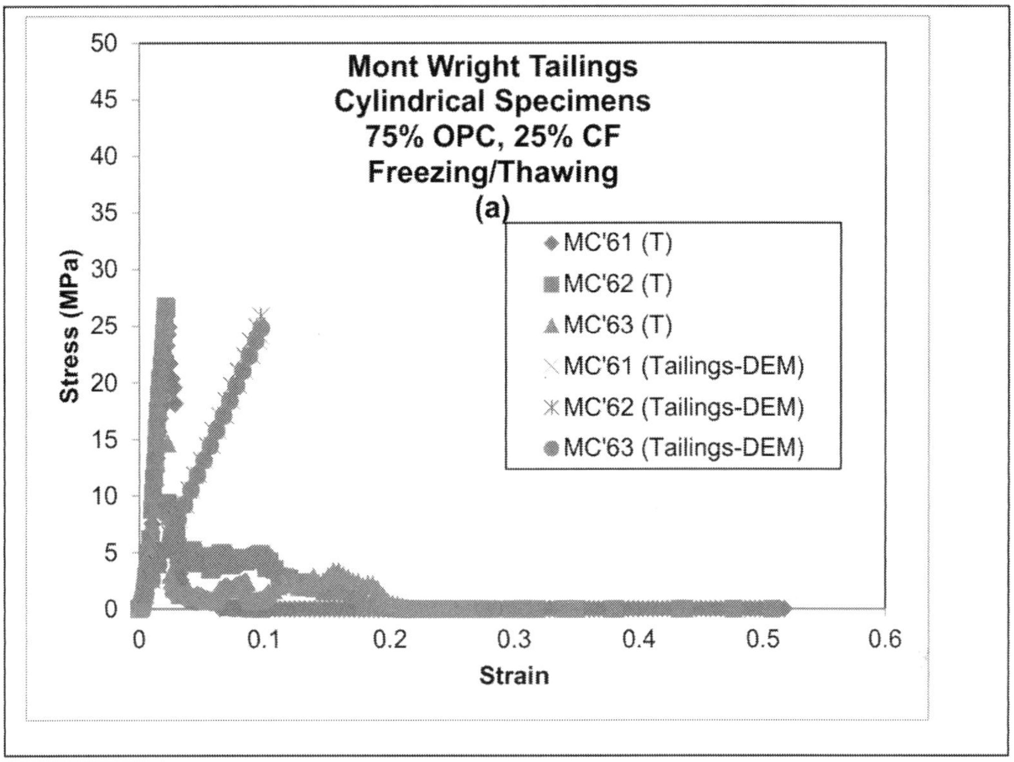

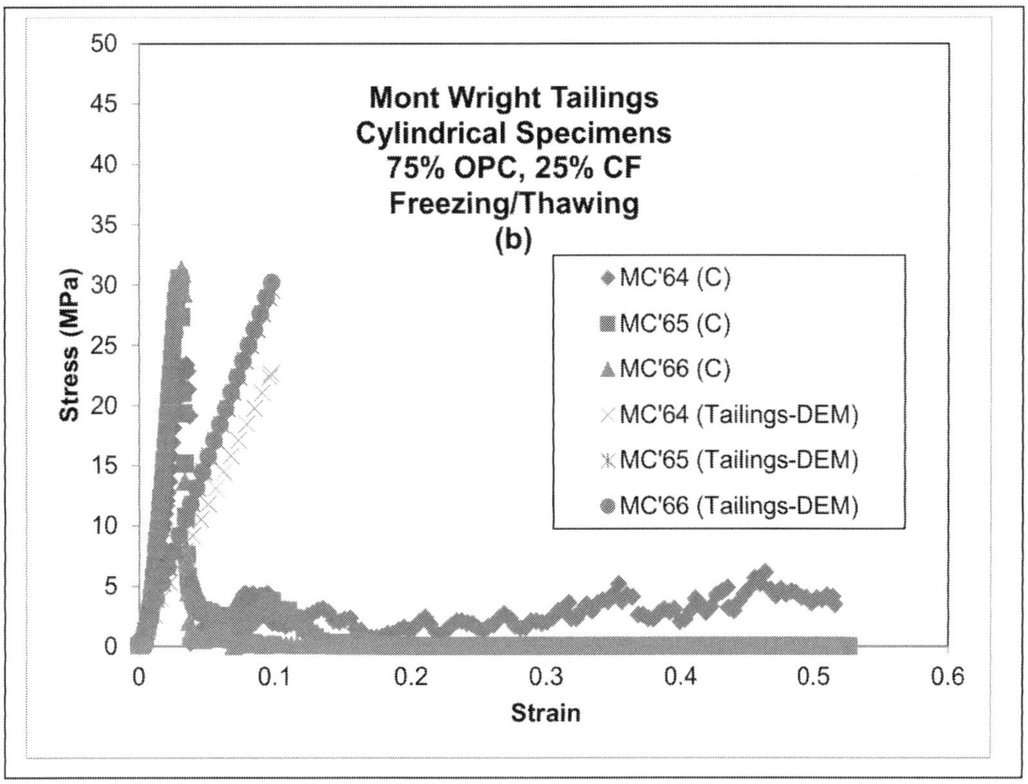

Figure 3. Computational and experimental UCS values for Mont Wright samples MC'61(a)-MC'66(b) after freezing/thawing

IOP Conf. Series: Journal of Physics: Conf. Series **1049** (2018) 012038 doi:10.1088/1742-6596/1049/1/012038

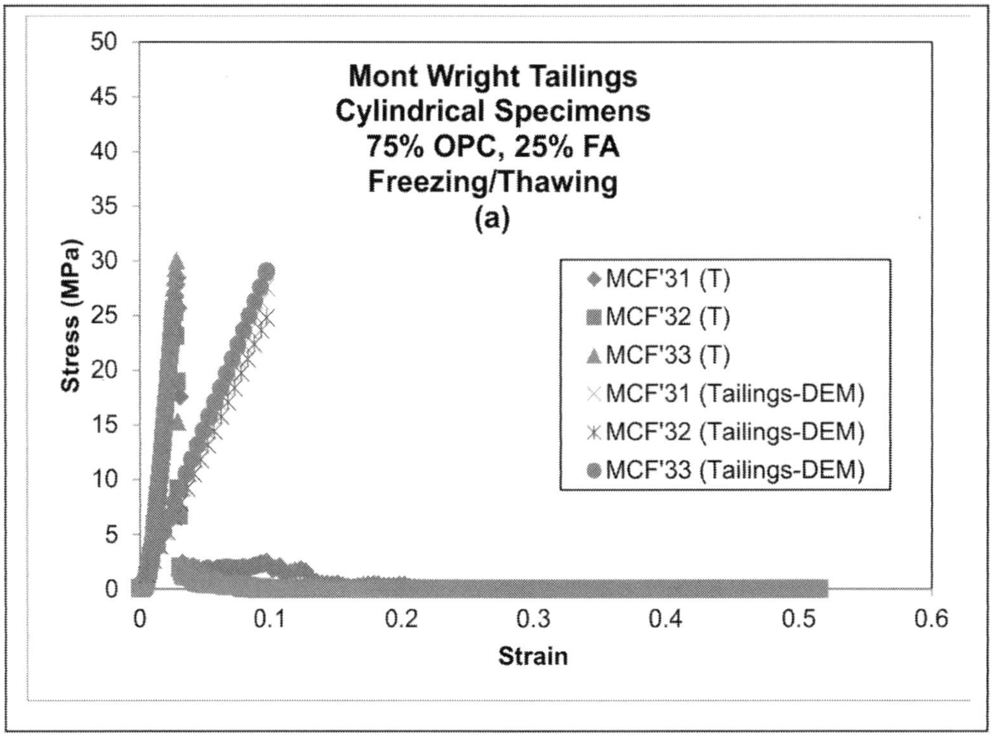

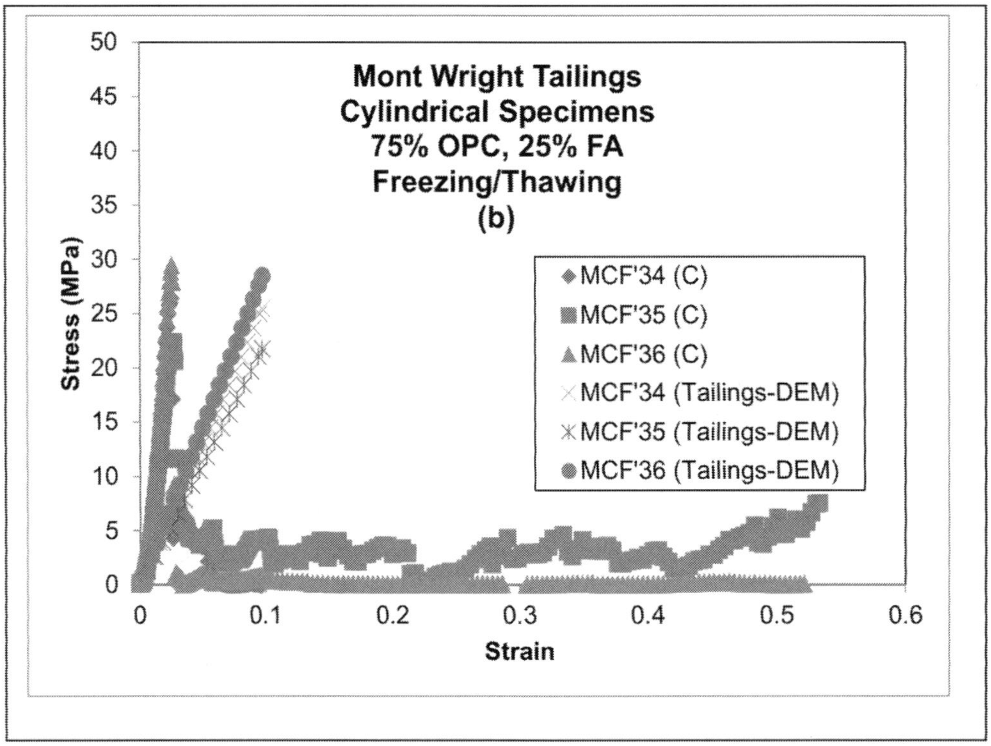

Figure 4. Computational and experimental UCS values for Mont Wright samples MCF'31(a)-MCF'36(b) after freezing/thawing

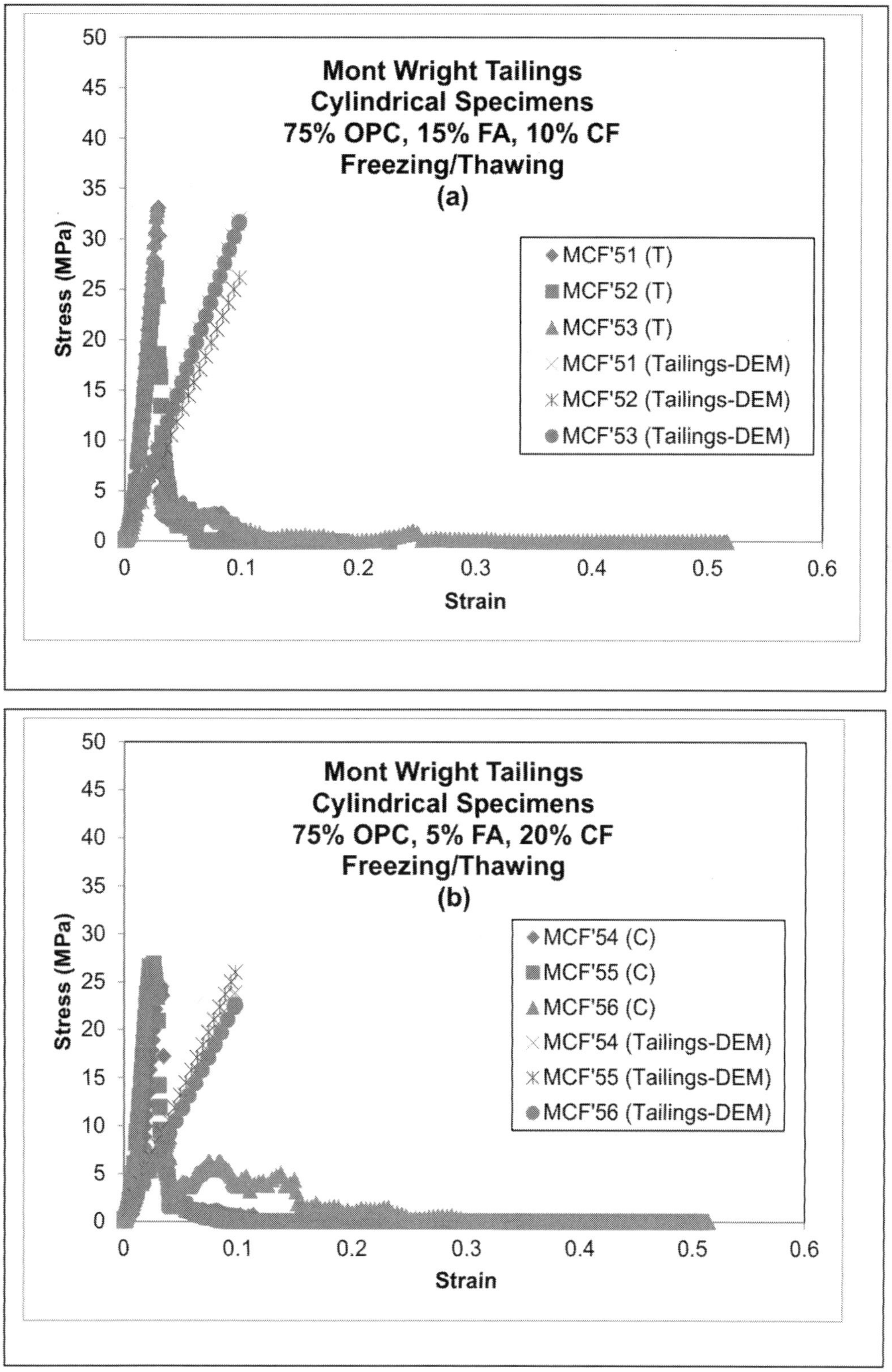

Figure 5. Computational and experimental UCS values for Mont Wright samples MCF'51(a)-MCF'56(b) after freezing/thawing

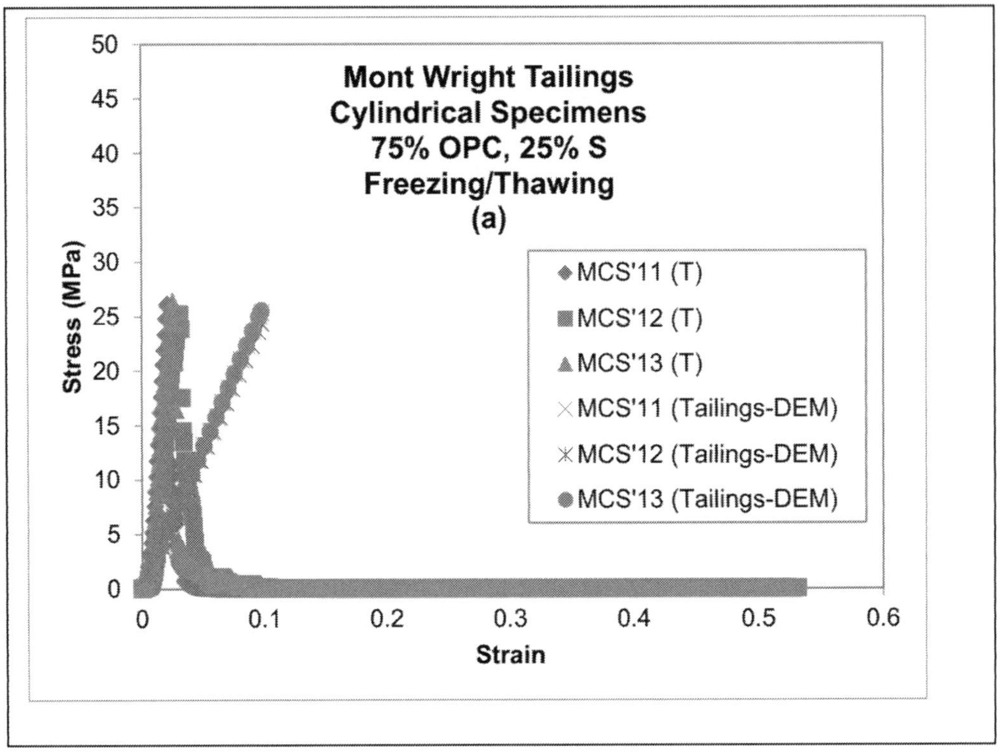

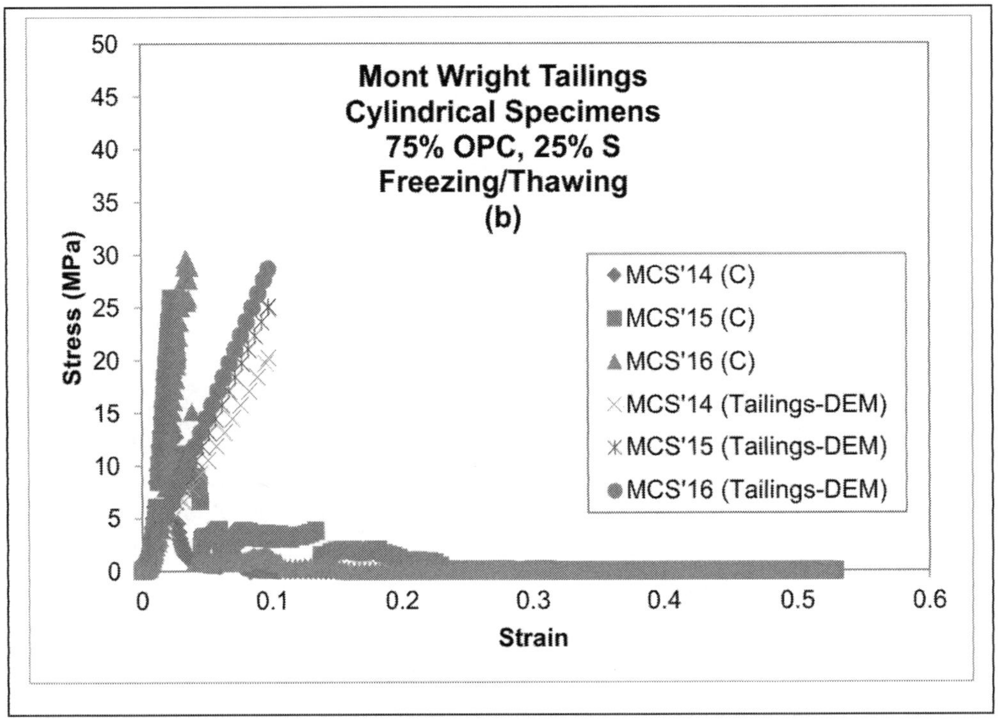

Figure 6. Computational and experimental UCS values for Mont Wright samples MCS'11(a)-MCS'16(b) after freezing/thawing

7. Discussion of Tailings-DEM™ strength modeling

Experimental and computational UCS values for the Mont Wright tailings matrices, for the cases shown, have similar types of curves indicating that computational modeling of this test was to a large extent accurate. Maximum values and slopes for the experimental and computational results followed closely. Marginal differences between the experimental and computational values could be attributed to several factors.

The marginal differences between the computational and experimental UCS results could be attributed, partially, to the experimental errors in the preparation of the specimens tested. It is known that soil-cement matrices suffer from result variability [28]. Furthermore, the presence of iron (Fe) has further complicated the outcome of the UCS experiments [28].

Another reason could be the assumptions used in defining the computational program Tailings-DEM™. The assumptions were based on modelling the particles as right-isosceles triangles. This assumption had affected the other assumption of contact mechanism. In reality, however, soil particles have varied shapes and smaller sizes and their contact mechanism is much more complicated. In addition, most soils contain different amounts of water, an affect that was not modeled in Tailings-DEM™.

It is thought that the main difference between the computational and the experimental results is the approximation to 2D modeling when writing the program. Also, binder constituents were not modeled in Tailings-DEM™, but were largely modeled through the UCS that was derived statistically [28]. In addition, the platen speed in the experimental tests was not modeled in the computational program.

Since Mont Wright tailings were predominantly sandy, it is thought that the program can be incorporated to model other types of aggregates, such as sands and gravely sands, with a particle size distribution close to Mont Wright tailings.

8. Conclusion

The DEM tool developed (Tailings-DEM™) is capable of simulating, with reasonable accuracy, the UCS strength values of Mont Wright tailings matrices. However, more physical and chemical characteristics, such as the binders' constituents and the loading platen speed need to be modeled to make the simulation more accurate.

It is thought that Tailings-DEM™ can be incorporated to model other types of aggregates with similar particle sizes, such as sands and gravely sands.

Acknowledgment
The authors would like to acknowledge with gratitude the support that Dr. Michelle Nokken and Dr. Catherine Mulligan of Concordia University (Montreal, Canada) have provided for this study.

9. References

[1] Simieritsch T Obad J and Dyer S 2009 Tailings plan review- an assessment of oil sands company submissions for compliance with ERCB Directive 074: tailings performance criteria and requirements for oil sands mining schemes Pembina Institute Box 7558 Drayton Valley Alberta T7A 1S7.

[2] Robinsky E 1978 Tailings disposal by the thickened discharge method for improved economy and environmental control Tailings Disposal Toda **2**, pp. 75-92 1979 by Miller Freeman Publications Inc. San Francisco California *In* Proceedings of the 2nd International Tailings Symposium May 1978 Denver Colorado.

[3] Robinsky E I 1999 Tailing dam failures need not be disasters-The Thickened Tailings Disposal (TTD) system CIM Magazine **92**(1028) 140-142.

[4] Wasayo S 2003 Suitability of mine tailings for shotcrete as a ground support. M.A.Sc. thesis Mining Engineering Dalhousie University Halifax Nova Scotia.
[5] Garand P, Vezina S and Bocking K 2000 Effects of flocculent on deposition of tailings sludge for upstream raise of impoundment dykes *In* Proceedings of the 53rd Canadian Geotechnical Conference Canadian Geotechnical Society 15-18 October 2000 Montreal Quebec pp. 633-640.
[6] Benzaazoua M, Belem T and Bussiere B 2002 Chemical factors that influence the performance of mine sulphidic paste backfill Cement and Concrete Research **32**(7) 1133-1144.
[7] Theriault J A, Frostiak J and Welch D 2003 Surface disposal of paste tailings at the Bulyanhulu gold mine Tanzania *In* Proceedings of Sudbury 2003 Mining and the Environment 25-28 May 2003 Sudbury Ontario.
[8] Demers B and Haile G 2003 Management of tailings stabilized by lime and cement at Canadian electrolytic zinc Valleyfield Quebec *In* Proceedings of Sudbury 2003 Mining and the Environment 25-28 May Sudbury Ontario.
[9] Zou D H and Sahito W 2004 Suitability of mine tailings for shotcrete as a ground support Canadian Journal of Civil Engineering **31**(4) 632–636.
[10] Celik O, Elbeyli I Y and Piskin S 2006 Utilization of gold tailings as an additive in Portland cement Waste Management and Research **24**(3) 215-224.
[11] Roy S, Adhikari G R and Gupta R N 2007 Use of gold mill tailings in making bricks: a feasibility study Waste Management and Research **25**(5) 475–482.
[12] Swami R K, Pundhir N K S and Mathur S 2007 Kimberlite tailings a road construction material Transportation Research Record: Journal of the Transportation Research Board **2**(1989) 131–134 Transportation Research Board of the National Academies Washington D.C.
[13] Fall M, Benzaazoua M and Ouellet S 2005 Experimental characterization of the influence of tailings fineness and density on the quality of cemented paste backfill Minerals Engineering **18**(1) 41–44.
[14] Fall M, Benzaazoua M and Saa E G 2008 Mix proportioning of underground cemented tailings backfill Tunnelling and Underground Space Technology **23**(1) 80–90.
[15] Benzaazoua M, Bussiere B, Demers I, Aubertin M, Fried E and Blier A 2008 Integrated mine tailings management by combining environmental desulphurization and cemented paste backfill: Application to mine Doyon,Quebec, Canada Minerals Engineering **21**(4) 330–340.
[16] Ercikdi B, Kesimal A, Cihangir F, Deveci H and Alp I 2009 Cemented paste backfill of sulphide-rich tailings: importance of binder type and dosage Cement and Concrete Composites **31**(4) 268–274.
[17] Fall M and Pokharel M 2010 Coupled effects of sulphate and temperature on the strength development of cemented tailings backfills: Portland cement-paste backfill Cement and Concrete Composites **32**(10) 819–828.
[18] Ercikdi B, Cihangir F, Kesimal A, Deveci H and Alp I 2010 Effect of natural pozzolans as mineral admixture on the performance of cemented-paste backfill of sulphide-rich tailings Waste Management and Research **28**(5) 430–435.
[19] Helinski M, Fahey M and Fourie A 2010 Coupled two-dimensional finite element modeling of mine backfilling with cemented tailings. Canadian Geotechnical Journal **47**(11) 1187-1200.
[20] Yang H, Chen C, Pan L, Lu H, Sun H and Hu X 2009 Preparation of double-layer glass-ceramic/ceramic tile from bauxite tailings and red mud Journal of the European Ceramic Society **29**(10) 1887–1894.
[21] Conner J 1990 Chemical fixation and solidification of hazardous wastes Van Nostrand Reinhold New York.
[22] ASTM C 989 2006 Standard specification for ground granulated blast-furnace slag for use in concrete and mortars Annual Book of ASTM Standards American Society for Testing and Materials Philadelphia Pennsylvania Vol. 04.02.

[23] Lafarge North America 2007 12950 Worldgate Dr. Ste. 500 Herndon Virginia 20170.
[24] NovaFrit International 2006 1200 Garnier Street Ville Ste-Catherine Quebec J0L 1E0.
[25] Haile G, O'Callaghan W B, Hartney T F and Cameron W 2000 Experience with thickened red mud tailings (paste) stacking tropical and temperate climates and applicability of the technology to other tailings *In* Proceedings of the 53rd Canadian Geotechnical Conference Canadian Geotechnical Society 15-18 October 2000.Montreal Quebec.
[26] Das Braja M 2000 Fundamentals of geotechnical engineering Brooks/Cole Publishing Company California.
[27] ASTM D 4842-90 1996 Standard test method for determining the resistance of solid wastes to freezing and thawing Annual Book of ASTM Standards American Society for Testing and Materials Philadelphia Pennsylvania Vol. 11.04.
[28] Mahmood Ali A 2012 Experimental and computational assessment of tailings binder matrices for construction purposes in cold regions Ph. D. thesis Concordia University Department of Building Civil and Environmental Engineering Montreal Quebec.

Influence of Rainfall Characteritics on Total Suspended Solid Concentration

N A Zuraini , N Alias, N Abd Rahman, S Harun Z Ibrahim; S Azman; and M Jumain

Department of Hydraulics and Hydrology, Faculty of Civil Engineering, Universiti Teknologi Malaysia, 81310 UTM Skudai, Johor, Malaysia

E-mail: noraliani@utm.my

Abstract. Urban stormwater quality has a significant impact to receiving water. Therefore, it is important to determine the quality of the urban stormwater from the university area as stormwater runoff will be transported to the receiving water. The objective of this study is to investigate the influence of rainfall characteristics on Total Suspended Solids concentration of stormwater runoff from the university area. The samplings of stormwater runoff and rainfall measurement were carried out in a small catchment in Faculty of Civil Engineering, Universiti Teknologi Malaysia Skudai, Johor, Malaysia. Twenty five samples from two storm events were analysed. The results show that the pattern of rainfall plays an important role of TSS concentration. First flush phenomenon was also observed from both events.

1. Introduction

Urbanisation process includes catchment development has significant effects on the quality of urban stormwater runoff (1). Urbanisation process changes the natural vegetation and physical characteristic of the catchment, thus reduced the infiltration capacity and increases the runoff velocity. During rainfall events, soil particles and debris from industrial and street, commercial, and residential areas are dissolved by rainfall–runoff and transported into any nearby water body, resulting in water pollution. Understanding storm-water runoff quality is required to develop effective urban storm-water runoff management. Urban stormwater quality is greatly influenced by point and non-point source pollution.

NPS pollution is influenced by many factors such as rainfall characteristics (rainfall intensity, rainfall duration, and rainfall depth), land use, number and distribution of dry days, and street cleaning (2). Total suspended solid measurement is an important water quality parameter for managing storm-water runoff because of its impact on receiving waters, sediment load computation, and potential application as an indicator of overall water quality parameters. Therefore, the objective of this study is to investigate the influence of rainfall characteristics on TSS concentration.

2. Materials and methods

2.1. Study Area

Stormwater samples were collected from M50, Faculty of Civil Engineering, an area in University Teknologi Malaysia, Johor Bahru, Malaysia. The catchment is 3-ha with 1150 m long drain. The slope

Content from this work may be used under the terms of the Creative Commons Attribution 3.0 licence. Any further distribution of this work must maintain attribution to the author(s) and the title of the work, journal citation and DOI.
Published under licence by IOP Publishing Ltd

ranges between 1.2% and 9%. The catchment drainage system separates stormwater system from the sewer system.

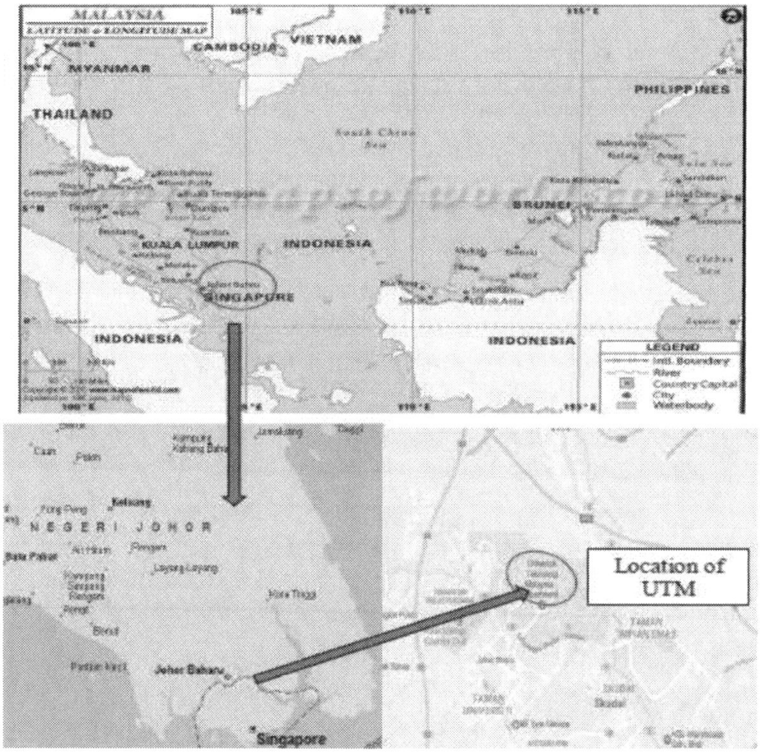

Figure 1. Location of the study site

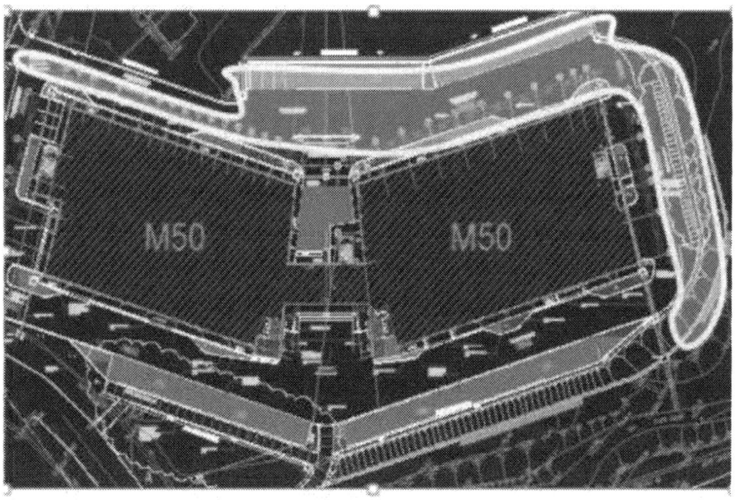

Figure 2. Catchment Area

2.2. Flow Measurement and Stormwater Sampling
Stormwater were collected manually using 1L polyethylene bottles. For every sample taken, the sampling time and water level were recorded. The bottles were labelled with date, sample number and time. The samples were collected from the beginning of storm water runoff flowing out from the drain

IOP Conf. Series: Journal of Physics: Conf. Series **1049** (2018) 012039 doi:10.1088/1742-6596/1049/1/012039

to the completion of 10 to 20 runoff samples, or the end of the storm event, which ever came first. Data was obtained from two observed rainfall events.

Samples were collected every two minutes, depending on the intensity of the storm. After the sampling process, the samples were brought to the Environmental Laboratory at Faculty of Civil Engineering, Universiti Teknologi Malaysia immediately for water quality analysis. The laboratory analysis was based on Standard Methods for the Examination of Water and Wastewater (3). In this study, only Total Suspended Solids (TSS) was the only water quality parameter analysed. This is due to TSS is related to the other pollutants as heavy metals, hydrocarbons and nutrients tend to adsorb to particulate surfaces (4, 5, 6). TSS wasthe dry-weight of particles trapped by a 0.45- m filter and does not include dissolved solids in the urban stormwater. TSS were measured by filtration, drying at 103–105 'C and weighing (7).

Figure 3. Catchment outlet

2.3 Rainfall Measurement
Tipping bucket rain gauge (ISCO) with volume resolution 0.01 inch/tip rainfall was used continuously to measure the rainfall depth. The rain gauge was set up on a levelled platform on a roof at M50 building to provide sufficient exposure with minimize obstruction. The gauge is connected to the central data acquisition system (data logger).

3. Results and Discussion

3.1 Characteristics of rainfall events
Two rainfall events were captured in this study. The first event with the duration of 46 minutes while the second event was 27 minutes duration. Table 1 presents the characteristics of rainfall obtained during the study period. It can be observed that antecedent dry period ranged from 13.5hour to 48.37hour, while the rainfall depths ranged between 12 mm and 12.72 mm.

Table 1. Rainfall characteristic

Event	Date	Storm duration (min)	Total Rainfall Depth (mm)	Rainfall Intensity (mm/hr)	Antecedent Dry Period (hr)
1	22/5/2017	46	12	16.59	13.5
2	24/5/2017	27	12.72	15.65	48.37

3.2 Pollutograph, hyetograph and hydrograph

Figures 4 and 6 present the hyetographs for Event 1 and Event 2 while Figure 7 and 7 shows the hydrograph and pollutograph of Event 1 and Event 2. It is observed that the rainfall depths range between 0.75 mm and 1.75 mm for Event 1 and 0.5 mm to 8.5 mm for Event 2. Event 1 has higher rainfall depth at the beginning of the event while Event 2 has higher rainfall depth in the middle of the event.

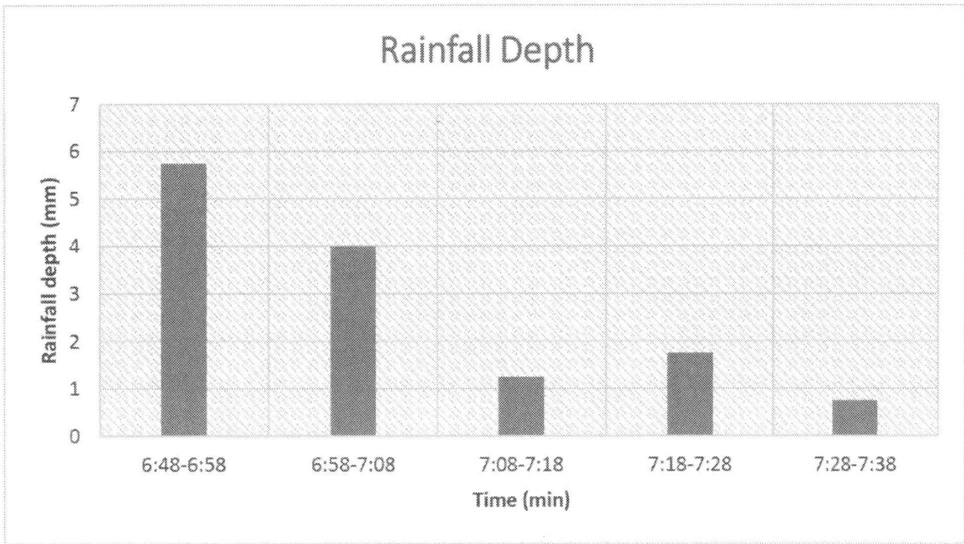

Figure 4. Rainfall Depth for Event 1

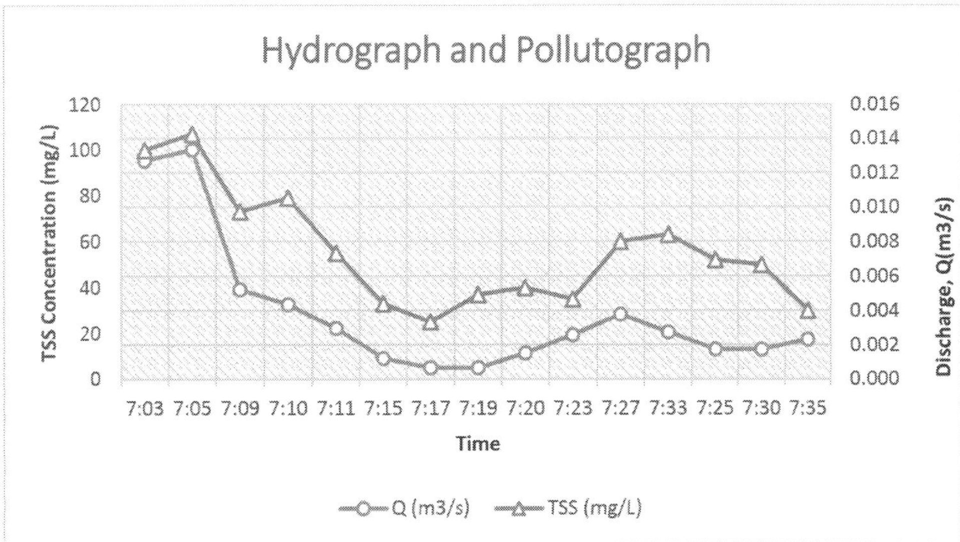

Figure 5. Pollutographs for TSS concentration and hydrographs for Event 1

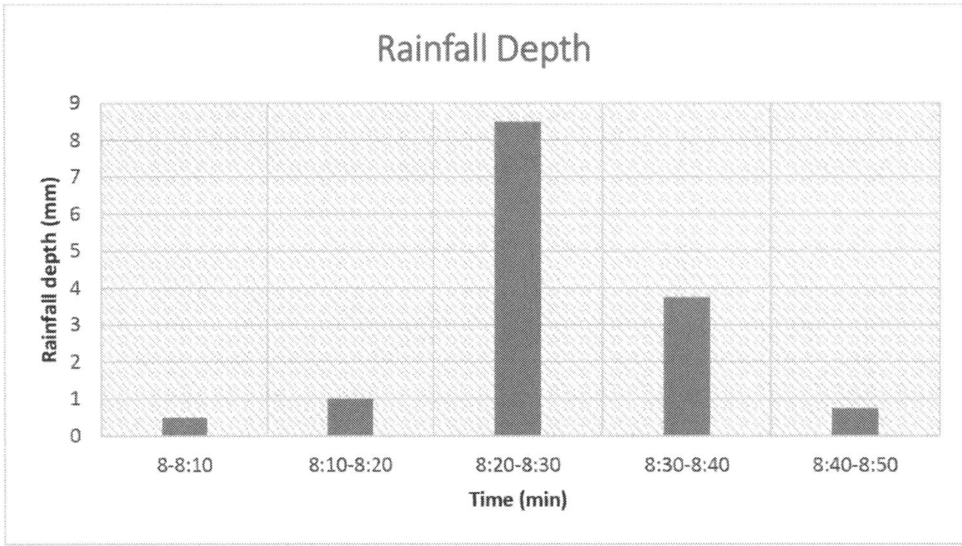

Figure 6. Rainfall Depth for Event 2

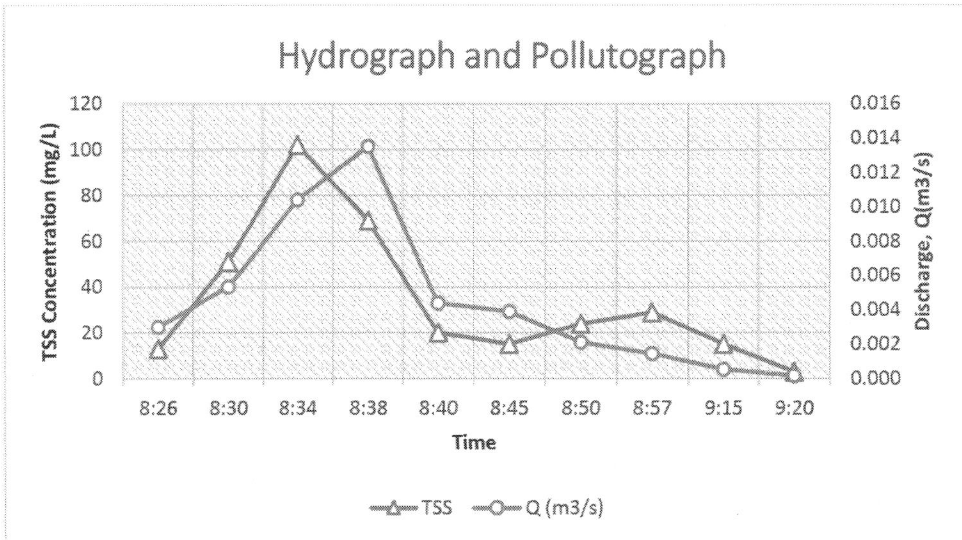

Figure 7. Pollutographs for TSS concentration and hydrographs for Event 2

As can be observed in Figures 5 and 7, the pattern of rainfall play an important role of the pollutographs' pattern. Event 1 has high rainfall intensity at the beginning of the event (between 6.48am and 7.05am), therefore, high concentration of TSS was observed in the beginning of the event as the first and second sample's concentration were not significantly different. In contrast, for Event 2, the concentration of TSS for the first sample (10mg/L) had significant different compared to second sample (100mg/L) and this is due to high rainfall intensity in middle of the rainfall event (8.20am-8.30am). However, it is observed that the concentration of TSS for Event 1 was higher compared to the TSS concentration of Event 2 as can be observed in Figures 6 and 8. This suggest that high rainfall intensity in the beginning of the event also will influence the wash-off process as higher TSS concentration was observed for Event

1 compared to Event 2. Rainfall characteristics are among the main influential factors in relation to urban stormwater quality (12, 13,14).

In terms of the pollutograph, it is observed that both events show similar pollutograph pattern, where higher concentration of TSS was observed at the initial part of the event and TSS concentration is decreasing until the end of the event. This suggests the occurrence of first flush phenomenon (8, 9, 10, 11). Therefore, both events proved that the first flush phenomenon occurred during the wash-off process.

4. Conclusion

This paper presents the influence of rainfall characteristics on TSS concentration. It was found that the pattern of rainfall characteristics influence the pattern of pollutograph. High rainfall intensity in the beginning of the event produces higher TSS concentration in the beginning of the event compared to the rainfall with high rainfall intensity in the middle of the event. First flush phenomenon was also observed for both events. Therefore, this study further prove the influence of rainfall characteristics on TSS concentration. These findings also useful as a basis to improve the design criteria and strategies for controlling stormwater pollution in urban areas, especially in university campus.

References

[1] Li, D, Wan, J, Ma, Y, Wang, Y, Huang, M, and Chen 2015 Stormwater Runoff Pollutant Loading Distributions and Their Correlation With Rainfall and Catchment Characteristics In a Rapidly Industrialized City. *PloS one, 10*(3), e0118776.

[2] Yoon, S W, Chung, S W, Oh, D G, and Lee 2010 Monitoring of non-point source pollutants load from a mixed forest land use. *Journal of Environmental Sciences, 22(6)*, 801-805.

[3] APHA, AWWA and WEF 1995 Standard Methods for the Examination of Water and Wastewater, 19th edn., American Public Health Association, Washington, *DC.Sze S M 1969 Physics of Semiconductor Devices* (New York: Wiley–Interscience)

[4] Ball J. E., Wojcik A. and Tilley J., (2000). Stormwater quality from road surfaces:monitoring of the Hume Highway at South Strathfield University of New South Wales.

[5] Dong A., G.Chester and G.V.Simsiman. (1983). Metal composition of soil, sediments, and urban dust and dirt samples from the Menomonee river Watershed, Wisconsin, U.S.A. *Water, Air and Soil Pollution, 22*, 257-275.

[6] Hengren L., Goonetilleke A. and Ayoko G. A. (2005). Understanding heavy metal and suspended solids relationship in urban stormwater using simulated rainfall. *Journal of Environmental Management, 76*, 149-158.

[7] Lee, J, Bang, K, Ketchum, L, Choe, J, Yu and M 2002 First Flush Analysis of Urban Storm Runoff. *Science of the Total Environment, 293(1),* pp 163-175.

[8] Sansalone J. J. and Cristina C. M. (2004). First Flush Concepts for Suspended and Dissolved Solids in Small Impervious Watersheds. *Journal of Environmental Engineering, 130*(11), 1301-1314.

[9] Thornton R. C. and Saul A. J. (1986). Some quality characteristics of combined sewer flows. *Public Health Engineer, 14*, 35-38.

[10] Acharyaa, A; Thomas C. Piechotaa, T.C and Acharya, K (2010). Characterization of First Flush Phenomenon in an Urban Stormwater Runoff: a Case study of Flamingo Tropicana Watershed in Las Vegas Valley *World Environmental and Water Resources Congress;* pp3366- 3376.

[11] Kang, J H, Kayhanian, M, Stenstrom and M K 2006 Implications of a Kinematic Wave Model For First Flush Treatment Design. *Water research, 40*(20), pp 3820-3830.

[12] Liu, A, Egodawatta, P, Guan, Y and Goonetilleke 2013. Influence of rainfall and catchment characteristics on urban stormwater quality. *Science of the total environment, 444*, pp 255-262.

[13] Schiff, K C, Tiefenthaler, L L, Bay, S M and Greenstein, D J 2016. Effects of Rainfall Intensity and Duration on the First Flush from Parking Lots. *Water, 8*(8), p 320.

[14] Nazahiyah, R, Yusop, Z, and Abustan, I 2007. Stormwater quality and pollution loading from an urban residential catchment in Johor, Malaysia. *Water science and technology, 56*(7), pp 1-9.

Characterization of Leachate at Simpang Renggam Landfill Site, Johor Malaysia

L W Mohamad Zailani[1], **N S M Zin**[1],

[1] Department of Water and Environmental Engineering, Faculty of Civil and Environmental Engineering, Universiti Tun Hussein Onn Malaysia, Malaysia
Corresponding author: nursya@uthm.edu.my

Abstract. Nowadays, the world facing a major problem in managed solid waste due to the increasing of solid waste. Malaysia, one of the country also involves in this matter which is 296 landfills are open to overcome this problem. Currently, the best alternative option to manage solid waste is by using landfilling method because it's have low costing advantages. The negatives side of landfill method, its might cause a pollution by producing leachate that will give an effect to the ground and surface water resources. This study focuses on analyzed the leachate composition at Simpang Renggam Landfill(SRL) site for seven parameter included COD, BOD, SS, turbidity, pH, BOD_5/COD, and ammonia (NH_3-N). All the data obtained were compared with previous researcher and MEQA (1974). From the result, SRL site was categorized as partially stabilized leachate with the parameter of $BOD_5/COD > 0.1$. SRL site are recommended to use a physical-chemical method for a better treatment because the leachate composition is classified as old leachate and aerated lagoon method are not satisfied to be use in treating the old leachate at SRL site.

1. Introduction

around the world become a major problem due to the large amount production of solid waste and might give a significant effect to manage solid waste. Furthermore, the situation become worse when the production of solid waste exceeded the limitation which is the factor causes from rapid growth of population, industrial field and economic activities [1]. The increasing of solid waste quantity gives negative effect to the environmental which can cause pollution to air, soil, and water [2].

Solid waste generates from different source such as industrial, institution, municipal service, construction and demolition and [3]. From the source stated, there a few methods are popularly used as an option to manage solid waste disposal such as sanitary landfill, gridding and discharge to sewer, open dump, incineration, compaction, milling, dumping, reduction and anaerobic digestion and composting [4]. Among these method, landfilling become popular method and have been applied for a long time [5]. Other than that, it has beneficial in lower cost, economic and using simple method to dispose the solid waste. However, using landfilling method, the production of leachate can cause a pollution in terms where the sediment releases the pollutant (leachate), the groundwater resources highly risk to be polluted and affected the water quality [6].

Content from this work may be used under the terms of the Creative Commons Attribution 3.0 licence. Any further distribution of this work must maintain attribution to the author(s) and the title of the work, journal citation and DOI.
Published under licence by IOP Publishing Ltd

Landfill leachate can be defined when the liquid (water from precipitation) absorb through or run out on surface of waste deposits at landfill site area [7]. Leachate contains high substances of organic matter and inorganic matter, consist of chemical substances, heavy metal such as ammonia, sodium, calcium, sulphur, iron, nickel, cadmium, lead and other [8]. Leachate have different characteristic due to the factor such as age of landfill, temperature, moisture content, oxygen content and solid waste composition [9]. From the factor stated, the leachate characteristic can be identified and categorized as young leachate or old leachate.

Many treatments can be used to treat leachate such as biological, chemical and physical treatment. Mostly young leachate has high BOD concentration and usually biological treatment are used to treat the leachate because it has efficiency in removal of high BOD concentration [10]. Some researcher stated that by using chemical-physical treatment along biogical treatment for old leachate increased the treatment efficiency. This treatment commonly been applied because its ability in removing difficult compound and non-biodegradable such as heavy metal, PCBs, AOXs, fulvic acid, humic acid from leachate [11].

In this study, seven parameter such as COD, BOD, SS, turbidity, pH, BOD5/COD, and ammonia (NH3-N) were measure and analysed from Simpang Renggam landfill site. The analysed data will be compare with Laws of Malaysia Environmental Quality Act MEQA 1974 and previous researcher data. The aim of this study is to provide a latest information about leachate characteristic from Simpang Renggam Landfill site, Johor. Furthermore, from this study, other researchers can use this information for better understanding in leachate characteristic and proposed new idea or solution to improve efficiency of leachate treatment.

2. Methodology

2.1 Location of Site

The location of SRL site is located at latitude 10 53'41.64 "22'34.68 North and 1030" East "in Kluang district, Johore, Malaysia. SRL site area cover around 6 hectare and aerated lagoon method are used as a treatment process for leachate treatment. SRL site now has been operation more than 12 years old and the total solid waste received by SRL site is around 250 tonnes per/day which is cover three areas namely, Simpang Renggam, Batu Pahat and Kluang. Furthermore, the SRL site had produces the leachate. The Fig 1 show site.

Fig 1: Simpang Renggam Landfill site

2.2 Leachate Sampling

A raw leachate samples were collected from SRL site by using HDPE ('high density polyethylene") container and all analyses and procedure of the experimental were conducted accordance to American Public Health Association (APHA 2008) Standard Method. Seven samples were collected and analysed during the period from January to April 2017. The sample were collected every two-week interval within four-month period. The analyses data of SRL leachate characteristic were compared with previous researcher and MEQA (1974). In this research, the leachate samples from SRL site were characterize into seven parameter representative landfill leachate included chemical oxygen demand (COD), Biological oxygen demand (BOD), BOD5/COD, suspended solid (SS), pH, ammonia (NH3 - N), and turbidity. These parameters were carried out at Environmental Laboratory, Faculty of Civil and Environmental Engineering (FKAAS), University Tun Hussein Onn Malaysia

3. Result and Discussion

Table 1 shows, the characteristics of landfill leachate versus the age of landfill [14] and Table 2 shows the characteristic of leachate samples from Simpang Renggam Landfill (SRL) site.

Table 1: Characteristics of landfill leachate versus the age of landfill [14].

Types of leachate	Young	Intermediate	Old
Age	<5	5-10	>10
pH	<6.5	6.5-7.5	>7.5
COD (mg/L)	>10,000	4000-10,000	<4000
BOD_5/ COD	0.5-1.0	0.1-0.5	<0.1
Organic Compounds	80% volatile fatty acid (VFA)	5%-30% VFA + Humic and fulvic acid	Humic and fulvic acid
Ammonia nitrogen (mg/L)	<400	N. A	>400
TOC/COD	<0.3	0.3-0.5	>0.5
Kjeldahl nitrogen (g/L)	0.1-0.2	N. A	N. A
Heavy metal (mg/L)	Low to medium	Low	Low
Biodegradability	Important	Medium	Low

Table 2: Characteristic of Leachate Samples from Simpang Renggam Landfill (SRL) Site

Parameter	Range	Average	MEQA (1974
COD (mg/L)	1836- 2150	1993	400
BOD_5 (mg/l)	110-230	170	20
SS (mg/L)	78-582	330	50
Turbidity (NTU)	172-190	181	-
pH	7.9- 8.3	8.1	6.0-9.0
Ammonia, (mg/L)	692-1272	982	5
BOD_5 / COD	170/1993	0.085	-

3.1 General Characteristic

3.1.1 Chemical Oxygen Demand, COD

The COD value were obtained from SRL site varied at average 1993 mg/L. According to Zawawi [12] in year 2013, recorded value of COD from SRL sites was at range 13166 – 13500 mg/L. Meanwhile, according to Fatihah [13] in year 2015, the value obtain at range 9839-15680 mg/L . Its show that the value of COD from year 2013 to 2017 decreasing due the age of landfill >10 years [14]. In this study, value of current COD value obtained were considerable low and categorized as old leachate because its characteristic such as age of the landfill more than 10 years as the limitation COD value of leachate was within <4000 [14]. While comparing with MEQA (1974), value COD obtained are exceeded the permissible limit 400mg/L for leachate discharge. Thus, proper treatment need to be consider for leachate discharge to avoid any pollution accour

3.1.2 Biochemical Oxygen Demand, BOD

The range BOD value obtained from this study was 170 mg/L. By comparing to the previuos data of BOD were recorded by Fatihah [13] in year 2015, average value was high at range 3183 mg/L. According to Tchobanoglous [8] its stated that for mature landfills, the BOD value are varies at range 100-200 mg/L. As the value BOD obtained from SRL are in range 100 -200 mg/l, the landfill are considered as mature landfills. However, The BOD value are greater that allowbble limit standrad by MEQA(1974), which is exceed 50mg/L and a better treatment need to be consider to reduced the BOD concentration

3.1.3 Suspended Solid, SS

From previous study, according to Zawawi [12] in year 2013, the range was 270-1200 mg/L. While in year 2015, previous data recorded by Fatihah[13] the SS value increases at range 1200-1240 mg/L. In 2017, value of suspended solid obtained at range 78-582 mg/L. Based on that, value of SS are decreasing from year 2013 to 2017. However, by comparing with MEQA (1974) , the value of suspended solid obtained are greater than allowabble limit which is exceed than 50mg/.

3.1.4 pH

Ph value obtained from SRL site varied at range 7.9 – 8.3. The pH value increasing due to the factor of age of the landafill and production of gas at landfill site [8]. Present data was recorded in year 2013 by Zawawi, ph value obtained was 8.31- 8.47 [12]. Leachate that has pH < 7.5 are categerizod as old leachate [14]. Leachate from SRL site are categorized as old leachate b referring to table 1. By comparing with MEQA (1974) the pH value obtained are acceptable in range 6.0- 9.0 by MEQA (1974), so leachate can be discharge without futher pH adjustment .

3.1.5 Turbidity

From previous recorded data by Zawawi in year 2013 [12], turbidity value was at range 2200-2300 NTU. Compared to the turbidity value obtained from SRL site in year 2017 are deacreasing at range 172-190 NTU

3.1.6 Ammonia

From the data obtained from SRL site show that ammonia concentaration between 692-1272 mg/L. Its summarized by Bikash [15], ammonia concentration at range 200 to 1000 mg/L have no adverse effect on anaerobic process. Thus, biological treatment is not advisable for SRL leachate From the previous data recorded by Zawawi [12] show that the higher concentaration ammonia between 755 to 2670 mg/L in 2013. However, ammonia concentration decrease due to the lack of organic matter in SRL site leachate which is ammonia generally produced from organic matter as stated by Bikash[15]. By comparing with MEQA 1974, ammonia value are higher than acceptable limit (5mmg/L). Ammnonia concentration at SRL leachate can be reduced in many method such as physical-chemical treatment to achive standard limit discharge by MEQA (1974).

3.1.7 BOD_5/COD

Generally, biodegradability of leachate used to ditermined the age of landfill through the BOD_5/COD ratio [15]. For SRL site, the biodegradabilty of SRL leachate are low with BOD_5/COD ratio 0.085. From BOD_5/COD ratio, the leachate are categerizod as stabilized old leachate with BOD_5/COD ratio less than <0.1 [14]. Its can be prove when the concentration of BOD and COD decrease throughout the increasing of age of landfill [15]. With biodegrabdabily characterizes (BOD_5/COD ratio) , its can be taken as best alternative for ditermined degree of bilogical and chemical decomposition that are taken place in the SRL site.

4. Conclusion

Leachate characteristic of Simpang Renggam Landfill site were analyzed and examined. For this study, seven parameters were analyzed on its composition. Generally, among these parameters such as BOD, COD show that the age of the landfill has a significant effect on the composition of leachate characteristic of SRL site. As a result, the composition of the leachate characteristic from year 2013 are change due the several factors. It can be summarized that the leachate from SRL site have contain low concentration of organic and inorganic constituent and based on biodegradability (BOD_5/COD ratio). Simpang Renggam Landfill site are categorized as stabilized old leachate. Upgrade treatment method need to be consider for leachate treatment due to the change of leachate composition.

5. Acknowledgement

This work was supported by Universiti Tun Hussein Onn Malaysia under grant scheme FRGS (Vot 1570) and GPPS (U794). The authors also wish to acknowledge Majlis Perbandaran Kluang for their assistance during the sampling process.

6. References

[1] Pervez Alam,Kafeel Ahmade, "Impact Of Solid Waste On Health And The Environment," Special Issue of International Journal of Sustainable Development and Green Economics (IJSDGE), Vols. : 2315-4721, V-2, I-1, 2,, 2013.

[2] D.Moeler, "Environmental Health(3rded.)," Cambridge, MA.Harvard University Press, 2005.

[3] Aziz, H. A., M.A.Yusoff, Aziz, S. Q., M.Umar & M.J.K.Bashir, "A leachate quality at Pulau Burung, Kuala Sepetang and Kulim landfills-A comparative study," Conference AWAM'09,Kuala Lumpur, Malaysia. USM., 2010.

[4] Zin, Nur Shaylinda Mohd, "Characterization of Leachate At Matang Landfill," Academic Journal of Science, vol. 1(2):317–322, 2012.

[5] Christopher O. Akinbile and Mohd S.Yusoff., "Environmental Impact of Leachate Pollution on Groundwater Supplies in Akure, Nigeria," International Journal of Environmental Science and Development,, Vols. Vol.2, No.1, 2011.

[6] A. Ikem, O. Osibanjo, M.K.C. Sridhar, and A. Sobande, "Evaluation of Groundwater Quality Characteristics near Two waste Sites in Ibadan and Lagos, Nigeria," Water, Air, and Soil Poll,Kluwer Academic Publishers, the Netherlands. , vol. 140: 307–333. , 2012.

[7] Healy, M.G., Morrison, L., Brennan, R.B.,Hynes, S., Norton, D. & Clifford, E., "Management of landfill leachate," The Legacy of European Union Directives Waste Management, vol. 55: 355–363. , 2016.

[8] Tchobanoglous, G, Hilary, T and Samuel, V., "Integrated solid waste management, Engineering Principles and Mangement Issues. McGraw-Hill INC.," p. 1993.

[9] Macbean, E.A., Rovers, F.A. Dan Farquhar, G.J., "Solid Waste Landfill Engineering and Design," 1st Ed, New Jersey: Prentice Hall Tr. Inc., 1995.

[10] Uygur, A. Dan Kargi, F., "Biological Nutrient Removal from Pre-Treatment Landfill Leachate in A Sequencing Batch Reactor," Journal Of Enviromental Managemen. , pp. Vol.71(1),9-12004.

[11] Abdulhussain A. Abbas, Guo Jingsong, Liu Zhi Ping,Pan Ying Ya and Wisaam S. Al-Rekabi , "Review on Landfill Leachate Treatment," American Journal of Applied Sciences, Vols. 672-684, 2009

[12 Zawawi Daud, Nur Fatihah Muhamad Hanafi, Halizah Awang, "Optimization of COD and Colour Removal From Landfill Leachate by Electro-Fenton Method, " Australian Journal of Basic and Applied Sciences,Vols. ISSN 1991-8178, pp. 263-268, 2013

[13] N. F. B. M. Hanafi, Landfill Leachate Treatment By Combination Of Electro Fenton And Sequencing Batch Reactor Method, Faculty of Civil & Enviromental Engineering,Universiti Tun Hussein Onn Malaysia, 2015, p. 80

[14] Li, W, Zhou, Q and Hua, T.," Removal of organic matterFrom landfill leachate by advanced oxidation process: Review.," International Journal of Chemical Enginering Vol .2010., p.2010

[15] Bikash Adhikari, Sanjay Nath Khanal, "Qualitative Study of Landfill Leachate from Different Ages of Landfill Sites of Various Countries Including Nepal," IOSR Journal of Environmental Science, Toxicology and Food Technology,Vols. Volume 9

[16] Renou, S, Givaudan, J.G, Poulain, S, Dirassouyan, F and Moulin, P., "Landfill leachate treatment: Review and opportunity.," Journal Hazardous Materials, pp.Vol.150 (3), pp.468-493., 2008.

IOP Conf. Series: Journal of Physics: Conf. Series **1049** (2018) 012041 doi:10.1088/1742-6596/1049/1/012041

Value Creating Activity from Aluminium Recycling on Metal Scrap Collector Site: Malaysia Case Study

Umol Syamsyul Bin Rakiman[1], Raja Zuraidah Binti Raja Mohd Rasi[2], Aryani Ahmad Latiffi[3], Muhammad Saeed Shahbaz[4]

[1] Department of Production and Operation Management, Faculty of Technology Management and Business, Universiti Tun Hussein Onn Malaysia (UTHM), 86400 Batu Pahat, Johor, Malaysia, avid8787@yahoo.com.
[2] Department of Production and Operation Management, Faculty of Technology Management and Business, Universiti Tun Hussein Onn Malaysia (UTHM), 86400 Batu Pahat, Johor, Malaysia, rzuraida.uthm.edu.my
[3] Department of Construction Management, Faculty of Technology Management and Business, Universiti Tun Hussein Onn Malaysia (UTHM), 86400 Batu Pahat, Johor, Malaysia, aryani.uthm.edu.my
[4] Department of Production and Operation Management, Faculty of Technology Management and Business, Universiti Tun Hussein Onn Malaysia (UTHM), 86400 Batu Pahat, Johor, Malaysia.

Email: msaeed.shahbaz@gmail.com

Abstract. This research is on the value creating activities from aluminium recycling on the metal scrap collector site. Less is known on how value is created from recycling. Value creation is important in order to generate profit. Metal scrap collector is one of the main component in the recycling chain which creates value for recycling. Understanding the value creation part of metal scrap collector will help to understand the value creation from recycling as metal scrap collector is one of the actors in the recycling industry. A case study was conducted in order to gather the data for this research. An indepth interview session was done with Managing Director of the metal scrap collector firm. A few value creating activities were identified consisting of two categories which is primary and supporting activities. Eventhough the separation between primary and supporting activities is in line with the value chain analysis by Porter. But the activities in the primary and supporting section differs slightly in a few section. This indicates that the value creation activities in the metal scrap collector differs from the original value chain analysis by Porter. Further research on other actors in the recycling value chain is needed in order to test the homogenousity throughout the recycling calue chain.

1. INTRODUCTION

Aluminium is the second most processed metal in the world after iron [1]. One of the primary aluminium production process is the Bayer process based on the extraction of bauxite. Aluminium is then obtained from alumina through electrolysis, a highly energy intensive process [2]. Nevertheless, unique from other materials, aluminium presents the advantage to be recycled indefinitely without any loss of its physical

Content from this work may be used under the terms of the Creative Commons Attribution 3.0 licence. Any further distribution of this work must maintain attribution to the author(s) and the title of the work, journal citation and DOI.
Published under licence by IOP Publishing Ltd

properties [3]. For this reason, it is estimated that 75 % of the total aluminium ever produced is still in use today and that the recycling of aluminium reduces emissions by over 90 % compared to primary production [4]. This advantage makes aluminium recycling more cost friendly than the production of aluminium from bauxite.

According to the report by [5], aluminium stock in the developed country is in rise from daily and industrial usage and in need to be recycled. With the high demand and cumulating amount of aluminium around the world, aluminium is important and valuable in the recycling industry. Some nations with high developing rate quickly adopt recycling as a business. China is the country with highest aluminium recycling rate being followed by Europe and Japan [6]. The high recycling rate of aluminium in many nations indicated aluminium is a promising business venture and being competitive is the way to remain on top of the game. This means for a developing country, it needs to prepare a good recycling approach to recycle aluminium.

The issue of recycling had become a major issue in many developing countries in the world. It is facing greater challenges in order to keep all of the waste in proper management. Aluminium recycling is one of the major components of recycling. It is not clear whether first recycling existed as a mean to save the environment or as a business. But after years of the recycling campaign, a chain of recycle has developed that serves as recycling and business opportunity at the same time.

Most of the recycling campaign mainly deals with the environmental benefits of recycling regardless of the materials. As an example, [7] handles the issue of recycling by manufacturer by putting emphasis on environmental effects. Environment is the main emphasise in recycling. But usually the value part of recycling in term of creation and monetary still lag. The creation of value and monetary perspective of recycling is important as we want to strive to the future and preserving the nature, often quoted as sustainability.

Aluminium recycling has become one of the green technologies which help to increase economy while at the same time save the environment. Aluminium recycling has never been standing as a single entity. It composed of many parties such as collectors, dismantler, metal merchants, shredders and many more [8]. These components of aluminium recycling created a chain. This chain creates value for the final product which is the recycled aluminium. It has to be evaluated in order to understand the value that can be created from the recycling chain [9].

In the case of aluminium, the main source of the aluminium recycling is transportation, building, engineering, packaging and others [10]. The industry which contributes towards the highest number of the aluminium waste consistently from 2007 until 2013 is transportation industry, followed by building industry and subsequently packaging, engineering and others industry [8,11-17].

According to [18,19], the main source of aluminium waste is the transportation industry not only in Europe but for the rest of the world. However this figure might differ from country to country based on the level and type of industry's activities. As an example even though Malaysia has its own transportation industry, the utilisation of aluminium in vehicle is not as extensive as in Europe. For the countries without an automobile industry or does not utilise it in their transportation manufacturing, the construction industry is probably the largest markets of aluminium, consuming some two and nine million tonnes of aluminium products per year in Europe and the world respectively [18,21]. Research by [21] has confirmed construction as a source of aluminium for recycling in Malaysia.

Once again, the level of aluminium waste may vary considerably from country to country due to the level and type of industry activities, which implies that the scrap generated at the end of life of the building will also be different. The total stored aluminium in the industry of building is the largest since the beginning of the industrial use of aluminium, amounting to nearly 170 Metric tons worldwide [18,20]. The main use of aluminium in this industry is to provide materials for roofing and cladding, and window and door frames, as well as small applications such as shutters, door handles, ceiling partitions [21].

Before the value part is investigated, the actors and activity which they performed should be identified and understood. Hence this research focuses on the waste or scrap of aluminium on the metal scrap collector site as metal scrap collector is one of the actors in the aluminium recycling industry. The main objective of this research is to trace what are the value creating activities on the metal scrap collector site.

IOP Conf. Series: Journal of Physics: Conf. Series **1049** (2018) 012041 doi:10.1088/1742-6596/1049/1/012041

2. LITERATURE REVIEW

Aluminium recycling is the process by which aluminium waste or scrap can be reused in products after its initial production. The process partly involves re-melting the metal, which is far less expensive and energy intensive than creating new aluminium through the electrolysis of alumina, which must first be mined from bauxite ore and then refined [2]. Aluminium recycling has a number of key environmental and economic benefits. Compared to other high volume materials, aluminium production has one of the largest energy differences between primary and secondary production: 186 MJ/kg for primary compared to 10-20 MJ/kg for secondary [22]. The International Aluminium Institute has came out with their ideal of well-developed aluminium recycling structure as shown in Fig. 1 as follows:

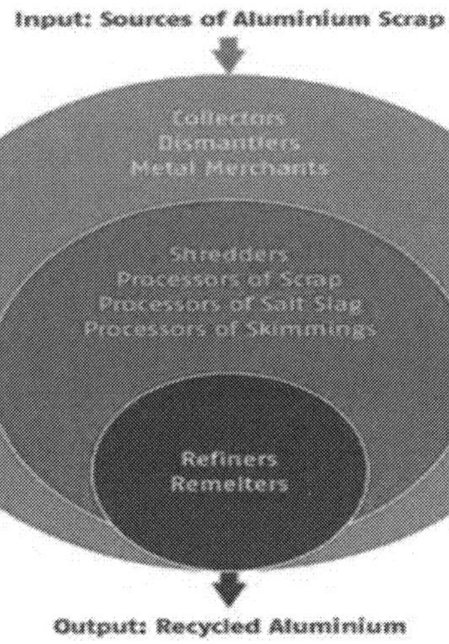

Fig. 1: Structure of a Well Developed Aluminium Recycling Industry [12]

Fig. 1 shows the required processes and parties in the aluminium recycling industry. As in Fig. 1, after the last process of recycling, how finally the recycled aluminium reaches the manufacturer again to be processed as a product is still unknown. Initially also, the aluminium product does not simply thrown away as a waste then someone collect it. The collecting phase might involves a few party especially if the aluminium is not a waste instead is a scrap in a manufacturing company such as a car factory or a construction site which uses aluminium as part of its input. All of these is important as a way to understand the value being created through the aluminium recycling process. Due to many benefits of recycling, it is a worthy area to be explored and understood. The value of the aluminium recycling industry should never be underestimated. The next section will be on the methodology on how this research is conducted.

3. METHODOLOGY

This study employs a case study method to identify the value creating activity of aluminium recycling on a metal scrap collector site. An in depth interview session was conducted with possible respondent in charge on site. The respondent was selected based on the use of aluminium and recycling activity in their facilities. This method is chosen due to the explorative nature of this study [22] since the activity might differ depending on country and main aluminium contributing industry in any particular country. The research begins by identifying metal scrap collector site that utilises aluminium as one of material being recycled. After that, it is validated whether they perform any aluminium recycling activity on site. This is done through observation and the interview itself. Person in charge and the value creating activity from the aluminium recycling activity being performed is the main concern in the interview. The next section is on the results of this research.

IOP Conf. Series: Journal of Physics: Conf. Series **1049** (2018) 012041 doi:10.1088/1742-6596/1049/1/012041

4. RESULTS

This case study was conducted with a metal scrap collector company. It is a metal recycling company operating in Johor. This company covers materials for recycling around Malaysia and oversea such as Singapore. The interview session was conducted with the founder of the company. Respondent holds two positions in the company which are the Chief Executive Officer (CEO) and the Managing Director. Aluminium, copper and magnesium are among the metals being recycled in this firm. Respondents has 20 years of experience in metal scrap business.

As the CEO of the company, the respondent founded the company and manages most of the recycling activities of the metal scraps. Roles of the respondent are to ensure the firm is appropriately organised and staffed and to have the authority to hire and terminate staff as necessary. The respondent also ensures expenditures of the firm are within the authorised annual budget. The respondent assesses the principal risks of the firm and to ensure that these risks are being monitored and managed. The respondent also has to ensure the firm has appropriate systems to enable it to conduct its activities both lawfully and ethically. The respondent also finds the supplier and buyer for the aluminium which the company has processed.

In this case study, the respondent has confirmed one of the material being purchased is aluminium. Based on the documents of the materials being purchased and processed; and researcher observation, aluminium is among the main materials in the compound of the firm. This firm processes aluminium to be further recycled into high purity aluminium. Supplier of the aluminium scrap are from construction site, manufacturing firm and petroleum processing company. This conformed this case study as an actor in the aluminium recycling industry. Table 1 is the description of the case study. This is an excerpt of the interview with the respondent.

"Before this aluminium scrap mostly came from the manufacturing firm and petroleum processing company. As we are going more green, even building use aluminium. So we have new source of aluminium scrap from the construction site" (Respondent).

Table 1: Description of Case Study

Type of firm	Metal scrap collector
Location	Johor, Malaysia
Position of respondent	Chief Executive Officer (CEO) and the Managing Director
Other identified actors	Construction site Manufacturing firm
Materials	Magnesium, copper and aluminium
Working experience	20 year experience in metal scrap business
Roles and Responsibilities	a. **Founder,** b. **Manager,** c. **Budgeting,** d. **Human resource management,** e. **Finds supplier and** f. **Buyer for the aluminium scrap**

Value creation can be understood by exploring the activities that impact the value of the intended material. It usually consists more than one activity to create value for a product. In the case of aluminium scrap, the activities that impact the value creation is still unclear. The aim of this part is to investigate and explore activities that impact value of the aluminium scrap being sold. This case study recycles aluminium scrap among other material such magnesium and copper. The aim of this part is to explore the activities that affect the value of the recycled aluminium.

There a few factors that affects the value of the aluminium scrap being sold. The first factor is warehousing. Warehousing is the activity to store the scrap aluminium after it arrives, waiting for processing and after processing. Warehouse is where the unprocessed and compressed aluminium scrap being stored. Two workers were assigned to take care of the warehouse. Compressed aluminium will not be sold directly if the price of the aluminium is down. When the price is down, then compressed aluminium scrap is kept in the warehouse. Once the price is high, then it will be sold. Respondent mentioned that once there was 6 months waiting for the right price to sell the compressed aluminium scrap. Sometimes when the warehouse is full, the respondent will have to rent another place to store the compressed aluminium scrap. When selling the processed aluminium scrap, the cost and selling price will be considered by the respondent.

The next value creating activity is worker. The number of worker affect the cost needed to transport and process the aluminium scrap. The company focuses on optimising the number of worker. Optimise means hiring and maintaining productive and skilful worker. For example, worker that can handle weighing process and then help with the compression process is regarded as productive and skilful. Emphasising on more than one skill and more than one work. During the interview time, respondent does not increase number of workers to process the aluminium scrap. This situation remain the same even if the amount of the aluminium scrap increases. If the amount of aluminium scrap increases, respondent will just increase the processing time. High number of worker is helpful when the amount of aluminium scrap is high. But when the number of aluminium scrap drops, there is no work to do, yet the payment is still on. Thus number of worker is helpful if it is accompanied with skill. Worker without skill and don't know how to do the work is ineffective. That is why the respondent try to maintain the existing worker to maintain production even if the time to process the aluminium scrap is prolonged.

In term of the recycling process, other than compressing machine, the respondent uses shredder machine. Compression machine is used to compress the aluminium scrap after it has been segregated by the worker. Magnet is used to segregate the aluminium scrap more easily. Sometimes the metal scrap is mixed with other metal. By having certain level of magnetism, it can help to attract different metal. Shredder machine functions especially for the colour coated aluminium. It will help to eliminate the colour on any coloured aluminium scrap such as aluminium can. Then the machine will shred and compress the can. After that, it can go directly into the furnace if there is. In this case study, the respondent does not have any furnace.

Transportation is the next factor that affect value. Transportation was used in three ways. The first one is to collect the aluminium scrap from the supplier and take it to the firm. The second one is to carry the processed aluminium scrap within the firm after compression. The last one is to send processed aluminium scrap to the customer after it has been purchased. Lorry and forklift are two types of transportation being used. Lorry is used to carry aluminium scrap into and out of the firm. While forklift is used to carry the aluminium scrap within the firm. Use of transportation requires driver and fuel. Both driver and fuel requires cost. This will ultimately affect the value of the aluminium scrap.

When the respondent has bought aluminium scrap from supplier, he will send workers to collect it. Worker also will be dispatched from the firm to send processed aluminium scrap whenever the customer paid for the item. This will be done upon request from the customer. Service affect the aluminium scrap value in two ways. First, if the supplier calls, then this company will send the lorry to collect the aluminium scrap. Second, sometimes the customer will call back and complain about the sold aluminium scrap. This happens when the item has problem. Then the respondent will send workers to confirm the situation. For example, if the sold aluminium scrap is not as specified, then it will be taken back to the firm. Magnesium physically looks like aluminium in term of the greyish colour. This situation sometimes causes aluminium scrap to be mixed with magnesium scrap in the form of compressed or contaminant in the compressed aluminium scrap. If the customer check the compressed aluminium scrap thoroughly, and magnesium is detected. The whole batch will be double checked. This requires worker to be sent to the customer's facility to do the checking. If the contaminant is found, then it will be brought back to the firm. This process will take time and cost.

The present of contaminant is an important issue when buying or selling the aluminium scrap. For example, if the bought aluminium scrap has a lot of oil. Then the weight will be reduced after certain discussion and consideration. The customer will estimate the weight of the oil and will be subtracted from the total weight. The same process also will be done for other contaminant. The aluminium scrap is the purchased item, not the contaminant. Observation is important in the buying process. Oil, waste, rust, soil, and other contaminant are among the factors to be considered during the buying process.

The cost for production is considered minor in determining the value of the aluminium scrap value. It is more on the volume and cost of purchase of the aluminium scrap. The respondent prefers supplier and customer that want high volume of aluminium scrap since it affects the cost of transportation. If the volume is too low, for example it does not even fill one lorry, it is not favourable. As the cost for the transportation and worker remains the same even if the volume is low. The net profit will be 5% to 10% from the total cost. Basically the issue of quantity and cost is the main consideration in determining the value of the aluminium scrap.

In term of selling, the respondent send the aluminium scrap that they sell to the buyer. Price of the bought aluminium scrap is taken into consideration before it is sold and transported to the customer. Sometimes when fast cash is needed, the respondent will sell the aluminium scrap to the trader which can give instant cash compared to the aluminium smelter. But usually the respondent calls all of the possible buyers to get the best price. Then the respondent sells to the buyer which offers the highest price.

The respondent mentioned on the trouble of making sure that the aluminium scrap that they buy is not a stolen item. This is important since it will cause a delay in processing the items since the interference by the authority like police is unavoidable when this thing occur. The respondent do not allow any entrance without permission. It is risky. The respondent might lessen the production of recycled aluminium due to lack of trusted workers. The price on the market will definitely affect the value but the challenge of keeping the recycled items when the price is high from being stolen is another challenge.

Another challenge is buying wrong materials. For an example, magnesium looks like aluminium. But aluminium is more valuable. If magnesium is mistaken for aluminium, then a loss will happen. The respondent is paying high for cheap item.

In term of improvement, there should be stricter regulation using manufacturing raw materials regardless whether it of primary or secondary source. This is important to control the rate of pollution. Even when recycling is exercised, pollution still take place. For an example, aluminium recycling can produce dross which cannot be further processed and need to be disposed. The government control the purchase and use of raw material from recycled source such as aluminium. The control can be in the form of a quota to control the amount of aluminium or materials used in the manufacturing process as this will cause pollution.

All of the aforementioned factors are considered value creating for the respondent. Some of it is internal factor while some of it is external factor. Table 2 is the summary of the factors that affect the value of the aluminium scrap in this case study together with the annotation from the respondent.

Table 2: Activities that Impact Value of the Recycled Aluminium

Value creating activities
a) Warehouse "When the warehouse is full then we sell it. Sometimes the place is not enough then we rent a place to store the items. The price for renting the place the store also will be considered when we want to sell the items".
b) Worker "Two workers will help to compress the aluminium scrap. One will help to put the scrap from the forklift. The other controls the compression machine".
c) Sale "I will consider a lot of buyer with high prices of course. When we need fast cash, I opt for trader".

d) Transportation "We use lorry to pick the aluminium scrap from the supplier. When we want to send the processed and sold aluminium scrap we also deliver it by lorry".
e) Service "Sometimes the customer will complain that our scrap has magnesium. We have to double check this claim by sending worker to check the claim".
f) Purchase "We send the worker to take the purchased aluminium scrap from Penang".
g) Volume "I really consider big volume of purchase. When you buy small volume, there will be empty space in the lorry. Very wasteful. Yet we still have to pay for the trip".
h) Contaminant "We don't want contaminants like oil to be in the product that we buy or sell. It reduce the selling price".
i) Government "We hope government could increase the use of recycled materials so that we can increase our production. Fortify the policy to recycle and make use of recycled items compulsory".

5. DISCUSSION
Table 2 is the value creating activities on the metal scrap collector site from the aluminium recycling. There are 9 activities taken place. Some of these activities creates value directly while some other affect value by affecting other activities. For example, warehousing affect value creation because the respondent needs to pay for the rent of the warehouse. Same goes with worker, sale, transportation, service and purchase. While other activity such as the present of contaminat, government policy and volume each preceed the activities mentioned above. This is because, if the aluminium is too contaminated, the aluminium scrap will not be purchased since it cannot be recycled any further. Same goes with the government policy and volume. Government policy to recycle will make people recycle and as a result sell the recyclable such as aluminium scrap to be recycled. While volume is also important in affecting value creation since if the volume is too small no this company will be reluctant to purchase the aluminium scrap. In combination, these will hamper the value creation process.

6. CONCLUSION
It can be concluded the combination of the nine activities affect value directly or indirectly. The dynamic of how they work together need more work to be understood fully. Just by collecting the aluminium waste, economic, social and environmental benefit can be achieved. It is not only save the environment but also economically profitable. Our economy can be sustained by recycling aluminium by maintaining our source of aluminium for future generation. Socially, by recycling it can avoid any safety and legal issue such as illegal dumping which is dangerous and can cause havoc to the society. Among many benefits of recycling is less pollution, saves energy and sustaining environment. This action also can ensure a sustainable future on the construction site. Rather than throwing away the aluminium scrap mindlessly and illegally which can be costly if get caught by the authority, we can get some of the money back by selling the scrap of aluminium.

7. FURTHER WORK
Since this research is only on exploring the activities that create value. The dynamic of how actually they work together as value creating factor is still in haze. But, the first step to identify the activities is rather important as it shows how many factors or activities will impact in just one actor from the whole aluminium recycling value chain. More activities from other actors should also be explored to investigate the whole aluminium recycling industry value creation.

8. ACKNOWLEDGEMENT
The authors would like to thank Ministry of Higher Education (MOHE), Research Management Center (RMC), UTHM, for supporting this research under the RACE grant Vote No. 1444, Centre for Graduate Studies (UTHM) and Faculty of Technology Management and Business, UTHM, for supporting this research.

REFERENCES
[1] Billy R. (2012) Material Flow Analysis of Extruded Aluminium in French Buildings Opportunities and Challenges for the Implementation of a Window-to-Window System in France, Norwegian University of Science and Technology.
[2] Schlesinger, M. E. (2014). Aluminum Recycling, Second Edition. Hoboken: Taylor and Francis.
[3] Jones, M. (2014). Sustainable Event Management: A Practical Guide (2nd ed.). Routledge.
[4] Martchek, K. J. (2006). Modelling More Sustainable Aluminium: Case Study. International Journal of LCA, Sustainable Management of Natural Resources - Aluminium, 11, pp. 34-37. Retrieved from http://www.alcoa.com/sustainability/en/pdfs/KMartchek_IJLCA_7772.pdf
[5] UNEP (2010), International Panel for Sustainable Resource Management. (2010). Metal stocks in society. Retrieved from United Nations Environment Programme website: http://www.unep.org/resourcepanel/Portals/24102/PDFs/Metalstocksinsociety.pdf
[6] IAI. (2017). Global Mass Flow Model Global Mass Flow Model - 2015 [2016 draft] (2017) International Aluminium Institute. Retrieved 30th July 2017 http://www.world-aluminium.org/media/filer_public/2017/05/08/2015_2016draft.xlsx
[7] Bülent Başaran, (2013) "What makes manufacturing companies more desirous of recycling?", Management of Environmental Quality: An International Journal, Vol. 24 Iss: 1, pp.107 - 122
[8] EAA. (2007). Activity Report 2007. European Aluminium Association, 1-28.
[9] Porter ME. Competitive advantage: creating and sustaining superior performance. New York: The Free Press; 1985
[10] EAA (2014). SUSTAINABLE DEVELOPMENT INDICATORS FOR THE ALUMINIUM INDUSTRY IN EUROPE 2012 Key Facts and Figures. European Aluminium Association, 1-8.
[11] EAA. (2008) Activity Report 2008. European Aluminium Association, 1-28.
[12] EAA. (2009). Activity Report 2009. European Aluminium Association, 1-30.
[13] EAA. (2010). Activity Report 2010. European Aluminium Association, 1-40.
[14] EAA. (2011). Activity Report 2011. European Aluminium Association, 1-40.
[15] EAA. (2012). Activity Report 2012 Engaging for Sustainable Growth. European Aluminium Association, 1-48.
[16] EAA (2013). 'Agenda for Action: Towards a Competitive and Sustainable Aluminium Industry in Europe'.
[17] IAI. (2017). Global Mass Flow Model Global Mass Flow Model - 2015 [2016 draft] (2017) International Aluminium Institute. Retrieved 30th July 2017 http://www.world-aluminium.org/media/filer_public/2017/05/08/2015_2016draft.xlsx
[18] Rakiman, U.S., Rasi, R.Z.R.M., & Latiffi, A.A. (2014). Value Chain Analysis: The Insight of Aluminium Recycling. Aust. J. Basic & Appl. Sci., 8(15): 377-382, 2014
[19] Luo, Z., & Soria, A. (2008). Prospective Study of the World Aluminium Industry. European Commission Joint Research Centre, 22951 EN, 1-79. Retrieved September 21, 2014, from http://ipts.jrc.ec.europa.eu/publications/pub.cfm?id=1556
[20] EAA. (2013). Activity Report 2013 Contributing to a Better Competitive Europe. European Aluminium Association, 1-36. Retrieved September 21, 2014, from http://www.alueurope.eu/wp-content/uploads/2011/09/EAA-AR-13-v14-REPRINT_medium-res.pdf
[21] Green JAS (2007). Aluminum recycling and processing for energy conservation and sustainability. Materials Park, OH: AS542 mail: drecklies@themanagement.de
[22] Andre, K. (2013.). The adaptation of supply chains to climate change (Unpublished master's thesis). Heriot-Watt University.

IOP Conf. Series: Journal of Physics: Conf. Series **1049** (2018) 012042 doi:10.1088/1742-6596/1049/1/012042

Hardness and wear rate of Al LM6 hollow cylinder fabricated using horizontal centrifugal casting

K Maximus[1], J Saifulnizan[2], A Ismail[2], N Hisyamudin[1], M N Khir[2], and A K Kamarul[2], M K Awang[2]

[1]Faculty of Mechanical Engineering & Manufacturing, Universiti Tun Hussein Onn Malaysia, 86400 Batu Pahat, Malaysia
[2]Crashworthiness and Collisions Research Group, Faculty of Mechanical and Manufacturing Engineering, Universiti Tun Hussein Onn Malaysia, 86400 Batu Pahat, Malaysia

E-mail: hd160094@siswa.uthm.edu.my

Abstract. Horizontal centrifugal casting is one of the methods to produce functionally graded materials (FGMs). The pouring temperature and rotational speed of the mold are among the main parameters affect the quality of casting. The effect of these two casting parameters on the hardness and wear properties of the Al LM6 hollow cylinder fabricated are investigated. The hardness value along the radial direction of the Al LM6 hollow cylinder fabricated is measured using Micro Hardness Vickers test. Meanwhile the wear rate of Al LM6 hollow cylinder fabricated is determined by measuring the mass loss after loaded with 5 N force and sliding distance of 75 m on silicon carbide grit paper (200 μm). The result after the testing shows that the Al LM6 hollow cylinder which is fabricated at pouring temperature 725°C and motor speed of 1500rpm produces the highest hardness sample and the cylinder fabricated at pouring temperature 690°C and motor speed of 1700rpm have the lowest wear rate value.

1. Introduction

Centrifugal casting is one of the methods in material processing where the quality of the final product are affected by the flow pattern of the molten metal [1]. This method also effective in producing a graded composite materials but there are still several process parameters that influence the characteristics of the product such as die wall temperature, pouring temperature, rotational speed of the mold, centrifugal force and many more [2]. The literature study of the fluid flow in centrifugal casting process is still inadequate. Theoretically, a true uniform cylinder should be able to produce when the mold is rotated at low speeds but practically it also depend the conditions of the molten metal where the mold need to be accelerated at a certain speed to obtain a uniform cylinder shape [3]. By using a higher speed, a uniform casting is formed due to the influence of liquid metal. To form a good casting, the molten metal must fill in the entire inner surface and then solidify. To achieve this, the melt must rotate in a high rotational speed at its low viscosity state. When the rotational speed increases, an irregular shape casting will be formed [3]. By means of the centrifugal force, the finer grain structures are produced to attain homogeneous and isotropic mechanical properties [4]. Functionally graded materials are produced through the segregation of particles due to centrifugal force during centrifugal casting process. The segregation occurs either at the outer or the inner periphery of the casting, depending on the difference in densities of the particles in the melt [5].

Content from this work may be used under the terms of the Creative Commons Attribution 3.0 licence. Any further distribution of this work must maintain attribution to the author(s) and the title of the work, journal citation and DOI.
Published under licence by IOP Publishing Ltd

Important variables which affect the rate of cooling of the melt are rotational speed of the mold during solidification process. When the speed of rotation increased, the centrifugal force are increased which will produce a strong convection in the liquid pool and produce a homogenous temperature in the bulk liquid. The fluid display a different flow patterns when the speed of the motor varies [2]. A rapid solidification of the cast shows the well distribution of fine grains while slow solidification rate of the cast shows coarse grain size [6].Solidification process in centrifugal castings is identical with other casting process where the state phenomena are changed in which governed by heat transfer. The optimal representation of solidification behavior in a horizontal centrifugal casting is to assume that crystal growth commences with the liquid metal in contact with the mold wall, and then proceeds right across the section until the last remaining liquid freezes uniformly at the inner surface [5].

Particulate reinforced aluminium alloy composites have shown a significant improvement in tribological properties, including sliding and abrasive wear resistance and seizure resistance. Thus, increased attention has been directed towards particulate metal matrix composites for tribological applications due to the advantages of MMCs such as good sliding wear resistance, high load carrying capacity and light weight. M Kok and K Ozdin [7] used three different Sic abrasive papers which are 600 grit, 360 grit and 240 grit under 2N and 5N loads to investigate the wear resistance of aluminium alloy using pin-on-disc abrasion test apparatus. The result shows that the wear volume loss of the matrix alloy and the composites almost increased linearly with increasing the sliding distance. The reinforcement of the 2024 Al matrix alloy with Al2O3particles significantly improved the abrasion wear resistance of all composites tested against all the abrasives used, and the wear resistance of the composites was much higher than that of the unreinforced 2024 aluminium alloy.G Chirita, D Soares and F S Silva [8] investigate the advantage of mechanical properties using vertical centrifugal casting compared with traditional gravity casting. Aluminium alloy AS12UN used in the study was melted at 670°C and it was poured inside the mold which was preheated at 130°C. The specimen was tested using Dartec tensile testing machine and the results shows that the mechanical properties shows a higher value for centrifugal casting process compared with traditional gravity casting. Besides that, the increment for centrifugal casting samples was 35% for rupture strength, 160% for rupture strain and 18% for Young's Modulus.

In this study, the effect of different casting parameter towards the hardness and wear resistance of the Al LM6 hollow cylinder fabricated is investigated. The Al LM6 is casted at three different pouring temperatures and three different speed of motor rotation. The hardness value and wear rate are measured using micro hardness Vickers test and wear test machine.

2. Methodology

Al LM6 hollow cylinders are fabricated using centrifugal casting method [9, 10, 11] with varying two casting parameters i.e. pouring temperature and motor speed. Three different pouring temperatures and motor speed set are 690, 710 and 725°C and 1300, 1500 and 1700 rpm, respectively.

The process starts by weighing small pieces of Al LM6 ingot. A total of 160 grams of ingot is required for each casting producing a hollow cylinder with volume of $5.5\text{x}10^{-5}$ m^3 as the density of Al LM6 is 2650 $kg.m^{-3}$. Then, the ingot is melted inside a crucible using induction furnace. After melted, the molten Al LM6 is poured inside the spinning mold with the desired motor speed. The mold is keep spinning for two hours to let the molten Al LM6 solidified. The solidified Al LM6 will be removed from the mold and left for 5 hours to let it cool down. Table 1 shows the casting conditions of Al LM6 hollow cylinders fabricated.

Hardness test and wear test are carried out to characterize the hardness and wear behaviour of Al LM6 hollow cylinder fabricated. The hardness test is carried out using Micro Hardness Vickers test and prepared as per ASTM E92-82. The hardness values at five points with distance of 2 mm each along the radial direction (from inner to outer) of the hollow cylinder fabricated are measured with a test load of 100 g (980.7 mN). Figure 1 shows the cross sectional area of the hollow cylinder fabricated, prepared for Micro Hardness Test.

Table 1. Parameter of centrifugal casting of Al LM6 hollow cylinder

Specimen	Material	Parameter	
		Motor Speed (rpm)	Pouring Temperature (°C)
1	AL LM6	1300 (57 G)	690
2			710
3			725
4		1500 (75 G)	690
5			710
6			725
7		1700 (100 G)	690
8			710
9			725

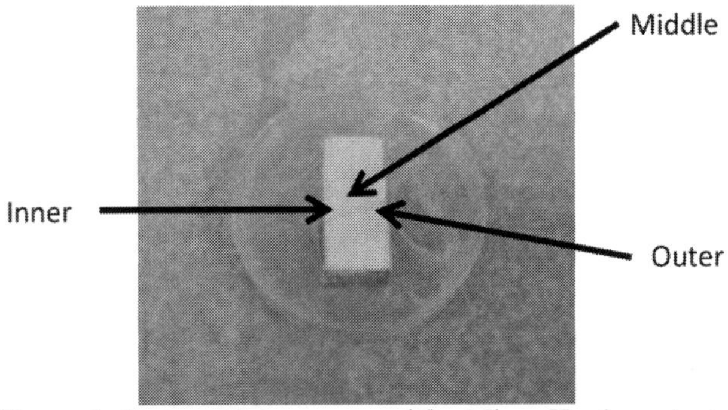

Figure 1. Cast specimen prepared for Micro Hardness test.

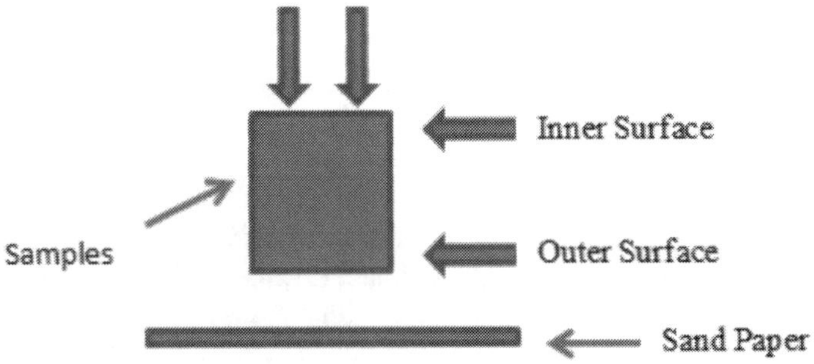

Figure 2. The position of sample during wear testing

The wear test is conducted by using pin and disc method under 5N loads using an abrasive grit size of 200 μm with a sliding distance of 75 m and 50 rpm speed [12]. The specimen size of 6 mm x 6 mm x 17 mm is cut from the hollow cylinder fabricated. The mass of the samples are weighed before and after the wear test and the test are conducted three times to obtain the average value of wear rate. Figure 2 shows the position of the samples for wear test. The wear rate obtained through the calculation using the wear rate test formula which is:

$$W = \frac{w}{\rho x S} \quad (1)$$

where W is the wear rate in mm^3/m, w is the weight loss in gram, ρ is the density of the material in gram/cc and S is the sliding distance in m.

3. Result and Discussion

Figures 3 to 5 shows the hardness values along the radial direction of the hollow cylinder fabricated. From the figures, it found that there are significant different in term of hardness value along the radial direction due to the different of Si content and size. As shown in our previous work [9, 10] the size of Si is the biggest at the inner part and causes the bonding of Al and Si particles becomes low. At a normalized radial distance between 2 to 6, it shows that all of the samples have a decreasing in hardness. This is because the cooling rate during solidification process is non-uniform inside the mold which might affect the distribution of Si particles at the middle part of the sample. Besides that, the inner surface of the samples is low due to the presence of gas porosity. This gas bubbles are thrown at the inner surface because it has low density and by the means of centrifugal force during casting process [10].

Averaging the five hardness value of each specimen shown that Sample 6 which is fabricated with 1500rpm motor speed and 725°C pouring temperature gives the highest value of hardness with the average value of 63.3 Hv while sample 2 that is fabricated with 1300 rpm motor speed and 710°C gives the lowest value with the average value of 55.6 Hv. This is due to the effect of viscosity and cooling rate of the molten metal during solidification process. Viscosity of the material is important parameter in order to move the molten along the mold.

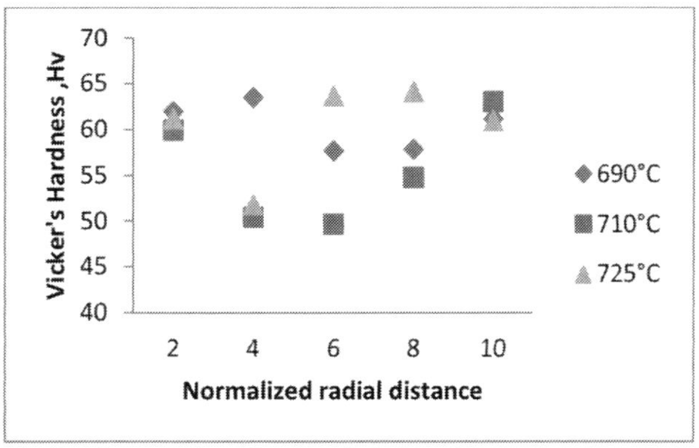

Figure 3. The hardness value along radial direction (motor speed of 1300 rpm).

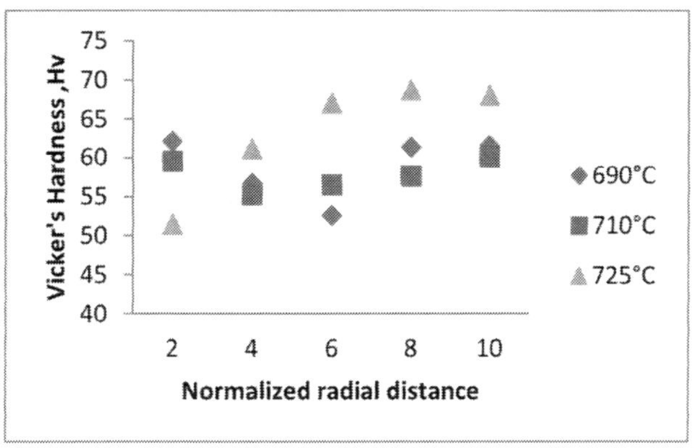

Figure 4. The hardness value along radial direction (motor speed of 1500 rpm).

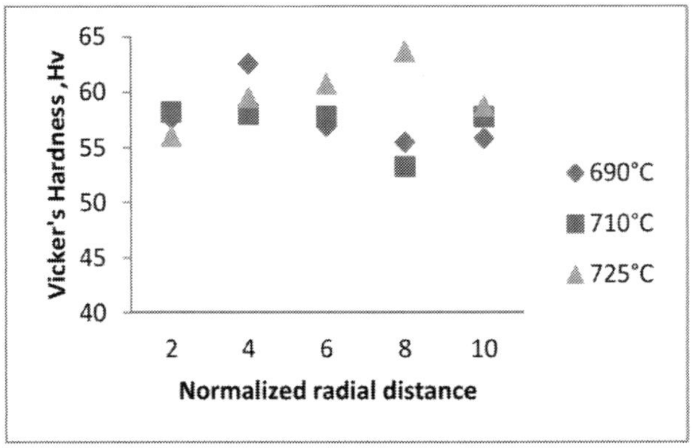

Figure 5. The hardness value along radial direction (motor speed of 1700 rpm).

Bonollo et al [12] states that a low melting temperature with a high viscosity causing the particles to move freely which will gives a better result. However, a suitable pouring temperature and a suitable speed must be obtained in order for the particles of Al and Si to distribute uniformly. A uniform distribution of Al and Si will gives a better hardness result.

Figure 6 below shows the wear rate for all of the hollow cylinders fabricated. From the figure, the wear rate for samples that is fabricated at 710°C increases as the speed of motor rotation increases. For the samples that are fabricated at 690°C, it shows that the wear rate is highest when it is fabricated at 1500rpm but for samples that are fabricated at 725°C the wear rate is lowest when it is fabricated at 1500 rpm. From the figure, it shows that sample 8 has the highest wear rate which is 2.803×10^{-4} mm^3/m while sample 7 that is fabricated with 690°C pouring temperature with 1700rpm motor speed has the lowest wear rate which is 1.605×10^{-4} mm^3/m. Lowest wear rate indicates that the sample has high resistance towards wear while highest wear rate indicates that the sample has low resistance towards wear.

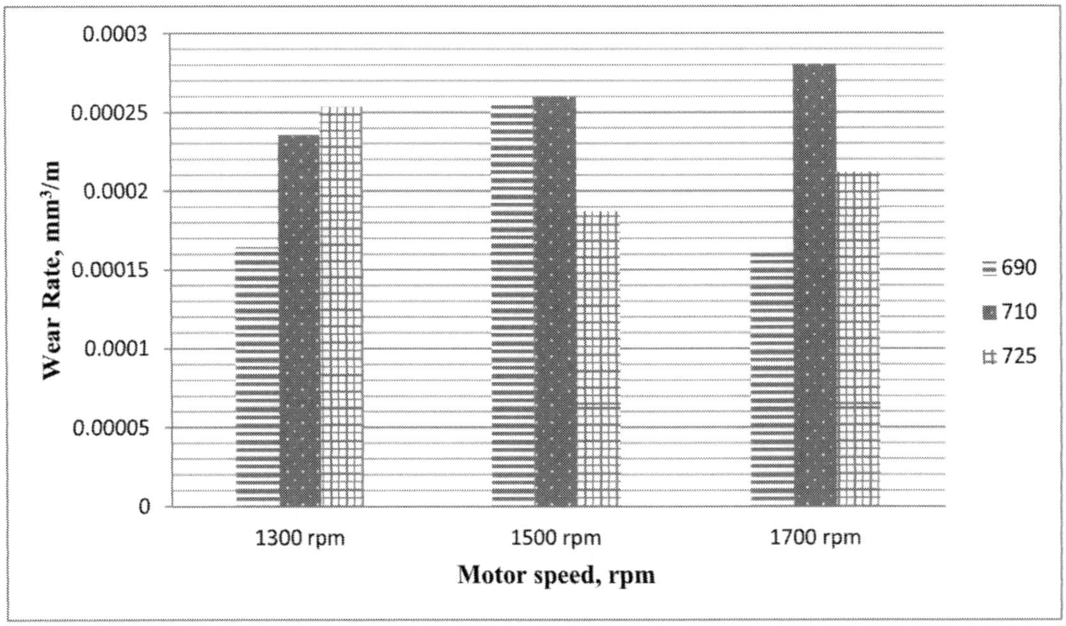

Figure 6. Wear rate for all of the samples.

At optimum speed, the formation of fine grains leads to better mechanical properties and hence improved wear resistance [3]. In Figure 6, it shows that sample 7 have the lowest wear rate hence the pouring temperature of 690°C with motor speed of 1700 rpm is at its optimum parameter. The next lowest wear rate is sample 1 with wear rate of $1.645x10^{-4}$ mm^3/m and pouring temperature of 690°C with motor speed of 1300 rpm followed by sample 6 with wear rate of $1.872x10^{-4} mm^3/m$ and pouring temperature of 690°C with motor speed of 1500 rpm.

The abrasive wear of a material mainly depends on its material characteristics, test parameters, and also environmental conditions. The material characteristics include properties of matrix material and reinforcement phase, hardness and fracture toughness. The major test parameters include applied load, sliding speed, time duration, sliding distance and wear track diameter. The grit size of abrasive particle and orientation of individual grit particles represent the nature of counter surface. It was noted that grit size of abrasive and hardness of material will have a direct influence on nature of wear mechanism and wear rate [13].

4. Conclusion

In conclusion, it can be conclude that this study have achieve its objectives to study the effect of pouring temperature and speed of motor rotation to produce an Al LM6 hollow cylinder by using horizontal centrifugal casting method.

Pouring temperature and the speed of motor rotation during casting are the important parameter to produce the best specimen. From the mechanical testing of hardness, sample 6 gives the best result due to the influence of viscosity and cooling rate which gives the parameter of pouring temperature 725°C and motor speed of 1500rpm produces the highest hardness sample and sample that is fabricated at pouring temperature 690°C with 1700motor speed gives the lowest result in hardness. The result from wear test shows that sample 7 gives the lowest value of wear rate and sample 5 gives the highest value of wear rate because the different pouring temperature and different motor speed influence the nature of wear mechanism and wear rate.

5. Acknowledgement

This research is funded by Universiti Tun Hussein Onn Malaysia (UTHM) under Graduate Research Grant (GPPS) with vot no. U801 and TIER 1 with vot. no U870.

6. References

[1] Shailesh R. A., Mahantesh S. T. And Shrikantha S. R 2014 Effect of Rotational Speeds on the Cast Tube During Vertical Centrifugal Casting Process on Appearance, Microstructure and Hardness Behavior for Al-2Si Alloy The Minerals, Metals & Materials Society and ASM International DOI:10.1007/s11663-014-0257-y.

[2] Madhusudhan, Narendranath S and Kumar G C M 2013 Properties of Centrifugal Casting at Different Rotational Speeds of the Die, International Journal of Emerging Technology and Advanced Engineering **3** 1

[3] Mukunda P G, Shailesh R A and Shrikantha S R 2010 Influence of Rotational Speed of Centrifugal Casting Process on Appearance, Microstructure and Sliding Wear Behaviour of Al-2Si Cast Alloy, Met. Mater. Int. **16** 137-143.

[4] Ping W S, Rong L D, Jie G J, Yun L C, Qing S Y and Zhi F H 2006 Numerical Simulation Of Microstructure Evolution of Ti-6al-4v Alloy in Vertical Centrifugal Casting, Materials Science And Engineering A 426 240–249.

[5] Kanaujiya K and Yugesh Mani Tiwari 2015 Analysis of Cooling Rate of Centrifugal Casting at Different Rotational Speed Based on Experiment, International Journal of Emerging Research in Management &Technology **4** 10

[6] Madhusudhan, Narendranaath S, Mohankumar G C and Mukunda P G 2010 Effect of mould Wall Thickness on Rate of Solidification of Centrifugal Casting, International Journal of Engineering Science and Technology **2(11)** 6092-6096.

IOP Conf. Series: Journal of Physics: Conf. Series **1049** (2018) 012042 doi:10.1088/1742-6596/1049/1/012042

[7] Kok M and Ozdin K 2007 Wear resistance of aluminium alloy and its composites reinforced by Al_2O_3 particles, Journal of Materials Processing Technology **183** 301–309
[8] Chirita G, Soares D and Silva F S 2008 Advantages of the centrifugal casting technique for the production of structural components with Al–Si alloys, Materials and Design **29** 20–27
[9] Maximus K, Saifulnizan J, Al-Emran I, Nik H M N, Mohd Khir M N, Kamarul A K 2017 Effect of Rotating Mold Speed on Microstructure of Al LM6 Hollow Cylinder Fabricated Using Centrifugal Method, Journal of Physics: Conference Series **914** 012039 1-10.
[10] Jamian S, Yutiman S & Hehsan H 2016 Al LM6 Hollow Cylinder Fabricated Using Centrifugal Casting, ARPN Journal of Engineering and Applied Sciences**11** (18) 11221-11225.
[11] Jamian S., Ayob S. N., Abidin R. Z., Muhd. Nur N.H. 2016 Fabrication of functionally graded natural fibre/epoxy cylinder using centrifugal casting method, ARPN Journal of Engineering and Applied Sciences **11**(4) 2327-2331.
[12] Bonollo F, Moret A, Gallo S, and Mus C 2004 Cylinder liners in aluminium matrix composite by centrifugal casting, La metallurgia italiana **6** 49-55
[13] Deuis R L, Subramanian C and Yellup J M 1996 Abrasive Wear of Aluminium Composites-A Review, Wear **201** 132–44

Cadmium and Chromium Accumulation in Cockles along the Estuary of Sungai Tampok and Sungai Sanglang

Ahmad Luqman Halit, Shamila Azman, Mohd Ismid Mohd Said, Noraliani Alias, Nazri Ali.
Faculty of Civil Engineering, Universiti Teknologi Malaysia, Malaysia
shamila@utm.my

ABSTRACT. Sungai Tampok and Sungai Sanglang located in Johor thrives in terms of aquaculture industry. Since most of the villagers are fisherman, the aim of this study is to determine heavy metals on cockles (*Anadara granosa*), and seawater. Cockles were collected from four stations at Sungai Tampok on 26th December 2016, 16th March 2017. Meanwhile, cockles from Sungai Sanglang were collected from three sampling stations on 11th January to 26th March 2017. The samples were then analysed for cadmium (Cd) and chromium (Cr). Water samples were also collected from the same location as cockle sampling stations. Cd and Cr in seawater were above Class 3 based on Marine Water Quality Criteria and Standard showing that the area is unsuitable for marine aquaculture, aquatic life and fishing activities. Accumulation of heavy metals on cockle also exceeded the Malaysian Food Regulation (1985).

Keywords: Cadmium; Chromium, Cockles (*Anadara granosa*).

INTRODUCTION

Cockle also known as *Anadara granosa* is a saltwater clam, a marine bivalve mollusk from the family of CARDIIDAE. Cockle has high economic value as food and it can be kept in aquaculture. Besides being consumed as a protein rich food, they are also used as a bio-indicator for detecting toxicity in aquatic communities (Tu, 2011). Cockles lives in low oxygen environment and ingest more viruses and bacteria unlike other varieties of clams that are safe to eat. Studies has been conducted on the toxicity of heavy metals in the tissue of cockles to ensure that cockles are safe for human consumption (Yap, 2003). Cockle sampling were conducted at Sungai Tampok and Sungai Sanglang which is located in Johor.
Human activities such as plantation, industrial and farming at the vincinity of Sungai Sanglang and Sungai Tampok, had affected water quality at the estuary. These factors have an important role to cause heavy metal accumulation in the area. Besides heavy metal accumulation, the distribution and quantity of the cockles are also affected by human activities along the areas. Polluted river can affect the cockles and spats, which resulted in high mortality (Neville, 2016). Both river areas serve as very important natural resources for the local people. It also helps the local people in term of economics where most of them depend on fishery as their income. The water quality at the area has significantly been affected due development. In 2005, Malaysia produced around 100,000 tonnes of cockles during its peak harvesting for local consumption and export. However, in 2015, the amount of cockles per tonnes produced was reduced to 16,000 tonnes. The harvested cockles in the industry were valued at RM160 millions for 2015. The impact are felt by consumers who have to pay between RM10 and RM 15 per kilogram in 2016 compared to just RM 2 and RM 3 a decade ago.

Heavy metals can be absorbed by living organisms because of their high solubility in the aquatic environments. Once they enter the food chain, large concentrations of heavy metals may accumulate in the human body. The metals that are ingested beyond the permitted concentration can cause serious health disorders (Babel, 2003). There are many ways or paths for the heavy metals to emit to the environment for example from the air during combustion, extraction and processing, to the surface waters (e.g. runoff and releases from storage and transport) and to the soil (groundwaters and crops). The greatest concern in terms of human health is the atmospheric emission. The quantities and widespread disperse along with the potential exposure are often ensues. The exposure does not result only from the presence of a harmful agent in the environment, infact contact is the key word to get the exposure (Berglund, 2001). The contact should be between the agent and the outer boundary of human body such as airways, skins and the mouth.

Cadmium compounds are mainly used in rechargeable nickel cadmium batteries such as our mobile phone. The emission has increase dramatically during the 20th century. Its products are often dumped together with household waste because the products are rarely recycled item. Cadmium is also present as a pollutant in phosphate fertilizers, sewage sludge to farm land and industrial emissions. These may lead to contamination of soils and increased cadmium uptake by crops and vegetables. Low pH will enhance the uptake process of soil cadmium by plants (Jarup, 1998). Cadmium exposure may cause kidney damage and its long term affects in high cadmium exposure may cause skeletal damage.

Chromium can be exposed through breathing, eating or drinking and through skin contact with chromium or chromium compounds. The level of chromium in air and water is generally low. In drinking water, the level of chromium is usually low as well, but contaminated well water may contain chromium (IV). The main human activities that increase the concentrations of chromium (III) are leather and textile manufacturing. The main human activities that increase chromium (VI) concentrations are chemical, leather and textile manufacturing, electroplating and other chromium (VI) applications in the industry. These applications will mainly increase concentrations of chromium in water. Through coal combustion chromium will also end up in air and through waste disposal chromium will end up in soils. Chromium is not known to accumulate in the bodies of fish, but high concentrations of chromium, due to the disposal of metal products in surface waters, can damage the gills of fish that swim near the point of disposal. In animal chromium can cause respiratory

Content from this work may be used under the terms of the Creative Commons Attribution 3.0 licence. Any further distribution of this work must maintain attribution to the author(s) and the title of the work, journal citation and DOI.
Published under licence by IOP Publishing Ltd

problems, a lower ability to fight disease, birth defects, infertility and tumor formation. The objectives of this study is to identify accumulation of cadmium and chromium in cockles and level of pollution along the estuary of Sungai Sanglang, Pontian and Sungai Tampok, Benut, Johor, Malaysia

METHODOLOGY

Samples were collected at 3 different sampling stations for Sungai Sanglang, whereas 4 sampling stations for Sungai Tampok. The cockles were collected manually at low tide from inter-tidal areas. The cockles were then cleaned externally by washing it thoroughly with clean water before being transferred into sampling bags. The collected samples were kept frozen in the refrigerator prior to heavy metal analysis. The samples from every station were grouped into 3 sizes i.e. small, medium and large. The sizing of the samples was based on its size range. Small cockle was around 2 cm and below, while medium and large sizes are around 2-3 cm and more than 4 cm respectively. The samples were prepared before analyzing the heavy metal concentration. The classified samples kept dried in the oven at 60°C for a day until its totally dried. The samples were then grinded. The grinded samples for each size were weighted for 1.0 g and added with 10 mL of HNO_3 diluted with deionized water in the ratio of 1:1. The sample was transferred into a beaker and covered with a watch glass and heated at 95°C for about 10-15 minutes using hot plate without being boiled. The lid of the beaker was kept closed to avoid the steam in the air to evaporate. After that, the sample was cooled and 5 mL of concentrated HNO_3 was added. The sample was refluxed for about 30 minutes with the lid opened and the solution was evaporated to 5 mL without boiling. The sample was then cooled and 2 mL of water and 3 mL of H_2O_2 was added. The lid of the beaker was then closed and warmed on the hot plate until the solution started the peroxide reaction. Samples were then heated until effervescence subsides and cool in the beaker. Then H_2O_2 was continually added in 1 mL with warming until the effervescence is decrease and sample appearance unchanged. However, please ensure the added H_2O_2 was not greater than 10 mL. The solution was added with 5 mL of concentrated HCL and 10 mL of water. The solution in the beaker was covered and returned to hot plate to reflux for 15 minutes without boiling. After cooled, sample was diluted with 100 mL of deionized water using volumetric flask. Atomic Absorption Spectrometer (AAS); Perkin Elmer AAS Pinnacle 900T was used for analysis of heavy metals i.e. chromium (Cr) and cadmium (Cd).

RESULTS AND DISCUSSION

Cadmium

Figure 1.0 shows the concentration of Cd in water for Sungai Tampuk and Sungai Sanglang. Sungai Sanglang has similar concentration for both sampling occasions except for S1which was slightly higher on 26th December 2016. Water samples at Sungai Tampuk have higher concentration on 11 Jan 2017 compared to 26 March 2017. The highest Cd concentration was recorded at S1-Sg Tampuk at 0.24 mg/L. Based on Malaysia Marine Water Quality Standards (MWQS), Sungai Tampok and Sungai Sanglang are classified as polluted with cadmium because the concentration recorded was more than 0.01 mg/L at Class 3.

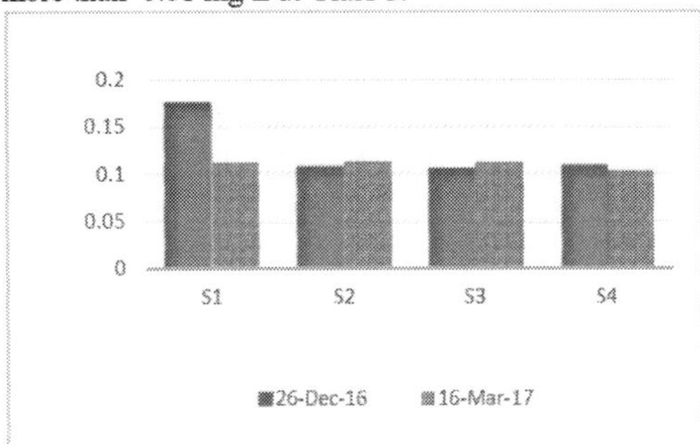

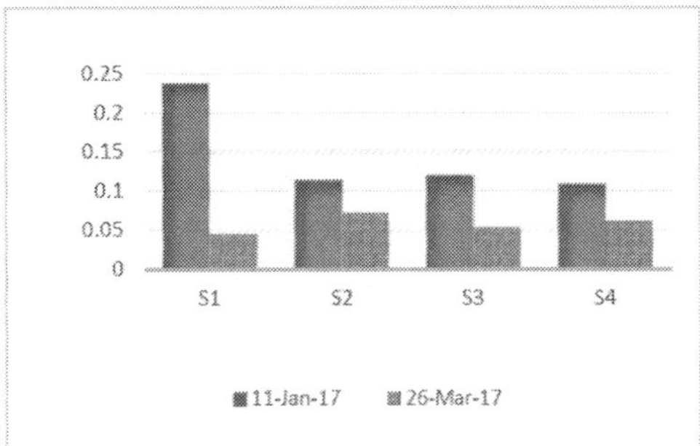

(a) **(b)**

Figure 1.0: Cadmium in water samples for (a) Sungai Tampok and (b) Sungai Sanglang

Sungai Tampok recorded concentration of Cd in cockles in the range of 0 to 4.28 µg/g as shown in Figure 2.0. The highest concentration of Cd was 4.28 µg/g at S4-Sg Tampuk for medium size. However Cd concentration at Sungai Sanglang was in the range of 0-3.85 µg/g. Based on Malaysian Food Regulation (1985), the limited concentration of Cd in seafood is 1.0 µg/g. It shows that most of the sampling stations for both Sungai Tampok and Sungai Sanglang exceeded the concentration limit. Cockles does not need cadmium in their growth, thus it means the amount of cadmium could be toxic to cockles and is not suitable for food sources.

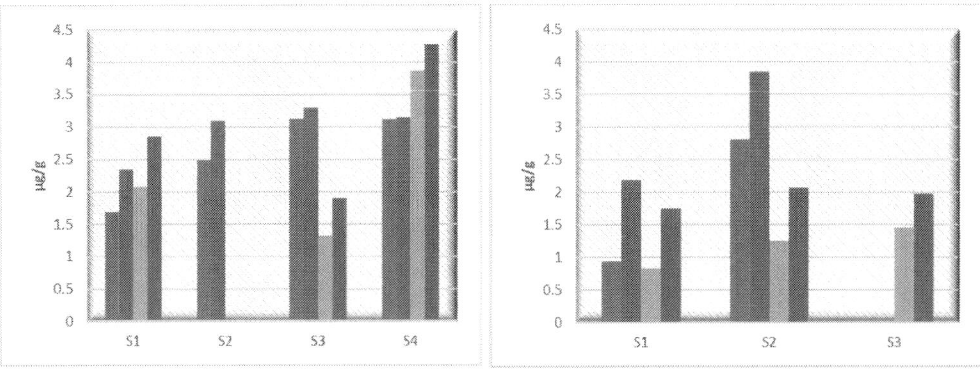

(a) **(b)**

Figure 2.0: Cadmium in cockles for (a) Sungai Tampok and (b) Sungai Sanglang

Chromium

Figure 3.0 shows the concentration of Cr in water with the highest concentration was recorded at sampling S2-Sg Tampuk on 16 march 2017. Meanwhile, concentration of chromium at Sungai Sanglang was highest at 1.177 mg/L at S1-Sg Sanglang. Based on MWQS, Sungai Tampok and Sungai Sanglang are considered polluted with chromium because the concentration was higher compared to class 3 at 0.048 mg/L.

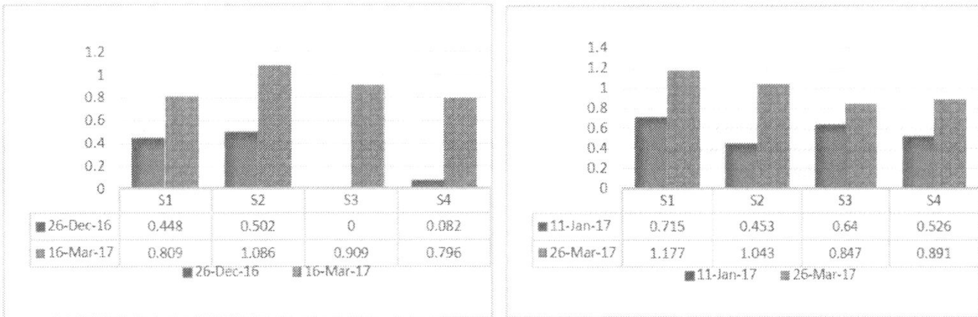

(a) **(b)**

Figure 3.0: Chromium in water samples for (a) Sungai Tampok and (b) Sungai Sanglang

The highest concentration of Cr in cockles was recorded at 98.3 μg/g at S2-Sg Tampuk in medium sizes (Figure 4.0). Cr concentration at Sungai Sanglang was in range of 32.34-89.30 μg/g. Malaysian Food Act 1983 and Regulation 1985 and WHO/FAO 2004 did not set any limit for Cr in bivalves. However, IAEA – 407 reported that a permitted level for Cr in the marine organism is 0.73 mg/kg.

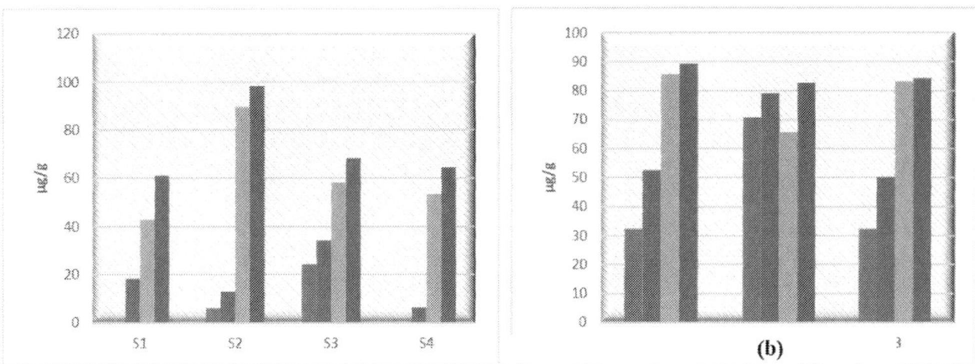

(b)

Figure 4.0: Chromium in cockles for (a) Sungai Tampok and (b) Sungai Sanglang

CONCLUSION

The concentration of Cd and Cr in seawater were above Class 3 based on Marine Water Quality Criteria and Standard showing that the area is suitable for ports, oil & gas field and unsuitable for marine aquaculture, aquatic life and fishing activities. Accumulation of heavy metals on cockle also exceeded the Malaysian Food Regulation (1985) for Cd. It shows that the area is polluted and is unsuitable for aquaculture and fishing. Sungai Sanglang has a recreational park near the estuary. Human attraction location may lead to increase of human density, thus the amount of littering is higher. Land-based activities such as development and agriculture increased might decrease the water quality levels.

REFERENCES

Annual States Fisheries Statistics. 1984 Kuala Lumpur, Malaysia, Department of Fisheries, Mm. of Agriculture.

Babel, S., Kurniawan, T.A., (2003). Various treatment technologies to remove arsenic and mercury from contaminated groundwater: an overview. In: Proceedings of the First International Symposium on Southeast Asian Water Environment, Bangkok, Thailand, 24 25 October, pp. 433–440.

Berglund M, Elinder CG, Järup L. (2001) Humans Exposure Assessment. An Introduction. WHO/SDE/OEH/01.3, 2001.

Jarup L, Berglund M, Elinder CG, Nordberg G, Vahter M., (1998). Health effects of cadmium exposure—a review of the literature and a risk estimate. Scand J Work Environment Health 1998; 24 (Suppl 1): 1–51

Neville,S.(2016, May 17). Cockles Trade Threatened. The Star. Retrieved from http://www.thestar.com.my/news/nation/2016/05/17/cockle-trade-threatened-production-down-to-16000 tonnes-from-100000-tonnes/

Tu N. P. C., Ha N. N., Agusa T., Ikemoto T., Tuyen B. C., Tanabe S., Takeuchi I., (2011). Trace elements in Anadara spp. (Mollusca: Bivalva) collected along the coast of Vietnam, with emphasis on regional differences and human health risk assessment. Fisheries Science 77(6):1033–1043

Yap C. K., Ismail A., Tan S. G., 2003 Heavy metal (Cd, Cu, Pb and Zn) concentrations in the green-lipped mussel Perna viridis (Linnaeus) collected from some wild and aquaculture sites in the west coast of Peninsular Malaysia. Food Chemistry 84:569-575.

Zhao, S., Feng, C., Quan, W., Chen, X., Niu, J., and Shen, Z. (2012). Role of living environments in the accumulation characteristics of heavy metals in fishes and crabs in the Yangtze River Estuary, China. Marine Pollution Bulletin, 64(6): 1163–1171.

IOP Conf. Series: Journal of Physics: Conf. Series **1049** (2018) 012044 doi:10.1088/1742-6596/1049/1/012044

3D Numerical Model of Soil-Tunnel Interaction Induced by Segment Joint Parameter

S N Jusoh[1], **H. Mohamad**[2], **A. Marto**[3], **R.A. Abdullah**[1], **N.M. Yunus**[1]
[1]Department of Geotechnics and Transportation,Universiti Teknologi Malaysia, Johor Bahru, Johor
[2]Department of Civil and Environmental Engineering, Universiti Teknologi PETRONAS, Bandar Seri Iskandar, Perak
[3]Department Environmental Engineering and Green Technology, Malaysia-Japan International Institute of Technology (MJIIT), Universiti Teknologi Malaysia, Kuala Lumpur

E-mail: snorafida@utm.my

Abstract. In a shallow tunnel, tunnel construction will triggers a chain of ground movements, resulting in settlements at the ground surface, which become more significant with the decrease in tunnel depth. Therefore, by considering behaviour of the segment tunnel joints (can affect the integrity of tunnel in both circumference and longitudinal directions), the surface settlement trough were investigated. Initial model of dual segmental tunnel lining was model and validated with laboratory test to gain nonlinear response of segment joint. Then, a soil-tunnel simulation model was developed with various segment joint parametric models for Singapore Circle Line Stage 3 (C852) project. Results showed that with the use of flexible segment joints that allow movement (hinge-nonlinear model) in staggered ring tunnel model, the ground deformation depicted higher resemblance of surface settlement pattern to the field data. Steady state settlement was observed after 10m of cutter head distance.

1. Introduction

Soil-tunnel interaction is uncertainty phenomenon. Tunnel excavation causes relaxation of in-situ stress, which is only partially restricted by the insertion of the tunnel support. In fact, it is not possible to create a void instantaneously and provide an infinitely stiff lining to fill it exactly. Hence, a certain amount of the deformation of the ground will take place at the tunnel depth, which triggers a chain of movements, resulting in settlements at the ground surface, which become more significant with the decrease in tunnel depth. In mean time, non-uniform ground pressures and joint interaction induced tunnel lining movement (*i.e.,* bending moment). Lots of researchers agreed that joints in a tunnel lining evidently affect the tunnel behaviour [1-7]. Therefore, segmental joint needs to be less stiff and shall possess more deformation than the main portion [4]. Yanzhi *et al.* [8] reported that the stiffness of the tunnel joints is comprehensively influenced by many factors such as; 1) thrust force exerted during tunnel advancement, 2) the remaining longitudinal force and 3) property of the packer material, etc. presenting a non-linear characteristic. Hence, Xiaochun *et al.* [9] specifically mentioned that the most influence factor impudence the bending moment is the rotational joint stiffness, K_θ. It is crucial to zoom into rotational stiffness as previous researchers have stated that each segment joint treated to have a unique value for the rotational stiffness but in realistic in verily change non-linearly [10].

[1] snorafida@utm.my

Content from this work may be used under the terms of the Creative Commons Attribution 3.0 licence. Any further distribution of this work must maintain attribution to the author(s) and the title of the work, journal citation and DOI.
Published under licence by IOP Publishing Ltd

Recent approach has led researchers to conduct more detailed investigations on the impacts of joint rotational to the global behaviour of the tunnel lining. Blom [10] presented a visualization of three stages behaviour of segment joints for tangential bending moments as a function of the rotation. Respective researcher proposed a direct closed solution by introducing reduction factor, ζ (EI_{eq} = $\zeta.EI_{full,hom}$) that considers the global influence of segment joint and is valid for single ring consideration with segment joint is in stage of closed joints. This solution is also accounted for non-linear segment joint conditions. Despite the proposed solutions are fit for linear and multi-linear models, however, Blom's solution is only ideal for single ring problems. In addition, the model of the solution was only based on the soil as radial spring and was derived only for the case of full slip soil-lining interfaces.

Therefore, it is important to carry out detail research on how much joint in lining affect the stress redistribution in the medium around the tunnel. This paper will discuss the development of segment joint model nonlinearly and tunnel model in order to improve soil-tunnel interaction. Nonlinear joint model and staggered ring tunnel model was simulated in the soil block to help investigate the global behaviour of ground. This is crucial because in-depth understanding of joint behaviour will contribute to an improved understanding of tunnel lining as a whole.

2. Simulation Model

Simulation model has been divided into two main phases; i) model development of dual-segment model (parametric analysis of joint model), ii) soil-tunnel model development.

2.1 Model Development

has been divided into two main phases; i) model development of dual-segment model (parametric analysis of joint model), ii) soil-tunnel model development. Dual-segment model consist of half segment tunnel lining jointed with parametric segment joint (hinge, spring and tie constrains) and verified with laboratory work [11, 12]. A series of dual segment model itself in early had been validated with empirical solution and single tunnel lining model, respectively. In soil-tunnel model, two types of segment's joint has been developed and embedded in a full tunnel lining together with soil tunnel model. Simulation model was developed based on the constructed tunnel profile of Singapore Circle Line Stage 3 (C852) project at Serangoon Interchange Station. Non-linear segment's joint model was adopted from Jusoh [11,12] by means hinge non-linear model and was compared to tie model (to represent continuous ring condition). Respective tunnel of 6.35 m outer diameter with 1.4 m width and 0.275 m thick segmental linings were modelled as a full ring and assigned with ring's joint and segment's joint connections. Five standard segments with opening angle of 67.5° were used to assemble the ring. Key segment was introduced to complete the whole ring. A series of three staggered ring model are used herein (follow [13]). Three successive staggered rings at angle of 11.25° rotated clockwise were developed followed by all-at-once tunnel model (3S+AAO) with the length of 22.4m in longitudinal axis (excavation length in y-direction). Total length of tunnel was equivalence to 19 rings. Another tunnel model consist of 19 rings tunnel model was also has been developed.

2.2 Mesh assignation

The entire soil and tunnel model was discretized. Solid element of 8-node brick linear hexahedron (C3D8P) with trilinear displacement and trilinear pore pressure was assigned to the soil model. In this case, the soil mesh was assigned to be well matched to the tunnel excavation length of 1.4 m in y-direction (which resembling the ring's width). Denser meshing was assigned at the respective axes so that the mesh of tunnel lining with segment's joint can be also accounted. A total of 9680 of solid elements with 10994 of total nodes were assigned. Pore pressure elements were used to model the soil (solid continuum element type) in geostatic analysis [14]. A total of 1522 linear quadrilateral shell elements of S4R were assigned to the tunnel lining model with the total number of nodes are 1832. Meshing of tunnel lining model initially had been validated with initial work in Jusoh [11,12].

IOP Conf. Series: Journal of Physics: Conf. Series **1049** (2018) 012044 doi:10.1088/1742-6596/1049/1/012044

2.3 Tunnel lining and soil properties

For the tunnel properties, the lining was simulated by using isotropic linear elastic model. The lining was assigned with Young Modulus, E_L of 33GPa and Poisson's ratio of 0.2. This full soil-tunnel model was designated with appropriate boundary condition and mesh. Figure 1 presents the soil-tunnel lining model developed in ABAQUS 6.10. Different colour schemes were used to represent the different type of soil. Selection of the developed model size (*i.e.,* boundary of model) is discussed in the next section while the details on the assigned soil properties are tabulated in table 1.

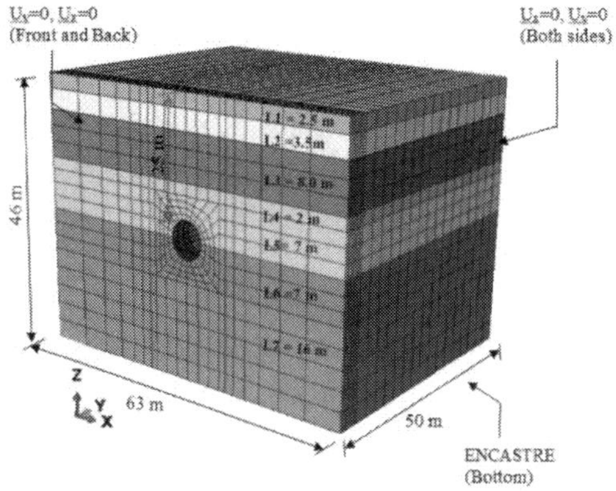

Figure 1. 3D model of soil-tunnel system with respective boundary condition and meshing

Table 1. Summary of the adopted soil properties

Soil layer	Soil type	Young Modulus, E_s (kPa)	Bulk density, γ (kN/m^3)	Poison's ratio, v_s	Angle of friction, ϕ(°)	Cohesion, c (kPa)
L1	Fill	7000	19	0.333	30	0.3
L2	Estuarine	3000	15	0.35	20	0.3
L3	Fluvial clay	3000	19	0.35	22	0.3
L4	Fluvial sand	7000	20	0.32	32	0.3
L5	Bukit Timah granite formation, G4 (VI)	59200	20	0.333	30	2
L6	Bukit Timah granite formation, G4 (v)	86400	20	0.3	35	2
L7	Bukit Timah granite formation G2 (III)	3500000	23	0.32	35	400

2.4 *Interaction analysis in full soil-tunnel model*

In full soil-tunnel model, three different interaction analyses were studied; i) ring's joint interaction, ii) segment joint interaction and iii) soil-tunnel interaction. Ring's joints were modelled as tie constraints at the periphery of both rings, assigned as surface to surface contact (see Figure 2(a)). Recent model of ABAQUS software allows for shell-solid surface to surface interactions [14]. Therefore, soil-tunnel coupling was accomplished by surface to surface contact algorithm by linking the contact interaction property and penalty friction of tangential behaviour adopted as 0.35 (Figure 2 (b)). The mesh tie constraints method called "master-slaves" was used to model the interaction between both lining and soil [14]. In this interaction analysis, the displacement and pore pressure of the "slaves" surface were set to be equal to the value of the master surface to which it is the closest. In general, the master surface is a surface of stiffer body or coarser mesh and vice versa for slave surface [15].

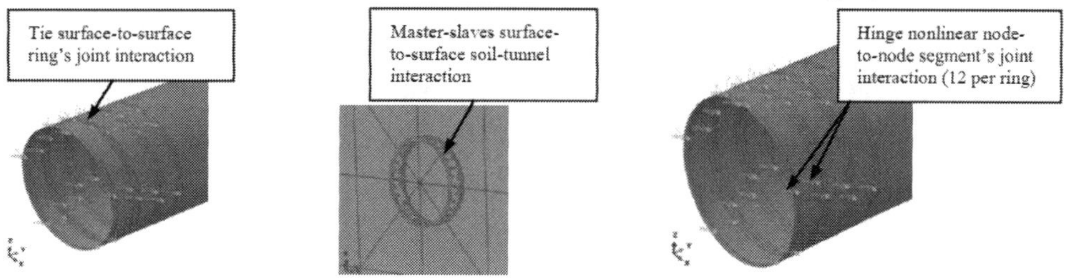

(a) (b) (c)

Figure 2. Interaction model, (a) Tie constrain of surface to surface type is assigned at the ring's joint, (b) Soil-tunnel interaction "master-slaves" formulation with surface to surface contact algorithm with contact interaction property, penalty friction of tangential behaviour; 'master' was represent by red

IOP Conf. Series: Journal of Physics: Conf. Series **1049** (2018) 012044 doi:10.1088/1742-6596/1049/1/012044

colour and 'slave' in purple and (c) Segment joint assigned with hinge nonlinear or tie constrain at two different wire link node-to-node position in tunnel lining

To have more certainty in model, a parametric model via tie constraints, spring and hinge type of interactions were investigated initially in dual-segment joint model to model the segment to segment interactions. The tie constraints of segment's joint were performed as tie (assigned as penalty type), representing the continuous ring condition. Tie and spring model did not require any stiffness joint data (with spring model allowed changes in boundary conditions), whereas for hinge model; rigid, linear and nonlinear joint stiffness properties shall be assigned (figure 4(c)). Non-linear hinge model was carried out with respective laboratory findings (the non-linear joint stiffness data from moment-rotation results)[11]. The obtained results were presented and discussed in next subsections.

3. Results and discussion

In this section, the results are discussed mainly based on the segment joint modelling in dual-jointed segment model and in fully soil-tunnel model.

3.1 Dual-jointed segment results

In simplified dual-jointed segment tests, tie constraints, spring and hinge type of interactions were investigated and verified with laboratory findings [16]. Hinge with non-linear angular joint stiffness properties (H-NL) showing the closest moment response in dual-jointed segments in which it was later selected to be adopted in the soil-tunnel model.

3.2 Soil-tunnel model results

After successfully model the interaction of simplified segment-to-segment model, the H-NL model was then adopted in the segment joint of a fully soil-tunnel numerical model. Tie constraints segment's joint were also adopted to show the comparison *(i.e.,* represent continuous ring condition). Instead of 3S+AAO model, a fully staggered 19 rings model had also been designated. Figure 3 presents the results of the surface settlement which were later compared with field data.

Surface settlement induced by tunnelling was validated and compared with data from a case study where the behaviour of full tangential bending moment tunnel was investigated. Field data showed sudden drop reading occurred at 38 to 40 m from tunnel start point due to construction process. This was due to nature of tunnelling process. As cutter head of tunnel construction is stopped at the final ring, different face pressure in the ground surrounding occurred, thus lead to obvious effect on settlement trough here. However, tunnel model via simulation possess slightly different in trend (lower initial surface settlement) but still obey general trend and gave good agreement to the cumulative Gaussian S-curve proposed by Peck [17]. This is due to the tunnel lining installation [18], effect of boundary fixities and material properties in tunnel lining.

When compared to the continuous ring model (tie-model), FEM-hinge model demonstrated well agreement in term of maximum settlement (suppose a steady state surface settlement) at a distance of 20m for about 32mm of surface settlement, almost similar to the steady state results obtained by Brillouin Optical Time Domain Reflectometer (BOTDR) or sometimes referred to as the Optical Fiber Strain Analyzer which has been measured in the tunnel ring of R540 in the case study and surface settlement observed from settlement marker of G2407 [19]. In the opposite, a simplified adopted joints of tie-model lead to unexpected heave measurement of surface settlement at the tunnel excavation range of 0 to 20 m. This unexpected heave due to due to segment's joint properties (*i.e.,* tie) act as continuous lining create buoyancy effect in lining. Another improved model with separated rings model (instead of all-at-once model) lead to higher resemblance of surface settlement pattern to the field data, with steady state settlement observed after 10m of cutter head distance. However, due to limited computer capacity, simulated model was abruptly stopped at ring 19, causing to monolithic (not a sudden drop reading) effect observed in field data. Nevertheless, one could see that this improved model (FEM-Hinge-model of 19 rings), presented more convincing results which depicted steady state surface settlement after 10 m at the distance of cutter head and ahead and shows almost similar pattern of settlement profile such as reported in G2407 [19].

4. Conclusions

This paper presents the development of segment joint modelling in order to improve soil-tunnel interaction. Various type of segment joints were modelled in initial dual-segment model and the effects of joints were observed. Hinge model that anticipated the non-linear moment-rotation reveals the results most matching to the field measurement. The respective model then extended to a full soil-tunnel simulation model. Ground deformation or commonly represented by a settlement trough at the ground surface as resulted from the effect of tunnel construction together with soil-tunnel, segment's and ring's joint interactions were measured parallel at the top of tunnel crown. Results showed that the flexible segment joint with non-linear properties induced higher resemblance of surface settlement pattern to the field data, with steady state settlement observed after 10m of cutter head distance.

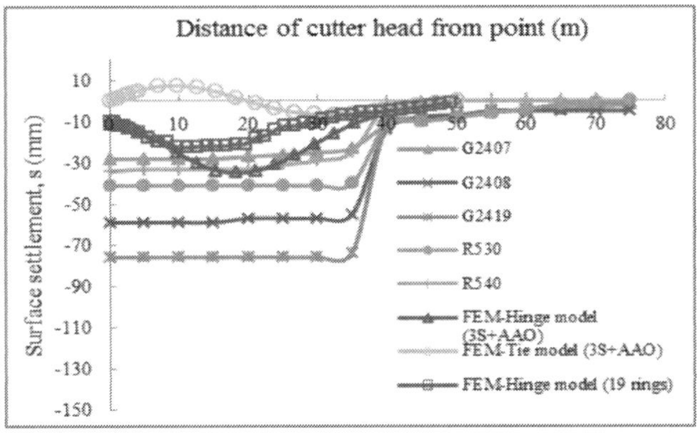

Figure 3. FEM results of three staggered and all-at-once tunnel (3S+AAO) of hinge-model, tie-model and FEM results of 19 rings hinge-model compared to field data of Outer bound tunnel (OT) at R530 –R540 of Singapore Circle Line Stage 3 (C852) project, (Note: G2407, G2408 and G2419 were surface settlement that measured by settlement marker on OT tunnel; and R530 and R540 were settlement reading at the specific rings; measured by BOTDR)

Acknowledgement

The authors also would like to thanks the financial support from UTM Research University Grant Scheme (RUGS), Vote #14J16.

References

[1] Hudoba, I. (1997) "Contribution to Static Analysis of Load-bearing Concrete Tunnel Lining Built by Shield-driven Technology". Tunnelling and Underground Space Technology, 12(1): 55–58.
[2] Blom, C. B. M., van der Horst, E. J. and Jovanovic, P. S. (1999) "Three-dimensional Structural Analyses of The Shield-Driven "Green Heart" Tunnel of The High-Speed Line South". Tunnelling and Underground Space Technology, 14(2): 217–224.
[3] Klappers, C., Grübl, F. and Ostermeier, B. (2006) "Structural Analyses of Segmental Lining – Coupled Beam and Spring Analyses versus 3D FEM Calculations with Shell Elements". Tunnelling and Underground Space Technology, 21(3-4): 254-255.
[4] Teachavorasinskun, S. and Chub-uppakarn, T. (2010) "Influence of Segmental Joints on Tunnel Lining". Tunnelling and Underground Space Technology, 25(4): 490–494.
[5] Arnau, O. and Molins, C. (2012) "Three-dimensional Structural Response of Segmental Tunnel Linings". Engineering Structures, 44, 210–221.
[6] Do, N. A., Dias, D., Oreste, P. and Djeran-Maigre, I. (2013a). "2D Numerical Investigation of Segmental Tunnel Lining Behavior". Tunnelling and Underground Space Technology, 37: 115–127.
[7] Do, N. A., Dias, D., Oreste, P. and Djeran-Maigre, I. (2013b). "Three-dimensional Numerical Simulation for Mechanized Tunnelling in Soft Ground: The Influence of The Joint Pattern". Acta Geotechnica, 9(4): 673–694.
[8] Yanzhi, Y., Weiwei, Z., Jianwei, W. and Zhihao, Y. (2014) "Three-dimensional Orthotropic Equivalent Modelling Method of Large-scale Circular Jointed Lining". Tunnelling and Underground Space Technology, 44: 33–41.

[9] Xiaochun, Z., Wei, Z., Zhengrong, H. and Yuewang, H. (2006) "Effect of Joint Structure on Joint Stiffness for Shield Tunnel Lining". Tunnelling and Underground Space Technology, 21(3-4): 407–408.

[10] Blom, C. B. M. (2002) "Design Philosophy of Concrete Linings for Tunnels in Soft Soils". Ph.D Thesis, Delft University, Netherlands.

[11] Jusoh, S. N. (2017) "Performance of Precast Bolted Tunnel Lining Through Physical And Numerical Modelling". Ph.D Thesis. Universiti Teknologi Malaysia, Johor Bahru.

[12] Jusoh, S. N., Marto, A. , Mohamad, H. Yunus, N.Z., Kassim, F., and Hezmi M.A. (2016).Numerical Modelling On The Mechanics Of Precast Segment Tunnel Lining, International Graduate Conference on Engineering, Science and Humanities (IGCESH2016), Universiti Teknologi Malaysia, Johor Bahru, Malaysia 15 -17 August 2016

[13] Do, N. A. (2014) "Numerical Analyses of Segmental Tunnel Lining under Static and Dynamic Loads". Ph.D Thesis. Civil Engineering, INSA de Lyon, English, Hanoi, Vietnam.

[14] Abaqus 6.10. (2010). User's manual. Hibbitt, Karlson and Sorenson, Inc.Pawtucket, RI, USA.

[15] Namazi, E., Mohamad, H., Hong, A. K. B., Hajihassani, M., Jusoh, S. N. and Abad., S. V. A. N. K. (2012) "Ground Behaviour Around A Tunnel Using Various Soil Models". Electronic Journal Geotechnics Engineering (EJGE), 17: 609-621.

[16] Siti Norafida Jusoh, Hisham Mohamad, Aminaton Marto, Nor Eliza Alias, Muhammad Azril Hezmi, Rini Asnida Abdullah, Nor Zurairahetty Mohd. Yunus (2016). Investigation on The Mechanics Of Precast Segment Tunnel Lining. Jurnal Teknologi. 78 (8-6) pp 61-66.

[17] Peck, R. B., Hendron, A. J. and Mohraz, B. (1972) "State-of-the-Art of Soft Ground Tunnelling". Proceedings of the Rapid Excavation and Tunnelling Conference, Chicago, IL, 1: 259–286.

[18] Vermeer, P., Bonnier, P. G. and Möller, S. C. (2002). "On a smart use of 3D-FEM in tunnelling." Numerical Models in Geomechanics Proceedings of the 8th International Symposium NUMOG VIII.

[19] Mohamad, H., Bennett, P. J., Soga, K., Mair, R. J., Lim, C. S., Knight-Hassell, C. K. and Ow, C. N. (2007). "Monitoring Tunnel Deformation induced by Close-proximity Bored Tunneling using Distributed Optical Fiber Strain Measurements". Proc. 7th Int. Symp. on Field Measurements in Geomechanics, Boston.

IOP Conf. Series: Journal of Physics: Conf. Series **1049** (2018) 012045 doi:10.1088/1742-6596/1049/1/012045

Fuzzy Conjoint Modelling in Studying User Willingness to Switch to Bicycle as Transportation in Ipoh City

Nik Mohd Iznan Tuan Yaakub[1], Muhamad Nazri Borhan[2], Muhamad Razuhanafi Mat Yazid[3], and Sharinatol Akmanida Jamaludin[4]

Jabatan Kejuruteraan Awam, Politeknik Ungku Omar, Jalan Raja Musa Mahadi, 31400 Ipoh, Perak[1,4],
Fakulti Kejuruteraan Awam & Alam Bina, Universiti Kebangsaan Malaysia, 43600 Bangi, Selangor,[1,2,3,4]

nimoizty@gmail.com[1], mnazri_borhan@ukm.edu.my[2], razuhanafi@ukm.edu.my[3]
akmanida@puo.edu.my[4]

Abstract. The increase of vehicles on the road contributed to the unsustainable city. The user dependence on motorised vehicles such as cars, motorcycles, vans, buses, and trucks should be reduced. Bicycle as transportation is the alternative methods that can be done as active transportation. The usage of bicycle can be reducing traffic, improve health, and reduce environmental pollution. As a model to assess the willingness of consumers to switch to bicycle transportation, total of 400 questionnaires were distributed in Ipoh city. The analysis uses fuzzy conjoint model as introduced by Zadeh and Tuksen to assess the extent to which consumers are willing to shift to cycling transport. The findings concluded, there are 6 strategies to improve the level of bicycle use special facilities for cyclists, a special crossing for cyclists, cycle parking provision, the application of the current cycle, associated facilities and bicycle users, and location of routes for cyclists.

1. Introduction

Latest town development and current climate change necessitate that we reconsider and analyse ongoing city development and management. Flash flood, landslide, thunderstorm, extreme greenhouse effect, global rising temperature and widespread air pollution are all impacts from significant climate change [1]. A sustainable city development in Malaysia revolves around 10 major criteria including culture & family, environmental administration, urban disaster, urban demography, ecological support, awareness & education, city services, natural resources management, city economics and public health. Motorised vehicles is the main contributing factor to an unsustainable environment besides manufacturing industry, development activities, power generation, land clearing and open air burning.

Ipoh is a city with a rich ecosystem and heritage that should be sustained. But now, city development is so rapid that the city sustainability is significantly affected. All main roads face serious congestion problem during peak hours. An effective way to reduce traffic congestion in big cities is the optimal use of bicycles. The use of bicycles as active transportation is popular in big cities around the world for instance in China, Belgium, Switzerland, Japan, Finland, Norway, Sweden, Germany, Denmark

Content from this work may be used under the terms of the Creative Commons Attribution 3.0 licence. Any further distribution of this work must maintain attribution to the author(s) and the title of the work, journal citation and DOI.
Published under licence by IOP Publishing Ltd

and Netherland. Moreover, in certain countries, the use of bicycles as a public transportation, or better known as bicycle sharing concept, has also been widely adopted. This concept doesn't require high cost and it is suitable for short distance travels not more than five kilometers. Japan for example, despite being acknowledged as a major automotive manufacturer – with Honda and Toyota being the most well-known brands, still adopts the culture of bicycle as a mode of transportation. Abdullah O.K. Rahmat [2] stated that this system was introduced to reduce pollution.

Figure 1. Ipoh city scenery

2. Research Methodology

To analyse the level of priorities placed by respondents with respect to the use of bicycle as a mode of transportation, researchers adapted the fuzzy conjoint method introduced by Turksen [3]. Furthermore, Turksen and Willson [4] had also used the fuzzy conjoint method in a group model based on the priority model. Indeed, the fuzzy conjoint model is often used in the business sector to identify consumers' preference level for a particular product. A study by Bass and Talarzyk [5] stated that preferences for a particular brand or product related to attitude based on beliefs in relative importance to each attribute. In another study, Dhar [6] said that user preference levels include subjective, fuzzy, doesn't persist, unstable or consistent.

Based on this model, a set of fuzzy R was designed to represent the hierarchy of responses against each specific characteristic. In this approach, the levels for each selected attribute was defined. To determine respondents' preference, the use of basic likert scale was identified as follows:

i. Strongly not preferred
ii. Not preferred
iii. Not sure
iv. Preferred
v. Strongly preferred

The fuzzy set theory was used to solve for formulation activities which are not precise, uncertain and fuzzy [7][8]. According to Baheri *et al.*[9] the fuzzy set theory is an extension of classic set theory and probability theory. Classic set theory only uses the values of 0 and 1, which is {0.1}, but fuzzy set theory uses the range of values between 0 and 1, which is (0,1). Fuzzy theory can be used to represent ambiguity or uncertainty that exist in the linguistic variables definition. Linguistic variables are represented by terms such as strongly not preferred, not preferred, not sure, preferred and strongly preferred. The figure below shows the concept of linguistic variables for beliefs.

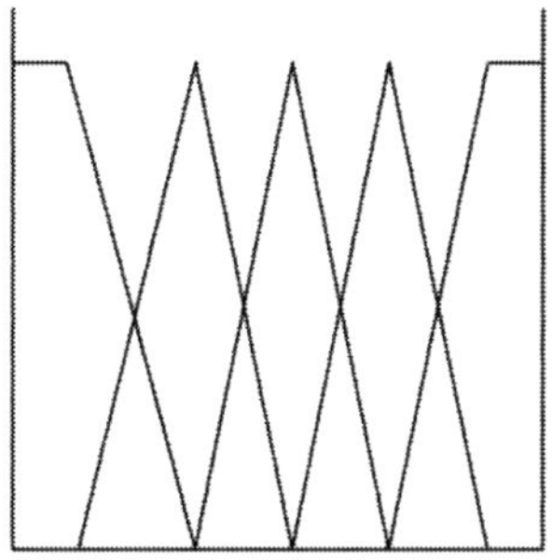

Figure 2. Linguistic variables values for beliefs in a fuzzy conjoint model.

The equation below is the fuzzy conjoint model adapted from Turksen (1992).

$$\mu_R\left(y_j, A\right) = \sum_{i=1}^{T} \frac{Wi}{\sum_{k=1}^{T} Wk} \mu_{Bi}\left(x_j, A\right)$$

Where:

Wi= Weightage that represent respondents' preference level / against all respondents in the study

A = Criteria being considered

$\mu_R\left(y_j, A\right)$ = Estimated degree of association for linguistic values R for all respondents towards item A based on linguistic values y_j = 1,2,3,….., k)

$\mu_{Bi}\left(x_j, A\right)$ = Degree of association for linguistic values B for respondent *i* towards item A based on linguistic values x_j = 1,2,3,…,k

K = The number of linguistic values used

N = The number of respondents

The table below shows the early definition for the linguistic values of the fuzzy set for variables preference level adapted from Yahaya and Mohamad [10]. This definition was based on Turksen and Willson [11].

Table 1. Linguistic variables and values.

Linguistic variables	Linguistic values
Strongly not preferred	{1/1, 0.75/2, 0.5/3,0/4,0/5}
Not preferred	{0.5/1,1/2,0.75/3,0.25/4,0/5}
Not sure	{0/1,0.5/2,1/3,0.5/4,0/5}

IOP Conf. Series: Journal of Physics: Conf. Series **1049** (2018) 012045 doi:10.1088/1742-6596/1049/1/012045

Preferred	{0/1,0.25/2,0.75/3,1/4,0.5/5}
Strongly preferred	{0/1,0/2,0.5/3,0.75/4,1/5}

The formula used to determine the degree of similarity between two fuzzy sets as follows:

$$S(F,M) = \frac{F \cdot M}{max\,(F \cdot F, M \cdot M)}$$

Where

$F = \mu_F\,(x_1),\ \mu_F\,(x_2), \ldots$
$F = \mu_M\,(x_1),\ \mu_M\,(x_2), \ldots$
F and M are vector X
$= (x_1, x2, \ldots.)$ '·' means dot product.

Table 2. Preference items for bicycle transportation.

NO.	ITEM
1	Bicycle lanes distinctly separate from roads for motorised vehicles
2	Bicycle lanes separated using blocks
3	Bicycle lanes by the roadside
4	Cyclists sharing the same roads with motorised vehicles
5	Covered bicycle lanes
6	Non-covered bicycle lanes
7	Lanes on the left and right of roads lined with trees for shade
8	Lanes lined with fragrant flower plants
9	The use of overhead bridge
10	The use of traffic lights
11	Absence of control
12	Involvement of authority such as the police
13	Covered bicycle parking
14	Multi-storey bicycle parking
15	Non-covered bicycle parking
16	Combination of covered and non-covered
17	Bicycle parking near shops
18	Separate parking for bicycles and cars by the roadside
19	Automated bicycle parking
20	Bicycle parking without security or locks
21	Complete cycling outfit including safety helmet and shoes
22	Everyday wear
23	Office wear
24	Traditional wear (*pelikat* sarong)
25	Rest and relaxation facilities (R&R)
26	Bicycle parking
27	Able to carry bicycle in public transportation
28	Separate lanes from motorised vehicles to avoid congestion
29	Bicycle routes nearby workplace
30	Bicycle routes nearby schools and higher education institutions
31	Bicycle routes nearby shops and malls
32	Bicycle routes nearby recreational areas

The following were the steps in fuzzy conjoint analysis after data collection process was conducted in Ipoh City. The first step was to determine the items that influence users' preference when using bicycles based on incentives and disincentives as listed in Table 2.

The second step was to define the value measurement scale for the five fuzzy linguistic value as introduced by Turksen and Willson [12]. Linguistic variables involved are as follows, where L_k = {strongly not preferred, not preferred, not sure, preferred, strongly preferred}.

$$L_1 = \left\{\frac{1}{1},\frac{0.75}{2},\frac{0.5}{3},\frac{0}{4},\frac{0}{5}\right\}$$

$$L_2 = \left\{\frac{0.5}{1},\frac{1}{2},\frac{0.75}{3},\frac{0.25}{4},\frac{0}{5}\right\}$$

$$L_3 = \left\{\frac{0}{1},\frac{0.5}{2},\frac{1}{3},\frac{0.5}{4},\frac{0}{5}\right\}$$

$$L_4 = \left\{\frac{0}{1},\frac{0.25}{2},\frac{0.75}{3},\frac{1}{4},\frac{0.5}{5}\right\}$$

$$L_5 = \left\{\frac{0}{1},\frac{0}{2},\frac{0.5}{3},\frac{0.75}{4},\frac{1}{5}\right\}$$

Next, in the third step, questionnaires distributed to 400 respondents were analysed. Respondent sets were marked as R = (R1, R2, R3, R4, R5.....R400). Preference set for the 32 questions by the 400 respondents were tabulated as 1 = (4, 4, 5 ...5) as shown in Table 3 below.

Table 3. Respondents' opinion on preference items.

	R1	**R2**	**R3**	**Rn**	**R400**	$\sum$
1	4	4	4	.	5	1926
2	1	5	5	.	4	1500
3	4	1	2	.	2	789
4	2	2	1	.	1	571
5	2	4	5	.	5	1649
n	.	.	.	.	.	.
32	2	2	4	.	4	1395

The fourth step was to compute the weightage for the evaluation of each respondent against the evaluation of all 400 respondents by dividing the value for the individual respondent against the sum of values for total respondents (refer to Table 4).

Table 4. Weightage scale for respondents' choices.

	W1	**W2**	**W3**	**Wn**	**W400**
1	4/1926	4/1500	4/789	.	5/1395
2	1/1926	4/1500	5/789	.	4/1395
3	4/1926	4/1500	2/789	.	2/1395
4	2/1926	4/1500	1/789	.	1/1395
5	2/1926	4/1500	5/789	.	5/1395
n	.	.	.	.	.
32	2/1926	4/1500	4/789	.	4/1395

IOP Conf. Series: Journal of Physics: Conf. Series **1049** (2018) 012045 doi:10.1088/1742-6596/1049/1/012045

The fifth step was to compute the degree of association of the linguistic value for the fuzzy set for each respondent. R value was derived by multiplying the weightage to linguistic value of the fuzzy set definition as the respondents' value as shown in Table 5.

Table 5. Degree of association for all 400 respondents towards linguistic values.

1	**L1**	**L2**	**L3**	**L4**	**L5**
R1	0.0000	0.0005	0.0016	0.0021	0.0010
R2	0.0000	0.0005	0.0016	0.0021	0.0010
.	.	.	.	.	.
R400	0.0000	0.0000	0.0013	0.0019	0.0026

The sixth and final step was to compute the linguistic value of the fuzzy set for the 400 respondents, which is L_i by summing the degree of association for each domain set value separately.

IOP Conf. Series: Journal of Physics: Conf. Series **1049** (2018) 012045 doi:10.1088/1742-6596/1049/1/012045

Table 6. Summation of degree of association.

QUESTION	L1	L2	L3	L4	L5
1	0.0010	0.0350	0.5343	0.7835	0.9304
2	0.0507	0.2765	0.7020	0.8775	0.5453
3	0.4563	0.8048	0.7117	0.3685	0.1204
4	0.6375	0.6756	0.6173	0.3091	0.1278
5	0.0400	0.1402	0.5879	0.7433	0.7841
6	0.4446	0.6059	0.6851	0.4418	0.1851
7	0.1300	0.3494	0.6652	0.7309	0.5397
8	0.1754	0.3566	0.6561	0.7055	0.5125
9	0.0092	0.0696	0.5558	0.7921	0.8792
10	0.0326	0.2200	0.6720	0.8740	0.6234
11	0.7622	0.7226	0.6006	0.1555	0.0366
12	0.3522	0.6872	0.7038	0.4704	0.2402
13	0.0027	0.0690	0.5662	0.8109	0.8650
14	0.0691	0.2378	0.6500	0.8121	0.6310
15	0.4050	0.6525	0.6969	0.4799	0.2013
16	0.3824	0.5897	0.6813	0.4895	0.2550
17	0.0644	0.2102	0.6249	0.7615	0.7350
18	0.1524	0.3900	0.6649	0.6792	0.6656
19	0.0581	0.1715	0.6047	0.7820	0.7636
20	0.7083	0.5312	0.3541	0.0000	0.0000
21	0.0175	0.0748	0.5411	0.7574	0.8938
22	0.0376	0.2437	0.6853	0.8769	0.5860
23	0.3447	0.6789	0.6981	0.4673	0.2426
24	0.6169	0.6814	0.6068	0.2822	0.1593
25	0.0460	0.1219	0.5618	0.7518	0.8303
26	0.0365	0.2199	0.6715	0.8528	0.6206
27	0.4734	0.6488	0.6625	0.3915	0.2017
28	0.3051	0.4955	0.6435	0.5368	0.4078
29	0.1446	0.2508	0.6032	0.6912	0.6490
30	0.1311	0.3424	0.6513	0.7101	0.5663
31	0.2634	0.4273	0.6424	0.5846	0.4519
32	0.0953	0.2043	0.6054	0.7484	0.6939

3. Results and Discussion

The first strategy proposed was customised facilities for cyclists in Ipoh City. For this section, eight questions were posed to respondents. Each respondent had given different responses. Two items had yielded the "strongly preferred" linguistic value result. The response recorded with weightage value of 0.9304 was for bicycle lanes to be separated from roads for motorised vehicles. Another item under "strongly preferred" was covered bicycle lanes with weightage value of 0.7841. Three items under the linguistic value "Preferred" were identified, which were bicycle lanes separated using blocks, lanes on the left and right of roads lined with trees for shade, and lanes lined with fragrant flower plants. The respective values for these items were 0.8775, 0.7309 and 0.7055. However, there were respondents who chose the linguistic value "Not sure" for the item non-covered bicycle lanes with weightage value of 0.6851. For the item bicycle lanes by the roadside, the linguistic value result was "Not preferred" with a weightage value of 0.8048. This was also the case for the item cyclists sharing the same roads with motorised vehicles, except with a lower weightage value of 0.6756.

The second strategy was providing dedicated crossing facilities for cyclists. Research finding only showed one item under the category "Strongly preferred" which was the use of overhead bridge. The weightage value was 0.8792. This was followed by the use of traffic lights with weightage value of 0.8740 under the category "Preferred". However, for the question posed about the involvement of authority such as the traffic police and Road Transport Department, the result was "Not sure". The weightage value for this item was 0.7038. For the item about crossing with the absence of control, the linguistic value found was "Strongly not preferred" with weightage of 0.7622.

The third strategy was providing bicycle parking. Researchers' initial perception was that bicycle users do not require covered bicycle parking, similar to motorcycle users who park their motorbikes anywhere regardless whether the space is covered or otherwise. Evidently, this premature assumption was unfounded. Based on research finding, the linguistic value for the item covered bicycle parking was "Strongly preferred" with a weightage of 0.8650. Four items were found to be under the category "Preferred" which were multi-storey bicycle parking, bicycle parking near shops, separate parking for bicycles and cars by the roadside, as well as automated bicycle parking. The weightage values for these items were 0.8121, 0.7615, 0.6792 and 0.7820 respectively. Items with linguistic value "Not sure" were non-covered bicycle parking and combination of covered and non-covered parking with their weightage at 0.6969 and 0.6813 respectively. Finally, there was only one item with linguistic value "Strongly not preferred" which was bicycle parking without security or locks with a weightage value of 0.7083.

The fourth strategy involved cyclist outfit when cycling. It is acknowledged that safety aspect is very important when cycling on busy roads. Road users are constantly exposed to risks of accidents which can be fatal. The respondents had concurred by choosing the linguistic value "Strongly preferred" for the item complete cycling outfit including safety helmet and shoes. The weightage value recorded for this question was 0.8938.

For everyday wear or normal clothing, the weightage value was found to be 0.8769 which was in the "Preferred" category. As for office wear or working attire, the item fell under the category "Not sure" with a weightage value of 0.6981. For the final question under this section, respondents were also asked about traditional wear such as pelikat sarong. The result was "Strongly not preferred" with a weightage value of 0.6814.

The fifth strategy proposed was providing facilities for the convenience of cyclists. In general, not many facilities are built specifically for cyclists in majority of countries around the worlds, except for developed countries with an established policy to encourage the use of bicycles as daily mode of transportation. Analysis results showed that most respondents had chosen the item rest and relaxation facilities for cyclists as the most important factor with a weightage value of 0.8303. A dedicated rest area was expected to be a need in an area with hot weather all year round. So it was not surprising that most respondents saw this facility as a requirement in Ipoh.

As for the item bicycle parking, the weightage value found in this research was 0.8528 which was "Preferred". Meanwhile, most respondents had selected "Not sure" for the ability to carry bicycles in public transportation, as well as for the item separate lanes from motorised vehicles to avoid congestion. The weightage values for these two factors were 0.6625 and 0.6435 respectively.

The sixth strategy involved the location of routes for the cyclists. Each item yielded a different weightage value. However, the results showed that the linguistic value "Preferred" was most frequently selected. The weightage value for bicycle routes nearby workplace was 0.6912, for bicycle routes nearby schools and higher education institutions 0.7101, and for bicycle routes nearby recreational areas 0.7484, all under the category "Preferred". On the other hand, there was one item with the linguistic value "Not sure" which was bicycle routes nearby shops and malls with a weightage value of 0.6424.

Evidently, the location of bicycle routes plays an important role to determine how likely the respondents in this research demography will switch to bicycle as a mode of transportation. Most respondents had given the "Preferred" linguistic value for bicycle routes nearby their workplace, regardless whether they were working in the civil service or in the private sector. Bicycle routes nearby schools and higher education institutions as well as recreational areas also proved to be of utmost importance.

Figure 3: Covered and non-covered bicycle parking.

4. Conclusion

Despite Japan's capability to produce various types of cars filling up roads all over the world, its own society is content and takes pride in using bicycles; a mode of transportation that is not only cheaper and more convenient, but is also more environmentally friendly, devoid of pollution impact which could threaten global temperature stability [13].

This research had contributed a significant implication towards discovery of ideas and new findings in transportation system, especially in the area of non-motorised vehicles in Ipoh City specifically, and in Malaysia generally. The findings and results ought to be considered by local authorities such as the City Council, the Department of Town and Country Planning, and the Public Works Department to upgrade the infrastructure for bicycle transportation system. This includes dedicated lanes for cyclists to experience a safer journey compared to sharing roads with motorised vehicles such as cars and buses. This will affirm that each road user has the rights to fair use of road facilities.

At the federal government level, a dedicated budget to upgrade bicycle transportation system need to be allocated adequately so that the relevant authorities will have sufficient funding to improve infrastructure in order to encourage the public to switch to bicycles as their means of transportation. In facing global challenges, there are no other unique ways to resolve them other than to increase user awareness, to be more concerned and conscientious towards their cities. According to Anable (2005), psychological factors such as perception, identity, social norms, and habits are increasingly being studied to understand travel behaviours.

Developed countries around the world such as Holland, Denmark, Finland, German, Sweden and Belgium had long adopted a policy which sides more towards non-motorised road users [14]. It is hoped that this change in stance can inspire and become a basis for Malaysians to switch to an established transportation system that supports a sustainable city.

ACKNOWLEDGEMENT

The authors would like to thank Universiti Kebangsaan Malaysia (UKM), Polytechnic Ungku Omar and the Ministry of Higher Education (MOHE), Malaysia for the financial support under research grants:GGPM-2017-059

REFERENCES

[1] Rosilawati Zainol, R, Mohamad@Ahmad, I, Ahmad, F & Nordin NA 2014, *Mesra pejalan kaki di bandar warisan*, Penerbit Universiti Malaya, Kuala Lumpur.
[2] Abdullah O.K. Rahmat, RA 2015, *Perancangan pengangkutan bandar,* Penerbit Universiti Kebangsaan Malaysia, Bangi.
[3] Turksen, IB 1992, 'Fuzzy expert systems for IE/OR/MS', *Fuzzy Sets and System,* vol. 51, no. 1, pp. 1–27.
[4] Turksen, IB & Willson, IA 1994, 'A fuzzy preference model for consumer choice, *Fuzzy Sets and Systems*, vol. 68, no. 3, pp. 253–266.
[5] Bass, FM & Talarzyk, WW 1972, 'An attitude model for the study of brand preference', Journal of Marketing Research, vol. 9, no. 1, pp. 93–96.
[6] Dhar, R 1995, 'New directions in behavioral decision theory: implications for consumer choice,' *Advance in Consumer Research*, vol. 22, pp. 203.
[7] Alriksson, S & Oberg, T 2008, 'Conjoint analysis for environmental evaluation. A review of methods and applications', *Environmental Science Pollution Research*, vol. 15, no. 3, pp. 244–257.
[8] Soutar, GN & Turner, JP 2002, 'Students' preferences for university: A conjoint analysis,' *International Journal of Educational Management*, vol. 16, no. 1, pp. 40-45.
[9] Baheri E, Dalvand MR, Ansarinejad A, Miri-Nargesi S, & Hatami-Shirkouhi L 2011, 'A fuzzy conjoint analysis approach for evaluating credit card services: A case study of Iranian bank,' *African Journal of Business Management*, vol. 5, no. 7, pp. 2753–2765.
[10] Yahaya, YH & Mohamad, N 2011, 'Designing software usability measurement using fuzzy set conjoint model', in *International Conference on Computer Communication and Management, Australia, 2-4*th *May 2011*, Sydney, Australia.
[11] Turksen, IB & Willson, IA 1995, 'A fuzzy set model for market share and preference prediction, *European Journal of Operational Research*, vol. 82, no. 6, pp. 39–52.
[12] Turksen, IB & Willson, IA 1995, 'A fuzzy set model for market share and preference prediction, *European Journal of Operational Research*, vol. 82, no. 6, pp. 39–52.
[13] Mat Yazid, MR, Jamaludin, SA, Tuan Yaakub, NMI & Borhan, MN 2015, 'Pendekatan Pemodelan Konjoin Kabur dalam Pengangkutan Aktif,' *Advanced Journal of Technical and Vocational Education,* vol. 1, no. 2, pp. 160–168.

IOP Conf. Series: Journal of Physics: Conf. Series **1049** (2018) 012046 doi:10.1088/1742-6596/1049/1/012046

Developmental research of sustainable technologies to minimise problematic road embankment settlements

N Ratha[1,2], T N H T Ismail[2,3], D C Wijeyesekera[2], I Bakar[2], A J L M Siang[2], M K Abu Talib[2] & A Zainorabidin[2]

[2]Faculty of Civil & Environmental Engineering, Universiti Tun Hussein Onn Malaysia, Johor, Malaysia

[3]Faculty of Engineering Technology, Universiti Tun Hussein Onn Malaysia, Johor, Malaysia.

E-mail: nurvin05@ymail.com

Abstract. Challenging, problematic and non-uniform ground conditions are a night mare to geotechnical engineers tasked with the design and construction of buildings and transport infrastructure. These often suffer undesirable structural settlements. Designing within the current understanding of geotechnics; settlement in peat and organic soils need to be recognised to include the known "primary and secondary consolidation characteristics" and the lesser known "tertiary consolidation phase". These eventually contribute cumulatively to the consequential uneven and hazardous "bumpy road" surfaces. Undulating flexible road pavements result primarily from the transference of the heavy self-weight of the embankment fill to yielding and non-uniform subgrade. The adoption of conventional design/repair methods such as pile, vertical drain, soil replacement and soil stabilisation are expensive and inappropriate in very soft ground conditions. These then lead to unjustifiably high and repetitive maintenance costs. There being no one quick fix solution for all; pragmatic research must necessarily identify the best/progressively improved practical and sustainable solution. A viable solution is to develop criteria and explore the concept of a "masonry arch bridge structure/lintel-column structure" and adopting sustainable materials through pragmatic searching for appropriate recyclable waste materials. This will lead to the basis for a sustainable, innovative, strong, stiff, permeable composite mat structure that can be used on soft and/or yielding ground conditions. Conceptual lightweight fill technology including the popularly used expanded polystyrene (EPS) and the innovative composite mats recently being developed by the research team are outlined.

1. Introduction

Current world news on natural geo environmental disasters of (somewhat predictable) hurricanes and (unpredictable in location and time) earthquakes attempt to question the credibility of engineering science when subjected to limiting conditions. On a similar note, but with a considerable time lapse, road users question their discomfort even in some newly constructed highways. On the basis of the extent of areal coverage and of all the engineering structures, highways/pavements leave the most

[1] To whom any correspondence should be addressed.

Content from this work may be used under the terms of the Creative Commons Attribution 3.0 licence. Any further distribution of this work must maintain attribution to the author(s) and the title of the work, journal citation and DOI.
Published under licence by IOP Publishing Ltd

dominant imprint on the natural landscape. Accordingly, rigid/flexible pavements contribute to the urban environments as a vital socio-economic pathway to a better quality of life. The pace of all infrastructure development in Malaysia and in other countries have led to the limited availability of "suitable" sites or the demands of highway routing necessitate the adoption of challenging sites bearing problematic ground conditions. Rigid pavements are increasingly desirable in highway construction due to the lesser maintenance needed when compared to flexible pavements. Nevertheless material sciences have researched into sustainable waste materials to reduce the compressive strength of the concrete in the rigid pavements while being able to increase the tensile or flexural strength of the normal concrete mix. Such composite materials combinations are bottom ash and waste foundry sand, steel slag and waste glass, and third combination being bottom ash and fly ash.

Under the 11th Malaysian Plan (2016-2020) [1], the government plans to spend a whopping amount of RM1.4 billion to build and upgrade 700 km of rural roads nationwide. A sum of RM200 million is provided for the upgrading of roads in Federal Land Development Authority (FELDA) settlements. FELDA lands, primarily being an agricultural hub are located in areas of clay and peat soil (in other words, soft soil). The development plan accordingly cites that this is where construction and road upgrading works would be carried out extensively. Thus, this government plan gives directed impetus to the significance of exploring new and innovative concepts with a sustainable approach in developing long term solutions in tackling problematic soft soils.

Historically, appropriate technologies were adopted as far back as in 2500 BC in Babylon. Figure 1 illustrates the construction technology, where hewn tree trunks (logs) were laid on the yielding soft ground to form a permeable, stiff but disconnected mat structure.

Figure 1. Road in Ur, circa 2500 BC (Adopted from [2])

2. Extent of soft soil in Peninsular Malaysia (Possible construction sites)

Figure 2 is a map of Peninsular Malaysia showing the general distribution of Quaternary Deposits including peat and soft soils, and implying the extent of the areas over which construction of infrastructure on soft soil can occur. Hence the prime outcome of this research will benefit many Malaysians. Problematic grounds other than the Quaternary peat or soft soil deposits but resulting from inland geo environmental conditions are not shown. The figure highlights the current road networks, especially the North-South Highway which traverses the western part of Peninsular Malaysia and is constructed extensively soft soil areas such as Alor Setar, Kedah, Georgetown, Penang and western regions is Johor. In an article of the Sunday Daily Mail, dated 18th June 2014, it mentions that Plus Malaysia Berhad (PLUS) spends some RM600 million or 20% of the company's annual toll collection for highway maintenance and upgrading works [3].

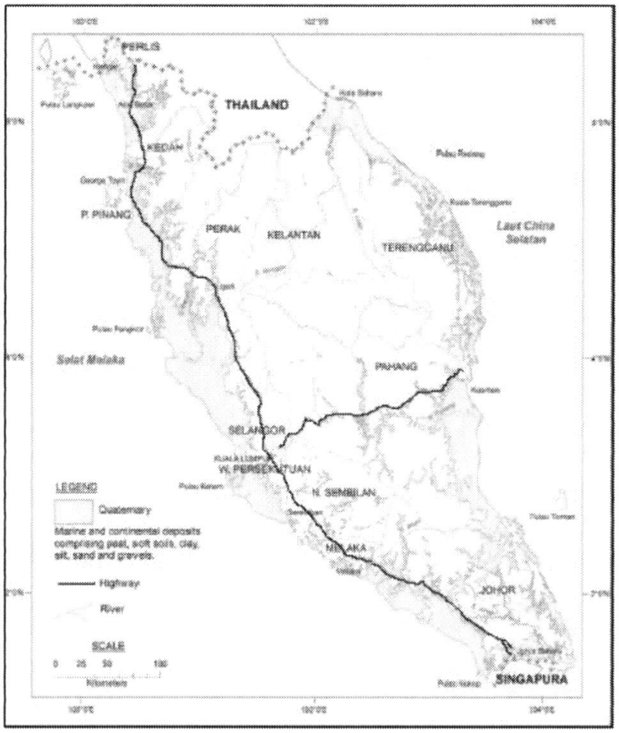

Figure 2. General Distribution of Quaternary Deposits including Peat and Soft Soils in Peninsular Malaysia [4]

All soil types are subjected to settlement due to compressive loads with consequent reduction of volume filled in the ground. The evaluation of settlement evolution should be carried out during the design process with the aim to limit the amount of residual settlement occurring at the end of construction [5]. The amount of total or differential settlement that a structure can undergo without resulting in unsightly building cracking, structural damage, or loss of intended use is dependent on the type of structure, structural system type, and planned use. The structural engineer provides the total and differential settlement tolerances for the structure. But these values should not be "zero," because every structure that imparts a load on the ground must cause some settlement, even if it is very small [6].

2.1 Case studies of soft soil turning uneven and creating hazardous infrastructures

Research conducted by personnel from the Public Work Department (JKR) in Malaysia [7], shows statistics of geotechnical forensic investigation on problematic projects carried out as summarized in table 1 since 2010 to 2015. There are 182 cases out of 252 forensic highway distress cases (approximately 72%) are related to the issue of ground settlement, the remaining 28% are caused by other factors such as vibration, erosion, foundation failures and so forth.

IOP Conf. Series: Journal of Physics: Conf. Series **1049** (2018) 012046 doi:10.1088/1742-6596/1049/1/012046

Table 1. Geotechnical forensic cases carried out by JKR [7]

	Year						
Contributing Factors	2010	2011	2012	2013	2014	2015	Total
Vibration				2	1	3	6
Soil erosion				3	2	4	9
Foundation failures		2	1	6	3		12
Retaining structure failures			1	4	2		7
Slope stability			6	3	12	4	25
Ground settlement	1	8	10	63	70	30	182
Water infiltration				4	4	1	9
Collapsible soil			1		1		2
Total	1	10	19	85	95	42	252

2.1.1 Settlement of bridge embankment in Parit Yaani.

A local and current case closely monitored by the authors in Jalan Sengkuang, Sri Bengkal, Parit Yaani is a site where the approach fill embankment to a bridge, built by JKR in 2010 has often undergone pavement distress due to settlements under the embankment loading. Here the sub grade soil largely consist of marine clay. After the completion of the bridge's construction and during its liability period, the edge/slant which is supported by the reinforced concrete bridge had undergone movement due to settlement in the horizontal direction and the resulting hazardous road surface deformations have had to be continuously repaired, every 7-8 months at a considerable road repair cost as shown in figure 3. To solve the settlement problem, JKR has repaired the embankments by adding fill materials on the settled area to make it even again, however this solution has not been effective as the replaced soil fill induces further settlement of the subgrade and the embankment becomes uneven again.

Figure 3. Chronology of pavement distress observed at Parit Yaani Bridge embankment (Oct 2015 – Jan 2016 – Oct 2016 – Jan 2017)

3. Innovative approaches in Highway Construction Materials

In civil engineering, the lightweight construction technology is adopted to describe particular methods for increasing economic and structural efficiency in reducing the weight of building structural constructions together with costs and construction lead time with the use of special construction materials. Highway 'failure' occurs due to settlement of subgrade affected by dead load, and not necessarily live load, which is momentary and has less effect on subgrade settlement, the awareness to use lightweight fill technologies as a means to develop more sustainable highways is crucial, [8]. Suitable for soft organic soil or peat soils that tends to settle unevenly due to its geological history or in man-made situations such as back filled inactive mining sites. There is no limit on the thickness of the lightweight fill to be adopted but developments in sustainable material sciences have widened the choice of such materials. Table 2, [8] shows some recent researches of alternative lightweight fills and their properties. RAP shown in the table is an example of sustainability in the form of recycling being adopted to form a flexible pavement. This is still heavy (compared to soil) and does not fully arrest the bumpy settlements that are caused when dead and live loadings are imposed on soft subgrades of variable thickness. As a consequence the roads demand frequent expensive maintenance.

Geofoam blocks in the form of lightweight expanded polystyrene (EPS) blocks have been popularly adopted in highway construction as a pioneering stiff and discontinuous mat structure (conforming to the brickwork concept described later in section 4 – and in contrast to the stiff linear structures used with logs in the road in Ur [2]. EPS is a manufactured material, and is affected by buoyancy, rodent attack and poor fire rating. Waste tyres have been bundled together to form irregular but not light mats are being used in Canada to make access roads over marshy soft ground conditions. Sustainable, stiff and light mat structures are being developed at UTHM and more information on these will be published as laboratory and field data becomes available. Ismail [9] reports on the performance of the use of such a product made of recycled plastic tubes.

Table 2. Recent Published Research on Alternative Lightweight Fill Material [8]

Alternative Composite Material & Production Method	**Functional use, Technical Properties & Approximate Costs**
a)Foamed recycled glass ; Low weight foamed recycled glass aggregates of up to 40 mm in size	***Non-structural fills in embankments, Retaining wall backfill and pipe bedding;*** Gap graded material Max. Dry density = 2.84 kN/ m3 LA Abrasion = 94%
b)Recycled Asphalt Pavement(RAP) – Fly ash(FA) Geopolymer A 30 % FA mixture and a 50:50 ratio of NaOH and Na2SiO3 is used at a curing temperature of 40 C.	***Stabilized pavement base course material;*** Compressive strength= 5.3 MPa Fly Ash 1 tonne – Price insignificant. Transporation cost and safety certificate required due to hazardous nature.
c)Cement-treated sand and expanded polystyrene (EPS) beads EPS beads are heated with steam to expand 50 times its original size. Cement is included to bind the cohesionless sand and EPS beads.	***Backfill material for use in highway embankments in soft soil areas*** Compression Strength = 800 KPa Unit Weight = 10 kN/m^3
d)Clays Modified with Recycled Crumb Rubber Kaolin clay and bentonite were used as soils for modification. From the weight of the soil, 2-4 % of rubber crumbs are used to give maximum strength.	***Soil modification*** Max dry density = 1296 kg/m^3 Max. Strength = 482 kPa
e)Blast furnace slag The slag in liquid for is cooled and then transported to a crushing and screening plant where it is further processed into various products including aggregates.	***As fill material for subgrade, sub-base or base.*** Particle Density= 2,550 — 2,650 kg/m^3 LA abrasion= 37 — 43% OMC= 8 — 12 % RM 50.10 /metric tonne
f)Geofoam Polymerization of styrene monomer is followed by impregnation of the polymerized polystyrene beads with a blowing agent. Styrene monomer and water is charged to the reaction kettle equipped with an agitator.	**As fill material for highways and embankments on peat or soft soil.** Min. apparent density= 15 kg m^3 Compressive stress=60 – 110 kPa Flexural strength= 60 – 300 kPa Shear strength= 80 – 130 kPa Tensile strength= 110 – 290kPa 1.2m x 1.8m x 0.6m ~RM 300.0

3.1Changing Rags to Riches (Recycling Waste Materials)

About 2 million tonnes of resins for the plastics industry are produced locally in Malaysia per annum, [10] and out of which 40% of the plastic resin are used for packaging. Polyethylene terephthalate (PET) where drinking water plastic bottles are made is the most widely recycled plastic in the world. PET is a plastic resin and a form of polyester that is formed by combining two monomers: modified ethylene glycol and purified terephthalic acid. PET plastic bottles, commonly small, portable 500ml and 1500ml sizes, are considered safe and reliable for food contact use. PET is used for numerous types of packaging for many foods, including everything from ketchup, peanut butter, soft drinks, and juices to beer, wine and spirits, [11]. In an online environmental awareness page, The World Counts, and under its Bottled Water Waste Facts [12], it mentions that more than 100 million plastic bottles are used up in the world every day and only 1 out of 5 bottles end up being recycled. Thus 80% of the plastic bottles end up being buried in landfills, and it takes approximately about 700 years for a PET plastic bottle to start decaying. Thus, realising the magnitude of the harm waste plastic bottles can cause to the environment leads innovating sustainable use of these materials in the construction industry. One of the sustainable mats being produced utilises the waste plastic bottles in block form.

4. Exploration of the masonry arch concept

According to Northern Architecture [13], masonry bridges in the United Kingdom has over 120 designs and most have already exceeded the Department of Transport's design life of 120 years and therefore must be considered as an archive of good practice and proportion. The masonry arch concept is based on an arch bridge shaped as a curved arch and supported on abutments at each end. This is illustrated in Figure 4. Arch bridges work by transferring the weight of the bridge and its loads as vertical abutment support reactions and also partially into a horizontal thrust restrained by the abutments at either side. This is also synonymous with the lintel and column construction in structures. Both structures remain stable without any settlement even though they traverse a void beneath them. Traditional masonry arches are generally durable, (by virtue of its unique fit to size construction technique) and somewhat resistant to settlement or undermining. However, relative to modern alternatives, such bridges are very heavy, requiring extensive foundations. They are also expensive to build wherever labour costs are high. Thus by replacing the heavier materials with lightweight fill material, the setbacks can be overcome.

Figure 4. Vertical bricks laid in the arch of the bridge and most bridges like the Erriff bridge are aged over 150 years [13]

5. Conclusions

Developing sustainable technologies to minimise problematic road embankment settlements is becoming more crucial, especially in finding a balance between having a long term solution, going green and constructing at an affordable cost. Understanding the measure of impact that construction designs can have on the settlement of embankments leads to an advance in the research of alternative materials and innovative design concepts. However, ideas and concepts should eventually crystallize, thus the research team is set on working to develop a sustainable technology based on the masonry arch concept.

6. References

[1] Unit E P Malaysia 2016 *11th Malaysian Plan* (2016-2020).
[2] Information available at http://sumerianshakespeare.com/mediac/450_0/media/3d3dfff71aa0f304ffff8169ffffe417 . Jpg [Accessed 12 Jan. 2014]
[3] Chelvy S T 2014 *Thesundaily*.my *Plus spends RM600 mil for annual maintenance.* [online] Available at: http://www.thesundaily.my/news/1086212 [Accessed 18 Sep. 2017].
[4] Construction Research Institute of Malaysia 2015 *Guidelines for Construction on Peat and Organic soils in Malaysia*, page 117-120.
[5] Marradi A, Pinori U, Betti G 2012 *The use of lightweight materials in road embankment construction*. Procedia-Social and Behavioral Sciences 53:1000-9
[6] Beltzer B, Gerhardt B R, Kubes K A & Smith S J 2013 *Goetechnical Engineering, Chapter 1 from Infrastructure from the Ground Up*. ISBN: 978-1-61438-763-3
[7] Mohamad N O, Razali C E, Hadi A A A, Som P P, Eng B C, RuslI M B, & Mohamad F R 2015 Challenges in Construction Over Soft Soil – Case Studies in Malaysia. *IOP Conf. Series: Materials Science and Engineering* 136 (2016) 012002 doi:10.1088/1757-899X/136/1/012002
[8] Wijeyesekera D C, Ratha N, Siang A J 2017 Critique on the Conceptual Developments of Innovative Practices with Lightweight Fills *In MATEC Web of Conferences* ***103***, p. 07015 EDP Sciences.
[9] Ismail T N H T 2017 *A critical performance study of innovative lightweight fill to mitigate settlement of embankment constructed on peat soil*, PhD thesis, University Tun Hussein Onn Malaysia, Johor, Malaysia
[10] Malaysian Plastics Manufacturer Association (MPMA) 2010 *Development of the Malaysian Plastics Industry and Its Contribution to the Economy*. Available online: http://jpspn.kpkt.gov.my/resources/index/user_1/Sumber_Rujukan/kajian/JPSPN%20Plastic%20Study%20-%20Final%20Report%20GESB%20-%20Softcopy%20English%20Ed2.pdf
[11] International Bottled Water Association (IBWA) 2014 *PET Facts | IBWA | Bottled Water*. [online] Available at: http://www.bottledwater.org/education/recycling/pet-facts[Accessed 18 Sep. 2017].
[12] The World Counts 2014 *Bottled Water Waste Facts*. [online] Available at: http://www.theworldcounts.com/stories/Bottled_Water_Waste_Facts [Accessed 18 Sep. 2017]
[13] Hesson R 2017 *Northern Architecture: Design of masonry arch bridges - Bridge Engineering* [online] Available at: https://www.northernarchitecture.us/bridge-engineering/design-of-%20masonry-arch-bridges.html [Accessed 18 Sep. 2017]

7. Acknowledgements

Different research grants (MyBrain15, MTUN, FRGS and even anonymous research funding) have collaborated and helped contribute to the research outlined in this paper.

IOP Conf. Series: Journal of Physics: Conf. Series **1049** (2018) 012047 doi:10.1088/1742-6596/1049/1/012047

Performance of Artificial Neural Network and Particle Swarm Optimization Technique based Maximum Power Point Tracking for Photovoltaic System Under Different Environmental Conditions

Said Zakaria Said[1,2], **Lamine Thiaw**[3]
[1]Department of Electrical and Electronic Engineering, Pan African University Institute for Basic Sciences Technology and Innovation, PAUSTI, Nairobi, Kenya.
[1]Department of Mechanical Engineering, Polytechnic University Institute of Mongo, Ndjamena, Chad.
[1]Department of Electrical Engineering, Ecole Superieure Polytechnique, Chekh Anta Diop University UCAD, Dakar, Senegal.

E-mail: djiddosaid@gmail.com

Abstract. Photovoltaic (PV) array may receive different levels of solar irradiance and temperature under different environmental conditions, such as partially shaded by clouds or nearby building. However, all PV systems have two major drawbacks: the efficiency of PV power generation is very low and the output power of a PV system is nonlinear, which depends closely on weather conditions, such as ambient temperature and the solar irradiance. Hence, tracking the maximum power of the PV arrays at real time is very important to increase the whole system performance. Multiple peak power points occur when PV module is under partially shaded conditions, which would significantly reduce the energy produced by PV without proper control. Therefore, a Maximum Power Point Tracking (MPPT) algorithm is used to extract maximum available PV power from the PV array. However, most of the conventional MPPT algorithms are incapable to detect global peak (GP) power point with the presence of several local peaks (LP). A hybrid Artificial Neural Network and Particle Swarm Optimization (ANN-PSO) algorithm is proposed in this report to detect the global peak power. A PV system which consists of PV array, DC–DC boost converter, a hybrid ANN-PSO Algorithm, and a resistive load, is simulated using MATLAB/Simulink. The simulation results are carried out, compared and discussed. The proposed algorithm should perform well to detect the Global Peak of the PV array under different environmental conditions.

1. Introduction

Photovoltaic (PV) is one of the most promising renewable sources due to its environmental friendliness and low maintenance cost [1,2]. However, there are two major challenges that need to be tackled for small-scale PV systems to be implemented: (1) high installation cost, and (2) low efficiency in PV conversion [3,4]. Moreover, the PV output characteristics are nonlinear as it always varies with solar irradiance and module temperature. Due to these characteristics, a maximum power point tracking (MPPT) controller is utilized to extract the maximum available power from PV array. The MPPT algorithm is used to control the duty cycle of the DC–DC or DC–AC converter which is inserted in between the PV modules and the load [2, 4].

Content from this work may be used under the terms of the Creative Commons Attribution 3.0 licence. Any further distribution of this work must maintain attribution to the author(s) and the title of the work, journal citation and DOI.
Published under licence by IOP Publishing Ltd

Over the years, numerous MPPT algorithms for PV array under uniform irradiance have been proposed [5]. They widely used techniques including Perturb and Observe (P&O) [5, 6], Incremental Conductance (IC) [6], Hill Climbing (HC) [7, 8], open-circuit voltage [8], and short-circuit current algorithm [8, 11]. Recently, several artificial intelligent methods, i.e., Fuzzy Logic Controller (FLC) [9, 10], Artificial Neural Network (ANN) [7, 9, 10] are explored. The above-mentioned conventional MPPT algorithms are not capable of tracking the true Maximum Power Point (MPP) if the PV array is partially shaded, such as covered by heavy clouds, falling tree leaves, birds' litters on the array or shaded by buildings [10] and so on. PV cells under low illumination (shaded condition) could be damaged by overheating problem (known as "hot-spot heating") attributed to a larger current that flows from other fully illuminated PV cells. This problem can be overcome by inserting bypass diode across the PV cells [11]. However, the insertion of bypass diodes creates multiple peaks, namely global peak (GP) and local peaks (LP), in which only the GP is the true MPP on P–V characteristic curve. The conventional algorithms are not intelligent enough to differentiate among the Global Peak and the Local Peak, where the control of operating point tends to linger around the LP [12].
Particle Swarm Optimization (PSO) is a global gradient less stochastic search method. It is used to search for continuous variable for optimization problems [11, 12]. Artificial Neural Network (ANN) is an information processing paradigm, which is based on the functional concepts of biological nervous systems. In this work, a performance of PSO and ANN technique is proposed to extract the GP under different environmental conditions. The ANN algorithm initializes the optimal voltage, initial value at the prevailing solar irradiance and temperature levels and aids the PSO algorithm to locate GP. This initial voltage is then fed into the PSO to reach the GP location. Therefore, PSO can reach the true GP in a shortest computational time. This avoids the operating point from lingering at LP and guarantees the reach of GP.

2. Particle Swarm Optimization and Artificial Neural Network technique based Maximum Power Point Tracking

2.1. Description of the ANN algorithm
The inputs of the ANN algorithm are the solar radiation **G**, the temperature **T** and the PV short-circuit current ***Ipv*** and the output of the algorithm is the voltage ***Vpv*** with thirty hidden neurons (layers) as shown in Figure 1.

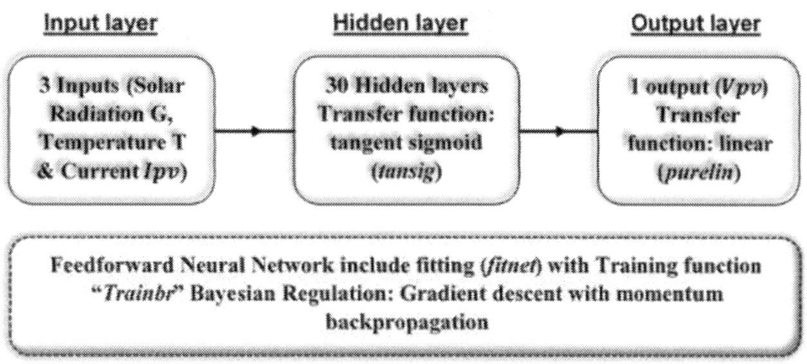

Figure 1. The block diagram of the specification of ANN algorithm in the simulation.

In Neural Network Algorithm (Figure 2), we use feedforward neural network with the most widely recommended, backpropagation method the "*fitnet*". The fitting networks are feedforward neural networks (*feedforwardnet*) used to fit an input-output relationship. The transfer functions for the hidden layer are the "*sigmoid function*" and the transfer functions for the output layer "*purelin function*".

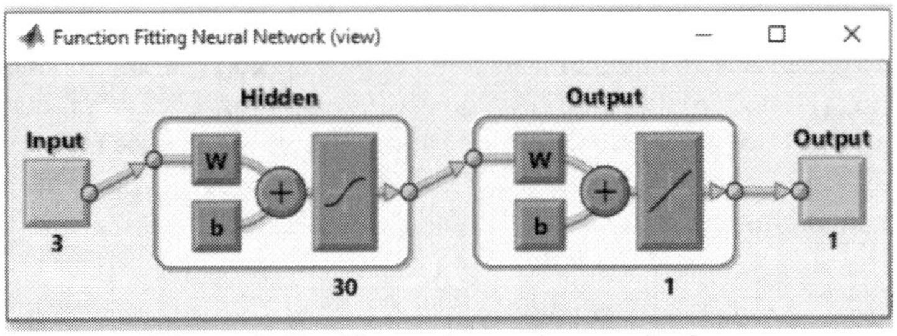

Figure 2. Feedforward neural network view.

The performance of ANN algorithm is defined by the graphs of the mean squared error (MSE) against number of epochs (Figure 3). It can be noticed that train set error, and validation set error have same characteristics as the blue line converges to the best for the 2655 epochs graph. Its best validation performance is 0.0073. Therefore, 2655 epochs are chosen for the ANN algorithm.

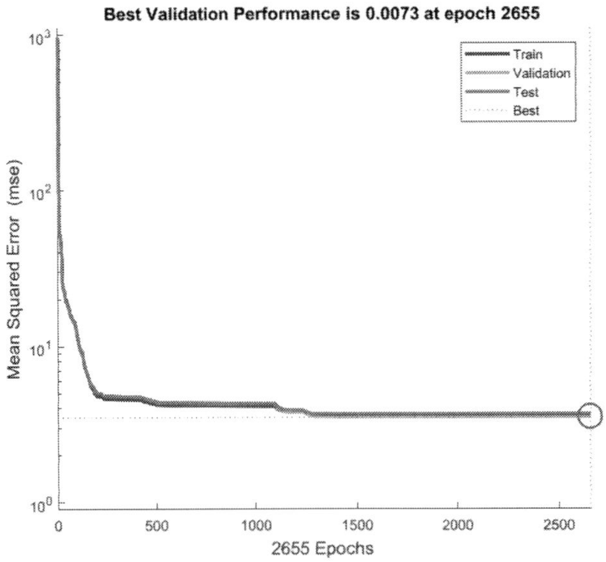

Figure 3. Best Regression for Feed-forward Back propagation with BR training function.

2.2. Particle Swarm Optimization Algorithm: Codes in MATLAB environment

2.2.1. Problem formulation of MPPT

The objective function (OF) required here must be a maximization problem. The maximum power point tracking of photovoltaic generators can be formulated as an optimization problem where the objective is to maximize the power by summing the operating voltages of all the photovoltaic generators. Therefore, the objective function (OF) is expressed as:

$$P(OF) = fitness(I_{pv}, G_1, T_1, G_2, T_2, G_3, T_3, G_4, T_4) \quad (1)$$
$$V_1 = NN\_PV([I_{pv}; G_1; T_1]) \quad (2)$$
$$V_2 = NN\_PV([I_{pv}; G_2; T_2]) \quad (3)$$
$$V_3 = NN\_PV([I_{pv}; G_3; T_3]) \quad (4)$$
$$V_4 = NN\_PV([I_{pv}; G_4; T_4]) \quad (5)$$

In the fitness function, for each PV module, the voltage *Vpv* is determined using the current value (the Position x) defined in the PSO Algorithm, Solar Radiation G and Temperature T of the PV Generators obtained from the model ANN of the latter. The objective function defined above is subjected to:

$$P_{ANN\_PSO} = max \sum_1^n V_{\text{pv,n}} \times I_{best} \tag{6}$$

$$V_{opt} = V_1 + V_2 + V_3 + V_4 \tag{7}$$

$$P_{ANN\_PSO} = V_{opt} \times I_{best} = I_{best} \times (V_1 + V_2 + V_3 + V_4) \tag{8}$$

Where P_{ANN_PSO} is the power at the MPP given by the hybrid algorithm, V_{opt} the output voltage of the PV array Generators as the sum of the voltages across all Generators in series and *NN_PV* is the neural network of the PV generators, V is the voltage of the module, I is the current of the PV array, G is the solar radiation and T is the temperature.

2.2.2. PSO Algorithm: Main program

The initial population (swarm) of size N and dimension D is denoted as $X = [X_1, X_2,...,X_N]^T$, where 'T' denotes the transpose operator. Each individual (particle) X_i $(i = 1, 2, ..., N)$ is given as $X_i=[X_{i,1}, X_{i,2},...,X_{i,D}]$. Also, the initial velocity of the population is denoted as $V=[V_1, V_2,..., V_N]^T$. Thus, the velocity of each particle X_i $(i = 1, 2, ..., N)$ is given as $V_i=[V_{i,1}, V_{i,2}, ..., V_{i,D}]$. The index i varies from 1 to N whereas the index j varies from 1 to D. The detailed algorithms of various methods are described below for completeness [11].

$$V_i^{k+1} = \omega V_i^k + c_1 r_1 \left(Pbest_i^k - X_i^k\right) + c_2 r_2 \left(Gbest_i^k - X_i^k\right) \tag{9}$$

$$X_i^{k+1} = X_i^k + V_i^{k+1} \tag{10}$$

In equation (9), $Pbest_{i,j}^k$ represents *personal best* j^{th} component of ith individual, whereas $Gbest_j^k$ represents j^{th} component of the *best individual* of population up to iteration k. Figure 2 shows the search mechanism of PSO in multidimensional search space [12,14].

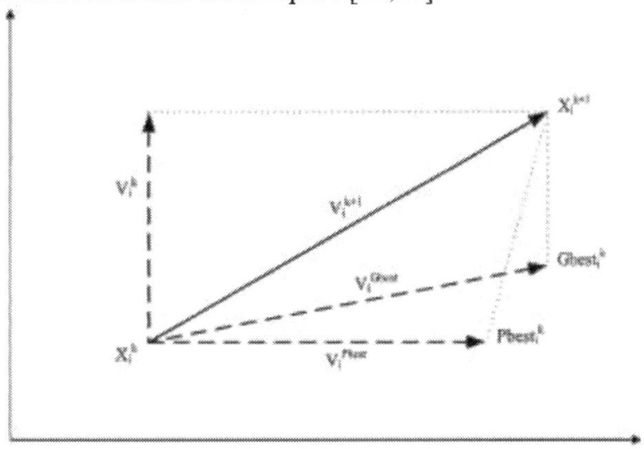

Figure 4. PSO search mechanism in multidimensional search space.

The different steps of PSO are as follows [11]:

Step 1: Set parameters W_{min} and W_{max}, $c1$ and $c2$ of PSO

Step 2: Initialize population of particles having positions $\boldsymbol{X}$ and velocities V

Step 3: Set iteration $k = 1$

Step 4: Calculate fitness of particles $F_i^k = f(X_i^k)$, $\forall i$ and find the index of the best particle b,

Step 5: Select $Pbest_i^k = X_i^k$ $\forall i$ and $Gbest^k = X_b^k$

Step 6: $\omega = \omega_{max} - \frac{\omega_{max} - \omega_{min}}{Iter_{max}} \times Iter$

Step 7: Update velocity and position of particles

$V_i^{k+1} = \omega \times V_i^k + c_1 rand() \times \left(Pbest_i^k - X_i^k\right) + c_2 rand \times \left(Gbest_i^k - X_i^k\right)$

$X_i^{k+1} = X_i^k + V_i^{k+1}$

Step 8: Evaluate fitness $F_i^{k+1} = f(X_i^{k+1})$, $\forall \boldsymbol{i}$ and find the index of the best particle $b1$,

Step 9: Update $Pbest$ of population $\forall i$,
If $F_{b1}^{k+1} > F_b^k$) then $Pbest_i^{k+1} = X_i^{k+1}$, else $Pbest_i^{k+1} = Pbest_i^k$,
Step 10: Update $Gbest$ of population,
If $F_{b1}^{k+1} > F_b^k$) then $Gbest^{k+1} = Pbest_{bi}^{k+1}$, else $Gbest^{k+1} = Gbest^k$,
Step 11: Print optimum solution as $Gbest^k$.
The parameters used in the PSO algorithm are considered as follows:

- Inertial weight: $\omega_{max} = 0.9$ to $\omega_{min} = 0.4$
- Acceleration factors (*c1* and *c2*): *c1=1.2* to *c2=1.6*
- Population size: *10* to *100*
- Maximum iteration (*MaxIt*): 50 to 1000
- Initial velocity: *10 %* of position

The block diagram which show the combination of the hybrid ANN and PSO Algorithm is illustrated in Figure 5.

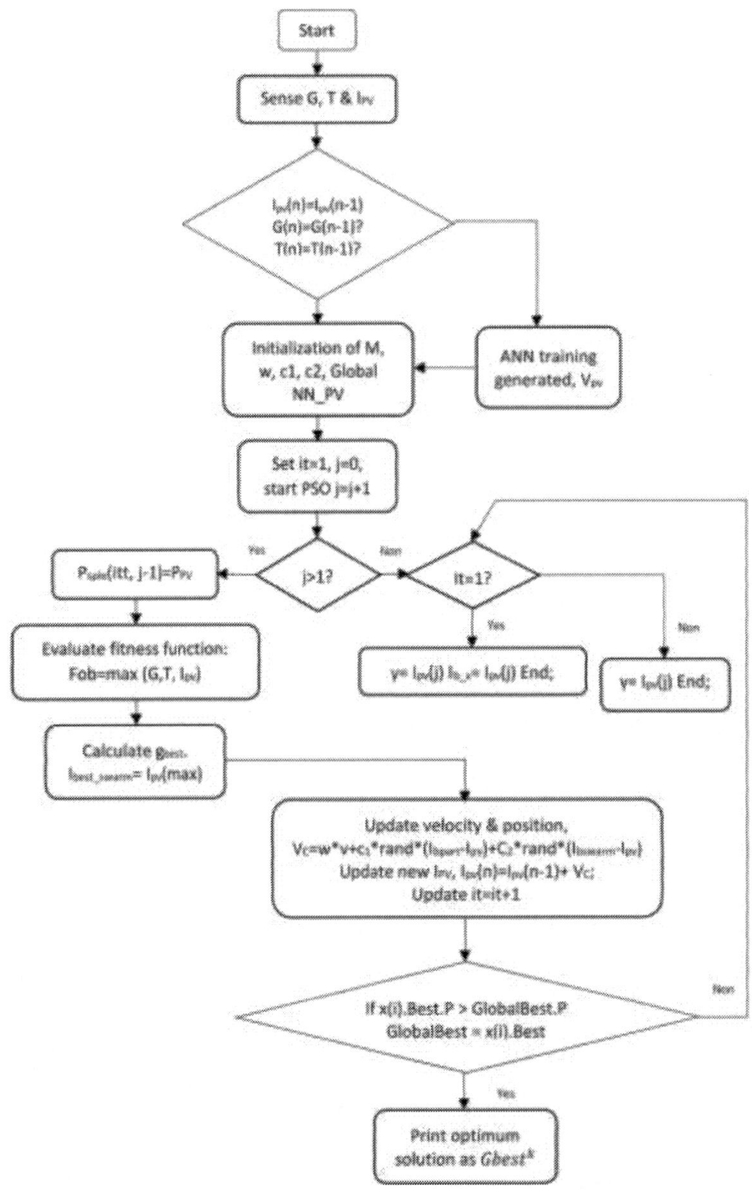

Figure 5. Flowchart of the proposed ANN-PSO Algorithm.

3. Modelling and simulation setup

3.1. Design and implementation process of scale solar photovoltaic

The simulation setup consists of four PV modules connected in series to form the PV array as shown in Figure 6. Each PV module is modelled using the single diode equations with bypass diodes connected across a group of PV cells based on the specifications of 1Soltech 1STH-215-Polycrystalline solar panel [11]. The specifications of the described PV module are tabulated in Table 1. The inputs of the PV array are the values of solar irradiance and temperature. Therefore, the inputs of the PV modules were varied for different environmental weather conditions.

Table 1: The specifications of 1Soltech 1STH-215-Polycrystalline PV module

Module data & Parameters	**Symbol**	**Typical value**
Open circuit voltage	Voc	36.9 V
Short circuit current	Isc	8.09 A
Maximum power voltage	Vmp	29.6 V
Maximum power current	Imp	7.58 A
Maximum Power	Pmax	224.368 W
Temp. coefficient of current	Ki	0.12 %/°C
Temp. coefficient of voltage	Kv	-0.36 %/°C
Number of solar cell	Ncell	60
Light generated current	IL	8.1056 A
Shunt resistance	Rsh	333.95 ohms
Series resistance	Rs	0.37039 ohms
Irradiance at STC	G	1000 W/m^2
Temperature at STC	T	25°C

As in Figure 6 shows the final model of PV array developed using Simulink includes irradiation and temperature as the inputs parameter and in voltage output *Vpv* and current output *Ipv* as the results. This model simulates a photovoltaic string for different environmental conditions and plot its I-V and P-V curves by changing the values of radiation and temperature. Here are the steps to follow to draw the I-V and P-V characteristic curves by setting the irradiance and temperature profile in the Solar Radiation Block and run Simulink.

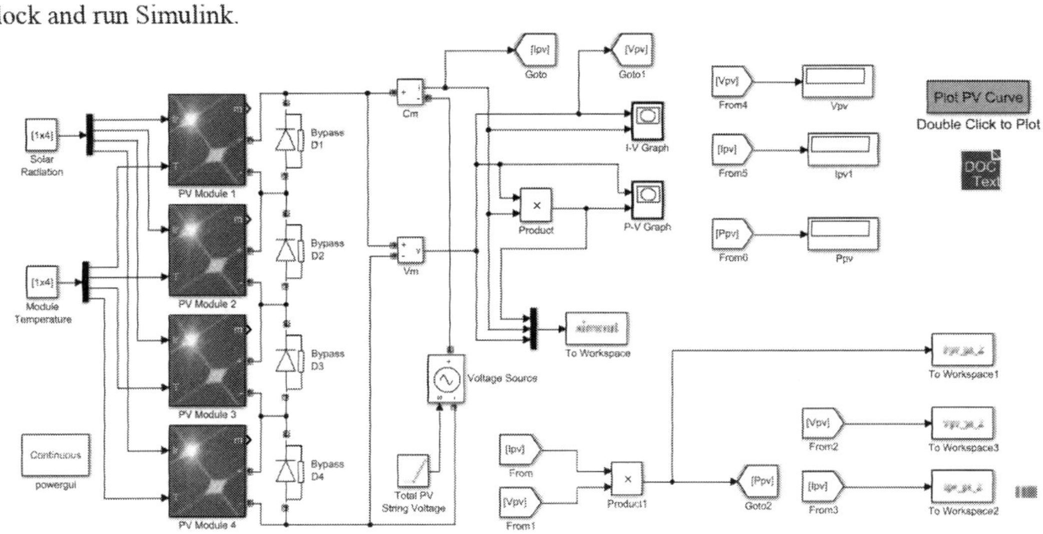

Figure 6. PV String Simulink Model.

3.2. Boost converter design and components parameters

The boost power stage is utilized in power supply because the required output voltage is always higher than the input voltage. The input current for a boost power stage is continuous and is the same as the inductor current. The output capacitor supplies the entire load current for the switching cycle. Selection of component parameters [12]:

Table 2: Boost power converter initial components parameters

Components parameters	Symbol	Typical value
Minimum input voltage	Vin	128 V
Switching frequency	fs	25 kHz
Initial output voltage	Vout	300 V
Power converter	Pc	897.313 W
Conversion efficiency of the boost	n	0.9
percent of the output voltage	Dv_percent	1

The boost converter parameter calculation and its design in Matlab Simulink.

1. D: Duty cycle

$$D = 1 - \frac{Vin x n}{Vout} \qquad \%D = 1 - \frac{128 \times 0.9}{300} = \%61.60 \tag{11}$$

2. di: input current

I_{ripple}: A good estimation for the inductor ripple current is 20% to 40% of the output current

$$di = I_{ripple} \times I_{out} \times \frac{Vout}{Vint} = 0.4 \times 3.33 \times \frac{300}{128} = 3.122\, A \tag{12}$$

Iout = 3.33 A

3. L: inductance(Henry)

$$L = \frac{Vin \times (Vout - Vin)}{(di \times fs \times Vout)} = \frac{128 \times (300-128)}{(0.9375 \times 25{,}000 \times 300)} = 0.00313\, H \tag{13}$$

4. dv: Output voltage ripple

$$\mathrm{dv} = \mathrm{Vout} \times \mathrm{dv\_percent}/100 = 300 \times 1/100 = 3\,\mathrm{V} \tag{14}$$

5. C: Capacitor (Farad)

$$C = \frac{Iout \times D}{fs \times dv} = \frac{Iout \times 0.616}{25{,}000 \times 3} = 0.0000273\, F \tag{15}$$

6. R: Resistive load

$$R = \frac{Vout}{Iout} = \frac{300}{3.33} = 90.10\, \Omega \tag{16}$$

7. The above equation is usually expressed as a ratio of the output voltage, V_{out}, to the input voltage, V_{int}, and is usually called M as shown:

$$M = \frac{Vout}{Vint} = \frac{1}{1-D} \times \frac{1}{1+\frac{R_L}{R \cdot (1-D)^2}} = \frac{300}{128} = 2.344 \tag{17}$$

D = 0.616; Vin_min = 128; fs = 25,000; L = 87.04×10^{-4} ; C = 1.11606×10^{-4} ; R = 90.10 ; M = 2.344.

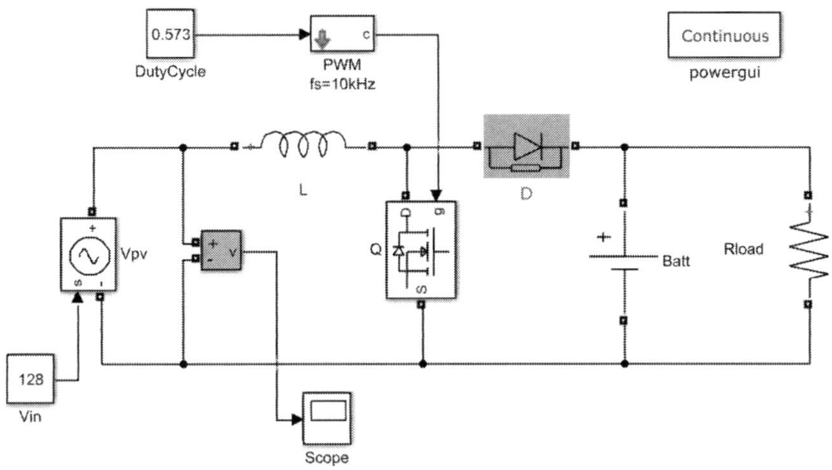

Figure 7. DC-DC boost converter Simulink model.

To reach the desired maximal voltage level, a DC–DC boost converter is used as the power stage interface between the PV system and the resistive load. The pulse width modulation (PWM) technique achieved by the MPPT duty cycle is applied to electronic switch of converter which is MOSFET(Q). The duty cycle is expressed by the following parameters.

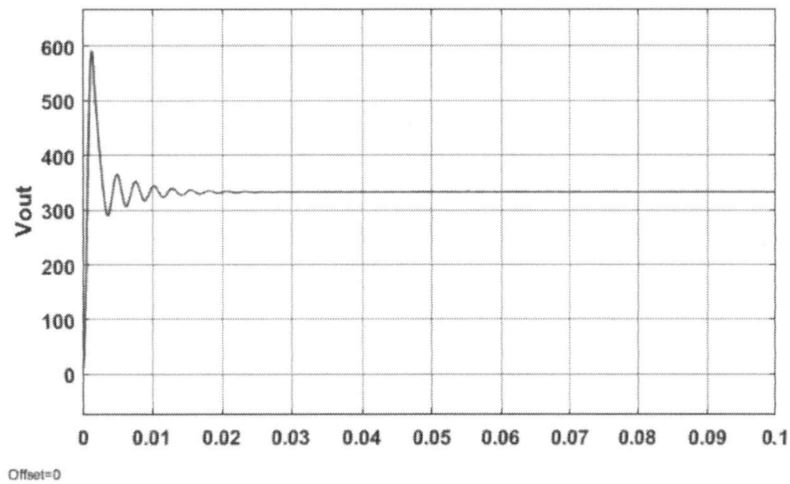

Figure 8. Output voltage signal of the boost converter.

3.3. Modelling of the MPPT PV system

The simulation setup consists of four PV modules connected in series to form the PV array as shown in Figure 9. The module data and model parameters of the described PV module 1Soltech 1STH-215-Polycrystalline solar panel are tabulated in Table 1. The inputs of the PV array are the values of solar irradiance and temperature.

The input of the DC-DC boost converter is the PV voltage which is generated by the PV array. The proposed hybrid PSO-ANN algorithm worked with PID controller to control the PWM block. The feedforward neural network algorithm consists of three inputs (solar irradiance G, Temperature T and the PV current $\boldsymbol{I_{pv}}$ for the four PV modules) and one output (PV voltage $\boldsymbol{V_{pv}}$).

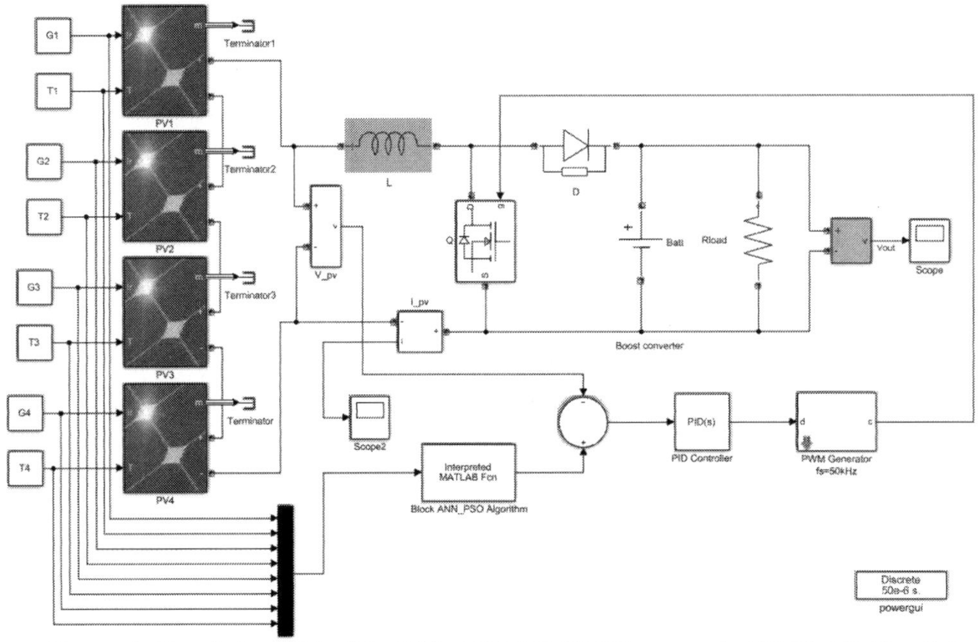

Figure 9. Simulink model of the PSO-ANN MPPT PV system in MATLAB/Simulink.

4. Results and discussion

4.1. Varying irradiation, fixed temperature

As the results based per pattern in the Table 2 are proven, the I-V and P-V characteristics with varying irradiation and temperatures are obtained.

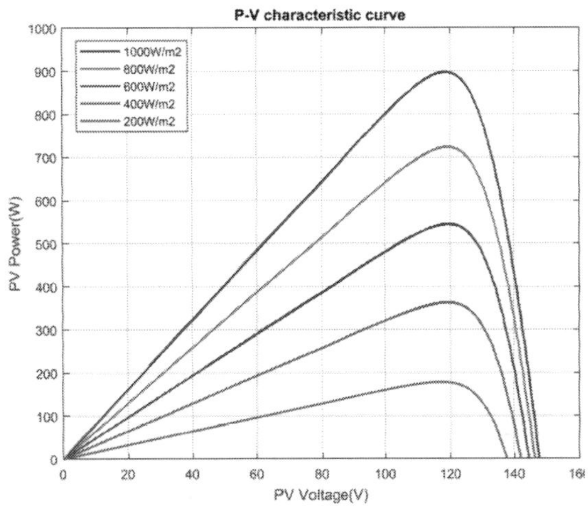

Figure 10. I-V characteristic varying irradiation, fixed temperature.

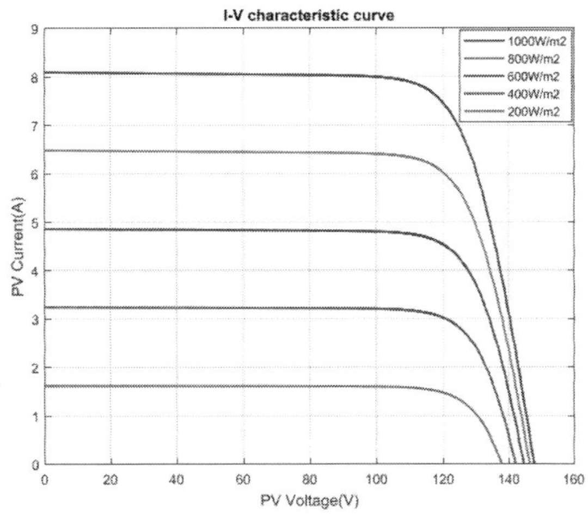

Figure 11. I-V characteristic varying irradiation, fixed temperature.

Figures 10 and 11 show the I-V and P-V output characteristic of PV module. The inputs for irradiation are varied by five different values of irradiation which are 1000W/m^2, 800W.m^{-2}, 600W.m^{-2}, 400W.m^{-2} and 200W.m^{-2} respectively. The temperature is fixed and it set as the value of PV panel reference temperature, Tr = 25°C. The tracking efficiency of the MPPT algorithm are evaluated [12] by:

$$E = P_{PSO\_ANN}/P_{max} \times 100 \tag{18}$$

Table 3: Varying irradiation G, fixed temperature Tr patterns for PV generators

	Ambient Condition: Tr = 25°C			
Case	**Irradiance**	**Vmpp(V)**	**Impp(A)**	**Pmpp(W)**
Pattern1	200W.m^{-2}	117.0750	1.5227	178.2681
Pattern2	400W.m^{-2}	119.1015	3.0436	362.5009
Pattern3	600W.m^{-2}	119.4375	4.5607	544.7186
Pattern4	***800W.m^{-2}***	***119.1015***	***6.0725***	***723.2465***
Pattern5	1000W.m^{-2}	118.4085	7.5781	897.3083

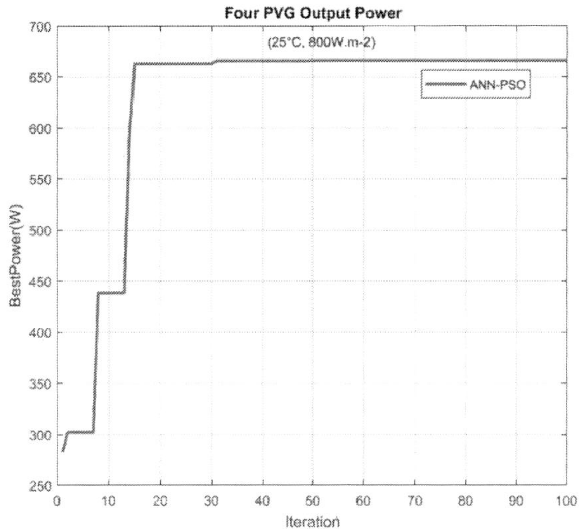

Figure 12. Performance of ANN-PSO output power at varying irradiation-fixed temperature.

$$E_1 = \frac{P_{PSO\_ANN}}{P_{max}} \times 100 = \frac{665.7782}{723.2465} = 92.05\% \quad (19)$$

4.2. Fixed irradiation, varying temperature

The irradiation input is set to 1000 W/m^2 and the temperatures are set as 15°C, 25°C, 35°C, 45°C and 55°C. This is to observe the PV characteristics output of the PV module when varying the irradiation shown in Figures 13 and 14. As for the increases in operating temperature, the output current increases while the voltage drops accordingly. This leads to decreasing value in PV power output.

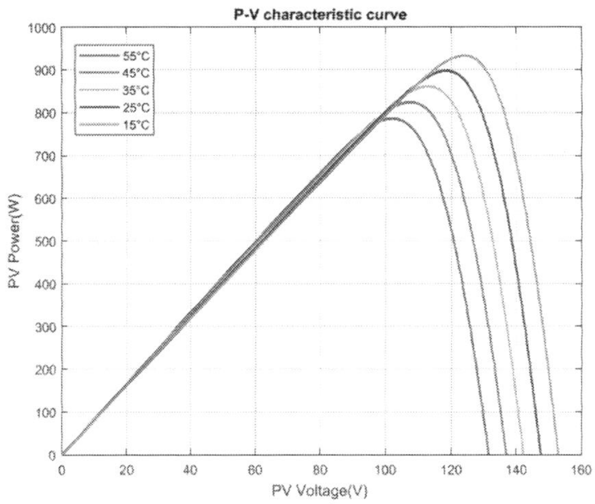

Figure 13. P-V characteristics fixed irradiation, varying temperature.

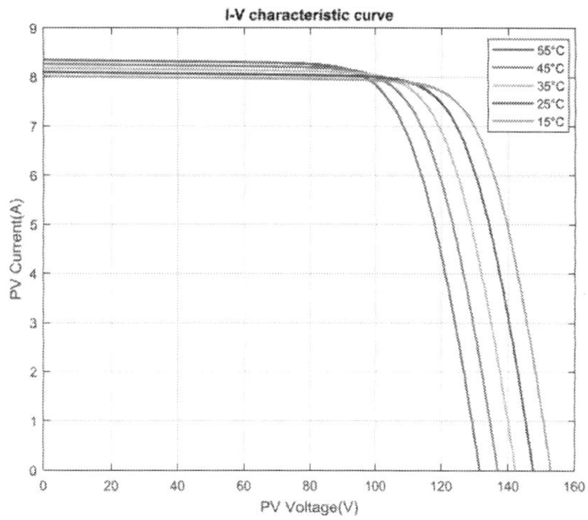

Figure 14. I-V characteristics fixed irradiation, varying temperature.

Table 4: Fixed irradiation G, varying temperature Tr patterns for PV generators

	Ambient Conditions: G=1000 W/m²			
Case	**Temperature**	**Vmpp(V)**	**Impp(A)**	**Pmpp(W)**
Pattern1	55°C	102.0705	7.6984	785.7795
Pattern2	45°C	107.4780	7.6636	823.6726
Pattern3	***35°C***	***112.9275***	***7.6231***	***860.8619***
Pattern4	25°C	118.4085	7.5781	897.3083
Pattern5	15°C	123.9105	7.5294	932.9775

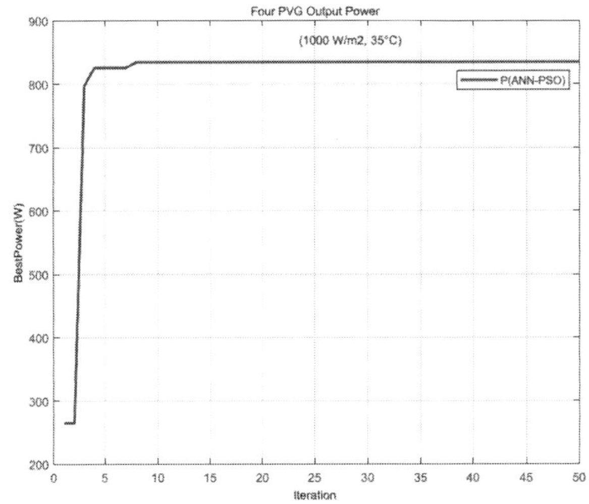

Figure 15. Performance of ANN-PSO output power at fixed irradiation-varying temperature.

$$E_2 = \frac{P_{PSO\_ANN}}{P_{max}} \times 100 = \frac{834.6977}{860.8619} = 96.96\% \quad (20)$$

4.3. Constant weather conditions

The output characteristic curves of the photovoltaic generators on the constant weather conditions test which means the temperature and the solar radiation are uniform in all the PV modules in series.

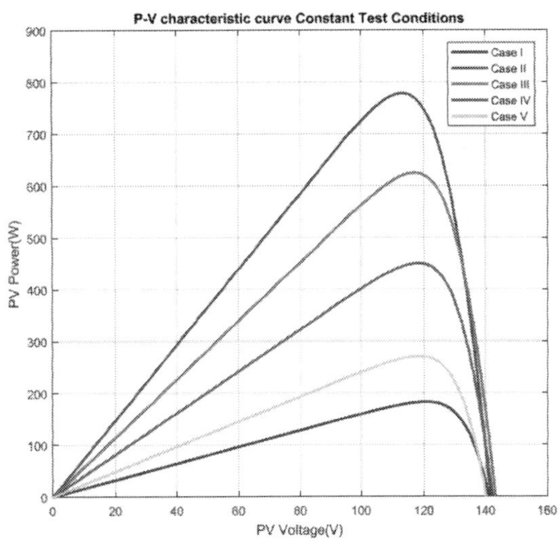

Figure 16. P-V characteristics curves of PVG at constant condition.

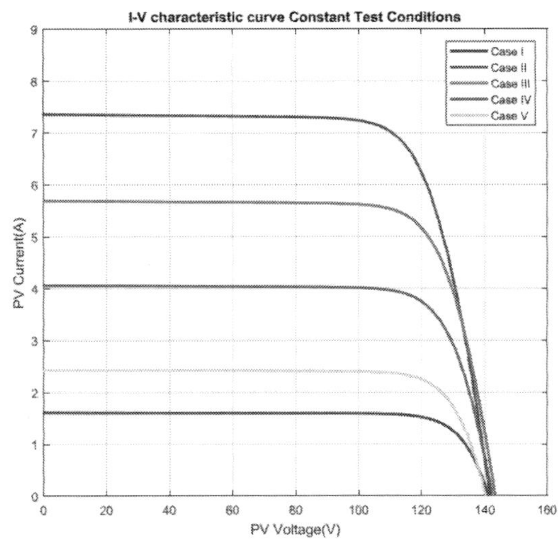

Figure 17. I-V characteristics curves of PVG at constant condition.

Table 5: Constant weather conditions patterns for series-connected PV generators

Case	Irradiance (W/m²)	Temp. (°C)	Voltage (V)	Current (A)	Power (W)
Pattern 1	200	20	120.5624	1.5164	182.8222
Pattern 2	900	35	113.3008	6.8666	777.9857
Pattern 3	700	29	117.1067	5.3308	624.2679
Pattern 4	500	27	118.2562	3.8079	450.3117
Pattern 5	***300***	***25***	***118.4302***	***2.2837***	***270.4569***

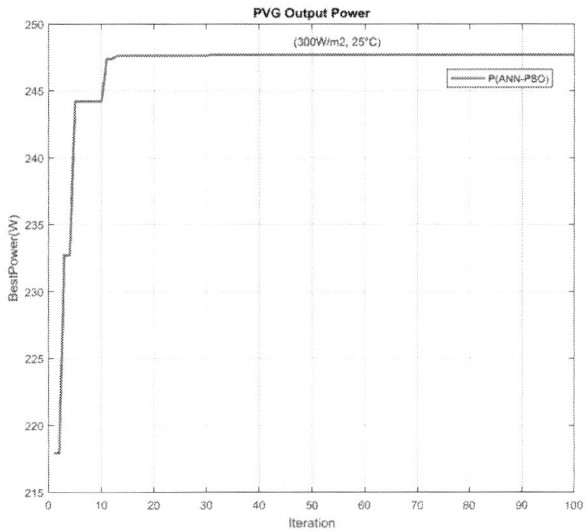

Figure 18. Performance of ANN-PSO output power of PVG at PSC.

$$E_3 = \frac{P_{PSO\_ANN}}{P_{max}} \times 100 = \frac{247.7043 \times 100}{270.4569} = 91.59\% \quad (21)$$

4.4. Effect Partial Shading Condition

Under a Standard Test Condition (STC), when the entire strings PV or the PV array receive a uniform irradiance and temperature, the typical P-V curve shows a single MPP, as illustrated in the curves of Figures 11 and 13 with different radiations and temperatures in each case.

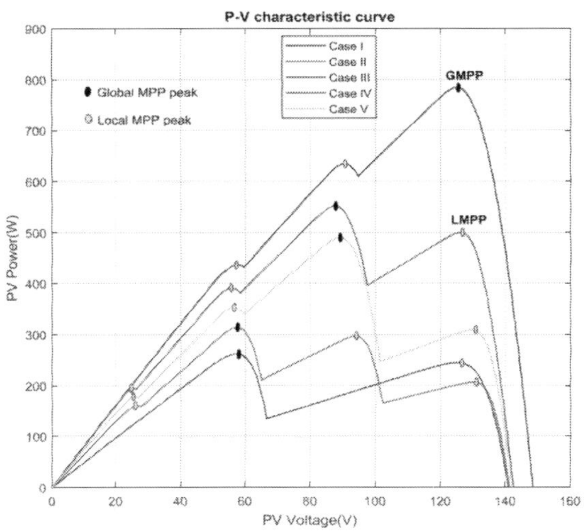

Figure 19. The *P-V* characteristics curves of PV generators at PSC.

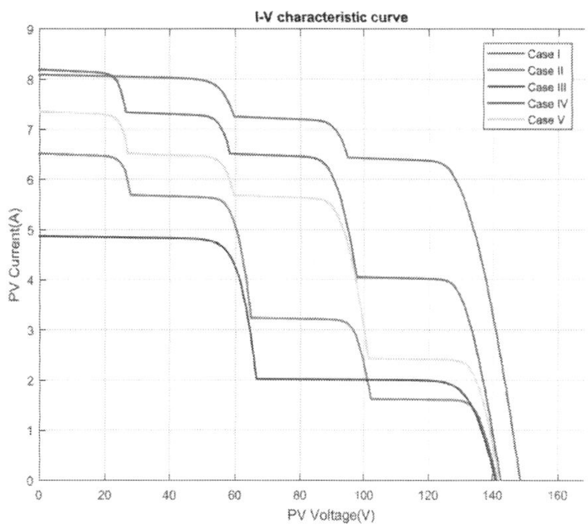

Figure 20. The *I-V* characteristics curve of PV generators at PSC.

Table 6: Shading patterns for series-connected PV generators

Case	Ambient Conditions				Global Peak
	PV module 1	**PV module 2**	**PV module 3**	**PV module 4**	**Pmax (W)**
Case I	***1000 W/m²*** ***25°C***	***1000 W/m²*** ***25°C***	***900 W/m²*** ***20°C***	***800 W/m²*** ***19°C***	***784.9260***
Case II	800 W/m² 32°C	700 W/m² 29°C	400 W/m² 26°C	200 W/m² 24°C	313.5360
Case III	600 W/m² 28°C	600 W/m² 28°C	250 W/m² 25°C	250 W/m² 25°C	261.7500
Case IV	900 W/m² 35°C	500 W/m² 27°C	1000 W/m² 40°C	800 W/m² 30°C	551.5869
Case V	300 W/m² 25°C	800 W/m² 31°C	900 W/m² 35°C	700 W/m² 27°C	490.3700

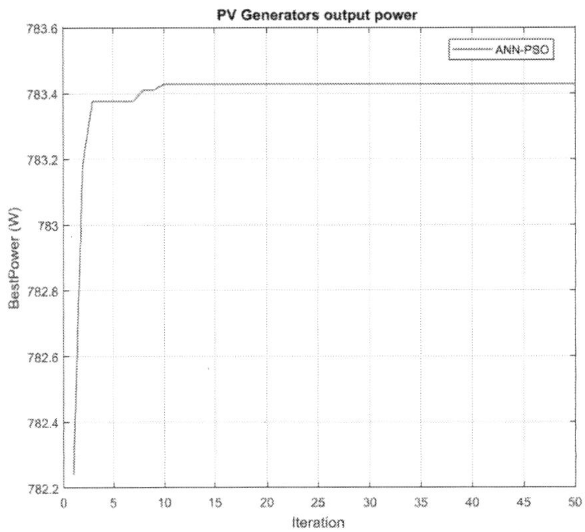

Figure 21. Performance of ANN-PSO output power of PVG at PSC.

$$E_4 = \frac{P_{PSO\_ANN}}{P_{max}} \times 100 = \frac{783.4215 \times 100}{784.9260} = 99.81\% \quad (22)$$

The PV system is simulated using Matlab/Simulink 2017 environment. The simulations are carried out and the proposed hybrid PSO-ANN algorithm is tested under different weather conditions: constant and uniform irradiation, rapidly varying irradiation and partially shading or non-uniform irradiation condition. Tracking the maximum power of the PV arrays at real time is very important to increase the whole system performance. The performances are evaluated in terms of the efficiency tracking and response time. The corresponding maximum power of PV generators for each conditions are recorded in tables.

4.5. Discuss on the limitation of the proposed system

To evaluate the performances of the proposed MPPT based on hybrid PSO and ANN algorithm controller during constant and non-uniform irradiance and temperature levels, four PV modules connected in series are partially shaded. Shading patterns for series-connected PV generators are simulated in five (05) cases with different ambient conditions (see Table 5). From the output P–V characteristic curves of the PV generator depicted, single and multiple power peaks are observed corresponding to the Local Maximum Power Point (LMPP) and the Global Maximum Power Point (GMPP). The proposed hybrid PSO and ANN algorithm controller based on MPPT has the ability to distinguish between the global peak (GMPP) and local peak (LMPP) through the scanning procedure.

Furthermore, it can be noticed that the method is able to track the GMPP on fully and partially illumination of PV generators as well as on partial shading conditions under no-uniform irradiance and temperature. The performances of the proposed hybrid ANN-PSO method maximum power point tracking of PV generators are computed and the proposed algorithm can always detect the maximum power even when the PV generators are partially shaded, partially of fully illuminated. So, the results show that the tracking efficiency of the PSO-ANN algorithm for each condition is in the range of 95.50%. Hence, the proposed algorithm can always reach to MPP if there is a change of solar irradiance and temperature.

5. Conclusion

The performance of the proposed particle swarm optimization with artificial neural network (ANN-PSO) is investigated when the PV generators perform under constant, standard test and partially shaded conditions. In this study, PV array is modeled based on four series connected PV modules. The developed hybrid ANN-PSO algorithm is tested under five different cases and its performances in optimizing the output power are given. The simulation results showed that the hybrid ANN-PSO algorithm is able to optimize the generation of PV system by tracking the Global Maximum Power Point (GMPP) when the ambient conditions (solar radiation and temperature) are changed. Furthermore, the proposed hybrid algorithm can control the PV system to perform at a more precise operating voltage. The hybrid ANN-PSO algorithm can always detect the Global Peak (GP), track the GMPP and provides better steady state during standard test and partial shading conditions. Power output of PV module is directly proportional to solar radiation. Hence, in order to select the best location to install PV module, the strength of solar radiation need to be considered. Besides that, environmental factors and local climate such as humidity, temperature and wind also need to be considered as it will affect the output power of PV module. These effects are shown clearly in this simulation result where the performances of PV module depend on the amount of solar radiation and also the temperature of surrounding.

6. References

[1] Kuperman A 2014 An analytical solution for tracking photovoltaic module MPP *IEEE J Photovoltaics* **4** 733

[2] Brito MAG, Galotto L Jr, Sampaio LP, Melo GA, Canesin CA 2013 Evaluation of the main MPPT techniques for photovoltaic applications *IEEE Trans Industr Electron* **603** 1156

[3] Thiaw, L., Sow, G., & Fall, S. 2014 Application of Neural Networks Technique in Renewable Energy Systems https://doi.org/10.1109/SIMS.2014.12

[4] L. Bouselham, B. Hajji, and H. Hajji 2015 Comparative Study of Different MPPT Methods for Photovoltaic System *3rd Int. Renewable and Sustainable Energy Conf.* p 1

[5] Jiang L, Maskell DL 2014 A simple hybrid MPPT technique for photovoltaic systems under rapidly changing partial shading conditions *40th Photovoltaic Specialist Conf.* p 0782

[6] C. S. Chin, Y. K. Chin, B. L. Chua, A. Kiring, and K. T. K. Teo 2012 Fuzzy logic based MPPT for PV array under partially shaded conditions *Proc. of the Int. Conf. on Advanced Computer Science Applications and Technologies* p 133

[7] K. L. Lian, J. H. Jhang, and I. S. Tian 2014 A maximum power point tracking method based on perturb-and-observe combined with particle swarm optimization *IEEE J. Photovoltaics* **4** 626

[8] M. Miyatake, M. veerachary, F. Toriumi, N. Fujii, and H. Ko 2011 Maximum power point tracking of multiple photovoltaic arrays: a PSO approach *IEEE Trans. Aerosp. Electron. Syst.* vol. **47** p 367

[9] Hartmann LV, Vitorino MA, Correa MBR, Lima AMN 2013 Combining model-based and heuristic techniques for fast tracking the maximum-power point of photovoltaic systems. *IEEE Trans Power Electron* **28** 2875

[10] A. A. Kulaksız, ve R. Akkaya 2012 Training Data Optimization for ANNs using Genetic Algorithms to Enhance MPPT Efficiencyof a Stand-alone PV System, *Turkish Journal of Electrical Engineering & Computer Sciences*, **20** p 241

[11] A. A. Kulaksız, ve R. Akkaya 2012 Training Data Optimization for ANNs using Genetic Algorithms to Enhance MPPT Efficiency of a Stand-alone PV System *Turkish Journal of Electrical Engineering & Computer Sciences* **20** 241
[12] Lyden S, Haque ME 2015 Maximum Power Point Tracking techniques for photovoltaic systems: a comprehensive review and comparative analysis. *Renew Sustain Energy Rev* **52** 1504

IOP Conf. Series: Journal of Physics: Conf. Series **1049** (2018) 012048 doi:10.1088/1742-6596/1049/1/012048

Investigation of Source of Distraction among the intrastate bus driver: *Focus Group Interview*

Muhammad Nur Annuar Mohd Yunos[1], Jalil Azlis-Sani[1], Musli Mohammad[1], Mohd Zamani Bin Ngali[1], Munzilah Rohani[2], Mohd Saiful Izwan Aazizi[3] and S.M. Sabri S.M. Ismail[4]

[1]Faculty of Mechanical and Manufacturing Engineering, Universiti Tun Hussein Onn Malaysia, 86400 Batu Pahat, Johor, Malaysia
[2]Faculty of Civil and Environment Engineering Universiti Tun Hussein Onn Malaysia (UTHM), 86400 Parit Raja, Johor, Malaysia
[3]Rapid Bus Sdn Bhd, Kompleks Rapid Bus Cheras Selatan, 71, Jalan Kb 2/15, Kampung Baru Balakong, 43300 Seri Kembangan, Selangor, Malaysia
[4]Strategy & Transformation Office, Prasarana Malaysia Berhad, Level 21, Tower B, Menara UOA Bangsar, No 5, Jalan Bangsar Utama 1, 59000 Kuala Lumpur, Malaysia

E-mail: muhdnurannuar@gmail.com

Abstract. Driver distraction have found as one of the main cause of accident all around the world. The driver distraction represents a significant problem to the personal transportation as well as the public transportation sector. However, factors of distraction especially in the context of public transport still has not widely explored. Thus, the aim of this research is to investigate the sources of distraction among the intrastate bus driver. This study will undergo focus group interview in order to achieve the objective of this research. 16 participants which are 8 participants from expert management and 8 participants from expert practitioner are involved in the focus group interview. From this research, there are 26 sources of distraction found in the focus group interview that overlap between both focus group. The result from this research can be used as guideline for improvement for the management of the intrastate bus services company and also for the future researcher as an aid to design the solution to overcome the sources of distraction.

1. Introduction

Distracted driving behaviour has been cited as one of the most dangerous error that can a driver make while driving a vehicle [1]. Many researchers had agreed that the driver distraction become the main cause of traffic accident all around the world [2]. Through the analysis from the United State Department of Transportation, it is found that the significant cause of total motor vehicle crashes is the driver distraction [3]. Besides, the Fatality Analysis Reporting System (FARS) in 2008, the driver distraction was identified as the factor of crashed at about 16% of all fatal crashes. Driver distraction also have been estimated at about 80% as the factors of traffic crashes which directly or indirectly contributed by the driver [4]. The report from The National Highway Traffic Safety Administration (NHTSA) state that the driver distraction contributes approximately 424,000 injuries which is 18 % of total injury crashes and 3,154 crashes related with fatalities which at about 10% of all fatal crashes [5].

Content from this work may be used under the terms of the Creative Commons Attribution 3.0 licence. Any further distribution of this work must maintain attribution to the author(s) and the title of the work, journal citation and DOI.
Published under licence by IOP Publishing Ltd

One of the main public transport chosen by the people in this country is the intrastate bus services. There are various factors of distraction of an intrastate bus driver which will have impact to the driver such as the number of hours driven, gender, experience and age [6]. An intrastate bus driver needs to drive the intrastate bus along a route which located in a high density-populated area, they will be facing a great number of external source of distraction as there will be more passengers riding the bus which resulting more frequent stops. Furthermore, there are over that 70% of distraction to be voluntary actions such as the use of mobile phone, adjusting the in-vehicle system as well as interacting with the passenger while driving the vehicle. In particular, only a few researches that had been discussed one the distraction of a bus driver and most of them were focused on the usage of mobile phone by the bus drivers [7]. Therefore, this study will identify the source of distraction among the intrastate bus driver.

2. Literature Review

Source of distraction are categorizing into two main categorizes which are internal and external distraction. Internal Distraction is referring to any distraction occur from the inside of the vehicle. It will make the driver's attention more to the activities inside the vehicle instead of focusing for the safety driving. The reaction of the driver for hazardous situation can reduce when the driver glance or the driver's attention more inside the vehicle [8]. Besides, the moments when the driver return their eyes from the glancing inside the vehicle to the roadway also will affect the reaction of the driver to the hazardous situation [9]. The glance of the driver away from the roadway must be less than two seconds in order to performing a safety driving procedure [9]. The longer the time for the drivers' eye off the road, the greater the risk of crash to the drivers [10]. Therefore, while undergo the task as intrastate bus driver, there have many factors inside the vehicle that are believed can be as distraction to the driver.

Firstly, technology as the factor of distraction. Technology is something that can always help human but sometimes technologies itself can be as distraction to the human. One of the common technology equipment own by the intrastate bus driver is mobile phone. The usage of mobile phone while driving can distract the intrastate bus driver and increase the risk of involving in accident [11]. In United States, mobile phone distraction resulted in 5% of injury and 18% of fatal crashed based on the police crash report [12]. The usage of mobile phone reported that it can reduce the awareness of the driver to the safety driving, increase the behaviour that is not safe, and increase the risk of accident for the pedestrian [13]. Besides, the handheld text message sending and receiving found that it required high mental demand, more frequent and longer glances away from the roadside which it can degraded the performance of the driver [14]. The driver sensitivity toward the upcoming events and associated perception and awareness of the road environment can be degraded when the driver simultaneously using hands-free mobile phone while driving [7]. The safety of the driver will be decreasing when they are texting while driving and having a negative impact on traffic operation [15]. The entertainment system in the vehicle also can give greatest negative impact on the driver performance [16].

Some of the intrastate bus driver in this country need to issue ticket to the passenger at the bus stop before the passenger enter the bus. It increases the workload of the intrastate bus driver and issuing ticket by the bus driver is classified as high risk as it is one of the distracting activities based on the previous research [17]. Besides, passenger also can be the source of distraction to the driver. Based on previous research, at about 10.9% of the distraction comes from the other occupants in the vehicles [18]. Besides, it is reported that talking to the passenger were found as one of the source of distraction towards the driver while driving the vehicles [19]. Interaction to the passenger also found as the source of distraction to the driver [20]. The other source of distraction found in the vehicle is adjusting the setting in the vehicle [21]. Eating and drinking inside the vehicle while driving also can be consider as one of the source of distraction to the driver [22]. So, it can be said that while the driver is undergoing the driving task, doing the secondary task can distract them [21]. Other than that, the personal of the driver also can be source of distraction to the driver for example health of the driver. Source of distraction that referring to the personal of the driver based on the salmon 2010 are fatigue, sickness and medication [23].

Distraction can occur from the activities not only inside the vehicle, also can happened from the outside of the intrastate bus driver [24]. The weather can distract the intrastate bus driver especially during the bright sunlight and heavy rain [23]. Besides, the infrastructure along the road need to be focus as it related with the driver distraction. One of the example of infrastructure is roadside advertisement and it is found that roadside advertisement can distract the intrastate bus driver [25]. Salmon 2011 also said that of the source of distraction related with the infrastructure is roadside advertising [23]. Lastly, the condition of the road can become one of the source of distraction to the driver.

3. Methodology

The collection data process for this research is focus group interview. The focus group interview is focusing to the expert which related with the research area. The main objective of this focus group interviews is to identify the source of distraction among the intrastate bus driver. For this research, there will be two focus group that which are participated by two group of experts which are management and practitioner. The management group is a group of experts who involve in the management team from the prominent intrastate bus company. Most of them are managers, depot managers, executive and also trainers from the bus training department. Besides, the practitioner's focus group is a group expertise which are the intrastate bus driver who experienced driving the intrastate bus for more than ten thousand hours or equal to 5 years' experience. The participant for the practitioner's focus group are chosen randomly by the prominent intrastate bus company in which the minimum required is their involvement driving the bus at least 5 years' experience. For this focus group interview, purposive sampling is used. Purposive sampling can be described as a selection of a group of individuals which have a basic basis information related with the research area. All the participants are chosen based on their expertise whether from the management or practitioner of an intrastate bus.

For this focus group interview, the participant for both group will give their opinion about the sources of distraction that distract the intrastate bus driver. The similar interview protocol is prepared for both group in order to ensure the reliability and consistency for both focus group interview. The participants are needed to brainstorm every source that distract the intrastate bus driver weather from inside and outside the intrastate bus. Two research assistants will be in each focus group interview who will be as moderator and another one will setup the recording equipment and also will write all the main point from the interview.

The focus group for both group are running simultaneously and are recorded audio and video for analysis purpose with permission from the participant. All the participants are informed about the recording and sign the consent letter before any discussion starts. Both group of participants were assisted by the interviewer about the general of this research all about before splitting into two different discussion room based on their expertise. At the end of this focus group interview, there will be a recorded audio and video and also a list of written of the main point about the sources that distract the intrastate bus driver from each group of the focus group interview. The recorded audio and video is transcribed into a written form and undergo content analysis. All the participants are coded during this focus group interview to ensure the confidentially of the information from the participants.

For data analysis, the interviews were transcribed into written form. The data from the interviews that is already transcribe is analysing by using content analysis techniques. Content analysis techniques can be defining as a technique that is abstracting and also categorizing information from the text. This technique is one of the research tool that are used to identify the information or certain point within a set of texts.

The analysis starts with two researchers who involve in this research read all the transcript several times to get familiar and understand the overall content of the interview data. Then, both researchers get a sense of the similar idea and noted the ideas into a preliminary codebook. This preliminary coding is done as to break the entire transcript into several meaningful segments. The two researchers then got together and compare their codebook as they create the preliminary codebook independently. To improve the reliability, both researchers review both their preliminary codebook and reach a consensus on creating a new codebook which containing the matched themes from their preliminary codebook. Several meetings between them which conducting the data rearrangement and mind mapping in order to create the most suitable themes from the matched themes of their preliminary codebook. In the analysis, there are 26 activities that related to the distraction among the intrastate bus driver and overlap between both group of experts.

4. Result and Discussion

This research is to investigate the source of distraction among the intrastate bus driver via the focus group interview. The participants of the interview are the managers, executive, trainers and experienced bus driver who already drove the bus at least 5 years' experience. The participants are divided into 2 group based on their expertise which are expert managements and expert practitioner. Both group of interviews are recorded audio and video for further analysis. After being transcribe and analysis, there are 26 sources of distraction that are overlap between each group.

4.1. Advertising video on the screen in the bus

This is the new source found as the source of distraction among the intrastate bus driver. Based on the respondent EM 7, he said that the distraction comes from the voice and sound from the advertising video that can distract the bus driver. Respondent EP 7 also said that sound from the video can distract the bus driver.

4.2. Technology devices for recording

The technology devices that being used for recording such as CCTV in the bus can distract the bus driver. Respondent EM 1 said that the CCTV in the bus is beneficial to record any accident happening in the bus but the driver tends to scared when they know that the CCTV is recording everything in the bus. The driver will be more cautious and always look at the CCTV. This source of distraction agreed by all other respondent in the Expert Management group. Based on the findings from the expert practitioner, the respondent EP 4 said that one the source of distraction is CCTV in the bus the respondent said that when there is CCTV recording all the activities, it is very helpful as it will make working as no one will stop working.

4.3. Technology equipment

This type of source of distraction include the uses of mobile phone, personal gadget and entertainment devices while driving the bus. EM 8 said that, the main source of distraction is the usage of the mobile phone while driving the bus. In addition, EM 6 said that the distraction involves the usage of communication device, power bank and personal gadget. Respondent EP 1 in expert practitioner said that, the usage of application such as WhatsApp and Instagram while driving can distract the bus driver.

4.4. Traffic congestion

Traffic congestion found as one of source of distraction to the intrastate bus driver. EM 4 said that, traffic congestion is one of source of distraction. Besides, EM 1 said that because of the traffic congestion, the intrastate bus driver might miss the bus stop especially during the peak hour. In addition, because of the traffic congestion, the trip sometimes can be up to 2-3 hours longer than usual duration which can lead to other problem which can distract the intrastate bus driver. EP 7 and EP 1 said that, when there are lots of other road users on the road which make the traffic congested, it will distract the intrastate bus driver.

4.5. Others road user

Other road user can be one of the source of distraction to the intrastate bus driver. EM 1 said that the other vehicles sometimes block the bus to reach the bus stop which make the intrastate bus driver stop the bus outside the specific point to drop and pick up the passenger. EP 7 also said that other road user can distract the intrastate bus driver. Besides, EP 1 said that the other road user will make the traffic slow as they are using application such as WhatsApp and Instagram while driving.

4.6. Identifying bus stop

Identifying bus stop can distract the bus driver. This situation happened based on a few situations. EM 1 said that the intrastate bus driver got distract to identify the bus stop when the passenger pressing bell repeatedly to stop although the distance to reach the next bus stop still 1 km away. EM 1 said that normally the next bus stop is about 250m, but when this situation occurs, the driver got distract when the driver stops the bus at the next bus stop, the passengers are still in the bus as their targeted bus stop is the next bus stop. Meanwhile based on EP 8 the intrastate bus driver need to identify random location to stop the bus as requested by the passenger although it is not the verified bus stop.

4.7. The effectiveness of the working schedule

Working schedule can distract the intrastate bus driver. EM 5 said that the working schedule might distract the intrastate bus driver especially during morning shift. EP 1 said that the working schedule can distract the intrastate bus driver. EP 1 also said that, the working schedule is distract the driver especially during the peak hours because of not enough staff. Besides, during the peak hour, the timing is too tight which make the driver has no rest. EP 4 said that, peak hour is usually in the morning during the office breaks and during evening after the working hours. EP 1 add, the peak hour also can be during the school break during the afternoon.

4.8. Changing the route

Changing the route can distract the intrastate bus driver. EM 1 said that it can be distract to the driver but it depends on the purpose and condition. EM 6 said that, the route for each driver to drive the bus is changing according to the situation and rotation. EP 2 said that the changing of route makes the intrastate bus driver distracted.

4.9. Personal issue

Personal issues are one of the source of distraction to the bus driver. It is referring to the job empowerment, internal stress, family problem, colleague and others. EM 5 said that, the distraction can occur to the bus driver because of the driver's personal issues that occur related to the family matters. EP 5 also said that, the personal issue that can distract the driver is about the far distant between their accommodation to the workplace.

4.10. Driver behavior

The driver's behaviour also can distract them while driving the bus. It includes the eating, drinking and smoking while driving. Besides, the personality of the driver such as hot-tempered person also can distract themselves while driving the bus. EM 1 said that, when the driver is angry because of the passenger behaviour, the driver tends to make emergency break in intention to make the passenger fall. Besides, EM 1 said that a trip usually around 30-40km and the driver have problem such as couldn't go to the toilet during the journey. This is due to the heavy traffic and it is not appropriate for him to stop the bus randomly along the road while there are so many passengers in the bus. EM 7 add that it will make the driver angry and get distract because of this situation. EM 4 said that, eating and drinking while driving can distract the intrastate bus driver. Finding from expert practitioner group, EP 1 said that sometimes there are overacting passengers who challenge the driver for multiple time which ended with the fighting between the driver and the passenger.

4.11. Less experienced driver

EM 1 said that, sometime the distance and the condition of the road might be varies every route. When a less experienced driver goes through a new route, it is difficult for them and it can distract them from performing safety driving procedure while driving the bus. Besides, EP 4 said that, if the driver is not familiar with the condition of the bus that they drove, it can distract them. EP 3 said that, issues happen when the management change the bus due to the medical leave of another driver and need another driver to replace. EP 3 add that when the driver less experience, there will be higher tendency for accidents.

4.12. Driver's salary

The driver's salary might distract the driver. EM 1 said that, the salary of the driver is low compared to the workloads. The EM 1 also said that the bus driver classified as unskilled work. That is why they have low salary. EM 1 inform that, nowadays the management try to increasing the salary of the bus driver by improve the level of skill from unskilled into semi-skill. Based on the expert practitioner group, EP 7 said that he realises that accidents usually happen after they got their salary. He said that, the accidents occur because of the driver need more money for their daily expense. Since the salary is low, then they need to think about doing some part time to get extra income to cover their daily expense. In addition, EP 7 said that this problem especially happened to the new driver as their salary is still low compared to the experienced driver who already have increment salary after working as bus driver for certain time.

4.13. Driver's welfare

Driver's welfare focusing on the temporary resting place. EM 1 said that, the management provide the temporary for the driver to rest. The use of the temporary resting place based on the EM 1, is a place that the driver can sleep when the finished working quite late or the driver's house is very far work the work place. EM 1 add, problem comes and might distract the bus driver is when the some of the depot don't have temporary resting place and some of the them misused which is a place to store things beside a place for the driver to rest. Based on EP 3, the company provided temporary resting place but it is not comfortable to stay as it is not suitable with the surrounding. EP 8 have opposition opinion with EP 3 which EP 8 said that the temporary resting place provide by the management is satisfying enough. This contradict opinion is accepted as the EP 8 and EP 3 are from different depot which each depot has their own temporary resting place based on the surrounding.

4.14. Driver's health

EM 2 said that the health of the driver might distract the intrastate bus driver. This issue arises when there are not enough drivers. The company having this problem because the drivers is not interested with the scheme provided by the company. When there is not enough driver, the existing driver need to work overtime. When the driver didn't have a good rest, the driver will be tired and sleepy. It spoiled the mood of the driver when coming to the work which can affect their performance. EP 8 said that the concentration and performance might be distracted because of the driver's health. The driver sleeps late last night and need to come back work early in the morning which will make the driver fatigue. EP 3 said because of this situation, the driver always sick and unwell when come to work. EP 5 said that when driver has not enough rest, it will affect their emotion.

4.15. Open trunked radio

EM 4 said that the trunked radio used to communicate with Bus Control Centre (BCC). The issues come when any communication between driver and BCC, all other driver will hear and it lead to questions and distract the driver. EM 1 said that because of the situation, driver tend to turn down the volume. Although the communication from BCC is not frequent unless if there are emergency cases, it is not good for the driver to turn down the driver, as the drivers are not getting any important information from BCC. Besides, based on EM 2, when the BCC is not communicate using trunked radio, the drivers are misuse the radio by interaction between them with jokes which will distract other driver who not related with the interaction. EP 3 said that the driver tends to turn down the volume because it found that the radio is distracting them while driving.

4.16. Comfortability of the driver's seat

EM 3 said that the driver's seat does not suit the height of the driver. Although the driver's seat can adjust based on the comfortability of the driver, but it still can distract the driver while driving the bus. EP 7 said that the driver's seat is not same every bus. Sometimes the driver familiar with their current bus, but when they need to change the bus, the seat is not suitable for them for example, the driver need to brake the bus by using the end of the toes although after adjusting the seat.

4.17. Condition of the bus

EM 6 said that based on his opinion, the speed brake sometimes is not well functioning. EM 1 response that nowadays the model from this company are more up-to-date hence generally they are all fine. EP 4 said that some of the bus is no comfortable to drive. When driving the bus, the driver knows the problem because the bus is not well maintained. It will distract the driving of the driver. EP 5 said that although the report has been made, the problem is still there. When the bus is not well managed, the duration of the bus will take some extra time. EP 1 said that, some of the depot have brand new buses, shiny and clean but some of the depot are not.

4.18. Systematicity of the management system

This source of distraction is the interaction between the management and the bus driver. EM 4 said that when the bus driver come in the morning, the management scold the driver on what had the driver done the day before. It will distract the mood of the bus driver along the day. EP 8 said that, every accident happened, the management will scold the driver and sometimes the bus driver face heavy punishment. EP 3 said that, sometime when the bus driver give opinion about the operation of the bus, it is not usually accepted. EP 7 said that, the managers don't usually willing to change what they have decided. Besides, EM 2 said that, the bus driver might distract because of the bus driver need to be multitasking hence increasing the tendency of accidents happening. This happen because of the management of the company decide to minimize their running budget by eliminate the conductors which the driver need to issuing the ticket also instead of only drive the bus.

4.19. Advertising along the roadside

The advertising along the roadside found as one of the source of distraction (Salmon et al., 2011). In this research, EM 4 said that the advertising billboard along the roadside distract the driver and all the participants in the expert management agree with the statement. In the expert practitioner group, majority of the participant agree that advertising along the roadside distract the driver and some of them have contradict opinion from majority. Majority said that if the billboard is interesting, then it can distract the driver. They also said that, if the advertisement shows car or woman, it can possibly distract the driver. EP 3 said that, the advertising especially the digital billboard is not distracting because the advertisement plays too fast and the driver can't even focus on it but EP 4 said that, the digital billboard can distract the driver especially during the night trip as the lighting from the billboard make the driver dazzle.

4.20. Special lane for bus

Special lane for the bus become one of the source of distraction among the intrastate bus driver. EM 6 said that, once there is specific lane for bus, the journey duration, and the recess time for the driver can be confirm which can prevent other source of distraction from occur. EM 1 said that, the special lane already has but not for all route. The special lane for bus supposed to be advantage to the bus company but the issues also happened because the lane is not well enforced. The lane is all occupied with car which make the bus difficult to move along the special lane for bus. EP 4 said that the operation of the bus better if there is special lane for bus. EP 3 said that, the special lane already has but only in the city not in the countryside. EP 4 add, the special lane should be empowered so that the lane only can be used for the bus driver driving the bus instead of the other vehicle.

4.21. Location of the bus stop

Location of the bus stop is found as one of the source of distraction. EM 4 said that the bus location of the bus stop can distract the bus driver and all other participants in the expert management agree with it. EP 4 said that, the company should increase the numbers of bus stop. Accident can occur by other vehicle hit back of the bus when the bus not dropping the passenger at the exact bus stop. The driver has to follow the instruction of the passenger otherwise, they will complaint to the management. EP 5 said that, the location should be in appropriate location. The EP 5 realise that the some of the location of the bus stop is not suitable. EP 1 said that, the bus stop should have enough space for the bus to stop and will not distract other road user if not, traffic congestion will be happened.

4.22. Communication between passenger and the driver

Communication between passenger and the driver found as one of the source of distraction among the intrastate bus driver. EM 1 said that, there are lots of elderly women who use the intrastate bus services. This is because, there are usually not working and they use the instate bus services to go to the market. EM 4 said that these elderly women usually very friendly and talkative. They would like to make a small talk throughout the journey. If the bus driver didn't respond, the elderly women will start asking questions to the driver. This can cause a distraction among the intrastate bus driver. Besides, the EP 7 said that, the communication between passenger and driver can be one of the source of distraction is because of the language barrier. Passengers of the intrastate are not only local, but also foreigner who from Indonesia, Bangladesh, Myanmar and others. Issues occur when the foreign passengers can't speak Bahasa Malaysia and also English. It will distract the driver when the driver wants to issue the ticket as they don't understand the destination that the foreign passenger wants to go. EM 7 said that, sometimes the foreign passengers communicate between each other very loudly which can cause distraction to the intrastate bus driver.

4.23. Distracted passenger behaviour

The intrastate bus driver facing thousands of passengers everyday with thousand variety behaviour. It is found that the passenger's behaviour can distract the intrastate bus driver. EM 1 said that, on of the issues usually occur related to the passenger behaviour is when the driver need to face the hot-tempered passenger. This type of passenger usually likes to complaining about the ticket priced and the delayed arriving time of the bus although the delay is because of the bus problem. They will scold the bus driver and it can cause a distraction to the driver. EP 7 said that, the passenger behaviour is related with the passenger attire. There is a few cases where the passenger wear impropriate will entering the bus which can distract not only the intrastate bus driver, the other passengers also can be distracted.

4.24. Smell from the passenger

Smell from the passenger found as one of the source of distraction among the interstate bus driver. EM 7 said that the smell from the body odour of the passenger especially from the foreigner can distract the intrastate bus driver. Besides, the smell can come from the goods that passenger bring into the bus. EM 5 said that some of the passenger use the intrastate bus services to go to the fish market. When the enter the bus from the market, they bring along the good such as fish, and other smelly raw food. The smell from the raw food will distributed all inside the bus which can distract the driver and also the passengers. It comes to worse when the bus uses the air-conditioning system which the smell from the raw food is trap inside the bus. EP 7 also said that the smell from raw food such as fish can distract the intrastate bus driver.

4.25. Under-age passenger

It is found that the under-age passenger is one of the source of distraction among the intrastate bus driver. Under-age passenger is referring to the passenger who is 18 years and below . EM 1 said that, the under-age passenger can distract the bus driver when they are pressing the bell as warning to stop at the next bus stop repeatedly. Besides, EM 1 also said that the under-age passenger focusing on the school kids, they tend to make vandalism inside the bus. When the driver advises them, they will provoke and sometimes it can create fight if they are in a group. Some of the school kids will report to their parents and complaint will be made by their parent to the management of the intrastate bus service. As result, the bus driver will be punished and it can distract the performance of the bus driver. EP 3 said that under-age passenger usually uses the intrastate bus services in group. The distraction comes when they enter the bus and the driver want to issue the ticket to them.

4.26. Troublesome passenger

Troublesome passenger can distract the intrastate bus driver. EM 1 said that, some of the cases when the passenger uses the bus service when they come to KL to attend some events and they get drunk. Then they will disturb other passenger and some of them even peeing in the bus, showing their private part to another passenger. This situation sometimes happened every day. EM 1 also said that there are many cases involving pick pocketer every day and some of them are threatening the driver. EM 4 said that, sexual harassments also always happened in the bus which can distract the bus driver. EM 1 add that there are passengers who paid invalid ticket as they paid for 2 zones but the passenger uses the bus until more than that. When the bus driver realizes, the bus driver stops the bus and call the passenger to pay for the next destination in which it will cause argument and can distract the bus driver. EP 1 also said that the drunk passenger will distract the bus driver. The driver only realizes the passengers in drunk when they enter bus and start disturbing another passenger.

5. Conclusion

As conclusion, this research found 26 sources of distraction that are very important for the management of the intrastate bus services company to be noticed in order to minimize the risk on the accident to occur especially related with the intrastate bus driver. Besides, the 26 sources of distraction also can be used as guideline for the future researcher who will continues the research related with this area.

Acknowledgment

The authors are supported by the Fundamental Research Grant Scheme (FRGS) Vot 1542, Office of Research, Innovation, Commercialization and Consultation (ORICC) Universiti Tun Hussein Onn Malaysia. The authors would also like to thank Rapid Bus Sdn Bhd on the contribution and participation for this study.

References

[1] Craft, R. H., & Preslopsky, B. 2009 *Driver Distraction and Inattention in the Usa Large Truck and National Motor Vehicle Crash Causation Studies. Driver Distraction and Inattention Conference,* 1–17.

[2] Ranney, T. a. 2008 *Driver Distraction: A Review of the Current State-of-Knowledge. National Highway Traffic Safety Administration, (April),* 1–32.

[3] NHTSA. *2010 Traffic Safety Facts - Distracted Driving 2009. US Department of Transportation, (September),* 1–8.

[4] Dingus, T. a., Klauer, S. G., Neale, V. L., Petersen, A., Lee, S. E., Sudweeks, J., Knipling, R. R. 2006 *The 100-Car naturalistic driving study phase II – Results of the 100-Car field experiment. Dot Hs 810 593, (April),* No. HS-810 593.

[5] Ranney, T. a., Garrott, W. R., & Goodman, M. J. *2000 NHTSA Driver Distraction Research: Past, Present, and Future. Driver Distraction Internet Forum,* **(233),** 9.

[6] Young, K. L., & Lenné, M. G. 2010 *Driver engagement in distracting activities and the strategies used to minimise risk. Safety Science,* **48(3),** 326–332.

[7] Treffner, P. J., & Barrett, R. 2004 *Hands-free mobile phone speech while driving degrades coordination and control. Transportation Research Part F: Traffic Psychology and Behaviour,* **7(4–5),** 229–246.

[8] Klauer, S. G., Guo, F., Simons-Morton, B. G., Ouimet, M. C., Lee, S. E., & Dingus, T. A. 2014 *Distracted Driving and Risk of Road Crashes among Novice and Experienced Drivers. New England Journal of Medicine,* **370(1),** 54–59.

[9] Borowsky, A., Horrey, W. J., Liang, Y., Garabet, A., Simmons, L., & Fisher, D. L. 2016 *The effects of brief visual interruption tasks on drivers' ability to resume their visual search for a pre-cued hazard. Accident Analysis and Prevention,* **93,** 207–216.

[10] Peng, Y., & Boyle, L. N. 2015 *Driver's adaptive glance behavior to in-vehicle information systems. Accident Analysis and Prevention,* **85,** 93–101.

[11] Törnros, J. E. B., & Bolling, A. K. 2005 *Mobile phone use - Effects of handheld and handsfree phones on driving performance. Accident Analysis and Prevention,* **37(5),** 902–909.

[12] Caird, J. K., Johnston, K. A., Willness, C. R., Asbridge, M., & Steel, P. 2014 *A meta-analysis of the effects of texting on driving. Accident Analysis and Prevention,* **71,** 311–318.

[13] Nasar, J. L., & Troyer, D. 2013 *Pedestrian injuries due to mobile phone use in public places. Accident Analysis and Prevention,* **57,** 91–95.

[14] Owens, J. M., McLaughlin, S. B., & Sudweeks, J. *2011 Driver performance while text messaging using handheld and in-vehicle systems. Accident Analysis and Prevention,* **43(3),** 939–947.

[15] Stavrinos, D., Jones, J. L., Garner, A. A., Griffin, R., Franklin, C. A., Ball, D., Fine, P. R. 2013 *Impact of distracted driving on safety and traffic flow. Accident Analysis and Prevention,* **61,** 63–70.

[16] Horberry, T., Anderson, J., Regan, M. A., Triggs, T. J., & Brown, J. 2006 *Driver distraction: The effects of concurrent in-vehicle tasks, road environment complexity and age on driving performance. Accident Analysis and Prevention,* **38(1),** 185–191.

[17] D'Souza, K. A., & Maheshwari, S. K. 2012 *Improving performance of public transit buses by minimizing driver distraction,* **128**, 281–293.

[18] Stutts, J., Reinfurt, D., Staplin, L., & Rodgman, E. 2001 *The Role of Driver Distraction. AAA Foundation for Traffic Safety, (May).*

[19] McEvoy, S. P., Stevenson, M. R., & Woodward, M. 2007 *The contribution of passengers versus mobile phone use to motor vehicle crashes resulting in hospital attendance by the driver. Accident Analysis and Prevention,* **39(6),** 1170–1176.

[20] Patel, J., Ball, D. J., & Jones, H. 2008 *Factors influencing subjective ranking of driver distractions. Accident Analysis and Prevention,* **40(1),** 392–395.

[21] Stevens, A., & Minton, R. 2001 *In-vehicle distraction and fatal accidents in England and Wales. Accident Analysis and Prevention,* **33(4),** 539–545.

[22] Stutts, J., Feaganes, J., Reinfurt, D., Rodgman, E., Hamlett, C., Gish, K., & Staplin, L. 2005 *Driver's exposure to distractions in their natural driving environment. Accident Analysis and Prevention,* **37(6),** 1093–1101.

[23] Salmon, P. M., Young, K. L., & Regan, M. A. *2011 Distraction "on the buses": A novel framework of ergonomics methods for identifying sources and effects of bus driver distraction. Applied Ergonomics,* **42(4),** 602–610.

[24] Edquist, J., Horberry, T., Hosking, S., & Johnston, I. 2011 *Effects of advertising billboards during simulated driving. Applied Ergonomics,* **42(4),** 619–626.

[25] Crundall, D., Van Loon, E., & Underwood, G. 2006 *Attraction and distraction of attention with roadside advertisements. Accident Analysis and Prevention,* **38(4),** 671–677.

IOP Conf. Series: Journal of Physics: Conf. Series **1049** (2018) 012049 doi:10.1088/1742-6596/1049/1/012049

Applying Psychomotor Domain for Competency Based Teaching in Vocational Education

Ahmad.A[1], Kamin, Y[1], Md.Nasir, A.N[1]

[1]Faculty of Education, Universiti Teknologi Malaysia

Email: a-adnan@utm.my

Abstract. Teaching is a teaching–learning activity between two different parties, namely, ,students and instructors. Teaching strategies are used by instructors in teaching sessions. Practical teaching requires paying attention to the application of all teaching strategies in the psychomotor domain. Psychomotor domains that need to be emphasized in teaching and learning processes should include seven elements, namely, perception, setup, controlled movement, mechanism, specific movement, settlement, and originality. This study aims to discuss the application of psychomotor domain practical teaching strategies adopted by teachers in vocational colleges. This study involves the observation of vocational college teachers who are experts in practical technology work. This study is expected to provide new ideas to develop instructors' skills in applying psychomotor elements in teaching methods in the form of demonstrations for teaching practical workshops.

1. Introduction

Technical and vocational secondary schools, which have been upgraded to vocational colleges, are assets that are enabling the technological advancement of Malaysia. Vocational curricula are changed and reviewed to adapt to the working environment. Vocational college curricula should consider every teaching need [1]. The teaching process should meet established standards. [2] explained that vocational trainers must be professional when selecting teaching strategies to realize learning goals. The selected teaching strategies must enhance knowledge (cognitive), skills (psychomotor), and attitudes (affective). Additional emphasis should be given to teaching psychomotor skills, specifically practical skills in workshops.

1.1. Background of study

Vocational subjects focus on facilitating learning through practical workshops rather than only through theoretical knowledge [3]. Practical teaching method is an effective classroom learning strategy [2]. Trainers need teaching and practical skills to ensure quality delivery of accurate information to students. Ethically, trainers should be accurate. Their teaching strategies should help students accurately apply all theories, teaching techniques, and program elements in workshops [4].The selected teaching methods must also be suitable for the lessons to be taught. [5]stated that appropriate teaching strategies are needed to develop skills in giving presentations and demonstrations, recalling, training, and performing practical projects, as well as improving students' performance. Teaching

Content from this work may be used under the terms of the Creative Commons Attribution 3.0 licence. Any further distribution of this work must maintain attribution to the author(s) and the title of the work, journal citation and DOI.
Published under licence by IOP Publishing Ltd

skills must involve the application of all aspects of psychomotor skills [6]. Methods for teaching psychomotor skills are required, especially for teaching workshops on practical skills [2]. Psychomotor domains that are emphasized in teaching and learning processes include perception, setup, response, controlled movement, mechanism, movement, settlement, and originality. Teaching and learning involving these psychomotor domains are coordinated and elevated by instructors by selecting proper teaching strategies. Selecting an effective teaching method can help improve the efficiency of the students in mastering practical skills in workshops. Practical teaching method is an effective form of demonstration [7]. The psychomotor domains of perception, setup, response, controlled movement, mechanism, specific movement, settlement, and originality should be applied in all teaching methods, teaching strategies, and practical workshops.

1.2. Simpson's Psychomotor Domain

Psychomotor in learning is present in physical skill with demonstrations of equipment or tools in the classroom, workshop, and laboratory. Figure 1 illustrates Simpson's psychomotor domain.

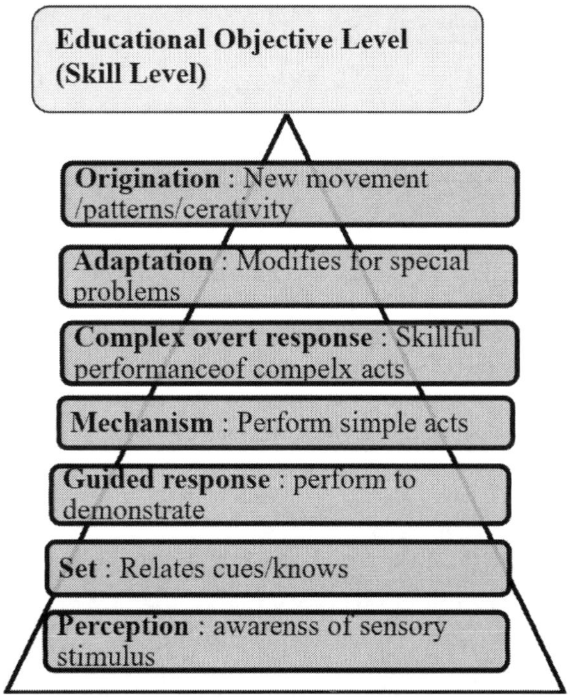

Figure 1 Simpson's psychomotor domain.

Seven categories in Simpson's domain are applied in this research [8].Perception is the ability to use sensory cues to guide skills or motor activity. The range is from sensory stimulation, cue selection, until translation. A set level is readiness to act, which includes mental, physical, and emotional aspects. This disposition is predetermined for different situations or mindsets. Guided response is the early stage in complex skill learning and includes imitation and trial and error. The performance will be achieved with practice. Mechanism is an intermediate skill, in which a learned response has become habitual and the movements can be performed with some confidence and proficiency. A complex overt response indicates that skillful performance involves complex movement patterns. This level includes performing without hesitation and automatic performance. The next level is adaptation, in which skills are well developed and individuals can modify movement patterns to meet special

IOP Conf. Series: Journal of Physics: Conf. Series **1049** (2018) 012049 doi:10.1088/1742-6596/1049/1/012049

requirements. The final level is origination, which involves creating new movements to fit particular situations or problems. This level emphasizes learning outcome based on highly developed skills.

2. Methodology

This study investigates how teachers apply the psychomotor domains in teaching practical workshops on the basis of Simpson's domain [8].The method used in this study is a type of unstructured observation; observation is made freely, and all aspects are considered important [9].The application of the psychomotor domains is analyzed through structured observation. The structured observation is more likely conducted through a survey, and respondents are asked the same set of questions. Researchers use purposive sampling to implement this observation method. Four teachers who teach automotive technology are selected for observation. [10] explained that researchers can use personal considerations to select the samples and that the sample can represent the same features of the entire sample. Peer structured observation was conducted (two observers participated) to make sure that the checklist is not biased to one observer. Observers stay in a teaching session conducted in a workshop, and all time allocated in each session for all teachers is the same because it is based on the time allocated for skill-based subjects. The aim of observation was to gain a better understanding of the teaching environment and the teaching method that was applied. The observers prepared a checklist based on the psychomotor domain criteria.

3. Results

The results on application of psychomotor domain in teaching strategies starting with respondents' responses in perception, set up, controlled movement, mechanism, specific movement, settlement and originality.

3.1. Perception (P1)

On the basis of observation, the psychomotor domain of perception (P1) is used in an observational strategy teaching method that involves memorizing the previous lesson and demonstration. When applying this strategy, teachers list the tools needed to perform practical tasks and ask students to classify and distinguish the use of each tool for practical sessions. Learning techniques for scan back are associated with the current situation that occurs in the environment and in the workshop. Students are asked to list the things that were taught in the classroom on that day and compare their practical performance with other specified manufacturing work. Instructors prefer the demonstration method in the perception (P1) aspect so that students will understand the procedure of the given task. Teachers initially show the cleaning of the equipment and machinery in workshops. Once a demonstration is completed, instructors ask all members of the class to perform the demonstrated practice. Teachers use methods to demonstrate the application of perception (P1) in teaching. Discussion is a teaching strategy in which instructors apply the psychomotor domain of perception (P1). Teachers ask students to discuss the relationship between the practical teaching implemented and other specified manufacturing work. The students who are asked to share with other students are selected from the discussions agreed between group members. Students can ask the trainers questions with regard to future concerns about practical workshops. Students and instructors discuss the problems jointly in the teaching and learning sessions.

3.2. Set up (P2)

A teaching strategy (i.e., discussions in applying the psychomotor domain setup) is chosen. For the application of the setup domain (P2), instructors ensure their mastery of practice prior to starting teaching and learning session involving practical questions. All existing problems are discussed jointly with teachers who teach other practical workshops. The students are ensured to fully focus on the things that are presented during the description of instructor practice. Students are asked to discuss existing problems if problems in the use of tools and machinery occur. This strategy prevents the occurrence of errors in handling machines and tools in pandemonium workshops.

3.3. Controlled movement (P3)

To develop competence in controlled movement (P3) in teaching strategies that shapes previous lesson is also applied in teaching sessions and learning workshops. Instructors ensure that students can prepare the workplace correctly without their assistance. Instructors remotely observe selected work of students. They remind students to revisit the preparation procedure of materials they have performed in the past. Controlled movement (P3) in teaching is also applied in the form of demonstration, wherein instructors show students the wrong way to use the machine and tells them the consequences of working without following the correct procedures and specifications. Instructors demonstrate the method by providing material specifications and showing the correct usage of the machine. This demonstration gives students an overview of practical learning beyond the applicability of controlled movement (P3).

3.4. Mechanism (P4)

To develop competence at the level of mechanism teachers, before starting a practical workshop, teachers ask students about setting and maintaining the correct tools. The trainers ensure that students can maintain the equipment safely and properly while performing without supervision. A teaching strategy is used by educators to apply mechanism as a psychomotor domain (P4). Trainers carry out teaching strategies in groups as they demonstrate to students during practical sessions. This process is only conducted with weak students during teaching sessions before a practice run. The instructors show students in advance and then ask students to open and install back the project when they have extra time during the practice run to increase student competency. For the application of mechanism (P4), trainers use video presentations. In the vocational college, instructors show an overview of the visual manufacturing process in the industry and associate the knowledge with practical workshops. Trainers also show students examples of right and wrong practices using techniques and procedures as guide during the practice run.

3.5. Specific movement (P5)

To achieve competence of specific movement (P5) also occurs in practical teaching strategies shaping scan back. Instructors highlight different criteria in each step in the workshops on the basis of the past experiences of practical students. Teaching strategies for demonstrations are conducted by trainers. Instructors show the use of each button on the machine so that the students become focused, incrementally upgraded to the machine, and cautious. The functions of each button are also described to the students. The application of a specific movement (P5) is included in teaching strategies involving demonstration.

3.6. Settlement (P6)

To achieve competence of (P6) teachers show variations that can be applied while performing practical work. Prior to practical sessions, instructors demonstrate techniques in installing and testing matches correctly. The practical sessions are carried out after instructors ask all the students gathered at the venue for updates. Teachers ask students to explain and demonstrate the return procedure implemented on the basis of tasks performed in actual projects. Instructors demonstrate techniques in installment and test if the work is correctly performed. The instructors demonstrate how to update the work to students in groups if the work material is inaccurate. Instructors request weak students to explain and demonstrate the return procedure implemented from start to finish to increase their understanding and skills. These efforts aid in the inculcation of adapted strategies for teaching demonstrations. The teaching strategies in the form of discussion and adaptation (P6) are also applied by instructors in teaching and learning practices in workshops. The teachers ask students to discuss and explain return procedures in implementing the practice that has been carried out. Instructors show students through media how to install and match materials correctly. The trainers also show variations of art that can be applied while developing the practice.

3.7. Originality (P7)

Lastly, to achieve the highest level of competence in the psychomotor domain teachers also occurs the originality (P7) is one of the domains in the question and answer session during learning activities. Teachers ask students to list the materials or other hand tools that can be used in practice if students lack handheld materials in the workshop.

4. Discussions

Observation or perception is applied in three teaching strategies, namely, scan back, demonstrations, and discussions. The setup domain is only applied in discussions. Controlled movement is applied in memorizing previous lesson , demonstrations, and video workshops. Mechanism is applied in teaching strategies such as question and answer sessions, demonstrations, and video workshops. Specific movement is applied in scan back and demonstrations. Adaptation is applied in teaching strategies such as discussions and demonstrations, particularly video demonstrations. Originality is applied in the question and answer sessions in workshops. [7] and [11] stated that teaching strategies involving demonstrations are effective for practical subjects. In the present study, demonstration-oriented teaching strategies adopted by teachers do not use all the psychomotor domains. Instructors do not apply setup and originality when using teaching strategies involving demonstrations. The curriculum structure contains four fundamental questions: What should be taught? How should it be taught? How should learning be measured? Who are the learners? [12] stated the components of a curriculum structure for students to know the standards and objectives, activities and experiences, test and assessments, and materials and equipment. Hence, this research focused on a related issues of curriculum structure and a proposed match with teaching delivery to achieve educational objectives. The revised Bloom's taxonomy also clarifies the intended outcome for educational objectives.

5. Conclusions

The contribution of this research focuses more in the vocational field. Experts' views were discussed based on aspects in the psychomotor domain, and the quality of teaching will be enhanced when teachers design their teaching strategies. Usually, teachers prepare their lesson plan based on the syllabus and taxonomy level to achieve their teaching goals. However, they do not emphasize the psychomotor domain even when teaching skill-based subjects. Therefore, the application of perception, setup, controlled movement, mechanism, specific movement, settlement, and originality require more focus and clarification in preparing lesson plans for skill-based subjects. The investigated psychomotor domain is based on the nature of automotive technology courses in vocational colleges. However, the methodology is also suitable for other skill-based subjects and laboratory modes as long the teaching and learning session involves more physical activities. When teachers plan how they should assess learning, the intersection of cognitive, psychomotor, and affective processes can facilitate the selection of learning activities to provide modeling and practices to others.

References

[1] Rojewski. J.W (2009). A Conceptual Framework for Technical and Vocational Education and Training. dlm. Rupert, M., David, W., Chris, C. *International Handbook of Education for the Changing World of Work, Bridging Academic and Vocational Learning.* Germany: Springer Science Business Media

[2] Tan, A. M. (2010). Pentaksiran Berasaskan Sekolah (PBS) Di Malaysia-Kesediaan Guru, Isu, dan Panduan Pelaksanaan. Universiti Kebangsaan Malaysia: Gerak Budaya Enterprise.

[3] Daniel, M. (2010). *Effective Teaching Evidence and Practice.* London: Sage Publication Asia Pacific Pte.Ltd.

[4] Laurie, B. dan Kerry, K. (2003). *Curriculum Construction.* 2nd ed. Boston: Pearson Education.

R. Nicole, "Title of paper with only first word capitalized," J. Name Stand. Abbrev., in press.

[5] Tiia, R. & Hants, K. (2011). *Teaching Strategies for Direct and Indirect Instruction in Teaching Engineering.* Estonian Centre for Engineering Pedagogy: Tallinn University of Technology.

[6] Harold, W. (2001). *Teacher Classroom Practices and Student Performance: How School Can Make A Difference.* New Jersey: Educational Testing Service Statistic and Research Division Princeton.

[7] Alessi, S. M., & Trollip, S. R. (2001). *Multimedia For Learning: Methods And Development 3rd Ed.* Boston MA: Allynn & Bacon, inc. ISBN 13:9780205376912

[8] Simpson, E. (1972). Taxonomy For The Psychomotor Domain in Snowman,

[9] Idris, N. (2010). *Penyelidikan Dalam Pendidikan.* Malaysia: McGraw-Hill.

[10] Jack, R. F., Norman, E. W., & Helen, H. H. (2011) . *How To Design And Evaluate Research In Education.*8th Ed. San Francisco: Mc Grow - Hill.

[11] Fishwick, P. A. & Hill, D.R.C (1999). Web-based simulation. Special Issue of the Simulation Journal, 4,July.

[12] Anderson, L. (2006, May). Revised Bloom's taxonomy. Paper presented at North Carolina Career and Technical Education Curriculum Development Training, Raleigh, NC.

IOP Conf. Series: Journal of Physics: Conf. Series **1049** (2018) 012050 doi:10.1088/1742-6596/1049/1/012050

Career Choice and Employability Skills for Vocational College Students

Siti Norain Jaafar[1], Normah Zakaria[1], Norulhuda Abd Rasheid[2]
[1]Universiti Tun Hussein Onn Malaysia, UTHM
[2]Universiti Teknologi Mara, UiTM

E-mail: ct_snj88@yahoo.com[1], norma@uthm.edu.com[1]

Abstract. This study is a quantitative research based on the survey conducted that aims to look at the relationship between employability skills and career choice for Vocational College students. The study involves 106 respondents consisting year three students from Batu Pahat Vocational College and Muar Vocational College using questionnaires as instruments. The results from the analysis show that the relationship between skills employability and career choice is high, with an ETA value of 0.742, which proves that the variables have a strong correlation. In conclusion, employability skills affect students in such a way that they tend to choose a career based on the course they take up.

1. Introduction

Employability skills are an important aspect in ensuring qualified graduates are employed. Most technical graduates depend solely on their academic credentials to get a job. They lack non-technical skill competency or better known as employability skills [1]. This situation creates dissatisfaction and becomes the main problem for employers to choose the right candidate for a certain post. However, the career choice by the student must be the right one and suitable because career is a work experience that affects the lifestyle and well-being [2].

Therefore, career choice is a very important decision for a student. It should be in line with the courses taken up by the students when they were studying at the Vocational College (VC) so that they are competent enough to carry out any tasks given to them when they are in the realm of work. Technical and Vocational Education training student need of emerging abreast of changes from time to time for industrial economy based on economy. Most employers now require students who will be employees who not only have technical skills but also requires employability skills to improve productivity and competitiveness.

Students with employability skills will be able to adapt to all types of work and versatility. Graduates students from educational institutions have no labour market because the students do not have soft skills and employability skills. Prospective employer are less optimistic about the graduates of the institutions, especially in the aspect of personality, appearance, self-confidence, lack of communication skills, the ability to make decisions is limited and not able to work in groups [3].

The study states that if the career choice is not taken seriously, the individual may be trapped in a career that they are not interested in, making the career less meaningful to them [4]. However, local graduates are advised not to be too choosy in finding jobs after they graduate because excellent careers

[1] Corresponder author: ct_snj88@yahoo.com

Content from this work may be used under the terms of the Creative Commons Attribution 3.0 licence. Any further distribution of this work must maintain attribution to the author(s) and the title of the work, journal citation and DOI.
Published under licence by IOP Publishing Ltd

can also be built from bottom level [1]. To obtain a desired career, students are required to have academic ability, technical skills and employability skills [5]. The RMK-10 also emphasizes a combination of soft skills and efficiency of the work force that is able to face the challenges of the 21st century (10th Malaysia Plan).

Therefore, VC graduates should be prepared to equip themselves with these skills and form self-esteem to engage in the realm of work [6]. Instead, any workplace requires workers with high technical skill couple with well-developed employability skills [7]. Employability skills are important and have been proven in helping people in adapting with changes and improve career opportunities in the workplace [8]. So, the objective of this research is to look the relationship between employability skills and career choice for Vocational College students.

1.1. Problem Statement
There are several persistent problems among some students who venture into this field who do not have the skills required by employers and hence disqualified for the job. The problem of career choice in this engineering field is when the graduates are over confident in choosing a career in the same field that they took up during their VC education without having the self-employed skills. Many technical graduates are still unemployed because they only rely on academic qualifications and the lack of employability skills required by employers [9].

However, the students are not entirely to be blamed as the students explained that the application of employability skills through the course content provided by schools or institutions is not sufficient to meet the demand of the industry [3]. There is one of barriers to successful employment that is a lack of employability skills [10]. There are issues surrounding the role of higher education, deficiencies in the classification of graduate destinations, the challenging of predicting the needs of employer of the future, and gaps between strategies, perceptions and the realities [11] and the results show in [11] study shows that the student will find their own unique responses to these challenges.

The potential of employability skills has been argued for increasing work-related outcomes and also help the people adapting with changes and improving career opportunities in the workplace [12]. Based on the problems presented, it clearly shows that in the past research done only focus on students from universities and graduates from technical and vocational fields that have been working in a particular sector. However, there is no research done on the Vocational College students yet.

1.2. Term of Definition
There is several terms use in this study to make readers understand more about career choice and employability skills among VC students.

1.2.1 Employability Skills
Employability skills are the quality of the individual that is required by the employer [13]. This skill can be applied in any field of work. While [14] defines the employability skills as the skills required by an employee to succeed in a job. [15] The employability skills list consists of several skills which includes communication skills, critical thinking and problem solving, teamwork skills, continuous learning and information management, entrepreneurship skills, ethics and moral and professional leadership.

1.2.2 Career
Career is defined as individual perception or a characteristic attitude and behaviour of a person related to work activity and experience in life [16]. However, [17] defines a career as a journey or the progress of an individual in a field of life that makes the choice of employment for daily living. [18] stated career refers to the first thing that students sought after graduation. Career is an occupation that

someone needs to have in their life to make an effort in their daily life. If they have a good job, it can make their life better and they can live in happily.

1.2.3 Career Choice

Career choice is an important aspect of individual life. This is because the chosen career will determine the role pattern of the individual in society. In addition, [19] concludes that there are three main reasons for difficulty in making career choices. Firstly, that person has interests in various fields. Secondly, the person could not gauge his/her ability and personality. Finally, the lack of knowledge and information about the profession to be chosen.

2. Methodologies

The study is quantitative research based on the survey conducted. The findings from the sample were then generalized to population studies. The selection of the design is in accordance with the main focus of this research as the relevant studies conducted on employability skills are often done on the technical and vocational students [20-22].

2.1. Population and Sample

This study was conducted at Vocational College located in Johor. There are three VC in Johor which have same courses that are Kluang Vocational College, Batu Pahat Vocational College and Muar Vocational College. Although only two VC have been involved in this study. Kluang VC is a pioneer for Johor, so their students had started a year earlier and was not in the same semester with two colleges yet. Therefore, the colleges selected for this study are Batu Pahat VC and Muar VC.

The selected factor for both VC because they have same courses of Construction Technology, Electronic Technology and Welding Technology and the students are also in the same semester. The population for this study is the final year students which should be 4th year students. However, both the new Batu Pahat VC and Muar VC operations in January 2013, then the first group students were admitted into the 3rd year. Therefore, the researcher decided to select third-year students who took Construction Technology, Electronic Technology and Welding Technology as the population of this study.

The population for this study was the third-year students of the Construction Technology, Electronic Technology and Welding Technology courses from Batu Pahat Vocational College and Muar Vocational College. There are 90 respondents from Batu Pahat Vocational College and 60 respondents from Muar Vocational College. The sample size determination for this study is based on the calculations proposed by [23] using the formula [24]. Based on the calculation of sample size, the total numbers of students required as a sample for the study are 106 people.

2.2. Research Instrument

The instruments used in this study was a set of questionnaires which contains four parts, namely part A, B, C and D. Part A is a demographic questionnaire, part B is a questionnaire adapted for self-efficacy of CDSE-SF from [25], part C is the employability skills questionnaire adapted from SCANS (2001) and part D is a career choice question. The reliability of the instrument is evaluated by experts in the Faculty of Technical and Vocational Education at University Tun Hussein Onn Malaysia (UTHM).

2.3. Data Analysis

Data analysis for this study uses descriptive and inferential analysis. In this study, descriptive analysis is used to describe information such as gender, course, course selection and respondents' aspirations. Inferential statistics involved in this study was Eta-squared (η^2). An eta-squared test is a measure of effect size for use in ANOVA used to identify the relationship between the different level of variables that are nominal and interval. It is used to identify the relationship between employability skills and career choice.

3. Results

The data obtained through this study were analysed using *Statistical Package for Social Sciences* (SPSS) version 20.0. Table 1 shows the total distribution of students based on gender. Table 2 shows the distribution of students based on three courses namely Construction Technology, Electronic Technology and Welding Technology.

Table 1. Distribution of students based on gender.

Gender	Frequency (n)	Percentage (%)
Male	77	72.6
Female	29	27.4
Total	**106**	**100**

Table 2. Distribution of students based on courses.

Course	Frequency (n)	Percentage (%)
Construction Technology	32	30.2
Electronic Technology	43	40.6
Welding Technology	31	29.2
Total	**106**	**100**

Table 3 shows the course selection rates either by the Vocational College or by their own choice. The results of analysis show that majority of the students choose their own courses. Only a small number of students is forced to take up courses set by that college. Table 4 shows the suitability of courses to suit ambition. The selection of the field is based on the student's ambition, but not on a small group whose selection is determined by the college.

Table 3. Course selection by students and Vocational College.

Course Selection	Frequency (n)	Percentage (%)
By Vocational College	20	18.9
Own Choice	86	81.1
Total	**106**	**100**

Table 4. The suitability of course selection to suit ambition.

Suitability	Frequency (n)	Percentage (%)
No	15	14.2
Yes	91	85.8
Total	**106**	**100**

Table 5 shows the descriptive analysis of the employability skills level of the Vocational College students. Basic skills show the highest average mean score of 4.05 and high interpretation mean. This indicates that the students master basic skills such as reading, writing and calculating well. Technology skills show the lowest average mean score which is 3.58 and moderate interpretation mean. This demonstrates the skills that involve the students' ability to seek out diverse information sources, assess existing data, analyse written data, computerize data and disseminate analysis results to others are not dominated by all students.

Table 5. Analysis descriptive of the employability skills level of Vocational College students.

Item	Average mean score		Interpretation mean
	Mean	Standard deviation	
Basic skills	4.05	.759	High
Thinking skills	3.86	.631	Moderate
Personal skills	3.79	.687	Moderate
Resource skills	3.76	.658	Moderate
Interpersonal skills	4.01	.681	Moderate
Information skills	3.71	.677	Moderate
Technology skills	3.58	.639	Moderate
Total average mean score	**3.82**	**.676**	**Moderate**

Table 6 shows the selection of career according to the course being pursued. Majority of the students are more confident to choose a career based on the field that is being followed. However, there are a few who do not choose a career according to the field, but they rather choose to venture into the government sector hoping for a guaranteed comfortable life.

Table 6. Career choice according to the course being pursued.

Career Choice	Frequency (n)	Percentage (%)
No	15	14.2
Yes	91	85.8

Total	**106**	**100**

Table 7 shows the value of the coefficient of correlation of the relationship between the employability skills and career choice. There is a strong and positive relationship with ETA reading value of .742. Students are confident of the skills they have and they will succeed in pursuing the same fields as their current course.

Table 7. The value of the coefficient of correlation for the relationship between employability skills and career choice.

Relationship	ETA Value
Employability skills VS Career choice	.742

4. Findings and Discussion

Based on the findings of the analysis carried out through the research studies for the employability skills, it is found that the VC students are at a moderate level. This is in accordance with the study of [15] where the response of employers to the employability skills of technical graduates employed shows that the employability skills of technical graduates as a whole stood at a moderate level and employers are quite dissatisfied with the skill level.

However, if the analysis on the skills of the employability skill is conducted separately, it is found that the mastery of basic skills and interpersonal skills are at high level. Research conducted [1] also demonstrates that the basic skills and interpersonal skills possessed by technical graduates are high compared to other skills. However, the VC students' employability skills should be enhanced through soft skills classes and programs to shape students' personalities in dealing with work environment situations [26].

In the other hands, [20] argues that the employability skills acquired by technical graduates still need to be further improved to meet industry requirement and enable graduates to work and adapt well in the industry environment. Subsequently, the findings for employability skills shows strong and positive relationship which mean that the employability skills of students affect their career choice. Besides that, [1] states that students who have high employability skills can work with others, have creative thinking and be more responsible in choosing a career.

This is because they believe that the skills that they have mastered will help them in carrying out any tasks given to them successfully. Students who lack of skills are hesitant to apply for job because they are worried that they won't be able to do work as expected by employers.

5. Conclusions

Through the study, it can be concluded that the employability skills possessed by students are inadequate and need to be improved. VC plays a role in ensuring their graduates dominate the employability skills as expected. By acquiring good employability skills, VC graduates will be much better. They indeed have been provided with good technical skills through the existing curriculum. Hence, the acquisition of high employability skills will make them the most demanding graduates who are much needed by employers. Researchers concluded that good employability skills help students to choose careers precisely because they have no problem to constantly adapt to the new working environment and they are confident to implement skills with their careers. So, this paper will contribute on student in making career choice in their life.

ACKNOWLEDGMENTS

IOP Conf. Series: Journal of Physics: Conf. Series **1049** (2018) 012050 doi:10.1088/1742-6596/1049/1/012050

This paper was partly sponsored by the Centre for Graduate Studies Universiti Tun Hussein Onn Malaysia.

References

[1] Rasul, Mohamad Sattar, et al. "Peranan institusi pendidikan teknikal dalam pemupukan kemahiran ‘employability’pelajar." *Sains Humanika* 50.1 (2009).

[2] Amla, M. S. "Pendidikan Kerjaya dan Pembangunan Modal Insan." (2010).

[3] Mohamed Rashid, N. B., and M. Mohd Rashahidi. "The Year 2004 Polytechnic Convocation Survey." Proceedings of National Seminar “The development of Technology And Technical-Vocational Education And Training In An Era of Globalization: Trend and Issues. 2005.

[4] Career Services, Rutgers University. Make a Decision (2007). Achieved May 17, 2015 from http://careerservices.rutgers.edu/OCAmakedecision

[5] Ahmad, Shaharuddin, et al. "Generic Competency Profile among Students in Institute of Higher Learning: A Case of Universiti Kebangsaan Malaysia (UKM), Malaysia." College Student Journal 44.3 (2010).

[6] Kamro, Maisarah. *Kesediaan menceburi kerjaya kejuruteraan binaan dalam kalangan pelajar jurusan binaan bangunan di Sekolah Menengah Vokasional*. Diss. Universiti Tun Hussein Onn Malaysia, 2012.

[7] Singh, Gurvinder Kaur Gurcharan, and Sharan Kaur Garib Singh. "Malaysian graduates' employability skills." *UNITAR e-Journal* 4.1 (2008): 15-45.

[8] Rasul, Mohamad Sattar, et al. "Development of employability skills assessment tool for manufacturing industry." *Jurnal Mekanikal* 30 (2010): 4-61.

[9] Zaini, Ahmad. "Students and employers as customers of multimedia college."*Proceedings of National Seminar “The development of Technology And Technical-Vocational Education And Training In An Era of Globalization: Trend and Issues”. Kuala Lumpur*. 2005.

[10] Ju, Song, Dalun Zhang, and Jacqueline Pacha. "Employability skills valued by employers as important for entry-level employees with and without disabilities." *Career Development and Transition for Exceptional Individuals*35.1 (2012): 29-38.

[11] Andrews, Georgina, and Marilyn Russell. "Employability skills development: strategy, evaluation and impact." *Higher Education, Skills and Work-Based Learning* 2.1 (2012): 33-44.

[12] de Guzman, Allan B., and Kyoung Ok Choi. "The relations of employability skills to career adaptability among technical school students." *Journal of Vocational Behavior* 82.3 (2013): 199-207.

[13] The Conference Board of Canada. Employability skills profile: The critical skills required of the Canadian workforce. *The Corporate Council on Education Center*. Ottawa Ontario United State: Wadsworth Thomson Learning. (2002).

[14] Rothwell, Andrew, and John Arnold. "Self-perceived employability: development and validation of a scale." *Personnel review* 36.1 (2007): 23-41.

[15] Ng, Poh Yen, Shamsul Kamariah Abdullah, and Fidella Tiew. "Employers' feedback on business graduates: an exploratory study in Curtin Sarawak."*International Review of business research papers* 5.4 (2009).

[16] Abdullah, A. B. *Kerjaya dalam bidang kejuruteraan*. PTS Professional. (2007).

[17] Rahim, Rohana Binti Abdul, and DOKTOR FALSAFAH. *Minat Kerjaya, Efikasi Kendiri Keputusan Kerjaya, Kemahiran Employabiliti, dan Pilihan Kerjaya Pelajar Vokasional Pertanian di Semenanjung Malaysia*. Diss. Tesis PhD. Universiti Putra Malaysia, 2010.

[18] Tadić, Ivana. *Career development of graduates in economics and business administration in Croatia*. Diss. Ekonomska fakulteta Ljubljana, Sveučilište u Ljubljani, 2005.

[19] Noah, Sidek Mohd. *Reka bentuk penyelidikan: falsafah, teori dan praktis: sebuah buku mesra pengguna*. Penerbit Universiti Putra Malaysia, 2003.

[20] Buntat, Yahya. "Integrasi kemahiran" employability" dalam program pendidikan vokasional pertanian dan industri di Malaysia." (2004).

[21] Ivan, H. (2007). *Perlaksanaan pendidikan dan latihan teknik-vokasional di German Malaysian*
[22] Rahim, Rohana Binti Abdul, and DOKTOR FALSAFAH. *Minat Kerjaya, Efikasi Kendiri Keputusan Kerjaya, Kemahiran Employabiliti, dan Pilihan Kerjaya Pelajar Vokasional Pertanian di Semenanjung Malaysia*. Diss. Tesis PhD. Universiti Putra Malaysia, 2010.
[23] Barlett, James E., Joe W. Kotrlik, and Chadwick C. Higgins. "Organizational research: Determining appropriate sample size in survey research."*Information technology, learning, and performance journal* 19.1 (2001): 43.
[24] Cohen, Jacob. "Statistical power analysis for the behaviour science." *New York, NY: Lawrence Erlbaum Associated. Dorfberger, S., Adi-Japha, E., & Kami, A.(2009). Sex differences in motor performance and motor learning in children and adolescents: An increasing male advantage in motor learning and consolidation phase gains. Behavioural Brain Research* 198 (1988): 165-171.
[25] Betz, Nancy E. "Self-efficacy theory as a basis for career assessment."*Journal of career Assessment* 8.3 (2000): 205-222.
[26] Rasul, Mohamad Sattar. Kajian tahap kemahiran kebolehdapatan kerja lulusan progran diploma kejuruteraan mekanikal Institusi Pengajian Tinggi di Malaysia. Tesis PhD.tidak diterbitkan: Universiti Putra Malaysia. (2010)

IOP Conf. Series: Journal of Physics: Conf. Series **1049** (2018) 012051 doi:10.1088/1742-6596/1049/1/012051

Exploring Pedagogical Constructs Based on 21st Century Learning Model for Technical-based Subject Delivery in School

Alias Masek[1], Nurul Ain Suhadi[2]

[1,2] Faculty of Technical and Vocational Education, Universiti Tun Hussien Onn Malaysia, Batu Pahat, Johor, Malaysia.

Email: aliasmasek@uthm.edu.my[1]

Abstract. This study aims to explore the teachers' pedagogical elements based on the 21st century learning model for delivery of the theoretical part of technical-based subjects in secondary schools. This study adopts the qualitative approach using the semi-structured interviews to gather data from three participants and expert teachers. The selection of teachers was based on these criteria: has more than ten years of teaching experience in any technical-based subjects; is a school's trainer for 21st century teaching and learning; and adopts the 21st century pedagogical approach in teaching and learning. The transcribed interview data were thematically analysed to identify various themes according to patterns and meanings. The analysis yielded this result: a total of 15 themes classified into two major parts, namely teachers' self-skills and teachers' pedagogical skills. Teachers' self-skills include technology agility and exploration, evaluating information, adopting latest applications, integrating multiple media, customising materials, and diversifying techniques. Teachers' pedagogical skills include interpreting visuals, interpreting information, guiding information access, gaining student reflection, stimulating student thinking, improving students' thinking, developing students' soft skills, motivating, nurturing values and providing information support. The implication is that these elements are useful for teachers to arouse the interest of 21st century students in learning and to sustain their focus during the time of content delivery; this is especially crucial for teaching the theoretical part of technical-based subjects in schools.

1. Introduction

The 21st century students look for a significant change in the delivery approach in order to stay alert during long hours of lecture. The conventional method using chalk and talk is most probably less effective nowadays; it is very hard to capture students' attention in the classroom without attractive and dynamic teaching materials. Teaching and learning based on texts, visuals, and audios might be the right solution to whet students' appetite for learning. Based on Cheryl (2002), three important types of literacy are essential for 21st century teaching, namely the technology, information and visual literacy. Therefore, well blended teaching materials using texts, visuals and audios could be the right tool for effective teaching of the 21st century students.

The literature has documented many studies of 21st century learning (i.e. Cheryl, 2002; Berry, 2012; Leavitt, 2012; Ling Koh & Chai, 2015), focus on students (Baru, Abdullah, Ali & Yusoff 2014), and use the right tools in 21st century teaching (Nussbaum and Sharples, 2014), and tapping into teachers' knowledge (Kereluik, Mishra, Fahnoe and Terry, 2013). Research that focuses on teachers' delivery approach has been lacking in concrete evidences, particularly pedagogical skills in the 21st century. Although most of the younger-generation teachers have embraced technology agility, the ability of knowledge delivery (i.e. pedagogy skills) using the recent technology is an issue that needs to be resolved. In addition, the necessity for a deeper understanding of the theoretical part of technical-based subjects is very crucial, before students start doing practical tasks in the schools' workshop. The use of technology depends on the teachers' pedagogical expertise in facilitating students' learning: the contents of the lesson can easily be understood, actions or basic skills taught can easily be imitated by students, and the delivery of teachers is in an attractive form that can capture students' attention.

Pedagogical skills form the most important criterion for a good teacher. Good pedagogical skills are able to draw students' attention and sustain their engagement in learning (Hashim, Yaakub & Ahmad (2003), besides having good knowledge (Kereluik, Mishra, Fahnoe and Terry, 2013). In the setting of the traditionally chalk and talk approach, good pedagogical skills stress several areas, for instance, good eye contact to attract students' attention and a dynamic vocal intonation to help students focus. The most important thing is that the teacher can deliver a good explanation on the content, and the teacher is able to pose a stimulating question to trigger students' thinking. For the technical part, explaining an action or skills is very difficult by using verbal descriptions; it is also extremely hard to demonstrate a movement by using static photos. Nevertheless, teaching and learning in the 21st century can be very different, since teaching and learning can occur by employing a wider spectrum of methods. These include the use of video and audio materials, electronic learning (e-learning) in the form of virtual classroom, mobile learning, and multiple approaches of student-centred learning. Although the teacher has a particular level of agility in using technology, facilitating learning using technology requires some new pedagogical skills so that students' learning can be more effective. For that reason, this study was conducted to explore the teachers' pedagogical elements, based on the 21st century learning model, particularly for teaching the theoretical part of the technical-based subjects in the secondary school.

1.2 The 21st Century Learning Model

The literature proposes many guidelines for teaching in the 21st century; these include the integrating model (Cheryl, 2002), framework (21st Century learning framework, 2017), proposing a work process in teaching (Ling Koh & Chai, 2015), and improving teaching using recent high-tech tools (Nussbaum and Sharples, 2014). For example, an enGauge 21st Century Skills model from Cheryl (2002), has been the most preferred model for teaching the 21st century students. The model was developed by the North Central Regional Education Lab (NCREL) and Metiri Group in 2003. The EnGauge 21st Century Skill Model consists of four components, namely Era-Digital Literacy, Inventive Thinking, Effective Communication and High Productivity. The 21st Century Skills Model has a deeper assessment of the technology, visuals and literacy information, which are included in the various constructions as a component of literacy in the digital era.

Similarly, Inside-Out School: The 21st Century Learning Model also provides guidance for teaching in the 21st century. This model is a follow-up to the 9 characteristics of the 21st Century Learning Model that was developed in 2009. This model focuses on the role of play, a variety of digital and physical media, and is designed to be interdependent between the society and school. Another model for the 21st century teaching is the 21st Century Learning Framework (2017). Unlike Inside-Out School: The 21st Century Learning Model, this model incorporates all the existing components of literacy by emphasising various features such as information literacy, accessing information, efficiency (time) and effectiveness (resources), evaluating information critically and

IOP Conf. Series: Journal of Physics: Conf. Series **1049** (2018) 012051 doi:10.1088/1742-6596/1049/1/012051

efficiently. Hence, the model can be seen almost the same as the approach used. However, the enGauge 21st Century Skills model is an option to conduct this study based on Figure 1.

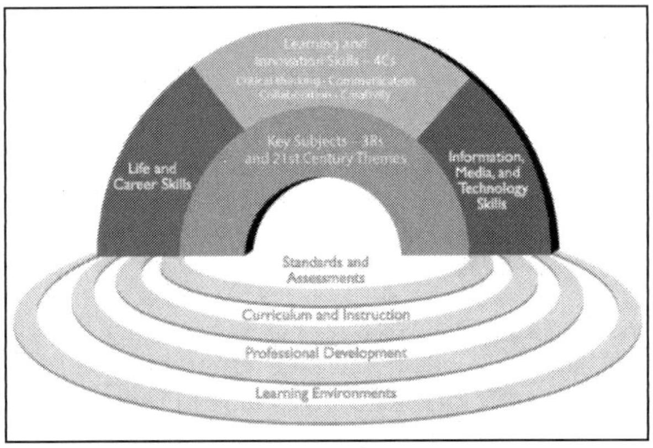

Figure 1: P21 Framework for the 21st Century Learning (source: Ministry Education)

2. Methodology

A qualitative study was conducted using the semi-structured interviews, in order to gain insights into and have an in-depth exploration of the phenomena, as according to Creswell (2003). The interview sessions were conducted face-to-face in the participants' office; the sessions were recorded using audio recorder software embedded within the mobile phone, so that the participants would feel at ease and comfortable. Three expert teachers were selected as participants. They met the selection criteria: is a teacher with more than ten years of teaching experience in any subjects; is a school's trainer for 21st century teaching and learning; and adopts the 21st pedagogical approach into his/her teaching and learning.

The first participant (P1) in this study is an experienced teacher, who has been a trainer for 21st century teaching and learning using frogs VLE application. The second participant (P2) is a teacher with ten years of teaching experience in the field educational technology. His task is to monitor teachers' digital literacy implementation in schools within the southern region. The third participant (P3) is a teacher with more than 10 years of teaching experience using the most recent technology in teaching and learning available in the school. The technical-based subjects include information technology, technology design, vocational technology, and living skills subjects.

The interview was guided by an interview protocol developed based on enGauge model of 21st century learning from Cheryl (2002). The questions were limited to three main constructs, namely technology literacy, information literacy and visual literacy. The interview protocol has been widely used in the qualitative research (Brayman & Bell, 2003) to achieve the study aims.

The audio files of the interviews were transferred into a computer, and they were immediately transcribed into an MS Word file. The data were analysed thematically according to several steps (Creswell, 2003): first, pre-coding was generated by selecting keywords (code); second, the axial coding categorised the coding that has the same character or meaning. Next, selective coding was performed by selecting the axial coding that corresponds to a theme/pattern studied. Then, definitions were drafted for the themes/patterns identified.

3. Result

The data analyses indicate a total of 15 themes, which can be divided into two main categories, namely teacher self-skills (i.e. skills a teacher should have before teaching/preparation for teaching), and pedagogical skills (skills for teaching). Teacher self-skills include technology agility and exploration, evaluating information, adopting latest applications, integrating multiple media, customising materials, and diversifying techniques. Teacher pedagogical skills include interpreting visuals, interpreting information, guiding information access, gaining student reflection, stimulating student thinking, improving student thinking, developing student's soft skills, motivating, nurturing value and providing information support.

Here is an insight into the themes: participants express the importance of a teacher to explore the technology beforehand; such preparation of teaching is necessary in the 21st century. Exploring the technology can help the teacher to prepare for a better teaching session. For instance,

"... knowledge about the use of technology, how it functions and its purpose in helping teachers ..." (P1)

Participants' also highlighted that teachers should be able to evaluate information before it can be disseminated to students. Seeking and accessing information form an essential skill, but evaluating the information is another crucial skill for teachers. Participants expressed these views:

"... actively seeking, accessing, acquiring and using information ..." (P3)

"... the skills used to find information, the ability to access and evaluate information ..." (P2)

In addition, a teacher should also adopt the latest application as a preparation for teaching in the 21st century classroom. A comprehensive technology is important to be adopted to arouse the students' interest in learning. Participants therefore highlighted that the use of online learning or electronic learning (e-learning) such as Celestia and Edmodo could be very helpful for the teaching process.

"... using Calestia learning applications ..., ... comprehensive learning applications such as 'Edmodo' ..." (P2)

Moreover, as a good 21st century skill for teaching preparation, a teacher should be able to integrate multiple teaching media and materials. Participants in the study stated that in visual literacy pedagogy, the objects used should match the topic; several media, including audio and video can be combined.

"... Using images, visuals, objects and videos that are relevant to the teaching topics ..." (P3)

"...Combine multiple media such as text, voice, image, animation and video in a software..."(P2)

Besides integrating media and materials, a teacher should have the ability to customise new materials from the internet that match the learning outcome. Some materials in the internet might be provided in a package; the teacher therefore needs to select only the relevant information to be used for a particular class. Here is a participant's opinion:

"... materials that are relevant to the teaching content can help students solve related questions." (P2)

In addition, a teacher with a good self-skill for teaching preparation should be able to diversify techniques to deliver the subject content. These techniques include forums, Quizzes and Gamutics.

"... an interesting quiz to challenge students' understanding ..." (P1)

IOP Conf. Series: Journal of Physics: Conf. Series **1049** (2018) 012051 doi:10.1088/1742-6596/1049/1/012051

"... insert the gamut element which students learn through play ..."
(P2)

For teachers' pedagogical skills, participants have highlighted that the most important pedagogical skills include visual and information interpretation. During the teaching process, a teacher should be able to help students interpret the meaning of, for instance, pictures, videos, objects, symbols, diagrams, and sketches that appear in the media. These pedagogical skills are important to avoid students misinterpreting visuals and information.

"... proficiency in interpreting materials using image, video or object ..." (P1)

"... Ability to interpret image and video use ..." (P2)

"... the ability to interpret visual acts, objects, or symbols ..."(P3)

In order to ensure students have multiple sources of information, a teacher should provide guidance on information access by giving them links, websites, or redirecting students to an individual for collaboration. Besides, students should be guided to acquire the right information.

"... guide students to find / access and obtain the necessary information using existing resources." (P3)

"... the ability to have the information evaluated and transmitted across multiple media." (P1)

"... the ability of the student in determining the exact information needed ..." (P2)

Helping students to reflect on the things they have learnt is one of the necessary pedagogical skills in the 21st century learning. Students' reflection improves their learning and sharpens the teacher's pedagogical skills. For instance, participants have shared experience that:

"... create a challenging yet fun learning atmosphere ..." (P2)

"... attracting student attention if such literacy is used ..." (P3)

In the 21st century, a teacher should also have the skills to stimulate and improve students' thinking. According to participants, teachers should relate the content of learning to real-world problems. Some real-world problems come in a complicated way that demands a complex problem-solving technique. Thus, it is important for teachers to also improve the students' thinking process.

"... Students associate the objects seen with daily life ..." (P1)

"... Enhances student thinking, especially in making decisions ..." (P2)

"... providing materials that can stimulate students thinking in learning in this 21st century ..." (P1)

Moreover, participants of the study stressed that teachers also need to have skills in nurturing students' soft skills, such as to encourage students to communicate and think critically. This also includes infusing a good value such as cooperating and competing in a healthy manner. The teacher should also encourage students to be more independent.

"... encourage students to communicate using materials provided ..." (P1)

"... encourage students to think critically ..." (P3)

"... nurturing the value of student collaboration in group discussions ..." (P1)

"... encourage students to get information using computers independently ..."(P2)

Motivating also appears as an important skill for teachers in the 21st century. During the teaching and learning process, teachers are encouraged to provide some tokens to students, especially for non-face – to-face learning using media. For example, a participant highlighted this gesture of encouragement:

"... in encouraging the students to continue to work, I'll give you some

money ..." (P3)

Teachers also need to have some skills in giving information support to ease the learning process. These include the skills in comparing, managing, compiling and combining information obtained by students. The Internet provides tons of information, but students sometimes need help in sorting out the more precise and meaningful information.

"... comparing, managing, compiling and combining information obtained ..." (P3)

5. Discussion

The study findings yielded a total of 15 themes, which are divided into two main categories, namely teachers' self-skills and teachers' pedagogical skills. Teachers' self-skills are those that require a teacher to make preparation for teaching, while teachers' pedagogical skills are those that a teacher uses to deliver content knowledge in a more effective way. Teachers' self-skills are important for teachers to integrate the 21st century technologies into teaching; these include technology agility and exploration, evaluating information, adopting latest applications, integrating multiple media, customising materials, and diversifying techniques. Teachers' pedagogical skills are important to arouse students' interest in learning, so that the content can be understood by students. These include skills in interpreting visuals, interpreting information, guiding information access, gaining students' responses, stimulating students' thinking, improving students' thinking process, developing students' soft skills, motivating, nurturing values and providing information support.

These findings of the elements of teachers' self-skills and pedagogical skills are similar to those of several previous authors. For instance, Reeves (1994) suggests 14 elements of pedagogical skills for teachers, three of which are in line with the outcome of this study, namely adopting latest applications, nurturing values and motivating. This result clearly suggests that values should remain as the most important part of education, and this element should be incorporated into teaching and learning of the 21st century. Similarly, in the Quinn (1996) study, two main skills are highlighted: first, teachers should be able to help students to reflect on the things they have learnt; second, teachers should also provide information support. Reflection is the most important element in learning; the process gives the students more impact in the learning process (Ryan & Ryan, 2013). On the other hand, information support is important so that students have the right information from credentialed sources, besides more alternatives of information access.

The 1996 study from Squires and Preece stresses the importance of teachers' skills in developing a student's soft skills. Soft skills development remains the most important domain of education even in the 21st century, although the teaching might not always occur in the face-to-face method. Traditionally, teachers infuse soft skills while teaching; for instance, teachers may comment on students' presentation to improve communication skills, which is face-to-face learning. However, if learning occurs through the social media, teachers should have the skills to inculcate in students the communication skills within that particular context as well.

In order to keep students interested in long hours of teaching, Albion (1999) has previously suggested that teachers should diversify the teaching techniques; the idea is similar to the findings of this study. In diversifying the teaching technique, teachers' self-skills are necessary. With the self-skills, teachers can design the teaching process using a more creative and innovative approach, such as online learning (Nussbaum and Sharples, 2014) and design thinking (Ling Koh & Chai, 2015). It is believed that the approach of diversifying techniques can also stimulate students' thinking. Students will have a different perception of content knowledge when the teacher uses multiple techniques for a similar learning outcome. Based on the findings of this study, stimulating students' thinking is one of teachers' pedagogical skills; this finding is similar to the study findings of Horila, Nokelainen, Syvinen and Overland (2002).

Besides, teachers should be able to motivate students to learn in the 21st century. Motivating skills are important for teachers since students are easily distracted from their learning, especially when learning through media such as social media (Horila, Nokelainen, Syvinen & Overland, 2002). Furthermore, based on the previous studies of Reeves (1994), and Horila, Nokelainen, Syvinen and Overland (2002), the authors of both studies agree that the motivating element is an essential part of pedagogy for teaching in the 21st century.

6. Conclusions

This study explores teachers' pedagogical elements for a new pedagogical construct relevant to the 21st century teaching and learning of the theoretical part of the technical-based subjects in the secondary school. Based on the three major constructs of the 21st century learning model from Cheryl (2002), there are 15 pedagogical elements. These elements are classified into two major groups of skills, which are necessary for teachers to teach the 21st century students. These elements are useful for teachers to arouse the interest of the 21st century students in learning and to sustain their focus and attention during the learning sessions, especially when delivering the theoretical part of the technical-based subjects.

7. References

[1] Albion, P. (1999) Heuristic evaluation of educational multimedia: From theory to practice. *Paper presented at the 16th Annual conference of the Australasian Society for Computers in Learning in Tertiary Education.*

[2] American Association of College of Teacher Education & Partnership for 21st Century Skills (2010). *21st Century Knowledge and Skills in Educator Preparation.* U.S: Pearson.

[3] Baru S. A., Abdullah, L, Ali, A. & Yusoff, H. (2014). "Permodelan Penerimaan Pelajar Terhadap Pembelajaran Maya (VLE)," *Journal of Business and Social Development*, , vol. 2, No. 2, pp. 36-47.

[4] Berry, W. (2012). The Inside-Out School: A21st Century Learning Model. Teachthought We Grow Teachers. Available from http://www.teachthought.com/the-future-of-learning/inside-outside-school-21st-century-learning-model/

[5] Cheryl, L. (2002). NCREL's *enGauge 21st Century Skills: Digital Literacies for a Digital Age*. Naperville, Ill.: North Central Regional Educational Laboratory.

[6] Creswell, J. W. (2003). *Research Design-Qualitative, Quantitative, and Mixed Methods Approaches*.2nd ed. London: SAGE Publications.

[7] Framework for 21st Century Learning (2017). Teachthought We Grow Teachers. Available from http://www.p21.org/about-us/p21-framework/349-media-literacyteachthought.com/the

[8] Hashim S., Yaakub R., & Ahmad M. Z. (2003). *Pedagogi: Strategi dan Teknik Mengajar Dengan Berkesan*. Pahang: PTS Publications.

[9] Horila, M., Nokelainen, P., Syvanen, A., & Overlund, J. (2002). *Criteria for the pedagogical usability, version1.0,* Hameenlinna, Finland: Hame Polytechnic and University of Tampere.

[10] Kereliuk, K., Mishra, P., Fahnoe, C. and Teryy, L. (2013). What knowledge is of most worth. *Journal of Digital Learning in Teacher Education,* 29(4), pp.127-140.

[11] Leavitt, M. (2012). *Learning from Visuals How well-Designed and Well-Used Visuals Can Help Students Learn:* An Academic Review. Arizona State University.

[12] Ling Koh, J.H. and Chai, C.S. (2015). Technological pedagogical content knowledge and design thinking: a framework to support ICT lesson design in 21st century learning. *The Asia-pacific Education Researcher,* 24(3), pp. 535-543.

[13] Nusbaum, M. & Sharples, M. (2014). E-learning in school education in the coming 10 years for developing 21st century skills: critical research issues and policy implications. *Journal of Education Technology and Society,* 17(1), pp. 70-78.

[14] Quinn, C. (1996). *Pragmatic Evaluation: Lesson from Usability,* retrieved November 24, 2017 from: http://www.ascilite.org.au/conferences/adelaide96/apaers/18.html

[15] Reeves, T. C. (1994). Evaluating what really matters in computer-based education. In Wild, M. & Kirkpatrick, D. (Eds.) *Computer education: New Perspectives,* Perth, Australia: MASTEC, 219-246.

[16] Ryan, M. and Ryan, M. (2013). Theorizing a model for teaching and assessing reflective learning in higher education. *Higher Education Research & Development, 32(2)*, pp. 234-251.

[17] Squires D., & Preece, J. (1996). Usability and Learning: Evaluating the Potential of Educational Sofware. *Computers & Education,* 27 (1), 15-22.

IOP Conf. Series: Journal of Physics: Conf. Series **1049** (2018) 012052 doi:10.1088/1742-6596/1049/1/012052

Development of Central Processing Unit Teaching Using Web-Based Learning

Norhasyimah Hamzah[1], Tamil Selvan Subramaniam[2], Nuraini Hassan[3], Arihasnida Ariffin, Siti Nur Kamariah Rubani
Fakulti Pendidikan Teknikal dan Vokasional,
Universiti Tun Hussein Onn Malaysia

E-mail: hasyimah@uthm.edu.my, tselvan@uthm.edu.my, ainihassan@gmail.com, hasnida@uthm.edu.my, kamaria@uthm.edu.my

Abstract. E-learning is a distance learning method that uses electronic and the Internet. The objective of this study is to help students from Faculty of Technical and Vocational Education of Universiti Tun Hussein Onn Malaysia (UTHM) to improve their understanding on the topic "Central Processing Unit." The students consisted of third-year students in their second semester of Bachelor of Vocational Education (Multimedia Creative) programme who were taking the course "Computer System and Network." The learning video was developed based on self-learning and contains syllabus from the topic "Network System in Communication." The development of this online learning video followed the Hannafin and Peck Model which was deemed suitable for online learning development. The Central Processing Unit learning video was developed using Weebly software as the main platform along with the other support software such as Adobe Flash Professional CS6, Adobe Photoshop CS6, and Camtasia Studio 8. Experts were called in to attest the result, and they verified that the information design, interaction design, and interface design were appropriate. Their suggestions were also considered and improvements were made to the platform The findings lead to some recommendations for future works and explain the multimedia software that has been developed.

1. Introduction

In the twenty-first century, information and communication technology (ICT) has expanded rapidly across the globe and reformed the education systems in various countries including Malaysia [1]. In 2016, the Ministry of Education Malaysia (MOE) outlined the programme "Toward Education Schools in Malaysia" to improve the quality of teaching and learning (T&L) with the use of information technology. Accordingly, the e-learning method has been adopted in various educational institutions to meet the evolving learning needs.

The e-learning method has proven to deliver several advantages, among which are fun learning experience and ability to increase students' understanding of a difficult topic. Other advantages noted by [2] are that this alternative learning process:

---

[1] To whom any correspondence should be addressed.
[2] To whom any correspondence should be addressed.
[3] To whom any correspondence should be addressed.

Content from this work may be used under the terms of the Creative Commons Attribution 3.0 licence. Any further distribution of this work must maintain attribution to the author(s) and the title of the work, journal citation and DOI.
Published under licence by IOP Publishing Ltd

1. offers unrestricted learning in that students can access information from any place with internet connection;
2. can attract students with engaging learning;
3. enables students to revisit weak areas; and
4. saves the costs of books and notes.

Given the above advantages, the e-learning thus has been perceived by the ministry as a means to produce highly skilled generation by the year 2020.

Modern technology particularly the Internet has expanded education beyond classrooms. According to [3] defines e-learning as an R & D process in which electronic technology is used as a medium to create learning experiences. Therefore, this e-learning method is in line with the Malaysian education system where this method follows the learning needs from time to time according to the circulation in this modernization cycle. The existence of e-learning makes the student more independent to understand in depth about a difficult topic. Additionally, students will not rely solely on educators to master a topic.

Technology has made Malaysia's education system no longer tied to conventional teaching methods and it provides a fun new experience for students to explore the latest technologies. To take on the challenges of the Y-generation Vision 2020, actively explore this sophisticated cyberspace. According to [4] to realize the goal of 2020, Malaysia needs a highly educated generation and are skilled in various fields to advance the nation.

1.1 Theory of Learning Behaviorism

Behaviourism views that the outcome of learning process can be measured by the observation of behaviour; it depends on the close relationship between students' attributes/attitudes (such as intelligence, social background, and ability) and learning outcomes [5]. After the learning, users are more focused on behaviourism learning theories which support the development of online-based learning related to the Central Processing Unit. As such, the behaviourism theory can ensure the behavioural change of students' responses to line-based learning.

Based on the exercises, notes, quizzes and videos developed by the researchers, this theory can support the change of students' perceptions about learning in front of a computer. For example, instructors can give credit to students who score their quizzes and exercises, which in turn, would motivate the students to improve their learning. In conclusion, the study of behaviourism theoretical study can occur due to the link between stimulus and response in the web-based T & L process.

2. Research Objectives

The objectives of this study are to

1. design a central processing unit for online teaching videos
2. develop a central processing unit for the online teaching videos, and
3. evaluate the functionality of the central processing unit for the online teaching videos.

3. Research Methodology

The design of the study was based on Hannafin and Peck [6] Model which illustrates the design and implementation of teaching materials as comprising three phases, namely (1) requirements assessment phase, (2) design phase, and (3) expansion phase and teaching implementation. All phases require an assessment and review process. The models also emphasise the ability of a class to receive media-based materials.

The first phase required the researcher to assess the characteristics of the target group and determine the hardware and software for the analysis. The target audience was identified to consist of (1) instructors and (2) students who were taking the subject "Computer and Network Systems." The hardware deemed appropriate for the study was the specification of computer systems, and the softwares used were Weebly, Adobe Flash CS6, Adobe Photoshop CS6, Adobe Illustrator CS4, and Camtasia Studio 8. These programmes were selected for their ability to facilitate the researcher to produce an online learning platform characterised by the five elements of text, graphics, audio, video and animation. Next, evaluations and checks were carried out to ensure that the hardware and software needs to develop web-based learning for Central Processing Unit Teaching was implemented to ensure that it is in line with the developer's needs. Once the needs were identified, the researcher assessed and review the platform to ensure that it meets all the requirements prior to development.

In the second phase, the developers provided storyboards to guide the creation of the teaching materials such as notes, exercises, videos, and quizzes. The contents of the material were based on the syllabus of "Computer and Network Systems" and the topic "Central Processing Unit" (CPU). Three designs–information design, interaction, and interfaces–were identified to guide the assessment of the specifications. The final phase–the expansion and implementation phase–involved translating the key elements of the design into a practical form of programming language system–a website. Formative and summative evaluations were carried out to test and review back-up web-based learning for central processing unit (CPU). Formative assessment was carried out throughout the media development and summative assessment was performed after the media were completed.

Questionnaires were distributed to the students with the aim to improve the quality of online learning and to meet the needs of consumers in information design, interaction, and interface. The respondents consisted of 30 third-year students of Vocational Education (Creative Multimedia) who were taking the course "Computer and Network Systems" for the second semester of 2016/2017. There are two sections in the questionnaire: Section A, which queries respondents' demographics, contains 4 items, and Section B contains 19 items with regard to information design, interaction design, and interface design. A five-point Likert scale was used with the options very disagreeable (STS), disagree (TS), disagree (KS), agree (S), and strongly agree (SS). The content and designs of the web-based learning were evaluated by experts in web-based development.

4. Research Findings

Three experts (lecturers) from the Creative Multimedia programme of the Faculty of Technical and Vocational Education (FPTV) commented that the font size of the teaching notes was too small. The experts also looked at the development of web-based learning for the topic "Central Processing Unit" (CPU). All these comments were considered and changes were incorporated into the finalised set of questionnaire. Table 1 shows the mean values for the level of user ratings on the aspects of information design.

Table 1. Mean for the Level of User Ratings on Aspects of Information Design

Item	Statement	Mean
1	The purpose of the learning content is clearly stated	4.47
2	The type of writing used in this line-based learning is appropriate	4.47
3	The writing size used in this line-based learning is appropriate	4.43
4	Information is communicated in a way that is easy to understand	4.43
5	The language presented in the form of text is easy to understand	4.47
6	These web-based learning information are not too shallow or deep	4.53
7	Information in teaching notes is easy to understand	4.30
8	The tutorial video view can be clearly understood	4.53
	Overall Mean	4.45

The highest mean value for information design is item 6 (web-based learning information is not too shallow or deep) (4.53). The lowest mean value is recorded for item 7 (information in teaching notes is easy to understand). The overall mean for information design is 4.45. Table 2 shows the mean values for the level of user ratings on the aspects of interaction design.

Table 2. Mean for the Level of User Ratings on Aspects of Interaction Design

Item	Statement	Mean
9	Users do not encounter any problems while using this web-based learning	4.40
10	Each button/icon menu in this web-based learning works well	4.53
11	Hyperlink /link works fine	4.50
12	Quiz and practice activities can be easily reached	4.53
	Overall Mean	4.49

The highest mean value of 4.53 obtained for the interaction design is for item 10 (each button/icon menu in this web-based learning works well). Item 9 (users do not encounter any problems while using this web-based learning) receives the lowest mean of 4.40 for interaction design. Table 3 shows the mean values for the level of user ratings on the aspects of interface design.

Table 3. Mean Values for the Level of User Ratings on Aspects of Interface Design

Item	Statement	Mean
13	The video tutorial size display used in this web-based learning is appropriate	4.50
14	The text display of teaching notes in this web-based learning is easy to read	4.50
15	The background music/audio used in the tutorial video is appropriate	4.33
16	The length of video display is appropriate	4.47
17	The graphical interface used is appropriate	4.47
18	Elements of animation are developed in harmony	4.40
19	The use of animated elements can attract users' attention	4.37
	Overall Mean	4.43

The highest mean value (4.50) for the interface design was noted for item 13 (video tutorial size display used in this web-based learning) and item 14 (the textbook display in teaching notes in this web-based learning is readable). The lowest mean value (4.33) was recorded for item 15 (the background music/audio used in the tutorial video is appropriate). The overall mean value for interaction design is 4.49, followed by information design (4.45) and interface design (4.43). This findings indicate that the users accepted the design of information, interaction, and interface of the web-based learning, and thus attesting that the platform meets all the three criteria.

5. Discussion

As a result, after evaluating experts has been commented on, developers have improved the development of web-based learning where developers have expanded the size of texts in the teaching notes so that they can be read by the users clearly. Additionally, the font type used has also been fixed so that it is appropriate to the user's age level and the topic presented. Furthermore, the use of text color has also been changed so that it does not affect the user's views in exploring the learning activities provided

For consumer assessment, item 6, which is online-based learning information, is not too shallow or deepest to record the highest mean for the information design of 4.53. This shows that the developed online learning is agreed upon that it meets the level of understanding of the students which is not too difficult to understand by students. According to [7], learning that meets the needs of students where it can attract students because the content presented is easily understood according to the level of students' understanding. It is also supported by content experts whose content has been reviewed by lecturers who teach subjects related to the Central Processing Unit topic specifying the appropriate content and according to the syllabus of the teaching set by the faculty.

Additionally, the mean of item 2 of the type of writing used in this web-based learning corresponds to the mean 4.47. This shows that the user agreed that the type of writing used is appropriate according to the user level. According to [8], majority students agree that the form of writing used in a multimedia application should be clear and easy to read so that it is appropriate to their learning.

6. Conclusion

In conclusion, the development of web-based learning is expected to have a positive impact on the students as it can be used at any time without being bound by the traditional T&L process which is in lecture alone. Furthermore, it is expected to help improve understanding after using web-based learning. Lecturers or teachers can also use this web-learning tool as a Teaching Assistant Tools as an alternative if lectures cannot be conducted face-to-face as students are able to browse online to understand any information related to this Central Processing Unit topic.

7. Acknowledgements

This research was supported by Short Term Grant (STG) at Universiti Tun Hussein Onn Malaysia (U647).

8. References

[1] Zainudin, Hassan., Mohd, Zaki. Mohamed. Som dan Nur, Azza. Abd Aziz. (2015) Jurnal Pendidikan Nusantara Indonesia. Perkembangan dan Reformasi Sistem Pendidikan Di Malaysia. 1(1): 1-14

[2] Santosh Kumar Behera. (2013). International Journal on New Trends in Education and Their Implications. E- And M-Learning: A Comparative Study. 4(3): 65-78.

[3] Horton, W. (2011). E-Learning by Design. 2nd Ed. Pfeiffer.

[4] Suradin, Ali and Salehuddin, Mohd Noraidil and Mustafa, Mohamad Zaid and Abd. Razzaq, Abdul Rasid and Madar, Ahmad Rizal and Ab Hadi, Mohd Yusop (2010). Persepsi Ibu Bapa Terhadap Pendidikan Vokasional. 1-6

[5] Zainal Abidin Zainuddin and Afrinaleni Suardi. (2010). Keberkesanan Kaedah Konstruktivisme Dalam Pengajaran dan Pembelajaran Matematik. Universiti Teknologi Malaysia

[6] Hannafin, M. J., Peck, L. L. 1998. The Design Development and Education of Instructional Software. New York: Mc. Millan Publ., Co

[7] Hartsell, T., & Yuen, S. (2006). Video Streaming in Online Learning. AACE Journal, 14(1), 31-43.

[8] Baharuddin Aris, Bilal Ali, Jamaluddin Harun dan Zaidatun Tasir (2001). Sistem Multimedia dan Aplikasinya. Kuala Lumpur: Venton Publisher.

IOP Conf. Series: Journal of Physics: Conf. Series **1049** (2018) 012053 doi:10.1088/1742-6596/1049/1/012053

Designing Tile Installation Tool for Floor Finishing Works

A Ariffin[1], M A Haron[2], T S Subramanian[3], N Hamzah[4] and S N K Rubani[5]
[1] Universiti Tun Hussein Onn Malaysia

E-mail: hasnida@uthm.edu.my

Abstract. This study is a research on product development of tile installation tool and focused on criteria that could help in finishing works for a floor that involved in small scale construction projects such as renovation of room, hall and others. The objective of this study is to design and develop tile installation tool in order to assist the finishing work for flooring installation. Besides, this study also examined the functionality of the tile installation tool in assisting the finishing work for the floor. Product evaluation was done through the approval of expertise from three experts in Akademi Binaan Malaysia (ABM) and also 2 professional contractors with the background of civil engineering. The material used in this study is by using Perspex (acrylic) equipped with a suction cup method. The test result shows that the tile can be lifted by using this tool could reach about 2.40 kg for the tile sized 2" x 2". The advantages of this product are the developed tool could save time and even more practical for the tile installation process compared to the complicated conventional method. In addition, by using this product, the quality of tile installation process could be controlled from the beginning till the end of process. Recommendations for product improvement in the future is by diversifying the types of tile used/installed, such as the tiles with rough or uneven texture.

1. Introduction

In Malaysia, one of the necessary packaging in a new building is the installation of tiles such as the installation of floor tile in the building. There are various types of floor tiles including ceramic tiles, porcelain tiles, granite tiles and these are based on user applications and requirements. Among the favorite tiles are ceramic tiles because of its easy maintenance, high load and reliable (Baharudin, 2007).

Tiles are generally used to protect floors, walls or other objects such as table surfaces and decorative chairs. Accordingly, Pieter (2006) states that some packaging products are required to protect concrete surfaces. In addition, tiles can also refer to the same unit using lightweight materials such as perlite, wood and mineral wool (fibers). In addition, the quality of the tile used should meet the specifications of the JKR and any details of the contract or the details of the tender (Specification of JKR, 1989). Not only that, tiles are also used to form walls and floor coverings. Its use is in interior areas such as kitchen, bathroom and garage space which are often made from materials such as glass, metal and stone. Installing tiles are important considering the size of the tile used, the shape of the area after completion, the general specification requirements for the area to be installed and position planning.

Content from this work may be used under the terms of the Creative Commons Attribution 3.0 licence. Any further distribution of this work must maintain attribution to the author(s) and the title of the work, journal citation and DOI.
Published under licence by IOP Publishing Ltd

The purpose of the Tile Installation Tool is to facilitate the installation of tile. This tool is also used for finishing work in the building. Besides that, educator can used Tile Installation Tool to teach finishing work because it's easy to show real practice to student. Typically, using this product, tile installation is done quickly without taking into precision using spirit level in every view. For the safety aspect, it can be harmful to the occupants in every building that is built. Additionally, the use of mortar between insufficient tile pieces makes the distance between these tile pieces to be too close. This is because tile craftsmen usually take easy steps for fast work and want to save the costs. Refer to specification of British Standard Institution (1987), which the use of mortar distance acceptable tiles is between 2 mm - 6 mm. Installing tiles require a high degree of precision and they can illustrate the position of the tile. CIDB Accredited Training (2006) records that the frequent damage to tiles is like cracking or fracture at the corners of the tile because the cement is not plastered equally and is not in the right position. In fact, the conventional way to install tiles is more tiring than using tile installation tools.

2. Problem Statement

Every year, the conventional tile installation employed by skilled workers in the construction industry is seen to be some other problems in which produce neat, clean and accurate installation is difficult. Lack of product development to facilitate tile installation work has caused construction workers to use conventional methods. Using this conventional method, it becomes a burden to new construction workers or unskilled workers in the use of equipment as the construction process is quite complicated and requires a long period of time to complete the tile installation process. Jahiman (2011) says that, finishing work is a rather complicated stage and requires high skill and competence to make sure installation tiles between 2 mm – 6 mm.

Additionally, the use of this manual installation method may also result in unsatisfactory installation quality. This illustrates that the installation of conventional tiles can result in losses if less skilled workers do so. Uneven and tedious installations can result in defects and cracks in the floor (CIDB Accredited Training, 2006). With this tile installation tool can also help to install more organized and neat. Even the risk of fracture and imprudence can be reduced and avoided.

3. Research Question

The study was conducted to answer the questions:

(a) How does a tile installation tool be designed and developed in the floor finish work?
(b) Is this tile installation function working properly on the floor finishing?

4. Research Methodology

The design of this study is to develop a tile installation tool to be used as a product to help floor finishing work. This study uses the Hanaffin & Peck Design Model (1988) as the basis for the development of product design in this study. Based on figure 1, Hannafin and Peck (1988) design models involve three phases, namely the need phase, the design phase and the implementation phase. Each phase is evaluated, tested and reviewed using an interview form instrument and expert legal form to see the effectiveness continuously.

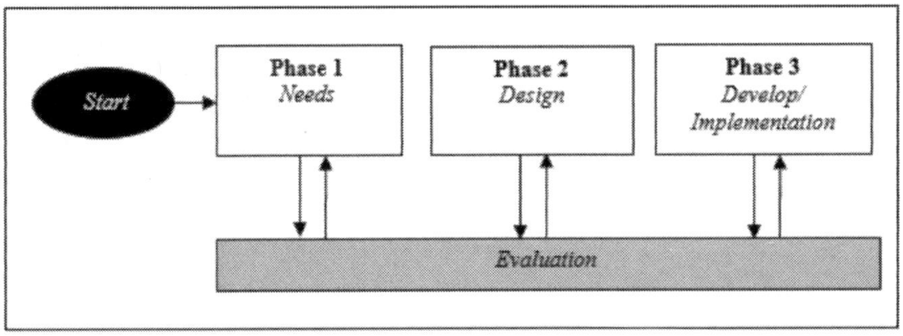

Figure 1: Hannafin and Peck (1988) Model

In the phase of their needs and phase designs, the sketcher draws initial sketches (Figure 2) to design tile installation tool products in assisting floor finishing work.

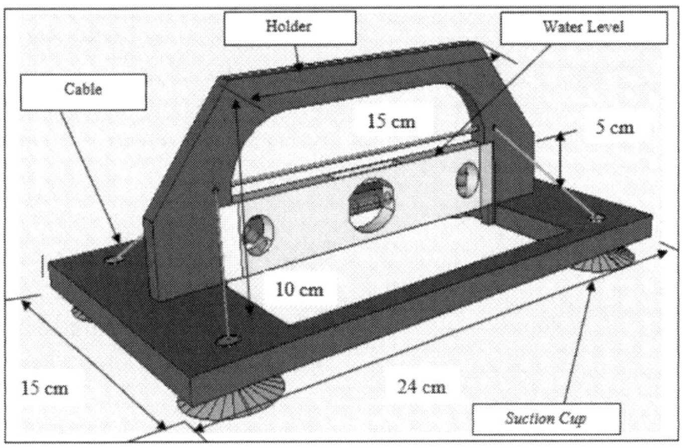

Figure 2: Design Tiles Installation Tool

Through the initial sketch, researchers also interviewed several experts to determine the size of the model and the appropriate material for developing the product. In the product development phase, researcher select the right material such as perspective, suction cup, spirit level, bolt and nut after a full sketch is made. Cutting work starts with making cuts of 11 cm x 24 cm to be used as a tool holder. The next work is a 25 cm x 14 cm cut-off work to serve as a base for tile-mounting tools. Cutting work is done using a drill machine to puncture holes in 1-inch.

For connecting work, the first connection is the installation of spirit water level on the tool holder. Then, the holder and the base are attached to the chlorophyll and the screw so that the installation will be stronger. Next is the installation of a suction cup using bolts and nut on each side of the tool set so that each piece will be joined to each other. After that, the spring will be connected to each suction cup that has been installed to be easily pulled. Finally, connecting the cables to four parts of the spring is connected which works as a pull trigger.

Figure 3: Tiles Installation Tool

5. Finding and Discussions

Analysis of loads involves several activities including the process of collecting information on how much tile load weight can be raised by this product. Weight measurement is done by using digital weighing. Tile load that is capable of lifting this product is as shown in table 1 below.

Table 1: Capable of lifting tiles

Weight	Function
Tiles 1"x1" (1.38kg)	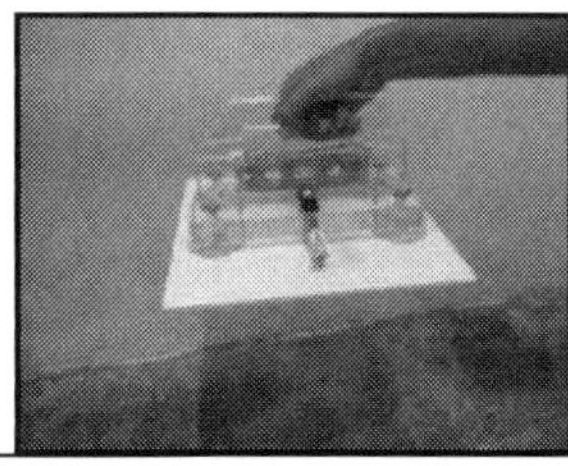
Tiles 2"x2" (2.40 kg)	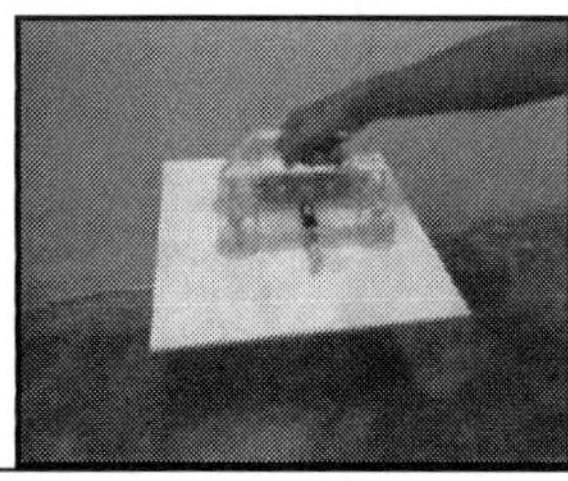

Referring to table 1, it can be concluded that the product has been tested in terms of the amount of load weights capable of lifting. Through this test it can be seen that the weight of the tile that can be lifted with the resulting product is up to 2.40 kg for 2 "x 2" size. This test can indirectly prove the suction cup resistance used for this product. The next analysis relates to the comparison of the time taken for the conventional tile installation and the installation of tiles by using the new product. The time taken is as shown in table 2 below.

Table 2: Comparison of the time taken

Size Tiles	Method	Times taken for installation tile $1m^2$ (minute)
Tile 1"x 1"	Conventional	15 minutes 52 seconds
	Tile Installation Tool	15 minutes 32 seconds
Tile 2"x 2"	Conventional	5 minutes 15 seconds
	Tile Installation Tool	5 minutes 3 seconds

Based on table 2, comparison of time taken for conventional tile installation and installation of tiles by using newly produced products. This analysis is aimed to prove that using this new product can save time and more practical for tile installation work compared to more complicated conventional methods. From the analysis, a short time was taken to bind $1m^2$ of tiles using new product methods rather than using conventional methods. Saving time using new products may not be as significant compared to conventional tile installation, but it is able to assist in the tile installation work.

6. Conclusion

Tile Installation Tool is a product that can help construction workers or educators to install tile more easily, save time and energy. This tool can also help to improve the quality of the tile installation and meet the JKR specifications. The main material used in this tool is suction cup which is a new method in the construction of products in the building industry. The use of suction cup is to grip and lift the tile and put it without damage. Users will find it easy to use, easy to buy if broken and increase productivity.

The tile installation tool is intended to assist the construction worker in floor finishing work to be effective as the functionality of the tool has been achieved. Jahiman (2011) says finishing work is a bit complicated and requires high skill and efficiency. This is because using conventional methods, it becomes a burden to new construction workers or unskilled workers in the use of equipment as the construction process is quite complicated. Conventional process requires a long period of time to complete.

The product development process has been carefully planned by researchers to avoid significant wastage and loss. Researchers are also able to develop this product at a minimal cost compared to existing equipment in the market. However, such products are still in the market and the development of this tool is seen to be potentially marketable and suitable for use in the industry.

Acknowledgements

This research was partially supported by University Tun Hussein Onn Malaysia, Johor using Short Term Grant (U646). We thank our colleagues from focus group who provided insight and expertise that greatly assisted the research.

References

[1] Abdullah, M. N (2003). *Teknologi Binaan*, Putrajaya: Sukatan Pelajaran Teknologi Binaan.

[2] Akademi Binaan Malaysia (2016). *Teori Tiling Level 1*. Malaysia: TLR (LP1).

[3] Bahadi, N. (2005). *Pembangunan manual elektronik latihan kerja amali : litar arus terus pengajian Kejuruteraan Elektrikal Tingkatan 4 di Sekolah Menengah Teknik. Kolej Universiti Teknologi Tun Hussein Onn*. Tesis Sarjana.

[4] Baharudin, M. A. (2007). *Development of Automatic Floor Tile Laying Machine*. Universiti Malaysia Perlis: Tesis Sarjana.

[5] Jahiman, B. (2011). *Teknologi Binaan Bangunan*. Edisi Kedua. Petaling Jaya, Selangor. IBS BUKU SDN. BHD.

[6] Kanapathy, V. (2006). *Migrant Workers In Malaysia. Paper presented at the workshop on an Asean Corperation Framework for Migrant Labour*. Migrant Workers, 1-2.

[7] Khan, A. (2011). *A Semi-Automated Floor Tiling Robotic System*. Univesiti Industri Selangor: Tesis Sarjana Muda.

[8] Masri, A. (2008). A *New Model For Examination Room Assignment Problem*: Case Study At University Kebangsaan Malaysia. (2), 1-6.

[9] Mat Amin, N. (2007). *Bahan Binaan Kejuruteraan* : Buku Diploma Kejuruteraan Awam.

[10] Wamuziri, S., and Smith, I. (2003). "Automated construction in Japan." Proceedings of the Institution of Civil Engineers - Civil Engineering.

IOP Conf. Series: Journal of Physics: Conf. Series **1049** (2018) 012054 doi:10.1088/1742-6596/1049/1/012054

Low Power Ultra-Wideband VCRO in 130nm CMOS Technology

Nabihah Ahmad[1,2*], Muhammad Khairul Khalis Bin Ibrahim [2], M. Hairol Jabbar[1,2]

[1] IC Design Research Group,
[2] Faculty of Electrical and Electronic Engineering,
Universiti Tun Hussein Onn Malaysia,
86400 Parit Raja, Batu Pahat, Johor, Malaysia.

*Email: nabihah@uthm.edu.my

Abstract. The development of voltage-controlled oscillator (VCO) that can operate at high performance within the ultra-wideband (UWB) frequency with low power consumption is useful for the future device implementation in communication system. This paper presents a design of low power ultra-wideband voltage-controlled ring-oscillator (VCRO) in 130 nm complementary metal-oxide-semiconductor (CMOS) technology with single-ended differential pair architecture for UWB application. The aim of this project is to optimize the VCRO with low power technique to ensure it consumes less power and able to operate within the UWB frequency. All design architecture, schematic and layout and simulation and analysis are implemented using Cadence Tool in 130 nm CMOS technology. Through simulation and analysis process, this VCRO achieved a wider frequency from 7.186 GHz to 9.651 GHz with lowest power consumption of 0.78 mW. This VCRO has tuning range of 25.54%. The output noise and figure of merit (FOM) at 1 MHz offset frequency is -153.1 dBc/Hz and -234.83 dBc/Hz while at a 10 MHz offset frequency, the output noise and FOM is -162.0 dBc/Hz and -223.73 dBc/Hz respectively. Through overall analysis, it can be conclude that this VCRO has better performance compared to previous work in terms of power consumption, output noise and frequency tuning range.

1. Introduction

In the rapidly growth of wireless communication, the device that capable to operate in high performance along with the long lifetime is very demanding. Thus, the development of voltage-controlled oscillator (VCO) with ultra-wideband (UWB) and at same time consumed low power could be useful for the future devices implementation [1].

The UWB technology had widely been developed in designation of wireless system. The UWB technology capable to transmit information within wide bandwidth over 3.1GHz to 10.6GHz [1], [2]. It also consumed low power within short-range data transmission and does not interfere with different wireless technology. These characteristic assure the UWB technology is suitable for short-distance application such as wireless personal area network (WPAN) and wireless body area network (WBAN) systems [2], [3].

Content from this work may be used under the terms of the Creative Commons Attribution 3.0 licence. Any further distribution of this work must maintain attribution to the author(s) and the title of the work, journal citation and DOI.
Published under licence by IOP Publishing Ltd

The UWB technology has very wide tuning range, it is a challenge to build a VCO with ultra-wideband frequency along with the low power consumption. Since many wireless devices are battery-powered, it is very important to minimize the power consumption.

In this modern world, it is high demand of portable device which is not only capable to be handle and easily remove, but also able to operate with high performance and long-lifetime. The high performance means it could able to operate effectively within short time. In terms of UWB application, the devices should be able to transfer the data with high speed and high bit rates within a short time.

Motivated by this demand, the design of VCO with ultra-wideband frequency and consume low power could help to amplify the performance in phase-locked loop (PLL) application for example. The existence of high performance of PLL will further the development of wireless application and make better future.

The VCOs have been designed in various method. Two common types that been used for design VCO are LC-based oscillator and ring-based oscillator [4]. LC-based has narrow tuning range of frequency but much better in term of temperature, noise frequency stability and power supply compared to ring-based. Alternatively, the narrow tuning range can be overcome by adding the switched inductor, switched capacitors and variable active inductors [5]. These alternatives may extend the tuning range but it will increase the complexity of circuit which can affect the high power consumption. While, ring-based oscillator gives wider tuning range and usually used in monolithic IC due to small chip area occupied. Thus, it is more cost effective compared to LC-based [4], [6]. VCO is the important part in communication system and mainly used in designing the frequency synthesizer using phase locked loop (PLL) [1]. PLL frequency synthesizer offer very high level of stability and accuracy in communication system. In order to get these offers, a VCO with high performance is needed and implemented into that system. Thus, it is necessary to design the VCO with wider tuning range but simplicity and ease of integration [1].

2. Single-Ended Differential Pair of VCRO architecture

The proposed VCRO architecture implemented using CMOS three-stages single-ended differential pair configuration of ring oscillator. Figure 1 shows the circuit of the proposed VCRO with three delay stages.

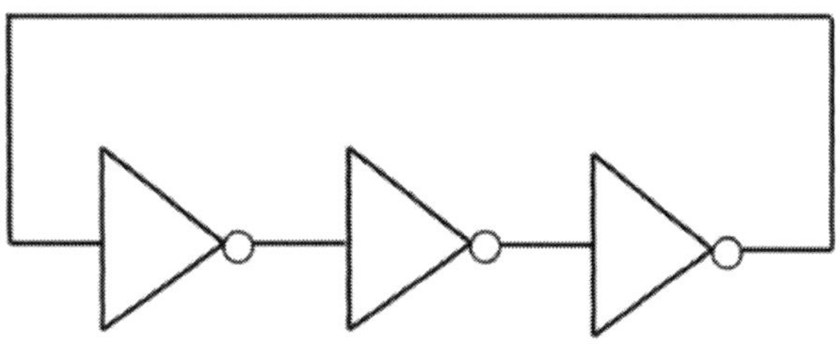

Figure 1. Proposed VCRO circuit architecture

The transistor level circuit of the proposed VCRO consists of eight transistors which the upper part of circuit consists of four PMOSs while the lower part of circuit consists of four NMOSs as shown in Figure 2 implemented using Cadence tool in 130 nm CMOS technology. The control part of the PMOS and NMOS are set with length/width of 0.13 µm/ 1 µm and 0.13 µm/ 3 µm. On the other hand, the three stages of delay cell that combined together become ring oscillator are set with length/width of 0.13 µm/ 2 µm for PMOS and 0.13 µm/ 3 µm for NMOS. The simulation was carried out with a 1.3V supply

IOP Conf. Series: Journal of Physics: Conf. Series **1049** (2018) 012054 doi:10.1088/1742-6596/1049/1/012054

voltage and voltage control varying between 0.5V to 2.5V with the load capacitance of 0.01pF. The frequency of 6 GHz with 1V of amplitude was applied for initial sinusoidal wave.

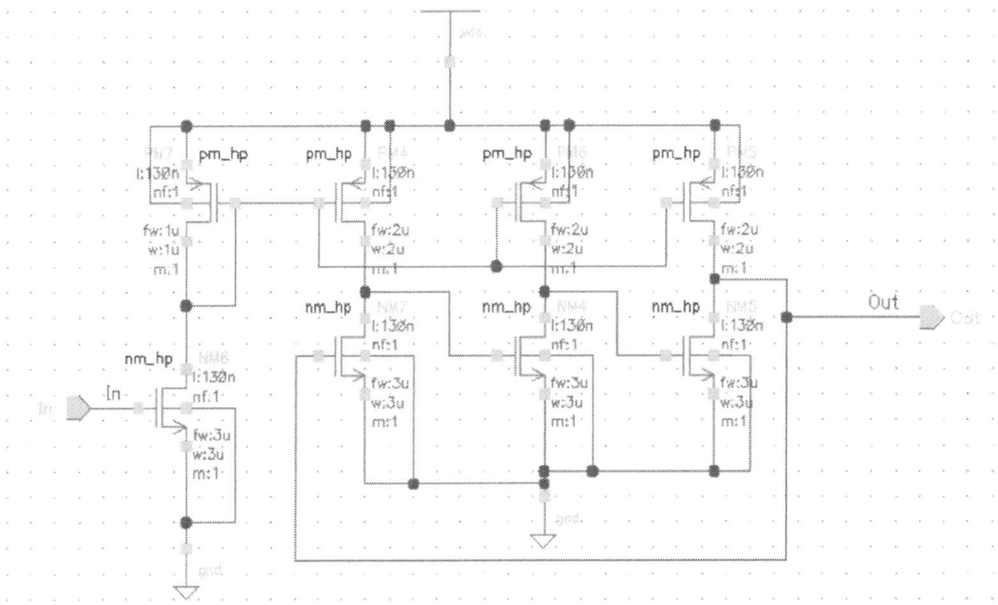

Figure 2. Proposed VCRO circuit implementation

3. Results and discussion

The setting of simulation was done by using Virtuoso Analog Design Environment with WaveScan Waveform Tool. The output waveform were evaluated and analysed by using WaveScan Waveform Tool. Figure 3 indicates the input of VCRO with power supply of 1.3V and voltage control of 2.5V at the initial sinusoidal frequency of 6 GHz and 1V amplitude. The simulation setting of transient analysis at the stop time of 20 ns. Figure 4 demonstrates the output of VCRO. Based on the output waveform, the oscillation frequency is calculated through WaveScan Calculator. The result shows the highest oscillation frequency is 9.651 GHz. Since the VCRO contains three stages of delay cell, the input pulse will be invert three times and gives the output with same shape but smaller voltage. Via manually calculation, the frequency also can be calculated using Equation 1

$$fosc = \frac{1}{T} \tag{1}$$

where T indicate the period of waveform.

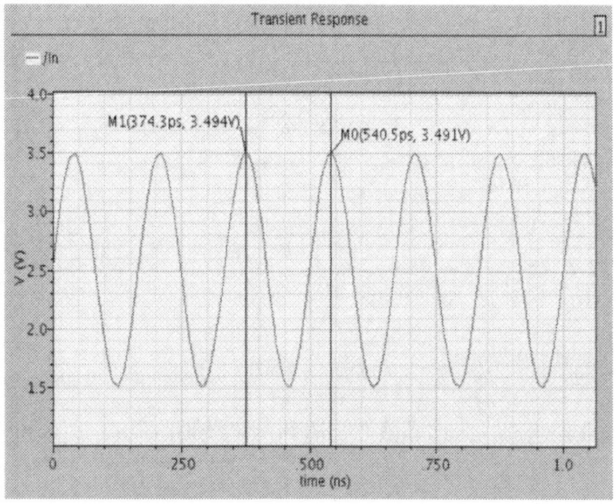

Figure 3. Input waveform of VCRO

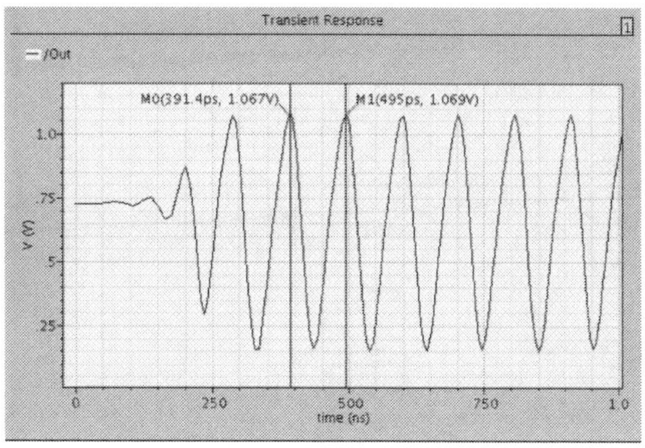

Figure 4. Output waveform of VCRO

The result of transient analysis and DC operating point at various control voltage between 0.5 V to 2.5 V with initial frequency and amplitude of 6 GHz and 1 V, respectively was shown on Table 1. The static power and dynamic power are be calculated by using Equation 2 and Equation 3, respectively.

$$P_{static} = I_{total} \text{ X } V_{DD} \quad (2)$$

where I_{total} indicate the total current of overall current and $V_{DD}$is the voltage of power supply.

$$P_{dynamic} = \text{Capacitor X } V_{DD}^{2} \text{ X Frequency} \quad (3)$$

Thus, the total power can be measured with addition of static power and dynamic power, as shown on Equation 4.

$$P_{total} = P_{static} + P_{dynamic} \quad (4)$$

Through investigation on Table 1, it was clearly that V_{ctrl} is the variable that changes through the analysis and producing result in term of oscillation frequency and overall current. When the V_{ctrl} is varied to 0.5 V, the overall current shows the lowest value which is 0.507 mA. Total power consumption by the circuit is the lowest which is 0.78 mW. In contrast, by setting the V_{ctrl} to 2.5 V, the overall current reached 1.489 mA. This overall current affects and increases the power consumption through the circuit to 2.099 mW. In terms of oscillation frequency, minimum and maximum frequency can be achieved when the V_{ctrl} is varied to 0.5 V and 2.5 V. From these values, the tuning range percentage and centre frequency can be determined through Equation (5) and (6). Based on the desired output oscillation frequency, tuning range can be measured using Equation .

$$TR (\%) = [(f_{max} - f_{min})/f_{max}] \text{ X } 100\% \tag{5}$$

where f_{max} indicate maximum frequency and f_{min} indicate minimum frequency of desired oscillation.

$$f_{centre} = [(f_{max} - f_{min})/2] + f_{min} \tag{6}$$

where f_{centre} referred to centre frequency of oscillation

Based on Equation 5, the tuning range of this VCRO is:

$TR (\%) = [(f_{max} - f_{min})/f_{max}] \text{ X } 100\%$
$TR (\%) = [((9.651 - 7.186) \text{ GHz})/9.651 \text{ GHz}] \text{ X } 100\%$
$TR (\%) = 25.54 \%$

Regarding to Equation 6, the centre frequency of oscillation is:

$f_{centre} = [(f_{max} - f_{min})/2] + f_{min}$
$f_{centre} = [((9.651 - 7.186) \text{ GHz})/2] + 7.186 \text{ GHz}$
$f_{centre} = 8.42 \text{ GHz}$

The VCRO tuning range is 25.54% while f_{centre} is 8.42 GHz as shown in Table 1, offer better tuning range to allow the VCRO to operate at variety of frequencies. On the other hand, the value of f_{centre} is used to find the figure of merit (FOM).

Table 1. Simulation result of proposed VCRO at different control voltage

V_{ctrl} (V)	**Frequency (GHz)**	**I_{total} (mA)**	**P_{static} (mW)**	**$P_{dynamic}$ (mW)**	**P_{total} (mW)**
0.5	7.186	0.507	0.659	0.121	0.780
1.0	8.179	1.456	1.893	0.138	2.031
1.5	9.283	1.482	1.927	0.157	2.084
2.0	9.632	1.488	1.934	0.163	2.097
2.5	9.651	1.489	1.936	0.163	2.099

Figure 5 demonstrates the output noise of the VCRO. The simulation setting was set with noise analysis. It can be investigated that the overall output noise of the VCRO is -153.1 dBc/ Hz at a 1MHz offset frequency and -162.0 dB/ Hz at a 10 MHz offset frequency.

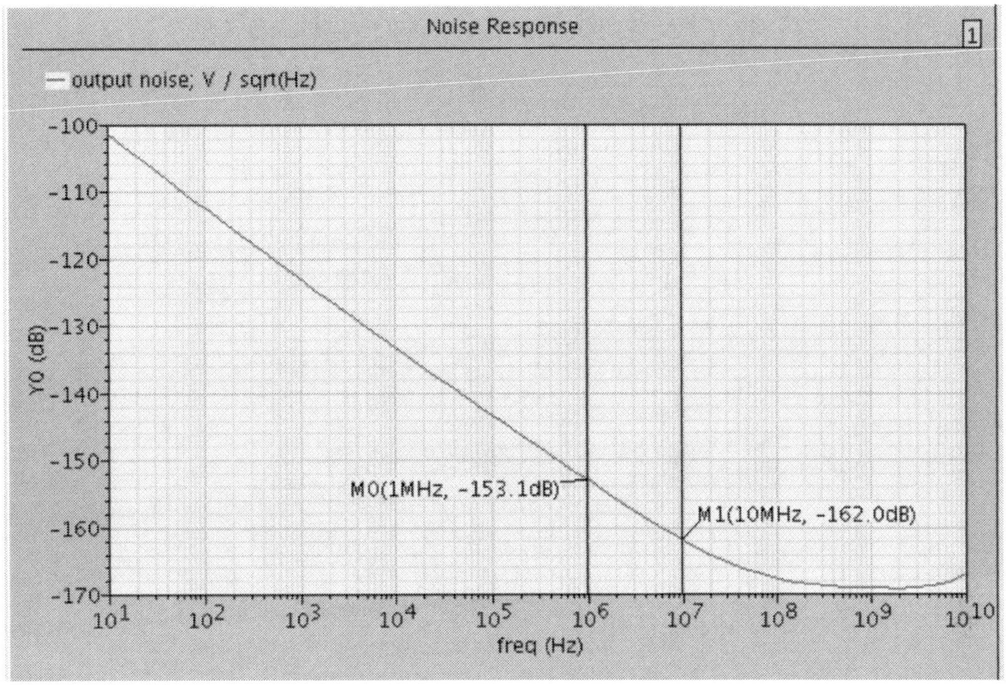

Figure 5. Output noise of VCRO

Through overall analysis, the FOM can be determined. FOM is used to estimate the performance of VCRO [1] and can be calculated using Equation 7.

$$\text{FOM} = P_{noise} - 20\,log\frac{f_{centre}}{f_{off}} - 10\,log\frac{P_{diss}}{1mW} \tag{7}$$

where P_{noise} indicates the output noise with an offset frequency, f_{centre} is centre frequency and f_{off} is offset frequency, and P_{diss} is the dissipated power by VCRO.

By using Equation (7), at a 1 MHz offset frequency, the FOM of the VCRO can be calculated as:

$$\text{FOM} = P_{noise} - 20\,log\frac{f_{centre}}{f_{off}} - 10\,log\frac{P_{diss}}{1mW}$$
$$\text{FOM} = -153.1\text{ dBc/ Hz} - 20\,log\frac{8.42\,GHz}{1\,MHz} - 10\,log\frac{2.099\,mW}{1mW}$$
$$\text{FOM} = -234.83\text{ dBc/ Hz}$$

Hence, by putting the entire variables in Equation (7), the FOM of this VCRO can be determined. At a 1 MHz offset frequency, the FOM is -234.83 dBc/ Hz and at a 10 MHz offset frequency, the FOM is -223.73 dBc/ Hz. The values of FOM is depends on the output noise The low value of FOM offers better performance of the VCRO. The results of the overall analysis are encapsulate in Table 2.

Table 2. Result summary of VCRO at different parameters

Result of Analysis	This Work
Frequency Maximum	9.651 GHz
Frequency Minimum	7.186 GHz
Maximum Power Dissipation	2.099 mW
Minimum Power Dissipation	0.778 mW
Tuning Range	25.54%
Centre Frequency	8.42 GHz
Output Noise @ 1MHz	-153.1dBc/ Hz
Output Noise @ 10MHz	-162.0dBc/ Hz
FOM @ 1MHz	-234.83 dBc/ Hz
FOM @ 10MHz	-223.73dBc/ Hz

Based on Table 3, this work is the third lowest power consumption when compared to [2] and [14] but achieved higher tuning range with3.32% increases. Comparing to research in [1], even though it has the widest tuning range, but this proposed work has better power consumption. In terms of output noise, this proposed work have better noise performance compared to other previous work with using the same technology with increasing of 80.54% from research in [1].

Table 3. Comparison of VCRO at different parameters with same technology

No.	Year	Technology (nm)	Power (μW)	Frequency (GHz)	Tuning Range (%)	Output noise (dBc/ Hz)
This work	2017	130	778	7.186-9.651	25.54	-153.1 @ 1MHz
[6]	2016	130	835	3.328-4.606	27.75	-74.5 @ 1MHz
[14]	2015	130	750	3.500-4.500	22.22	-79 @ 1MHz
[5]	2013	130	50	3.500-4.500	22.22	-79 @ 1MHz
[1]	2008	130	7020	3.400-7.630	55.43	-84.8 @ 1MHz

Figure 6 shows the layout circuit of three-stage VCRO with single-ended differential pair configuration. The three-stage ring oscillator of PMOS and NMOS were set with width of 2um and 3um respectively. Meanwhile, the control PMOS and NMOS were having width of 1um and 3um, respectively. The total layout area size of yield VCRO without the bonding pad is 11.80 μm X 9.9 μm (116.82 μm^2)

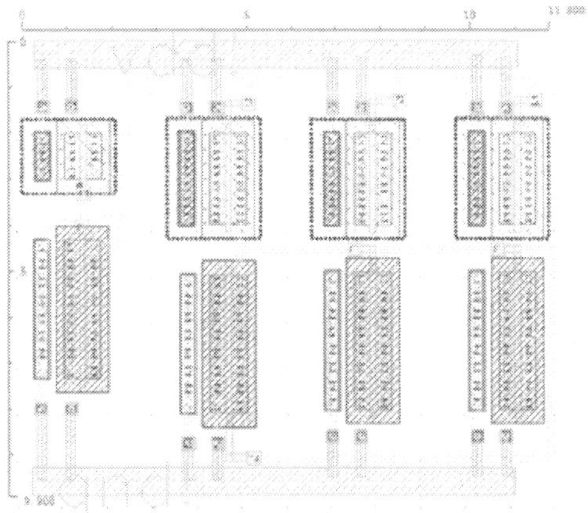

Figure 6. Layout circuit of three-stage VCRO

This VCRO has successfully passed DRC and ERC verification without any error occurred as the design followed all the rules and can be proceeding to next physical verification. The LVS was also done and pass successfully.

4. Conclusion

The proposed three stage of voltage-controlled ring-oscillator (VCRO) with single –ended differential pair architecture have been designed that able to operate in the range of frequencies of 7.186 to 9.651GHz at power supply of 1.3V implemented in 130nm CMOS technology. It has wide tuning range which is 25.54% with the power consumption of 0.78mW.

Acknowledgement

This work was financial supported by FRGS Grant Vot Number 1538.

References

[1] N. Rashidi, A. Nabavi, and M. Mehrabian 2008 7GHz voltage control ring oscillator for UWB applications *Proc. Int. Conf. Microelectron* p 429
[2] M. Ali, H. Shawkey, and A. Zekry 2013 Ultra-low power 3.5-4.5GHz FM-UWB transmitter *2013 Saudi Int. Electron. Commun. Photonics Conf.* p 5
[3] M. Ali, M. Sawan, H. Shawkey, and A. Zekry 2016 FM-UWB transmitter for wireless body area networks: Implementation and simulation *Proc. - IEEE Int. Symp. Circuits Syst.* **2016–July** 2395
[4] M. Hsieh and G. E. Sobelman 2006 Comparison of LC and Ring VCOs for PLLs in a 90 nm Digital CMOS Process *Proceedings, Int. SoC* p 19
[5] Y. Ito, K. Okada, and K. Masu A tunable wideband frequency synthesizer using LC-VCO and mixer for reconfigurable radio transceivers *J. Electr. Comput. Eng.* **2011** 1
[6] B. S. Patro, J. K. Panigrahi, and S. K. Mandal 2012 A 6-17 GHz linear wide tuning range and low power ring oscillator in 45nm CMOS process for electronic warfare *Proc. - 2012 Int. Conf. Commun. Inf. Comput. Techno.* p 17–20
[7] Malviya, Rahul 2011 Low Power Current Starved VCO in 70nm CMOS Technology.
[8] B. Li, X. Fan, and Z. Wang 2012 A wideband LC-VCO with small VCO gain variation and adaptive power control *J. Semicond.* **33** 105008

[9] S. Chauhan and R. Mehra 2016 CMOS Design and Performance Analysis of Ring Oscillator for Different Stages *Int. Journal of Eng. Trends and Technology* **32** 234

[10] S. Suman and P. B. P. Singh 2012 Design of Temperature Sensor Using Ring Oscillator *Int. Journal of Scientific & Engineering Research* **3** 1

[11] N. Gargouri, D. Ben Issa, A. Kachouri, and M. Samet 2015 A performance comparison of single ended and differential ring oscillator in 0.18 μm CMOS process *Int. J. Sci. Res. Eng. Technol.* **3** 123

[12] P. S. Athare, S. P. Agnihotri, and Y. A. Salame 2016 Design and Implementation of low power Ring VCO Using Tail Transistor *Int. Journal of Industrial Electronics and Electrical Engineering* **6** 3

[13] B. Goyal, S. Suman, and P. K. Ghosh 2016 Design and Analysis of Improved Performance Ring VCO based on Differential Pair Configuration *Int. Conf. on Electrical, Electronics, and Optimization Techniques* p 3468

[14] M. Ali, H. Shawkey, and A. Zekry 2015 A 750μW 3.5–4.5GHz FM-UWB transmitter," *J. Electr. Syst. Inf. Technol.* **2** 133

[15] N. Kim and J. M. Rabaey 2015 RMO2A-3 Reduction Technique for UWB-based Cognitive Radios p 67–70

[16] Gagandeep Singh, Mandeep Singh Angurana 2014 Design of Wide Tuning Range and Low Power Dissipation of VCRO in 50nm CMOS *Int. Journal of Advanced Research in Electrical, Electronics and Instrumentation Engineering* 3 9675

[17] I. Abdo and K. Abugharbieh 2013 A 14.7GHz wideband 28nm CMOS ring VCO *Proc. IEEE Int. Conf. Electron. Circuits, Syst.* p 909

IOP Conf. Series: Journal of Physics: Conf. Series **1049** (2018) 012055 doi:10.1088/1742-6596/1049/1/012055

Dimensions of Learning Styles among Engineering Students

A. R Jamali[1], M. M Mohamad[1]
[1]Faculty of Technical and Vocational Education
Universiti Tun Hussein Onn Malaysia
Parit Raja 86400, Batu Pahat Johor Malaysia

Email: *alirazabinismail@gmail.com*

Abstract. Current study is conducted to examine the types of learning styles among engineering students. Felder Silverman learning style model (FSLSM) was used to measure the types of learning styles represent engineering students. FSLSM contains four dimensions that intended to identify learning preferences of students with respect to process information, perceive information, input information and understand information. Each dimension consisted of two sub-dimensions such as in processing dimension learner can be either active learner or reflective learner, this pattern follows in all four dimensions, perceive dimension distinguishes sensing or intuitive learner, input dimension distinguishes visual learners or verbal learners, understand distinguishes sequential or global learners. This study applied survey research method. Index of learning styles (ILS) was used as a tool to collect data. A total of 315 engineering students participated from Universiti Tun Hussein Onn Malaysia. Finding revealed that engineering students are tending to be visual learners.

1. Introduction

Learning style plays an energetic role in engineering education that portrait the techniques in which learners generally obtain, retain and get back information. It helps students to increase their cognitive capacity and to deal with the learning difficulties which successively improves their academic performance [1]. Every student has different learning style preference depends on their likings and disliking. They acquire knowledge best by auditory, listening, observing, and doing [2]. Certain external environmental inputs like seeing, hearing, reflecting and acting helps students in learning. These inputs support students as learning process is a way of reflecting and acting, reasoning logically and intuitively, memorizing and visualizing [3].

Students gain knowledge well in a class when teaching styles matches to their preferred learning styles [4]. Such as, the learning styles are the preferences and characteristics by which individual receive and process information [5]. For instance, a classroom is comprised of dissimilarities in terms of educational background, preferred learning style, cultural influence and cognitive ability from a social perspective.

Furthermore, mismatches between teaching methods with students preferred learning styles may lead them towards poor academic performance [6]. Thus, to understand students learning preferences which accomplish their needs and enhance their academic achievements, it requires to determine the best feasible learning style. In higher education, third and fourth year students are supposed to be matured enough to deal with their ability of understanding assignments and subjects by own **[7]**. However, majority students who fails examination usually blame their failure to external stimulus such as mismatch teaching methods or lack of academic standards [8].

Teachers are main facilitating and guiding sources in classroom for that reason they must possess the ability to understand that how well students acquire information and what ways [9]. Therefore, it is necessary that educators must adjust their teaching methods according to students learning styles preferences as discrepancies in teaching and learning styles can create difficulties to understand subjects

Content from this work may be used under the terms of the Creative Commons Attribution 3.0 licence. Any further distribution of this work must maintain attribution to the author(s) and the title of the work, journal citation and DOI.
Published under licence by IOP Publishing Ltd

and assignments in class. Hence, there is a need to deliver an effective teaching that must be comprised of teaching style and preferred learning style in their teaching activities particularly taking cognitive and intellectual activities into consideration [10].

Moreover, A continuous process of learning styles may lead students' higher performance [11]. Felder opinions that strong preference of any student for a particular learning style may trouble if teaching style do not match with student learning style [12]. Every student has different mental approach, learning styles, speed of pickup any information, passion, and motivation to learn however teaching methods and academic activities are different. So, the learning styles try to find out individual thinking skills, motivation, and preferred ways of acquiring knowledge to enhance students' performance. Thus, providing educational facilities in a classroom; educators must understand categories of students to enhance the learning abilities so that they can meet expected educational goals [13].

1.1 Defining learning styles

There are numerous definitions of learning styles can be found in literature. For instance, some definitions are mentioned below.

Dunn (1993) defined learning style as the technique in which each learner initiates to concentrate on, process and maintain new information [14].

Felder (1996) describes learning styles are individuals take in information by means of preferences and strengths in their preferred ways of learning [15].

Vermunt (1996) expresses learning style as a logical attention of learning activities that scholars generally employ in their educational orientation and intellectual learning activities [16].

Keefe (1985) defines learning style as the combined characteristic cognitive, affective and physiological behaviors that help as relatively constant signs of how students observe, interrelate with, and react to the educational environment [17].

1.2 Felder-Silverman Learning Style Model (FSLSM)

The Felder-Silverman Learning Styles Model (FSLSM) was developed by Richard Felder and Linda Silverman and was first published in 1988 [18]. The model was originally developed to address learning differences amongst engineering education. The model categorizes students into four dimensions. The first-dimension process information distinguishes learners into active learners and reflective learners. Active learners learn best by applying, and trying things out and working with learning materials. Also, they tend to be more interested in communication with others and prefer to learn by working in groups where students can discuss the learned material. While reflective learners like better to think about materials and reflect on the materials. The second dimension of FSLSM model is perceive information that consisted of sensing and intuitive learners. Sensing learners prefer to learn facts and concrete material also they are considered more sensible and realistic.

Moreover, they like to relate the learned material with real world. While intuitive learners like better abstract learning resources for instance theories and their fundamental meanings. Intuitive learners like to discover relationships and possibilities and they have a tendency to be more innovative and creative than sensing learners. The third dimension is input information that differentiates learners into visual learners or verbal learners. Visual learners learn best by pictures, diagrams, flowcharts or by demonstrations while verbal learners learn best through written and spoken materials. The dimension of FSLSM model is understand information that categorized learners in two styles sequential and global learners. Sequential learners learn best by small incremental steps, step by step paths in finding solution and from small parts to whole. Whereas, global learners learn best by using holistic thinking process and learn in large leaps.

In addition, the main focus of Felder's work is learning style preferences of students. To measure the learning style preferences based on FSLSM model, Index of Learning Styles (ILS) questionnaire was designed by Felder and Soloman [19]. Index of learning style questionnaire is based on Felder Silverman learning style model that consisted of 44 items. ILS items are divided according to FSLSM model dimensions, each dimension contains 11 items with two answers "a" and "b". Where "a" denotes sub scales active, sensing, visual and sequential respectively and "b" denotes reflective, intuitive, verbal and global respectively. With regard to the psychometric qualities of instrument ILS, various studies illustrated that the ILS deals predictive value and more consistency than other normally used learning

IOP Conf. Series: Journal of Physics: Conf. Series **1049** (2018) 012055 doi:10.1088/1742-6596/1049/1/012055

style instruments [11]. The index of learning items was grouped based on semantic similarities as described in FSLSM model. Table 1 shows the ILS items grouped according to semantic similarities [7].

Table 1. Semantic groups of Index of learning style

Learning Style	Semantic group	ILS questions (answer a)	Learning Style	Semantic group	ILS questions (answer b)
Active	Trying something out Social oriented	1a 5a 9a 13a 17a 21a 25a 29a 33a 37a 41a	Reflective	Think about material Impersonal oriented	1b 5b 9b 13b 17b 21b 25b 29b 33b 37b 41b
Sensing	Existing way Concrete material Careful with details	2a 6a 10a 14a 18a 22a 26a 30a 34a 38a 42a	Intuitive	New ways Abstract material Not careful with details	2b 6b 10b 14b 18b 22b 26b 30b 34b 38b 42b
Visual	Pictures, diagram, graphs, flowchart, demos	3a 7a 11a 15a 19a 23a 27a 31a 35a 39a 43a	Verbal	Spoken & Written words Difficulty with visual style	3b 7b 11b 15b 19b 23b 27b 31b 35b 39b 43b
Sequential	Detail oriented From parts to the whole	4a 8a 12a 16a 20a 24a 28a 32a 36a 40a 44a	Global	Overall picture Relations or connections	4b 8b 12b 16b 20b 24b 28b 32b 36b 40b 44b

2. Problem Statement

Knowledge is refined when importance is given on learners leaning style preferences as it helps in achieving necessary academic achievements rather also improves students' cognitive ability. The learning process is a way of communication between students, teachers, and teaching resources. For better outcomes, students learning process should always be given importance [13]. Preferably, teacher's teaching methods should be matched with learners' preferred learning style. Mismatches in instruction and preferred learning style of students often leads them towards lower academic grades [20]. For instance, in education learning style is playing a major role, teachers must not neglect students learning preferences in classes to enhance their cognitive and academic performance.

2.1 Objective of study

- To identify the types of learning styles among engineering students.
- To identify the learning style differences among engineering courses.

3. Methodology

Present study adopted quantitative survey research as this method gives better accuracy and consistency of research findings [21]. Survey method delivers possible information regarding the population and data collection from respondents about their educational, financial, and social backgrounds [22]. The purpose of this research is to explore the types of learning styles of engineering students represent. There were 315 final year engineering students participated from three undergrad courses mechanical, electrical and civil at Universiti Tun Hussein Onn Malaysia. Random sampling method was applied in this study. The index of learning style questionnaire was distributed among participants for data collection. This study aimed to determine types of learning style represent engineering students and learning style differences among the courses. For the first objective, calculation method was adopted from Felder and Spurlin. And for second objective ANOVA (Analysis of variance) was applied to analyze the differences among the courses.

4. Results

- ***To identify the types of learning styles among engineering students***

Index of learning styles by Felder and Solomon was used to categorize students based on their preferences. The following is the scoring method of index of learning style questionnaire was described by Felder and Spurlin [23]. Students learning style scores were calculated by allocating a value of one to each of the 11 questions in the four learning style dimensions. The smaller number was then subtracted from the larger one and the letter from the larger number was taken over, leading to a value between 1 and 11 as well as a letter for the learning style dimension. For example, if under Active/Reflective, the learner had 2 "a" and 9 "b" responses, the dominant learning style is 7b (9b-2a=7b). That means the learner is a reflective learner. For each dimension, if learner's score on a scale is 1-3, student has a mild preference for the one or other dimension. If the learner's score on a scale is 5-7, learner has a moderate preference for one dimension of the scale and will learn more easily in a teaching environment which favors that dimension. If the score on a scale is 9-11 then student has a strong preference for one dimension of the scale and may have difficulty learning in an environment which does not support that preference. The learning style scores were transferred to an Excel spreadsheet to analyze the data.

Figure 1 reveals that respondents had a preference for the Active and reflective learning styles with mild, moderate and strong preferences, the 114 students are mild active with percentage of 36.19%, 91 students with percentage of 28.88% have moderate preference on active learning style and 16 learners with 5.07% percentage have strong preference on Active learning style. While on reflective dimension, 70 students with percentage of 22.22% are mild-reflective, 23 students with 7.30% percentage are categorized as moderate preference and only one student have strong preference on reflective dimension with percentage of 0.31%.

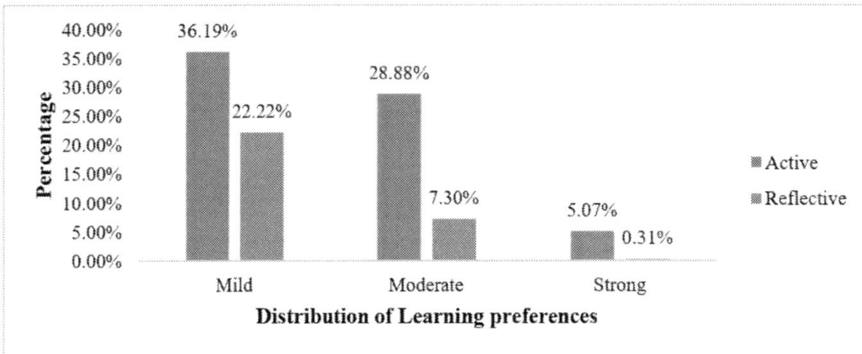

Figure 1. Distribution of Active-Reflective Learning Styles

Figure 2 shows the distribution of sensing and intuitive learning styles with mild, moderate and strong preferences. 116 (36.82%) students have mild preference on sensing dimension, 56 (17.77%) students are moderate on sensing information and 19 learners with percentage of 6.03% have strong preference on sensing dimension. While on Intuitive dimension, 89 (28.25%) students are mild-intuitive preference, 31 (9.84%) have moderately intuitive preference and 4 students with percentage of 1.26 have strong intuitive preference.

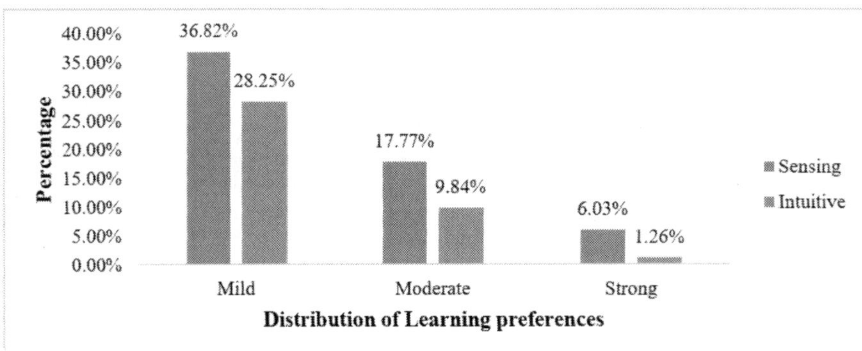

Figure 2. Distribution of Sensing-Intuitive Learning Style Scores

Figure 3 categorized visual learners and verbal learners with mild, moderate and strong preferences. 76 students with percentage of 24.12% have mild-visual preference, 113 (35.87%) students are moderately visual preference and 107 (33.96%) have strong visual preference. While on verbal dimension, 15 (4.76%) students are mild verbal learners, 4 students with percentage of 1.26%

moderately verbal learners and 0 student means no any students have strong preference on verbal dimension.

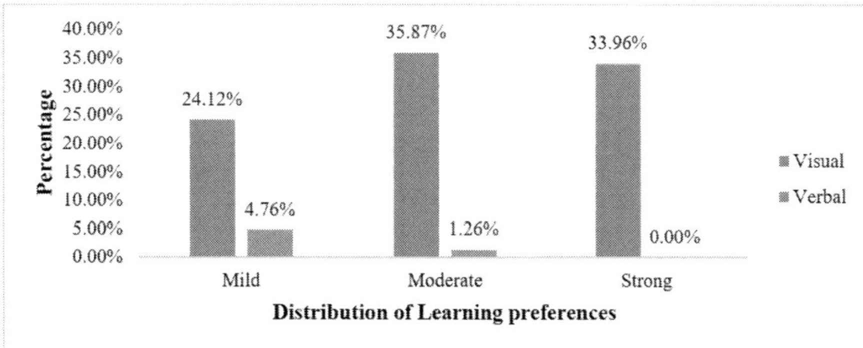

Figure 3. Distribution of Visual-Verbal Learning Style Scores

Figure 4 categorizes the sequential and global learning style preferences with mild, moderate and strong preferences. On this dimension 125 (39.68%) students have mild-Sequential preference, 57 (18.09%) students have moderate sequential preference and 14 students with percentage of 4.44% have strong preference on sequential dimension. While on global learning style there are 79(25.07%) students are mild, 32 (10.15%) have moderate level preference and 8 students with 2.53% have strong level preference.

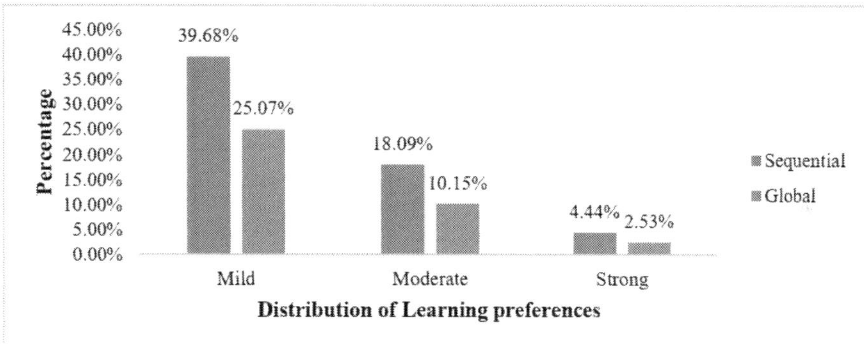

Figure 4. Distribution of Sequential-Global Learning Style Scores

- ***To identify the learning style differences among engineering courses***

The one way ANOVA (Analysis of Variance) was used to determine if any difference exists or not among the learning styles between electrical, mechanical and civil degree students. There is no any difference found between engineering courses mechanical, electrical and civil as the "p" value is greater than 0.05 that is 0.357. For the assumption statistically significant difference the "p" value must be less than 0.05. This shows that all three engineering students from mechanical, electrical and civil courses share similar learning style preferences. Furthermore the preferred learning styles of all engineering students was found visual learning style. The results of ANOVA analysis is shown below in Table 2.

Table 2. Differences between learning styles among courses

	Sum of Squares	df	Mean Square	F	P (Sig.)
Between Groups	.969	2	.484	1.033	.357
Within Groups	146.279	312	.469		
Total	147.248	314			

5. Discussion and Conclusion

Knowing the leaners preferred learning styles may support to increase the quality of learning and teaching. As Sabine Graf mentioned that identifying preferred learning style of learners is road to improve learning and teaching in classroom [20]. The findings of this study are supported by many previous researches. Constant has conducted a study at Iowa State in educational field where respondents preferred learning style was visual preference with percentage of 85% [24]. Likewise, Sabine Graf has conducted a study on students of Engineering and Information Management from Massey University, New Zealand and the University of Vienna, Austria that resulted with 87% visual learning preference [20]. Furthermore, Tee in 2015 has found majority of students preferred visual learning with percentage of 90% in Business Management and Hospitality programs at Vocational college of Malaysia [25]. Most of learners in different social science, natural science, and engineering have visual preference, however, in every course the information presented is verbal, such as written texts on the board, spoken words in classroom lectures [23]. In this study majority of engineering students have strong preferred learning style found to be visual learning.

Besides, engineering courses mostly based on designs diagrams but some subjects are based on theories. Educators need to reform their teaching style based on students' preferences so that better academic performance can be obtained. As in this study engineering students showed their strong preference on visual learning. This implies that institutes should focus more on visual leaning environment so that learners can easily enhance their academic achievement. This shows a serious concern that mismatching teaching and learning can cause problem to students' academic performance. Additionally, learning styles and teaching styles are closely connected to each other. Several studies found that when educators teach courses according to students preferred learning styles, students perform well, and they improve their academic achievements. Through this match between learning styles and teaching styles will give a good impact on learners' academic achievements.

The findings of second objective of this study revealed that there is a no any difference exist between the engineering courses. This implies that engineering courses are different by nature but students' preferences for learning is identical. Therefore, teaching methods should be made flexible so that better outputs can be gained. It argues that curriculum and teaching methods should be made according to courses. Based on study findings, the researcher suggests that lectures, teaching materials should practice more visual graphical tools such as demonstration, presentation using software PowerPoint etc. in classroom teachings. That will suit according to visual learning style preferences. Moreover, educators should also generate a mix teaching atmosphere to support active learning between the learners. Activities doings like discussion, cooperative learning, demonstration, brainstorming, problem based learning and project based learning are strongly suggested. By nature, mostly engineering subjects are involved in procedures. Due to this reason educators should deliver instructions systematically like from easy steps to difficult steps.

Finally, educators can start a lecture by highlighting on details then follow by hands on doings to support the learner's view on new information. At the time, delivering examples will assist the learners to relate class work with real work condition outside the tutorial room.

6. Acknowledgment

This research funded by Fundamental Research Grant Scheme;1501 Ministry of Higher Education and Universiti Tun Hussein Onn Malaysia.

7. References

[1] Mohamad, M. M. B., Mei Heong, Y., & Tze Kiong, T. (2014). Conceptions of learning through learning styles and cognitive dimension in vocational education. *Journal of Technical Education and Training*, 6(1).

[2] Graf, S., & Kinshuk. (2008). *Learner modelling through analyzing cognitive skills and learning styles*. Handbook on Information Technologies for Education and Training, 179–194.

[3] Yee, M. H., Yunos, J. M., Othman, W., Hassan, R., Tee, T. K., & Mohamad, M. M. (2015). Disparity of Learning Styles and Higher Order Thinking Skills among Technical Students. *Procedia - Social and Behavioral Sciences, 204(November)*, 143–152.

[4] Mohamad, M. M., Sulaiman, N. L., Sern, L. C., & Salleh, K. M. (2015). Global Journal of Business and Social Science Review the Composite of Students' Characteristic and Cognitive Dimension in Vocational Education, 1(March), 164–175.
[5] Letele, M. J., G. Alexander, & Swanepoel, Z. I. (2013). Matching/Mismatching of Teaching and Learning Styles in Rural Learning Exologies of Lesotho: Does it Enhance Academic Achievement. *Journal of Human Ecology*, 41(3), 263–273.
[6] Graf, S., Viola, S. R., & Leo, T. (2007). In-Depth Analysis of the Felder-Silverman Learning Style Dimensions. *Journal of Research on Technology in Education*, 40(1), 79–93.
[7] Romanelli, F., Bird, E., & Ryan, M. (2009). Learning styles: A Review of theory, application, and best practices. *American Journal of Pharmaceutical Education*, 73(1).
[8] Mohamad, M. M., Yusof, Y., Muhammad Hanafi, N., Yee, M. H., & Tee, T. K. (2013). Connecting learning styles and cognitive dimension in building construction education. *In Proceeding of the International Conference on Social Science Research*, ICSSR (e-ISBN 978-967- 11768-1-8). 4-5 June 2013, Penang, Malaysia.
[9] Koh, Y. Y., & Chua, Y. L. (2012). The Study of Learning Styles among Mechanical Engineering Students from Different Institutions in Malaysia. Procedia - Social and Behavioral Sciences, pp.636–642.
[10] Alias, N. A., and Zainuddin, A. M. (2005). Innovation for Better Teaching and Learning: Adopting the Learning Management System. *Malaysian Online Journal of Instructional Technology*, 2 (2), 27-40.
[11] Felder, R. M., Brent, R., & Prince, M. J. (2011). Engineering instructional development: Programs, best practices, and recommendations. Journal of Engineering Education, 100(1), 89–122.
[12] Felder, R. M., and Soloman, B. A. (1997). Index of Learning Styles Questionnaire. Retrieved 1st July 2016, from http://www.engr.ncsu.edu/learningstyles/ilsweb.html.
[13] Eishani, K. A., Saa'd, E. A., & Nami, Y. (2014). The Relationship between Learning Styles and Creativity. Procedia-Social and Behavioral Sciences, 114, 52-55.
[14] Dunn, R. (1993). Learning Styles of the Multiculturally Diverse. *Emergency Librarian*, *20*(4), 24-32.
[15] Felder, R. M. (1996). Matters of style. ASEE Prism, 6(4), 18-23. Retrieved from: www4.ncsu.edu/unity/lockers/users/f/felder/public/PapersILS-Prism.htm.
[16] Vermunt, J. D. (1996). Metacognitive, cognitive and affective aspects of learning styles and strategies: A phenomenographic analysis. *Higher education*, *31*(1), 25-50.
[17] Keefe, J. W. (1985). Assessment of learning style variables: The NASSP task force model. *Theory into practice*, *24*(2), 138-144.
[18] Felder, R. M., & Silverman, L. K. (1988). Learning and teaching styles in engineering education. Engineering Education, 78(7), 674–681.
[19] Felder, R. M., and Soloman, B. A. (1997). *Index of Learning Styles Questionnaire*. Retrieved from http://www.engr.ncsu.edu/learningstyles/ilsweb.html.
[20] Graf, S., Liu, T. C., Chen, N. S., & Yang, S. J. (2009). Learning styles and cognitive traits–Their relationship and its benefits in web-based educational systems. *Computers in Human Behavior*, 25(6), 1280-1289.
[21] Creswell, J. W. (2008). Educational research, planning, conducting, and evaluating quantitative and qualitative research (3rd Ed.). Upper Saddle River, NJ: Pearson Education.
[22] Krosnick, J. a. (2010). Survey Research. Research Design and Data Collection, 159–185.
[23] Felder, R. M., & Spurlin, J. (2005). A validation study of the Index of Learning Styles. *International Journal of Engineering Education*, 21(1), 103–112.
[24] Constant, K. P. (1997). Using Multimedia Techniques to Address Diverse Learning Styles in Materials Education, *Journal of Materials Education*, 19.
[25] Tee, T. K., Yunos, J. M., Kuppusamy, B., Yee, M. H., Mohamad, M. M., Othman, W., & Hanapi, Z. (2015). The Pattern of Learning Styles among Second Year Students in Business Management and Hospitality Programs at One of the vocational college in Northern Zone. *Procedia-Social and Behavioral Sciences*, *204*, 62-72.

IOP Conf. Series: Journal of Physics: Conf. Series **1049** (2018) 012056 doi:10.1088/1742-6596/1049/1/012056

TVET lecturer empathy and student achievement

S Dahri[1], Y Yusof[1] and C Chinedu[1]
[1]Faculty of Technical and Vocational Education, Universiti Tun Hussein Onn Malaysia (UTHM), Malaysia

E-mail: gb160007@siswa.uthm.edu.my

Abstract. This study aims to highlight the importance of empathy in TVET lecturers to develop healthy lecturer-student relationship. Because healthy lecturer student relationship is crucial for better student learning, achievement and lecturer's own well-being. Therefore, this article is important in helping TVET lecturers to understand the importance of empathy and enhance their ability to empathize with their students. This is an effort to make realize the importance of empathy towards creating strong bonds with students for their better learning and ultimately producing good human beings.

1. Introduction

Most of the educational research nowadays is in search of the utmost effective factors in teaching and learning. In this search, researchers' main focus have been on lecturers as a factor of effective learning. Recently, emotional abilities of a lecturer are considered to be highly important for effective teaching and learning. Although, it is believed that lecturers just need brain power to undergo through all the tasks, yet there is a growing evidence that they also need emotions as the compulsory fuel for the higher functioning of the brain [1-5]. Hence, emotional abilities are increasingly regarded as an ingredient that contributes to the effectiveness of lecturers.

Among several emotional abilities, empathic ability of a lecturer is often cited as an important characteristic of an effective lecturer [6]. Because for an effective teaching and learning, developing healthy lecturer-student relationship is very critical. And to establish it, empathy is considered as fundamental component by many researchers [7,8]. It is therefore important to ensure that lecturers have the ability to empathize especially in the field of TVET. TVET requires lecturers not only with technical skills yet an ability to have strong bonds with students and can create a safe classroom environment. Hence, this study will therefore highlight the role of lecturer empathy towards developing healthy lecturer-student relationship.

2. Background

TVET is an educational field that emphasizes on skills, mastery and competence of its individuals. Thus, TVET lecturers need to ensure that students are mastering the skills and knowledge efficiently. A study reveal that TVET lecturers are less committed towards their jobs, lack dedication, showing carelessness towards their duty and are lacking of ability [9]. Which can affect the reputation of teaching profession.

---

[1]gb160007@siswa.uthm.edu.my.

Content from this work may be used under the terms of the Creative Commons Attribution 3.0 licence. Any further distribution of this work must maintain attribution to the author(s) and the title of the work, journal citation and DOI.
Published under licence by IOP Publishing Ltd

The way lecturer behaves with students and act in classroom puts a long lasting effect on students' learning and their lives. If the lecturer cannot empathize with students or cannot create a better learning environment, that lecturer will make a new generation of lame [9]. Thus, it is very crucial for a lecturer to have the ability to empathize.

3. Empathy

The term empathy, started to be used in 1880 by a German psychologist named, Theordore Lipps who created it from the term Einfuhlung which means, in feeling. It is also believed that it was conceptualized by both Lipps and another psychologist Edwards Titchener [10]. This ability of a person allows him/her to get aware of how others feel and understand as well as contextualize others thoughts, emotions, feelings and actions [11]. It recognizes other's feelings, points out their possible causes and share their emotional experience from outside.

Empathy consists of two components, namely, cognitive and affective [11]. Cognitive component of empathy is the mental activity that accurately infer, what others thinking or feeling [12]. Whereas affective empathy is the person's emotional response to the affective state of the other person [11]. In other words we can say that the former involves understanding people's thoughts and feelings and the latter involves a sensation resulting from people's feelings or thoughts.

4. LECTURER EMPATHY

Lecturer empathy is the fundamental teaching skill for the promotion of positive learning atmosphere to students. Lecturer empathy is the ability of a lecturer to show concern and tune into the perspective of a student [13]. These are the lecturers that become a role model for their students by developing strong bonds with them. These strong bonds and positive interaction with students, make students self-efficacious and motivate them towards their studies [14]. Furthermore, these lecturers have high moral standards, can communicate with the students by keeping their emotional states in mind and also encourage them communicate effectively with others [8].

A research study got the views of college students about their best and poorest lecturers with the help of student evaluation [15]. The lecturers that they reported as best lecturers were the ones who exhibited empathy towards them. Another similar study also had the same findings as mentioned above [16]. On the other hand, a mixed mode study conducted by Ceylan, on 216 lecturers found that lecturers with high empathic ability also possess high self-esteem [17]. Thus, empathy is an ability of a lecturer that cannot be overlooked, instead it should be encouraged in lecturers.

5. Lecturer Empathy and Lecturer-Student Relationship

Every student is different from one another based on their demographics, nature and the experiences that they have in their lives. For TVET students to be more motivated towards their education and academic achievement, they need to believe that their lecturer can relate to them. This can happen when TVET lecturers will have empathy towards their students.

In any educational field, the role of lecturer empathy cannot be ignored. If lecturers become empathic to their students, it will result into students who will feel more connected to lecturer, and when lecturer takes the perspective of students, it inspires the lecturer to make the learning more relevant to the students [18]. Increasing lecturer's ability to be empathic leads towards understanding and responding their students' needs appropriately [19]. It is a professional asset for lecturers, social workers, physicians and everybody who works with people6.

Quality in teaching and learning is often considered to be related with the quality of lecturer-student relationship. As Muhammad and his fellows, while discussing qualities of TVET lecturers put emphasis on creating strong bonds with TVET students as this is one of the main qualities of a professional TVET lecturer [9]. According to Vygotsky, the cognitive development of students is at highest level when they are rooted in social relations [20]. These relations and the environment in which they are rooted are crucial for students to feel valuable and make meaning to the learning. Empathic TVET lecturers act as a model of morality by engaging their students in positive interaction that enhances the quality of learning and teaching. The empathy of lecturer could enhance lecturer-student relationship and having positive lecturer-students relationship leads towards improving student performance [21]. However, there has been very limited literature found that emphasize on TVET lecturers being empathic, which is research gap in the field of TVET. The following studies will therefore discuss the role lecturers' empathy in general, not specifically on TVET, due to the lack of studies in the context of TVET. Nevertheless, the following studies do give an idea that lecturers' empathy is important regardless of their particular educational field.

Peck and fellows studied the role of empathy in partnership between lecturer and student. The lecturers who were more empathetic described their relationship with students in a positive way and the lecturers who reflected critical view, showed dissatisfaction with their relationship to students [22]. On the contrary, Barr examined the relationship between lecturers' empathy and perceptions of their school culture. School Culture Scale and Interpersonal Reactivity Index were used for this purpose. Out of the other findings, one of the finding was that empathy was not related to student-lecturer relations [19]. It takes more than empathy to deal with the complex relationship of student-lecturer. However, this study was the first attempt to examine the relationship between these two variables. Lecturer empathy is yet a term which is not studied largely; also this particular study was conducted on a small sample, so it cannot be generalized. Thus, lecturer empathy may play an important role in improving school culture.

Lecturer empathy also facilitates learning by creating a psychologically safe classroom environment for student. Coffman found that when lecturers show empathy towards their students, it creates an empathic classroom environment. He also found that the students who thought their lecturer as empathic had higher grade expectations [23]. However, the methodology of the study is questionable as the researcher used a convenience sampling that is the participants were researcher's and his colleagues students, also the instrument he developed to measure empathy, the validity and reliability of the instrument was not defined. Instead of the fact, it is noticeable that this study does provide a support that empathic lecturers create an empathetic classroom environment that enhances students' learning. On the other hand, Cooper also claims that lecturer's role is significant in creating a learning environment, being responsive to the needs of students and also the concerns of their students [24]. Similarly, Muhammad and his fellows argue that TVET lecturers' role is holistic and comprehensive, he/she has to establish an atmosphere where their students feel free to search all aspects of itself. Further they add that TVET lecturer should be caring and friendly with the students in order to create a positive learning environment [9].

In developing relationships between TVET lecturer and student, understanding the perspective of students is required for TVET lecturer. Empathy makes a strong bond between lecturer and student and enhances care. A caring, empathetic TVET lecturer helps students emotionally, academically and socially by building connections with them.

In a qualitative research, novice and experienced lecturers were asked about the important characteristics of a lecturer they believe should be. The findings indicated that creating a caring relationship with students was common response of all the responses. After creating five overall categories that were divided into sub categories as "effective" and "ineffective" they verified the reliability of their results by having two independent scores classify 100 of the statements. That yielded in 97% of the agreement rate [25]. This research provides good evidence that the ability of lecturers to create caring relationship with students is considered as effective lecturers.

Similarly, Collier state that empathy is the source of effective lecturer student relationship, it builds trust and friendship between lecturer and student. And this trust leads students towards better learning [26]. Kordi and his colleagues concluded several factors that contribute in the development of relationship with students. First, as lecturer's willingness to give attention or concern to the goals and needs of the students or being sensitive to the needs and moods of the students was considered as a contributor in developing caring relationship. Second, the positive atmosphere of the class created by the lecturer by showing supporting and helpful attitude in the class enhances the attention and learning of the students. Third, spending plenty of time with students and designing activities for them. Fourth when they explain the confusing and ambiguous points patiently to students [27]. Thus empathy is critical for TVET lecturers to develop relationship with students and create a healthy classroom climate.

Hence, empathy is the characteristic that can help TVET lecturers in promoting strong bonds with students, better learning environment, higher student achievement and a caring attitude towards students.

6. Lecturer Empathy and Student Achievement

Currently, there are few researches that have attempted to see the correlation between lecturer empathy and student achievement. Nevertheless, they do give an idea that lecturer empathy is vital for student achievement.

Muller analyzed the students' and lecturers' perception of one another and the effects of those perceptions on student achievement. He used National Longitudinal Study of 1988 (NELS) and taking public school lecturers and students as population. He found that the students who perceive their lecturers as caring spend more effort in their academic achievement. And when lecturers perceive their students as they are making an effort to expend their academic performance, there is positive lecturer student relationship [28].

On the other hand, Bostic studied the relationship between lecturer empathy and students' scores on a standardized test. A correlational research design was used to see the relationship between lecturer empathy and students score on a standardized test. To measure lecturer empathy, he used a subscale of perspective taking from Interpersonal Reactivity Index (IRI) [18]. Although, the study indicated no relationship between lecturer empathy and students' scores on standardized test. However, it is arguable here that the scores on standardized test possibly may not be liable lecturer characteristics. Also, the standardized test scores might be measuring just the small amount of student achievement and that the research findings have threats to its generalizability.

McAllister and Irvine did a qualitative research from a diversity perspective. They conducted a training program on diversity for the lecturers to train them about being empathetic to their students from different cultures. They found that this training program, gave a deeper understanding to the lecturers about the diverse cultures and this understanding lead them to be empathetic to their diverse students. Also, it helped them to be more effective in the classroom. Above all, the lecturers realized that being empathetic towards their students enhanced their interaction with them and that resulted in more positive learning environment [29].

Nordina and colleagues state that there a need for TVET lecturers to develop strong relationship with their students in order to improve student achievement. TVET lecturer should make sure that he/she is friendly, has a trust worthy relationship with the students and is able to understand their students [30]. The main focus of TVET education is to develop mastery skilled and competent workforce and for that it is very crucial for lecturers to make sure that students are satisfied with the lecturer and are learning skills and knowledge effectively [9]. To achieve this goal creating a psychologically safe classroom environment for learning is important. Which can be developed by having empathic TVET lecturers.

7. Research Design and Method

The research design adapted for the study is correlational research design. Correlational research design is used when a researcher is interested in the extent to which two or more variables are correlated, that is, when changes in one variable are reflected in changes in the other [31]. The researcher has chosen this research design because, this research will examine the relationship between TVET lecturers' empathy students' achievement.

Data collection will be done by distributing a Likert scale questionnaire of 30 items. The questionnaire will be measuring empathy of TVET lecturers. The researcher is using questionnaire in the study because it is considered suitable when the study is intended to investigate the perspectives, opinions and attitudes of respondents regarding certain characteristic or problem [31].

This study will have two groups of respondents as the target populations for the study. They include TVET final year degree students and the lecturers who are teaching those students in the Faculty of Technical and Vocational Education (FPTV) and Faculty of Technology Engineering (FTK) from University Tun Hussein Onn Malaysia (UTHM). There are eight faculties in the UTHM, however only two faculties are chosen. They are chosen purposely for this study because both of them share the similar degree programmes that is mechanical, electrical and civil engineering. Since they share similar degree programmes, the researchers intend to see the difference in the lecturers' empathy sharing similar programmes yet with different educational background. Which are FPTV lecturers with education background and FTK lecturers with non-education (technical) background. As based on the previous research findings, it shows that individuals who have education background tend to have better understanding of empathy than the individuals who come from different educational background and then choose teaching as a profession [32].

T-test will be used to find out the difference between the level of empathy of TVET lecturers in FPTV and FTK from lecturers' perspective and from students' perspective. Then, after getting the perception of lecturers and students from both the faculties, they will be correlated by using Pearson correlation to find out the relationship between the lecturers' and students' perception. Finally, Pearson correlation will also be used to find out the relationship between lecturer's empathy and students' achievement in both the faculties.

8. Conclusion

Overall, empathy of a TVET lecturer needs to be encouraged and enhanced for creating healthy relationship with students. Which makes students self-efficacious, motivated towards their studies, positively affect their academic outcomes and ultimately shapes a good human being. In addition, lecturer empathy not only help students but also is good for lecturer's own well-being and keeps them determined towards this noble task.

Acknowledgment
Centre for graduate studies, Universiti Tun Hussein Onn Malaysia.

10. References

[1] Day C., & Qing, G. (2009). Teacher emotions: Well being and effectiveness Advances in teacher emotion research (pp. 15-31): Springer.
[2] Goroshit, M., & Hen, M. (2014). Does Emotional Self-efficacy Predict Teachers' Self-efficacy and Empathy? Journal of Education and Training Studies, 2(3), 26-32.
[3] Madalinska-Michalak, J. (2015). Developing Emotional Competence for Teaching. Hrvatski časopis za odgoj i obrazovanje, 17(Sp. Ed. 2), 71-97.
[4] Shanmugasundaram, U., & Mohamad, A. R. (2011). Social and emotional competency of beginning teachers. Procedia-Social and Behavioral Sciences, 29, 1788-1796.
[5] Thomas, M., Francis, P., Shahid, S. A. M., & Jani, S. H. M. (2015). The Need for Emotional Intelligence Skills among Knowledge Workers at the Tertiary Level. International Journal of Social Science and Humanity, 5(4), 347.
[6] Stojiljković, S., Djigić, G., & Zlatković, B. (2012). Empathy and teachers' roles. Procedia-Social and Behavioral Sciences, 69, 960-966.
[7] Cornelius-White, J. (2007). Learner-centered teacher-student relationships are effective: A meta-analysis. Review of educational research, 77(1), 113-143.
[8] Jennings, P. A., & Greenberg, M. T. (2009). The prosocial classroom: Teacher social and emotional competence in relation to student and classroom outcomes. Review of educational research, 79(1), 491-525.
[9] Muhammad, S. N. H., & Jaafar, S. N. (2015). TVET Teacher Professionalism in Leadership Personality Formation. Journal of Education and Practice, 6(1), 143-147.

[10] Torres, L. H., Bonilla, R. E. B., & Moreno, A. K. A. (2016). Empathy in future teachers of the Pedagogical and Technological University of Colombia. NEW APPROACHES IN EDUCATIONAL RESEARCH, 5(1), 30-37. doi: 10.7821/naer.2016.1.136
[11] Baron-Cohen, S., & Wheelwright, S. (2004). The empathy quotient: an investigation of adults with Asperger syndrome or high functioning autism, and normal sex differences. Journal of autism and developmental disorders, 34(2), 163-175.
[12] Swan, P., & Riley, P. (2012). " Mentalization": A Tool to Measure Teacher Empathy in Primary School Teachers. Australian Association for Research in Education (NJ1).
[13] Tettegah, S., & Anderson, C. J. (2007). Pre-service teachers' empathy and cognitions: Statistical analysis of text data by graphical models. Contemporary Educational Psychology, 32(1), 48-82.
[14] Cooper, B. (2004). Empathy, interaction and caring: Teachers' roles in a constrained environment. Pastoral Care in Education, 22(3), 12-21.
[15] Greimel-Fuhrmann, B., & Geyer, A. (2003). Students' evaluation of teachers and instructional quality--Analysis of relevant factors based on empirical evaluation research. Assessment & Evaluation in Higher Education, 28(3), 229-238.
[16] Slate, J. R., LaPrairie, K., Schulte, D. P., & Onwuegbuzie, A. J. (2009). A mixed analysis of college students' best and poorest college professors. Issues in Educational Research, 19(1), 61-78.
[17] Ceylan, R. (2009). An examination of the relationship between teachers' professional self-esteem and empathic skills. Social Behavior and Personality: an international journal, 37(5), 679-682.
[18] Bostic, T. B. (2008). Teacher Empathy and Its Relationship to the Standardized Test Scores of Diverse Secondary Students.
[19] Barr, J. J. (2011). The relationship between teachers' empathy and perceptions of school culture. Educational Studies, 37(3), 365-369.
[20] Cooper, B. (2011). Empathy in education: Engagement, values and achievement: Bloomsbury Publishing.
[21] Hassan, R. (2010). Strenghtening technical and vocational education (TVET)-producing innovative TVET teachers for 21st century students. my science work.
[22] Peck, N. F., Maude, S. P., & Brotherson, M. J. (2015). Understanding Preschool Teachers' Perspectives on Empathy: A Qualitative Inquiry. Early Childhood Education Journal, 43(3), 169-179.
[23] Coffman, S. L. (1981). Empathy as a relevant instructor variable in the experiential classroom. Group & Organization Studies, 6(1), 114-120.
[24] Cooper, B. (2010). In search of profound empathy in learning relationships: Understanding the mathematics of moral learning environments. Journal of Moral Education, 39(1), 79-99.
[25] Walls, R. T., Nardi, A. H., von Minden, A. M., & Hoffman, N. (2002). The characteristics of effective and ineffective teachers. Teacher education quarterly, 29(1), 39-48.
[26] Collier, M. D. (2005). An ethic of caring: The fuel for high teacher efficacy. The Urban Review, 37(4), 351-359.
[27] Kordi, L., Hasheminejad, S., & Biria, R. (2012). Do we care? Investigating how a caring relationship might influence comments and responses in EFL writing classes. Theory and Practice in Language Studies, 2(6), 1249.
[28] Muller, C. (2001). The role of caring in the teacher-student relationship for at-risk students. Sociological inquiry, 71(2), 241-255.
[29] McAllister, G., & Irvine, J. J. (2002). The role of empathy in teaching culturally diverse students: A qualitative study of teachers' beliefs. Journal of teacher education, 53(5), 433-443.
[30] Nordina, A. S., Hassan, R., & Hasan, A. (2015). Teaching of Professional Trainers at Vocational Colleges in Malaysia.
[31] Creswell, J. W. (2012). Qualitative inquiry and research design: Choosing among five approaches: Sage publications.

[32] Darling-Hammond, L. (2000). How teacher education matters. Journal of teacher education, 51(3), 166-173.

IOP Conf. Series: Journal of Physics: Conf. Series **1049** (2018) 012057 doi:10.1088/1742-6596/1049/1/012057

Eliminating shyness through co-curricular activities towards enhancing the career development of engineering students

Normah Zakaria[1], **Masadliahani Masduki** [2], **Nur Nadiah Ismail** [3]

Faculty of Technical and Vocational Education, Universiti Tun Hussein Onn Malaysia, 86400 Parit Raja, Johor, Malaysia [1,3]
Sekolah Kebangsaan Seri Bali, 85400 Chaah, Johor, Malaysia [2]

Email: adliq@yahoo.com

Abstract. The purpose of this study is to find the most effective strategies for eliminating shyness amongst engineering students through co-curricular activities. As shyness causes negative impacts towards students' development, it becomes a major concern for lecturers and student affairs departments to find strategies that can solve the problem. However, shyness cannot be eliminated solely through the formal classroom learning. Ironically, co-curricular activities provide a good platform to develop students' potentials, but previous research failed to analyse the specific co-curricular activities that can decrease the impacts of shyness. This paper seeks to address this gap. 102 articles were reviewed and 6 articles were used to suggest co-curricular activities that can reduce the impact of shyness. Finding shows that co-curricular activities such as debate, choir, sports team, physical activities and exercise, career and technical student organizations and entrepreneurship clubs and societies are found to be the core activities that can enhance the social and communication skill, participation in the teaching and learning, self-esteem and career development of shy students. The recommendation activities can be used by the student affairs departments to eliminate shyness amongst engineering students by ensuring their active participation in the respective clubs.

1. Introduction

In developing a career, engineering students need to have a high level of self-confidence and not being shy to communicate. According to [1], communication aspects such as oral presentations represent one of the tasks required in the career development of an engineer. A professional engineer needs to familiarize himself and allocate nearly 60% of their time in oral communication, especially for project presentations where they need to work with various types of agencies [1]. Thus, in building a career, engineering students need to develop good interpersonal communication skills so that they dare to express their ideas effectively. This is because communication is the most anticipated generic skill by many potential employers. According to [2], engineering students have the strength in terms of technical and theoretical skills, but they are lacking in social and communication skills. However, conventional learning methods have little to assist in improving their interpersonal skills and decrease shyness among students [3].

Content from this work may be used under the terms of the Creative Commons Attribution 3.0 licence. Any further distribution of this work must maintain attribution to the author(s) and the title of the work, journal citation and DOI.
Published under licence by IOP Publishing Ltd

Enhancing shy engineering students' self-esteem in the classroom setting seems impossible because lecturers need to meet their teaching objectives in certain period given. Furthermore, [4] pointed out that there are many negative cultures in a classroom that cause negative influences towards shy students. In this sense, out-of-class programs could be the most appropriate solution to be applied for shy engineering student at the universities. Students' participation in the co-curricular activities able to give positive impacts on academic development and to shape the students' self-esteem towards developing their career [5]. This is important because highly skilled workers with good social and communication skills are the characteristics of graduates required by current employers [6].

1.1 Engineering students and the problem of shyness
Self-confidence is pivotal in everyone's life. Individuals with a high self-esteem can accomplish their desired goals and not easily give up in achieving their success. However, not all individuals have positive thinking and high self-esteem. There are shy individuals with low self-esteem and always think negative. Most scholars relate shyness with social and communication skills deficiency, in which shy individuals feel discomfort in social communication situations, fear of speaking and tendency to be silent, passive or irresponsive to others [7, 8, 9]. Shyness is a universal problem experienced by all individuals, from children to adults. Shyness among students causes negative impacts towards their career development. Engineering students must fit in the world of work, so, each of the student needs to be prepared with the skills needed to meet the requirements of today's industrial market. However, previous research found that among the shortcomings of engineering graduates is the lack of social and communication skills as demanded by the employer in addition to the competitive job market [10]. These skills are crucial to ensure that students are able to cope with any problems in their study, especially when they do not understand certain topics, or have something to be argued [11, 12]. Students need to give opinions and ask questions, join the discussions and participate in activities during the learning sessions to develop their generic skills [13, 14]. In order to enhance the generic skills among engineering students, many of them face the difficulties to participate actively in social situations, especially during the classroom learning and also during their practical session due to shy behaviour [13].

Numerous studies have considered co-curricular activities as the holistic approach to develop students' skills in general, but only limited research looking at the benefits for each type of the co-curricular activities it can offer. Although international studies have emphasized various ways to overcome shyness, they are generally focused on shy children and very few studies conducted to cater shyness among college students. Furthermore, many studies have focused on the benefits of some extra-curricular activities for students who are at the risk of academic failure or dropout [15], emphasizing the program as a possible way to enhance engagement, but ignoring to analyze its impacts towards reducing or eliminating shyness. Thus, there is a vital need to investigate the type of effective co-curricular activities in eliminating shyness among engineering students who play an important role in the development of human capital. This paper seeks to address this gap by reviewing the literature on co-curricular practices to propose specific activities that able to eliminate or reduce shyness among engineering students to achieve the full potential of student development.

2. Methodology
The effects of students' shyness were identified according to analyses of selected publications. Then, a small-scale literature review of published articles was undertaken to identify the most relevant type of co-curricular activities to eliminate shyness. To carry out the search, online literature search engines were used to find academic papers and other resources using few keywords like; co-curricular and self-esteem, co-curricular and communication skill, co-curricular and shyness, co-curricular and career development, and, college students and club. A total of 102 articles were reviewed but only six were relevant and selected as the main references in constructing the suggested co-curricular activities to eliminate shyness and achieve the maximum potential of students' career development (see Table 1).

This paper first deliberates the effects of shyness towards engineering students. An overview is given to the effects of shyness towards students in terms of communication skill, social skill, participation in the teaching and learning, self-esteem and career development. Then, the most relevant types of co-curricular activities that able to reduce the impacts of shyness were discussed.

Table 1. Co-curricular activities to eliminate shyness and enhance career development

Types of co-curricular activities	Researcher(s)	Findings
Debate club	Lawhorn [16]	Self-confidence, workplace skill, presentation skill, analytical and logical reasoning skills, ability to think and speak spontaneously
Choir	Jacob, Guptill, & Sumsion [17]	Sense of community, social bonding, accomplishment, expressing personality
Sports team	Edwards, Edwards, & Basson [18]	Enhances mental health, social, psychological well-being, self-esteem
Physical activities and Exercise	Page & Hammermeister [19], Ekeland, Heian, Hagen, & Coren [20]	Lower score of shyness related to longer time doing physical activities, higher self-esteem
Career and technical student organizations	Lawhorn [16]	Academic, interpersonal skill, career skill, leadership skill, improve self-esteem, encourage civic involvement
Entrepreneurship clubs and societies	Pittaway, Rodriguez-Falcon, Aiyegbayo, & King [21]	Motivation, enhance social, increase self-efficacy, self-esteem, gaining practical experience & networking

3. The impact of shyness

Shyness affects engineering students in various parts of their development. In this part, first, the impact of shyness towards students' communication skill were discussed, followed by the impacts on their social skills, participation in the teaching and learning and self-esteem. Finally, the impact of shyness towards students' career development were discussed.

3.1 The impact on communication skills

Communication is vital for an engineer. However, shyness causes engineering students fail to share ideas and opinions, refuse to ask questions for clarification, inactive in group discussions, ineffective in the classroom, and unable to deliver presentations effectively. Past studies have identified shyness is a strong and stable trait that brings negative impacts to oral communication among students [22, 23], [24, 25]. Shyness inhibits one's interpersonal goals and prevents students to start a friendship because they are not confident to communicate [12]. Shyness makes a student become less talking and feels awkward or behave like completely wrong when interacting [26]. Shy students are associated with the low perception of communication efficiency [27, 28]. Shy students rarely express their ideas or thoughts. They prefer to keep silent when being asked and refuse to be participative in classroom's activities. These students are unable to talk fluently in front of others like when delivering a presentation due to the lack of self-confidence.

Shyness in students is closely related to the difficulty of communicating because they are too concerned about other person or people's negative judgments against them [26]. Shy students often considered as unfriendly and unable to communicate effectively [28, 29] also reported that shy students have low communication competence and unable to speak fluently and confidently in a social situation. Students who are shy will have problems in both verbal and non-verbal communication [26]. According to [28], shy students face the difficulties to start a conversation because they find it hard to initiate and structuring the conversation. Because of this, shy students speak less and take more time to give a response [26]. Shy students exhibit poor reciprocity smiling behaviour and lack of verbal competencies [30]. Shy students often considered as having an introverted personality, thus they have a low level of verbal responding [24]. Shy engineering students will have the difficulties to throw their ideas and do presentations in front of others as they have a low level of communication skill.

3.2 The impacts on social skills

Shy students feel more comfortable with computer-mediated communication, compared with the face to face communication, especially when it involves interaction with the opposite gender [31]. [28] listed down some of the consequences of shyness among students, which are they tend to have negative emotions such as loneliness, stress, and isolation because they face the difficulty to find new friends. According to [29], shyness was correlated to social competencies of a student whereby he or she will avoid the social situations and demonstrate awkwardness during social interaction.

Shy students have a lower social acceptance [32] and do not dare to start a conversation in a group setting or doing any social action in group activities [31]. The study on the effects of shyness by [33] focused almost the exclusive correlation between shyness and the difficulties in social interactions. The researcher found that shy students experienced more loneliness and have a smaller support network compared to the non-shy students. Shy students have lack interaction with others and would demonstrate low social skill, speak less, lack eye-contact, sit far from others, less calm and less friendly than non-shy students [29]. Shy engineering students will try to avoid meeting other people, not active in the organization and avoid going to conferences. Thus, they have lack of networking and experiences that are important for their future career.

3.3 Participation in the teaching and learning

According to [34], university lecturers have identified that there is a decrease of student involvement in the classroom. Shyness leads to an increase of unasked questions where students feel afraid to ask for clarifying things they are not sure. Students are shy to speak and ask questions during lectures due to fears of being ridiculed or underestimated by others. Shy students are also difficult to express opinions and ideas [28]. The same situation occurs in the study conducted by [11]. The psychologist found that most students refused to talk and did not participate actively in the class due to shyness.

Among the effects of shyness in the classroom are that students rarely start discussions, they avoid to raise new topics, do not challenge lecturers, rarely ask for clarification and do not respond voluntarily [11]. The researcher argued that due to the lack of interaction in the classroom, the passive learning atmosphere has been unnoticed. In the study of [35], he found that among the consequences of shyness are students afraid to grab the opportunity to stand out like competing to answer questions. The researcher added that in group activities, especially during discussions, shy students do not give many contributions, and just giving irrelevant comments if requested. Shy students are afraid to ask questions because they have a low self-esteem and rarely communicate or interact with lecturers, especially if they were asked to answer the questions or being asked to present their work individually or in groups [36]. Shy students are introverted and have a high level of communication apprehension. This makes them become unnoticed by the lecturers and friends due to the lack of participation in the classroom [37]. Shy engineering students are not active in the classroom, do not dare to speak up and get less attention from lecturers. These affect their academic achievements and the shortcoming limits their chance to improve their interpersonal skill.

3.4 The impacts on self-esteem

Individuals with a high level of self-esteem will also have a high level of self-confidence and could achieve their career goals. However, shy students do not have positive thinking and high self-esteem. Shyness causes students to have negative thoughts and negative self-belief [38, 29]. Shyness has a significant and negative correlation towards students' self-esteem [23]. Shyness causes students to have lack self-esteem, hence, they feel tense when interacting with others. This makes them feel isolated and uncomfortable in any social situation [27]. According to [39], shy students have a low self-esteem and they tend to blame themselves for their social failure that causes them to feel lonely. The inability of a student to contribute to their social and the expectation that they will be looked down by others can contribute to shyness, thus lowering the self-esteem and decrease in self-confidence [12]. Shy students believe that they are less competent than their peers, thus this belief lowers down their self-esteem [26]. Shy students with low self-esteem will have lack of confidence in social situations [37] and this makes them become inactive in the classroom. Shyness causes engineering students to have a low level of self-esteem that they always think they are incapable of doing challenging tasks. This limits their chances to improve their skills.

3.5 The impacts on career developments

Shyness causes negative impacts towards career development in which shy students fail to grab the opportunities available to them because they have lack of self-confidence and not active in the effort to prepare themselves for the career challenges [40]. Additionally, shyness causes students to achieve lower employment status and delay in the career stability [41]. According to [42], conceptually, shyness creates the barrier in clarifying vocational self-concept because the development of self-concept involves social interaction. [40] examine the notion that shyness is inversely related to some behavior of early career development. The researchers reported that shy undergraduate students are less likely to find information about career, do not know to make career decisions, and lack of interest in the field of interpersonal careers. This agreed by [43] who argued that the personality of shyness affects the career choices of the students in which they tend to involve less in the social career. Shy students are less involved in the career exploration because shyness prevent social interactions that make it difficult for them to get information as they need to interact with people [44, 39].

Vocational self-concept is the level of self-clarity pertaining to the career in terms of attitudes, values, interests, needs, and abilities [45]. Shy students always face problems in the development of self-concept and matching it to the appropriate career [42]. Thus, shy students are sceptical in determining career options [46, 43], assuming they have no talent and have a potential job that only fits their shy ability [40]. Shy students have no career maturity, less exploring and planning, and are not sure to be in the world of work [42]. Shy students have a weak vocational self-concept, so if they are employed they are less likely to be a leader, have lack of promotion opportunities and often feel dissatisfied with various aspects of their work [40]. In developing the career, students need to build career management competencies that include generic skills [47]. However, shyness affects the generic skills of the students. [48] reported that 62.3% of the technical graduates are unemployed due to lack of generic skills. [6] found that communication skills are the most important generic skills in the working industry, but are least mastered by the engineering graduates. The study was supported by [49], who reported that the most significant weakness of engineering students when they are in the industry was the lack of communication skills, especially during oral presentations, due to the high level of shyness. Thus, it is clear that shyness prevent engineering students to develop their career as they are lacking in social and communication skill, vocational self-concept and generic skills.

4. Discussion

Shy engineering students are less confident to talk and often not contributive in discussions. Shy students can be helped to reduce their shyness by providing them the platform to enhance their self-esteem and self-confidence through their involvement in specific co-curricular activities. Such activities will not only develop their social and communication skill but help them in enhancing their vocational self-concept and generic skills towards positive career development.

4.1 Debate club

[50] suggested that the theme-based approach in drama and debate could integrate various skills like listening and speaking skill, reading and writing skill, and thinking skill. This would progressively help students to further develop their self-confidence. As engineering students, it is crucial for students to learn how to speak up in front of people to throw their comments or thoughts. Ironically, the public speaking skills can only be learned through a lot of practices during debate clubs. Thus, students who join debate clubs will get used to participate actively during the teaching and learning sessions in their classroom. The support and cooperation from group members provide debate activity a positive atmosphere that motivates shy students to speak confidently, besides improving their social and communication skills. Through practices in debates, students become brave and confident to talk, able to deliver their ideas confidently and reduce communication apprehension. Debate enables students to improve their oral communication, teamwork and critical thinking skill [51, 52] besides developing their confidence to fight their stage fright [52]. This skill is very much required for engineering students to improve their interpersonal skill and enhance their self-esteem to make them able to speak confidently. Participation in a debate club also enhances shy students' workplace skill, especially the presentation skill, in which engineering students will be needed to present their works in front of their superiors and co-workers. Besides that, engineering students may develop their analytical and logical reasoning skills, together with the ability to think and speak spontaneously. According to [16], these skills benefit students in college and will be looked by future employers. For engineering students, they can have the engineering related themes for the debate activities, such as sustainable development and engineer-entrepreneur.

4.2 Choir

The choir is another type of communication activities that could decrease shyness in students by enhancing their self-esteem and increasing their confidence level because it encourages both oral and non-oral expressions in front of audiences. Choir and orchestra could enhance shy students' self-esteem. Although students felt nervous to perform on the stage, the good thing is they gain a lot of benefits like cooperative and teamwork skills, social skill, new friendship, and increase their self-esteem [15]. These enhance their self-confidence and interpersonal skill, yet able to eliminate shyness. By participating in the choir club, social anxiety among engineering students could be decreased gradually as they will be interacting in social interactions. The challenges students face in the choir club act as the platform to motivate them. Students could feel the sense of belonging, yet decrease their shyness because they feel related to others, and experienced a sense of camaraderie. These enhance their teamwork skill and benefit their career development.

A shy student would have an introverted personality [53]. According to [54], an introverted person would choose a career that does not involve too many social interactions so that he or she does not need to speak too often due to the communication apprehension. However, an engineering student should not avoid social interactions due to the nature of the work that needed them to communicate with clients and team members [1]. The interactions between shy students and their team members in the choir club, also the interactions between them with the audiences and the communities help to decrease their communication apprehension. These would enhance their social and communication skills and develop their career as engineers.

4.3 Sports team

Besides the oral performance-based clubs, shyness can also be reduced through sports-based co-curricular. Previous studies proved that sports can improve social skills and enhance the self-esteem of

IOP Conf. Series: Journal of Physics: Conf. Series **1049** (2018) 012057 doi:10.1088/1742-6596/1049/1/012057

shy students [55, 18, 56]. The involvement in sports would gradually reduce social anxiety among shy students because they interact with group members and with the spectators. Sports team like hockey promotes mental health and psychological well-being that also provide social interactions for students in a meaningful way which can eliminate their shyness. Sports team fosters co-operation between team members. This could enhances their teamwork and leadership skills that automatically improve the communication skill in which these are the generic skills required by the employers [47].

4.4 Physical activities and exercise
There is evidence that higher rate of shyness and loneliness is associated with not exercising frequently [19]. Shy students avoided physical activity due to social anxiety. However, through the involvement in physical education, students can be avoided from being humiliated and enhance their self-esteem [20]. According to [19], students who exercise 7 times a week earn shyness and loneliness scores that are much lower than other groups due to the increased of heartbeat and interpersonal factors during exercise with their friends. Rugged physical activity and involvement in sports teams decrease shyness in students [56]. When students involve themselves in physical activities, this makes them become fit and helps them to become active during the classroom interactions. Furthermore, by exercising or doing physical activities, students would automatically involve themselves in social interaction in a leisure setting and this would slowly decrease their social anxiety that increases their social skill. When involved in physical activities, engineering students practice their motor skills and this helps them improve their technical skill. [57] argued that engineering graduates should master both technical and generic skill so that they can fit in the working industries.

4.5 Career and technical student organizations
In career and technical student organizations, there are many specific categories of vocational education that is not only developing shy students' technical skills but also enhance their communication skills in the informal setting [58]. This is due to the goals of the organization to improve students' self-esteem and civic development [16]. Students could develop their social and communication skill and explore their career opportunities when they are participating actively in the community. These will increase shy students' self-esteem and when they involve in the activities organized by the organization during the co-curricular program, they will also increase their competencies in both technical and generic skills. When self-esteem is enhanced, students will become more confident and would involve actively in the discussions and activities during formal classroom learning [37]. Besides enhancing interpersonal communication, career and technical student organizations provide students a platform to practice their public speaking skill that develops their self-confidence to eliminate shyness, while developing most workplace skills needed for their career development.

4.6 Entrepreneurship clubs and societies
Activities conducted in entrepreneurial clubs help to reduce shyness and increase students' self-esteem to face and communicate with others. Students are also able to build and operate their own company, as well as not fear to participate in any competition because their self-confidence is built during the entrepreneurship activities. The challenges that students face during joining the club increase their motivation level and self-efficacy that enhance their social skill. Students found that this club changed their personal behaviour, like from introverted towards more extroverted students [21]. Besides gaining practical experiences, students also learn communication skill, team-work skill and gain networking that benefits their career development. [21] also reported that students could enhance their self-esteem and the activities conducted had improved their public speaking skill.

5. Conclusion
Shyness causes negative impacts towards engineering students' development in terms of their social and communication skill, participation during the teaching and learning session, self-esteem and their

career development. These impacts of shyness can be decreased through the involvement of engineering students in one of the specific co-curricular activities which are; debate, choir clubs, sports team, physical activities and exercise, career and technical student organizations and entrepreneurship clubs and societies. These co-curricular activities are believed as the platform to boost engineering students' confidence level, lower down their communication apprehension and decrease the social anxiety that eliminates shyness and enhance the self-esteem of engineering students towards a positive career development. When students' communication skill is improved, it will help them in the career exploration because they will be able to get into social interactions and get information regarding their career path. Thus, it is pivotal for student affairs divisions to provide support, opportunities, and encouragement for engineering students to achieve the maximum potential of development.

Acknowledgments

This research work is supported by the Contract Grant RAGS and R065 supported by Universiti Tun Hussein Onn Malaysia.

References

[1] E. Bhattacharyya and A. Z. Zainal. What do students and engineers have to say about communicative competence in technical oral presentations?, *Pertanika J. Soc. Sci. Humanity.*, vol. 23, no. August, pp. 123–142, 2015.

[2] D. R. Woods *et al.*, Developing Problem Solving Skills: The McMaster Problem Solving Program, *J. Eng. Educ.*, vol. 86, no. 2, pp. 75–91, 1997.

[3] M. S. Tahar. Rekabentuk pengajaran kejuruteraan (Sistem TPU) satu terjemahan konsep pengajaran visual dan kinestetik, *J. Kejuruter.*, pp. 1–9, 2000.

[4] H. Gazelle. Class climate moderates peer relations and emotional adjustment in children with an early history of anxious solitude: A child x environment model., *Dev. Psychol.*, vol. 42, no. 6, pp. 1179–1192, 2006.

[5] G. Tchibozo. Extra-Curricular Activity and the Transition from Higher Education to Work : A Survey of Graduates in the, *High. Educ. Q.*, vol. 61, no. 1, pp. 37–56, 2007.

[6] F. B. Hairi, M. N. Ahmad Toee, and W. Razzaly. Employers' perception on soft skills of graduates : A study of Intel Elite soft skill training, *Int. Conf. Teach. Learn. High. Educ.*, no. 2011, 2011.

[7] J. M. Cheek and A. K. Watson. The definition of shyness: Psychological imperialism or construct validity?, *J. Soc. Behav. Pers.*, vol. 4, no. 1, pp. 85–95, 1989.

[8] J. C. Mccroskey and V. P. Richmond. Communication apprehension and shyness: Conceptual and operational distinctions, *Commun. Stud.*, vol. 33, no. 3, pp. 458–468, 1982.

[9] P. G. Zimbardo. Shyness: What is it and what to do about it., *MA Addison-Wesley Publ. Company.*, 1977.

[10] A. Esa, S. Padil, and S. Hassan. Employability skills in engineering programs: A case study in Malaysian polytechnics, *Adv. Sci. Lett.*, vol. 21, no. 7, p. 2442–2445(4), 2015.

[11] F. E. Anderson. The enigma of the college classroom: Nails that don't stick up, in *A handbook for teaching English at Japanese colleges and universities*, P. Wadden, Ed. Oxford: Oxford University Press, 1993, p. 101–110.

[12] J. B. Asendorpf. Shyness and adaptation to the social world of university, in *Shyness: Development, Consolidation and Change*, W. R. Crozier. Ed. London: Routledge, 2000, pp. 103–120.

[13] J. Rimmer, J. Good, E. Harris, and M. Balaam. Class participation and shyness: Affect and learning to program, in *Proceedings of the 23rd Annual Psychology of Programming Interest Group Conference*, 2011, pp. 1–9.

[14] A. Masek. An appropriate technique of facilitation using students' participation level measurement in the PBL environment, *Int. J. Eng. Educ.*, vol. 32, no. 1, pp. 402–408, 2016.

[15] S. E. Pitts. Anything goes: A case study of extra-curricular musical participation in an English secondary school, *Music Educ. Res.*, vol. 9, no. 1, pp. 145–165, 2007.

[16] B. Lawhorn. Extracurricular activities : The afterschool connection, *Occup. Outlook Q.*, vol. 9, no. 2008, pp. 16–21, 2008.

[17] C. Jacob, C. Guptill, and T. Sumsion. Motivation for continuing involvement in a leisure-based choir: The lived experiences of university choir members, *J. Occup. Sci.*, vol. 16, no. 3, pp. 157–162, 2009.

[18] D. J. Edwards, S. D. Edwards, and C. J. Basson. Psychological Well-Being and Physical Self-Esteem in Sport and Exercise., *Int. J. Ment. Health Promot.*, vol. 6, no. 1, pp. 25–32, 2004.
[19] R. M. Page and J. Hammermeister. Shyness and loneliness: Relationship to the exercise frequency of college students, *Psychol. Rep.*, vol. 76, no. 2, pp. 395–398, 1995.
[20] E. Ekeland, F. Heian, K. Hagen, and E. Coren. Can exercise improve self esteem in children and young people? A systematic review of randomised controlled trials, *Br. J. Sports Med.*, vol. 39, no. 11, pp. 792–798, 2005.
[21] L. Pittaway, E. Rodriguez-Falcon, O. Aiyegbayo, and A. King. The role of entrepreneurship clubs and societies in entrepreneurial learning, *Int. Small Bus. J.*, vol. 29, no. 1, pp. 37–57, 2011.
[22] R. L. Duran and L. Kelly. The cycle of shyness: A study of self-perceptions of communication performance, *Commun. Reports*, vol. 2, no. 1, pp. 30–38, 1989.
[23] S. Koydemir and A. Demir. Shyness and cognitions: an examination of Turkish university students., *J. Psychol.*, vol. 142, no. 6, pp. 633–644, 2008.
[24] M. R. Leary. The Conceptual Distinctions Are Important: Another Look at Communication Apprehension and Related Constructs., *Hum. Commun. Res.*, vol. 10, no. 2, pp. 305–312, 1983.
[25] J. C. McCroskey. Avoiding communication, *Shyness, reticence, Commun. apprehension*, pp. 13–38, 1984.
[26] A. Arroyo and J. Harwood. Communication competence mediates the link between shyness and relational quality, *Pers. Individ. Dif.*, vol. 50, no. 2, pp. 264–267, 2011.
[27] J. M. Cheek and A. H. Buss. Shyness and sociability., *J. Pers. Soc. Psychol.*, vol. 41, no. 2, pp. 330–339, 1981.
[28] P. A. Pilkonis. The Behavioral Consequences of Shyness, *J. Pers.*, vol. 60, no. April, pp. 31–46, 1996.
[29] R. S. Miller. On the nature of embarrassability: Shyness, social evaluation, and social skill., *J. Pers.*, vol. 63, no. 2, pp. 315–339, 1995.
[30] J. B. Evans, M. A., Rubin, K. H., & Asendorpf. Communicative competence as a dimension of shyness, in *Social withdrawal, inhibition, and shyness in childhood*, New York: Psychology Press, 1993, pp. 189–212.
[31] A. Afshan, I. Askari, and L. S. S. Manickam. Shyness, Self-Construal, Extraversion-Introversion, Neuroticism, and Psychoticism: A Cross-Cultural Comparison Among College Students, *SAGE Open*, vol. 5, no. 2, 2015.
[32] R. R. McCrae and O. P. John. An Introduction to the Five-Factor Model and its applications, *J. Pers.*, vol. 60, pp. 175–215, 1992.
[33] P. Zimbardo, P. Pilkonis, and R. Noorwood. The silent prison of shyness, California, 1977.
[34] H. T. Holbrook. The quiet student in your classroom., *Lang. Arts*, vol. 64, no. 5, pp. 554–557, 1987.
[35] P. G. Friedman. Shyness and reticence in students, *Educational Resources Information Center (ERIC)*. National Education Association, Washington, DC, p. 49, 1980.
[36] W. . Kurniawan, Y; Abdullah, T; Zakaria, K; Wan Abdullah. Mengapa Pelajar Takut Bertanya dalam Kuliah?: Sebuah Kajian di Universiti Malaysia Kelantan (Why Students Are Afraid to Ask during Lectures?: A Study in University Malaysia Kelantan), *J. Pers. Pelajar*, vol. 15, pp. 13–22, 2012.
[37] J. Phillips, F. Sponsor, and B. L. Morgan. Please Don't Call on Me: Self-esteem, communication apprehension, and classroom participation, *J. Undergrad. Res. Univ. Wisconsin–LaCrosse 4*, pp. 81–87, 1998.
[38] R. stephen Briggs. Shyness: Introversion or Neuroticism?, *J. Res. Pers.*, vol. 307, pp. 290–307, 1988.
[39] J. Zhao, F. Kong, and Y. Wang. Self-esteem and humor style as mediators of the effects of shyness on loneliness among Chinese college students, *Pers. Individ. Dif.*, vol. 52, no. 6, pp. 686–690, 2012.
[40] S. D. Phillips and M. A. Bruch. Shyness and dysfunction in career development, *J. Couns. Psychol.*, vol. 35, no. 2, pp. 159–165, 1988.
[41] A. Caspi, G. H. Elder, and D. J. Bem. Moving against the world: Life-course patterns of explosive children., *Dev. Psychol.*, vol. 23, no. 2, pp. 308–313, 1987.
[42] R. J. Hamer and M. A Bruch. Personality factors and inhibited career development: Testing the unique contribution of shyness, *J. Vocat. Behav.*, vol. 50, no. 3, pp. 382–400, 1997.
[43] B. C. Myers. An inhibition from being shy: Shyness and its effects on career preferences of college students, *Am. J. Psychol. Res.*, vol. 1, no. 1, pp. 60–70, 2005.
[44] S. . Philips and M. Bruch. Shyness and dysfunction in career development, *J. Couns. Psychol.*, vol. 35, no. 2, p. 159, 1988.
[45] Q. Weng and J. C. McElroy. Vocational self-concept crystallization as a mediator of the relationship

IOP Conf. Series: Journal of Physics: Conf. Series **1049** (2018) 012057 doi:10.1088/1742-6596/1049/1/012057

between career self-management and job decision effectiveness, *J. Vocat. Behav.*, vol. 76, no. 2, pp. 234–243, 2010.

[46] J. M. Cheek and L. A. M. M. Carpentieri, Shyness and self-concept, in *Perception of Self in emotional disorder and psychotherapy*, 11th ed., Springer US, 1986, pp. 113–131.

[47] R. Bridgstock. The graduate attributes we've overlooked: Enhancing graduate employability through career management skills, *High. Educ. Res. Dev.*, vol. 28, no. 1, pp. 31–44, 2009.

[48] Z. Hanapi, A. Kamis, T. K. Tee, and M. H. Hanapi. Jurang integrasi kemahiran employabiliti di Malaysia : Satu kajian empirikal graduan kejuruteraan Kolej Komuniti Integrated employability skills gaps in Malaysia : An empirical study of Community College graduates, *Malaysian J. Soc. Sp. 12*, vol. 3, no. 3, pp. 145–153, 2016.

[49] R. Mustapha, M. Y. Husain, S. A. M. Syed Mohamad, and Seri Bunian Mokhtar. Persepsi Majikan Kejuruteraan Terhadap Tahap Kemahiran Empolyabiliti Pelajar Kejuruteraan Institusi Teknikal: Satu Kajian Kes, *J. Sci. Math. Technol.*, vol. 1, no. 2, pp. 41–55, 2014.

[50] M. Sasaki and M. Kono. Integrating drama performance and debate-style speech performance in English Language teaching: Paralleled collaborative teaching, 2007.

[51] M. M. Garrett, L. M. Schoener, and L. M. Hood. Debate: A teaching strategy to improve verbal communication and critical-thinking skills, *Nurse Educ.*, vol. 21, no. 4, pp. 37–40, 1996.

[52] P. Zare and M. Othman. Students' perceptions toward using classroom debate to develop critical thinking and oral communication ability, *Asian Soc. Sci.*, vol. 11, no. 9, pp. 158–170, 2015.

[53] P. T. Costa and R. McCrae. Normal personality assessment in clinical practice: The NEO Personality Inventory, *Psychol. Assess.*, vol. 4, no. 1, pp. 5–13, 1992.

[54] J. L. Holland. A personality inventory employing occupational titles., *J. Appl. Psychol.*, vol. 42, no. 5, pp. 336–342, 1958.

[55] C. J. Blomfield and B. L. Barber. Brief report: Performing on the stage, the field, or both? Australian adolescent extracurricular activity participation and self-concept, *J. Adolesc.*, vol. 32, no. 3, pp. 733–739, 2009.

[56] R. M. Page and E. P. Zarco. Shyness, physical activity, and sports team participation among Philippine high school students, *Child Study J.*, vol. 31, no. 3, p. 193, 2001.

[57] A. Rugarcia, R. M. Felder, D. R. Woods, and J. E. Stice. The Future of Engineering Education: Part 1. A Vision for a New Century., *Chem. Eng. Educ.*, vol. 34, no. 1, pp. 16–25, 2000.

[58] K. M. McNally and M. W. Harvey. Career and technical student organizations: A perfect path to self-determination and successful transition, *Prev. Sch. Fail.*, vol. 45, no. 3, pp. 114–118, 2001.

IOP Conf. Series: Journal of Physics: Conf. Series **1049** (2018) 012058 doi:10.1088/1742-6596/1049/1/012058

The Significant Effects of Communication Activities in the Co-Curricular Towards Reducing Shyness Amongst Elementary School Children

Masadliahani Masduki [1], **Normah Zakaria** [2], **Nur Nadiah Ismail** [3]

Sekolah Kebangsaan Seri Bali, Jalan Kg Jawa Baru, 85400 Chaah, Johor, Malaysia[1]
Faculty of Technical and Vocational Education, Universiti Tun Hussein Onn Malaysia, 86400 Parit Raja, Johor, Malaysia [2, 3]

E-mail: adliq@yahoo.com

Abstract. Shy children tend to have verbal communication problems and low self-esteem. However, the confidence and communication skills of these children could not be developed solely through classroom instructions. Ironically, the co-curricular program provides a good medium for communication activities that can improve their confidence and communication skills. Therefore, this study examined children's communication skills and communication activities during the co-curricular program at their school. The areas of study were the children's level of shyness, the implementation and the effectiveness of communication activities in the co-curricular program. The quantitative study adopted a non-experimental descriptive research design. The respondents were 357 Grade Six students from elementary schools in Segamat, Johor which was selected from a random sampling. Results showed that most children were having a medium level of shyness and the communication activities were moderately implemented in the co-curricular program at their schools. Almost all respondents viewed communication activities in the co-curricular program as very effective to reduce shyness, enhance their self-esteem and communication skills. Hence, communication skills of shy children can be enhanced through appropriate communication activities conducted during the co-curricular program. Children should be encouraged and trained to speak in public by participating in the debate, public speaking and theatre activities during the co-curricular to reduce their shyness.

1. Introduction

Shyness is a universal problem happens to everyone's life. In order to develop self-efficacy among children, issues like shyness and low self-esteem are major problems faced by them in their classrooms, especially if they were asked to express their personal opinions individually. They are silent when being asked, and would typically refuse to talk much in groups or class discussions. Even if they do speak out, their voices are not clear and they are unable to construct a complete set of sentence structures. At the same time, they would be demonstrating several signs of nervousness through their body gestures, for example, shaky hands, sweating and trembling and constant stuttering. Shyness is a personality trait that gives a negative self-evaluation towards the behaviour of an individual [1]. [2] defined shyness as the reaction during social interactions that cause individuals to

Content from this work may be used under the terms of the Creative Commons Attribution 3.0 licence. Any further distribution of this work must maintain attribution to the author(s) and the title of the work, journal citation and DOI.
Published under licence by IOP Publishing Ltd

experience discomfort, overthinking about something, that makes them feel extremely ashamed of their own behaviours. Shy students could be described as having low self-esteem with an anxious feeling in social situations [3] Shy or inactive students demonstrate poor participation, quiet, lack of group skills, and lack of confidence [4].

Shy children tend to feel lonely [5, 6] as they have lower social acceptance together with shortcomings in physical appearance not active in the classroom [7, 8] and also outside the classroom [8]. [9] holds the view that shyness is a negative trait that lowers one's self-esteem. [10] argued that the success of a classroom instruction largely depends on the activities of 'talking'. In this case, shy children are typically found to face difficulties in delivering good talk, and having a negative feeling about themselves; the main reason for them to avoid talking [11]. This opinion is similar to the study of [12] on shy students' participation in a seminar. The researcher found that the problem of shyness among the students was due to several reasons; (i) no idea to express their thinking and feelings, (ii) students feel nervous to speak up, and (iii) students are afraid if they accidentally utter something silly. This affects children's academic achievements since oral communication is pivotal in the classroom learning [7]. Furthermore, shyness affects the cognitive, affective, and behaviour of a person [13].

2. Literature Review

Previous research discovered that 13% to 15% children are shy. [10] identified 13.3% children as extremely shy with low self-esteem. These findings are almost similar to one finding by [14] who revealed that 14.4% children are shy. [1] found a prevalence of lifetime shyness in 13.3% from their studied population. With regards to the attitude of shy children, [15] argued that they tend to have (i) verbal communication problems, (ii) take a longer time to offer verbal comments, and (iii) avoid joining a conversation [16]. Shy children speak less in conversation because they have poor language skills with a limited amount of vocabularies [17]. This is a worrying scenario because if the children do not utilise the opportunity in such social interaction, their language abilities are at risk and affect their academic learning as well as social competencies [18]. The researchers argued that shyness and poor communication skills lower the self-esteem of an individual. In this case, self-esteem could only be developed through one's positive experience and enhanced by positive communication activities [11].

2.1. Shyness and communication skill

Communication skill is one of the employability skills in which students need to master during both of their school and university years [19]. However, previous studies found that most of the students today are shy and this affects their communication ability. Communication plays an important role towards achieving goals, and past studies found a close relationship between shyness and communication ability [20]. While most researchers proved that poor communication skills are the primary effect of children's shyness, some of them also found that poor language skills and fear to communicate might result in students' shyness [21]. Shy children prefer to just observe, rather than joining the activities in the classroom. When shy children are not active in the class, the learning process would become less effective and teachers face problems to achieve their teaching and learning objectives. Shy children often fail to answer questions, unable to share their thoughts and opinions, or simply ask questions to clarify things that they are not sure of.

However, the level of confidence and communication skills among these shy children could not only be developed through classroom instructions as there are certain negative factors influencing them to remain shy in the class [22]. If no action is taken to overcome this situation, they might not be performing well in their academic. Language, communication, and attention skills are perceived as the root of shyness in children whereby teachers able to observe shy behaviour in them through the way they interact in a formal classroom setting [17]. Lacking in communication skills causes negative self-perception and contributes to the shyness itself [23]. This means that shyness and communication skills are closely related, acting as a causal-effect factor. Because of this, shy children's social and emotional development could be well developed through communication activities.

[24] stressed that the behavourial problem and skill deficiency of an individual need to be treated. This means that teachers need to eliminate shyness in children at the first place before enhancing their communication skill. In contrast, [25] argued that shyness can be treated through communication activities since the activities may enhance children's self-esteem and in the same time improve their communication skill. This will then eliminate or at least decrease their level of shyness. Teachers are highly encouraged to integrate beneficial communication activities into a form of language tasks, especially for the purpose of teaching and learning of shy children [26]. Children should be supported through interactive educational activities, which could be further supported by a comprehensive set of co-curricular program, aimed at enhancing their self-esteem and socio-communicative skills [26, 27, 28]. This would gradually help children to progress further into a better state of self-reflection and self-acceptance. So, it is clear that shy children would have a low level of communication skill in which this problem can be treated through communication activities, especially during the co-curricular program.

2.2. Communication apprehension (CA)

Previous authors have reported that shyness is strongly related to communication apprehension [29] [30, 31, 32]. Communication apprehension (CA) is an experience of anxiety when individuals are communicating and have the feeling of fear in social situations [33]. McCroskey added that a high level of CA causes problems such as extreme anxiety and avoidance of direct communication whereby the negative thoughts could make a person feels restless, prevented him or her from speaking and tend to have low self-esteem [24]. Individuals with a high level of shyness will also have a higher level of CA. This makes them feel bad and nervous when communicating in front of others. Thus, shy children would also have the problem of communication apprehension that causes them to feel fear and nervous to communicate in front of others.

2.3. Communication activities

Besides developing a healthy relationship with other individuals that mainly help children to cope with their problem of shyness, the theories of emotion and self might also be created through the execution of educational language activities [26]. According to [34], shyness could be reduced through efforts taken to develop communication skills by having sufficient practices in educational language activities. Encouraging children to speak in certain situations helps them to be more open up in other situations. They would gradually realise and embrace the fact that there is nothing that they should be scared of [10]. Cooperative learning is a type of learning approach that encourages the spirit of togetherness or cooperation. This approach is essential in developing children's communication and social skills to overcome the shyness related problem. Cooperative learning could be achieved through the involvement in clubs and societies at the school. Cooperative learning, along with collaborative learning and interactions are three common pillars of communication activities [35]. It is also used as the platform to provide mutual support, enhance students' speaking abilities, improve their communication skills and foster students' motivation [36]. Cooperative learning is more structured and more directive whereby children learn how to work together in groups. This way, it will be easier for the children to eliminate their shyness, especially when they are getting the chances to interact and communicate with group members.

There are several types of cooperative learning that are suitable to be integrated into the co-curricular program such as drama, theatre, choral speaking, discussion, debate, choir and public speaking or speech presentation. The cooperation and support from other members are important for speech presentation [37, 38]. This will create an encouraging ambiance that stimulates shy children to speak confidently. Besides speech, debate is typically used by facilitators to improve students' verbal communication and critical thinking skills [39, 40], enhance teamwork skill and confidence level, and also adding their knowledge and overcoming their stage fright [40]. [41] recommended drama as an effective communication activity for shy students because it incorporates spontaneous verbal and non-verbal expressions which stimulate both cognitive and affective domains. Therefore, the drama could maximize the process of speaking and thinking. Through drama, students' motivation is able to be stimulated whereby it fosters self-awareness, self-esteem, and self-confidence. [42] examined the effectiveness of the Pyramid After-School Club in improving children's socio-emotional health and self-esteem. Children's confidence was developed through series of social and communication activities such as ice-breaking, morning circle, simulations, and cooperative games. Therefore, the activities might enhance children's communication skills and self-esteem

[43] argued that teachers play a vital role and responsibility to provide self-concept support, encouragement, and opportunities to develop shy children's confidence and improve their communication skills. In response to this situation, [44] stressed that communication activities and skills could be conducted and acquired everywhere at any time, without necessarily being classroom-bounded. Co-curricular program is therefore a good alternative to help improve students' language and communication skills. This program had proven to be successful in decreasing the level of negative emotions within children [45]. [11] in his study highlighted that the schools need to plan co-curricular activities, especially for shy and emotional students. This deems important to develop self-esteem. Interesting co-curricular activities might result in active children's participation and contribute to a high level of motivation, which then increases their self-confidence and improve self-esteem. Co-curricular activities are believed as the program that can eliminate shyness amongst students, due to the reason that the activities carried out during the program include various types of communication activities that help the students bring down the level of their communication apprehension and shyness. Keeping in view with the damaging impact of shyness, the present study was undertaken to;

a) Identify the level of shyness among year six students.
b) Identify the implementation's level of communication activities in the co-curricular program.
c) Identify the effectiveness of communication activities in the co-curricular program towards shy children.

3. Methodology

3.1. Sample
Respondents in the present study were 357 Grade Six students (156 boys, 201 girls). The sampling procedure included a random selection from elementary schools in Segamat, Johor. Since there is no debate, public speaking, choir or theatre club for elementary schools, these participants were involved in the clubs that have the highest chances for the implementation of communication activities during the co-curricular program. Thus, participants were involved in English Club (107), Malay Language Club (116) and Culture Club (134) at their respective school.

3.2. Measures
A questionnaire consists of 34 items had been developed and approved by four experts in terms of construct validity and face validity. Students were asked to rate the questionnaire based on 5-point Likert Scale (very low = 1, very high = 5). The questionnaire is divided into three part items; the level of shyness (10 items), the implementation of communication activities in the co-curricular program (14 items), and the program's effectiveness (10 items).

4. Results and Findings

4.1. Level of Children's Shyness
Level of shyness: Results indicated that 27.03% children are having a high level of shyness. 48.65% children are having the medium level of shyness and only 24.32% children reported the low level of shyness. Overall, children have a moderate level of shyness ($M = 3.49$, $SD = 0.76$). "*Shy during individual oral presentation*" showed the highest mean score items, whereas "*shy to get involved in group activities*" is the lowest mean score result. Table 1 below shows the level of children's shyness.

Table 1. The level of children's shyness

Level of Shyness	**Mean (SD)**
shy during individual oral presentation	3.89 (0.99)
shy to deliver ideas	3.59 (0.90)
speak in low voices	3.47 (0.89)
shy to answer when being asked	3.46 (0.87)
shy to give a try	3.46 (0.90)
shy to go to the front	3.46 (1.19)
shy to raise a hand	3.43 (0.93)
shy to ask a question	3.41 (1.07)
passive during discussions due to shy	3.38 (1.01)
shy to get involved in group activities	3.35 (0.86)
	3.49 (0.76)

4.2. Implementation of communication activities in the co-curricular program at schools

The result from the current study indicated a medium overall mean score *(M = 3.50, SD = 0.67)* for the implementation of communication activities in the co-curricular program at schools. Overall, students reported that among the communication activities conducted, communicative games are the most implemented activities at the schools during a co-curricular program (M = 4.24, SD = 0.72), compared to speeches (M = 2.92, SD = 0.95) which showed the lowest score result. Table 2 shows the implementation of communication activities in the co-curricular program at the respective schools.

Table 2. Communication activities in the co-curricular program

Communication activities	**Mean (SD)**
communicative games	4.24 (0.72)
singing	4.20 (0.66)
group activities	4.06 (0.48)
discussions	3.90 (0.62)
on stage activities	3.81 (0.62)
story telling	3.78 (0.83)
poems	3.50 (0.77)
debates	3.45 (0.42)
on stage competitions	3.27 (0.73)
involved all students, including the quiet ones	3.02 (0.40)
acting	2.97 (0.87)
simulation	2.95 (0.63)
presentation	2.93 (0.39)
speeches	2.92 (0.95)
	3.50 (0.67)

4.3. The effectiveness of communication activities in the co-curricular program

The main goal of this analysis was to explore whether communication activities in the co-curricular program bring good impacts to reduce shyness in children. Majority of the respondents found that communication activities in the co-curricular program are very effective in reducing their level of shyness. "*Raise a hand more frequently*" item shows the highest mean score among all items, followed by *"become braver to try"*. Item *"answer questions voluntarily"* shows the lowest mean score result. Table 3 shows the effectiveness of communication activities in the co-curricular program.

IOP Conf. Series: Journal of Physics: Conf. Series **1049** (2018) 012058 doi:10.1088/1742-6596/1049/1/012058

Table 3: The effectiveness of communication activities in the co-curricular program

Impact of communication activities in the co-curricular	Mean (SD)
raise a hand more frequently	4.22 (0.63)
become braver to try	4.16 (0.87)
become more active	3.95 (0.62)
become more confident	3.89 (0.57)
often ask questions	3.89 (0.77)
involved actively in group activities	3.86 (0.48)
able to do the individual oral presentation	3.86 (0.79)
easily mix with others	3.84 (0.60)
give more response to discussions	3.81 (0.62)
answer questions voluntarily	3.78 (0.63)
	3.93 (0.51)

5. Discussion

The result indicated that children experienced the highest level of shyness when they need to do oral presentation individually. This supports the theory from [24] who defined CA as having fear in a social situation, resisting individuals to orally communicate with others, especially in a classroom setting. This is due to the action of internalising certain unavoidable negative thoughts as well as the fear of making mistakes and being dishonoured. These have resulted to low self-esteem [46] holds the view that in any learning situation, the attitude of a student is related to the emotion that would determine his or her behaviour. Shyness at schools also demonstrates that children have a high level of *neuroticism* and low level of *extraversion* which cause them to become emotional, sad, inhibited, quiet, inactive, shy and refuse to participate in any discussions and classroom activities. Children were having the lowest level of shyness when they need to get involved in group activities. This is because they already knew their peers, so that they feel more comfortable, especially in a smaller group setting. This analysis supported by the study of [47] who found that students were having critical problems in a few situations such as when they need to deliver a speech or presentation and when they are in a large group, compared to a small group setting. Furthermore, students feel shy mostly when they are the focus of attention, whereby this is normally happens in a large group setting [47].

Next, findings from this research indicated that beneficial communication activities are not being widely implemented during the co-curricular program. According to [48], co-curricular activities are structured activities under adult supervision which focus on developing skills, have scheduled meetings, and have their own rules and expectations. In this case, each organisation or club and societies is shown to have its own specific activities and goals [48, 49] which do not include communication activities in specific. However, communicative games are found to be the most implemented activities in this study due to the activities do not take much time to be planned by the teachers. [50] argued that language activities are highly motivating and children naturally get their full satisfaction, pleasure, and excitement from communicative games. Communicative games are also meaningful, playful, fun, and able to increase children's interest and curiosity [50]. Besides that, shy children would not feel scared of making mistakes through communicative games as well as willing to volunteer and eager to talk more [45].

Meanwhile, it has been previously agreed that co-curricular program brings good impacts for children development. Results from this study do not differ from the previous research. Communication activities in clubs and societies provide formal and informal opportunities for children to develop their interests and talents, practice new skills, and get a sense of personal recognition or groups which could ultimately increase their confidence level and enhance their self-esteem [49]. Children tend to be more motivated to learn and this results in the increase of self-efficacy and participation, causing children to be more active both in class (for academic purpose) and during co-curricular activities [51]. These would automatically able to overcome their shyness. Children in this study found that communication activities in the co-curricular are less effective in motivating them to answer questions voluntarily. This might be due to the reason that when answering questions, a student will be the focus of attention. When this happens, the student will be evaluated by the teachers and friends. This makes them become shy and afraid that they will give wrong, inaccurate or stupid answers [52, 53, 12].

6. Conclusion

The level of shyness in children is at a medium level and the communication activities are not widely implemented during the co-curricular at schools. However, children found that communication activities in the co-curricular program benefit them in terms of increasing their self-esteem and reducing their shyness. Teachers should encourage all children, including the quiet ones to talk in front of others by planning various communication activities during the co-curricular program such as speech, debate, and drama. Finally, this research is among the first to explore the role of communication activities in the co-curricular program for shy school children. It is hoped that the findings will help school administrators in planning the appropriate communication activities during the cocurricular program. Future studies should consider a more specific communication activity in the co-curricular program, for instance, debate club in order to further explore this issue.

7. Acknowledgement

This research work is supported by the Centre for Graduate Studies, Universiti Tun Hussein Onn Malaysia.

References

[1] L. Henderson, P. Zimbardo, and B. Carducci. Shyness, *The Corsini Encyclopedia of Psychology*. John Wiley and Sons, Inc., 2010.
[2] A. Afshan, I. Askari, and L. S. S. Manickam, Shyness, self-construal, extraversion-introversion, neuroticism, and psychoticism: A cross-cultural comparison among college students, *SAGE Open*, vol. 5, no. 2, pp. 1–8, 2015.
[3] S. Koydemir and A. Demir. Shyness and cognitions: an examination of Turkish university students., *J. Psychol.*, vol. 142, no. 6, pp. 633–644, 2008.
[4] A. Masek. An appropriate technique of facilitation using students' participation level measurement in the PBL environment, *Int. J. Eng. Educ.*, vol. 32, no. 1, pp. 402–408, 2016.
[5] A. Arroyo and J. Harwood. Communication competence mediates the link between shyness and relational quality, *Pers. Individ. Dif.*, vol. 50, no. 2, pp. 264–267, 2011.
[6] B. A. S. Gökhan. An investigation of the relationship between shyness and loneliness levels of elementary students in a Turkish sample, *Educ. Sci.*, vol. 2, no. 2, pp. 419–440, 2010.
[7] B. J. Carducci, *The psychology of personality: Viewpoints, research, and applications*, 2nd ed. Malden, MA, 2009.
[8] R. J. Coplan, A. DeBow, B. H. Schneider, and A. A. Graham. The social behaviours of inhibited children in and out of preschool, *Br. J. Dev. Psychol.*, vol. 27, no. 4, pp. 891–905, Nov. 2009.
[9] P. A. Cowden. Communication and conflict: Social anxiety and learning, in *Proceedings of the Academy of Organizational Culture, Communications and Conflict*, 2009, 14, no. 2, pp. 16–18.
[10] S. Balda and K. Duhan. Social inhibition in preschool children : Causes and coping strategies, *J. Psychol.*, vol. 1, no. 2, pp. 79–83, 2010.
[11] A. Q. Chaudhry. Self esteem of shy and emotional students at elementary Level, *J. Educ. Pract.*, vol. 4, no. 28, pp. 121–125, 2013.
[12] W. R. Crozier and Crozier W.R. Shyness and students' perceptions of seminars, *Psychol. Learn. Teach.*, vol. 4, no. 1, pp. 27–34, 1985.

[13] L. J. Nelson, L. M. Padilla-Walker, S. Badger, C. M. Barry, J. S. Carroll, and S. D. Madsen. Associations between shyness and internalizing behaviors, externalizing behaviors, and relationships during emerging adulthood, *J. Youth Adolesc.*, vol. 37, no. 5, pp. 605–615, 2008.
[14] K. Rubin, L. Both, and M. Wilkinson. The Waterloo longitudinal project: Correlates and consequences of social withdrawal in childhood, *Infanc. y Aprendiz.*, vol. 13, no. 49, pp. 73–90, 1990.
[15] D. M. Buss. Evolutionary biology and personality psychology. Toward a conception of human nature and individual differences, *Am. Psychol.*, vol. 39, no. 10, pp. 1135–1147, 1984.
[16] W. R. Crozier. Shyness, self-perception, and reticence, in *Self-Perception: Interpersonal Perspectives on Individual Differences*, no. 2, Westport, Conneticut: Greenwood, 2001, pp. 53–76.
[17] K. M. Rudasill, A. Prokasky, X. Tu, S. Frohn, K. Sirota, and V. J. Molfese. Parent vs. teacher ratings of children's shyness as predictors of language and attention skills, *Learn. Individ. Differ.*, vol. 34, pp. 57–62, 2014.
[18] R. Wadman, K. Durkin, and G. Conti-Ramsden. Self-esteem, shyness, and sociability in adolescents with specific language impairment (SLI),"*J. Speech. Lang. Hear. Res.*, vol. 51, no. 4, pp. 938–952, 2008.
[19] E. Bhattacharyya. Communicative competence in technical oral presentation: Perspective of ESL educators and professional engineer, *Pertanika J. Soc. Sci. Humanit.*, vol. 22, no. March, pp. 1–16, 2014.
[20] K. Durkin. Language, shyness and social contexts: Commentary, *Infant Child Dev.*, vol. 18, no. 3, pp. 291–298, 2009.
[21] M. Byrne, B. Flood, and D. Shanahan. A qualitative exploration of oral communication apprehension, *Account. Educ.*, vol. 21, no. 6, pp. 565–581, 2012.
[22] H. Gazelle. Class climate moderates peer relations and emotional adjustment in children with an early history of anxious solitude: A child x environment model, *Dev. Psychol.*, vol. 42, no. 6, pp. 1179–1192, 2006.
[23] K. L. Bierman and W. Furman. The effects of social skills training and peer involvement on the social adjustment of preadolescents, *Child Dev.*, vol. 55, no. 1, pp. 151–162, 1984.
[24] J. C. McCroskey. Avoiding communication, *Shyness, reticence, Commun. apprehension*, pp. 13–38, 1984.
[25] S. L. Tominey and M. M. McClelland. Red light, purple light: Findings from a randomized trial using circle time games to improve behavioral self-regulation in preschool, *Early Educ. Dev.*, vol. 22, no. 3, pp. 489–519, 2011.
[26] S. L. Bosacki, R. J. Coplan, L. Rose-Krasnor, and K. Hughes. Elementary school teachers' reflections on shy children in the classroom, *Alberta J. Educ. Res.*, vol. 57, no. 3, pp. 273–287, 2011.
[27] R. J. Coplan and M. Armer. Talking Yourself Out of Being Shy: Shyness, Expressive Vocabulary, and Socioemotional Adjustment in Preschool, *Merrill-Palmer Quarterly*, vol. 51. pp. 20–41, 2005.
[28] R. J. Coplan and M. Weeks. Shy and soft-spoken: Shyness, pragmatic language, and socio-emotional adjustment in early childhood, *Infant Child Dev.*, vol. 18, no. 3, pp. 238–254, 2009.
[29] W. R. Crozier. Shyness and self-esteem in middle childhood, *Br. J. Educ. Psychol.*, vol. 65, pp. 85–95, 1995.
[30] R. L. Duran and L. Kelly. The cycle of shyness: A study of self-perceptions of communication performance, *Commun. Reports*, vol. 2, no. 1, pp. 30–38, 1989.
[31] V. Martin and M. C. Thomas. The Journal for Specialists in Group Work A model psychoeducation group for shy college students A Model Psychoeducation Group for Shy College Students, *J. Spec. Gr. Work*, vol. 25:1, no. December 2014, pp. 79–88, 2000.
[32] S. D. Phillips and M. A. Bruch. Shyness and dysfunction in career development, *J. Couns. Psychol.*, vol. 35, no. 2, pp. 159–165, 1988.

[33] McCroskey, Oral communication apprehension: A summary of recent theory and research, *Hum. Commun. Res.*, vol. 4, no. 1, pp. 78–96, 1977.

[34] J. B. Evans, M. A., Rubin, K. H., & Asendorpf. Communicative competence as a dimension of shyness, in *Social withdrawal, inhibition, and shyness in childhood*, New York: Psychology Press, 1993, pp. 189–212.

[35] R. Oxford. Cooperative learning, collaborative learning, and cnteraction: Three Communicative strands in the language classroom, *Mod. Lang. J.*, vol. 81, no. 4, pp. 443–456, 1997.

[36] M. H. Chou. The influence of learner strategies on oral presentations: A comparison between group and individual performance, *English Specif. Purp.*, vol. 30, no. 4, pp. 272–285, 2011.

[37] J. Hommes, B. Rienties, W. de Grave, G. Bos, L. Schuwirth, and A. Scherpbier. Visualising the invisible: A network approach to reveal the informal social side of student learning," *Adv. Heal. Sci. Educ.*, vol. 17, no. 5, pp. 743–757, 2012.

[38] Josefa J Mardijono. Humanistic strategies in the EFL speaking class, *kta*, vol. 3, no. 2, pp. 92–108, 2001.

[39] M. M. Garrett, L. M. Schoener, and L. M. Hood. Debate: A teaching strategy to improve verbal communication and critical-thinking skills, *Nurse Educ.*, vol. 21, no. 4, pp. 37–40, 1996.

[40] P. Zare and M. Othman. Students' perceptions toward using classroom debate to develop critical thinking and oral communication ability, *Asian Soc. Sci.*, vol. 11, no. 9, pp. 158–170, 2015.

[41] A. Maley and A. Duff. *Drama techniques in language teaching: A resource book of communication activities for language teachers*, 2nd ed. New York: Cambridge University Press, 1983.

[42] Á. E. McKenna, T. Cassidy, and M. Giles. Prospective evaluation of the pyramid plus psychosocial intervention for shy withdrawn children: An assessment of efficacy in 7- to 8-year-old school children in Northern Ireland, *Child Adolesc. Ment. Health*, vol. 19, no. 1, pp. 9–15, Feb. 2014.

[43] J. Brophy. ERIC Digest., *ERIC Clearinghouse on Elementary and Early Childhood Education Urbana IL.* 1996.

[44] S. Guo. Impact of an out-of-class activity on students' English awareness, vocabulary, and autonomy, *Lang. Educ. Asia*, vol. 2, no. 2, pp. 246–256, 2011.

[45] M. Xiao, L., & Luo. English co-curricular activities: A gateway to developing autonomous learners, in *CamTESOL Conference on English Language Teaching*, 2009, vol. 5, no. Selected papers, pp. 239–251.

[46] M. Ghaddafi. Attitude towards Bahasa Melayu among the speakers in the east coast of Peninsular Malaysia, *Pertanika J. Soc. Sci. Humanit.*, vol. 24, no. S, pp. 167–180, 2016.

[47] P. G. Zimbardo, "Shyness: What is it and what to do about it., *MA Addison-Wesley Publ. Company.*, 1977.

[48] J. L. Mahoney, R. W. Larson, and J. S. Eccles. *Organized activities as contexts of development: Extracurricular activities, after-school and community programs*, vol. 21. New Jersey: Lawrence Erlbaum Associates, Inc., Publishers, 2005.

[49] J. Roth, J. L., & Brooks-Gunn. What exactly is a youth development program? Answers from research and practice, *Appl. Dev. Sci.*, vol. 7, no. 2, pp. 94–111, 2003.

[50] Y.-H. Wang. Using communicative language games in teaching and learning English in Taiwanese primary schools, *J. Eng. Technol. Educ.*, vol. 7, no. 1, pp. 126–142, 2010.

[51] P. Akos. Extracurricular participation and the transition to middle school, *Res. Middle Lev. Educ.*, vol. 29, no. 9, pp. 1–9, 2006.

[52] M. R. Leary. The Conceptual Distinctions Are Important: Another Look at Communication Apprehension and Related Constructs., *Hum. Commun. Res.*, vol. 10, no. 2, pp. 305–312, 1983.

[53] S. Vahedi. The factor structure of the revised cheek and buss shyness scale in an undergraduate university sample, *Iran. J. Psychiatry*, vol. 6, no. 1, pp. 19–24, 2011.

IOP Conf. Series: Journal of Physics: Conf. Series **1049** (2018) 012059 doi:10.1088/1742-6596/1049/1/012059

A Low Power CMOS Phase Frequency Detector in High Frequency PLL System

Fadhilah Binti Noor Al Amin[2,*], Nabihah Ahmad[1,2] and M. Hairol Jabbar[1,2]

[1] IC Design Research Group
[2] Faculty of Electrical and Electronic Engineering,
Universiti Tun Hussein Onn Malaysia,
84600 Batu Pahat, Parit Raja, Johor

*E-mail: fadhilahamin90@gmail.com

Abstract. The design of low power consumption for the different application and electronic product has become one of the challenges in high performance of Very Large Scale Integration (VLSI) design today. Therefore, many techniques have been introduced to minimize the power consumption and one of the technique is by using Multi-threshold Complementary Oxide Semiconductor (MTCMOS) power gating technique. This paper aimed to design a High Speed Phase Frequency Detector (HSPFD) using MTCMOS power gating technique implemented in 130 nm CMOS technology by using Cadence Virtuoso Tool. This design achieved a power consumption of 0.116 μW and frequency of 1.26 GHz. The result shows that the power consumption of HSPFD with MTCMOS power gating technique achieved about 67.14% smaller than conventional HSPFD while the frequency improved about 26%. The total area of the proposed HSPFD is 1423 μm^2 (61.1100 μm x 23.290 μm). It can be conclude that, this design has better performance compared to previous work and it is suitable for applications like wireless communication system that need low power blocks to have long life battery.

1. Introduction

The rapid growth of electronic system has contributed to the use of Phase Locked Loops (PLL) as one of the inevitable necessities in modern days. It has been used as a part of digital system and communication system like wireless and wire-line [1]. PLL is widely used in many application such as clock generation and clock recovery in microprocessor, networking, communication system and frequency synthesizers. PLL basically consists of Phase Frequency Detector (PFD), Charge Pump (CP) and Loop Pass Filter (LPF) and Voltage Control Oscillator (VCO). Figure 1 shows the block diagram of PLL system where in the PFD, the phase of feedback signal from VCO with the input reference signal is compared and then it generate up or down output according to the phase difference. While for CP, it charges or discharges the capacitor of the low pass filter according to up or down signal from PFD and the VCO will increases or decreases the output frequency according to the control voltage that produced by the CP.

Content from this work may be used under the terms of the Creative Commons Attribution 3.0 licence. Any further distribution of this work must maintain attribution to the author(s) and the title of the work, journal citation and DOI.
Published under licence by IOP Publishing Ltd

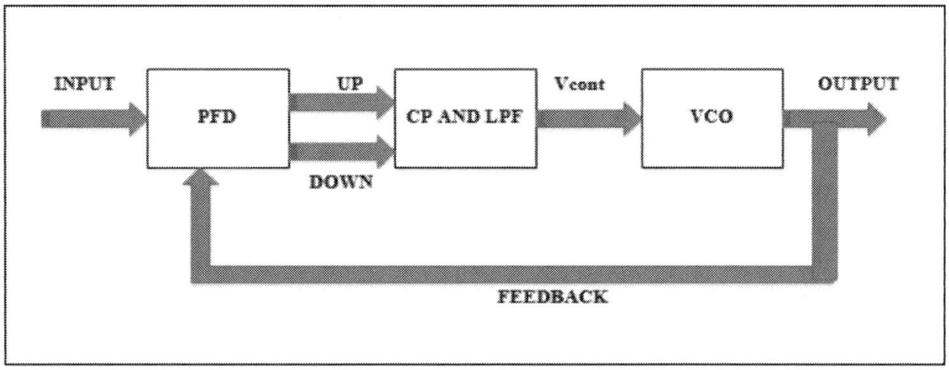

Figure 1. Block diagram of PLL system [1].

Complementary Metal Oxide Semiconductor (CMOS) technology offer a PFD design in smaller scale design with low power consumption and low area which able to perform in higher frequency with a lower power consumption. This paper focus on the Phase Frequency Detector (PFD) where it is one of the important block in PLL system. The PFD have two inputs that are clock reference (CLK_{REF}) and clock from voltage control oscillator (CLK_{VCO}) where the first input is from the reference clock and second input is from the feedback signal that coming from VCO. PFD also have two outputs known as UP and DOWN signal according to the phase and frequency difference of the input signals. The result of the PFD output were connected to the charge pump to generate the related control signal for VCO. Figure 2 shows a block diagram of PFD with its inputs and outputs.

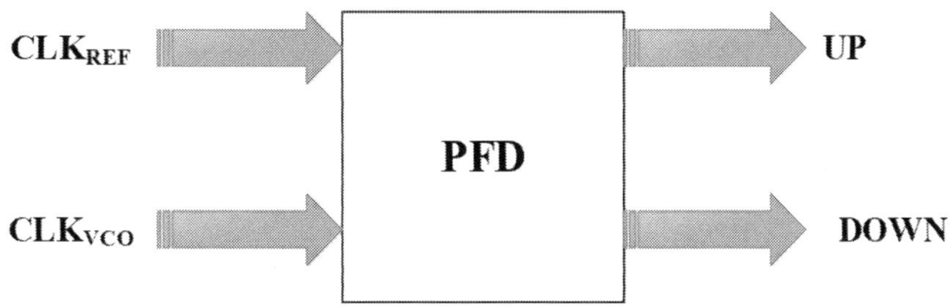

Figure 2. Phase Frequency Detector [1].

This paper is divided into four sections and organized as follow. In Section 2, the HSPFD MTCMOS with power gating architecture is discussed. In Section 3, the simulation results and discussion for the findings is presented. Finally, the conclusion is in Section 4.

IOP Conf. Series: Journal of Physics: Conf. Series **1049** (2018) 012059 doi:10.1088/1742-6596/1049/1/012059

2. HSPFD MTCMOS with power gating architecture

The conventional HSPFD architecture consists of D Flip-flop, two inverter and NAND gate are used to reduce the power consumption and achieved the high speed by detecting the rising and falling edges of the input signals. The proposed HSPFD is using the MTCMOS power gating technique where two sleep transistors are applied between the supply voltage and ground line known as power gating. The MTCMOS technique provides a simple and effective power gating structure by utilizing high speed, low threshold voltage (L_{Vt}) transistors for logic cells and low leakage, high threshold voltage (H_{Vt}) devices as sleep transistors. Sleep transistors disconnect logic cells from the supply and/or ground to reduce the leakage in the standby mode [20]. This technique is based on disconnecting the L_{Vt} logic gates from the power supply and the ground line via cut-off H_{Vt} sleep transistor. The transistor that having low threshold voltage are used to isolate the low threshold voltage transistors from supply and ground during standby (sleep) mode to prevent leakage dissipation. While in the active mode, sleep transistor are turned on and the logic that consists of L_{Vt} transistors can operate with high speed and low switching power dissipation. When the circuit is in sleep mode, the H_{Vt} transistors are turned off causing isolation of L_{Vt} transistors from supply voltage and ground and therefore it reduced the sub-threshold leakage currents [12]. Figure 3 shows the HSPFD MTCMOS with power gating technique.

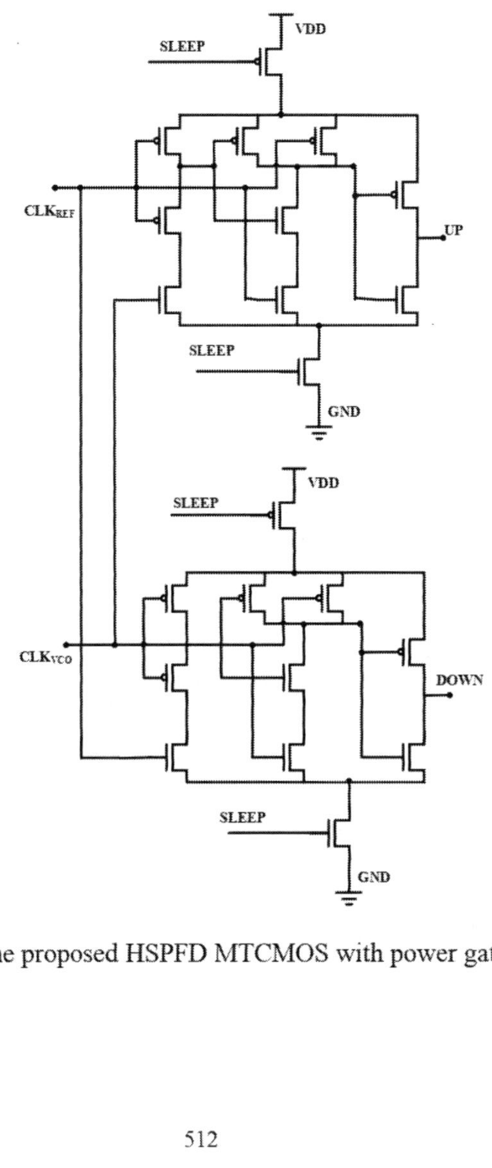

Figure 3. The proposed HSPFD MTCMOS with power gating technique.

IOP Conf. Series: Journal of Physics: Conf. Series **1049** (2018) 012059 doi:10.1088/1742-6596/1049/1/012059

3. Result and Discussion

The simulation of HSPFD MTCMOS with power gating technique circuit was carried out in Cadence Virtuoso with supply voltage of 1.2 V and load capacitance of 1 pF.

The power consumption of the proposed HSPFD circuit is 0.116 μW with the frequency of 1.26 GHz. The peak-to-peak high and bottom peak were measured in these three condition that are in lock condition, when CLK leading condition and when CLK_{REF} leading condition. In Figure 4(a), during lock condition, the value of peak-to-peak high and bottom peak measured are same that is 1.193 V. During CLK_{REF} leading condition in Figure 4(b), the value of peak-to-peak high and bottom peak of UP output is slightly higher compared to DOWN output which is 1.229 V and 1.205 V while peak-to-peak high for DOWN output is 1.212 V. In CLK leading condition shown in Figure 4(c), the result of peak-to-peak high and bottom peak measured is vice versa from the CLK_{REF} condition where the DOWN output has higher value of peak-to-peak high and bottom peak that is 1.229 V and 1.205 V while the UP output has the value peak-to-peak high measured is 1.212 V. This is happen due to the active and inactive mode of sleep transistor that has been explained in section 2. The layout of proposed circuit is 1423μm^2 (61.1100 μm x 23.290 μm) is shown in Figure 5.

Table 1 tabulated the comparison of different low power technique together with the proposed design done that is HSPFD MTCMOS with power gating technique in this project. Based on the table, the proposed design measured the least power consumed in the HSPFD among other design. About 67.14% of power consumption of the HSPFD with MTCMOS using power gating technique is improved compared to the conventional HSPFD technique [17]. The frequency also improved about 26% compared to the conventional HSPFD technique [17] that achieved about 1 GHz only. This also proves that with smaller CMOS technology scale, lower power consumption can be achieved.

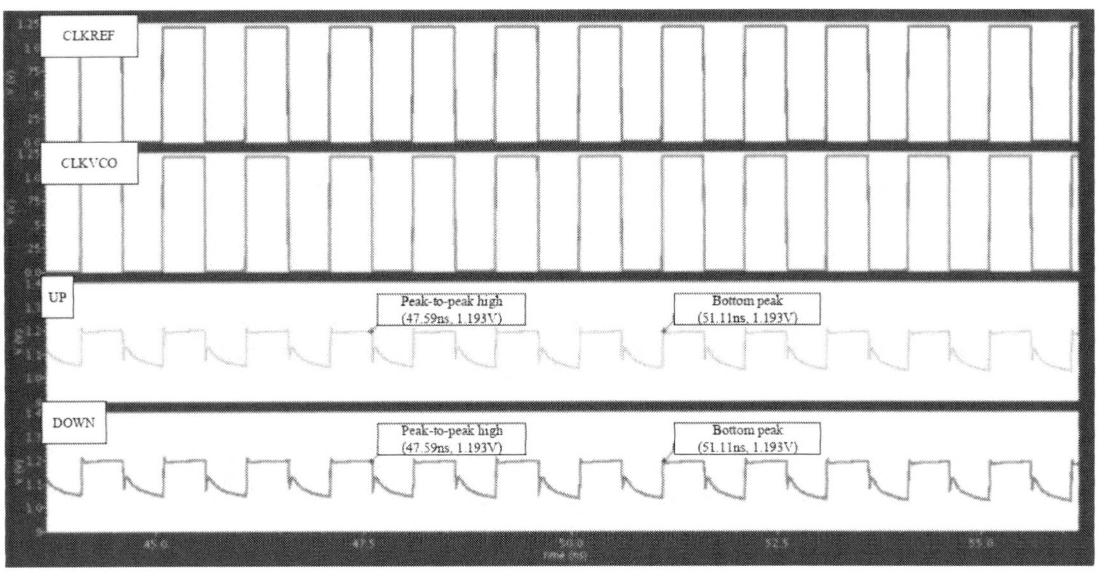

Figure 4 (a). HSPFD MTCMOS with power gating waveform at lock condition.

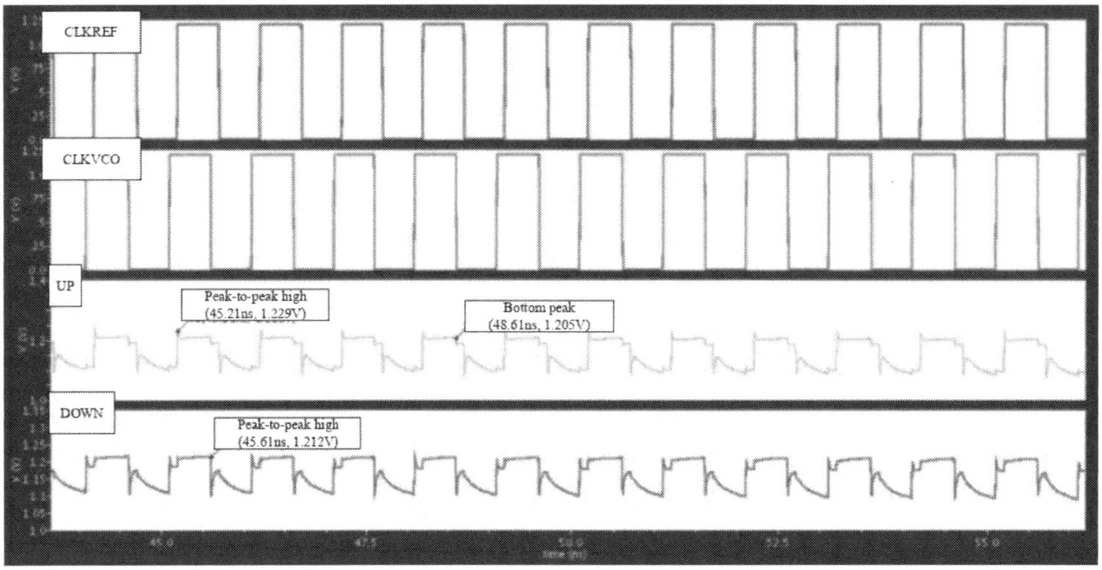

Figure 4 (b). HSPFD MTCMOS with power gating waveform at CLK_{REF} leading condition.

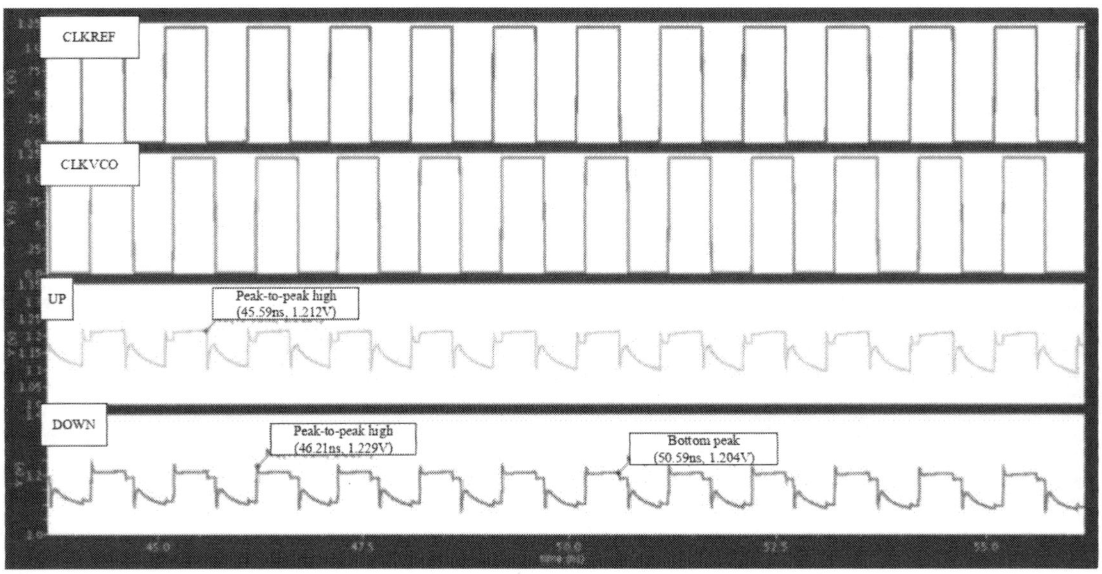

Figure 4 (c). HSPFD MTCMOS with power gating waveform at CLK_{VCO} leading condition.

Table 1. Summarization of proposed HSPFD with previous HSPFD technique.

Authors	**Techniques**	**Technology (nm)**	**Power (μW)**	**Frequency (GHz)**
[17]	Conventional HSPFD	0.13	0.353	1
[16]	HSPFD	0.18	8.513	1
[18]	Conventional HSPFD	0.35	0.39	1
[14]	Conventional PFD	0.18	33.5	0.5
Proposed Design	HSPFD MTCMOS	0.13	0.116	1.26

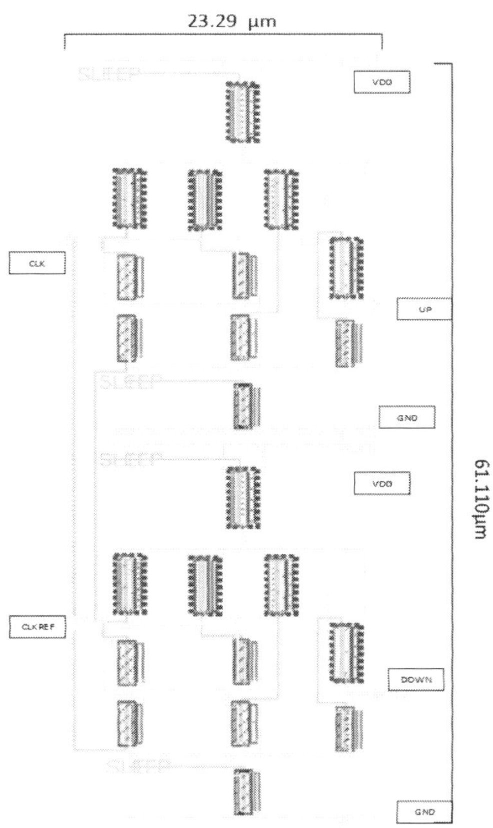

Figure 5. Layout of HSPFD MTCMOS with power gating technique.

4. Conclusion

Minimization of low power consumption is essential for high performance of VLSI systems. A HSPFD MTCMOS with power gating technique has been described in this paper where it is proven that the design was able to get low power consumption of 0.116 μW and high frequency of 1.26 GHz that implemented in 130nm CMOS technology. Therefore, this is proved that the MTCMOS with power gating technique was able to improve overall performance in PFD design.

Acknowledgement

This work was financial supported by FRGS Grant Vot Number 1538

Reference

[1] Anshul Agrawal and Rajesh Khatri 2015 Design of Low Power, High Gain PLL using CS- VCO on 180nm Technology *Int. Journal of Computer Applications* **122** 18

[2] Luiz C. Moreira, WAM Van Noike and C. Dionisio 2009 A pulse generator UWB ultra wide band using PFD phase frequency detector in 180 nm CMOS technology *SBMO/IEEE MTT-S Int. Microwave and Optoelectronic Conference Proceedings p* 239-43

[3] Kashyap K Patel and Nilesh D Patel 2013 Phase frequency detector and charge pump for DPLL using 0.18 μm CMOS Technology *Int. Journal of Emerging Technology and Advanced Engineering (ISSN 2250-2459, ISO 9001: 2008 Certified Journal* 3.

[4] Belorkar, Ujwala A. and SA Ladhake 2010 Design of low power phase locked loop (PLL) using 45nm VLSI Technology *International journal of VLSI design & Communication Systems (LSICS)* **1** 2

[5] Kim, Kyung-Ki, and Yong-Bin Kim 2007 A 32nm and 0.9 V CMOS phase-locked loop with leakage current and power supply noise compensation *JSTS: Journal of Semiconductor Technology and Science* **7** 1

[6] Minhad, K. N., M. B. I Reaz, and J. Jalil 2014 A low power 0.18-μm CMOS phase frequency detector for high speed PLL *Elektronika ir Elektrotechnika* **20** 9

[7] Thakore, Kruti P., Harikrishna C. Parmar, and N. M. Devashrayee 2011 Low power and low jitter phase frequency detector for phase lock loop *Int. Journal of Engineering Science and Technology* **3** 3

[8] Sampath, Parthasarathy, and Vikas Choudhary 2011 High speed phase frequency detector *U.S. Patent No. 7,940,088.*

[9] Gupta, Bipin, and Sangeeta Nakhate 2012 Transistor gating: A technique for leakage power reduction in CMOS circuits *International journal of emerging technology and advanced Engineering* **2.4** 321

[10] Zhou, Qiang, et al. 2009 An MTCMOS technology for low-power physical design. *Integration, the VLSI Journal* **42.3** 340

[11] W. Bo, Z. Jun and T. Kim Tony 2012 Maximization energy efficiency utilizing MTCMOS technology *IEEE 4th Asia Symposium on Quality Electronic Design* p 35.

[12] Nirmal, Uma, Geetanjali Sharma, and Yogesh Misra 2011 A low power high speed adders using MTCMOS Technique *IJCEM Int. Journal of Computational Engineering & Management* **13**.

[13] Tonk, Anu 2015 A comparative review and analysis of different phase frequency detectors for phase locked loops *International Journal of Engineering Research and General Science* **3** 6

[14] Supraja Batchu, Jayachandra Prasad Talari and Ravi Nirlakalla 2015 Analysis of low power and

high speed phase frequency detectors for phase locked loop design *Procedia Computer Science* p 57

[15] Nandini Raj, and Himadri Singh Raghav 2013 Comparison of low power phase frequency detectors for delay locked loop *International Journal of Engineering Research and Technology* **2.1** ESRSA Publications.

[16] Arshak, K., O. Abubaker, and E. Jafer 2004 Design and simulation difference types CMOS phase frequency detector for high speed and low jitter PLL *Devices, Circuits and Systems, Proc. of the Fifth IEEE International Caracas Conference on.* **1**.

[17] Rashmi, S. B., and Siva S. Yellampalli. 2012 Design of phase frequency detector and charge pump for high frequency PLL *Int. Journal of Soft Computing and Engineering* **2** 88

[18] Thakore, Kruti P., Harikrishna C. Parmar, and N. M. Devashrayee 2011 Low power and low jitter phase frequency detector for phase lock loop *Int. Journal of Engineering Science and Technology* **3** 3

[19] Rajani, H. P., and Srimannarayan Kulkarni 2012 Novel sleep transistor techniques for low leakage power peripheral circuits *Int. Journal of VLSI Design & Communication Systems* **3.4** 81.

[20] S. Kim, S.V. Kosonocky, Stephen, and D.R. Knebel 2003 Understanding and minimizing ground bounce during mode transition of power gating structures *Proc. Int'l Symp. Low Power Electronics and Design* p 22

IOP Conf. Series: Journal of Physics: Conf. Series **1049** (2018) 012060 doi:10.1088/1742-6596/1049/1/012060

Charge Pump and Loop Filter for Low Power PLL Using 130nm CMOS Technology

Nabihah Ahmad[1,2*], Nur Atikah Binti Ishaimi[2], M. Hairol Jabbar[1,2]

[1] IC Design Research Group,
[2] Faculty of Electrical and Electronic Engineering,
Universiti Tun Hussein Onn Malaysia,
86400 Parit Raja, Batu Pahat, Johor, Malaysia.

*Email: nabihah@uthm.edu.my

Abstract. A tri-state charge pump circuit and second order low pass filter circuit were designed to be used in Phase Lock Loop (PLL) system. The proposed design reduces the non-ideal effects such as a current mismatch and charge sharing. Therefore, it can be minimized by providing an equal value for the two switches UP and DOWN. While the charge pump output determines the output condition of the low pass filter. The proposed design have been simulated by using 130nm Complementary Metal Oxide Semiconductor (CMOS) technology in Cadence Tools. The simulation also includes the parameters for tri-state charge pump and second order low pass filter using voltage supply of 1.2 V. The power consumption of the design is 2.07 mW with the output voltage swing from 288 mV to 413.8 mV. The frequency achieved from the proposed design is 4.7 GHz. The total area of the layout that have been measured is 31.4 μm x 22.6 μm (0.7096 mm^2). Thus, the proposed design able to achieve the scope of low power consumption and high frequency in smaller technology.

1. Introduction

Phase Lock Loop (PLL) is a system that consists of three major parts; Phase Frequency Detector (PFD), Charge Pump and Loop Filter, and Voltage Controlled Oscillator (VCO). A PLL is highly preferred because it is a feedback system that compares the output frequency from the input frequency and can survive in a single chip. A PLL is normally used in well-timed clock generator, recovery of signal from noisy communication channel and high performance wireless with additional application in PLL's parts [1]. Figure 1 shows the basic block diagram for a PLL system.

The charge pump circuit is connected with loop filter and located within PFD and VCO. Charge pump is functioning as a converter for the logic states of the PFD into an analog signal in order to control the VCO. The frequency of the VCO is controlled by the output signal of the charge pump circuit. The output voltage of the charge pump circuit must be held at a constant voltage, when PLL locks in some frequency. The charge pump consists of two switched current source that pump charge in or out of the loop filter according to two logical inputs.

Loop Filter circuit is also important to the performance of the PLL because it removes high frequency noise of the detector, influences the hold and captures ranges, and influences the switching speed of the loop in lock [2]. Loop filter convert the output signal of phase frequency detector to control voltage and reduce the ripple from the charge pump. A charge pump circuit along with low pass filter is used to minimize the disturbances at the input of current-starved voltage controlled oscillator (CS-VCO) and to get a sharper and smoother signal at the CS-VCO output.

Content from this work may be used under the terms of the Creative Commons Attribution 3.0 licence. Any further distribution of this work must maintain attribution to the author(s) and the title of the work, journal citation and DOI.
Published under licence by IOP Publishing Ltd

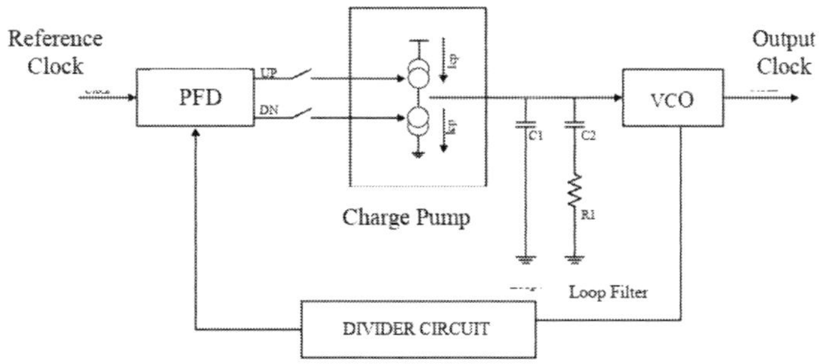

Figure 1. PLL Block Diagram

2. Charge Pump Circuit

A charge pump circuit consists of two symmetrical current sources that is controlled by the PFD output, and it will convert the sequential logic state of PFD into analog signals that are suitable to control VCO [5]. It also carries the phase error between the reference and the divided signal into analog quantities or current pulses. There are several types of charge pump, for example, conventional charge pump, current steering, different input with single-ended output and high voltage charge pump.

From previous studies, most of the charge pump circuit that have been used are basic charge pump with two switches, ideal charge pump, Dickson charge pump and so on. Figure 2 shows an ideal charge pump that gives the output frequency of 1.5 GHz and input voltage of 90 mV. Table 1 shows the operation of ideal charge pump.

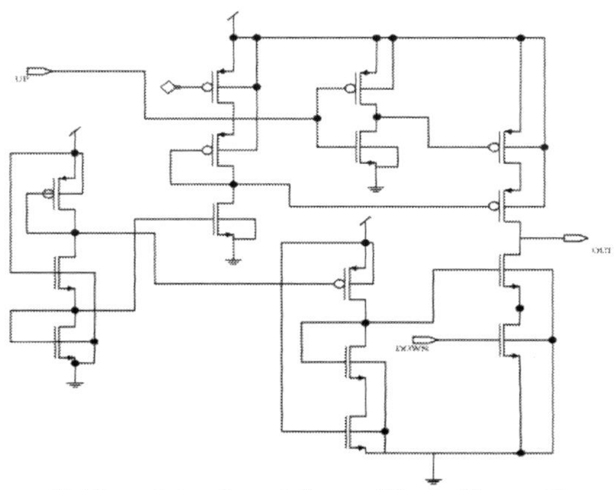

Figure 2. Transistor Level for an Ideal Charge Pump [1]

Table 1. Operation of Ideal Charge Pump [1]

UP Signal	DN Signal	Condition	Note
1	0	Charging	I_{cp} flows into filter
0	1	Discharging	I_{cp} flows out from filter
0	0	Vout constant	$I_{cp} = 0$
1	1	Vout constant	I_{cp} = not 0

Charge pump circuit used in [6] is a Dickson charge pump that causes the output voltage to be higher than the input voltage as shown in Figure 3. It is consists of capacitors and MOS switches. The two-stage charge pump is used as a step-up converter to prevent problems at the threshold voltage. By using 3.3 V input voltage, this charge pump produces 6.3 V of output voltage and more than 100 mA of load current.

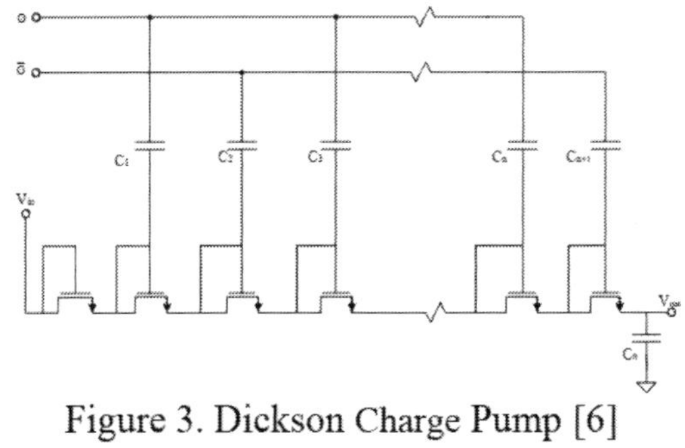

Figure 3. Dickson Charge Pump [6]

Tri-state charge pump also have been used in PLL system. There are a few architectures for charge pump. For example, a conventional tri-state that consist of three topologies of switch in source, gate and drain. This type of charge pump consumes lower current depending on the frequency of the PFD. Figure 4 shows the transistor level for conventional tri-state charge pump with the three topologies. After one of the topologies have been properly biasing, the output voltage gained in the range of 0.5 V to 2.5 V with 3 V supply.

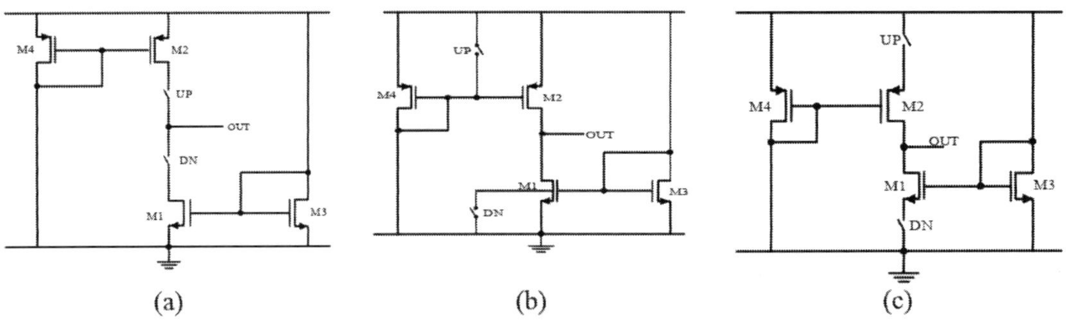

Figure 4. Conventional tri-state charge pump with different topologies. (a) Switch in drain (b) Switch in Gate (c) Switch in Source.

3. Loop Filter Circuit

Loop filter needs to be chosen properly with the correct value to prevent inappropriate values that may either lead the loop to oscillate for long without reaching the locked state or small variations in the input data may cause the loop to unlock. From the previous studied, there are several types of loop filters hab been designed. For example, passive and active low pass filter [1], inductor-capacitor (LC) filter [10], first order passive lag filter [4], simple resistor and capacitor (RC) low pass filter [6]. Figure 5 shows the second order loop filter while Figure 6 shows the lag filter from the previous studies.

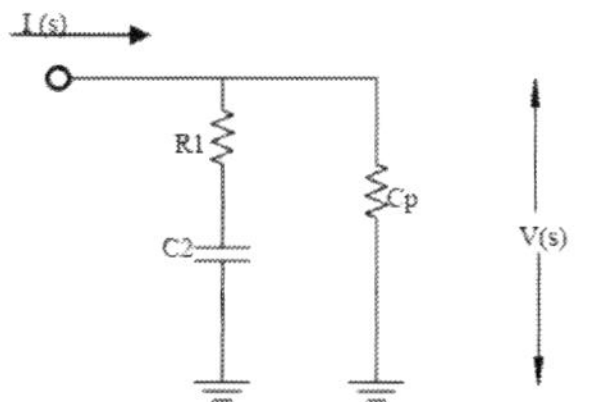

Figure 5. Second Order Loop Filter [1]

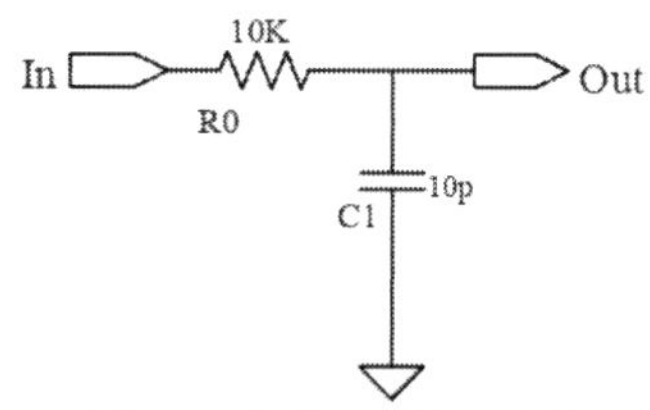

Figure 6. Lag Filter [4]

4. Proposed Design Architecture

The Charge Pump in a PLL design is constructed in an integrated circuit technology consisting of pull-up and pull-down transistor and on-chip capacitor. The output from the PFD provides UP and DOWN gating signals that work as switches so that when UP is ON then DOWN is OFF. It only works on either way. Low Pass Filter is a filter that builds with a resistor and capacitor that pass through the low-frequency signals and also blocking high frequency signals. The capacitor is a reactive device that offers very high resistance to low frequency signal and vice versa. Since the current always takes the path with lowest resistance, the capacitor that represents a low resistance in a circuit for high frequency signal will build its path. Figure 7 shows the block diagram for charge pump and loop filter.

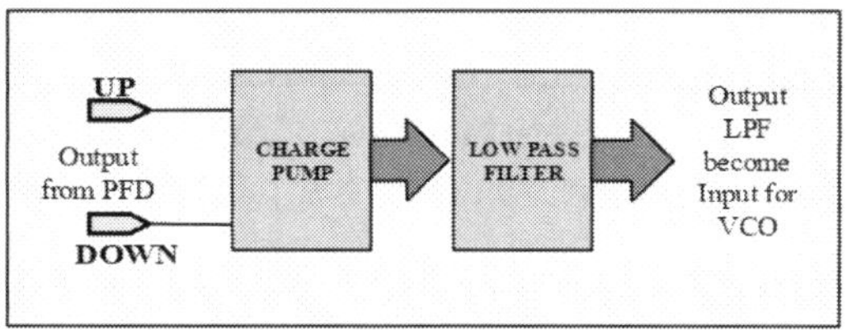

Figure 7. Block Diagram for Charge Pump and Loop Filter.

4.1 Tri-state Charge Pump Circuit

Tri-state charge pump are commonly known as the low current consumption depends on the frequency of the PFD. This tri-state charge pump consists of one inverter, one amplifier and a total of five transistors. When the input is low, the current pump into filter also low. Thus, an amplifier was added as shown in Figure 8 where the amplifier was placed in between switch UP and the inverter to increase the amplitude of the voltage and the current signal of the circuit. The equal values for switch UP and DOWN are provided to prevent the current mismatch for the proposed charge pump circuit. For this 130 nm CMOS technology, this circuit is using 1.2 V supply voltage. The delay given to switch UP and DOWN is 100 ps and 500 ps respectively as shown in Table 2.

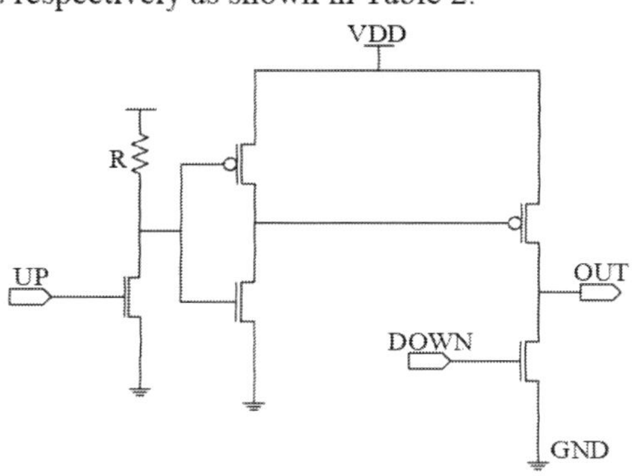

Figure 8. Transistor level for proposed tri-state charge pump.

Table 2. Properties for Switch UP and DOWN

	UP	DOWN
DC Voltage (V)	1.0	1.0
Voltage 1 (V)	200m	200m
Voltage 2 (V)	1.0	1.0
Delay (s)	150p	150p
Rise Time (s)	10p	10p
Fall Time (s)	10p	10p
Pulse Width (s)	5n	5n
Period (s)	1n	1n

4.2 Second Order Low Pass Filter Circuit

Loop Filter for a PLL system needs to be chosen wisely to prevent the locked loop to be unlocks due to the inappropriate value and small variations in the data input [5]. The main function of the loop filter used, second order low pass filter, which converts the current generate from the charge pump into a controlled voltage signal for controlling the VCO after it filters out the noise. Loop filter is used to overcome the noise and high frequency signal component from the charge pump and stabilize the loop. Figure 9 shows the schematic circuit for second order low pass filter.

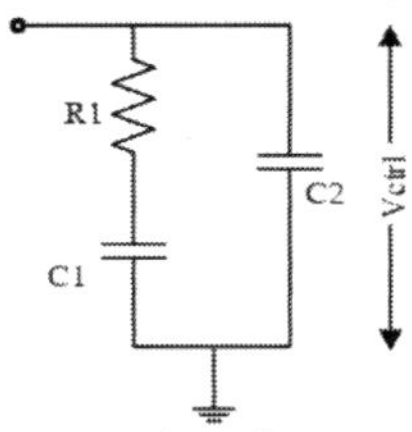

Figure 9. Second Order Low Pass Filter

By referring to the method introduced by Ken Holladay [9], the parameter values for the loop filter must be defined in order to design the basic requirement. For this PLL design, the frequency range is 3.1 GHz to 5.1 GHz and the VCO sensitivity, K_{VCO} are assumed at 10 kHz. The charge pump current, I_{CP} is at 5 μA while the PLL loop bandwidth, BP_{PLL} is 100 kHz.

$$F_{step} = F_{osc\ max} - F_{osc\ min} \tag{1}$$

$$N = \frac{F_{osc\ max}}{F_{step}} \tag{2}$$

The frequency step, F_{step}, is obtained by subtracting the maximum oscillator frequency and minimum oscillator frequency and the value of N is obtained when the $F_{osc\ max}$ is divided by F_{step}. The F_{step} obtained from the calculation is 2 GHz and the value of N is 2.55. Considering the damping factor, ξ, is typically at 0.707 therefore equation (3) will calculate the natural frequency, F_N, that is equal to 30.01 kHz.

$$F_N = \frac{2 \times BP_{PLL}}{2\pi \times \left(\xi + \frac{1}{4\times\xi}\right)} \tag{3}$$

From Equation (4), (5) and (6), the value of the capacitor, C_1 is defined by calculating the value of C_2 and R_1. Thus, the value of C_2 and R_1 is 551.49fF and 19.35kΩ respectively. The value of C_1 is equal to 55.1 fF.

$$C_2 = \frac{I_{CP} \times K_{VCO}}{N \times (2\pi \times F_N)^2} \tag{4}$$

$$R_1 = 2 \times \xi \times \sqrt{\frac{N}{I_{CP} \times K_{VCO} \times C_2}} \tag{5}$$

$$C_1 = \frac{C_2}{10} \tag{6}$$

Equation (7) is the equation to calculate, T_s, the time desired for the filter to step to the new frequency. The frequency of the carrier within the desired time after a step, F_a, is normally in 1000 Hz. Thus, the value of T_s is equal to 54.42 μs or 50 μs.

$$T_S = \frac{-1 \times \left(\ln \frac{F_a}{F_{step}}\right)}{F_N \times 2\pi \times 2\xi} \tag{7}$$

From the value of R1, C1 and C2 calculated above, the frequency cutoff, f_c, for this proposed design are calculated by using Equation (8). Thus, frequency cutoff is 6.53 GHz.

$$f_c = \frac{1}{2\pi\sqrt{R_1 C_1 C_2}} \tag{8}$$

5. Results and Discussion

This proposed design of charge pump and loop filter was designed in CMOS 130 nm technology and operates at 1.2V power supply. Figure 10 shows the combined circuit of the tri-state charge pump and second order low pass filter. The charge pump output determined the condition of the capacitor in low pass filter is charged or discharged.

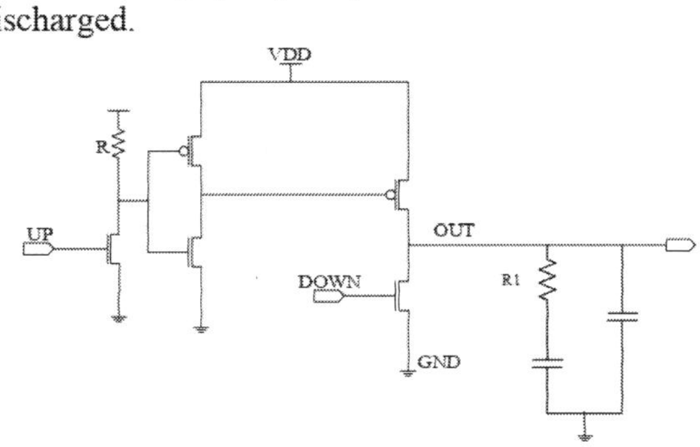

Figure 10. Combined Circuit of Charge Pump and Loop Filter

The result gained from the tri-state charge pump and second order low pass filter has been plotted as shown in Figure 11. The waveform plotted is almost in ideal behavior with a glitch or noise in it. From the waveform, by using the Result Browser in WaveView in Cadence Tool, the power consumption was

IOP Conf. Series: Journal of Physics: Conf. Series **1049** (2018) 012060 doi:10.1088/1742-6596/1049/1/012060

plotted at 2.07mW with voltage supply of 1.2V as shown in Figure 12. The power consumption gained is within the range of the scope wanted.

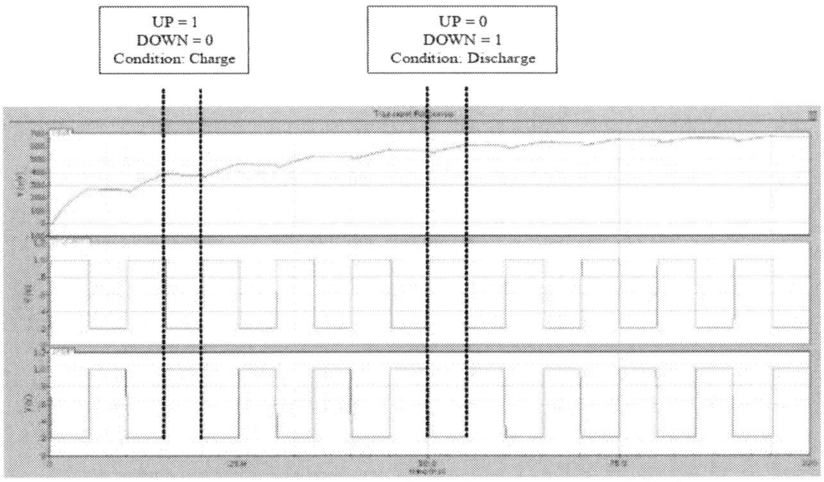

Figure 11. The output waveform for the proposed design.

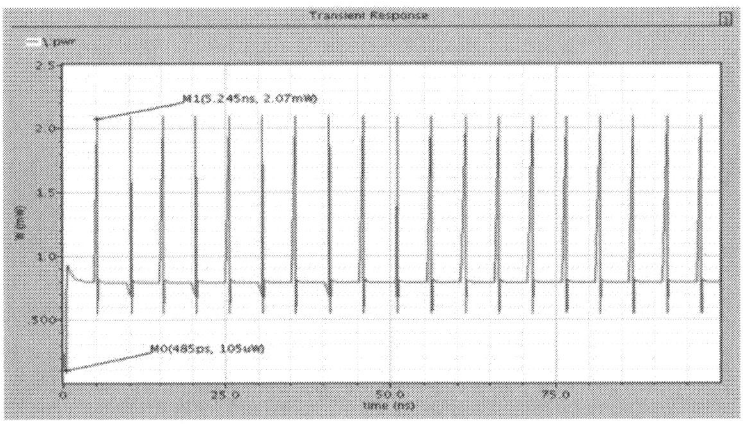

Figure 12. The power consumption of the proposed design is 2.07 mW.

This circuit also has been tested in different voltage supply which is 1.0 V, 1.2 V, and 1.8 V. The different usage of power supply can produce different power consumption as shown in Table 3. By using voltage supply of 1.0 V, smaller power consumption was consumed compared to 1.2 V, where the power consumption is 2.07 mW. Meanwhile, Figure 13 shows the plotted frequency that is measured at 4.7 GHz by using Calculator in the Analog Design Environment. Figure 14 shows the layout for tri-state charge pump and low pass filter that have been arranged. From the figure, the charge pump area is smaller than loop filter area. This is because the loop filter is using analog components, which is a resistor and capacitor. The total area of the layout that have been measured is 31.4 μm x 22.6 μm (709.6 μm^2). The impairment that Figure 15 is the comparison of power consumption and frequency that have been plotted.

Table 3. Comparison of Different Values of Voltage Supply.

Voltage Supply (V)	**Average Power (μW)**	**Power Consumption (mW)**

IOP Conf. Series: Journal of Physics: Conf. Series **1049** (2018) 012060 doi:10.1088/1742-6596/1049/1/012060

1.0	1.72	1.10
1.2	1.956	2.07
1.8	4.0	5.14

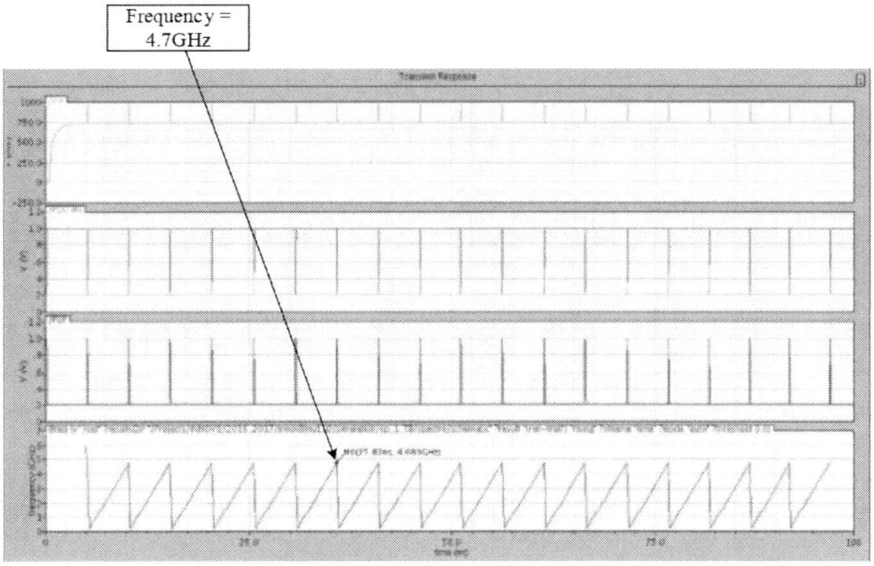

Figure 13. The plotted frequency of the proposed design.

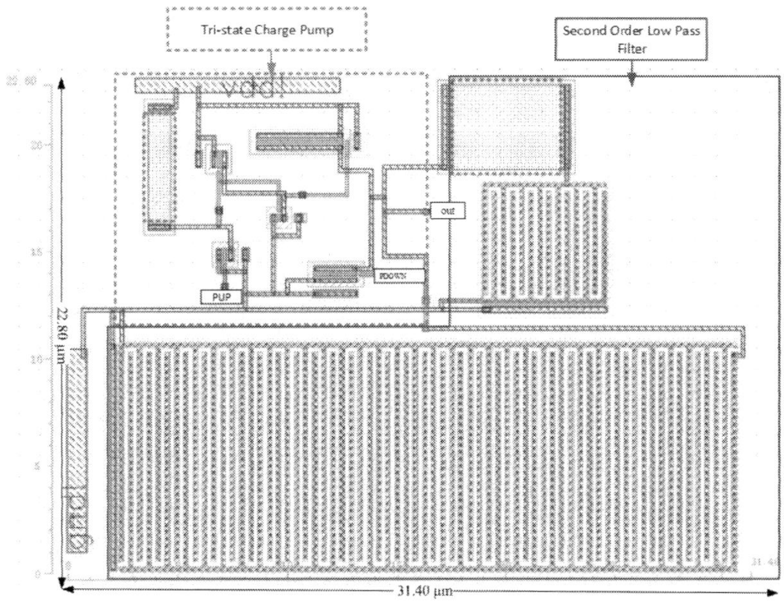

Figure 14. Layout of Tri-State Charge Pump and Low Pass Filter

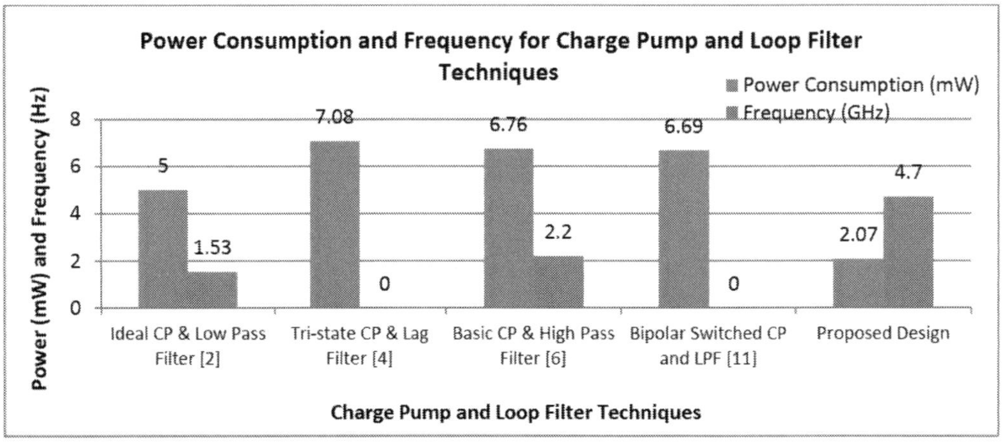

Figure 15. Power Consumption and Frequency for Charge Pump and Loop Filter Techniques

There is a lot of technique used by charge pump and loop filter whether in PLL system or not. But, most of the techniques used 180nm CMOS technology and above. Power consumption of the proposed design is improved about 58.6% compared to ideal charge pump and low pass filter. Basic charge pump and high pass filter improved about 69.38% when compared to the proposed design and also compared to other. Last but not least, bipolar switched charge pump and loop filter only focus on the power consumption of the whole system. The difference is about 69.05% compared to the proposed design.

6. Conclusion

A tri-state charge pump and second order low pass filter has been designed by using Cadence Tools in 130nm CMOS technology. For the proposed designed, the aim of the power consumption and frequency required are achieved. The power consumption achieved is 2.07 mW with the range of frequency is from 3.1 GHz to 5.1 GHz and frequency achieved is 4.7 GHz. From the final design, the area measured for the whole tri-state charge pump and low pass filter is 31.4 µm x 22.6 µm (709.6 µm2).

Acknowledgement

This work was financial supported by FRGS Grant Vot Number 1538.

References

[1] K. K. Patel, N. D. Patel, K. P. Thakore 2013 Charge Pump, Loop Filter and VCO for Phase Lock Loop Using 0.18 µm CMOS Technology *IOSR Journal of VLSI and Signal Processing* **2** 21

[2] P.E Allen. 2016 Lecture 120-Filters and Charge Pumps (Reading: 4, 6, 9, and 10). [Online]. Available FTP: http://users.ece.gatech.edu/jskenney/L120-Filter&CPs(2UP).pdf Accessed on 20 September 2016.

[3] K. K. Patel and N. D. Patel 2013 Phase Frequency Detector and Charge Pump For DPLL Using 0.18 µm CMOS Technology *Int. Journal of Emerging Technology and Advanced Engineering* **3** 4

[4] A. Mishra 2014 Performance Analysis of Power Optimal PLL Design Using Five-Stage CS-VCO in 180nm *2014 Int. Conf. on Signal Propagation and Computer Technology* (ICSPCT) p 764

[5] I. Toihria 2013 Design of an Effective Charge Pump-Phase Locked Loops Architecture for RF Applications *Int. Journal of Computer Applications* **74** 38

[6] J. W. Park 2013 Integrated high voltage boost converter with LC filter and charge pump *Microelectronic International* **31** 54-60
[7] Y. Sen, N. Jain 2014 Design and Implementation of Phase Locked Loop Using Current Starved Voltage Controlled Oscillator *Advance in Electronic and Electric Engineering* **4** 637
[8] A. Agrawal, R. Khatri 2015 Design of Low Power, High Gain PLL using CS-VCO on 180nm Technology *Int. Journal of Computer Application* **122** 26
[9] K. Holladay and F. Microelectronics 2000 Design a PLL for a specific loop bandwidth *EDN Network* p 173
[10] B. Goyal, S. Suman, and P. K. Ghosh 2016 Design of Charge Pump PLL Using Improved Performance Ring VCO *Int. Conf. on Electrical, Electronics, and Optimization Techniques* p 3254
[11] M. Bhardwaj 2016 Design of a Wide Output Range and Reduced Current Mismatch Charge Pump PLL with Improved Performance *2016 Int. Conf. on Advances in Computing, Communications and Informatics* p 2644
[12] Abdul Majeed K.K, Binsu J. Kailath 2015 CMOS Current Starved Voltage Controlled Oscillator Circuit for a Fast Locking PLL *IEEE INDICON* p 1
[13] Cadence Tools. [Online]. Available FTP: https://www.cadence.com/. Accessed on 20 April 2017.
[14] D. Clien, G. Shimokura, Technical Contributor 1999 Layout Design in CMOS IC Layout: Concepts, Methodologies, and Tools, Woburn MA p 22-67
[15] S. Tiwari, S. Katiyal, P. Parandkar, R. Malviya 2011 Low Power Current Starved VCO in 70nm CMOS Technology Sri Ram Institute of Technology Jabalpur
[16] J. Gupta, A. Sangal, H. Verma 2011 High Speed CMOS Charge Pump Circuit for PLL Applications using 90nm CMOS Technology *IEEE* p 346
[17] M. Zucchelli, L. Colalongo, A. Richelli, Z. M. K. Vajna 2016 Dickson Charge Pump using Integrated Inductors in Complementary Metal-Oxide Semiconductor Technology," *The Institution of Engineering Technology* **9** 553

IOP Conf. Series: Journal of Physics: Conf. Series **1049** (2018) 012061 doi:10.1088/1742-6596/1049/1/012061

Green Practice Knowledge Profiling among the Technical and Vocational Education Masters Programme Students

M.F.Lee, N.S. Muhtar & C.S.Lai
Faculty of Technical & Vocational Education,
Universiti Tun Hussein Onn Malaysia, 86400 Parit Raja, Batu Pahat, Johor

E-mail: mflee@uthm.edu.my

Abstract. Global warming is utmost problem because of the negative impact to all creatures of the earth. The best solution for this issue should through education. Therefore, this study was conducted to identify the green practice knowledge among the Technical and Vocational Education Masters' students in Universiti Tun Hussein Onn Malaysia. Green practice knowledge refer to National Green Technology Policy four key areas by KeTTHA which is energy sector, building sector, water and waste water management sector, and transportation sector. A total numbers of 87 students were selected as respondent using stratified sampling technique in this study. A survey method using questionnaire was employed as research design for this study. The gathered data were analyse using sum, mean score, standard deviation, Kruskall Wallis test and Mann Whitney U test. The findings found that the level of green practice knowledge was at the medium level. The findings also shown that there was no significant in the level of green practice knowledge from the aspect of gender and background of study among the respondents. However, there was a significant difference in green practice knowledge among the respondents in terms of age. As a conclusion, the global warming impact can be minimize by increasing the green knowledge, awareness and practising among the communities in their lifestyle.

1. Introduction

The global warming triggered extreme climate phenomena that resulted in various natural disasters such as snowstorms and floods throughout the world. If this phenomenon is not being controlled, there will be major catastrophes that will hit the world's inhabitants. According to Wan Omar (2009), the impact of global warming has been felt in Malaysia only in moderate capacity compared to other countries. He also noted that the rise in global temperatures can cause changes in nature, including rising sea levels and extreme weather such as floods, droughts, heat waves, typhoon and tornado. In addition, the phenomenon of global warming has also resulted in the occurrence of long, dry, and dry droughts. This situation will also result in the occurrence of forest fires that can kill lives, destroying property, crops and pets. The fire also invites the occurrence of haze that can affect human health and vision. Overall, the issue of global warming is a world issue and not just involving some countries. Malaysia's contribution to global warming may not be so great, but as a global contribution of Malaysians, it is necessary to seek an end to the day-to-day activities that will not contribute to the ever increasing global temperature. Sulaiman and Junus (2005) state that extreme weather changes are heavily influenced by human activity. Therefore, Malaysians should practicing green practices in their daily lives as a must to fight the issue of global warming.

Content from this work may be used under the terms of the Creative Commons Attribution 3.0 licence. Any further distribution of this work must maintain attribution to the author(s) and the title of the work, journal citation and DOI.
Published under licence by IOP Publishing Ltd

While the results of the study conducted by Muhd Hussin (2007) found that the environmental awareness and understanding of the students on green concepts were at a satisfactory level. However, assessments of student attitudes and practices on green are at a moderate level. The only way to overcome the issue of global warming is through education, and the teachers will play an utmost important role. When educators are knowledgeable about green practices in everyday life, they can influence their students and the people surrounding to participate in practising green during everyday life. Malaysia has been prominent as a developing country. Malaysia is facing a huge challenge in ensuring country development can be sustained. Among the environmental issues that are often discussed today are, urban air quality, river water quality, forest destruction, household waste and waste and chemicals from the plant are examples of environmental issues occurring in Malaysia. Although green practices were introduced in the early 2000s, the study on green practices in Malaysia is remain under investigate. Therefore, this study aims to determine green practise knowledge among the students of Master of Technical and Vocational Education, University Tun Hussein Onn Malaysia (UTHM). The research questions of this study are as follows:

- What is the level of knowledge of Green practice among students?
- Is there a significant difference in green practices knowledge level from the aspect of gender?
- Is there a significant difference in green practices knowledge level from the aspect of age?
- Is there a significant difference in green practices knowledge level from the aspect of courses?
- Is there a significant difference in green practices knowledge level from the aspect of mode of study?

2. Green Practice Knowledge
The green practices emphasized in this study are the concept of vehicle sharing, energy saving and water resources, the reduction of use of disposable items and the use of green technology among the students. These elements are based on National Green Technology Policy Pillars (KeTTHA, 2011) as follow:

- Energy - Seek to attain energy independence and promote efficient utilisation;
- Environment - Conserve and minimize the impact on the environment;
- Economy - Enhance the national economic development through the use of technology; and
- Social - Improve the quality of life for all.

According to KeTTHA (2011), the significant progress and major improvement must be done in the following four (4) key areas to achieve National Green Technology Policy Pillars. The first key area is energy sector and energy utilization sector, which includes application of Green Technology in power generation and in the energy supply side management, including co-generation by the industrial and commercial sectors; as well as application of Green Technology in all energy utilization sectors and in demand side management programmes. Second key area is buildings sector that focus on adoption of Green Technology in the construction, management, maintenance and demolition of buildings. The third key area is water and waste water management sector that emphasize on adoption of Green Technology in the construction, management, maintenance and demolition of buildings. Fourth key area is transportation sector that cater on incorporation of Green Technology in the transportation infrastructure and vehicles, in particular, biofuels and public road transport.

3. Methodology
The research design for this study was a survey using an achievement test as instrument. A total number of 87 Technical and Vocational Education Master students was involved in this study. The sample was selected using stratified sampling technique. An achievement test was developed to measure the samples' green practice knowledge based on National Green Technology Policy four key

areas by KeTTHA which is energy sector, building sector, water and waste water management sector, and transportation sector. The reliability test for this achievement test using alpha Cronbach was .916. The gathered data were analyse using sum, mean score, standard deviation, Kruskall Wallis test and Mann Whitney U test. In order to determine the green practice knowledge level among the samples, the total scores of them was categorize into three group, namely low, medium, and high level.

4. Results

Table 1 shows the score of respondents in green practise knowledge. Findings indicate that the level of knowledge in green practice among students of Master of Technical and Vocational Education, was at a moderate level. The average score among of them was 11.07 with standard deviation = 1.66. This means that the students only can answered correctly 11 from the 16 questions given. The minimum score was 6 and the maximum score was 15.

Table 1. Green Practise Knowledge Score among the students.

Mean Score	Standard Deviation	Minimum score	Maximum score	Level
11.07 / 16	1.66	6/16	15 / 16	Medium

Table 2 shows the score of green practise knowledge between male and female students. Findings determine that male students (M=11.13; SD=2.21) tend to have higher mean score than female students (M=11.06, SD=1.45). However minimum score for female was 8 while male was 6 and the maximum score for female was 14 while male was 15. Even though male and female tend to have different mean score but Mann Whitney U test in Table 3 and Table 4 show that there are no significant mean ranks of male (448.11) and female (42.61) on green practise knowledge, $U = 624.5$, $p = .364$.

Table 2. Green Practise Knowledge Score between Male and Female students.

Gender	n	Mean Score	Standard Deviation	Minimum Score	Maximum Score
Male	22	11.13	2.21	6	15
Female	65	11.06	1.45	8	14

Table 3. Ranks

	Gender	N	Mean Rank	Sum of Ranks
Score	Male	22	48.11	11058.50
	Female	65	42.61	2769.50
	Total	87		

Table 4. Test Statistics[a]

	Score
Mann-Whitney U	624.500
Wilcoxon W	2769.500
Z	-.907
Asymp. Sig. (2-tailed)	.364

a. Grouping Variable: Gender

IOP Conf. Series: Journal of Physics: Conf. Series **1049** (2018) 012061 doi:10.1088/1742-6596/1049/1/012061

Table 5 shows the score of green practise knowledge among the students with different ages. Findings determine that students with < 25 year (M=12.31; SD=1.49) tend to have highest mean score than others age students. Kruskal Wallis Test in Table 6 and Table 7 show that there are a significant mean ranks of among the students with different ages on green practise knowledge, $\chi^2 = 8.909$, $p = .031$. The mean rank in descending order are less than 25 year (61.69) > 31-35year (46.54) > 26-30 (40.53) > more than 36 year (36.41) in green practise knowledge.

Table 5. Students' Green Practise Knowledge Score from age aspect.

Age	n	Mean Score	Standard Deviation	Minimum Score	Maximum Score
< 25 year	13	12.31	1.49	10	15
26 – 30 year	51	10.84	1.52	6	14
31 – 35 year	12	11.25	1.48	9	14
> 36 year	11	10.45	2.07	7	13

Table 6. Ranks

	Age	N	Mean Rank
Score	< 25 years	13	61.69
	26-30 years	51	40.53
	31-35 years	12	46.54
	> 36 years	11	36.41
	Total	87	

Table 7. Test Statistics[a,b]

	Score
Chi-Square	8.909
df	3
Asymp. Sig.	.031

a. Kruskal Wallis Test

b. Grouping Variable: Age

Table 8 shows the score of green practise knowledge among the students from different background of study. Findings determine that others course students (M=11.50; SD=1.77) tend to have highest mean score compare to others. Kruskal Wallis Test in Table 9 and Table 10 show that there are no significant mean ranks of among the students from different background of study on green practise knowledge, $\chi^2 = 7.985$, $p = .092$. The mean rank in descending order are others course (50.46) > civil engineering (50.17) > electric & electronic engineering (36.17) > commerce (35.35) > mechanical engineering (33.78) in green practise knowledge.

Table 8. Students' Green Practise Knowledge Score among Students from different Background of Study

Background of Study	n	Mean Score	Standard Deviation	Minimum Score	Maximum Score
Commerce	10	10.50	1.72	8	13
Mechanical	9	10.11	2.37	6	13
Electric & Electronic	18	10.72	0.96	9	13
Civil	12	11.41	1.08	9	13
Others	38	11.50	1.77	7	15

Table 9. Ranks

	Background of Study	N	Mean Rank
Score	Commerce	10	35.35
	Mechanical	9	33.78
	Electric & Electronic	18	36.17
	Civil	12	50.17
	Others	38	50.46
	Total	87	

Table 10. Test Statistics[a,b]

	Score
Chi-Square	7.985
df	4
Asymp. Sig.	.092

a. Kruskal Wallis Test
b. Grouping Variable: Background of Study

Table 11 shows the score of green practise knowledge between full time and part time students. Findings determine that full time students (M=11.12; SD=1.73) tend to have higher mean score than part time students (M=11.02, SD=1.61). However, minimum score for full time was 6 while part time was 7 and the maximum score for full time was 15 while part time was 14. Even though full time and part time tend to have different mean score but Mann Whitney U test in Table 12 and Table 13 show that there are no significant mean ranks of full time (45.06) and part time (43.05) on green practise knowledge, $U = 899.5, p = .704$.

Table 11. Green Practise Knowledge Score between Full Time and Part Time students.

Mode of Study	n	Mean Score	Standard Deviation	Minimum Score	Maximum Score
Full time	41	11.12	1.73	6	15
Part time	46	11.02	1.61	7	14

Table 12. Ranks

	Mode of Study	N	Mean Rank	Sum of Ranks
Score	Full time	41	45.06	1847.50
	Part time	46	43.05	1980.50
	Total	87		

Table 13. Test Statistics[a]

	Score
Mann-Whitney U	899.500
Wilcoxon W	1980.500
Z	-.380
Asymp. Sig. (2-tailed)	.704

a. Grouping Variable: Mode of Study

5. Conclusion

The knowledge on green practice among the students is at medium level. This condition needs to be improved because some of them are in-service teacher whereas some of them are the future teacher. They must aware that they are one of the individuals responsible for disseminating knowledge about of green practice to future generations in order to achieve the aims of sustainability development. Comparison of the green practice knowledge level from the aspect of gender, age, courses and mode of study showed that only students with different ages tend to have significant difference. The senior students need more exhortations and remind on green practise. Efforts to raise awareness about the importance of green practices in everyday life is utmost urgent. Additionally, future work on 5Rs (Reuse, Reduce, Reuse, Recycle, Repair and Recover) practising among the students is under investigation. As a citizen in the earth, they must remember their responsible with their waste throughout the 5Rs practising.

6. References

[1] Wan Omar, W. M. R. (2009). Minda: Ancaman Serius Pemanasan Global. Harian Metro. Retrieved from http://www.met.gov.my/pdf/newsArticle/HMetro6_26May2009.pdf

[2] Sulaiman, K. Dan Junus, L. (2005). Fenomena Ekuinoks dan Cuaca Panas. Utusan online. Retrieved from http://utusan.com.my/utusan/info.asp?y=2005&dt=0226&pub=utusan_malaysia&sec=rencana&pg=re_05.htm&arc=hive&arc=hive

[3] Muhd Hussin, S. H. (2007). *Penilaian Prestasi Alam Sekitar: Dalam Aspek Kesedaran Alam Sekitar dan 3R Di Kalangan Pelajar Universiti Malaysia Sabah.* Universiti Malaysia Sabah: Tesis Sarjana Muda.

[4] KeTTHA (2011) National Green Technology Policy Main Areas. Retrieved from http://www.kettha.gov.my/portal/index.php?r=kandungan/index&menu1_id=3&menu2_id=75&menu3_id=95#.Wgphw1uCzIU

[5] KeTTHA (2011) National Green Technology Policy Pillars. Retrieved from http://www.kettha.gov.my/portal/index.php?r=kandungan/index&menu1_id=3&menu2_id=75&menu3_id=93#.Wgpky1uCzIU

Acknowledgments
The publication of this work was supported by the PPG grant (No. Vot: V012) awarded by Ministry of Education Malaysia (MOE) and the Office for Research, Innovation, Commercialization and Consultancy Management (ORICC) of Universiti Tun Hussein Onn Malaysia. The authors would also like to thank to those who graciously gave their time to participate in this study.

IOP Conf. Series: Journal of Physics: Conf. Series **1049** (2018) 012062 doi:10.1088/1742-6596/1049/1/012062

Continuous Monitoring of Blood Pressure Based on Heart Pulse Analysis

Valerie Tan[1], Audrey Huong[1], Xavier Ngu[1]
[1] Faculty of Electrical and Electronic Engineering
University Tun Hussein Onn Malaysia

E-mail: audrey@uthm.edu.my

Abstract. This paper aims to demonstrate the use of a finger-cuff system in the approximation of one's systolic and diastolic value based on the acquired heart pulse data. The prediction of the required systolic and diastolic value of blood pressure is via the use of two-element Windkessel Model. The results are compared for three groups of individuals of different clinically measured blood pressure range. A comparison with the result given from clinically available arm-cuff blood pressure monitor revealed a relatively low mean difference of 1.6 ± 3 mmHg predicted for the systolic values, while a considerably larger difference of 25.3 ± 6.3 mmHg is found for the estimated diastolic values. This work concluded that the proposed strategy that used heart pulse in its analysis may be used as an alternative means for continuous measurement of one's blood pressure, but a system with high sampling resolution may be required to improve the accuracy of the predicted value.

1. Introduction

Continuous measurement of blood pressure is vital especially for aged population, pregnant women and individuals who either have just undergone surgery or with critical medical conditions. It is crucial to continuously monitor the blood pressure of these groups so that prompt action can be taken following the detected sudden change in their blood pressure as the latter may result in death. The conventional sphygmomanometer technique, which required the use of a blood pressure cuff and a stethoscope, is inconvenient for continuous measurement of blood pressure. The arm-cuff blood pressure monitor is a more convenient alternative as it provides continuous measurement of the value and it does not require the use of a stethoscope. Nonetheless both of these approaches may potentially cause harm especially to those who suffered from hypertension and hypotension. The exertion of pressure on the arm of these populations via the use of a cuff narrows the size of local blood vessels; this increases the chances of one getting stroke due to a possible insufficiency in the supply of oxygen to the brain. Therefore, cuffless Photoplethysmography (PPG) [1] and the use of a finger-cuff [2] were previously introduced to overcome these problems. Both of these techniques are wearable technologies that do not require the occlusion of artery in their measurement. While the implementation of the latter technique is limited to fingertip, the former may include a wide range of wearable devices such as wrist-watch, ring and clothing [3, 4].

PPG is a non-invasive optical technology that measures relative blood volume changes in vessels close to the skin surface. Many works have been done in the past to predict one's blood pressure using the PPG data and via the use of Windkessel model [5, 6], wherein it was reported that the accuracy of value predicted using this model increases with the order of the element used. Meanwhile other study

Content from this work may be used under the terms of the Creative Commons Attribution 3.0 licence. Any further distribution of this work must maintain attribution to the author(s) and the title of the work, journal citation and DOI.
Published under licence by IOP Publishing Ltd

[7] emphasized the importance of PPG signals measurement from the wrist. According to this study, only blood pressure predicted from the wrist radial artery can be validated via the sphygmomanometer. Many of these inventions allow portable and non-invasive measurement of blood pressure. It is the objective of this work to explore the feasibility of using heart pulse collected from the developed circuit for continuous and real time prediction of one's blood pressure value via the measurement from fingertip.

2. Materials and Methods

2.1. Finger-cuff blood pressure system

The finger-cuff system developed in this study worked in transmission mode for real-time measurement of heart pulses from one's fingertip. The schematic diagram of the constructed circuit is shown in figure 1, while figure 2 showed the overall operation diagram of the system. This work used infrared (IR) ray in the wavelength of 940 nm for the illumination of the selected skin site. The detection of the PPG signals was via the use of market available finger pulse sensor (HRM-2511E), which wrapped and held in position around the selected finger during the experiment, to minimize the possibility of motion artifacts. The measured analog PPG signals were converted to digital pulses via a 10-bit analog to digital (A/D) converter (MCP3008). These pulses were processed in real time for the prediction of the required systole and diastole blood pressure; these values were then sent to a computer via Raspberry Pi 3 for online monitoring of one's blood pressure value.

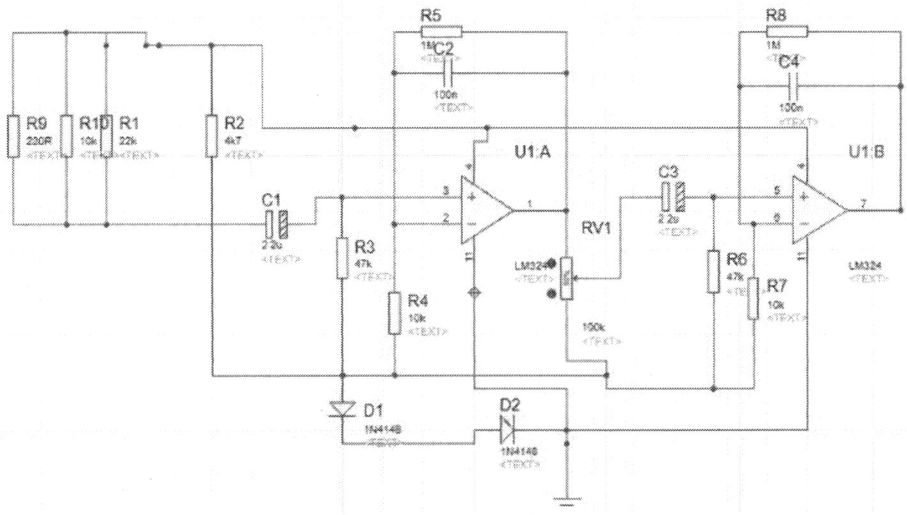

Figure 1. Heart pulse detection circuit diagram

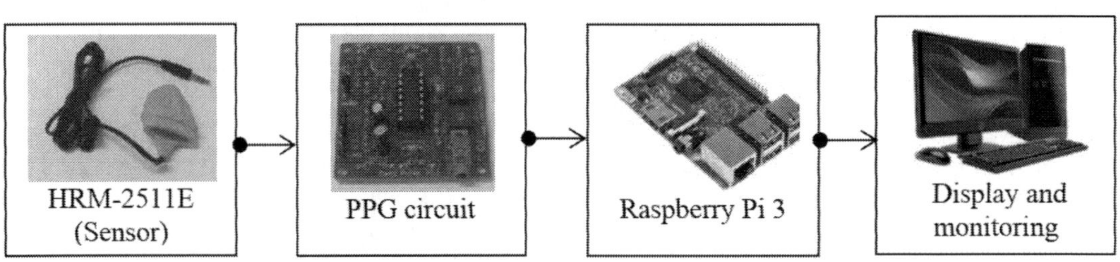

Figure 2. The operation diagram of the constructed finger-cuff PPG system

2.2. Two-element Windkessel model

The Windkessel model was previously used in the past to describe the *in-vivo* arterial system with respect to the blood flow and blood pressure. The use of this model in the prediction of one's blood pressure was extensively demonstrated in the work by Choudhury *et al.* [5]. This study employed a two-element Windkessel model shown in equation (1) due to the simplicity of the model. This model has an analogy of an electrical circuit, wherein a resistor was placed in parallel with a capacitor. It was reported that the resistance, R, represents total peripheral resistance while capacitance, C, represents the arterial compliance. Both of these values were taken to be 1 [8].

$$\frac{P(t)}{R} + C\frac{dP(t)}{dt} = I(t) \tag{1}$$

where I and P represent blood flow and blood pressure, respectively. Solving for equation (1) yields an expression of blood pressure during systolic and diastolic cycle. This is viable with the assumption that I is a sinusoidal wave during systole with peak blood flow represented by I_o, and is zero during diastole. The expression of I_o from ventricles to artery is given by [5]:

$$I_o = \frac{C_o T_c}{60\int_0^{T_s}\sin\left(\frac{\pi t}{T_s}\right)dt} \tag{2}$$

where C_o and T_c represent cardiac output and cardiac time, respectively. The cardiac output is defined as the product of stroke volume and average heart rate in a minute. The stroke volume in the unit of millilitre was assumed to have the same value as the body weight of the investigated individual in kilogram. In this study T_c was taken as the total duration of the high of the pulse as shown in figure 3.

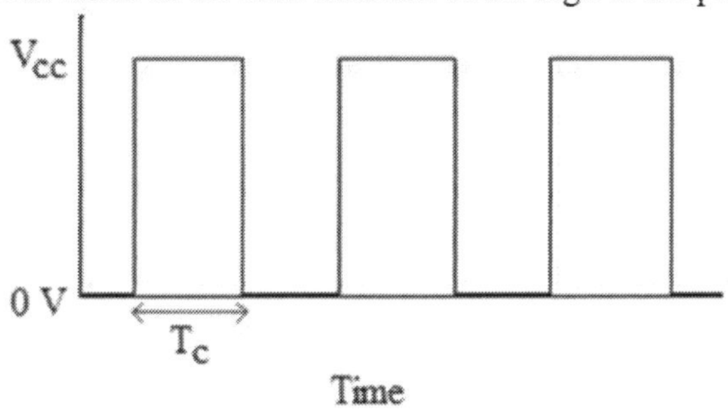

Figure 3. Measurement of cardiac time from a pulse

It was also assumed that the systolic time, T_s, is one-third of the total cardiac time, while the diastolic time, T_d, is two-third of the total cardiac time as followed:

$$T_s = \frac{1}{3}T_c \tag{3}$$

$$T_d = \frac{2}{3}T_c\,. \tag{4}$$

From equation (3) and (4), the systolic blood pressure value is expressed as:

$$P_s = P_{ts}\exp\left(\frac{-T_s}{RC}\right) + \frac{I_o T_s C\pi R^2}{T_s^2 + C^2\pi R^2}\left(1 + \exp\left(\frac{-T_s}{RC}\right)\right) \tag{5}$$

while the diastolic blood pressure is given by:

$$P_d = P(t|t = T_d) = P_{td}\exp\left(\frac{-T_d}{RC}\right). \tag{6}$$

where P_{ts} and P_{td} is the value of P_s and P_d calculated from its preceding pulse. In this work the initial value of P_{ts} is set as 80 mmHg; the P_s value calculated from equation (5) is then used as the initial value of P_{td} in the calculation of P_d from equation (6).

2.3 Experimental subject and method

This study recruited three groups of individuals (aged between 20 and 30 years) from University Tun Hussein Onn Malaysia (UTHM) for a comparison of their blood pressure value measured from the conventional cuff technique and the developed approach. The device used for the cuff technique, which was used as the gold standard in this study, is OMRON arm-cuff blood pressure monitor (model no. HEM 907). These volunteers were grouped according to their blood pressure value measured from the arm-cuff technique. There were ten subjects from each group of blood pressure value range shown in Table 1. All self-declared healthy with no known underlying illness, and the recruited females were not pregnant during the study. These recruits gave their consent prior to the study.

During the experiment, the subjects were told to relax in a chair and breathe normally for five minutes before blood pressure measurement was collected from either their right or left arm using the arm-cuff blood pressure monitor. This is repeated for measurement using the finger-cuff system on their index finger.

Table 1. Blood pressure value range defined in this study

Blood Pressure range	Systolic Value (mmHg)	Diastolic value (mmHg)
Low	< 115	< 75
Normal	< 140 and > 115	< 90 and > 75
High	> 140	> 90

3. Results

The systolic and diastolic blood pressure value measured for the recruited subjects of different blood pressure range tabulated in table 1 using arm-cuff blood pressure monitor and the developed system is shown in table 2. From table 1 the mean and standard deviation of systole pressure value measured using cuff blood pressure monitor and the developed system, respectively, for groups of individuals with high, normal and low blood pressure is calculated as 151.4 ± 4.51 mmHg, 128.5 ± 7.82 mmHg, 104.9 ± 6.72 mmHg and 151.3 ± 5.59 mmHg, 128.6 ± 5.93 mmHg, 106.7 ± 6.36 mmHg. Meanwhile the mean and standard deviation of the diastole pressure value measured using cuff blood pressure monitor and the developed system for these groups is given by 90.9 ± 6.41 mmHg, 81.9 ± 3.54 mmHg, 67.7 ± 3.83 mmHg and 123.9 ± 3.84 mmHg, 105.1 ± 4.77 mmHg, 87.3 ± 5.2 mmHg, respectively. The average of the mean and standard deviation of absolute difference in the systolic and diastolic value given from these systems is calculated as 1.6 ± 3 mmHg and 25.3 ± 6.3 mmHg, respectively.

4. Discussion

From table 2, it can be clearly seen that the results for systolic value measured from the developed system for all groups of volunteers are similar to that from the arm-cuff blood pressure monitor with the average mean error of 1.6 ± 3 mmHg. Nonetheless, a comparatively larger mean difference was observed for diastolic measurement using these two techniques given by 25.3 ± 6.3 mmHg. This is likely due to the poor measurement of the cardiac time shown in figure 3 where the T_d and T_s value used in equation (5) and (6) were deduced. The heart pulses used in this study were obtained following the conversion of analog PPG signals to digital pulses via a 10-bit converter, this converter provides a fast sampling rate at the price of a low resolution, and this poor sampling may misrepresent the shape of the PPG. Moreover, the foot of PPG pulse wave signals, which contains the latter part of the diastole, is likely masked by noise and may go undetected. This yields an underestimation of the T_d value calculated from equation (4), the erroneous in this value propagates through equation (6) resulting in an overestimation of the diastolic blood pressure. This error is then manifested itself in the following diastole blood pressure values through the equation (6). This leads to a relatively consistent difference in the diastolic blood pressure value predicted using the arm-cuff blood pressure monitor and the developed system with the mean standard deviation given by 6.3 mmHg.

This work demonstrated that the information deduced from a heart pulse may potentially be used to approximate one's systole and diastole blood pressure value. This approach allows a real time processing of the signals as it placed a lower demand on the system required for processing and sampling of the signals. However, an A/D conversion system with high sampling resolution may be needed to allow a more accurate measurement of the cardiac time, and to more accurately predict the required systole and diastole blood pressure.

Table 2. The measured diastolic and systolic value from gold standard (arm-cuff blood pressure monitor) and the developed finger-cuff system

Blood pressure range	High Blood pressure	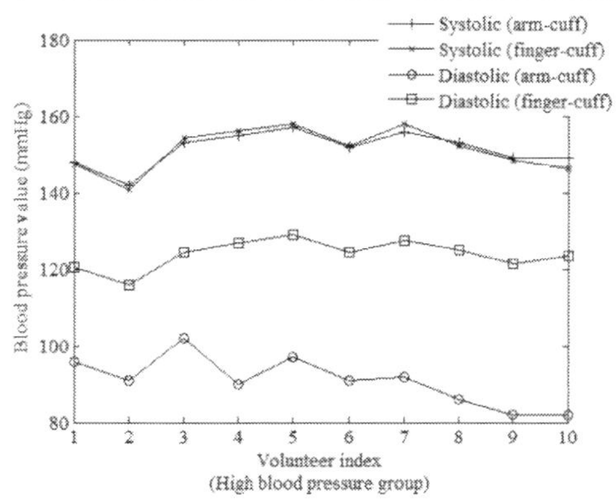
	Normal blood pressure	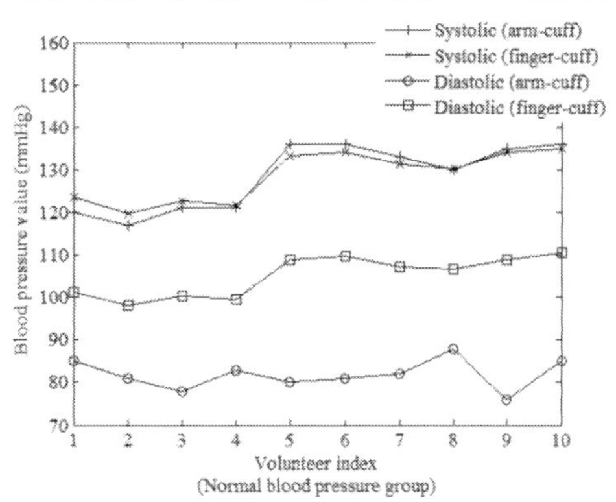

	Low blood pressure	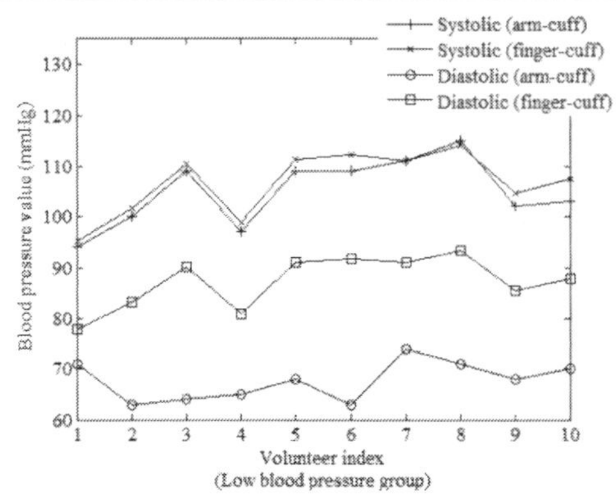

Conclusion

This study showed that the developed finger-cuff system is able to provide an online prediction of one's systolic and diastolic value based on the analysis of the heart pulse. The results showed high consistency in the systolic value given from the market available arm-cuff blood pressure monitor and the developed system for all the recruited groups. Nonetheless a relatively large difference is observed in the measured diastolic value due to a lack in the A/D conversion efficiency. This work concluded that the systolic and diastolic blood pressure value may be approximated from heart pulse signals through an accurate measurement of cardiac time and with the use of a two-element Windkessel model.

Acknowledgement

The authors would like to thank all volunteers who participated in this study. The authors are also grateful to Nooreis Shadilla Jarkasi from Department of Electronic Engineering, UTHM, for her help during data collection. This work was financially supported in part by the Ministry of Education Malaysia under FRGS grant no. 1581.

References

[1] Y Zheng, C C Y Poon, B P Yan and J Y W Lau 2016 *Journal of Medical System* **40** 1
[2] K Matsumura, T Yamakoshi, P Rolfe and K Yamakoshi 2016 *IEEE Transactions on Biomedical Engineering* **64** 1131
[3] C C Y Poo, Q Liu, H Gao, W H Lin and Y T Zhang 2011 *Journal of Computing Science and Engineering* **5** 246
[4] Y T Zhang, C C Y Poon, C H Chan, M W W Tsang and K F Wu 2006 *3rd IEEE/EMBS International Summer school on medical devices and Biosensors*
[5] A D Choudhury, R Banerjee, A Sinha and S Kundu 2014 *Conf Proc IEEE Eng Med Biol Soc* p 4567
[6] E Maksuti, N Westerhof, B E Westerhof, M Broome, N Stergiopulos 2016 *Plos one*
[7] S H Song, J S Cho, H S Oh, J S Lee and I Y Kim 2009 *IEEE Explore: Computer in Cardiology* 741
[8] M Hlaváč, "Windkessel 2004 *Proc 2004 Student Electrical Engineering, Information and Communication Technologies* 1

IOP Conf. Series: Journal of Physics: Conf. Series **1049** (2018) 012063 doi:10.1088/1742-6596/1049/1/012063

The Development of Spectroscopy System for Non-Invasive Measurement of Skin Blood Oxygen Saturation

H. L. Kam[1], Audrey Huong[1*], Nur Anida Jumadi[1], N. Ibrahim[1], Wan Mahani Hafizah Wan Mahmud[1], Wan Nur Hafsha Wan Kairuddin[1]
[1] Faculty of Electrical and Electronic Engineering
Universiti Tun Hussein Onn Malaysia

E-mail: audrey@uthm.edu.my

Abstract. This paper presents the use of the developed spectroscopic system for noncontact measurement of one's skin blood oxygen saturation (S_tO_2). This study employed light diffusion model in the prediction of the required value using light source of wavelength 488 nm, 522 nm and 684 nm. The performance of this system was tested on ten volunteers of different skin tone to investigate the robustness of the system. The mean and standard deviation of S_tO_2 value calculated for the recruited subjects is given by 71 ± 4.7 %. The results revealed differences in the predicted value with the recruits' skin tone. This work concluded that the developed system may potentially be used in clinical settings to determine skin oxygen saturation and to monitor wound healing progress.

1. Introduction

Hemoglobin is a protein in red blood cells that carries oxygen to one's organs and tissues, and at the same time transports carbon dioxide gas from organs and tissues back into the lungs to be expelled from the system. The percentage of hemoglobin is commonly referred to as the blood oxygen saturation. Oxygen has a significant role in wound healing [1-2], this element is essential to help in providing the additional energy source needed for tissues repairing. Under conditions of systemic hypoxia, there would be a decrease in wound oxygen level, henceforth affect the healing outcomes. Healthy people would have a normal range of arterial blood oxygen saturation, S_aO_2, fluctuating from 95 to 100 %; a drop in this value to 90 % is sometimes common and does not raise an alarm.

Hemoglobin level test is a blood test that provides information on the effectiveness of red blood cells in carrying oxygen to cells [3]. This test is important because the amount of oxygen readily available to the tissues depends upon the concentration of hemoglobin in red blood cells and local perfusion of the tissues. In the events of hemoglobin insufficiency, tissues would be deprived of oxygen, both heart and lungs would have to work harder to compensate for the demand.

Some commercial and readily available instruments include pulse oximeter and fiber-optic pulmonary artery catheter oximeter [4]. Both of these equipments are commonly used for *in-vivo* blood oxygen saturation measurement. Pulse oximeter is a non-invasive technology that uses a contact probe for measurement of one's S_aO_2 from well perfused fingertip. Nonetheless, the use of a probe is not practical if burnt or wound injuries were located on the extremities where the probe would be used. Besides, pulse oximeter failed in its prediction of the S_aO_2 value under medical situations such as hypovolemia, hypothermia, hypoxia, hypoxemia and hypoperfusion as it cannot distinguish between

Content from this work may be used under the terms of the Creative Commons Attribution 3.0 licence. Any further distribution of this work must maintain attribution to the author(s) and the title of the work, journal citation and DOI.
Published under licence by IOP Publishing Ltd

different forms of hemoglobin [5]. Meanwhile blood gas analyzer is another means to measure the oxygen saturation in blood. Even though this method is able to provide a more accurate estimation of blood oxygen saturation, it is an invasive approach. This work aims to develop and demonstrate the use of a non-invasive and non-contact spectroscopic system in the measurement of one's blood oxygen saturation with the assumption that oxyhemoglobin (HbO_2) and deoxyhemoglobin (Hb) were the only absorbers present. It is also the objective of this study to propose the use of this invention for future clinical use.

2. Material and Method

2.1. Experimental system and subjects

The in-house developed optical system shown in figure 1 consisted of a Light-Emitting Diode (LED model no. SMD 5730), a reflective diffraction grating (model no. GR50-1205), a plano-convex lens and an imager (model no. CW835). The LED was aligned at an angle of 45° from normal to illuminate the surface of the selected skin site. Light interacted with the skin would be reflected onto the diffraction grating; light diffracted from this element produces a spectrum of visible light. A slit was placed in front of the diffraction grating to gate the light. Only light of the selected wavelength would pass through the slit. This single wavelength light passed through a lens before it was focused onto the imager. The latter was placed at a distance of approximately 150 mm from the skin site. The selection of the wavelength is via the rotation of the diffraction grating. This work chose light beam of wavelength 488 nm, 522 nm and 684 nm for illumination of the skin owing to the large variability in the hemoglobin absorption at these wavelengths. The search of these selected wavelengths is via the change in the diffraction angle of the diffraction grating from normal as followed:

$$n\lambda = d(\sin i + \sin i') \qquad (1)$$

where n is the order of diffraction, λ is the diffracted wavelength, d is the grating constant (the distance between successive grooves), i is the angle of incidence measured from the normal and i' is the angle of diffraction measured from the normal. This study used the first order diffraction beam, $n = 1$ and d is calculated as 833 μm.

In this work, experiment was conducted on the palm of the right hand of ten healthy volunteers (aged between 23-27 years) of both genders with different skin tone to investigate the range in the predicted blood oxygen saturation, S_rO_2. These subjects gave their consent to participate in the study. For each subject, three data were consecutively collected from the same skin site.

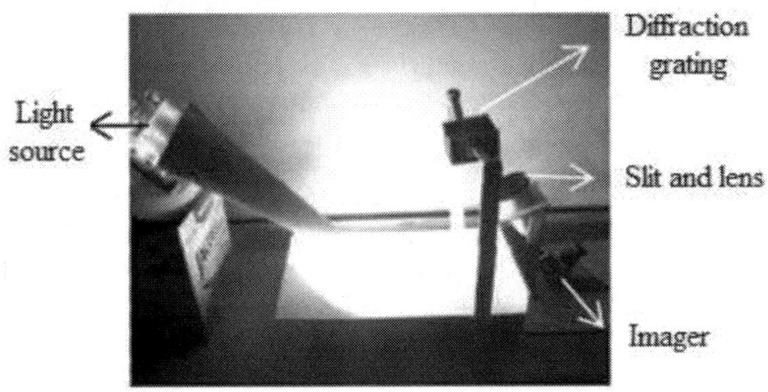

Figure 1. Non-contact optical spectroscopy system

2.2 Analytical model and iterative fitting process

Each image collected from the imager was calculated for its overall mean value to speed up the processing time. An algorithm was written in MATLAB to calculate the light attenuation of each wavelength given by:

$$A = \log\left(\frac{I_w}{I_s}\right) \tag{2}$$

where I_s is the light intensity reflected from the palm of the hand while I_w is the light intensity reflected from the surface of a white reference sheet. The prediction of the required health parameter value is via the wavelength, λ, dependent light diffusion approximation model defined as followed [6]:

$$A(\lambda) = \sqrt{\frac{3\mu_a(\lambda)}{\mu_s'(\lambda)}} \tag{3}$$

where the relationship between light absorption, μ_a, and S_tO_2 is given by

$$\mu_a(\lambda) = \left(\left(\varepsilon_{HbO_2} - \varepsilon_{Hb}\right)S_tO_2 + \varepsilon_{Hb}\right)T\,. \tag{4}$$

The wavelength dependent extinction coefficient value, ε, of the hemoglobin components can be found in the work by Zijlistra [7]. The symbol T in equation (4) denotes total blood concentration and is taken as 268 μmol/L[8]. Meanwhile the reduced scattering coefficient, μ_s', shown in equation (3) is expressed as:

$$\mu_s'(\lambda) = 2 \times 10^4 \lambda^{-1.5}. \tag{5}$$

The search of the required S_tO_2 value shown in equation (4) is via an iterative fitting algorithm using the *fminsearch* function in MATLAB. The optimal S_tO_2 value is assumed to have been found when the absolute difference between the experimentally measured value and the light attenuation calculated from equation (3) is less than 10^{-20}.

3. Results

This study conducted experiments on ten volunteers; among these volunteers five were female (volunteer index number 1, 3, 6 ,7 and 8 in figure 2) while others were male. The mean and standard deviation of the S_tO_2 value predicted for each recruited volunteer using three consecutively collected data are plotted in figure 2. The average of S_tO_2 value is calculated as 74 ± 3.3 % and 68 ± 6.1 %, respectively, for the recruited females and males. Meanwhile the overall average of the S_tO_2 shown in figure 2 is calculated as 71 ± 4 %. The difference in the S_tO_2 value predicted for the two genders was statistically evaluated using an unpaired two-tailed student t-test in SPSS software (SPSS 22, Inc., Chicago, Illinois) with confidence level of 95 %. This statistical test revealed a significance value, p, of 0.752.

4. Discussion

Figure 2 showed that the skin blood oxygen saturation value predicted for majority of the recruited volunteers fluctuated in between 70 – 80 %, with the overall mean value is calculated as 71 %. This range of blood oxygen saturation was also observed in other related works [9, 10] for measurement on the palm of the hand where abundance of arteriovenous anastomoses (AVAs) can be found. The AVAs provides a direct shunting path between venules and arterioles. The measured data reflected from skin is, therefore, taken as the average of blood oxygen saturation information carrying signals from local capillaries, veins and arteries. Meanwhile the calculated significance value, ρ, of 0.752 indicates a high similarity in the value predicted for both genders.

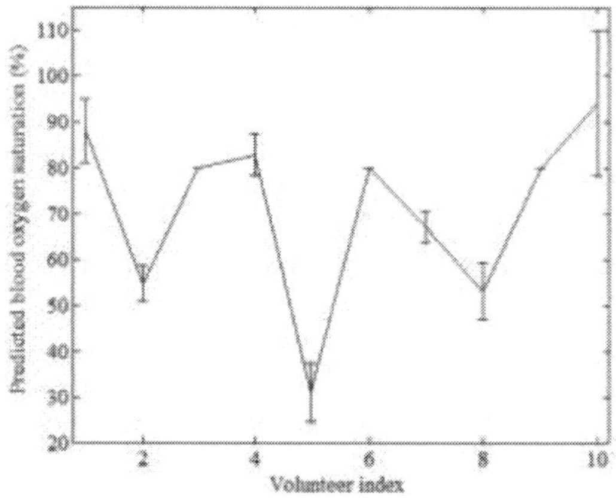

Figure 2. The calculated mean and standard deviation of the S_tO_2 predicted for the recruited volunteers.

This work found a considerably low S_tO_2 value predicted for volunteer index 5 with mean and standard deviation value calculated as 30 ± 6 %. This is owing to the fact that the corresponding volunteer has a comparatively darker skin tone compared to others. This showed that the employed analytical technique is sensitive to one's skin pigmentation. This can be overcome by taking into consideration changes in light attenuation with skin melanin absorption via equation (4), as is demonstrated in the work of [11]. Nonetheless it must also be mentioned that differences in one's skin thickness, health status, age and diet would also affect the predicted value [12].

The obtained results revealed the developed spectroscopy system and the employed analytical strategy may potentially be further improved in the estimation of one's S_tO_2. In addition, the use of light of wavelength 488 nm, 522 nm and 684 nm as the illumination beam is shown suitable for use in developing a skin blood oxygen saturation monitor. Among the possible clinical applications, following improvement of this system, include non-invasive evaluation of wound healing progress, and in post-operative care to monitor the success of skin graft.

5. Conclusion

This work demonstrates the use of a spectroscopy system to non-invasively monitor one's skin blood oxygen saturation using light of wavelength 488 nm, 522 nm and 684 nm. The obtained results revealed an overall mean S_tO_2 of 71 ± 4 %, while no significant difference was observed for the value predicted for the two genders. This work concluded that the performance of the analytical strategy may be improved by including the effects of skin pigmentation on the total light attenuation defined in the employed analytical model.

Acknowledgement

The authors would like to thank all volunteers who participated in this study. This work was financially supported in part by the Ministry of Education Malaysia under FRGS grant no. 1581.

References

[1] M H Chen, C W Kerechanin, D G Greenspan, T B Criss, S C Franckowiak, J A Vincent and R S Pattay 2005 *Johns Hopkins Apl Technical Digest* **26** 67

[2] A Huong, S Philimon, and X Ngu 2017 *Journal of Innovative Optical Health Sciences* **10** 17500041
[3] D Nicoll, C M Lu, M Pignone and S J McPhee 2008 *Pocket Guide to Diagnostic Tests* (The McGraw-Hill Companies)
[4] S Whitener, R Konoske, and J B Mark 2014 *Best Practice & Research Clinical Anaesthesiology* **28** 323
[5] T Slusher, A Bjorklund, H T Aanyu, A Kiragu and Christo Philip 2017 *J Pediatr Intensive Care* **6** 066
[6] A K C Huong 2012 *Spectroscopic analysis of scattering media via different quantification techniques* (The University of Nottingham)
[7] W G Zijlstra, A Buursma and O W van Assendelft 2000 *Visible and near infrared absorption spectra of human and animal hemoglobin: determination and application*: VSP
[8] A Huong and X. Ngu 2014 *Journal of Innovative Optical Health Sciences* **7** 1
[9] S Philimon, A Huong and X Ngu 2017 *IOP Conference Series: Materials Science and Enginering*, **226** 1
[10] S Philimon, A Huong and X. Ngu 2016 *APRN Journal of Engineering and Applied Sciences* **11** 3951
[11] A Huong and X. Ngu 2016 *Proceedings of the 3rd International Conference on Biomedical and Bioinformatics Engineering* 46
[12] S De Spirt, H Sies, H Tronnier and U Heinrich 2012 *Skin Pharmacol. Physiol.* **25** 2

IOP Conf. Series: Journal of Physics: Conf. Series **1049** (2018) 012064 doi:10.1088/1742-6596/1049/1/012064

VANET Routing Protocols: Review, Implementation and Analysis

Muhammad Rizwan Ghori, Ali Safa Sadiq and Abdul Ghani

Faculty of Computer Systems and Software Engineering, University of Malaysia Pahang, Gambang Campus, Kuantan, Malaysia

rizwanghori@outlook.com, alisafasadiq@ump.edu.my, abdulghani@ump.edu.my

Abstract. Wireless technology is developing very fast. Most of the researchers are working in the field of wireless communication. VANET is an evolving technology in the field of wireless communication and with the advancement it will contribute more to the smart transportation system in days to come. VANET gives a communication framework that has enhanced the traffic service and helped in reducing the road accidents. Data sharing in this system is time sensitive and require quick and vigorous network connection forming. VANET is serving the said purposes but there are some issues and challenges like efficient handling of fast handovers for video streaming applications. Therefore, in this paper we have reviewed and discussed several studies related to the routing protocols to judge which one is the best for video applications in VANET. Moreover, after studying different systems made by the researchers, we have critically analyzed them and found advantages and disadvantages for the future works. Also, simulation is performed to check the delays and throughput comparisons between the routing protocols. Furthermore, with the experiments we have proven that the AODV performance is better than the other ad-hoc protocols in VANET environment.

1. Introduction

Vehicular Ad-hoc NETwork (VANET) has become the key research study due to increasing demand of road safety and management. VANET is a subclass of Mobile Ad-hoc NETwork (MANET) which belongs to a family of Wireless Ad-hoc NETwork (WANET). Talking of MANET, it is basically a self-organizing communication system that is not dependent on any infrastructure. Mostly it is used by the military. But nowadays it is becoming common. In simple way, MANET uses the same basic communication methodology just like blue tooth adhoc network used for data sharing between the computers. The basic principle of VANET also is same like MANET. VANET system comprises of mobile nodes which are the sensors embedded in the vehicle, fixed infrastructure consisting of Road Side Units (RSU) (Shakyawar and Tiwari 2016). RSU are the permanently installed units that works as a gateway for connection to the server or internet for getting information. The most vital service provided by the said network is the driving safety as road accidents are the 9th major cause of deaths. Moreover, according to the survey, most of the accidents can be avoided if driver gets the warning half a second before an accident. VANET is serving the said purpose by sharing road safety and traffic analysis information through internet.

According to Yong et al. (2016), the VANET architecture consist of vehicle and infrastructure components. Vehicle workings consists of the On Board Unit (OBU) and the application that will be working for OBU to enable it to communicate. Moreover, infrastructure components consists of RSUs commonly connected to the internet. Subsequently, Figure 1 is showing a VANET system architecture.

In VANET, there are mainly two kinds of communication such as Vehicle To Vehicle (V2V) and Vehicle To Infrastructure (V2I) (Vijayakumar et al. 2016). V2V is a wireless communication between

Content from this work may be used under the terms of the Creative Commons Attribution 3.0 licence. Any further distribution of this work must maintain attribution to the author(s) and the title of the work, journal citation and DOI.
Published under licence by IOP Publishing Ltd

the vehicles where as V2I is the communication between the vehicle and infrastructure. VANET is quite different from other ad-hoc networks in terms of features such as high mobility, abrupt changes of topology, time critical, high computational ability etc. Besides its good features and applications there are some challenges also linked to the said network such as security, scalability, quality of service, power control etc.

Over the years there has been many researches on developing applications and usage models for VANET type of communication. As more people spend time on the road, hence, more requirement of internet connection to communicate with each other, to receive real time news, traffic information and weather reports etc. Moreover, some of the latest applications developed related to VANET are online file sharing, real time video updates and entertainment via connection to the internet through RSUs or V2V type of connections. Moreover, the VANET applications are categorized as safety and comfort applications (Hassan et al. 2016).

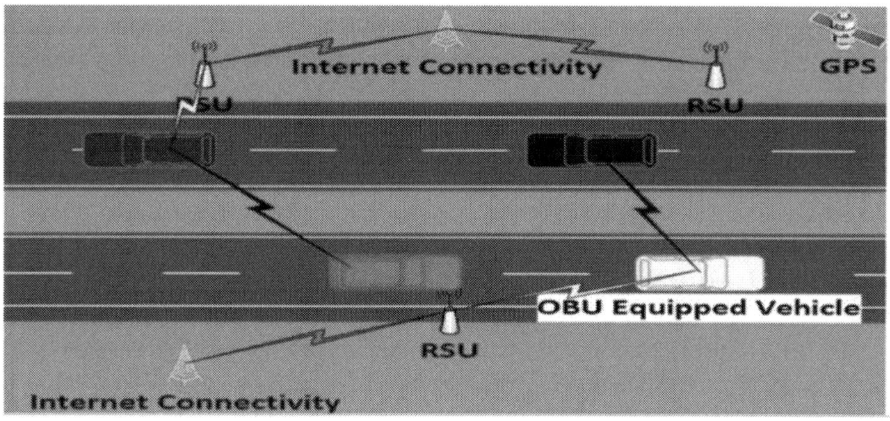

Figure 1: VANET System Architecture

The rest of the paper is organized in a manner that in Section 2, the routing protocols used in VANET are discussed while Section 3 will discuss the related works followed by in Section 4, we talked about video streaming in VANET. In Section 5, we discussed about the OPNET implementation steps. Moreover, Section 6 will show the simulation results and with Section 7, we conclude the paper.

2. Routing Protocols

Due to the high mobility feature of VANET, the use of correct routing protocol is of great concern. The packets in the network are send from vehicle to vehicle that are moving with speed and also the density of vehicles is increasing and decreasing which increases the challenges related to routing protocols. Due to highly challenging nature of VANET, the researcher came out with different types of protocols which will be explained in following sections. Moreover, Figure 2 is showing the classification of routing protocols.

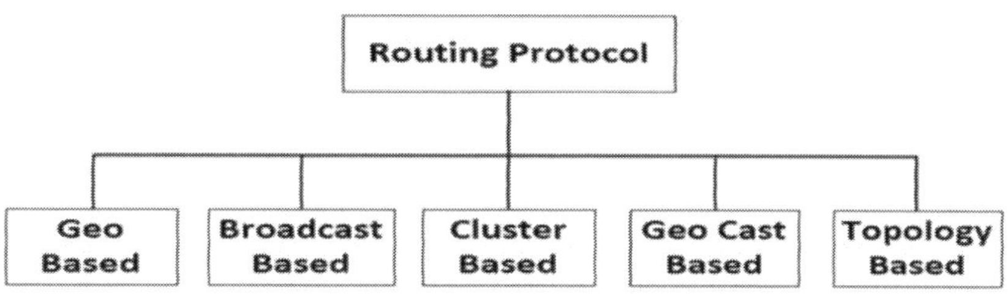

Figure 2: Classification of Routing Protocols

2.1 Geo Based Protocols

In these protocols, a source will communicate with the destination by using geographical positions as well as with its network address. Load Balancing Routing Protocol (LBRP) calculate as well as configure the route as per information based on the location of nodes. Therefore, there is no such need to build the routing tables. The protocol consists of three components such as beaconing, location and forwarding services. The drawback of this protocol is that it requires the Global Positioning System (GPS) assistance to get the vehicles location. Moreover, the satellite signals get weak when the vehicle goes in the area like tunnel. But as far as highway environment is concerned it gives the best performance. Additionally, its advantages include efficient in high mobility environment. Example of these protocols are Greedy Perimeter Stateless Routing and Distance Routing Effect Algorithm for Mobility (DREAM).

2.2 Broadcast Based Routing Protocols

This protocol floods the data packet to the entire VANET to all the available nodes in the broadcast domain. The said protocol is used whenever the destination node is out of the range of the source node. Mostly these protocols are used with application that are concerned with the safety such as road and weather condition warning, emergency warning messages etc. Example of broadcasting routing protocols includes Distributed Vehicular Broadcast Protocol (DV-CAST), Position Aware Reliable Broadcasting Protocol (POCA) and Density Aware Reliable Broadcasting Protocol (DECA). The positive point of the said protocols is its reliability. But these types of protocols consume more bandwidth and many duplicate packets reaches the node which is not good for the protocol performance.

2.3 Cluster Based Routing Protocols

In this protocol, the vehicle with same characteristics such as the speed, direction etc. are combined in one cluster. Moreover, if a vehicle node needs to communicate with the node within the cluster then the data will follow a direct path as it is considered to be a local communication. Moreover, if the vehicle node needs to communicate the node which is outside the cluster than it requires the help of its cluster head for reaching the destination. The scalability factors make it a good choice for the network designers. But traffic delays is its drawbacks. Clustering for Open Inter Vehicular Communication Network (COIN) is the prime example of this protocol.

2.4 Geo-cast Routing Protocols

Geo cast protocol consists of two main zones i.e. Zone Of Relevance (ZOR) and Zone Of Forwarding (ZOF). ZOR is that area dedicated for the nodes of that region. The main goal of this protocol is to make communication possible between the vehicle present in ZOR. Moreover, if the source vehicle wants to communicate with the vehicle that is not in ZOR of that vehicle than the vehicle will become part of ZOF and any vehicle that comes in ZOF has the responsibility of forwarding the information to other ZORs. Due to the frequent changes in the zones, the connection disconnection can take place regularly and the said point comes in the bracket of drawbacks.

2.5 Topology Based Routing Protocols

The focus of this paper is on topology based protocols. The said type consists of two kinds of protocol such as proactive and reactive. Moreover, the classification of routing protocols used in ad-hoc environment is shown in Figure 3.

IOP Conf. Series: Journal of Physics: Conf. Series **1049** (2018) 012064 doi:10.1088/1742-6596/1049/1/012064

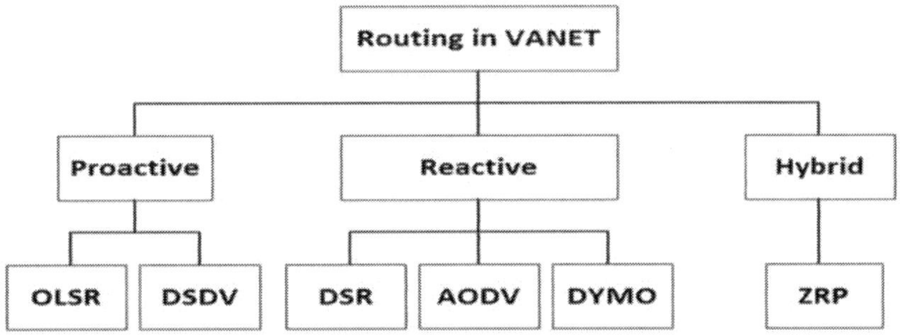

Figure 3: Classification of Ad-hoc Routing Protocols

2.5.1 Proactive Protocols

The protocols keep on computing the route regularly and therefore, the routing table changes or updates frequently. These protocols use Bellman Ford Algorithm in which all the nodes keep the information related to the next node. The one of the advantage of the said protocol is that the route will be known whenever a packet wants to send a data. Example of these protocol includes Optimized Link State Routing (OLSR) and Destination Sequenced Distance Vector (DSDV) protocol.

OLSR is basically a refined version of link state protocol. The working of link state protocol is such that any change in a topology will be broadcasted to all the nodes in a network which will increase the network overhead. OLSR handles with two kinds of messages like hello and a message to control the topology. Hello messages are used to find the data about the connection status. While topology control message is used to broadcast its own neighbour information with the help of multi point relay (MPR) selected list. Because of the use of MPR, the overload has reduced as it was in the case of pure link state protocol.

DSDV protocol is a modification of Bellman Ford Algorithm. This algorithm resolved the problem of looping in routing by maintaining the sequence number information of each node.

2.5.2 Reactive Protocols

This type of protocols does not possess the information about all the nodes (Zhang and Sun 2016). It only keeps the information of the nodes that comes in the route. The example of reactive protocols is an Ad-hoc On-Demand Distance Vector (AODV), Dynamic Source Control Routing (DSR) and Dynamic Manet on Demand (DYMO) protocol.

AODV is the most commonly used protocol in VANET. It possesses the information of next-hop for the destination nodes. Moreover, each routing table lasts for certain period. If there is no route demand within the specified time than route will get expired and a new route will be defined on demand. As per AODV, whenever a source node wants to send a data to the destination node, it will check the route in its routing table. If the route information exists in the table, then packet will be forwarded to the destination. Otherwise the source node will broadcast the route discovery request to the neighbours. AODV is a mobile ad-hoc protocol that is suitable in a high mobility traffic. The said protocols have reduced the broadcasting overhead. Moreover, route discovery process will be done on demand (Sallam and Mahmoud 2015).

DSR is a productive protocol for routing. It is basically made for multi-hop WANET. It is a self-organizing and configuring protocol that does not require any administration. The two main functions of the said protocol are route discovery and maintenance. The said functions work with each other for node discovery and route maintenance

DYMO is another on demand protocol designed after AODV. DYMO routing protocol can be implemented as proactive as well as reactive (Spaho et al. 2012). Moreover, route discovery methodology is on demand whenever required.

2.5.3 Hybrid Protocols
Hybrid is a composite of proactive and reactive routing protocols which reduced the overhead and delays occurrence due to the periodic sharing of topology information (Jain and Jeyakumar 2016). With the hybrid approach, the efficiency and scalability feature of network has improved. On the other hand, the drawback of hybrid approach is high latency for navigating new routes. The common protocol based on hybrid approach is Zone Routing Protocol (ZRP).

3. Related Works
With the advancement of wireless communication, there are different kind of networks developed according to the people requirement. The use of wireless devices has increased tremendously. New networks are being introduced and among all VANET is one of them. Moreover, in VANET the packet routing is the main challenge to handle. In view of aforementioned, some works related to VANET routing is discussed in this section.

Prokop (2011) worked on the analysis of routing protocol and proved DYMO protocol to be the best choice. Moreover, Spaho and Ikeda (2014) made a system to analyse the protocols such as DSDV and DYMO and concluded that DYMO outperformed the DSDV as DYMO can work as proactive as well as reactive protocol. Gupta and Chuba (2014), Zhang and Sun (2015) developed an analysis that among all reactive and proactive protocols, AODV gave the best results in the high mobility environment. Kaur and Malhotra (2015) made the performance analysis of AODV, OLSR and ZRP and came out with the result that hybrid protocol works well in VANET environment due its dual nature. Moreover, Sallam and Mahmoud (2015) came out with the conclusion that AODV is far better than OLSR as it gives very good result with high speed vehicles. Another performance evaluation of topology based routing protocol (DSDV, AODV and LSGR) is done by Kumar, Baghel and Mishre (2016) and reported AODV as a good protocol. Lengliz and Slama (2016) worked on the improvement of AODV protocol and came out with the version AODV-RLT in which they enhanced the life time of the route. Another research done on the comparison of protocols such as Q-AODV and GPSR by Mouhib, Ouadghiri and Naanani (2016). As per the study, authors proved that Q-AODV better than its competitor. Subsequently, another study done by Salman et al. (2016) on topological routing protocols by considering the packet drop ratio and end to end delay. The authors came up with an idea that reactive protocols work well in ad-hoc environment. Setiabudi et al. (2016) made the comparative study between GPSR and ZRP. The author concluded that hybrid protocols always work well in ad-hoc environment and specially with high speed nodes. Kumar and Baghel (2016) made an analysis of AODV and LSGR protocols and as per the authors LSGR is more robust for VANET environment.

According to the comparative study done in Table I, it is proved that AODV is the most suitable protocol for use in VANET environment. Moreover, according to the research, new protocol can be designed by combining the features of AODV with ZRP to enhance the performance.

IOP Conf. Series: Journal of Physics: Conf. Series **1049** (2018) 012064 doi:10.1088/1742-6596/1049/1/012064

TABLE I
Summary of Comparative Study

References	Protocols	Performance Results
Salman et al. 2016	Topology Based Routing Protocols	Reactive Protocols works better
Setiabudi et al. 2016	GPSR and ZRP	Hybrid Protocol (ZRP) works better
Kumar and Baghal 2016	AODV and LSGR	LSGR proved better
Sallam and Mahmoud 2015	AODV and OLSR	AODV is better in more dense and high mobility area
Kaur and Malhotra 2015	AODV, OLSR and ZRP	ZRP proved to be the best in terms of node density and packet sizes
Zhang and Sun 2015	AODV and DSR	AODV performance is better with high mobility nodes then DSR
Gupta and Chaba 2014	AODV, DSR, OLSR and DSDV	AODV is among the best with high mobility nodes
Spaho and Ikeda 2012	DSDV and DYMO	DYMO has better performance than DSDV
Prokop 2011	AODV, DSR and DYMO	DYMO proven best AODV can be used as second choice

4. Video Streaming in VANET

The challenge other than routing is the provision of good quality video streaming in VANET. In VANET, the nodes move very fast. There are many applications that are running on the VANET. But, the application like video streaming is supposed to be a difficult task in VANET environment as the hand offs are very fast. So, with such quick changes, maintaining a quality of service for the aforesaid application is a challenging job (Ennaciri et al 2016). In VANET, there is frequent change in the topology which result in link breakages that causes heavy loss of packets.

Some works has been done to resolve the problems in video streaming. Mammeri, Boukerche and Fang (2016) enhanced the existing internet protocol i.e. real-time transport protocol (RTP) used for the live video streaming. RTP basically based on the principle of user datagram protocol (UDP) in which if packet will loss than there will no retransmission of it. Subsequently, Sarakis et al. (2016) explored VANET performance with running the video streaming application. The author performed the tests by using IEEE 802.11 p and ETSI ITS protocol stack with varied size videos and different vehicle speeds. Moreover, they prove that with the use of proper available communication and computational parameters a good performance can be achieved. Joshi et al. 2016 presented an analysis for video streaming in VANET with the use of fast hand off techniques. According to the authors, the rate of packet loss increases with the increase of video streaming bit rate. Layered Cooperative Cache Management Video Content-Centric Network (LCC-VCCN) was proposed by Wei et al. (2016). Moreover, authors proved with the help of simulation results that the proposed caching strategy performed well as compared to the existing strategies.

After studying different work done, we have come up with the conclusion that there is a need to improve the existing protocols that are being used in VANET for video streaming. In our view, instead of using the UDP, different TCP flavours can be embedded in the protocols to reduce the packet loss in video streaming.

Moreover, as per the review conducted by us in section 2, AODV proved to be the best protocol for the VANET. So, in next section we will implement the VANET scenario with the help of simulation software by using AODV and DSR.

5. OPNET Implementation

To monitor different performance matrices related to AODV and DSR protocols in VANET environment, we have simulated some scenarios with the help of OPNET modeler 14.5. This section will explain in detail the scenarios and parameters used for the simulation.

5.1 Simple VANET Scenario

This scenario consists of two parts having 20 and 40 nodes (mobile stations) respectively and given them the random trajectories for the movement as shown in Figure 4 and 5. Moreover, for the application designation we have included the application config and profile config to set the applications (ftp and http) used by the nodes. Additionally, we have added the mobility config node to set the profile related to mobility of all the nodes. Subsequently, we changed the routing protocol of all the nodes to AODV.

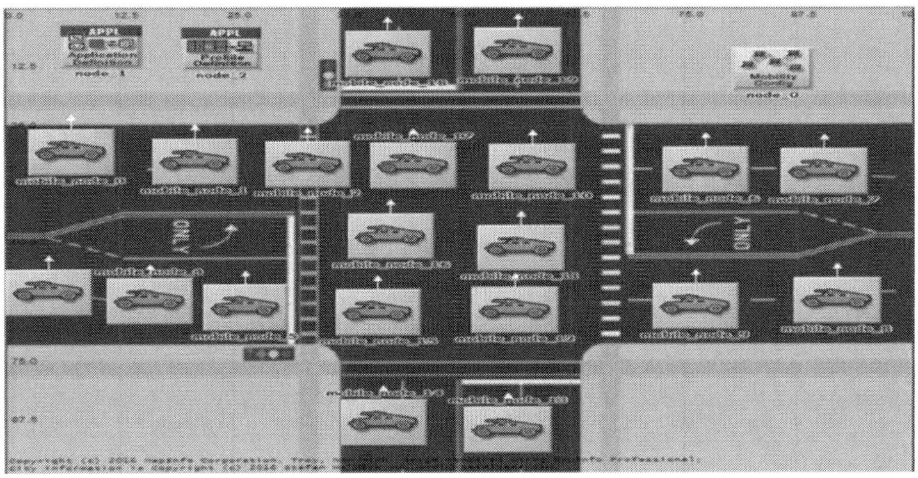

Figure 4: Simple Scenario with 20 Nodes

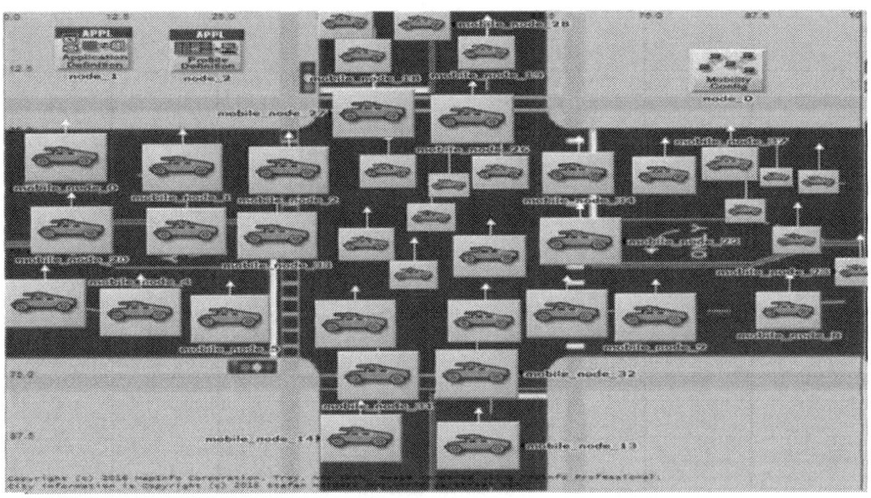

Figure 5: Simple Scenario with 40 Nodes

Moreover, we repeated the same procedure as AODV one except for routing protocol. In this scenario, we used the DSR protocol to monitor the difference of performance as compared to AODV.

5.2 Complex VANET Scenario

In the said scenario, we have added the road side units (RSU) along with 20 nodes. We have used the same applications and settings like previous scenario except that we put four RSUs along the road. Moreover, we have given different BSS identifiers to all RSUs. Additionally, each RSU has designated nodes associated with it and the associated node will have the same BSS identifier as of the RSU which is bound to give the services as shown in Figure 6. Also, we have assumed that there are two different VANET networks and the RSU that are in one network are linked with PPP 28k for their communication. Subsequently, this scenario implemented with the AODV protocol. Similarly, the same scenario is repeated by using the DSR protocol for the comparisons. In Fig. 6, the red blocks are showing different VANETs. Moreover, in every VANET as discussed, all the nodes will have same identifiers related to BSS.

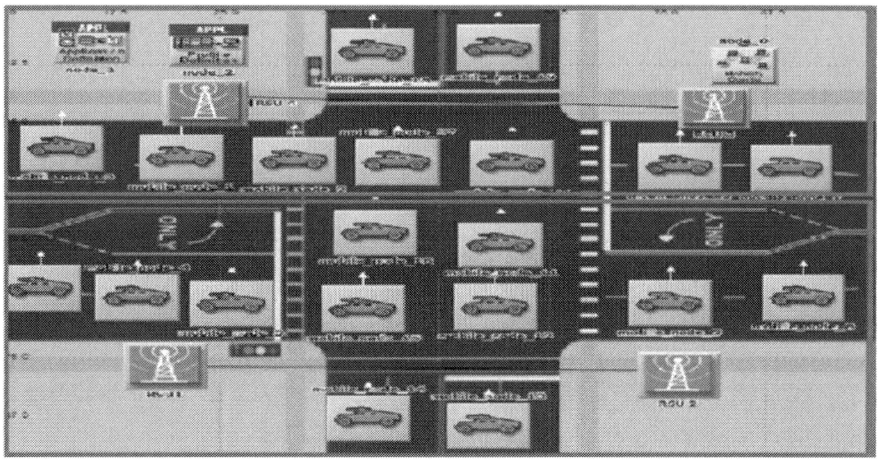

Figure 6: Complex Scenario

5.3 Simulation Parameters and Performance Matrices

For any simulation, there must be some parameters that has to be defined. Same is the case in our simulation. Our simulation parameters are shown in Table II.

TABLE II
Simulation Parameters

Maximum Simulation Time	1 Hour
Environment Size	100 x 100 km
Number of Nodes	20, 40
Mobility	Random way point
Speed	10 km / hour
Type of Traffic	FTP and HTTP
Routing Protocol	AODV and DSR

6. Simulation Results

This section will present the simulation results achieved from the scenarios described in the previous section for different scenarios.

6.1 Simple VANET Scenario Results

Figure 7 and 8 is showing the through put and delay comparison respectively between the AODV and DSR with 20 vehicles in simple scenario.

According to the Figure 7 and 8, the performance of AODV in terms of throughput and delay is far better than the DSR protocol.

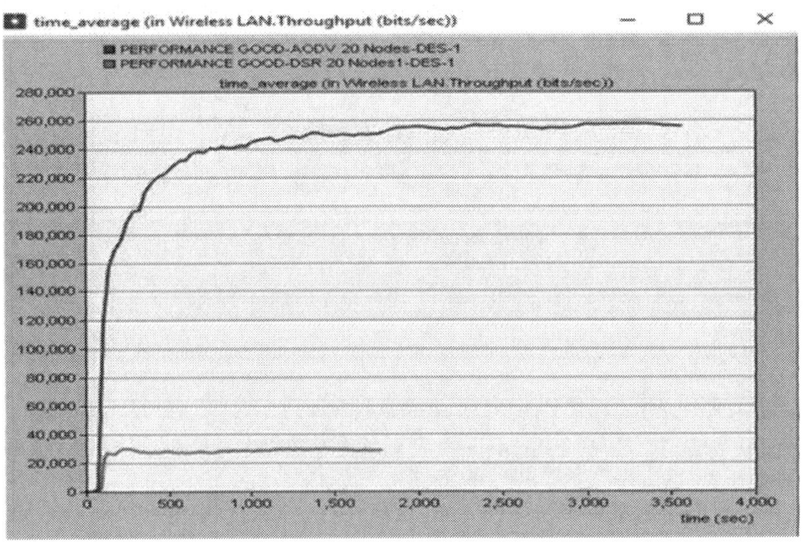

Figure 7: Throughput Comparison AODV and DSR with 20 Nodes

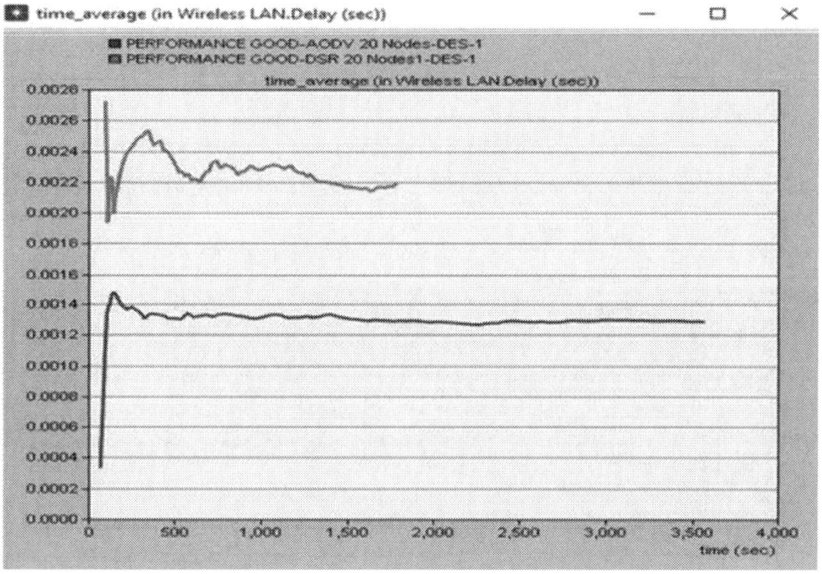

Figure 8: Delay Comparison AODV and DSR with 20 Nodes

Moreover, Figure 9 and 10 are showing the throughput and delay comparison with 40 nodes. As per the Figure 9 and 10, the AODV is better than DSR in terms of throughput but showing more delay as compared to DSR.

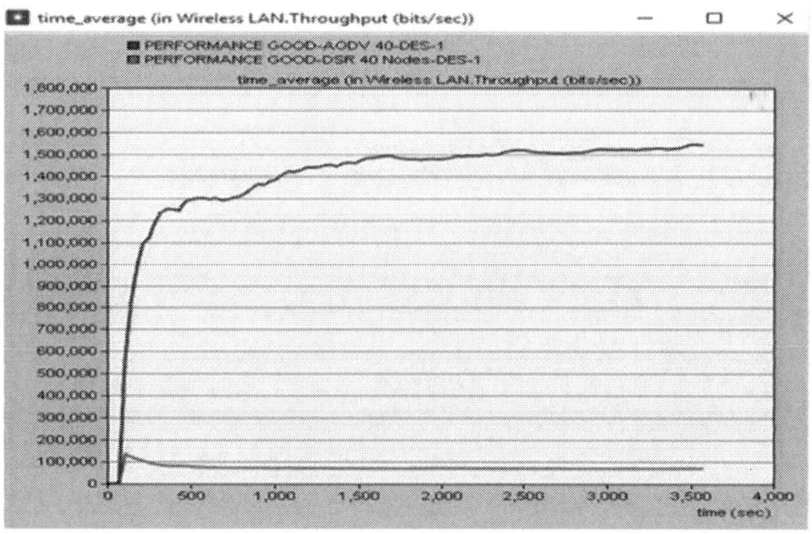

Figure 9: Throughput Comparison AODV and DSR with 40 Nodes

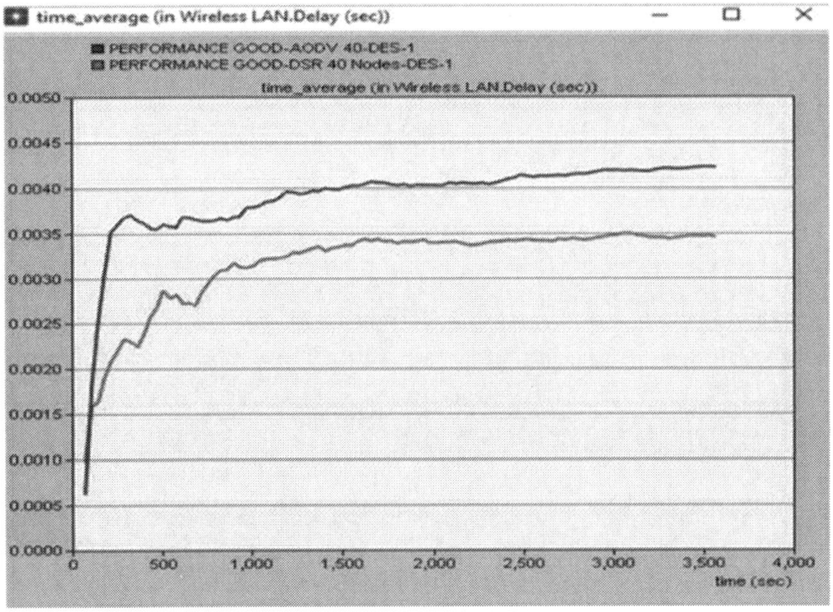

Figure 10: Delay Comparison AODV and DSR with 40 Nodes

6.2 Complex VANET Scenario Results

In this scenario, we have kept 20 mobile nodes and did comparison by using AODV and DSR routing protocols. According to Figure 11 and 12, like previous scenarios AODV performed good as compared to DSR in terms of throughput as well as delay.

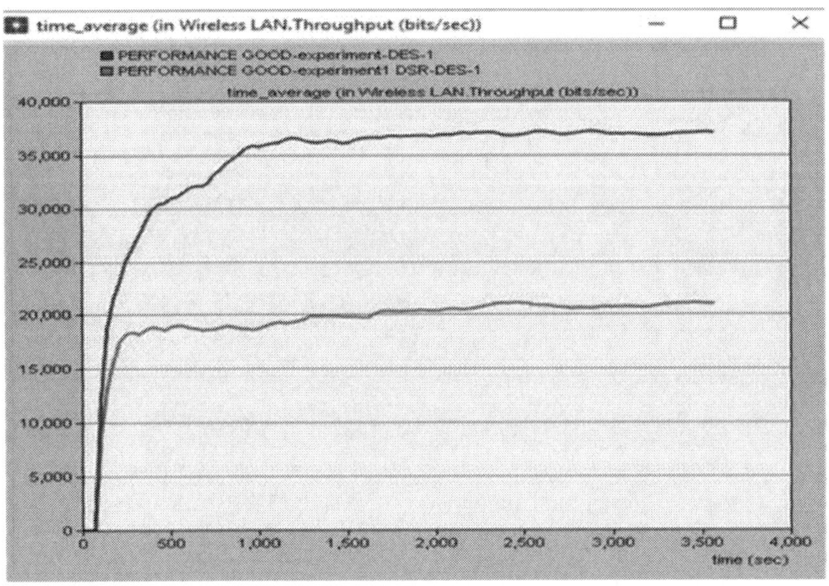

Figure 11: Throughput Comparison AODV and DSR with RSUs

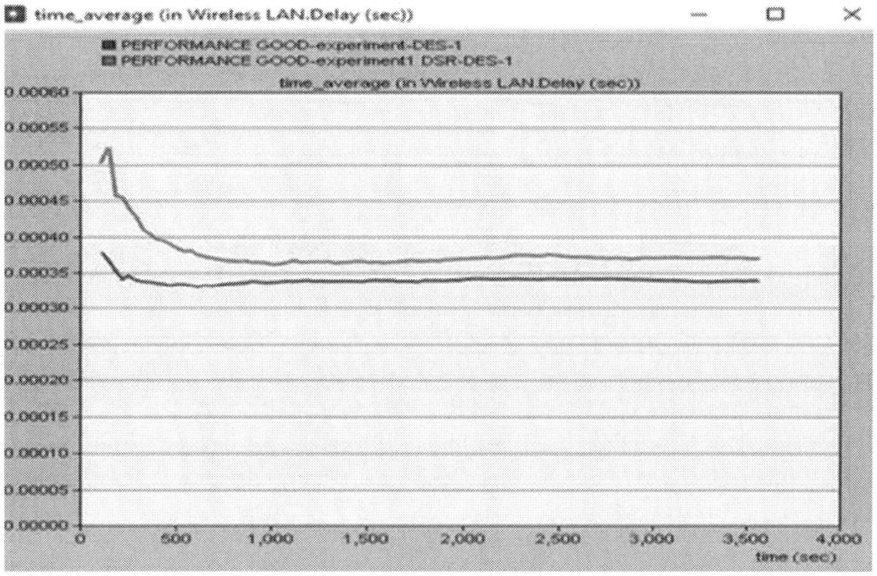

Figure 12: Delay Comparison AODV and DSR with RSUs

7. Conclusion

In this paper, we have reviewed many studies related to routing protocols. As per the research completed, AODV proved to be the best routing protocol in VANET environment. Moreover, video streaming is another challenging task to achieve in VANET. According to the study, we have concluded that lot of effort is required to get a good quality video streaming in VANET and AODV again is proven to be better for video applications. For the aforesaid we have done simulation using two scenarios that is the simple and complex. In simple we only kept mobile nodes and took the performance results of AODV and DSR. Also, in complex scenario, we added the RSUs to have more comprehensive results. Both the aforementioned situations have shown that AODV performance is far better than the DSR. Moreover, for future works we recommend that a new protocol can be designed by combining some flavors of TCP with UDP.

References

[1] **Ennaciri, A**. et al, 2016. Performance analysis of streaming video over vehicular ad-hoc, in Proc. of 13th International Conference on Computer Graphics, Imaging and Visualization (CGiV), pp. 375-380.

[2] **Gupta, P**. and Chaba, Y., 2014. Performance analysis of routing protocols in vehicular ad-hoc networks for cbr applications over udp connections, in Proc. of International Journal of Engineering and Computer Science, vol. 3, pp. 6418-6421.

[3] **Hassan, A.S.A**., Hossain, M.S and Atiquzzaman, M., 2016. Security threats in vehicular ad-hoc networks, in Proc. of International Conference on Advances in Computing, Communications and Informatics (ICACCI), pp. 404-411.

[4] **Joshi, J**. et al., 2015. COMIP: Cluster based overlay and fast hand off mobile internet protocol system for video streaming in VANETs, in Proc. of 2015 IEEE 3rd International Conference on Smart Instrumentation, Measurement and Applications (ICSIMA), pp. 1-6.

[5] **Kaur, A**. and Malhotra, J., 2015. On the selection of quality of service provisioned routing protocol through realistic channel for VANET, in Proc. of International Journal of Scientific and Technology Research, vol. 4, issue. 07.

[6] **Kumar** et al., 2016. Performance Evaluation of DSDV, AODV and LSGR Protocol in Ad Hoc Networks, in Proc. of 2016 International Conference on Electrical, Electronics, and Optimization Techniques (ICEEOT), pp. 4261-4266.

[7] **Lengliz, I**. and Slama, A., 2016. Enhancing AODV and DSR with the RLT strategy for efficient routing in VANETs, in Proc. of 2016 International Symposium on Networks, Computers and Communications, pp. 1-6.

[8] **Mammeri, A**., Boukerche, A. and Fang, Z., 2016. Video streaming over vehicular ad hoc networks using erasure coding, in Proc. of IEEE Systems Journal, vol. 10, pp. 785-796.

[9] **Mouhib, I**. et al., 2016. Network as a service for smart vehicles: A new comparative study of the optimized protocol Q-AODV and GPSR protocol, in Proc. of 2016 International Conference on Engineering and MIS (ICEMIS), pp. 1-5.

[10] **Prokop, M**., 2011. Routing protocol evaluation development of a fully functional simulation environment for vehicular ad hoc networks, Msc. Thesis, Rochester Institute of Technology Kate Gleason College of Engineering, New York, USA.

[11] **Sallam, G**. and Mahmoud, A., 2015. Performance evaluation of OLSR and AODV in VANET cloud computing using fading model with SUMO and NS3, in Proc. of International Conference on Cloud Computing (ICCC), pp. 1-5.

[12] **Salman, A**. et al., 2016. Analysis of topology based routing protocols for VANETs in different environments, in Proc. of 2016 IEEE International Multidisciplinary Conference on Engineering Technology (IMCET), pp. 27-31.

[13] **Sarakis, L**. et al., 2016. Providing entertainment application in VANET environment, in Proc. IEEE Wireless Communication Conference, pp. 30-37.

[14] **Setiabudi, A**. et al., 2016. Performance comparison of GPSR and ZRP routing protocols in VANET environment, in Proc. 2016 IEEE Region 10 Symposium (TENSYMP), pp. 42-47

[15] **Spaho, E**., Ikeda, M., Barolli, F., Xhafa, F., Younas, M and Takizawa, M., 2013. Performance evaluation of OLSR and AODV protocols in a VANET, in Proc. of IEEE 27th International Conference on Advanced Information Networking and Applications (AINA), pp. 577-582.

[16] **Vijayakumar, P**., Azees, M. and Deborah, J., 2016. Dual authentication and key management techniques for secure data transmission in Vehicular Ad Hoc Networks, in Proc. of International Journal of IET Intelligent Transport Systems, vol. 17, pp. 1015-1028.

[17] **Wei, Y**., Xu, C., Wang, M. and Guan, J., 2016. Cache management for adaptive scalable video streaming in Vehicular Content-Centric Network, in Proc. of 2016 International Conference on Networking and Network Applications (NaNA), pp. 410-414.

[18] **Yong, X**., Libing, W.U., Yubo, Z. and Jian, S., 2016. Efficient and secure authentication scheme with conditional privacy preserving for VANETs, in Proc. of Chinese Journal Electronics, vol. 25, Issue. 5.

[19] **Zhang, J**. and Sun, Z., 2016. Assessing multi-hop performance of reactive routing protocols in wireless sensor networks, in Proc. of IEEE International Conference on Communication Software and Networks (ICCSN), pp. 444-449

IOP Conf. Series: Journal of Physics: Conf. Series **1049** (2018) 012065 doi:10.1088/1742-6596/1049/1/012065

Ergonomic Assessment in Small and Medium Enterprises (SMEs)

Voon Yen Siong, Jalil Azlis-Sani, Nik Hisyamudin Muhd Nor, Muhammad Nur Annuar Mohd Yunos, Jacquelyne Anne Boudeville, Syazwan Ismail

Faculty of Mechanical Engineering & Manufacturing Universiti Tun Hussein Onn Malaysia (UTHM), 86400 Parit Raja, Johor.

E-mail: azlis@uthm.edu.my

Abstract. Small and Medium Enterprises (SMEs) industry plays important role in economic growth in Malaysia especially food production industries. Ergonomics in SMEs food industries are one of the most common issues being neglected among people. In this study, the objective is to identify the possibility of an adverse health effect, to conduct hazard and risk assessment and lastly to evaluate the ergonomics problem at SMEs. Hazards identification was done through interview sessions and Industrial Accidental Prevention Association (IAPA) checklist. Checklist based on direct observation on working activities is being evaluated in SMEs. Hazards recorded down are mainly focus on potential ergonomics problems. Then, Hazard Identification, Risk Analysis and Risk Control method (HIRARC) is used to further analyse the hazards. Results show that majority of the employees are unfamiliar with the concept of ergonomics which easily expose themselves to ergonomic injuries hence, it is indeed important to increase awareness among the employees and employer of the company to emphasize on the safety of the instruments and machines used. Lastly, recommendations on each hazard of working activity are given to improve the overall condition and production of the company. This study would serve as a guidance and reference related to this field to eliminate the potential ergonomics issues.

1. Introduction

Small and Medium Enterprises (SMEs) plays an increasingly crucial role in economic growth in Malaysia. SMEs are one of the significant industries in Malaysia besides running as an important source in developing a rural area and maximising the resources allocation by processing the raw part into products. SMEs in Malaysia, for instance in machinery industries, food supplies industries, and heavy workload industries are facing several problems such as they are still performing traditional technology [1]. Mikami and Tanaka (2008) shown that the food processing industry is one of the most important industry in developing economic growth[2].

Small and Medium Enterprises (SMEs) are playing huge and significant contribution to the economic development that can includes business activities in our country regardless of urban or rural area. The economy of our country is changing to a more knowledge-based economy rather than the industrial based economy to achieve the vision of 2020 [3].

Normally built in a rural area, people in those areas get their job to make their monthly wages for their living. Beck (2013) stated that SMEs were defined as an engine of innovation and growth and they help decrease poverty[4]. Ergonomics related problems are one of the most common issues being

Content from this work may be used under the terms of the Creative Commons Attribution 3.0 licence. Any further distribution of this work must maintain attribution to the author(s) and the title of the work, journal citation and DOI.
Published under licence by IOP Publishing Ltd

neglected or ignored often among the industries. Ergonomics issues are mostly addressed by working posture and design of workstations or tools. Punnett L and Wegman (2004) also stated that risk factors for MSDs are rapid work pace and repetitive motion patterns, insufficient recovery time, heavy lifting and forceful manual exertions, non-neutral body postures, mechanical pressure concentrations, and body vibration [5].

In this rapidly globalizing world, safety performance is a key issue for the industries to become a world-class competitor. Occupational accidents may lead to permanent disabilities or deaths and/or economic losses or both [6]. In order to develop a good safety culture, the attitude of the workers needs to be changed by adopting best practices, good housekeeping, and changes in work culture and work practices.

The major contributing factors include physical factors, environmental factors, individual factors and work organization factors. Ilman et al., (2012) has implied the need for employers and employees to take more active roles in addressing issues related to work postures [7]. When employees are exposed to a combination of these contributing factors, the chances of having work-related injuries increase. Thus, identifying all contributing factors is crucial in reducing potential ergonomics issues.

The objectives of this study are to identify the possibility of an adverse health effect, to conduct hazard and risk assessment and to evaluate the ergonomics problem at the small and medium enterprises (SMEs). The research study has a scope that only limited to focus on the potential ergonomics problem among the workers at the SMEs and observation on working activities of the workers in small and medium food enterprises (SMEs) industry in Parit Raja, Johor.

2. Methodology

Methodology for this study is divided into 3 main components which are direct observations, ergonomics checklist, interview session and Hazard Identification, Risk Analysis and Risk Control (HIRARC) method.

The HIRARC method is to ensure that any potential risk to 3 three safety, health, environment and business aspects of any operation is minimized. To identify actual injury hazards, it is proposed that greater attention is given to exposure assessment by systematically analyzing injury cases and case series [8]. As for risk control, hazard should be controlled at their source. In order to reduce the accident or incident level, it is important to ensure that safe working practice is being observed and controlled [9]. The checklist is used in this study of assessing the potential ergonomic problems at the SMEs to evaluate the effectiveness in reducing ergonomic injuries and disorders in small and medium food enterprise industry. A form is very vital because observations have shown that most of managers or supervisors do not carry safety form or checklist when supervising their workers in the factory [10].

Firstly, interview questions are being prepared and applied during industries visit to further understand the condition of the company. Ergonomics questions are being asked during the interview. Then, Industrial Accidental Prevention Association Checklist also known as IAPA checklist is used to assess or evaluate the instruments the employees used at workstation and machines they operate. Ergonomics assessment is made based each category and comments were given as well.

Photos on each workstation are taken to further observe the condition of the working environment. After that, Hazard Identification, Risk Analysis and Risk Control method is used to provide a more systematic and objective approach ways to control risks where risk is calculated based severity and likelihood of hazard occurrence. Recommendations are given based on each working process.

Two crisps production industries were chosen and visited in this study which named as Company A and Company B. The crisps are mainly made from cassava roots or 'Ubi Kayu' in Malay language. Cassava root is also named scientifically as Manihot Esculenta. From the interview sessions, Company A and Company B have the same number of employees which is twenty persons. Company A has nine foreigners and eleven local workers whereas company B has four foreigners and sixteen locals.

3. Data Collection

According to the interview session, Company A has an industry accident record of five injuries over the past three years. Company B has only one injury case happened in three years. Majority of the employees in Company A and B have no knowledge of the concept of ergonomics. The process of making chips or crisps includes peeling, washing, slicing, deep-frying, colouring, and flavouring, weighing and packaging and lastly marketing.
Industrial Accidental Prevention Association checklist or IAPA checklist is used to identify hazard of each working process. The criteria of the checklist include records of injuries, employee comments, physical demands of work tasks, layout and condition of the workplace or workstation, characteristics of objects handled, environmental conditions, work clothing and PPE or personal protective equipment and characteristics of the organization of the work. A Yes or No answer is ticked based on the criteria through direct observation. Table 1 shows the results of the checklist in company A and company B.

After identified hazards of each working process, HIRARC method is used to deliver a more systematic approach to assess hazards and their respective risks based on each identified hazard as well as provide a method to control the risk. Hazard identified of work activities are divided into five which are cassava roots skin peeling, slice cassava roots with cutting machine, transfer buckets of cassava roots from places to places, stirring cassava roots in frying oils, deep-frying and packaging.

Table 2 illustrates the hazard identification of company A and B. Risk Analysis is also being calculated and ranked from the highest risk to the lowest risk. Table 3 shows the risk analysis of Company A and Company B. From the results of risk analysis acquired from company A acquired, stirring and deep-frying have the highest risk, 20. Cassava roots slicing with cutting machine has the second highest rank in risk,12 followed by buckets of cassava roots transferring from places to places which rank the third, 10. As for the fourth in ranking, cassava roots skin peeling has a risk of 8 whereas the least risk out of all work activities is 3, packaging.

Table 1. Hazard Identification

No.	Work Activity	Hazard		Effect	
		Company A	Company B	Company A	Company B
1	Cassava roots skin peeling	Non-ergonomic seats		Backpain	
2	Slice cassava roots with cutting machine	Cutting blades		Amputation, fingers get caught into the machine	
3	Transfer buckets of cassava roots from places to places	Heavy loads		Fatigue and back injuries	
4	Stirring cassava roots in the frying oil, Deep-frying	Extremely high temperature of frying oil, splatter of oil		Heat exhaustion, fatigue, Upper limb injuries and legs injuries	
5	Packaging	Heavy loads		Muscular strain and backpain	

Table 2. Risk Identification

	HAZARD IDENTIFICATION	A	B
Records of Injuries	Are there records of injuries or accidents to indicate a risk of adverse health effects due to ergonomic factors in the task evaluated?	Y	Y
Employee Comments	Are there employee comments to indicate a risk of adverse health effects due to ergonomic factors in the job or task being evaluated?	Y	Y
Physical Demands	Is forceful physical handling such as carrying, lifting, lowering, pushing, pulling a part of the job?	Y	Y
	Are there contact forces exerted on to the body?	N	N
	Does the employee have difficulty gripping on object or tool, which has a smooth, slippery surface?	Y	N
	Are object handled with a pinch grip?	N	N
	Does the work involve repetitive motions or many similar movements?	Y	Y
	Is the work fast-paced or controlled by a machine or process?	Y	Y
	Are employees required to sit or stand continuously for more than two hours or in total for more than three hours in the shift?	Y	Y
	Does the task require that any part of the body be maintained in a static posture?	N	N
	Does the task require the employee to work with any body part in an awkward position instead of a neutral one?	Y	Y
Layout and Condition of the Workplace or Workstation	Do working heights or reaches cause employees to bend or reach beyond a comfortable range?	N	N
	Does the workplace layout require awkward or extreme movements?	Y	N
	Does the layout of the workstation restrict movements of the body, for example, by limiting leg room?	N	N
	Do observations indicate problems with the design of seating?	Y	Y
	Are employees unsure of how to adjust their workstations?	Y	N
	Do floors or sloped surfaces such as ramps pose a risk of slipping, cause problems for employees who stand on them for long periods, or cause problems for employees who stand on them for long periods, or cause problems for pushing or pulling objects?	Y	N
Characteristics of Objects Handled	Are there problems handling an object due to its size, shape or weight?	N	N
	Are there problems handling an object due to its condition? For example, is the object fragile, unbalanced, or non-rigid?	N	N
	Are handles on containers an inappropriate size or shape, or not strong enough for the weight and size of the object?	N	N
	Are the handles for tools or equipment in size, shape or height?	N	N
	Is vibration from the tool or equipment transmitted to the operator's hand/arm?	Y	Y
	Is the palm or base of the hand used like a hammer for striking?	N	Y
	Do objects, tools or parts of the station with hard, sharp or uneven surfaces put pressures on any body part?	N	N

Table 2. Risk Identification (con't)

Environmental Conditions	Is the employee exposed to extreme temperatures?	Y	Y
	Are any parts of the body exposed to cold from exhaust air, cold liquids or other objects?	N	N
	Do employees assume awkward postures to overcome problems associated with glare, in adequate lighting or poor visibility?	N	N
	Is the employee's whole body exposed to vibration for significant portions of the work shift?	Y	Y
Work Clothing and PPE	If the employee wears gloves, do the gloves hinder gripping or restrict movement?	N	N
	Do records, employee comments or observations indicate fatigue or postural problems from the use of personal protective equipment?	N	N
Characteristics of the Organization of the Work	Are there indication of excessive fatigue or pain, or symptoms of adverse health effects due to extended work days or overtime?	Y	Y
	Are there indications of excessive fatigue or adverse health effects due to shift work or piecework?	Y	N
	Is there build-of fatigue or a risk of adverse or a risk of adverse health effects due to insufficient rest periods or task variety?	Y	Y
	Are tasks in a job rotation program similar to one another, and therefore not providing a variation in movements?	N	N
	Do peak workloads or sudden increases in pace occur with the task?	Y	Y

However, in Company B, deep-frying and stirring has also the highest risk, 16. Cassava roots slicing with cutting machine is the second risky activity. Next, transfer buckets of cassava roots from places to places and cassava roots skin peeling has the third and the fourth rank in risk, 8, followed by the least risk in rank, packaging. Company B has a lower risk in average compared to company A. Lastly, control measures are suggested to improve the condition of each work station.

Comparisons between companies are shown in Table 4. Layout design of company A and company B are sketched but they are not the exact dimension and scale. It clearly shows that Company B has a better layout design than company A. Suggestion and modification is also given to have a better layout design which is shown in Figure 1.

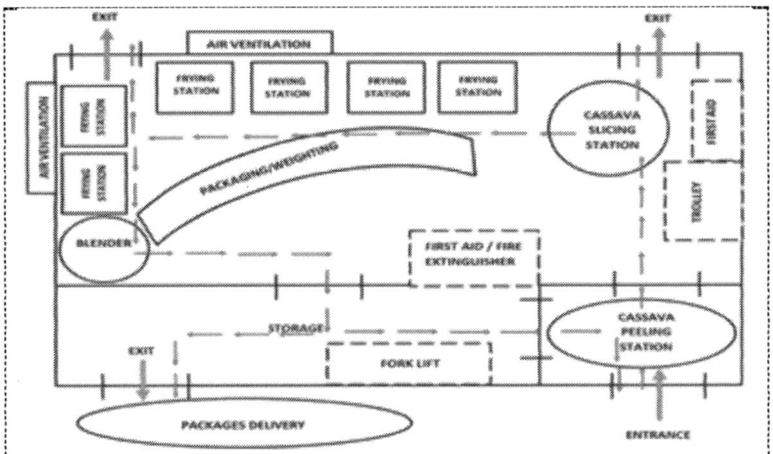

Figure 1. Figure Suggested and modified layout design.

Table 3. Risk Analysis

Risk Analysis		Company A			Company B		
Work Activity	Existing Risk Control (if any)	Likeli-hood	Severity	Risk	Likeli-hood	Severity	Risk
Cassava roots skin peeling	Nil	4	2	8 (Medium)	4	2	8 (Medium)
Slice cassava roots with cutting machine	Safe work practice	4	3	12 (Medium)	3	3	9 (Medium)
Transfer buckets of cassava roots from places to places	Usage of trolley	5	2	10 (Medium)	4	2	8 (Medium)
Stirring cassava roots in the frying oil, Deep-frying	Safe work practice, safety boots	5	4	20 (High)	4	4	16 (High)
Packaging	Manual lifting procedure	3	1	3 (Low)	1	1	1 (Low)

Table 4. Comparison between companies

Company	Differences A	B
Design of the industry	Opened space industry	Closed space industry
Floor condition	Slippery floor	Dry and safe floor
Layout Design	Disorganised layout of the stations	Systematic arrangement of the stations
Comfortability of the working condition	Cool working condition	Hot working condition
Similarities between company A & B		
Organizational practice - No job rotation work practice		

4. Conclusion

Based on this study that had been done, it can be conclude that it is indeed important to increase awareness among the employees and employer of the company to emphasize on the safety of the instruments and machines used. Level of ergonomics problem should also be highlighted to reduce all type of potential injuries.

Recommendation is given based on each work station. However, employee involvement is also the key to the working environment. Open communication between management or employer and the workers throughout the ergonomic improvement process allows for a flow of information critical to identify contributing factors and solving problems. Once the communication channel is in place, the next step is to form an ergonomics team whose job is to look for task that may lead to potential ergonomics injuries. Having a team who administrates ergonomics can benefits the company because they can always come out with suggestions that can be easily implemented with minimal cost and mechanical intervention.

This research serves to identify possibility of adverse health effect as well as to evaluate the ergonomics problem of small and medium enterprises (SMEs). This could ensure the safety of the employees and the management of the company as well as to provide a more systematic layout design in order to increase the effectiveness and efficiency of the production. This study is also essential in recommending suitable approach at workplace so as to assist and provide employees a better working surrounding. Through this ergonomics assessment, risk measures and several recommendations were given to ease the employees at work and to avoid any unnecessary potential ergonomics issues.

IOP Conf. Series: Journal of Physics: Conf. Series **1049** (2018) 012065 doi:10.1088/1742-6596/1049/1/012065

Acknowledgment
The authors are supported by the Fundamental Research Grant Scheme (FRGS) Vot 1542, Office of Research, Innovation, Commercialization and Consultation (ORICC) Universiti Tun Hussein Onn Malaysia. The authors would also like to thank SMEs on the contribution and participation for this study.

References
[1] Senik, G. 2010 *Small-scale food processing enterprises in Malaysia.*

[2] Mikami, K., & Tanaka, S. 2008 *Food processing business and agriculture cooperatives in Japan: Market Power and Asymmetric Information. Asian Economics Journal,* ***22*****(1)**, 83–107.

[3] Ong, J. W., Yeap, P. F. , & H. 2010 *Malaysian small and medium enterprises: the fundamental problems and recommendations for improvement. Journal of Asia Entrepreneurship and Sustainability,* ***6*****(1)**

[4] Beck, T. 2013 *Bank financing for SMEs - Lessons from the Literature. National Institute Economic Review,* **16**.

[5] Wegman DH., & Punnett L. 2004 *Work-related musculoskeletal disorders: The epidemiologic evidence and the debate. Journal of Electromyography and Kinesiology,* **14(1)**, 13-23

[6] Dağdeviren M., & Yüksel İ. 2008 *Developing a fuzzy analytic hierarchy process (AHP) model for behaviour-based safety management.* ***178*****,** 1717-1733.

[7] Ilman, A. Y., & Y. Helianty. 2012 *Rancangan perbaikan sistem kerja dengan metode Quick Exposure Check (QEC) di bengkel sepatu X di Cibaduyut. Jurnal Online Institut Teknologi Nasional,* **2(1)**, 121-129.

[8] Robert M.Park. 2002 *Hazard identification in occupational injury;Reflections on standard epidemiological methods. International Journal of Occupational.*

[9] Radlinah Kunju Ahmad 2000 *Developing a proactive safety performance Measurement Tool(SPMT) For Construction Sites .Loughborough University. Degree of Doctor of Philosophy.*

[10] Nedumaran, B. 2004 *Industrial safety and risk management. Sri Venkateswara College of Engineering. India.*

IOP Conf. Series: Journal of Physics: Conf. Series **1049** (2018) 012066 doi:10.1088/1742-6596/1049/1/012066

Visualization on Colour Based Flow Vector of Thermal Image for Movement Detection during Interactive Session

Nabilah Ibrahim[1]**, Mohamad Nurazmi Raman, Lina Farhana Mahadi, Wan Nurshazwani Wan Zakaria**
Faculty of Electrical and Electronic Engineering,
Universiti Tun Hussein Onn Malaysia, Batu Pahat, Johor MALAYSIA

E-mail: nabilah@uthm.edu.my

Abstract. Recently thermal imaging is exploited in applications such as motion and face detection. It has drawn attention many researchers to build such technology to improve lifestyle. This work proposed a technique to detect and identify a motion in sequence images for the application in security monitoring system or outdoor surveillance. Conventional system might cause false information with the present of shadow. Thus, methods employed in this work are Canny edge detector method, Lucas Kanade and Horn Shunck algorithms, to overcome the major problem when using thresholding method, which is only intensity or pixel magnitude is considered instead of relationships between the pixels. The results obtained could be observed in flow vector parameter and the segmentation colour based image for the time frame from 1 to 10 seconds. The visualization of both the parameters clarified the movement and changes of pixel intensity between two frames by the supportive colour segmentation, either in smooth or rough motion. Thus, this technique may contribute to others application such as biometrics, military system, and surveillance machine.

1. Introduction

Recent advances in thermal image digital storage and video hardware has led to new applications such as surveillance, driver license application systems, biometrics, and virtual reality. These applications have attract the attention of researchers in motion detection system to be applied in current lifestyle [1]. In applications such as outdoor surveillance, where the background temperature is largely different from humans, thermal imaging can be employed in identifying and tracking people. One of the advantage of thermal imaging the ability to unrecognized shadow or light illumination, which is usually hold up on most systems based on motion detection system. Furthermore, contour of the heat signatures is the main concern in motion tracking because the contour have the physical properties that been used as subject to track down all the movement. The shadow region can cause false information and by avoiding this region it helps to improve the sensitivity of the motion detection. Using this information, thermal images are capable to detect the motion of the targeted person.

In addition to detect motion in thermal image, some may suggest the thresholding method to distinguish the motion subject and the stationary subject. However, the major problem when using thresholding method is that only the intensity is considered instead of relationships between the pixels. Extraneous pixels could easily be included that are not part of the desired region which lead to missing

[1] nabilah@uthm.edu.my

Content from this work may be used under the terms of the Creative Commons Attribution 3.0 licence. Any further distribution of this work must maintain attribution to the author(s) and the title of the work, journal citation and DOI.
Published under licence by IOP Publishing Ltd

isolated pixels within the region especially near the boundaries of the region. Since the pixels intensity do not represent the normal intensity in the region, these effects become worse as the noise become higher [2]. Another problem with global thresholding is that changes in illumination across the scene may cause some parts appear brighter in the light and darker in shadow. Thus, this project employed canny edge detection method to resolve the problem by extracting the edges in a thermal image. The aim is to find points in an image where discontinuities or sharp changes in intensity occur [3]. An effective edge detector reduces a large amount of data but still keeps most of the important feature of the image [4]. By using Lucas Kanade algorithm, motion detection based on pixel can be used by using the edges created in edge detection process. Pixel in the region of the contour is monitored in every frame to find any movement. Every pixel that move out of the region or going inside of the region in every angle in the picture can be identified as movement.

2. Data Acquisition

Figure 1 shows the procedure to identify motion in thermal image. Firstly, video of thermal image in AVI format as an input was recorded for 30 seconds during the interactive session among student in a room by using FLUKE VT04 Visual IR Thermometer camera. Figures 2 (a) and (b) show the frame extracted from recorded video and the frame of thermal image, respectively. In this part, a lot of interaction occur between subject and student may contribute to the thermal heat signature. During the recording time, student A who act as subject was move around and interact with other student to create a motion in thermal image. When interaction between the subject and object is surrounding, background and foreground are mixed together. This contour is used as the maximum edge to differentiate one object from another. Besides that, the contour used in Canny edge detection method is to identify the important feature of the image and motion when images in sequence is combine into one frame.

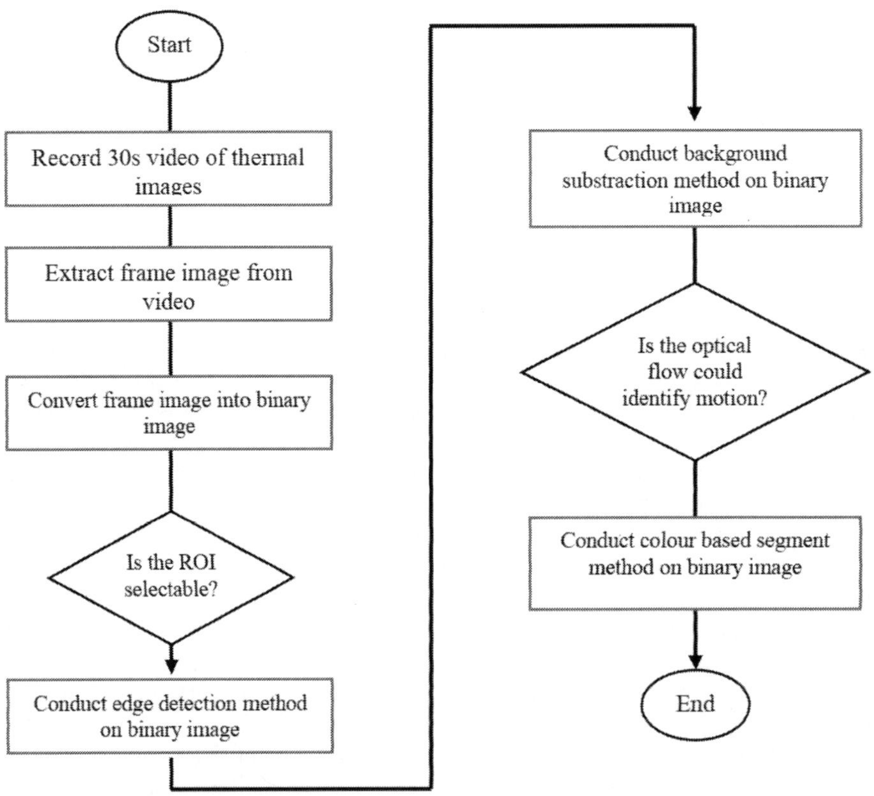

Figure. 1 Steps of motion identification in thermal image

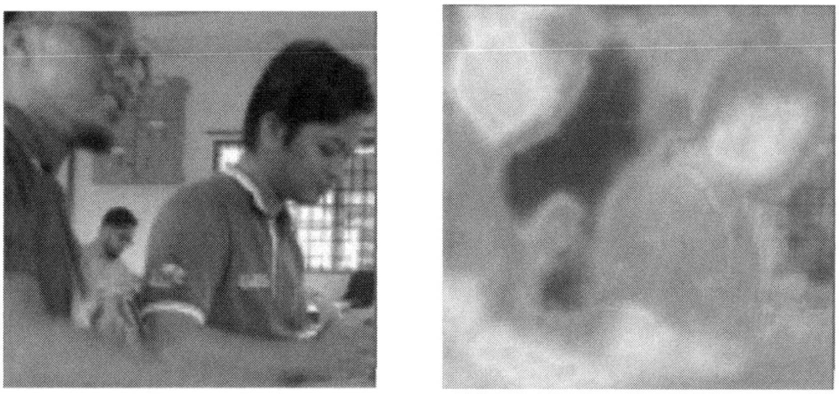

(a) (b)

Figure. 2 a) Output of video after extracted into frames and b) Frame in thermal image

3. Data Analysis

3.1 Canny Edge Detector

To identify the edges, the threshold plays a huge role where the intensity of the pixel above the minimum threshold are kept as edges and the rest are be removed to avoid false information. Discontinuity of edges are cropped and removed and all smooth and sparse images might have enough intensity that met threshold requirement are kept. Canny edge detector used to smooth the edges while derivative function used to calculate the maxima and minima [5]. It applied the second derivative zero crossing point to consider the strong and weak edges. Two thresholds are used to detect strong edges and weak edges [3]. The intensity image contains an edge only when the edge exists precisely in the colour image. Figures 3(left) and (right) show the grayscale image obtained from the converted thermal image and the applied Canny edge detector image on the grayscale image, respectively. Canny operator is able to detect all borders in the image (right), while the grayscale image (left) is not able to detect any edges.

Figure. 3 Grayscale image and applied canny edge detector image

To obtain smooth edges, each point of pixel is convolve with:

$$Gx = [-10-10]\ ,\ Gy = [11-1-1] \qquad (1)$$

Gx = vertical edge response
Gy = horizontal edge response

The magnitude and orientation of the gradient are computed as:

$$M[i,j]=\sqrt{Gx[i,j]2+Gy[i,j]2} \quad (2)$$
$$\theta[i,j]=\tan-1(Gy[i,j],Gx[i,j]) \quad (3)$$

M[i,j] = magnitude of the gradient
θ[i,j] = direction of the gradient

3.2 Optical Flow Method
This method called Lucas-Kanade method, identifies the presence of moving object by considering the difference between two consecutive frames and velocity of each movement [6]. By subtracting second image from the first image frame, operator in consecutive frame obtains the desired output. This method is efficient in detecting the intensity of gray level changes between images. As part to achieve the aim of this work, the optical flow is essential to estimate the intensity of object movement. This method was applied on the greyscale image shown in Figure 3 (left). This method divided into 3 parts:

i. Reference or background image. Reference image can be selected as a frame which is temporally adjacent image from a dynamic sequence.
ii. Arithmetic subtraction operation algorithm.
iii. Selection of suitable threshold.

Based on the obtained flow vector, Hornn Schunck algorithm translates the vector into colour based for each flow depending on changes in the value of the block [6]. The Hornn Schunck technique is to estimate the optical movement that presents the constraint of smoothness in resolving the aperture problem. The algorithm segmented the images into colour based on optical flow vector data which each colour represents different vector and magnitude. However, this calculation depends on a differential method figured by utilizing a gradient constraint with a whole picture smoothness to acquire an expected velocity field. The brightness of every pixel is effect along its movement direction in the picture sequences.

4. Results and Discussions
This section shows and discusses the results obtained as represented in figure 5. Figure 5 is the result of flow vector and segmented colour images analysed during the time frame of 1, 5, 10, and 15 seconds. Flow vector images (middle) indicate every movement changes in the image which calculated between the pixels of two time frames. It clearly can be observed that the high magnitude of vector is measured at the edges of image. For the segmented colour image (right), each colour represents different vector and magnitude. Blue colour represents a smooth movement in 8x8 region of neighborhood pixel, while red colour represents movement of pixel outside 8x8 region neighborhood pixel which can be false information because the pixel is moved to another far region from their original place. In other hand, pink and green colours are the optical flow constraint that lead to the unable estimation of the movement in the region. Other color were the optical constraint and aperture problem which has not enough information to recover the optical flow at one point of the image intensity.

Based on figure 4(b) (right) during T = 5s, the image result represented a lot of blue colour that indicates the tiny movement between two frames where the pixel was move only in 8x8 region pixel. The algorithm minimizes both the optical flow constraint and the magnitude of the variations of the flow field. This can be visualized on the flow vector on image that have a very smooth vector (figure 4(b) (middle). Meanwhile, figure 4(c) (right) during T = 10s, showed a bundle of movement in frame which produce a lot of red region. In the presence of large displacements, it was complicated to estimate the vector when the gradient of the image is not a smooth. The flow vector result as observed

IOP Conf. Series: Journal of Physics: Conf. Series **1049** (2018) 012066 doi:10.1088/1742-6596/1049/1/012066

in figure 4(b) (middle), the vector is rough and long which unable to estimate and allocate the new pixel into right location.

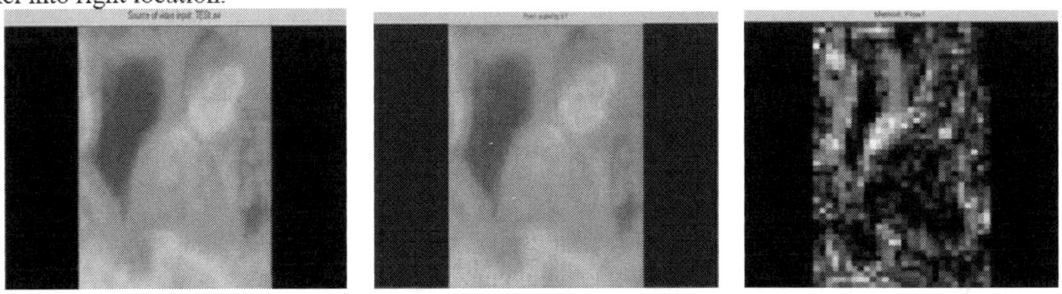

(a) Image analysis at $T = 1$ s

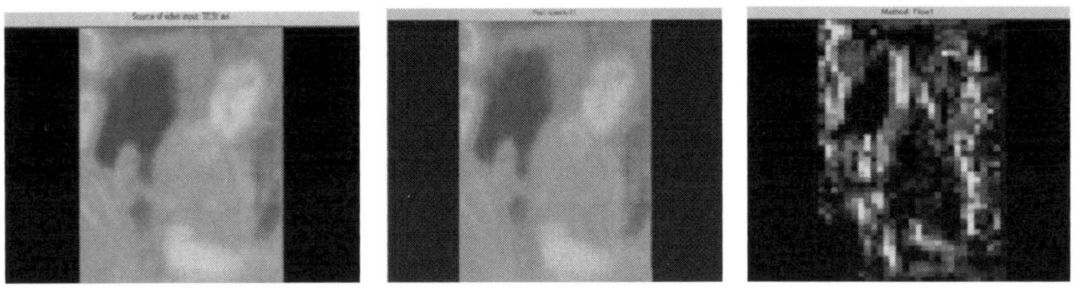

(b) Image analysis at $T = 5$ s

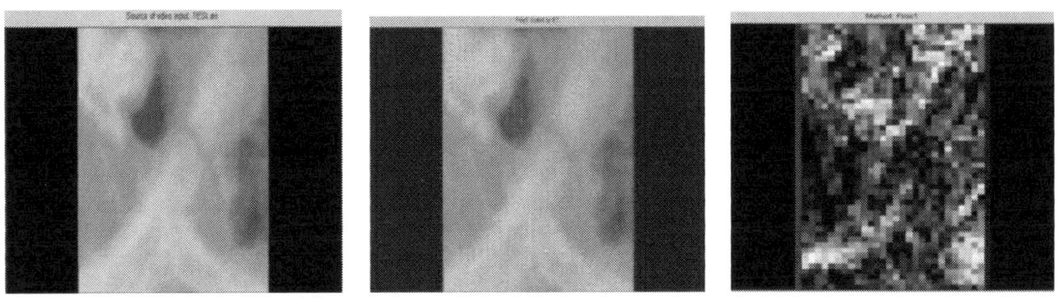

(c) Image analysis at $T = 10$ s

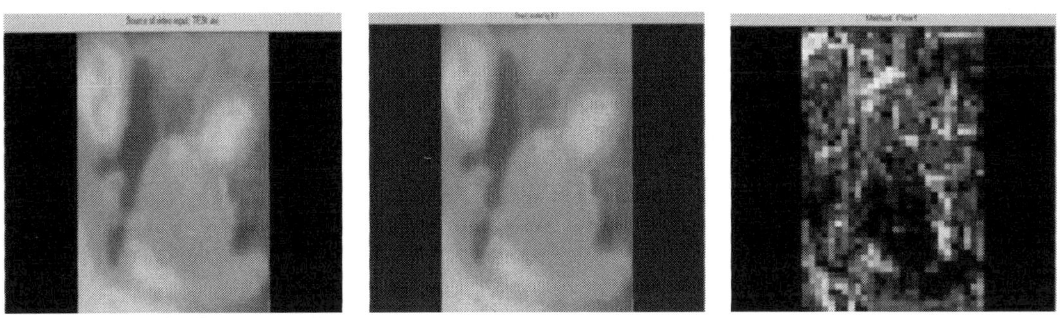

(d) Image analysis at $T = 15$ s

Figure 4. Result of (left) grayscale image, (middle) flow vector on image, and (right) segmented colour image.

5. Conclusion

The objective of the work was to identify motion in thermal images by detecting the image edges from the thermal image, and applying the flow vector before segmenting the colour on the analysed image. By implementing the Canny edge detection, the smoothing and the gradient calculation are done globally, but the thresholding in the hysteresis is done locally within each block so that even the weak edges could be detected. The method presented promising the possible steps to be applied in current technology such as surveillance and security monitoring system to improve a healthy lifestyle.

References

[1] TB Moeslund, E Granum 2001 *Computer Vision and Image Understanding* **81** 231.
[2] R Medina-Carnicer, A Carmona-Poyato 2010 *IEEE Trans Image Process* **19** 165.
[3] Shaveta Malik, Tapas Kumar 2016 *Communications on Applied Electronics* **5** 2.
[4] Suhaila Sari, Sri Erna Ervinna Asahrori, Hazli Roslan and Nabilah Ibrahim 2015 *ARPN Journal of Engineering and Applied Sciences*, **10** 8812.
[5] Mat Harun, Noor Hafizzatul 'Izzah and Ibrahim, Nabilah and Aziz, Nur Shazilah 2015 *ARPN Journal of Engineering and Applied Sciences* **10** 8635.
[6] Mehmet Sezgin, Bulent Sankur 2004 *Journal of Electronic Imaging* **13** 146.

Acknowledgments

The authors would like to thank Universiti Tun Hussein Onn Malaysia (UTHM) and Malaysia Government for the support and of this study (FRGS Vote. No:1583)

IOP Conf. Series: Journal of Physics: Conf. Series **1049** (2018) 012067 doi:10.1088/1742-6596/1049/1/012067

Preliminary Design of a Robotic Exoskeleton for Arm Rehabilitation

Abdul Malik Mohd Ali[1], Radzi Ambar[2], Kushairy Abdul Kadir[1], Mohd Norzali Haji Mohd[2], Mohamad Reyasudin Basir Khan[1], Sabilah Abdul Halim[1], Wan Adila Wan Sharon[1] and Nurul Najihah Aznizam[1]

[1] Electrical Section, Universiti Kuala Lumpur British Malaysian Institute Bt. 8, Jalan Sungai Pusu, 53100 Gombak, Selangor, Malaysia

[2] Computational Signal, Imaging and Intelligence (CSII) Research Group, Department of Computer Engineering, Faculty of Electrical and Electronic Engineering, Universiti Tun Hussein Onn Malaysia, 86400 Parit Raja, Batu Pahat, Johor, Malaysia.

amalik@unikl.edu.my

Abstract. This research paper presents the design of a low-cost and easy-to-use 2 degree of freedom (DOF) robotic exoskeleton for arm rehabilitation. The developed exoskeleton consists of a 2 DOF robotic arm attached on a chair. Force sensitive resistors are also utilized in the design of the device to measure muscles activities during rehabilitation process. Kinovea software is used to analyse the performance of the patient during exercise via video capture. The measured data hopefully can assist physicians and caregivers in designing suitable rehabilitation process for stroke patient. The proposed design provides a novel tool towards upper limb stroke rehabilitation process. Although there are many exoskeleton robotics arms which are commercially available, however, due to the disadvantages such as weight, high-cost and complex mechanisms, this paper proposed new ideas on solving these problems by designing an exoskeleton which is functional, low-cost and users friendly

1. Introduction

The objective of developing a robotic exoskeleton or robotic arm is to replicate or imitate sensory-motor capabilities of the human hand [1]. Development of new methods and sensors for various application contributes in the development of novel techniques for biomechanical applications that promote studies on human motion analysis and rehabilitation processes [1, 2]. Furthermore, due to the advancement of robotic technologies, it has changed the method of utilizing grippers with only two rigid fingers, and no phalanges, to the development of human-like hands with at least three to five functional fingers, each with two to three phalanges [3].

A master-slave robotic system is a popular tool in application related to rehabilitation and remote handling operation [4, 5]. This system enables the personnel to maintain safe working distance from hazardous work environments. Moreover, in the field of healthcare such as tele-surgery and rehabilitation, remote handling tools involving the usage of robotic hands are also employed to improve human limb function. There are many types of five fingers robotic hand that has been developed. The robotic hand, involving innovative mechanisms or myoelectric control systems are an example of the advanced types. Vinet et al. develop a five finger adult sized anthropomorphic hand called the Montreal hand with passive adaptive capabilities by means of a clutch, a cable system, and a

Content from this work may be used under the terms of the Creative Commons Attribution 3.0 licence. Any further distribution of this work must maintain attribution to the author(s) and the title of the work, journal citation and DOI.
Published under licence by IOP Publishing Ltd

spring-loaded pulley mechanism [6]. Rajiv et al. developed a multiple motor and sensory feedback robotic hand which can grasp objects [7]. From both cases, the robotic hand gave a more human-like finger function. But there are many setbacks mainly due to oversize, overweight and costly. It is proven that imitation is an element which is important in proper improvement of social and communicative skills [8]. Mirror Visual Feedback (MVF) therapy or mirror therapy is an imitation method introduced back in early 1990s, which is based on mirror illusion to help patient's limb practice due to cerebral vascular accident (CVA) injuries, post-stroke or amputated [9]. Motivated by this study, the authors have developed a master-slave robotic hand that substitute the paralyzed hand in MVF therapy to aid in recovery process of patient's upper limb function [10].
The objective of this study is to develop a low-cost, modular and easy-to-use robotic exoskeleton that can be utilized as a tool for arm rehabilitation. This paper describes the preliminary design of the robotic exoskeleton that consists of a robotic arm attached on a chair and flex sensors to record muscle activities during movement. Preliminary experimental result to show the usefulness of the device are demonstrated at the end of the paper.

2. Design Description

Figure 1(a) shows the 3D design and schematic of the developed planar two (2) degree of freedom (DOF) exoskeleton with coordinate frames of the base and each joint that allows movement at the shoulder and elbow. The planar exoskeleton that is fixed on a portable adjustable seat chair. As seen in the figure, the mechanism of the exoskeleton is positioned above the user's arm, where all actuators are located on top of the user's arm and aligned with the user's elbow and shoulder joints positions. In the current design, the joints movements are directly actuated by DC motors which are fixed on two (2) links (Link 1 and Link 2) made using polymer material which is not only solid but also lightweight. Furthermore, as shown in Figure 1(a), Link 2 is fixed with an arm carriage that supports user's arm. Figure 1(b) shows the image of the actual exoskeleton.
As shown in Figure 1(a), Joint 1 represents the shoulder joint and its axis of motion is along z_0 axis. This joint provides a rotational angular motion around z_0 axis in x_0-y_0 plane. On the other hand, Joint 2 represents the elbow joint and its axis of motion is along z_1 axis. This joint provides a rotational angular motion around z_1 axis in x_1-y_1 plane.

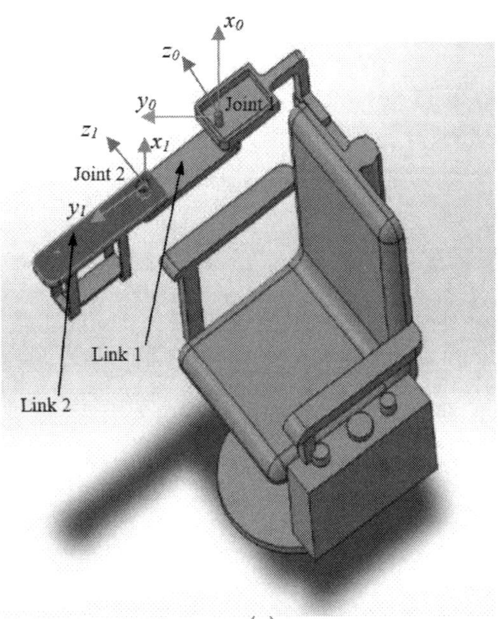

(a) (b)

Figure 1. (a) Schematic of the proposed exoskeleton, (b) 3D design of the proposed exoskeleton

Figure 2(a) to (d) show CAD drawings of the proposed device demonstrating the range of motions for shoulder and elbow joints. Figure 2(a) and (b) show the initial position and 45deg inward movement of the shoulder joint, respectively. Figure 2(c) and (d) show the elbow joint inward motions from 45deg and 90deg, respectively. The range of motions for both shoulder and elbow joint are similar, which is [-90deg, 90deg].

(a) (b) (c) (d)

Figure 2. Movement of robotic arm: (a) Shoulder and elbow joints initial angles; (b) Shoulder inward 45deg; (c) Elbow joint 45deg inward movement at shoulder joint initial angle; (d) Elbow joint 90deg inward movement at shoulder joint initial angle.

2.1. Hardware Design

Figure 3 shows images of the developed two DOF exoskeleton for assisting patient in arm rehabilitation process. Modular design of the exoskeleton that can be reconfigurable to be suited for the use of either left or right arm will be developed in the future. However, in the current preliminary design, the exoskeleton is developed for upper right hand only.

As shown in the figure, the exoskeleton is fixed on an office chair via a 20cm x 10cm x 5cm base made from metal sheet. This base is attached with a DC power window motor with 1:10 gear ratio. This DC power window motor drives the shoulder joint which is connected to Link 1 and the lower arms that consists of the elbow joint and Link 2. The DC power window motor is covered by a semi-transparent box as a cover. Furthermore, the links (Link1 and Link 2) of the robotic arm are made of polymer material which is strong and lightweight. Technically, the structure of the shoulder is designed so that it is directly attached to the elbow joint's actuator.

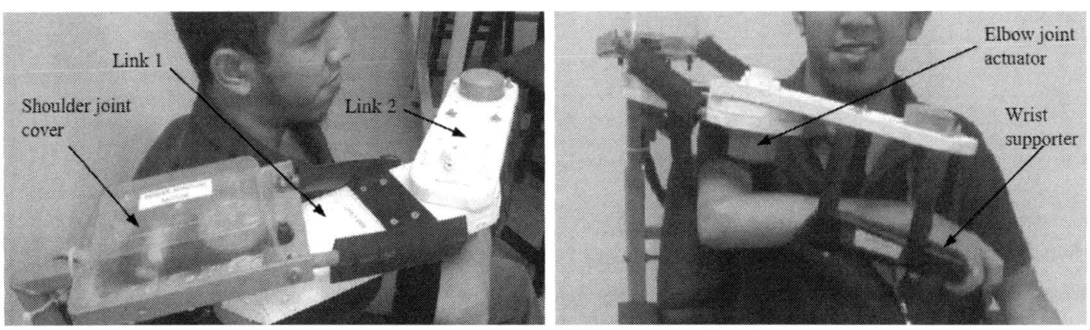

Figure 3. Exoskeleton for arm rehabilitation

On the other hand, the rotational joint of the elbow uses a low-cost and compact-sized 20W SPG20 DC brushed motor with 50:1 gear ratio as shown in Figure 3. This DC brushed motor consists of a 4mm diameter D-shaft. One of the most critical components of this devices is the motor shaft. Each actuator's shaft on the shoulder and elbow joint is positioned with a twelve teeth motor shaft. In this work, the design of the gear system mechanism is crucial to control the exoskeleton to mimic the movement of human arm. Both actuators are powered by 12V power supply. Each actuator is a self-contained drive unit with a built-in motor, controller, amplifier, and communication interface. Table 1 shows the specification of the torque produced by each actuator and its angle range of motion. The rotational movement of the elbow and shoulder joints are limited from 0deg to 90deg only. A stopper spring mechanism is designed on each joint to limit the movement for only up to 90deg.

An Arduino microcontroller controls the overall system of the device, which communicates with motor driver and interfacing between hardware via serial communication for data processing. Each joint of the exoskeleton is controlled by a motor driver. The shoulder joint is also passive, manually positioned units that use friction or a lock to hold their position.

Table 1. Joint range of motion (ROM) and torque value

No	Joint	Angle (deg)	Torque (NM)
1	Shoulder	0 to 90	20
2	Elbow	0 to 90	20

2.2. Force Sensitive Resistor

A force sensitive resistor (FSR) is a type of resistor that composed of thick polymer film. The FSR works by changing its resistance when physical pressure applied to the active surface. Figure 4 shows the relation between the sensor's resistivity against physical force on its active surface. When there is no pressure, the sensor looks like an infinite resistor (larger than 1MΩ), while the resistance may reduce to several kΩ when the pressure on the active surface increases. FSR as shown in Figure 4 is very ideal for body-worn sensors and robotic applications since it require less power and easy interface, very lightweight and relatively sensitive to small physical force. In this work, the bicep and forearms of the user are attached with FSRs to record muscles excitation during rehabilitation.

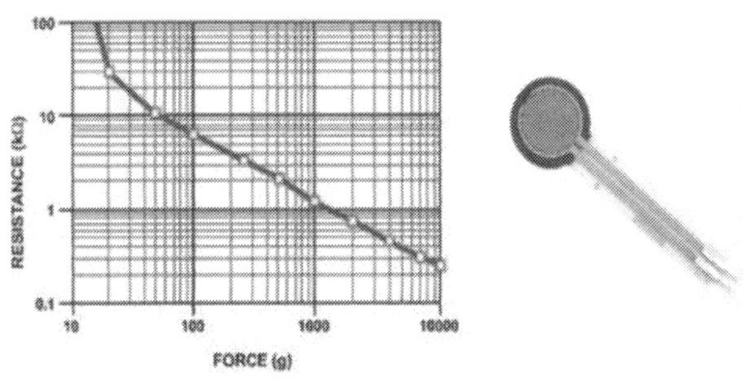

Figure 4. Force sensitive resistor

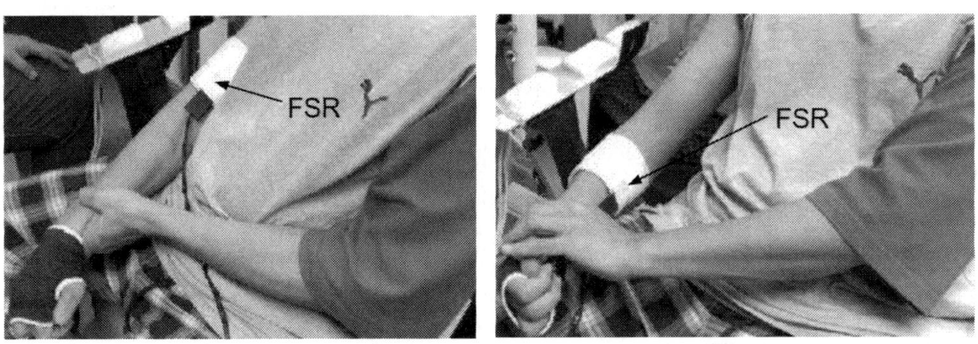

(a) (b)

Figure 5. FSR attached on user's (a) bicep and, (b) forearm.

IOP Conf. Series: Journal of Physics: Conf. Series **1049** (2018) 012067 doi:10.1088/1742-6596/1049/1/012067

3. Preliminary Experiments and Results

This section describes the preliminary experiment to verify the usefulness and usability of the developed device.

In this preliminary experiment, a patient with arm disability (level 3) was asked to do a simple elbow inward bending movement with the support of the exoskeleton robotic arm as shown in Figure 6. During the movement the arm, the muscle activities on bicep and forearm were recorded using the FSR explained in previous chapter. Then, the raw data from FSR were processed by Kinovea software. Kinovea is a free and open source video analysis software for viewing and analysing techniques of human motions [12].

For the purpose of analysis, the recorded data are divided into three (3) analysis criteria, where each of the criteria follows Modified Ashworth Scale (MAS) rehabilitation grading system for clinical applications. MAS measures resistance during passive soft-tissue stretching. FSR sensor was determined by measuring the latency of bicep and forearms extensor reflex force relative to the onset of muscle reflex myoactivity.

Figure 7(a) to (b) show the result of the preliminary experiment based on Kinovea software. Figure 7(a) shows the result of force exerted on the FSR attached on the shoulder. Figure 7(b) shows the result of force exerted on the FSR attached on the bicep. Both results show different peaks which is the position where the strength of the robotic arms is at the maximum use. From this signal, it shows that the Kinovea recorded the robotics arms and human arms kinematics in different angle. With respect to the way the bend sensors are utilized, it makes sense to take advantages from a step wise linear behaviour.

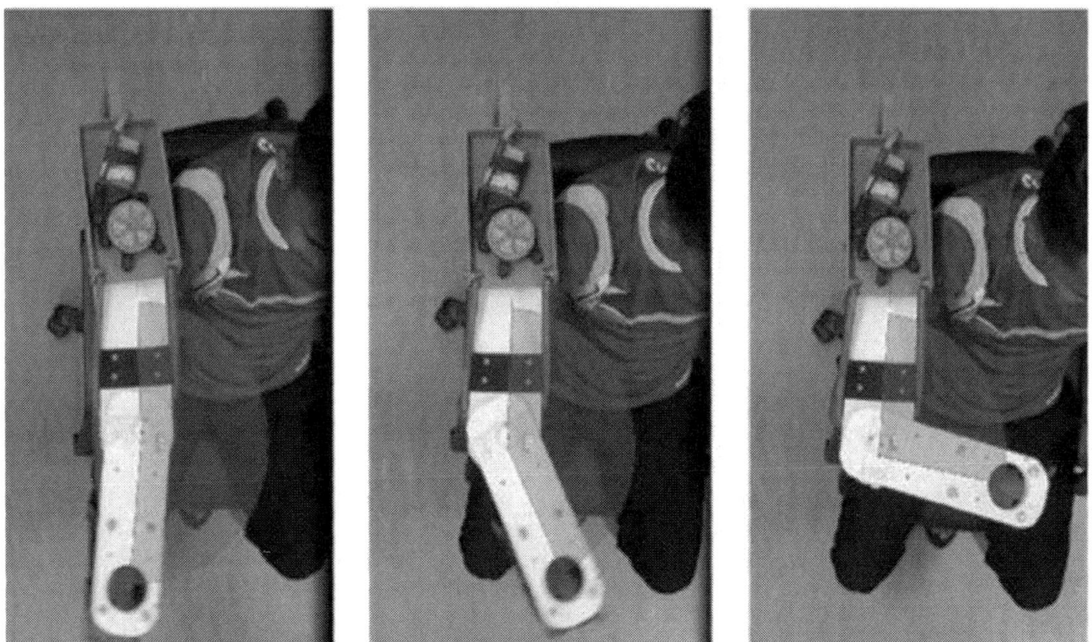

(a) Initial condition (b) 30deg arm bend (c) 90deg arm bend

Figure 6. Preliminary experiment involves simple arm bending movement to record muscle activity

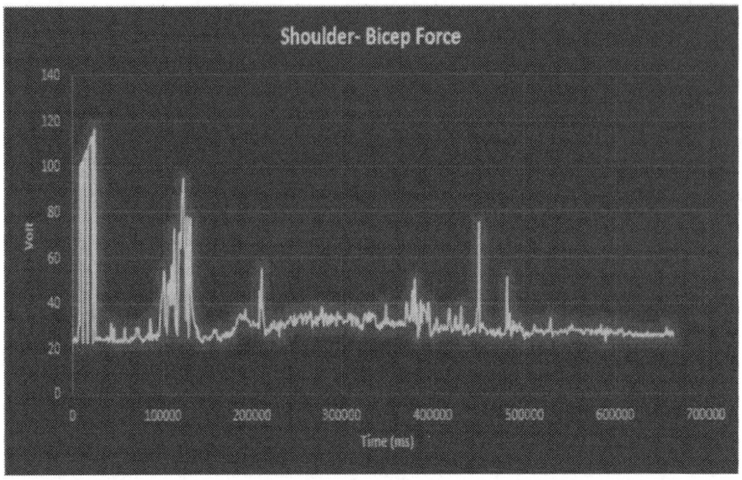

(a) Force exerted on shoulder

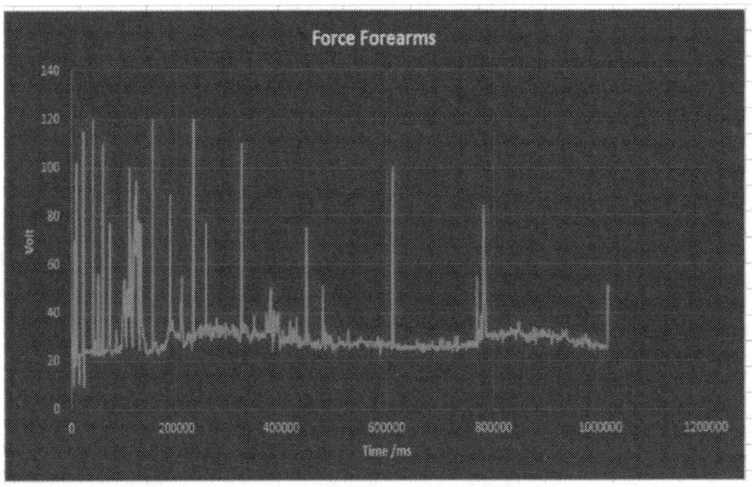

(b) Force exerted on bicep

Figure 7. Results show muscle activities during experiment while using the developed device.

4. Conclusion

This paper presents preliminary design and development of a robotic exoskeleton for arm rehabilitation. The design of this device is for right hand use only. However, taking into account modular design, the device will be able to be used for either hand in the future. As this device is still in developing stages, continuous efforts are being done to make the device useful in real application. Preliminary experimental result shows the device equipped with force sensitive resistor can provide information on how the muscle respond during robot assisted rehabilitation. Acknowledgement

The author would like to take opportunity to express her heartfelt appreciation to her respectful supervisor for his supervision, encouragement, contradictive ideas patience, guidance and invaluable advice enabling the author to produce this paper.

5. Acknowledgement

The researchers would like to thank Ministry of Higher Education Malaysia and UTHM for the research grant under the Incentive Grant Scheme for Publication (IGSP) Vot U685 for supporting this research work.

6. References

[1] Degeorges R et al. 2005 "Three-dimensional rotations of human three-joint fingers: an optoelectronic measurement," *Preliminary results Radiol. Anat.* **27** pp 43-50

[2] Nizam Y, Mohd M N, Abdul Jamil M M 2017 "Biomechanical application: exploitation of Kinect sensor for gait analysis," *ARPN Journal of Engineering and Applied Sciences* **12**(10) pp 3183-3188

[3] Beth T et al 2003 "Characteristics in human motion – from acquisition to analysis," *Humanoids 2003*

[4] Placidi G 2007 "A smart virtual glove for telerehabilitation," *Computers in Biology and Medicine* **37** pp 1100-1107

[5] Ambar R and Sagara S 2015 "Development of a master controller for a 3-link dual-arm underwater robot," *Artif. Life and Robotics* **20** pp 327-335

[6] Vinet S et al 1995 "Design methodology for a multifunctional hand prosthesis," *J. Rehab. Res. Dev.* **32** pp 316-324

[7] Rajiv D et al 1998 "The design and development of a gloveless endoskeletal prosthetic hand," *Journal of Rehabilitation Research and Development* **35** pp 388-395

[8] Ludovic D et al 2008 "HandCare: A Cable-Actuated Rehabilitation System to Train Hand Function After Stroke," *IEEE Transactions on Neural Systems And Rehabilitation Engineering* **16** pp 582-291

[9] Ramachandran V S and Alstchuler E L 2009 "The use of visual feedback in particular mirror visual feedback in restoring brain function," *Brain* **132** pp 1693-1710.

[10] Ali A M M, Ambar R, Wahi J M and Jamil M M A 2011 "Development of Master-Slave Robotic Hand for Mirror Visual Feedback (MVF) Therapy," *Journal of Engineering Technology* **1** pp 27-32

[11] Kinovea https://www.kinovea.org/ [Accessed on 15 September 2017]

[12] Kaya T et al 2011 "Inter-rater reliability of the Modified Ashworth Scale and modified Modified Ashworth Scale in assessing poststroke elbow flexor spasticity," *Int J Rehabil Res* **34** pp 59-64

Computational and Experimental Study of Gradiometer Coil Sensor Design for Single-Sided Magnetic Particle Imaging

Abdulkadir Abubakar Sadiq[1], **Nurmiza Binti Othman**[1], **Muhammad Mahadi Abdul Jamil**[1]

[1] Department of Electronics Engineering, Faculty of Electrical and Electronics Engineering, Universiti Tun Hussein Onn Malaysia, 86400 Parit Raja, Johor, Malaysia
E-mail: engrsadiqgam@gmail.com

Abstract. This paper presents the design and construction of an induction coil sensor, suitable for single-sided Magnetic Particle Imaging (MPI) scanner. Tomographic images of the magnetic nanoparticles (MNPs) used as tracer need to be formed from the voltage induced in the pickup coil as a result of the particles' magnetization response when excited by a superimposed magnetic field from the excitation and selection fields. The coil sensor is looped in the form of the solenoid (Maxwell arrangement) and inserted into the coaxial arrangement of the excitation and selection field coils to form the single-sided scanner geometry. A first-order gradiometer is realized by winding two identical coils, each coil having 208 turns, with a baseline distance of 30 mm. Simulation and experimental results revealed that the induced voltage is directly proportional to the excitation field, and inversely proportional to the distance away from the excitation coil. Results obtained showed that the designed gradiometer pickup coil is sufficient for single-sided MPI scanner, to be applied in the sentinel lymph node biopsy (SLNB) for breast cancer detection.

1. Introduction

Magnetic particle imaging (MPI) is an innovative imaging method that utilizes the non-linear magnetization response of superparamagnetic iron-oxide nanoparticles (SPIONs) employed as tracer materials to quantify their local concentration [1]. MPI scanners comprise of excitation coils, selection field coils as well as pickup coils. Excitation coils provide the periodic excitation magnetic field to which the magnetic tracers are exposed to, which they become saturated and responds by giving out another periodic magnetic field to be recorded by the pickup coil in the form of the voltage signal. The periodic magnetization response is confined to a small region in space called the field-free point (*FFP*) with the help of the selection field coils [2].

This paper looks into the design and construction of a highly sensitive receiver coil in gradiometer form, for recording MPI signals in the form of voltage. The gradiometer like other induction coil sensors works based on the fundamental Faraday's law of electromagnetic induction, having a transfer function of $V = f(B)$.

$$V = -N\frac{d\emptyset}{dt} = -NA\frac{dB}{dt} = -\mu_0 NA\frac{dH}{dt} \quad (1)$$

where N is the number of turns, A is the cross-sectional area of the coil, $\emptyset$ is the magnetic flux passing through the coil, B is the magnetic flux density and H is the magnetic field strength. Other unique

[1] engrsadiqgam@gmail.com.

Content from this work may be used under the terms of the Creative Commons Attribution 3.0 licence. Any further distribution of this work must maintain attribution to the author(s) and the title of the work, journal citation and DOI.
Published under licence by IOP Publishing Ltd

types of induction coils sensors have been reportedly used in MPI research ranging from Rogowski coil, flux ball sensor, and the needle sensor [3]. Advantages of the gradiometer include its linearity, sensitivity, simplicity, and suitability with the single-sided MPI scanner topology [4].

2. Principle of Gradiometer Pickup Coil

The principle of operation of a gradiometer pickup coil is shown in Figure 1. H_{ext} stands for the external magnetic field, H_{x1} is the magnetic field sensed by the first coil and H_{x2} is the magnetic field sensed by the second coil. The external magnetic field is generated by a large source, for instance from the excitation coil as in this case. It could be the earth's magnetic field as well which makes the field uniform [3].

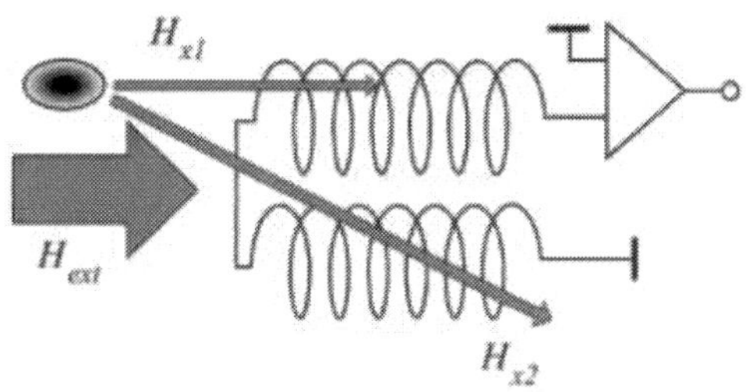

Figure 1. Gradiometer coil schematic [3]

If a gradiometer is inserted in the excitation field, the two coils will detect the same magnetic field (this shows that the influence of the excitation field is removed). In another case, if a small magnetic field source is present together with the excitation field (for example a magnetic field from a magnetic tracer material), the magnetic field detected by the first coil near the source is greater than in the second coil. The output of the gradiometer coil will be the difference between the two fields detected, called the gradient of the field. With this mode of measurement, it is likely to detect a quite small magnetic field in the occurrence of a very high magnetic field [4].

Typically, a gradiometer coil is arranged in vertical, planar or asymmetric mode, which shows zero reading when inserted into a uniform magnetic field when properly designed. Two or more coils can be employed to form higher order gradiometers [4]. For example, two coils are used to form first order gradiometer, three loops to form second order gradiometer and so on.

3. Methodology

Different methods and coil winding arrangements for magnetic nanoparticle imaging were reported by [6]. Coils are designed based on the application of the scanner to be constructed. Gradiometer pickup coils are usually designed in Maxwell arrangement, Helmholtz arrangement or in other forms of induction coil sensors like the Rogowski coil [7].

The pickup coil conceived in this research is for single-sided MPI scanner to be used for SLNB; an overview of the scanner is given by [8]. Therefore, the excitation coils as well as the selection field coil to be used in the scanner design need to be considered. In the single-sided scanner, all coils are assembled coaxially on one side. This means that the gradiometer has to be designed in such a way that it fits into the excitation coil and the excitation coil also fits into the selection field coil. Excitation coil design is reported in details by [9]. Figure 2 shows the MPI signal generation from the excitation coil to the pickup coil. The excitation coil generates the periodic excitation magnetic field, superimposed on the selection field from the selection field coil (for signal spatial encoding), together they saturate the MNPs. The particle's magnetization response is recorded in the form of the voltage signal by the gradiometer pickup coil, designed to cancel interferences. This voltage signal is finally passed through a resonant circuit, designed for tuning signals at the second harmonic frequency. Therefore a series *RLC* resonant circuit aims to serve as second harmonic frequency selection circuit.

IOP Conf. Series: Journal of Physics: Conf. Series **1049** (2018) 012068 doi:10.1088/1742-6596/1049/1/012068

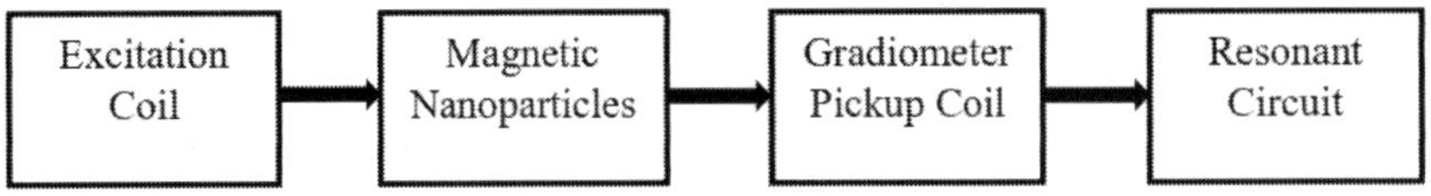

Figure 2. MPI signal generation

3.1. Particle magnetization

The magnetization response of the MNPs is approximated by the Langevin equation of superparamagnetism, as given in Equation (2).

$$M_D(t) = m_s C\,(\coth(\xi) - 1/\xi) \tag{2}$$

The magnetization of the particles distribution with diameter D is described by M_D, while

$$m_s = 1/6\pi D^3 M_s \tag{3}$$

denotes the magnetic moment at saturation. The saturation magnetization M_S is a material constant. The particle concentration is given by c, and the Langevin parameter reads:

$$\xi = (m_s \mu_0 H(t))/(K_B T_P) \tag{4}$$

where $H(t)$ is the external magnetic field, k_B is is the Boltzmann constant, T_P is the temperature of the particles, and μ_0 gives the permeability of vacuum [10]. Applying a sinusoidal external magnetic field, called drive field H_D, with frequency f_0, a non-sinusoidal oscillation of the particle magnetization M is induced due to the non-linearity of the Langevin function. In Fourier space, this deformation can be detected as higher harmonic frequency components [11].

3.2. Voltage induced in pickup coil

According to Faraday's law of electromagnetic induction, a voltage is induced in the coil through electromagnetic induction [12]. Therefore, if the frequency of the second harmonic is f, magnetic field signal generated by the MNPs is B_s, number of coil winding is N, magnetic flux interlinking the pickup coil is Φ , and A is the cross-sectional area of the coil, induced voltage signal V_s can be calculated using Equation (1) as:

$$V_s = N \times (d\Phi)/dt = N \times A\,(dB_s)/dt = NA\,d/dt\,(\sqrt{2}B_{s\,rms}\cos\omega t) \tag{5}$$

$$V_s = NA\omega B_{s\,rsm} = 2\pi NAfB \tag{6}$$

Equation (6) is simulated in MATLAB with the detection frequency of *45.6 kHz* and a cross-sectional area of $5.03 \times 10^{-3}\ m^2$. The number of turns is varied from 50 to 400 turns while the magnetic field is varied between *-3 nT to +3 nT*. Table 1 summarizes the constructed coil parameters.

Table 1. Gradiometer pickup coil specifications.

Parameter	Value
Number of turns, N	208
Cross-sectional area, A	$5.03 \times 10^{-3} m^2$
Internal diameter, D	5 mm
External diameter, d	15 mm
Height, H	30 mm
Baseline distance, b	30 mm
Core material	Polylactic acid
Wire material	Copper wire

4. Results and Discussion

The constructed gradiometer pickup coil is simulated in MATLAB environment, so as to have an idea of the voltage to be induced when the coil is exposed to a certain magnetic field, before finally constructing the coil. The built coil as shown in Figure 3, is then tested and experimented, and the simulation results are presented with the experimental results obtained.

Figure 3. Constructed gradiometer coil

4.1. Simulation results

Equation (6) is simulated in MATLAB environment by varying the number of turns (N) of the coil, as well as the magnetization response (B) of the MNPs (by assuming a range of $-3\ nT$ *to* $3\ nT$). The results obtained are depicted in Figure 4.

Figure 4(a) illustrates the induced voltage versus the magnetization response of the MNPs while Figure 4(b) shows the induced voltage with varying harmonic frequencies. The relationship between the induced voltage on the pickup coil and the magnetic field is linear, as shown in Figure 4(a), with nT range magnetization. The higher the magnetic field around the coil the higher the voltage induced and the wider the range of voltage induced in the coil. Woods et al. (2011) designed a gradiometer for the frequency range of 20 Hz to over 500 kHz; the open-circuit coil-terminal induced voltage is in μV at a magnetic field strength B in pT range [13]. Consequently, a magnetic field strength in pT range will induce a voltage in μV range. This supports our findings as shown in Figure 5(a).

Next, the voltage induced on the pickup coil when the harmonics of the fundamental excitation frequency are varied is presented in Figure 4(b). With 22.8 kHz as the excitation frequency, the second, third and fourth harmonics are 45.6 kHz, 68.4 kHz, and 91.2 kHz respectively. It can be deduced from Figure 4(b) that the higher the frequency, the lower the voltage induced [14].

In addition, the magnitude of the excitation signal is varied so as to see the effect on the voltage induced. Figure 5(a) shows the induced voltage with varying amplitude, which shows a linear relationship. It is also evident from Figure 5(b), that the higher the number of turns of the gradiometer pickup coil the higher the voltage induced, as given in Equation (6) [15]. Other parameters such as the surface area of the coil, the excitation frequency, and the magnetic nanoparticle's magnetization also obey the linear relationship of Equation (6). The results of induced voltage obtained revealed its relationship with the magnetic excitation field, harmonics of the fundamental frequency and the number of turns of the coil. These findings have given us an insight into the construction of a pickup coil that will sufficiently record the voltage to be produced from a region of the uniform magnetic field.

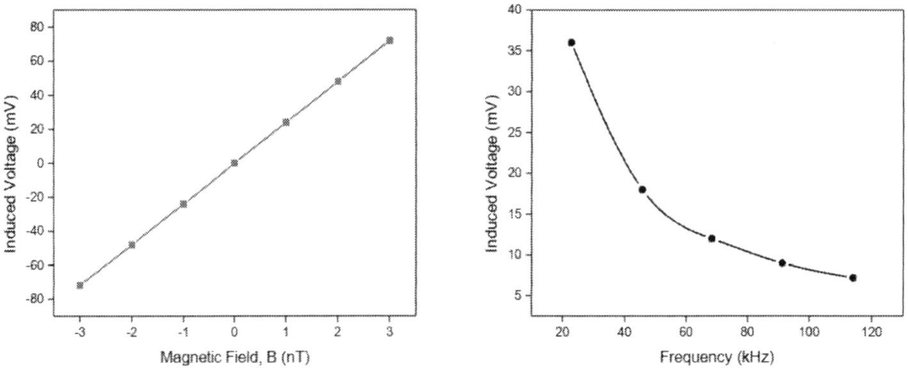

(a) (b)

Figure 4. Induced voltage with varying (a) magnetic field and (b) harmonic frequency

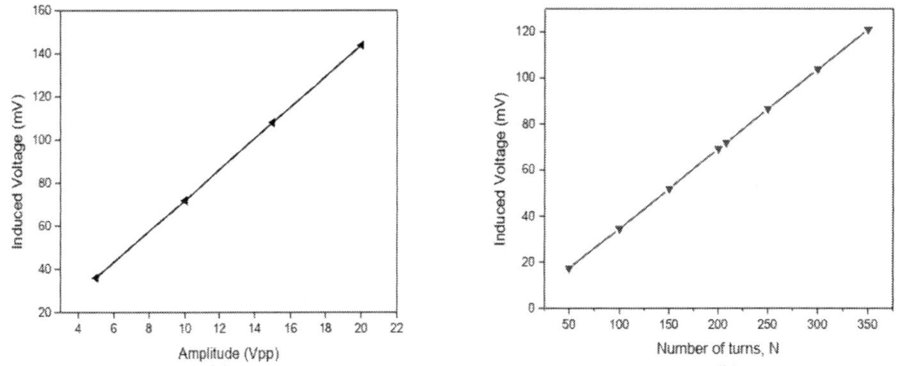

(a) (b)

Figure 5. Induced voltage against (a) amplitude of excitation field, and (b) number of turns of pickup coil

4.2. Experimental Results

The constructed gradiometer pickup coil is inserted coaxially into the excitation and selection field coils (experimental set-up) as depicted in Figure 6(a). *A* indicates the gradiometer pickup coil, while *B* is the excitation coil and *C* is the selection field coil. This is the experimental set-up assembled to measure the voltage induced by the magnetic fields generated by the excitation and selection field coils. Details on the excitation coil and selection field coil design is reported by Sadiq *et al.*, 2016 [9].

Initially, only the excitation coil (*EC*) is used together with the pickup coil, to differentiate the voltage produced when the selection field is added. The induced voltage recorded when only the excitation magnetic field and the combination of the superposition of the excitation and selection field (*EC* + *SFC*), are depicted in Figure 6(b). It is evident from the figure that the excitation frequency is directly proportional to the induced voltage. The trends tend to be non-linear when the excitation frequency is above 20 kHz. The excitation field alone induces voltage slightly higher than the superimposed field, because the selection field creates the *FFP*. At the *FFP*, only the effect of the excitation field is sensed, due to the signal spatial encoding. In this work, *EC* is generated with 22.8 kHz of excitation frequency, and 10 mV amplitude while *SFC* is generated with a DC voltage within the range of 0-5 mV.

(a)

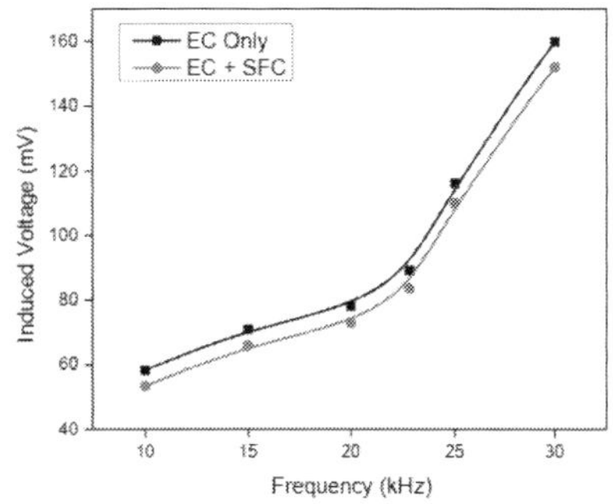

(b)

Figure 6. (a) Experimental set-up (b) Induced voltage from varied frequencies of excitation and selection fields

Next, the position of the pickup coil is varied with respect to the position of the combined excitation and selection field coils. Hence, the pickup coil is placed at the center of the excitation coil and moved at distances of 5 mm, 10 mm, 20 mm and 30 mm away from the excitation coil. Figure 7(a) shows the voltages induced on the pickup coil when the distances are varied from the center of the excitation coil. It is evident that the voltage induced when the pickup coil is at the center somewhat doubled that at 5 mm (almost at the surface of the excitation coil). This is the theory behind magnetic field induced by a coil which states that the magnetic field is concentrated in a nearly uniform field at the center of a long solenoid [16].

Finally, the induced voltage is recorded when the excitation field is generated by a fixed frequency of 22.8 kHz, and the selection field is generated by DC voltage source of 5 mV. Measurements were taken using two types of coils arrangement. Initially, the excitation coil (*EC*) and the pickup coil are used. Then, the combination of excitation coil and selection coil (*EC + SFC*), together with the pickup coils are used. The results obtained as shown in Figure 7(b), revealed that the amplitude of the excitation field, just like the frequency, has a linear relationship with the induced voltage.

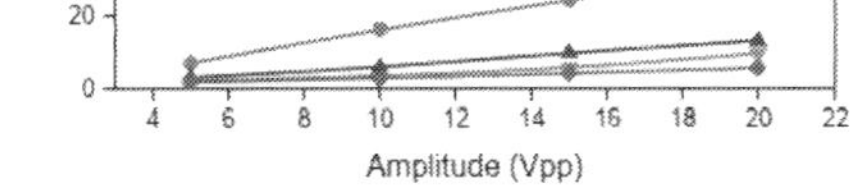

(a)

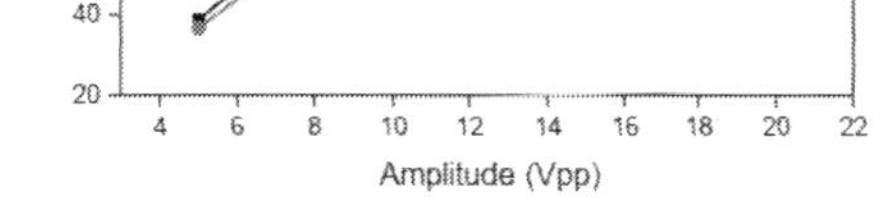

(b)

Figure 7. (a) Induced voltage with varying distance of excitation coil (b) Induced voltage with fixed excitation frequency

5. Conclusion

In conclusion, a gradiometer pickup coil is successfully designed, numerically simulated, and constructed to demonstrate an induction coil sensor suitable for single-sided magnetic particle imaging scanner. Maxwell arrangement type of coil winding is adopted in this research, for its suitability to the scanner geometry and smooth implementation. The induced voltage recorded follows the linear relationship of the theoretical equations governing the design of a coil sensor, with respect to the excitation frequency, its amplitude and the distance away from the excitation field. The *EC* and *SFC* were found to generate the required fields for the creation of *FFP* to be employed for MNPs signal spatial encoding. Thus, the gradiometer pickup coil presented will be suitable for single-sided MPI device to be used in SLNB application.

Acknowledgments
This research was conducted with financial support by RAGS Vot Number R067 and Postgraduate Research Grant (GPPS) Vot Number U589, supported by Universiti Tun Hussein Onn Malaysia and Ministry of Higher Education Malaysia (MOHE).

References

[1] Gleich B and Weizenecker J 2005 *Nature* **435** 1214
[2] Rahmer J E, Weizenecker J and Gleich B US Patent 9,215,990, US Patent and Trademark Office Washington DC USA
[3] Tumanski S 2016 *Handbook of magnetic measurements* (Florida: CRC Press)
[4] Tumanski S 2007 *Measurement Science and Technology* **18** R31
[5] Vrba J, Haid G J, Lee W M, Taylor B R and Tillotson M A 1997 *Inventors; Ctf Systems Inc., assignee* United States patent US 5,657,756
[6] Knopp T, Sattel T F, Biederer S, Rahmer J, Weizenecker J, Gleich B and Buzug T M 2010 *IEEE Transactions on Medical Imaging* **29** 12
[7] Fiorillo F 2010 *Metrologia* **47** S114
[8] Sadiq A A, Othman N B and Abdul Jamil M M 2015 *ARPN Journal of Engineering and Applied Sciences* **10** 8556-8561
[9] Sadiq A A, Othman N B and Abdul Jamil M M 2016 *Proc. Of the 2016 IEEE Asia-Pacific Conference on Applied Electromagnetics* 364
[10] Buzug T M, Bringout G, Erbe M, Gräfe K, Graeser M, Grüttner M and Haegele J 2012 *Zeitschrift für Medizinische Physik* **22** 323
[11] Othman N B, Tsubaki T, Yoshida T and Enpuku K 2011 *Bulletin of the Graduate School of Information Science and Electrical Engineering* (Fukuoka: Kyushu University Press) p 51
[12] Thong W M and Gunstone R 2008 *Research in Science Education* **38** 31
[13] Woods R C, Baker R M, Li F, Stephenson G V, Davis E W and Beckwith A W 2011 *Journal of Modern Physics* **2** 498
[14] Wu K, Wang Y, Feng Y, Yu L and Wang J P 2015 *Journal of Magnetism and Magnetic Materials* **380** 251
[15] Jiles D 2015 *Introduction to magnetism and magnetic materials* (Florida: CRC Press)
[16] Cullity B D and Graham C D 2011 *Introduction to magnetic materials* (New Jersey: John Wiley & Sons)

IOP Conf. Series: Journal of Physics: Conf. Series **1049** (2018) 012069 doi:10.1088/1742-6596/1049/1/012069

Resonant Tunnelling Diode Design for Oscillator Circuit

Halimatus Saadiah[3], Warsuzarina Mat Jubadi[1,3*], Nabihah Ahmad[1,2,3], M. Hairol Jabbar[2,3]

[1] Nano Simulation Research Group, FKEE, UTHM
[2] IC Design Research Group, FKEE, UTHM
[3] Faculty of Electrical and Electronic Engineering,
Universiti Tun Hussein Onn Malaysia,
86400 Parit Raja, Batu Pahat, Johor, Malaysia.

*Email: suzarina@uthm.edu.my

Abstract. Resonant Tunneling Diode (RTD) was introduced by Tsu and Esaki in 1973 and it is a p-n junction device that exhibits Negative Differential Resistance (NDR). Due to the NDR and its characteristics, RTD is suitable for high speed application device especially in the oscillator circuit. The fast-acting negative differential resistance of RTDs make them excellent devices for very high-frequency electronic applications. An NDR region is an important region at which an oscillation will occur. Therefore, this study perform a simulation of RTD design for oscillator circuit at a target frequency of 500 – 600 GHz. In this research, the simulation of I-V characteristics for RTD device and oscillator characteristics are performed in MATLAB. The C-V characteristics are identified and simulated in order to define the frequency value. It can be observed that a smaller capacitance value is required in order to obtain higher frequency oscillation. The target frequency oscillation at around 500-600 GHz have been achieved through the simulations of the RTD design.

1. Introduction

Today's technology has been developing very fast. Because of that, new and advanced devices that offer a different application at high frequency and function are produce every year. Resonant Tunneling Diode (RTD) was discovered about two decades ago and it gives an advantage in developing of electronic device. Its special characteristics lies on its structure, operation and the Negative Differential Resistance (NDR) region which make RTD very significant for high speed application; for example, oscillator circuit [1]. Besides, RTD is suitable for high switching speed because of its small capacitance and high current density and effectively functional at the room temperature [2].

Tunneling diodes can be very compact and are also capable of ultra-high-speed operation because the quantum tunneling effect through the very thin layers is a very fast process. Because of that, RTD structures is used as an oscillator with operating frequencies as high as 712 GHz and a mixer with operating frequency of 2.5 THz [2]. The oscillations are due to the Direct Current (DC) instabilities

Content from this work may be used under the terms of the Creative Commons Attribution 3.0 licence. Any further distribution of this work must maintain attribution to the author(s) and the title of the work, journal citation and DOI.
Published under licence by IOP Publishing Ltd

caused by the RTD inductance value, the shape of the RTD current-voltage characteristic and the bias voltage [1].

Oscillator circuit is very important in communication system such as in radio, and computer. An oscillator functions to generate a signal in computer or electronic devices. If this oscillator generates a signal with higher frequency, so electronic devices will have faster performance. Hence, oscillator circuit is designed by using RTD due to its excellent performance at very high frequency. The fast-acting negative differential resistance of RTDs make them admirable devices for very high-frequency electronic applications [1]. Therefore, the purpose of this study is to perform the simulation of resonant tunneling diode which is designed for oscillator circuit at around 500-600 GHz.

2. Circuit Design

2.1 RTD Circuit and Characteristics

RTD is an undoped quantum well, sandwiched between two undoped layers of barrier and the doped region are collector and emitter. Besides, RTD is operated in electron tunneling mechanism or known as quantum mechanical tunneling [5]. The phenomenon of resonant tunneling happen when the electron transmission coefficient is drastically increasing in equivalent of particular energy values [6]. The passing of electrons through a barrier which is very thin compared to the electron wavelength is known as quantum tunneling process [5].

There are two characteristics of RTD; C-V and I-V characteristics respectively. These characteristics are depending on the circuit that has been chosen. Figure 1 displays two types of small signal equivalent circuit for RTD; the G-C equivalent circuit and parallel-inductance equivalent circuit. Figure 1(a) shows the simplest circuit of RTD which consists of a parallel combination between conductance capacitance (G-C equivalent circuit). However, this circuit is not suitable for RTD at high frequency due to the real and imaginary parts of RTD admittances. The RTD conductance and capacitance does not change with frequency, instead of having a constant value [2]. Second type of circuit is shown in Figure 1(b) which is parallel-inductance equivalent circuit and it is chosen as the RTD circuit.

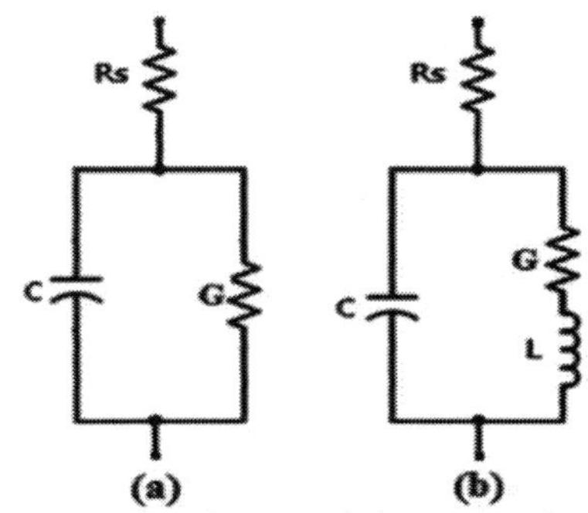

Figure 1. Small signal of equivalent circuit of RTD [3]

Figure 2 shows the graph of C-V characteristics for RTD. A capacitance value has a relationship with the voltages whereas the voltage further increase above 0.2 V, the capacitance will drop drastically as parabolic curve. It indicates the decreasing of charge density [10]. It means that voltage value will affected the capacitance value in RTD. Besides, the graph shows that the higher frequency of RTD i.e., 30 GHz will have smaller capacitance value as compared to frequency of 10 GHz. The graph in Figure 3 demonstrates the real part of the input admittance (conductance) versus the bias voltage at different frequencies. It has the same curve with capacitance graph but inversely to the shape of graph in Figure

IOP Conf. Series: Journal of Physics: Conf. Series **1049** (2018) 012069 doi:10.1088/1742-6596/1049/1/012069

2. Hence, it confirm that the frequency is dependence on the differential conductance, G and capacitance, C especially in the NDR region [2].

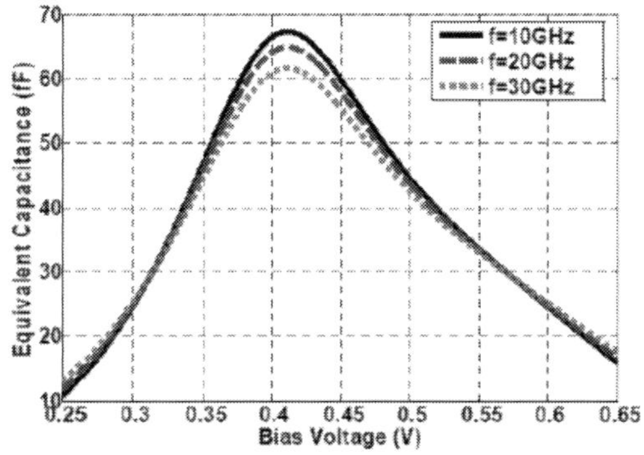

Figure 2. Equivalent capacitance versus the bias voltage at different [2]

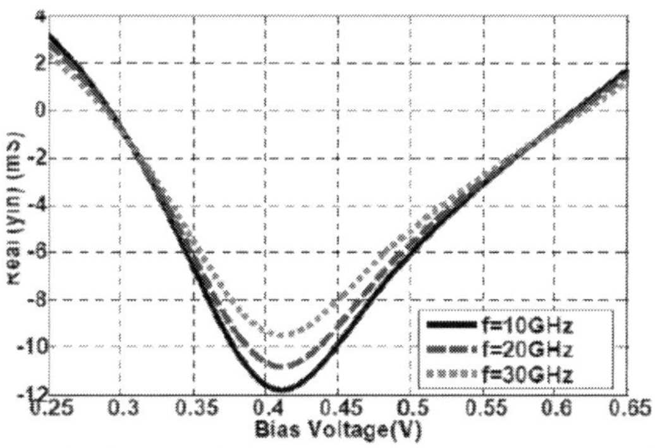

Figure 3. Real part of the input admittance (conductance) versus the bias voltage [2]

2.2 Oscillator Circuit Based on RTD

An oscillator circuit consists of a voltage supply, a serial inductor, an RTD, and a capacitor are required in designing of oscillator as shown in Figure 4. Furthermore, DC instabilities will cause oscillations due to the shape of the RTD current-voltage characteristic and bias voltage. The rising and failing time of the oscillator voltage edges will be determined by capacitance values. However, the capacitances have minor effect on the frequency as compared to the inductance [1].

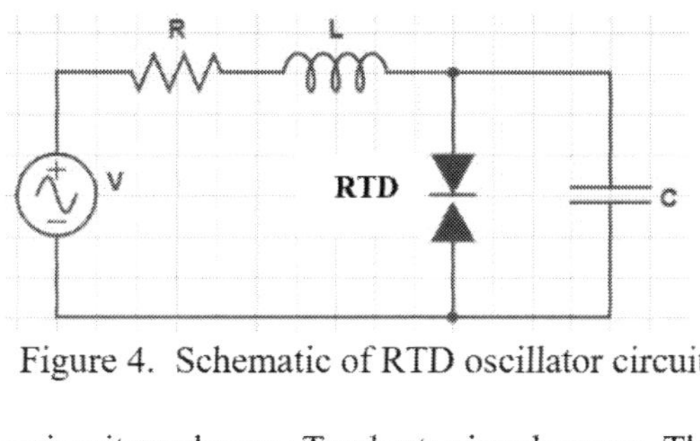

Figure 4. Schematic of RTD oscillator circuit

The RTD based oscillator circuit produces a Terahertz signal source. This oscillator will have highest maximum oscillation frequency (f_{max}) of the inherent Negative Differential Conductance (NDC) due to the resonant tunnelling among the electronic devices [7]. The improved performance of practical operation in microware frequency is achieved by RTD oscillator topologies demonstrated in [7]. Besides, the size of RTD based oscillator circuit is small and this reduces power loss of devices [6].

3. Simulation Parameters

The tunneling current for RTD device can be derived from equation (1) - equation (3). It is calculated based on integrating the transmission electron velocity over the electron distribution in k space [5]. Besides, these equations are also used for modelling RTD. Meanwhile, figure 2.10 illustrates the result after the derivation of tunneling current. This figure show the current density based on sample presented in WM Jubadi [5].

$$J = \frac{em*kT}{4\pi^2 h^3}\int_0^\infty dET(E,V).\,ln\left[\frac{1+e^{(E_F-E)/kT}}{1+e^{(E_F-E-eV)/kT}}\right] \quad \text{(Eq.1)}$$

$$T(E,V) = \frac{\left(\frac{\Gamma}{2}\right)^2}{\left[E-\left(E_r-\frac{eV}{2}\right)\right]^2+\left(\frac{\Gamma}{2}\right)^2} \quad \text{(Eq.2)}$$

$$J = \frac{em^*kT}{4\pi^2 h^3}.\,ln\left[\frac{1+e^{(E_F+E_r+^{eV}/_2)/kT}}{1+e^{(E_F-E_r+^{eV}/_2)/kT}}\right].\left[\frac{\pi}{2}\right]+tan^{-1}\left(\frac{E_r-\frac{eV}{2}}{\frac{\Gamma}{2}}\right) \quad \text{(Eq.3)}$$

Where Γ, Er, EF, $m*$, h are the resonance width, the energy in the $In_{0.8}Ga_{0.2}As$ well resonance level, the energy of the Fermi level in two contacts, the effective mass and the reduced Plank constant, E, is the energy of the resonant level relative to the bottom of the well at its center [5].

In this research, Equation (4) - Equation (7) were obtained from Abdallah, R. M., Dessouki, A. A., & Aly, M. H. [8] to model the I-V characteristics of RTD device. Therefore, these equations are used for fitting the graph of current density of the RTD device. The parameter of A, B, C_1, D_1 and E_1 in this equation indicates different I-V characteristics of RTD device. Besides, parameter C_2 is used to control the height of the added inverse tangent term. D_2 is used to control the slope of this function, and consequently, can change the slope of the negative differential conductance. Lastly, E_2 controls the voltage of the middle point of the tangent function [8].

$$J_{exponential}(V) = A \times \left(e^{(B\times V)} - e^{(-B\times V)}\right) \quad \text{(Eq.4)}$$

$$J_{Gaussian}(V) = C_1 \times \left(e^{(-D1(V-E1)^2)} - e^{(-D1(V+E1)^2)}\right) \quad \text{(Eq.5)}$$

$$J_{mod}(v) = C_2 \times (\tan^{-1}(D_2(V-E_2)) + (\tan^{-1}(D_2(V+E_2))) \quad \text{(Eq.6)}$$

$$I(V) = area \times \left(J_{exponential}(V) + J_{Gaussian}(V) + J_{mod}(v)\right) \quad \text{(Eq.7)}$$

4 Results and discussion

4.1 I-V Characteristics of RTD

In this study, four simulations of I-V characteristics are illustrated in one graph as show in Figure 5. Based on the graph, it shows the different value of peak and valley for current and voltage when the parameter of fitting are varied. Therefore, all these value have been summarized in Table 1. Based on the table, it is obvious that simulation 2 has higher Peak to Valley Current Ratio (PVCR) value but lower peak current. However, simulation 4 has higher peak current value compare to other simulation. So, it is indicated that simulation 4 is better simulation because the higher peak current value can be affected the oscillator frequency. It is evidence that parameter A, C_1 and C_2 are significant in order to achieve different peak current value. The pattern of I-V characteristic graph is the same with the previous study in [5]. Therefore, it is proven that the fitting value is tolerable.

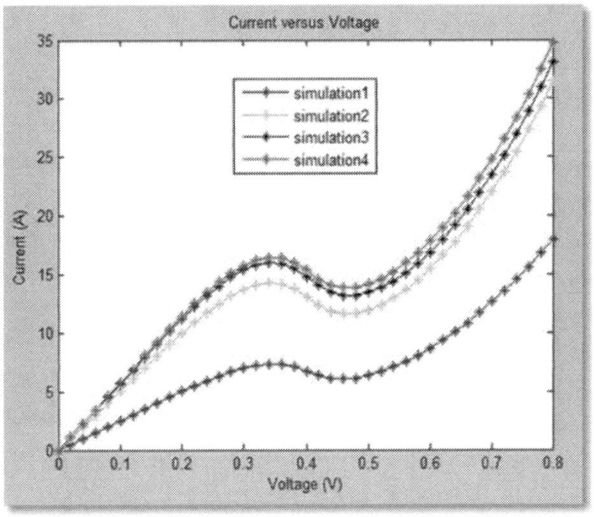

Figure 5. Graph I-V Characteristics for combination of simulation 1-4

Table 1: Result of peak and valley value for current and voltage.

Simulation	Current(A)		Voltage(V)		PVCR
	I_p	I_v	V_p	V_v	
1	7.315	6.042	0.34	0.46	1.21
2	14.27	11.57	0.34	0.46	1.23
3	16.00	13.20	0.34	0.46	1.21
4	16.46	13.86	0.34	0.46	1.20

4.2 C-V Characteristics

In this study, C-V and G-V value for 4 simulations are plotted in one graph as illustrated in Figure 6. Besides, Figure 7 shows the reciprocal of inductance versus voltage that will be used in the oscillator circuit in Figure 4. However, it will be omitted for the calculation at the higher frequency because the graph shows that inductance have very a small value and less significant [8-10]. Based on Figure 6(a) and Figure 6(b), one simulation point of conductance and capacitance value in each simulation has been chosen which is at V=0.34V. The findings are simplified and tabulated into the Table 2 in order to relate with the frequency of oscillation. This voltage value is chosen because it is around NDR region which reside between at the peak voltage and valley voltage.

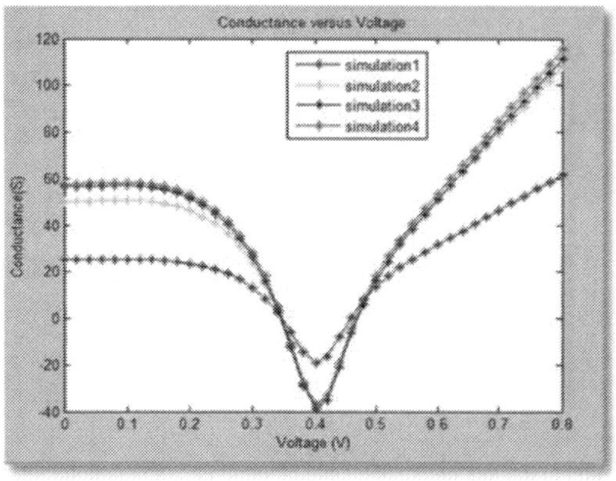

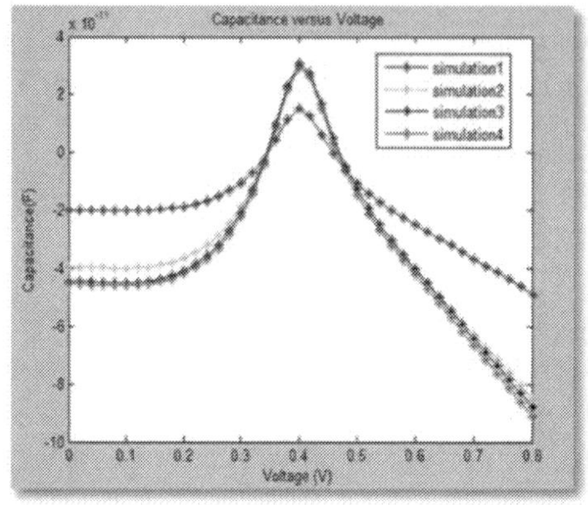

(a) (b)

Figure 6. (a) G-V Characteristics and (b) C-V Characteristics for combination of simulation 1-4

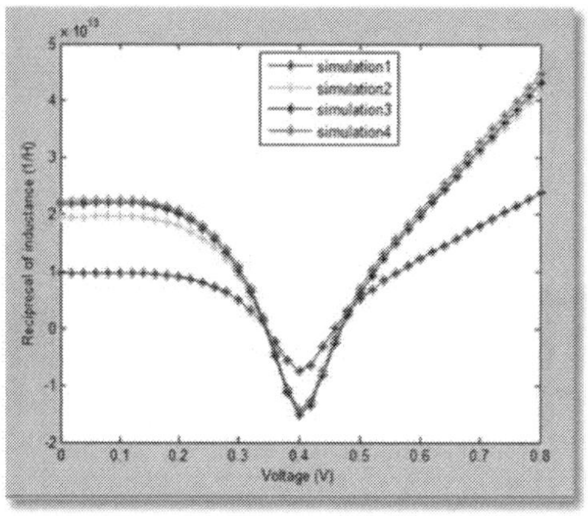

Figure 7. Graph reciprocal of inductance for combination of simulation 1-4

Table 2: Component values for oscillation frequency calculation

Simulation	G(S)	C (F)	L (H)
1	2.2853	-1.7761p	1.1290p
2	2.7231	-2.1220p	0. 94744p
3	3.6792	-2.8773p	0.70124p
4	5.2921	-4.1514p	0.48752p

The value of conductance and capacitance tabulated in Table 2 are used in order to calculate frequency for oscillator. Therefore, Figure 8 demonstrates the frequency versus voltage for various parameter values define in Eq (1) – Eq. (4). The graphs shows that the frequency value is dependent on the conductance value as it has the same G-V characteristics. The oscillation will occur at the NDR region where it is defined at V=0.34 V. Then, the frequency value from the simulations are summarized into Table 3. Based on the analysis, the capacitance and conductance had influenced the frequency value of oscillator circuit [11-12]. The frequency is higher when the conductance value is higher and lower capacitance value .It is observed that the capacitance value should be kept lower in order to get a higher oscillation frequency. Therefore, the best RTD design for oscillator is simulation 4 because it is achieved the targeted frequency which is 557.13 GHz.

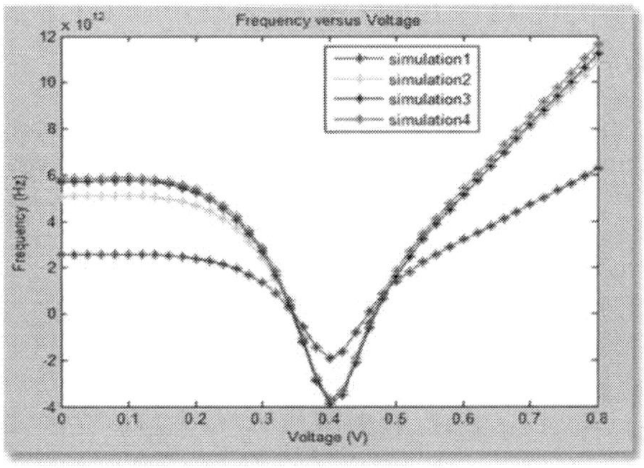

Figure 8. Frequency versus Voltage.

Table 3 Capacitance and Frequency value at V=0.34V (NDR of RTD device)

Simulation	G (S)	C (pF)	F(GHz)
1	2.28	-1.78	254.48
2	2.72	-2.12	298.51
3	3.68	-2.88	394.74
4	5.29	-4.15	557.13

5 Conclusion

Resonant Tunneling Diode (RTD) has been successfully design for oscillator circuit with targeted frequency which is 500-600 GHz. In order to achieve the target frequency, a few simulations have been conducted. The best result is illustrated in simulation 4 obtained by using A, B, C_1, C_2, D_1, D_2, E_1 and E_2 parameter value of 29.6695x1011A/m2, 2.3545 V-1, 25.1813x1011 A/m2, -18.9879x1011 A/m2, 15.5466 V-2, 15.5470 V-1, 0.3227 V and 0.4028 V respectively. Besides, simulation 4 has a higher peak current value as compared to other simulations which is 16.46A and the PVCR is 1.20. Based on simulation, the frequency is increased when conductance value increases and meanwhile the capacitance value is decreases. Therefore, the conductance and capacitance value of 5.2921S and -4.1514 pF to achieve an oscillation frequency of 557.13 GHz. In conclusion, oscillator circuit based on RTD is successfully designed based on this value.

Acknowledgement

This work was financial supported by FRGS Grant Vot Number 1538.

References

[1] Li, B., Zhang, W., Xu, C., & Li 2006 The Oscillator Circuit Based on Resonant Tunneling Diodes *WSEAS Transactions on Circuits and Systems* **5** 872

[2] Dessouki, A. A. S., Abdallah, R. M., & Aly, M. H 2014 A Simplified Analytical Technique for High Frequency Characterization of Resonant Tunneling Diode *Advances in Electrical and Computer Engineering* **14** 87

[3] Sun, J. P., Haddad, G. I., Mazumder, P., & Schulman, J. N. 1998 Resonant Tunneling Diodes: Models and Properties *Proceedings of the IEEE* **86** 641

[4] Zhao, P., Cui, H. L., Woolard, D. L., Jensen, K. L., & Buot, F. A. 2001 Equivalent circuit parameters of resonant tunneling diodes extracted from self-consistent Wigner-Poisson simulation *IEEE Trans. on Electron Devices* **48** 614

[5] W.M. Jubadi, M A Md Zawawi, M.Missous 2014 Simulative Study on Physical Modelling of Submicrometer Highly-strained $In_{0.8}Ga_{0.2}As/AlAs$ Resonant Tunnelling Diode *IEEE Transactions on Electron Devices* **978** *4799*

[6] Buccafurri, Emanuela 2010 Analytical modeling of Silicon Based Resonant Tunneling diodes for RF oscillator application *Ph.D. Dissertation, National Institute of Applied Sciences*

[7] Lee, J., Kim, M., & Yang, K. 2016 A 1.52 THz RTD Triple-Push Oscillator with a-Level Output Power *IEEE Transactions on Terahertz Science and Technology* **6** 336

[8] Abdallah, R. M., Dessouki, A. A., & Aly, M. H. 2014 A Simple Approach to Extract the Small Signal Model Circuit Elements for RTD *Proc. of Information Science, Electronics and Electrical Engineering (ISEEE), 2014 International Conference on* **3** 1747

[9] Q.Liu,A.Seabaugh,P. Chahal, and F.Morris 2004 Unfied ac model for the resonant tunneling diode *IEEE Transactions on Electron Devices* **51** 653

[10] Mizuta, H. and Tanoue, T. 2006 The Physics and Applications of Resonant Tunneling Diodes *Cambridge University Press* **2**

[11] S. Diebold et al. 2016 Modeling and Simulation of Terahertz Resonant Tunneling Diode-Based Circuits in *IEEE Transactions on Terahertz Science and Technology* **6** 716

IOP Conf. Series: Journal of Physics: Conf. Series **1049** (2018) 012070 doi:10.1088/1742-6596/1049/1/012070

A Preliminary Study on Malay Vowel Formant for Young Adult and Aged People

Norezmi Jamal[1], **Shahnoor Shanta**[1*], **Fadhilah Rosdi**[2], **Nabilah Ibrahim**[1],
[1]Faculty of Electrical and Electronic Engineering, Universiti Tun Hussein Onn Malaysia, Batu Pahat, Johor, Malaysia.
[2]Faculty of Information Science and Technology, Universiti Kebangsaan Malaysia, Selangor, Malaysia.

E-mail: shahnoor@uthm.edu.my

Abstract. This paper presents a study on Malay vowel formants of different age group between young adults and aged people. About 12 of the available young adult speech (6 males and 6 females; mean age 23 years) and 4 of the recorded aged people speech (2 males and 2 females; mean 67.3 years) were examined for analysis of formants to study the effect of age and gender. This work is significant in studying the hypothesis that the formant of aged people is lower than young adult population. The findings obtained shows that the formant of aged group is not critically affected by their age and male groups produces lower formant frequencies compared to the female groups either young adult or aged people. Formant frequency values also subject to variation due to the influence of surrounding phonemes.

1. Introduction

In recent years, the numbers of people suffering from stroke have increased gradually. In Malaysia, as many as six people experience a stroke every hour [1] and it is one of the top five leading causes of death [2]. Most of them are aged people. An aged person can be defined as a person whose age is above 60 years old [3]. They might suffer from communication disorders following a stroke. In communication disorder, a stroke survivor may have language impairment and articulation problem, which is known as aphasia [4]. To assess and diagnose this disorder, most clinicians and researchers use acoustic analysis tools [5]. The advantage of this analysis is to differentiate the pathologic and normal voice with the given healthy population.

Formant is one of the essential acoustic properties that will be considered in this study. It is known as the resonating frequency of the vocal tract [6]. In spoken language, vowels play an important role by carrying the linguistic information about any spoken word or sentence. Without vowel, the sound of word might not occur. Although some research has already been carried out on the acoustic property analysis of vowels, especially the formants, the point to note here is that it is normally incomparable if resulting from the cross-language analysis of phonetic units such as English, Swedish, Russian, Turkish, Bangla and Mandarin [5, 7-9]. For example, Kuhl et. al. claimed that the acoustic forms of speech are highly inconstant, changing with few reasons that include speaker gender, speaking rate, and the phonological context of the sound, which makes the task of sorting ambient language sounds into phonetic categories a complex one [7].

Demirhan et. al. in their study on acoustic voice analysis of young Turkish speakers, reported that the fundamental frequency, F0 of vowels (/a/, /i/ and /u/) for female were greater than male [5].

Content from this work may be used under the terms of the Creative Commons Attribution 3.0 licence. Any further distribution of this work must maintain attribution to the author(s) and the title of the work, journal citation and DOI.
Published under licence by IOP Publishing Ltd

The authors investigated the F0, perturbation parameters, and noise-to-harmonics ratio (NHR) for three sustained vowels in a group of young Turkish speaking adults. The perturbation values were compared in the study, which significantly changed for some pairs of vowels. On the other hand, there was no significant difference between NHR values of three vowels in men. The study is conducted, where the acoustic analysis is only limited to adult Turkish speakers with a small sample data [5]. Moreover, Ghosh et. al. reported the distinct Bangla vowels exhibiting different formant frequencies [8]. While, Liu and Ng studied the formant characteristics associated with the vowels produced by esophageal and laryngeal speakers of Mandarin [9]. The results indicated that the esophageal speakers were associated with significantly higher formant frequencies and a significantly diminished vowel space circumscribed by the three corners vowels (/a/, /i/, and /u/) compared to laryngeal speakers.

A considerable amount of literature has already been published on Malay vowel based acoustic properties [10, 11]. For example, Ting et. al. and Zourmand et. al. claimed that Malay acoustic properties of female is higher than male irrespective of young adult or children, especially considering the fundamental frequency and the formants of vowels. However, there are significant differences between the vowels reported for all perturbation measures in Malaysian Malay males. Rosdi et. al. also investigated the effect of speech rate on formant frequency of Malay vowels [12]. The results showed that there is a small variant, where the changing of second formant, F2 values from vowels spoken at a fast rate as compared to those spoken at a normal rate.

In short, formants can differ depending on the configuration of the vocal tract for articulating different types of voiced sounds. It can be noted that normally vocal tract of female is shorter than male [13]. Due to this, female formant is higher than the male one. Besides, the effect of age also can cause the difference on the formant as discussed in the age-related changes in acoustic characteristics of adult speech article [14]. The focus of the article was the aged population and English language. Still, there is not much study reported on Malay aged people formant analysis. Due to this, the aim of this work is to discuss the comparison of Malay aged people and young adult formant analysis. This paper is organized as follows. In section 2, the previous related works on the formant of vowels are discussed. Section 3 provides the methodology of formant analysis from a given audio file. In section 4, the results and discussion are presented. Finally, our work of this paper is summarized in the last section together with the future direction relevant to this work.

2. Related Works on Analysis of Formants Resulting from Malay Vowels

Basically, the different vowels could be represented with distinct characteristics formants due to the different phonetic sounds. The formant frequencies are the oscillation of frequencies that produces high amplitude of sounds (resonances) along the distinct cross- sectional area of the vocal tract [15]. In simple terms, vowels are the most significant sounds that often get affected. This is because; the vowels are produced from the excitation of fixed vocal tract shape and the vibration of the vocal cord [15]. The vowels sounds are also primarily based on the positions of tongue, jaws, lips and velum [15]. There are three typical formants, namely as the first formant (F1), the second formant (F2), and the third formant (F3).

A few researches have been carried out, examining the effect of different age, gender and ethnicity on the formant of vowels [16-17]. Table 1 shows the reported baseline formant values (F1, F2 and F3) of vowels (/a/, /u/ and /i/) for male and female speakers which were obtained by different authors in their respective research articles. The baseline value is calculated from the average of the research findings on the formant analysis of Malay native speakers considering different age and ethnicity. It can be seen that the formants of each vowel for male speaker is lower than that of female speaker, where F1, F2 and F3 were reported. Ting et. al. investigated the first three formant frequencies for the control Malaysian Indian young adult groups [16] and Malaysian Malay children group [17]. For Malay Indian young adults, the F1, F2, and F3 of average vowels for male speakers were reported at 359 Hz, 1457 Hz and 2799 Hz respectively [16]. While F1, F2 and F3 of average vowels for Malaysian Indian females were reported at 485 Hz, 1707 Hz, and 3081 Hz respectively

[16]. Malaysian Indian females had significant higher formant frequencies in all vowels than those of Indian male speakers. It is similar for the children speakers, where the results showed that, Malay female children had higher formant frequencies compared to male speakers [17].

Table 1. The baseline formant values of vowels based on the previous related works.

Authors	Details	Vowels	Male Speakers			Female Speakers		
			F1	F2	F3	F1	F2	F3
[16]	Indian Speakers	/a/	445	1155	2703	769	1458	3045
		/u/	320	853	2763	369	872	2853
		/i/	263	2376	3193	324	2763	3521
[17]	Children Speakers	/a/	749	1454	3225	787	1530	3247
		/u/	408	950	3012	436	981	2997
		/i/	353	2621	3573	361	2428	3584
	Baseline (Average)	**/a/**	**597**	**1304**	**2964**	**778**	**1494**	**3146**
		/u/	**364**	**902**	**2888**	**403**	**927**	**2925**
		/i/	**308**	**2499**	**3383**	**343**	**2596**	**3553**

3. Methodology

In this study, 16 of the native Malay speakers were categorized into two large groups, which are: (i) young adult (YA) and (ii) aged people (AP). For young adult group, the audio files were used from the available Malay digit speech corpus (databases) [18]. 12 of the recorded speech signals were examined including 6 male (YM) and 6 female speakers (YF). Their mean ages are 23 years. And, for aged people, 4 speakers including 2 males (AM) and 2 females (AF) ranging in age from 61 to 79 years were evaluated. Aged people speech data were recorded using a high performance Keystudio microphone attached with the pop filter at a clean environment. The details of the recorded audio data are; mono channel, 16-bit resolution and a sampling rate of 16.0 kHz respectively. All the speakers were given a task to say Malay digit words (*empat, tujuh* and *sembilan*) for three times. The vowels (/a/, /u/ and /i/) of three selected Malay digit words as shown in Figure 1 were analysed for acoustic properties. All the recorded data were imported into Praat speech analysis software. The formants frequencies (F1, F2 and F3) of Malay vowels were analysed by highlighting the desired vowel area. Formants of each Malay spoken digits were identified by averaging the value of formants from three different sets of similar spoken words. The resulting waveform for the word *empat* is shown in Figure 1 (a). While the word *tujuh* and *sembilan* are respectively illustrated in Figure 1 (b) and 1 (c).

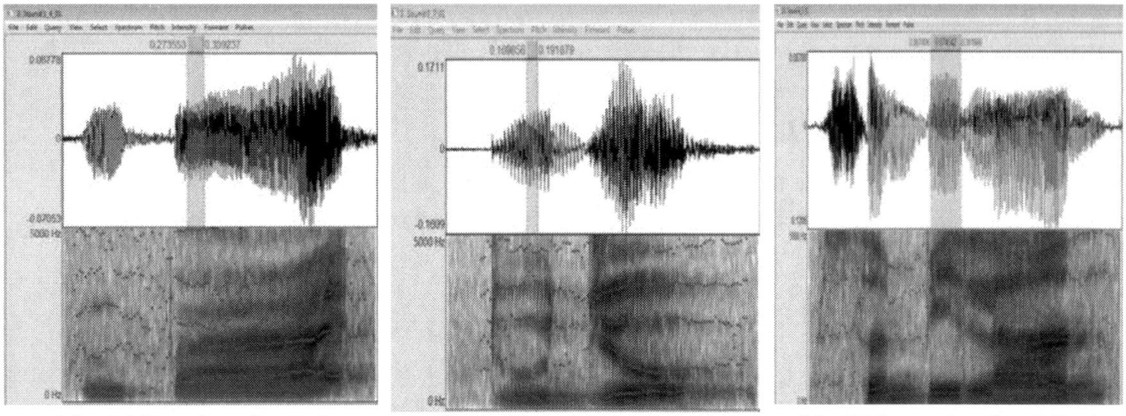

(a) /a/ from four (*empat*) (b) /u/ from seven (**tu**juh) (b) /i/ from nine (sembi**la**n)

Figure 1. Waveform and its spectrogram of Malay spoken digits

4. Results and Discussion

The formant values of each vowels (/a/,/u/,/i/) were compared for young adult (YA) and aged people (AP) speakers. Figure 2 shows the distribution of formants (F1, F2 and F3) for different vowels. For the vowel class /a/, young male (YM), young female (YF), aged male (AM) and aged female (AF) are clearly grouped in the same class. As shown in Figure 2, the AF group that is represented by red star legend are found to have significantly more formant values of /a/ vowel than the other three groups. In contrast to earlier findings in the age-related changes in acoustic characteristics of adult speech for English vowels reported in article [14], however, no evidence for the age of female population, affecting the formant values was detected. The authors reported that young adult female produced higher formants of /a/ vowel among other groups [14].

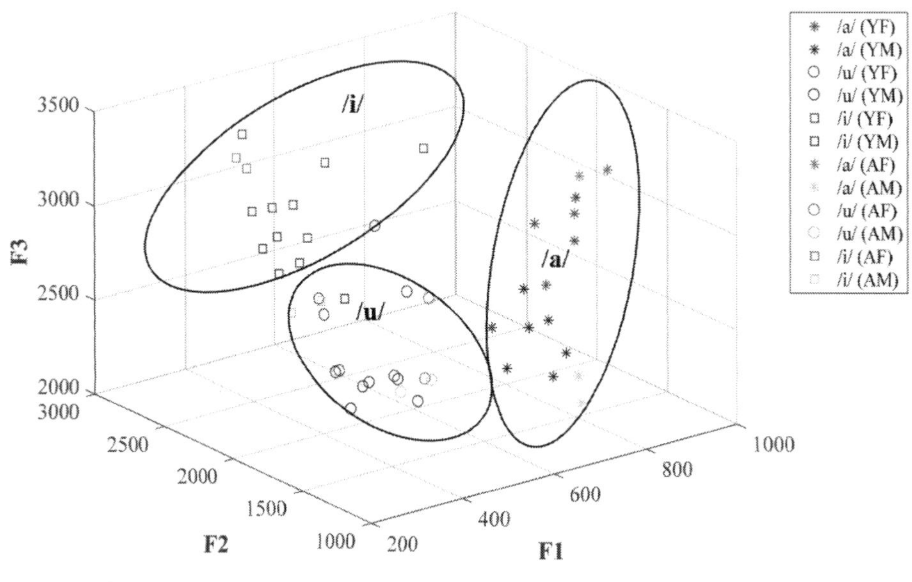

Figure 2. The distribution of formant values for different vowels from 16 Malay native speakers

The results obtained, also showed the outliers for /i/ and /u/ of the vowel class. The formant value of /i/ vowel for aged male was categorized into the /u/ vowel class. The overlapping as seen in the results obtained might be related to the gender and age differences and also due to the co-articulation [19]. Another reason might be that the aged male group might have slower speaking rates than the younger one. This unexpected result could be attributed to the fact that the formant values of /i/ and /u/ vowels keep continuously changing due to the influence of the neighbouring phonetic sounds as shown in the spectrogram of Figure 1 (b) and Figure 1(c) respectively. Table 2 shows the mean of the first (F1), second (F2), and third (F3) formant frequencies for each vowel (/a/, /i/, /u/) produced by young adult male (YA-M), young adult female (YA-F), aged people male (AP-M), and aged people female (AP-F) obtained from the preliminary analysis of data in Figure 2. Obviously, the mean formant value (F1, F2 and F3) of each vowel for male is lower than those of female groups. For example, for /a/ vowel, young adult and aged people male speakers respectively produced 683 and 680 Hz of F1 which are lower compared to the young adult and aged people female speakers. So, the results obtained demonstrated the similarity with other findings as discussed in section 2. The reason behind this is that the gender significantly affects the formant values; mainly because of the mean size of male adult vocal tract is normally about 17 cm, longer than that of the female and children population [13].

Table 2. Mean first (F1), second (F2), and third (F3) formant frequencies of vowels for four different groups of speakers

Groups	Vowels	F1 (Hz)		F2 (Hz)		F3 (Hz)	
		M	F	M	F	M	F
YA	/a/	683	793	1387	1663	2528	2845
	/u/	331	366	1382	1765	2503	2792
	/i/	338	444	2063	2231	2967	3226
AP	/a/	680	868	1254	1753	2245	3033
	/u/	329	373	1093	1162	2609	2808
	/i/	342	470	1930	2432	2937	3301

5. Conclusion

The study investigated the first three formant frequencies of /a/, /u/ and /i/ vowels for normal Malay young adult and aged people speakers. The results found, demonstrates that, generally, Malay female groups have higher formant frequencies compared to male groups. Moreover, the aged people group have lower formant values due to their slower speaking rate. Besides, it is also found that, there is formant pattern ambiguity of vowel sounds for /i/ and /u/. This is because of the formant values of /i/ and /u/ vowels keep continuously changing due to the influence of the neighbouring phonetic sounds. As a result, the findings show that formants are very useful features for speech recognition or early diagnosis for any speech disorder, but they have not been widely used because of the difficulty in estimating them. So, Mel Frequency Cepstrum Coefficient (MFCC) analysis or other feature extraction techniques could be a possible avenue for future research.

Acknowledgement

This work is financially supported by the Short Term Grant (STG), vot number 530 under Universiti Tun Hussein Onn Malaysia. Also, highly appreciated to Microelectronics & Nanotechnology - Shamsuddin Research Centre (MiNT-SRC), Universiti Tun Hussein Onn Malaysia for providing research work space.

IOP Conf. Series: Journal of Physics: Conf. Series **1049** (2018) 012070 doi:10.1088/1742-6596/1049/1/012070

References

[1] O W Mohamad, Z Taib M and Jailani R 2014 *Systems, Process and Control (ICSPC)* 2014 157
[2] Loo K W and Gan S H 2012 *International Journal of Stroke* **7** 165
[3] Ismail N N N N, Lokman A M and Redzuan F 2016 *e-Learning, e-Management and e-Services (IC3e)* 133
[4] Jamal N, Shanta S, Mahmud F and Sha'abani M 2017 *AIP Conference Proceedings* 020028.
[5] Demirhan E, Unsal E M, Yilmaz C, and Ertan E 2016 *Journal of Voice* **30** 378
[6] Rabiner L R and Schafer R W 2011 *Pearson Upper Saddle River* (NJ) 64
[7] Kuhl P K, Andruski J E, Chistovich I A, Chistovich L A, Kozhevnikova E V, Ryskina V L, Stolyarova EI, Sundberg U and Lacerda F 1997 *Science* **277** 684
[8] Ghosh T, Saha S and Ferdous A I 2016 *An International Journal (SIPIJ)* **7**
[9] Liu H and Ng M L 2009 *Journal of voice* **23** 255
[10] Ting H N, Chia S Y, Hamid B A and Mukari S Z M S 2011 *Journal of Voice* **25** 305
[11] Zourmand A, Mahmoudi A, Marimuthu J and Ting H N 2012 *Intelligent Systems, Modelling and Simulation (ISMS)* 379
[12] Rosdi F, Salim S and Mustafa M 2014 *Advanced Science, Engineering and Medicine* **6** 127
[13] Huang X, Acero A, Hon H W and Reddy R 2001 *Spoken language processing: A guide to theory, algorithm, and system development* (Prentice hall PTR)
[14] Torre P and Barlow J A 2009 *Journal of communication disorders* **42** 324
[15] Rabiner L R and Juang B H 1993 *Fundamentals of speech recognition*
[16] Ting H and Lee C 2013 *World Congress on Medical Physics and Biomedical Engineering* 375.
[17] Ting HN, Zourmand A, Chia S Y, Yong B F and Hamid B A 2012 *Journal of Voice* **26** 664
[18] Al-Haddad S A R, Samad S A, Hussain A and Ishak K A 2008 *American Journal of Applied Sciences* **5** 714
[19] Maurer D, D'heureuse C, and Landis T 2000 *International Journal of Neuroscience* 39

IOP Conf. Series: Journal of Physics: Conf. Series **1049** (2018) 012071 doi:10.1088/1742-6596/1049/1/012071

130 nm Low Power CMOS Analog Multiplier

Ahmad Safuan Abu Naim[1], **Siti Hawa Ruslan**[1,2*]

[1] Faculty of Electrical and Electronic Engineering,
[2] NanoSIM Research Group,
Universiti Tun Hussein Onn Malaysia,
86400 Parit Raja, Batu Pahat, Johor, Malaysia.

*Email: sitihawa@uthm.edu.my

Abstract. Processing analog signal often involves analog multiplier and the multiplier is part of system on chip (SoC). Designing such system with a low power consumption is crucial nowadays. It is very important to increase the system battery lifetime. The design also must be smaller in size. In order to reduce the power consumption of the multiplier, an architecture that require smaller current must be designed and the approach is to use a design that is able to function at a low voltage supply. This project has designed the analog multiplier with a low power consumption using Silterra 130 nm Complementary Metal Oxide Semiconductor (CMOS) technology. A four quadrant technique is applied in the design. The scaling of transistor will help in reducing the size of the analog multiplier, and the proposed circuit architecture has produced a compact multiplier. Cadence electronic design automation (EDA) Tools is used to design the circuit. The schematic, layout, physical verification and parasitic extraction with post layout simulation are done to verify the multiplier circuit is functioning. The analog multiplier is operated with 1.2 V voltage supply and the power consumption is 98 μW. At 1 V, the power consumption is 32 μW. The total area for the design is 99 μm².

1. Introduction

Analog circuit deals with continuous time signal that are used to generate, detect, measure, amplify, attenuate or filter signal. In analog circuits such as in adaptive filters, frequency doublers, and modulators [1], [2], analog multiplier plays an important part in it. It is used extensively in the field of telecommunication, control, instrumentation, measurement, and signal processing [3]. The output of analog multiplier, v, is a linear product of two continuous input signals v_1 and v_2. Therefore $v = k\, v_1\, v_2$ where k is a multiplication constant or gain of suitable dimension [4]. Figure 1 shows the basic idea of an analog multiplier.

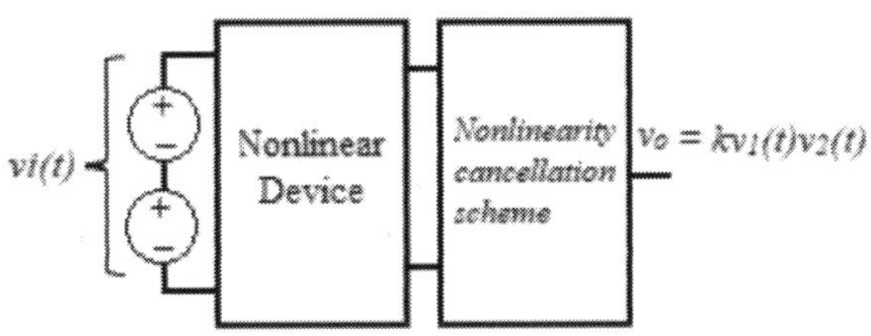

Figure 1. Basic idea of multiplier [5]

In portable applications, analog multiplier has to operate with a low voltage to improve power efficiency [6]. Hence, analog multiplier must be designed in the low voltage environment too.

2. Analog Multiplier

An analog multiplier had been designed in [7], and the design is claimed to be very compact. However, the circuit has an extra voltage reference terminal connected between the resistive loads. The extra voltage reference makes the circuit consumed more power and the circuit complexity is high [8]. Thus the analog multiplier circuit cannot be called compact. In [9], a fully differential four quadrant multiplier

Content from this work may be used under the terms of the Creative Commons Attribution 3.0 licence. Any further distribution of this work must maintain attribution to the author(s) and the title of the work, journal citation and DOI.
Published under licence by IOP Publishing Ltd

IOP Conf. Series: Journal of Physics: Conf. Series **1049** (2018) 012071 doi:10.1088/1742-6596/1049/1/012071

was designed with supply voltages of ± 1 V. The circuit consumed 0.326 mW of static power and was designed in 0.25 μm CMOS technology. A proper biasing is needed in this circuit in order to produce a correct output. A 180 nm analog multiplier is designed in [10] with a power consumption of 0.43 mW. The circuit works well with ± 0.8 V supplies but the power consumed is considered high. Another 180 nm analog multiplier is designed in [11] at the supply voltage of 1 V. The power consumed by the circuit is very low which is 1.2 μW, but the circuit has to work in weak inversion region. Hence a proper biasing is critical in this circuit. As for the multiplier circuit in [12], the technology is 130 nm and with a supply of 1 V, the power consumed is 25.14 mW.

The analog multiplier in this paper is designed to eliminate extra voltage reference in order to produce a compact circuit. With the help of current CMOS technology which has been scaled down over the years, a smaller transistor size of 130 nm is used which will yield higher device density, higher speed and reduced power consumption. For the operating supply of the circuit, a 1.2 V or 1 V can be applied. Cadence EDA Tools is used to design and simulate the analog multiplier in this project.

Figure 2 illustrates the proposed circuit of the analog multiplier that is designed in this project. The proposed analog multiplier comprises of a pair of common source amplifier with input transistors m1 and m2. This will give output currents in terms of squaring functions of input voltages (V1 and V2). A suitable value of resistance for resistors R in the circuit has been determined to ensure all the transistors in the circuit are working in the proper region. The basis for the circuit is a four quadrant multiplier circuit as mentioned in [1], [8], [13] and [14].

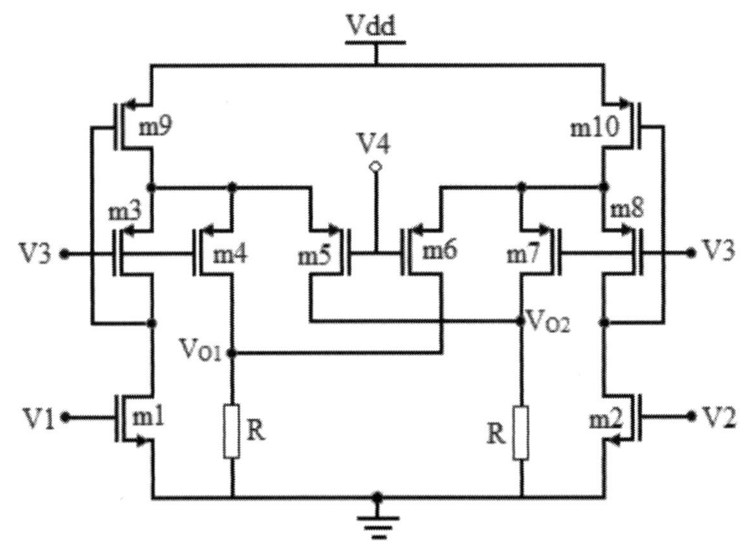

Figure 2. The proposed analog multiplier circuit

3. The Designed Circuit

Figure 3 shows the schematic circuit of the analog multiplier designed using Cadence software in 130 nm CMOS technology. The circuit contains 10 transistors consist of 8 PMOSs and 2 NMOSs. The circuit also includes 2 resistors, R1 and R2 with the value of 2.5 kΩ. Transistor m1 and m2 are the input transistors for the input voltage V1 and V2 respectively. These two transistors give output currents in terms of square functions of the two input [8]. Transistors m3-m10, serve as a square root circuit which act as non-linear cancellation path. The output currents of the input transistors are directed into the square root block to produce a differential output current of the overall circuit and resulting in a multiplication function of two input signal V12 and V34. Table 1 shows the transistor dimension used to design the analog multiplier.

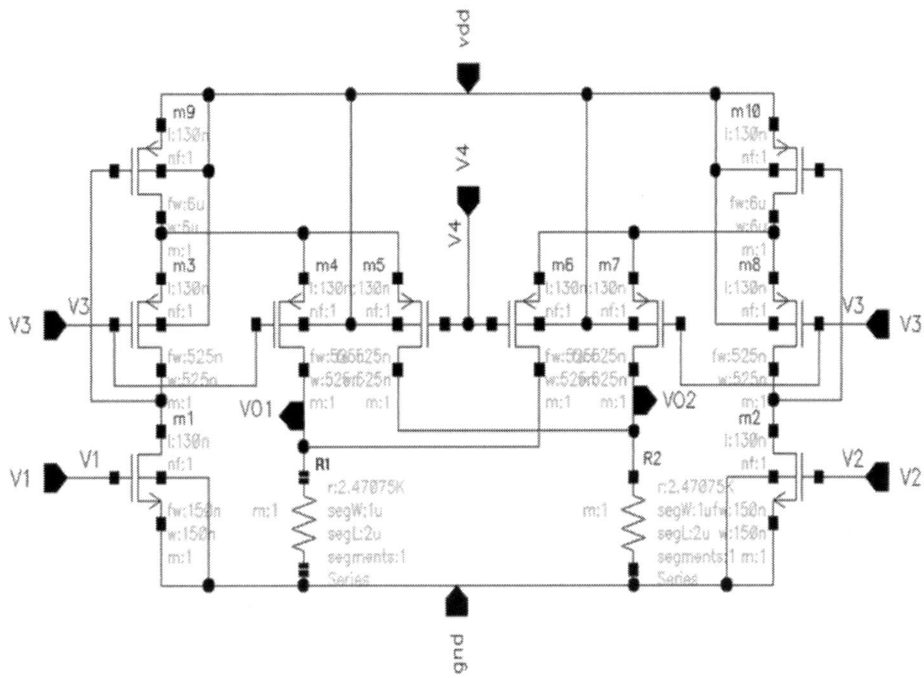

Figure 3. The schematic circuit of analog multiplier

Table 1. Transistor Sizing

Transistor	Width, W (nm)	Length, L (nm)
m1-m2	150	130
m3-m8	525	130
m9-m10	6000	130

4. Result and Discussion

The analog multiplier circuit in Figure 3 was designed and simulated using Cadence with 130 nm CMOS technology with two different supply voltages of 1.2 V and 1 V. The amplitude of V1 and V2 are varied with its sinusoidal carrier signal frequency of 25 kHz multiplied with V3 and V4 which are also varied with sinusoidal modulating signal frequency of 1 kHz. Figure 4 shows the simulated circuit with its input amplitude and frequency.

4.1 Transient Analysis

This analysis includes time versus voltage. The differential input voltage V12 is defined as V1-V2 while the differential control voltage V34 is equal to V3-V4. The output for this circuit is the product of the carrier signal V12 (Figure 5 (a)) with the modulating signal V34 (Figure 5 (b)) and the produced waveform is shown in Figure 5(c). The result obtained in Figure 5(c) is from Vout = V_{O1}-V_{O2}. The result is correct and according to the theory. The result obtained is similar to the result in [8] and [14], thus proved the analog multiplier circuit has been successfully designed.

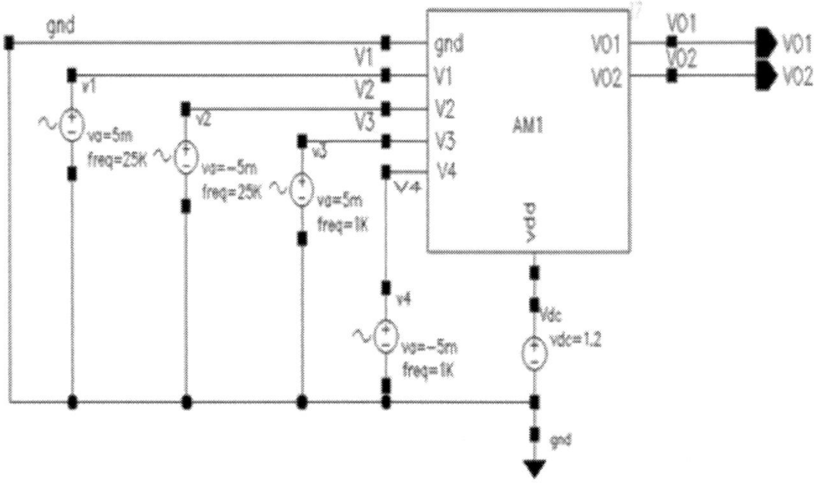

Figure 4. Simulation circuit

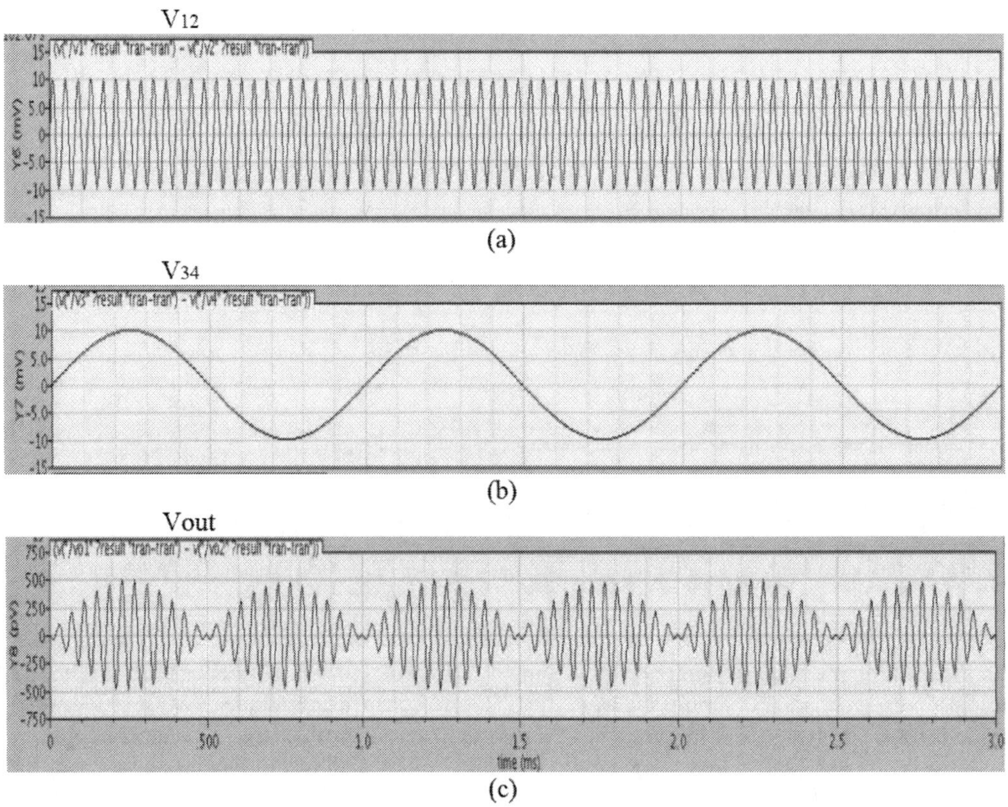

Figure 5. Output waveform from simulation (a) V_{12} sinusoidal carrier signal, (b) V_{34} sinusoidal modulating signal, (c) output waveform of analog multiplier

The circuit is also simulated with different voltage supplies to investigate the maximum and minimum input voltages that can be applied to the circuit. The simulation result listed in Table 2 are for 1.2 V and 1 V. It is found that with 1.2 V voltage supply, a higher amplitude of input signal can be applied. The power consumption of the proposed analog multiplier at 1.2 V is 98 μW while at 1 V the power consumption is 32 μW. A 1 V power supply consumed less power but the range of the input signal is lower. A supply voltage of lower than 1 V cannot be applied to the circuit because the output waveform is incorrect. A higher voltage than 1.2 V can be applied but the power consumed will be higher too. Nevertheless, the proposed analog multiplier has been shown to consume less power as compared to the design in [9], [10] and [12].

Table 2. Comparison of different voltage supply

Voltage Supply (V)	Input Voltage (V)		Output voltage (V)	Power Consumption (W)
1.2	Max	14.0m	968.1p	98μ
	Min	0.6m	1.71p	
1.0	Max	8.0m	596.0p	32μ
	Min	0.3m	0.8052p	

4.2 Design Rule Check (DRC) and Layout Versus Schematic (LVS) Test

DRC is a step imposed on the layout. It verifies the physical layout to ensure that none of the design rule is violated. Figure 6 show the layout for the proposed design. Transistors M9 and M10 occupied more area as compared to other transistors in the circuit. The total physical area for the analog multiplier is 99 μm². The area taken is smaller if compared to the circuit in [7] and [8]. This result is expected since these circuits were designed using a bigger CMOS technology.

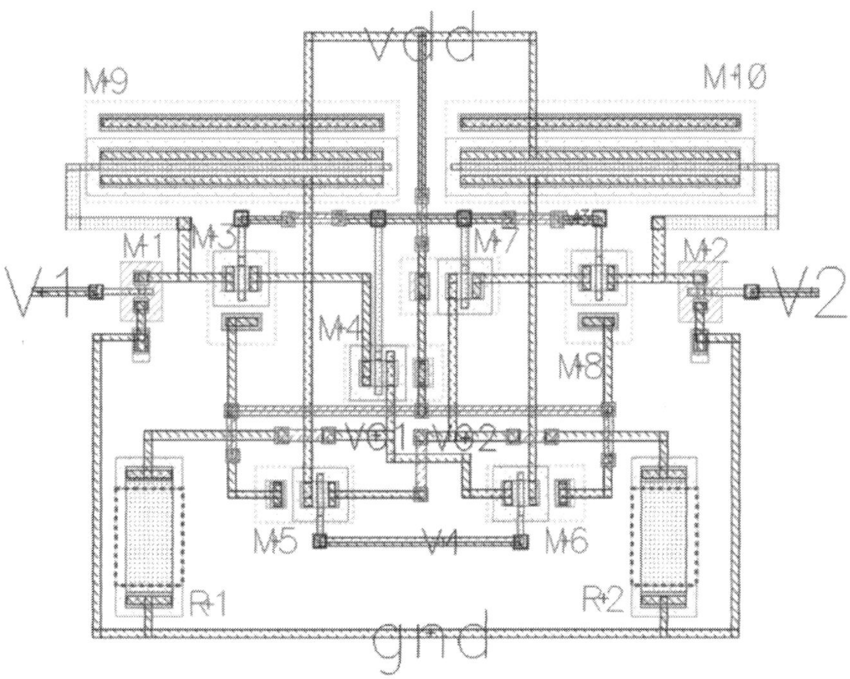

Figure 6. Layout design of the multiplier

LVS physical verification is vital tool to verify the connectivity comparisons between the layout and schematic. The result from the LVS test done on the multiplier circuit has shown the schematic and layout are equivalent.

5. Conclusion

An analog multiplier circuit has been constructed using 130 nm CMOS technology with an area of 99 μm^2, the circuit is more compact than previous work. A smaller transistor size had been used in this design. The circuit can be used with a voltage supply of 1 to 1.2 V with a power consumption of between 32 to 98 μW. The circuit has been simulated and performed well as an analog multiplier.

Acknowledgement
This work was supported by Faculty of Electrical Engineering (FKEE), Universiti Tun Hussein Onn Malaysia (UTHM), Silterra and Collaborative Microelectronic Design Excellence Centre (CEDEC) for Innovate Malaysia Design Competition 2017.

References

[1] Bult K and Wallinga H 1986 A four-quadrant analog multiplier *IEEE J. Solid-State Circuits* **21** 430
[2] Liu S I 1994 Low voltage CMOS four-quadrant multiplier *Electronics Letters* **30** 2125
[3] Flandre D, Raskin J P and Vanhoenacker-Janvier D 2001 SOI SMOS transistor for RF and microware application *International Journal of High Speed Electronics and Systems* **11** 1159
[4] Gibert B 1968 A precise four-quadrant multiplier with subnanosecond respons *IEEE J. Solid-State- Circuits* **3** 365
[5] Nandini A S, Madhavan S and Sharma C 2012 Design and implementation of analog multiplier with improved linearity *International Journal of VLSI Design & Communication Systems* **3** 93
[6] Patel D S and Amin G D 2014 Simulation and analysis of wideband and low power CMOS analog multiplier in deep submicron technology *International Journal of Advance Engineering and Research Development* **1** 1
[7] Sawigun C and Demosthenous A 2006 Compact low voltage CMOS four quadrant analog multiplier *Electronics Letters* **42** 1149
[8] Kiatwarin N, Ngamkham W and Kiranon W 2007 A compact low voltage CMOS four-quadrant analog multiplier *ECTI International Conference* 17
[9] Mahmoud S A 2008 CMOS fully differential CMOS four-quadrant analog multiplier 2008 *International Conference on Microelectronics* 27
[10] Ettaghzouti T, Hassen N and Besbes K 2016 High performance low Voltage low power voltage mode analog multiplier Circuit *7th International Conference on Sciences of Electronics, Technologies of Information and Telecommunications* 527
[11] Nikseresht S, Azhari S J and Danesh M 2017 High bandwidth four-quadrant analog multiplier *25th Iranian Conference on Electrical Engineering* 210
[12] Modi N B and Gandhi P P 2013 Four quadrant analog multiplier with VCVS in deep-submicron technology *Proceedings of 2013 IEEE Conference on Information and Communication Technologies* 1091
[13] Tijare A, Dhondse S, Nitnaware S and Pawar M. 2014 Low-voltage, high frequency CMOS analog multiplier *International Journal of Engineering and Computer Science* **3** 5004
[14] Saatlo A N and Ozoguz I S 2016 Design of a high-linear, high-precision analog multiplier, free from body effect *Turkish Journal of Electrical Engineering & Computer Sciences* **24** 820

Investigation of gender and prandial state as factors affecting skin bilirubin level

F. Y. Choong, Audrey Huong[1*]
[1] Faculty of Electrical and Electronic Engineering
University Tun Hussein Onn Malaysia

E-mail: audrey@uthm.edu.my

Abstract. This paper demonstrates the use of the developed non-invasive optical system to measure differences in the mean concentration of skin bilirubin $C_{bil(m)}$ among individuals of different gender and at different prandial state. This system employed the use of color filters to allow an illumination of the selected skin site using light beams of centre wavelength 420 nm, 440 nm and 460 nm. The prediction of the required skin bilirubin concentration value is via Modified Lambert Beer law. This study conducted experiments on the back of the hand of twelve recruits. The results revealed a higher mean $C_{bil(m)}$ value of 0.25 ± 0.007 µmol/L for males as compared to 0.14 ± 0.0086 µmol/L measured for females. In addition, this work observed a decrease in the mean $C_{bil(m)}$ value for the ten recruits of mixed gender from 0.23 ± 0.06 µmol/L at preprandial phase to 0.18 ± 0.04 µmol/L during the postprandial period. This study concluded that the developed system is able to detect changes in the skin bilirubin concentration level with one's gender and prandial state as reported in literature. However, this system may be further improved for future clinical application on jaundiced patients.

1. Introduction

In the medical arena, non-invasive devices able to provide comprehensive information of one's health status are highly sought after. This includes information of one's bilirubin concentration, which is a waste product produced during the breakdown of aged red blood cells. This orange-yellow pigment yields yellow stains in skin and other tissues (jaundice) if its concentration in blood is high. The poor functional efficiency of one's liver in eliminating aged or damaged red blood cells would lead to the accumulation of the latter, and hence an excessive production of bilirubin. This medical condition is often found either in individuals with liver diseases or newborns whose liver have yet fully developed.

In addition to the underlying health status, previous research revealed that one's blood bilirubin concentration would also be affected by the change in their dietary intake [1]. The corresponding work showed a decrease in this value for populations who consumed either carbohydrate and calorie-rich foods or diet containing high fat and carbohydrate as compared with the standard normal diet (ρ < 0.001) [1, 2]. This same pattern of results was also reported by [3] for frequent caffeine consumers. Meanwhile other study concluded that gender is a factor that may contribute to the differences in the bilirubin concentration level [4].

Many types of bilirubin detector are commercially available in the market; they can be categorized as invasive and non-invasive approach. The former is a laboratory blood test that required the drawing

Content from this work may be used under the terms of the Creative Commons Attribution 3.0 licence. Any further distribution of this work must maintain attribution to the author(s) and the title of the work, journal citation and DOI.
Published under licence by IOP Publishing Ltd

of a small amount of blood in the measurement of total and direct bilirubin concentration value. This process is time consuming, laborious and there may also be a risk of infection at the sampling site; hence it is not a preference for many. In view of this, non-invasive technique is often considered as the more favourable approach to use. This technique measured the reflected light signals upon interaction of light with the investigated skin site. The required value would then be translated from the measured signals using a lookup table. The existing non-invasive devices include Bilirubinometer and Bilitec [5]. These devices provide harmless and quick prediction of the required transcutaneous bilirubin concentration value. Nonetheless there were reports on inconsistency in the predicted value depending on the measured site and skin pigmentation [6]. This is likely owing to the fact that the good performance of the employed lookup table may be limited to certain groups of population.

For this reason, other work [7] reported the use of an analytical model as a more suitable approach in the estimation of the required health parameter value. The operating principle of this technique is based on analysis of the measured light spectrum to interpret the variation in light absorptivity underneath skin. Therefore, it is the aim of this study to develop and propose a multi-color filter system to provide non-invasive measurement of one's skin bilirubin level without the use of a look-up table. It is also the objective of this study to investigate the variation in this value predicted for individuals of different gender and at different prandial state.

2. Material and Method

2.1. Experimental system and subjects

The optical system developed for non-contact monitoring of skin bilirubin concentration value is shown in Figure 1. The employed light source in the system is a white Light Emitting Diode (model no. SMD 5730). Light reflected from the selected skin site would pass through a color filter to allow only the detection of light of the chosen wavelength by a Complementary Metal Oxide Semiconductor (CMOS part no. CW835). The latter was located at a distance of 400 mm and an angle of approximately 30° from normal. In this work, light balancing filter (model no. LB120 from Hoya Optics), blue filter (B-440, Hoya Optics) and green filter (G-533, Hoya Optics) were used as the color filter to pass light of centre wavelength 420 nm, 440 nm, and 460 nm, respectively, through them. The reason of the selection of these wavelengths is due to the high bilirubin absorptivity across wavelength range of 400 – 500 nm [8]. These color filters were mounted on a makeshift rotating filter wheel controlled by a stepper motor (17PM-J).

In this study, experiments were conducted on twelve non-smoking Asian volunteers (six males and six females) aged between 23 and 27 years. All volunteers gave their consent to participate in the study, they declared healthy and were not suffered from hepatitis. Experiments were conducted at the back of the right hand of these individuals seated in a chair. The same procedure was repeated for a group ten volunteers of mixed gender during pre-meal (fasting for over 12 hours) and post-meal (postprandial) session. The validation of the obtained value is via a comparison study among volunteers of different genders, and during pre and post-meal intake. This is following the previous reports [2,4], which cited these as the factors to the differences in the bilirubin concentration value. Meanwhile the reason of the choice of this skin site is to investigate the consistency in the pattern of results for individuals with different skin tone using the proposed quantification strategy.

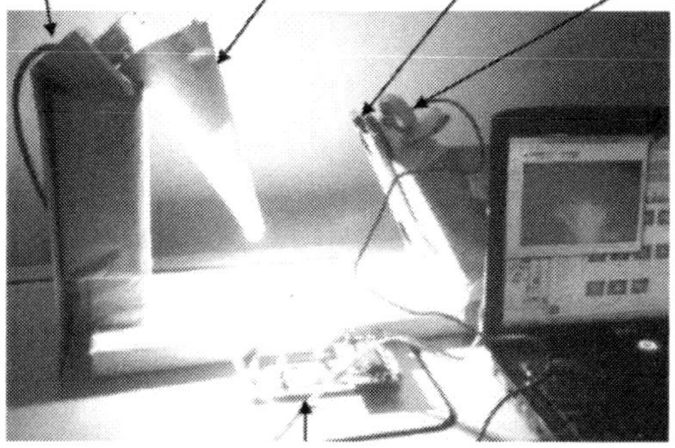

Figure 1. Non-contact multi-color filter system

2.2 Modified Lambert Beer model

In this work, Modified Lambert Beer (MLB) model shown equation (1) is used in the estimation of the skin bilirubin concentration. This model is commonly used to relate changes in light attenuation, A, with the wavelength dependent total absorptivity, μ_a, of the medium as followed:

$$A(\lambda) = G + \mu_a(\lambda)d. \tag{1}$$

The A is given from the measurement of reflected light intensity and is defined as:

$$A(\lambda) = \log\frac{I_W}{I_S} \tag{2}$$

where I_W and I_S represent average light reflectance from a white sheet and from the selected skin site, respectively. The wavelength dependent μ_a is given by:

$$\mu_a(\lambda) = \varepsilon_{bil}(\lambda)C_{bil}. \tag{3}$$

where $\varepsilon_{bil}(\lambda)$ and C_{bil} denote wavelength dependent extinction coefficient and concentration of skin bilirubin, respectively. The symbol d in equation (1) represents light mean pathlength, and is taken here as the reciprocal of the medium's reduced scattering coefficient, μ_s'. Since the tissues μ_s' value is approximately 10 cm^{-1} at light wavelength of 450 nm [9], the parameter d is calculated as 1 mm. The effects of light scattering on light attenuation is notably dominant at the shorter wavelength, so the scattering dependent attenuation offset, G, is assumed here as A (λ= 420 nm). Therefore, substituting equation (3) into (1), and rearranging the equation gives

$$C_{bil} = \frac{A(\lambda_x) - A(\lambda = 420\ nm)}{\varepsilon_{bil}(\lambda_x)d} \tag{4}$$

where λ_x represents light wavelength of 440 nm and 460 nm. The $\varepsilon_{bil}(\lambda_x)$ at these wavelengths is given by 52109 $cm^{-1}M^{-1}$ and 53869 $cm^{-1}M^{-1}$, respectively [10]. The C_{bil} calculated from equation (4) for light wavelength 440 nm and 460 nm is then averaged to give the mean bilirubin concentration, $C_{bil(m)}$.

3. Results and Discussion

Figure 2 showed the boxplot of skin bilirubin concentration value, $C_{bil(m)}$, predicted for the recruited females and males. The results revealed a mean and standard deviation of $C_{bil(m)}$ given by 0.25 ± 0.007 µmol/L and 0.14 ± 0.0086 µmol/L for males and females, respectively. Meanwhile a comparison in the $C_{bil(m)}$ value estimated for volunteers during pre and post-meal session, which changes one's blood glucose level, is shown in Figure 3. A test of correlation between the value predicted for volunteers during fasting and postprandial period was via a paired, two tailed sample *t*-test in SPSS software (SPSS 22, Inc., Chicago, Illinois). This statistical test revealed significant value, $\rho = 0.000$, using the confidence level of 95 %.

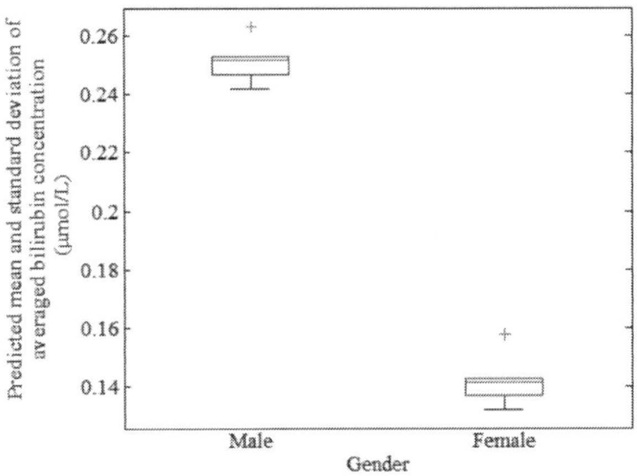

Figure 2. The predicted mean and standard deviation of the averaged skin blood bilirubin concentration value, $C_{bil(m)}$, for volunteers of different genders

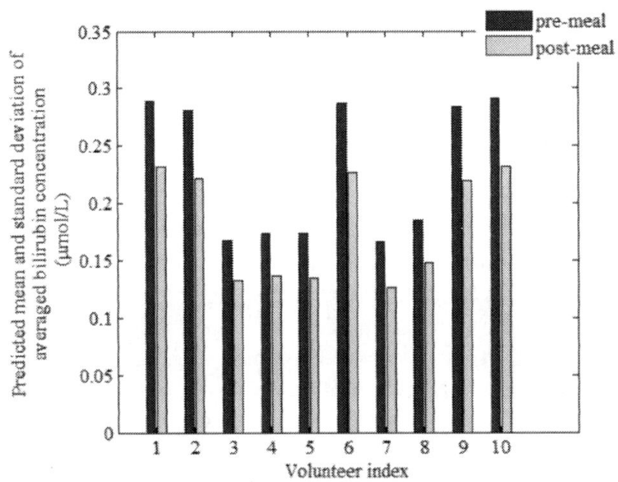

Figure 3. Mean skin blood bilirubin concentration value, $C_{bil(m)}$, predicted for the recruits during fasting and postprandial session

The results shown in Figure 2 for the two genders revealed a mean difference of 0.11 μmol/L in the calculated $C_{bil(m)}$, wherein the value estimated for males is predominantly higher than that of its counterpart gender. This observation is agreeable in the work reported by Zucker *et al.* [4], who observed a mean difference of 0.2 mg/dL in the value measured using *in-vitro* laboratories technique. This discrepancy may be contributed by factors such as differences in physical activities, body mass index and diet between females and the males as discussed in the work of Tanaka *et al.* [11]. The relatively low standard deviation in the values predicted for males and females given by 0.007 μmol/L and 0.0086 μmol/L, respectively, may indicate the robustness of the proposed quantification approach towards differences in skin pigmentation.

In addition, this work found a decrease in the predicted skin bilirubin concentration value following glucose load as shown in Figure 3. This value dropped from mean value of 0.23 ± 0.06 μmol/L measured during fasting state to 0.18 ± 0.04 μmol/L after meal. The notable difference between these values was further confirmed with the calculated $p = 0.000$. This decrease in the value was likely to be contributed by an increase in the liver stiffness after food intake [12]. These changes in the liver function would, therefore, modify one's bilirubin concentration in blood. These preliminary findings showed that the developed optical system that used light beam of wavelength 420 nm, 440 nm and 460 nm is able to detect changes in the predicted C_{bil} in the same pattern as that reported in the literature. However, this system may be further improved via the use of diffused laser beams as the light source to allow illumination of the selected skin site using a narrower light spectrum, and hence a more precise measurement of light attenuation value at each wavelength.

4. Conclusion

This study demonstrated the use of the developed multi-color filter system to provide a non-invasive and non-contact monitoring of one's skin bilirubin level without the use of a look-up table. This work found a variation in the predicted C_{bil} value depending on genders, and a negative association between the estimated value and the blood glucose level. The results revealed a small variation in the value predicted for individuals of the same gender, which may suggest the robustness of the employed strategy towards differences in the skin pigmentation. This work concluded that the reasonable good performance of this system may imply the possibility of its application in clinical settings for the diagnosis of neonates jaundice.

Acknowledgement

The authors would like to thank all volunteers who participated in this study. This work was financially supported in part by the Ministry of Education Malaysia under FRGS grant no. 1581.

References

[1] P D Loprinzi and S E Mahoney 2015 *Angiology* **66**
[2] B N Limketkai, M Choe, S Patel, N D Shah and V Medici 2017 *Current Nutrition Reports* **6**
[3] A O Chinwel, N Johnkennedy, O Hope, O Constance and U Helen 2013 *Journal of Current Research In Science* **1** 104
[4] S D Zucker, P S Horn and K E Serman 2004 *Hepatology* **40**
[5] P Bechi, F Pucciani, F Baldini, F Cosi, R Falciai, R Mazzanti, A Castagnoli, A Passeri and S Boscherini 1998 *Dig Dis Sci* **38** 1297
[6] N Polley, S Saha, S Singh, A Adhikari, S Das, B R Choudhury and S K Pal 2-15 *J. Biomed. Opt.* **20**
[7] A Huong and X Ngu 2014 *Journal of Innovative Optical Health Sciences* **7**
[8] P E Ong, A K C Huong, W M Hafizah, K G Tay and S P Philimon 2016 *IEEE EMBS Conference on Biomedical Engineering and Sciences (IECBES)*
[9] S L Jacques 2013 *Physics in Medicine and Biology* **58**
[10] G Agati and F Fusi 1990 *J. Photochem. Photobiol.* **7**
[11] M Tanaka, S Budhathoki, A Hirata, M Morita, S Kono, M Adachi, H Kawate, K Ohnaka and R Takayanagi 2013 *BMC Endocrine Disorders* **13**
[12] E A Rewisha, A K Helal, A Eljaky, M M Ahmed and T A Salman 2017 *Journal of American Science* **13**

Low Power 130 nm CMOS Johnson Counter with Clock Gating Technique

Nur Syuhadah Amran[1], **Siti Hawa Ruslan**[1,2*]

[1]Faculty of Electrical and Electronic Engineering, Universiti Tun Hussein Onn Malaysia, 86400 Parit Raja, Batu Pahat, Johor, Malaysia.
[2]NanoSIM Research Group, Universiti Tun Hussein Onn Malaysia, 86400 Parit Raja, Batu Pahat, Johor, Malaysia.

*Email: sitihawa@uthm.edu.my

Abstract. In a very large scale integration (VLSI) of integrated circuit (IC) nowadays, digital circuit with low power design is the target of the IC designer. This is to prolong the battery life of the circuit especially if it is meant for wearable devices. In most of the digital circuits, counters are used widely and these counters consumed a lot of power. Therefore in this project the reduction of power consumption of Johnson Counter by using clock gating technique is presented. Johnson Counter is used extensively to generate particular data and shift the data synchronously as per the output sequence of the counter. To ensure the power consumption is reduced, a clock gating technique is incorporated to the Johnson Counter. This counter is implemented in Cadence software using 130 nm Complementary Metal Oxide Semiconductor (CMOS) technology. The design is observed by comparing the design of a 4 bit Johnson Counter using clock gating technique and another 4 bit Johnson Counter without using the clock gating technique. The result shows the power consumption of the Johnson Counter using the clock gating technique is 21.22 μW while the regular Johnson Counter consumed 67.09 μW. Thus the power consumption is reduced by about 68.3% when a clock gating technique is used.

1. Introduction

The circuitry and power consumption of digital system are increased day by day. More functions are integrated into electronic devices. This is achievable as technology is growing fast whereby chip size is getting reduces, speed become increases and device has become as small as possible. Since many tasks can be done in one time, power consumption has become an important part in design parameter. In a digital system, all arithmetic, logic and memory operations will be performed synchronously according to the system clock. Therefore, efficient design of sequential circuit is very important [1].

A Johnson Counter is a counter which provide data sequence and is used in many applications such as digital to analog (D/A) converter [2]. This counter provides a very specific loop synchronously which is very useful for other logic design [3]. Johnson Counter requires only a half number of flip-flops as compared to the conventional ring counter. Since the Johnson Counter is controlled synchronously by a clock, the circuit power consumption is very high. All the flip-flops in the counter will be triggered even though there is no change in the output of a certain flip-flop blocks. Thus to reduce the power consumption of the counter, clock gating technique is used to exclude those avoidable clock transitions at which no logic change is required by the block. By doing this process,

Content from this work may be used under the terms of the Creative Commons Attribution 3.0 licence. Any further distribution of this work must maintain attribution to the author(s) and the title of the work, journal citation and DOI.
Published under licence by IOP Publishing Ltd

the power can be minimized for that sequential circuit [4]. Figure 1 shows the basic process of clock gating technique. It is a logic circuit that controlled the input of the clock to the sequential circuit.

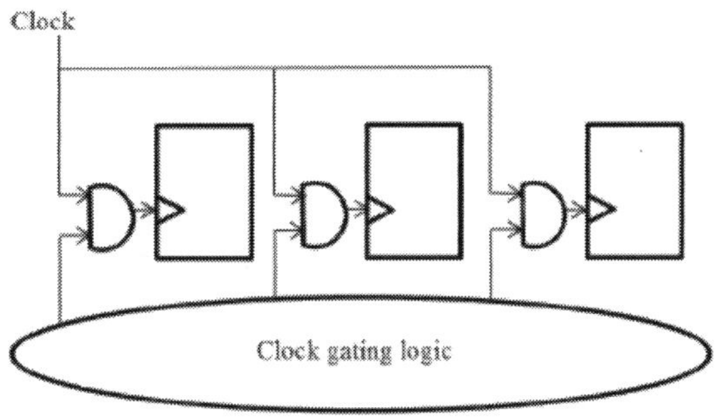

Figure 1. Clock gating technique [2]

2. Clock Gating

Several methods and circuits have been proposed and designed by previous researchers in order to improve performance, achieve higher speeds and lower the power consumption of sequential circuits over the past decades [5-7]. However, there is still a challenge to provide a more low power consumption with the latest design for modern applications and with the advancement of small scale CMOS technology. This project is conducted in 130 nm CMOS technology to improve the power consumed by the Johnson Counter by incorporating a clock gating technique to govern the counter clock management system. With this, a greater device density, higher speed and low power consumption are obtained.

In clock gating technique the clock to the flip-flops in the Johnson Counter are coming from the master clock (clk_f) that is controlled according to equations given in equations (1) to (4). These equations are for 4 bit Johnson counter. Clock clk_{f1} and clk_{f2} are the clock master that had been divided by two and four respectively.

$$Clk_{Q0} = clk_f \,.\, clk_{f1} \,.\, clk_{f2} \qquad (1)$$

$$Clk_{Q1} = clk_f \,.\, \overline{\mathrm{clk}_{f1}} \,.\, clk_{f2} \qquad (2)$$

$$Clk_{Q2} = clk_f \,.\, clk_{f1} \,.\, \overline{\mathrm{clk}_{f2}} \qquad (3)$$

$$Clk_{Q3} = clk_f \,.\, \overline{\mathrm{clk}_{f1}} \,.\, \overline{\mathrm{clk}_{f2}} \qquad (4)$$

3. The Designed Circuit

Two Johnson Counters are designed in this project by using Cadence EDA Tools. The first one is the conventional Johnson Counter and the other is the proposed Johnson Counter with the clock gating circuit. The operating supply of the circuits is 1.2 V which is smaller than other designs referred. Figure 2 and Figure 3 illustrate the conventional and proposed circuit of the Johnson Counter with clock gating technique designed in this project. A 4 bit Johnson Counter using J-K flip-flop is used in both designs. The four J-K flip-flops are used to shift the data synchronously according to the data sequence of Johnson Counter. The clock of conventional Johnson counter is not controlled in producing the output. As for the proposed Johnson counter, the flip flops will only get the clock pulse when they are needed to toggle the value stored in it. Otherwise the flip-flops are not triggered. Thus power consumption can be minimized by turning off the unused J-K flip flop.

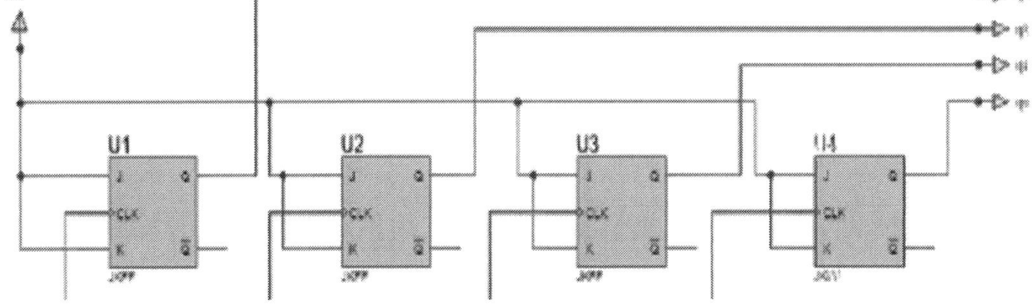

Figure 2. The conventional Johnson Counter

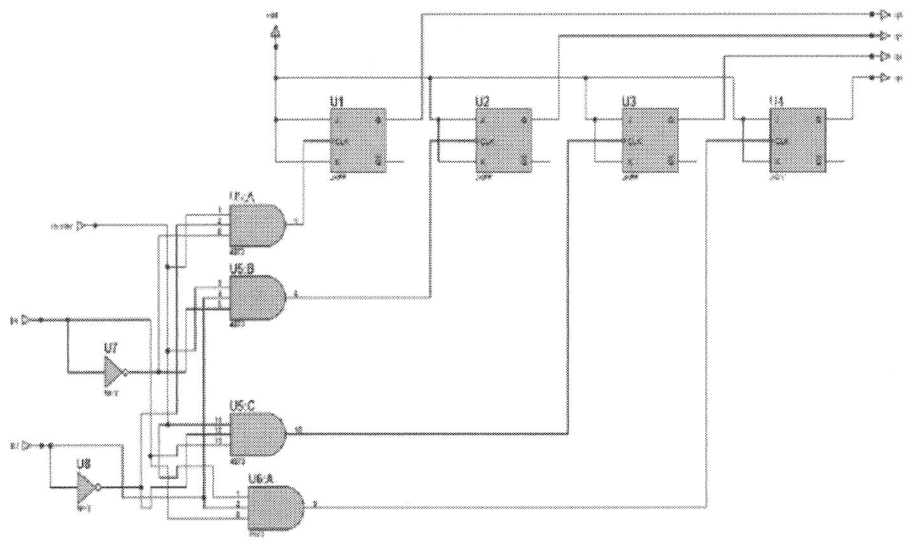

Figure 3. The proposed Johnson Counter with clock gating technique

Figure 4 shows the schematic circuit of the conventional Jonson Counter designed using four J-K flip-flops, while Figure 5 shows the schematic circuit of the Johnson Counter with clock gating technique. The J-K flip-flop used in this design is in toggle mode and is built by using INVERTER and AND gates. The clock gating circuit is constructed also using a combination of INVERTER and AND gates according to the equations mentioned earlier to ensure the clock pulses to each of the flip-flop will be delivered only when it is required.

IOP Conf. Series: Journal of Physics: Conf. Series **1049** (2018) 012073 doi:10.1088/1742-6596/1049/1/012073

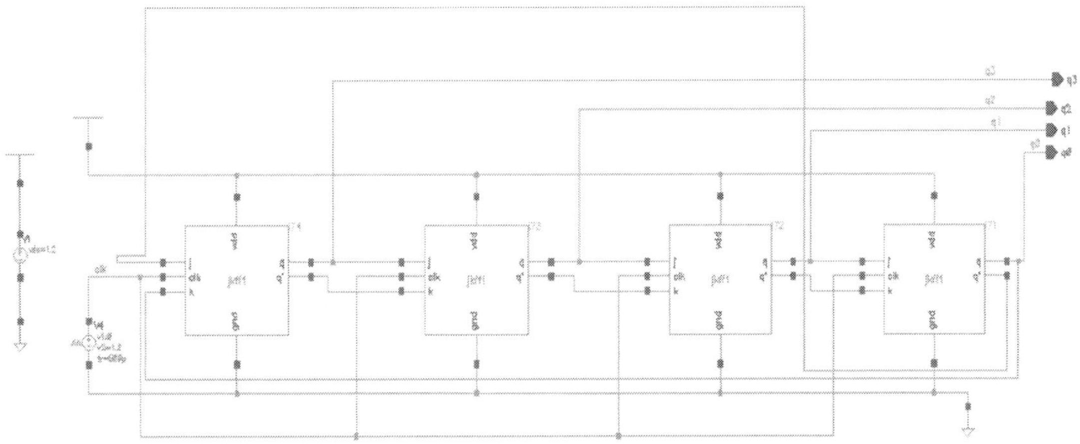

Figure 4. The schematic circuit of conventional Johnson counter

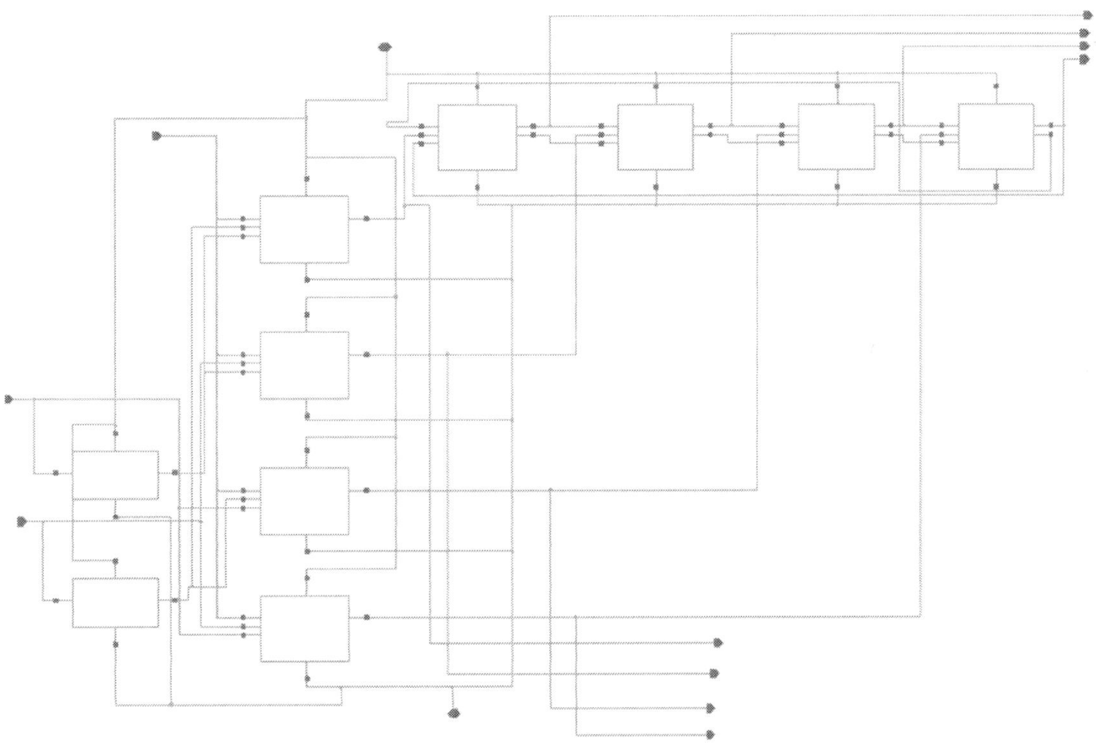

Figure 5. The schematic circuit of Johnson counter with clock gating

4. Result and Discussion

The Johnson Counter with clock gating technique circuit in Figure 6 was designed and simulated using Cadence with 130 nm technology and supply voltage of 1.2 V. The input for clk_{f2}, clk_{f4} and master is connected with different pulse input. Figure 6 shows the simulated circuit with its input pulse and frequency.

IOP Conf. Series: Journal of Physics: Conf. Series **1049** (2018) 012073 doi:10.1088/1742-6596/1049/1/012073

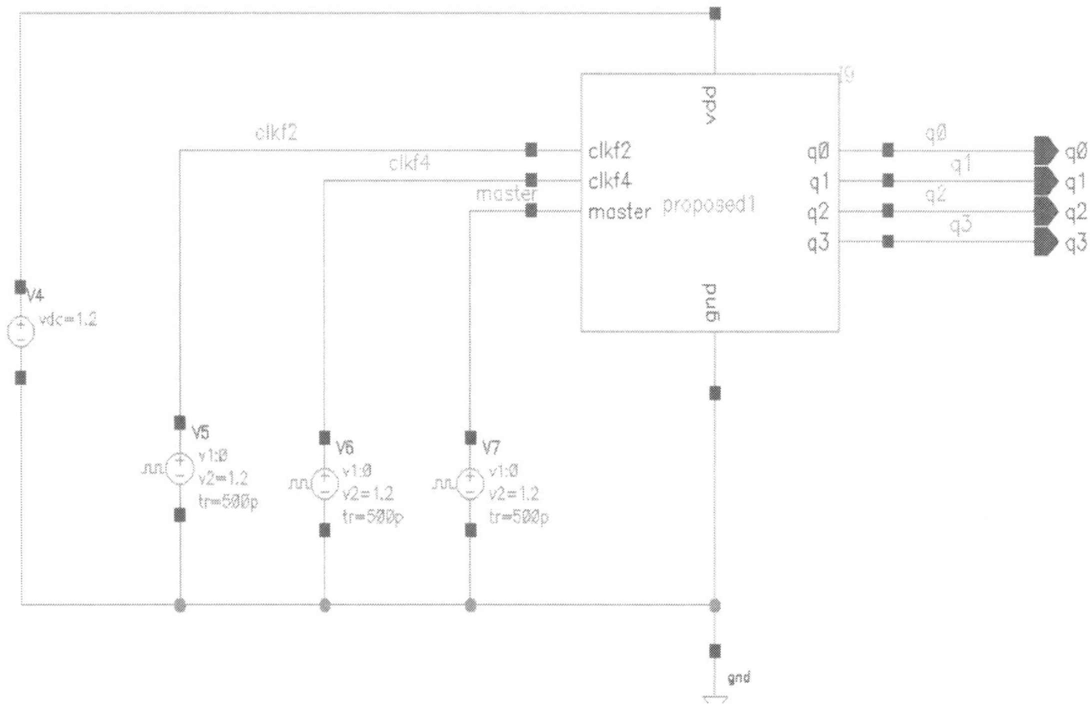

Figure 6. Test bench circuit

4.1 Transient Analysis

Figure 7 shows the simulation result of the conventional counter while Figure 8 shows the simulation result of the counter with clock gating circuit. It is shown that the counter with clock gating technique is getting a pulse when it is needed to toggle the output. Each flip flop is receiving a controlled clock sequence generated by the combinational logic circuit, master clock and the available masking clocks. Hence the power consumption of the system can be controlled efficiently. From the simulation, the power consumption for conventional Johnson Counter circuit is 67.09 μW while the power consumption for Johnson Counter with clock gating technique circuit is 21.22 μW. There is a reduction around 31% or about 45.87 μW when clock gating technique is used.

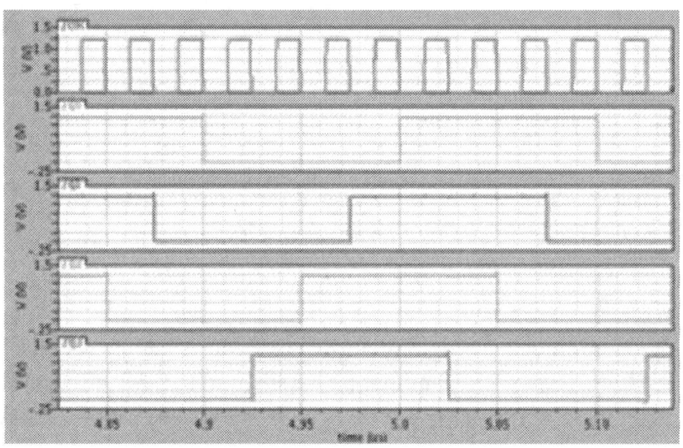
Figure 7. Output waveform of conventional counter

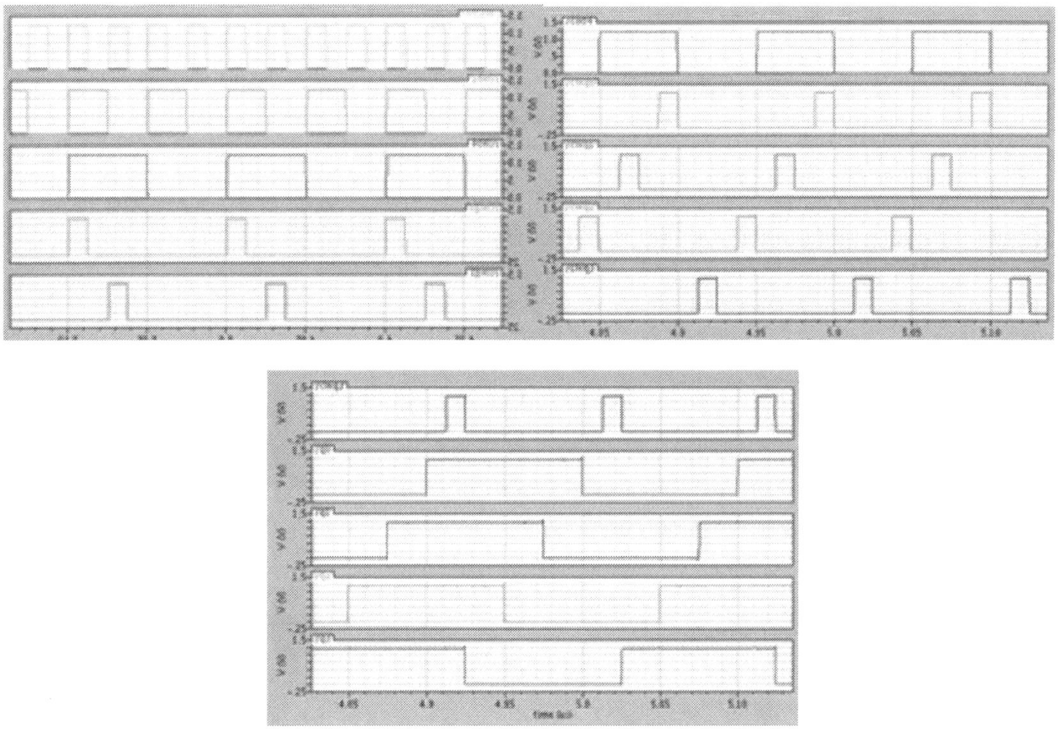

Figure 8. Output waveform of Johnson Counter with clock gating

4.2 Design Rule Check and Layout versus SchematicTest

Design Rule Check (DRC) is a physical design process to determine if the circuit layout drawn, satisfies a number of rules as defined by the semiconductor manufacturer. Figure 9 shows the layout for the proposed design and the result of the DRC for the layout is successful. The total physical area for the proposed analog multiplier is 129 μm².

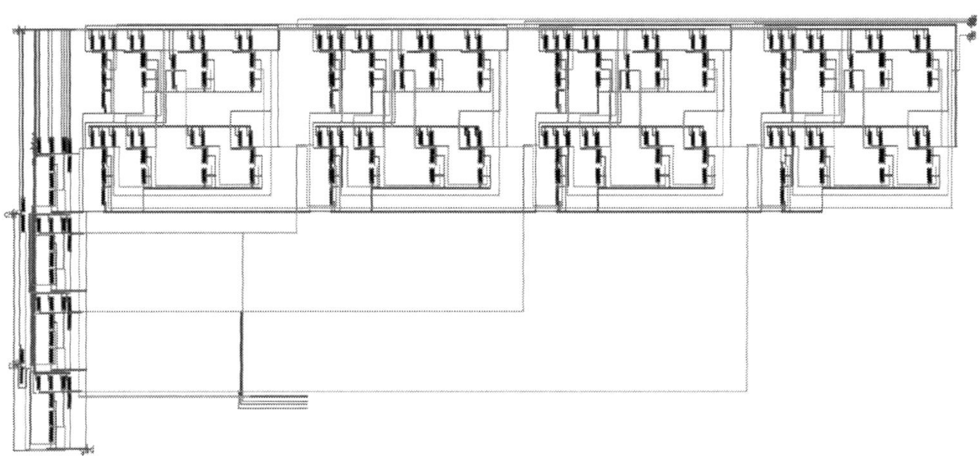

Figure 9. Layout design of Johnson Counter with clock gating technique.

The layout versus schematic (LVS) is done to verify whether the layout really representing the schematic circuit. In this test, a comparisons between the layout and schematic is done to check the circuit connectivity. The result from the LVS test has shown the schematic and layout are equivalent.

5. Conclusion

A 4 bit Johnson Counter with a clock gating technique are designed in this project. The clock gating technique is used as a method to reduce the power consumption of the Johnson Counter. The circuit is designed utilizing 1.2 V power supply using Cadence EDA tools with 130 nm CMOS technology. The simulation result shown that the clock gating systems is effective in reducing the power consumption if Johnson Counter.

Acknowledgement

Special thanks to Nano Simulation Group (NanoSIM) and MiNT-SRC, Universiti Tun Hussien Onn Malaysia (UTHM) for providing the platform and the laboratory to conduct the research. This work was supported by Faculty of Electrical Engineering (FKEE), Universiti Tun Hussein Onn Malaysia (UTHM), Silterra and Collaborative Microelectronic Design Excellence Centre (CEDEC) for Innovate Malaysia Design Competition 2017.

References

[1] Sujatha B V and Venkateswarlu V T 2015 An efficient design for reduction of power dissipation in Johnson counter using clock gating *International Journal of Science and Research* **4** 2723
[2] Ismail S M, Rahman A B M S and Islam F T 2012 Low power design of Johnson counter using clock gating *IEEE 15th International Conference on Computer and Information Technology* 510
[3] Ismail S M, Rahman S, Rahman M M and Ferdous N S 2012 A design scheme of toggle operation based Johnson counter with efficient clock gating. *International Conference on Computational Intelligence, Modelling and Simulation* 393
[4] Bansal E and Bajaj N 2016 A review on power efficient clock gated Johnson counter *Imperial Journal of Interdisciplinary Research* **2** 1284
[5] Pandey S, Uniyal S and Goel N 2015 An improved low power Johnson counter design with clock gating *International Journal of Electronics and Communication Technology* **6** 9
[6] Shelke M and Humane B 2015 Low power VLSI design using clock gating technique. *International Journal of Advanced Research in Computer Engineering and Technology* **4** 3270
[7] Wu Q, Pedram M and Wu X 2000 Clock gating and its application to low power design of sequential circuits *IEEE Transactions on Circuits and Systems 1: Fundamental Theory and Applications* **47** 415
[8] Benini L and Micheli G D 1995 State assignment for low power dissipation *IEEE Journal of Solid State Circuits* **30** 258
[9] Pedram M, Wu Q and Wu X 1998 A new design of double edge triggered flip-flops *Proceedings of ASP-DAC* 417
[10] Wu X, Wei J and Pedram M 2000 Low-power design of sequential circuits using a quasi-synchronous derived clock *Proceedings of ASP-DAC*

IOP Conf. Series: Journal of Physics: Conf. Series **1049** (2018) 012074 doi:10.1088/1742-6596/1049/1/012074

Preliminary study of image processing techniques for the detection of varicocele based on 2D ultrasound images

Mohd Asyraf Abd Rahim[1], Nurmiza Othman[1], Wan Mahani Hafizah Wan Mahmud[1]
Department of Electronic Engineering,
Fakulti Kejuruteraan Elektrik dan Elektronik, UTHM Johor, Malaysia.

E-mail: wanmahani@uthm.edu.my

Abstract. Varicocele is quite common among male population but can only be confirmed and treated through surgery which may affect the fertility. While ultrasound (color Doppler technique) can be used to diagnose the varicocele, the need to perform the Valsalva maneuver technique when performing the color Doppler ultrasound scanning might make patient become uncomfortable. Therefore, this preliminary study has been done to see whether any image processing techniques can be used to enhance the 2D ultrasound image to help the diagnosis. The purpose of this study is to enhance the ultrasound image of 2D varicocele testicle by using image enhancement and edge detection. The images used are the retrospective images of the normal testicle, and testicle with varicocele and severe varicocele (dilated vein). The image enhancement was done using median filter while the edges were detected using Sobel operator. Based on the analysis, sobel technique does an excellent job to show the varicocele region especially the dilated vein. To summarise, further image processing methods could be applied using different image processing techniques as the analysis could be helpful in giving the urologist the second opinion during diagnosis of varicocele before the surgery.

1. Introduction

Varicocele is defined as a scrotal abnormality or swollen because of elongated, dilated and tortuous veins that cause the testicle drain. Usually, varicoceles can occur in 15% among male population, and it might be an evolutionary consequence of men's upright posture. More than 80% cases of varicocele occur on the left side and the remainder on both sides. Generally they tend to happen during puberty for male population [1].

There are no symptoms for varicocele disease but it can cause uncomfortable condition and scrotal pain. It can happen while sitting for a long period, doing vigorous activity or standing in a wrong way and it could be relieved by lying down. It also could be uncomfortable before sleeping after having a hectic day with full of activities. The pain will go away after resting for the night. Usually, when a person got the pain he will feel just like tooth ache, dull, and congestive but the pain do not spread to other parts of the body [2]. However, a person who gets varicocele pain can hardly aware of it until the scrotal wall become thick and contracted. The most danger for this case is it can be associated with male fertility and there are no urination issue and erectile dysfunction. When the dilated vein becomes

[1]wanmahani@uthm.edu.my

Content from this work may be used under the terms of the Creative Commons Attribution 3.0 licence. Any further distribution of this work must maintain attribution to the author(s) and the title of the work, journal citation and DOI.
Published under licence by IOP Publishing Ltd

larger, the varicocele could be clumpy and the person could feel like a "bag of worms" in the scrotum and could give uneasy feeling to the person as well [3].

Currently, the best way to confirm and treat varicocele disease is through surgery. This surgery is a minor surgery and patients can leave the hospital on the same day as their surgery took place. During this procedure, patients will receive anesthesia treatment which is several type of numbing medication. Then, the urologist will make a cut at the lower abdomen part then the abnormal veins will be tied off to prevent the blood flow and as the result, the blood will flow into normal veins area. For reducing the swelling, an ice pack is placed on the area by the patient within 24 hours after surgery [4]. However, it is very important to examine and diagnose this disease before doing the surgery to avoid male infertility.

Ultrasound imaging can be used to produce image of a man's testicles and surrounding tissues. In helping to determine abnormality of the testicles, epididymis and scrotum ultrasound imaging is the main or primary method. Moreover, ultrasound does not use ionizing radiation so it is safe and noninvasive [3]. Color doppler ultrasound (CDUS) is one of the technique that could be used to examine varicose vein [5]. However, the patient needs to perform the Valsalva maneuver technique each time during the diagnosis, thus making the patients feel uncomfortable.

Therefore, this study proposes to use the processing of the 2D ultrasound images, to see whether is it possible to diagnose the varicocele using 2D ultrasound (without the need to perform the Valsalva maneuver as in CDUS) with the help of the pre-processing techniques to those images.

2. Material and Methods

This study uses the 2D ultrasound images of varicocele from the male subjects. The study was performed to the retrospective images thus details on the patients could not be retrieved. Nevertheless, all DICOM images used are still in good condition and the diagnosis of the images has been confirmed and evaluated by specialist. For this study, all images used are roughly in the size of 300x240 before cropping. Figure 1 shows the example of the original images of (a) healthy testicle, (b) varicocele region of the left testicle, (c) testicle that potentially has varicocele and (d) larger dilated vein that shows the severity of the varicocele pain.

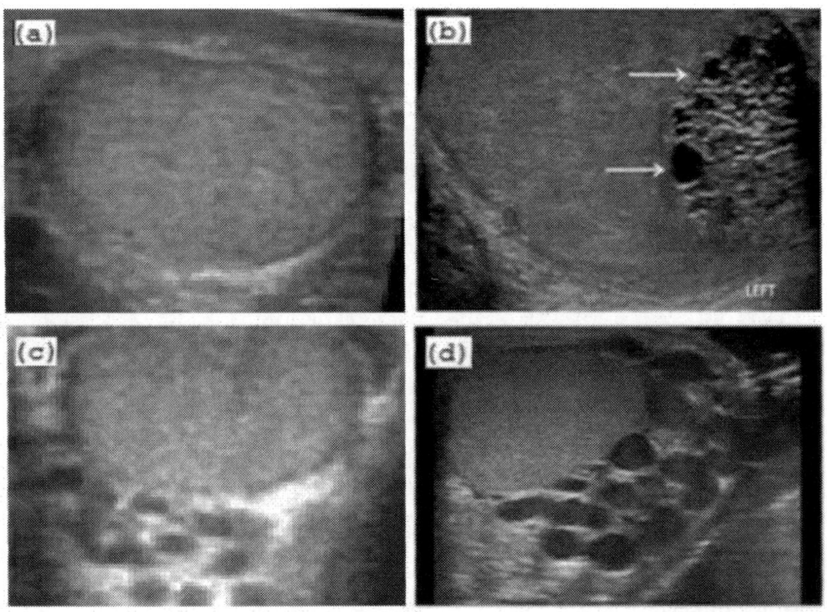

Figure 1. Ultrasound image of; (a) healthy testicle, (b) varicocele region, (c) potential varicocele, (d) larger dilated vein.

IOP Conf. Series: Journal of Physics: Conf. Series **1049** (2018) 012074 doi:10.1088/1742-6596/1049/1/012074

For this study, general pre-processing techniques were applied to the images as in block diagram in Figure 2; consist of data collection, image cropping, image filtering and edge detection. The data is gathered from a local hospital and categorized into 3 groups for analysis purposes. Group 1 consists of normal testicle for comparison with those with varicocele. Group 2 are the data with the varicocele and Group 3 are those with the severe varicocele (dilated vein).

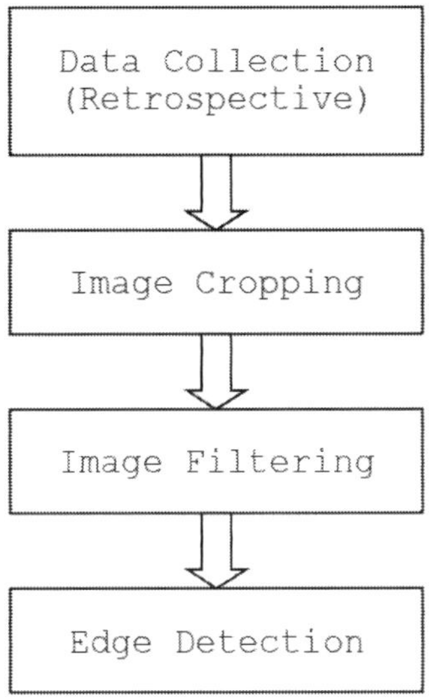

Figure 2. Process applied for this study

After gathering all the retrospective data needed, all images were cropped so that only the testicle region would be examined. For images with vericocele, the part that only have those vericocele were selected. The example of the cropped window is as in Figure 3.

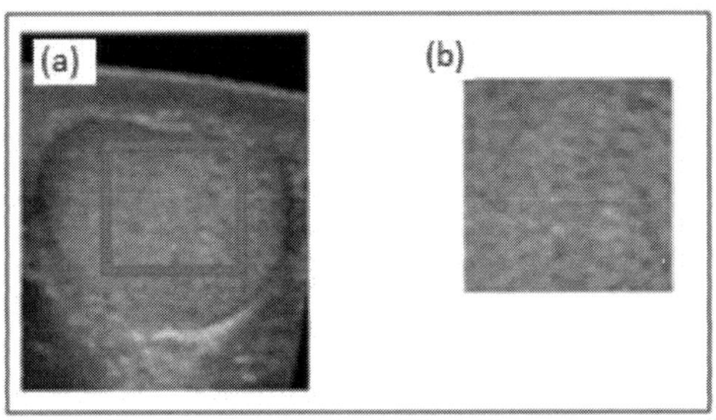

Figure 3. Example of cropped image; (a) original images showing the selected area (red square) and (b) the cropped image

Then, the image filtering was performed using median filter. Median filter is a non-linear filter and is also known as rank filter [6]. Overall, it can reduce the salt and pepper noise and at the same time maintain the sharp edges of the images. The median filter will consider every pixel in the image and based on its nearby pixel, it will decide whether or not it represents the surroundings [9]. The median can be calculated by performing few main steps;

IOP Conf. Series: Journal of Physics: Conf. Series **1049** (2018) 012074 doi:10.1088/1742-6596/1049/1/012074

Step 1: Arrange the neighborhoods pixels into the numerical order.
Step 2: Find the middle pixel value.
Step 3: Replace all the neighborhoods pixels value to the middle pixel value.

Based on previous researches, median filter has shown to give a good result with high value of power signal to noise ratio (PSNR), low value of mean squared error (MSE), while still preserving the important edges when using it with ultrasound images [6-9]. Therefore, in this preliminary study, median filter were applied to the images. The example of the result when filtered using the median filter is as in Figure 4. From the figure, it shows that the images has smoother surface as compared to their original versions.

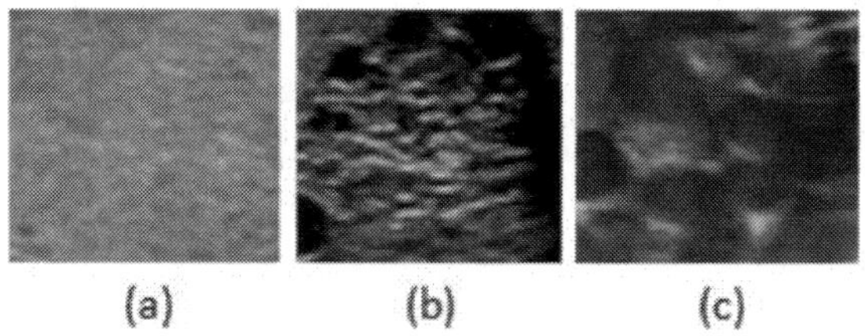

Figure 4. Result after median filtering for; (a) Group 1, (b) Group 2, (c) Group 3.

Lastly, this study also performs edge detection. Edge detection is a process of finding and locating image edges. Edge detection of an image is a very important step towards understanding image features because edges may contain features that could be significant and meaningful [10]. There are various types of edge detection operators [12], but the one that was chosen for this study is the Sobel operator. This operator measures the 2-D spatial gradient of the image and then emphasizes those regions with the high spatial. This operator has two 3x3 convolution kernels which were 90° to each other. This is important as it can optimize the respond to edges running vertically and horizontally relative to the pixel grid, where one kernel represents a certain orientation and another kernel represents its perpendicular orientation [12]. Sobel operator has been chosen to be applied as it offers two main advantages as below [11]:

1. It has smoothing effect to the random noise in the image because of the average factor introduced in the operator.
2. The edge seems thick and bright as the elements of the edge on both sides have been enhanced (differential of two rows or two columns).

3. Result and Analysis

The Sobel technique does its job to show the varicocele region especially the dilated vein. Some observations have been made when performing the Sobel operator to the filtered images. As for healthy testicle in Figure 5, the white spots around the testicles are in sequence shape to indicate no dilated vein. Furthermore, size of the white spot also smaller meaning that the vein in normal condition without any enlargement. There are also no significant size differences between left and right testes for the healthy testicle.

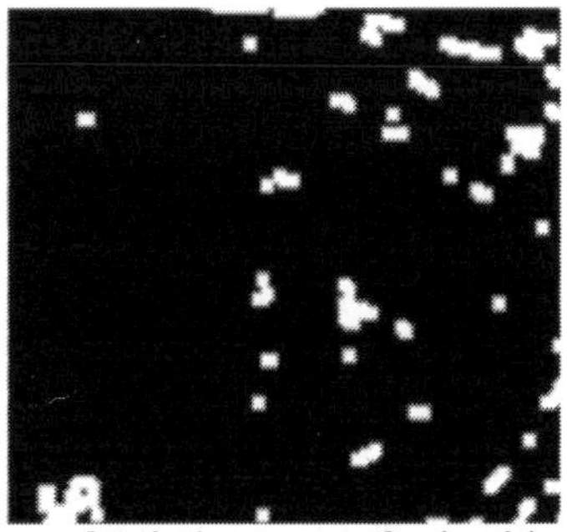

Figure 5. The output of Sobel operator for healthy testicle (Group 1)

For cases with varicocele (Group 2), the white spot coagulate at the testicle. It is shown there is dilated vein and cause the size of the vein increases. The size of the white spot little bit larger than healthy testicle. Then, the present of the dilated vein indicate there is varicocele pain as in Figure 6.

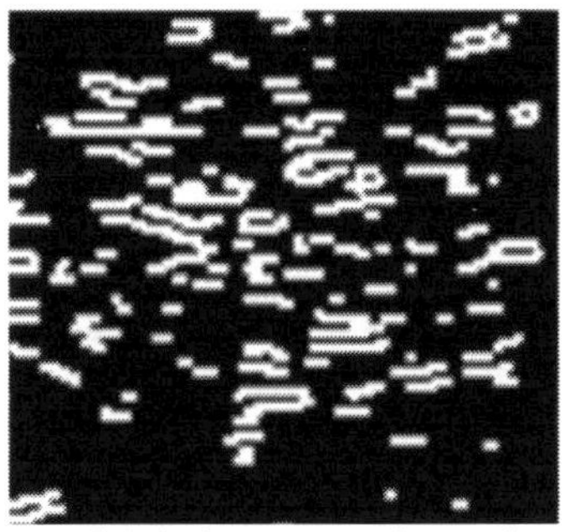

Figure 6. The output of Sobel operator for vericocele testicle (Group 2)

Figure 7 shows the output of Sobel operator for dilated vein (Group 3). The white spots are bigger and coagulated at certain parts. It is shown there is big dilated vein that increase the severity of the varicocele.

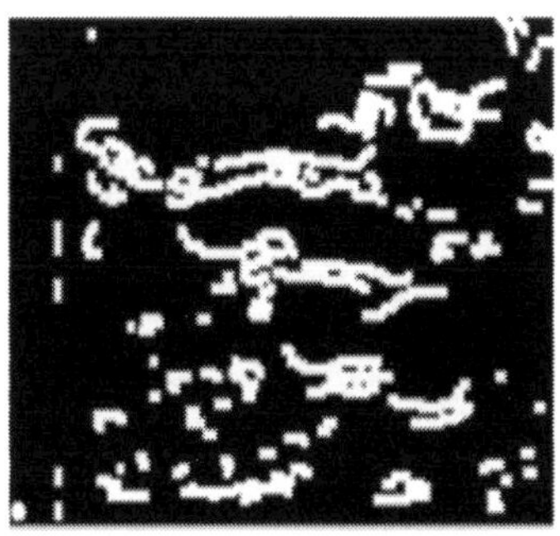

Figure 7. The output of Sobel operator for testicle with dilated vein (Group 3)

Based on the result, it can be seen that pro-processing techniques is helpful in enhancing the important features of the 2D ultrasound images. As in the result, we can see that that Group 1 has only few white spots. While as the varicocele become severe, the white spots will become more complicated and coagulated.

4. Conclusion

From this research, the sobel technique shows clearly the dilated vein in the form of white spots. User also could observe the size and condition of the dilated vein. This result can be helpful for urologist to diagnose the severity of varicocele. However, detail testing on different image filtering techniques, and different edge detection could be performed in order to get a better result.

Acknowledgments

Authors would like to acknowledge assistance from colleagues, special work by the specialist and financial support from RAGS Grant R067.

References

[1] Wolverson M K, Houttuin E, Heiberg E, Sundaram M and Gregory J 1983 *AJR Am J Roentgenol* **141** 775
[2] Rajeev K and Rupin S 2005 *J Obstet Gynecol India* **5** 505
[3] Williams D H, Karpman E and Lipshultz L I 2006 *Can. J. Urol* **13** 13
[4] Yamamoto M, Katsuno S, Yokoi K, Hibi H and Miyake K 1995 *Nag. J. Med. Sci.* **58** 47
[5] Pauroso S, Di Leo N, Fulle I, Di Segni M, Alessi S and Maggini E 2011 *J. Ultrasound* **14** 199
[6] Mohan K and Aroquiaraj I L 2014 *Int. J. Comp. Intel. & Inf.* **4** 207
[7] Ambardar S and Singhal M 2014 *Int. J. Emerg. Tech. & Adv. Eng.* **4** 824
[8] Razman N R, Mahmud W M H W and Shaharuddin N A 2015 *Proc. IEEE Stud. Conf. R&D* 402
[9] Hafizah W M and Supriyanto E 2011 *Int. J. Comp. Apps.* **21** 15
[10] Vincent O R and Folorunso O 2009 *Proc. Inform Sci. & IT Edu. Conf.* 97
[11] Gupta S and Mazumdar S G 2013 *Int. J. Comp. Sci. & Mgmt. Res.* **2** 1578
[12] Krishna R, Pragya G, Kapil D T and Vipul G. 2017 *Int. J. Res. Appl. Sci. & Eng. Tech.* **5** 534

IOP Conf. Series: Journal of Physics: Conf. Series **1049** (2018) 012075 doi:10.1088/1742-6596/1049/1/012075

A 1.2V, 7.5ppm/°C Bandgap Reference with Start-up and Power Down Circuit

Siti Idzura Yusuf[1*], Suhaidi Shafie[1], Hasmayadi Abdul Majid[2]

[1] Faculty of Engineering,
Universiti Putra Malaysia,
43400 UPM Serdang, Selangor Darul Ehsan, Malaysia.
[2] MIMOS Berhad Technology Park Malaysia,
57000 Bukit Jalil, Kuala Lumpur, Malaysia.

*Email: idzura@uthm.edu.my

Abstract. A reference voltage is expected to be stable against process variation, voltage supply and temperature. The architecture of bandgap reference circuit in this paper consists of start-up and power down circuit to minimize power dissipation in system on chip application. The circuit is simulated using Cadence tools in 0.18μm Silterra CMOS technology and operated with 1.8V±8% supply voltage. Based on the post layout simulation, the bandgap reference circuit generated 1.202V stable voltage with temperature coefficient of 7.5ppm/°C over temperature range −40°C to 100 °C and -78.9dB power supply rejection ratio at DC frequency. The simulation results show that the circuit is well function to supply a stable input voltage to another circuit application such as Analog Digital Converter.

1. Introduction

Reference circuit is an important block to provide a constant DC voltage independent and insensitive from process changes, supply voltage and temperature (PVT). It is important that the reference circuit must be precising functioning for example as an input for Analog Digital Converter (ADC) circuit. The resolution of the bit converter is affected by the accuracy of the stability voltage input and temperature range [1].

The reducing transistor size has also stimulated the trend of system integration into a full-chip Silicon-on-Chip (SoC). As the technology growth fast, system integration SoC encourages into many applications. To maintain the operation of the system, the battery frequency replacement will be taken into consideration. The power consumption requirement is a critical issue that every circuit design must be prepared as a main factor for circuit design specification [2]. To reduce the power use in a system, the circuit should have low current voltage and also Power Down (PD) system. This reference circuit will provide start up and power down circuit to minimize battery operation and also fulfills the specification of current consumption, with minimum temperature coefficient, wide temperature range and acceptable Power Supply rejection ratio (PSRR) across corners simulation.

2. Bandgap Reference Architecture

The topology of Bandgap Reference (BGR) has been introduced since 1971 and was followed by Brokaw in 1974. This Bandgap reference circuit designed based on Brokaw topology by using

Content from this work may be used under the terms of the Creative Commons Attribution 3.0 licence. Any further distribution of this work must maintain attribution to the author(s) and the title of the work, journal citation and DOI.
Published under licence by IOP Publishing Ltd

characteristic of base-emitter voltage of two Bipolar Junction Transistor (BJT) biased at unequal current densities can provide a temperature independent voltage [3].

2.1 Bandgap Reference Core

A stable BGR voltage with temperature is obtained by Voltage Complementary to Absolute Temperature (VCTAT) derived from the BJT base-emitter voltage (VBE) compensated by the Voltage Proportional to Absolute Temperature (VPTAT) by the difference of VBE voltages of *n* number BJT operated under different current densities. Figure 1 shows a basic principle of BGR and can be represented as equation 1. VCTAT is the VBE biased with constant current that has a negative temperature coefficient (TC) of -2 mV/ C and VPTAT is ΔVBE proportional to the thermal voltage, (VT) that has a positive temperature coefficient of 0.086 mV/ C . By appropriately scaling the value of resistance ratio and the numbers of BJT between two difference voltages the stable 1.2V reference voltage can be produced [4].

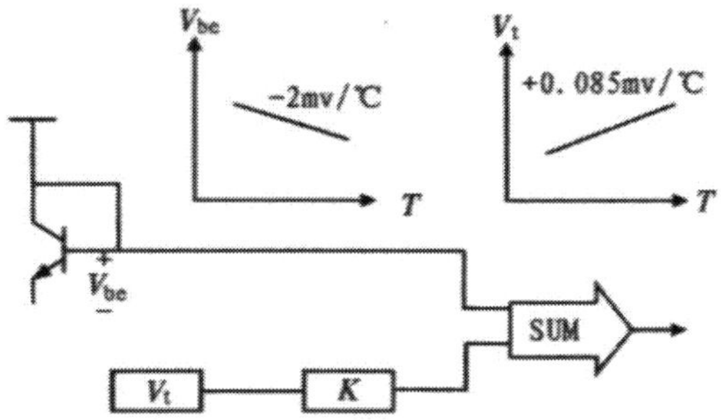

Figure 1: Basic principle of bandgap reference [5]

$$VBG = VBE + V_T \ln n \left(1 + \frac{R2}{R1}\right) \tag{1}$$

The architecture of bandgap reference is shown in Figure 2. To produce proportionate temperature of VCTAT, five BJT transistors of Q2 are chosen and 9.375 value of resistance ratio, (R2/R1) has been determined to increase the current densities of VPTAT. The two-stages op-amp differential amplifier was used in this design. The loop gain of the op-amp is 70dB resulting the output that sufficient to drive the R1 and R2 resistors. The first stage differential amplifier contains of transistors M8-M14 and transistors M15 is a common-source amplifier transistor. By placing C1 capacitor as a Nested-Miller compensation technique, it can improve BGR circuit phase margin for better stability and faster settling time [6].

IOP Conf. Series: Journal of Physics: Conf. Series **1049** (2018) 012075 doi:10.1088/1742-6596/1049/1/012075

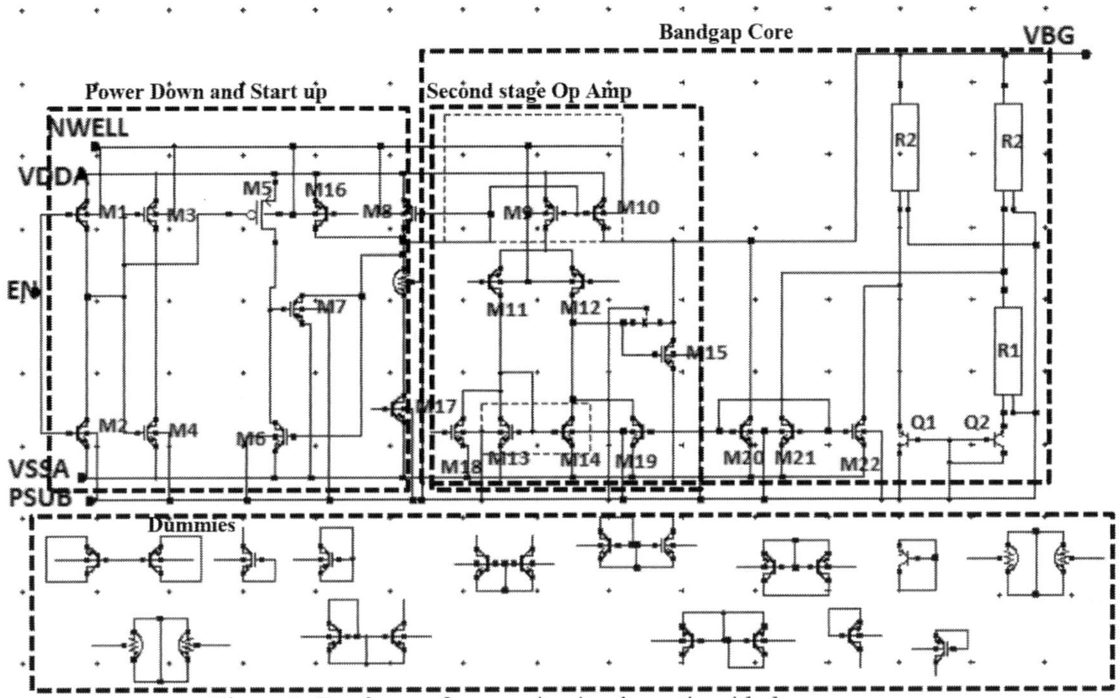

Figure 2. Bandgap reference circuit schematic with dummy

2.2 Start-up Circuit and Power Down

To solve the power dissipation of the BGR, this design will use start-up and power down circuit. The start-up circuit will avoid unnecessary operation current of the BGR after stable voltage was reached. Meanwhile, the power down will disable the operation when the BGR circuit is not in use. The enable (EN) input as shown in Figure 2 would power up the circuit when the input is high. The start-up circuit consists of series of diode-connected transistor, M5 and gradually switch on when EN =1 and increase the Vg voltage of M7. With that increasing voltage, transistor M7 will turn ON and the current flows through M7 giving signal to start-up the bandgap core circuit to the stable voltage. Besides that, the current also flows through transistor M6 that pulls down the Vg voltage of M7 and turn it OFF. Hence, the start-up circuit will turn off and the BGR circuit operates.

During the power down mode, EN = 0, the transistors M1-M4 work as switches and transistor M16-M20 will pull down every node of bandgap core transistor to be isolated from the power supply. This technique will reduce the power dissipation during power down mode.

3. BGR Layout design

The layout of BGR has completed with zero Design Rule Check (DRC) and without Layout Versus Schematic (LVS) errors as shown in Figure 3. The post layout simulation results including the parasitic extraction using Calibre view are verified. The layout also consists of common centroid technique for matching purpose and including dummies. The active silicon area of the BGR circuit is 0.027mm^2 smaller than previous designed [12].

IOP Conf. Series: Journal of Physics: Conf. Series **1049** (2018) 012075 doi:10.1088/1742-6596/1049/1/012075

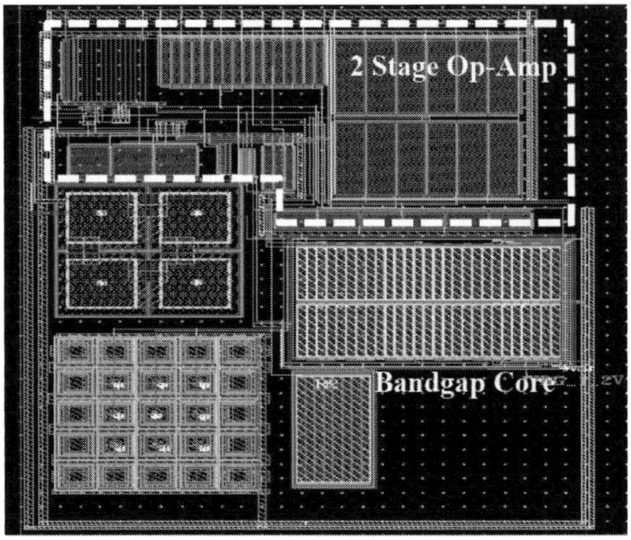

Figure 3: Bandgap reference layout design

4. Result and Discussion

The BGR circuit was designed by using 0.18μm Silterra CMOS process with the supply voltage of (VDDA) 1.8V with ±8% rate of change and separated bulk source to prevent the short channel effect result. The post-layout result was simulated through five process corners (TT,SS, SF,FS and FF) with temperature variation ranging from -40 to 100°C. There are nine matrix conditions consist of three conditions of supply voltage (1.65V, 1.8V and 1.95V) and three temperature conditions (-40°C, 27°C, 100°C).

4.1 Temperature Coefficient Analysis

The DC analysis gives bandgap reference voltage, VBG as a function of temperature. Figure 4 shows the VBG variation of typical corners, while Figure 5 shows VBG variation for five corners analysis. The variation of VBG value in typical process is 1.2658mV, maximum variation is 1.3008mV (FF, 27°C, 1.65V) and minimum is 1.2373mV (SS, -40°C, 1.95V). By using equation in (2), the temperature coefficient of typical corners is 7.52ppm/°C and for the most extreme corners is 7.73ppm/°C. This result shows low temperature coefficient result compared with previous design [9][10][11][12].

$$ppm/°C \frac{VBG,max - VBG,min}{VBG,typ\ (Tmax - Tmin)} . 10^6 \qquad (2)$$

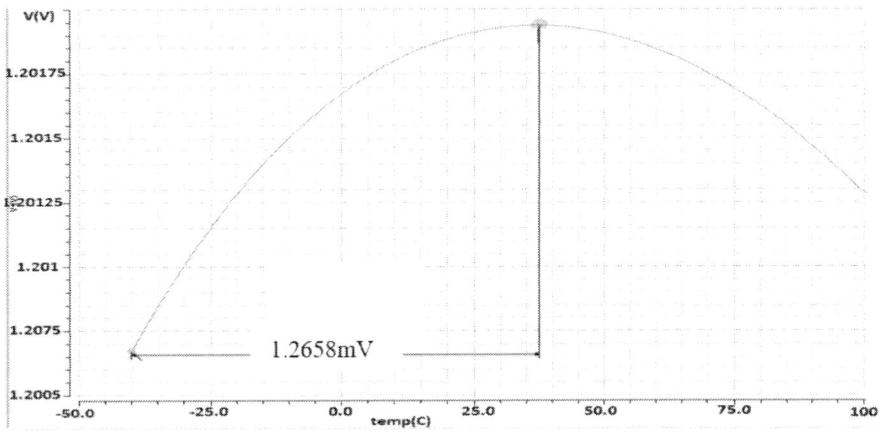

Figure 4: Typical corners of temperature dependence

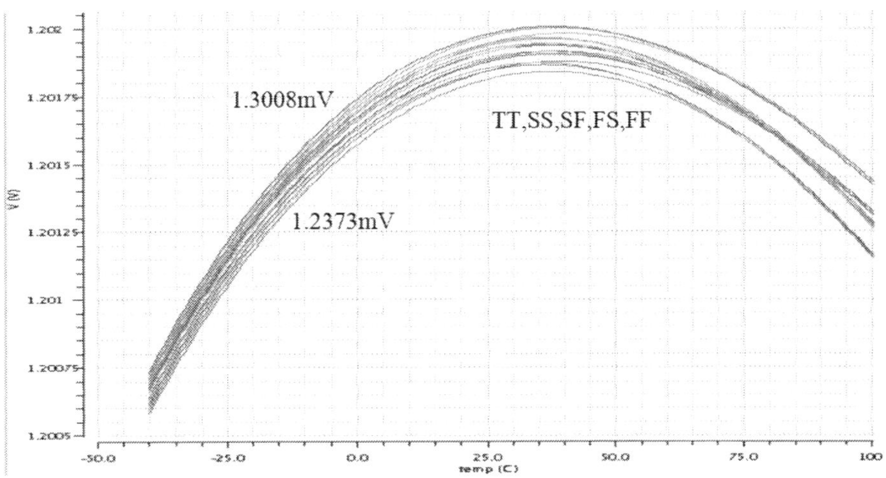

Figure 5: Voltage variation over five process corners

4.2 Bandgap Reference Transient Analyses

Transient simulation analyses in Figure 6 shows the settling time of BGR output. It is found that the circuit acquired 35μs to produce stable output when the supply voltage ramping up from 0V to 500ns. The enable signal from Figure 7 shows the VBG output when EN signal is high and in power down mode. It can be observed during active mode the current is 53.54μA, and 252.5pA in power down mode for typical condition. Therefore, the power consumption of this circuit is 96.4μW and consume minimum power dissipation during power down mode compare with design in [7]. The value of VBG for typical process is 1.20192V while the maximum value is 1.20199V and minimum value is 1.20058V, thus the constant voltage (CV) for the VBG is 0.135%. The close up graph in Figure 8 shows the stable output of BGR across corners can be observed.

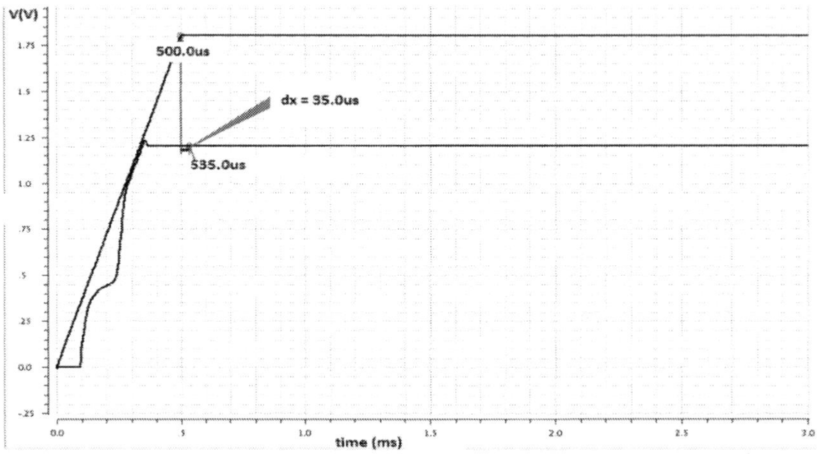

Figure 6: VBG settling time

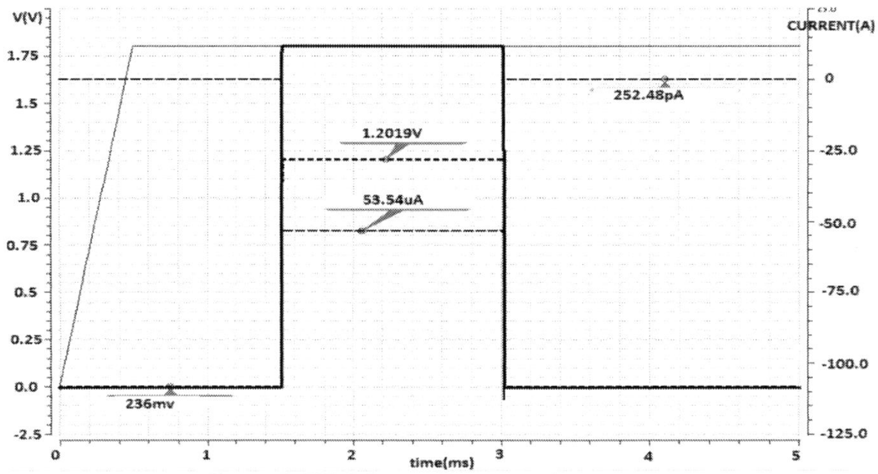

Figure 7: VBG Power Down analyses

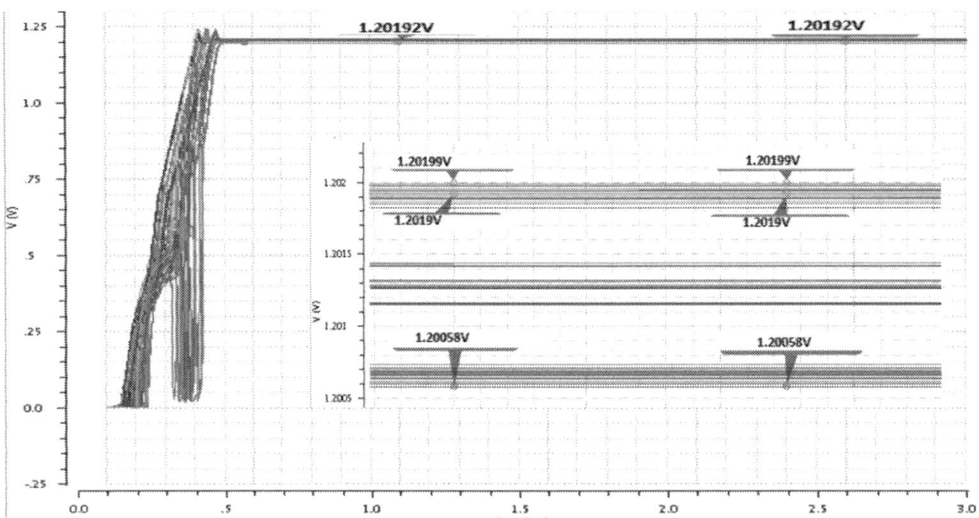

Figure 8: VBG output over five process corners

4.3 Bandgap Reference Power Supply Rejection Ratio (PSRR)

The PSRR value of the circuit is -78.86dB at typical corners at 27C DC as in the PSSR Vs frequency plot in Figure 9. The maximum value of PSSR is -119.5dB (SF, 27°C, 1.95V) and the minimum value is -71dB (SF, 40°C, 1.65V). The result shows the PSSR values are acceptable for every corners process. The performance metric comparison between propose bandgap reference circuit and a set of other state of the art are shown in Table 1.

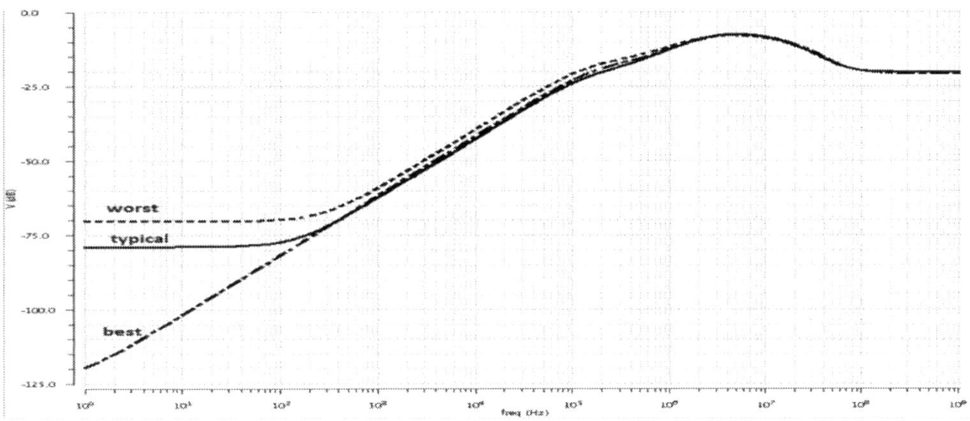

Figure 9: PSRR value of VBG

Table 1. Performance comparison with previous Bandgap reference circuits

Parameter	[7]	[8]	[9]	[10]	[11]	[12]	This work
CMOS process (μm)	0.15	0.18	0.18	0.18	0.18	0.18	0.18
Power supply(V)	3.3	1	1.25-2	1.1-2.2	3.3	1.7 -1.9	1.8±8%
VBG (V)	1.25	0.7136	0.536	0.8	1.25	1.176	1.202
Temperature Range(°C)	-55-125	-40-120	-40-85	-40-125	-40 - 125	-40 -140	-40 -100
Total Current(μA)	704.84	49.3	0.24	13.2	84	50	53
TC (ppm/°C)	0.758	3.23	19.302	9	11.6	13	7.5
Area(mm^2)	NA	0.03	0.0077	0.04	0.014	0.05	0.027
PSRR(dB) @DC	-82	-57@1kHz	-55@10Hz	-108	-115	NA	-78.9

5. Conclusion

A 0.18μm CMOSS bandgap reference with start-up and power down circuit was designed and the post layout simulation was verified. The circuit was simulated for five process corners (TT,SS,FS,SF,FF) by using 1.8V ±8% supply voltage over temperature of -40°C to 100°C. A 1.202V stable voltage reference has been produced with TC 7.5ppm/°C and PSRR at DC frequency -78.9dB in typical corners. It can be concluded that the BGR circuit is compatible for system on chip application because it can be operated over wide range of temperature, low temperature coefficient and also high PSRR ratio in all corners. Besides that, with the power down circuit designed, it is well suit for low power dissipation and also occupied layout area of 0.027mm^2.

Acknowledgement

This work is supported by UPM via PUTRA IPI 9443300 and IPS 9513700 and collaboration with MIMOS Berhad.

References

[1] Oljaca M and Baker B 2009 *Analog Appl J.* **3Q** 13
[2] Asyaei M 2018 *Integr VLSI J.* **60** 263
[3] Brokaw AP 1974 *IEEE J. Solid-State Circuits* **SC-9(6)** 388
[4] Banba H, Shiga H, Umezawa A and Miyaba T. 1999 *IEEE J. Solid-State Circuits* **34(5)** 670
[5] Khot P and Shettar RB 2015 *201519th Int Symp VLSI Design and Test* (Amedabad,India:IEEE) p1
[6] Palmisano G and Palumbo G 1995 *IEEE Trans. Circuits Syst. I, Fundam. Theory Appl.* **42(3)** 178
[7] Calvillo JP, Guilherme J and Horta N 2017 *20171st New Gen. of Circuit and System* (Genova:IEEE) p197
[8] Jun L, Yiqiang Z and Junfeng G 2012 *Chin J. Semicond.* **33** 025014
[9] Singh S and Halonen K. 2015 *Microelectronics J.* **46(1)** 36
[10] Wang L, Zhan C, Zhao S, Chai G, Liu Y, Huang Q and Li G 2016 *IEEE Int. Conf. Electron Devices Solid-State Circuits* (The University of Hong Kong: IEEE) p410
[11] Zhu Y, Liu F, Yang Y, Huang G, Yin T and Yang H. 2014 *Proc IEEE 2014 Cust Integr. Circuits Conferance* (San Jose,CA: IEEE) p3
[12] Hanhart M, Wunderlich R and Heinen S 2017 *2017 13th Conf PhD Res in Microelectronics and Electronics* (Giardini Naxos–Taormina, Italy:IEEE) p269

Groundwater Exploration in Granitic Rock Formation Using Electrical Resistivity and Induced Polarization Techniques

Saiful Azhar Ahmad Tajudin[1,2], Mohd Jazlan Mad Said[1,2], Aziman Madun[1,2], Mohd Hazreek Zainalabidin[1], and Mohd Zainizan Sahdan[2]

[1]Faculty of Civil and Environmental Engineering,
Universiti Tun Hussein Onn Malaysia, Batu Pahat, Johor, Malaysia.
[2]Preston GeoCEM Sdn. Bhd, Batu Pahat, Johor, Malaysia.

Email: saifulazhar870@gmail.com

Abstract: Electrical resistivity and induced polarization surveys have been conducted for groundwater exploration at granitic formation. This study discusses the earth materials resistivity and chargeability of granitic for groundwater exploration at IIUM campus. In this study Terrameter model LS2, cables, electrodes, cable connectors, battery and remote cable were used as the tools for measurement. The spacing between electrodes is 5 m and maximum length of spread line is 400 m by using Pole-Dipole protocol. By comparing the resistivity values and the chargeability values, it would be able to provide better interpretation for ground water exploration for granitic rock. The result also shows the important of chargeability to verify the resistivity value for locating the groundwater position.

Keywords: Groundwater, Geophysics, Electrical resistivity, Induced Polarization

1.0 Introduction

There are two (2) source of waters can be obtained in the earth, such as surface water and groundwater. In many urban areas, the most common source of water is surface water which provided by the water supply department of government or private sectors. There are certain situations happen lately when the surface water is not sufficient to supply for the locals especially during drought season and worst case scenario during floods due to water contamination. Therefore, source of groundwater is essential as a secondary source or alternative for local usage. Groundwater source exists beneath the subsurface in soil pore spaces and the fractures in rock formations [1]. It can be obtained for the use of the people for domestic or agricultural use [2]. The underground water contains and transmits in aquifer, which are characterized either unconfined aquifer or confined aquifer. The occurrence of this aquifer can be determined by electrical technique survey, one of the geophysical applications for underground survey [3,4,5]. The electrical survey is commonly adopted by using resistivity and induced polarization methods. By combining these two methods, subsurface can be modelled and the groundwater bearing can be detected. Electrical resistivity and induced polarization methods provide resistivity and chargeability values of the subsurface for entire survey line [6,7]. The principal of electrical resistivity technique is in term of how current is opposed to flow between two electrodes. Meanwhile induced polarization is measuring the time of the earth material can store the charges. This paper discusses the induced polarization characteristics between two distinct type rock formations and compare it with resistivity values for groundwater exploration at Universiti Islam Antarabangsa Malaysia, Gombak, Selangor, Malaysia.

Content from this work may be used under the terms of the Creative Commons Attribution 3.0 licence. Any further distribution of this work must maintain attribution to the author(s) and the title of the work, journal citation and DOI.
Published under licence by IOP Publishing Ltd

2.0 Literature Review

2.1 Electrical resistivity method

The electrical resistivity survey is the oldest application of geophysical survey techniques used for determining water existence underground [8]. This main purpose of this technique is to determine the subsurface resistivity distribution by making measurements on the ground surface [9]. According to Hazreek et. al [10], this technique is sensitive to the variations of earth materials with the existence of water. By injecting electrical current to the ground through electrode [11], it provides estimated resistivity value of ground which is affected by ground parameters such as the minerals, fluid content, porosity and degree of water saturation in the rock [7,11]. Asry et.al [12] stated that the resistivity can be used to investigate boundary between crystalline and sedimentary rocks, compact quartzite rocks with schist or phyllite where differentiation between rocks can be intrepreted in horizontal direction meanwhile the maximum resistivity value can change in vertical direction. The electrical resistivity method has followed the fundamental physical law of Ohm's Law and determined the value of resistivity in Ohm meter (Ωm) [13].

2.2 Induced Polarization Method

The induced polarization (IP) method is recently development of multiple electrode data acquisition and has been used to identify changeability of subsurface materials. This method used the same survey configuration as resistivity method. Loke *et. al* [10] stated that the resistivity measurement are conjunction with induced polarization measurement for complex mineral exploration. By using recent geophysical electrical tools of terrameter, both resistivity and induced polarization method can be performed simulteneously. The induced polarization method uses parameters of time and frequency domain to show the induced polarization effect [6,11]. Induced polarization effects are caused by the two main effects such as the membrane polarization and electrode polarization effects. Membrane polarization is mainly caused by the existence of clay mineral in the sediment or rock. Meanwhile, electrode polarization is caused by conductive minerals in rocks where electrical current flow through partly electrolytic (groundwater) and partly electronically (conductive mineral). Induced polarization have two types of measurement taken in time-domain or frequency domain where most often measurement is taken using latest terrameter LS2 are in time-domain. The induced polarization effects were measured by the residual decay voltage after the current was switch off, this is called induced polarization measurement taken in time-domain and reading in milivolt per volt (mV/V) or in milliseconds (ms).

2.3 Geological Setting

Geology formation in Universiti Antarabangsa Islam Malaysia is classified as an intrusive rock of granite as shown in Figure 1. It is near boundary between Chert-Argillite formation as known as Gombak Chert and Selut Schist formation which consist of quartz-mica schist and some metaquartzite. The Gombak Chert is an olistostromal block of interbedded chert, mudstone and siltstone within group near Genting Sempah. The study location is also near to the Kuala Lumpur fault line where may have a broad fracture rock zone that may contain groundwater.

IOP Conf. Series: Journal of Physics: Conf. Series **1049** (2018) 012076 doi:10.1088/1742-6596/1049/1/012076

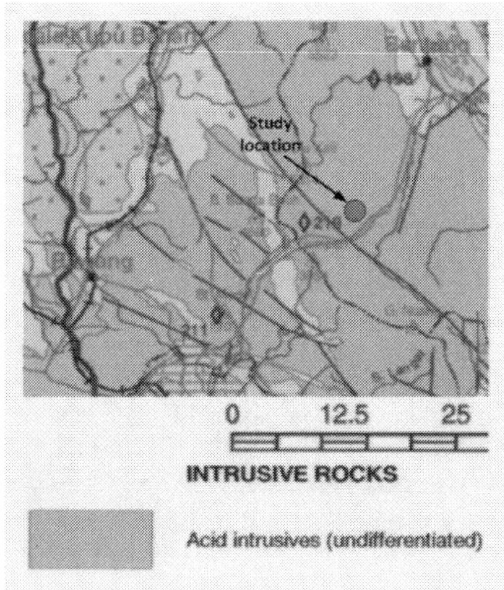

Figure 1. Geology map of Universiti Islam Antarabangsa Malaysia, Gombak [14].

3.0 Methodology

3.1 Equipment Setup

For this study, the latest Terrameter LS2 manufactured by ABEM was used for measurement including several equipments such as 4 units of multi-purpose cable, 64 units of jumper cable, 61 units of stainless steel electrode, 2 units of cable connector, 1 unit of 12 volt battery and 1 unit of remote cable. Figure 2 shows the equipment used for electrical resistivity and induced polarization survey whereas, Figure 3 shows the equipment arrangement for electrical methods survey. The spacing between electrodes is 5 m and maximum length of profile line is 400 m. Both study locations used Pole-Dipole protocol with remote cable setup about perpendicular from spread line with a distance of 300 m. During data acquisition, the LS2 terrameter is configured to take resistivity and induced polarization measuremens at the same time.

Figure 2. The ABEM Terrameter LS2 resistivity meter and supported equipment.

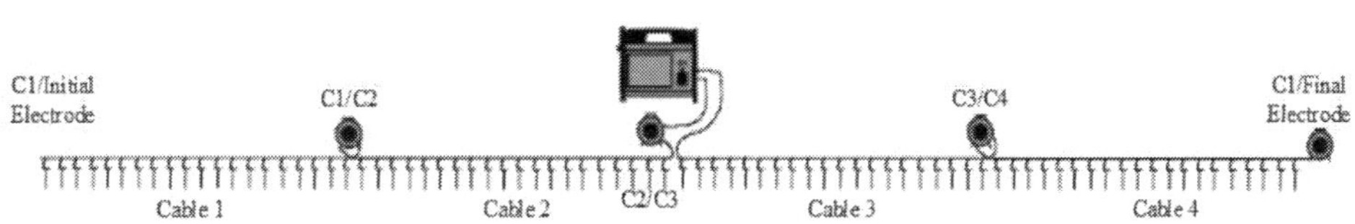

Figure 3. The arrangement of ABEM Terrameter LS2 resistivity meter and other equipments.

IOP Conf. Series: Journal of Physics: Conf. Series **1049** (2018) 012076 doi:10.1088/1742-6596/1049/1/012076

3.2 Interpretation Technique

RES2DINV program was used to convert raw data in the extension of the DAT format. The earth materials resistance measurements are reduced to apparent resistivity values by inversion process. RES2DINV used a least-squares inversion scheme to determine the appropriate resistivity value so that the calculated apparent resistivity values are in agreement with the measured values [13,15]. It will produce a smooth boundary which is recommended to differentiate between soil material and rocks. The inversion process is carried out to obtain three types of resistivity section which consist of calculated apparent resistivity, measured apparent resistivity and inverse model resistivity. The misfit between measured and calculated apparent resistivity produce root mean square (RMS) values. The resistivity contour value is adjusted based on geological information that fit the resistivity range with different colors [16].

4.0 Result and Discussion

For groundwater exploration, the resistivity and induced polarization parameters are suggested to utilize thus would be able to give realistic subsurface model for groundwater interpretation. In resistivity measurement, it is recommended to differentiate fresh groundwater based on resistivity value from 10 to 100 Ohm.m. Meanwhile, in induced polarization measurement, chargeability for water is 0 ms. There are three (3) profiles lines have been conducted electrical survey around study location. Table 1 shows the resistivity and chargeability values for Universiti Islam Antarabangsa Malaysia (UIA) for ease the interpretation. Figure 4 shows an overview of three (3) profile lines of results on google map at Universiti Islam Antarabangsa Malaysia. Figure 5, Figure 6 and Figure. 7 show the 2-dimensional tomography of resistivity and induced polarization at each profile lines.

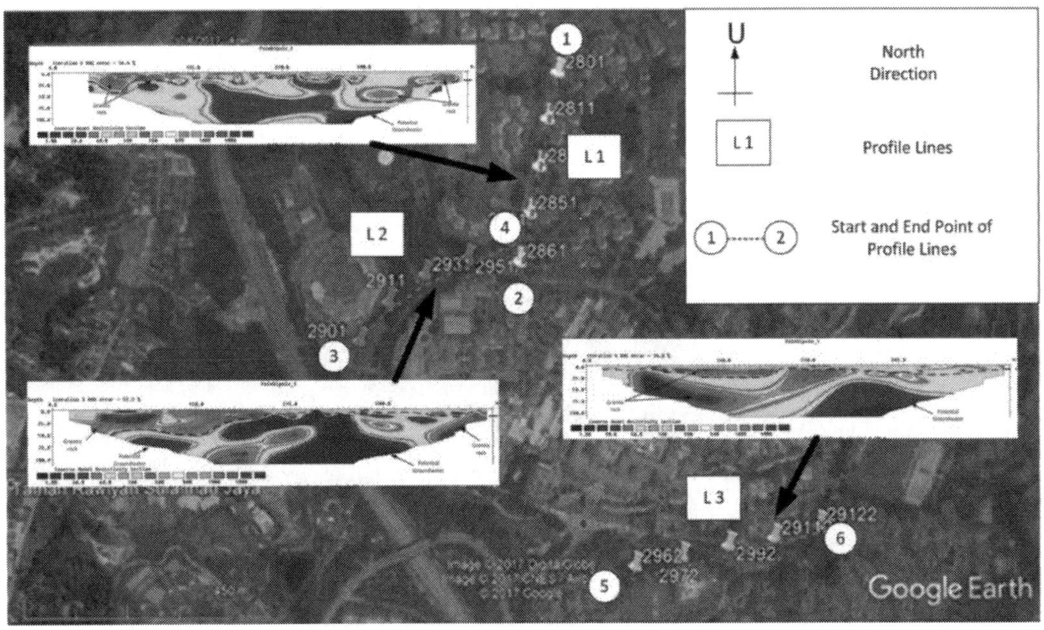

Figure 4. An overview of resistivity and induced polarization results at UIA.

IOP Conf. Series: Journal of Physics: Conf. Series **1049** (2018) 012076 doi:10.1088/1742-6596/1049/1/012076

Table 1. The resistivity and chargeability value at Universiti Islam Antarabangsa Malaysia and its interpretation.

Resistivity value (Ωm)	Resistivity legend	Interpertation
1 – 100	1.00 30.0 60.0 100	Fractured rock saturated with water
> 100	600 800 1000 4000	Fractured to Fresh solid rocks
Chargeability value (millisecond, ms)	Chargeability Legend	Intepertation
0.0 - 1.0	0.00 0.300 0.700 1.00	Fractured rocks contain fresh water
1.0 – 12.0	3.00 5.00 9.00	Fractured rocks zone to solid rocks
>12.0	12.0	Fresh solid rock zone

The 2-dimensional tomography at Permian granite rock formation at Universiti Islam Antarabangsa Malaysia was separated based on the variation of resistivity values i.e. less than 100 ohm.m and above 100 ohm.m. Below 100 ohm.m is the zone which can be considered as saturated fractured rock with groundwater. Meanwhile, above 100 ohm.m is suggested as a zone of dense rock. The chargeability from induced polarization tomography is separated between the value of below 1.0 ms and above 1.0 ms. It is expected at granitic rock mass has the value above 1.0 ms due to its ability to retain the electrical charges. Meanwhile, in water zone, the value is 0 ms chargeability due to its inability to retain the electrical charges. Therefore, by combining resistivity and chargeability values, the groundwater bearing zone can be well predicted and detected

Based on Figure 5, Figure 6 and Figure. 7, all the profile lines have shown potential of groundwater around the study areas based on the contrast of resistivity value and chargeabilty value. For profile line 1, the potential groundwater zones are very large within the distance from 134 to 340 m with the depth from 10 m until more than 80 m at fracture zone. The most potential groundwater zone is at profile line 2. The groundwater is mainly located at rocks formation which indicated volume of water are less than in fractured zone at distance of 155 to 320 m. Figure 7 shows potential groundwater zone at profile line 3 at the distance from 180 to 350 m and the depth more than 25 m.

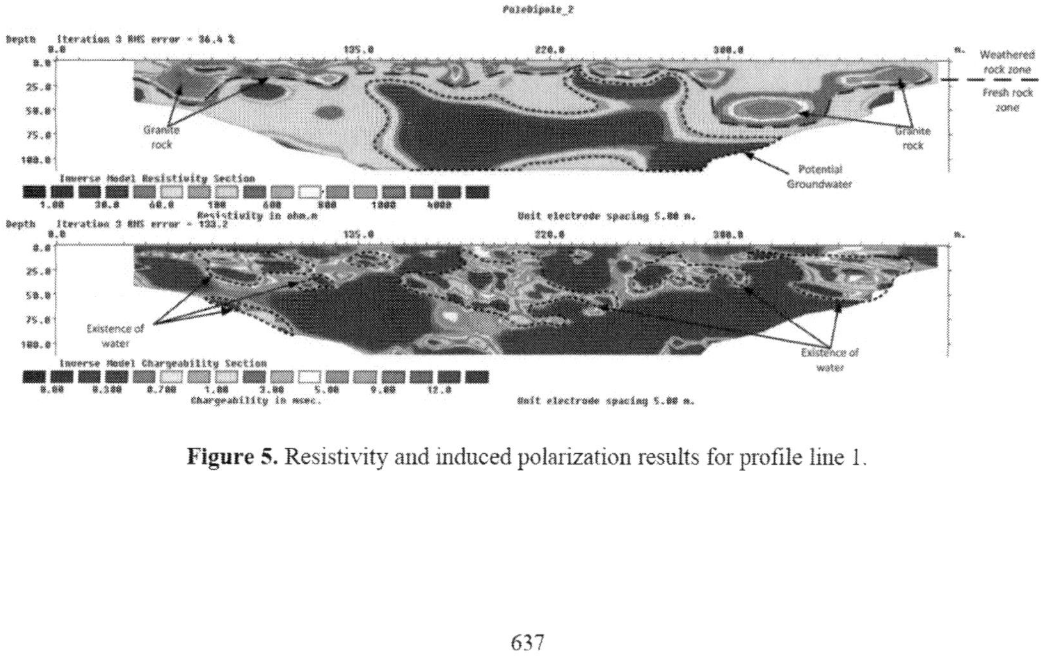

Figure 5. Resistivity and induced polarization results for profile line 1.

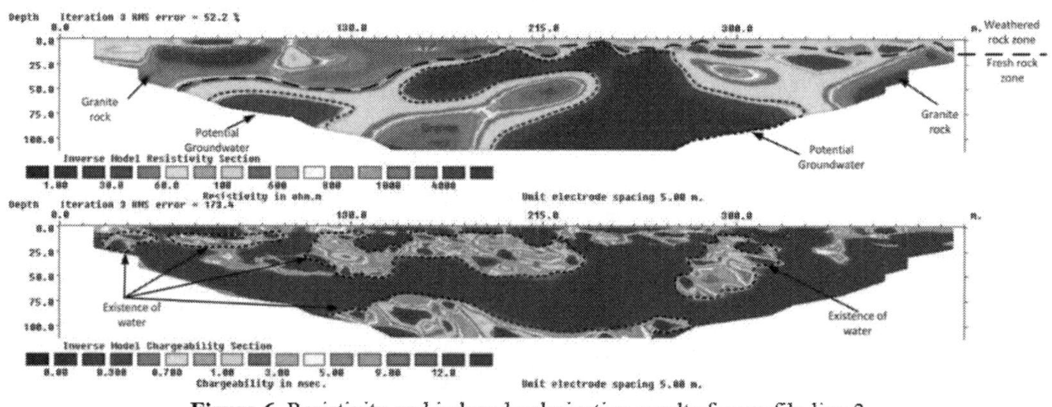

Figure 6. Resistivity and induced polarization results for profile line 2.

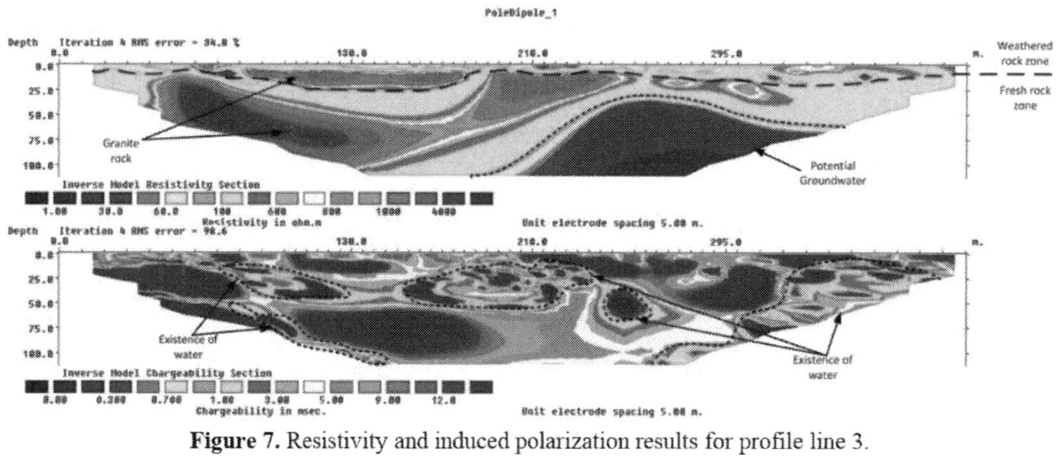

Figure 7. Resistivity and induced polarization results for profile line 3.

Resistivity measurement could determine the location of groundwater in older rock formation in distinct between fractured or solid rocks. Through induced polarization method, the results affected by the electrode polarization effect which caused by conductivity of rocks material thus the current will flow in partial electrolytic (groundwater) and partly electronic (conductive mineral). By comparing resistivity method and induced polarization method, existense of groundwater can be predicted and detected either the water is exists in fractured zone or in dense rocks zone. The high chargeability value (more than 12 ms) in the potential groundwater zone has shown their existense of water in dense rock zones in the resistivity result. This is probably the indication of low volume of groundwater. While, the chargeability value between 0 – 9 ms are the evidence of the existense of water in fractured rock zones which have higher concentration of groundwater.

5.0 Conclusion

This study used geophysical electrical survey to determine the groundwater by combining resistivity and induced polarization techniques in granitic formation. There were three (3) profiles lines of measurements have been conducted at Universiti Antarabangsa Malaysia and all the results have shown potential groundwater zones. From the results, there are potential groundwater zones of 10 – 120 Ohm.m at dense rock formation with chargeability more than 12 ms. Meanwhile, the fractured rock formation has the chargeability value ranging between 0 – 9 ms.

Acknowledgemnet

The authors would like to thank to Ministry of Higher Education and Universiti Tun Hussein Onn Malaysia for their financial support on FRGS vot. 1455.

References

[1] Tyson, A.N. 'Georgia's Groundwater Resources U.S Geological Survey Bulletin'. (1993) pp. 1096:1-11.

[2] Dongmei, H., Matthew, J.C., Guoliang, C., Benjamin, H. Alterations to groundwater recharge due to anthropogenic landscape change, *In Journal of Hydrology,* Volume 554 (2017), pp 545-557.

[3] Rayner, S.F., Bentley, L.R., Allen, D.M., 2007. Constraining aquifer architecture with electrical resistivity imaging in a fractured hydrogeological setting. *Journal of Environmental and Engineering Geophysics*, Volume 12 (4), (2007), pp.323-335.

[4] Gélis, C., Revil, A., Cushing, M.E., Jougnot, D., Lemeille, F., Cabrera, J., De Hoyos, A., and Rocher, M. Potential of electrical resistivity tomography to detect fault zones in limestone and argillaceous formations in the experimental platform of Tournemire, France. *Pure and Applied Geophysics* Volume 167 (11), (2010), pp. 1405-1418.

[5] Schütze, C., Vienken, T., Werban, U., Dietrich, P., Finizola, A., Leven, C., 2012. Joint application of geophysical methods and direct push-soil gas surveys for the improved delineation of buried fault zones. *Journal of Applied Geophysics* Volume 82, (2012) pp. 129-136.

[6] Telford, E., Geldart, W.M., Sheriff, R.E., Applied Geophysics. *Cambridge University Press*, UK, (1990).

[7] Innocent, M., Hlatywayo, D.J., Nel J.M., Chuma C. Electrical resistivity survey for groundwater investigations and shallow subsurface evaluation of the basaltic-greenstone formation of the urban Bulawayo aquifer, *Physics and Chemistry of the Earth.* Volume 50–52, (2012), pp 44-51.

[8] Loke, M.H., 2011. Electrical resistivity surveys and data interpretation. in Gupta, H (ed.), Solid Earth Geophysics Encyclopaedia (2nd Edition) "Electrical & Electromagnetic" Springer-Verlag, 276-283.

[9] Loke, M.H. Geoelectrical Imaging 2-D & 3-D. GEOTOMOSOFT SOLUTIONS. Manual books. (1995-2017).

[10] M Hazreek Z.A., M Faizal T.B., Mohd Hafiz Z., Nor'aishah M.A., Aziman M., Saiful Azhar A.T. Groundwater Seepage Mapping using Electrical Resistivity Imaging. *Applied Mechanics and Materials* Vols. 773-774 (2015.) pp 1524-1534

[11] Loke, M.H., Chambers, J.E., Rucker, D.F., Kuras, O. Wilkinson, P. B. (2013) Recent developments in the direct-current geoelectrical imaging method. *Journal of Applied Geophysics* Vol. 95 (2013) 135–156.

[12] Zeinab A. Abdul Rahim S., Wan Zuhairi Y. and Jasni Y. Groundwater investigation using electrical resistivity imaging technique at Sg. Udang, Melaka, *Malaysia. Bulletin of the Geological Society of Malaysia*, Volume 58, (2012), pp. 55 – 58

[13] Reynolds, J. M., 1997. An Introduction to Applied and Environmental Geophysics. England:Wiley

[14] Geology Map of Peninsular Malaysia, General Mineral and Geoscience Department Malaysia (1985).

[15] M Hazreek Z.A., Rosli S., Chitral W.D., Fauziah A., S Azhar A.T, M Aziman M. and B Ismail B. Soil Identification using Field Electrical Resistivity Method. *Journal of Physics Conference Series* (2015).

[16] Loke M.H.,Acworth I. andDahlin T., 2003.Acomparison of smooth and blocky inversion methods in 2D electrical imaging surveys. Exploration Geophysics 34, 182–187.

IOP Conf. Series: Journal of Physics: Conf. Series **1049** (2018) 012077 doi:10.1088/1742-6596/1049/1/012077

VLSI Implmentation of a Fast Kogge-Stone Parallel-Prefix Adder

Lee Mei Xiang, Muhammad Mun'im Ahmad Zabidi, Ainy Haziyah Awab, Ab Al-Hadi Ab Rahman

Faculty of Electrical Engineering, Universiti Teknologi Malaysia, 81310 Skudai, Malaysia

E-mail: l.mei.xiang@hotmail, raden@fke.utm.my, ainy.haziyah@gmail.com, hadi@fke.utm.my

Abstract. This paper presents a VLSI implementation of a high speed Kogge-Stone adder (KSA) using 0.18μm process technology. The adder is known to be one of the fastest adder architectures, and this is validated through a comparison with other adder architectures including the standard ripple carry adder and the carry look ahead adder. Furthermore, our KSA adder is also compared with a default optmized adder from the Artisan standard cell library. The adders are compared for bit widths of 8, 16, and 32. The adders are designed using Verilog and synthesized using both front-end and back-end tools, with complete validation and verification stages, including analysis for performance, power, and area. Results show that in terms of performance, KSA results in the lowest propagation delay with almost constant delay for all bit widths, with up to 70% less delay as compared to all other architectures. Area and power penalty is found to also increase by roughly 59%. In terms of energy usage, the KSA adder results in up to 64% less. In the case when speed and energy are critical, this fast and energy efficient KSA adder can be readily integrated into custom VLSI designs.

1. Introduction

Adder is one of the fundamental building blocks in a digital system. It is used in microprocessors, as well as many signal processing operations such as filtering and convolution. These basic blocks are also used in larger signal processing systems such as in image and video compression, object recognition, modulation, etc. Accelerating adders used in these systems therefore enables many applications to speed up. In terms of hardware implementation for VLSI, many different types of adder architectures have been reported, each with their own strenghs and weaknesses.

The main aim of this paper is to implement one of the fastest adder architectures using VLSI technology, which is the Kogge-Stone adder, and validate its performance, power, and area as compared to other adder architectures, including the ripple carry adder, carry look ahead adder, and the default adder from the standard cell library. The objective is to have a fast VLSI adder that can be integrated into larger systems and improve its performance.

Some of the works that implement the KSA include [1] where an adaptive power gating is applied on a 32-bit KSA design. However, the technique is only applicable to partially idle circuits and not efficient for general addition operation. The work in [2] implements the KSA on FPGAs, but may not be as efficient in VLSI since the implementation is mainly on gate-level. The work in [3] is most similar to the work in the present paper, where the KSA is implemented using 0.18μm process technology. However, the focus was mainly on Monte-Carlo simulation

Content from this work may be used under the terms of the Creative Commons Attribution 3.0 licence. Any further distribution of this work must maintain attribution to the author(s) and the title of the work, journal citation and DOI.
Published under licence by IOP Publishing Ltd

IOP Conf. Series: Journal of Physics: Conf. Series **1049** (2018) 012077 doi:10.1088/1742-6596/1049/1/012077

and temperature analysis than benchmarking with other adder architectures. The work in this paper proposes to implement in VLSI the standard KSA adder and benchmark against some of the popular and commonly used adder architectures.

2. Background

The adder is a fundamental building block and known to be an important element in many applications especially in digital signal processing and microprocessors. The most basic adder architecture is known as the Ripple Carry adder (RCA), where the longest path is the carry delay. The path becomes longer as the number of bits are increased.

Extensive research was carried out to improve this long carry delay path. In 1958, Weinberger and Smith proposed using propagate and generate blocks to handle the carry issue. This design is today known as the Carry Look Ahead adder (CLA) [4]. It consists of the propagator and generator, and the carry generator block as shown in Figure 1 and 2. The top-level 8-bit CLA is shown in Figure 3. It shows that the PG runs parallel for each bit. However, the process in PPC depends on the C_{out} from previous bit to complete. Hence, the time delay is not solved completely by CLA [5].

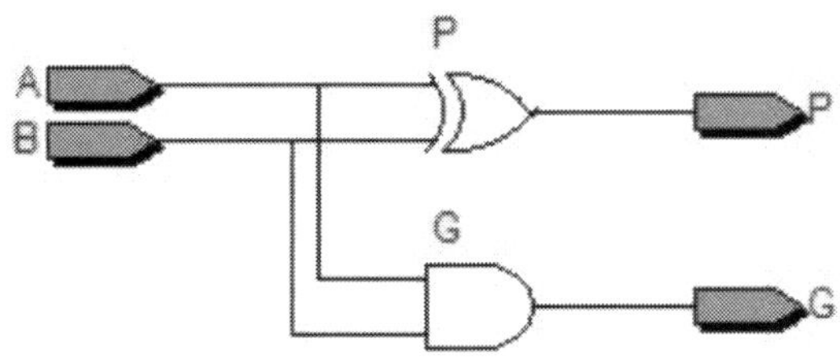

Figure 1. Architecture of Propagator and Generator Block.

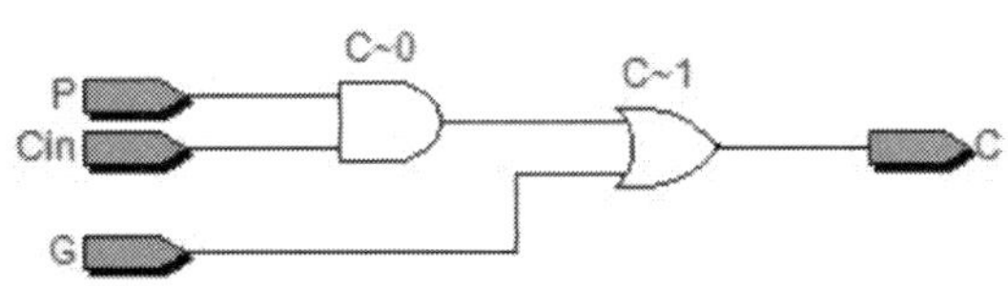

Figure 2. Architecture of Various Carry Generator Block.

IOP Conf. Series: Journal of Physics: Conf. Series **1049** (2018) 012077 doi:10.1088/1742-6596/1049/1/012077

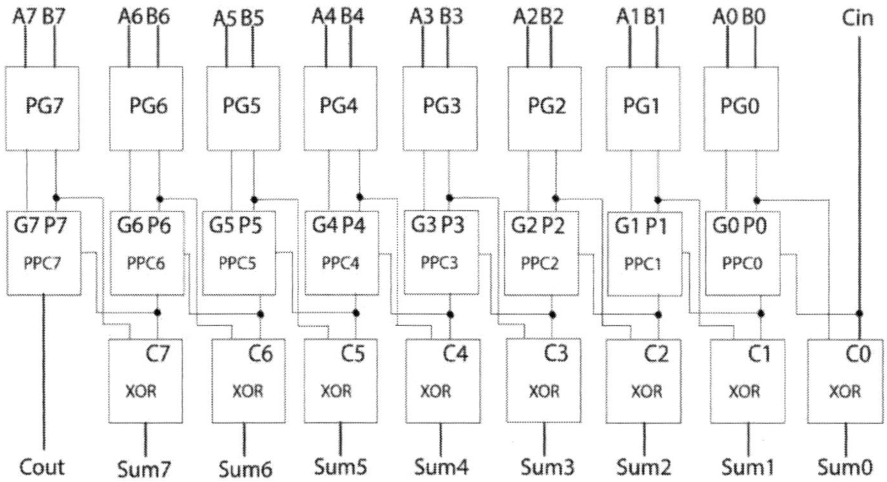

Figure 3. Architecture of 8-bit Carry Look Ahead Adder.

Parallel Prefix Adder was proposed, and expected to have better performance than previous adder. In Parallel Prefix adder, binary addition usually presents in terms of Generation Signal (G_i), Propagation Signal (P_i), Carry Signal (C_i) and Sum Signal (Sum_i). Each bit position follows the range $n \leq i \leq 1$. All of these signals can be obtained from the equation below [6]:

$$G_i = A_i + B_i \tag{1}$$

$$P_i = A_i \oplus B_i \tag{2}$$

$$C_i = \begin{cases} G_i \\ G_i + P_i C_{i-1} \end{cases} \tag{3}$$

$$Sum_i = P_i \oplus C_{i-1} \tag{4}$$

The sum for Parallel Prefix Adder can be obtained by three stages as shown in the block diagram in Figure 4. [7].

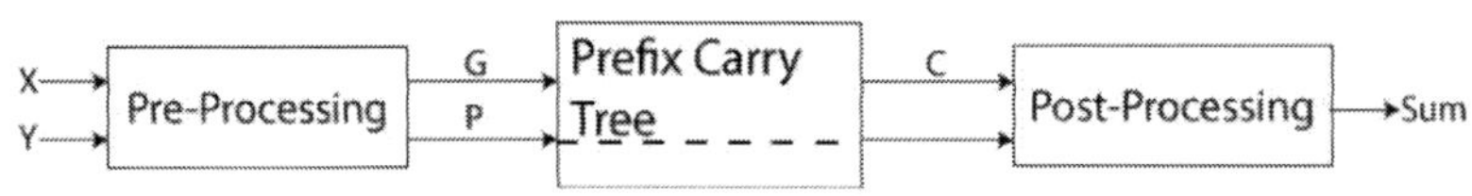

Figure 4. Block Diagram in getting the sum for Parallel Prefix Adder.

P_i and G_i generated from pre-processing block based on equation (1) and (2). The signals then proceeds to the next stage, Prefix Carry Tree to generate C_i. This stage is the part that differentiates the performance of an adder used. On the other hand, the final stage is aimed to get the final adder result Sum_i by following equation (4).

There are a lot of Parallel Prefix adders that have been developed starting from 1960s. For example J.Sklansky conditional adder from 1960, Kogge-Stone adder from 1973, Ladner-Fisher adder from 1980 and etc [8].

IOP Conf. Series: Journal of Physics: Conf. Series **1049** (2018) 012077 doi:10.1088/1742-6596/1049/1/012077

The concept of KSA was developed by Peter M. Kogge and Harold S. Stone, published in a seminal paper in 1973 [9]. It was introduced to solve a large class of recurrence problems in parallel computer such as the Illiac IV. Nowadays, KSA is known as the special and fast adder in VLSI. The architecture consists of three blocks, which are pre-processing, carry generator and post processing blocks. Figure 5 shows the architecture of KSA.

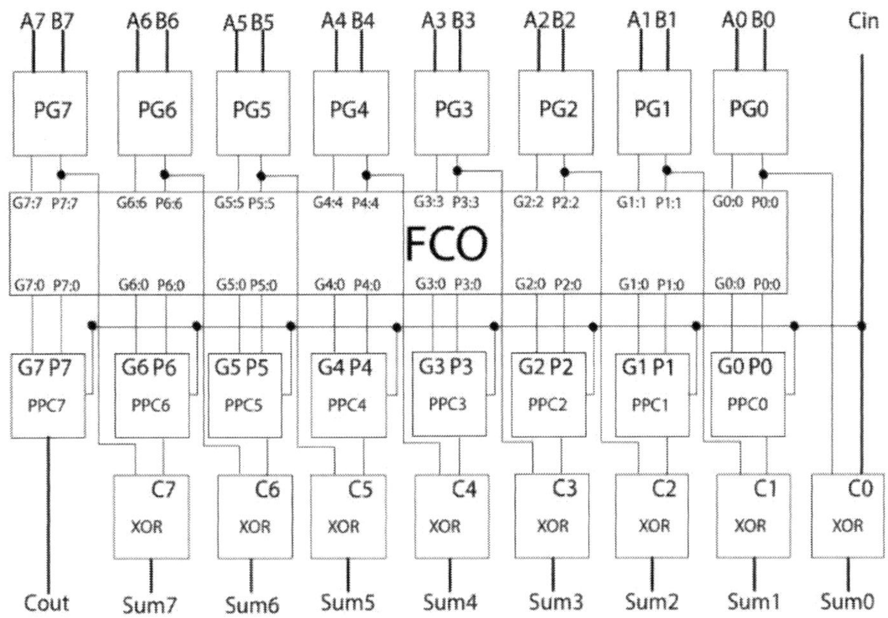

Figure 5. Architecture of 8-bit Kogge Stone Adder.

The architecture of KSA is quite similar to CLA. However, the fundamental carry operator (FCO) block lies between the PG blocks and PPC blocks. This block processes the propagator and generator bits in pre-processing structure. Figure 6 shows the architecture of FCO used in KSA.

In one bit KSA circuit, it requires 2 gate delays. However, it can be used to combine any P_i and G_i signals that are generated from PG block, or combine some P_i and G_i signals that are already combined. The combination does not affect the previous bits but only the level.

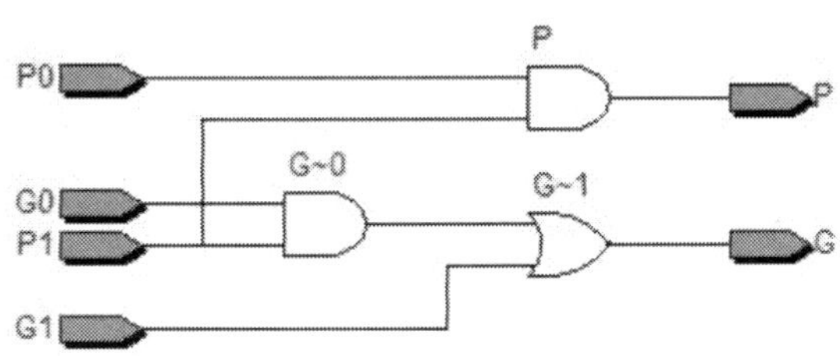

Figure 6. Architecture of FCO Block.

In addition, The PPC block in KSA is slightly different compared to CLA. The input to each bit of PPC block in KSA is the C_{in} which is decided by end-user or external factors. Unlike CLA, the input in PPC block requires the Cout from previous bit. With this architecture, the time delay in KSA will be in the optimum stage.

IOP Conf. Series: Journal of Physics: Conf. Series **1049** (2018) 012077 doi:10.1088/1742-6596/1049/1/012077

3. Methodology

The basic ripple carry adder, CLA, and KSA as explained in the previous section are implemented in Verilog and sent for both front-end and back-end digital IC design flow, including full validation and analysis for performance, power, and area. The adder from the standard cell library is instantiated directly using the Verilog code below:

```
module Default_adder (A, B, Cin, Sum, Cout);
parameter N=7;
input [N:0] A, B;
input Cin;
output [N:0] Sum;
output Cout;
assign (Cout, Sum) = A + B + Cin;
endmodule
```

3.1. EDA Tools Flow

The Synopsis and Mentor Graphics EDA tools are used for VLSI implementation and verifications of the adder architectures. Synopsys Design Vision and PrimeTime are used for front-end synthesis and timing analysis, while the Synopsys ICC is used for backend layout generation. Simulation is performed using Synopsys VCS, while DRC, LVS, and PEX are performed using Mentor Graphics Calibre. Figure 7 shows our tool flow.

4. Results and Discussion

This section presents results for the VLSI implementation of the adder architectures. Each adder architecture has been designed using Verilog and synthesized using full ASIC flow to obtain an IC layout. Figure 8 shows an example of a VLSI layout for the 32-bit KSA. All the results have also been analyzed post-layout for performance, area, and power. Registers are also inserted at input and output of each adder in order to obtain valid timing analysis data.

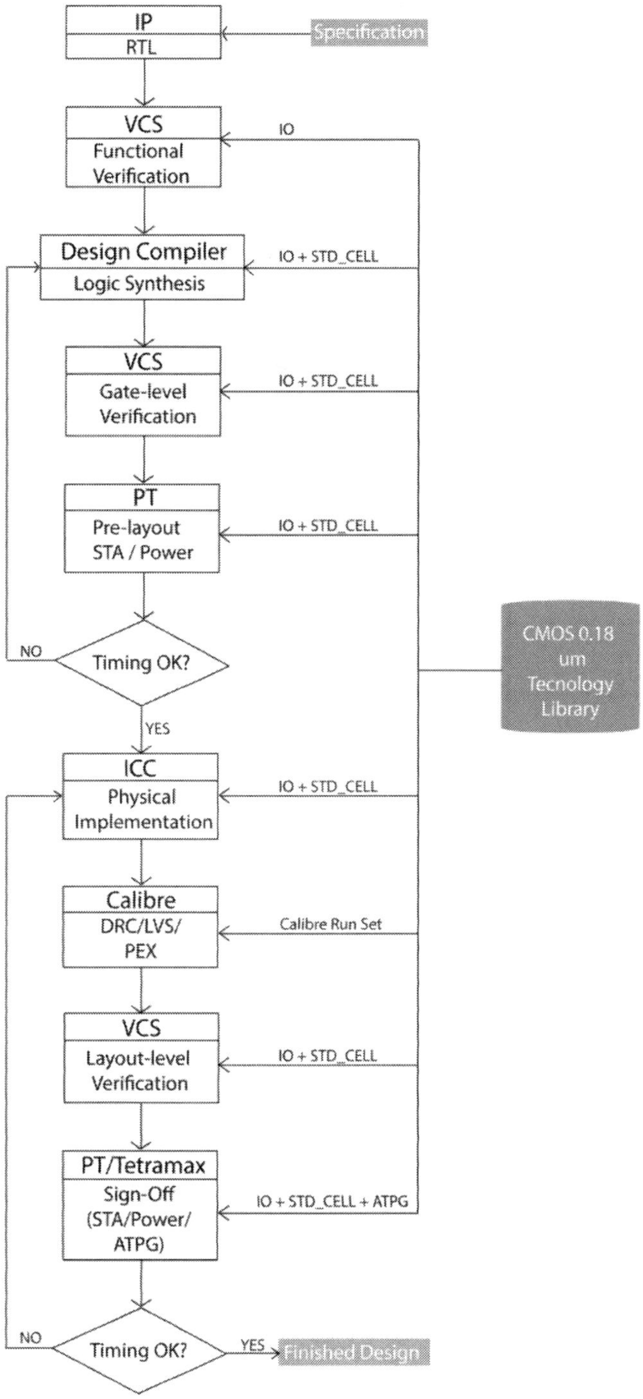

Figure 7. Synopsis design flow with Mentor Graphics tools for verifications

4.1. Propagation Delay

Propagation delay information is obtained from post-layout timing analysis. The required frequency is set to 100MHz. The slack value obtained is then used to estimate the actual

IOP Conf. Series: Journal of Physics: Conf. Series **1049** (2018) 012077 doi:10.1088/1742-6596/1049/1/012077

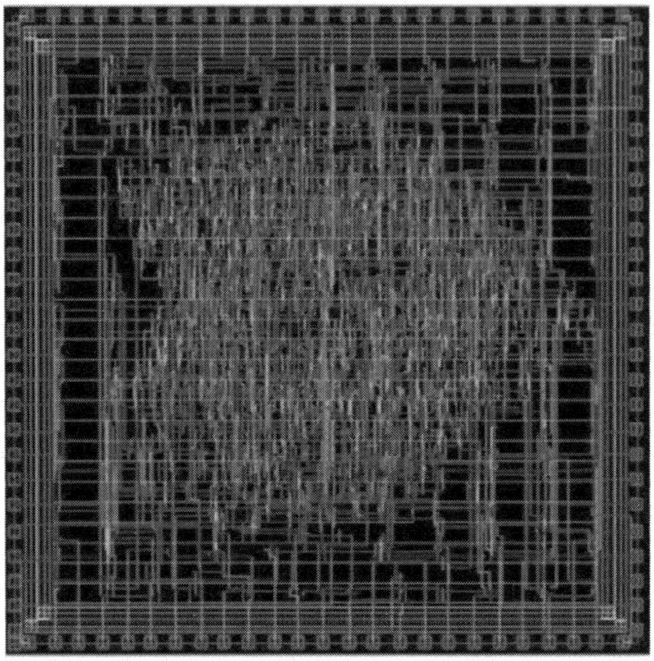

Figure 8. VLSI layout of a 32-bit KSA.

arrival time, which is then normalized to propagation delay. As shown in Figure 9, the KSA results in the lowest with almost constant delay from 8-bit to 32-bit. The higher the number of bits required, the further the distance between KSA and others adders. KSA is rougly 37% less delay in 8-bit and 70% less delay in 32-bit as compared to the worst case. In terms of maximum frequency, the KSA can run at an estimated maximum speed of $1/3.02ns = 331MHz$ for 8-bit KSA. As explained in section 2, the logic depth of KSA is constant regardless of the number of bits, which conforms to our experimental data.

IOP Conf. Series: Journal of Physics: Conf. Series **1049** (2018) 012077 doi:10.1088/1742-6596/1049/1/012077

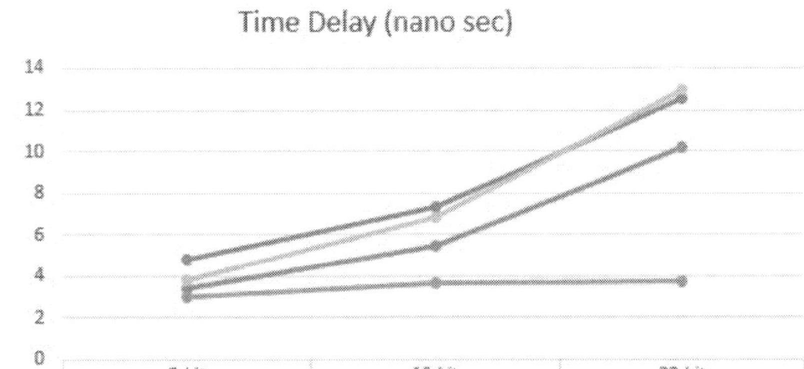

	8-bit	16-bit	32-bit
RCA	4.83	7.35	12.5
CLA	3.44	5.45	10.15
KSA	3.02	3.61	3.75
adder	3.78	6.83	12.91

Figure 9. Time delays for each adder.

4.2. Cell Area
According to the results in cell area as shown in Figure 10, KSA requires the highest cell area. This is expected since extra FCO blocks are required in order to generate the sum out and C_{out} results in parallel. The area increase is up to 59% for 32-bit adders.

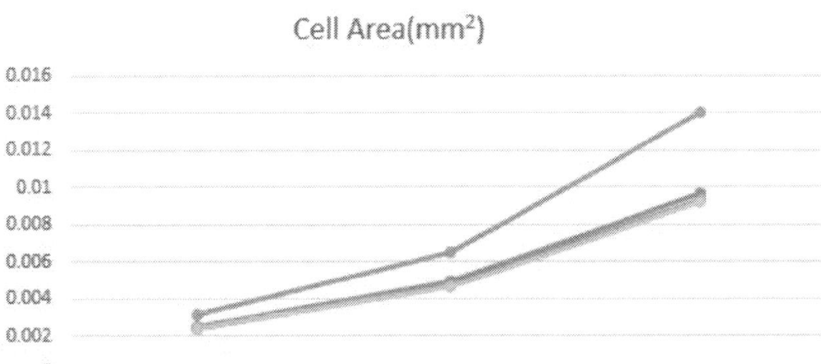

	8-bit	16-bit	32-bit
RCA	0.0024	0.0047	0.0093
CLA	0.0025	0.0049	0.0097
KSA	0.0031	0.0065	0.014
adder	0.0024	0.0046	0.0092

Figure 10. Cell Area coverage of each adder.

Cell count information for various adder architectures is given in Table 1.

4.3. Total Power
Power consumption information is obtained from Quality of Result (QoR) report for post-layout. It is taken as the sum of static and dynamic power. Results show that maximum increment is

Table 1. Cell count for various adder architectures.

Type of Adder	Cell Count (8-bit)	Cell Count (16-bit)	Cell Count (32-bit)
RCA	34	66	134
CLA	51	99	199
KSA	68	148	330
Default	28	52	104

up to 24% for 32-bit adder as compared between RCA and KSA. The total power is found to be almost the same for all adder architectures in 8-bit and 16-bit. However, the power consumption for KSA increases sharply when going to 32-bit. This is shown in figure 11.

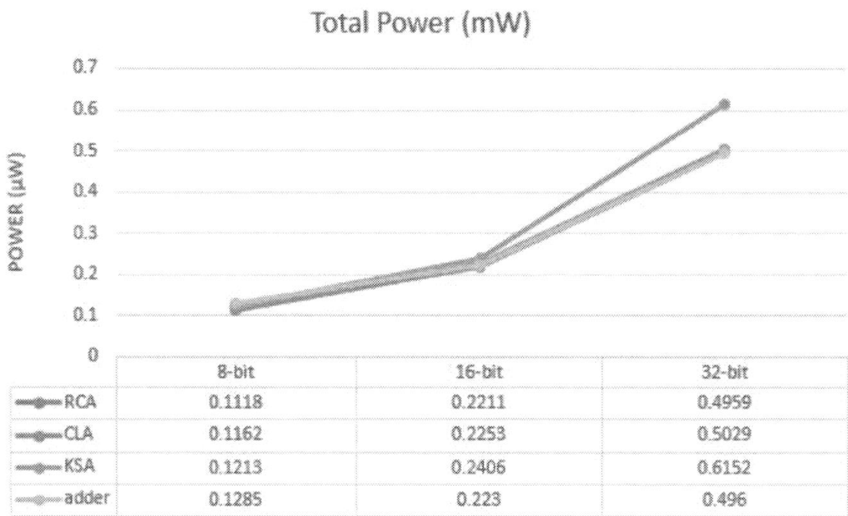

Figure 11. Power Consumption of each adder.

Another interesting criteria to evaluate is the energy efficiency. Energy can simply be defined as follows:

$$E = Pt \tag{5}$$

Where P is the total power and t is the time delay. Table 2 summarizes this information, where it can be seen that the KSA uses significantly less energy compared to other adders, with up to 64% less in the best case at 32-bit.

Table 2. Energy efficiency of various adder architectures.

Type of Adder	Energy (pJ) (8-bit)	Energy (pJ) (16-bit)	Energy (pJ) (32-bit)
RCA	0.54	1.63	6.19
CLA	0.40	1.23	5.10
KSA	0.37	0.87	2.31
Default	0.49	1.52	6.40

5. Conclusion

In this paper, we have designed a fast Kogge-Stone adder and implemented using 0.18um process technology. The high speed adder has been validated and compared against other adder architectures including the ripple carry adder, carry look-ahead adder, and the default adder from the standard cell library. All of these adder architectures are implemented using Verilog, and synthesized using both front-end and back-end tools from Synopsys to obtain the ASIC layout. The designs have also been validated using tools from Mentor Graphics. Results show a very significant improvement in time delay for the KSA as compared to other architectures, at the cost of relatively larger area and power consumption. This adder can be readily applied to speed critical applications in VLSI. Currently, work is ongoing to further improve the adder by custom VLSI design.

Acknowledgments

The authors would like to acknowledge UTM for providing the facilites for research, and also the Ministry of Education for providing the funds (Vot No 4F659).

References

[1] Bezati E, Yviquel H, Raulet M and Mattavelli M 2011 A unified hardware/software co-synthesis solution for signal processing system *Proceedings of the 2011 Conference on Design and Architectures for Signal and Image Processing (DASIP)* p1
[2] Xilinx 2017 ZedBoard Technical Specifications Internet: http://zedboard.org/content/zynq-fpga-fabric-clock-speed
[3] Lin Z and Chow C 2013 Zcluster: A zynq-based hadoop cluster *Field Programmable Technology (FPT) International Conference on. IEEE.* p450
[4] Powell A and Silage D 2015 Statistical performance of the ARM cortex A9 accelerator coherency port in the xilinx zynq SoC for real-time applications *ReConFigurable Computing and FPGAs (ReConFig) International Conference on* p1
[5] Sugimoto N and Amano H 2014 Hardware/software co-design architecture for Blokus Duo solver *Field-Programmable Technology (FPT) International Conference on* p358
[6] Ahmed K E and Farag M M 2015 Hardware/software co-design of a dynamically configurable SHA-3 System-on-Chip (SoC) *Electronics, Circuits, and Systems (ICECS)* p617
[7] ISO/IEC 2006 Information technology âĂŤ MPEG video technologies âĂŤ Part 2: Fixed-point 8x8 IDCT and DCT transforms ISO/IEC 23002-2
[8] Xilinx 2017 FPGA coprocessing for C/C++ programmers (part II) Internet: http://xillybus.com/tutorials/vivado-hls-c-fpga-howto-2
[9] Miano J 1999 Compressed Image File Formats âĂŞ JPEG, PNG, GIF, XBM, BMP Addison Wesley Longman Inc USA
[10] Chiper D F 2016 A new VLSI algorithm for a high-throughput implementation of type IV DCT *International Conference on Communications (COMM)* p17
[11] Arai Y, Agui T and Nakajima M 1988 A Fast DCT-SQ Scheme for Images *Transactions of IEICE* E71 1095
[12] Kovac M and Ranganathan K 1995 JAGUAR: A Fully Pipeline VLSI Architecture for JPEG Image Compression Standard *Proceedings of the IEEE* 83 247
[13] Bhaskaran V and Konstantinides K 1999 Image and Video Compression Standards Algorithms and Architectures âĂŞ Second Edition Kluwer Academic Publishers USA
[14] Zabidi N N and Ab Rahman A A 2017 VLSI Design of a Fast Pipelined 8x8 Discrete Cosine Transform *International Journal of Electrical and Computer Engineering (IJECE)* 7 1430
[15] Xilinx 2013 Xilinx Zynq-7000 All Programmable SoC Overview Xilinx Inc

Determining the Elements of TVET Teachers Competency for Nigerian Higher Learning Institutions

Lai Chee Sern, Mohamad Hamisu, Kahirol Mohd Salleh
Fakulti Pendidikan Teknik dan Vokasional, Universiti Tun Hussein Onn Malaysia, Batu Pahat, 86400 Johor, Malaysia.

E-mail: lcsern@uthm.edu.my

Abstract. This paper determined the suitable elements of competency framework to be applied for Nigerian higher learning institutions. Lack of competent TVET teachers in Nigerian higher institutions has led to several problems such as low quality graduates and unemployment. Competency is a vital element for assessing the quality of technical and vocational education and training (TVET) teachers. Therefore, this research determined the suitable elements of competency model to be applied to Nigerian TVET higher institutions based on Malaysian Human Resource Development Practitioners (MHRDP) competency model for workplace learning and performance (WLP). The study was fully quantitative and 218 cross sectional survey instrument was systematically distributed to TVET teachers from five higher institutions and 205 questionnaires were retrieved based on the stratified sampling technique. Exploratory factor analysis (EFA) was conducted to analyse the data. The research discovered that there are 19 elements relevant to the competency framework for TVET teacher Nigerian TVET higher institutions. A similar research should be undertaken to extend the result to reflect other Non-TVET institutions in Nigeria.

1.0 Introduction

Nigeria as developing country in the world, needs to have competent technical and vocational education and training (TVET) teachers for educational development and socioeconomic advancement that would lead to egalitarian society full of opportunities. In order to be among the successful countries of the world, higher learning institutions need to provide competent teachers with diverse technical know-how and advanced skills to meet the challenges of real time in higher learning institutions [1]. A lot of challenges have been taking place in the world of work including Nigerian higher institutions [2, 3, 4]. This has resulted in the need for continuous learning and updating competencies of employees or teachers across all ages. Therefore, competency needs are categorized in many dimension of life endeavour such as organisational competency, thinking competency and application competency possessed by employees in order to perform a task effectively and efficiently [5, 6]. It is more valuable for employees to develop and improve their work ability, capability and skills thoroughly most especially for TVET teachers [7]. Lack of competency in the higher institutions of learning can bring more challenges which required new ways to accomplish teaching and learning processes [8]. Therefore, it is understandable that the challenges occur in different ways usually affect the institutions worldwide specifically Nigerian higher institutions of learning. In order to minimize these unconditional challenges the competency framework need to be restructured.

Content from this work may be used under the terms of the Creative Commons Attribution 3.0 licence. Any further distribution of this work must maintain attribution to the author(s) and the title of the work, journal citation and DOI.
Published under licence by IOP Publishing Ltd

Competency framework can be regarded as an instrument or tool that identifies the needed competencies in order to minimize the challenges that are currently existing in the institutions [9]. Consequently, many institutions are adapting and adopting competency based model or framework in order to achieve their objectives [10]. Therefore, this research aimed at determining the suitable elements of competency framework to be applied for Nigerian TVET higher institutions of learning. In specific, this paper focused on organisational, thinking and application competencies based on MHRDP competency model.

2.0 Literature review
The following section provides some information concerning the definition of competency and competency needs for TVET teachers.

2.1 Competency
Whenever we talking about knowledge, skills, attitude or behaviour. We are referring to quality. This word quality is regarded as "competence". The word of "competence" comes from the Latin word "competere" which means "competent" that is a person's ability to possess adequate skill, knowledge, attitude or behaviour to perform certain activity. Thus quality, ability and skills to do some certain activity competently.

The word competency was initially discussed and assessed by David McClelland in the early 1970s as a real features of individuals which they possess such as skills, knowledge and ability of worker performance which can be predicted, measured and assessed. Although, the first research of competency that came out with human resource development was done by McLagan in 1989 who believed competency is an area of knowledge, attitude and skills of individuals to produce vital key output.

For more than three decades, the term 'Competency' is defined by different researchers with different perceptions and perspectives and in different ways. According to McLagan and Suhadolnik [11], competency can be defined as a centre for knowledge, attitude and skills that are vital in producing key outputs. Likewise, Boyatzis [6] defined competency as the capability and ability of individual skills, knowledge and behaviour to complete the task assigned to them. While Lee [12] defined competency as set of related skills, abilities knowledge and behaviour which can influence the main aspect of workers job. Another researcher, Abel [13] conceptualised competency as to apply some knowledge into practice within a specific context. In addition, according to Sulaiman et al [14], competency refers to the workers or individuals or practitioners performance as related to the organisational performance in doing task or job that can be evaluated or measured. Taken together, competency in the context of current research is defined as the ability of an individual to apply the knowledge, skills and attitude to complete the given task that can be evaluated.

2.2 Competencies of TVET Teachers
Competencies of TVET teachers as illustrated by Andersson and Köpsén [15] and Arifin and Rasid [16] are exactly those skills, knowledge, attitudes, values, tasks and appreciations that are related to teaching and training within the sphere of TVET. Globally, TVET teachers should be competent in managing classroom and workshop, handling teaching aids, assessing and evaluating students' performance, applying various teaching methods, recognising students' learning styles, meeting the needs of students in the classroom, and more importantly, imparting the necessary technical knowledge and vocational skills efficiently and effectively [2, 17, 18].

According to German Council of Education outline two major areas of teacher's competencies, namely educational competencies and specialized competencies [19]. The individual's capability, ability or behaviour to organize around underline concept or intent, that should be their competencies [6]. It is a worker or individual performance as related to organisational performance in doing task or job [8]. Therefore, from the above mentioned points we should know what kinds of competencies are require by TVET teachers in order to perform their duties as competent teachers.

Subsequently there are different TVET teacher's competences by different scholars in different perspectives. Yunos, Esa, Jamil, and Rosli [20] and Ali [21] pointed out six different TVET teachers' competences that need to be embedded in teacher training program such as skills, knowledge, ethics and professionalism, social process, social accountability and entrepreneurship. In addition, Deitmer and Heinemann [22] outlined five TVET teacher competencies, namely, professionalism, management, analytical, social and communication and knowledge about framework condition. Likewise, Setiabudhi [23] claims that it is compulsory for the TVET teachers to acquire competency related to teaching methodology, knowledge at workplace, research management and implementation and analyse of work process.

3.0 Methodology

The research was carried out in Bauchi State, north eastern Nigeria. Five TVET higher learning institutions in Bauchi were involved in this research, namely, Abubakar Tafawa Balewa University (ATBU), College of Education (COE) Kangyere Bauchi, Abubakar Tatari Ali Polytechnic (ATAP), College of Education (COE) Azare and Bauchi Institute of Education and Islamic Studies (BIEIS). Bauchi State is chosen because it the only state in the north east that has a high number of TVET higher institutions of learning in the region as mentioned above

3.1 Research sample

A total of 427 Nigerian TVET teachers from five higher institutions of learning (University, Polytechnic and Colleges of education) were targeted in this research. The stratified sampling technique was applied. A sum of 218 questionnaires were distributed and 205 completed questionnaires were returned. The return rate was 94.04%. A total of 37.6% of the respondents were University TVET lecturers, 25.1% from Colleges of Education, and 27.3% from Polytechnics. Of the total respondents, 81% (166) were males with the remaining (19% or 39) being females.

3.2 Research instrument

Data were collected through the use of structured questionnaire which consists of questions on demographics and 25 elements of competencies from three research constructs: organisational competency (10 elements), thinking competency (8 elements) and application competency (7 elements) each was measured on a 5 point Likert scale, ranging from 1 = not important to 5 = extremely important. These items were adapted from Salleh, (2012). The reliability coefficient of the instrument was large, $\alpha = 0.88$.

3.3 Administration of Instrument

A total of 218 questionnaires were distributed to five higher institutions of learning based on systematic sampling distribution of 51% to each institution which consist of University with 84 questionnaires, while Polytechnic with 57 questionnaires and three Colleges of education (COE) with 77 questionnaires (COE Azare = 30 questionnaires, COE Kangyere = 25 questionnaires and COE BIEIS = 22 questionnaires). Out of 218 set of questionnaires distributed, 205 questionnaire were returned.

4.0 RESULTS

The outcomes of data analysis are presented as follows:

4.1 Data screening

The data was screened for univariate outliers. The screening analysis did not detect any data with extreme value. The minimum quantity of data for EFA was fulfilled, with 205 as final sample size.

IOP Conf. Series: Journal of Physics: Conf. Series **1049** (2018) 012078 doi:10.1088/1742-6596/1049/1/012078

4.2 Data Analysis: Exploratory factor analysis (EFA)

The use of EFA in this research was to determine the suitable elements of competency model to be applied for Nigerian TVET higher institutions. Firstly, the adequacy of sample size was checked. The Kaiser-Meyer-Olkin coefficient was 0.604, which was slightly above the recommended cut-off value of 0.6. The Bartlett's test of sphericity was significant, χ^2 (171) = 5104.33, $p < .00$.

The 25 measurement items used in the instrument were then subjected to the principal component analysis which showed the presence of three components with eigenvalues more than 1. The initial eigen values indicated that the factor one explained 26.43% of the variance, whereas the factor two and factor three explained 16.00% and 11.78% of the variances respectively.

The outcomes of EFA indicated that six elements have to be eliminated due to poor factor loadings (<0.4). See Table 1. Of the six elements, two elements belong to Thinking, Organisational, and Application Competencies.

Table 1 Rotated Component Matrix

	Components			**Cronbach Alpha**
Code	1	2	3	
THI1	.961			
THI3	.960			
THI4	.957			.957
THI6	.885			
THI7	.864			
THI8	.819			
THI2	.367 (removed)			
THI5	.341 (removed)			
ORG2		.843		
ORG4		.834		
ORG5		.581		
ORG6		.556		
ORG7		.518		.741
ORG8		.487		
ORG9		.455		
ORG10		.406		
ORG3		.368 (removed)		
ORG1		.319 (removed)		
APP1			.784	
APP2			.765	
APP6			.601	.847
APP4			.595	
APP7			.448	
APP3			.354 (removed)	
APP5			.335 (removed)	

Extraction Method: Principal Component Analysis
Rotation Method: Varimax with Kaiser Normalization

IOP Conf. Series: Journal of Physics: Conf. Series **1049** (2018) 012078 doi:10.1088/1742-6596/1049/1/012078

a. Rotation converged in 4 iterations
b. THI = Thinking competency, ORG=organisational competenc
APP= application competency

The rotated component matrix in table 4 shows a meaningful loading of three factors with the Cronbach's Alpha coefficient. Factor one ranges from .961 to .816 with α=.957, factor two (2) ranges from .843 to .434 with α= .741, factor three ranges from .833 to .488 with α= .847 respectively.

4.3 Determinants of TVET teacher's competencies in Nigerian tertiary institution

The factors with description of elements and factor loadings are presented in the table 6 six elements were found to load under first construct which was labelled as thinking competency. Under second construct eight elements were found to load which was also labelled as organisational competency and five elements onto third construct which was labelled as application competency. All the constructs were labelled in line with their elements that loaded onto them. Therefore, all the elements that were used during the EFA analysis were reduced to 19 elements that are relevant to Nigerian TVET tertiary institutions. The factor loading matrix is presented in Table 2.

Table 2 Determinants of TVET teacher's competency

Items	Thinking Competency	Organisational Competency	Application Competency
Workplace	.961		
Facilitation	.960		
Standard identification	.957		
Model building	.890		
Analytical thinking	.862		
Leadership	.816		
Communication		.843	
Work environment		.833	
Goal implementation		.591	
Buy in / Advocacy		.585	
Consulting		.515	
Negotiating /Contracting		.487	
System thinking		.450	
Vision		.434	
Staff selection theory and application			.833
Training theory and application			.825
Career development theory and Application			.544
Reward system theory and application			.537
Process consultation			.488

Six elements were loaded onto thinking competency, namely, workplace, facilitation, standard identification, model building, analytical thinking and leadership. These are the most appropriate elements of thinking competency to be applied to Nigerian TVET higher institutions based on the respondents' feedbacks.

Similarly, eight elements were loaded onto organisational competency which includes communication, work environment, goal implementation, buy in / advocacy, consulting, negotiating/contracting, system thinking and vision which are the most suitable elements of competency to be applied to Nigerian TVET higher institutions based on the answers of the respondents.

Lastly, five elements were found loaded in the application competencies, namely, staff selection theory and application, training theory and application, career development theory and application, reward system theory and application and process consultation.

5.0 Discussion

The research found that, out of the 25 elements of competency, 19 elements emerged to be the most suitable elements of competency applicable to Nigerian TVET higher institutions. This findings is supported by studies of Dubois and Rothwell [24] suggested that competency model has been shown to undergo several reviews over the years since its development. Similarly, Berge et al., (2002) mentioned that many institutions are adapting and adopting competency model in order to develop and achieve their needs and objectives.

Within the context of this research, organisational competency, thinking competency and application competency are the main area of concern, there are only 19 elements relevant to these competency component that is suitable for TVET teachers who serve in TVET higher institutions of learning. It means that six elements are not related to the three competency components based on TVET teachers' perspective.

In organisational competency, there are two elements that are not relevant, namely, critical entrepreneurial issues and group dynamic. The irrelevance of identification of critical entrepreneurial issues to organisational competency might be attributed to business competency construct because identifying the critical business issues it has do with determining the key business issues and force for change and applying that knowledge to performance improvement strategies which more related to business competency construct, while group dynamic is likewise attributed to leadership competency construct because group dynamic is a way of assessing how group of people function or working together, as they seek to meet the needs of members of public which required a leader who can lead them to achieve their goals as both indicated in ASTD competency model developed by Rothwell et al. [25]. Therefore, this might be the reason why these elements do not belong to the organisational competency because this research focused only on organisational, thinking and application competencies.

In thinking competency, two elements that is competency identification and questioning are not applicable to thinking competency. This as a result of inappropriateness of competency identification to thinking competency, might be associated or linked with analytical competency construct in ASTD model because competency identification is a process of identifying the skills, knowledge and attitude which are required by employees to perform their duties effectively and efficiently. Therefore, this element is more related to analytical competency construct as indicated in ASTD. While, questioning is similarly attributed to technical competency construct because questioning is a method of collecting information via pertinent questions asked during surveys, interviews and focus group for the purpose of performance analysis which also more related to the technical competency construct also in ASTD model [26]. This might be the reason why these elements do not belong to the thinking competency because this study focused only on thinking, organisational and application competencies.

In the same instance, two elements are not relevant to application competency, namely feedback and institutional development. The feedback is more relevant to technical competency construct since feedback provides useful performance information to the appropriate people. While the irrelevance of institutional development to application competency might be attributed to analytical competency construct because institutional development theory and application is a way of comprehending the theories, techniques and appropriate application of institutional development interventions by employers as they are used for performance improvement as pointed out in ASTD competency model. Therefore, this might be the reason why the two elements are not relevant to the application competency.

6.0 Conclusion

The 25 elements of competency components were later analysed in EFA thus, organizational, thinking and application competency, the result revealed 19 elements of competency from three constructs were also found to be suitable for Nigerian TVET higher institutions of learning. Thus, organisational competency with 8 elements, thinking competency with 6 elements and application competency with 5 element.

References

[1] Ismail, S., & Mohammed, D. S. (2015). employability skills in tvet curriculum in Nigeria federal universities of technology. *Procedia-Social and Behavioral Sciences, 204*, 73-80.
[2] Oni, C. S. (2007). Globalization and its implications for vocational education in Nigeria. *Essays in Education, 21(1),* 30-34.
[3] Berg, S. A., & Chyung, S. Y. (2008). Factors that influence informal learning in the workplace. *Journal of workplace learning, 20*(4), 229-244.
[4] Paloniemi, S. (2006). Experience, competence and workplace learning. *Journal of Workplace Learning, 18(7/8),* 439-450.
[5] Salleh, K. M. (2012). *Human resource development practitioners perspectives on competencies: An application of American society for training and development (astd) workplace learning and performance (wlp) competency model in Malaysia*. Colorado State University: PhD. Thesis.
[6] Boyatzis, R. E. (2008). Competencies in the 21st century. *Journal of Management Development, 27(1),* 5-12.
[7] Billett, S. (2001). *Learning in the workplace: Strategies for effective practice*. Allen & Unwin, PO Box 8500, St Leonards, 1590 NSW, Australia.
[8] Salleh, K. M., Khalid, N. H., Sulaiman, N. L., Mohamad, M. M., & Sern, L. C. (2015). Competency of adult learners in learning: Application of the iceberg competency model. *Procedia-Social and Behavioral Sciences, 204*, 326-334.
[9] Fogg, C. D. (1999). *Implementing your strategic plan: How to turn" intent" into effective action for sustainable change*. Washington DC: AMACOM.
[10] Berge, Z., de Verneil, M., Berge, N., Davis, L., & Smith, D. (2002). The increasing scope of training and development competency. *Benchmarking: An International Journal, 9(1),* 43-61
[11] McLagan, P. A., & Suhadolnik, D. (1989). *Models for human resource development practice:The research report*. Washington, DC: ASTD
[12] Lee, Y. (2009). Competencies needed by Korean HRD master's graduates: A comparison between the ASTD WLP competency model and the Korean study. *Human Resource Development Quarterly, 20(1),* 107-133.
[13]Abel, M.H. (2008). Competencies management and learning organizational memory. *Journal of Knowledge Management, 12(6),* 15-30.
[14] Sulaiman, N. L., Salleh, M. K., Mohamad, M. M., & Sern, L. C. (2015). Technical and vocational education in malaysia: Policy, leadership and professional growth on Malaysian women. *Asian Social Science, 11, 24,* 153-161.

[15] Andersson, P., & Köpsén, S. (2015). Continuing professional development of vocational teachers: participation in a Swedish national initiative. *Empirical Research in Vocational Education and Training, 7*(1), 1-20.

[16] Arifin, M. A., & Rasid, R. M. (2017). The competent vocational college teacher: A proposed model for effective job performance. *International Journal of Academic Research in Business and Social Sciences, 7 (2)* 2222-6990.

[17] Okoye, K. R., & I. M. O. (2015). Enhancing technical and vocational education and training in Nigeria for sustainable development: Competency based training approach. *Journal ofEducation and Practice, 6(29),* 66-69.

[18] Oluwasola, A. J. (2014) Professional competence of technical teachers: A factor analysis of the training needs of technical college teachers in ekiti state. *Journal of European Education, 3(5)* 281-293.

[19] Bauer,W., & Gollmann, P. (2008).*Vocational Training Research for the Professionalisation of Vocational School Teacher*. Bremen: ITB, Uni-Bremen.

[20] Yunos, J., Esa, A., Jamil, M., & Rosli, D. (2010). Transnational standards design framework for tvet teacher training program. *Proceedings of International Conference on Education for Sustainable Development in Technical and Vocational Education and Training,* Manila, Philippines. *pp* 227-237.

[21]Ali, M.(2015). *Developing the knowledge-based human resources that support the implementation of the national dual training system (NDTS): Evaluation of tvet teacher's competency at mara training institutions*. Universiti Tun Hussein Onn Malaysia: PhD Thesis.

[22]Deitmer, L., & Heinemann, L. (2009). Evaluation approaches for workplace learning partnerships in vet: Investigating the learning dimension. *Towards integration of work and learning* (*pp.* 137-151). Springer Netherlands.

[23] Setiabudhi, J. D. (2013). Post study pre-service practical training program for tvet teacher students. *The 2nd UPI International Conference on Technical and Vocational Education and Training*. Indonesia: Bandung, *pp*. 36-46.

[24]Dubois, D., & Rothwell, W. (2004). *Competency-based human resource management: Discover a new system for unleashing the productive power of exemplary performers*. Nicholas: Brealey Publishing.

[25]Rothwell, W. J., Sanders, E. S., & Soper, J. G. (1999). *ASTD models for workplace learning and performance: Roles, competencies, and outputs*. Washington DC: American Society for Training and Development.

[26]Rothwell, W. J. (2002). *Beyond training and development: State-of-the-art strategies for enhancing performance*. New York: AMACOM.

IOP Conf. Series: Journal of Physics: Conf. Series **1049** (2018) 012079 doi:10.1088/1742-6596/1049/1/012079

Groundwater Responses on Pumping Analysis for Clayey Aquifer

S Musa[1], N A Zakaria[2], T L Lau[2], M A Ahmad Nazri[1], R. Buhari[3].
[1,3]Faculty of Civil and Environmental Engineering, Universiti Tun Hussein Onn Malaysia, 86400, Batu Pahat, Johor, MALAYSIA
[2]*River Engineering and Urban Drainage Research Centre (REDAC)*,
Engineering Campus, Universiti Sains Malaysia, Seri Ampangan, 14300 Nibong Tebal, Penang, MALAYSIA

E-mail: sabariah@uthm.edu.my

Abstract. The relationship between water quantity and clayey layer shows that water storage was very limited to an infiltration into the ground layer which is still retains as a surface water during recharge by precipitation. This paper consists of pumping analysis by using AQTESOLV model in a groundwater response for clayey area to identify the capability of water quantity actions during the discharge and recharge flows. Based to pumping analysis, including type of pump used, discharge rate, water table response, recharge rate, type of soil, and remaining time shows that this area was unconfined aquifer with has low hydraulic conductivity below than 9.6 mm/hr. Therefore, this simulation shows that the quantity and capacity of water intake able to withdrawal more than 40 m^3/day by using 1hp pump and above with a constant rate after 4 days pumping continuously. The capability of storage and intake should be considered by natural (rainfall) or man-made recharge to show the water balance in the system whiles the process of pumping continuously. Therefore, this analysis able to predicts the capacity of groundwater storage at the clayey characteristics while it needs to improve any strategy to generate more water intake.

1. Introduction

In the area of Sembrong, Johor, sand occurs as isolated layer within the extensive clay layers and has limited capacity as an aquifer. Some other parts of the investigation area have only a few meters thick sand layers [1]. This situation is a contributor to the cause in conducting the correct investigation in order to find the best solution for the area. This analysis was performed based on recharge well built in system (REWES) model for subsurface model at Research of Soft Soil Centre Malaysia (RECESS) area. This subsurface model consists of recharge and discharge well (RW) and four monitoring well between 4 m to 35 m from RW. There are extra well (EW) at 4 m, monitoring well 1(MW1) at 10 m, monitoring well 2 (MW2) at 20 m and old well (OW) at 35 m distances.

This study area is generally covered by clay that makes up more than 60% of the total layers up to 35 m depth [2]. Sandy layer is limited in the extent both vertically and laterally in layers and it is often admixed with sandy clay. A survey on the sub-surface profile using the resistivity technique at the Universiti Tun Hussein Onn Malaysia (UTHM) campus shows that the sand exists as isolated lenses

Content from this work may be used under the terms of the Creative Commons Attribution 3.0 licence. Any further distribution of this work must maintain attribution to the author(s) and the title of the work, journal citation and DOI.
Published under licence by IOP Publishing Ltd

within the extensive clay layers and therefore the capacity of aquifer is limited [3]. Therefore, the location of this study indicates that the sand discovery has a wide lateral continuity in the sub surface as its presence is only a few meters thick and almost invisible.

Study on soil permeability has been conducted in the area of UTHM [4]. The top layer of soil in this area mostly consists of silty clay with very low rate of infiltration. Study by[5] shows that most parts in the Sri Gading area are flat (low laying area) and around 1.0 m to 1.5 m above mean sea level. This topographic condition results in the river being unable to drain the surface runoff directly to the downstream or into the sea effectively.

Generally, a top layer of brownish, peaty clay covers most of the area. Below the top covering is a layer of greyey clay, which is frequently mottled and locally contains a lateritic redness. Greenish, shell-bearing clay, probably of marine origin, is also extensive, particularly near the coast. Sand is limited in extent, both vertically and laterally, and is often admixed with a little clay [6].

The report by [7] also shows the similar result of an average 20 m depth of clay and silt cover the top layer. Only two points of the borehole (BH3 and BH10) show a sandy layer on the top surface as shown in Table 1. A total of 48 undisturbed soil samples were obtained from a very soft clayey layer in BH-1, BH-2, BH-4, BH-5, BH-7, BH-8, and BH-9. Nevertheless, this study investigates a deeper recharge well over 40 m deep, which found limestones one of the porous layer at this area as potential groundwater resources [8]. Therefore, meteorological and geo-hydrological/ geophysical information about the catchment area, the existing water table and its fluctuations, water demand, availability of runoff water, and socio-economic condition are considered as fundamental input for aquifer characteristics studies [9].This investigation in the study area on the subsurface lithological and hydrogeological conditions aims to determine the amount of groundwater available for domestic consumption and/or industrial/agricultural use.

Table 1. Types of soil for boreholes around study area [3]

		Thickness of layer (m)		
Borehole no.	BH depth (m)	Clay	Silt	Sand
1	36.3	0-27	27-36.30	-
2	31.93	0-23	23-31.93	-
3	39.15	1.10-25	0-1.10 , 28.2-39.15	25-28.2
4	34.8	0-23.6	23.6-34.8	-
5	34.62	0-22	22-34.62	-
6	34.62	0.8-24.8	0-0.8, 24.8-34.62	-
7	31.87	0-24.4	24.3 - 31.87	-
8	28.65	0-21.8	21.8- 28.65	-
9	37.77	0-23.2	23.2-37.7	-
10	45.3	0.8-11.2 , 13-19	0-0.8 , 19-45.3	11.2 - 13

2. Method analysis

Pump test analysis using software models is usually used to obtain good precision and simulations. The reading results obtained from fieldwork are used as the primary data source for analyzing the findings for the purpose of simulating and predicting a series of data in certain cases. In addition, it is also used to test data and verify the completion of the analysis through relevant derivation in other cases. A solution using AQTESOLV model is suitable for analyzing the situation in a groundwater response and the procedure as summarized in Figure 1. The combination methods between vertical drains (infiltration basins) and REWES models are used to analyze the effectiveness of detecting water contained in the area without disturbing the ecology and the sustainability of the study area.

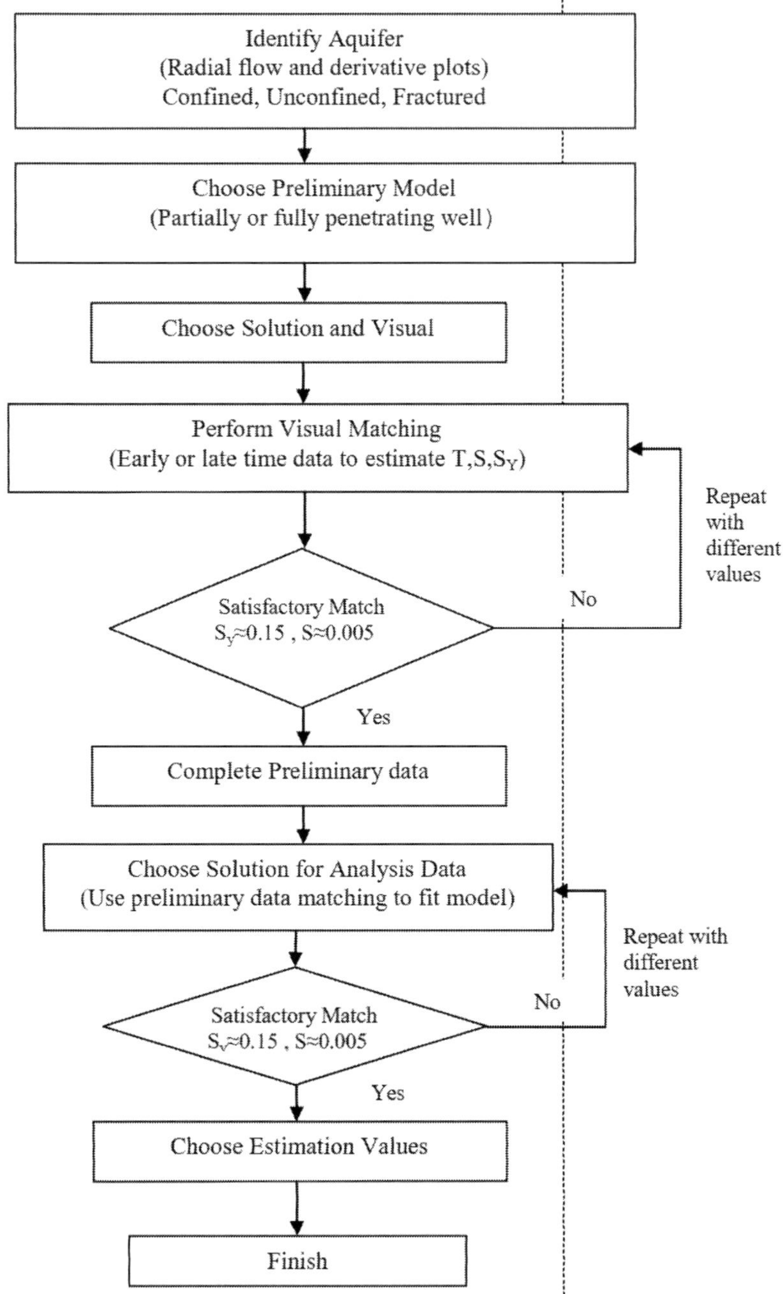

Figure 1. Steps analysis for aquifer characteristics

To run this program, several sets of data are required for the assessment in generating an aquifer parameter output with a minimum error. Table 2 shows the pumped well and the monitoring wells parameters related to the preliminary (set) input data. The model is available to evaluate the well characteristics based on the nearest well requirement by generate the results automatically.

Table 2. Data set values for wells test

	Output		
Parameter	RW	OW	EW
Head of water level, H	0.59m	0.3m	0.44m
Depth of aquifer, b	45m	45m	45m
Ratio of hydraulic conductivity, k_v/k_h	1	1	1
Coordinate for x direction, x	100m	62m	102m
Coordinate for y direction, y	100m	90m	103.5m
Depth of well without screen, d	11.5m	5.7m	11.5m
Depth of screen well, L	30m	39m	30m
Radius of inside well, r (c)	0.074m	0.074m	0.049m
Radius of outside well, r (eq)	0.075m	0.075m	0.05m
Radius of outside well, r (w)	0.075m	0.075m	0.05m
Transmissivity, T	1440 m^2/day	1440 m^2/day	1440 m^2/day
Storage, S	0.001	0.001	0.001
Specific yield, S_y	0.1	0.1	0.1
Beta or water compressibility, β	0.1	0.1	0.1

Note: RW – recharge well, OW- old well, EW- extra well.

3. Result estimation

The data set for the REWES test is selected during dry and wet (short term) conditions through the pumping process for more than 4 hours. Some outputs for displacement (water level) and discharge flows were produced for this solution.

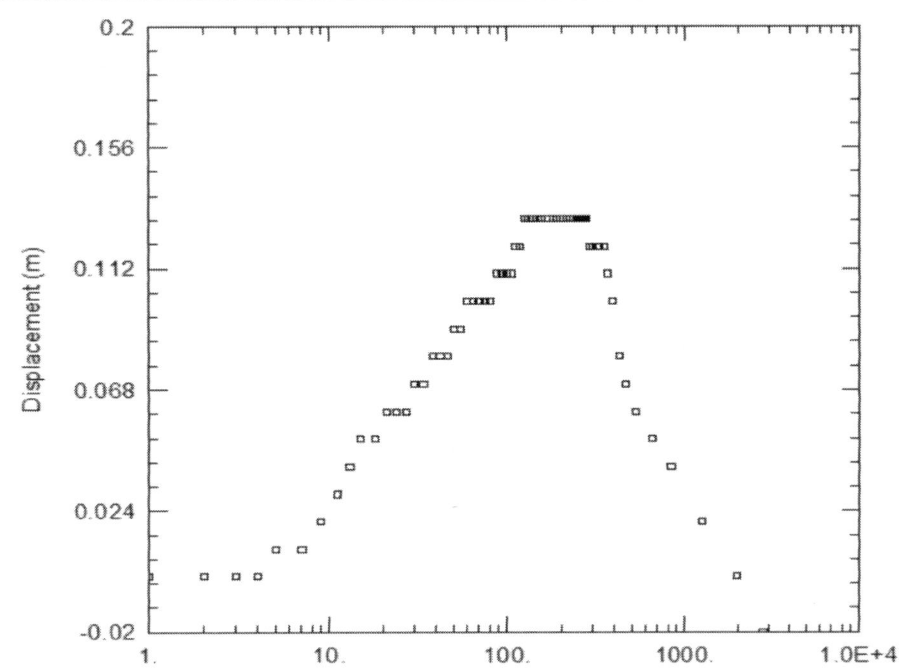

Figure 2. Radial flows for pumping and recovery activities.

A RW data test based on this software analysis was conducted to evaluate the type of aquifer in the areawhich is done by examining radial flow distributions plot. These estimations and simulations were conducted in the aquifer test, which were basically outputs from REWES model. Predictive data is also included in order to be able to predict the discharge record values for long-term needs. Figure 2 shows the confirmation of the assumption of radial flow in an unconfined aquifer with delayed gravity response until 10 minutes. Although the wells in this case was built as penetrate partially, thus, the Moench method [10] was employed to obtain first-cut estimates of aquifer properties.

This pumping analysis showed that the radial flows within 4 hours were achieved around 0.135m with rapid recovery time compared to pumping time. Meanwhile, the water table in this area remains high while the pumping was discharge more than 4 hours continuously.

The second step in choosing the preliminary model is to identify whether a well is a partially or a fully penetrating well and a partially penetrating well, the Moench method is applied. The solution for an unconfined aquifer type was selected based on the required parameters of the program. The visual curve been examined in order to update the initially defined parameter. The match curve is manually drawn on the distribution data to fit the output. The updated parameter on the right side is then verified to determine the suitable values for *S* nearest to 0.005. This is to ensure the accuracy in the visual curve simulating. Data analysis is continued with matching curve manually with the Neuman (1974) solution for an unconfined aquifer with delayed gravity response as shown in Table 3 as an output validation.

Table 3. Preliminary outputs for testing analysis

AQTESOLV	RW	
Analysis	Test	Simulation
Method	Moench	Moench
Aquifer Model	Unconfined	Unconfined
Transmissivity, T	1482 m^2/day	1482 m^2/day
Storativity, S	0.557	0.004955
Beta or water compressibility, β	0.1	0.1
Specific yield, S_y	0.1	0.1
Hydraulic conductivity, K = T/b	32.93 m/day	32.93 m/day
Specific storage, S_s = S/b	0.001238 1/m	0.001101 1/m
Alfa, α	1×10^{30} m^{-1}	1×10^{30} m^{-1}
Radius of outside well, r(w)	0.075 m	0.075 m
Radius of inside well, r(c)	0.074 m	0.074 m

Preliminary data simulation is employed to fit the selected model. Beta value of 0.1, early time (type A) curve matching for Neuman method is used to find the values of *T* and the *S*, and late time (type B) curve matching is adopted to update *T* and S_Y. This process is repeated until a requirements accurate value are satisfied. The diagnostic details are shown in Table 4.

Table 4. Outputs for RW analysis

AQTESOLV Diagnostic	RW Simulation
Method / Aquifer Model	Nueman / Unconfined
T	3.121 m^2/day
S	0.1
β	1×10^{-5}
S_y	0.5
K_z/K_r	0.001246
K = T/b	0.06937 m/day
S_s = S/b	0.00222
Root Sum Square, RSS	37.81 m^2
variance	0.5816 m^2
standard deviation	0.7627 m

3.1 Forward Solution.
The discharge evaluation for long-term pumping is shown in Figures 3 and 4 for RW and OW, respectively. Simulation for pumping forecasting was conducted at dry time to ensure that the minimum values (40 m^3/day) were produced at the discharge rate. For example, discharge at the RW with 1 horsepower pump was 40 m^3/day compared with OW, which was 16 m^3/day more (58 m^3/day), with twice the horsepower of pump for RW. These values reflect the average for discharges over 40 m^3/day for both wells over a seven-day rolling for daily average discharge rate. These forecast values were analyzed to determine the potential or minimum discharge values to forecast for long term pumping activities. Therefore, the quantity and capacity of water intake has shown that this area able to withdrawal more than 40 m^3/day with a constant rate after 4 days pumping continuously.

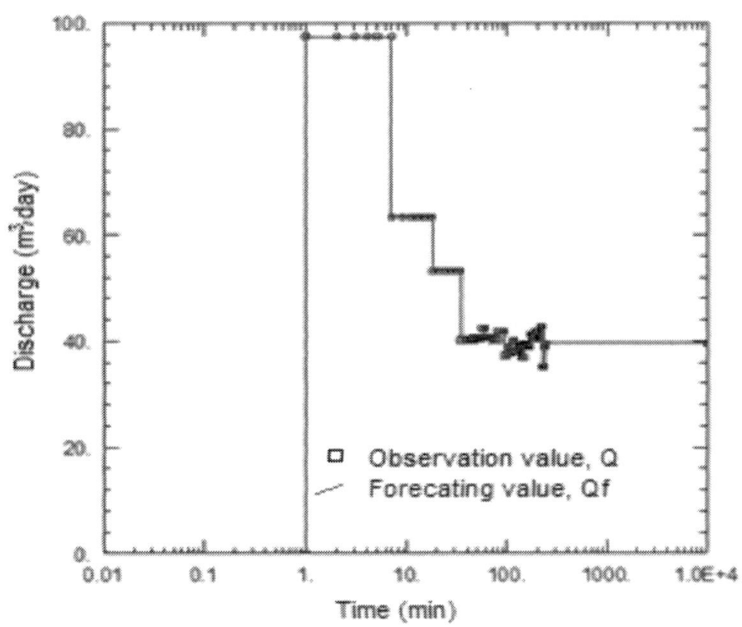

Figure 3. Discharge forecast for recharge well during dry condition in a week-long evaluation

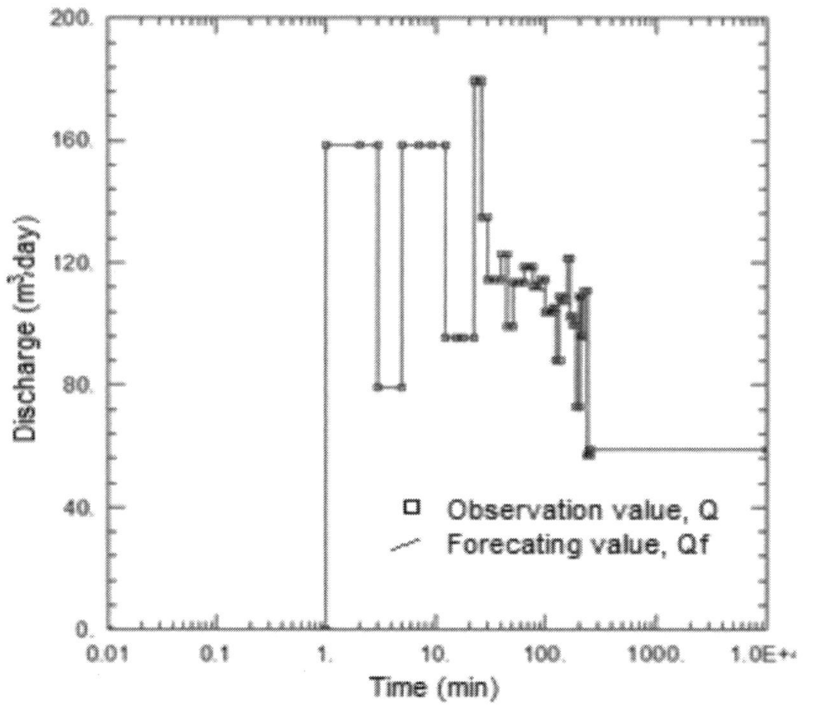

Figure 4. Observed forecast for OW discharge during dry condition in a week

In response to the wet condition, the flow rate of ground water recharge increases because of the increase in the water table and the mounting pressure. Hydraulic conductivity and discharge also have increased values. For every pumping set, two monitoring wells evaluate the aquifer type in a clay point. Almost all parameters have similar values, except for the anisotropy ratio (*Kz/Kr* = ratio of the hydraulic conductivity in horizontal and vertical flows) values, which always vary depending on flow movement during discharge time. The anisotropy ratio (*Kz/Kr*) shows the relationship between the aquifer type and the distance from the pumped well effluences the flows during the pumping activities.

The ratio become bigger with nearest monitoring well compared to long distance. For example, water withdrawal at RW has an anisotropy ratio of 12.46 at EW, with 4 m of a monitored well, compared with OW, which has an anisotropy ratio of 0.1312 with 35 m from RW for the same values in a condition. Also, response to a pumped well at OW, which is monitored in respective monitor wells, covers a 35 m to 39 m range and a ratio between 0.104 and 0.1312. This ratio shows that flow rate movement is capable of increasing the values based on the nearest distance and water content.

The pumped well from RW, the nearest monitored well is EW with 4 m of distance while pumped well from OW, the nearest monitored well is RW with 35 m of distance. Therefore, the nearest monitored well and analysis more accurate outputs for analysis use. The EW outputs were accepted for available record and use for both conditions for RW and RW are for OW analysis.

3.2 Long term estimation.

One pumping response for 8 hours at RW during the dry condition shows a delayed response in monitoring records after 30 minutes of pumping. The comparison pumping analysis in Figure 5 is a continuous step for long duration was performed to estimate the available capacity at dry condition to support the water supply later. It also available use for evaluate the aquifer characteristics in unconfined type because the observation values has good data recorded and profiles of water level of pumped well and monitored wells. Monitoring level decreased depending on the distance. For example, at EW located at only 4 m, the level is reduced after 30 minutes of pumping. MW1, located at approximately 10 m, the pumping can reduce the level after 480 minutes of pumping. This natural response is related to coverage width and to the condition of an aquifer. The minimum distance for this clay area was built 4 m from pumped well and maximum distance for monitoring well is 39 m to covers the observations data.

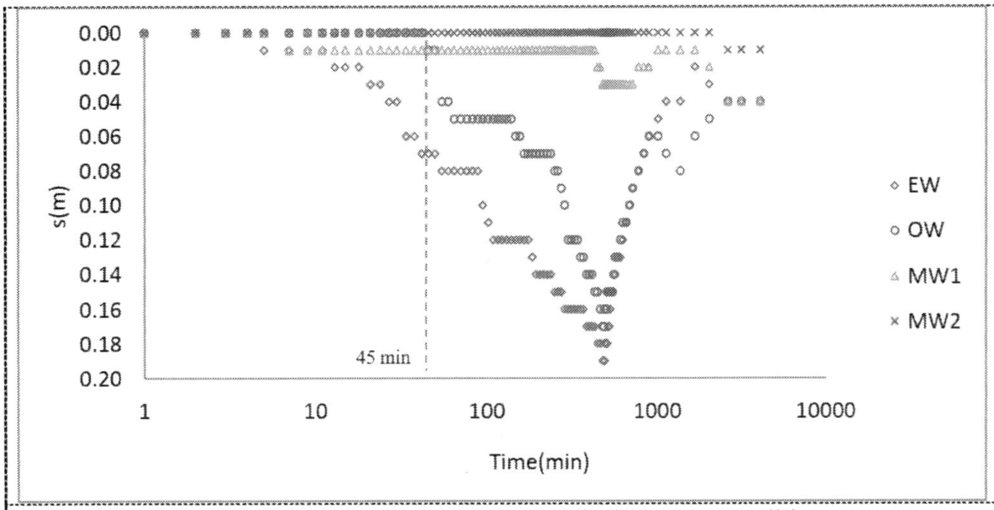

Figure 5. Long-term monitoring wells during dry condition

3.3 Clayey responses. In a clayey aquifer, there is delay response to distance. A clayey aquifer also has physical factors that reduce the level in a short period. Monitoring at OW covers 35 m from a pumped well, thus reducing the level after 45 minutes of RW pumping. This condition implies that the accuracy level (same level of construction well) is applicable if the monitoring wells achieve functionality within the aquifer or in the same level as the pumped well.

Figure 6 is plotted based on real site data to formulate functionality of discharge capacity and flow record in the clayey area. Then, the type of soil and flow rate under the ground influences the hydraulic conductivity values. It also depends on the pumping activities, including type of pump used, flow rate of discharge, water table response, recharge rate, type of soil, and remaining time. Equation (1) has good response in order to manage the time constrain to withdrawal capacity for any duration. It means that, the drawdown occurs at 5.7 minutes for 1 m distance of monitoring. However, the actual drawdown for a clayey unconfined aquifer in this area is assumed as;

$$D_t = -0.00001 d_m^{\ 2} + 0.034 d_m + 5.647 \tag{1}$$

where;

D_t = Drawdown duration in time (min)

d_m = distance from pumped well (m)

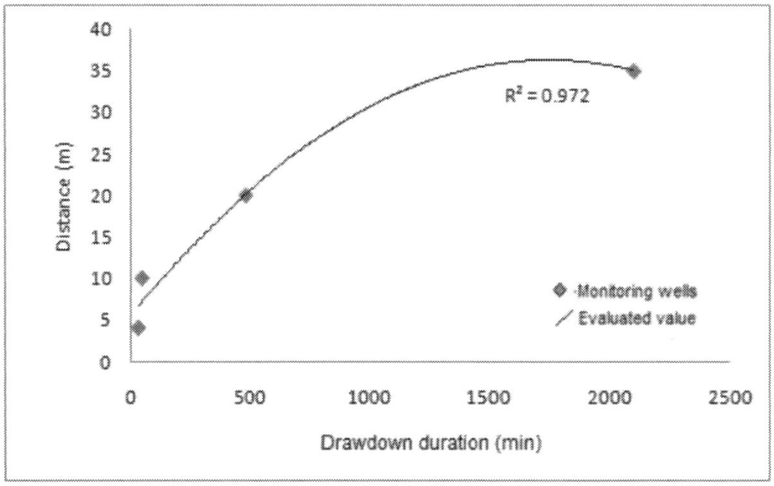

Figure 6. Polynomial relationship between duration time and distance observation at recharge well during long-term pumping at dry condition

Then, linear concepts are used in the coverage of side effects to predict the affected discharge area. This groundwater extraction needs to be controlled to ensure that its excessive consumption will not contribute to the negative effects of natural hydrology. Extraction flow rate could be controlled based on area level decrease (drawdown) and rate of recharge. Therefore, the water table control can be used in this case based on power of pump flow rate as;

$$A = 1000 p_w + 2200 \tag{2}$$

where;
A = coverage area (m^2)
p_w = horse power of pump (w)

3.4 Potential and perspective aquifer responses.
Lateral pumping and recovery techniques are widely used to estimate aquifer properties in a simple solution. Many methods are currently used for basic aquifer estimation. The results are very encouraging to the findings of the researcher because of the ability of this technique in identifying the characteristics and types of aquifers. This study also focuses on multi-step of pumping analysis that could be used to obtain the natural response from ground properties and supported factors. Difference steps analyze has real response to the system. These multi-step analyze applied to the system in a clayey aquifer or in an unconfined layer, provide good feedback as well as potential responses to drainage system management.

4. Conclusion
This linearity concept was useful to formulate the estimation capacity and applicable for any power of pump. But, the recovery was limited the water quantity attracted because the clay area has low hydraulic conductivity below than 9.6 mm/hr. Natural (rainfall) or man-made recharge is considered to show the water balance in the system. Recharge quantity in clayey soil is not the same as discharge amount at RW because recharge works in gravity and separately with natural flow in the system, is commonly used in long-term travel, especially in the case of clayey soil. Therefore, this analysis shows that water well can be applied at clayey area for groundwater extraction in terms of domestics and small industries use.

References

[1] Musa S, Zakaria N A, Lai S H, and Tjahjanto D,2010. The Drainage Exchange of Stormwater Potential in Flat Area Problems *J. of Environmental Science and Engineering*, ISSN 1934-8932 **4** pp1-8.

[2] Musa S, Zakaria NA, Tjahjanto D, and Lai SH,2009. The Potential of Recharge Well System in Flat Area with Low Infiltration Rate, *Proceedings of the International Conference on Water Resources* Langkawi Malaysia pp 5.

[3] Musa S, Zakaria N A, Hin L S, and Md. Ali Z,2010.An Authentication the Soil Type of Top Surface Using Tension Infiltrometer, *World Engineering Congress 2010*, Conference on Engineering and Technology Education, Kuching, Sarawak, Malaysia.

[4] Tjahjanto D, Musa S, and Ridzuan M B,2008.A Study on Artificial Recharge Well as a Part of Drainage System and Water Supply in UTHM.*National Seminar on Sustainable Development: Ecological, Economical and Social Aspects, Sabak Bernam, Malaysia*; pp. 106-111.

[5] Tjahjanto D, Baharuddin M F T, Madun A, and Ab. Latiff, AA,2006. Water Availability and Storm Water Control Using Artificial Recharge on Ground Water. *National Conferences on Water for Sustainable Development Towards, Port Dickson, Malaysia*, pp. 5.

[6] Logananthan, Kiat A N, and Sulaiman Z A,1987. Groundwater Reseources Study of the Sri Gading Area, Johor and Rembau-Tampin-Gemas Area, Negeri Sembilan, Fourth Malaysia Plan (1981-1985) Project, *Geology Survey Depertment*, Ministry of Primary Industries.

[7] Maju Teknik Kota Sdn. Bhd. 2007. Laporan Kerja-kerja Pembinaan Telaga Tiub UTHM 1 di Universiti Tun Hussein Onn Malaysia (UTHM) Parit Raja, Batu Pahat,Johor.

[8] Mohammed T A, and Ghazali A H,2009. Evaluation of Yield and Groundwater Quality for Selected Wells in Malaysia. Pertanika *Journal Science and Technology*. Vol. 17, No 1, pp. 33-42.

[9] Kamra S K,2004. Enhancing Groundwater Recharge in North- West Indo-Gangetic Plains, Groundwater Use in North-West India–Workshop Papers. *Centre for Advancement of Sustainable Agriculture*, New Delhi, pp. 202.

[10] Moench A F,1997. Flow to a Well in a Water-table Aquifer: An Improved Laplace Transform Solution, *Ground Water*, Vol. 34, No. 4, pp. 593-596.

IOP Conf. Series: Journal of Physics: Conf. Series **1049** (2018) 012080 doi:10.1088/1742-6596/1049/1/012080

2D Platform-based Game of Arithmetic Game-based Learning

Zoe Yee Tze Yun[1], Noorhaniza Wahid[1], Norhanifah Murli[1], Rahayu A Hamid[1], Muhammad Fakri Othman[1]

[1]Department of Multimedia, Faculty of Computer Science and Information Technology, Universiti Tun Hussein Onn Malaysia, Parit Raja, 86400 Batu Pahat, Johor, Malaysia

E-mail:nhaniza@uthm.edu.my

Abstract. Mobile games have attracted attention among game developers and users. Throughout the years, mobile games have developed to become one of the domination of the digital world. They have evolved from just a non-coloured with few animated dots and lines to a coloured and more realistic three-dimensional gameplay. Besides, mobile games do not only act as an entertainment but also as a medium for improving mental arithmetic skill if the content is designed accordingly. However, most of mobile games nowadays have been developed for leisure purpose only. There is less game on IQ test and education while more on skills and action alternatively. Hence, arithmetic game-based learning application is developed as an alternative to be included in the variety of gameplay or content of the game. The game created based on platform genre with endless running style. Arithmetic questions are provided in the gameplay as one of the obstacles to both provide challenges and IQ test for the players. This game is implemented in mobile platform. Technology Acceptance Model (TAM) is adopted to measure the game quality based on the evaluation of user acceptance level towards the gameplay, functionality and playability, and overall performance of the game. Overall, 50% of respondents agreed that the gameplay is satisfying. In addition, 40% of respondents strongly agreed that the functionality and playability of the game is stable and the overall performance of the game functions perfectly.

1. Introduction

A mobile game is a game which characterized as game played in handheld devices with or without network functionality [1]. Besides, it can come in preloaded and also can be downloaded by users from any available store such as App Store and Google Play or the internet. Nowadays, mobile phones specifically smartphones have a broad series of connectivity features like Bluetooth, Wi-Fi and 3G. These type of technologies allow users to even engage in wireless multiplayer games which can involve two or more players. There are several types of mobile game genres available such as adventure, arcade, platform, puzzle, strategy, shooter, role-playing, simulation, sports, survival, and hybrids [2]. Among them, platform-based game genre is one of the most popular game genre in market place. This game genre involves the movement of the character through a series of platforms to avoid obstacles while collecting items if there is any available. Traditionally, platform-based games are side-on 2-dimensional (2D) in perspective and very popular on earlier gaming platforms. Playing digital

Content from this work may be used under the terms of the Creative Commons Attribution 3.0 licence. Any further distribution of this work must maintain attribution to the author(s) and the title of the work, journal citation and DOI.
Published under licence by IOP Publishing Ltd

games has become a popular pastime amongst children. On average, children aged 8–18 spend 50 min per day playing digital games [3]. Playing digital games also reported as the second most common activity after finishing schoolwork [4].

However, mobile games have indirectly become an issue as the games developed are more violent-based [5]. Some of the games in store have also caused feelings of aggression and violence to exist within an individual. The simulated violence of certain genre of games may affect a player's thoughts, feelings and physical arousal. Hence, this in turn develops a short term and probably a long term effect on a single person's judgment of an aggressive act [6]. Not only that, there are people who get addicted to the games too easily once played repetitively. In addition, other games were mainly developed simply for leisure purpose with no values added. Therefore, a 2D platform-based mobile game-based learning application, called Defeatobot is developed as an alternative to be included in the variety of gameplay or content of the game. In addition, the gameplay of Defeatobot is a simple one-button game in which players have to drive a character to pass the obstacles. The game contains arithmetic questions as one of the obstacles in which player has to answer in a fast and precise way. Thus, the gameplay is hoped to enhance students' arithmetic skills.

The rest of the paper is presented as follows: related work discusses about mobile games as learning strategy in arithmetic. Next, the methodology describes the mobile development process, followed by the section that discusses the results. Finally, the concluding remarks are presented.

2. Related Works

Mobile games have played a crucial part in human life since the existence of it. However, mobile games did not obtain wide recognition even though they have existed for a long time until the appearance of new colour handsets with better usability, sound and Java technology. Not only that, mobile games have received a lot of attention due to the popularity of smart devices since the improvement in the design and features of the current games. Mobile games have been widely applied as an alternative approach in learning strategy. Shin *et al.* [7] investigate the effect of game technology on elementary students learning mathematics. In this study, students who played technology-based arithmetic games outperformed students who played paper-based arithmetic games. Research also found that, playing games can lead to gain in accuracy and calculation speed and improvement in attitude towards mathematics in 10 to 11 years-olds [8]. Findings in [8] also confirmed by [9] where playing an educational games could help second grade pupils improve their arithmetic skills as well as can increase their accuracy compared to paper exercises or no exercise. Moreover, the mathematics games can increase mental calculation speed in a similar way as an equivalent number of paper exercises. In terms of affective outcomes, playing the game was reported to be more enjoyable than filling out the paper exercises [9].

3. Methodology

Defeatobot mobile game application is developed in Android based operating system. Game Development Life Cycle (GDLC) [10] is applied to develop the game application. There are six phases involved in the development of Defeatobot game: initiation, pre-production, production, testing, beta and release. Detail explanation of each phase will be discussed in the following topic.

3.1. Initiation phase
Initiation phase involves creating a rough concept of what kind of game that will be created. The output of initiation is the game concept and a simple game description. Table 1 shows a sample of the game plan document produced for Defeatobot game.

Table 1. Defeatobot game plan document

Item	Description
Game Concept	2D side-scrolling, Platform Game, endless game style
Game Theme	Educational Game
Educational Content Value	Drill and practice about arithmetic
Character Name	Robot
Obstacles	Mechanical gearbox, arithmetic questions (randomly generated)
Other features	Boosters and game life
Environment	Famous landmarks and skyscrapers of Malaysia
Storyline	Defeatobot is a game that requires player to navigate the Robot past the obstacles (mechanical gearbox) while answering the arithmetic questions throughout the gameplay. In the beginning, the player is given 3 lives. If the player obtained certain score, power-ups booster will come out to help player get rid of the obstacles as well as increasing the score if the player managed to answer the next arithmetic questions correctly.

3.2. Pre-production phase
Pre-production involves game design and creating game prototype. Game design focuses on defining and preparing game genre, gameplay, mechanics, storyline, characters, challenges, fun factors, technical aspects, and game design document (GDD) or storyboard. Defeatobot game implements single playing mode that is offline playable. The setting of this game focuses on local content with the famous landmarks and skyscrapers in Malaysia, which are displayed in a more futuristic style by using a friendly robot as the main character of the game. Those landmarks are presented as game scene background. The graphics creation was completely in 2D graphics design as it is of a parallax scrolling style of gameplay. Thus, it does not require the game to be in three dimensional environment. In terms of gameplay strategy, the game structure is designed as an endless running where player has to play until it reaches zero live before they can start a new gameplay. In this game, the player must consistently control the character to fly upward. Player has to drive the character through the game by passing the obstacles while gathering as many collectibles as they can. As for the obstacles, this game uses gearbox and arithmetic questions which appear at random position to obstruct the player. The player needs to escape the arithmetic obstacles by choosing the correct answer before it reaches the character. Each question has four options of answer. Once the player chooses the correct answer, the obstacle itself will disappear and thus, the player can continue with the gameplay. Any correct answer will gain 10 marks and vice versa. The purpose of applying arithmetic obstacles is to activate player's mental arithmetic while increasing their calculation speed. Besides having obstacles, this game also provides power-up booster to player in order to accomplish the mission. Meanwhile, there are two iterations happened in this phase: first iteration produced a form of prototype's foundations and structure; second iteration produced a complete prototype.

3.3 Production phase
Production phase is the main activity which involves the assets creation, source codes creation, and the integration of both elements. The output of this phase is a formal and refined prototype. Production phase involves the game development starts by preparing all kinds of game assets for example, text, graphics, buttons, characters and audio. Those assets are created using specific software. The game assets are arranged according to the planning done in the storyboard. The assets such as character, props and buttons are scaled to the appropriate sizes and organized correctly in particular arrangement

of the game application. After completing the assets creation, integration of all assets are performed to produce complete scenes as shown in Figure 1 and Figure 2.

The game consists of seven scenes such as introduction, main menu, instruction, high score, about game, credits, and gameplay. Each scene holds a certain amount of buttons and functions. For example, the main menu scene has four buttons namely play, high score, credits and about game, as shown in Figure 1. They are arranged according to the standard arrangement of game main menu. As for the gameplay scene, it has two buttons which include pause and 'fly' buttons as in Figure 2. They are organized at the positions where it can be visible to player. Figure 2 depicts the gameplay when player meets the arithmetic obstacle, multitasking action should be performed. The player needs to control the character by tapping the 'fly' button and at the same time answer the arithmetic question correctly in a shortest time. At the end of the game, a high scoreboard of the game is displayed. Only the top five player will be listed on the scoreboard.

Figure 1. Main menu interface of Defeatobot game

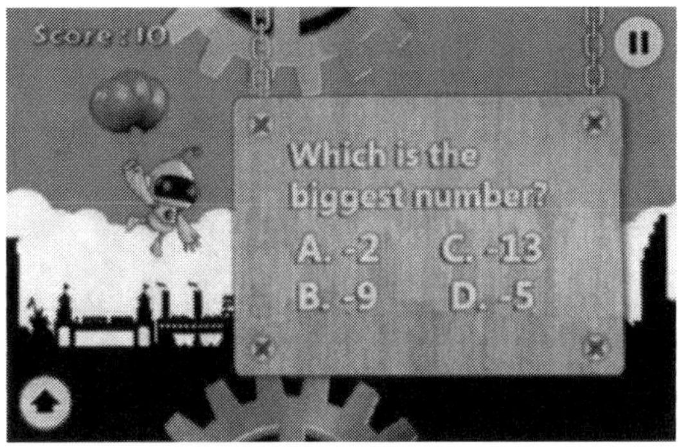

Figure 2. Gameplay of Defeatobot game

```
public void FlyUp()
{
  if (canFly = true) {
    rigidhody2D.AddForce (Vector2.up * flyingSpeed);}
}
void FixedUpdate() {
  temp = transform.position;
  temp.y = Mathf.Clamp (temp.y,-10f,2.69f);
  transform.position = temp;
  if (lives > 0) {
    transform.Translate(new
Vector3(1,0,0)*Time.deltaTime*forwardSpeed);}
  if (Input.GetKeyDown ("space")) {
    rigidbody2D.AddForce (Vector2.up * flyingSpeed);}
}
```

Figure 3. Code segment for player movement

```
public class MovingBG : MonoBehaviour
{
  public float scrollspeed = 1f;
  private float offset = 0f;

  void Update () {
      offset += CalculatingVelocity.dispSpeed *
      scrollspeed;
      renderer.material.SetTextureOffset ("MainTex",new
      Vector3 (offset, 0));
  }
}
```

Figure 4. Code segment for parallax scrolling of background

Next, single source code for the prototype is created by using C# programming language. Figure 3 depicts the code segment which allows character to be controlled by player in the game. The character can only move to the right continuously, or fly up or fall down during the gameplay. Besides that, the background of the gameplay was integrated by using parallax scrolling of a game effect as shown in Figure 4. After completing the source code creation, the assets and scripts are integrated in Unity software to execute the application in the correct way. Unity 3D software is used as game engine that enable user to interact in mobile platform. C# programming language is chosen to be used in Unity to implement the game application. Android Application Package (*.apk*) file is needed for the Android platform installation. This script linked the objects and scenes together, and allowed the player to interact with the game and navigate through the application.

3.4 Testing phase
In this phase, formal details testing (alpha testing) are performed using playtest by developer internally to assess the game functionality and playability. Game functionality testing involves game stability, integration of game assets. Any bugs, loopholes, or dead-ends during playtesting are discovered and fixed. Meanwhile, game playability testing involves testing the ease and correctness of tapping method used to control the main character and navigate through the game environment.

3.5 Beta phase
Beta phase is to conduct external tester called beta testing. Technology Acceptance Model (TAM)[11],[12] is adopted to measure user acceptance level in terms of gameplay, functionality and playability, and overall performance. The game is released to a limited number of users in online and offline form. Online beta testing was carried out by sending the *.apk* file to 20 target users via Google Drive. The respondents downloaded the file into their Android device and answered the questionnaire in Google form. Offline beta testing was conducted by 10 target users within the organization.

3.6 Release phase
Release phase is the time when the game build has reached final stage and ready to be released to public. However, Defeatobot game application is released internally within the organization for educational purpose.

Table 2. Technology Acceptance Model for evaluating game quality

Construct	Evaluation variables	Measured item
Perceived of usefulness (**PU**)	Gameplay (arithmetic obstacles)	**PU1**: I found that this game is fun and interesting to be played **PU2**: I found that this game is challenging to be played **PU3**: I found that this game is able to activate my arithmetic skill
Attribute of usability (**AU**)	Functionality and playability	**AU1**: The game is stable when played in mobile device **AU2**: The control buttons work perfectly in the game **AU3**: The integration of game assets is implemented perfectly
User satisfaction (**US**)	Overall performance	**US1**: The navigation of the game is easy to understand and control **US2**: The screen design is suitable with the theme featured in the game **US3**: The words and illustrations are suitable in each screen display **US4**: The presentation of the information and instruction is clear **US5**: The aesthetic/art value in the game is good **US6**: The icon buttons are compatible with the their respective function **US7**: The overall game application functions perfectly

4. Results and Discussion

The game application was successfully built with the size of 1280 x 800 pixels resolution and less than 40MB memory size. User testing was carried out to evaluate three variables based on TAM: users' acceptance level towards the gameplay approach (arithmetic obstacles), functionality and overall performance. 30 respondents of the age of 16-34 were involved in the testing. Table 2 shows the construct and measured items adopted from TAM to evaluate the game quality.

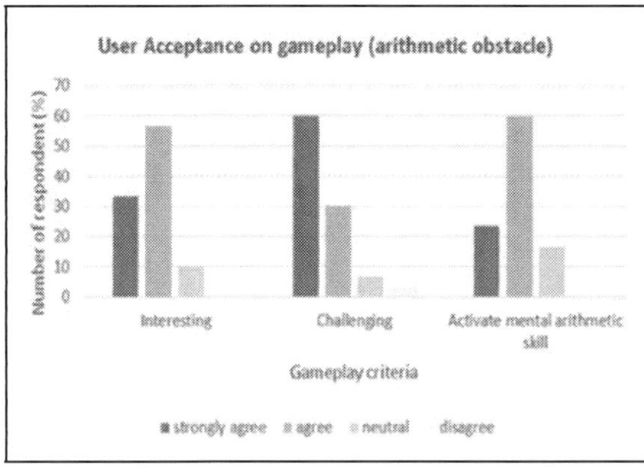

Figure 5. Gameplay approach with arithmetic obstacle

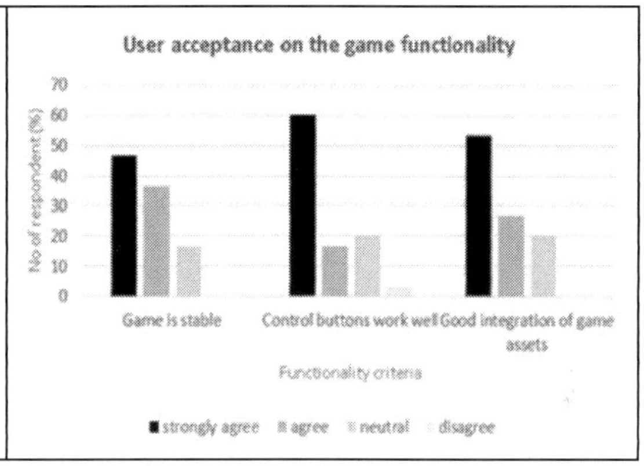

Figure 6. Game functionality and playability

Figure 5 shows that 57% respondents agreed while 33% strongly agreed that the arithmetic obstacles are interesting. Furthermore, 60% respondents agreed and strongly agreed the game is challenging and activate mental arithmetic skill, respectively. Conversely, only 3% respondents disagreed that the game was challenging. In terms of game functionality and playability as in Figure 6, more than 40% respondents strongly agreed that the game is stable. Interestingly, more than 50% respondents strongly agreed that the game has good button control and integration of game assets. Conversely, 3% respondents disagreed that the game has good button control. This is due to unfamiliar game control using respective control button. Meanwhile, Figure 7 shows that more than 40% respondents strongly agreed that overall game performance criteria worked well and have been achieved the game objectives.

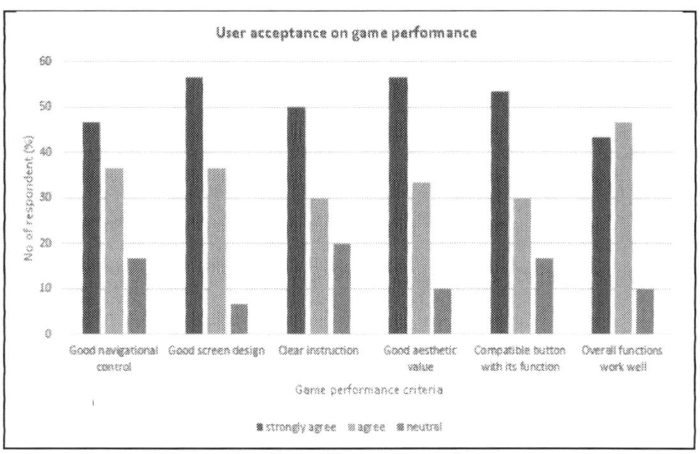

Figure 7. Overall game performance

5. Conclusion and Future Works

Overall, it can be concluded that the implementation of arithmetic obstacles as the gameplay in Defeatobot mobile game application has potential to be used as an alternative way of learning approach. Compare to other casual mobile game, Defeatobot is able to stimulate user to think and having fun while playing the game. Future enhancement is needed to improve user acceptance criteria in terms of its gameplay criteria as well as increasing performance level criteria for more convenient and user friendly mobile game application.

Acknowledgement

This study has been supported in part of the Short Term Grant (STG) Vot U367 under the Malaysia Ministry of Higher Education (MOHE) and Universiti Tun Hussein Onn Malaysia (UTHM). This research also supported by GATES IT Solution Sdn. Bhd. under its publication scheme.

References

[1] E. J. Jeong and D. J. Kim, Definitions, Key Characteristics, and Generations of Mobile Games Mobile computing: concepts, methodologies, tools, and applications, Information Science Reference, 2009.

[2] T. Stahl, "Video game genres", 2005, Available at http://www.thocp.net/ software/ games/ reference/ genres.htm

[3] D. J. Miller and D. P. Robertson, "Educational benefits of using game consoles in a primary classroom: a randomised controlled trial," British Journal of Educational Technology, Vol. 42, No. 5, pp 850–864, 2011.

[4] F.o.A. Scientists, Harnessing the Power of Video Games for Learning, in: Summit on Educational Games, 2006.

[5] D. A. Gentile, "The effects of violent video game habits on adolescent hostility, aggressive behaviors, and school performance," Journal of Adolescence, Vol. 27, No. 1, pp 5-22, 2004.

[6] T. A. Kooijmans, "Effects of video games on aggressive thoughts and behaviors during development", 2004, Available at http://www.personalityresearch.org/papers/kooijmans.html

[7] Shin N et al., "Effects of game technology on elementary student learning in mathematics," British Journal of Educational Technology, Vol. 43, No. 4, pp 540-560, 2011.

[8] S. Livingstone and L. Haddon, EU Kids Online: Final Report LSE: EU Kids, 2009.

[9] E. N. Castellar et al., "Improving arithmetic skills through gameplay: Assessment of the effectiveness of an educational game in terms of cognitive and affective learning outcomes," Information Sciences, Vol. 264, pp 19–31, 2014.

[10] R. Ramadan and Y. Widayani, "Game Development Life Cycle Guidelines," Proc. International Conference on Advanced Computer Science and Information Systems (ICACSIS), 2013.

[11] H. M. Abu-Dalbouh, "A Questionnaire Approach Based on the Technology Acceptance Model for Mobile Tracking on Patient Progress Applications," Journal of Computer Science, Vol. 9, No. 6, pp 763-770, 2013.

[12] C. S. Wang, Y. M. Huang and K.S. Hsu, "Developing a mobile game to support students in learning color mixing in design education," Advances in Mechanical Engineering, Vol. 9, No. 2, pp 1–6, 2017.

IOP Conf. Series: Journal of Physics: Conf. Series **1049** (2018) 012081 doi:10.1088/1742-6596/1049/1/012081

Fortune Run: A Mobile Game Showcasing Cultural Celebration in Malaysia

Chong Yong Khong[1], Rahayu A Hamid[1], Noorhaniza Wahid[1], Norhanifah Murli[1], Muhammad Fakri Othman[1]

[1]Department of Multimedia, Faculty of Computer Science and Information Technology, Universiti Tun Hussein Onn Malaysia, Parit Raja,86400 Batu Pahat, Johor, Malaysia

E-mail: rahayu@uthm.edu.my

Abstract. Mobile games have become one of the most popular entertainment medium nowadays. Mobile games with wide variety of genres can be found and downloaded easily through digital marketplace such as Google Play and Apps Store. Elements of cultural festival has also been developed as mobile games. However, the amount of existing mobile games that contain Malaysia's local cultural content are limited. Malaysia is a unique country consists of multiple ethnics and cultures. Therefore, a game has been developed in this project to introduce the unique elements of Malaysia's culture to the user. The game is created based on the theme of Chinese New Year celebration in Malaysia. Game Development Life Cycle (GDLC) was used as a methodology. The game is implemented in mobile platform and in English language. The game uses a 2D side scrolling design approach where swiping method is used to control the game character. Technology Acceptance Model (TAM) is adapted to measure user acceptance level in terms of game design, functionality and overall game performance. Results of user acceptance test shows that 50 % of respondents agreed that the developed mobile game able to provide general idea of how Chinese New Year is celebrated in Malaysia.

1. Introduction

Mobile game is a video game played on mobile devices such as featured phone, smartphone and tablet computer [1]. However, game features designed for mobile devices are not as rich as Personal Computers (PCs) or consoles platform because mobile devices have limited system resources [1]. Various game approaches have been applied in mobile games development. One of the most popular game approaches is 2-dimensional (2D) side scrolling where user can move the character from one side of the screen to the other through side camera view. Although games are created with the goal to provide entertainment, it has implicitly encourage its players to learn [2]. Nowadays, learning about cultural heritage [3] and festivals has also been made available through mobile games. Nevertheless, only a few of these mobile games contain Malaysia's local content.

Malaysia is a multicultural country with many unique elements. Festivals celebration is one of the unique elements that unite all races in Malaysia. In Malaysia, Chinese New Year is not only celebrated by the

Content from this work may be used under the terms of the Creative Commons Attribution 3.0 licence. Any further distribution of this work must maintain attribution to the author(s) and the title of the work, journal citation and DOI.
Published under licence by IOP Publishing Ltd

IOP Conf. Series: Journal of Physics: Conf. Series **1049** (2018) 012081 doi:10.1088/1742-6596/1049/1/012081

Chinese, but also celebrated by other races together. This unique element of Malaysia maybe known by majority of the Malaysians, but there is still a possibility of some Malaysians or even foreigners have no knowledge about it. Among the downloadable 2D side scrolling mobile games, only few of them contain Malaysia's local cultural content. Therefore, Fortune Run mobile game features Malaysia's local content that can provide some general ideas about what Malaysia looks like. Moreover, the theme of Chinese New Year celebration should be able to introduce the environment of Chinese New Year festival in Malaysia to the user.

The rest of the paper is presented as follows: related work discusses about mobile game development and comparison of existing games with similar game approach. Next, the methodology describes the mobile development process, followed by the section that discusses the results. Finally, the concluding remarks are presented.

2. Related Work

Chinese New Year, or also known as 'Spring Festival' is an important festival celebrated by the Chinese all around the world. The first day of the Chinese New Year falls on the first day of the first month in Chinese calendar and the celebration last for 15 days. Each year of the Chinese calendar is named after one of the 12 animals of the Chinese Zodiac [4]. In Malaysia, Chinese New Year is not only celebrated by the Chinese but also celebrated together by other races. During the festival, people of other races will visit and greet their Chinese friends or neighbours.

Three 2D side scrolling mobile games which are similar are analysed in Table 1. The developed application is named as (c) Fortune Run, is compared with 2 other existing applications, which are (a) Chhota Bheem Jungle Run [5] and (b) Nasi Lemak Delivery [6]. From the table, we can identify that Fortune Run contains more game features compared with the other two similar mobile games. The features in Fortune Run mobile games consist of boosters, virtual game shop, four levels with different game environments and three character lives. Besides that, Fortune Run uses swiping control method where player can control the character by swiping up or down on the mobile screen. Most importantly, Fortune Run features the combination of Chinese New Year theme and Malaysia's local content.

Table 1. Comparison between 2D side-scrolling games

Function	**Chhota Bheem Jungle Run**	**Nasi Lemak Delivery**	**Fortune Run**
Goals	Control the character to travel as far as possible, collect coins and avoid obstacles.	Control the character to deliver Nasi Lemak, collect coins and energy, and avoid obstacles.	Control the character to collect red envelopes, complete in-game objective and avoid obstacles.
Malaysia's Local Content	Not available	Kampung scene	Chinese New Year celebration
Boosters	Available	Not available	Available
Virtual game shop	Available	Not available	Available
Game level	One level	Ni ne levels	Four levels
Game environment	Jungle	Village	Village, city, residential area and in-house
Number of character lives	One	One	Three

3 Methodology

Fortune Run game application is developed in Android based operating system using GDLC methodology [7]. There are six phases involved in the development of Fortune Run game: initiation, pre-production, production, testing, beta and release. Details of each phase will be discussed in the following topic.

3.1 Initiation

In the initiation phase, a rough concept of what kind of game that will be created is first identified.
The output of initation is the game concept and a simple game description. Game description for the Fortune Run game can be referred to in Table 1.

3.2 Pre-production

Pre-production involves the creation of game prototype. The prototype functions to assess the game design and the whole idea. Game design focuses on defining game genre, gameplay, mechanics, storyline, characters, challenges, fun factors, technical aspects, and storyboard. Fortune Run is created as a single player offline mode. The setting of the game depicts the celebration of Chinese New Year in four different game environments in Malaysia. The game uses graphics created in 2D. Fortune Run was developed as an action game as it contains series of actions such as jumping and running. The game also requires fast responses from the user to control the game character. 2D side scrolling approach was implemented in this game development. In addition, the game uses swiping control method for character controlling where player can control the character by swiping up and down on the mobile screen. Fortune Run consists of four game levels; each with its own environment and associated obstacles. The game environments can be referred to in Table 1. The list of obstacles includes chicken and rock (village), traffic cone and road sign (city), cat and water splash (residential area) and beverage cans and mandarin peel (in-house). In the game, the player needs to collect as many red packets as they can and in doing so need to avoid the obstacles. Once the player has collected the required amount of red packets, they will unlock and proceed to the next level. The game finishes in the fourth level with the setting of firecrackers.

3.3 Production

Production is the main process which involves the assets creation, source codes creation, and the integration of both elements. The output of this phase is a formal and refined prototype. The game consists of eight scenes which are main menu, level selection menu, tutorial, game shop and game levels 1 until 4. Each scene contains game assets that are related to the theme of Chinese New Year. Figure 1 shows a few scenes from the Fortune Run game. The game assets created in this phase included graphical arts and game audio. The graphical arts developed in this project include game character, game environments, obstacles, collectables and buttons. All of the graphical arts were developed using Adobe Flash CS6. The graphical arts was painted using colours that match the theme of Chinese New Year. After the game assets were created, they were imported into Unity. The graphical assets were dragged into the scene and scaled to suitable size using the scaling tool in Unity. Then, the graphical assets were arranged and aligned in appropriate position according to the storyboard. Figure 2 shows the assets arrangement of main menu using Unity.

Integration between game assets and scripts is a process of attaching the script files to the game assets through game engine. In this project, Unity was used to integrate the game elements together. Scripting in Unity is implemented using C# script language to execute series of actions or commands. In the game, different scripts were created to execute commands such as to control character, generate game props and calculate score. After the scripts were created, they were then attached to the corresponding object in Unity. Figure 3 shows the attachment of script to the game character. Figure 4 shows the partial code in character control script that used to control the character movements by swiping up and down on the screen.

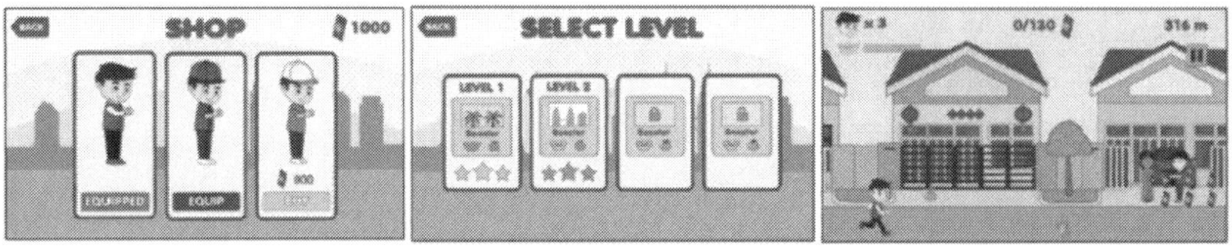

(a) (b) (c)

Figure 1. Fortune Run game: (a) game shop, (b) level selection menu, (c) level 3 game environment

Figure 2. Graphical Assets Arrangement of Main Menu using Unity

Figure 3. Character Control Script Attached to Game Character

IOP Conf. Series: Journal of Physics: Conf. Series **1049** (2018) 012081 doi:10.1088/1742-6596/1049/1/012081

```
if (Input.touchCount > 0)
{
    Touch touch = Input.GetTouch(0);
    if (touch.phase == TouchPhase.Began) {
        startPos = touch.position;

    } else if (touch.phase == TouchPhase.Ended)
    {
        endPos = touch.position;

            Vector2 distance = endPos - startPos;
            if (Mathf.Abs (distance.x) < Mathf.Abs (distance.y)) {

                if (distance.y > 0 && grounded) {
                    characterRigidBody.velocity = new Vector2(characterRigidBody.velocity.x, jumpForce);
                    jumpSound.Play();
                    Debug.Log ("Up");
                }

                if (distance.y < 0 && !grounded) {
                    characterRigidBody.velocity = new Vector2(characterRigidBody.velocity.x, -jumpForce/2);
                    Debug.Log ("Down");
                }
            }

    }
}
```

Figure 4. Partial Code Used to Control Game Character.

Game building is the last step of production. Firstly, the publish platform was set to Android and all the game scenes were loaded into the build list. Then, platform settings such as resolution, orientation, game icon and title were set. Finally, the game installer with Android Package Kit (APK) format was created.

3.4 Testing phase

Testing phase is the phase where alpha testing was conducted to test the game functionality and playability. A formal details testing plan were created for the purpose. Testing are conducted using features playtest to assess the game functionality and playability. Functionality involves testing whether the game assets such as main character, boosters and buttons functions correctly. Game playability involves testing the ease and correctness of swiping method used to control the main character and navigate through the game environment. Any bugs, loopholes, or dead-ends discovered during playtesting are documented in the testing plan and fixed.

3.5 Beta phase

Beta phase is where the beta test of a game is conducted. Beta testing still using the same testing methods as the previous testing method. The purposes of beta testing are to find any bugs remained in the game and collect satisfaction feedback from the testers. In this project, Technology Acceptance Model (TAM) [8] is adopted to measure user acceptance level in terms of game design, functionality and overall game performance. Therefore, a closed beta test was conducted among 20 selected target users. The testers were required to provide feedback about the game by answering a questionnaire. Questionnaires that consist of evaluation on user acceptance, functionality and design of the mobile game were created and distributed to the testers.

3.6 Release

Release phase is the time when the game build has reached final stage and ready to be released to public. However, Fortune Run game application is released internally within the organization for educational purpose

IOP Conf. Series: Journal of Physics: Conf. Series **1049** (2018) 012081 doi:10.1088/1742-6596/1049/1/012081

4. Results and Discussion

User testing was carried out to evaluate three variables based on TAM users' acceptance level towards the game design, functionality and overall game performance. 20 respondents of the age of 16-23 were involved in the testing. Table 2 shows the construct and measured items adopted from TAM to evaluate the game quality.

Table 2. Technology Acceptance Model for evaluating game quality

Construct	Evaluation variables	Measured item
Perceived of usefulness (**PU**)	Game design	**PU1**: I found that the game gives general idea of Chinese New Year celebration in Malaysia. **PU2**: I found that the game design matches the theme of Chinese New Year celebration in Malaysia. **PU3**: I found that the colours used matches the theme of Chinese New Year celebration in Malaysia. **PU4**: I found that the graphical arts used matches the theme of Chinese New Year celebration in Malaysia.
Attribute of usability (**AU**)	Functionality and playability	**AU1**: The main character is easy to control. **AU2**: The movement of main character is smooth. **AU3**: The game has a smooth navigation.
User satisfaction (**US**)	Overall performance	**US1**: The game is fun to play with. **US2**: The game is easy to be understood. **US3**: The game runs smoothly on Android device **US4**: The buttons functions correctly **US5**: The game instruction are clear.

Figure 5 shows that more than 50% of the respondents agreed and the rest strongly agreed that the game design provides general idea of how Chinese New Year is celebrated in Malaysia. Similarly more than 80% of the respondents strongly agreed that the game design and the use of colours and graphical arts are suitable for the theme of Chinese New Year celebration. From the game environment, respondents were able to identify and associate graphical assets used with the festival celebration.

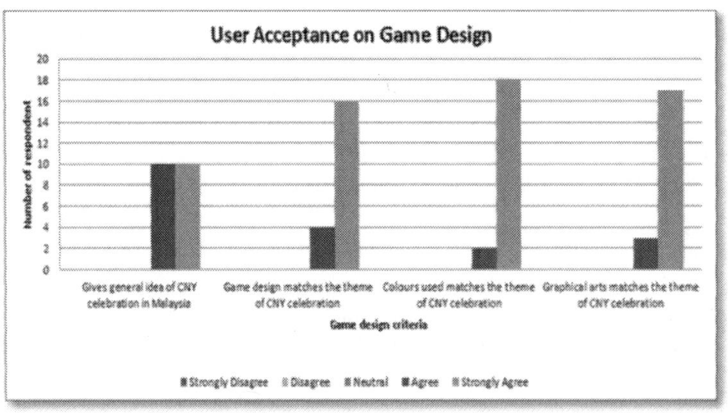

Figure 5. Game design

Results of user acceptance test on functionality (Figure 6) shows that 65% of respondents agreed that the game character is easy to be controlled. However, 15% of the respondents have neutral opinion on the difficulty of character control. This is probably due to unfamiliarity of the swiping method used in the game, as compared to the normally used tapping method. Nevertheless, more than 60% of respondents agreed that both character movement and game navigation is smooth. Meanwhile Figure 7 shows the results of the game's overall performance whereby more than 40% of respondents strongly agreed that the game is fun to be played and can be easily understood. Moreover, the respondents were satisfied with the game as it runs smoothly on the device, have fully functional buttons and clear game instruction.

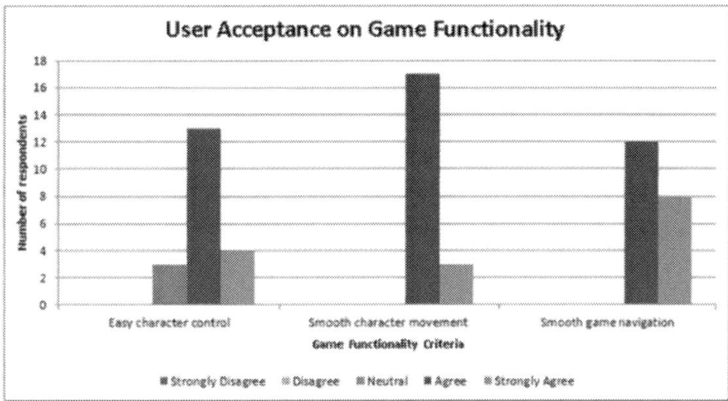

Figure 6. Game functionality

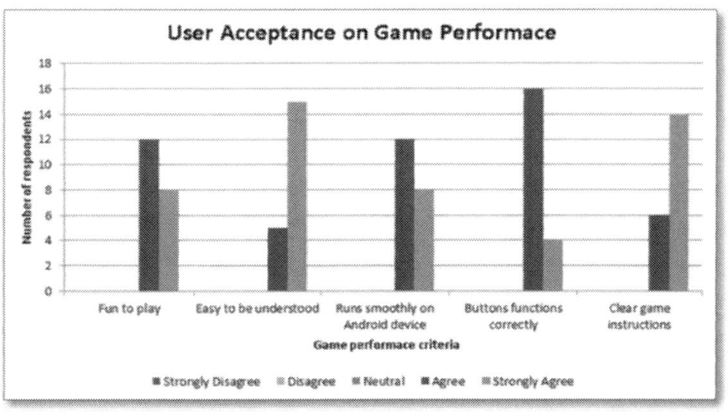

Figure 7. Overall game performance

5. Conclusion and Future Works

Fortune Run mobile game has been successfully developed to showcase the celebration of Chinese New Year in Malaysia. Compared to other locally theme games, Fortune Run captures the spirit and environment of the occasion by incorporating important elements such as red packets, mandarin and fire crackers in the game. As the game was developed as an offline single player mode, future works would include adding online scoreboard so that user can compete with other players in real time.

Acknowledgement

This study has been supported in part by the Short Term Grant (STG) Vot U367 under the Malaysian Ministry of Higher Education (MOHE) and Universiti Tun Hussein Onn Malaysia. It is also supported by GATES IT Solution Sdn. Bhd. Under its publication scheme.

References

[1] C. Janssen. "Techopedia Explains Mobile Games". 2013, Available at: http://www.techopedia.com

[2] J.P. Gee, "What Video Games Have to Teach Us About Learning and Literacy". Computers in Entertainment, - Theoretical and Practical Computer Applications in Entertainment Vol. 1, No. 1, pp, 20-20 DOI=http://dx.doi.org/10.1145/950566.950595

[3] J. Barwick, J. Dearnley, and A. Muir. "Playing games with cultural heritage: a comparative case study analysis of the current status of digital game preservation". Games and Culture, Vol. 6, No. 4, pp. 373-390, 2011.

[4] History.com Staff (2010). "Chinese New Year". 2010, Available at: http://www.history.com/topics/holidays/chinese-new-year

[5] Nazara Technologies (2015). "Chhota Bheem Jungle Run". 2015, Available at https://play.google.com/store/apps/details?id=com.junglerunnazara.com.junglerun.nazara&hl=en

[6] Y. P. Chong, "Nasi Lemak Delivery". Final Year Report Bachelor Degree.Universiti Tun Hussein Onn Malaysia, 2015.

[7] R. Ramadan and Y. Widayani, "Game Development Life Cycle Guidelines," Proc. International Conference on Advanced Computer Science and Information Systems (ICACSIS), 2013.

[8] C. S. Wang, Y. M. Huang and K.S. Hsu, "Developing a mobile game to support students in learning color mixing in design education," Advances in Mechanical Engineering, Vol. 9, No. 2, pp 1–6, 2017.

IOP Conf. Series: Journal of Physics: Conf. Series **1049** (2018) 012082 doi:10.1088/1742-6596/1049/1/012082

Voice Converter on Android

W Y Lee, N Murli, N Wahid, R A.Hamid, M F Othman
Department of Multimedia, Faculty of Computer Science and Information Technology, Universiti Tun Hussein Onn Malaysia, Parit Raja, 86400 Batu Pahat, Johor, Malaysia
E-mail: hanifah@uthm.edu.my

Abstract. Voice Converter is an application developed for Android mobile platform that can help user to change his voice the way he likes. Existing voice changer mobile application provides the functions for users to change their voice into the voice of limited preset templates, this cannot see to the needs of film or animation producers, because the templates are limited and the uniqueness of the production cannot be secured. With the Voice Converter mobile application, film or animation producers can easily have access to a number of sound models that can be used to convert audio files which can be used in production later. Besides that, film makers can custom the sound of the conversion by inputting desired values of playback rate to change the audio sound in the way they would like to. The application uses speaker and microphone of the phones to play and record audio. For the recording process, the 'MediaRecorder' APIs is used to allow the device to capture audio, save the audio and play it back. The recordings captured can be converted into various sound models and also a model with custom parameters. The SoundPool Library is adopted to alter the playback rate of the audio. A mobile application named Voice Converter which is compatible for Android platforms is successfully developed. The functions of the application include recording and saving audio, converting input audio into existing template sound and converting recorded audio with custom parameter

1. Introduction

Sound can be referred to as vibration that propagates through a transmission medium such as gas, liquid or solid matter and can be heard by human's or animal's ear. It is generated from a sound source, where it creates vibrations and these vibrations travels away from the source to the destination in the form of sound wave. Voice in the other hand is the sound uttered through the mouth of living creatures, especially human being in speaking, singing, and shouting. The mechanism for generating human voice can be divided into three parts: the lungs, the vocal cord within the larynx, and the articulators. In the lungs, air is exhaled to create an airstream in the trachea and across the larynx, which is often called the voice box.

When air passes through, the vocal cords vibrate very quickly to produce sound [1]. The articulators act as filters that to some degree can interact with the airflow to control the strength of the sound source. Every human being has their unique voice due to a number of factors. The frequency of the vibration of the vocal cords, the length and width of the vocal cords, the shape and size of the resonance chambers, are among the factors that make a person's voice unique [2]. A voice changer program is a program or an application that alters a person's voice to either make them sound like someone else or disguise their voice. Voice changer applications work by changing the tone or pitch, to change the nature of the voice into that of another.

Film and animation production involve various post-production activities, and one of them is the voice dubbing process. With the aid of the Voice Converter mobile application in which audio file can be converted into model with custom parameters, the work will be made easy without having to worry much about the voice actor's condition.

Content from this work may be used under the terms of the Creative Commons Attribution 3.0 licence. Any further distribution of this work must maintain attribution to the author(s) and the title of the work, journal citation and DOI.
Published under licence by IOP Publishing Ltd

This paper is organized as follows. In Section 2, the related works are presented. In Section 3, the development of the Voice Converter is discussed. In Section 4 discusses the implementation of voice converter application and finally, future works and conclusion of this project is summarized in Section 5.

2. Related Works

Voice plays an important role in movie and film production, which makes up a big part of entertainment for people around the world. In some drama series or animated series (anime), dubbing is a core post-production process. However, problem arises as drama or anime series filmed over years, and the children characters in the series remain children, while the voice actor has already aged by a few years and may have undergone puberty and have their vocal line changed. In these cases, a new voice actor with similar vocal line is needed but in most cases the new voice actor will not provide a voice identical to the previous voice actor and more or less will affect the quality of the series.

Existing voice changer mobile application provides the functions for users to change their voice into the voice of very limited pre-set templates such as Voice Changer Mobile Application [3], Voice Changer with Effects Mobile Application [4] and Best Voice Changer Mobile Application [5]. This cannot fulfil the needs of professional film or animation producers, because the templates are limited and the uniqueness of the production cannot be secured. With this voice changer application, users are allowed to record and convert human voice into a custom sound. Some software on the computer provides similar functions in professional approaches but users have to be in a particular location to access the hardware. This is inconvenient when the voice actors have difficulties to present on the spot. The Voice Converter mobile application however provides the convenience to the users where they can access the function anytime and anywhere.

The following section will discuss about the concepts and techniques that is employed in the development of the Voice Converter mobile application.

2.1. MediaRecorder API

MediaRecorder is a set of Application Program Interface (API) for Android Mobile Platform. The set of programming algorithms allows the recording of media such as video and audio and for us to access the microphone and storage on an Android device. The Voice Converter mobile application will adopt this algorithm as the basic technique to allow recording of media on Android devices as it is simple to use and is compatible on most Android mobile devices.

2.2 SoundPool Library

SoundPool library is a Java class that manages and plays audio resources for applications. It is a collection of samples that can be loaded into memory from a resource inside the APK. The SoundPool library uses the MediaPlayer service to decode the audio into a raw 16-bit PCM mono or stereo stream. This allows applications to ship with compressed streams without having to suffer the CPU load and latency of decompressing during playback. SoundPool is capable of setting the audio to loop and also change the playback rate of an audio when playing it. The SoundPool class allows loading a number of sound clips into a device's memory where they can be played on demand. It even allows playing a number of sound clips at the same time.

3. Methodology

The development of this application consists of 3 phases: system design, interface design and application development. The process model adopted in the development of the Voice Converter mobile application is the Waterfall Process Model. It allows for departmentalization and control of the project. A schedule can be set with deadlines for each stage of development to ensure that a phase is completed before moving to the next. Table 1 shows the phases and activities involved in the development of the project.

3.1 System Design

The Voice Converter mobile application consists of only two modules, which are the "Convert" and "Custom" interfaces. The "Convert" module provides the function of recording the voice of user and the conversion of the recorded voice into another voice, while the "Custom" module allows user to change their voice with the tempo and pitch-shift values input by themselves. Figure 1(a) shows the interface of the Convert Module – Audio Input. This interface provides functions for user to record audio, playback the recorded audio, and redo the recording if they would like to before moving to the next interface. This allows user to record their voice for many attempts before they proceed with the one they are satisfied with. Figure 1(b) shows the second interface of Convert Module Interface – Model Selection. This interface provides different sound models for the users to choose from (e.g. Slow-mo, Speed up, Helium, Witch, Ghost, Rhino, Chipmunk, Male, Female, Demon, or Robot). By pressing on the button, the recorded audio will be altered and played. Figure 1(c) shows the record interface of the Custom module. This is the first interface of the module that allows the user to record and playback their voice. Figure 1(d) shows the second interface of Custom module. In this interface, users are required to input the playback rate of the audio. Users then press the convert button to convert and play the audio.

(a) (b) (c) (d)

Figure 1. (a) Convert Module Interface – Audio Input, (b) Convert Module Interface – Model Selection, (c) Custom Module Interface – Audio Input; and (d) Custom Module Interface – Convert

3.2 Implementation of Convert Module

The Convert Module allows users to record their voice, to layback, to convert their voice into a list of sound models, and to save and to playback the converted audio. The recording function is implemented in Android Studio using the MediaRecorder API which is a Java-based API. It allows the application to access the device microphone to record audio. Shown in the appendix are the code segments for the recording function in Android.

For the purpose of declaration, the recorder bit-rate is set to 16-bit and the recordings format is set to .wav. A recording temporary file will be created during the recording process. The sample rate and channels of the recordings are also declared. Figure 2 shows the code segment for the implementation of the startRecording function method, in which a new MediaRecorder is introduced. Next the recorder function is called to start the recording and the state of recording is set to True. Finally, the writeAudioDataToFile function is called to write recorded audio data to storage, and to save the recorded audio into the device. Figure 3 shows the code segment of stopRecording function, the recording state is first set to False, the recorder is called to stop recording and the whole recorder function is called to End. In the next function interface of the "Convert" module as shown in Figure 1(b), user can choose from a number of sound models that he or she wants to convert to. A SoundPool library is also introduced to the class, so that the alteration of playback rate of audio can be implemented. The default value for the parameters of soundpool explosion is set to 0. The buttons for different sound models are declared with onclicklisterner that allows them to implement the functions when it is clicked. A new SoundPool function is then declared and the function is set to load the recorded audio that is saved to the device. Figure 4 shows the code segment of the core method of the audio alteration. For different sound models, different parameters for the SoundPool function are defined. The declaration of SoundPool explosion method allows a few parameters to be manipulated when an audio is played. First are the left and right volume for the audio file when played. 1.0 is the maximum value for the playback volume. Next the priority of audio file can be set, with 0 is the lowest priority; the higher the number, the higher the priority. Followed by is the loop value of the audio file: 0 indicates no loop and -1 will set the audio to loop continuously. And finally the playback rate or frequency of the audio. For instance 0.65f indicates a playback rate of 65% of the original audio for the Demon sound model.

```
private void startRecording(){
   recorder = new AudioRecord(MediaRecorder AudioSource.MTC, RECORDER_SAMPLERATE,
      RECORDER_CHANNELS, RECORDER_AUDIO_ENCODING, bufferSize);

   int i = recorder.getState();
   if(i==1)
         recorder.startRecording();

   isRecording = true;
   recordingThread = new Thread(new Runnable(){
         @OverRide
       public void run(){
         writeAudioDataToFile();
         }
   },"AudioRecorder Thread");
recordingThread.start();
}
```

Figure 2. Code Segment for startRecording Method

```
Private void stopRecording(){
   if (null != recorder){
      isRecording = false;

      int i = recorder.getState();
      if(i==1)
         recorder.stop();
      recorder.release();

      recorder = null;
      recordingThread = null;
   }
}
```

Figure 3. Code Segment for startRecording Method

```
public void onClick(View v){
   switch (v.getId()){
      case R.id.imageButton_demon:
         sp.play(explosion, 1, 1, 0, 0,
            .65f);
      break;
      case R.id.imageButton_chipmunk:
         sp.play(explosion, 1, 1, 0, 0,
            2.2f);
      break;
   }
}
```

Figure 4. Code Segment for stopRecording Method

3.3 Implementation of Custom Module

The recording function of the Custom module is implemented using the same codes and functions as in the Convert module. Therefore, this section will discuss about mainly the implementation of converting audio into a custom model. An EditText symbol is used to get the frequency value from the user. A new SoundPool function is then declared and the function is set to load the recorded audio that is saved to the device. In Figure 5 we can see that the SoundPoolfunction to alter the frequency of audio is implemented with a button click. The parameters are predefined except for the playback rate, it is defined by the value input by the user in the application.

IOP Conf. Series: Journal of Physics: Conf. Series **1049** (2018) 012082 doi:10.1088/1742-6596/1049/1/012082

```
public void onClick(View v){
    switch (v.getId()){
        case R.id.buttonConvert:
            float freq = 0.01f*Float.parseFloat(editFreq.getText().toString());
            sp.play(explosion, 1, 1, 0, 0, freq);
        break;
    }
}
```

Figure 5. Code Segment for SoundPool explosion function

3.4 Development of Sound Models

The development of sound models in the Voice Converter mobile application is implemented using Java programming language in Android Studio. The playback rate or frequency of the input audio is altered with different values so that different sound models can be produced. When we increase the playback rate, to 200% of its original value, the audio file will be shortened to half its length and the speech will sound two times faster, the audio will also sound higher by a certain value, which means that the pitch of the audio will also be altered at the same time. In the Voice Converter mobile application, a total of 11 sound models are produced, each of them provides different sound effects and has different parameters. Table 1 below shows the sound models available with their respective playback rate alterations

Table 1. Sound Models with playback rate alteration values

No	Sound Model	Playback rate (% of original audio)
1	Speed-up	200
2	Slow-mo	60.5
3	Helium	180
4	Male	70
5	Robot	120
6	Female	142
7	Demon	65
8	Ghost	40
9	Witch	78
10	Chipmunk	220
11	Rhino	35

4. Results and Discussions

The testing was carried at Batu Pahat mall targeting open users of age 13-30. A total of 30 respondents were involved regardless of their gender and race. Questionnaires were distributed among them to get feedback on their experience of using the application. The data and feedback collected are then statistically analyzed based on the three categories of questions that are user acceptance, usability and design. The results are tabulated using bar charts as shown in Figure 6 – 8 below. As in Figure 6, almost all respondents agree and strongly agree that the application is fun and easy to use and gives good experience to them. In terms of the overall design (Figure 7), all respondents find that the sound models produced are attractive (100%), the use of buttons are appropriate (90%) and the selection of colours are good and appropriately chosen (70%). And finally, referring to Figure 8, the functionality level of the application is rather at a satisfying level. Most users responded positively, i.e. they strongly agree and agree that the application run wells on Android devices and the buttons function properly. Most of them also agree that the navigation is simple (80%), the application provides quality conversions (70%) and the sound models are fun and useful (60%).

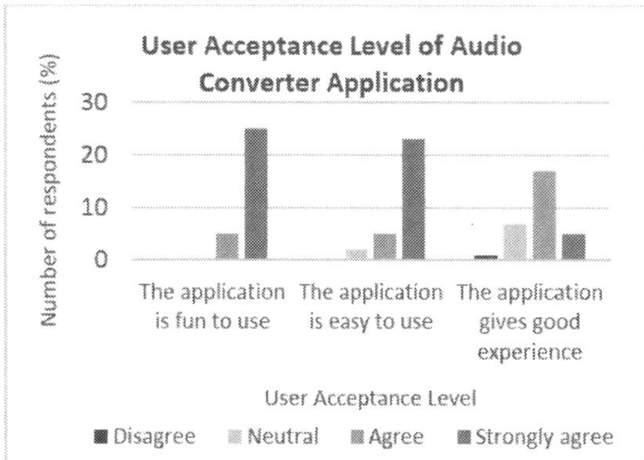

Figure 6. User acceptance level of Audio Converter Application

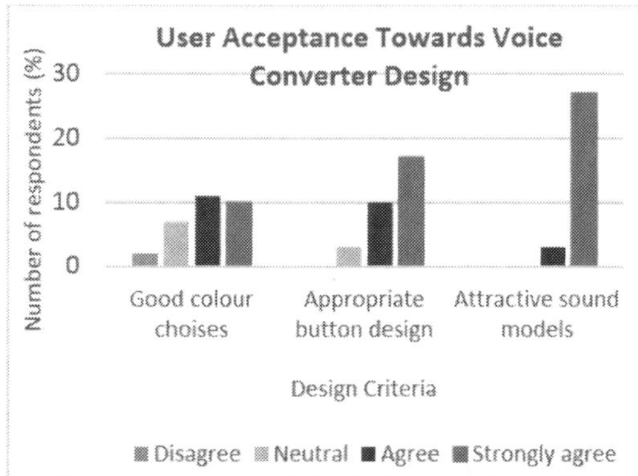

Figure 7. User Acceptance Towards Voice Converter Design

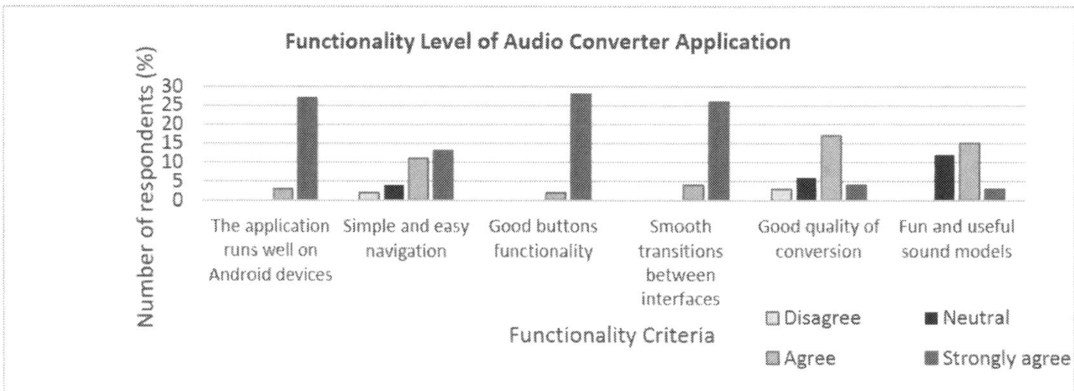

Figure 8. Functionality level of Audio Converter Application

5. Conclusion and Future Works

The Voice Converter mobile application possess some advantages compared to similar application as the user can input the value to alter the frequency of the audio themselves. On the other hand, the application has also some critical limitations, one of them is that the application does not allow users to save the converted audio file to the device.

The future works of the mobile application involve solving limitations and providing extra features. The Voice Converter mobile application is a production developed in a short timeframe and under limited resources, which gives it a big room for improvement in the future. Main functions like saving the converted audio to the device and also importing existing audio file to be converted should be allowed instead of having to record every time they wish to convert. Instead of having user to input the frequency, sliders shall be created to ease users and avoid invalid input. The application also shall be developed to operate on iOS platform so that it can be made available to more users and larger market. More manipulating properties for the audio other than playback rate, for example altering the pitch and tempo, or adding echo to the audio, this can help to produce more sound models and ensure the application continuously provide fun and usefulness to the users.

Acknowledgement
This study has been supported in part of the Short Term Grant (STG) Vot.U367 under the Malaysia Ministry of Higher Education (MOHE) and Universiti Tun Hussein Onn Malaysia (UTHM). This research also supported by GATES IT Solution Sdn. Bhd. under its publication scheme.

References

[1] The Propagation of sound Retrieved June 2015 from http://pages.jh.edu/~virtlab/ray/acoustic.htm
[2] Kampwirth K 2013 What Determines What Your Voice Sounds Like (Retrieved September 2016) from http://mentalfloss.com/
[3] Voice Changer Mobile Application (Retrieved September 2016) from https://play.google.com/store/apps/details?id=com.androidrocker.voicechanger
[4] Voice Changer with Effects Mobile Application (Retrieved September 2016) from https://play.google.com/store/apps/details?id=com.baviux.voicechanger
[5] Best Voice Changer Mobile Application (Retrieved September 2016) from https://play.google.com/store/apps/details?id=com.scoompa.voicechanger

IOP Conf. Series: Journal of Physics: Conf. Series **1049** (2018) 012083 doi:10.1088/1742-6596/1049/1/012083

Investigation of Circular Patch Metasurface (MS) on Inverted Suspended Circular Polarized Antenna

H. A. Bakar, M. Z. A. Abd. Aziz, B. H. Ahmad and N. Hassan
Center for Telecommunication Research and Innovation (CeTRI), Faculty of Electronics and Computer Engineering (FKEKK), Universiti Teknikal Malaysia Melaka (UTeM), 76100 Durian Tunggal, Melaka, Malaysia

E-mail: hamizan421@yahoo.com

Abstract. In this paper, an analysis of different stage of circle patch metasurface (MS) on basic structure of inverted suspended circular patch with square slot antenna design is proposed to investigate its effect toward the basic antenna performance. The basic antenna was designed based on inverted suspended circular patch with air gap separation between the patch substrate and copper ground plane at a distance of 10 mm (Basic design). Then, different design of MS structure (Design A, B, C & D) consisting of 5 x 5 unit cell of homogeneous periodic element is place in between the air gap of the basic antenna. All of MS design is located at a fixed position to study the effect of different MS shape to the Basic antenna performance. The proposed antenna and MS structure had been designed and simulated by using Computer Simulation Technology (CST) Software. Target application for this antenna design is for Wireless Local Area Network (WLAN) at operating frequency of 2.4 GHz. Simulation performances in term of return loss, resonant frequency, bandwidth, realized gain, directivity, axial ratio, total efficiency and radiation pattern at the design frequency are investigated and discussed. From observation on simulation result, resonant frequency of Basic design is shifting to a lower frequency from 2.4 GHz to 2.19 GHz (Design C and D) with the integration of the MS structure. Design C obtained the widest axial ratio bandwidth among all of other design with 1380 MHz covered from 1.31 GHz to 2.69 GHz. Directivity for Basic designs at 2.4 GHz increases from 7.94 dBi to 8.17 dBi, 8.16 dBi, 8.15 dBi and 8.06 dBi (Design A, B, C and D) respectively.

1. Introduction

Antenna is one of the most important equipment in a wireless communication system which enables the transmission and reception of signals through air as medium. Nowadays, with the great requirement for compact wireless devices, a small effective antenna with common structures continues to be a subject of great attraction [1-3]. Other than that, antenna with circularly polarized (CP) [4-6] also getting more attention because they offer flexibility in the orientation angle between the transmitter and receiver. In recent year, a lot of efforts have been made by researchers to study and modified simple antenna design to be integrated with MS structure in order to represent as circular polarization antenna [7,8], compact [9,10] the structure and miniaturize [11,12] the size of planar antenna for wireless communication system application. Metasurface (MS) is a 2-D planar equivalent of metamaterial structure and has receiving researchers interest in the last few years for planar antenna performance enhancement such as wide bandwidth and high gain [13-17]. Other than that, as in

Content from this work may be used under the terms of the Creative Commons Attribution 3.0 licence. Any further distribution of this work must maintain attribution to the author(s) and the title of the work, journal citation and DOI.
Published under licence by IOP Publishing Ltd

[18,19] it demonstrate that the performances of patch antenna can be enhanced simultaneously by the addition of MS at the top of the antenna design.

In this work, it is proposed to use different stage of circle patch unit cell of MS to study the effect of MS structure on Basic antenna behaviour in term of return loss, resonant frequency, bandwidth, return loss bandwidth (RLBW), realized gain, directivity, axial ratio, axial ratio bandwidth (ARBW), total efficiency and radiation pattern. A Basic design of inverted suspended circular patch with square slot (Basic) is proposed to be integrated with the MS structure. All of MS design consist 5 x 5 unit cell of homogeneous periodic element. Design A consist of unit cell of one circle patch, Design B and C consist two combination of circle patch and Design D consist of three combination of circle patch. The configuration of Basic design and design with integration of MS structure is demonstrated as in Figures 1.

2. Antenna Design

In this paper, the basic antenna and the MS structure was designed by using FR4 substrate with thickness, h = 1.6 mm, dielectric constant, ε_r of the substrate = 4.4, tangent loss and tan δ = 0.019. Thickness of copper, t = 0.035 mm is used for the ground plane and as the conductive material printed antenna and MS. The basic circular polarized antenna is an inverted suspended circular patch with square slot as shown in Figure 1 (a). Where the antenna feedline and circular patch with square slot are printed at the bottom sided (back) of the antenna substrate as shown in Figure 1 (c) without any copper patch at other sided (front) of the substrate as shown in Figure 1 (b). The inverted suspended circular patch antenna and its ground are separated by 10 mm air gap as illustrated in Figure 1 (a) side view. The feedline was fed by using 50 Ohm SMA coaxial probe connector thus represent as an inverted suspended L-probe technique for the antenna input port.

Figure 1 (d) side view shows the configuration of basic antenna with the integration of MS in between the antenna air gap. From the perspective view illustrated three different layer of the antenna design where layer 1 consisting of inverted suspended circular patch with square slot and feedline, layer 2 consisting of inverted 5 x 5 unit cell of different MS structure design (Design A, B, C & D) and layer 3 consisting of copper ground plane. The width and length of layer 1, layer 2 and layer 3 is equal that is 80 x 80 mm respectively. The MS position of Design A, B, C and D is fixed as shown in side view of Figure 1 (d) where Air gap 1 = 3 mm and Air gap 2 = 5.4 mm. All of MS design has the same gap (g) that separating the unit cell element of MS between each other centre with 16 mm distance as shown in Figure 1 (e), (f), (g) and (h). The radius (r) of all circle patch MS design is the same that is 5 mm respectively as shown in Figure 1 (i), (j), (k) and (l).

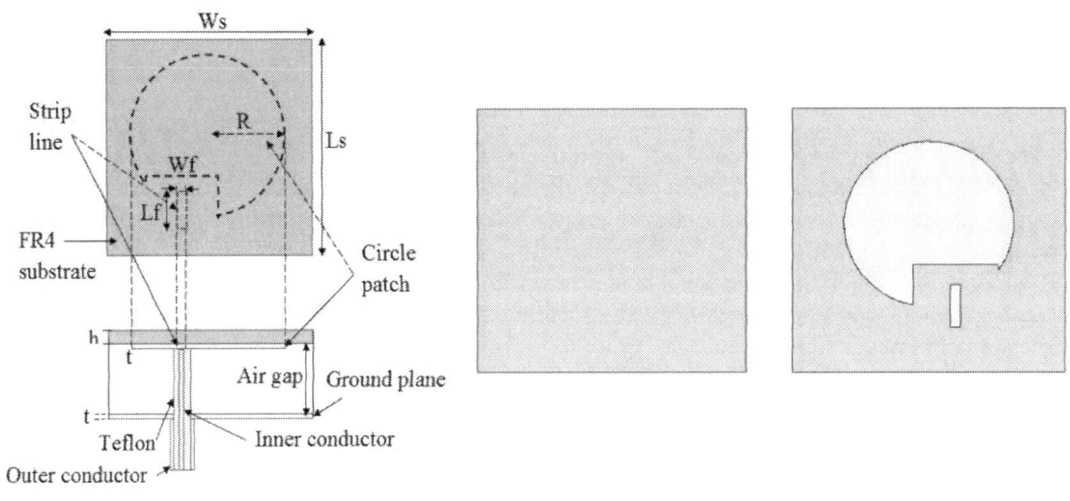

(a) (b) (c)

IOP Conf. Series: Journal of Physics: Conf. Series **1049** (2018) 012083 doi:10.1088/1742-6596/1049/1/012083

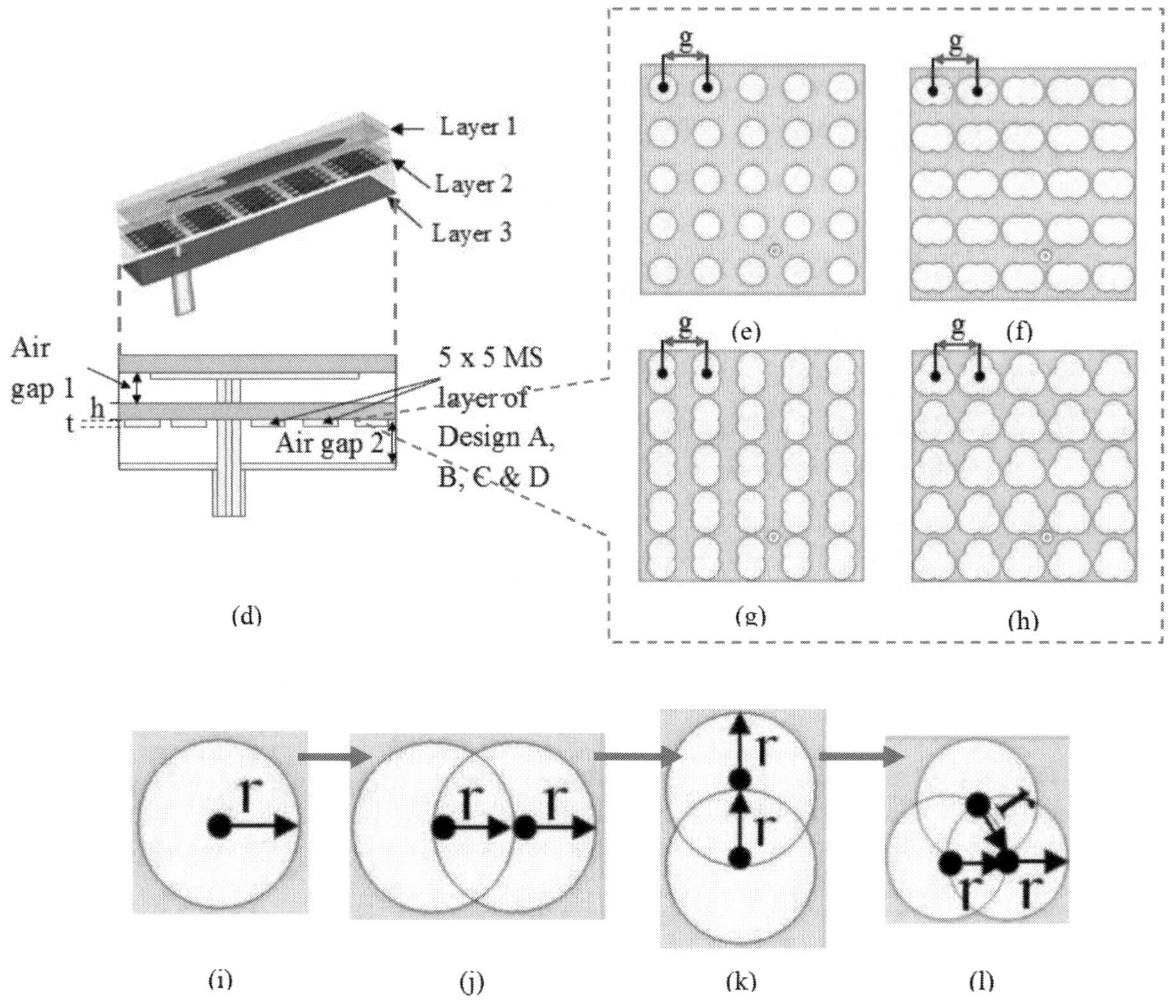

Figure 1. Configuration of Basic design and design with MS structure (a) Basic design top and side view (b) Basic design front view (c) Basic design back view (d) Basic design with MS perspective and side view (e) 5 x 5 unit of Design A (f) 5 x 5 unit of Design B (g) 5 x 5 unit of Design C (h) 5 x 5 unit of Design D (i) single unit MS Design A (j) single unit MS Design B (k) single unit MS Design C (l) single unit MS Design D

The MS is designed as inverted facing toward copper ground plane where the 5 x 5 element unit cell is printed only at the bottom (back) of a single-sided substrate. All of MS design consists of x 5 homogeneous pattern of circle patch MS where each of the unit cells are printed periodically along the x-axis and y-axis. Unit cell of Design A MS has only one circle patch while Design B and Design C has 2 combination of unit circle Design A that periodically in horizontal and vertical position as can be seen in Figure 1 (j) and (k). Design D MS is a combination of 3 unit circle of Design A MS with respect to the radius of circle patch as in Figure 1 (l). The optimize dimension of basic antenna at the frequency of 2.4 GHz and basic antenna with integration of MS are tabulated as in Table 1.

Table 1. Optimize dimension of Basic antenna with and without MS

Design Parameter	Dimension (mm)	Description
Air gap	10	Air gap basic antenna
Air gap 1	3	Air gap MS
Air gap 2	5.4	Air gap MS
h	1.6	Thickness of substrate
t	0.035	Thickness of copper
L_s	80	Length of substrate
L_f	13	Length of feedline
W_s	80	Width of substrate
W_f	3.1	Width of feed
R	25	Radius of basic antenna circle patch
r	5	Radius of metasurface circle patch
g	16	Gap between metasurface

3. Result and discussion

In this section, simulation antenna parameter which comprised of return loss (RL), resonant frequency (f_r), bandwidth (BW), return loss bandwidth (RLBW), realized gain, total efficiency, directivity, axial ratio, axial ratio bandwidth (ARBW) and radiation pattern are analysed and discussed at 2.4 GHz frequency. The antenna performance results of Basic design, Design A, Design B, Design C and Design D from frequency range of 1.2 GHz to 3.5 GHz are compared and demonstrated as in Figure 2 (a) and (b), 3 (a) and (b) and 4. The comparison antenna parameters at 2.4 GHz of all design are tabulated as in Table 2.

3.1. Return loss (RL), Resonant frequency (f_r), bandwidth (BW) and Realized gain

The comparison of simulation S11 and realized gain for Basic design, Design A, B, C and D are illustrated as in Figure 2 (a) and (b). Based on Figure 2 (a), the resonant frequency of Basic design is at 2.398 GHz with -35.64 dB. With the addition of MS, resonant frequency clearly can be observed shifted to a lower frequency at 2.26 GHz (Design A), 2.24 GHz (Design B) and 2.19 GHz (Design C & D). Design A has the most lower resonant frequency with return loss of -39.80 dB compared with other antenna design, while Design D has only -23.16 dB at its resonant frequency. The bandwidth of Basic design is 556 MHz covering from 2.21 GHz to 2.76 GHz. With the addition of MS structure Design A, B, C and D, the bandwidth decreases to 424 MHz, 340 MHz, 334 MHz and 324 MHz respectively.

According to Figure 2 (b), the realized gain of Basic design, Design A, B, C and D at 2.4 GHz are 7.76 dB, 7.74 dB, 7.54 dB, 7.35 dB and 7.15 dB respectively. Even though the resonant frequency are shifting with the addition of MS, the gain of all MS design still more than 7 dB at 2.4 GHz. Maximum gain of Basic design is at 2.5 GHz with 7.96 dB, while maximum gain with the addition of all MS design are at the same frequency which is at 2.3 GHz. All of gain results at 2.3 GHz for MS design are more than 7.7 dB. For a comparison of gain for MS design at 2.3 GHz, Design B has the highest gain with 7.87 dB.

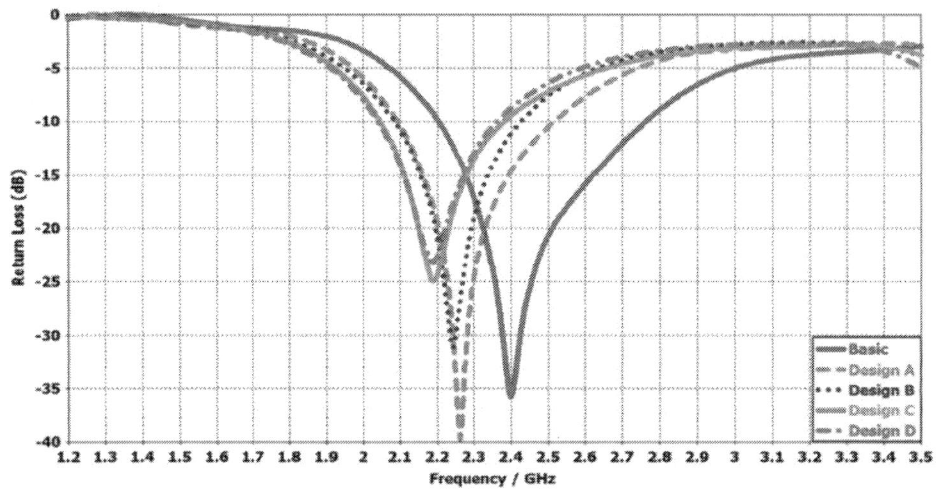

(a)

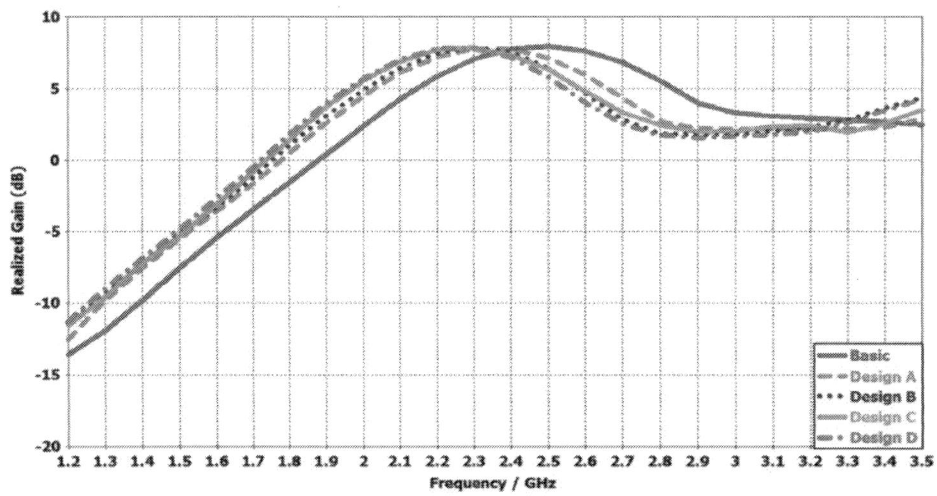

(b)

Figure 2. Comparison simulation result of Basic design, Design A, Design B, Design C and Design D (a) return loss (b) realized gain

3.2. Total efficiency and directivity

Figure 3 (a) and (b) illustrated the simulation view of the comparison result of MS Design A, B, C and D with Basic antenna design for total efficiency and directivity versus frequency. Based on Figure 3 (a), maximum total efficiency of Basic design is at 2.4 GHz frequency which achieved -0.18 dB. With addition of MS Design A, maximum efficiency is shifted to 2.3 GHz with -0.25 dB. Design B, C and D has the same maximum total efficiency frequency, which is shifted 100 MHz from Design A maximum efficiency at 2.2 GHz with -0.26 dB, -0.23 dB and -0.24 dB respectively. Total efficiency above 50 % of Basic design covering frequency range of 2 GHz to 3.3 GHz. While, total efficiency above -3 dB with addition of MS design ranging around 1.85 GHz to 2.98 GHz respectively. Total efficiency of all design at 2.4 GHz is less than -1 dB.

According to Figure 3 (b), the simulation result of directivity for Basic designs at 2.4 GHz increases from 7.94 dBi to 8.17 dBi, 8.16 dBi, 8.15 dBi and 8.06 dBi (Design A, B, C and D). With the addition of MS, maximum directivity of Basic design shift from 2.5 GHz (8.19 dBi) to 2.4 GHz (Design A and B) and 2.3 GHz (Design C and D). The maximum directivity for Design C and D are 8.26 dBi and 8.29 dBi respectively. For a comparison of maximum directivity of all design, Design D obtained the highest directivity. From observation, even though using different MS designs, the performance result between each MS design is quite similar in term of realized gain, total efficiency and directivity.

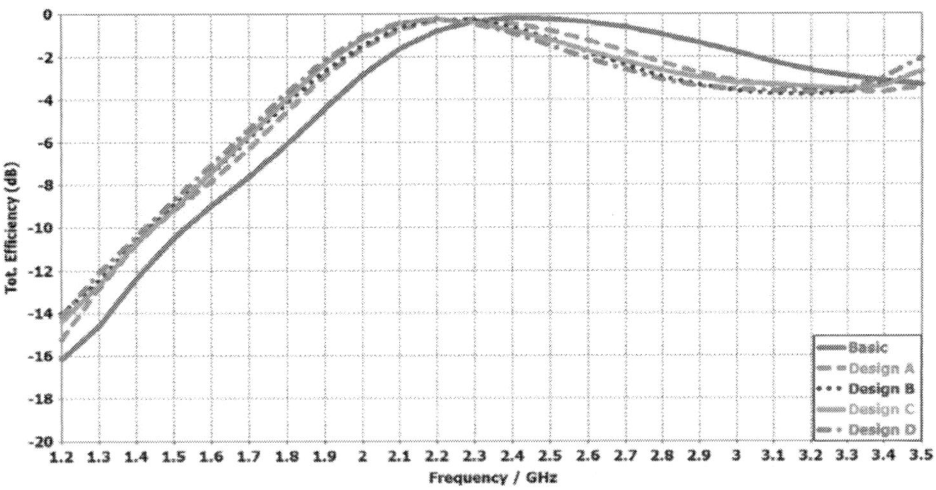

(a)

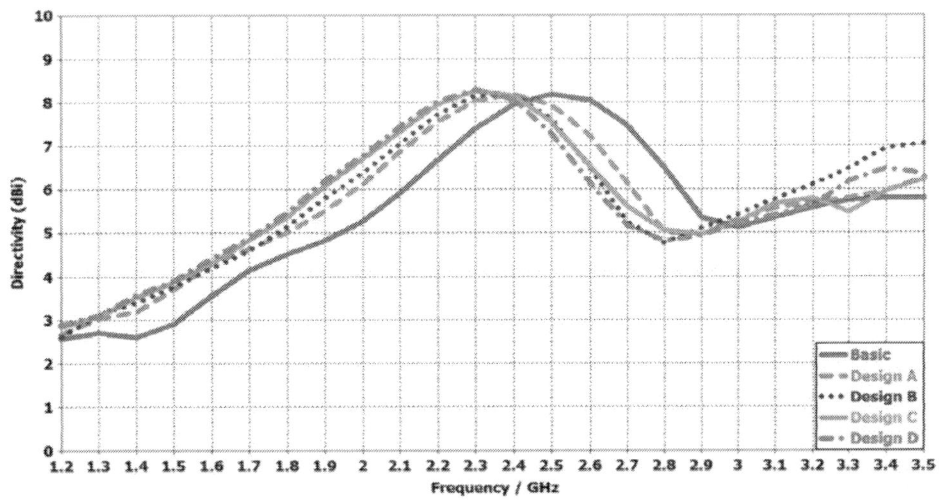

(b)

Figure 3. Comparison simulation result of Basic design, Design A, Design B, Design C and Design D (a) total efficiency (b) directivity

3.3. Axial ratio, Axial ratio Bandwidth (ARBW)

As demonstrated in Figure 4 is the simulation of axial ratio for Basic design and design with the addition of all different MS structure. Based on Figure 4, the minimum axial ratio for Basic design is at 2.1 GHz with 0.07 dB. While at 2.4 GHz, the axial ratio is 0.62 dB. With addition of all MS design, the axial ratio found to be more than 1 dB at 2.4 GHz. Where Design A has 3.13 dB (linear polarization), Design B, C and D remain below 3 dB which is 2.94 dB, 1.98 dB and 2.29 dB respectively. Axial ratio bandwidth (ARBW) of Basic design is 1120 MHz which covered from frequency range of 1.37 GHz to 2.49 GHz. While Design A and Design B only covered from 1.29 GHz to 2.40 GHz with 1110 MHz and 1.31 GHz to 2.41 GHz with 1103 MHz. However, Design D covered frequency range of Design A and B which is from 1.29 GHz to 2.45 GHz respectively. In comparison of ARBW for all design, Design C and Design D obtained a wider ARBW. Where, Design C obtained the widest axial ratio bandwidth among all of other design with 1380 MHz covered from 1.31 GHz to 2.69 GHz.

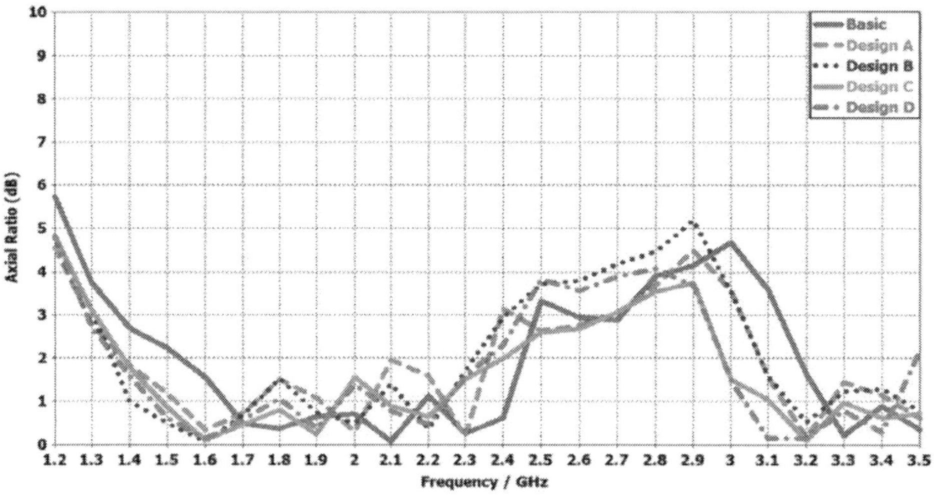

Figure 4. Comparison axial ratio simulation result of Basic design, Design A, Design B, Design C and Design D

3.4. Radiation pattern

Comparison of simulated radiation pattern for phi = 0°, phi = 90° and theta = 90° of Basic design, Design A, B, C and D at 2.4 GHz frequency are demonstrated as in Figure 5. The shapes of radiation pattern of all design are quite similar to each other. As demonstrated in Figure 5, at phi = 0° and phi = 90°, the orientation of main lobe direction are toward 2° and 0°. While at theta = 90°, main lobe direction change toward 129° respectively. Phi = 0° have one back lobe, phi = 90° have two back lobe while theta = 90° has omnidirectional lobe of radiation pattern. From observation, Design D has the biggest lobe of radiation pattern and Basic design has the smallest lobe of radiation pattern in comparison with other design.

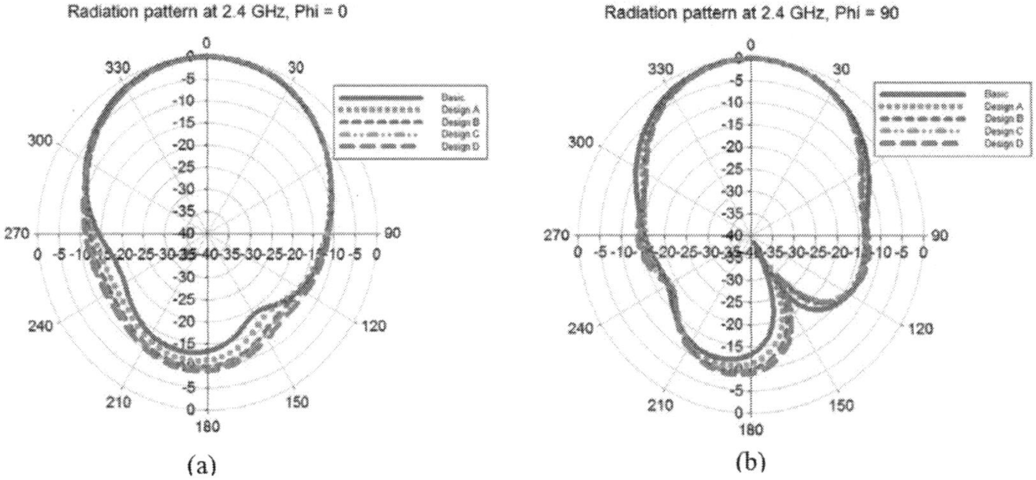

(a) (b)

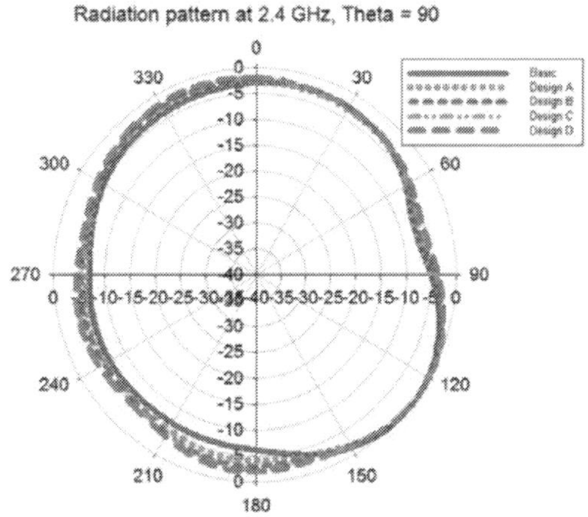

(c)

Figure 5. Comparison simulation result of radiation pattern for Basic design, Design A, Design B, Design C and Design D (a) phi = 0° (b) phi = 90° (c) theta = 90°

The comparison simulation antenna parameter result of Basic design, Design A, Design B, Design C and Design D at 2.4 GHz are tabulated as in Table 2. Based on Table 2, the result of return loss (RL), realized gain, axial ratio (AR), total efficiency and directivity are demonstrated.

Table 2. Comparison of simulation antenna parameter results performance at 2.4 GHz for Basic design, Design A, Design B, Design C and Design D

Antenna parameter	Basic	Design A	Design B	Design C	Design D
Return loss (dB)	-35.64	-14.62	-11.12	-9.44	-8.72
Realized gain (dB)	7.76	7.74	7.54	7.35	7.15
Axial ratio (dB)	0.62	3.13	2.94	1.99	2.29
Total efficiency (dB)	-0.18	-0.44	-0.63	-0.79	-0.91
Directivity (dBi)	7.94	8.18	8.16	8.15	8.06

4. Conclusion

In this paper, using four different MS design (Design A, B, C & D) to investigate its effect on inverted suspended circular polarized antenna (Basic) has been proposed in this study. The MS which consisting of 5 x 5 unit cell is located in between the basic antenna air gap at a fixed position. Based on the results, it shown with the integration of MS, the basic antenna resonant frequency shifted to a lower frequency from 2.4 GHz to 2.19 GHz (using Design C and D). Other than that, antenna bandwidth and gain (at 2.4 GHz) also decreases gradually with the addition of MS Design A, B, C and D to basic antenna. Total efficiency of all design at 2.4 GHz is less than -1 dB. Directivity for Basic designs at 2.4 GHz increases from 7.94 dBi to 8.17 dBi, 8.16 dBi, 8.15 dBi and 8.06 dBi (Design A, B, C and D). Moreover, Design C obtained the widest axial ratio bandwidth among all of other design with 1380 MHz covered from 1.31 GHz to 2.69 GHz. At 2.4 GHz, only Design A axial ratio is more than 3 dB which is 3.13 dB. Radiation pattern of all design at 2.4 GHz are quite similar. With the integration of MS structure to the basic antenna, size reduction can be achieved by reducing the radius of circular patch. In order to assemble as more compact and smaller sized antenna, for the next step that may be taken is miniaturizing the antenna substrate size while maintaining a similar antenna performance. Antenna with compact in size, large return loss bandwidth (RLBW) and axial ratio bandwidth (ARBW) and can operate as circular polarization can represent as a good candidate for compact and broadband circular polarization antenna for WLAN application at targeted frequency of 2.4 GHz.

Acknowledgments
Authors wishing to acknowledge Universiti Teknikal Malaysia Melaka (UTeM) for supporting in this research work under the grant PJP/2017/FKEKK/HI13/S01540. Other than that, special thanks to the Government of Malaysia which provide MyBrain15 program for financial support of this work under the research grant TRGS/1/2014/FKEKK/02/1/D00001 from Ministry of Higher Education (MOHE).

5. References

[1] Sun X L, Cheung S W and Yuk T I 2014 *Microwave Opt Technol Lett* **56** 1360
[2] Li L, Sing Wai C and Tung Ip Y 2014 *IET Microwaves Antennas Propag* **8** 1021
[3] Sun X L, Cheung S W and Yuk T I 2013 *Microwave Opt Technol Lett* **55** 1765
[4] Mohammadi P and Rafii V 2013 *Progress In Electromagnetics Research Letters* **43** 105
[5] Xu P, Yan Z H, Zhang T-L and Yang X-Q 2014 *Progress In Electromagnetics Research Letters* **44** 125
[6] Fan F, Wang W, Yan Z-H and Tan K-B 2014 *Progress in Electromagnetics Research C* **47** 47
[7] Zhu H, Chung K L, Sun X L, Cheung S W and Yuk T I 2012 *IEEE Antennas and Propagation Society, AP-S International Symposium (Digest)* p1
[8] Malik J, Patnaik A and Kartikeyan M V 2017 *Asia-Pacific Microwave Conference Proceedings (APMC)* p1
[9] Yue T, Jiang Z H and Werner D H 2016 *IEEE Transactions on Antennas and Propagation* **64** issue 5 1595
[10] Majumder B, Kandasamy K, Mukherjee J and Ray K P 2015 *Electronics Letters* **51** issue 17 1310
[11] Dong Y, Toyao H and Itoh T 2012 *IEEE Transactions on Antennas and Propagation* **60** issue 2 (PART 2) 772
[12] Zhu H L, Cheung S W and Yuk T I 2015 *Microwave and Optical Technology Letters* **57** issue 9 2050
[13] Nasimuddin N, Chen Z N and Qing X 2016 *IEEE Antennas and Propagation Magazine* **58** issue 2 39
[14] Chung K L and Chaimool S 2012 *Microwave and Optical Technology Letters* **54** issue 2 529
[15] Hu P F and Pan Y M 2017 *Asia-Pacific Microwave Conference Proceedings* 1
[16] Zhao Y 2016 *AIP Advances* **6** issue 5 1
[17] Zhou C, Cheung S W, Li Q and Li M 2017 *Applied Physics Letters* **110** issue 21 1
[18] Chung K L and Chaimool S 2011 *In: Proceedings of the 5th European Conference on Antennas and Propagation* p48
[19] Zhu H L, Cheung S W, Liu X H, Cao Y F and Yuk T I 2014 *In: 8th European Conference on Antennas and Propagation* p2575

IOP Conf. Series: Journal of Physics: Conf. Series **1049** (2018) 012084 doi:10.1088/1742-6596/1049/1/012084

Implementation of an 8x8 Discrete Cosine Transform on Programmable System-on-chip

[1]Bryan Chan Jia Ching, [2]Ab Al-Hadi Ab Rahman, [3]Nabihah Ahmad

[1,2] Faculty of Electrical Engineering, Universiti Teknologi Malaysia, 81310 Skudai, Malaysia

[3] Faculty of Electrical and Electronic Engineering, Universiti Tun Hussein Onn Malaysia, 86400 Parit Raja, Malaysia

E-mail: icebryanchan@gmail.com, hadi@fke.utm.my, nabihah@uthm.edu.my

Abstract. This paper presents experimental results that compares between a full software (SW) implementation and a software/hardware (SW/HW) co-design implementation of a DSP algorithm on a Xilinx Zynq programmable System-on-chip (SoC). The case study being used is the 8x8 two-dimensional discrete cosine transform (2D DCT), present in the popular JPEG and MPEG4-AVC encoder. The full SW design is implemented on a hardcore ARM processor on the FPGA SoC. The SW/HW co-design utilizes both the ARM processor and the Configurable Logic Blocks (CLB) of the FPGA SoC, where the communication channel is implemented using the Xillybus FIFO buffers, implemented as an external DRAM. In this case, the core 2D DCT operations are executed on the CLB, while data initialization and transfers are implemented on the processor. Results show that SW implementation is faster compared to SW/HW implementation for data inputs of less than 0.48 megapixels. As data inputs get larger, SW/HW implementation shows better performance, with up to 2x faster for 2 megapixels data input size. This study proves the viability of implementing the 2D DCT operations as dedicated hardware accelerator in multimedia encoders.

1. Introduction

The trend in systems design is gearing towards the use of heterogenous system-on-chip (SoC) where a general purpose processor (GPP) is integrated as a single chip with custom hardware accelerators. Modern FPGAs today provide the capability to program the SoC for both the custom hardware parts and the software part running on the GPP. In such case, the FPGA is commonly referred to as a programmable SoC. Major FPGA vendors from Intel (formerly Altera) and Xilinx today also provide specialize FPGA chips that consist of hardcore ARM processors with configurable logic blocks (CLB) in a single chip.

Compared to classical full hardware implementation on FPGAs, programmable SoCs give flexibility to designers to implement non speed critical or very complex parts of the design in SW, while allowing inherently parallel parts of the code to be implemented as custom logic for increased performance. As designs today become more complex, full hardware implementation may not be feasible as time to market becomes shorter. HW implementation is known to be difficult and more challenging compared to SW implementation, but usually achieves better performance. Programmable SoCs give a nice balance for designers in terms of performance and flexibility.

Content from this work may be used under the terms of the Creative Commons Attribution 3.0 licence. Any further distribution of this work must maintain attribution to the author(s) and the title of the work, journal citation and DOI.
Published under licence by IOP Publishing Ltd

IOP Conf. Series: Journal of Physics: Conf. Series **1049** (2018) 012084 doi:10.1088/1742-6596/1049/1/012084

This paper explores SW/HW Co-design methodology on a programmable SoC. Some part of the code runs on SW, while some others on HW. Both of the GPP and CLB resources are used in the FPGA chip. Co-design methodology overcomes the complexity faced from full HW implementation. However, the challenge in co-design is establishing the handshake between HW and SW, as well as managing the data transfer between them. One of the major issues in co-design is the communication cost between the two platforms, whereby the throughput of the system may be limited. Co-design is often suggested to be used in large data processing or those with intensive calculation requirements and may not be suitable for systems that require large and continuous data transfers [1].

This paper presents experimental results using the Zedboard from Digilent [2]. The board contains a single programmable SoC called the Zynq-7000 that consists of two ARM CPUs and the CLB for custom HW. There are several works reported in literature on the use of Zynq SoCs, but none validated on a streaming signal processing application, as proposed in this work. The work in [3] presents an implementation of a hadoop cluster for data clustering; [4] profiles the performance of the ARM cortex A9 on the Zynq; [5] and [6] present a co-design architecture for Blokus Duo Solver and SHA-3 encryption/decryption respectively.

The application proposed for the co-design implementation is the two dimensional discrete cosine transform, usually referred to as the 8x8 2D DCT. The algorithm is implemented in C code using the ISO/IEC 23002-2 algorithm [7]. Xilinx Vivado HLS is used to convert the C code to Verilog for HW implementation. The ARM CPU on the Zynq is used mainly for data initialization, transfers, and validation, while the core operation is implemented on the CLB. Data transfer buffer between SW and HW is implemented using Xillybus [8]. This co-design methodology is then compared with a full SW implementation running only on the GPP in terms of performance.

2. Background

2.1. 2D Discrete Cosine Transform

Discrete Cosine Transform (DCT) is a mathematical function that is mostly used in image and video compression, i.e. JPEG, MPEG-4 AVC/H.264, and HEVC encoders. DCT transforms data from time domain into frequency domain for removing data redundancies in a lossy compression, as down sampling and quantization is irreversible [9]. A fast DCT is necessary for encoding or decoding images with large resolutions or videos with high frame rate [10].

The 2-D DCT calculation has a high degree of computation complexity [11]. There are many algorithms to compute the 2-D DCT coefficients [12] [13] and the algorithm chosen in this paper is based on the algorithm defined by ISO/IEC 23002-2 standard [7]. This algorithm has been simplified in [14], and shown in Figure 1. In an 8x8 input matrix or input block, the first 1-D DCT is running row-wise while the second 1-D DCT is running column-wise based on the outputs from the first 1-D DCT. This simple pipelined algorithm reduces the complexity of the calculation factor by four [13].

3. SW and SW/HW Implementation Methodology

3.1. Zedboard and Xilinx Zynq-7000

Xilinx Zynq-7000 Programmable SoC consists of one embedded chip of a dual-core ARM Cortex-A9 MP Core microprocessor and a FPGA Fabric. At the board level, it can be divided into two parts, namely Processor System (PS) and Programmable Logic (PL). The dual-core ARM processor runs at a maximum frequency of 677MHz while the maximum frequency of PL or FPGA Fabric is determined by the design or by default it is 100MHz [2]. PS is made up of ARM microprocessor, SDRAM, flash memory controllers, peripherals as well as general-purpose input/output (GPIO) . Zedboard can be used to connect with a host machine through Universal

IOP Conf. Series: Journal of Physics: Conf. Series **1049** (2018) 012084 doi:10.1088/1742-6596/1049/1/012084

```
x0 = ind0 + ind7;
x1 = ind0 - ind7;
x4 = ind1 + ind6;
x5 = ind1 - ind6;
x2 = ind2 + ind5;
x3 = ind2 - ind5;
x6 = ind3 + ind4;
x7 = ind3 - ind4;

xa = pmul_1_2(x3);
x3 = pmul_1_1(x3);
xb = pmul_1_2(x5);
x5 = pmul_1_1(x5);

x3 = x3 + xb;
x5 = x5 - xa;

xa = pmul_2_2(x1);
x1 = pmul_2_1(x1);
xb = pmul_2_2(x7);
x7 = pmul_2_1(x7);


x1 = x1 - xb;
x7 = x7 + xa;
xa = x1 + x3;
x3 = x1 - x3;
xb = x7 + x5;
```

```
x5 = x7 - x5;
x1 = xa + xb;
x7 = xa - xb;
xa = x0 + x6;
x6 = x0 - x6;
xb = x4 + x2;
x2 = x4 - x2;
x0 = xa + xb;
x4 = xa - xb;

xa = pmul_3_2(x2);
x2 = pmul_3_1(x2);
xb = pmul_3_2(x6);
x6 = pmul_3_1(x6);

x2 = xb + x2;
x6 = x6 - xa;

outd0 = x0;
outd1 = x1;
outd2 = x2;
outd3 = x3;
outd4 = x4;
outd5 = x5;
outd6 = x6;
outd7 = x7;
```

```
pmul_1_1( x){
    return ( x - (x>>3) - (x>>7) );

}

pmul_1_2( x){
    return ( (x>>3) - (x>>7) +  (((x>>3) - (x>>7) - (x>>11))>>1 )   );
}

pmul_2_1( x){
    return ( (((x >> 9) - x)>>2) - ((x >> 9) - x));
}

pmul_2_2( x){
    return (x>>1);
}

pmul_3_1( x){
    return ( ((x + (x>>5))>>2) + (x>>4) );
}

pmul_3_2( x){
    return ( (x + (x>>5)) - (( x + (x>>5))>>2) );
}
```

Figure 1. Fast 1D DCT algorithm defined in the ISO/IEC 23002-2 standard

Asynchronous Receiver/Transmitter (UART) or through a memory card with bitstream file generated by Vivado software [15].

3.2. HW/SW Co-design using Xillybus

Xillybus IP is used to connect AXI ports and FIFOs with I/O devices on linux based operating system on the Zynq called Xillinux. In a SW/HW co-design methodology, when an application

IOP Conf. Series: Journal of Physics: Conf. Series **1049** (2018) 012084 doi:10.1088/1742-6596/1049/1/012084

logic is called, the processor passes the configuration header and messages to the hardware accelerator to compute the message.

3.3. Xillybus IP Core

Xillybus is a Direct Memory Access (DMA)-based IP core for data transport in between a FPGA and a host machine. In other words, Xillybus is a bridge for Processor System (PS) and Programmable Logic (PL) for data transfer. Xillybus supports wide range of products and board including Altera and Xilinx devices. FPGA application logics connects to the Xillybus through standard FIFOs. The Xillybus IP core then connects directly to the PS. Data transfers are made simple since PS handles the files in and out from PL through the Xillybus. The operation is represented in Figure 2. On host machine or PS side, there is no need for software library, any programming language can be used to access Xillybus streams without any new specific extensions. As shown in the figure, the PS is interfaced to the PL via PCIe bus connected to the Xillybus IP core. The core in turns are interfaced to application FIFO, where by default are implemented as external DRAM. The FIFOs connect directly to the PL.

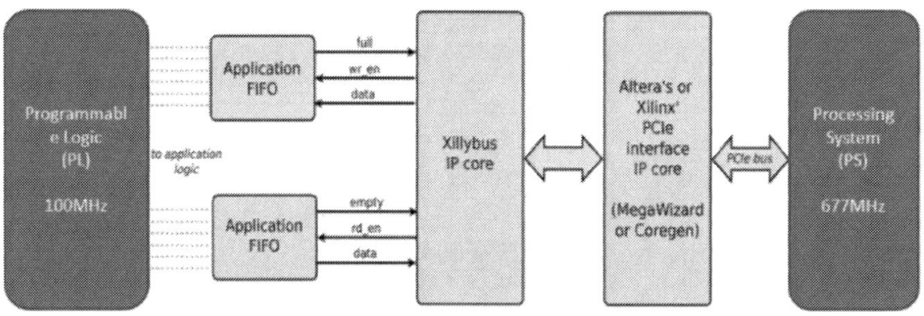

Figure 2. Xillybus IP core

3.4. Xillinux by Xillybus

Xillinux is a Linux distribution with additional feature like FGPA code kit. Just like any other Linux distribution, Xillinux allows Xilinx products, including but not limited to Zedboard, to run a graphical desktop with the support of keyboards, mouse, monitor screen and many other input output devices. The main feature or purpose of this Linux distribution is to make the integration between Processing System (PS) and Programmable Logic (PL) simple and easy. Xillinux uses SD card as its hard disk and it comes with a built-in PL development kit. The development kit allows Xillybus to be piped with the PS (ARM processors for Zedboard). Application logic can be connected to the FIFOs in PL and plain files can be sent in and out using simple operations from Xillinux.

3.5. Wrapper Function for the PL

Wrapper function is responsible for the interface between the PS and the PL. Inside the wrapper function, pragma directives are used to interface with the Xillybus IP core. Wrapper function works like the main function in a simple C program which will later make calls to the calculation function which is called the synthesized function. Programmer or user needs to decide how a normal program will separate into software part and hardware part. For the hardware part of a program, it will be included in the synthesized function. This part of the program will be running in the PL as the hardware part of the software-hardware co-design. In this work, the whole 2-D DCT will be put inside the synthesized function to maximize the performance. The wrapper function together with the synthesized function will then be debugged and synthesized by Vivado HLS. After that, the synthesized hardware code will then be exported into RTL in dcp format which will later be combined with Verilog code of the provided firmware for Xillinux, in order to generate a bitstream file. The bitstream file be will copied into the SD card where the Operating System of Xillinux is installed. When the Xillinux is booted up with the new bitstream file, the hardware code is ready to be used. Any changes to be made on the hardware part of the code must be done starting from the step of designing the wrapper and synthesized function.

3.6. Host Program for the PS

After the PL is ready with the desired synthesized function, a host program is written and compiled in PS to communicate and work with the function inside the PL. The PS and PL are interconnected with two standard FIFOs, one for sending data from PS to PL and another is vice versa. Therefore, it is wise to make use of basic parallel programming which includes a simple forking in the host program. The host program is forked into two processes, which are writing and reading data. This will prevent the processing from stalling due to lack of data to process or output data to be cleared up. Arrays are also used to send blocks of data into PL for processing and receive processed blocks of data from PL. These approaches will reduce but not totally eliminate the I/O overhead as well as the impact of software-hardware latencies.

The host program written in this work reads a text file containing the input blocks and then sends to PL for DCT before outputting the resulting blocks into a new text file. The host program or the software part of the program is responsible for reading inputs and writing outputs in the software environment. A detailed flowchart is shown in Fig. 3.

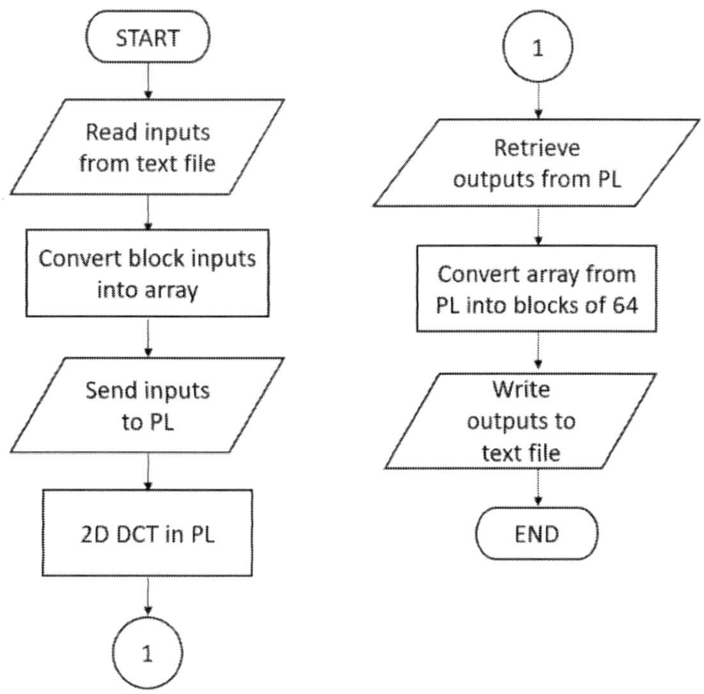

Figure 3. Host program interfacing with the PL

3.7. In-code Performance Profiling

In order to compare the performance between the SW/HW co-design program and the full software designed program, a c program with exactly the same functions is written and compiled in the Xillinux environment.

In this work, in code performance profiling is used to ensure the accuracy of the result because an external profiler might cause a heavy load on the operating system. Therefore, to ensure an unbiased profiling, the timing interval must be logical.

Since both of the programs are having the same functionality, it is safe to put the profiling interval as the beginning of the main function and just before the main function ends. The resulting time taken for executions is then noted down for further analysis.

3.8. Results and Analysis

Both the full software program and SW/HW co-design programs are executed for different numbers of 8x8 blocks. Numbers of blocks are set to 100, 1000, 2000, 3000, 4000, 5000, 6000, 7000, 8000, 9000, 10000, 12000, 14000, 16000, 18000, 20000, 22000, 24000, 26000, 28000, 30000, 32000 and 34000 respectively. Each execution is repeated for at least 100 times in order to obtain the average execution time for each number of blocks processed.

The results are plotted in the graph as shown in Figure 4. Note that the graph is not using a constant scale for the x-axis.

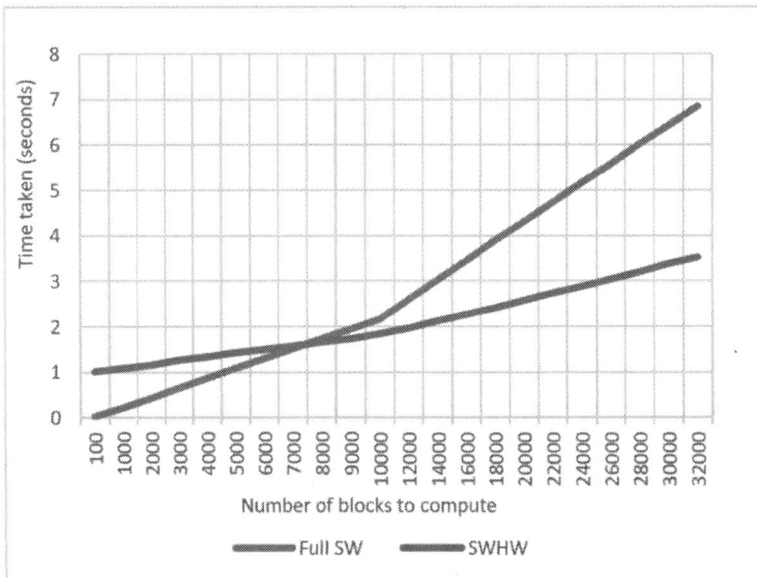

Figure 4. Graph of time taken to execute the DCT for different number of 8x8 blocks

As shown by the graph, both programs show linear relationship between the number of blocks processed and the time taken to complete the given tasks. As the number of blocks increase, the time taken to complete the DCT of the blocks increases linearly.

Initially, when the number of blocks fed into the DCT for both programs are below 7500 blocks, the co-design program shows relatively slower performance up to 1 second difference than that of full software designed program. This is mainly due to the input/output (I/O) communication time is relatively large when handling small number of blocks. Besides, an idle function of 1 second is included in the co-designed program in order to prevent the Xillybus devices from stalling and it is necessary to keep the program functioning as expected.

However, both of the line graphs crosses over at approximately 7500 of blocks where the performance of co-design program has caught up on the performance of software program. At this number of blocks processed, the time taken for I/O has become more insignificant compared to the performance of Programmable Logic.

Both line graphs continue to increase linearly with a constant gradient respectively after exceeding 7500 blocks of data processed. However, the performance difference between both programs is becoming bigger. The gradient of full software program execution time (0.213 milliseconds/block) is greater than that of software-hardware co-designed program (0.077 milliseconds/block). This means when the amount of input or blocks in this case increases, the advantage of using software-hardware co-design is becoming more significant.

Thus, from the result shown, co-design is suitable for programs that deal with large amount of data. In this work, less than 7500 blocks of data are considered as small amount and the DCT design should be done just by using full software design.

However, nowadays with the advanced growing technology in image processing and cameras, even smartphone cameras are generally above 8 megapixels. This means JPEG encoder nowadays requires the capability to handle more than 125,000 blocks of data if a block contains 8x8 of pixels.

Figure 5 illustrates clearly how the co-design program outperforms the full software designed program. The full software designed program shows a constant throughput of 0.3 megapixels per second despite the increase of pixels to be processed. However, the throughput of co-design program increases linearly over the increase of pixels to be processed up to 0.58 megapixels/second at 2 megapixels. Figure 5 also shows that the throughput of co-design program outperforms that of full software designed program once the processed pixels exceed 480,000 pixels or 0.48 megapixels.

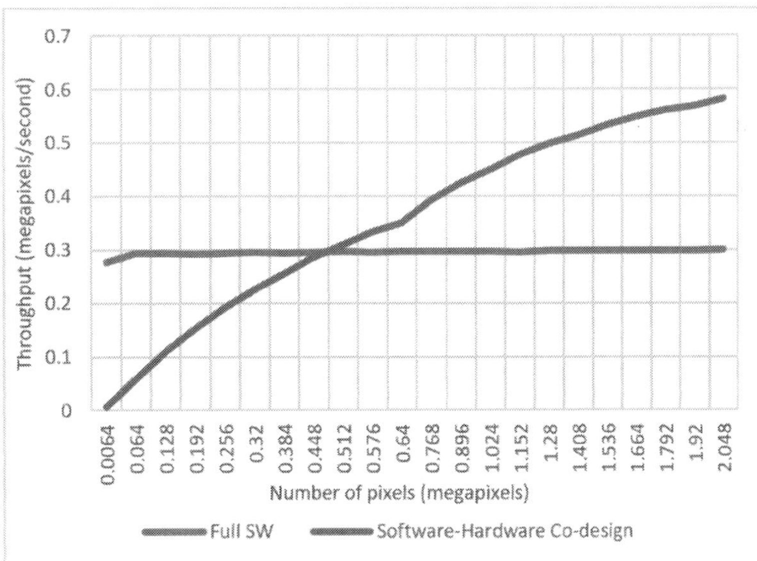

Figure 5. Graph of throughput vs number of pixels.

Resource utilization on the PL side is shown in Figure 6. As can be seen, the resource utilization is very low since the design size is relatively small. The power consumption can be expected to be low as well due to low resource usage.

Name	BRAM_18K	DSP48E	FF	LUT
DSP	-	-	-	-
Expression	-	-	0	73
FIFO	-	-	-	-
Instance	-	2	1125	1449
Memory	5	-	36	24
Multiplexer	-	-	-	178
Register	-	-	144	-
Total	5	2	1305	1724
Available	280	220	106400	53200
Utilization (%)	1	~0	1	3

Figure 6. Resource utilization on the PL side.

4. Conclusion

In this paper, we provide experimental results and insights on SW/HW co-design versus full SW implementation of a 8x8 DCT on Xilinx Zynq programmable SoC. The DCT is a very important process in image and video compression technologies such as in JPEG and MPEG, and known to be the most compute intensive part. In order to speed up the process compared to full SW implementation, the C code of the DCT has been synthesized to Verilog and implemented on HW, while data initialization, transfers, and validation on SW. Results show that due to communication cost, SW/HW implementation outperforms full SW implementation for data size larger than 0.48 megapixels. Results also reveal performance improvement of roughly 2x for 2 megapixel input size for DCT. Future work is to implement the full JPEG encoder using the SW/HW co-design methodology and validate the viability of using the CLB for acceleration.

Acknowledgment

The authors would like to thank UTM for providing the facilites, and the Ministry of Education for providing the funds for this research (Vot No. 4F659).

Reference

[1] Bezati E, Yviquel H, Raulet M and Mattavelli M 2011 A unified hardware/software co-synthesis solution for signal processing system *Proceedings of the 2011 Conference on Design and Architectures for Signal and Image Processing (DASIP)* p1

[2] Xilinx 2017 ZedBoard Technical Specifications Internet: http://zedboard.org/content/zynq-fpga-fabric-clock-speed

[3] Lin Z and Chow C 2013 Zcluster: A zynq-based hadoop cluster *Field Programmable Technology (FPT) International Conference on. IEEE.* p450

[4] Powell A and Silage D 2015 Statistical performance of the ARM cortex A9 accelerator coherency port in the xilinx zynq SoC for real-time applications *ReConFigurable Computing and FPGAs (ReConFig) International Conference on* p1

[5] Sugimoto N and Amano H 2014 Hardware/software co-design architecture for Blokus Duo solver *Field-Programmable Technology (FPT) International Conference on* p358

[6] Ahmed K E and Farag M M 2015 Hardware/software co-design of a dynamically configurable SHA-3 System-on-Chip (SoC) *Electronics, Circuits, and Systems (ICECS)* p617

[7] ISO/IEC 2006 Information technology âĂŤ MPEG video technologies âĂŤ Part 2: Fixed-point 8x8 IDCT and DCT transforms ISO/IEC 23002-2

[8] Xilinx 2017 FPGA coprocessing for C/C++ programmers (part II) Internet: http://xillybus.com/tutorials/vivado-hls-c-fpga-howto-2

[9] Miano J 1999 Compressed Image File Formats âĂŞ JPEG, PNG, GIF, XBM, BMP Addison Wesley Longman Inc USA

[10] Chiper D F 2016 A new VLSI algorithm for a high-throughput implementation of type IV DCT *International Conference on Communications (COMM)* p17
[11] Arai Y, Agui T and Nakajima M 1988 A Fast DCT-SQ Scheme for Images *Transactions of IEICE* E71 1095
[12] Kovac M and Ranganathan K 1995 JAGUAR: A Fully Pipeline VLSI Architecture for JPEG Image Compression Standard *Proceedings of the IEEE* 83 247
[13] Bhaskaran V and Konstantinides K 1999 Image and Video Compression Standards Algorithms and Architectures âĂŞ Second Edition Kluwer Academic Publishers USA
[14] Zabidi N N and Ab Rahman A A 2017 VLSI Design of a Fast Pipelined 8x8 Discrete Cosine Transform *International Journal of Electrical and Computer Engineering (IJECE)* 7 1430
[15] Xilinx 2013 Xilinx Zynq-7000 All Programmable SoC Overview Xilinx Inc

IOP Conf. Series: Journal of Physics: Conf. Series **1049** (2018) 012085 doi:10.1088/1742-6596/1049/1/012085

Perception of Mathematics Game's Design for Primary School: *Based on Teachers' Opinions*

Nur Faizura Ahmad Fuadi[1], **Muhammad Fakri Othman**[2] , **Norhalina Senan**[3]
Universiti Tun Hussein Onn Malaysia, 68400 Parit Raja, Batu Pahat, Johor.

E-mail: faizurafuadi@gmail.com[1]fakri@uthm.edu.my[2], halina@uthm.edu.my[3]

Abstract. Unmistakable methods can be used for learning, and they can be looked at in a few viewpoints, particularly those identified with learning results. In this paper, we introduce an examination with a specific end goal to think about the design adequacy and development's requirement of a game based learning (GBL) approach that is about to be used in LINUS screening for mathematics subject in primary school. The approach includes multiple interaction forms regarding addition and subtraction operation in mathematics based on LINUS constructs. Ten teachers from three different school located in Batu Pahat have participated in the study. The investigations involving survey activity by using questionnaire as the instrument. While breaking down the results, the outcomes demonstrated that the kids observed the amusement to be all the more fulfilling if there are less levels and more colours. Since the survey were conducted to a very common type of school in Malaysia, we believe game that is about to be built based on opinion gained could be utilized as an effective instrument in primary schools to strengthen pupils' lessons.

1. Introduction

Computerized strategy may not exist to totally supplant the conventional technique but rather to supplement the current technique by giving the clients convenience. Kids frequently play either separately or cooperatively for entertainment and pleasure only. Game gives way for kids to learn and investigate in a more significant manner. It might be stated, game is an autotelic and wilful act that requires regular reasons [1]. Notwithstanding, when set with regards to work or training, play is regularly observed as a trivial action. Regardless of such feedback, game was appeared to add to children's social, intellectual, and passionate improvement [2]. Also, battles that the improvisational ability of game in making people more flexible in methods for getting things done and more open to switch, together with showing alternatives practices and contemplations [3].A decent UI outline, as it allows the client to do errands, empowers a simple, normal, and connecting with communication between a client and a framework [4].

2. Literature Review

As complex systems, games show children strategic problem solving but, even more vitally in a systemic thinking on how one individual's activities can influence the whole framework [5]. By joining instructive recreations with pre and post-tests, numerous examinations demonstrate it to be in any event as viable as different strategies for evaluation, and understudies favour them over near techniques [6]. Instructive advancements that are very much outlined could supplement and bolster the

Content from this work may be used under the terms of the Creative Commons Attribution 3.0 licence. Any further distribution of this work must maintain attribution to the author(s) and the title of the work, journal citation and DOI.
Published under licence by IOP Publishing Ltd

learning condition [7]. This "21st-century preparing stage" has risen as one approach to address the numerous difficulties of instructing children [8][9][10]. With whenever, anyplace figuring, new developments might be constructed that are moderate, convenient and that can change our thoughts of data get to and learning. The significance of the component of happiness has been generally worried in learning and training. Absence of delight and weariness can be connected to withdrawal and disappointment in learning [11]. Primary school students who were presented a mobile game were discovered with eagerness contribute extra vitality clinched alongside realizing, which incited expanded learning.[12], this statement can be concluded that technologies do attract children to learn.

In concurrence with other specialists', proposal that help gave before playing a diversion may fortify learning, added to the game's guideline videos, which kept going around 3 min each. In these videos somebody plays the game while thinking accordingly presents normally how the amusement is played and which techniques can be utilized [13].In various conditions, remarkably in the academic setting, engagement sparklesgood enthusiastic experiences including fun, vitality, and duty in regards to persistent work [14].

Most computer games and other instructive programming now being utilized as a part of grade school Mathematics training centre around the initial two angles: number truth learning and operation aptitudes [15]. Besides, games encourage players to try, do mistakes, and after that try again without feeling ashamed [16]. Due to this harm-free condition and the quick denigration gave by the computer system, players are animated to investigate and analyse [17].

3. Method

The content of the game is based on 12 LINUS constructs used in the courses of the standard 1 until standard 3 children in Malaysia. The main objective of the developing game is to strengthen children's knowledge regarding mathematics. In this game, the children should be able to learn about addition operation and subtraction operation. Apart from the addition and subtraction operation, the children also learn how to solve problem in mathematical expression.

Each school participating in the study was acknowledged as an instance investigation. The examination construct itself with respect to a subjective technique for it was trusted this was the most reasonable way to deal with investigate inadequately contemplated issues or issues exhibiting numerous obscure variables. Thus, it might have been attempted that 10 teachers who represented these schools stated in table 1 and answered the survey would be related to Mathematics in schools through the development of syllabus with kids in standard 1 until standard 3.

Table 1. Respondents involved

Name of School	**Number of respondents**
Sekolah Kebangsaan Pintas Raya	10
Sekolah Kebangsaan Pintas Puding	10
Sekolah Kebangsaan Rejo Sari	10

After few trials of reaching the targeted school by email, an approach was made through a meeting with its representatives. The study concentrated on Sekolah Kebangsaan because at the moment, these schools are strong expansion in context of computing according to the technologic plan

IOP Conf. Series: Journal of Physics: Conf. Series **1049** (2018) 012085 doi:10.1088/1742-6596/1049/1/012085

of education and by the Ministry of Education [18]. After a few minutes of explaining the objectives of study and its aim to all teachers involved, they were given some times to ask any questions to ensure their understanding in this study. A set of questionnaire being used in this study to get the result. 10 questionnaires were distributed simultaneously to all respondents and they had taken approximately 5 to 10 minutes to finish answering all 10 questions given.

4. Result

The data for this study were then analysed using SPSS version 11.5 to get the reliability value of the questionnaire. The reliability obtained is $\alpha = 0.966$. This reliability value shows that the questionnaire is good and can be used for this study. This is because the instrument which has a coefficient value of less than 0.6 can be considered as having low value of reliability [19]. Teachers reported a multitude of fun experiences answering the questionnaire as the questions are all easy for them to understand.

The teachers would like a game that has less than 5 levels. More than 60% of them responded if there are more than 5 levels, they expected the children would be engaged to the game and given efficient effect. Another 30% think that they would be boring if there are more than 5 levels. As the pupils might be struggling to finish those levels, efficiency level might drop as well.

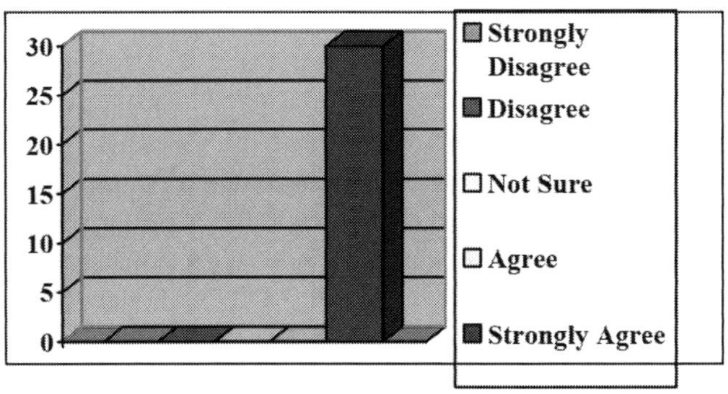

Figure 1. Question 5 responses

Statement in question number 5; "Colours play important roles in educational game?" have 100% of respondents responded "Strongly Agree" and that showed they really want the game to be colourful and fun enough to be enjoyed. Almost all of the teachers suggesting a display of points gained by players as a reward for the children when they had completed a level, displaying what they had accomplished.

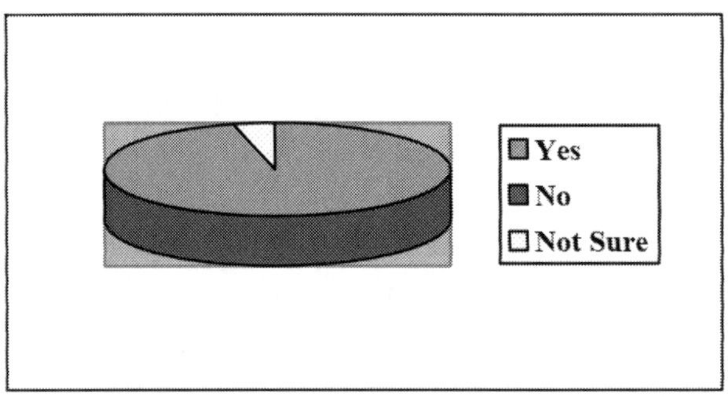

Figure 2. Question 10 responses

Regarding the theme, question number 10 which is "In your opinion, if thereis a game that can help the kids to enhance their numeracy skill, would you like it to be classroom-like theme or other than that?" got 96% of "Yes" answer from the respondents. This kind from claiming actual reproduction will be possibility will empower school's environment to be joined with students' physical, existed universe. [20]. Formative assessment and game-based learning have an inalienable suspicion of being incorporated into basic learning situation [21].

5. Discussion

A game model that can achieve a unique equilibrium is preferred during game theoretical modelling and analysis. Below is the characteristics that researchers listed based on the opinion given by the respondents:

- Efficiency (E): It is hard to give an exact meaning of efficiency as this requirement is related to the goals of models. This requirement can make sure that the game is designed with not only effectiveness but also high proficiency so it can be generally utilized. By only using game, teachers can generate kids' interests towards Mathematics subject.

- Feasibility (F): A feasible model refers to a circumstance that the result elements of players are characterized by the possible data.

- Scalability (SC): This prerequisite guarantees that a composed amusement model can be broadened to an extensive scale situation with more players taking an interest in and less confinements on trial conditions. A scalable model can fit into other application scenarios easily without too much alteration. In many ways, the game could also be 2 players game which would enhance the enjoyment and increase the eagerness in players to compete with each other.

- Individual Rationality (IR): It is unreasonable that a player has motivation to participate in a diversion where it can just acquire negative utility. In a non-cooperative game, all players pick systems from the perspective of augmenting their own benefits, consequently this prerequisite is fulfilled in each non-helpful diversion. We say an agreeable diversion is singular reasonability when the players in fantastic coalition can acquire a larger number of benefits than it can get alone (without participation with any other person). Personally, by playing game based learning (GBL), the kid himself could be motivated to get highest points they could in the game without anyone knowing.

Other than that, we decided the game to have only 3 levels to match the respondents' opinions. The children have to solve every questions given in every levels. Based on the answers given by the respondents, between each levels in the game, explanations should be showed, explaining the rules and goals to achieve a particular goal in the upcoming level. The game would be started with a guidance, which introduced the players and help them with clue for them to solve the problem given in next level. As in LINUS screening itself, children still need help from the teachers to guide them throughout the screening. At the age of 7 until 9 year-old children, we agreed to the fact that they still needed help and guidance to the new condition especially regarding technology use.Garris et al. suggested that the learning outcomes occur outside of the game during reflection anddebriefing (figure 3) [22], showing that game's characteristics is one of the important things in achieving the expected outcomes from certain GBL.

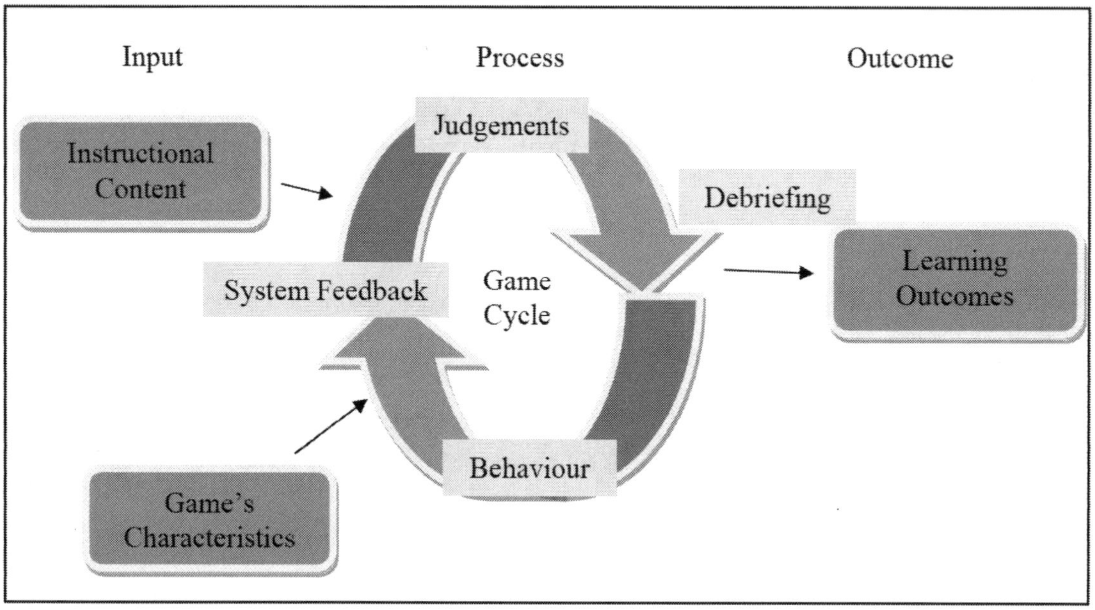

Figure 3. Learning in GBL [22]

6. Conclusion

For future works, the study may focuses on the process to achieve learning outcomes in GBL. Besides, they might also study whether the gender and age of kid's influences their achievements as there might be certain ages that the kids not even know how to use the game and some other ages very eager in playing game. Other subjects in primary school can also use game as a helping tool in classroom to increase the kids' interests in learning.

Acknowledgement
The research publication is seconded by Universiti Tun Hussein Onn Malaysia (UTHM) Short Term Grant (STG) Vot.U367 and is also supported by Gates IT Solution Sdn. Bhd. publication scheme.

Reference

[1] Feezell, R.2010 A pluralist conception of play. Journal of the Philosophy of Sport, 37(2), 147-165. http://dx.doi.org/10.1080/00948705.2010.9714773
[2] Fisher, K. R., Hirsh-Pasek, K., Golinkoff, R. M., & Gryfe, S. G. 2008 Conceptual split? Parents' and experts' perceptions of play in the 21st century. Journal of Applied Developmental Psychology, 29(4), 305-316. http://dx.doi.org/10.1016/j.appdev.2008.04.006.
[3] Brown, S. L. 2009 Play: How it shapes the brain, opens the imagination, and invigorates the soul. New York: Penguin Group.
[4] Muhamad Fairus Kamaruzaman et al. 2016Developing user interface design application for children with autism. Procedia - Social and Behavioral Sciences 217, 887 – 894.

[5] Squire, Kurt. 2011 Videogames and learning: Teaching and participatory culture in the digital age.New York: Teachers College Press.
[6] Delacruz, G. C., Chung, G. K., & Baker, E. L. 2010 Validity evidence for games as assessment environments. CRESST Report, 773, National Center for Research on Evaluation, Standards, and Student Testing (CRESST) Retrieved from http://www.cse.ucla.edu/products/reports/R773.pdf
[7] Morali, I. U. and S. 2010 A Short View on the Relationship of Mathematics and Game for Literature Context and Concept of the Educational Mathematics Game. World Applied Sciences Journal, 9(3), pp. 314-321.
[8] Jones, M., & Marsden, G. 2006 Mobile interaction design. NY: John Wiley & Sons.
[9] Papanikolaou, K., & Mavromoustakos, S. 2006 Critical success factors for the development of mobile learning applications. Paper presented at the 24th IASTED International Conference on Internet and Multimedia Systems and Applications, Innsbruck, Austria.
[10] Seong, D. S. K. 2006Usability guidelines for designing mobile learning portals. Paper presented at the Third International Conference on Mobile Technology, Applications & Systems (Mobility)
[11] Shernoff, D. J., Csikszentmihalyi, M., Schneider, B., & Shernoff, E. S. (2003). Student engagement in high school classrooms from the perspective of flow theory. School Psychology Quarterly, 18(2), 158e176. http://dx.doi.org/10.1521/scpq.18.2.158.21860.
[12] Sandberg, J., Maris, M., & De Geus, K. 2011 Mobile English learning: An evidence based study with fifth graders. Computers & Education, 47, 1334–1347
[13] Leemkuil, H., & De Jong, T. 2004 Games en gaming (Games and gaming). In P. Kirschner (Ed.), ICT in het onderwijs: The next generation. Katern bij onderwijskundig lexicon, uitbreiding editieIII. Alphen aan de Rijn, Netherlands: Kluwer.(pp. 41–63)
[14] Niemi, H., Harju, V., Vivitsou, M., Viitanen, K., Multisilta, J., & Kuokkanen, A. 2014 Digital storytelling for 21st-century Skills in virtual learning environments. Creative Education, 5, 657-671
[15] Mullis, I. V. S., Martin, M. O., Foy, P., & Arora, A. 2012 TIMSS 2011 international results in Mathematics. Chestnut Hill, MA: TIMSS & PIRLS International Study Center.
[16] Gee, J. P. 2005 Good video games and good learning. Phi Kappa Phi Forum, 85(2), 33–37.
[17] Kirriemuir, J. 2002 The relevance of video games and gaming consoles to the higher and further education learning experience. Techwatch report. http://tecnologiaedu.us.es/nweb/htm/pdf/301.pdf Retrieved on 10.08.17.
[18] Ministério da Educação. 2007 Despacho nº 16 149/2007 de 25 de Julho, Diário da República, 2ª Série, nº 142, p. 21077.
[19] Mohd Salleh dan Zaidatun Tasir. 2001 "Pengenalan kepada Analisis Data Berkomputer: SPSS 10.0 forWindows." Kuala Lumpur: Venton Publishing.
[20] Squire, K., & Klopfer, E. 2007 Augmented reality simulations on handheld computers. The Journal of the Learning Sciences, Lawrence Erlbaum Associates. 16(3), 371–413
[21] Broussard, M.J.S. 2012 Using Games to Make Formative Assessment Fun in the Academic Library, The Journal of Academic Librarianship, http://dx.doi.org/10.1016/j.acalib.2012.12.001
[22] Garris, R., Ahlers, R., & Driskell, J. E. 2002 Games, motivation, and learning: A research and practice model. Simulation & Gaming, 33(4), 441-467

IOP Conf. Series: Journal of Physics: Conf. Series **1049** (2018) 012086 doi:10.1088/1742-6596/1049/1/012086

Empirical Analysis on Sales of Video Games: A Data Mining Approach

Amar Aziz[1], Shuhaida Ismail[2], Muhammad Fakri Othman[1], Aida Mustapha[1]

[1]Faculty of Computer Science and Information Technology, Universiti Tun Hussein Onn Malaysia, 86400 Parit Raja, Batu Pahat, Johor, Malaysia.

[2]Faculty of Science and Technology, Universiti Tun Hussein Onn Malaysia, 84600 Pagoh, Muar, Johor, Malaysia.

E-mail: shuhaida@uthm.edu.my, fakri@uthm.edu.my, aidam@uthm.edu.my

Abstract. This paper studies factors that make the sales of video games becomes a blockbuster. The dataset used is collected from an online database maintained by VGChartz.com. Using the dataset, the Rapid Miner tool is used to select the features or factors and produce efficient estimation of the data. The techniques used in this project included the k-Nearest Neighbour (k-NN), Random Forest and Decision Tree. The factors and differences in the results are deliberated and discussed.

1. Introduction

In technology world, people from various age ranging from small kids to adults play the video games. This has lead to the spike in the sales of video games nowadays. Video games are released by major publishers across many popular hardware platforms. It provides the only real experience of interactive entertainment offered by the modern technologies. It also provides a rapidly growing form of entertainment and is being used for educational as well as business purposes. In the previous decade, several major video gaming releases have raised the bar of conventional entertainment goods in terms of revenues earned. There are several types of blockbuster video games on sale today such as Grand Theft Auto IV' by Rock Star Games and Call of Duty' Series by Activision. These types of video games have produced a series of annual records for revenues over the course of a three-year period.

Predictive modelling has long been the goal of many individuals and organizations. This science has many techniques, with simulation and machine learning at its heart. Aside from the potential of simulations, machine learning techniques are known for their ability to uncover hidden data trends. While the choice of algorithms used in each study may differ, they all had one common similarity, they give choices that human track experts made and are able to use the data to create arbitrage opportunities [7]. For example, [5] used Back-Propagation Neural network with Principal Component Analysis to predict the weekly video games sales. Other than that, transformation of classification trees into a decision-analytic model has also been used in solving the value-maximizing game development policy. The results showed that the compact predictive models created by data mining algorithms can help to make decision-analytic feed-forward control feasible, even for large, complex problems [6].

Content from this work may be used under the terms of the Creative Commons Attribution 3.0 licence. Any further distribution of this work must maintain attribution to the author(s) and the title of the work, journal citation and DOI.
Published under licence by IOP Publishing Ltd

IOP Conf. Series: Journal of Physics: Conf. Series **1049** (2018) 012086 doi:10.1088/1742-6596/1049/1/012086

The main purpose of this study is to find out the contributing factors that lead video game sales becoming blockbuster. The collected dataset consisting of approximately 1,800 observations relating to individual video games titles released across a various platforms. This study uses a unique dataset of video game title to estimate the effect of dependent variables that used to empirical the lifetime unit sales of games released in the US in September 2010, which are collected from online database maintained by VGChartz.com. Then, the Rapid Miner tool is used to produce efficient estimation of the data. Aside from finding the factors that lead video game sales becomes blockbuster, the experiment also estimate the effect sales of video games becoming blockbuster.

The remaining of this paper proceeds as follows. Section 2 presents the methodology of the project including the machine learning algorithms and the dataset, Section 3 presents the experimental setup, Section 4 presents and discusses the results, and finally Section 5 concludes the work.

2. Methodology

In this research, a benchmark dataset which was originally collected and analyzed by Dr. Joe Cox is used. The dataset was collected from two Portuguese secondary schools during 2005-2006 by using two different methods which are mark reports and questionnaire. Decision Tree is used to predict and to find the correlation between features and as before for pre-processing the dataset. Data cleaning is performed in order to remove noise and to correct the inconsistencies exits in the data. The next step is merging two different data sets by using data integration technique. Classification is performed to analyse the statistical value between pair or two items. For this research, techniques such as k-NN [2], Random Forest [3] and Naïve Bayes [4] will be used to identify the difference results and outputs. The different and descriptions of techniques used during this project were documented in following section. Figure 1 shows the steps taken in the methodology to perform the experiment.

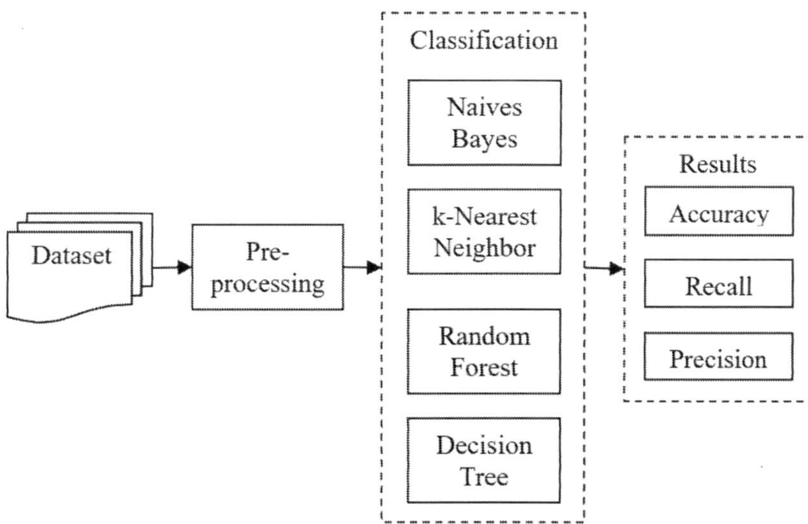

Figure 1: Methodology

A. Machine Learning Algorithms

There are four machine learning algorithms used in the experiments, which are the naïve Bayes, k-Nearest Neighbor, Random Forest, and Decision Tree.

- Naïve Bayes: This technique is used to create a Bayesian model that predicts the value of a target attribute (often called class or label) based on several input attributes of the dataset. The naive Bayes technique was used by previous work in [5].

IOP Conf. Series: Journal of Physics: Conf. Series **1049** (2018) 012086 doi:10.1088/1742-6596/1049/1/012086

- k-Nearest Neighbor: By comparing a given test example with training examples that are similar, the k-Nearest Neighbour algorithm is learning based on the analogy. The k-nearest neighbour algorithm is also one of the simplest of all machine learning algorithms.
- Random Forest: The Random Forest operator is used to generate a set of random trees. The random trees are generated in exactly the same way as the Random Tree operator generates a tree. The resulting forest model contains a specified number of random tree models. The number of trees parameter specifies the required number of trees.
- Decision Tree: The technique used to create a classification model that predicts the value of a target attribute (often called class or label) based on several input attributes of the dataset.

B. Dataset

Rapid Miner studio is used as the tool to import the data set and the attribute and type for the data is assigned before into the process. The attribute and the data type are shown in Figure 2. The data is reviewed and normalization is performed accordingly. Normalization is a pre-processing technique that used to rescale attribute values to fit in a specific range. Data normalization is very important especially when dealing with attributes that have different units and scales. For example, some data mining techniques use the Euclidean distance. Therefore, all attributes should have the same scale for a fair comparison between them. In other words, normalization is a technique used to level the playing field when looking at attributes that widely vary in size as a result of the units selected for representation. This operator performed normalization of selected attributes. Four normalization methods are provided. These methods are explained in the parameters.

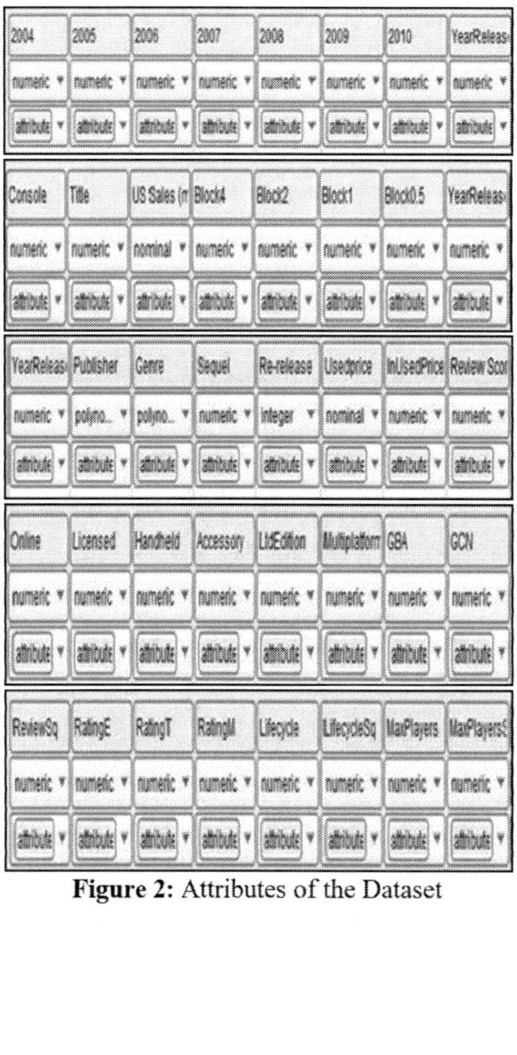

Figure 2: Attributes of the Dataset

3. Experiments

This study uses a unique dataset of video game title to estimate the effect of dependent variable that used to empirical the lifetime unit sales of games released in the US in September 2010, which are collected from online database maintained by VGChartz.com. RapidMiner Studio Basic is use as a tool to run the experiment. RapidMiner Studio Basic is a code-free environment for designing advanced analytic processes with machine learning, data mining, text mining, predictive analytics and business analytics [1].

A. Data Pre-Processing

Data pre-processing technique that has been used in this experiment is normalization. Normalization technique is generally used to rescale the attribute values in the dataset as shown in Figure 1. Other than normalization, data cleaning and data integration are also use in order to remove noise, correct the inconsistencies and merge two different data sets respectively.

B. Data Transformation

For our work, cross-validation technique is used to estimate the statistical performance of the learning operator and to estimate the accurate of model performance in training and testing phase.

C. Operators Parameters

Based on the dataset, suitable operators are chosen in order to produce the result and to fix error occurs during the experiment. Table 1 shows the details and the parameter of the operators used in this work.

D. Assessment Criteria

The results of each technique, which are Naïve Bayes, Decision Tree, K-NN and Random Forest were classified into three parameters which are accuracy, recall and precision. Assessments criteria accuracy is calculated by taking the percentage of correct predictions over the total number of examples, while the weighted recall is the result of an average of recall of every class. The weighted precision on the other hand, is calculated by taking the average of precision of every class.

Table 1: Operators and Parameter

No	Type	Parameter
1	Frequency-based Discretization	Filter type: regular expression Regular expression: Usedprice Number of bin: 2 Range name type: long
2	Decision Tree	Criterion: gain ratio Maximal depth: 20 Confidence: 0.25
3	k-Nearest Neighbor (k-NN)	k: 1 Measure types: Mixed Measures Mixed Measure: Mixed Euclidean Distance
4	Random Forest	Number of Tree: 10 Criterion: gain ratio Maximal depth: 20 Confidence: 0.25
5	Naïve Bayes	Laplace Correction: yes
6	Performance Classification	Main criterion: First Accuracy: Yes Weighted mean recall: Yes Weighted mean precision: Yes

4. Results and Discussion

The accuracy, recall and precision of the data were gathered by based on four machine learning algorithms; Naïve Bayes, Decision Tree, k-NN and Random Forest. The result of the data by using these techniques were documented in the following section. Frequency-based discretization was used to convert the selected numerical attributes into nominal attributes by changing the numerical attributes into a user-specified number of bins. Table 2 shows the final results for Naïve Bayes, Decision Tree, K-NN and Random Forest.

Table 2: Results for Naïve Bayes, Decision Tree, k-NN and Random Forest

Techniques	Accuracy	Recall	Precision
Naïve Bayes	81.58%	43.00%	44.12%
Decision Tree	99.55%	86.61%	86.20%
K-NN	24.86%	15.07%	13.97%
Random Forest	26.89%	3.45%	0.93%

The results in Table 2 showed that the percentage of accuracy for all method using Decision Tree produced the same percentage, which is 99.55% and the percentages of recall and precision for all method produced 86.61% and 86.20% respectively. Meanwhile the highest accuracy, recall and precision is obtained using z-Transformation method for k-NN technique with the values of 24.86%, 15.07% and 13.97% respectively. The highest accuracy, recall and precision using Random Forest technique are recorded at 26.89%, 3.45%and 0.93%, respectively. Besides, the highest accuracy, recall and precision percentage using Naïve Bayes technique is the Interquartile Transformation which produced 81.58%, 43.00% and 44.12% respectively. The experiments also showed that the most contributing factors to the blockbuster game sales include the year that the game is released, the genre, the price, and the review score.

5. Conclusion

In conclusion, from dataset there has been several technique perform to compare the result to find the outcome of what makes blockbuster video game. The techniques are Decision Tree, K-NN Result and Random Forest. The results are then compared between three criteria, which are accuracy, recall and precision. The results showed that Naïve Bayes technique using Interquartile Transformation have much closer accuracy, recall and precision compared to decision tree technique. It is proven that this method is more suitable to calculate the accuracy, recall and precision of the work compared to others.

Acknowledgement

The research publication is supported by Universiti Tun Hussein Onn Malaysia (UTHM) under Short Term Grant (STG), Vot.U367 and is also seconded by Gates IT Solution Sdn. Bhd.

References

[1] RapidMiner Studio 6.5. (2016). Retrieved on 27 November 2016 from http://docs.rapidminer.com
[2] K-NN (k-Nearest Neighbor). (2016). Retrieved on 27 November 2016 from http://docs.rapidminer.com/studio/operators/modelling/predictive/lazy/k_nn.html.
[3] Random Forest. (2016). Retrieved on 27 November from http:// docs.rapidminer.com/studio/operators/modelling/predictive/trees/parallel_random_forest.html
[4] Naïve Bayes. (2016). Retrieved on 27 November 2016 from http://docs.rapidminer. com/studio/operators/modelling/predictive/bayesian/naive_bayes.html
[5] Marcoux, J., & Selouani, S. A. (2009). A hybrid subspace-connectionist data mining approach for sales forecasting in the video game industry. In Computer Science and Information Engineering, 2009 WRI World Congress on (Vol. 5, pp. 666-670). IEEE. Brydon, M., & Gemino, A. (2008). Classification trees and decision-analytic feedforward control: a case study from the video game industry. Data Mining and Knowledge Discovery, 17(2), 317-342.

[6] Schumaker, R. P., Solieman, O. K., & Chen, H. (2010). Predictive modelling for sports and gaming. In Sports Data Mining (pp. 55-63). Springer US.
[7] Cox, J. (2014). What makes a blockbuster video game? An empirical analysis of US sales data. Managerial and Decision Economics, 35(3):189-198.

IOP Conf. Series: Journal of Physics: Conf. Series **1049** (2018) 012087 doi:10.1088/1742-6596/1049/1/012087

Design of Circle Defected Waveguide with Monopole Antenna

S. J. Chin, M. Z. A. Abd. Aziz, M. R. Ahmad
Center for Telecommunication Research and Innovation (CeTRI), Faculty of Electronics and Computer Engineering (FKEKK), Universiti Teknikal Malaysia Melaka (UTeM), 76100 Durian Tunggal, Melaka, Malaysia

E-mail: shujia2605@gmail.com

Abstract. In this paper, the circle DWS with monopole antenna is designed and studied. From previous research studies, there are design of waveguide with metamaterial by using Electromagnetic Bandgap (EBG) and Frequency Selective Surface (FSS). Yet, Defected Waveguide Structure (DWS) is not familiar to design with waveguide which is the motivation of this paper. Meanwhile, the design of waveguide with monopole antenna is not explored widely. The DWS is designed in circle shape with Ultrawideband (UWB) monopole antenna. Several factors are studied on the size, inter element spacing and number of circle DWS towards waveguide. Circle DWS with monopole antenna are analyzed through antenna parameters such as return loss, efficiency, gain and directivity. In overall, the circle DWS with monopole antenna able to operate in wide bandwidth of 8.66GHz which is suitable for UWB applications. It also achieves higher efficiency which is more than -3dB. Highest gain of 7.29dB and directivity of 8.62dBi are achieved at 9GHz and 9.5GHz respectively. The use of DWS improves the gain and directivity of low profile monopole antenna but at the same time still maintaining the its wide bandwidth.

1. Introduction

A great deal of attention is dedicated to metamaterial in recent year. This is because it consists of material properties that is made artificially by man [1]. There are a few examples of metamaterial such as Electromagnetic Bandgap (EBG), Frequency Selective Surface (FSS), Defected Ground Structure (DGS) and Defected Microstrip Structure (DMS) [2-5]. Defected structure which are designed at ground plane and microstrip feed line will be called as DGS and DMS. In the past, defected structure is designed in patch or slot to disturb the normal current distribution. For example, DGS in [4] enhances bandwidth while DMS in [5] introduces higher gain. Despite there are defected structure on the ground plane and microstrip feed line, no researches on the Defected Waveguide Structure (DWS) is obtained.

For the past, waveguide is one the earliest types of transmission line used to transmit signal [6]. Nowadays, it has been focused on the applications especially in antenna and filter site [7-8]. Besides that, it also has been designed with metamaterial to determine its reflection and transmission characteristics [9]. There is theoretical modal used to study the field distribution of waveguide with anisotropic walls [10]. The anisotropic walls are referred to walls with metamaterial. From the previous studies, there are types of metamaterial such as FSS, EBG [9, 11] are designed with waveguide but no DWS design is found with waveguide.

There is another hot research topic which is Ultrawideband (UWB) technology [12-13]. This is because people request for the wide coverage for high date rate and high resolution. Thus, UWB

Content from this work may be used under the terms of the Creative Commons Attribution 3.0 licence. Any further distribution of this work must maintain attribution to the author(s) and the title of the work, journal citation and DOI.
Published under licence by IOP Publishing Ltd

technology requires wideband antenna for operation. Monopole antenna is of great interest for its potential of low profile and low cost [13-15]. There are researches are done for the modifications on monopole antenna by using metamaterial by internally or externally [15-16]. For instance, the DGS improves the bandwidth of monopole antenna by introducing L-shaped DGS slot at ground plane in terms of internally [15]. For external, FSS reflector is placed underneath monopole antenna achieves enhancement in gain [16]. However, there is no DWS design with monopole antenna is found.

In this paper, the circle DWS waveguide is designed with monopole antenna. The monopole antenna is a microstrip-fed with circular radiating patch which is worked for UWB frequency range 3.1GHz-10.6GHz. The circle DWS waveguide is designed externally to the monopole antenna. Antenna parameters such as return loss, efficiency, gain and directivity are used to observe its performance.

2. Antenna Design

The design and simulation are constructed in CST Microwave Studio. Monopole antenna is designed in circular radiating patch and microstrip feed line. This antenna is designed to support UWB applications at 3.1GHz-10.6GHz. The Equations (1) and Equations (2-4) are referred for the fundamental of circular radiating patch and microstrip feed line. There are two copper layers at front and back of substrate where circular radiating patch and feed line at the front while a ground plane at the back as shown in Figure 1. The material used for antenna design is FR4 substrate with dielectric constant 4.4, tangent loss 0.019 and thickness 1.6mm.
Radius of circular radiating patch can be determined by using equations as below [17]:

$$R = \frac{F}{\sqrt{1 + \frac{2h}{\pi \varepsilon_r F}\left[ln\left(\frac{\pi F}{2h}\right) + 1.7726\right]}} \tag{1}$$

where R = actual radius

$F = \frac{8.791 \times 10^9}{f_r\sqrt{\varepsilon_r}}$

f_r = resonant frequency

h = substrate height

ε_r = substrate dielectric constant

The feed line can be also calculated by using the microstrip feed line equations as below [6]:

Characteristic impedance, Z_o

$$Z_o = \frac{60}{\sqrt{\varepsilon_{reff}}} ln\left(\frac{8h}{W_f} + \frac{W_f}{4h}\right) \; for \; \frac{W_f}{h} < 1 \tag{2}$$

$$Z_o = \frac{120\pi}{\sqrt{\varepsilon_{reff}}\left[\frac{W_f}{h} + 1.393 + 0.667ln\left(\frac{W_f}{h} + 1.444\right)\right]} \; for \; \frac{W_f}{h} > 1 \tag{3}$$

where W_f = line width

Effective dielectric constant, ε_{reff}

$$\varepsilon_{reff} = \frac{\varepsilon_r + 1}{2} + \frac{\varepsilon_r - 1}{2}\frac{1}{\sqrt{\left(1 + 12\frac{h}{W_f}\right)}} \tag{4}$$

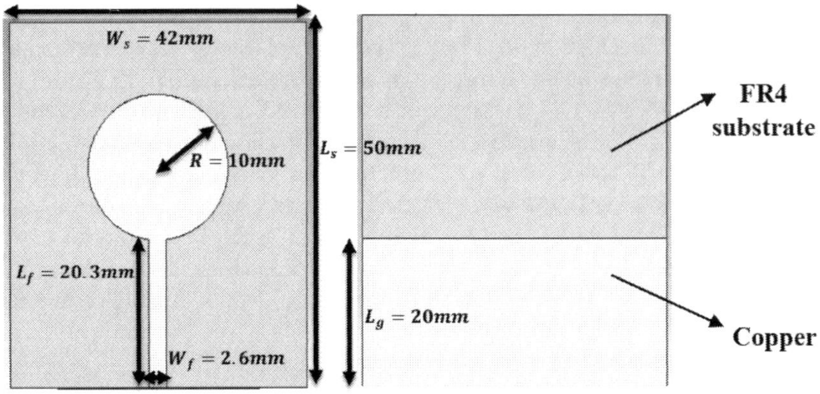

Figure 1. The front and back view of monopole antenna

Typically, dimension of waveguide (width x height) is important to determine cut off frequency as shown in Equation (5). Equation (5) shows the relationship between width (a) and height (b) with cut off frequency (f_c) [6]. The larger the a and b, the smaller the f_c can be propagated through waveguide where m and n represent the modes of waveguide. However, the rectangle waveguide with full copper at inner walls is designed based on the dimension of monopole antenna as shone in Figure 2. The waveguide width (a=44mm) and height (b=52mm) which include dimension and small vacuum gap of 1mm to avoid both waveguide and antenna surface are contacted. Two waveguide ports are set at both open-ended surfaces of waveguide to measure the transmission coefficient (S21).

$$f_c = \frac{1}{2\pi\sqrt{\mu\varepsilon}}\sqrt{\left(\frac{m\pi}{a}\right)^2 + \left(\frac{n\pi}{b}\right)^2} \tag{5}$$

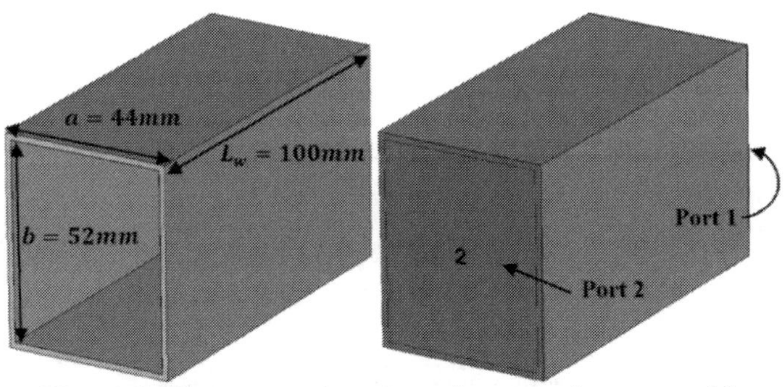

Figure 2. The perspective view of rectangular waveguide

In this paper, the DWS waveguide in circle shape is focused instead of waveguide with full copper. The copper layer at the inner walls of waveguide are removed to have circle DWS. The simple basic circle shape is utilized due to ease of fabrication. Throughout the designs, radius refers to the radius of circle while space refers to the inter element spacing. Meanwhile, variable *X*, *Y* and *Z* refer to number of circle element according *X*, *Y* and *Z* directions respectively at top bottom and both sides walls. Initially, the circle DWS are designed with *Radius=3mm*, *Space=1mm*, *X=2*, *Y=3* and *Z=15* as shown in Figure 3.

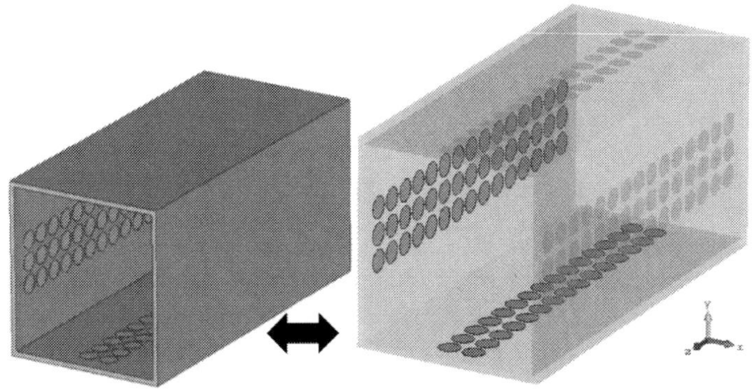

Figure 3. The perspective view of circle DWS waveguide

Before the circle DWS waveguide is inserted with monopole antenna, a few parameter studies are done to determine effects of DWS towards waveguide. The parameter studies are done on the *Radius*, *Space* and number of circle element for *X*, *Y* and *Z* directions. The circle DWS waveguide is designed with full circle DWS to insert monopole antenna as shown in Figure 4. In addition, the number of circle DWS in *Z* direction is *12* to ensure a full wave cycle for propagation of longest wavelength in waveguide. Their performances are observed and analyzed through S-Parameters which are reflection coefficient (S11) and transmission coefficient (S21). As seen in Figure, the monopole antenna is then acted as input to the circle DWS waveguide. Antenna parameters in terms of return loss, efficiency, gain and directivity is observed for the circle DWS with monopole antenna.

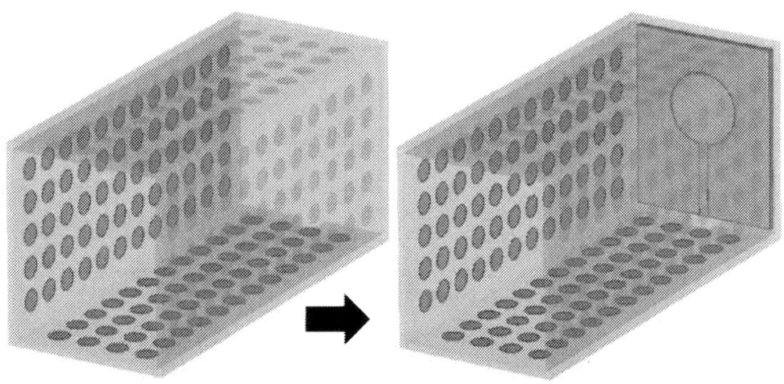

Figure 4. Insert monopole antenna in circle DWS waveguide

3. Result and discussion

The parameter studies of the circle element are done on its size and internal circle element spacing as shown in Figure 5-6. The size of circle element is varied in terms of radius from *3.0mm-6.6mm*. It is observed that the S-Parameters results are heavily depended on the size of circle element. Smaller size of circle element with *radius=3.0mm* can achieve widest bandwidth at 3GHz-9GHz for transmission as shown in Figure 5. For convenience in description, the S-Parameters results is enlarged and divided into two groups in terms of transmission and reflection. The bigger the size of circle element, the narrower the operating band and the higher the reflection at higher frequency.

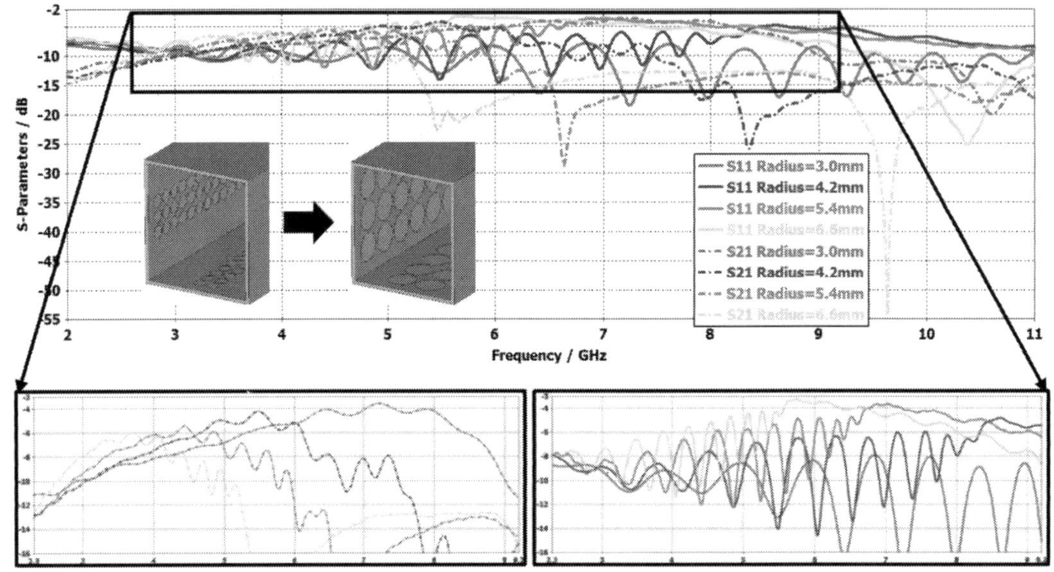

Figure 5. S-Parameters results by varying size of circle DWS

Less significant changes are observed through the S21 results by varying the inter element spacing in Figure 6 compares to size. They vary tardily but significant changes only occur at higher frequency range at 9GHz onwards. The optimized inter element spacing for circle DWS is *3mm* with widest bandwidth of 6.5GHz. In Figure 6, almost similar curve is achieved for S11 below 5.5GHz. Above 5.5GHz, the S11 with bigger space fluctuate more rapidly to have more cycles with higher phase changes compare to S11 with smaller space.

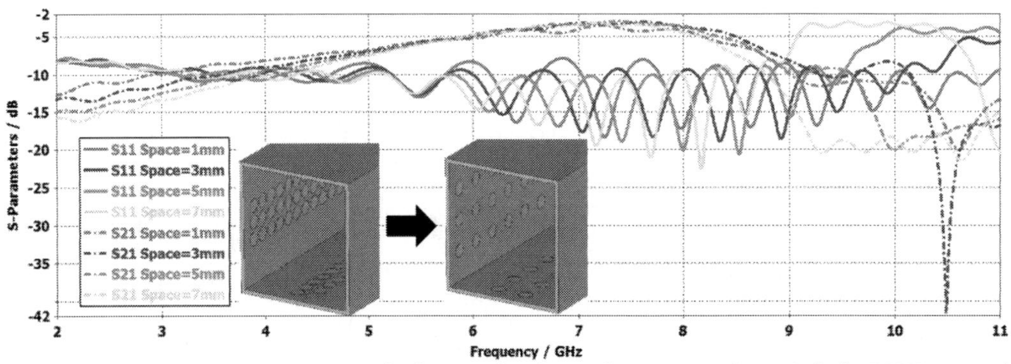

Figure 6. S-Parameters results by varying inter element spacing of circle DWS

Meanwhile, the number of circle DWS is also varied in *X*, *Y* and *Z* directions. The number of circle DWS in *Z* direction still maintain in *Z=15*. While varying the number of circle DWS in *X* direction, the number of circle DWS in *Y* direction is designed to occupy fully at the walls with maximum number of *Y* and vice versa. It is noticed that the variance of number in *X* direction has no impact on the reflection and transmission of waveguide as shown in Figure 7. However, Figure 8 shows there is an obvious change for S21 at 8GHz and onwards for the variance of number in *Y* direction. With no circle DWS at both sides walls, the waveguide can achieve high pass at 4GHz with higher transmission and low reflection. Besides that, fluctuated S11 with biggest different in 25dB is achieved for *Y=0*. While adding the number of circle DWS, the S11 is changed faster in terms of phase to have narrower cycle of S11. However, there is improvement on transmission at lower frequency range when adding the number of circle DWS in direction *Y*.

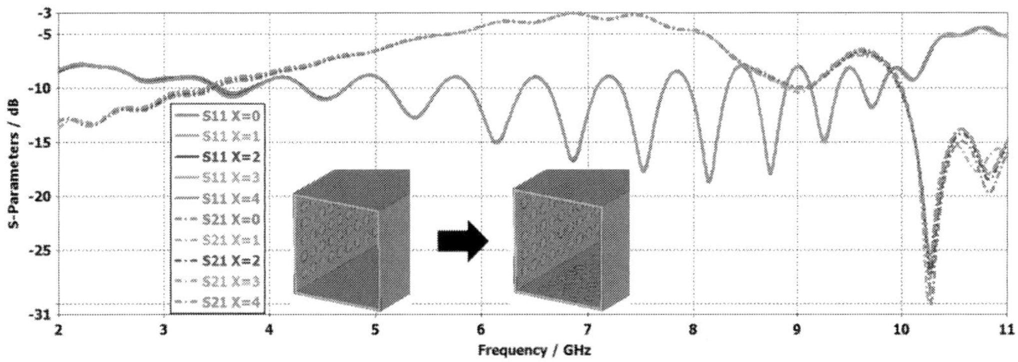

Figure 7. S-Parameters results by varying number of circle DWS in *X* direction

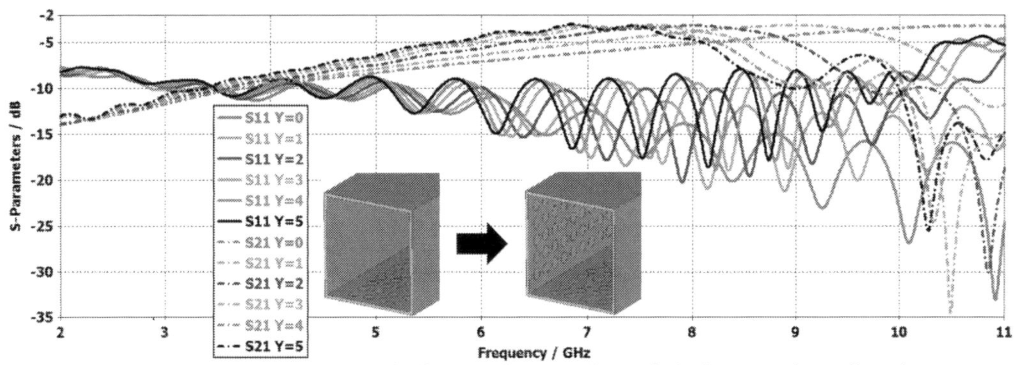

Figure 8. S-Parameters results by varying number of circle DWS in *Y* direction

For the length of circle DWS waveguide, the number of circle DWS in *Z* direction is varied from number 5-30. As seen in Figure 9, the more the number of circle DWS in *Z* direction, the lower the transmission at higher frequencies. It is understood that with the increasing of number in *Z* direction, the longer the circle DWS waveguide, more losses are occurred in transmission especially at high frequencies. In zoomed view of Figure 9, the shorter waveguide with *Z=5* shows biggest difference of fluctuated S11 with 25dB. By adding the number of *Z*, the S11 results tend to have one more cycle but smaller change in magnitude dB. For S21, shorter waveguide with least number in Z direction shows higher transmission and wider working band at higher frequency. In Figure 9, the S11 in phase is showed at frequency 5GHz-10GHz. It is observed that almost similar position but different in degree of phase are achieved for all designs at frequencies 6.02GHz and 8.08GHz. The shortest waveguide demonstrates lowest in phase at both frequencies.

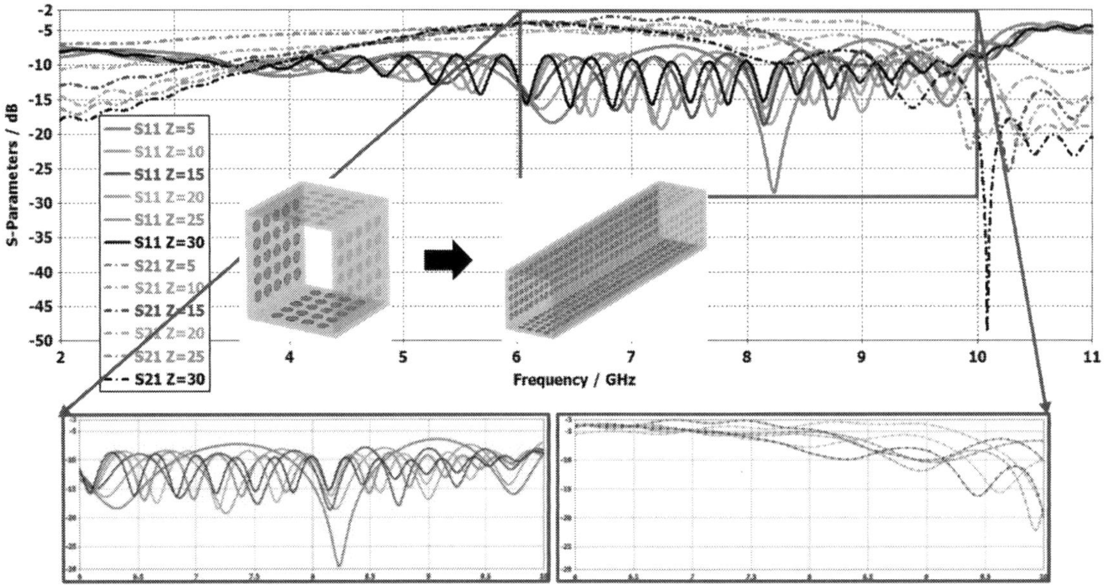

Figure 9. S-Parameters results by varying number of circle DWS in *Z* direction

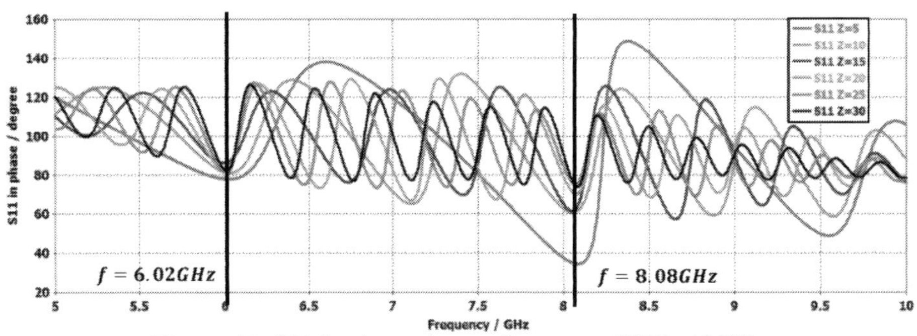

Figure 10. S11 in phase view at frequency 5GHz-10GHz

Design of circle DWS with monopole antenna is observed through return loss, efficiency, gain and directivity as shown in Figure 11. For return loss, it achieves wide bandwidth of 8.66GHz at 2.34GHz-11GHz with two resonances at 3.386GHz and 8.624GHz. Besides that, the efficiency is more than -3dB to ensure more than 50% of the signal can be radiated successfully. Furthermore, it can achieve more than 2dB and 2dBi for gain and directivity respectively. Both directivity and gain show almost similar curve with highest directivity and gain are 8.62dBi at 9.5GHz and 7.29dB at 9dB. Almost high directivity and gain are achieved at lowest resonant frequency 8.624GHz.

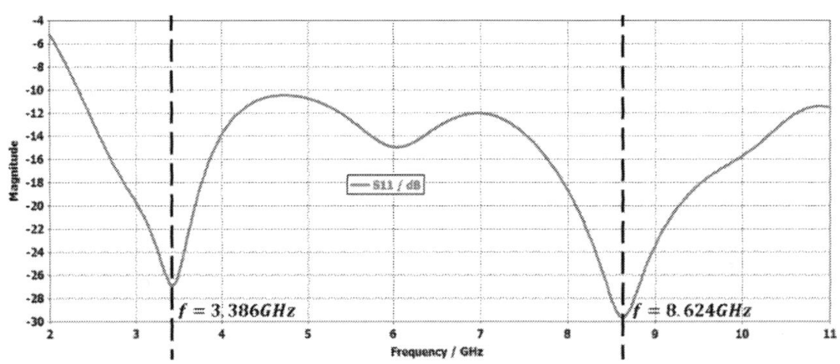

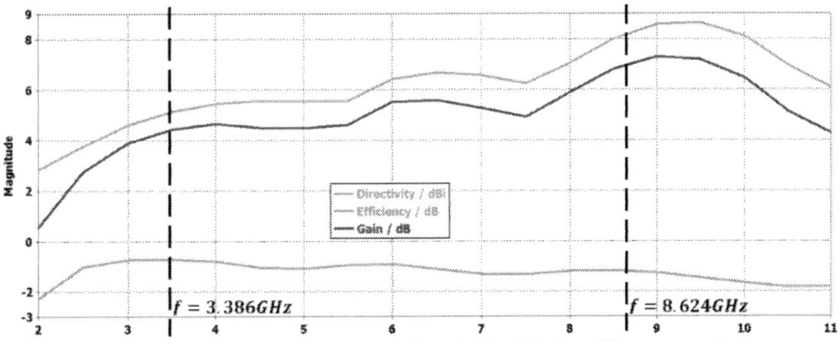

Figure 11. Antenna parameters results circle DWS with monopole antenna

The surface current distribution is also observed at a few interested frequencies. They are plotted at *phase=90* degree in Figure 12. It is noticed that bottom wall has less concentration of surface current compares to side wall. From the side wall, the transmission of wave can be observed. At 2GHz, the distribution of surface current is only occurred near monopole antenna due to longest wavelength in propagation. At first resonant frequency 3.386GHz, higher concentration of surface current is achieved at side wall. This show the propagation of wave is occurred through circle DWS waveguide. Surface current is also distributed at the side wall but with less concentration due to high reflection at 4.637GHz. Meanwhile, at second resonant frequency 8.624GHz, the surface current distribution is occurred at both side and bottom walls. Higher concentration of surface current still can be promised due to lowest reflection and high gain at higher frequency.

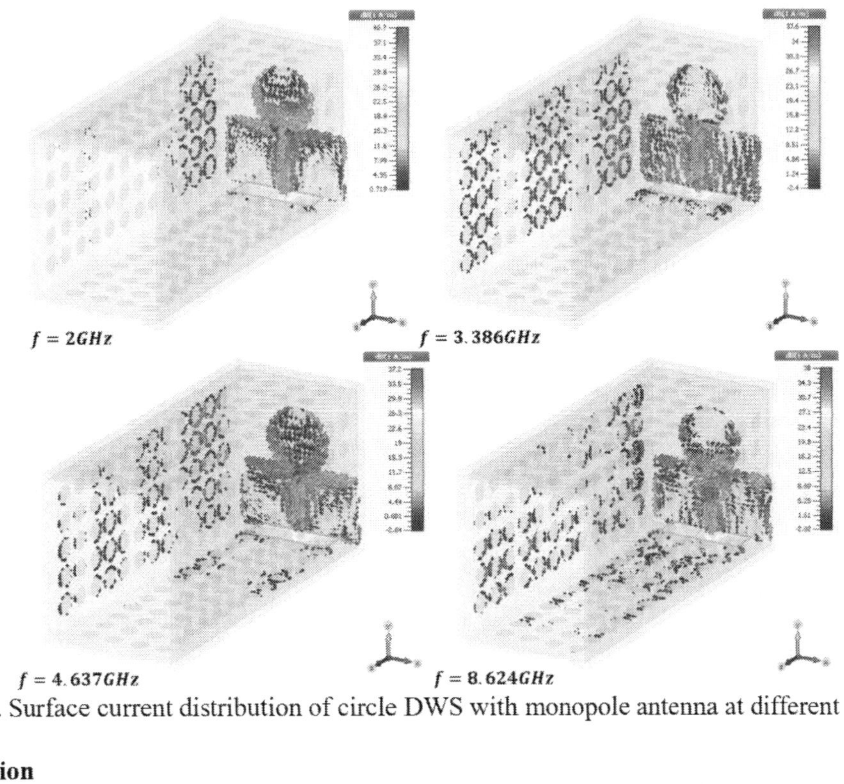

Figure 12. Surface current distribution of circle DWS with monopole antenna at different frequencies

4. Conclusion

This paper presents and studies the circle DWS with monopole antenna. Basically, a few factors for circle DWS towards waveguide are studied which includes size, inter element spacing and number of circle DWS. The design of circle DWS with monopole antenna still able to work in wide bandwidth of

IOP Conf. Series: Journal of Physics: Conf. Series **1049** (2018) 012087 doi:10.1088/1742-6596/1049/1/012087

8.66GHz at 2.34GHz-11GHz. In addition, the efficiency is higher than -3dB to ensure the satisfactory radiation. Furthermore, it can achieve higher gain and directivity with more than 2dB and 2dBi respectively. Highest gain and directivity are occurred at higher frequency range. In future, this circle DWS with monopole antenna will be fabricated for verification purpose. Besides that, others shape will be also used to replace the circle DWS to study the effects towards waveguide.

Acknowledgments
Authors wishing to acknowledge support from Universiti Teknikal Malaysia Melaka (UTeM) and Ministry of Higher Education (MOHE) during the project work. Other than that, a special thank is also given to the UTeM Zamalah Scheme for funding this project.

5. References

[1] Gangwar K, Paras and Gangwar R P S 2014 *Advance in Electronic and Electric Engineering* **4** 97-106
[2] Liu Q, Xi Y Z, Reniers A C F and Smolders A B 2016 *10th European Conf. on Antennas and Propagation* (Davos) 1-4
[3] Zahir Joozdani M and Khalaj Amirhosseini M 2017 *IEEE Transactions on Antennas and Propagation* **65** 705-712
[4] Upadhyay D, Acharya I and Gupta A K 2015 *Int. Conf. on Communications and Signal Processing* (Melmaruvathur) 0176-0179
[5] Oudayacoumar S, Karthikeyan T and Hariprasad V 2014 *Int. Conf. on Advances in Electrical Engineering* (Vellore) 1-4
[6] Pozar D M 2012 *Microwave engineering* 4th edition, John Wiley & Sons (New Jersey)
[7] Zhu H L, Ding C, Wei G and Guo Y J 2016 *Int. Symp. on Antennas and Propagation* (Okinawa) 196-197
[8] Zou T, Zhang B and Fan Y 2016 *IEEE 9th UK-Europe-China Workshop on Millimetre Waves and Terahertz Technologies* (Qingdao) 219-221
[9] Fu Y G and Yuan N C 2005 *Asia-Pacific Microwave Conf. Proc.* 2
[10] Byrne B, Raveu N, Capet N, Le Fur G and Duchesne L 2016 *IEEE Int. Symp. on Antennas and Propagation* (Fajardo) 1091-1092
[11] Cucini A, Caiazzo M, Bennati P and Maci S 2004 *IEEE Antennas and Propagation Society Symp.* **1** 807-810
[12] Bukkawar S and Ahmed V 2017 *Int. Conf. on Nascent Technologies in Engineering* (Navi Mumbai) 1-4
[13] Nikolaou S and Abbasi M A B 2017 *IEEE Transactions on Antennas and Propagation* **65** 2063-2067
[14] Suraj P and Aslam R 2017 *Int. Conf. on Inventive Systems and Control* (Coimbatore) 1-5
[15] Sharma M K, Kumar M, Saini J P, Sukla S and Bhati N 2016 *3rd Int. Conf. on Computing for Sustainable Global Development* (New Delhi) 1545-1549
[16] Acharya I and Ramesh R 2015 *Int. Conf. on Communications and Signal Processing* (Melmaruvathur) 0388-0391
[17] Kwaha B J, Inyang O N and Amalu P 2011 *Int. Journal of Recent Research and Applied Studies* **8** 86-95

Prediction of Photovoltaic (PV) Output Using Artificial Neutral Network (ANN) Based on Ambient Factors

Siti Amely Jumaat[1], Flora Crocker[2], Mohd Helmy Abd Wahab[3], Nur Hanis Mohammad Radzi[4], Muhammad Fakri Othman[5]

[1,2,4]Green and Sustainable Energy (GSEnergy), Faculty of Electrical & Electronic Engineering, Universiti Tun Hussein Onn Malaysia (UTHM), 86400, Batu Pahat, Johor, MALAYSIA

[3]Computational, Signal, Imaging and Intelligence (CSII), Faculty of Electrical & Electronic Engineering, Universiti Tun Hussein Onn Malaysia (UTHM), 86400, Batu Pahat, Johor, MALAYSIA

[5]Department of Multimedia, Faculty of Computer Science and Information, Technology, Universiti Tun Hussein Onn Malaysia (UTHM), 86400, Batu Pahat, Johor, MALAYSIA

E-mail: sitiamely1979@gmail.com[1], flora.wa90@gmail.com[2], helmy@uthm.edu.my[3], nurhanis@uthm.edu.my[4], fakri@uthm.edu.my[5]

Abstract. This paper is proposed an artificial neural network (ANN) to apply in the system of prediction of power output from photovoltaic (PV) panel system. In order to test the efficiency and reliability of a proposed ANN model experimental output will be comparing with mathematical equation. The objectives of this project are to develop the ANN model that capable of predicting power output. The activation functions using for the hidden layer is hyperbolic tangent. The training algorithm is used Levenberg-Marquardt backpropagation. The meteorology data as input data was obtained from RET screen database in the period from 1st January 2015 until 31st August 2016. There were two locations in Malaysia to be the subject test; Melaka and Kuala Lumpur. From the result, for Melaka, Malaysia the outputs V_m (MAPE = 0.0003% and RMSE = 8.5%) and I_m (MAPE = 4.3% and RMSE = 26.8%). Then, for Kuala Lumpur, Malaysia have a less error and good correlation with V_m (RMSE = 0.2%) and I_m (MAPE = 0.008% and RMSE = 0.3%). Hence, average power output was high level in January to March for both locations. The conclusion shows that the performance of power output is depending on the level of solar radiation on the day.

1. Introduction

The three main important issues that is facing by worldwide power generation are increasing oil prices, the depleting of fossil fuel day to day and increasing the worst issues of environmental [1]. The environmental issues in reducing greenhouse gases (GHG), the utilization of renewable energy sources (RES) is now growing rapidly and widely accepted an alternative power supply [2], [3]. The PV systems provide the direct method to convert solar energy into an electric energy. Therefore, modelling and simulation are important role in the development and investigate the performances of PV modules as well to design of PV system. The PV module performances are relied to provide output

[1] sitiamely1979@gmail.com

Content from this work may be used under the terms of the Creative Commons Attribution 3.0 licence. Any further distribution of this work must maintain attribution to the author(s) and the title of the work, journal citation and DOI.
Published under licence by IOP Publishing Ltd

prediction for proposed and existing PV system under the diverse of ambient environmental condition [4]. With the forecasting and prediction method it will easy to control the operation of harvesting sun energy and energy management and planning [5-6]. Nowadays, ANN technique is popular as use to forecasting due to could reduce the cost, achieving higher processing speed and simpler implementations [7].

There are various methods to implement ANN as the solar radiation as a forecaster. Birinchi Bora *et al* in [8] present Artificial Neural Network method based modeling of PV module to predict the output. İlhan Ceylan *et al* in [9] discussed an artificial neural network that used for prediction of photovoltaic module temperature. There are ambient factor involves to evaluate the PV performance. S.S. Priya in [10] studying the feasibility of an Artificial Neural Network (ANN) based method to estimate and predict global solar radiation (GSR). James Mubiru *et. al* in [11] using ANN could be used to predict monthly typical daily direct solar radiation. In paper [12] presented using the Artificial Neural Network (ANN) model to predict the output. The ANN models create using a neural network fitting toolbox (nftool) and the Levenberg-Marquard (LM) algorithm is used in this analysis. A. Saberian *et al* [13] discussed on modeling and prediction of the photovoltaic power output using artificial neural networks. There two neural network structures use for this model are general regression neural network (GRNN) and feedforward back propagation (FFBP). Valerio Lo Brano *et al* in [14] discussed about predicting the power output of a PV panel by using the ANN model. M. Benghamen *et al* [15] discussed about the ANN model that use for estimating and modeling of the daily global solar radiation. They are used four inputs built six different ANN model using different combination inputs. They are chosen feed-forward neural network (FFNN) with back propagation training algorithms. M.A.Behrang *et al* [16] presented a prediction daily global solar radiation (GSR) on horizontal based on meteorological data using ANN method. They are implemented multi-layer perceptron (MLP) and radial bias function (RBF). In [17] S.A.Jumaat *et. al* proposed ANN to be applied in the system of prediction as fast method used for forecasting output of solar radiation for PV module. The purpose of this study is to predict the solar radiation that receives by PV module. The multilayer perceptron (MLP) with back propagation is used as a network topology.

In this research a create an ANN for forecaster data of PV module for location in Malaysia. The multilayer perceptron (MLP) is selected as network topology for forecast model. Then, the Levenberg-Marquard (LM) algorithm is used in this analysis. This forecast model are divide into two model; Model I to predict amount of solar radiation, Model II to predict current and voltage generated by a PV panel system. In order to test the efficiency and reliability of a proposed ANN Model, the experimental output will compare with mathematical equation.

2. Artificial Neural Network (ANN)

Simon Haykin in [18] a neural network is a massively parallel distributed processor which is the unit with a simple process, has ability to sorting experiential knowledge and making it available to use. It is two aspect neural network look like a human brain: (i) the knowledge or input is learned by the network through a learning process. (ii) the inter-neuron connection strengths is used as storage the knowledge, which known as synaptic weights. Artificial Neural Network (ANN) is a field of study for human brain system, which is expected to be applied in real life situation. ANN is able to act like a human brain, which to solve the problem by learning method compared to conventional computing which is pre-programmed. ANN has a processing unit to process information and responds to the input given before giving the output desired. Hence, it can be recognized that ANN is used as a method for replacement human brain as a computing device which is far more powerful than conventional logic algebra computation [11] and basic method to optimization technique and minimizes error [19] and [20].

2.1 Multilayer Perceptron (MLP)

MLPs are the most common type of feed-forward networks use for prediction data. The MLP which consist of three types of the layers: an input layer, a hidden layer and an output layer. As known that, bias weight is represented the extra weight for each unit which fixed to 1 and it can be changing by adjusted same value with weight in the network. The important to get the best training performance for MLP is by choosing architecture for giving input parameter. Most an architectural example used is MLP with two-layer perceptron, which with sigmoid transfer function (non-linear). These models can approximate arbitrary accuracy the problem solving with choosing one hidden layer units. Model selection is depending on complexity of the problem that MLP model needed to be solved. However, if there are too many hidden unit will cause the model overfitting and effect the interpolation of test set become poor quality. Then, at least ten network different with different hidden layer should be trained. The initial of the weight is in small random of value near zero and can be both positive and negative. Instead that, the learning rate is considerable to the generalization performance of the network, then reasonable value is in the range 0.05 to 1.0 as default. Before training process, the data should to be partitioned into three types; train, test and validate set. Input data usually need to normalized according to the desired range (0 to 1) or (-1 to 1). The pattern from training set will present the network pattern and one epoch of training is set. It can be hundred epochs occur during the learning process. Nevertheless, this learning process has a stop criterion when validation stop decreases. The validation set is used to decide stopping training. Hence stop criterion should reach the minimum point of error on validation set in order to get good performance [21], [22]. In this paper, there are two ANN Models are created in order to predict solar radiation as for ANN Model I and current and voltage for ANN Model II as shown in Figure 1(a). Figure 1(b) is illustrated developing of ANN model using MLP network topology with Levenberg-Marquardt (LM) algorithm neural network.

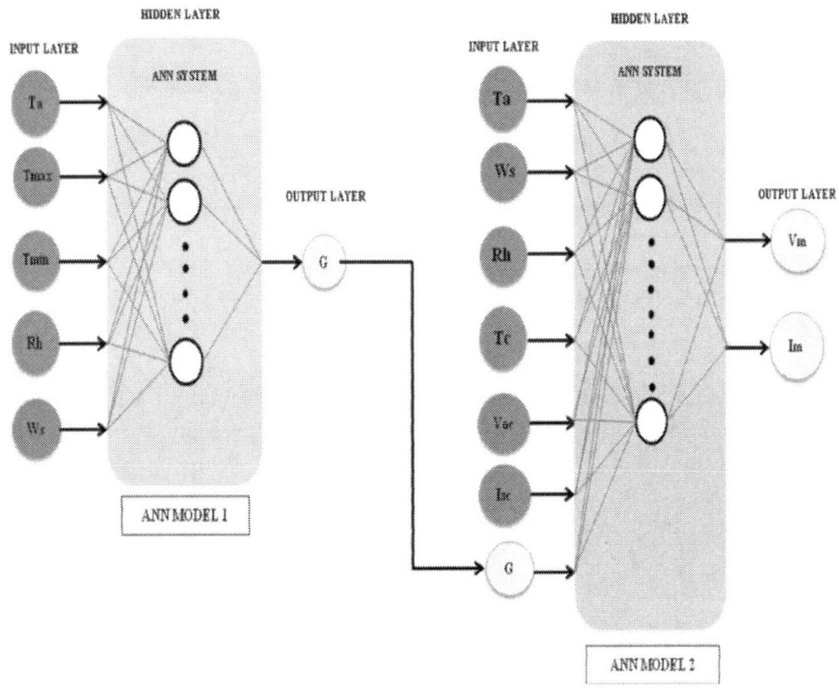

(a)

IOP Conf. Series: Journal of Physics: Conf. Series **1049** (2018) 012088 doi:10.1088/1742-6596/1049/1/012088

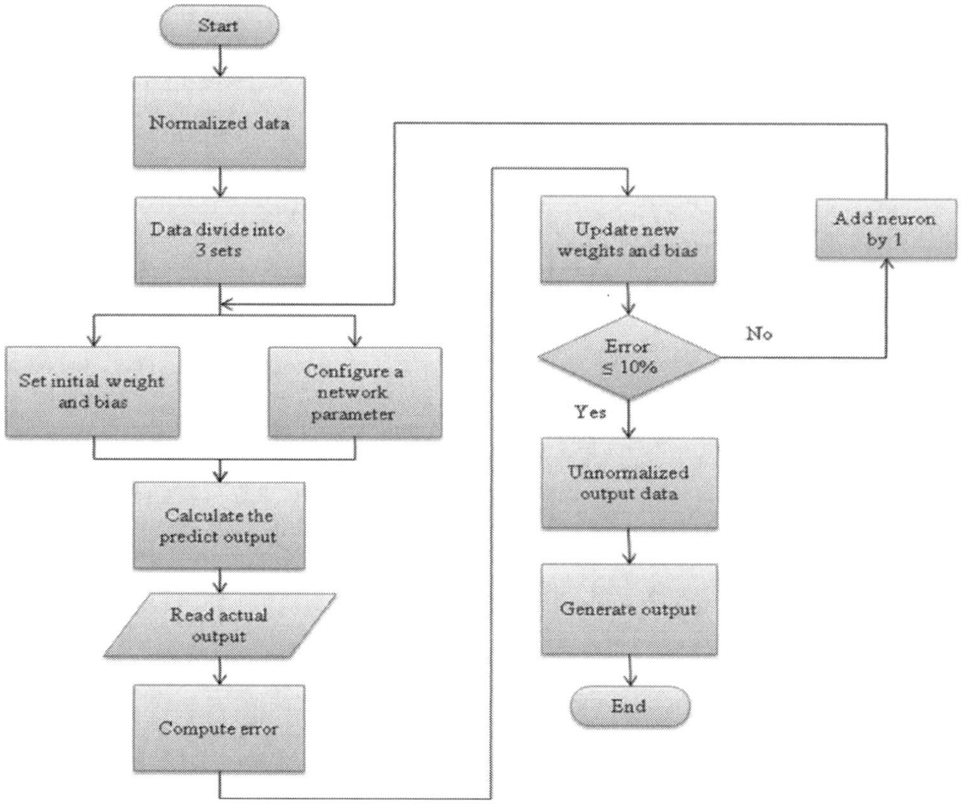

(b)

Figure 1 (a): ANN architecture network (b): The flow chart for developing MLP network

2.2 Input Data Vector For System

To occupy and train an ANN Model, there is the large database of specific data which represents the analyzed parameter of system is required. The proposed Equations (4) and (5) are used in calculating the value of voltage and current as actual output for proposed the ANN Model 2. However, for the power value is the product of voltage and current, so power as in Equation (6) use common Ohm Laws [23]. Hence, the cell temperature is used to obtain the current and voltage in consider of using PV panel parameter. Based on ambient temperature, relative humidity, wind speed, and global irradiance the use to calculate cell temperature is consider. The intended equation to predict the module temperature formula of Tc (°C) adopted from [24] and expressed in Equation (1). The physical data used for the training of the ANN Model were evaluated by using following equations [14], [23], [25 - 26];

i. Air temperature, T_a (°C)
ii. Relative humidity, R_h
iii. Wind speed, W_s (m/s)
iv. Maximum temperature, T_{max} (°C)
v. Minimum temperature, T_{min} (°C)
vi. Solar radiation, G (W/m^2)
vii. Cell temperature, T_c (°C)
viii. Open circuit voltage, V_{oc} (V)
ix. Short circuit current, I_{sc} (A)

These three last parameter are important in order to improve the evaluation of the current and voltage that generated by PV panel. Their values were evaluated by expressions as follow;

$$T_c(°C) = 0.954T_a + 0.03G - 1.629W_s + 0.088R_h + 3.9 \quad (1)$$

$$I_{sc} = I_{sc\cdot ref}(G/G_{ref}) + \mu I_{sc}\,(T_c - T_{ref}) \quad (2)$$

$$V_{oc} = V_{oc,ref} + \mu V_{oc}\,(T_c - T_{ref}) \quad (3)$$

In this paper, the electrical parameters for experiment data were obtained from a 175W PV panel [14] system, which utilized in developing the ANN Model. The system specification was shown in Table 1 as follow.

Table 1: Data Sheet of Kyocera KC175GH-2 polycrystalline panel

Maximum Power, P_{max} [W]	175
Maximum Voltage, V_{mpp} [V]	23.6
Maximum Current, I_{mpp} [A]	7.42
Open circuit voltage, V_{oc} [V]	29.2
Short circuit current, I_{sc} [A]	8.09
V_{oc} thermal coefficient, μ_{Voc} [V/°C]	-0.109
I_{sc} thermal coefficient, μ_{Isc} [mA/°C]	3.18

3. Experimental Setup

There are five basics steps used to designing ANN model such as; (1) Data collection, (2) Pre-processing data, (3) Creating network, (4) Training network and (5) Test performance of the model. These basic steps are to ensure the designing model follow a systematic procedure and will easy to troubleshoot the system.

3.1 Data collection

The RET screen database was used as data collection to obtain the climate data for subject test location such as Melaka and Kuala Lumpur. All data were collected started from 1st January 2015 until 31st August 2016 and accumulate for the further calculations and comparisons. In fact, Malaysia was experiencing two distinct monsoons there were from October to March was a rainy season (northeast monsoon) and June to September was a dry season (southwest monsoon). Data collection was a first step to collect and prepare the input data for ANN Model. Hence, present study investigates the data of these 20 months start from 1st January 2015 until 31st August 2016. There will be 609 samples of data for each input and output parameter. All input data was meteorology data which obtain from RET screen database. The output data were current and voltage that generated from PV panel system which calculated with Equations (4) and (5) as shown. This parameter will be calculating using equation which adopted from [14], [23], [25], [26] as follow;

$$V_m = V_{oc}\,[1 - (b/U_{oc}) \ln a - R_s\,(1 - a - b)] \quad (4)$$

$$I_m = I_{sc}\,(1 - a - b)] \quad (5)$$

Where

$a = U_{oc} + 1 - 2.U_{oc}.Rs$, $b = a/(1+a)$, $Uoc = Voc/Vt$ and $V_t\,(V) = 0.025[(Tc\,(°C) + 273)/300]$

$$P = Vm \times Im \quad (6)$$

3.2 Pre-processing data

Pre-processing data also known as normalization, this step was important before train the ANN Model in order to obtain the network more efficient. This data normalization was essential as there were varies of variables for a different unit. The normalization step was applied to both the input and the output in the data set. In this way, the network output always falls into a normalized range. The network output can then be reverse transformed back into the units of the original output data when the network was put to use in the field. The min-max technique was used in the present study to normalize the input data as in Equation (7). The range of data normalization was within the range -1 to 1 [13].

$$y = (y_{max} - y_{min}) * (x - x_{min}) / (x_{max} - x_{min}) + y_{min} \quad (7)$$

3.2.1 Creating network

Creating a network was the stage that includes setting network parameter such the number of the neuron for hidden layer, transfer function, training function, weight and bias learning function, and performance function. During the training process, the parameter like the number of training epochs, learning rate, and the number of hidden layer could be changed depend on the error performance of ANN output. This study, multilayer perceptron (MLP) network was used as it has high accuracy to predict and forecast output compared to other networks as discuss in Section 2. The several numbers of the neuron for hidden layer, the number of epochs and learning rate were approaches in order to create the finest network. The number of neurons was set at range 1 to 10, the learning rate will be set in a range from 0.001 to 1.0 and the number of epoch set from 10 to 1000 [22]. Since the MSE value is very small and correlation coefficient, R is approaching to 1, then it could be considering as the selected network. The lower MSE value was showing a good training of the network [25].

3.2.2 Training network

The next step was to train the input data using Levenberg-Marquardt (LM) back propagation. The weight and bias value were set into initial value before training process. Then, out of 609 samples of data will divided into 70% used for training only 427 sample data taken. The rest for testing and validation both were 15% of left data samples. During training process, the algorithm was selects the data randomly from the data set, so every time of train process will got varies of MSE value [27] that depend on 70% of input data was chosen.

3.2.3 Test performance of the model

Then, the next step was testing the performance of the developed ANN Model. In order to evaluate the ANN model proposed in this study, three error statistics used which were; the mean squared error (MSE), the mean absolute percentage error (MAPE), and root mean squared error (RMSE). The MSE represents the error between the actual and prediction output. Then, the MAPE was as accuracy indicator for the neural network. RMSE was used indicate the efficiency of the develop ANN Model in prediction. A large positive RMSE indicates that there was a big deviation in the predicted data from the measured data and the lower RMSE show the accurate of the prediction output. It also represents the measurement of the variation of the predicted data around the measured data [27]. All the errors formula expressed in a percentage as defined as follow:

$$MSE = [0.5(Actual - Predict)^2] \times 100\% \tag{8}$$

$$MAPE = \left[\frac{1}{N} \sum \frac{Actual - Predict}{Actual}\right] \times 100\% \tag{9}$$

$$RMSE = \left[\sqrt{\frac{1}{N} \sum (Predict - Actual)^2}\right] \times 100\% \tag{10}$$

4. Result and Discussion

The trial of investigate the number of neuron, training epochs and learning rate were evaluated to find the MSE of the MLP network. Tables 2, 3 and 4 were represents the percentage of MSE for training and testing and the R value for different number of neuron, training epochs and learning rate. After considering the value of MSE testing at neuron 7 among the other neuron was the lowest compared with MSE training, while the best fit for testing was the finest among MSE training was approaching 1 means a positive linear relationship [25]. It can decide that most suitable number of the neuron was at neuron 7. Additional, insufficient number of the neuron will cause the network overfitting or over training. Aspect to be considered are the MSE value for training data must be smaller than MSE value of testing data and the best fit of training data must be greater than the best fit of testing data [22], [25]. From result experiment, the suitable number of training epochs was 600 and the suitable learning rate was 0.05. If learning rate was set to highest, the algorithm can oscillate and become unstable. If learning rate was lowest, the algorithm takes too long to converge. Then, the MLP network topology for the ANN Model 1 was 5-7-1 and the ANN Model 2 was 7-7-2.

Table 2: MSE for different number of hidden node

No. of hidden node	MSE training	MSE testing	R value
1	0.13	0.22	0.99800
2	0.00252	0.00163	0.99995
3	0.00069	0.00102	0.99999
4	0.00033	0.00078	0.99999
5	0.00086	0.00556	0.99997
6	0.00038	0.00202	0.99999
7	0.00031	0.00023	1.00000
8	0.00051	0.00089	0.99999
9	0.00009	0.00059	0.99999
10	0.00041	0.00036	0.99999

Table 3: MSE for different number of epoch

Epochs	MSE training	MSE testing	R value
100	0.0005	0.0010	0.99999
200	0.00046	0.00103	0.99999
300	0.00009	0.00260	0.99999
400	0.00003	0.00158	1.00000
500	0.00002	0.00001	1.00000
600	0.00001	0.00006	1.00000
700	0.00004	0.00009	1.00000
800	0.00010	0.00362	0.99999
900	0.00005	0.00055	1.00000
1000	0.00031	0.00029	1.00000

Table 4: MSE for different learning rate

Learning rate	MSE training	MSE testing	R value
0.001	0.00001	0.00015	1.00000
0.05	0.000004	0.00003	1.00000
0.2	0.00150	0.00597	0.99995
0.4	0.00073	0.00112	0.99999
0.6	0.00009	0.00248	0.99999
0.8	0.00002	0.00134	1.00000
1.0	0.00012	0.00062	0.99999

The two locations in Malaysia were selected in order to test and check the efficiency of the developed ANN Model. The chosen locations are Melaka and Kuala Lumpur. The first develop ANN Model was predict the solar radiation, G (W/m^2). Moreover, second ANN Model was predicting voltage and current at maximum point in order to forecasting the power generate by PV panel system in different climate condition.

4.1ANN Model I: Solar Radiation

IOP Conf. Series: Journal of Physics: Conf. Series **1049** (2018) 012088 doi:10.1088/1742-6596/1049/1/012088

The model was performed well in this study but it cannot forecast the solar radiation values same as actual very well. For this case, it was obtained low coefficient correlation R in the experiment. The correlation coefficient, R for Melaka was 58%, and Kuala Lumpur was 51%, respectively. Figures 2 and 3 were shows the actual against predict values for selected location. After considering the ambient factor, experiment result for solar radiation that underestimates and overestimates for all location could be affected by these factors; rainy, dry and heavy rainfall. Hence, the findings suggest that the count of an ambient factor as input variable does insufficient to provide improvement for the forecasting model [28]. Therefore is recommended to consider the other variables like latitude, longitude, altitude and sunshine duration, sunshine ratio, mean daily solar irradiance, mean daily air temperature, day and month, sunshine accuracy and mean average temperature, atmospheric data and insolation data [5]. However, the findings obtained here show that the ambient factor does not provide significant precision to forecasting results as support by K.A.Baharin *et al* [28] find that ambient temperature not enough to predict solar radiation. From the experimental result can conclude that the high of solar radiation was recorded in January until March for Melaka (117.87-263.14W/m^2) and Kuala Lumpur (123-274.73W/m^2). The harvesting electrical energy is a suitable period during January until March for Melaka, and Kuala Lumpur.

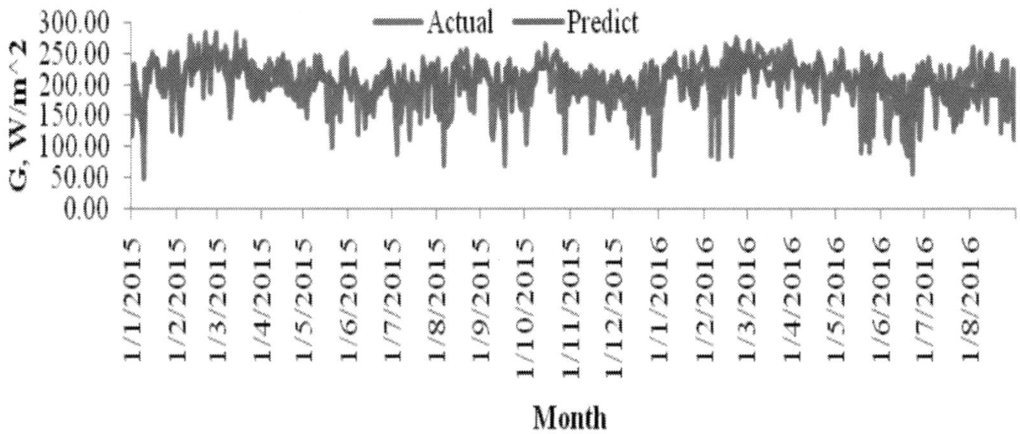

Figure 2: Actual vs. predict value of solar radiation (Melaka)

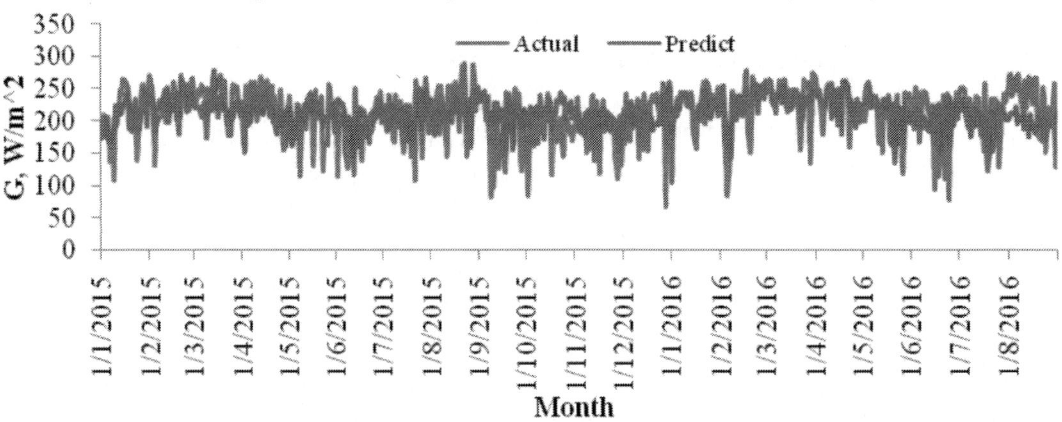

Figure 3: Actual vs. predict value of solar radiation (Kuala Lumpur)

4.2ANN Model II: Voltage and Current

The average actual and estimated voltage values for Melaka were 22.06V – 22.08V and 22.18V–22.94V. Figure 4 show the voltage pattern for estimated was similar with actual value. However, there was some error different between estimated and actual value. The voltage value was decreasing trend on March and May both years due to probably current value fluctuated. Others fact could be a location itself probably have a presence the disturbance such as cloud cluster that could cause the variation in weather that affects the solar radiation. The voltage values produce high on January to February in all probability this location no experience the rainy season. Moreover, the average actual and estimated current values for Melaka were 0.4A – 2.23A and 0.98A–2.12A. Figure 5 show the current pattern for estimated was similar with actual value. However, there was big gap different between estimated and actual value. The current value was decreasing trend in January was rainy season and June to August due to the fact there was dry season condition. Hence, the location probably has a presence the disturbance such as cloud cluster that could cause the variation in weather. The current values produce high on February-March in all probability this location no experience the rainy season. In addition, Kuala Lumpur was experiencing the average value of actual and estimated voltage at 22.19V–23.04V and 22.21V–23.01V respectively.

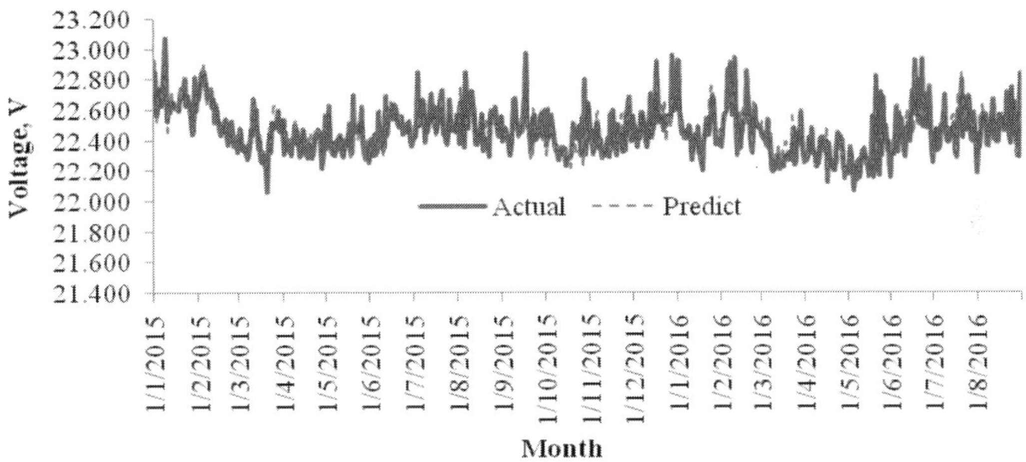

Figure 4: Result for actual and predict data voltage for Melaka

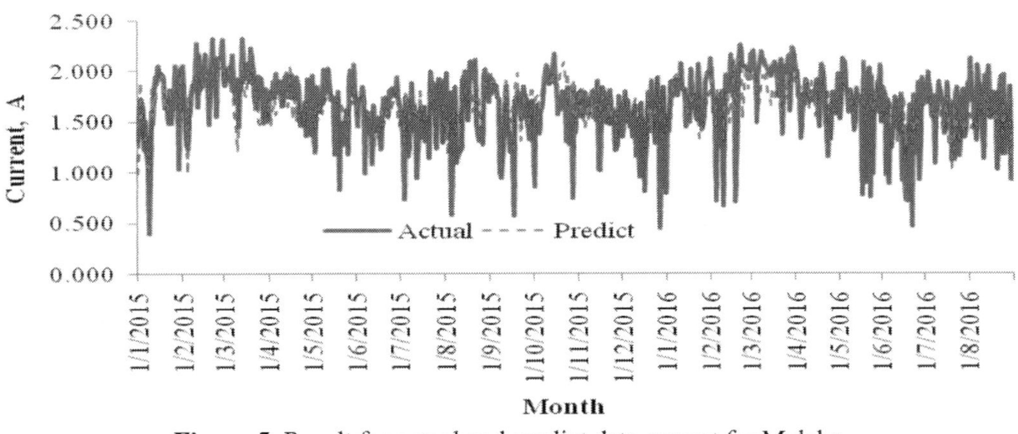

Figure 5: Result for actual and predict data current for Melaka

Figure 6 show the actual and estimated voltage was a similar trend with less error different for both values. The voltage was descending trend on March to May at the rainy season and heavy rainfall. The voltage was an upward trend on February and July and same pattern for both years at this location. However, Kuala Lumpur was experiencing the average value of actual and estimated currently at 1A–2.25A and 1.07A–2.22A respectively. Figure 7 show the actual and estimated current was a similar trend with less error different for both values. The current was a downward trend on February at rainy season, May and July of 2016 at dry season. This could be affected by the solar radiation which it

IOP Conf. Series: Journal of Physics: Conf. Series **1049** (2018) 012088 doi:10.1088/1742-6596/1049/1/012088

directly depends on value on parameters such as like wind speed, relative humidity, and temperature. This was agreement by A. Babatunde *et al* [30] the parameters which were solar radiation, temperature, humidity and wind speed shows related with each other, especially for solar radiation and temperature. Figure 8 show the power output that generated by PV panel system. Overall from result experiment, the power output produces for polycrystalline PV panel almost certainly up to 25%. This could be a fact due to characteristics of the polycrystalline panel itself which the module efficiency is about 10%-11%. Besides, the meteorology data also could influence the power output for PV panel likes ambient temperature, solar radiation, humidity, or wind speed. However, the most important parameter affect the power output was solar radiation. This could the lower solar radiation value which below 400 W/m^2, the power output of PV produces also at low values. This was agreement by A. Ghazali *et al* [30]that polycrystalline show higher power output efficiency at the high level of average solar radiation. This shows that the performance of polycrystalline is depending on the level solar radiation. Hence, from the experimental result, the high potential of producing electrical energy using PV panel for selected locations are in January until March.

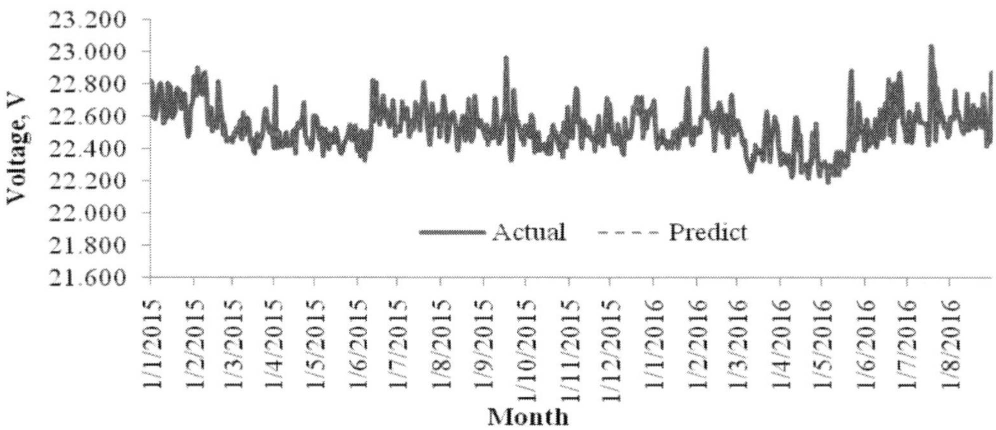

Figure 6: Result for actual and predict data voltage for Kuala Lumpur

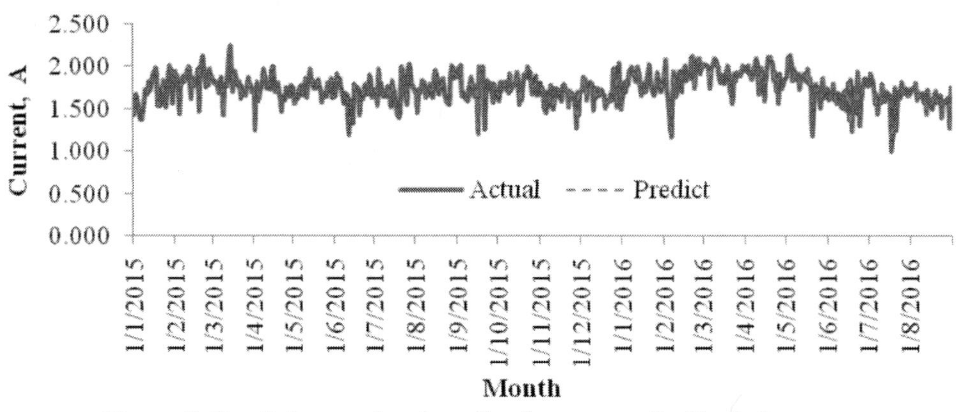

Figure 7: Result for actual and predict data current for Kuala Lumpur

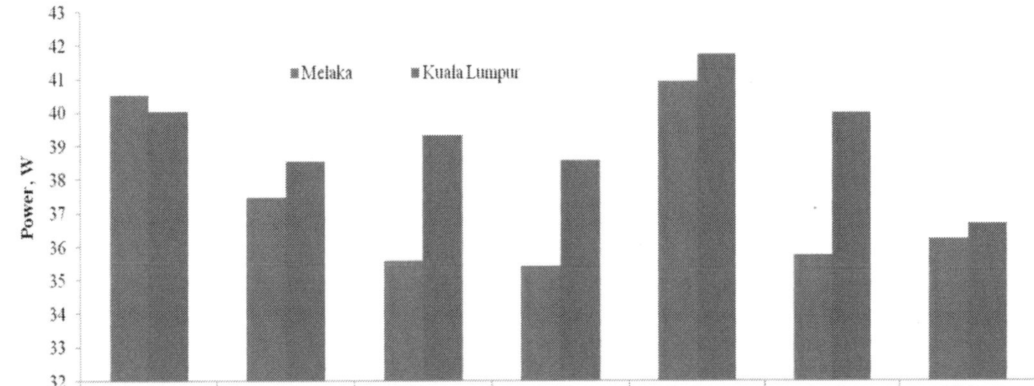

	Jan-Mar	Apr-Jun	Jul-Sept	Oct-Dec	Jan-Mar	Apr-Jun	Jul-Aug
Melaka	40.523	37.461	35.599	35.442	40.923	35.751	36.227
Kuala Lumpur	40.051	38.537	39.333	38.574	41.743	39.99	36.698

Figure 8: Average value of power in Watt for predict data in Melaka and Kuala Lumpur regions

Figures 9 (a) and (b) shows the best curve fit and the MSE performance in order to predict voltage and current for Melaka of ANN Model 2. MSE training of this model was lower than MSE testing and MSE validation. It means the data of model have been learning very well during training.

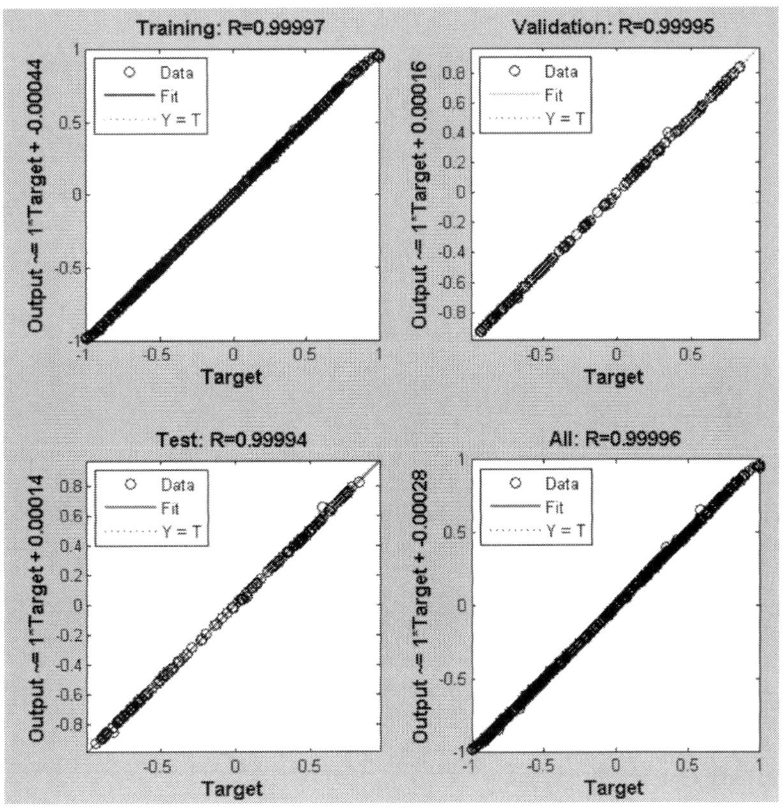

(a)

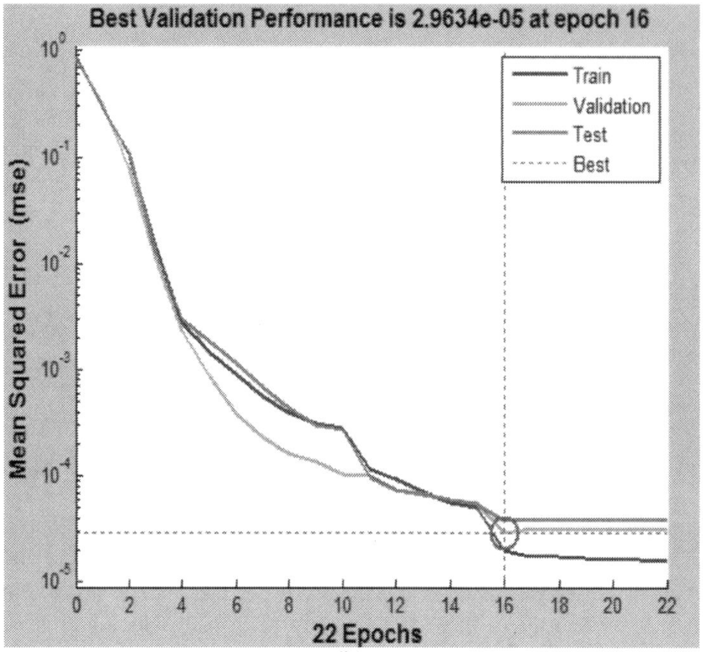

(b)

Figure 9: (a) The best curve fit and (b) MSE performance of ANN Model for Melaka

Figures 10 (a) and (b) shows the best curve fit and the MSE performance in order to predict voltage and current for Kuala Lumpur of ANN Model 2. The best performance for MSE testing of this model was lower than MSE training and MSE validation. The input data of network was learning well during training and testing.

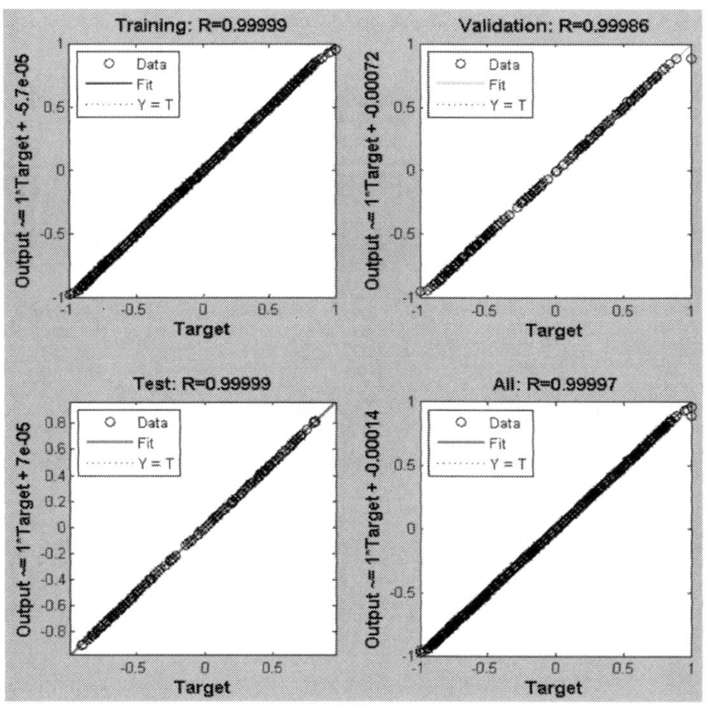

(a)

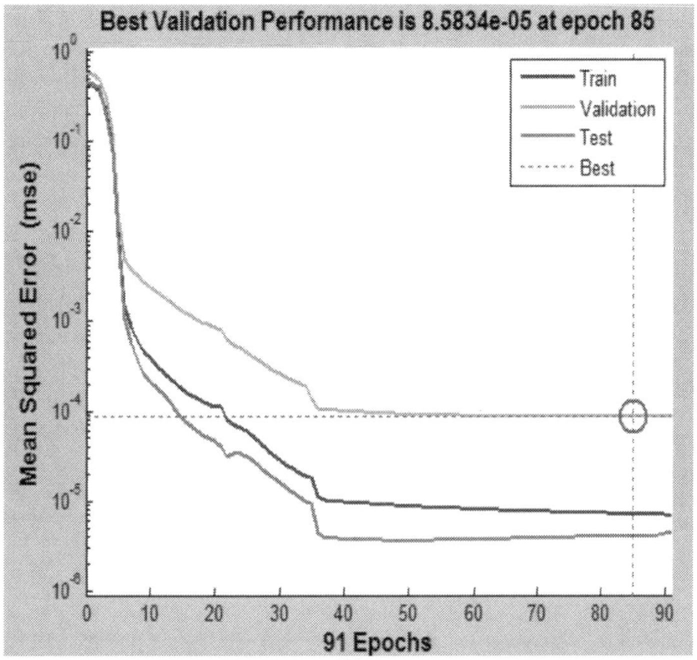

(b)

Figure 10: (a) The best curve fit and (b) MSE performance of ANN Model for Kuala Lumpur

Then, Table 5 has listed the MSE value in percentage and total response of system for each locations. As can be concluded from this table, for location of Kuala Lumpur has 0.0007% train and 0.0004% error. It has good correlation between actual and prediction output. As the MAPE is represent as an accuracy indicator for the neural network. RMSE is represented the efficiency of the proposed model in prediction. Hence, from experiment result in Table 5, the current accuracy recorded for two locations were Melaka = 95.7% and Kuala Lumpur = 99.9%. Then, the voltage accuracy recorded for both locations were 99.9% with a small number of the error. Thus, the efficiency of ANN Model 2 were to predict Vm and Im for both location were considered as lower for Melaka because of RMSE was higher than Kuala Lumpur.

Table 5: MSE, R-value, MAPE and RMSE for selected location

Location	**MSE Train (%)**	**MSE Test (%)**	**R**	**R^2**	**MAPE (%)**		**RMSE (%)**	
					V_m	**I_m**	**V_m**	**I_m**
Melaka	0.002	0.004	0.99996	0.99992	0.0003	4.3	8.5	26.8
Kuala Lumpur	0.0007	0.0004	0.99997	0.99994	0	0.008	0.2	0.3

5. Conclusion

This paper represents the two design of ANN Model were used for predicting the solar radiation, voltage, Vm and current Im that presented in MLP neural network topology. However, the output of ANN Model 1 was training well but not has good correlation between output and target. Besides, the value of regression is not fit to MLP's theory which it should get regression most near to 1. There still have many errors between training set and the target. The model develops as to predict the performance of PV module based on changing ambient factors affect the solar radiation. Then, the output of ANN Model 2 represents the relationship between amounts of ambient factor that affect PV panel output in term of V_m and I_m at the both location. The performances of evaluating ANN Model 2 were using MSE, MAPE and RMSE value which presented the accuracy and efficiency of the system itself. Moreover, this model of the network is successfully developed and predicts the V_m and I_m output. By using this neural network method are easy and fast to predict the output of the system and to solve the problem. The significant of this study it could help out the electrical production for PV panel system owner to control, estimate and planning the electrical bill.

Acknowledgement

This research was supported by Office of Research, Innovation and Commercialization Center (ORICC), UTHM.

6. References

[1] M. Bouzguenda, T. Salmi, a. Gastli, and a. Masmoudi, "Evaluating solar photovoltaic system performance using MATLAB," *2012 First Int. Conf. Renew. Energies Veh. Technol.*, pp. 55–59, 2012.

[2] J. Eynard, S. Grieu, and M. Polit, "Wavelet-based multi-resolution analysis and artificial neural networks for forecasting temperature and thermal power consumption," *Eng. Appl. Artif. Intell.*, vol. 24, no. 3, pp. 501–516, 2011.

[3] Ayu Wazira Azhari, Kamaruzzaman Sopian, Azami Zaharim, "A New Approach For Predicting Solar Radiation In Tropical Environment Using Satellite Images – Case Study Of Malaysia," *Environ. Dev.*, vol. 4, no. 4, pp. 373–378, 2008.

[4] F. B. S. Mekhtoub, "PV Cell Temperature / PV Power Output Relationships Homer Methodology Calculation," *Ipco*, vol. 2, no. 1, pp. 1–12, 2014.

[5] H. G. Chua, B. C. Kok, and H. H. Goh, "Solar Insolation Forecast Using Artificial Neural Network For Malaysian Weather," no. April, pp. 4–9, 2012.

[6] N. Mahamad, M. Kamal, and B. M. Amin, "Power Prediction Analysis using Artificial Neural Network in MS Excel," *Isbn*, pp. 54–58.

[7] F. M. Dias, A. Antunes, and A. M. Mota, "Artificial neural networks: A review of commercial hardware," *Eng. Appl. Artif. Intell.*, vol. 17, no. 8, pp. 945–952, 2004.

[8] B. Bora, O. S. Sastry, and R. Dahiya, "Artificial neural network based modeling of PV module to predict the output," no. December, 2013.

[9] I. Ceylan, O. Erkaymaz, E. Gedik, and A. E. Gurel, "The prediction of photovoltaic module temperature with artificial neural networks," *Case Stud. Therm. Eng.*, vol. 3, pp. 11–20, 2014.

[10] S. S. Priya and Mohammad Hashif Iqbal, "Solar radiation prediction using Artificial Neural Network," *Int. J. Comput. Appl.*, vol. 116, no. 16, pp. 28–31, 2015.

[11] J. Mubiru, "Using Artificial Neural Networks to Predict Direct Solar Irradiation," *Adv. Artif. Neural Syst.*, vol. 2011, pp. 1–6, 2011.

[12] A. K. Yadav and S. . Chandel, "Artificial Neural Network based Prediction of Solar Radiation for Indian Stations," *Int. J. Comput. Appl.*, vol. 50, no. 9, pp. 1–4, 2012.

[13] A. M. Saberian, H. Hizam, M.A.M Radzi, M.Z.A Ab Kadir, and Mirzaei "Modelling and Prediction of Photovoltaic Power Output Using Artificial Neural Networks," *Int. J. Photoenergy*, vol. 2014, 2014.

[14] V. Lo Brano, G. Ciulla, and M. Di Falco, "Artificial Neural Networks to Predict the Power Output of a PV Panel," *Int. J. Photoenergy*, vol. 2014, p. 12, 2014.

[15] M. Benghanem, A. Mellit, and S. N. Alamri, "ANN-based modelling and estimation of daily global solar radiation data: A case study," *Energy Convers. Manag.*, vol. 50, no. 7, pp. 1644–1655, 2009.

[16] M. A. Behrang, E. Assareh, A. Ghanbarzadeh, and A. R. Noghrehabadi, "The potential of different artificial neural network (ANN) techniques in daily global solar radiation modeling

based on meteorological data," *Sol. Energy*, 2010.
[17] Siti Amely Binti Jumaat, Flora Crocker, Mohd Helmy Bin Abd Wahab, Nur Hanis Binti Mohammad Radzi, Investigate the Photovoltaic (PV) Module Performance by Using Artificial Neural Network (ANN), 2016 IEEE Conference on Open Systems (ICOS), 10-12 Oct 2016 pp. 59-64
[18] S. Haykin, *Neural Network: A Comprehensive Foundation*, Second. India: Person Prentice Hall, 2005.
[19] B. Vasumathi and S. Moorthi, "Implementation of hybrid ANNPSO algorithm on FPGA for harmonic estimation," *Eng. Appl. Artif. Intell.*, vol. 25, no. 3, pp. 476–483, 2012.
[20] R. Pino, J. Parreno, A. Gomez, and P. Priore, "Forecasting next-day price of electricity in the Spanish energy market using artificial neural networks," *Eng. Appl. Artif. Intell.*, vol. 21, no. 1, pp. 53–62, 2008.
[21] L.Tarassenko, *A Guide Neural Computing Application*. New York, 1998.
[22] S. Sumathi, S. N. Sivanandam, and S. N Deepa, Neural Network Using Matlab 6.0, Second. New Dehli: Tata McGraw-Hill Education, 2006.
[23] W. Zhou, H. Yang, and Z. Fang, "A novel model for photovoltaic array performance prediction," *Appl. Energy*, 2007.
[24] G. Tamizhmani, L. Ji, Y. Tang, L. Petacci, and C. Osterwald, "Photovoltaic Module Thermal/Wind Performance: Long -Term Monitoring and Model Development For Energy Rating," 2003.
[25] S. S. B. Jamal, "An Application of Artificial Neural Network for DC servo Motor," Universiti Teknologi Malaysia, 2009.
[26] K. A. Baharin, H. Abdul Rahman, M. Y. Hassan, and C. K. Gan, "Short-term forecasting of solar photovoltaic output power for tropical climate using ground-based measurement data," *J. Renew. Sustain. Energy*, 2016.
[27] K. Abhishek, A. Kumar, R. Ranjan, and S. Kumar, "A rainfall prediction model using artificial neural network," *2012 IEEE Control Syst. Grad. Res. Colloq.*, no. Icsgrc, pp. 82–87, 2012.
[28] A. M. Ameen and J. Pasupuleti, "Sensitivity Of Artificial Neural Network Based Model For Photovoltaic System Actual Performance," pp. 241–244, 2014.
[29] K. A. Baharin, H. Abdul Rahman, M. Y. Hassan, and C. K. Gan, "Short-term forecasting of solar photovoltaic output power for tropical climate using ground-based measurement data," *J. Renew. Sustain. Energy*, vol. 8, no. 5, p. 53701, 2016.
[30] A. Babatunde and O. Olusola, "Regional Estimation of Solar Radiation using Routinal Meteorological Variables," *Pacific J. Sci. Technol.*, vol. 5, no. 17, pp. 147–154, 2015.
[31] A. Ghazali M. and A. M. Abdul Rahman, "The Performance of Three Different Solar Panels for Solar Electricity Applying Solar Tracking Device under the Malaysian Climate Condition," *Energy Environ. Res.*, vol. 2, no. 1, pp. 235–243, 2012.

Effect of Rainfall Patterns on Concentration Of CO_2, Soil Temperature And Matric Suction For Acidic Barren Soil

M I S Fazlina[1], A T S Azhar[1*], M E Abdullah[1], N M Sunar[2], Z Embong[3] and M Aziman[1]

[1] Faculty of Civil and Environmental Engineering, Universiti Tun Hussein Onn Malaysia, Johor, Malaysia

[2]Faculty of Engineering Technology, Universiti Tun Hussein Onn Malaysia, Johor, Malaysia

[3]Faculty of Science, Technology and Human Development, Universiti Tun Hussein Onn Malaysia, Johor, Malaysia

Corresponding author: saifulazhar870@gmail.com*

Abstract. This study is based on field monitoring of a soil-atmosphere toward on acidic barren soil. CO_2 in the cavity are dependent on climatic parameters, mainly on the outdoor temperature during rainfall. Observations were taken over a longer period of time. Maximum CO_2 values in the barren slope are reached in the warmest months and are in accordance with soil CO_2 values. The maximum CO_2 concentration in the barren slope is 3022 ppm on average, while the minimum is 2438 ppm. To describe the field findings, CO_2 production and diffusion experiments related to the soil behavior were developed. The results showed that the soil CO_2 production increases as the soil temperature and matric suction increase. The relationship between the gas diffusion coefficient and soil water suction is necessary data (i.e water evaporation or CO_2 evolution above the soil surface).

Keywords: matric suction, slope, CO_2, rainfall, temperature.

1.0 Introduction

The surface soil layer plays a key role in all these processes [1]. On the one hand, it is the responsible for the CO_2 production, which depends on weather conditions and soil physical, chemical and biological properties [2], such as organic matter content, the abundance of microorganisms, type and density of vegetation, etc. On the other hand, soil constitutes the primary interface through which liquid and gas pass between the outside atmosphere and the underground environment. This fluid transfer (i.e., gas and liquid movement) occurs through the soil pore network and is affected by the physical properties of soil, such as density, texture, pore network, mineral composition and organic matter content, among others. In addition, fluid transfer is also regulated by the degree of moisture in the soil, which in turn depends on the infiltration processes whereby water enters the soil and adds to the total soil moisture [3].

2.0 Materials and methods

2.1 Site description

The cut slope was located at the Research Centre for Soft Soil which is a new Research and Development (R & D) initiative by Universiti Tun Hussein Onn Malaysia. The soil sample was collected from Ayer Hitam barren acidic soil [4]5]. From that, the area of soil embankment was constructed with the length of 5 m and the width of 5 m at the bottom. Meanwhile, the small area at the top section which has 1 m length and 1 m width was built for installation of monitoring

Content from this work may be used under the terms of the Creative Commons Attribution 3.0 licence. Any further distribution of this work must maintain attribution to the author(s) and the title of the work, journal citation and DOI.
Published under licence by IOP Publishing Ltd

instrument and as a working platform . The height of embankment is 1.2 m whereas the angle of the slope is 30°. The diagram of the embankment for barren soil is shown in Figure 1.

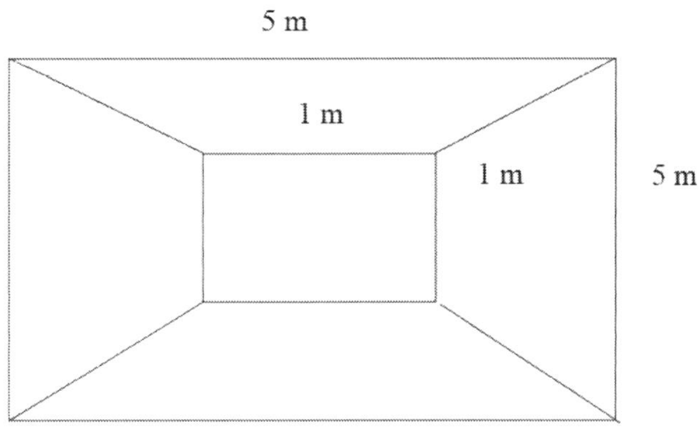

Figure 1. Schematic diagram of scale embankment from the top view

2.2 Field instrumentation monitoring program

To monitor the CO_2 and temperature changes in the slope, the Watermark Soil Moisture Sensor was installed at 20 cm depth(Figure 2). Since rain gauges were used in this study, rainfall data were monitored using a Data-Logging Rain Gauge, WatchDog 1120 (Figure 3).

Figure 2. Watermark Soil Moisture Sensor are installed.

Figure 3. Data-Logging Rain Gauge.

2.3 Soil properties

A series of laboratory tests were conducted in determining the soil properties by which the soil type and geotechnical properties can be characterized as presented in Table 1. The main physical index properties of the soil investigated in this study is soil classification. It depends on several factors such as the moisture content, permeability test, in situ density test (Core Cutter Method), specific gravity, Particle Size Distribution, and Atterberg Limit of the soil.

Table1. The properties of soil material in this study area [4]

Composition	Value
Moisture content (%)	28
Permeability (m/s)	7.45x10^{-7}
Bulk density, ρb **(Mg/m^3)**	1.517
Dry density, ρd **(Mg/m^3)**	1.426
Specific gravity, Gs	2.70
Liquid Limit, LL	46
Plastic Limit, PL	27
Plasticity Index, PI	19
Void Ratio,e	0.756
Porosity,n	0.431
Dry unit weight, γd **(kN/m^3)**	15.08
Saturated unit weight, γsat **(kN/m^3)**	19.3

3.0 Results and discussions

Results were analyzed and tabulated to point out brief respiration of the barren soil physically and mechanically.

3.1 Effect of rainfall intensity on the CO2 content and temperature of the soil.

Data from the barren soil were collected from August 13th to August 20th, 2017 (Figure 4). During this period, the total amount of rainfall was 36.2 mm. Immediately after the first rainfall episode, the CO_2 concentration has increased (from 2637 to 2700 ppm in 17 h) in the soil. According to Pla *et al.*, [6] when the rainfall episode started, the soil takes time to become saturated since it has already contained a considerable amount of water. Therefore, transportation of water is faster initially (Figure 4). This result was supported by Kabwe *et al.*[7] They showed that the CO_2 gas efflux was dramatically reduced after heavy rainfall events, however, the impact was of relatively short duration.

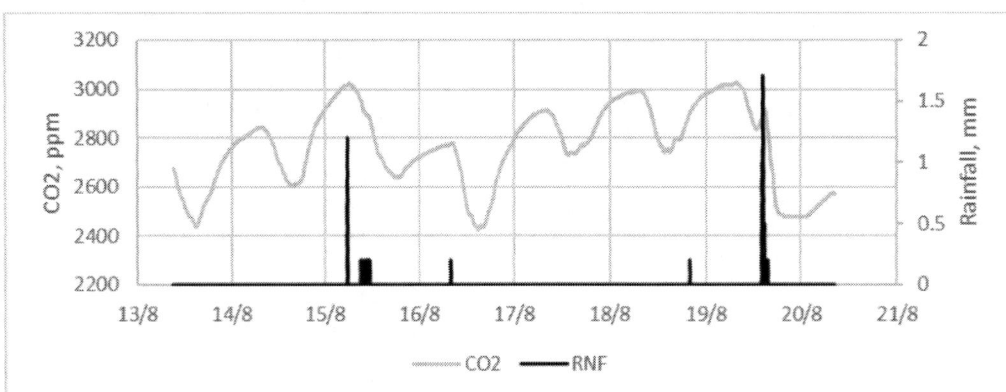

Figure 4. Soil conditions on CO_2 during the rainfall events

On August 16th, in response to the rainfall episode, soil CO_2 has changed from 2458 to 2462 ppm. On the following days, the amount of water in soil was substantially higher than the mean annual value due to the continuous influx of water, which also affected the soil temperature[8]. The last and most significant rainfall event (27.5 mm) was occurred on August 19th. After the occurance, the CO_2 concentration in the soil decreased until August 21st. Under other circumstances, i.e., without rain, the natural CO_2 trend would have decreased during this period of time [9].

Figure 5 shows a typical soil temperature plot. As depicted in this figure, the temperature fluctuates from 21°C to 31°C. Although the plot seems to show a considerable scatter but the general trend is that there is a decrease in temperature due to the rainfall [9]. Likewise, there is a decrease in soil temperature due to the rainfall.

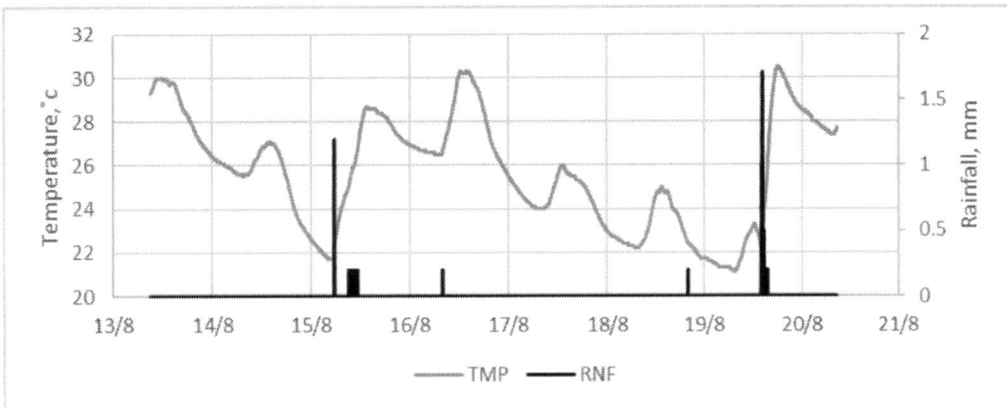

Figure 5. Temperature - time plot

3.2 Relationship of temperature and CO_2 on matric suction

Figure 6 below shows the relation between matric suction on CO_2 and temperature of the soil. The reading was taken after the rainy event occurs for one hour. It showed that an increase in temperature increases the matric suction. The pattern was supported by Nguyen *et al.*, [9] where the magnitude of the temperature rise depends on the amount of water in the porous ceramic matrix, and in turn depends on the matric suction. In other words, the temperature rise is a function of matric suction with the relation of $R^2 = 0.983$. In addition, during dry seasons matric suction increases due to the net loss of water. In contrast, the soil has a net gain of water during the wet season and therefore the matric suction decreases [11]. At the temperature of 22.2°C, the measured matric suction is about 108 kPa. The suction rises to about 117.7 kPa when the temperature reached 24.5°C.

Regarding the relationship between suction and CO_2, the matrix suction is directly proportional to CO_2 as a function of $R^2 = 0.9621$. Because the soil is close to saturation point, almost all the pore space is filled with water, and thus the gas fluxes (e.g., CO_2 and O_2) are expected to be reduced significantly [7]. It should be noted that the free diffusion coefficient of CO_2 is about four orders of magnitude larger in air than in water [12]; diffusive transport in the water-filled pores is much slower than that in the air-filled voids.

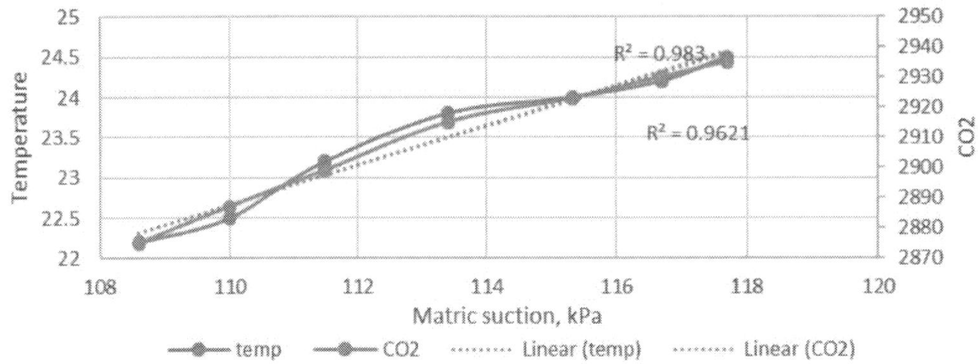

Figure 6. Graph of matric suction versus temperature and CO_2

Hashimoto, in 2004 [13], stated that the gas diffusion coefficient changes with air porosity and soil water content, which in turn is affected the soil water suction. The relationship between the gas diffusion coefficient and soil water suction are of necessary data for various simulations (i.e water evaporation or CO_2).

3.3 Relationship of CO_2 and soil temperature

Higher temperatures lead to higher levels of CO_2, and vice versa as shown in Figure 7. It is predicted that the increase in the atmospheric [CO_2] will lead to increases in air and soil temperatures [14]. Possible high temperature will promote soil respiration, that is to say, the release of CO_2 to the atmosphere [15]. According to Hashimoto[13], soil CO_2 profile depends on gas diffusivity and CO_2 production within the soil. CO_2 increase withi increasing temperature. In addition, the increment of the CO_2 gas gradient with increasing temperature is caused by the increment of CO_2 production in soil. The CO_2 in soil is strongly controlled by soil temperature and moisture [16].

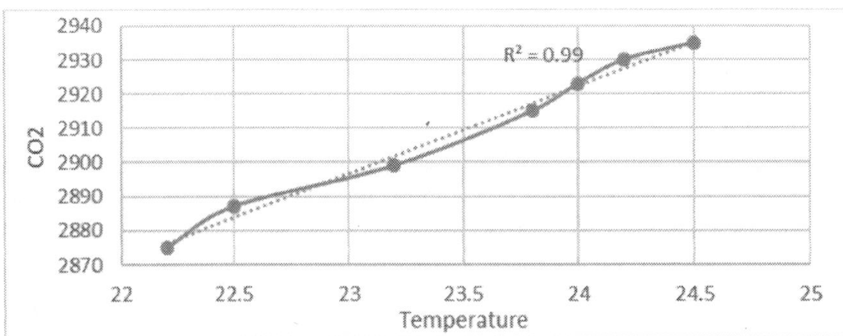

Figure 7. Relationship between soil temperature and CO_2

4.0 Conclusion

In the current study, the effects of heavy rainfall events on the near surface water contents and the resultant CO_2 efflux were investigated. Results showed that the CO_2 gas efflux was dramatically reduced after heavy rainfall events but the impact was of relativly short duration. To quantify the total CO_2 emission, all information about the underlying controls upon respiration from different soil and vegetation were important (e.g. soil disturbs and rainfall intensity). On the other hand, the main factor influencing the soil respiration is soil temperature, because the soil respiration followed the same pattern as the temperature, while rainfall only caused a brief disturbance in the soil respiration. Rainfall seems to contribute only by creating favorable conditions for a quick decrease in temperature and consequently the respiration followed the physical effect of soil water percolation. Soil CO_2 diffusion is also strongly linked to the soil pore space, which is responsible for the gas movement that depends on the water saturation. Soil temperature and moisture directly affect soil CO_2 production although different patterns have been found for different time scales. While daily CO_2 soil cycles are mainly controlled by the soil temperature, for the longer CO_2 time scale, the direct relationship between the soil temperature and the soil CO_2 disappears.

Acknowledgment

The authors would like to acknowledge to the Ministry of Higher Education and Universiti Tun Hussein Onn, Malaysia for supporting this research under Research Grant Contract Vot. U561.

References

[1] E. P. S. Cañete, P. S. Ortiz, A. S. Kowalski, C. Oyonarte, and F. Domingo, "Subterranean CO2 ventilation and its role in the net ecosystem carbon balance of a karstic shrubland," vol. 38, pp. 8–11, 2011.

[2] R. M. Carbonell-bojollo, M. A. R. De Torres, A. Rodríguez-lizana, and R. Ordóñez-fernández, "Influence of Soil and Climate Conditions on CO 2 Emissions from Agricultural Soils," pp. 3425–3435, 2012.

[3] S. X. Yao, C. C. Zhao, T. H. Zhang, and X. P. Liu, "Response of the soil water content of mobile dunes to precipitation patterns in Inner Mongolia , northern China," *J. Arid Environ.*, vol. 97, pp. 92–98, 2013.

[4] A. T. S. Azhar, M. I. S. Fazlina, Y. M. F. , M Aziman, K. Azman, and Z. A. M, and Hazreek, "Preliminary Analysis on Matric Suction for Barren Soil," pp. 2–10, 2016.

[5] J. Norashira, E. Zaidi, M. Aziman, and A. T. S. Azhar, "Observation of pH Value in Electrokinetic Remediation using various electrolyte ($MgSO_4$, KH_2PO_4 and $Na(NO_3)$) for Barren Acidic Soil at Ayer Hitam, Johor, Malaysia," *IOP Conf. Ser. Mater. Sci. Eng.*, vol. 136, p. 12022, 2016.

[6] C. Pla *et al.*, "Role of soil pore structure in water infiltration and CO 2 exchange between the atmosphere and underground air in the vadose zone : A combined laboratory and field approach," *Catena*, vol. 149, pp. 402–416, 2017.

[7] L. K. Kabwe, G. W. Wilson, and J. M. Hendry, "field measurements of surface gas fluxes and surface- water conditions for mine waste rock management 1," pp. 866–884, 2006.

[8] C. Pla, S. Cuezva, N. Fusi, and G. B. Crosta, "Role of soil pore structure in water infiltration and CO2 exchange between the atmosphere and underground air in the vadose zone : a combined laboratory and field approach," vol. 149, no. Part 1, pp. 402–416, 2017.

[9] F. Kitz, K. Gerdel, A. Hammerle, T. Laterza, F. M. Spielmann, and G. Wohlfahrt, "In situ soil COS exchange of a temperate mountain grassland under simulated drought," *Oecologia*, vol. 183, no. 3, pp. 851–860, 2017.

[10] Q. Nguyen, L. Samarasekera, D. G. Fredlund, B. L. Marjerison, and S. Highways, "Presentation of long-term matric suction measurements in a highway subgrade using thermal conductivity sensors," pp. 487–492, 2001.

[11] T. Robin, "Matric suction on total stability evaluations," 2015.

[12] K. A. Birkedal, L. P. Hauge, A. Graue, and G. Ersland, "Transport Mechanisms for CO2-CH4 Exchange and Safe CO2 Storage in Hydrate-Bearing Sandstone," pp.

4073–4095, 2015.
[13] S. Hashimoto, "Studies on carbon dioxide production and its transport in forest soil," 2004.
[14] IPCC, "Climate Change 2014 Synthesis Report Summary Chapter for Policymakers," 2014.
[15] J. I. L. Morison and D. W. Lawlor, "Interactions between increasing CO 2 concentration and temperature on plant growth," vol. 44, pp. 659–682, 1999.
[16] S. Reth, M. Reichstein, and E. Falge, "The effect of soil water content , soil temperature , soil pH-value and the root mass on soil CO 2 efflux – A modified model," pp. 21–33, 2005.

Investigation of Combustion Process of Biodiesel Derived from Waste Cooking Oil, Jatropha, Palm Oil and Algae using Rapid Compression Machine

Mohd Irham Anas[1], Amir Khalid*[1,2,a], Norrizam Jaat [1], Norshuhaila Mohamed Sunar[1], Faisal Hushim[1], Azahari Razali[2], Azwan Sapit[2]
[1]Automotive and Combustion Synergies Technology Group, Advanced Technology Centre (ATC), Faculty of Engineering Technology, Universiti Tun Hussein Onn Malaysia, Parit Raja, Batu Pahat 86400 Johor, Malaysia
[2]Combustion Research Group (CRG), Centre for Energy and Industrial Environment Studies (CEIES) Faculty of Mechanical and Manufacturing Engineering, Universiti Tun Hussein Onn Malaysia, 86400 Batu Pahat, Johor, Malaysia

*Corresponding Author: amirk@uthm.edu.my[a]

Abstract. Biodiesel essentially comes from plants and animal fats which is safe and can be used in diesel engines with few or no modification needed toward existing engine. Higher of the biodiesel blending ratio reflects to the higher viscosities that influences to the fuel-air premixing, ignition process and combustion characteristics. The objective of this research is to investigate the effect of different blending ratio on ignition delay, ignition process, combustion characteristics and emission production using Rapid Compression Machine (RCM). Rapid Compression Machine is used to simulate combustion behaviour similar with the real combustion for internal combustion engines. The standard petroleum diesel was blend with biodiesel derived from waste cooking oil, algae and jatropha with blending ratio from 2vol% to 40vol%. The biodiesel were tested using 130 MPa of injection pressure with constant ambient temperature of 950 K. Increased of the blending ratio of biodiesel significantly influences the longer of ignition delay. Changes of biodiesel properties thus influence to the fuel-air mixing, ignition process and combustion characteristics. In addition, the emission of NO_x and CO_2 increased when increasing biodiesel blending ratio.
Keywords: Biodiesel, Rapid Compression Machine, Ignition delay, Emission, Waste cooking oil, Algae, Jatropha

1. Introduction

Fossil fuel depletion and global warning issues have strongly motivated research in fuel production from the biomass. Biodiesel is one of alternative fuel manufactured from animal fats or vegetable oils [1-5]. In this study, the effect of different biodiesel blending ratio towards ignition delay and emission was investigated in rapid compression machine (RCM). Many studies conducted using rapid compression machine by previous researcher. Some of them require continuation and improvement from the previous research. The RCM used to simulate the occurrence of combustion process [6-9]. Besides, the RCM used is a single cylinder, free moving piston and direct injection of fuel and air. This system allowance of studying about ignition of fuel-air mixture in a controlled condition compared to standard compression ignition engine [10-12]. Ignition delay in a diesel engine was the time interval between the start of injection and the start of combustion [13-14]. The ignition delay process divided to two process which are physical process and chemical process. Physical delay

Content from this work may be used under the terms of the Creative Commons Attribution 3.0 licence. Any further distribution of this work must maintain attribution to the author(s) and the title of the work, journal citation and DOI.
Published under licence by IOP Publishing Ltd

process involves fuel-air mixture formation which contributes to atomization of the fuel. Meanwhile, chemical delay process more refer to premixed combustion [15-16]. The injection parameter and fuel properties have been studied to improve combustion characteristics. Emission of hydrocarbon and particulate matter gas formation due to incomplete fuel-air mixing occur in ignition delay. The ignition delay plays an important role for controlling the emission [16-18].

In this research, the ignition delay and emission are investigated with constant injection pressure and constant ambient temperature. The blending machine used for fuel preparation. This experiment was performed in Rapid Compression Machine (RCM). The experiment from Palm-Oil biodiesel, Algae biodiesel, Jatropha biodiesel were compared with standard diesel (petroleum based) in order to understand the effects of variant biodiesel properties on ignition process and ignition delay.

Figure 1 illustrates the blending machine while Figure 2 illustrates the schematic diagram of blending machine. The different biodiesel with different blending ratio used to study the effect of ignition delay and emission. Blending machine system used to prepare the fuel. The temperature of boiling water was set up for 70°C. The biodiesel was stirred about 2 hour with standard diesel.

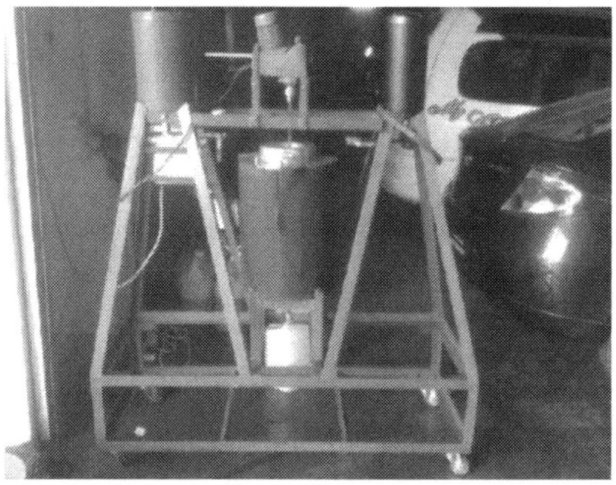

Figure 1: Blending machine.

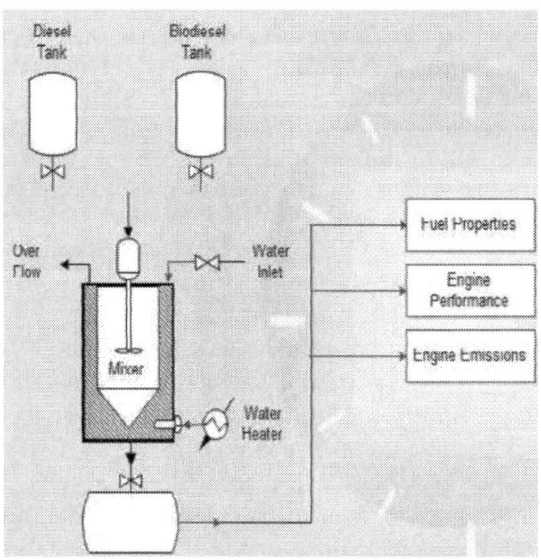

Figure 2: Schematic diagram of blending machine.

In this research, the Rapid Compression Machine was used to stimulate the combustion process as real as compression ignition engine. The injection pressure used was 130 MPa with constant ambient temperature of 950 K. Figure 3 illustrate the Rapid Compression Machine and Figure 4 illustrate the schematic diagram of RCM with research area. The rapid compression was heating up until desired temperature, gasket and a free piston during installation. Tighten all the connection parts to avoid leakage. Once the valve is opened, the nitrogen gas was supply for pneumatically pushed piston forward to stopper installed in front of cylinder liner. The single shot commonrail system injects the fuel into the combustion chamber for combust. Pico Scope 3000 series used to plot a graph for the process.

Figure 3: Rapid Compression Machine

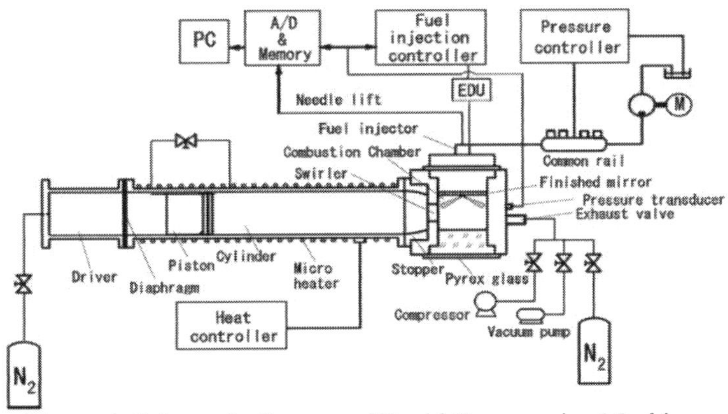

Figure 4: Schematic diagram of Rapid Compression Machine.

Furthermore, Table 1 shown the fuel properties of biodiesel derived from Waste Cooking Oil, Crude-Palm Oil, Algae and Jatropha. Table 2 shows the fuel properties used in this experiment.

Table 1: Experimental condition

Fuel	Injector type	6 holes, Ø = 0.16 mm
	Fuel type	WCO,CPO, Algae, Jatropha
	P_{inj} [MPa]	130
	q_i [ml]	0.04
Ambient	T_i [K]	950
	r_s [m/s]	19
	ρ [kg/m^3]	16.6
	O_2 [vol%]	21

Table 2: Fuel properties

Fuel Type	Properties	
	Density (g/m^3)	Kinematic Viscosity (Cp)
A0	0.860	4.4
A5	0.831	3.6
J0	0.878	4.3
J10	0.838	3.6
B5	0.837	3.0
B10	0.838	2.9
B15	0.840	3.0
Wco5	0.889	3.0
Wco10	0.890	3.18
Wco15	0.892	3.2

RESULT AND DISCUSSION

The effect of biodiesel blending ratio on ignition delay is investigated for Waste Cooking Oil Biodiesel (Wco5, Wco10 and Wco15). The experiment condition used injection pressure of 130 MPa with constant ambient temperature of 950K. The other constant experiment condition was fuel quantity injection which was 0.04 ml. Then, chamber pressure with 0.1 MPa. Besides, air density used was 16.6 kg/m^3.

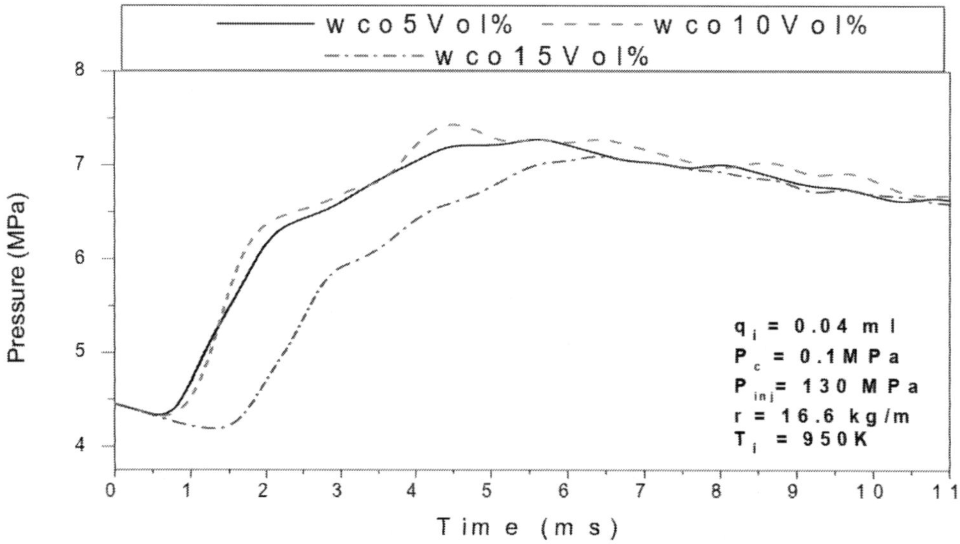

Figure 5: The variation of waste cooking oil biodiesel on ignition delay.

Figure 5 clearly demonstrate the ignition delay versus biodiesel blending ratio. In addition, the Pressure axes demonstrate the patterns of combustion occur in combustion chamber and time axis demonstrate the duration of ignition delay until end of combustion. As seen in Figure 5, biodiesel fuel Wco5 has shortest ignition delay compared to Wco10 and Wco15. The ignition delay duration for Wco5 is 0.7 ms. Meanwhile, the duration for Wco10 and Wco15 are s 0.8 ms and 1.7 ms, respectively. From figure 5, the increasing blends of biodiesel increase the ignition delay. This was due to increasing viscosity of the fuels. Higher viscosities affect the poor atomization and reduce the efficiency of the mixture formation. Then, the results shown that Wco15 has longer ignition delay compared to Wco5.

Figure 6 illustrate the different biodiesel derived from Waste Cooking Oil (Wco5, Wco10 and Wco15) and Palm-Oil (B5, B10 and B15). The experiment condition used injection pressure of 130 MPa with constant ambient temperature of 950K. Based on Figure 6, Palm-Oil biodiesel has same results with Waste Cooking Oil biodiesel. The increasing in blending ratio effects the increasing in ignition delay of the corresponding fuel.

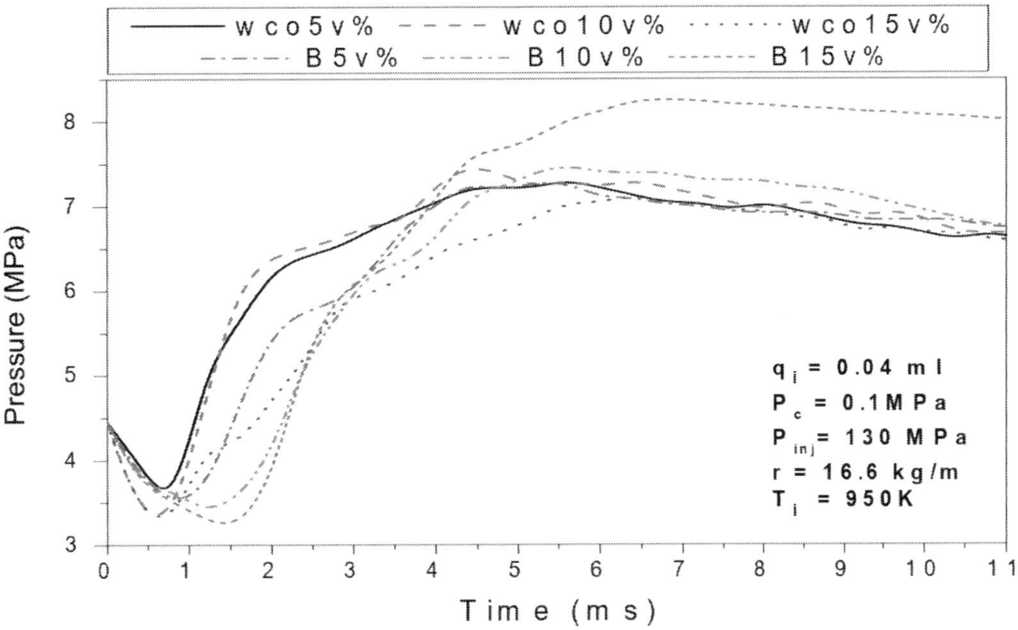

Figure 6: Comparison of WCO and CPO on ignition delay.

Then, the ignition delay for biodiesel fuel derived from Jatropha had discussed. The blending ratio varies at 5%, 10% and 15%, biodiesel fuel J5 has shortest ignition delay compared to J10 and J15. The ignition delay duration for J5 was 2.14ms. Meanwhile, the duration for J10 was 2.12 ms and J15 was 2.8 ms respectively. From Figure 7, the increasing blends of biodiesel increase the ignition delay. This was due to increasing viscosity of the fuels. Higher viscosities affect the poor atomization and reduce the efficiency of the mixture formation. Then, the results shown that J15 has longer ignition delay compared to J5 and J10.

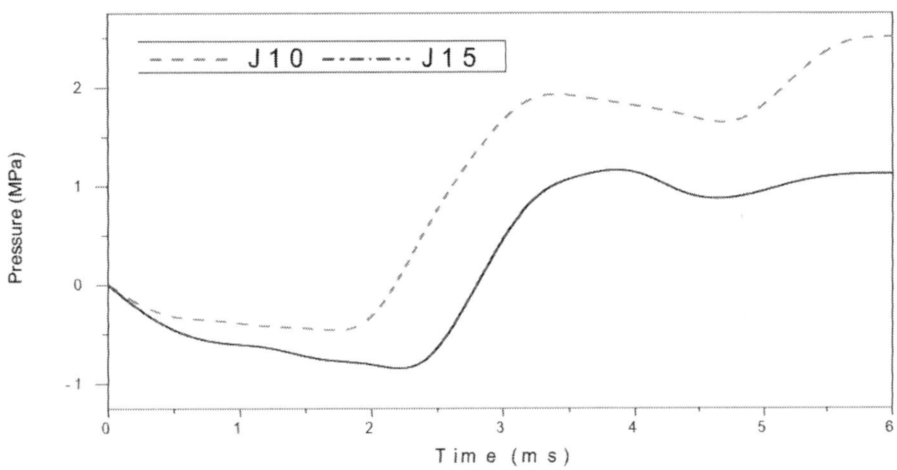

Figure 7: Ignition delay for Jatropha biodiesel.

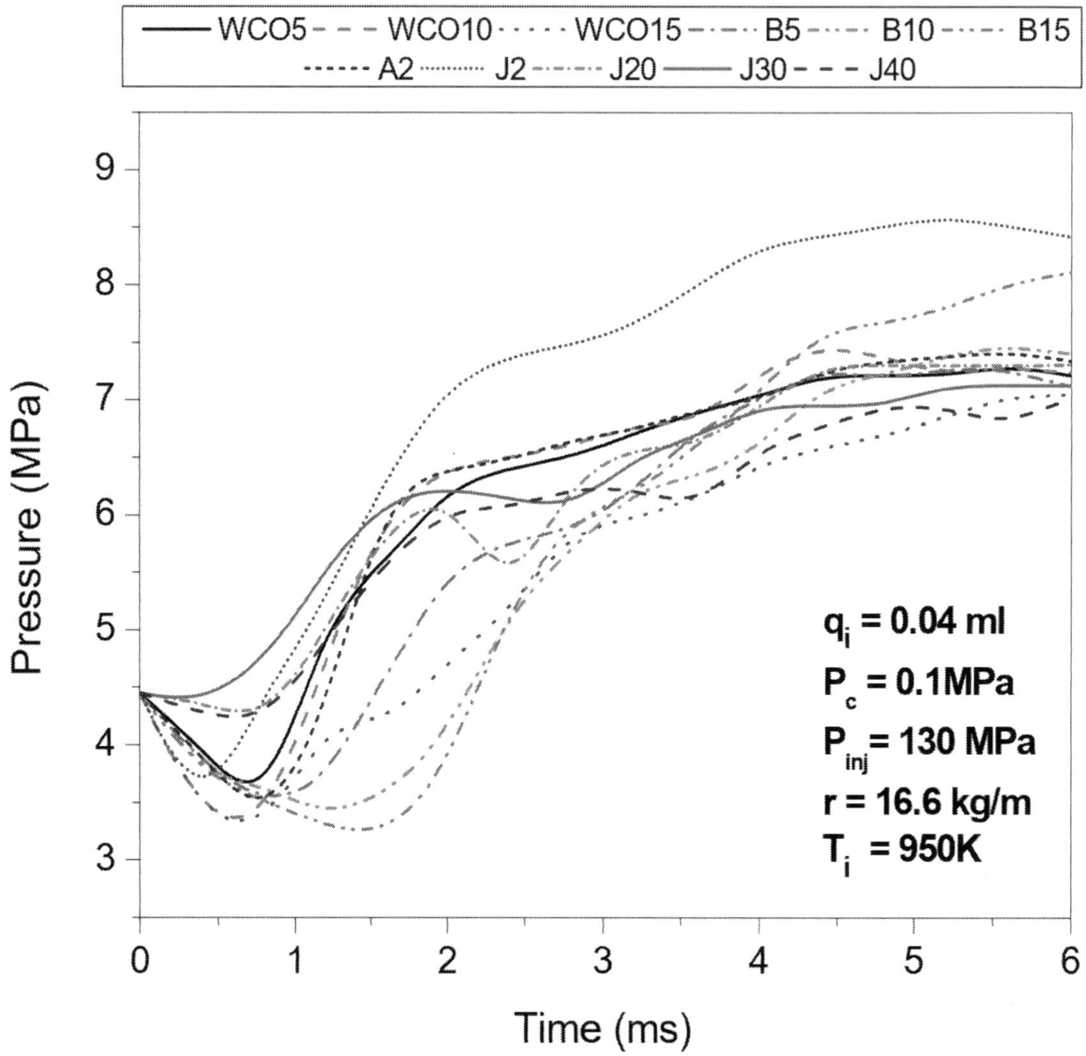

Figure 8: Comparison of different biodiesel blending ratio on ignition delay.

According to Figure 8, the ignition delay for A2 biodiesel is 1.0 ms. The duration is shorter compared to B5, B10 and B15 because of Algae biodiesel (A2) blending ratio consists of 2% of Algae fuel and 98% of diesel fuel. Moreover, Algae biodiesel (A2) has low fuel viscosity and high ignitibility to improve the decreasing in ignition delay duration. Then, Jatropha biodiesel (J2) was compared with previous biodiesel. Jatropha biodiesel (J2) contain same percentage blend with Algae biodiesel (A2). The ignition delay for J2 is 0.7ms. Meanwhile, the ignition delay for J20 is 0.7ms, J30 is 0.8ms and further increase to 0.9ms. This is because high molecular weight of Jatropha breaks down due to chemical reaction at high temperature results shorter ignition delay with low blending ratio.

Then, the effect of biodiesel blending ratio on the exhaust emission discussed in Figure 9. The experiment used the variation of biodiesel fuel which was Waste cooking Oil Biodiesel (WCO5, WCO10 and WCO15), Palm-Oil biodiesel (B5, B10 and B15), Algae biodiesel (A2) and Jatropha Biodiesel (J2, J5, J10, J15, J20, J30 and J40). The experiment condition used injection pressure of 130 MPa with constant ambient temperature of 950K. Figure 9 below illustrates the effects of biodiesel blending ratio on the emission. From Figure 9, the emission of hydrocarbon (HC) and soot formation decreased with the increasing of biodiesel blending ratio. This was due to higher blending fuel contains higher amount of organic condensates and volatiles particles likes oxygen. In addition, Algae biodiesel has low hydrocarbon emission compared to Jatropha due to Algae fuel was lower volatile than Jatropha fuel. In contrast, the percentage of carbon dioxide (CO_2) and NO_X emission increase constantly as the increasing in percentage of biodiesel blending ratio. More oxygen in higher blending of the fuel helps in combustion process to be more complete. High oxygen content contributes to increase the NO_X emission. Next, the emission of carbon monoxide (CO) decreases constantly with higher biodiesel blending ratio. Due to combustion, more carbon converted to carbon dioxide rather than carbon monoxide. Algae biodiesel has low carbon monoxide emission compared to Jatropha biodiesel because Algae contain more oxygen molecules.

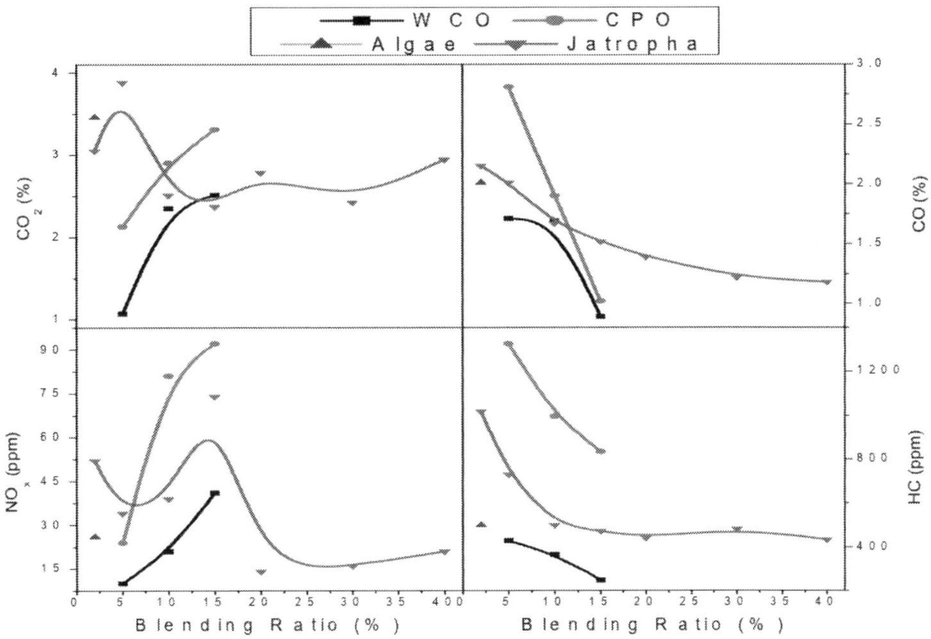

Figure 9: Effect of different biodiesel blending ratio on emission.

CONCLUSION

In this paper, effects of different biodiesel blending ratio had been discussed. The experiment used the variation of biodiesel fuel which was Waste cooking Oil Biodiesel (WCO5, WCO10 and WCO15), Palm-Oil biodiesel (B5, B10 and B15), Algae biodiesel (A2) and Jatropha Biodiesel (J2, J5, J10, J15, J20, J30 and J40). The experiment condition used injection pressure of 130 MPa with constant ambient temperature of 950K. Increased of the blending ratio of biodiesel significantly influences the higher fuel viscosities and longer of ignition delay. It seems that the changes of biodiesel properties thus influence to the fuel-air mixing, ignition process and combustion characteristics. In addition, the emission of NO_x and CO_2 increased when increasing biodiesel blending ratio.

ACKNOWLEDGEMENTS

The authors also would like to thank the Universiti Tun Hussein Onn Malaysia for the Short Term Grant (STG) VOT.0997 and Fundamental Research Grant Scheme (FRGS) vot.1054.

REFERENCES

[1] Meher LC, Sagar DV, Naik SN. Technical aspects of biodiesel production by transesterification – a review. Renew Sustain Energy Rev 2006;10:248–pg 68.
[2] Khalid, A., Hayashi, K., Kidoguchi, Y., Yatsufusa, T., "Effect of air entrainment and oxygen concentration on endothermic and heat recovery process of diesel ignition", SAE Technical

Papers, 2011, Society of Automotive Engineers of Japan, Inc. and SAE International, DOI: 10.4271/2011-01-1834

[3] Khalid, A., Yatsufusa, T., Miyamoto, T., Kawakami, J., Kidoguchi, Y., "Analysis of relation between mixture formation during ignition delay period and burning process in diesel combustion", SAE Technical Papers, 2009, SAE International

[4] Khalid, A. (2013). Effect of ambient temperature and oxygen concentration on ignition and combustion process of diesel spray. Asian Journal of Scientific Research, 6(3), 434-444. doi:10.3923/ajsr.2013.434-444

[5] Agarwal, A. K., & Dhar, A. (2010). Experimental Investigation of Preheated Jatropha Oil Fuelled Direct Injection Compression Ignition Engine Part 1: Performance, Emission, and Combustion Characteristics. Journal of ASTM International (JAI), 7(6).

[6] Wilbur, A. A. Ignition Studies- The Determination of Auto Ignition Temperatures of Hydrocarbon Fuels. Washington DC: Naval Research Laboratory. 1974.

[7] Lyn,W and Valdmanis " The Effect of Physical Factors on Ignition Delay" 1968.

[8] Khalid, A., Mudin, A., Jaat, M., Mustaffa, N., Manshoor, B., Fawzi, M., Razali, M.A., Ngali, Z. (2014), Effects of biodiesel derived by waste cooking oil on fuel consumption and performance of diesel engine, (2014) Applied Mechanics and Materials, 554, pp. 520-525. DOI: 10.4028/www.scientific.net/AMM.554.520

[9] J. M. Luján, B. Tormos, F. J. Salvador, and K. Gargar, "Comparative analysis of a DI diesel engine fuelled with biodiesel blends during the European MVEG-A cycle: Preliminary study (I)," Biomass and Bioenergy, vol. 33, no. 6–7, pp. 941–947, Jun. 2009.

[10] M. Canakci, "Performance and emissions characteristics of biodiesel from soybean oil," Proceedings of the Institution of Mechanical Engineers, Part D: Journal of Automobile Engineering, vol. 219, no. 7, pp. 915–922, Jan. 2005.

[11] M. J. Haas, K. M. Scott, T. L. Alleman, and R. L. Mccormick, "Engine Performance of Biodiesel Fuel Prepared from Soybean Soapstock : A High Quality Renewable Fuel Produced from a Waste Feedstock|," no. 2, pp. 1207–1212, 2001.

[12] Khalid, A., Jaat, N., Sapit, A., Razali, A., Manshoor, B., Zaman, I., Abdullah, A.A. Performance and emissions characteristics of crude jatropha oil biodiesel blends in a diesel engine (2015) International Journal of Automotive and Mechanical Engineering, 11 (1), pp. 2447-2457. DOI: 10.15282/ijame.11.2015.25.0206

[13] Khalid, A., Nursal, R. S., Tajuddin, A. S. A., & Hadi, S. A. (2016). Performance and emissions characteristics of alternative biodiesel fuel on small diesel engine. ARPN Journal of Engineering and Applied Sciences, 11(12), 7424-7430

[14] H. Xie, L. Song, Y. Xie, D. Pi, C. Shao, and Q. Lin, "An Experimental Study on the Macroscopic Spray Characteristics of Biodiesel and Diesel in a Constant Volume Chamber," Energies, vol. 8, no. 6, pp. 5952–5972, Jun. 2015.

[15] A. K. Agarwal, A. Dhar, J. G. Gupta, W. Il Kim, C. S. Lee, and S. Park, "Effect of fuel injection pressure and injection timing on spray characteristics and particulate size – number," Appl. Energy, vol. 130, no. October, pp. 212–221, 20

[16] Khalid, A., Mustaffa, N., Manshoor, B., Zakaria, H., Alimin, A. J., Leman, A. M., & Sadikin, A. (2014). The comparison of preheat fuel characteristics of biodiesel and straight vegetable oil, (2014) Applied Mechanics and Materials, 465-466, pp. 161-166. DOI: 10.4028/www.scientific.net/AMM.465-466.161

[17] Anbarasu Augustine, Loganathan Marimuthu, Saravanan Muthusamy, Performance and Evaluation of DI Diesel Engine by using Preheated Cottonseed Oil Methyl Ester, Procedia Engineering,Volume 38, 2012, Pages 779-790.

[18] Lee, S., Park, S., & Daisho, Y. (2005). An experimental study of the effects of combustion systems and fuel properties on the performance of a diesel engine. Proceedings of the Institution of Mechanical Engineers, Part D: Journal of Automobile Engineering, 218(11), 1317–1323. http://doi. org/10.1243 /0954407 042579987.

Influences of Storage Duration on the Fuel Properties of Biodiesel derived from Jatropha and Waste Cooking Oil

Nadiarulah Nanihar[1], Amir Khalid*[1,2,a], Fathul Hakim[1], Norshuhaila Mohamed Sunar[1], Bukhari Manshoor[2], Izzuddin Zaman[2]
[1]Automotive and Combustion Synergies Technology Group, Advanced Technology Centre (ATC), Faculty of Engineering Technology, Universiti Tun Hussein Onn Malaysia, Parit Raja, Batu Pahat 86400 Johor, Malaysia
[2]Combustion Research Group (CRG), Centre for Energy and Industrial Environment Studies (CEIES) Faculty of Mechanical and Manufacturing Engineering, Universiti Tun Hussein Onn Malaysia, 86400 Batu Pahat, Johor, Malaysia

*Corresponding Author: amirk@uthm.edu.my[a]

Abstract. Biodiesel is an alternative fuels that have competitive properties as petroleum diesel. Biodiesel is usually can be derive from any fatty acid or oil by transesterification process. This research is to investigate the effects of storage duration and indoor ambient condition on biodiesel properties and characteristics. There are three types of blending which is 5vol% blends (5vol% plant oil 95vol% diesel), 10vol% blending (10vol% plant oil and 90vol% diesel) and 15vol% blending (15vol% plant oil and 85vol% diesel) each called CPO5 (crude palm oil 5vol%), CPO10 (crude palm oil 10vol%),CPO15 (crude palm oil 15vol%), JO5 (jatropha oil 5vol%), JO10 (jatropha oil 10vol%), JO15 (jatropha oil 15vol%) WCO5 (waste cooking oil 5vol%), WCO 10 (waste cooking oil 10vol%), and WCO15 (waste cooking oil 15vol%) respectively. All samples are store at indoor condition and he changes of fuel properties such as acid value, water content, flash point, density and viscosity were measured until 1960 hours periods. The result showing and increasing value is the water content and density. But for the acid value and viscosity the data was showing a constant value throughout the time period. In flash point case the data was varied by the type of oil used, for jatropha group it was showing and increasing value and for another group was showing a decreasing result. Under 1960 hours of storage duration, the effect of degradation was happen although the effect is not significance because the changes are still in acceptable ranges.

Keywords: Biodiesel, Storage, Acid Value, Water Content, Flash Point, Density, Viscosity

1. Introduction

Biodiesel were known as Fatty Acid Methyl Ester FAME) is the most discussed among researchers a decade ago. It can be used as a pure fuel or blended with petroleum based fuel with different percentage but the standard storage and handling procedures used for biodiesel are the main issue due to the biodiesel fuel specifications [1-3]. This topic became an epidemics among the researchers because the increasing of the awareness about our renewable energy and alternative fuel. In this research, the characteristic of various sample [4-6]. Oxidative degradation undergo the complex secondary reactions further oxidize into acids, leading to an increase in acid value cause by the formation of hydrogen peroxide [2]. Increasing rate of acid value depends on storage duration [3–5]. High temperature of storage can cause the increasing value for the result [5]. Bouaid et al has done the

Content from this work may be used under the terms of the Creative Commons Attribution 3.0 licence. Any further distribution of this work must maintain attribution to the author(s) and the title of the work, journal citation and DOI.
Published under licence by IOP Publishing Ltd

research using different storage condition (expose and not expose to sunlight). Sunlight give higher increasing rate of acid value to biodiesel storage [4]. M. Shahabuddin et al stated that acid value, density and viscosity is increasing throughout the time [5]. A. Obadiah et al. found that biodiesel without antioxidant gave a significant increase in kinematic viscosity after 50 weeks. The sample with antioxidant still maintain under control for kinematic viscosity. Antioxidant gives more stability to biodiesel to maintain its properties. Water content in sample would make the kinematic maintain and decrease because the kinematic viscosity is measure of resistance of substance to flow. From Khalid et al shows that the Crude Palm Oil Blending will decrease the emissions was recorded by using burner system [9-10]. So this could be the answer for the future, greener fuel.

In this study, six ratio of biodiesel blended from crude palm oil (CPO) based were using for this research, CPO5, CPO10, CPO15, JO5, JO10,J015,WCO5, WCO10 andWCO15. All of the samples were stored in a liter transparences glass bottles in two condition; indoor condition (22-24°C) and outdoor condition (28-30°C). Sample were stored for 1960 hours and tested every week. Five types of properties testing were tested for this study; water content, density, kinematic viscosity, flash point and acid value. For these testing, European standard testing procedures were used. In this study, biodiesel is be produced by using jatropha oil, waste cooking oil and palm oil were involve a number of processes including titration process, transesterification, and investigate the physical properties of biodiesel. First, the titration was run to determine the content of FFA in the JO. If the value is higher FFA content, JO then must go through to reduce the amount of FFA esterification under 2%. If the FFA content of less than 2%, it will involve this transesterification process [8-10].

Biodiesel is fuels that can be compose from mono-alkyl esters of long chain fatty acids originates from plant or animals that will need to fulfil requirement of ASTM D6751 fuel specification. The advantages of using extraction biodiesel is it can use either co-products or by-products [8]. Biodiesel is very environmental friendly and non-toxic. If compare with petroleum based diesel, biodiesel is the best in term of combustion emissions. It capable to reduce the emissions of carbon monoxide, particulate matter and unburned hydrocarbons [11]. Carbon dioxide produced form biodiesel can recycled by photosynthesis and will minimize the effects of greenhouse effects [12]. The comparison of diesel and biodiesel is described as Table 1;

Table 1 : The Comparison of Standards For Diesel And Biodiesel Based On American Society For Testing And Material (ASTM) [13].

Fuel Property	Diesel ASTM D975	Biodiesel ASTM D6751
Lower Heating value, BTU/gal	129,050	118,170
Kinematic viscosity @ 40°C, cSt	1.3-1.4	4.0-6.0
Specific gravity @ 60°C, g/cm3	0.85	.88
Carbon, wt%	87	77
Hydrogen, wt%	13	12
Oxygen, by dif. wt%	0	11
Sulphur, ppm	500	0
Boiling Point, °C	180 to 340	315 to 350
Flash point, °C	60 to 80	100 to 170
Cloud point, °C	~15 to 5	~3 to 12
Pour point, °C	~35 to ~15	~15 to 10
Cetane number	40-55	48-65
Lubricity (HFRR), µm	300-600	<300

Jatropha Curcas

Jatropha (Jatropha curcas) is one of such non-edible oils, which has an estimated annual production potential of 200 thousand metric tons in India and it can be grown in waste land [14]. The oil content for jatropha kernel is 63.16% [15] and higher linseed, soybean, and palm kernel which is 33.33%, 18.35% and 44.6%, respectively [16]. Hence jatropha will be the perfect biodiesel because it more economical in term of oil contains. Table 2 shows the fuel properties of jatropha oil, jatropha biodiesel, and diesel; Table below will show the fuel properties of jatropha oil, jatropha biodiesel, and diesel [17];

Table 2: Fuel properties of jatropha oil, jatropha biodiesel, and diesel.

S.no	Fuel Blended	Density (Kg/m3)	CV (kJ/kg)	Viscosity @ 40 °C (cSt)	Flash Point (°C)
1	Diesel	850	44000	2.87	76
2	JO20	852	43759.5	3.02	88
3	JO50	857	42673	3.59	113
4	JO100	873	43690	4.23	148
Testing Procedure		ASTM D4052	ASTM D240	ASTM D445	ASTM D93

2.0 Storage Method and Parameter

In this study, density, viscosity, acid value, water content and flash point of biodiesel blend were observed under ambient condition of range of 28^{0}C and 31^{0}C and stored up to 1960 hours of storage duration (10 weeks). Figure 3.1 shows the flowchart of the process involved in this study. Biodiesel samples were stored in glass containers and were monitored on a weekly basis through the test properties[18].

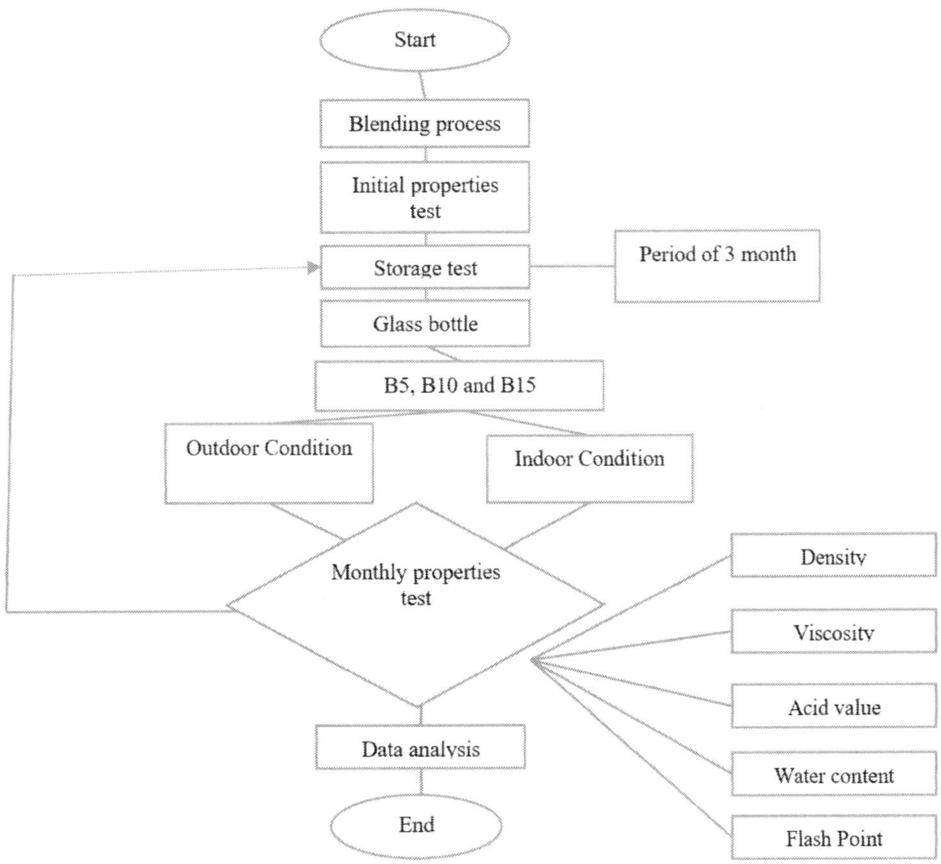

Figure 1: Experiment Flow Chart

This process was observed at a certain temperature and mixing them to produce biodiesel in the presence of methanol and base catalyst. Physical characteristics have been considered including acid value, viscosity, density, water content, and flash point.

In this study there are 4 samples are taken which is 5% blends of jatropha oil (JO 5), 10% blends of jatropha oil (JO 10), 15% blends of jatropha (JO 15). A block diagram of blending process and schematic diagram that used in the mixing process were shown in Figure 2 and Figure 3. In addition, Table 3 describes the blending machine components and its function during the blending process. The purified jatropha oil methyl ester was blended with commercial diesel. The mixer was scale at 60°C and the mixture was stirred at 70°C for 1 hour. The blade speed were maintained at 270rpm[9].

Figure 2: The Blending Machine

Table 3: The Component of Blending Machine.

Parts Number	Component	Function
1	Diesel Tank	Storage and level the diesel
2	Motor	Transmit motion stir
3	Biodiesel Tank	Storage and level the biodiesel
4	Stir	Stir the mixture and biodiesel
5	Water Inlet Pipe	Use to inlet at the around
6	Water Outlet Pipe	To take out the water
7	Mixture Tank	Mixture diesel and biodiesel
8	Switch Box	On the devise and set the temperature
9	Mixture Outlet Pipe	To take out the mixture

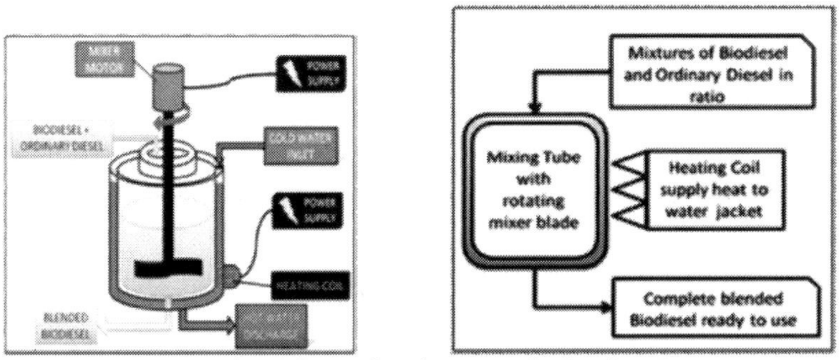

Figure 3: The Schematic Diagram of Blending Process

Properties Test

All of the samples is kept on two different conditions, first is inside a cupboard, and another is on a field. Then the sample is undergo five monthly properties test, which is density, viscosity, water content, acid value and lastly flash point test. Density is an important property which influences the overall standard of a biodiesel. The density of biodiesel blend was measured according to, EN ISO 3675 European Standard. By taking approximately 30 ml of sample it was poured into a beaker and left to cool down to 15°C inside a refrigerator. We measured the weight of an empty 10.104 ml pycnometer by using a weighing machine and level the weight indicator. The cooled sample was poured into the pycnometer until it was completely filled it. The pycnometer was measured again on a weighing machine and the weight shown this time would be the weight of the sample inside the pycnometer. Hence, the density of the sample could be calculated as follow:

Volume = 10.104ml

$$Density = 1000 \times weight\ (g)/volume \qquad (1)$$

Viscosity is an important property which indicates the ability of a material to flow. For the viscosity test, EN ISO 3104 European Standard was followed. The sample needed to be heated up to the temperature of 40°C. Heated sample was then poured into a graduated cylinder. A solid stainless steel sensor of the viscometer (Viscolite 700) was then immersed into the sample. The sample level had to be higher than the line on the sensor and the side of the viscometer sensor must not be touching with the graduated cylinder. The reading of the viscometer result was in centipoises (cPs). The left-over of sample on the solid stainless steel sensor was wiped cleanly after used.

The acid value analysis was based on the titration process of the sample through the use of sodium hydroxide, phenolphthalein and biodiesel sample. In the process of preparing the alkaline reagent, sodium hydroxide (NaOH) 0.1M, 2 g of sodium hydroxide (NaOH) in powder or pellets form were mixed with 500 ml of distilled water inside a volumetric flask. The alkaline mixture was mixed well by shaking the flask thoroughly. The mixture was poured into a burette until it was fully filled. In order to ensure no bubble was trapped inside the burette, few drop of solution was titrated out until the bubble was gone.

For the acid value test, EN 14104 European Standard was followed. The sample was heated up to around 60 – 70 °C. By taking 4 g of the heated sample was mixed with 50 ml of 2-propanol inside a conical flask. The mixture was heated up again to 40°C and then 5-7 drops of phenolphthalein is added in. It was titrated with the sodium hydroxide mixture until the first pink colour appeared and lasted for around 30 seconds. The volume of alkaline solution used for the reaction was recorded.

$$Volume\ of\ alkaline\ solution\ used,\ ml = Final\ volume - Initial\ volume \qquad (2)$$

$$acid\ value, \% = \frac{ml\ of\ alkaline \times 0.1 \times 56.1}{weight\ of\ sample} \qquad (3)$$

For the water content test, EN ISO 12937: 2000 European Standard was followed. Coulometric Karl Fischer Titrators. It was suitable for small water volumes such as biodiesel blend in ppm unit. A syringe was used to draw approximately 5 ml of biodiesel sample and was weighted by using a weighing machine. The weighing machine was then levelled. The start button on the machine was pressed. The sample was injected into the machine solution through the inlet until a keypad screen appeared at the monitor. It was important that the sample is injected slowly so that no excess sample was injected. The remaining sample in the syringe was weighted again. The weight appeared was the amount of sample injected and was inserted into the machine through the keypad appeared. The result of total water content in the sample was shown in a few seconds. Flash point is a parameter of practical importance because as the flash point increases, the higher is the safety level during handling, transportation and storage. For flash point test, EN ISO 2719: 2002 European Standard was followed. By taking approximately 60 ml of sample was poured into the brass cup until the inner line had reached. Every parts of the flash point tester were assembled back carefully and the estimation of boiling point of the sample was inserted into the machine which is around 80°C. The machine was switched on and the sample was heated up. The overall process was comprised of heating and cooling down of the sample. The duration of the overall process was done in 20 minutes.

3.0 Results and Discussions

Effect of the storage duration of biodiesel blending ratio from jatropha oil (JO) under room temperature was investigated at the base STD (just after blends) for biodiesel JO 5, JO 10 and JO 15 respectively. The average storage temperature for room temperature and outdoor temperature are 24°C and 33°C, respectively. Figure 4 shows the flash point and water content with variant storage duration for standard diesel, JPO, WCO and CPO. For flash point the Crude Palm Oil Biodiesel (CPO) and Standard Diesel showing a steady trend but increasing trend for all of JO 5, JO 10 and JO 15.

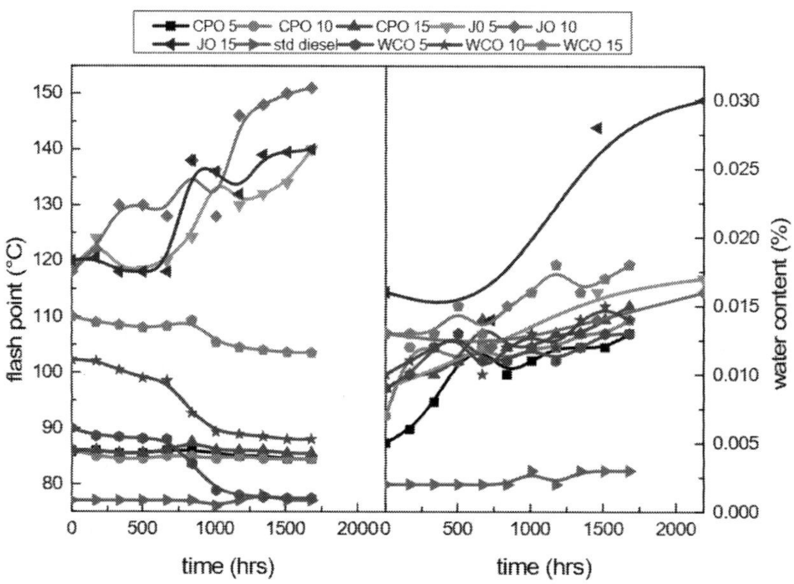

Figure 4: Flashpoint and Water content in Indoor Condition

For jatropha derivation the lowest flashpoint recorded is at 120°C but the data keep increasing until JO10 reach peak value which is 150°C. Higher value of flash point is good for safety of a machine because diesel is self-ignition fuel so it can reduce the change of explosion during handling it. It shows that water content is increased with the increasing of time duration. The highest value is 0.0029% from JO 15 at 2190 hours. The lowest value is standard diesel at 1176 for 0.002% water content. Next Figure 5 elaborates the effects of CPO, JO, WCO and standard diesel for acid value, viscosity and density. The water content is closely related to acid value due to the hydrolysis of methyl ester that causes the increasing of water content value.

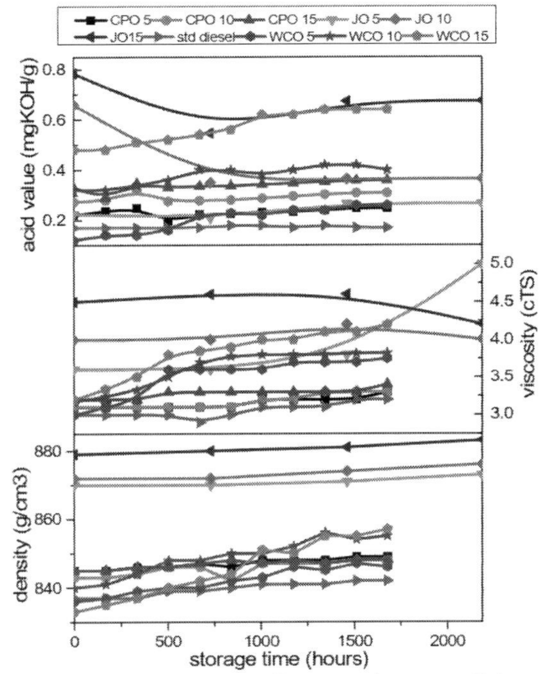

Figure 5: Different Blending at Indoor Condition

The acid value content is slightly increased for CPO, JO5 and standard diesel. For JO10 and JO15, there were showing a decreasing value. The highest acid value recorded at early state which is 7.9 mgKOH/g and the lowest is at 0.17 mgKOH/g. For the viscosity CPO, JO5 and standard diesel increase in value but for JO10, and JO15 the value is decreasing. The highest viscosity was recorded at 2190 hours, 4.6 cTs for JO5. The lowest viscosity is at 672 hours for 2.9 cTs. For density the JO show us a very distinct data which starting from 845 g/cm^3 and the lowest is 835 g/cm^3.

4.0 Conclusions

This research investigate the effects of ambient condition and storage time on flash point, water content, acid value, density and viscosity of CPO (crude palm oil, JO (jatropha oil) and WCO (waste cooking oil). The summary as follows: This study shows that low changes in the physical properties of biodiesel mixed samples when stored at ambient temperature or low temperature. Indoor storage conditions which are less susceptible to delayed break light chain fatty acids. Biodiesel storage at low temperatures is suitable and more advantageous because the impact on the physical properties is minimal and beneficial to slow down the degradation of biodiesel and storage. The physical properties majorly affected by the blending ratio and type of derivation used.

5.0 Acknowledgements

The authors also would like to thank the Universiti Tun Hussein Onn Malaysia for the Short Term Grant (STG) VOT.0997 and Ministry of Higher Education, Malaysia for supporting this research under COE-MTUN Grant Scheme VOT.C009

References

[1] M. Mittelbach and S. Gangl, "Long storage stability of biodiesel made from rapeseed and used frying oil," J. Am. Oil Chem. Soc., vol. 78, no. 6, pp. 573–577, 2001.

[2] Khalid, A., Nursal, R. S., Tajuddin, A. S. A., & Hadi, S. A. (2016). Performance and emissions characteristics of alternative biodiesel fuel on small diesel engine. ARPN Journal of Engineering and Applied Sciences, 11(12), 7424-7430

[3] Khalid, A., Mustaffa, N., Manshoor, B., Zakaria, H., Alimin, A. J., Leman, A. M., & Sadikin, A. (2014). The comparison of preheat fuel characteristics of biodiesel and straight vegetable oil, (2014) Applied Mechanics and Materials, 465-466, pp. 161-166. DOI: 10.4028/www.scientific.net/AMM.465-466.161

[4] A. Obadiah, R. Kannan, A. Ramasubbu, and S. V. Kumar, "Studies on the effect of antioxidants on the long-term storage and oxidation stability of Pongamia pinnata (L.) Pierre biodiesel," Fuel Process. Technol., vol. 99, pp. 56–63, 2012.

[5] M. Shahabuddin, M. A. Kalam, H. H. Masjuki, M. M. K. Bhuiya, and M. Mofijur, "An experimental investigation into biodiesel stability by means of oxidation and property determination," Energy, vol. 44, no. 1, pp. 616–622, 2012.

[6] M. Berrios, M. A. Martín, A. F. Chica, and A. Martín, "Storage effect in the quality of different methyl esters and blends with diesel," Fuel, vol. 91, no. 1, pp. 119–125, 2012.

[7] C. Pattamaprom, W. Pakdee, and S. Ngamjaroen, "Storage degradation of palm-derived biodiesels: Its effects on chemical properties and engine performance," Renew. Energy, vol. 37, no. 1, pp. 412–418, 2012.

[8] Khalid, A., Hayashi, K., Kidoguchi, Y., Yatsufusa, T., "Effect of air entrainment and oxygen concentration on endothermic and heat recovery process of diesel ignition", SAE Technical Papers, 2011, Society of Automotive Engineers of Japan, Inc. and SAE International, DOI: 10.4271/2011-01-1834

[9] Khalid, A. (2013). Effect of ambient temperature and oxygen concentration on ignition and combustion process of diesel spray. Asian Journal of Scientific Research, 6(3), 434-444. doi: 10.3923/ajsr.2013.434-444

[10] L. Natalie, "Study of hydrogenation derived renewable diesel as a renewable fuel option in North America," pp. 1–78, 2012.

[11] C. Y. Lin, H. A. Lin, and L. B. Hung, "Fuel structure and properties of biodiesel produced by the peroxidation process," Fuel, vol. 85, no. 12–13, pp. 1743–1749, 2006.

[12] S. M. Hafis, M. J. M. Ridzuan, R. N. Farahana, A. Ayob, and S. Syahrullail, "Paraffinic mineral oil lubrication for cold forward extrusion: Effect of lubricant quantity and friction," Tribol. Int., vol. 60, pp. 111–115, 2013.

[13] Khalid, A., Mudin, A., Jaat, M., Mustaffa, N., Manshoor, B., Fawzi, M., Razali, M.A., Ngali, Z. (2014), Effects of biodiesel derived by waste cooking oil on fuel consumption and performance of diesel Engine, (2014) Applied Mechanics and Materials, 554, pp. 520-525. DOI: 10.4028/www.scientific.net/AMM.554.520

[14] N. Kaushik, K. Kumar, S. Kumar, N. Kaushik, and S. Roy, "Genetic variability and divergence studies in seed traits and oil content of Jatropha (Jatropha curcas L.) accessions," Biomass and Bioenergy, vol. 31, no. 7, pp. 497–502, 2007.

[15] E. Akbar, Z. Yaakob, S. K. Kamarudin, M. Ismail, and J. Salimon, "Characteristic and Composition of Jatropha Curcas Oil Seed from Malaysia and its Potential as Biodiesel Feedstock Feedstock," Eur. J. Sci. Res., vol. 29, no. 3, pp. 396–403, 2009.

[16] M. H. Gordon, Chemistry of oils and fats – sources, composition, properties and uses, vol. 94. 2006.

[17] P. K. Sahoo and L. M. Das, "Combustion analysis of Jatropha, Karanja and Polanga based biodiesel as fuel in a diesel engine," Fuel, vol. 88, no. 6, pp. 994–999, 2009.

[18] Khalid, A., Jaat, N., Sapit, A., Razali, A., Manshoor, B., Zaman, I., Abdullah, A.A. Performance and emissions characteristics of crude jatropha oil biodiesel blends in a diesel engine (2015), International Journal of Automotive and Mechanical Engineering, 11 (1), pp. 2447-2457. DOI: 10.15282/ijame.11.2015.25.0206.

IOP Conf. Series: Journal of Physics: Conf. Series **1049** (2018) 012092 doi:10.1088/1742-6596/1049/1/012092

Review of Access Control Mechanisms in Cloud Computing

Muhammad Rizwan Ghori, Abdulghani Ali Ahmed

Faculty of Computer Systems and Software Engineering, University of Malaysia Pahang, Gambang Campus, Kuantan, Malaysia

rizwanghori@outlook.com, abdulghani@ump.edu.my

Abstract. Cloud Computing is an evolving technology in the field of IT. People are using this technology vastly as it reduces the storage and other services burden of the users as they use the services provided by the cloud. There are many advantages related to this technology like unlimited storage, cost effectiveness but lacking the strong authentication process. Many researchers are working on the security issue of cloud like user confidentiality, integrity and authenticity etc. However, in this paper we have reviewed and discussed several papers that are related to the authentication in cloud environment. Moreover, after studying different authentication system made by the researchers, we have critically analysed them and found advantages and disadvantages along with some suggestions for the future work.

1. Introduction

Cloud Computing (CC) is a growing field nowadays as it is being used commonly and is generally provided by the third party. The most prominent feature is accessibility of the data. Data kept in the cloud can be retrieved any time and at any place if there is a network access. The benefit of using this technology is that you get worry free related to equipment purchase, data storage capacity, accessibility speeds etc. as these are the concerns of service providers. In view of aforesaid, sharing of data has become more convenient with cloud storage. Moreover, the companies or industries are also taking advantages of the cloud for data management tasks as it becomes cheaper for the companies instead of purchasing the expensive hardware's and software's.

SPI is the most common CC service model for Software, Platform and Infrastructure. Software as a service (SaaS) is a software model in which applications are provided by the vendor or service provider on the requirement of the user on the internet. Platform as a service (PaaS) is a standard for providing the operating system to the customer on the internet without downloading or installation. Infrastructure as a service (IaaS) is provision of storage, hardware etc. facilities to the customer [1-3]. CC is basically providing two major services like data and computational service. With data being the most popular service as all the data management and storage tasks are performed by the cloud service providers. Further the tasks of computations are also managed by the cloud computational services [4-5].

Regardless of CC advantages, it has also increased the risk of the attacks and unauthorized access to the data as it is distributed and stored in different location. Moreover, as storage is shared among many users so there are chances of the illegal user access to the data which can be due to many reasons like device problems, bad intention of some user etc. To overcome these issues the data security must be provided. In network, the protection to the information can be provided with the help of encryption

Content from this work may be used under the terms of the Creative Commons Attribution 3.0 licence. Any further distribution of this work must maintain attribution to the author(s) and the title of the work, journal citation and DOI.
Published under licence by IOP Publishing Ltd

of the data or information. There are different aspects of security in CC such as data confidentiality, integrity and availability which is also known as CIA triad [5-12].

Data confidentiality can be called as the privacy. It is basically to prevent the sensitive information from reaching the wrong person. As said above that data encryption is the standard method for provision of data security. Other include the user id and password protection authentication. Moreover, biometric verification constitutes the other method for certification. Data integrity means that data should not be changed during transmission and steps must be taken to ensure that data cannot be altered by illegal intent of any person [2]. Availability can be defined as the system should always be available for the use. It can be best assured with timely hardware equipment maintenance and keeping up the system with upgrades to avoid any conflict. Authentication and access control is done with the matching of the credential provided by the user with the stored one in the relevant server.

With the advent of CC, user can custom the storage or service without having any idea of the location of the infrastructure. For example, if you are in country A and using the CC service for data storage. It might be possible that your data is being stored in the CC server which is in country B. Thus, keeping your data secure in such huge environment is difficult as CC service can store the data regardless of the location. CC being a convenient service have some disadvantages as well in terms of security as discussed. As the term security is quite vast, the major issue nowadays in CC is the management of access controls and authentications for users related to data sharing etc. [13]. Data authentication and access control in CC is the latest ongoing research topic. In this review, we have discussed and reviewed different aspect related to cloud data security in terms of authentication in detail.

The rest of the paper is organized in a manner that in Section 2, the data authentication and access controls in CC along with the classification of authentication methods are discussed while Section 3 will discuss in detail the comparative studies related to CC and at last with Section 4 we conclude the paper.

2. Data Authentication and Access Control in Cloud Computing

People are using CC commonly these days. Many small as well as large organizations use CC for managing their large amount of data as well as for using different applications etc. As far as customers are concerned data security is the main worry. There are many aspects of security but as told above the authentication and access control of data is now a debating topic among the IT professionals. As the hackers are becoming more advanced, the normal user id and password authentication is never remained reliable [22]. Additionally, security can be compromised when a member of Cloud Service Provider (CSP) can steal information for some financial benefits. Moreover, if you are using biometrics for the security checks it can also be stolen by the aforesaid staff. So, there is a strong need of data authentication and access control techniques to be implemented in CC to fully secure the cloud user. Subsequently, Fig.1 is showing the classification of authentication methods.

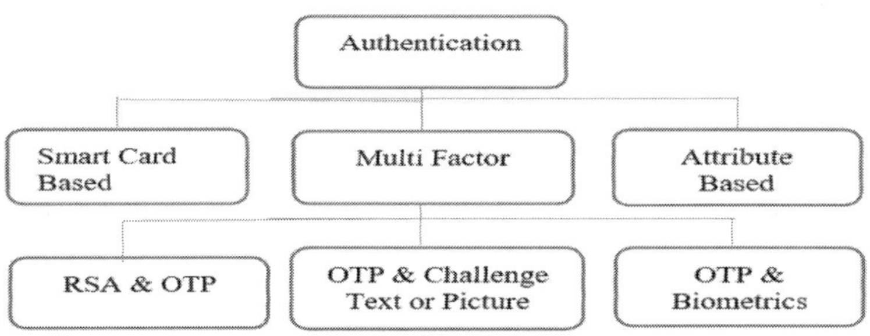

Fig.1: Classification of Authentication Methods

In [14], the author presented a hierarchical authentication model by modifying Anti Statistical Block Encryption (ASBE). The system consisted of five parties i.e. CSP, data owners, data consumers, domain authority and trusted party. In the system data owners, can encrypt the file and will store it in the cloud. Consumers can use the file stored in cloud by decrypting it. Data owners are managed by the respective domain authorities and domain authorities are administered by the trusted parties. The Fig.2 is showing the hierarchical view of system model in which the same process is being followed as discussed.

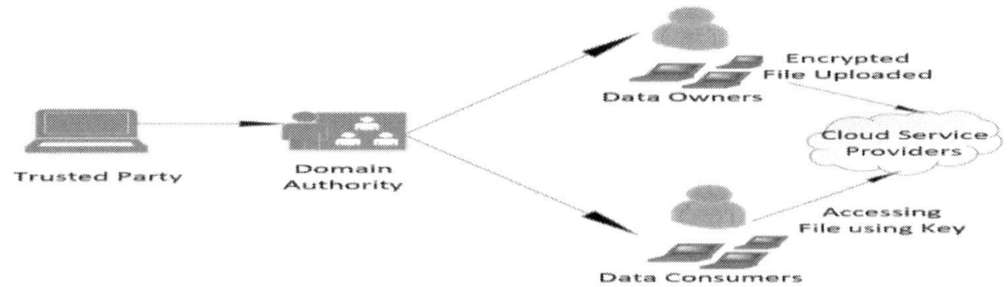

Fig.2: Hierarchical Authentication Model (Z. Wan et al. 2012)

Basically, according to the author, the domain authorities are represented by the organization or department and the user and consumers are under the aforesaid authority. With the permission of domain authority user or consumer can upload and download the file to and from the cloud respectively. As per system shown in Fig.2, it is lacking the proper authentication for user's security. According to the author, the users are dependent only on the trusted party, if the said party plays bad then there will be no security.

Another scheme was proposed by [15], that used Smart Card and Single Sign on (SSO). In the scheme, a combination of smart card and third party authentication was used to attain SSO in inter cloud environment. Moreover, [16] also presented the same model. The drawback of these models is that they have used a special smart card reader for the authentication purpose.

In [17], the writer has given a model scheme in which two factor authentications, Ron, Ravist and Adi (RSA) and one time password (OTP) for CC accessibility has been used. The connection establishment to the cloud consisted of three phases such as setting up for the connection, registration and finally the authentication phase. According to the author, first two steps will be performed by the user only for once. But the authentication step is mandatory every time for the user to access the cloud services. In this scheme, no extra devices have been utilized such as special purpose smart card readers etc. The scheme presented was good for authentication but [19] used the elliptic curve cryptography which is more faster and having smaller key sizes than RSA.

In [18], the system was proposed consisting of the Data Owners, CSPs and the Reputation Centers (RCs). Data owners were the users that request or use the services provided by the CSPs. RC was the trustworthy party with a good reputation and a system consisted of multiple RCs. Moreover, as per the system, CSP will pass stored data to the user on the directives of RC as it will be always available for the authorization of the data access. But according to the system proposed, RC is not given permission to access the data stored in the cloud. The system is shown in Fig.3.

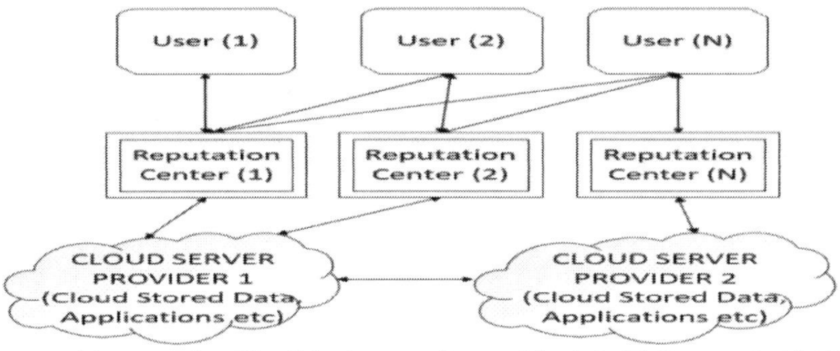

Fig.3: Architecture for Flexible Access Control in CC (Z.Yan et al. 2015)

The system for authentication was developed by [19] using A3 algorithm. According to the author, first step is the establishment of the connection between the cloud server and the user. For registration, the user must fill in the form with personal details. After the registration, Elliptic Curve Diffie Hellman (ECDH) equivalent key exchange will be established with the server. Furthermore, the server will create One Time Password (OTP) which will be send to the user according to the given details by using A3 algorithm. With the help of ECDH and OTP user id will be generated using A3 algorithm. Moreover, after every log in, the new password will be generated for the use. The model presented by the author is efficient as ECDH is faster and have small key sizes. Moreover, according to [22] multifactor authentication with biometrics proven more authenticated.

Moreover, [20] made an authentication system based on attributes for Cloud Server (CS). Attributes Based Signature (ABS) is a technique in which signer is identified by a set of attributes rather having a single attribute defining the signer identity. In this paper, the author has proposed an effective and dynamic signature generation scheme for ABS in which CS provides the half of the signature for the user and other half a user will generate and with less computational cost. Cost effectiveness along with security has made the ABS an emerging technique in the field of authentication and proving itself.

A study was made by [21] on factors that boosted privacy concern in CC environment. He collected 340 cases from the student population and concluded that CC security has to be improved to overcome the privacy concerns of the users.

In [22], the multifactor authentications such as user id, password, OTP and biometrics have been used. As per the system, the user id and password judges the knowledge of the user, fingerprints shows the users identity and OTP along with nonce, master and session keys used for the verification of user identity to the server and vice versa.

Additionally, for management of session keys, station to station Diffie-Hellman key exchange has been used for exchanging one time session key. Furthermore, the said keys will not be stored in server etc. for security concerns. Nonce has basically been used for handshaking for the avoidance of any undependable server encounters.

In addition, for preservation of privacy related to user identifications, the hashed credentials have been used for the verification by cloud server in place of original keys. Also, for symmetric encryption and decryption of the data communication between the user and server Advanced Encryption Standard (AES) has been implemented. The Fig.4 is showing the system scenario depicting that client will register with the third-party authentication server and all the server involved including CS and Authentication Server (AS) must get registered with each other by sharing a secret key.

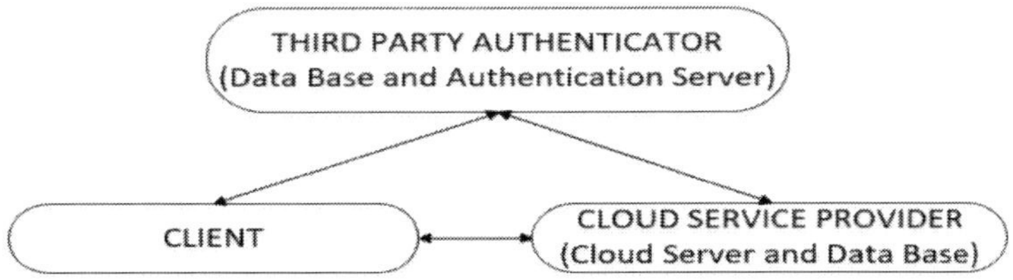

Fig. 4: Multi-Factor Authentication for CC (S.Nagaraju and L.Parthiban 2016)

Cloud Cat tool was developed by [23] for flexible, protected and effective management of multiple user accesses in cloud environment. The aforesaid model was based on the combination of Cyphertext Policy-Attribute-Based-Encryption (CP-ABE) and access control model based on roles (RBAC).

The model consisted of four modules such as Authentication Manager responsible to control access to the data in the cloud database According to the model, Authorization Manager controls the administration and implementation of security policies. Moreover, policy deployment can only be done by data owners or administrators. Another module included in the system is the User Access Control Management which provides access to the data and the last is Collaborative Ciphertext Policy Attribute-Role Based Encryption Model C-CP-ARBE access control algorithm which is used to provide the cryptographic processes for the users.

Also, in healthcare systems the authenticity is of great concern. In view of aforementioned, [24] has made a framework for Electronic Health Record (EHR) system over cloud. The system was made in a manner that patient report can be shared through cloud with the doctors in such a way that privacy of the patient should not be compromised. In addition, the requested EHR encryption is done using symmetric keys for the communication between the two doctors. The model was made in a way that an individual EHR server was placed on each health care provider side and finally they were connected to the cloud shared environment.

The Fig.5 is showing the EHR cloud system proposed by the author. For instance, two health servers are shown. As per the system, the doctors can share the information through the cloud server with more authentication as it is dedicated cloud for the EHR. Moreover, this system is typically for the health care, not very much reliable for the sensitive data.

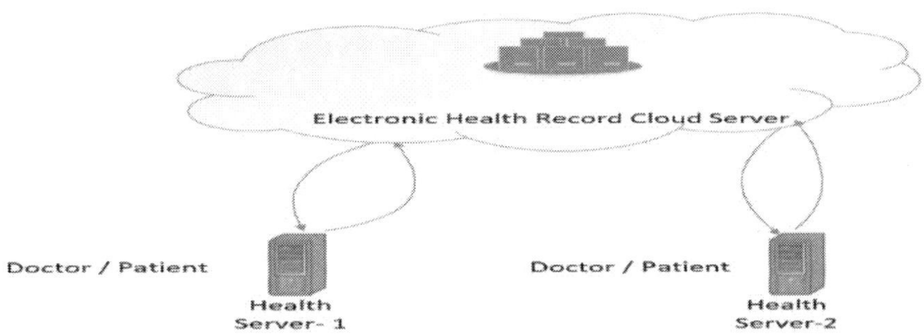

Fig. 5: Framework for Sharing EHR over Clouds (A. Ibrahim et al. 2016)

In other study, [25] proposed a security architecture for CSP to check the data integrity with user authentication system. For the said purpose, the authors used One Time Password (OTP) and modified SHA-2 hash function for authentication and integrity checks respectively. The system architecture along with steps is shown in Fig.6.

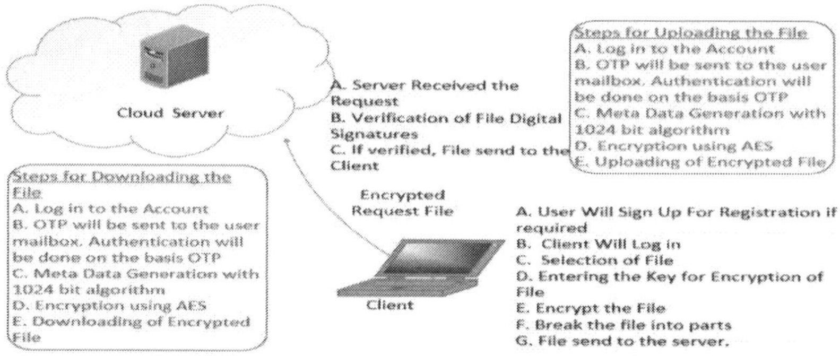

Fig. 6: Security Architecture for CSP (N.R.Patil and R.Dharmik 2016)

The model in Fig.6 has used single authentication method in which only OTP was required for the file encryption and decryption.

In [26], another authentication scheme was produced consisting of pass generator. The system was designed in such a way that user must pass through different phases like registration, login, image update as shown in Fig.7.

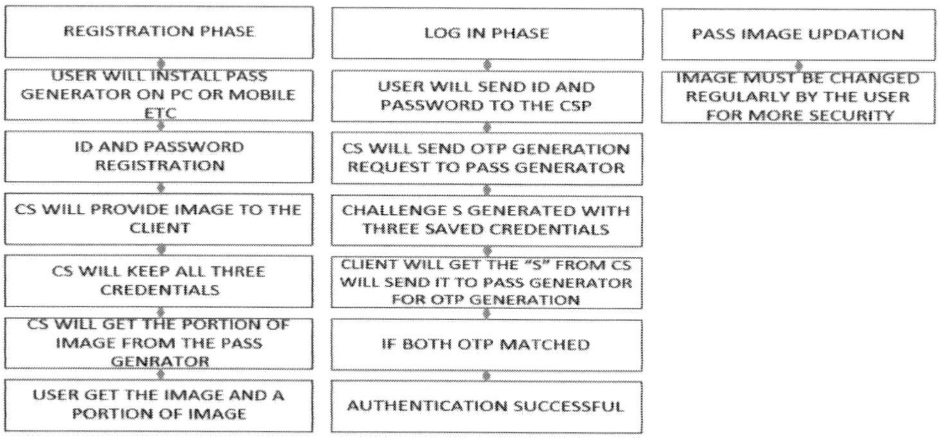

Fig. 7: Strong Password Generator Model (A.Abdellaoui et al. 2016)

According to the Fig. 7, for registration, the user must install the pass generator application in his/her PC or any other device. Then user has to register the id and password for the said phase and that will finally be saved in the Cloud Server (CS). According to the system, CS will assign an image to the client and will save all of the three credentials for the later use. Moreover, for more security the CS will get the image and a portion of image from the pass generator and will send it to the user who will save the aforesaid identifications. In the login phase the user will send the id and password to the CSP. After the authentication, CS will send the OTP generation request to the cloud password generator. In the next step, challenge S will be generated from the previously saved credentials which were id, password and the image. The client will get the challenge S from the CS and will send it to its password generator application which will then generate an OTP for the client. If OTP generated by the client and CS matches, then the client will be allowed to access the cloud for the services. Moreover, in the third phase, the image should be updated or changed by the user often for more security. The aforesaid system is shown in Fig.6 depicting the two-factor authentication with OTP and picture challenge. The stated system provided good authentication. More so, there is a need of a model that should support more layers of authentication for the sensitive data accessing. Moreover, use of additional layers of security can make system a bit slow but security in some cases gets the priority.

Additionally, more layers should be designed in such a way that it should not disturb the systems effectiveness.

3. Critical Analysis

The Table I is showing the latest work done in the field of authentication in CC along with the narration of pros and cons.

TABLE I Summary of Comparative Study

Ref	Year	Advantages	Disadvantages	Suggestions
[22]	2016	It is vigorous and efficient	Device theft can cause a security breach.	Multifactor can be combined with the more challenge so that it becomes stronger
[23]	2016	In Comparison, Secure for the users	Old Technique. No strong password encryption is used	Data and password must be encrypted more strongly
[24]	2016	Secure System for medical use	Not suitable for sensitive kind of data.	With the help of more advanced encryption algorithm it can get more secure and efficient
[25]	2016	Used One Time Password (OTP) Authentication. Easy to Implement	Only one layer of security i.e. OTP	Need to propose more secure technique for the users
[26]	2016	Authentication is done based on Challenge image. Difficult to hack	Time consuming.	Biometrics challenge can also be included for authentication

All the authors in Table I have done well but security and more so authentication in CC requires more research. Because hackers are becoming day by day stronger and equipped. The design of authentication should be in such a manner that to steal an information, a person should pass through different steps to get the information. A system with the combination of multifactor authentication along with the integration of biometrics [22] and a third layer must be implemented like challenge picture [26] can make an authentication system more vigorous and stronger. Moreover, this additional third layer of security will be for the people who are interested in fool proof security for the protection of the data. As far as normal users are concerned OTP and two-layer authentication is good enough.

4. Conclusion

In this paper, we have reviewed several works related to the authentication in CC. Researchers have done well in proposing their work by using different techniques. Still the gaps are there as we did not yet get the perfection in authentication system for cloud CC. After reviewing all the material related with pros and cons, we concluded that a system with multifactor authentication along with biometrics verification and integration of some challenge question for a user can make the CC system more vigorous as far as authentication processes are concerned. Integration of the OTP with biometrics along with the challenge text or image will add a third layer of authenticity for the users who require more security. Moreover, for future works, biometrics of all ten fingers of user hand should be in CS database. Whenever user want to log in he or she will be passed through multi factor authentication and can be asked for specific finger authentication for example when a user log in for the first time the server can ask for the biometric of the left thumb impression. Moreover, as far as data encryption is concerned, AES can be adopted as it is the strongest encryption technique. Additionally, there is no system yet made for authentication which can bypass the service provider staff involvement because they can steal any body's credentials any time. So, there is a need to make such a fool proof authentication system in which involvement of CSP staff should be less.

Acknowledgments

*Funding support was provided by the Ministry of Higher Education in Malaysia (No. FRGS/1/2016/ICT03/UMP/02/1), and the fund of COMSTECH-TWAS joint research grants the program for young scientists in OIC countries No. 14-340 RG/ITC/AS_C.

References

[1] R. Sravan Kumar and A. Saxena, Data integrity proofs in cloud storage. in communication systems and networks (COMSNETS), in Proc. of 2011 Third International Conference on, 2011.

[2] S. Nepal et al., DIaaS: Data Integrity as a Service in the cloud, in Proc. of IEEE International Conference on Cloud Computing, 2011.

[3] S. Subashini and V. Kavitha, A survey on security issues in service delivery models of cloud computing, in Proc. of Journal of Network and Computer Applications, pp. 1-11, 2011.

[4] C. Ning et al., Privacy-preserving multi-keyword ranked search over encrypted cloud data, in Proc. of IEEE Conference, 2011.

[5] F. Jun et al., Analysis of integrity vulnerabilities and a non-repudiation protocol for cloud data storage platforms, in Proc. of 39th International Conference on Parallel Processing Workshop, 2010.

[6] S.M. Khan and K.W. Hamlen, Anonymous Cloud: A data ownership privacy provider framework in cloud computing, in Proc. of IEEE 11th International Conference on Trust, Security and Privacy in Computing and Communications (TrustCom), 2012.

[7] F. Sebe et al., Efficient remote data possession checking in critical information infrastructures in Proc. of IEEE Transactions on Knowledge and Data Engineering, vol. 20, pp. 1034-1038, 2008.

[8] A. Juels and J. Burton S. Kaliski, Pors: proofs of retrievability for large files, in Proc. of 14th ACM conference on Computer and communications security, pp. 584- 597, 2007.

[9] G. Ateniese et al., Provable data possession at untrusted stores, in Proc of 14th ACM conference on Computer and communications security, pp. 598- 609, 2007.

[10] Z. Tan, Z. Tang, R. Li, A. Sallam, and L. Yang, Research on trust based access control model in cloud computing, in Proc. of 6th IEEE Joint International Information Technology and Artificial Intelligence Conference (ITAIC), vol.2, pp. 339- 344, 2011.

[11] S. Yu, C. Wang, K. Ren, and W. Lou, Achieving Secure, Scalable, and Fine-grained Data Access Control in Cloud Computing, in Proc. of IEEE INFOCOM, pp. 1-9, 2010.

[12] J. Ilanchezhian, V. Varadharassu, A. Ranjeeth, and K. Arun, To improve the current security model and efficiency in cloud computing using access control matrix, in Proc. of Third International Conference on Computing Communications and Networking Technologies(ICCCNT), pp.1-5,2012.

[13] Y.Jin, C.Tian, H.He and F.Wang, A secure and lightweight data access control scheme for mobile cloud computing, in Proc. of IEEE Fifth International Conference on Big Data and Cloud Computing, pp. 172179, 2015.

[14] Z.Wan, J. Liu, and R.Deng, A hierarchical attribute-based solution for flexible and scalable access control in cloud computing, in Proc. of IEEE Transactions on Information Forensics and Security, vol. 7, pp. 743-754, 2012.

[15] M.Hwang and T. Hung Sun, Using smart card to achieve a single sign on for multiple cloud services, in Proc. of IETE Technical Review, September, 2013.

[16] J.W. Tsaur, H.Li and B.W.Lee, An efficient and secure multi-server authentication scheme with key agreement, in Proc. of Journal of Systems and Software, vol. 85, pp. 876882, 2012.

[17] A. Yassin, H.Jin, A.Ibrahim, W.Qiang, and D.Zou, Cloud authentication based on anonymous one-time password, in Proc. of International Conference on Information Technologies and Applications, Netherlands, 2013.

[18] Z.Yan, X.Li, M.Wang and A.V. Vasilakos, Flexible data access control based on trust and reputation in cloud computing, in Proc. of IEEE Transactions on Cloud Computing, vol. pp, issue. 99, pp. 1-1, 2015.
[19] S.Singh and V.Kumar, Secured users authentication and private data storage- access scheme in cloud computing using elliptic curve cryptography, in Proc. of 2nd International Conference on Computing for Sustainable Global Development, pp. 791-795, India, 2015.
[20] Z.Liu, H.Yan and Z.Li, Server-aided anonymous attribute-based authentication in cloud computing, in Proc. of International Journal on Future Generations Computer System, Vol. 52, pp. 61-66, 2015.
[21] Z.Asadullah and O.Oyefolahan, Factors Influencing Information Privacy Concern in Cloud Computing Environment, in Proc. of International Symposium on Mathematical Sciences and Computing Research (iSMSC), pp. 238-242 , 2015.
[22] S.Nagaraju and L.Parthiban, Provably secure multi-factor authentication for the cloud computing systems, in Proc. of Indian Journal of Science and Technology, 2016.
[23] S.Fugkeaw and H. Sato, A collaborative access control tool for data outsourced in cloud computing, in Proc. of 10th International Conference on Digital Information Management, pp. 243-248, 2015.
[24] A.Ibrahim, B.Mahmood and M.Singhal, A secure framework for sharing electronic health records over clouds, in Proc. of IEEE International Conference on Serious Games and Applications for Health (SeGAH), pp. 1-8, 2016.
[25] N.R.Patil and R.Dharmik, Secured cloud architecture for cloud service provider, in Proc. of World Conference on Futuristic Trends in Research and Innovation for Social Welfare (Startup Conclave), pp. 1-4, 2016.
[26] A. Abdellaoui, Y.I. Khamlichi and H. Chaoui, A novel strong password generator for improving cloud authentication, in Proc. of International Conference on Computational Modelling and Security (CMS), vol. 85, pp. 293-300, 2016.
[27] K. Chauhan, A. Sanger and A. Verma, Homomorphic encryption for data security in cloud computing, in Proc. of International Conference on Information Technology, pp. 206-209, 2015.
[28] R. Mishra, D. P. Mishra, A. Tripathy, and S. K. Dash, A privacy preserving repository for securing data across the cloud, in Proc. of 3rd Int. Conf. Electron. Comp. Tech, pp. 6–10, 2011.
[29] M. S, E. Daniel, and N. Vasanthi, Survey on Various Data Integrity Attacks in Cloud Environment and the Solutions, pp. 1076–1081, 2013.
[30] K. Khan and M. Shaheen, Exploring data security in cloud without encryption, in Proc. of IEEE 16th International Conference on Computational Science and Engineering, pp. 9–14, 2013.
[31] S. Hohenberger, B. Waters, Attribute-based encryption with fast decryption, in Proc. of Public-Key Cryptography, pp. 162–179, 2013.

IOP Conf. Series: Journal of Physics: Conf. Series **1049** (2018) 012093 doi:10.1088/1742-6596/1049/1/012093

Investigating the Utilisation of Plastic Bottle as Aggregate Replacement for Concrete Block

Mardiha Mokhtar[1], Nor Baizura Hamid[2], Masiri Kaamin[3], Muhammad Nur Aiman Adnan[4], Muhammad Amzar Othman[5], Muhammad Shahruzi Mahadzir[6], Mah Mudin Amin[7]

[1,2,3,4,5,6,7] Department of Civil Engineering, Center for Diploma Studies, Universiti Tun Hussein Onn Malaysia (UTHM), 86400 Parit Raja, Batu Pahat, Johor, Malaysia.

E-mail: Mardiha@uthm.edu.my

Abstract. Concrete block plays an important role in the construction industry. Concrete blocks have been used in the making of various buildings, retaining wall and others. In this modern era, various technologies have been presented and invented in order to upgrade the life quality of mankind. Unfortunately, there are some technologies that bring bad consequences to the environment. For instance, the production of cement will contribute to a high emission of bad gases such as carbon dioxide that may be harmful to human's health. The manufacturing of cement is also emits hazardous fumes and dust that potentially can reduce the quality of air and lead to many respiratory problems. Therefore, the main objective of this paper is to search for a new alternative way to cope with this environmental problem. Plastic bottles such as Polyethylene terephthalate (PET) was used as the main component in construction of concrete blocks. After curing the blocks for 28 days, compression test and flexural test were prepared to test the ultimate strength of these concrete blocks. From both tests, it shows that bottle blocks were qualified to replace concrete blocks in walls as the maximum stress achieved by bottle blocks were higher than the minimum permissible compressive strength and flexural value. From these results, it can be concluded that plastic bottle has a potential to replace the current concrete blocks and further study about its other properties, such as lifespan and water cement ratio should be carried out.

1. Introduction

In this modern era, the innovation in science and technology are getting more competitive through new researches. As the result, the production of equipment in medical, industry, agriculture, information technology will be cost-saving and at the same quality or even better. Plus, the technology of polymer-based products are increased resulting from the new researches thus making the production of plastics more effective and good quality compared to the usual materials.

Nowadays, plastic-based products are getting more demand in many industries to produce their own products. Apart from that, the disposal of plastics is also getting serious and there are still no solutions yet to cope with this matter.

Therefore, it is essential for us to find and search for new alternatives to reuse plastics in a good way to prevent the environment from pollution. Plastic bottles need a large portion of area to undergo

Content from this work may be used under the terms of the Creative Commons Attribution 3.0 licence. Any further distribution of this work must maintain attribution to the author(s) and the title of the work, journal citation and DOI.
Published under licence by IOP Publishing Ltd

decomposition which means a larger site or landfill is needed. However, the price for gaining a property is way too expensive and because of that, plastic bottles are recommended to be reuse rather than been thrown away or disposed.

Disposed plastic bottles can harm the environment and not to forget, humans' health. Plus, plastic bottles disposed in the drain or small channels potentially can clog the channels thus resulting in flash floods that may cause loss of life and damaging valuable properties. Apart from that, the disposal of plastic bottles in rivers and sea can danger the natural ecosystem especially in the water.

The reuse and recycle programs in terms of plastic bottles have been actively organized by many societies as these are the most cost-saving and relatively effective. Besides that, recyclable materials can be used in many industries to produce something new with new and various functions as well as saving the energy. There are many materials that can be recycled for instance, old tires, rubber, waste products and others. This paper will prioritized the usage of plastic bottles for recycle and reuse in the production of concrete in construction.

2. Background of study

Polyethylene terephthalate (PET) is one of the most common thermoplastics resin of polyester used for various type of products mainly packaging. PET recycling consists in the transformation of bottles through two different kinds of processes which firstly is mechanical process which the most employed physical process to recycle PET. It consist three stages that include separation, washing, and grinding of containers. With this process, PET flakes produced can be directly employed without the need of being reprocessed as pellets in the creation of products by injection or extrusion.

Secondly, PET is recycled using chemical process consists of the separation of the basic components or monomers. The methanolysis, glycolysis, and hydrolysis are the elemental processes to achieve this transformation. Both ways are expensive and there are other alternative way to recycle PET materials which is using them as substitute of concrete aggregates. Due to demands of technological development in the construction area, the possibility for generating alternative materials that can be applied with increasing functionality, low costs, and better physical, chemical, and mechanical properties than conventional materials [1].

Concrete is a composite materials which mainly use in various type of construction. It made of cement, coarse and fine aggregate, sand and water. There are various type of admixture that can be use in concrete to increase its strength or to speed up productivity. The quality of concrete depends on type of materials use, mixture, transportation and compaction [2].

Generally blocks applied with PET, due to its lightweight structures compared to normal concrete blocks it will give less stress or load to the foundation thus engineer can designed more economical foundation due to less stress or load. Moreover, blocks applied with PET reduce volume of mixture which less labor needed because the mixture is lighter and also the cost of transportation can be reduce. Another feature for blocks applied with PET are good sound insulation and better shock absorption [3].

3. Materials and method

Firstly, 80 pieces of 500mL plastic bottles are filled with sand and compacted using iron rod. Formworks were cleaned and brushed with oil or grease to prevent the concrete from stick at the formwork surface during extraction. The mixture of concrete was poured into the formwork with three layers and compressed using a steel rod at 25 times. Cube and concrete cuboid was left for 24 hours before extraction from the formwork.

The size of concrete block was 440mm x 215mm x 100mm. Concrete blocks were tested for compression and deflection test after curing in water for 28 days. Figure 1 shows samples of concrete blocks after 28 days of curing period. There are 12 blocks produced for this study as shown in Table 1.

IOP Conf. Series: Journal of Physics: Conf. Series **1049** (2018) 012093 doi:10.1088/1742-6596/1049/1/012093

Figure 1. Concrete blocks after 28 days of curing period.

Table 1. Type of block concrete produced

Type of block	Description	Diagram (mm)
A	Single vertical	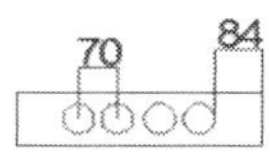
B	Double vertical	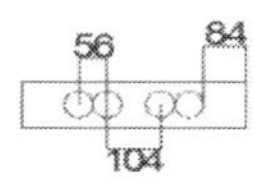
C	Horizontal	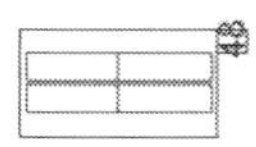
S	Standard	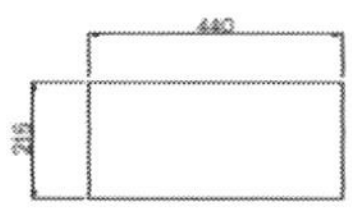

Curing is the process to avoid the water content in the concrete evaporate quickly by hot weather. The concrete should be put in a damp at least 14 days after it harden. Rather the purpose is to show that high-performance concrete has to be water cured as early as possible, late curing is of partially no value, but is still better than no curing at all, total shrinkage can be drastically reduced by appropriate early water curing and no water curing at all can be catastrophic [4].

Compressive strength test was carried out to obtain the strength of the concrete sample. This test is performed by BS 5628: Part 1: 1992 [5]. Sample used are concrete sample that had reached the age of 28 days. The main factors affecting the strength of concrete is the number of honeycomb are left in the concrete, the strength will reduced if the concrete more compressible. Therefore, it is imperative that concrete is made as dense as possible. Concrete strength also increases with age, but the rate increases is greatly affected by proper curing methods. Ideally concrete stored in humid conditions to allow the cement hydration occurs properly. Compressive strength is a more highly utilized parameter in the cement and concrete industries than are elastic modulus. [4].

IOP Conf. Series: Journal of Physics: Conf. Series **1049** (2018) 012093 doi:10.1088/1742-6596/1049/1/012093

4. Results and discussion

Table 2 showed the results from compression and deflection tests. The ordinary concrete block is acted as control sample to be compared with the bottle concrete blocks. From the compression strength test, the maximum stress that can be achieved by control sample, single vertical bond, double vertical bond and horizontal bond were 12.2 N/mm^2, 6.4 N/mm^2, 9.4 N/mm^2 and 4.3 N/mm^2 respectively. Among the three bottle blocks, the double vertical bond have the highest value of stress that can be achieved.

According to BS EN 771-3 [6], the minimum permissible average compressive strength shall be 7.3 N/mm^2 for 440mm x 215mm x 100mm solid concrete blocks. Therefore, this can prove that bottle blocks are qualified as wall materials to replace standard masonry as its compressive strength exceed the minimum permissible compressive strength. This prove that bottle concrete blocks are potentially in replacing standard blocks or masonry in wall.

Besides that, according to ASTM C293 [7], the standard flexural strength of concrete block is 5.5 N/mm^2. The single vertical bond, double vertical bond and horizontal bond can exert 5.02 N/mm^2, 4.21 N/mm^2 and 5.7 N/mm^2 of flexural strength respectively. In terms of flexural strength, only horizontal bond has exceed the standard value. Therefore, this can prove that bottle blocks are qualified as wall materials to replace standard masonry as its compressive strength exceed the minimum permissible compressive strength.

Table 2. Results from compression and deflection tests

Types	Compressive Strength (N/mm^2)	Flexural Strength (N/mm^2)
A	6.4	5.02
B	9.4	4.21
C	4.3	5.7
S	12.2	5.5

5. Conclusions

In a nutshell, the application of plastic bottles filled with sand as retaining wall or other wall structures to replace bricks and concrete blocks in the construction industry is acceptable as the strength is over the minimum permissible strength of British Standard and the flexural strength was according to ASTM C293. In terms of compressive strength, the double vertical bond have the highest value of strength. In the terms of flexural strength, the horizontal bond have the highest value of flexural strength. On the other hand, the plastic bottle have the ability to replace standard concrete blocks in Malaysia's buildings in the terms of saving budget and conserve the natural environment. The water cement ratio was 0.50 and the moisture content was 0.30. Besides that, the ratio of concrete mixture is 1:2:4 and the results are obtained. Hence, a research on the optimum ratio of the mixture should be carried out to obtain the strongest concrete.

Acknowledgements
The author would like to thank the Office of Research, Innovation, Commercialization and Consultancy (ORICC), UTHM, for supporting this research.

References

[1] L. Á. Córdoba, G. M. Barrera, C. B. Díaz, F. U. Nuñez, and A. L. Yañez 2013 *Effects on Mechanical Properties of Recycled PET in Cement-Based Composites* (*International Journal of Polymer Science*)
[2] N. A. Abdulla 2014 *Effect of Recycled Coarse Aggregate Type on Concrete* (*J. Mater. Civ. Eng.*)
[3] S. Chowdhury, A. T. Maniar and O. Suganya 2013 *Polyethylene Terephthalate (PET) Waste as Building Solution* (International Journal of Chemical, Environmental & Biological Sciences (IJCEBS)), vol 1
[4] P. Klieger and J. F. Lamond 1994 *Significance of tests and properties of concrete and concrete-making materials* (American Society For Testing And Materials (ASTM): Philadelphia)
[5] British Standard Institution 2003 *BS 5628 Part 1, Structural use of unreinforced masonry* (London-BSI)
[6] British Standard Institution 2003 *BS EN 771-3, Spesification for mansory units: Aggregate concrete masonry units* (UK-BSI London)
[7] ASTM 2007 *ASTM C293/C293M–10: Standard Test Method for Flexural Strength of Concrete* (American Society for Testing and Materials)

IOP Conf. Series: Journal of Physics: Conf. Series **1049** (2018) 012094 doi:10.1088/1742-6596/1049/1/012094

Effects of Vibration on Intrastate Bus Drivers

A F Azenan[1]**, Jalil Azlis-Sani**[2]
[1,2] Faculty of Mechanical Engineering & Manufacturing Universiti Tun Hussein Onn Malaysia (UTHM), 86400 Parit Raja, Johor.

E-mail: azlis@uthm.edu.my[2]

Abstract. *Lower back pain is a common effect of whole body vibration on bus drivers. Therefore, this study was carried out to identify effects of whole body vibration on intrastate bus drivers including lower back pain. A total number of 161 respondents in Putrajaya Sentral and 92 respondents in Larkin Sentral were selected and the data were analysed using Statistical Package for Social Science (SPSS). A total of 5and 7 bus drivers from Putrajaya Sentral and Larkin Sentral respectively were selected for the measurement. HVM 100 Larson Darvis and a seat pad accelerometer were being used. Based on the data collected, measurements of whole body vibration of two locations were conducted to investigate on the exposure level. Several pain areas faced by the drivers were identified at both terminals i.e; neck, upper back, lower back and shoulder. Based on the correlation analysis, upper back and lower back pain were identified as significant effect of the vibration. The results showed positive, strong and significant relationship between daily vibration exposure, A(8) with upper and lower back pain. In conclusion, the vibration exposed to the drivers at Larkin Sentral were significantly higher and exceeds the exposure limit compared to those at Putrajaya Sentral.*

1. Introduction

Presently, there are varieties of public transports in Malaysia especially in Kuala Lumpur such as bus, light rapid train and also monorail. Public bus operation has play an important role across the world for commuting passengers [1]. Vibration has become an important consideration in engineering as the development of industries and vehicles become aggressive [2]. Investigations have shown that bus drivers are exposed to high-intensity vibrations [3]. There are two types of human vibration exposure which are hand arm vibration (HAV) and whole body vibration (WBV). This study will be focussing on WBV on a person is supported on a vibrating surface. It can be seen while driving a bus, the seat will receive or expose to vibration and then the whole body of the drivers will expose to WBV. WBV is an occupational issue of concern due to adverse health effect or simple discomfort and annoyance [4] and can be measured in three orthogonal axes which are x-fore and aft, y-lateral and z-vertical [5]. The amount of vibration exposures depends on a number of factors, including the type and design of the vehicle, the speed at which the vehicle is travelling, the body posture, and the environmental conditions [6]. This sector (bus) also could lead to an issue such as musculoskeletal disorder apart from contributing to the economic growth and job opportunities [7].WBV can cause so many health problem such as back pain problem. One group of workers that have been recorded as being at an increased risk for lower back pain (LBP) is occupational drivers such as bus drivers [8]. Furthermore,

Content from this work may be used under the terms of the Creative Commons Attribution 3.0 licence. Any further distribution of this work must maintain attribution to the author(s) and the title of the work, journal citation and DOI.
Published under licence by IOP Publishing Ltd

there are four primary harmful effects of the vibration are motion sickness, impaired comfort, impaired activities and degraded health [9].

Efficient and proper process are needed in this study as to gain a better and accurate result. In order to do that, there are several objective in this study that have been focused which is to identify the level of whole body vibration among bus drivers and evaluate whether the vibration exposure have been already exceed the exposure limit based on ISO 2631-1(1997). Then, correlation analysis will be done between two variables which are daily vibration exposure, A(8) from vibration measurement and pain areas caused by whole body vibration from questionnaire as to investigate the effects of whole body vibration instead of lower back pain.

2. Methodology

This research was a cross-sectional study conducted among bus drivers in Rapid Bus Sdn Bhd (Putrajaya Central) and Syarikat Pengangkutan Maju (Larkin Central). The first method of this study was by using vibration instruments which were human vibration meter analyzer (HVM100) and a Tri-axial seat pad accelerometer. For the measurement of whole body vibration, there were 5 bus drivers from Putrajaya Central and 7 bus drivers from Larkin Central were selected to measure the whole body vibration exposure. While performing their normal driving condition, drivers must not be disturbed including researcher in order to get an accurate measurement of whole body vibration. The vibration instruments needed to be calibrated before collecting data in order to get an ideal sensitivity. The value for both weighting- frequencies (Wd and Wk) was (Wd=1.4wx, Wd=1.4wy and Wk=1.0wz) and the sensitivity in this study was (x=1.040 mV/g, y=1.024 mV/g and z=1.016 mV/g) with gain 20 dB each axes. HVM 100 could only recorded a data or register up to only 4 minutes and the next 4 minutes will be recorded in another register. Importantly, the seat pad accelerometer must in proper position (axes).

After getting the data, it will be transferred to Blaze software for further report, analysis on whole body vibration exposure which was correlation analysis for identifying the effects of vibration and calculated daily vibration exposure A(8) and vibration dose value (VDV) by using formulas given.

$$A_W = [\frac{1}{T}\int_0^T a_w^2(t)dt\,]^{\frac{1}{2}} \tag{1}$$

Where, Aw (t) is the vibration data and the average weighted vibration in m/s^2.
T is the time measurement in second (s).

$$VDV = [\int_0^T a_w^4(t)dt\,]^{\frac{1}{4}} \tag{2}$$

Where,
VDV is the vibration dose value in $m/s^{1.75}$
T is the time of measurement in second (s)

Measurement time for each drivers in Putrajaya Central was 36 minutes while for Larkin Central was 40 minutes. By using the Blaze software, the four minutes of each register were merged together and the value of A(8) in a merged file were converted to an 8-hour daily value by using HSE calculator to calculate the A(8) and VDV. Then, the results needed to be compared with Exposure Action Value (EAV) and Exposure Limit Value (ELV) for both A(8) and VDV based on Health Guidelines Caution Zone (HGCZ). The details was shown in Table 1.

Second method of this study was by using Modified Nordic Questionnaire. All selected respondents were required to fill up the questionnaires. The Modified Nordic Questionnaire consists of three major parts; personal details, drivers comfort and drivers health with objectives to screen the musculoskeletal disorders in an ergonomics context and for occupational health care service [10]. Each of the drivers were briefed nicely and wisely about the questionnaire and the importance of getting actual data from each of them. The data collected from the questionnaires were keyed in into SPSS for further statistical analysis such as the correlation analysis. The purpose of the analysis was to identify the relationship between two variables while the Pearson correlation was selected for the statistical analysis. Generally, there are three conditions of the analysis that were positive or negative relation, strength of the relation (-value) and significant of the relation (p-value). The -value and p-value were shown in Table 2.

Table 1: EAV and ELV details

EAV	ELV
EAV for A(8)	0.50 m/s^2
ELV for A(8)	1.15 m/s^2
EAV for VDV	9.1 $m/s^{1.75}$
ELV for VDV	21.0 $m/s^{1.75}$

Table 2: α-value and p-value

α-value & p-value	Strength and Significance
α-value < 0.5	Moderate-weak
α-value > 0.5	Strong-very strong
Sign (p-value) < 0.05	Very significant
Sign (p-value) > 0.05	Not significant

3. Results and Discussion

Based on Figure 1, the painful area caused by vibration on intrastate drivers in Putrajaya Central were presented and the most painful area was neck with percentage of 81.4%. The rest followed by lower back, shoulder, upper back, knee and hips. On the other hand, painful areas experienced by bus drivers in Larkin Central were more compared to Putrajaya Central. The highest percentage of painful areas in Larkin Central was lower back with 87% of 92 respondents. This finding was in lined with previous study on the relationship between back disorders and vehicle operation with vibration exposure [11]. Researchers agreed and recognized that low back pain is the most leading pain due to vibration. The other painful areas having by the bus drivers in Larkin Central were upper back, shoulder, neck, knee, hips, thigh and ankle.

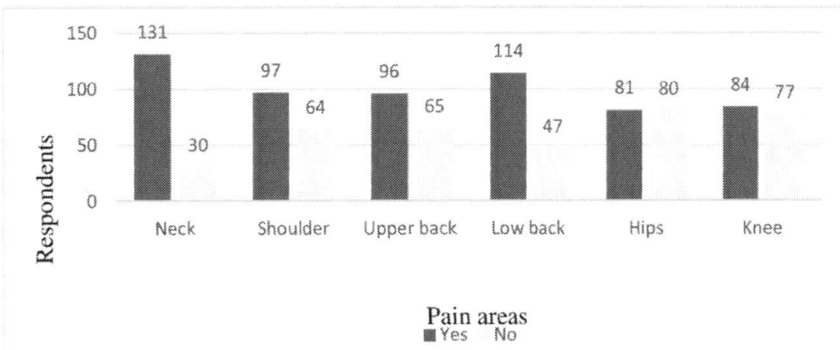

Figure 1: Painful area caused by vibration in Putrajaya Central

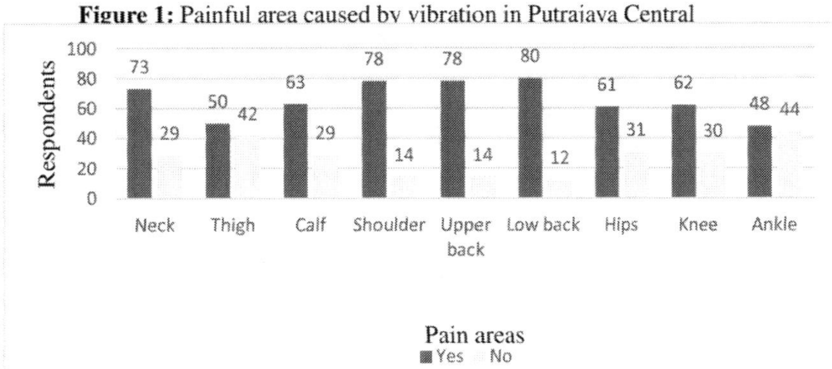

Figure 2: Painful area caused by vibration in Larkin Central

According to Figure 2 shows the A(8) for bus drivers In Putrajaya Central was obtained from Blaze Software and after calculating by using HSE calculator. Yellow line represented the value of EAV which was 0.5 m/s^2. Based on the figure below, there were no bus drivers who reach the limit of EAV. Only two bus drivers almost reach the EAV limit which was bus driver C and D with equal values of daily vibration exposure that were 0.48 m/s^2. The rest were bus driver B, E and the lowest was driver A (0.42 m/s^2, 0.40 m/s^2 and 0.34 m/s^2). Otherwise, the rest was safe to be used and might not harm the bus drivers, but safety precautions must be taken by the employer.

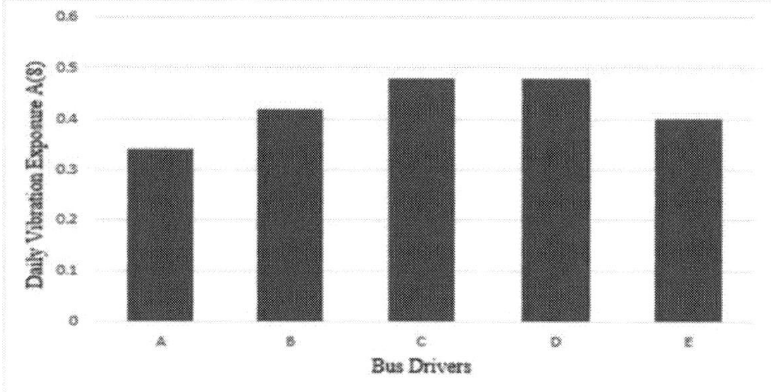

Figure 3. Daily vibration exposure, A(8) in Putrajaya Central.

Daily vibration exposure, A(8) in Larkin Central was shown in Figure 4 below. As been mentioned earlier, the yellow line represented the value of EAV while red line shows the value of ELV which was 1.15 m/s^2. Based on the figure, daily vibration exposure, A(8) of all these 7 different bus drivers surpasses the breaking point of EAV. The highest A(8) is from driver G which was 1.14 m/s^2 where it needed only 0.01 m/s^2 to reach the limit of ELV. Generally, the employer should consider all these buses to undergo maintenance as the buses are not fit for driving. Based on ISO 2631-1, the employer need to implement the health monitoring and also control measures. Otherwise, continuous monitoring and assessment on the buses and drivers should be done by employer to overcome the problem.

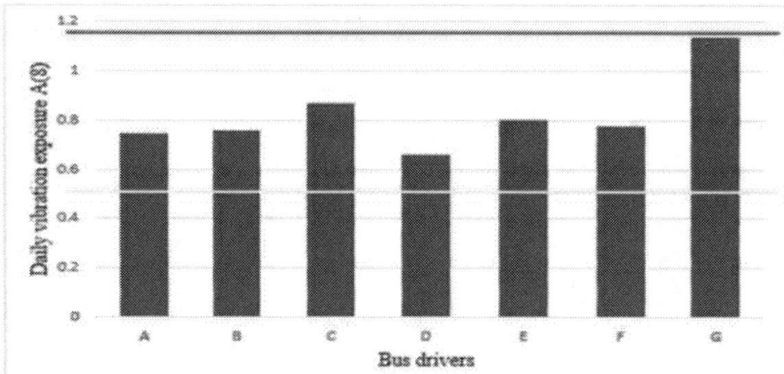

Figure 4. Daily vibration exposure, A(8) in Larkin Central.

VDV was the use of the time integrated fourth power of acceleration. The VDV analysis will be considered as there have shocks and impacts on these bus drivers. Based on ISO 2631-1, when crest factors are above 9, VDV analysis is recommended for evaluation when impulsive exposures are present [4]. Based on Figure 5, the drivers in Putrajaya Central were already exceed the EAV daily VDV exposure level which is 9.1 $m/s^{1.75}$. The value of VDV exposure level in highest and lowest order for drivers E, D, A, C and B were 10.5 $m/s^{1.75}$, 10.2 $m/s^{1.75}$, 10.1 $m/s^{1.75}$, 10.0 $m/s^{1.75}$ and 9.6

$m/s^{1.75}$. According to the VDV exposure level, there were several action needed to be done by the employer as being mentioned in A(8) analysis. Employee need to assess exposure levels, health surveillance for operators exceeding the action level or caution zone and training for management and operators [13].

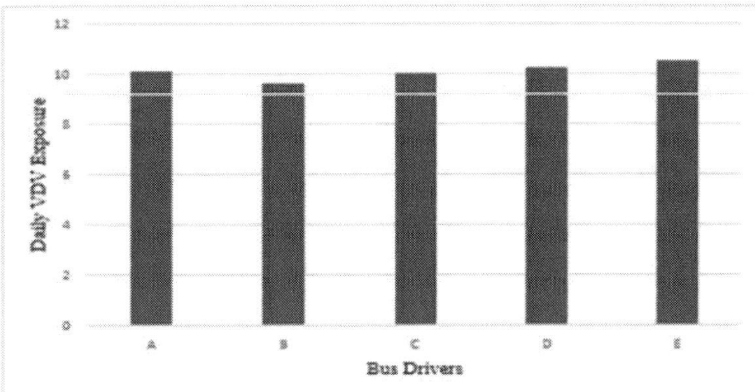

Figure 5. Daily VDV exposure (m/s1.75) in Putrajaya Central.

Refering to Figure 6 below, the daily VDV exposure of these 7 drivers in Larkin Central had exceeded the EAV level, meanwhile, there were three drivers(C, E and G) which were now surpassing the ELV level (21.0 $m/s^{1.75}$) with value of VDV exposure level 21.6 $m/s^{1.75}$, 21.0 $m/s^{1.75}$ and 28.1 $m/s^{1.75}$. The rest are driver B with 20.1 $m/s^{1.75}$, driver F with 19.2 $m/s^{1.75}$, driver A with 17.5 $m/s^{1.75}$ and driver D with 17.2 $m/s^{1.75}$. The VDV exposure level in Larkin Central was very dangerous and in critical situation. Several requirements were compulsory to be done such as full assessment of vibration levels and engineering to lower down the exposure levels.

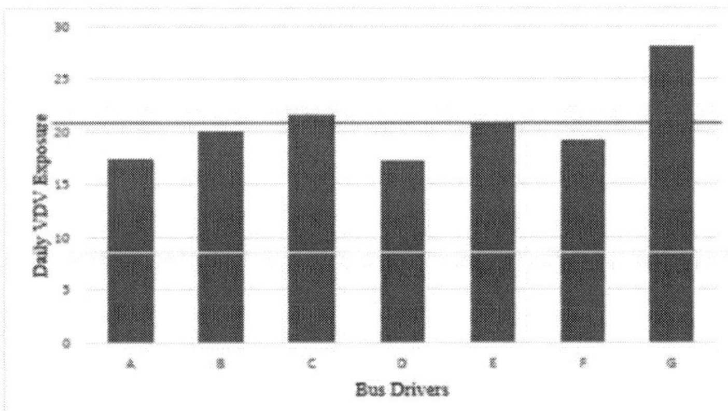

Figure 6. Daily VDV exposure (m/s1.75) in Larkin Central

Table 3 shows the statistical analysis which was correlation analysis. There was no significant correlation analysis that can be done in Putrajaya Central because the daily vibration exposure, A(8) was on the best level. Even though the Pearson correlation showed positive and strong correlation but it was not significant. Effects of vibration on intrastate bus drivers were identified from the Pearson correlation analysis in Larkin Central. The correlation showed positive, strong correlation and significant at p-value was less than 0.05 for Larkin Central. Therefore, the effects of vibration on intrastate bus drivers were identified in Larkin Central which were upper back pain and low back pain.

Table 3: Pearson correlation daily vibration exposure, A(8) with upper and low back pain in Putrajaya Central and Larkin Central

	Sample Size (N)	Upper back pain	Lower back pain
Putrajaya Central	5	.720	.728
		.155	.163
Larkin Central	7	.807*	.745*
		.028	.050

*Correlation is significant at the 0.05 level (2-tailed)

Scopes compared were daily vibration exposure, A(8), discomfort and correlation analysis. In Putrajaya Central, no drivers exceeded the EAV or ELV, which meant the overall operation was good in terms of maintenance progress, bus condition and road condition. Meanwhile, all the drivers in Larkin Central surpassing the EAV and one of them was very close to exceed the ELV. Thus, the locations should be identified, marked out, indicated and engineering measures should be taken to control vibration. There were many discomforts faced by the drivers in Larkin Central compared to Putrajaya Central. These discomforts were due to the driving technique, improper manual handling, road surface, suitability of vehicle and also driver age. On the other hand, there were no significant correlation analysis that was done in Putrajaya Central because the daily vibration exposure, A(8) was on the best level. Even though the Pearson correlation showed positive and strong correlation but it was insignificant. However, the Pearson correlation showed positive, strong correlation and significant for Larkin Central. Therefore, the effects of vibration on intrastate bus drivers were identified in Larkin Central which were upper back pain and low back pain.

4. Conclusion

In conclusion, there were several causes to vibration such as speed, design of the vehicles and environmental conditions. Furthermore, the posture of the driver body also was one of the caused that lead to WBV. Awkward postures combined with vibration will amplify the WBV exposure. Other causes such as vehicle maintenance, suspension, seat design and also driver`s skills were known as the caused that lead to WBV. In addition, the results of daily vibration exposure A(8) in Putrajaya Central shown that the drivers did not exceed the exposure action value while the drivers in Larkin Central were surpassing the exposure action value. Meanwhile for VDV exposure, all the drivers surpass the EAV and three drivers in Larkin Central which were C, E and G exceeds the ELV. According to Blaze and SPSS software analysis, the effects of WBV on bus drivers were upper and lower back pain. Several control measures can be implemented by employer to prevent unnecessary exposure by assessing the vibration, review the safe working system previously as far as practicable and organize training to impart knowledge and awareness and to implement health monitoring. Engineering control like proper maintenance or regular servicing of equipment may minimize exposure levels. Hence, this study and future study will unbox perceptions on the importance of lowering down the level of vibration and knowing the effects of it because there are risks and consequences if they neglect it and have no awareness regarding this matter.

5. References

[1] Rohani, M. M., Wijeyesekera, D. C., & Karim, A. T. A. 2013 *Bus Operation, Quality Service and The Role of Bus Provider and Driver. Procedia Engineering*, **53**, 167–178

[2] Katu, U. S., Desavale, R. G., & Kanai, R. A. 2005 *Effect Of Vehicle Vibration On Human Body – RIT Experience*, 1–9.

[3] Sekulić, D., Dedović, V., Rusov, S., Šalinić, S., & Obradović, A. 2013 *Analysis of vibration effects on the comfort of intercity bus users by oscillatory model with ten degrees of freedom. Applied Mathematical Modelling*, **37(18-19)**, 8629–8644

IOP Conf. Series: Journal of Physics: Conf. Series **1049** (2018) 012094 doi:10.1088/1742-6596/1049/1/012094

[4] Blood, R. P., Ploger, J. D., Yost, M. G., Ching, R. P., & Johnson, P. W. 2010 *Whole body vibration exposures in metropolitan bus drivers: A comparison of three seats. Journal of Sound and Vibration*, **329(1)**, 109–120

[5] Okunribido, O. O., Magnusson, M., & Pope, M. H. 2006 *Low back pain in drivers: The relative role of whole-body vibration, posture and manual materials handling. Journal of Sound and Vibration*, **298(3)**, 540–555

[6] Cheung, A. N. and B. 2006 *The effects of vibration frequencies on physical, perceptual and cognitive performance physical, perceptual and cognitive*, (October).

[7] Taib, M. Y., Ismail, A. R., & Yusoff, A. R. 2010 *A Review on Whole Body Vibrations towards Human*, (December), 597–602.

[8] Okunribido, O. O., Shimbles, S. J., Magnusson, M., & Pope, M. 2007 *City bus driving and low back pain: A study of the exposures to posture demands, manual materials handling and whole-body vibration. Applied Ergonomics*, **38(1)**, 29–38

[9] Kang, T.-H., & Kaizu, Y. 2011 *Vibration analysis during grass harvesting according to ISO vibration standards. Computers and Electronics in Agriculture*, **79(2)**, 226–235

[10] Kuorinka, I., Jonsson, B., Kilbom, a., Vinterberg, H., Biering-Sørensen, F., Andersson, G., & Jørgensen, K. 1987 *Standardised Nordic questionnaires for the analysis of musculoskeletal symptoms. Applied Ergonomics*, **18(3)**, 233–237

[11] Granlund, J., & Brandt, A. *Bus Drivers ' Exposure To Mechanical Shocks Due To Speed Bumps*, **8**, 1–10.

[12] Foster, G., & Leeuwen, M. Van. 2007 *Queensland Mining Industry Health and Safety Conference 2007 Managing Vibration Exposure More recent Standards for WBV exposure*

IOP Conf. Series: Journal of Physics: Conf. Series **1049** (2018) 012095 doi:10.1088/1742-6596/1049/1/012095

Practical Knowledge of Prospective Teachers' In Pedagogy

Siti Nur Kamariah Rubani[1], Tamil Selvan Subramaniam[2], Arihasnida Ariffin[3], Nurhasyimah Hamzah[4] and Maisarah Bidin[5]

Faculty of Technical and Vocational Education,
Universiti Tun Hussein Onn Malaysia.

E-mail: kamaria@uthm.edu.my[1]

Abstract. Practical knowledge in pedagogy is a summary of the procedures taught during practical work that needs to be mastered by a teacher who wanted to carry out teaching and learning in the laboratory. Therefore, the aim of this study was to examine the practical knowledge in pedagogy in terms of safety, handling machines and hand tools. This study uses a survey method. Questionnaires were distributed to a total of 475 prospective teachers with equal division of 31 respondents for each of the areas of specialization offered. Data were analyzed by descriptive statistic. The findings indicate that the scope of practical knowledge in pedagogy of the prospective teachers in the Faculty of Technical and Vocational Education is high (mean = 3.32). These findings may provide guidance to prospective teachers to prepare themselves to be an educator in the field of vocational with skills and practical knowledge in pedagogical.

1. Introduction

Practical knowledge in pedagogy is a knowledge of education related to the scientific study of the principles and methods of teaching, encompassing the optimum teaching process in the laboratory [1]. As pedagogical knowledge changes from time to time, the pedagogical concept also changes according to its development. Currently, pedagogical concepts are more accurately described as concepts involving teaching methods based on educational research with the aim of obtaining effective teaching outcomes [2]. Generally, pedagogical knowledge is a course that must be followed by potential teachers as a guide for teaching in both classroom and laboratories.

In addition to be exposed to science on field skills, prospective teachers at FPTV UTHM are also educated with pedagogical knowledge to ensure the graduates will have a high quality of knowledge and profession while undergoing teaching and practical and real work [3]. In order to produce highly educated and proficient teachers, they must mastered the scope of practical pedagogical knowledge that has become a must-have for a prospective vocational education teacher at FPTV UTHM [4]. The

1 Faculty of Technical and Vocational Education, Universiti Tun Hussein Onn Malaysia.
2 Faculty of Technical and Vocational Education, Universiti Tun Hussein Onn Malaysia.
3 Faculty of Technical and Vocational Education, Universiti Tun Hussein Onn Malaysia.
4 Faculty of Technical and Vocational Education, Universiti Tun Hussein Onn Malaysia.
5 Faculty of Technical and Vocational Education, Universiti Tun Hussein Onn Malaysia.

Content from this work may be used under the terms of the Creative Commons Attribution 3.0 licence. Any further distribution of this work must maintain attribution to the author(s) and the title of the work, journal citation and DOI.
Published under licence by IOP Publishing Ltd

scope of practical pedagogical knowledge is focused on safety and handling of machine tools [5]. According to [6], the excellence of students in mastering practical learning depends on the teacher's strategy in applying pedagogical knowledge. Thus, high quality and excellent students to develop a Malaysian nation can be born [7]. The location of teaching and learning of vocational education involves the demonstration of theory and practice teaching. The teaching plan of any course and vocational education training that is managed should involve campus environments and relevant employment environments based on the conditions within the Malaysian vocational qualification framework.

2. Problem statement

Safety aspects in pedagogical knowledge include pre-teaching activities such as student readiness checks, job assignment briefings, safety practice in laboratories and compliance with general safety procedures [8]. According to [8], students who are not concerned about the safety and the indifferent attitude towards work or work can pose a danger to themselves and their peers. This may be due to a teacher who does not prepare daily teaching instruction in teaching, especially in the laboratory [9]. Although teachers will be exposed to pedagogical knowledge, there are also some who cannot apply the knowledge learned in the teaching and learning process [9].

Teaching and learning in the lab also involve the handling of related machinery and hand tools. Machine handling covers things such as reading and abiding by manual usage, proper and accurate maintenance of machines such as cutting, drilling, welding, grinding, and similar hardware operated by [10]. While handling hand tools includes such as maintaining cleanliness, labelling and storing properly, perfect equipment usage procedures for proper work and safety knowledge of workpieces, work hardware, self and others nearby at work stations [8].

In addition to mastering the theories in the classroom, potential teachers need to master the practical lessons learned in order to give students an optimal understanding. However, in the handling of machines and hand tools, there are teachers who provide less clear explanations and bring problems to the practice of the workshop [11] This is because teachers feel less confident, less prepared and less practical skills to carry out relevant practical lessons [12]. Thus, students are difficult to understand the concepts taught by teachers and only memorize their learning and do not know how to apply them in everyday life [9]. Thus, a study has been conducted to give more understanding and guidance to future teachers in preparing to conduct better teaching and practical exercises in the lab better and effectively.

3. Research Question

This study is conducted with reference to the following questions:

(i) What is the practical knowledge in pedagogy of prospective teachers in terms of safety aspects?

(ii) What is the practical knowledge in pedagogy of prospective teachers in terms of machine control and hand-held equipment in the laboratory?

4. Methodology Research

A total of 475 respondents was selected through purposive sampling method. The respondents consisted of prospective teachers in Faculty of Technical and Vocational, Universiti Tun Hussein Onn Malaysia (FPTV, UTHM). This study utilized a survey design method. This study was conducted using a survey method. Using survey methods research can provide feedback by analyzing the data obtained from questionnaires. The study was started with the collection of raw data through questionnaire distribution. The questionnaire was divided into three parts: Part A (safety aspect) and Part B (machine control and hand-held equipment in the laboratory). A questionnaire form developed using five-point Likert scale that is very agreeable, agreeable, and uncertain, disagree and strongly disagree. After data collection was created, descriptive analysis was conducted to obtain a population attribute description of the central tendency value to determine the direction of diversion. In addition, descriptive analysis is also conducted to obtain the mean value. With the analysis of the data, the translation of the values obtained can assess the practical knowledge in pedagogy for the prospective teachers in FPTV UTHM.

5. Finding

This chapter will be discussed about the findings of the study. The data obtained from the questionnaires that have been distributed and the detailed data analysis shown to answer the research question of the study.

Part A in the questionnaire is safety aspects of the practical knowledge in pedagogy. This section contains 10 items. Table 1 shows the findings of the mean score for the respondent's.

Table 1. Safety aspects

Items	**Safety aspect**	**Mean**
1.	I have first aid knowledge	3.13
2.	I always use protection tools when doing practical work in the lab	3.29
3.	I know safety measures when using hand tools	3.37
4.	I know the safety when handling a machine	3.37
5.	I always keep a good ventilation when doing practical work	3.48
6.	I always keep safety during practical work to avoid accidents	3.56
7.	I make sure the hand tools are in a safe condition before use	3.52
8.	I make sure the machine is safe before use	3.46
9.	I know how to use a fire extinguisher in the laboratory	3.17
10.	I know there are laboratory regulations are patched in the laboratory	3.53
	Average	**3.38**

Based on Table 1, the findings for the safety aspects of the practical knowledge in pedagogy show the highest mean score is item 6 which is 3.56. Meanwhile, the lowest item is in item 1 which is the mean score of 3.13. Overall, the findings for the safety aspects of the practical knowledge of pedagogy have a higher mean score with an interpretation mean score of 3.38. This means prospective teachers' in FPTV, UTHM have practical knowledge in pedagogy in terms of safety.

Part B in the questionnaire is machine control and hand-held equipment in the laboratory for the practical knowledge in pedagogy. This section contains 10 items. Table 2 shows the findings of the mean score for the respondent's.

Table 2. Machine control and hand-held equipment in laboratory

Item	**Machine control and hand-held equipment in laboratory**	**Mean**
1.	I can handle hands tools very well	3.26

2.	I can handle the machine very well	3.29
3.	I know the function of each equipment in the laboratory	3.26
4.	I know the function of each machine in the laboratory	3.29
5.	I know how to choose the right equipment for the practical work	3.33
6.	I know how to choose the suitable machine for practical work	3.30
7.	I can explain how to use a proper hands tools	3.29
8.	I can explain how to use the right machine	3.29
9.	I'm good at using hands tools	3.29
10.	I'm good at handling the machine in the laboratory	3.25
11.	I know how to maintain the machine	3.02
12.	I know how to maintain the hands tools	3.07
	Average	**3.25**

Based on Table 2, the findings for machine control and hand-held equipment in laboratory for the practical knowledge in pedagogy show the highest mean score is on item 5 which is 3.33. Meanwhile, the lowest item is in item 11 which is the mean score of 3.02. Overall, the findings for the machine control and hand-held equipment in laboratory for the practical knowledge in pedagogy have a higher mean score with an interpretation mean score of 3.25. This means prospective teachers' in FPTV, UTHM have practical knowledge in pedagogy in terms of machine control and hand-held equipment in laboratory.

6. Conclusion

As a conclusion of the study, safety is the key that needs to be emphasized in a teaching and learning session in the laboratory. As such, the appropriate teachers should master the safety procedures to ensure the safety of the workshop is guaranteed by complying with the law set out in the laboratory. Based on the analysis of the study the practical knowledge in pedagogy of prospective teachers has a high mean value in term of safety aspect. In terms of safety, teachers should know how to use first aid, handling fire extinguishers and the personal protective equipment and so on. Thus, it can be stated that the prospective teachers in FPTV, UTHM are very concerned about safety during teaching and learning in laboratories.

For machine operation, it involves reading and compliance a manual for handling machine and maintenance. Based on the findings of the study, the practical knowledge in pedagogy has high mean values. Thus, it can be said that prospective teachers' in FPTV, UTHM are the individuals who are practically has a knowledge in handling a machines and hand tools during do their project in the laboratories.

7. Acknowledgements

The authors would like to thank the Universiti Tun Hussein Onn Malaysia for supporting this research under the Short Term Grant (STG) U645.

8. References

[1] Mok, S. S. 2008 *Pedagogi untuk Pengajaran & Pembelajaran.* 1st Ed. (Selangor Darul Ehsan: Penerbitan Multimedia Sdn Bhd.)

[2] Smith, P. L. & Ragan, T. J 2005 *Instructional Design.* 3rd Ed. (USA: John Wiley & Sons)

[3] Maksmovic, J., Petrovic, J. & Osmanovic, J. 2015 Professional Competencies of Future Pedagogues. *Article from Pedagogical Pluralism as The Basis of Education Strategy.* 179036 (2011-2014)

[4] Abdun, Z. 2008 *Amalan Pengurusan dan Penyelenggaraan Bengkel dalam Kalangan Guru-guru Kemahiran Hidup Bersepadu di 12 Buah Sekolah Menengah di Perlis.* (UTM: PSM Thesis)

[5] Wan Husain, W. L 2011 *Tahap Pengetahuan Guru Kemahiran Hidup Dalam Pengurusan Bengkel di Sekolah Menengah di Zon Skudai, Johor.* (UTM: PSM Thesis)

[6] Loughran, J., Berry, A., & Mullhall, P 2006 *Understanding and developing science teachers' pedagogical content knowledge.* (The Netherlands: Sense Publishers)

[7] Zainal, N., Ibrahim, N. H., Surif, J., Mokhtar, L. E., dan Abd Rahman, N. F 2013 Amalan Pengetahuan Pedagogi Kandungan (PPK) Dalam Kalangan Pensyarah Kimia Universiti Teknologi Malaysia. *2nd International Seminar on Quality and Affordable Education (ISQAE 2013).*

[8] Abd Samad, N., Hussein, A., Awang, H., Abd Baser, J., Hasan, A., Harun, H., Masek, A.,Idris, M. S., Amiruddin, M. H. & Ab Ghani, M. A. (2015). *BBT 30603 Modul Pembelajaran Kendiri Pengurusan Makmal dan Bengkel.* 3rd Ed. Malaysia: Penerbit UTM

[9] Md Hanafiah, Z 2008 *Faktor-Faktor Yang Mempengaruhi Keberkesanan Pengajaran Guru Pelatih UTM Dalam Menghadapi Latihan Mengajar.* (UTM: PSM Thesis)

[10] Sharudin, S. A 2008 *Faktor Yang Mempengaruhi Keberkesanan Pengajaran dan Pembelajaran di dalam Bengkel Vokasional Di Dua Buah Sekolah Menengah Teknik di Negeri Sembilan.* (UTM: PSM Thesis)

[11] Wearmounth, J 2005 *Perspectives on student behavior in school-exploring theory and developing practice.* (UK: Routledge ltd.)

[12] Jamaludin, J 2014 *Kesediaan Guru Kolej Vokasional Dalam Pengajaran Amali Teknologi Elektronik di Negeri Pahang.* (UTM: Sarjana Thesis)

IOP Conf. Series: Journal of Physics: Conf. Series **1049** (2018) 012096 doi:10.1088/1742-6596/1049/1/012096

Bivariate SPC Chart Pattern Recognition Using Modular-Neural Network

Nurul Adlihisam Mohd Sohaimi[1], Ibrahim Masood[1], Danial Md Nor[2]

[1]Faculty of Mechanical and Manufacturing Engineering,
Universiti Tun Hussein Onn Malaysia,86400 Parit Raja, Batu Pahat, Johor, Malaysia
[2]Faculty of Electrical and Electronic Engineering,
Universiti Tun Hussein Onn Malaysia,86400 Parit Raja, Batu Pahat, Johor, Malaysia

E-mail: ibrahim@uthm.edu.my

Abstract. In statistical process control (SPC), monitoring and identifying unnatural variation in manufacturing process is challenging when dealing with two correlated quality variables (bivariate). The conventional multivariate SPC charts were designed only for triggering unnatural variation but it is not provide information towards diagnosis. In recent years, several SPC chart pattern recognition schemes were proposed to provide information towards diagnosis based on various category of unnatural variation. In this paper, the modular neural-network scheme was developed to identify nine category of bivariate SPC chart patterns. The success factor for the scheme is outlined as a new perspective in realizing accurate monitoring-diagnosis in quality control.

1. Introduction

The advances in manufacturing technology and manufacturing systems involve processing method, material handling, and quality control. In quality control, selection for an effective statistical process control (SPC) scheme for monitoring and identifying unnatural variation is challenging when dealing with two correlated quality variables (bivariate). In monitoring aspect, the selected scheme should be capable to identify process condition whether in statistically in-control or out-of-control with minimum false alarm. In diagnosis aspect, the selected scheme should be capable to correctly identify the types of out-of-control condition.

The conventional multivariate SPC charts such as Chi-square, multivariate cumulative sum (MCUSUM), and multivariate exponentially weighted moving average (MEWMA) were designed based on monitoring approach to detect the presence of unnatural variation and it were constantly being improved to be more sensitive for triggering small shifts.

Later, the capability of diagnosis was improved through pattern recognition method. An early study on pattern recognition scheme using neural network (NN) recognizer was reported in 2003 [1]. NN is a massive parallel-distributed processor that is capable to learn, recall, and generate knowledge [2]. The properties that are required to recognize and classify the data are often contaminated with noise, unknown distribution and incomplete [3-4]. The other attributes for properties and capabilities to NN in pattern recognition is a non-linearity, input and output mapping, adaptability and fault tolerance, among other. One of the advantages using NN, it can control the noisy measurement and require the

Content from this work may be used under the terms of the Creative Commons Attribution 3.0 licence. Any further distribution of this work must maintain attribution to the author(s) and the title of the work, journal citation and DOI.
Published under licence by IOP Publishing Ltd

assumption of statistical distribution in monitored data [5]. Based on supervised learning approach, NN shows the ability to recognize and classify patterns directly using identified series of process data streams.

In the related study, many reported researches focused on monitoring and diagnosis of bivariate process mean and/or variance using NN [6-7]. NN has been successfully utilized as a pattern recognizer in classifying unnatural patterns and in estimating the shifts of quality variables [8]. NN training required sufficient data, typically in a large amount that is quite difficult to obtain from real manufacturing process. In most cases, synthetic SPC samples were model and simulated mathematically [9]. Several key success factors have been suggested in the reported researches such as the design of neural network structure, selection of training algorithms, training strategy, and design of input data representation, among others [10].

2. Framework of modular neural-network

The design of modular neural-network scheme consists of four phases as shown in Figure 1.

Phase I: Problem Identification

- Classification for bivariate SPC chart patterns. The process variation can be divided to nine categories as follows: Normal (N00), US01, DS01, US10, DS10, US11, DS11, USDS and DSUS. Symbols U and D respectively represent upward shifts and downward shifts. US10 shows there is upward shifts at the first quality variable (V1), while the second quality variable (V2) is remain at in-control condition. On the other hand, US01 shows there is upward shift at V2, while V1 is remain at in-control condition. In other case, US11 shows there are upward shift at both quality variables (V1, V2).
- The desired of target vector. In this research, the size of input vector was five (5) corresponding for shift. The number of output nodes in this research was set corresponding to the number of pattern classed.

Phase II: Recognizer and input representation, Training and Pre-Testing

- Design of pattern recognizer. A modular-NN recognizer with the multi-layer perceptron's (MLP) model was designed as shown in Figure 3 and Figure 4. The MLP model was applied since it has been proven effective for classification tasks [11].
- Design of input representation. In many cases, input representation for NN was utilized in the form of raw data or original SPC samples [12-14], features-based such as shape features or statistical feature [15-16] and the combination between raw data and features-based [17]. In this paper, a new statistical features set was selected using design of experiments analysis.
- Training and Testing. Target performance for training was determined at $\geq$ 95% for normal and shifts patterns.

Phase III: Validation test

The monitoring and diagnosis capability of the scheme can be judged based on the detection speed, rate of false alarm, and the classification accuracy.

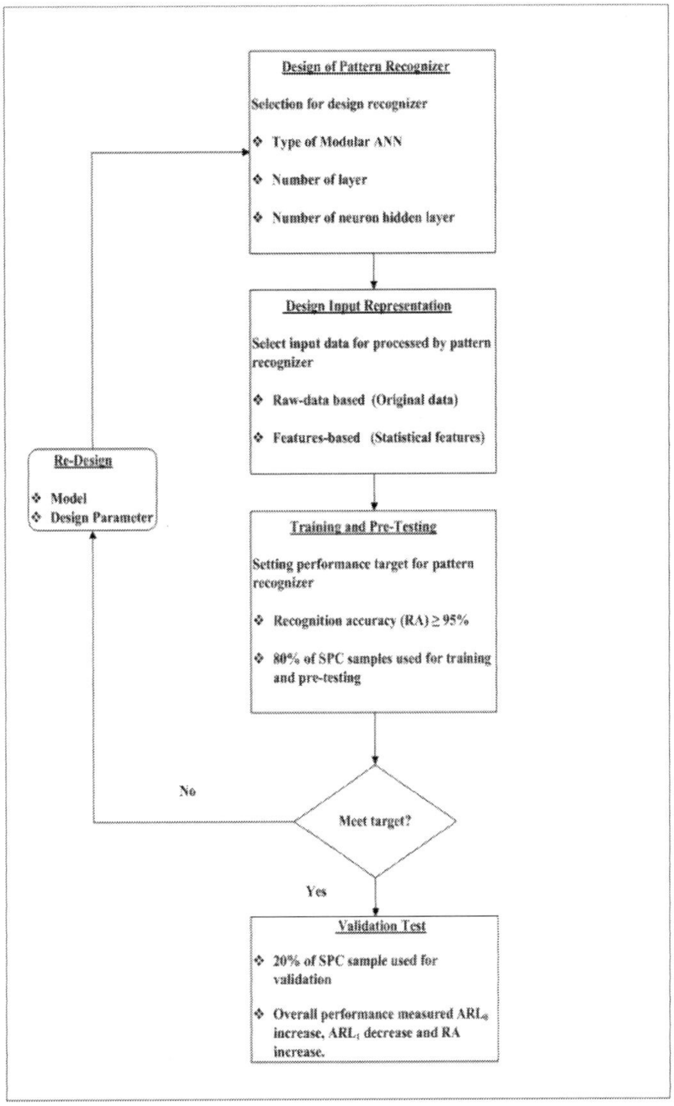

Figure 1. Process flow in designing the modular-NN scheme [18]

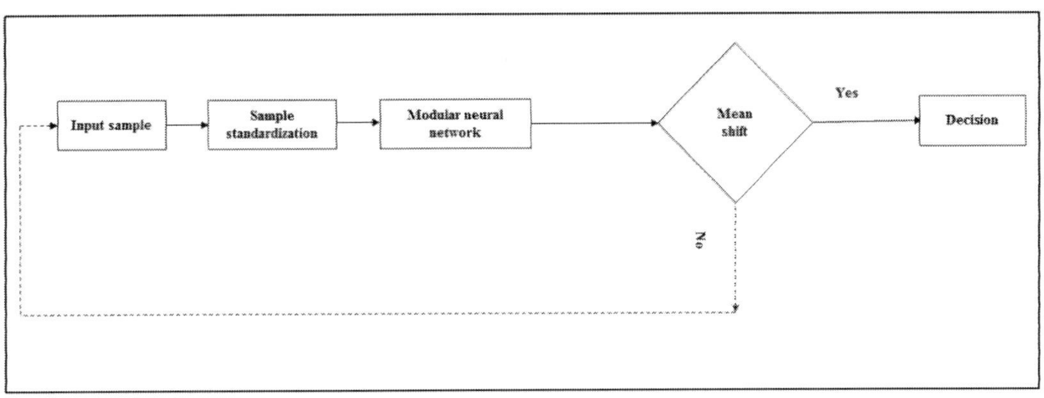

Figure 2. Conceptual diagram for the modular-NN scheme

The operating procedure of the modular-NN scheme is represented in Figure 2. In practice, input samples or standardized samples will be taken from real SPC samples. In this paper, simulated data in the form of dynamic SPC samples were used for analysis. The data start from in-control (normal) process and it was changing gradually to out-of-control (shift) process. In order to obtain effective decision in QC, the expected outcome of the pattern recognizer is that it should be able to detect an out-of-control process as fast as possible with minimum false alarm.

Figure 3 shows the model of modular-NN recognizer. It consists of a generalized-MLP and four specialized-MLP model. The generalized-MLP model functions to classify patterns based on five categoires, i.e., normal, shifts at first variable (US1), shifts at variable 2 (DS1), shifts at both variables (UD1) and shift at opposite direction (DSU). The specialized-MLP functions to classify patterns based on two categories as follows: US1 to classify US10 and DS10 patterns, DS1 to classify US01 and DS01 patterns, UD1 to classify US11 and DS11 patterns, and DSU to classify USDS and DSUS patterns.

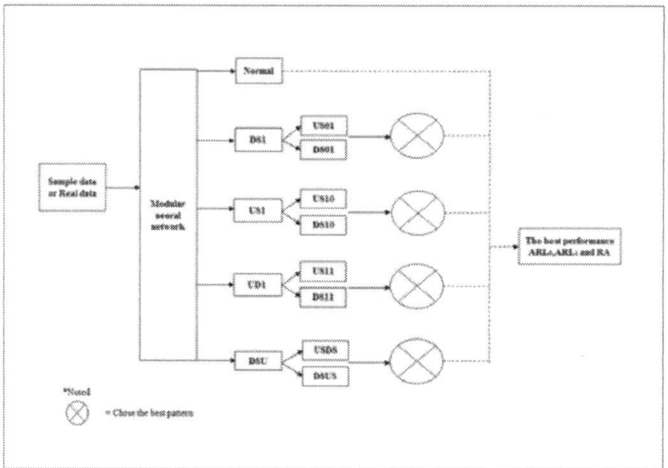

Figure 3. Framework for modular-NN recognizer

3. Results and Discussion

The monitoring and diagnosis performances of the proposed scheme were evaluated based on average run length (ARL_1) and recognition accuracy percentage (RA) as shown Table 1. The performance results involve various ranges of correlation as follow:

- US10, US01, DS10 and DS01 – correlation between two variables are low (0.1 to 0.3)
- US11 and DS11 – positive and high (0.5 to 0.7)
- For USDS and DSUS – negative and high (-0.7 to -0.5)

The results were represented by average ARL_1 and RA for different class of shifts pattern. In monitoring aspect, it can be observed that the smaller the mean shift, the longer the ARL_1. Based on this trend, it can be concluded that an out-of-control process with small variation would be more difficult to be detected due to its similarity to an in-control process. In opposite, it would be easy to detect an out-of-control process with large variation due to its difference to an in-control process.

In diagnosis aspect, it can be observed that the stronger the correlation between two variables, the higher the RA. Inversely, size of mean shifts does not change the values of RA. Based on this finding, it can be concluded that an unnatural variation of weak correlation condition would be more difficult to be classified due to mixed properties between in-control and out-of-control process. In opposite, an unnatural variation with strong correlation would be easy to be classified due to clear properties of out-of-control process.

Table 1. Testing results of modular-NN scheme

Mean Shift (std dev)	ARL_1 and RA	US10, US01, DS10, DS01	US11, DS11	USDS, DSUS
		0.1 / 0.3	**0.5 / 0.7**	**- 0.5, - 0.7**
1.0	ARL_1	10.55 / 10.42	7.94 / 8.58	8.12 / 8.68
	RA	**90.6 / 91.2**	**95.6 / 98.2**	**90.8 / 96.2**
1.5	ARL_1	5.91 / 5.72	4.91 / 5.25	5.04 / 5.20
	RA	**92.7 / 91.4**	**99.2 / 99.8**	**96.8 / 98.6**
2.0	ARL_1	4.35 / 4.18	3.63 / 3.77	3.66 / 3.78
	RA	**93.0 / 91.6**	**99.7 / 99.6**	**98.4 / 99.3**
2.5	ARL_1	3.47 / 3.33	2.90 / 3.04	2.94 / 3.08
	RA	**91.6 / 91.6**	**99.2 / 98.8**	**98.3 / 98.3**
3.0	ARL_1	2.91 / 2.81	2.47 / 2.65	2.45 / 2.60
	RA	**89.8 / 89.2**	**99.1 / 99.2**	**98.8 / 99.1**

4. Conclusion

Proper design of NN recognizer is important in the development of pattern recognition scheme. In this study, it has become more challenging to monitor and diagnose nine categories of bivariate SPC chart patterns with fast detection and high accuracy. This aim can be realized using a modular – NN recognizer. The utilization of smaller structure of a generalized-MLP and several specialized-MLP was able to achieve efficient recognition. As a result, this finding will be useful to improve quality in manufacturing.

5. Acknowledgement

The authors wish a highly gratitude to Universiti Tun Hussein Onn Malaysia (UTHM) for funding this research (Contract-Industry Grant Scheme, U524).

References

[1] Zorriassatine, F., Tannock, J.D.T (1998). A Reviewof Natural Network for Statistical Process Control, *Journal Intelligent Manufacturing* **9**, pp.209-224.
[2] Haykin, S. (1999) Neural Network: A Comprehensive Foundation, Prentice-Hall, Englewood Cliffs, NJ.
[3] Schalkoff, R.J., 1997. Artificial neural network, *McGraw-Hill*, New York.
[4] Haykin, S. (1999) Neural Network: A Comprehensive Foundation,2nd ed., *Prentise Hall*, New Jersey.

[5] Bag, M. and Gauri, S.K. (2012). An Expert System for Control Chart Pattern Recognition. *International Journal Advanced Manufacturing Technology*, pp. 291 – 301.

[6] Chen,L.H., Wang, T.Y.,2004. Artificial Neural Network to Classify Mean Shift From Multivariate X^2 Chart Signals. *Computer and Industrial Engineering* **28**, pp 195-205.

[7] Yu, J.B., and Xi, L.F., 2009. A Neural Network Ensemble-Based Model for On-Line Monitoring and Diagnosis of Out-of-Control Signal in Multivariate Manufacturing Processes, *Expert System with Applications* **36,** pp. 909-921.

[8] Guh, R.S. and Tannock, J.D.T. (1999). A Neural Network Approach to Characterize Pattern Parameters in Process Control Charts. *Journal of Intelligent Manufacturing* **10,** pp.449 – 462

[9] Masood I, Hassan A (2010) Issue in development of artificial neural network-based control chart pattern recognition schemes. *European Journal of Scientific Research* **39**(3), pp.336-355.

[10] Cheng, C.s., 1997. A Neural Network Approach for the Analysis of Control Chart Pattern, *International Journal of Production Research* **35,** pp. 667-697.

[11] Gauri, S.K. and Chakraborty, S. (2008). Improve Recognition of Control Chart Patterns Using Artificial Neural Network. *International journal of Advanced Manufacturing Technology*, pp.1191-1201.

[12] Pham, D.T., and Oztemel, E., 1993. Control Chart Pattern Recognition Using Combinations of Multilayer Perceptrons and Learning Vectro Quantization Neural Networks, *Proc. Instn. Mech. Engrs.* **207**, pp. 113-118.

[13] Zorriassatine, F., Tannock, J.D.T. and O'Brien, C. (2003). Using Novelty Detection to Identify Abnormalities Caused by Mean Shifts in Bivariate Processes. *Computer and Industrial Engineering*, pp.385 – 408

[14] Niaki, S.T.A and Abbasi, B. (2005). Fault Diagnosi in Multivariate Control Chart Using Artificial Neural Networks. *Quality and Realibility Engineering International*, pp. 825-840.

[15] Pham, D.T and Wani, M.A., 1997, Features-based controls chart pattern recognition. *International Journal of Production Research*, **35**, 1875-1890.

[16] Gauri, S.K. and Chakraborty, S. (2006). Feature-Based Recognition of Control Chart Patterns. *Computer and Industrial Engineering*, pp.726 – 742.

[17] Masood, I. and Hassan, A. (2014). Bivariate Quality Control Using Two-Stage Intelligent Monitoring Scheme. *Expert Systems with Applications* **41,** pp.7579 - 7595.

[18] Sohaimi, N.A.M., Masood, I., Mohammad, M. and Hassan, M.F. (2017). Design Methodology of Modular-ANN Pattern Recognizer for Bivariate Quality Control. *Journal of Telecommunication, Electronic and Computer Engineering* **9**(3−2)**,** pp.31 - 34.

IOP Conf. Series: Journal of Physics: Conf. Series **1049** (2018) 012097 doi:10.1088/1742-6596/1049/1/012097

Evaluation of Working Posture on Bus Traffic Controllers

Farhan Najmi Ferdinand Bakri[1], Jalil Azlis-Sani[1], Mohd Zamani Ngali[1], Muhammad Nur Annuar Mohd Yunos[1], Jacquelyne Anne Boudeville[1], Syazwan Ismail[1], Munzilah Rohani[2], Nor Aziati Abdul Hamid[3], Mohd Saiful Izwan Aazizi[4] and S.M. Sabri S.M. Ismail[5]

[1]Faculty of Mechanical Engineering & Manufacturing, Universiti Tun Hussein Onn Malaysia (UTHM), 86400 Parit Raja, Johor, Malaysia
[2]Faculty of Civil and Environment Engineering, Universiti Tun Hussein Onn Malaysia (UTHM), 86400 Parit Raja, Johor, Malaysia
[3]Faculty of Technology Management and Business, Universiti Tun Hussein Onn Malaysia (UTHM), 86400 Parit Raja, Johor, Malaysia
[4]Rapid Bus Sdn Bhd, Kompleks Rapid Bus Cheras Selatan, 71, Jalan Kb 2/15, Kampung Baru Balakong, 43300 Seri Kembangan, Selangor, Malaysia
[5]Strategy & Transformation Office, Prasarana Malaysia Berhad, Level 21, Tower B, Menara UOA Bangsar, No 5, Jalan Bangsar Utama 1, 59000 Kuala Lumpur, Malaysia

E-mail: azlis@uthm.edu.my

Abstract. This study aimed to evaluate the working posture of bus traffic controllers. The Musculoskeletal Disorders (MSDs) prevalence and the exposure to the ergonomic risks factors (ERFs) among bus traffic controllers were determined. There were 26 respondents (16 male and 10 female) who were bus traffic controllers (BTC) at a prominent bus service provider were selected as the subjects. Data for Musculoskeletal Disorders (MSDs) prevalence among the bus traffic controllers were collected using Nordic Musculoskeletal Questionnaire (NMQ) and direct observation and assessment of ergonomic risks factors were conducted via Rapid Office Strain Assessment (ROSA). By using ROSA, the BTCs working posture and work environment were checked and measured in order to quantify the risks to ERFs. According to the assessment, more than half (15/26 respondents) of the bus traffic controllers were found to have an extreme posture with the ergonomic risk level was high (more than 4) (action level 3) which required a further investigation and changes of work habits soon. The research showed that most of the bus traffic controllers (BTC) were exposed to high ergonomics risk for MSDs development. To avoid MSDs, the personal working behaviours and the design of the computer-based workstations must be improved based on ergonomic principles.

1. Introduction

In contemporary years, rapid development and use of computers has affected the workers and their workplace [1]. Introduction of computers into the workplace has changed the working environment, work organization, appearance of new risk factors, and a different use of worker physical and mental potential. These changes may have a bad effect on their health [2].

Content from this work may be used under the terms of the Creative Commons Attribution 3.0 licence. Any further distribution of this work must maintain attribution to the author(s) and the title of the work, journal citation and DOI.
Published under licence by IOP Publishing Ltd

Malaysia is one of the developing countries having musculoskeletal disorders (MSDs) as a major occupational health issue for workers [3]. Many aspects or factors including personal aspects, work related factors (WRF) and psycho-social factors (PSF) can lead to musculoskeletal disorders (MSDs). Generally, musculoskeletal disorders (MSDs) are caused by static body positions, repetitive movements, prolonged muscular contraction, and the use of force. Musculoskeletal disorders (MSDs) consist of minor physical disabilities as this term is used to describe various conditions that affect the muscles, bones, and joints. Work-related musculoskeletal disorders (MSDs) can affect shoulders, arms, elbows, wrists, hands, back, legs and feet [4].

Computer and office-based work, such as customer service work, is a growing occupation that is correlated with an increase in Work Related Musculoskeletal Disorders (MSDs) [5]. Based on a survey in Malaysia, it was found that 50% of 136 computer users from a sample of population of university students and office staff suffered some lower back pain did not have an adjustable backrest on their seats. Most of the users had higher Rapid Upper Limb Assessment (RULA) scores of the wrist and neck as it showed an increased risk of developing occupational overuse syndrome [6].

Studies have reported a high prevalence of musculoskeletal disorders among call center workers who sit daily for a prolonged length of time. The workers spent most of their time working in a workspace that uses a computer desk and chair. The musculoskeletal disorders were caused by the awkward posture from the use of fingers, wrists and arms in repetitive posture and always looking on the screen while working using computer [7]. Along with musculoskeletal disorders, they might suffer from eye, ear, and throat related health problems [8].

Therefore, a study was conducted among a prominent bus service provider workers to evaluate the working posture of bus traffic controllers at the control room which is a computer-based workstation. Their daily working activities and working postures were observed in order to identify the possibilities of musculoskeletal symptoms and their exposure to the ergonomic risk that lead to musculoskeletal disorders.

This study of evaluation of working posture on bus traffic controllers could provide better understanding regarding ergonomics. The identification of musculoskeletal disorders or symptoms and the exposure level to the ergonomic risks among bus traffic controllers who works in a computer-based workstation can be determined and this could develop better understanding regarding their working activity, posture and its effect. This research would help for improvements in terms of ergonomics which involved the working posture, working habits of the bus traffic controllers and the design of the workstation. The safety of the workers and the working environment can be improved to ensure the quality and efficiency of the bus control services.

2. Methodology

In this research, two methods were used for data collection. Structured interview sessions using Nordic Musculoskeletal Questionnaire (NMQ) was carried out in order to determine the MSDs among bus traffic controllers. Besides, direct observation using Rapid Office Strain Assessment (ROSA) was conducted to assess the ERFs among bus traffic controllers.

2.1. Subjects and Task Description

This study has been conducted among 26 bus traffic controllers (BTC) at the control room of a prominent bus service provider which is categorized as a computer based workstation. In this study, each BTC has to work at their workstation which during that time, their working posture or responsiveness and interaction with equipment was observed. Included in this study were the BTC who had to use the computer for their normal workday. The daily work tasks performed by the bus traffic controllers are such as key in the data using computer, letter writing, dealing with telephone and email enquiries.

2.2. Nordic Musculoskeletal Questionnaire (NMQ)

The Nordic Musculoskeletal Questionnaire [9] has been widely used to identify the possibilities and severity of self-rated musculoskeletal symptoms. The questionnaire includes questions asking about the experience of musculoskeletal problems in nine body areas which are neck, shoulders, elbows, wrists/hands, upper back, lower back, hips/thighs, knees, and ankles/feet over the past week and over the past year.

NMQ consists of three parts. Part A was about the demographic items including gender, age, height, weight, hand dominance, working experience, and hours of work per week. Part B is concentrated on trouble for locomotive organs which included questions about trouble during last 12 months regarding neck, shoulders, elbows, wrists/hands, upper/lower back, hips/thighs, knees, ankles and feet. The last part was Part C which concentrated on common area of musculoskeletal symptoms which are lower back, neck and shoulder. There were specific questionnaires on the details for the musculoskeletal symptoms on each body parts.

2.3. Rapid Office Strain Assessment (ROSA)

Rapid Office Strain Assessment (ROSA) is a picture based posture checklist which is designed to quantify exposure to risk factors in an office work environment. It is modelled after other picture based checklist and chart scoring systems such as RULA and REBA [10]. But ROSA is more specific for computer usage risk assessment compared to RULA [11].

ROSA consists of three sections which are Section A, Section B and Section C. Section A are scoring section for office chair which combined the height of the chair, its pan depth, the armrests and back support. Section B consists of sub-section for monitor and telephone score while Section C combined the scores for mouse and keyboard. When conducting an assessment by using ROSA, the observer will choose the most suitable scores based on the posture of the worker that being observed. From ROSA, there are 4 risk assessment score levels which are low (1-2 points), medium (3-4 points), high (5-7 points) and very high (8-10 points).

2.4. Data Collection

In this research, the data was collected using two methods that have been explained in previous sections which were Nordic Musculoskeletal Questionnaire (NMQ) and Rapid Office Strain Assessment (ROSA). The subjects were 26 bus traffic controllers (BTC) from a prominent bus service control room.

Firstly, a face to face structured interview was conducted among bus traffic controllers by using Nordic Musculoskeletal Questionnaire (NMQ). The questionnaire consists of three parts which are A, B and C. For Part A, they were asked for some demographic items such as gender, age, height, weight, hand dominance, years of work and hours of works per week. As for Part B, the questionnaire is concerning on the trouble with the locomotive organs. The BTCs were asked on information about the musculoskeletal symptoms frequency of pain and the intensity of the pain. The 12 months prevalence, 7 days point prevalence and intensity of musculoskeletal troubles were identified and questioned from the workers. The last part which was Part C is where the workers were questioned about the details area of musculoskeletal symptoms which involved lower back, neck and shoulder.

In addition, an observation on the working activities and working posture of the bus traffic controllers was conducted. A direct observation on the BTC's working posture using Rapid Office Strain Assessment (ROSA) in the computer based control room was performed in order to assess the ergonomics risk factors [12]. Posture and work environment were checked step by step and included sitting on chair, monitor and telephone used, mouse and keyboard respectively. Also considered was the estimated work duration which was reported by the computer user (BTC). This value was used as the assumed duration of use for all components of the workstation. Their working postures and work environment score were recorded and evaluated.

2.5. Data Analysis

From the collected data of Nordic Musculoskeletal Questionnaire (NMQ) which contain demographic variables, number of self-reported symptoms and reported lower back, neck and shoulder trouble, descriptive statistics was used to analyse the data such as percentages, mean and standard deviation. While for the result of ROSA obtained which included the scores for office chair (Section A), scores for monitor and telephone (Section B), scores for mouse and keyboard (Section C) and ROSA Grand scores were analysed as mean, standard deviation and action level. Results of both Nordic Musculoskeletal Questionnaire and Rapid Office Strain Assessment (ROSA) were carried out using Microsoft Excel and SPSS Software Version 23.0. Figure 1 shows the flowchart of data analysis.

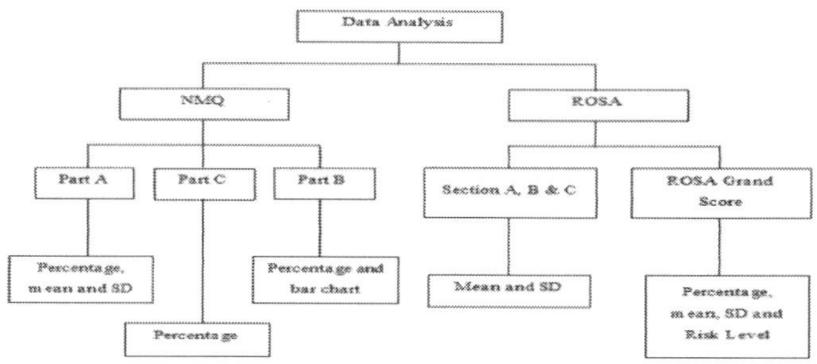

Figure 1. Data Analysis Flowchart

3. Results & Discussion

3.1. Nordic Musculoskeletal Questionnaire (NMQ)

3.1.1. Demographic Variables

Table 1 is a presented summary of the demographic variables. Based on the analysis, there was more male bus traffic controllers compared to female. Majority of the bus traffic controllers are aged between 21 to 40 years old and only 3 of them are 40 years old and over with the mean score for age was 30.31. Their total year of working experience was 2.89 years on average. Most of the bus traffic controller's working experiences were between 1 to 5 years with only 3 of them with 6 years and more working experience.

All of 26 of them spent more than 51 hours in the computer- based control room every week with the mean score for the weekly working time was 55 hours a week. From the questionnaire, it was found that 14 of the BTCs had normal weight for their Body Mass Index (BMI) while 9 of them were overweight. Besides, 1 of the BTC was underweight and 2 were obesity. The average body mass index for the BTC was 25.09 kg/m2 and only 2 of them were left handed.

Table 1. Demographic Variables of Bus Traffic Controllers (n=26)

Characteristic		N	%	Mean	SD
Gender	Male	16	61.54	-	-
	Female	10	38.46		
Age	≤20	0	0.00	30.31	7.70
	21-40	23	88.46		
	≥41	3	11.54		
Working Experience	1-5	23	88.46	2.89	2.15
	≥6	3	11.54		
Working hours in control room	40-50	0	0.00	55.00	0.00
	≥51	26	100.00		
Body Mass Index (BMI)	Underweight	1	3.85	25.09	4.10
	Normal Weight	14	53.85		
	Overweight	9	34.62		

	Obesity	2	7.69		
Hand Dominance	Right-handed	24	92.31	-	-
	Left-handed	2	7.69		

3.1.2. Number of Self-reported Symptoms

The numbers of self-reported symptoms among bus traffic controllers that occurred in the last 12 months were shown as percentage in Figure 2. The results of the research showed that different gender of the BTC had different prevalence of musculoskeletal symptoms on the same body region. The highest prevalence of body part trouble in last 12 months for bus traffic controllers was low back (84.62%), followed by upper back, neck, shoulder and knees while the least body part trouble were hips/thighs and ankles/feet both at 15.58% followed by elbows and wrists/hands which both was at 11.54%.

Female bus traffic controllers were more likely to get injured from their work activity as they had higher injury prevalence for all body parts compared to male bus traffic controllers. Their highest prevalence was upper back which suffered by all of the female bus traffic controllers. Meanwhile the least injury prevalence for female bus traffic controllers were elbows, wrist/hands, hips/thighs and ankles/feet as the score for all of the body parts were 20%. It was similar compared to previous study by Jensen et al (2002), as it showed that the same body region of MSDs suffered by female call-centre workers were neck with 53% prevalence of symptoms, shoulder (42%) and upper and lower back (30%) symptoms.

As for male bus traffic controllers, they had highest prevalence for low back injury which was at a rate of 81.3% as the least prevalence of body part trouble was elbows and wrists/hands which both scored 6.3%.

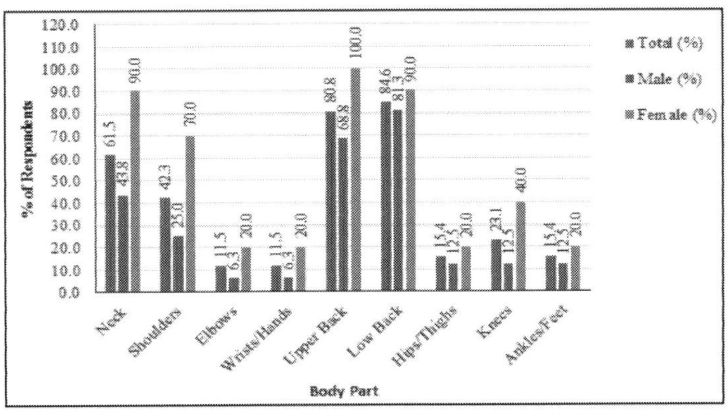

Figure 2. Percentage of NMQ Respondents with Body Part Trouble in Last 12 months

3.1.3. Reported Lower Back, Neck & Shoulder Trouble

There were additional results for lower back, neck and shoulder trouble which are related to accidents, work or leisure activities limitations, and treatment are constructed in Table 2. During the past 12 months, there were 5 of bus traffic controllers sought a professional for trouble regarding their neck, followed by 4 for lower back and 2 for shoulder trouble. Some of the bus traffic controllers claimed that they had to change their job or duty due to the trouble which there were 4 of them who had to change their duty due to lower back trouble and 2 due to neck trouble.

It was found that 22 out of 26 BTC suffered lower back trouble, 16 suffered neck trouble and 11 suffered shoulder trouble. In the last 12 months, most of the bus traffic controllers suffered lower back trouble for about 1 to 7 days followed by 15.38% for 8 to 30 days and 7.69% for more than 30 days of trouble. For neck trouble, the percentage was 46.15% for 1 to 7 days and 7.69% for both 8 to 30 days and more than 30 days. As for shoulder trouble, the percentage was 30.77%, 7.69% and 3.85% respectively. Those body parts trouble had negative effect on the bus traffic controllers work activity.

IOP Conf. Series: Journal of Physics: Conf. Series **1049** (2018) 012097 doi:10.1088/1742-6596/1049/1/012097

From the research, it was found that lower back symptom was severe enough to reduce bus traffic controllers' work activity and leisure activity during last 12 months with both at 26.9% followed by those who had neck trouble, 19.23% had to reduce work activity and 23.08% had to reduce their leisure activity. As for shoulder trouble, the percentage was 15.38% for both type of activity. Besides, 11.54% bus traffic controllers reported that they were prevented from work for 1 to 7 days while 3.85% of them were prevented from work for 8 to 30 days and more than 30 days due to lower back trouble. Furthermore, there were 7.69% bus traffic controllers prevented from work for 1 to 7 days, none for 8 to 30 days and 3.85% for more than 30 days due to neck trouble. Not to mention, there were also 3 bus traffic controllers prevented from work for 1 to 7 days due to shoulder trouble and only 1 for both 8 to 30 days and more than 30 days.

Similar to the result of Nordic Musculoskeletal Questionnaire for this study which found that 84.62% of BTC suffered lower back trouble, 61.54% suffered neck trouble and 42.31% suffered shoulder trouble, there were studies that reported high prevalence of MSDs among call centre workers. According to a study among call centre by neck pain, upper back pain and lower back pain were the top three ranked as highest prevalence during the last 12 months. Another study by Rocha et al (2005), reported that most of the call centre workers have high prevalence on neck symptom and shoulder symptom at 43.0% for both of the body region. The call centre workers were observed sitting daily for a prolonged length of time in their computer-based workstation performing several tasks such as key in the data using computer, letter writing, dealing with telephone and email enquiries which were similar to the task performed by bus traffic controllers (BTC).

Symptoms	Lower Back, % (n)	Neck, % (n)	Shoulder, % (n)
Any trouble ever	84.62 (22)	61.54 (16)	42.31 (11)
Ever had in accident	0 (0)	7.69 (2)	0 (0)
Change of job or duty	15.38 (4)	7.69 (2)	0 (0)
Total time with trouble last 12 months			
0 days	0 (0)	0 (0)	0 (0)
1-7 days	61.54 (16)	46.15 (12)	30.77 (8)
8-30 days	15.38 (4)	7.69 (2)	7.69 (2)
More than 30 days, but not every day	7.69 (2)	7.69 (2)	3.85 (1)
Every day	0 (0)	0 (0)	0 (0)
Reduce of Activity			
Reduce of work activity last 12 months	26.92 (7)	19.23 (5)	15.38 (4)
Reduce of leisure activity last 12 months	26.92 (7)	23.08 (6)	15.38 (4)
Total time prevented work last 12 months			
0 days	65.38 (17)	50.0(13)	23.08 (6)
1-7 days	11.54 (3)	7.69 (2)	11.54 (3)
8-30 days	3.85 (1)	0 (0)	3.85 (1)
More than 30 days	3.85 (1)	3.85 (1)	3.85 (1)
Sought a professional/Hospitalized	15.38 (4)	19.23 (5)	7.69 (2)
Trouble in last 7 days	42.31 (11)	34.62 (9)	34.62 (9)

Table 2. The Prevalence of Musculoskeletal Symptoms by Body Parts among Bus Traffic Controllers, (n=26)

Anatomical Region	Any trouble last 12 months, % (n)	Prevented from normal work, % (n)	Trouble last 7 days, % (n)
Upper Extremity			
Shoulders	42.31 (11)	19.24 (5)	34.62 (9)
Elbows	11.54 (3)	0 (0)	3.84 (1)
Hands/Wrists	11.54 (3)	0 (0)	7.69 (2)
Lower Extremity			
Hips/Thighs	15.38 (4)	0 (0)	0 (0)
Knees	23.08 (6)	3.84 (1)	3.84 (1)
Ankles/Feet	15.38 (4)	0 (0)	0 (0)
Axial Skeleton			
Neck	61.54 (16)	11.54 (3)	34.62 (9)
Upper back	80.77 (21)	15.38 (4)	42.31 (11)
Lower back	84.62 (22)	19.24 (5)	42.31 (11)

Table 3. 12-month prevalence of musculoskeletal symptoms and work interference by anatomical region

Besides having high prevalence of musculoskeletal symptoms which prevented them from doing their normal work, some of them had the musculoskeletal trouble in the last 7 days. For those having lower back symptom, 42.31% claimed to have the trouble during last 7 days. The percentage was the same for both who had neck and shoulder trouble as 34.62% reported to have the trouble during the last 7 days.

Referring to table 3, there were 3 anatomical regions which involved upper extremity, lower extremity and axial skeleton. For upper extremity, shoulders were the most affected region as there were 11 (42.31%) bus traffic controllers reported to have trouble. There were 3 (11.54%) of them who had trouble on their elbows and hands/wrists. For lower extremity, the most affected region were knees with percentage of respondent having the trouble was 23.08% followed by both hips/thighs and ankles/feet at 15.38%. Not to mention, the highest prevalence of musculoskeletal symptom among bus traffic controllers was the axial skeleton which involved lower back (84.62%). The results also showed that most of the bus traffic controllers were prevented from normal work due to shoulders(19.24%), neck(11.54%), upper back (15.38%) and lower back (19.24%). It was also found that most of the bus traffic controllers had trouble at those four regions during last 7 days with percentage of 34.62% for shoulders and neck, and 42.31% for upper back and lower back.

3.2. Rapid Office Strain Assessment (ROSA)

3.2.1. Reported Lower Back, Neck & Shoulder Trouble

In order to assess the ergonomic risks that can lead to musculoskeletal disorders (MSDs) in working environment among bus traffic controllers Rapid Office Strain Assessment (ROSA) method was applied. ROSA was designed as a tool to identify and quantify the factors related to discomfort due to computer jobs. The assessment included sitting posture, work station which includes chair (chair height, pan depth, armrest and back support), computer (monitor, mouse and keyboard) and the usage of telephone. Besides, the working duration was also considered. All the bus traffic controllers work for 11 hours per day and had continuous work using computer for more than 4 hours per day.

Based on the results, it was found that most of the BTCs had appropriate or suitable chair height which their knees were at 90° (73.1%). There were also BTCs who had their knees too low (knee angle < 90°) and too high (knee angle> 90°) which the proportions were 23.1% and 3.8% respectively. For chair pan depth, 80.8% of the bus traffic controllers had approximately 3 inches of space between their knee and the edge of seat. Unfortunately, there were some of them having less than 3 inches of space (15.4%) and more than 3 inches of space (3.8%). In terms of design of work station, the chair used have non-adjustable pan (seat) at 100%.

Majority of the BTCs had relaxed shoulder during their working period which their elbows were supported in line with their shoulder (69.2%) and 30.8% of them were observed to be in inappropriate posture with their shoulder shrugged and arm's unsupported. Not to mention, the design of the chair that have non-adjustable armrests (100%) will increase the ergonomic risks which might lead to MSDs. There were 34.6% of BTCs who had good sitting posture with adequate lumbar support which their chair reclined between 95° to 110°. Besides, it was observed that most of the bus traffic controllers (65.4%) have a habit to lean forward (no back support) when using computer.

The bus traffic controllers sat in arm's length distance (45-75cm) from monitor with screen at eye level and only 3.8% had low monitor/screen which was below 30°. All of them had no holder for document which forces them to flex and twist their neck in order to view papers positioned on the desk. When dealing with calls or using telephone, most of the bus traffic controllers had one hand on phone and neutral neck posture (76.9%). Unfortunately, there were 11.5% of them who had a habit to hold the telephone using their neck and shoulder. Some had to reach to the telephone with greater distance than 30cm (23.1%) and there were 88.5% who had no hands-free option for their phone.

All of the bus traffic controllers used their computer mouse in line with their shoulder but 100% of them had their mouse and keyboard on different surfaces. Majority of BTCs observed to have relaxed shoulder and straight wrists when using keyboard (65.4%) but there were also 34.6% who used their keyboard on positive angle (>15°) which lead to wrist extension. Their bad working habit and the deficiency of the workstation design will increase the ergonomic risk level that will lead to MSDs.

3.2.2. Section A – Chair Score

Section A which was chair section consists of 4 smaller sub-sections involving chair height, seat pan depth, the position of armrest and the position of back support. Table 4 shows the score obtained for each risk factors for Section A.

3.2.3. Section B – Monitor & Telephone Score

Section B was a score section which involved monitor and telephone. In this section, the length and position of monitor for all workstation for each of the bus traffic controllers were assessed. The angle of neck twist, glare on screen and documents holder were also considered to be risk. As for telephone, the position of the telephone from the bus traffic controllers was measured and assessed. The posture of neck and shoulder when using the telephone was observed as an addition score for the risk factor.

Table 5 shows the score obtained for Section B. Mean score for monitor was 3.04 (SD = 0.20) and the score was 1.35 (SD = 0.85) for telephone. The final score for Section B was 2.31 (SD = 0.62).

Table 4. Section A Score for Office Chair (n = 26)

Risk Factors	Score	
	Mean	SD
Chair height	1.23	0.43
Pan depth	2.15	0.37
Armrest	2.31	0.47
Back support	1.65	0.49
Duration	1.00	0.00
Final Score	4.00	0.80

Table 5. Section B Score for Monitor & Telephone (n = 26)

Risk Factors	Score	
	Mean	SD
Monitor	3.04	0.20
Telephone	1.35	0.85
Final Score	2.31	0.62

3.2.4. Section C – Mouse & Keyboard Score

Section C involved the score for mouse and keyboard. It was a score section related to shoulder, hand and wrists postures of the bus traffic controllers while doing their work. The position of mouse and keyboard were assessed as it may lead to abducted postures of the shoulder, hand and wrists. During the assessment, it was found that all the workstation had mouse and keyboard placed on different surfaces. It was also observed that some of the bus traffic controllers used the keyboard with positive angle (>15°) which led to wrist extension. Table 6 shows the score for mouse and keyboard.

Table 6. Section C Score for Mouse and Keyboard (n=26)

Risk Factors	Score	
	Mean	SD
Mouse	4.00	0.00
Keyboard	2.35	0.49
Final Score	4.35	0.49

Table 7. ROSA Grand Score and Level of Action (n = 26)

Score	Score	
	Mean	SD
Section A	4.00	0.80
Section B	2.31	0.62
Section C	4.35	0.49
ROSA Grand Score	4.58	0.50

3.2.5. ROSA Grand Scores

Based on the analysed data, the risk level for the BTCs working in the computer-based bus control room was only at 2 risk levels which were medium and high. Majority of the bus traffic controllers was at high risk level of 57.7% which the mean score was 5.0 while the medium risk level was 42.3% with 4.0 mean score. The range for medium level was from 3 to 4 points but in this case study, all of the bus traffic controllers scored 4 point risk level with frequencies at 42.3%. Similar to high risk level, the bus traffic controllers only scored 5 point risk level at frequency of 57.7% although the range for high risk level was 5 to 7 point.

Table 7 shows the summarized results for each section which are chair section, monitor and telephone section and mouse and keyboard section. All the section's score were analysed in order to get ROSA Grand Score. As discussed in the subsection 4.3.2, 4.3.3 and 4.3.4, the final score for Section A, Section B and Section C were 4.00 (SD = 0.80), 2.31 (SD = 0.62) and 4.35 (SD = 0.49) respectively. The mean score for ROSA grand score for 26 bus traffic controllers was 4.58 (SD = 0.50). According to the assessment, more than half (15/26 respondents) of the bus traffic controllers were found to have an extreme posture with the ergonomic risk level was high (5 to 7) (action level 3) which required a further investigation and changes of work habits soon. Not to mention, the rest of them (11/26 respondents) were at medium risk level with score from 3 to 4.

Similarly to this study, according to a study among call centre by Poochada & Chaiklieng (2015), neck pain (61.1%), upper back pain (55.7%) and lower back pain (53.1%) were the top three ranked as highest prevalence during the last 12 months. Another study reported that most of the call centre workers have a high prevalence on neck symptom and shoulder symptom at 43.0% for both of the body region [13]. Arvidsson et al (2008) [14] found that work-related musculoskeletal disorders (WRMSD) were frequently reported among computer operators and most of them focused on their neck and shoulder. The findings of previous studies found that MSDs among the call centre workers were related to the ergonomic factors which include personal working behaviours and the design of their workplace.

Based on ROSA, there were 3 sections of the assessment, Section A (Chair section), Section B (Monitor and Telephone) and Section C (Mouse and Keyboard) which related to the development of MSDs among bus traffic controllers. In this research, the ergonomics risk level was not very high. The assessment by using ROSA showed the final grand score for ergonomic risk level among bus traffic controllers was 4.58 (SD=0.50) which was considered as medium risk level. The reason might be that there was modern equipment used which can be adjusted to adapt to the individual in order to prevent MSDs [15] in the control room.

From the observation of the BTCs working posture related to the chair, it was found that most of the BTCs have a habit of leaning forward (no back support) when using the computer. Some of them also had their shoulders shrugged and their arms unsupported while working and had a habit of sitting in an awkward position with low (<90°) and high (>90°) knee angle. These behaviors along with static body positions and prolonged sitting will lead to MSDs such as back, shoulders, arms, elbows, wrists, legs and feet. In terms of workstation equipment and design, the non-adjustable pan depth and armrests of the chair will increase the risk level to MSDs.

From the assessment of the usage of monitor and telephone, it was observed that the deficiency of workstation design forced the BTCs to have non-neutral posture which then lead to symptoms of neck. Although the computer screens were designed to be at arm/s length distance (40-75cm) and at eye level, there was no document holder which forced the BTCs to flex and twist their neck in order to view papers positioned on the desk while using computer. Document holders to keep or hang printed materials such as papers are needed during tasks involving computers, and it must be placed close to the user and the monitor [17]. Previous studies proved that when using either a lateral or an in-line document location while working on a computer, needs both minimum neck movement and muscle activity than once they were placed on the desktop [16]. Besides, there were also telephones which placed too far to reach (outside 30cm) from the BTCs and did not have hands-free options which result in awkward postures of neck and shoulder when using telephone. Not to mention, prolonged conversation and the habit of bus traffic controllers to use the telephone by neck and shoulder hold (pinch) will increase the risk of having MSDs along their neck and shoulder.

Based on the observation on the usage of mouse and keyboard, the workstations were designed to have a mouse and keyboard on different surfaces. The keyboard was placed on the keyboard tray which was lower than the mouse located on the desk. This design will result in reaching for the mouse and the arms unsupported while using the mouse. Besides, some of the BTCs used their keyboard on a positive angle while typing (>15°) which forced to wrist extension. Working with the situation for a prolonged time and repetitive motion which forced the shoulder and arm to be in stress may lead to musculoskeletal disorders. Performing keying tasks with these awkward postures may lead to musculoskeletal disorders of the elbows, shoulders, hands, and wrists.

4. Conclusion

In a nutshell, the objectives of the research were successfully achieved, which the working posture of the bus traffic controllers was evaluated, the prevalence of musculoskeletal disorders (MSDs) and the level of ergonomic risks among bus traffic controllers were determined. By using Nordic Musculoskeletal Questionnaire (NMQ), it was found that the highest prevalence of body part trouble in last 12 months for bus traffic controllers was low back (84.62%) while the least body part trouble elbows and wrists/hands which both was at 11.54%.

As for male bus traffic controllers, they had highest prevalence for low back injury which was at a rate of 81.3%. Female bus traffic controllers were more likely to get injury from their work activity as they had higher injury prevalence for all body parts compared to male bus traffic controllers. Their highest prevalence was upper back (100%) which suffered by all of the female bus traffic controllers.

From the research, it is proved that the use of computer has been identified as a risk related in developing work-related musculoskeletal disorders (WMSDs). The risk factors from using the computer include prolonged awkward posture of the shoulder, lower back, elbow, hands, wrists, head and neck affected the bus traffic controllers. The MSDs among the bus traffic controllers were related to the ergonomic factors which includes their personal working behaviours and the design of their workstation.

Acknowledgment
The authors are supported by the Fundamental Research Grant Scheme (FRGS) Vot 1542, Office of Research, Innovation, Commercialization and Consultation (ORICC) Universiti Tun Hussein Onn Malaysia. The authors would also like to thank Rapid Bus Sdn Bhd on the contribution and participation for this study.

Reference

[1] Moom, R. K., Sing, L. P., & Moom, N. 2015 *Prevalence of Musculoskeletal Disorder among Computer Bank Office Employees in Punjab (India): A Case Study.*
[2] Ortiz-Hernández, L., Tamez-González, S., Martínez- Alcántara, S., & Méndez-Ramírez, I. 2003 *Computer use increases the risk of musculoskeletal disorders among newspaper office workers. Archives of Medical Research*, **34(4),** 331–342.
[3] Maakip, I., Keegel, T., & Oakman, J. 2016 *Prevalence and predictors for musculoskeletal discomfort in Malaysian office workers: Investigating explanatory factors for a developing country. Applied Ergonomics, 53, 252–257.*
[4] Sharan, D., Parijat, P., Sasidharan, A. P., Ranganathan, R., Mohandoss, M., & Jose, J. 2011 *Workstyle risk factors for work related musculoskeletal symptoms among computer professionals in India. Journal of Occupational Rehabilitation,* **21(4),** 520–525.
[5] Robertson, M. M., Ciriello, V. M., & Garabet, A. M. 2013 *Office ergonomics training and a sit-stand workstation: Effects on musculoskeletal and visual symptoms and performance of office workers. Applied Ergonomics,* **44(1),** 73–85.
[6] Cho, C.-Y., Hwang, Y.-S., & Cherng, R.-J. 2012
[7] Spyropoulos, P., Papathanasiou, G., Georgoudis, G., Chronopoulos, E., Koutis, H., & Koumoutsou, F. 2007 Prevalence of low back pain in Greek public office workers. Pain Physician, **10(5),** 651.
[8] Subbarayalu, A. V. 2013 *Occupational health problems of call center workers in India: A cross sectional study focusing on gender differences. Journal of Management,* **1(2),** 63–70.
[9] Kuorinka, I., Jonsson, B., Kilbom, A., Vinterberg, H., Biering-Sørensen, F., Andersson, G., & Jørgensen,
[10] Liebregts, J., Sonne, M., & Potvin, J. R. 2016 *Photograph-based ergonomic evaluations using the Rapid Office Strain Assessment (ROSA). Applied Ergonomics,* **52,** 317–324.
[11] Poochada, W., & Chaiklieng, S. 2015 *Ergonomic Risk Assessment among Call Center Workers. Procedia Manufacturing*, **3,** 4613–4620.

[12] Sonne, M., Villalta, D. L., & Andrews, D. M. 2012 *Development and evaluation of an office ergonomic risk checklist: ROSA–Rapid office strain assessment. Applied Ergonomics*, **43(1),** 98–108.
[13] Rocha, L. E., Glina, D. M. R., Marinho, M. de F., & Nakasato, D. 2005 *Risk factors for musculoskeletal symptoms among call center operators of a bank in Sao Paulo, Brazil. Industrial Health,* **43(4),** 637–646.
[14] Arvidsson, I., Hansson, G.-Å., Mathiassen, S. E., & Skerfving, S. 2008 *Neck postures in air traffic controllers with and without neck/shoulder disorders. Applied Ergonomics,* **39(2),** 255–260.
[15] Putz-Anderson, V., Bernard, B. P., Burt, S. E., Cole, L. L., Fairfield-Estill, C., Fine, L. J., Hurrell Jr, J. J. 1997 *Musculoskeletal disorders and workplace factors. National Institute for Occupational Safety and Health (NIOSH).*
[16] Goostrey, S., Treleaven, J., & Johnston, V. 2014 *Evaluation of document location during computer use in terms of neck muscle activity and neck movement. Applied Ergonomics*, **45(3),** 767–772.
[17] Occupational Safety and Health Administration of United States, OSHA 2010 *Computer Workstation Ergonomics.*

PUPILS PERCEPTION TOWARDS THE IMPLEMENTATION OF ENVIRONMENTAL EDUCATION ACROSS CURRICULUM IN MALAYSIA PRIMARY SCHOOL

Nurul Hidayah Liew Abdullah[1], Hashima Hamid[1], Haryati Shafii[1], Seow Ta Wee[1] and Jamilah Ahmad[2]

[1]Fakulti Pendidikan Teknikal dan Vokasional
[1]Fakulti Pengurusan Teknologi dan Perniagaan, Universiti Tun Hussein Onn Malaysia (UTHM), 86400 Parit Raja, Johor, Malaysia
[2]Fakulti Pendidikan, Univesiti Teknologi Malaysia (UTM), Skudai, Johor, Malaysia.

Email: hidayahl@uthm.edu.my, hashima@uthm.edu.my, haryati@uthm.edu.my, tawee@uthm.edu.my & jamilah_ahmad@utm.my

Abstract. Education is an important element in enhancing pupils' awareness on the environment. This is because education plays an important role in shaping and transforming society. In Malaysia, environmental education (EE) across curriculum has long been established. However, to date there is no specific assessment done to determine the successful implementation of these elements. Thus, the perception of pupils is needed to identify the needs of the implementer of the suitability and significance of EE across the curriculum on the pupils' environmental awareness. This study aims to identify the grade 6 pupils' perception toward the implementation of environmental education across curriculum in Malaysia primary school. The results showed that the pupils' perception of the environment was overall high and the location of the school does not significantly affect the pupils' perceptions of the environment. The majority of students defines the concept of the environment as objects. In addition, the percentage of pupils who give the concept of environmental rationally is also high. The perceptions of pupils on the need to protect the environment illustrate the majority agree on the benefits of future generations. The environmental concepts are presented somewhat rational and show the relationship between the environment and human beings. In conclusion, students' perceptions of the environment are positive and can determine it rationally and know the nature of natural disasters is a result of human action.

1. INTRODUCTION

Education is required to sow and instilled environmental knowledge and values through the learning and teaching process in both classroom and outdoors so students can positively react to the environment. This situation led to education being adopted as one of the most appropriate mediums to apply the values of the environment is not overwhelming as almost all the people go through it. Since the

Content from this work may be used under the terms of the Creative Commons Attribution 3.0 licence. Any further distribution of this work must maintain attribution to the author(s) and the title of the work, journal citation and DOI.
Published under licence by IOP Publishing Ltd

Intergovernmental Conference on Environmental Education in USSR in 1977, education has been accepted and appointed as the most appropriate medium in the effort to regulate the environment and sustainable development, environmental education and education for sustainability is the most effective way of dealing with environmental issues [1, 2]. This means that the school has been entrusted to deliver not only useful academic knowledge, but also the knowledge of a very important environment for humanity to survive. Moroye schools are also a key way to address environmental issues. Next, it can build a more perfect human life by raising awareness of the environment [3]. However, over the past decade, the scope of EE has traditionally been found to be limited in influencing the attitudes needed to reduce environmental pollution [4].

EE was debated issue internationally since the Stockholm Human and Environmental Conferences held in 1972. Figure 1 shows a series of international environmental issues discussions that are always looking for solutions to address the increasingly tenuous environmental quality.

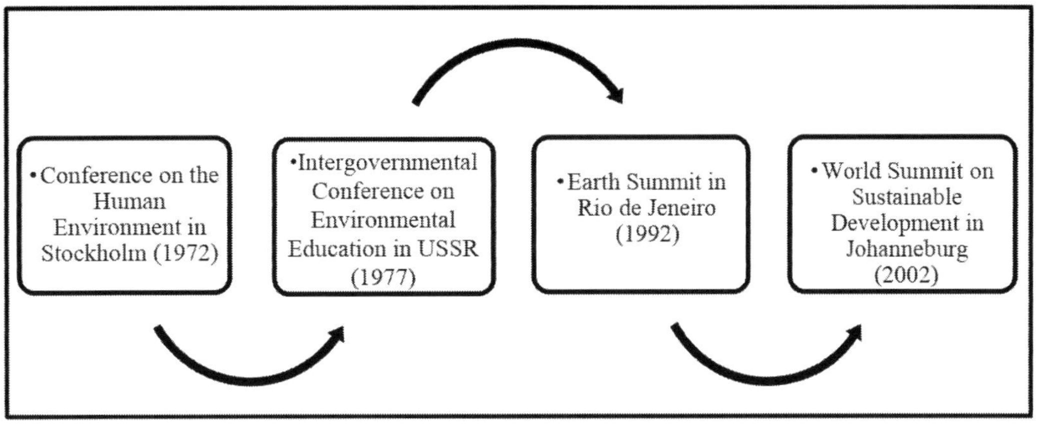

Figure 1: International EE discussions [5-7]

In line with that, in 1972, Malaysia began participating in environmental talks at the United Nations Conference on Human Environment when environmental issues arose as a global concern [8]. The role of education was seen as the most important agent at the Summit of Earth in Rio de Janeiro in 1992. The most important thing discussed in the conference was on Sustainable Development until there was Chapter 36 in Agenda 21 and among its resolutions had set out the dissemination of information on sustainable development through EE program. This is in line with the primary task of education, to distribute legitimate knowledge to the community [9]. The knowledge gained through the education process can produce knowledgeable human capital and the potential to advance the nation from various aspects. This is because human capital development of a country is very important to drive the development of a civilization [10].

Learning process is a continuous process that will usually cause a change in human behaviour in a positive direction. In other words, education is a highly influential agent in shaping human identity and mold to manage the environment and shaping the future. Meanwhile, EE is also an important element in improving the quality of the environment by nurturing and instilling awareness about the importance of environmental preservation and conservation. As a result, the preservation and conservation of nature and natural resources can contribute to the improvement of the quality of life [11].

Perception of the environment refers to an individual's perception or view of the environment and its surroundings. Therefore, different individual perceptions are usually different because each one views something from a different point of view. Human perception of the environment is influenced by various factors, whether external or internal. One's existing experience can also affect human perception to create a new situation. This is because the formation of perceptions, attitudes and relationships with the environment is a key goal in childhood and facilitates a variety of positive environmental experiences from one child to another [12]. Children begin to make a perception of the world at a very young age as early as they get formal education. Preliminary perception is formed through observation and personal experience in the local environment and media coverage of global issues [13]. Likewise, perceptions of environmental issues need to be emphasized because through perceptions will drive towards the behaviour or actual actions of individuals.

Perception is a person's view or response to something like a student's perception of environmental issues. Perception is a dependent variable that will change according to the influence of the environment, friends, attitudes, knowledge, parents and other related factors. In this study, the perception is chosen as one of the variables because through the perspective of an individual it will lead him to act either to bring about positive or otherwise positive effects on the environment.

2. METHODOLOGY

A total of 81 respondents was involved in this study using a qualitative approach through partial structured interviews conducted in selected schools. Meanwhile, 685 respondents were involved using a quantitative approach. The respondents' choice in this method was conducted randomly by the school according to the three levels of student achievement in academic, namely excellent, medium and weak in Kulai, Johor, Malaysia. Triangulation is done to determine the validity of the decisions obtained from both approaches and to strengthen the results of the study. Qualitative methods can be used to identify and obtain additional information that can support and reinforce the findings of quantitative methods. According to Azizi et al., this method provides clear confidence in the validity, decision and fact [14]. This interview was conducted in full compliance with the protocols of the interviews provided.

Interviews were analyzed manually by making transcriptions right after the completion of each interview session. The transcription is prepared based on the questions posed. Then, the analysis of the transcription is done by looking at the suitability of a particular theme to the answers given by the respondents in accordance with the process suggested by Bogdan and Biklen [15]. Even though there is software that can be used for analysis process but researchers prefer the manual way because computer software is not able to replace the process of interpreting the significance of data requires the identification of meaningful categories and determine the relationship between each of these categories. Meanwhile, the quantitative data was analyzed used t-test, percentage and mean.

3. Instrument

The research instrument applied in this study was a set of questionnaires and interview protocols. Among the information obtained through the questionnaire are demographic and 15 statements aimed at measuring pupils' perceptions of the environment. The items are in the form of a statement and students are asked to choose according to the 5 points Likert scale which is very disagreeable (STS), disagree (TS), uncertainly (TP), agree (S) and strongly agree (SS). The Likert scale was chosen in this study because the scale was easily administered, analyzed, more consistent and had high reliability. The given perceptual items are more geared towards the students' existing experiences. This allows the pupil to give perceptions of what is seen, experienced and what has been done before.

Then, the instrument was developed to three lecturers who are experts in this field of study for validation. The validation of the experts in the field studied is necessary to determine the accuracy and validity of the set of questionnaires that have been prepared [16]. Evaluation of instruments submitted to experts in this field takes between one to two months to be returned to the researcher. After review by experts, some questions and answers have been purified such as terms and sentences to facilitate student understanding.

Pilot study should be carried out before the developed instrument is used on the actual sample so that the reliability of the instrument can be ascertained. This step should be implemented to ensure the reliability of the research instrument can be performed on other respondents. Chua stated usually about 30 people who have the same characteristics as the respondents of the study are sufficient [17]. An important pilot study was undertaken to give a preliminary picture of the probability of the results of the study to be obtained in the actual study and the reliability of the instrument can be ascertained. However, the results of the study were not able to ensure the success of the actual study but can help in identifying the feasibility or the realities of the actual studies to be carried out.

For this purpose, pilot studies were conducted in a primary school in Kulai district, which was not involved in the actual study. The researchers themselves administered this study with the help of teachers in the school. Before respondents answered the questionnaire, the researchers first explained the background and the purpose of the study. The administrative procedures of this pilot study are also explained so that respondents can respond well. The findings of the pilot study were analyzed to obtain the reliability of the instrument. The results of the pilot study show that the Cronbach α coefficient is 0.787 (n = 31). Therefore, the researcher still maintains the number of items (15 items) for this part because the reliability value obtained is quite high.

The interview protocol consists of 10 questions relating to the environment, including several questions relating to waste management. The questions relate to the pupils' view of the environment, the pupils' efforts in environmental preservation, the application of pure environmental values and the adequacy of environmental knowledge delivered by the grade 6 science teacher. Interviews were conducted on a group 3 people at a time. The time for each interview session lasts between 14 and 18 minutes. The interview session was recorded with the sound recording device. In this context, the questions presented have been reviewed by three experts and experienced lecturers in this field of study. The purpose of the student interview protocol is to obtain an in-depth feedback on the questions raised. This section was also implemented to complement and support the results of this study.

4. RESULT AND DISCUSSION

The results of the study will be discussed and refined to determine the respondents' perceptions of the environment. Perception is an important component in determining the attitudes and behaviours of humans towards the environment as these components are mutually influential.

4.1. Respondents background

A total of 81 respondents (26 male, 55 female) was involved in this study using a qualitative approach through partial structured interviews conducted in selected schools. Meanwhile, 685 respondents (309 male, 376 female) were involved using a quantitative approach. Figure 2 shows the number of respondents by gender. All respondents are grade 6 pupils from national school, Chinese school and Tamil school. Respondents in this study are 12 years old and had 6 years experiences studying in primary school. There are 345 respondents from schools situated in an urban area and 340 respondents from school situated in rural area.

The results showed that the respondents' perception of the environment was overall high or positive (76.9%) with a mean score of 3.9225 (mean score = 3.67-5.00). Perceptions of 158 respondents were moderate or neutral (mean score = 2.34-3.66), while no respondents showed low perception (mean score = 1.00-2.33). Table 1 shows the level of perceptions. The results of this study are in line with past studies on children's views on the future, showing deep concern for social and environmental issues [18-20]. A high level of perception of the environment is a positive indicator of respondents or perceptions of respondents to the environment. The similarity of perceptions shows that education about the environment is important so that time and environment change does not change the positive perception of the environment.

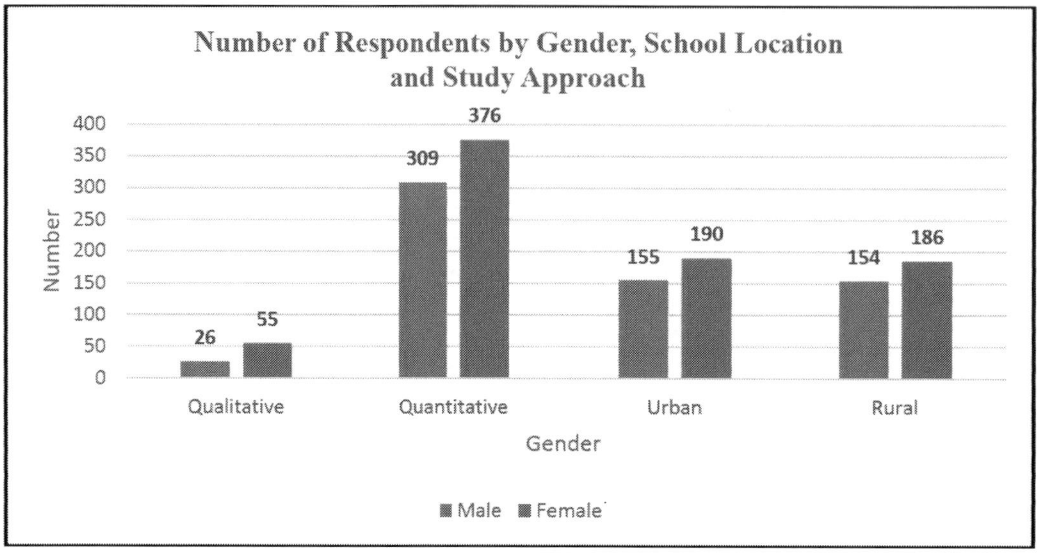

Figure 2: Number of respondents by gender, school location and study approach

Table 1: Level of Perception

Level	Perception	
	Number	%
High	**527**	**76.9**
Moderate	158	23.1
Low	0	0
Total	685	100.0

This study also intends to examine whether there is a difference between the school location and the respondents' perception of the environment. The analysis of the study using t test to achieve that goal has resulted $t = -1.487$ and $p = .137$ ($p > .05$). The results showed that respondents' perception of rural students was slightly higher (mean = 3.9506) compared to urban respondents (mean = 3.8949). The study concludes that there is a slight difference in perceptions of respondents between rural and urban zone schools, which is slightly better than urban students, but these differences are not significant. The results of this study are supported by the study of Roslee et al.found that there was no difference in environmental awareness between urban and rural students [21]. This means that the position or location of the school does not significantly affect the respondents' perceptions of the environment because of technological sophistication and informational channels at the fingertips. The findings of the previous study show that humans evaluate environmental problems from different angles as there are various factors that affect the perception and assessment of the problem [22, 23]. Rural or urban communities have easy access to the Internet but their use depends on the willingness and desire of the individual.

The respondents' perceptions of high levels of environment can drive the true action of this group positively towards the environment. However, there are other factors that affect the human action of the environment. Research on the data as a whole has found that respondents' perceptions of the environment are at high or positive levels. Alp et al. found that pupils had a higher inclination to translate feelings toward the environment, but did not show such a desire for efforts to act eco-friendly [24]. This is because students do not have enough information about the causes and consequences of their actions on the environment.

4.2. Environmental Concept

The diverse concepts of the environment brings different meanings to the existing experience, knowledge, local culture and environmental influences. In this study for interview approach, seven categories as pioneered by Loughland, Reid and Petocz have been adopted to determine the environmental concept by respondents' views [25]. The category is a place (surrounding area), a place with living things, an object (water, tree), a place where humans and others in the surroundings depend on each other for a living, God's creation, a place that humans are responsible for looking after and a place that supports human survival. Aini et al. also used the same category in their studies of secondary school pupils [26].

The results showed that the majority of respondents (n = 81) define the environment with object concept (38.3%). 39.5% of the respondents gave a rational concept about the environment, namely "A place that humans are responsible for looking after" and "a place that supports human survival". However, two respondents were unable to define them and 41 respondents responded to defining the environment towards the concept of objects (places and objects). The results also found that no respondents associate the environment with "A place where humans and others in the surroundings depend on each other for living". This situation may be caused by the viewpoints or understanding of respondents at this age (12 years old) to the unreached environment, still in the process of understanding or not being exposed by teachers and any parties. Table 2 shows the differences in concepts to the environment by respondents.

Table 2: Differences concepts to the environment by respondents

Environment concepts	Number	%
A place (surrounding area)	10	12.3
A place with living thing	3	3.7
An object (water, tree)	31	38.3
A place where humans and others in the surroundings depend on each other for living	0	0
God's creation	3	3.7
A place that humans are responsible for looking after	25	30.9
A place that supports human survival	7	8.6
No response	2	2.5
Total	81	100

4.3. The Need to Protect the Environment

This study also examines pupils' opinions on the need to protect the environment holistically and sustainably as well as the reasons the respondents think so as in Table 3. The reasons given are detailed to summarize pupils' responses to the environment in the context of environmental importance to human beings from all aspects. The findings show that the importance of protecting the environment according to the most frequent respondents' views is "For future generation" is 22.2% as in Table 3. Followed by "Avoid pollution" by 14.8% and "Ensuring health" of 10 people. The reason for avoiding pollution here refers to all types of contamination such as water, air, land and so on.

The study concludes that respondents are aware of the importance of protecting the environment for the benefit of future generations and preventing various pollution to ensure human health. However, there is a respondent who cannot express the importance of protecting the environment. Only two respondents stated that the importance of protecting the environment is to "attracting tourist" to Malaysia and this statement is linked to the national economy. Meanwhile, three responded to "comfortable living". The findings show the diversity of respondents' perceptions of the importance of protecting the environment.

Table 3: The need and the reasons to protect the environment

The need to protect the environment	Total	
	Number	%
For future generations	18	22.2
Ensuring health	10	12.4
Avoid pollution	12	14.8
Avoid global warming	6	7.4
Supply oxygen	7	8.7
Avoid the extinction of nature	6	7.4
Attracting tourist	2	2.5
Comfortable living	3	3.7
The environment is important for living	6	7.4
Natural disasters	4	4.9
Others	6	7.4
No response	1	1.2
Total	81	100

4.4. Perception on environmental issues

Respondents' perception to environmental issues, especially those involving feelings are also identified in this study. The results of the analysis show that respondents expressed sadness, disappointment, anger, dislike and others (dissatisfaction and dislikes) on environmental issues that occurred either locally or internally as in Table 4. Respondents are so heart-wrenching that most environmental issues occur as a result of human action.

The findings of the study show that the majority of respondents (44.4%) is saddened by the existence of environmental issues that usually have a negative impact on humans and the environment. Besides sadness, respondents also expressed disappointment (19.8%) against the irresponsible attitude of the environment without thinking that the environment is a common right.

Table 4: Feelings of respondents on environmental issues and human actions towards the environment

Feeling	Number	%
Sadness	36	44.4
Anger	22	27.2
Disappointment	16	19.8
Dislike	4	4.9
Others	3	3.7
Total	81	100

There were some respondents (27.2%) who expressed anger over irresponsible human activities and damaging the environment. 4.9% of respondents expressed dislike as well as 3 respondents have feelings of dissatisfaction and hate towards the actions of those who cause harm to the environment. To get a clearer picture, respondents are also asked to state why they feel such feelings. After conducting research on the data, the study found that the majority of respondents (50.6%) made irresponsible people as their sole cause of concern as in Table 5. Such a review because the environment is the right of everybody, but there is an irresponsible person doing damage, so it is harmful not only to the environment but also to the detriment of humans especially in terms of social, economic and political.

Table 5: Respondents comment on feeling about environment damage

Comment	Number	%
Irresponsible	41	50.6
Do not love the environment	15	18.5
No awareness	7	8.6
The environment belongs to everyone	4	4.9
Can bring disease	4	4.9
Uncomfortable life	3	3.7
No comment	3	3.7
Others	4	4.9
Total	81	100.0

Table 6: Effort that have been and will be respondents do to preserve the environment

Effort	Number	%
Do not throw trash everywhere	18	22.2
Recycle	14	17.3
Collaborative cleaning practices	10	12.3
Do not throw trash everywhere and recycle	9	11.1
Separate garbage and recycle	8	9.9
Admonish and advise friends	6	7.4
Collect trash and throw it into the trash	5	6.2
Plant a flower	3	3.7
Recycle and plant trees	2	2.5
Saving natural resources	2	2.5
Drawing posters for people's awareness	1	1.2
Others	2	2.5
Not specified	1	1.2
Total	81	100.0

There is no sense of affection for the environment as well as justification for respondents why feelings of sadness, disappointment, anger, dislike and others are caused by those who cause environmental damage. Research on data found that most respondents stated "Do not throw waste everywhere" as a pure effort to conserve the environment (18 respondents). This is in line with the willingness and ability of this group based on the age of about 12 years. Table 6 shows the efforts the pupils have and will do to preserve the environment.

Many of the respondents responded that recycling and collaborative cleaning practices were undertaken for environmental conservation with 14 and 10 respondents respectively. There are also other efforts undertaken by respondents to support and preserve the environment such as planting flowers, treating and advising friends, collecting garbage and throwing them into the garbage bin, saving natural resources and garbage isolation before being discarded. However, a respondent did not state his efforts on the preservation of the environment in this study. Overall, respondents are trying to maintain the environment.

4.5. Perception of natural disasters and their impact on human and nature

In this study, respondents' view of natural disasters was identified to determine the level of respondents' knowledge to environmental issues in general. The focus of this research is on the occurrence of natural disasters. Overall, all respondents stated that the incidence of natural disasters was caused by human action towards the environment. The reasons mentioned are divided into several aspects as shown in Figure 3.

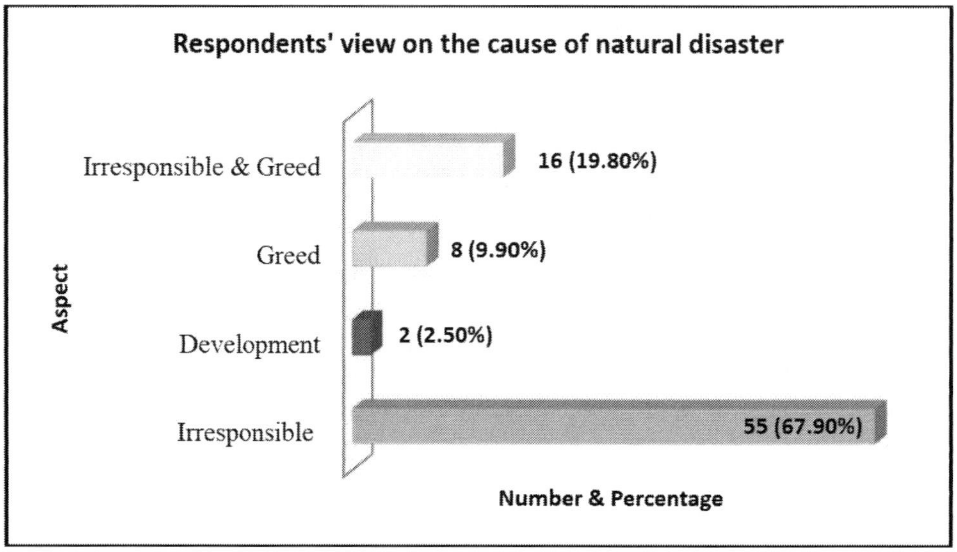

Figure 3: Respondents' view on the cause of natural disaster

Figure 3 shows that the "irresponsible" aspects of human beings are the main cause of natural disasters (55 respondents). Commentary from respondents shows that this aspect takes place from the smallest of individual behaviour such as trash throwing to the maximum extent of individual or collective behaviour such as uncontrolled forest exploration. 6 respondents combine irresponsibility and human greed when interacting with the environment. This aspect suggests that people are too profitable without thinking about the disasters they face. Meanwhile, 2 respondents who view that natural disasters occur due to the ongoing development process throughout the country. Unplanned development can end up with natural disasters and not for the benefit of all. 8 respondents stated that "Greed" is human greed as contributors to natural disasters.

For impact on human and nature, the majority of respondents believe that the impact of natural disasters on humans is affecting health (43.2%), leading to death (13.6%). Other effects can be seen in the summary of findings as in Table 7.

Table 7: Impact of natural disasters to humans and nature

Impact of natural disasters	Number	%
Impact on humans		
Impaired health	35	43.2
Loss of property	3	3.7
Basic facilities are affected	5	6.2
Death	11	13.6
Uncomfortable life	7	8.6
Impaired health and loss of property	6	7.4
Impaired health and homelessness	3	3.7
Death and loss of property	2	2.5
Others	6	7.4
Not specified	3	3.7
Impact on nature		
Destroyed	21	25.9
The trees are getting decrease	4	4.9
Extinction	16	19.8
Nature is polluted	29	35.8
Others	5	6.2
Not specified	6	7.4
Total	81	100.0

Most respondents (35.8%) think that the impact of natural disasters on nature is contaminated nature. 25.9% of respondents believed that the environment would be destroyed and 19.8% of respondents said that the extinction would occur in the environment. The view of the respondents leads to the destruction of the environment at the end of which is to start the contaminated nature, then cause extinction and ultimately total destruction of nature. There are respondents who view that the effects of natural disasters cause the trees to decrease (4.9%) and others (6.2%), while (7.4%) respondents do not express their views on this issue.

The findings of the perception study show that respondents have various concepts to define the environment. Respondents' perceptions of the importance of protecting the environment show that all parties should work together to create a harmonious and balanced environment without thinking of self-interest for the benefit of future generations. Generally, the environmental awareness of respondents is modest and can be improved and preserved by various parties.

5. CONCLUSION

In conclusion, respondents have various environmental concepts through their perceptions based on experience, knowledge, everyday life and environment. The findings of this perception study show that the environmental concept of respondents is more conceptualized. However, the stated environmental concept was related to the environment and human and rational. In addition, the importance of protecting the environment is more focused on the benefits of future generations and ensuring health. Therefore, the environment needs to be maintained sustainably so that pollution can be avoided. In this context, all parties need to be responsible and cooperate to protect the environment to improve the quality of the environment. Hence, educational institutions such as schools need to play an active role in spreading environmental knowledge to preserve it. Respondents' awareness of the importance of the environment can also be identified through a sense of emotion over environmental issues. Respondents also know that all parties have the right to the environment and should not be manipulated arbitrarily. Incidents of natural disasters will occur if human beings act extravagantly on the environment resulting in a negative impact not only on nature but also to the loss of universal humanity. In general, respondents are aware that the destruction of the environment is due to irresponsible, greed and human action that negates nature's pursuit for self-interest.

6. Acknowledgement

This research is funded by the PPG Grants (VOT V032), Universiti Tun Hussein Onn Malaysia.

References

[1] Huckle, J. (1983). Environment education. In Huckle, J. (eds.) *Geographic Education: Reflection and Action*. Oxford, UK: Oxford University Press. 99-111.

[2] Fien, J. (1993). *Education for Environment: Critical Curriculum Theorising and Environment Education*. Geelong, Victoria, Australia: Deakin University Press.

[3] Moroye, C.M. (2005). Common ground: An ecological perspective on teaching and learning. *Curriculum and Teaching Dialogue, 7(1/2)*: 123-139. http://www.highbeam.com/doc/1P3-1040187551.html

[4] Taylor, N., Nathan, S. & Coll, R.K. (2003). Education for sustainability in Regional New South Wales, Australia. *International Research in Geographical and Environmental Education*, 12(4): 291-311.

[5] Palmer, J. A. (1998). *Environmental Education in the 21st Century: Theory, Practice, Progress and Promise*. London and New York: Routledge.

[6] Mohamad Suhaily Yusri. C. N. & Mohmadisa. H. (2005). Pembangunan dan pendidikan alam sekitar di Malaysia: Isu dan Cabaran. In Mohmadisa. H & Mohamad Suhaily Yusri. C. N. (eds). *Pembangunan dan Alam Sekitar di Malaysia*. Tanjong Malim: Penerbit Universiti

Pendidikan Sultan Idris.
[7] Haryati, S., Sharifah Meryam, S. M. & Siti Sarawati, J. (2011). Pendidikan alam sekitar ke arah meningkatkan kualiti hidup masyarakat Malaysia: Isu dan cabaran. In Haryati, S. (ed.) *Isu-isu Kualiti Hidup Masyarakat di Bandar*. Parit Raja: Penerbit UTHM.
[8] Chan, N. W. (2012). Awareness of environmental issues and reality: A Malaysian perspective. Kertas kerja dibentangkan dalam *International Environment and Health Conference*. Vistana Hotel: Pulau Pinang. 6-7 Jun. XXV-XLV
[9] Gough, N. (1987). Learning with Environment. In Robottom, I. (ed.) *Environmental Education: Practice and possibility*, Geelong, Victoria: Deakin University Press.
[10] MOE. (2008). Modal Insan Cemerlang: Minda Kelas Pertama. Putrajaya: KPM.
[11] Mumtazah, O. (2011). Penggunaan lestari: Konsep dan pengenalan. In Mumtazah Othman & Nurizan Yahaya (eds.). *Penggunaan Lestari: Bagaimana Tingkah Laku Remaja?* Serdang: UPM. 1-11.
[12] NAAEE. (2010). *Early Childhood Environmental Education Programs: Guidelines for Excellence*. USA: Washington. IBSN 978-1-884008-23-8.
[13] Cullingford. C. (1995). Children's attitudes to their local environment. In G. Harris & C. Blackwell (eds.). *Monitoring change in education, 2: 21-36. Environmental Issues in Education*. Aldershot: Ashgate.
[14] Azizi, Y., Shahrin, H., Jamaludin, R., Yusof, B. & Abdul Rahim, H. (2007). *Menguasai Penyelidikan Dalam Pendidikan: Teori, Analisis & Interpretasi Data*. Kuala Lumpur: PTS Professional Publishing Sdn. Bhd.
[15] Bogdan, R. C. & Biklen, S. K. (2003). *Qualitative Research for Education: An Introduction to Theories and Method.* (4th ed). New York: Pearson Education Group, inc.
[16] Tay, M. G. (2003). Satu tinjauan terhadap sikap dan masalah guru Bahasa Melayu di Sekolah Rendah Jenis Kebangsaan. *Jurnal Penyelidikan Maktab Perguruan Batu Lintang Sarawak*, 4: 78-85.
[17] Chua, Y. P. (2006). *Kaedah dan Statistik Penyelidikan: Kaedah Penyelidikan. Buku 1*. Kuala Lumpur: The McGraw-Hill.
[18] King, L. D. (1995), *Doing Their Share to Save the Planet. Children and Environmental Crisis*. New Jersey: Rutgers University Press.
[19] Hicks, D. & Holden, C. (1995). *Vision of Future. Why We Need to Teach for Tomorrow*. London: Trentham Books.
[20] Barraza, L. (1999). Children's drawings about the environment. *Environmental Education Research*, 5(1): 49-66.
[21] Roslee, T., Dg. Norizah, A. K., Soon Singh, B. S. & Muniandy, S. (2012). Pendidikan alam sekitar dan hubungannya dengan pengetahuan dan kesedaran alam sekitar dalam kalangan pelajar sekolah menengah bandar dan luar bandar di daerah Tawau. In Azizi, M., Mohmadisa, H., Mohamad Suhaily Yusri, C. N. & Nasir, N. (eds.). *Pendidikan Alam Sekitar: Melestarikan kesedaran masyarakat di Malaysia*. Tanjong Malim: Penerbit Jabatan Geografi dan Alam Sekitar. 131-152.
[22] Dunlap, R. E. & Jone, R. E. (2002). Environmental concern: Conceptual and measurement issues. In Dunlap, R. E & Michelson, W. (eds.). *Handbook of Environment Sociology*. Greenwood Press, Westport, CT, 482-524.
[23] Milfont, T. L., & Gouveia, V. V. (2006). Time perspective and values: An exploratory study of their relations to environmental attitudes. *Journal of Environmental Psychology*, 26: 72-82.
[24] Alp, E., Ertepinar, H., Tekkaya, C. & Yilmaz, A. (2008). A survey on Turkish elementary school students' environmental friendly behaviours and associated variables. *Environmental Education Research,* 14(2): 129-143. http://dx.doi.org/10.1080/13504620802051747
[25] Loughland, T. Reid, A. dan Petocz, P. (2002). Young people's conceptions of environment: A phenomenographic analysis. *Environmental Education Research*, 8(2): 187-197.
[26] Aini, M.S., Nurizan, Y. & Fakhrul-Razi, A. (2007). Environmental comprehension and practices

of Malaysian youth. *Journal of Environmental Education Research*, 13 (1): 17-31.

IOP Conf. Series: Journal of Physics: Conf. Series **1049** (2018) 012099 doi:10.1088/1742-6596/1049/1/012099

The effect of crumb rubber on the physical and rheological properties of modified binder

Siti Khatijah Abu Bakar[1], Mohd Ezree Abdullah[1], Mazlina Mustafa Kamal[2], Raha Abd Rahman[1], Rosnawati Buhari[1], Ramadhansyah Putra Jaya[3], Suryati Sabri[1] and Kabiru Abdullahi Ahmad[1]

[1]Faculty of Civil and Environmental Engineering, Universiti Tun Hussein Onn Malaysia, 86400 Parit Raja, batu Pahat, Johor, Malaysia
[2]Malaysian Rubber Board, Technology Centre, RRIM Research Station Sg. Buloh, 47000 Selangor, Malaysia
[3]Faculty of Civil Engineering, Department of Geotechnics and Transportation, Universiti Teknologi Malaysia, 81310 UTM Johor Bahru, Malaysia

Corresponding author: ezree@uthm.edu.my

Abstract. Bitumen is widely used in highway pavement construction. The highway intensity of traffic causes overloading and subsequently contributes to pavement distress i.e. rutting, cracking and disintegration. Hence, to minimize the pavement distress and enhance its durability, conventional bitumen properties need to be improved. An effort has been made by modifying the conventional bitumen with crumb rubber. The physical and rheological properties of crumb rubber modified binder (CRMB) were study by conducting fundamental physical test (including penetration, softening point, ductility, and storage stability) and Dynamic Shear Rheometer (DSR) test. Results indicated that, penetration decreased, and softening point increased, for unaged and short term aged samples, respectively. Temperature susceptibility of CRMB also improved as PI value increased. Moreover, rheology testing exhibited an improvement of $G^*/\sin\delta$ value and increased bitumen's performance grade (PG) by one level. These properties enhancement indicates greater resistance to rutting.

1. Introduction

Asphalt pavement provides the necessary stiffness and strength to increased driving comfort and flexible maintenance and repair action [1]. Over the years, many studies have been conducted to improve the conventional bitumen for better performance and prolonged pavement life that benefits road authorities and road user by ensuring a comfortable ride and lower maintenance cost in the long run. A major concern of Malaysia highway industry is the excessive of permanent deformation while pavement is still in service. Premature failures resulting to the frequent repetition of heavy axle load and worsen with the presence of excessive moisture in pavement, particularly in wet climate. Hence, in order to reduce the pavement premature failure and improve its performance, addition of polymers is a very effective way [2-4]. However, since polymers are rather expensive, using waste polymer to act as alternative may help to improve the pavement performance as well as to solve the waste disposal problem [5-7].

Widely used waste in modifying bitumen is crumb rubber, which is scrapped from waste tires. Crumb rubber can be incorporated into base bitumen using two different methods, which are dry and wet

Content from this work may be used under the terms of the Creative Commons Attribution 3.0 licence. Any further distribution of this work must maintain attribution to the author(s) and the title of the work, journal citation and DOI.
Published under licence by IOP Publishing Ltd

IOP Conf. Series: Journal of Physics: Conf. Series **1049** (2018) 012099 doi:10.1088/1742-6596/1049/1/012099

process [8]. Dry process consists in the production of asphalt mixture with the addition of small ground rubber quantity from discarded tyres as a substitute for a part of the mineral aggregate. Meanwhile, wet process involves modification of bitumen with rubber to produce a new binder which captures beneficial engineering properties of both base material [9]. During the process, rubber and bitumen are allowed to blend together for a designate period of 45 to 60 minutes to promote an interaction between rubber and bitumen [10,11]. The blending temperature normally used is between 176 to 204°C [10]. Zanetti et al. [11] suggested that rubber is thoroughly mixed with bitumen at temperature of 175 to 225°C and subsequently kept in agitation at temperature of 150 to 215°C.

An effective modified binder should be able to increase rutting resistance at high temperature without making it too viscous during mixing or too brittle at low temperature. Sufficient content of rubber is crucial to ensure its compatibility with base bitumen to avoid phase separation during storage, transportation, application, and service [5].

This paper presents is intended to study the effect of crumb rubber on physical and rheological bitumen properties incorporating wet process.

2. Material and method

A conventional bitumen Penetration Grade (PEN) 80/100 was used in this study as virgin bitumen. Meanwhile, fine crumb rubber (40 mesh) was used as modifier to produce crumb rubber modified binder (CRMB). The preparation of CRMB involved the addition of different percentages of crumb rubber namely 2, 4, 6 and 8% by weight of bitumen. Prior to blending process, 500g of base bitumen was heated at 170 ± 5°C. The required amount of crumb rubber was then added and mixed with hot bitumen using high shear mixer at 1200 rpm for approximately 60 minutes. Subsequently, the CRMB was exposed to high temperature (163°C) in Rolling Thin Oven Film (RTFO) to prepare a short term aged binder.

3. Result and discussion

3.1. *Physical testing*

The fundamental physical tests such as softening point, penetration, ductility, penetration index (PI) and storage stability conducted on both conventional and modified bitumen were presented in Table 1. As can be seen from the table, penetration and ductility were decreasing and softening point was increasing with the increasing crumb rubber content. The result also showed an increment of PI to positive value with the increasing of crumb rubber content. Greater PI indicates the binder less susceptibility to temperature.

Meanwhile, the potential CRMB to segregation during storage was measured by storage stability test. The difference of softening point between top and bottom section less than 2.2°C can be considered as stable. As can be seen, increasing of crumb rubber content affected the storage stability result as the difference between top and bottom value increased. It was also found that, better stability can be seen in binder contains less than 4% of crumb rubber. Moreover, crumb rubber addition more than 4% could lead to binder instability. Consequently, rubber particles may separate gradually and assemble at the top during storage.

Table 1. Physical properties of CRMB

Testing	Control	2-CRMB	4-CRMB	6-CRMB	8-CRMB
Softening Point (°C)	48.3	51.9	54.0	57.2	59.2
Penetration (d-mm)	81.9	70.7	68.1	62.2	56.3
Penetration Index (PI)	-0.40	0.12	0.54	1.00	1.16
Ductility (cm)	142.4	32.4	28.3	26.4	24.5
Δ Softening point (°C)	1.0	1.9	1.6	8.0	14.4

3.2. High temperature performance grade

The effective high temperature performance grade (PG) for a given binder determined at G*/sin δ value greater than 1.0 kPa (unaged) and 2.2 kPa (short term aged). Table 2 showed the high service temperature as well as failure temperature obtained for control and CRMB. Results revealed that the increasing content of crumb rubber in bitumen resulting in higher service temperature compared to control binder. As expected, control binder recorded high temperature performance at PG 64 as failure temperature obtained were 65.6 and 64.6°C for unaged and short term aged condition, respectively. Likewise, modifying conventional bitumen with 2 and 4% crumb rubber produced a modified binder with PG 64. Meanwhile, adding crumb rubber at 6% and more, successfully improved the performance grade from PG 64 to PG 70. The upgrading of PG indicated by failure temperature recorded above 70°C for 6 and 8% at both unaged and short term aged conditions.

Table 2. High temperature performance of CRMB

Binder type	Failure temperature (°C)		Performance Grade (PG)
	Unaged (G*/sin δ > 1.0 kPa)	Short term aged (G*/sin δ > 2.2 kPa)	
Control	65.6	64.5	PG 64
2-CRMB	67.7	64.0	PG 64
4-CRMB	68.9	66.5	PG 64
6-CRMB	71.5	71.9	PG 70
8-CRMB	73.7	74.6	PG 70

3.3. Rutting resistance

Effect of crumb rubber content on rutting resistance (G*/sin δ) at 70°C was depicted in Figure 1. As can be seen, the addition of crumb rubber positively affected the binder's rutting resistance. Generally, G*/sin δ value improved with the increasing of crumb rubber content. 8% crumb rubber has the higher G*/sin δ value among all modified binder for unaged and short term aged conditions. In comparing to control binder, results recorded G*/sin δ value for unaged and short term aged binder increased by 60 and 69%, respectively, with the presence of crumb rubber up to 8%. It is believed that the improvement of G*/sin δ value is due to the interaction between bitumen and rubber particles during wet process causing rubber particles to swell. The swelling of rubber particles resulting in binder hardening and enhances its stiffness.

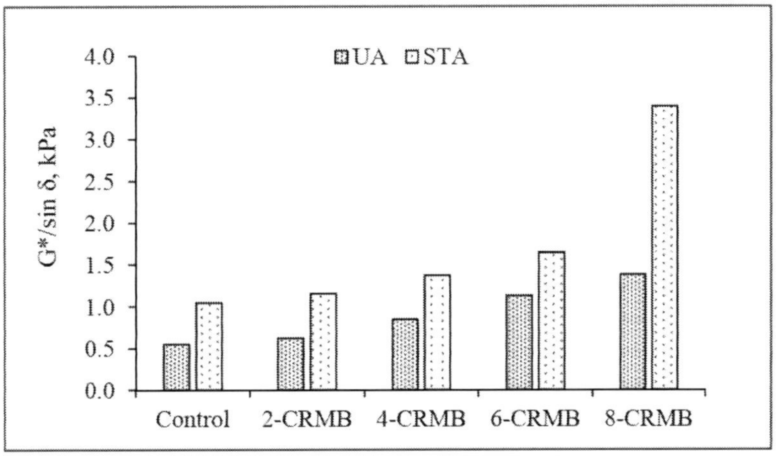

Figure 1. G*/sin δ of unaged condition and short term aged binder at 70°C

4. Conclusion

Based on the result, associating crumb rubber enhances the physical properties of conventional bitumen. The improvement of penetration and softening point, as well as PI indicates that bitumen response to temperature is improved. This improvement is supported by Dynamic Shear Rheometer (DSR) result, which shown an upgrading high service temperature performance grade by one level (from PG 64 to PG 70). It is also found that G*/sin δ value increases with the presence of crumb rubber. However, storage stability result reveals that, adding crumb rubber more than 4% causing binder instability during storage. Taken together, modifying conventional bitumen with appropriate content of crumb rubber does produce a modified binder that has high potential in improving rutting resistance and has high stability during storage, hence, could reduce asphalt pavement permanent deformation.

Acknowledgement

Research work presented in this paper was supported by Ministry of Higher Education Malaysia (MOHE), under MyBrain15 programme and IGSP grant (U669), Universiti Tun Hussein Onn Malaysia. Financial support for this study is highly appreciated.

References

[1] Butt AA, Birgisson B, and Kringos N (2012). Optimizing the highway lifetime by improving the self healing capacity of asphalt. *Procedia - Social and Behavioral Sciences,* **48** 2190–2200

[2] Ahmad J (2010). A study on moisture induced damage and rutting of hot mix asphalt. Ph D thesis, Universiti Teknologi MARA, Malaysia

[3] Moghadas Nejad F, Arabani M, Hamedi GH, and Azarhoosh AR (2013). Influence of using polymeric aggregate treatment on moisture damage in hot mix asphalt. Constr.and Build. Mater., **47** 1523–1527

[4] Ahmad J, Yusoff NIM., Hainin MR., Rahman MYA, and Hossain M (2014). Investigation into hot-mix asphalt moisture-induced damage under tropical climatic conditions. Constr. and Build. Mater., **50** 567–576

[5] Kalantar ZN, Karim MR, and Mahrez A (2012). A review of using waste and virgin polymer in pavement. Constr. and Build. Mater., **33** 55–62

[6] Aziz MMA, Rahman MT, Hainin MR, and Bakar WAWA (2015). An overview on alternative binders for flexible pavement. Constr. and Build. Mater. **84** 315–319

[7] Yan K, Xu H, and You L (2015). Rheological properties of asphalts modified by waste tire rubber

IOP Conf. Series: Journal of Physics: Conf. Series **1049** (2018) 012099 doi:10.1088/1742-6596/1049/1/012099

and reclaimed low density polyethylene. Constr. and Build. Mater. **83** 143–149
[8] Ganiron TU (2014). Waste Tire as an Asphalt Cement Modifier for Road Pavement. Int. J. Serv. Sci. Technol. **7** 181–194
[9] Lo Presti D, Airey GD, and Partal P (2012). Manufacturing Terminal and Field Bitumen-Tyre Rubber Blends: The Importance of Processing Conditions. Procedia - Social and Behavioral Sciences, **53** 485–494
[10] Hicks RG (2002). Asphalt rubber design and construction guidelines volume I – Design guidelines. Northern California Rubberized Asphalt Concrete Technology Center (NCRACTC) – California Integrated Waste Management Board (CIWMB)
[11] Zanetti MC, Fiore S, Ruffino B, Dalmazzo D, and Lanotte M (2015). Characterization of crumb rubber from end-of-life tyres for paving applications. Waste Management, **45** 161-170

IOP Conf. Series: Journal of Physics: Conf. Series **1049** (2018) 012100 doi:10.1088/1742-6596/1049/1/012100

Influence of Ageing Process on Rheological Characteristic of Waste Natural Rubber Latex Modified Bitumen

Siti Khatijah Abu Bakar[1], Mohd Ezree Abdullah[1], Mazlina Mustafa Kamal[2], Raha Abd Rahman[1], Ramadhansyah Putra Jaya[3], Rosnawati Buhari[1], and Kamarul Arifin Hadithon[2]

[1]Faculty of Civil and Environmental Engineering, Universiti Tun Hussein Onn Malaysia, 86400 Parit Raja, batu Pahat, Johor, Malaysia

[2]Malaysian Rubber Board, Technology Centre, RRIM Research Station Sg. Buloh, 47000 Selangor, Malaysia

[3]Faculty of Civil Engineering, Department of Geotechnics and Transportation, Universiti Teknologi Malaysia, 81310 UTM Johor Bahru, Malaysia

Corresponding author: ezree@uthm.edu.my

Abstract. Bitumen is widely used in highway pavement construction. Changes of bitumen properties occurs during the modification and ageing process. The ageing process does consequences in road deterioration; however, it is also benefits in determination of optimum properties during bitumen modification process. Thus, this study was conducted to evaluate the changes of bitumen (unmodified and modified) properties as it was exposed to artificial ageing process. In order to achieve the objective, base bitumen (PEN 60/70) was modified with waste natural rubber (NR) latex. Subsequently, the modified bitumen was subjected to a laboratory ageing process by the aided of rolling thin oven film (RTFOT) and pressure ageing vessel (PAV). Its rheological characteristic was evaluate using dynamic shear rheometer (DSR). It was found that, associating waste NR latex with bitumen improved its properties during short term and long term ageing, indicated by $G^*/\sin\delta$ and $G^*.\sin\delta$. The performance grade (PG) also successfully improved to PG76 with the addition of 5% waste NR latex.

1. Introduction

Bitumen ageing behaviour has a crucial impact on durability of pavement, subsequently resulting to the deterioration of asphalt pavements [1]. Basically, bitumen ageing is induced by chemical and physical changes that occurs during storage, mixing, transport and construction process as well as during its service life [2,3]. During production and construction process (short term ageing), bitumen ageing or hardening primarily associated with loss of volatile components. Meanwhile, progressive oxidation occurs during asphalt pavement service life (long term ageing) [4-7]. Both short term and long term ageing resulting in an increase of binder's viscosity, therefore, stiffen the mixture. Subsequently, the mixture become hard and brittle, and susceptible to disintegration and cracking failures. Moreover, ageing may render the mixture's durability, in term of wear resistance and moisture susceptibility. However, ageing is not necessarily negative phenomenon, since some ageing may help a mixture to achieve its optimum properties [7].

Content from this work may be used under the terms of the Creative Commons Attribution 3.0 licence. Any further distribution of this work must maintain attribution to the author(s) and the title of the work, journal citation and DOI.
Published under licence by IOP Publishing Ltd

To simulate bitumen ageing condition, laboratory ageing method has been conducted to quantitatively determine the ageing of bitumen. Simulation of field ageing entails increasing the temperature, decreasing bitumen film thickness, increasing oxygen pressure, or using combinations of these factors [8 -10]. The most utilized test to simulate the age hardening occurring during plant mixing and laydown is Rolling Thin Film Oven Test (RTFOT). Meanwhile, Pressure Ageing Vessel test (PAV) is used to simulate long term ageing during in-service after 5 to 10 years [11-13].

In this paper, changes of NR latex modified bitumen properties due to short term and long term aging was performed using Dynamic Shear Rheometer (DSR).

2. Material and method

All the modified bitumens were prepared using a high shear mixer a shearing speed of 1500 rpm. The mixing procedure began with heating the approximately 500g base bitumen (PEN 60/70) to 145 ± 5°C. Subsequently waste NR latex (3 -7% by weight) was gradually added into the hot bitumen under stirring condition for 45 minutes. Caution should be taken as the addition of latex into hot bitumen could produce frothing and foaming.

As specified in Superpave binder specification, short term and long term laboratory ageing of both unmodified and modified bitumen were performed using Rolling Thin Oven Film test (RTFOT, AASHTO T240) and Pressure Ageing Vessel test (PAV, AASHTO R28), respectively. During the ageing simulation, bitumen was exposed to high temperature of 163°C for 85 minutes during RTFOT; and afterwards, RTFOT residue was subjected to high pressure of 2.1 MPa for 20 hours at 100°C during PAV test. The unaged and aged bitumen were evaluated using Dynamic Shear Rheometer (DSR) at a frequency of 10 rad/s.

3. Result and discussion

3.1. Isochronal plot

Isochrones plot of complex modulus, G* and phase angle, δ at constant frequency of 10 rad/s in original state (unaged) and RTFO aged state (short term aged) were shown in Figure 1. Isochronal plot of the rheological properties of bitumen illustrates bitumen's stiffness and viscoelastic behaviour with the changes of temperature [5]. With increasing of NR latex content, G* value of modified bitumen in original state was promoted. Moreover, G* values of modified bitumen were more pronounced compared to base bitumen within temperature ranges of 46 to 58°C. An obvious increment of G* was clearly seen when NR latex content was increased up to 7% resulting in improvement in binder's stiffness. However, 7% NR latex showed a sharp increase in slope of G* isochrones with the increasing of temperature indicating a rapid reduction of G*.

During RTFO aged state, within temperature 46 and 52°C, base bitumen and 3% NR latex modified binder exhibit greater G* amongst all modified bitumen. However, both base bitumen and 3% NR latex modified bitumen recorded a reduction in G* values, therefore, reducing its stiffness and subsequently increased the temperature susceptibility. At high temperature (58 to 76°C), 5% and 7% NR latex modified binder resulted in greater G* indicating better temperature susceptibility compared to base bitumen.5% and 7% NR latex modified binder resulted in greater G* indicating better temperature susceptibility compared to base bitumen.

Meanwhile, isochronal plot for phase angle, δ illustrated that as temperature increased, δ for base bitumen approaching 90°, therefore, predominantly showed viscous behaviour. When rapid reduction of stiffness occurred, bitumen loss its elastic component and becoming a viscous liquid at high temperature [5]. With the increasing of NR latex content, δ value reduced towards elastic behaviour. 7% NR latex modified bitumen exhibited the lowest δ amongst all modified bitumen. As temperature changes from medium to high temperature, both unmodified and modified bitumen slowly behave toward viscous behaviour in response to the increasing of δ value and loses of elastic component.

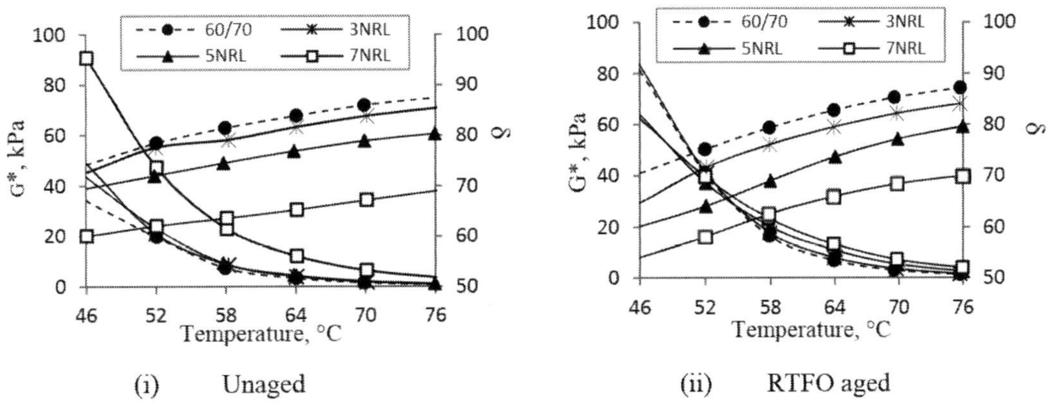

Figure 1. Rheological properties of (i) unaged and (ii) RTFO aged bitumen versus changing temperature

3.2. G/sin δ*

In order to evaluate rutting resistance characteristic (G*/sin δ), bitumen with higher G* and lower δ are preferred. Rutting resistance at high temperature was determined by limiting the minimum value of G*/sin δ are 1.0 kPa for original and 2.2 kPa for RTFOT residue, in accordance with Superpave binder specification. Figure 2 presented G*/sin δ values for original and RTFO aged state at 76°C. Obviously, bitumen associated with NR latex exhibited higher capability to resist rutting compared to base bitumen. As NR latex content was increased, G*/sin δ value was also gradually increased. It was recorded that 7% NR latex has the highest G*/sin δ value amongst all modified bitumen. Highest G*/sin δ value revealed that rutting resistance of bitumen was improved by the addition of NR latex. The changes in G*/sin δ of RTFO aged with response to NR latex content showed similar trend with original state bitumen. The increase of G*/sin δ could be attribute to the bitumen-rubber interaction during wet process. The interaction occurs as the light component of bitumen absorb into rubber particles causing binder hardening.

Moreover, the increasing in G*/sin δ value contributed to the improvement of bitumen's performance grade (PG), as shown in Table 1. Despite of the improvement in G*/sin δ, the addition of 3% NR latex remained the same performance with control bitumen (PG70). However, incorporating more than 5% NR latex successfully improved PG by at least one level (PG76). According to JKR/SPJ/2008-S4 requirement, polymer modified binder should achieve PG 76. Thus, 5 and 7% concentration satisfy the high service temperature and fulfil the JKR/SPJ/S4-2008 requirement.

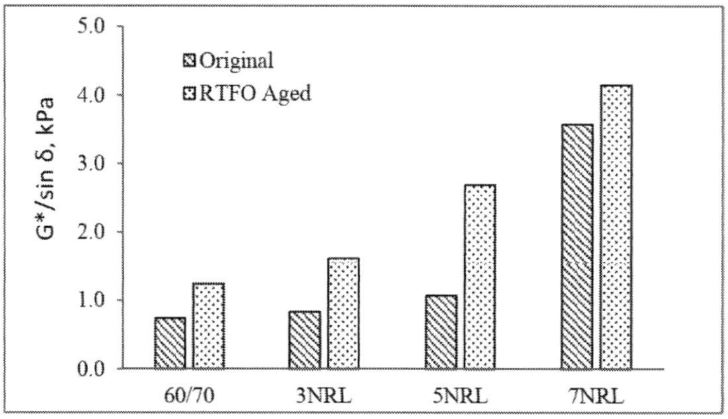

Figure 2. Rutting resistance (G*/sin δ) of modified binder at 76°C

Table 1. Rutting resistance properties of unmodified and modified bitumen before and after RTFO aged

Binder type	Temperature (°C)	Unaged (G*/sin δ > 1.0 kPa)	Short term aged (G*/sin δ > 2.2 kPa)	Performance Grade (PG)
60/70	70	1.53	3.02	PG 70
	76	0.74	1.37	
3NRL	70	1.81	3.58	PG 76
	76	0.89	1.71	
5NRL	70	2.24	5.44	PG 76
	76	1.19	2.70	
	82	0.68	1.41	
7NRL	70	6.89	7.68	PG 82
	76	3.80	4.16	
	82	2.24	2.39	

3.3. G.sin δ*

G*.sin δ value is measured in order to determine the characteristics of binders in terms of fatigue ability resistance. According to Superpave binder specification, G*.sin δ should have a maximum value of 5000 kPa. Low G* and δ are considered desirable attributes from the standpoint of resistance to fatigue cracking. In other word, the lower value of G*.sin δ at 10 rad/s are generally considered as good indicator for fatigue cracking resistance [12 (Ali, Mashaan, & Karim, 2013) (Abdullah, 2014).

Fatigue life of NR latex modified bitumen were tested on RTFO+PAV residue at 25°C and the results were depicted in Figure 3. In general, the higher content of NR latex contributed to lower G*.sin δ values of modified bitumen. The reduction in G*.sin δ was about 10%, 23% and 48% for NR latex content of 3%, 5% and 7%, respectively. From Table 2, both unmodified and modified bitumen recorded G*.sin δ values lower than 5000 kPa at 25°C. As the testing temperature was lowered to 22°C, G*.sin δ values exceeded the maximum value except for 7% NR latex which still recorded lower than 5000 kPa. It meant that, the increasing of NR latex content has a positive effect on modified bitumen during long term aging condition. Also, incorporating more than 5% NR latex could improve the resistance to fatigue cracking.

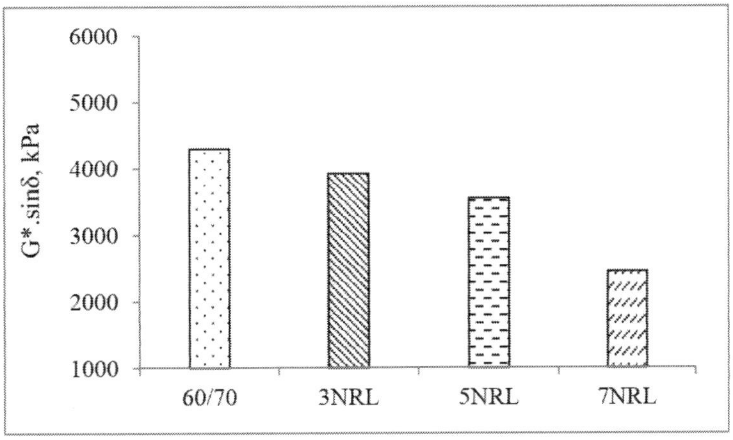

Figure 3. G*.sin δ of PAV residue at 25°C

Table 2. Fatigue cracking properties of modified bitumen after RTFO+PAV aged

Binder	G*.sin δ (kPa)		
	28°C	25°C	22°C
60/70	3002	4420	6739
3NRL	2769	3993	5949
5NRL	2488	3425	5109
7NRL	1706	2309	3425

4. Conclusion

Based on the result obtained, associating bitumen with NR latex has successfully improved its rheological properties during ageing condition (short term and long term ageing). It was also found that, an increasing of NR latex content enhances the bitumen's stiffness and elasticity properties, which indicated an improvement in temperature susceptibility and bitumen's performance at high temperature. In term of long term ageing, the higher NR latex content has positively affected on fatigue cracking resistance as lower G*.sin δ was recorded. Taken together, results from this study showed that the use of NR latex as bitumen modifier successfully improved at least one level of PG. It is also can benefits in alleviating permanent deformation and fatigue cracking problems that later affected the asphalt pavement performance during its service life.

Acknowledgement

Research work presented in this paper was supported by Ministry of Higher Education Malaysia (MOHE), under MyBrain15 programme, and Fundamental Research Grant Scheme number 1458 Universiti Tun Hussein Onn Malaysia. Financial support for this study is highly appreciated.

References

[1] Steiner D, Hofko B, Hospodka M, Handle F, Grothe H, Füssl J, Eberhardsteiner L, and Blab R (2015). Towards an optimised lab procedure for long-term oxidative ageing of asphalt mix

specimen. *International Journal of Pavement Engineering,* 1-7

[2] Lu X, and Isacsson U. (2002). Effect of ageing on bitumen chemistry and rheology. *Construction and Building Materials*, **16** 15–22

[3] Si Bachir D, Dekhli S, and Ait Mokhtar K. (2016). Rheological evaluation of ageing properties of SEBS polymer modified bitumens. *Periodica Polytechnica Civil Engineering*, **60** 397–404

[4] Sengoz B, and Isikyakar G. (2008). Analysis of styrene-butadiene-styrene polymer modified bitumen using fluorescent microscopy and conventional test methods. *Journal of Hazardous Materials*, ***150*** 424–432

[5] Peralta EJF. (2009). *Study of the interaction between bitumen and rubber*. Universiti de Minho. Retrieved from http://repositorium.sdum.uminho.pt/handle/1822/10557

[6] Airey GD, Rahman M, and Collop AC. (2003). Absorption of bitumen into crumb rubber using the basket drainage method. *International Journal of Pavement Engineering*, **4** 105–109

[7] Hofford S. (2015). Survey of fatigue resistance quantification of asphalt mixture. *International Journal of Advanced and Applied Sciences*, **2** 42–49

[8] Peterson JC, Branthaver JF, Robertson RE, Harnsberger PM, Duvall JJ, and Ensley EK. (1993). Effects of physicochemical factors on asphalt oxidation kinetics. *Transport Res Record*, **1391** 1–10

[9] Johansson LS, Lu X, and Isacsson U (1998). *Ageing of raod bitumen - State of the art*. Sweden

[10] Lu X, and Isacsson U (2002). Effect of ageing on bitumen chemistry and rheology. *Construction and Building Materials*, **16** 15–22

[11] Zhang J, Wang J, Wu Y, Wang Y, and Wang Y. (2009). Evaluation of the improved properties of SBR/weathered coal modified bitumen containing carbon black. *Construction and Building Materials*, **23** 2678–2687

[12] Ali AH, Mashaan NS, and Karim MR (2013). Investigations of physical and rheological properties of aged rubberised bitumen. *Advances in Materials Science and Engineering*, **2013** 1–8

[13] Ji J, Suo Z, and Xu, Y (2013). Laboratory Evaluation on the Long-term Aging Characterization of Warm Modified Binders. *Procedia - Social and Behavioral Sciences*, **96** 1640–1647

[14] Abdullah ME. (2014). Performance of warm mix asphalt (WMA) mixture using nanoclay modified asphalt binder. Universiti Tun Hussein Onn Malaysia. PhD Thesis.

Parameter Optimization for CO_2 Laser Cutting of Wood Polymer Composite (WPC)

M. Ibrahim*, M. Kesevaan

Advanced Manufacturing and Materials Center, Faculty of Mechanical and Manufacturing Engineering, Universiti Tun Hussein Onn Malaysia, 86400, Parit Raja, Batu Pahat, Johor.

Email: mymustaffa@gmail.com, kesevaanmorgan@gmail.com,

Abstract. Wood polymer composites (WPC) are being used widely in various industries due to its wood-like appearance and durability which makes it a suitable substitute for wood in internal decorations. This study involves cutting the WPC (20% wood) of 8mm thickness using CO_2 laser cutting machine and the best parameter is obtained in order to get the best cut. The laser cutting machine used in this experiment has a power rating of 50W and the experimental procedures involves variable power setting from 60% - 80%. The nozzle height was also varied ranging from 4mm to 6mm and the cutting speed from 2 – 4 mm/s. The cut material was analysed by taking the reading of top kerf width. The best parameter is then used to cut decorative wood panels which imitate traditional Malay wood carvings (*papan pemanis*) to be installed at homes. Through the analysis, there are total 10 solutions being generated which will produce good cut quality.

1. Introduction

This study was done to adapt a new technology for an old tradition of wood carving which has been a very popular internal and external house decoration (papan pemanis) [1] as shown in Figure 1. Previously, pure wood blocks carved manually by skilled craftsmen were used to make repetitive designs which will be then installed on walls or railings. The idea of this study is to replace traditional wood blocks with wood polymer composites (WPC) and it would be machined using a CO_2laser cutting machine.

There are many previous studies [2] which mentions about CO_2 laser cutting and its parameter optimization and therefore those techniques can be adopted to study the process parameter of cutting WPC using CO_2 laser cutting machine. The most common parameters studied are nozzle design, variations in shield gas velocity, cutting speed, type of shield gas and focal configuration [2].

Content from this work may be used under the terms of the Creative Commons Attribution 3.0 licence. Any further distribution of this work must maintain attribution to the author(s) and the title of the work, journal citation and DOI.
Published under licence by IOP Publishing Ltd

Figure 1: Multiple pieces of wood ornaments installed at a local mosque

2. Methodology

The total number of parameters usually being studied for laser cutting varies across different research and this can be limited due to available resources. For this study, only three factors are taken into consideration, which are laser power, cutting speed and nozzle height. The technique used for this experiment has been adopted from previous researches which studies individual effects and also explores the relationship between several variable factors and this technique is called response surface method (RSM). It uses statistical analysis to deduce the optimum values of factors to get the best response according to requirement.

2.1. Laser cutting parameters

Laser cutting technology has been developing to cater for various industries because of its efficiency and user friendly mechanism. It is being used to cut from metals as well as timber materials like medium density fibreboard (MDF) [2]. When a different industry adopts this machining technique, proper attention would be given to the laser cutting parameters and this further extends the usage of laser cutting machining. Previous studies have identified importance of process parameters like laser power, cutting speed, and shield gas to determine the cut quality for different materials [3]. Other than that, the nozzle design and the shield gas velocity can also be an important factor for certain materials like MDF. There are also some advanced factors that should be looked on for intricate details like beam spot diameter, thermal conductivity, reflectivity of workpiece material and interactions between various factors [5]. However, in this study, we would be focusing only on laser power, cutting speed and nozzle height.

2.1.1. Laser power. Laser power has direct effect on the surface quality due to the extreme heat produced while machining. One of the functions of assist gas is to remove the molten material before solidification takes place but this has to work simultaneously with suitable laser power. Previous studies has shown that the surface roughness generally increases with higher laser power [3, 6]. Improper setting of laser power would also increase the kerf width, cause a larger hole diameter as well as more irregular striation lines.

2.1.2. Cutting speed. For a given thickness of material, there would be a range of cutting speed including an optimum value based on cut quality. Cutting speed is also an important economic variable as a higher cutting speed would deduce a lower cycle time and therefore reduce the unit cost. It is shown that the energy loss at the cut zone can be decreased with an increase in the cutting speed, resulting in a more efficient process [3]. Similarly, when a thicker material is to be cut, the cutting speed has to be reduced, causing more energy to be wasted at the cut zone. The levels of thermal energy loss from the cut zone in most cases, rises rapidly with increasing material thickness coupled with the reduction in cutting speed [7].

2.1.3. Nozzle height. Many types of nozzle design has actually been adapted from industries that involves gas dynamics. A proper nozzle design has to provide a shock-free flow as well as ease of manufacturing. There are nozzles which have shaped contracting part and a smooth throat part that can produce a completely shock-free flow, but are impractical because of their high cost as no mass manufacturing technique exist as of the year 2008 [6]. Nozzle height is also important as an inappropriate height will cause formation of burr towards the lower part of the workpiece [8].

2.2. Material

Wood polymer composite (WPC) is a relatively new product being used extensively compared to commercial wood and timber in various industries [9]. WPC proves to be a better alternative as it is more environmental friendly and requires lesser maintenance. This quality has caused a demand for companies that specializes in manufacturing WPC for home application. WPC is made by mixing ground wood particles with selective thermoplastic resin and then extruded to produce end product with desired dimension and design. The main objective of this study is to produce 'papan pemanis', a traditional wood ornament using WPC and being machined using CO_2 laser cutting machine.

2.3. Design of experiment

The design of experiment used for this experiment was central composite design with alpha 1 and face-centered. Through this DOE there were total 20 number of runs due to 3 variable factors and 1 individual response. Table 1 shows the factors and response affiliated with experimental terms.

Table 1: Process variables and experimental design levels

No	Statistical term	Experimental term	Levels		
			-1	0	+1
1	Factor A	Laser power (%)	60	70	80
2	Factor B	Cutting speed (mm/s)	2	3	4
3	Factor C	Nozzle height (mm)	4	5	6
4	Response	Top kerf width			

3. Results and Discussion

All the data that were recorded after the cutting process were recorded in a table before analysis using a statistical software in a method called response surface method (RSM). The analysis involved the study of relationship among various parameters, namely cutting speed, laser power and nozzle height. The relation between individual parameter with the cutting quality were initially studied before obtaining the joined relation of multiple parameters with cutting quality.

All the raw data were initially included in Table 2 before proceeding with data analysis.

Table 2: Data extracted after 20 runs using CO_2 laser cutting machine

Runs	**Factor A**	**Factor B**	**Factor C**	**Response**	**Runs**	**Factor A**	**Factor B**	**Factor C**	**Response**
1	60	2.00	6.00	1.269	11	70	4.00	5.00	1.121
2	70	3.00	5.00	0.699	12	60	2.00	4.00	1.352
3	80	4.00	4.00	1.048	13	80	2.00	6.00	1.502
4	80	3.00	5.00	0.689	14	70	3.00	5.00	0.653
5	70	3.00	5.00	0.813	15	70	3.00	6.00	0.818
6	80	4.00	6.00	1.111	16	70	2.00	5.00	1.441
7	70	3.00	5.00	0.855	17	80	2.00	4.00	1.460
8	70	3.00	5.00	0.615	18	60	3.00	5.00	0.504
9	70	3.00	4.00	0.665	19	60	4.00	4.00	0.720
10	60	4.00	6.00	0.715	20	70	3.00	5.00	0.727

3.1. Analysis

For the beginning, from the fit summary analysis, the suggested highest order polynomial is quadratic. Next step is the analysis of variance (ANOVA), from which we can see the linear and quadratic effect of each factors involved in this experiment. Through this analysis, we can also identify if the factors are significant for the response or not by looking at the P value. Any value below 0.05 is considered to be significant and vice versa. However only the significant factors will be used in the general equation by the end of this analysis as the factors which are not significant does not have much effect on the response generally. Figure 2 shows the revised ANOVA table without the factors which are not significant.

Source	Sum of Squares	DF	Mean Square	F Value	Prob > F	
Model	1.85	4	0.46	73.51	< 0.0001	significant
A	*0.16*	*1*	*0.16*	*24.81*	*0.0002*	
B	*0.53*	*1*	*0.53*	*84.65*	*< 0.0001*	
A^2	*0.057*	*1*	*0.057*	*9.12*	*0.0086*	
B^2	*0.97*	*1*	*0.97*	*153.98*	*< 0.0001*	
Residual	0.094	15	6.298E-003			
Lack of Fit	*0.052*	*10*	*5.189E-003*	*0.61*	*0.7641*	*not significant*
Pure Error	*0.043*	*5*	*8.517E-003*			
Cor Total	1.95	19				

Figure 2: Revised ANOVA table (partial sum of squares)

From this table, we can see that the factors does not interact or effect the data through interaction. Factor C, nozzle height, does not have effect on top kerf width for this material. Factor A and B does have significant effect and they also have quadratic effect on top kerf width. Figure 3 shows graphically on how the laser power and cutting speed effects the top kerf width for WPC.

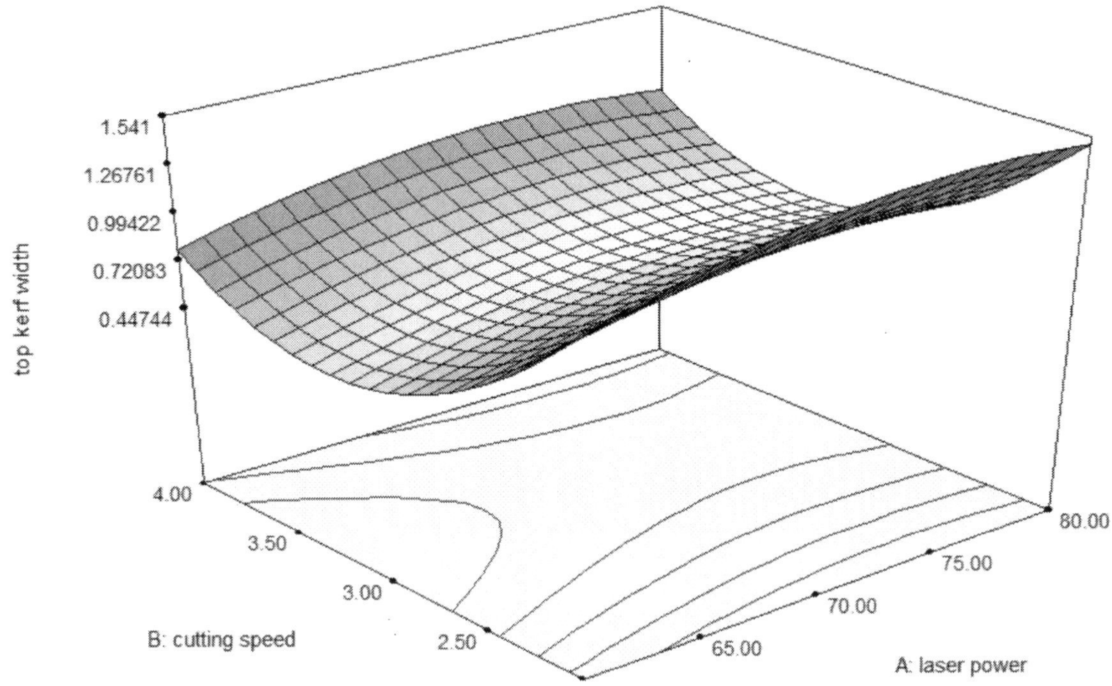

Figure 3: Effects of laser power and cutting speed on top kerf width of WPC

From this we can now obtain the general equation which can be used to estimate the value of top kerf width by considering only the factors which are significant. Therefore, the final equation in terms of coded factors is given by:

$$R1 = 0.73 + 0.13A - 0.23B - 0.13A^2 + 0.55B^2$$

where R1 is response (top kerf width), A and B are factors A and B which are laser power and cutting speed respectively.

3.2. Optimization

With all the analysis done, we can now generate the numerical optimization of factors which will theoretically ensure an optimum cut quality for the material WPC using this CO_2 laser cutting machine. There has been 10 solutions being generated which shows the values of factors. Table 4 shows all the possible solutions generated.

Table 3: Solutions with a total 10 numbers of starting points

No.	Laser power (%)	Cutting speed (mm/s)	Nozzle height (mm)
1	77.08	3.84	4.36
2	65.50	2.29	4.50
3	74.28	3.44	5.60
4	77.67	2.59	5.39
5	65.51	3.18	5.44
6	79.17	2.65	5.17
7	68.15	3.18	4.39

8	61.67	2.67	4.56
9	69.60	3.73	5.99
10	71.60	3.85	5.96

4. Conclusion
From this study, we can conclude that it is possible to machine WPC using CO_2 laser cutting machine when careful attention is given to the parameters involved in machining. Through this parameter optimization process, we are able to identify the best parameter that has to be set in order to get the best cut quality, which enables us to further increase the usage of WPC in the industry of internal house decorations. More specifically, we can now produce the traditional Malay wood carving, popularly known as 'papan pemanis' which was previously made using wood through intricate process which needs high level human skills.

References

[1] N. Utaberta, S. D. M. Sojak, M. Surat, A. I. Che-Ani and M. M. Tahir. International Journal of Civil, Architectural Science and Engineering Vol:6 No:7, 2012

[2] K.C.P. Lum, S.L. Ng, I. Black, CO2 laser cutting of MDF 1. Determination of process parameter settings. Optics & Laser Technology 32 (2000) 67-76

[3] Avanish Kumar Dubey, Vinod Yadava. Laser beam machining – A review. International Journal of Machine Tools & Manufacture 48 (2008) 609-628

[4] John C. Ion (2005). Laser processing of Engineering Materials (1st ed., p. 589). Oxford, UK: Elsevier Butterworth-Heinemann.

[5] Flaviana Tagliaferri. Study on the influences of laser parameters on laser asisted machining processes. Procedia CIRP 8 (2013) 170-175

[6] KJ. Muralidhara (2008) A study on CO2 laser cutting of mild steel and optimization of process variables.

[7] Wandera, C., Salminen, A., Olsen, F.O., & Kujanpaa, V. (2006). Cutting of stainless steel with fiber and disk laser. In 25th International Congress on Application of Lasers and Electro-Optics (pp. 211-220). Laser Institute of America

[8] K. Huehnlein, K. Tschirpke, R. Hellmann. Optimization of laser cutting processes using design of experiments. Physics Procedia 5 (2010) 243-252

[9] Denise Ortigosa Stolf, rancisco Antonio Rocco Lahr. (2004) Wood-polymer composite: Physical and mechanical properties of some wood species impregnated with styrene and methyl methacrylate. Materials Research, Vol. 7, No. 4, 611-617, 2004

IOP Conf. Series: Journal of Physics: Conf. Series **1049** (2018) 012102 doi:10.1088/1742-6596/1049/1/012102

The Integrated Academic Information System Support for Education 3.0 in Higher Education Institution: Lecturer Perspective

H P Utomo[1,2], A T Bon[1] and M Hendayun[2]

[1] Faculty of Technology Management and Business, Universiti Tun Hussein Onn Malaysia, Johor, Malaysia
[2] Faculty of Engineering, Universitas Langlangbuana, Bandung, Indonesia

hp150078@siswa.uthm.edu.my

Abstract. Education 3.0 has been implemented in many higher education institutions (HEIs). Education 3.0 has been directed the institution toward better educational experience. But on the other hands, the implementation of Education 3.0 also caused some problems. Previous research has found administrative problem experienced by the lecturer. This research explores deeper from the lecturer and suggested the solution from lecturer perspective, combined with information technology capabilities owned by the HEIs. The research used a case study as the method and conducted a qualitative research with a semi-structured interview. The interview analysis has found that the increase of the administrative processes is caused by online and offline administrative activities. The online activities are from e-learning and the offline activities are from traditional learning (face-to-face). The administrative processes also involved the academic information system (AIS). Simplified all of the administrative processes are more preferred. To overcome the problems, integrating the AIS and e-learning become necessary. This research suggests transforming the existing AIS into an integrated AIS and hopes the solution can simplify the administration process.

1. Introduction

The implementation of Education 3.0 has been directed the educational institutions toward better educational experience [1]. They can implement the characteristics of Education 3.0 one-by-one based on their needs. One of the characteristics is lecturing method. The lecturing characteristic in Education 3.0 more connected with technology. To provide and sharing the knowledge, e-learning often used as their technology. There are many e-learning technologies such as Moodle and Edmodo. Those applications are ready-to-use product. Many HEIs implemented e-learning with those applications [2]. Some of them developed their own e-learning application. The applications can support learning activity and administrative activity. In Table 1 below is a listed of whole characteristic of Education 3.0.

Table 1. The characteristics of Education 3.0 [3]

Content from this work may be used under the terms of the Creative Commons Attribution 3.0 licence. Any further distribution of this work must maintain attribution to the author(s) and the title of the work, journal citation and DOI.
Published under licence by IOP Publishing Ltd

	Education 3.0
Meaning is ...	Constructed in social community
Technology is ...	Everywhere, digital utilization
Lecturing is done ...	Face-to-face and people-technology-people
Institutions are located ...	Anywhere in the "creative society"
Parents view education institutions as ...	Creating knowledge place for their children and provide support for parent involvement
Lecturers are ...	Everybody, wireless assisted, technology provide raw material for knowledge production
Hardware and software in schools ...	Available in affordable cost and maximized for the effectiveness of the learning activities
Industry views graduate as ...	As entrepreneurs candidate and co-workers who can support the industrial involvement

Somehow, the implementation of Education 3.0, especially the lecturing characteristic caused some problems. Education 3.0 using dual lecturing method, the online method used e-learning and the offline/traditional method used conventional face-to-face learning. From [4] study, the increasing of administrative processes causes the lecturer often late to submit the final results. Their administrative activities are from e-learning, traditional learning and academic information system (AIS). A study by [5] also has found administrative problems when the e-learning and AIS are implemented in separated application. This research will explore deeper from the lecturer and the AIS capability. Finally, the aim is to find the solution from lecturer perspective, combined with information technology capabilities owned by the HEI.

2. Literature review

2.1. Academic information system

An information system (IS) often used in an educational institution is the academic information system (AIS). Many higher education institutions (HEIs) have been implemented the AIS [6]. Most of them implemented the AIS for helping the administrative processes in the academic affair [7]. They also used information from the AIS to make the decisions and policies [6]. With the current capabilities, the AIS can help the implementation of the Education 3.0.

According to [7], an academic information system (AIS) is application to support procedure the academic affair in an educational institution. Generally, AIS used to manage the student academic data, such as registration, courses selection, and provide the study results. Most of the AIS connected to the intranet for local used. The internet used when user need to access the AIS from outside the institution. The AIS described by [7] used only for the administrative purpose. The users are only the students and the administration staff.

From [6] study, AIS is a system used to manage and process the information resources in an HEI. The information output from this system will provide information to the decision makers or the executive leaders. AIS in [6] study used to help make a decision for institution development, instead for administration only. The AIS used by executive leader, lecturer, student, and administration staff.

The AIS has to accommodate to the needs of all users [8]. The [8] study compares the AIS in three different HEIs. The study found, the procedure and function of the AIS in the HEIs are similar. Therefore, the AIS must be flexible to development. The flexibility will help the system to stay up to date, and follow the needs of the users with better functionality over the latest technology.

From the facts above, most of AIS in the HEIs used for administrative process and support the decision maker. The current users also limited to executive leader, lecturer, student, and administration staff.

2.2. Education 3.0

Education world has a new paradigm called Education 3.0 [9]. The characteristics of Education 3.0 have extended the learning process. Education 3.0 also changed the academic stakeholder roles. Education 3.0 forced usage of advanced technology and extended the learner status [10]. The existing system must improve to support the characteristics.

Education has five (5) elements. There are learner, lecturer, context, subject, and evaluation. All the elements integrated to achieve the high level of meaningful learning. The achievement will emerge the empowerment in responsibility and commitment [11]. The collaboration among lecturer, student, parent, and administrator is a compulsory. The collaboration also need proffer technology to achieve the level.

According to [12], new ideas in lecturing and learning always had been awash the education concept. The administrator and lecturer got many reformation suggestions. They forced to use the new lecturing strategies and assessments in the new curricula. In the digital age, technology can support the all processes [13].

Education 3.0 characterized with multi-institutional and multi-cultural which the learners can be as knowledge creator and shared the knowledge into social community. The implementation of technology made distinction become blurred and no limitation for space and time [10]. The new approach offered by Education 3.0 mentioned by [11] and [13] where education enters the digital age and can emerging the empowerment. Those approach support by the integrated technology with wide access for learner and lecturer anytime and anywhere. With those, they can elaborate the knowledge in the system around the world [9].

In Education 3.0, students are not only consumed the knowledge but also empowered to produce it. [14]. As in Figure 1, Education 3.0 is made from Education 2.0 which is internet-enabled learning, and by centuries of experience with memorization in Education 1.0. Education 2.0 begins the transition to a new educational paradigm based on knowledge production and innovation production, the appropriate engines for viable 21st-century economies. Education 3.0 is qualitatively different incarnations that build upon Education 2.0 information sourcing capabilities and, to a lesser extent the memorization habits of Education 1.0 [3].

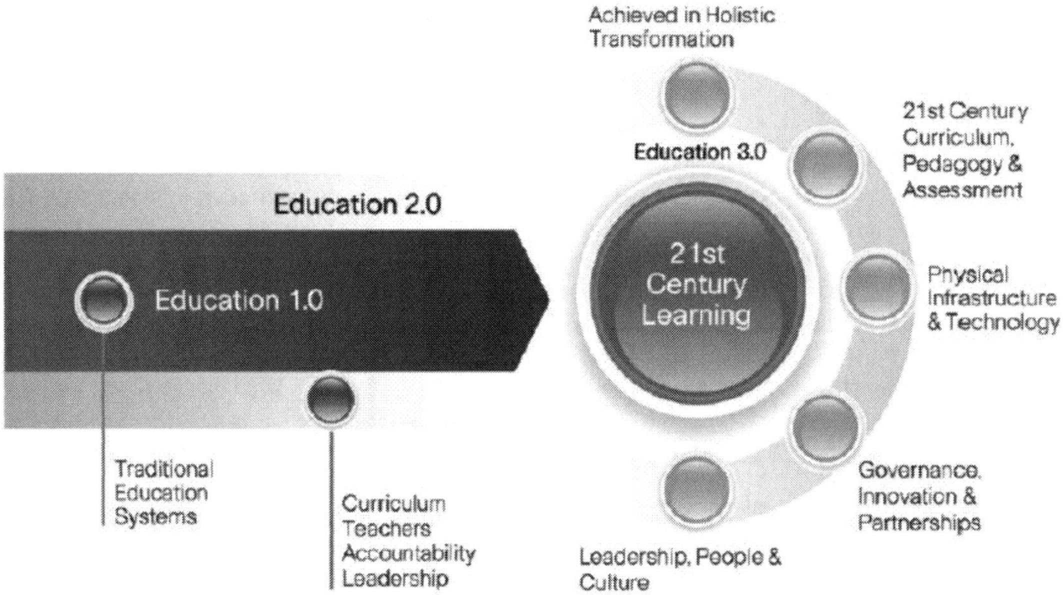

Figure 1. Education 3.0 paradigm [9]

3. Methodology

The case study method used in this qualitative research. Semi-structured interview conducted with lecturers who implemented Education 3.0 concept in their lecturing activities. They are also users of Sistem Informasi Akademik Universitas Langlangbuana (SIAk UNLA) in Bandung, Indonesia. Case study method suitable for investigating a contemporary phenomenon in depth when the boundaries between phenomenon and context are not clearly evident [15]. In this research, the Education 3.0 is a contemporary phenomenon and the AIS in the HEI is a context. This research starts the interviews with 10 (ten) lecturers. According to [16] and [17], the number of respondents in case study depends on the research itself. Their study suggests to do data analysis may follow or overlap with data collection. overlapping data collection and analysis has the advantage of adjusting the data collection process based on themes emerging from data analysis, or to further probe into these themes. The collecting data may stop if there is no longer anything new or already reach saturation level. After raw analysis, the data from respondents already saturated. The collected data will be analyzed and combined with AIS capability to provide the most suitable solution.

4. Analysis and Discussion

4.1. Analysis of interview data

From the interview with lecturers, they shared some problems when used e-learning and AIS in the separate application. They also used social media and instant messaging application to enrich their learning process. Because of those, they must spare more time to calculate the final results because they must combine and recap from electronic-based and paper-based marks. When all marks recapped, then they must key-in the results into the AIS. They also often late to submit the final results when handle more subjects or classes. The on time accomplishment of the final results is one of the performance indicators in higher education institution [18]. If many lecturers late to submit the final results, it will be affected to the institution's accreditation.

Besides the above problem, the interviews also discover the comforting usage from e-learning application with mobility support. They can access the e-learning from their mobile device anywhere and very enjoyed with the push notification function. But when they accessed the AIS with their mobile device, they felt the lack of function on it. They lost some information and function because the AIS did not support mobile platform yet. This also happens to the users that used their own institution's e-learning without mobile platform support. Furthermore, [19] found accessibility problems when the application did not support mobile platform. Besides that, the mobile platform provides an ideal avenue for the transfer of knowledge which will improve the learning capacity when collaboration is used in the learning process [20].

Based on the facts above, for the HEIs that have implemented Education 3.0 characteristics in their learning process are suggested to support those characteristics with their ICT capability. Application integration becomes necessary to simplified administration process and the application also must support mobile platform access.

4.2. Analysis of SIAk UNLA

As we know AIS is an information system for the academic purpose. The components are used in AIS will be same with IS components. The components of IS are hardware, software, database, procedures, people, and network. The hardware is a device such as a processor, monitor, keyboard, and printer. Together, these devices accept data and information, process them, and display them. The software is a program or collection of programs that enable the hardware to process data. A database is a collection of related files or tables containing data. A network is a connecting system (wire-line or wireless) that permits different computers to share resources. Procedures are the set of instructions about how to combine the above components in order to process information and generate the desired output. People are those individual who use the hardware and software, interface with it, or use its output [21, 22].

Based on the observation and documentation, the current state of SIAk UNLA from information system (IS) perspective can be shown below. The current state of SIAk UNLA based on IS components is given in Table 2.

Table 2. SIAk UNLA state based on IS components

IS Components	**SIAk UNLA**
Hardware	For both server-side and client-side already met the TIA-942 hardware standard
Software	Using open source OS and applications
Database	Using open source database (MySQL)
Procedures	AIS usage only for academic administration
People	Users are lecturer, students and administration staff
Network	Connected to the intranet and internet

The IS components can be used to relate the AIS and the Education 3.0. Research by [4] has found that all the characteristics of Education 3.0 can be supported by AIS. The characteristic of lecturing in Education 3.0 is lecturer to student, student to student, student to lecturer, people-technology-people (co-construction of knowledge). To support this characteristic, HEI must backup it with the code of conduct to avoid miscommunication and misunderstanding between lecturer and student. Code of conduct can clearly state the rights and obligations between lecturer and student [23]. From IS components overview, the support will come from procedures.

5. Conclusion

Based on analysis and discussion, this research has strengthened the problems faced by the lecturer when the HEI implement Education 3.0 in their learning process. Empowering ICT resources in the HEI can be a solution to solve the problems. AIS has capabilities to support the characteristics of Education 3.0 [4]. This research suggests AIS to transforms and enhancing by integrating with e-learning and becomes the integrated AIS with mobile platform supports. The integrated AIS hopes can decrease the administrative processes and makes lecturer more productive with less-administrative works.

References

[1] Lengel J G 2013 *Education 3.0: Seven steps to better schools* (New York: Teachers College Press)
[2] Light D 2012 Principals for web 2.0 success: 10 ways to build vibrant learning communities with the read/write web *Learn. Lead. with Technol.* **39** 18-20
[3] Harkins A M 2008 Leapfrog principles and practices: core components of education 3.0 and 4.0. *Futur. Res. Qual.* **24** 19-31
[4] Utomo H P, Bon A T and Hendayun M 2017 Academic Information System Support in the Era of Education 3.0 *IOP Conf. Ser.: Mater. Sci. Eng.* **226**
[5] Badu S Q, Kamuli S, Laisa Z, Dungga A and Nurdin M I 2012 E-leaning administrative simplification through integration with academic information system of higher education. *Proc. Int. Symp. on Open, Distance and Elearning.* (Jakarta: Cente for ICT Education, Ministry of Education and Culture)
[6] Indrayani E 2013 Management of academic information system (AIS) at higher education in the city of bandung *Proc - Soc. Behav. Sci.* **103** 628-636
[7] Utami E and Raharjo S 2014 Database security model in the academic information system *Int. J. Secur. its Appl.* **8** 163-174
[8] Alyoubi B A and Arif M J 2014 A comparative study between the academic information system of King Abdulaziz university and other Saudi Arabia Universities *Life Sci. J.* **11** 261–275
[9] Stevenson M, Cevenini P, Temple I, Halkett R and Patton R 2011 [cited 20 March 2016]. Available from https://www.cisco.com/c/dam/en_us/solutions/industries/docs/education/transforming_edu_wp.pdf
[10] Keats D and Schmidt J P 2007 The genesis and emergence of Education 3.0 in higher education and its potential for Africa *Fst. Mndy.* **12**
[11] Novak J D 2011 A theory of education: meaningful learning underlies the constructive integration of thinking, feeling, and acting leading to empowerment for commitment and responsibility. *Mean. Learn. Rev.* **1** 1-14.
[12] Wilson S M and Piterson P L 2006 *Theories of Learning and Teaching: What Do They Mean for Educators?* (Washington: NEA)
[13] Jones C 2015 Theories of learning in a digital age. *Networked Learning, Research in Networked Learning.* (Switzerland: Springer International Publishing)
[14] Gerstein J 2014 Moving from education 1.0 through education 2.0 towards education 3.0. *Educational Technology Faculty Publications and Presentations.* (Boise: Department of Educational Technology, BSU)
[15] Yin R K 2009 *Case study research: Design and methods* (London: Sage publications)
[16] Marshall B, Cardon P, Poddar A & Fontenot R 2013 Does sample size matter in qualitative research?: A review of qualitative interviews in IS research. *Journal of Computer Information Systems*, 54(1), 11-22.
[17] Baker S E, Edwards R & Doidge M 2012 How many qualitative interviews is enough?: Expert voices and early career reflections on sampling and cases in qualitative research. National Centre for Research Methods, Southampton.
[18] BAN-PT 2010 *Akreditasi Program Studi Sarjana* (Jakarta: Buku VI Badan Akreditasi Nasional-

Perguruan Tinggi)

[19] Lee G H, Talib A Z, Zainon W M and Lim C K 2014 Learning history using role-playing game (RPG) on mobile platform. *Adv. Comput. Sci. Its Appl.* **279** 597-602

[20] Ho R C and Chua H K 2015 The influence of mobile learning on learner's absorptive capacity: a case of bring-your-own-device (BYOD) learning environment. *Taylor's 7th Teach. Learn. Conf. 2014 Proc.* (Singapore: Springer Singapore)

[21] Turban E, Volonino L and Wood G 2013 *Information Technology For Management: Advancing Sustainable, Profitable Business Growth* (New Jersey: John Wiley & Sons, Inc)

[22] Laudon K C and Laudon J P 2016 *Essentials of Management Information Systems 12th Edition* (New York: Pearson)

[23] Barrett D E, Casey J E, Visser R D and Headley K N 2012 How do teachers make judgments about ethical and unethical behaviors? Toward the development of a code of conduct for teachers *Teaching and Teacher Education* **28** 890-898

IOP Conf. Series: Journal of Physics: Conf. Series **1049** (2018) 012103 doi:10.1088/1742-6596/1049/1/012103

Freezing/Thawing Weathering Resistance of Solidified Mine Tailings Matrices

Ali A Mahmood[1] and Maria Elektorowicz[2]

[1]Associate, Ash System International, 930 Av. Van Dyck, Brossard, Quebec, J4W 2E6, Canada
[2] Professor, Dept. of Building Civil and Environmental Engineering, Concordia University, Rm: EV-6.213, 1455 De Maisonneuve Blvd. W., Montreal, Quebec, H3G 1M8, Canada

Email: alimahmood45@yahoo.ca, mariae@encs.concordia.ca

Abstract. Mine tailings have been the source of water resources contamination over the years. This paper presents a novel approach for the solidification and stabilization of these mine tailings using Calsifrit binder and compares it to conventional Portland cement, slag, and fly ash. The resulting mine tailings matrices represent new materials with the potential of using them in construction processes in cold climates. Experimental results showed that the tailings binder matrices of concern passed the freezing/thawing durability tests. Furthermore, developed computer code Tailings-DEM™ was able to model successfully matrices' freezing/thawing characteristics.

1. Introduction

In cold climates repeated cycles of freezing and thawing can cause physical deterioration of the solidified waste matrices thus exposing them to contact with water and leaching mediums. This can cause serious environmental consequences, e.g. underground water resources and living environment contamination. Mine tailings have been the origin of several environmental calamities [1][2]. Therefore to prevent such disasters from taking place, solidification of mine tailings is investigated using several binders and additives, to reduce as much as possible their detrimental environmental impact. In this study, a novel binder called Calsifrit™ was applied in comparison to traditionally used Portland cement, fly ash, and slag to solidify 2 different types of mine tailings from Eastern Canada, namely, Mont Wright and Musselwhite. The objective of this study is to assess the response of the binders-tailings matrices to freezing and thawing effects in order to evaluate their potential use as construction materials in cold climates. This will include an evaluation of the potential benefits of using Calsifrit compared to Portland cement and traditional additives. These experimental results will be further used to evaluate the applicability of the developed computer program Tailings-DEM™ in numerically modeling these freezing/thawing experimental results. The objective of this study is to quantify the response of the new tailings matrices to freezing and thawing cycles.

Content from this work may be used under the terms of the Creative Commons Attribution 3.0 licence. Any further distribution of this work must maintain attribution to the author(s) and the title of the work, journal citation and DOI.
Published under licence by IOP Publishing Ltd

2. Methodology

Another incentive for this research is the fact that in 2011, the Quebec provincial government has instated an economic stimulus plan (Plan Nord) to invest C$80 billion in the northern region of the Province. This plan will be carried out for a period of about 25 years and is projected to create 20,000 jobs [3]. A big part of this economic stimulus will be directed towards the mining industry in that region, which, so far, has been active for many years.

Traditionally mine tailings have been treated as a waste material that requires landfilling inside specially created tailings dams. This approach has been practiced for many decades until these tailings dams started breaking up and causing substantial environmental damages, coupled in several instances with human fatalities [1][2].

In recent years there has been a new trend towards "managing" these tailings rather than treating them as a waste that requires dumping. Several investigators have attempted to re-use mine tailings in an environmentally sustainable manner. Examples include the works of Garand et al.[4], (Benzaazoua et al. [5], Theriault et al. [6], Roy at al. [7], Helinski et al. [8], Zhang [9], Ugama et al. [10] and Sabat et al. [11].

In order to assess the applicability of using mine tailings as construction materials, an experimental and computational program was designed to find their index properties and to assess their engineering properties with respect to the proposed function.

Investigations are composed of two phases: experimental and computational. The Experimental Phase 1 is divided into 3 stages: characterization of materials, formulation of tailing matrices and assessment of the matrices' usefulness as road construction materials. Phase 2 consists of 3 stages: development of an adequate computer program predicting matrices usefulness as road materials, verification of the program for strength and weathering tests.

This study focuses on the freezing/thawing environmental tests. Details of other tests and the computational phase can be found elsewhere [12].

2.1 The tailings

Bench scale and pilot tests performed by Alcan International Ltd. on copper and gold tailings had shown the applicability of the thickening technology to a variety of tailings [13], in addition to the alumina tailings that were investigated by the latter authors. Therefore in this study, two different mine tailings were selected for evaluation; Musselwhite tailings from Placer Dome (gold) mine:, in Musselwhite, Ontario, and Mont Wright tailings from Quebec Cartier mineral company, from Mont-Wright (iron) mine, in Quebec.

These two types of tailings were chosen due to their availability in the northern territories of Eastern Canada; close to the area where Plan Nord will be employed. The adverse environmental impact that tailings cause to underground water resources and the environment warrant a detailed investigation into the potential of reusing them sustainably in a construction application. Table 1 shows their physical properties. Where: D_{10} = diameter corresponding to 10% finer, D_{50} = diameter corresponding to 50% finer, D_{60} = diameter corresponding to 60% finer, D_{30} = diameter corresponding to 30% finer, C_u = uniformity coefficient = D_{60} / D_{10}, C_z= coefficient of gradation = $D_{30}{}^2 /(D_{10}\, D_{60})$, $P_{4.75mm}$ (%) = percentage passing sieve no.4, $P_{0.075mm}$ (%) = percentage passing ASTM sieve # 200, SW-SM = well graded sand with silt, SM = silty sand, SP = poorly graded sand, USCS [14] = unified soil classification system.

The mineralogical composition of Mont Wright tailings is: Sandstone (quartzite), Mica schist, Amphibolite, Gabroïque Granite, Gabroïque Feldsphate, Specular Hematite, Specular Magnetite, Quartz, Diopside, Tremolite, Actinolite and Gruenerite. While Musselwhite is composed of Quartz (SiO_2), Birnessite-syn (MnO_2), Calcium manganese oxide hydrate ($Ca_2Mn_{14}O_{27}.xH_2O$) and Dannemorite $(Fe,Mg,Mn)_7Si_8O_{22}(OH)_2$.

IOP Conf. Series: Journal of Physics: Conf. Series **1049** (2018) 012103 doi:10.1088/1742-6596/1049/1/012103

Table 1. Physical properties for the tailings

Type of Tailings	Musselwhite	Mont Wright
D_{10} (mm)	0.0045	0.147
D_{50} (mm)	0.016	0.255
D_{60} (mm)	0.023	0.3
D_{30} (mm)	0.012	0.2
C_u	5	2.04
C_z	1.39	0.91
$P_{4.75mm}$ (%)	100	100
$P_{0.075mm}$ (%)	74.9	2.05
Initial moisture content (%)	30.15	4.27
USCS	SM	SP

2.2 Binding materials

The binding materials for the mine tailings, which have been tested in this study, consist of Type I ordinary Portland cement (OPC), fly ash, slag and Calsifrit.

Unconfined compressive strength of fly ash lies in the range 1.38 to 6.895 MPa with permeability between 10^{-5} to 10^{-8} cm/s [15]. The fly ash was obtained from St. Laurence Cement Company in Longueil, Quebec.

Ground granulated blast furnace slag used in this study was obtained from Lafarge North America Cement Company in Montreal, Quebec under the trade name NewCem®. It is a light grey odorless powder produced in accordance with ASTM C 989 [16]. It is composed of (30-50) % Calcium Oxide, (0-20) % Magnesium Oxide with less than 1% Crystalline Silica. Its specific gravity is between 2-3, water pH 8-11. Its boiling point is above 1000 °C [17].

Calsifrit™ is a totally amorphous siliceous material, a matrix of calcium and sodium fluoro-aluminosilicate. This homogeneous solid substance has a blackish grey color, possesses a high reactivity potential and shows cementitious properties when finely ground. This product was obtained from the manufacturer NovaFrit International. Table 4 shows its chemical composition. Calsifrit has a pH (1% solution in water) of 6.5-7.5, a melting point of 800 °C, a specific gravity of 1.3 with a moisture content of less than 10% [18].

2.3 Testing Procedure

A procedure for measuring the solidified waste resistance to freezing and thawing was employed to determine its resistance to these environmental conditions [19]. This test is intended for the evaluation of the freezing/thawing resistance of monolithic solidified wastes.

As such, small cylindrical specimens were molded in metallic molds measuring 44 mm diameter by 74 mm height. After casting the solidified tailings matrices in these molds, they were left to thaw and solidify in the moisture chamber for a period of 43 days. These specimens were, then, subjected to 12 cycles of freezing at -20 degrees Celsius for 24 hours, followed by thawing in water at room temperature for 24 hours. The weight loss of each specimen was measured and compared with that of a control specimen. The change in weight was used to determine the state of the structural integrity of the specimen. Six identical specimens were used for each mixture; three test (denoted by T) and three control specimens (denoted by C).

After removing the specimens from the freezing cabinet and the moisture chamber, 240 milliliters of distilled chilled water at 4 °C was added to them. To the control specimens was added 240 milliliters of distilled water at room temperature (22 °C) before plastic wrap was placed on the beakers and all water covered specimens were stored at room temperature (22 °C) for a period of approximately 23 hours.

Afterwards each specimen was transferred to another dry beaker prepared as before, taking into account that all loosely attached particulates were removed by spraying with distilled water from a wash bottle to the surface of the specimen. Water was left to drain into the original beaker [19]. Then the specimens' weight loss was determined by measuring the amount of the solid residue in these beakers by evaporating water in the oven. A total of 12 cycles was performed in the prescribed manner. Figures 1, 2, 3, 4 show the experimental results of these tests.

3. Results and Discussion

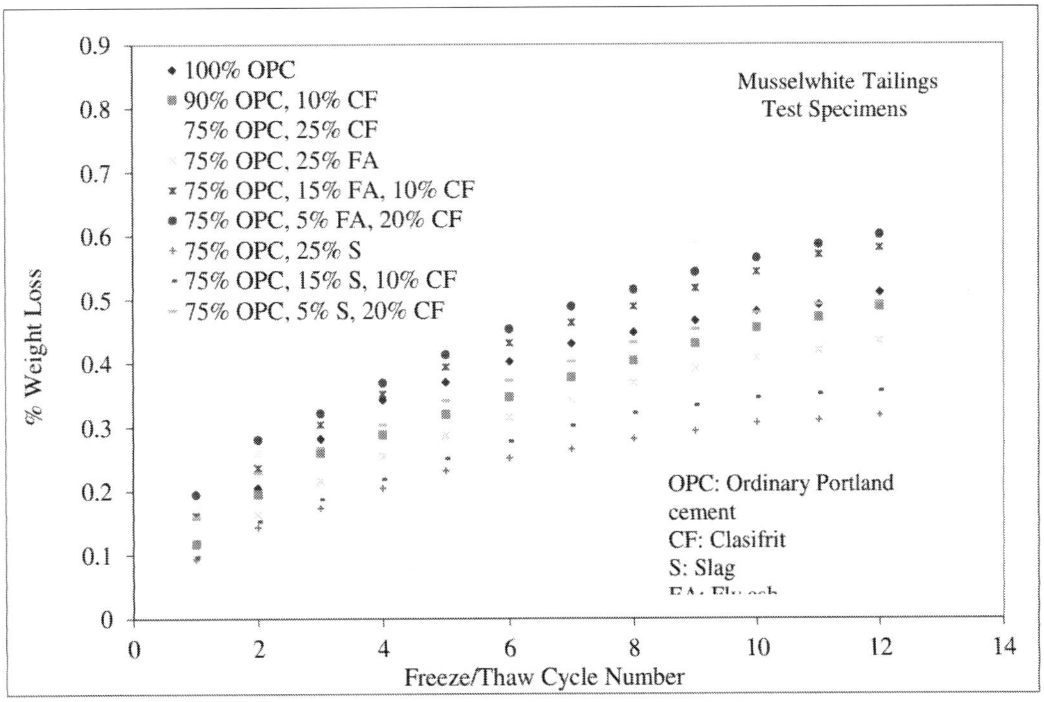

Figure 1. Weight loss after freezing and thawing test for Musselwhite test matrices

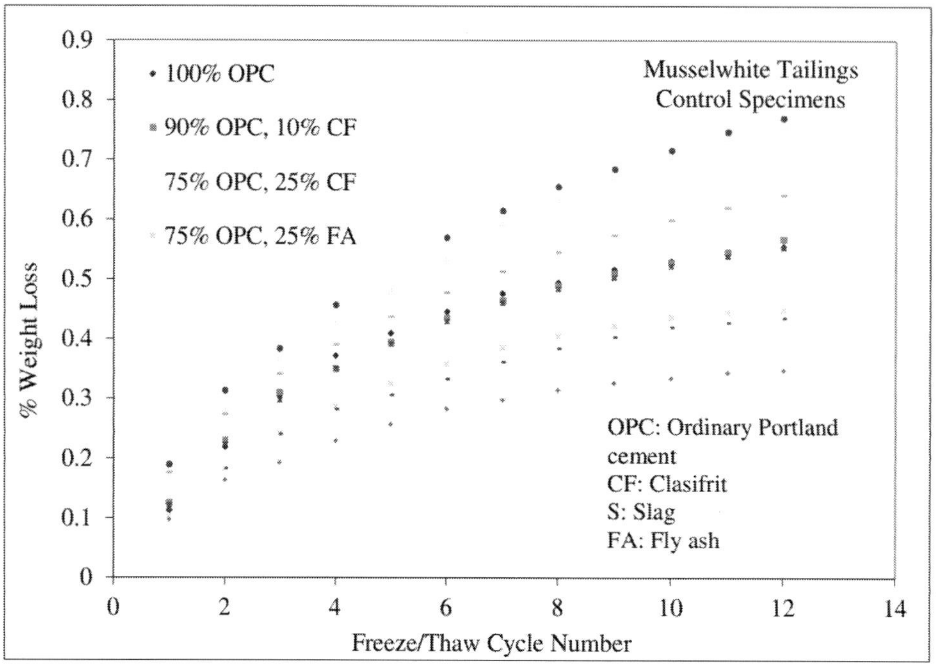

Figure 2. Weight loss after freezing and thawing test for Musselwhite control matrices

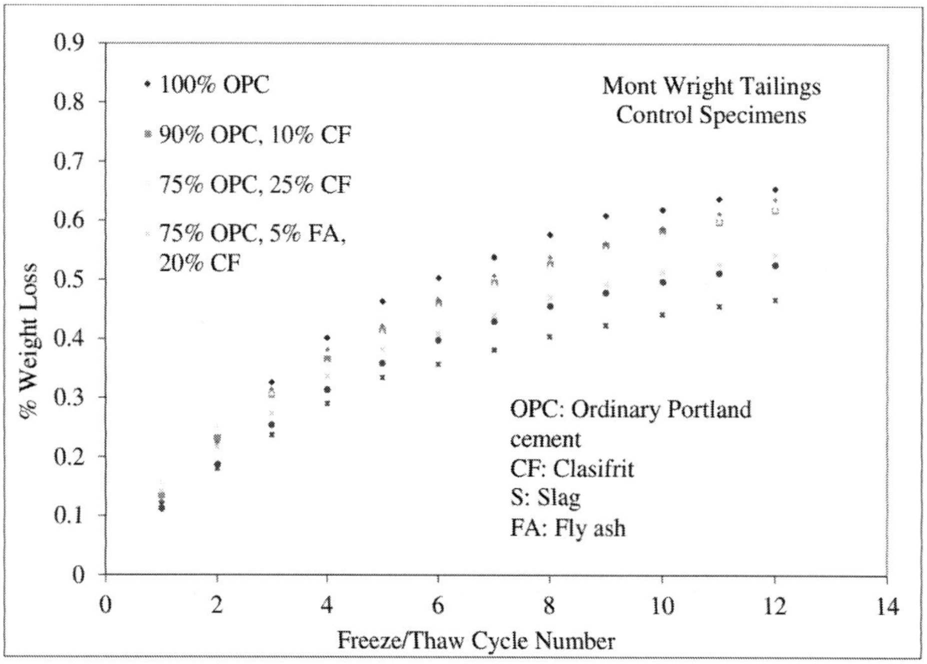

Figure 3. Weight loss after freezing and thawing test for Mont Wright control matrices

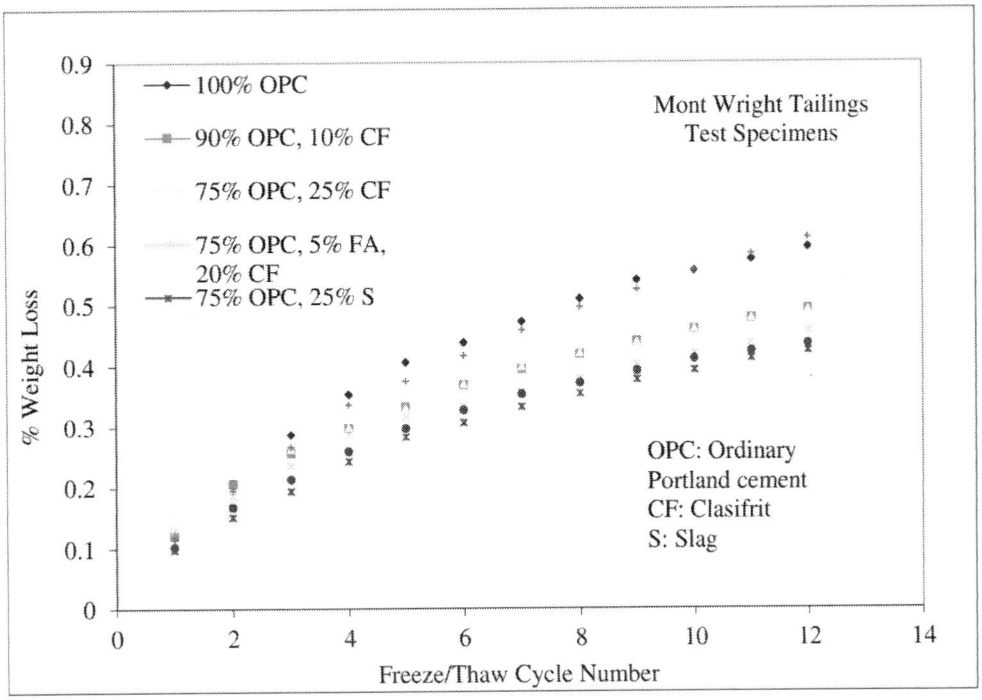

Figure 4. Weight loss after freezing and thawing test for Mont Wright test matrices

ASTM stipulates that for a sample to fail the freezing and thawing test it has to lose 30% of its weight during or after the completion of this test [19]. Accordingly, it can be seen from Figures 1 to 4 that all matrices tested for both Mont Wright and Musselwhite tailings fared well in this test as the maximum cumulative weight loss did not exceed 0.8% during the 12 cycles of freezing and thawing.

Figures 1-4 show that test and control specimens had similar weight loss values and characteristics, which is clear proof of rigidity of the mixtures and samples. They also show that adding Calsifrit to the cement-only specimens had a positive effect on durability for the Mont Wright specimens (testing and control) with weight loss decreasing with the addition of 10% and 20% Calsifrit to the cement-only mixtures. The same mixtures had less effect on the Musselwhite specimens. On the other hand slag-OPC matrices faired the best with the least weight loss out of all tailings matrices tested. The 5% Fly ash -20% Calsifrit -75% OPC matrix had less weight loss than the same matrix with slag for Mont Wright tailings. These mixtures showed an opposite effect on Musselwhite matrix specimens.

The increase in durability of Mont Wright specimens with addition of Calsifrit could be attributed to the improvement in the bond between the hydrated cement matrix and the tailings particles. This is due to the conversion of calcium hydroxide, which tends to form on the surface of the aggregate particles, to calcium silicate hydrate [20]. Musselwhite tailings on the other hand had an amount of Zn and Pb that retarded hydration as previously discussed, preventing the creation of a stronger bond between the aggregate particles and the hydrated cement matrix. This could be the reason that Musselwhite tailings matrices, generally, showed more weight loss than Mont Wright matrices.

4. Numerical modeling of particulate media

Traditionally, continuum based numerical analysis methods have been used qualitatively and quantitatively to analyze particulate and granular media. The most notable example in soil mechanics is the Finite Element Method and its various hybrids.

Although the finite element method was used to study discontinuous problems, it has several disadvantages. For example this method is continuous in nature, which means that it cannot efficiently model the discontinuity property. Also, as the accuracy of the model is as good as the assumptions used, reproducing complex behavior with continuous methods requires complex constitutive models, containing sometimes dozens of parameters and internal variables in order to capture discontinuous behavior. This can improve results but on the other hand they are computationally intensive and time consuming [21].

Due to their particulate nature, soils can experience significant local deformations and bifurcations. Continuum based methods cannot compensate for such anomalies without incremental analysis involving nonlinear elasto-plasticity based constitutive relations. Even so this approach is lacking in adequately modeling the most rudimentary behavior of soils, including non-linear deformations and local yielding [22].

4.1 The Discrete Element Method program Tailings-DEM™

In 1979 Cundall and Strack developed the Discrete Element Method [23]. This method models particulate media as a discrete collection of particles. This random collection of particles interacts through their contact forces. The method calculates the displaced positions and rotations of these particles at discrete time intervals. The DEM-simulation is started by first generating a model, which results in the random orientation of the particles with assigned initial velocities. Forces and moments acting on each particle are computed form the initial data and the relevant physical laws and contact models. Generally, the simulation consists of three steps: the initialization, the explicit time stepping and post processing.

In the DEM the microstructure of the system is modeled rather than using constitutive laws or complicated elements. Using DEM we can capture changes in microstructure, change in shape and deformations, dynamics and forces within the system in real time and in detail. Compared to conventional continuum methods, the DEM uses fundamental and fewer parameters when modeling discontinuous behavior [21].

Tailings-DEM™ is a Discrete Element Method computer program written for the evaluation of the strength and durability characteristics of these tailings binder matrices [12]. The program was written using the computer programming language C++ and the compiler Code Blocks. General assumptions employed when using Tailings-DEM™ are explained elsewhere [12].

Thermal stresses resulting from the change in temperatures of the freezing and thawing test [19], are calculated according to the following equation by Pytel and Singer [24]:

$$\sigma = E\,\alpha\,\Delta t \qquad (1)$$

Where:
σ = thermal stress (MPa)
E = modulus of elasticity (MPa)
α = coefficient of thermal expansion (m/m °C)

The United States Federal Highway Administration (FHWA) employs a value range of (18-20) x 10^{-6}/ °C for the coefficient of thermal expansion of saturated cement pastes [25]. The average for this range, 19 x 10^{-6}/ °C, is used in calculating the thermal stresses in this study.

The modulus of elasticity (E) for each tailings matrix is assumed to be equal to the slope of the linear portion of the stress versus strain diagram of that matrix.

Equation (1) was used to compute the thermal stresses in the matrix resulting from the change in temperatures during its exposure to freezing/thawing cycles. Matrix weight loss was used as the indicator of matrix response to the freezing/thawing environment.

The thermal stress resulting from the change in the specimen temperature is found according to Equation (1) shown above. The coefficient of thermal expansion in this equation is a measure of the materials expansion or contraction with temperature. Due to the relatively small length changes associated with thermal expansion, it is usually expressed in terms of micro strains per unit temperature change. The United States Federal Highway Administration employs a value range of (18-20) x 10^{-6}/ °C for the coefficient of thermal expansion of saturated cement pastes [25]. The average of this range, 19 x 10^{-6}/ °C, is used in this study for calculating the thermal stresses.

The modulus of elasticity (E) for each tailings matrix is assumed equal to the slope of the linear portion of the stress versus strain experimental diagram of that matrix. This value of E is used in Equation (1) to find the thermal stress.

Once the thermal stress is calculated according to Equation (1), it is multiplied by the cross sectional area of the cylindrical specimen to get the maximum thermal load. Dimensions of the cylindrical specimen used are: 44 mm diameter and 74 mm height. These values represent the actual dimensions used experimentally.

Subsequently, this load is used in finding the change in length (ΔL) associated with computational results, which represents the length response of the specimen to the vertical load applied to it during the UCS test. This change in length divided by the initial specimen length is assumed here to represent the weight loss from freezing and thawing thermal stresses expressed in units of MPa. The change in length (ΔL) is found by graphically reading off the displacement corresponding to the thermal load found using Equation (1) for the tailings matrix in question. The computational weight loss for the tailings matrix equals this displacement divided by the initial length of the specimen (74 mm).

Figures 5, 6, 7 show the computational percentage weight loss found using the procedure explained and its comparison with the experimental weight loss results for the freezing/thawing tests on the tailings matrices. The percent difference is found by subtracting the computational weight loss from the experimental and dividing the result by the experimental weight loss. Then, the factor of safety is found by dividing the experimental over the computational weight loss. Table 2 shows the specimen descriptions as used here.

As explained above, the matrix was previously exposed to the freezing/thawing cycles test following ASTM D 4842 [19]. Temperature range used in the computation was according to the laboratory test; i.e. from room temperature (22 °C) to the freezing temperature of-20 °C. Results show that all tailings matrices tested fared well in this test as the maximum value of the cumulative mass loss did not exceed 0.8% [12].

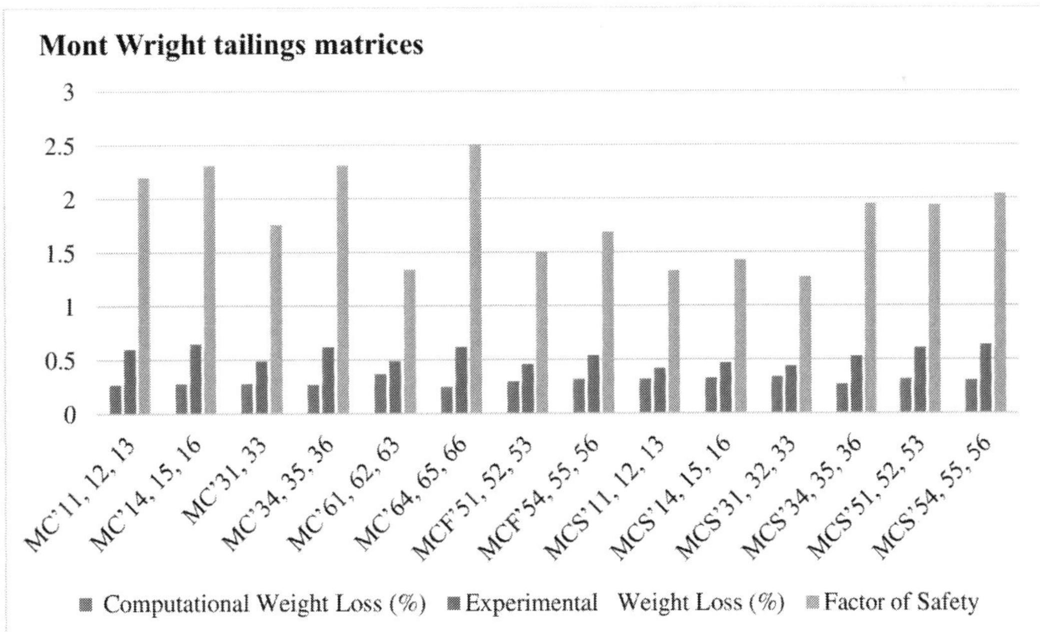

Figure 5. Weight loss comparison between the computational and experimental for Mont Wright matrices after freezing and thawing

IOP Conf. Series: Journal of Physics: Conf. Series **1049** (2018) 012103 doi:10.1088/1742-6596/1049/1/012103

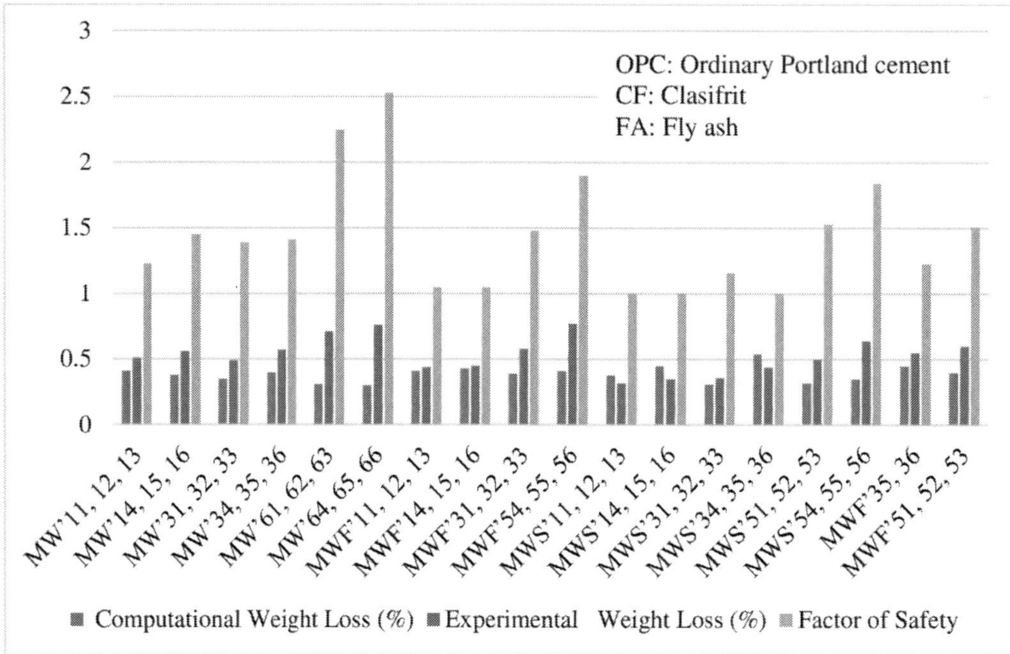

Figure 6. Weight loss comparison between the computational and experimental for Musselwhite matrices after freezing and thawing

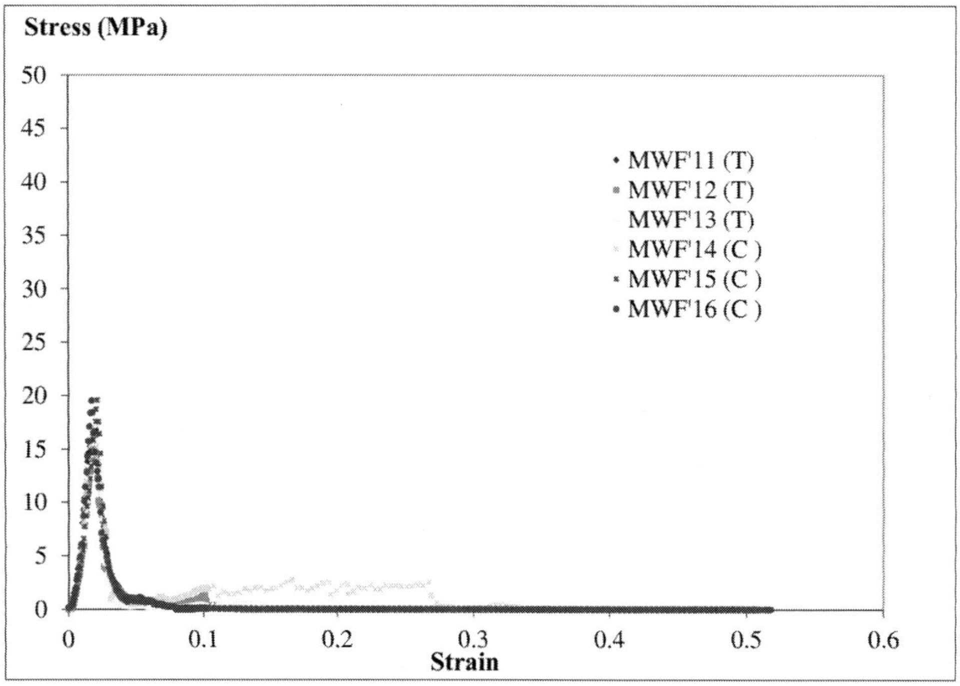

Figure 7. Compression strength for Musselwhite samples MWF'11-MWF'16 after freezing/thawing (Musselwhite tailings, Cylindrical Specimens, 75% OPC, 25% FA, Freezing/Thawing)

IOP Conf. Series: Journal of Physics: Conf. Series **1049** (2018) 012103 doi:10.1088/1742-6596/1049/1/012103

Table 2. Tailings matrices specimen codes

Tailings Type	Weathering Type	Code	Specimen	Binder/ Tailings	OPC/ Binder (%)	Calsifrit/ Binder (%)	Fly ash/ Binder (%)	Slag/ Binder (%)
Mont Wright	Wetting/ Drying	MC1	1, 2, 3, 5, 6	0.5	100	0	0	0
		MC3	1, 2, 3, 4, 5, 6	0.5	90	10	0	0
		MC6	1, 2, 3, 4, 5, 6	0.5	75	25	0	0
		MCF1	1, 2, 3, 4, 5, 6	0.5	75	0	25	0
		MCF3	1, 2, 3, 4, 5, 6	0.5	75	10	15	0
		MCF5	1, 2, 3, 4, 5, 6	0.5	75	20	5	0
		MCS1	1, 2, 3, 4, 5, 6	0.5	75	0	0	25
		MCS3	1, 2, 3, 4, 5, 6	0.5	75	10	0	15
		MCS5	1, 2, 3, 4, 5, 6	0.5	75	20	0	5
Musselwhite		MW1	1, 2, 3, 4, 5, 6	0.375	100	0	0	0
		MW3	1, 2, 3, 4, 5, 6	0.375	90	10	0	0
		MW6	1, 2, 3, 4, 5, 6	0.375	75	25	0	0
		MWF1	2, 3, 4, 5, 6	0.375	75	0	25	0
		MWF3	1, 2, 3, 4, 5, 6	0.375	75	10	15	0
		MWF5	1, 2, 3, 4, 5, 6	0.375	75	20	5	0
		MWS1	1, 2, 3, 4, 5, 6	0.375	75	0	0	25
		MWS3	1, 2, 3, 4, 5, 6	0.375	75	10	0	15
		MWS5	1, 2, 3, 4, 5, 6	0.375	75	20	0	5
Mont Wright	Freezing / Thawing	MC'1	1, 2, 3, 4, 5, 6	0.5	100	0	0	0
		MC'3	1, 3, 4, 5, 6	0.5	90	10	0	0
		MC'6	1, 2, 3, 4, 5, 6	0.5	75	25	0	0
		MCF'3	1, 2, 3, 4, 5, 6	0.5	75	10	15	0
		MCF'5	1, 2, 3, 4, 5, 6	0.5	75	20	5	0
		MCS'1	1, 2, 3, 4, 5, 6	0.5	75	0	0	25
		MCS'3	1, 2, 3, 4, 5, 6	0.5	75	10	0	15
		MCS'5	1, 2, 3, 4, 5, 6	0.5	75	20	0	5
Musselwhite		MW'1	1, 2, 3, 4, 5, 6	0.375	100	0	0	0
		MW'3	1, 2, 3, 4, 5, 6	0.375	90	10	0	0
		MW'6	1, 2, 3, 4, 5, 6	0.375	75	25	0	0
		MWF'1	2, 3, 4, 5, 6	0.375	75	0	25	0
		MWF'3	1, 2, 3, 4, 5, 6	0.375	75	10	15	0
		MWF'5	1, 2, 3, 4, 5, 6	0.375	75	20	5	0
		MWS'1	1, 2, 3, 4, 5, 6	0.375	75	0	0	25
		MWS'3	1, 2, 3, 4, 5, 6	0.375	75	10	0	15
		MWS'5	1, 2, 3, 4, 5, 6	0.375	75	20	0	5

IOP Conf. Series: Journal of Physics: Conf. Series **1049** (2018) 012103 doi:10.1088/1742-6596/1049/1/012103

5. Discussion of Tailings-DEM™ modeling of freezing/thawing

Based on the results shown in Figures 5, 6, 7 it is evident that differences in the freezing/thawing weathering results are within an average factor of safety of 2. The experimental values were higher than the computational for all Mont Wright and most Musselwhite matrices. The Musselwhite matrices with the following binder mixtures had higher computational weight loss values than experimental: 75% OPC and 25% slag (testing and control) and 75% OPC, 15% slag, 10% CF (control).

The reason behind this discrepancy could be attributed to the relatively high ratio of particles smaller than 75 micrometer that Musselwhite tailings contain as revealed in the physical properties shown in Table 1. Additionally, slag's average specific gravity of 2-3 may have resulted in more matrix voids filled by slag particles during the mixing stage rendering the matrix more resistant to freezing and thawing weathering effects. It is also shown that when the binder content contained 25% of slag, weight loss was lower for all Musselwhite and Mont Wright matrices, for both testing and control specimens. This is clearly evident in Figures 1, 2, 3, 4. Musselwhite tailings matrix with 25% slag had 1 mg/L Fe content according to the TCLP test [12]. Compared to this, no Fe content was detected with the equivalent Mont Wright tailings matrix. Al-Otaibi [26] who used steel mill (iron oxides) as fine aggregate in cement mortars, found that drying shrinkage was lower when using steel mill scale. He also showed that replacing 40% of sand with steel mill gave the highest increase in compressive strength and increased flexural strength. This may have contributed to Musselwhite matrices having less percent weight loss than their Mont Wright counterparts.

The US Federal Highway Administration cite research by Malhotra [27] who did freeze thaw tests on concrete incorporating 25-65% slag. Test results indicated that regardless of the water / (cement + slag) ratio, air entrained concrete slag specimens performed excellently in freeze thaw tests, with relative durability factors greater than 91% [25].

There is not much deviation between the experimental and computational results for the Musselwhite tailings matrix when the binder content is composed of 75% OPC and 25% fly ash as can be seen in Figures 5 and 6. This is explained by the fact that the weight loss was found computationally based on the experimental stiffness coefficient. It can be seen in Figure 5 that all 6 specimens of this matrix combination had very close strength and stiffness values, hence closing the gap between experimental and computational results.

Other reasons for the differences between computational and experimental results can be attributed to the triangular particle shape used in the Tailings-DEM™ program. Changing the assumed particle shape to a pentagon or a more circular shape may lead to the results being more closely related. Reducing their size and increasing their number may also enhance the results.

The above suggests that Tailings-DEM™ can be used with confidence to account for and predict freezing and thawing weathering effects, as described by the procedure in Mahmood [12].

However, it should be used with caution when the matrices contain a detectable Fe content coupled with slag binder content between 10% - 25%.

6. Conclusions

- It can be seen that Mont Wright and Musselwhite tailings matrices passed the freezing/thawing weathering tests with a maximum weight loss of 0.8%, which indicates that these matrices can be used as a construction material sustainably in cold regions,
- Adding Calsifrit to the cement-only specimens had a positive effect on durability for the Mont Wright testing and control specimens with weight loss decreasing with the addition of 10% and 20% Calsifrit to the cement-only mixtures. Although the same mixtures had less effect on the Musselwhite samples, it can be concluded that Calsifrit is a viable Portland cement replacement, within the limits shown above,
- Simulating the vulnerability of Mont Wright and Musselwhite tailings binder matrices to cold climate, through the freezing/thawing test, using the Discrete Element Method was successful,

- Differences between experimental and computational freezing/thawing weathering results are within an average factor of safety of 2.

Acknowledgment
The authors would like to acknowledge with gratitude the support that Dr. Michelle Nokken and Dr. Catherine Mulligan of Concordia University (Montreal, Canada) have provided for this study.

References

[1] Fourie A B and Papageorgiou G 2001. Defining an appropriate steady state line for Merriespruit gold tailings Canadian Geotechnical Journal **38**(4) 695-706.
[2] Fourie A B, Blight G E and Papageorgiou G 2001. Static liquefaction as a possible explanation for the Merriespruit tailings dam failure Canadian Geotechnical Journal **38**(4) 707-719.
[3] Government of Quebec, Societe du Plan Nord 2011 https://plannord.gouv.qc.ca/en/.
[4] Garand P, Vezina S and Bocking K 2000 Effects of flocculent on deposition of tailings sludge for upstream raise of impoundment dykes *In* Proceedings of the 53rd Canadian Geotechnical Conference Canadian Geotechnical Society 15-18 October 2000 Montreal Quebec pp. 633-640.
[5] Benzaazoua M, Belem T and Bussiere B 2002 Chemical factors that influence the performance of mine sulphidic paste backfill Cement and Concrete Research **32**(7) 1133-1144.
[6] Theriault J A, Frostiak J and Welch D 2003 Surface disposal of paste tailings at the Bulyanhulu gold mine, Tanzania *In* Proceedings of Sudbury 2003 Mining and the Environment 25-28 May 2003 Sudbury Ontario.
[7] Roy S, Adhikari G R and Gupta R N 2007 Use of gold mill tailings in making bricks: a feasibility study Waste Management and Research **25**(5) 475–482.
[8] Helinski M, Fahey M and Fourie A 2010 Coupled two-dimensional finite element modeling of mine backfilling with cemented tailings Canadian Geotechnical Journal **47**(11) 1187-1200.
[9] Zhang L 2012 Recycling and utilization of mine tailings as construction material through Geopolymerization *In* Proceedings U.S. EPA Hardrock Mining Conference 2012: Advancing Solutions for a New Legacy April 3-5 2012 Denver Colorado.
[10] Ugama T I, Ejeh S P and Amartey D Y Effect of tailings on the properties of concrete **6** (10) 2014 IISTE.
[11] Sabat V, Shaikh M, Kanap M, Chaudhari M, Suryawanshi S and Knadgouda K Use of iron ore tailings as a construction material **3**(2) Aug. 2015 International Journal of Conceptions on Mechanical and Civil Engineering.
[12] Mahmood Ali A 2012 Experimental and computational assessment of tailings binder matrices for construction purposes in cold regions Ph. D. thesis Concordia University Department of Building Civil and Environmental Engineering Montreal Quebec.
[13] Haile G, O'Callaghan W B, Hartney T F and Cameron W 2000. Experience with thickened red mud tailings (paste) stacking tropical and temperate climates and applicability of the technology to other tailings *In* Proceedings of the 53rd Canadian Geotechnical Conference Canadian Geotechnical Society 15-18 October 2000 Montreal Quebec.
[14] Das Braja M 2000 Fundamentals of geotechnical engineering. Brooks/Cole Publishing Company California.
[15] Conner J 1990. Chemical fixation and solidification of hazardous wastes Van Nostrand Reinhold New York.
[16] ASTM C 989 2006 Standard specification for ground granulated blast-furnace slag for use in concrete and mortars Annual Book of ASTM Standards American Society for Testing and Materials Philadelphia Pennsylvania Vol. 04.02.

[17] Lafarge North America 2007 12950 Worldgate Dr Ste. 500 Herndon Virginia 20170.
[18] NovaFrit International 2006 1200 Garnier Street Ville Ste-Catherine Quebec J0L 1E0.
[19] ASTM D 4842-90 1996 Standard test method for determining the resistance of solid wastes to freezing and thawing Annual Book of ASTM Standards, American Society for Testing and Materials, Philadelphia Pennsylvania. Vol. 11.04.
[20] Toutanji H, Delatte N, Aggoun S, Duval R and Danson A 2004 Effect of supplementary cementitious materials on the compressive strength and durability of short-term cured concrete Cement and Concrete Research **34**(2) 311–319.
[21] Matar M I 2005 Modeling of Montmorillonite clay-water interactions with particle subdivisions using three dimensional Discrete Element Method Ph.D. thesis North Dakota State University Fargo North Dakota.
[22] Khwaja M 1996. Discrete Element Method: micro-mechanical and large scale modeling M.Sc. thesis University of Massachusetts Lowell Massachusetts.
[23] Cundall P A and Strack O D L 1979 A discrete numerical model for granular assemblies Geotechnique **29**(1) 47-65.
[24] Pytel A and Singer F L 1987 Strength of materials Harper and Row Publishers Inc. New York.
[25] FHWA 2011 US Federal Highway Administration website info, http://www.fhwa.dot.gov 1200 New Jersey Ave SE Washington DC 20590.
[26] Al-Otaibi S 2008 Recycling steel mill scale as fine aggregate in cement mortars. European Journal of Scientific Research **24**(3) 332-338.
[27] Malhotra V M 1983 Strength and durability characteristics of concrete incorporating a pelletized blast-furnace slag ACI special publication SP-79 The use of fly ash, silica fume, slag and other mineral by-products in concrete Editor V M Malhotra 892 - 921 and 923—31 American Concrete Institute Detroit Michigan.

IOP Conf. Series: Journal of Physics: Conf. Series **1049** (2018) 012104 doi:10.1088/1742-6596/1049/1/012104

The computer program Tailings-DEM™ modeling the strength properties of Musselwhite tailings matrices

Ali A Mahmood[1] and Maria Elektorowicz[2]

[1]Associate, Ash System International, 930 Av. Van Dyck, Brossard, Quebec, J4W 2E6, Canada
[2]Professor, Dept. of Building Civil and Environmental Engineering, Concordia University, Rm: EV-6.213, 1455 De Maisonneuve Blvd. W., Montreal, Quebec, H3G 1M8, Canada

Email: alimahmood45@yahoo.ca, mariae@encs.concordia.ca

Abstract. Mine tailings are the waste materials from mining operations. They are traditionally stored in tailings ponds that have been the subject of several environmental catastrophes that resulted in the contamination of the environment and underground water resources. This paper presents the development of a numerical tool, called Tailings-DEM, for simulating the unconfined compressive strength of Musselwhite tailings, which are sandy materials. This tool is developed based on the Discrete Element Method and the programming language C++. It is shown here that Tailings-DEM™ can be utilized to model Musselwhite tailings matrices and sands with a particle size distribution close to Musselwhite tailings. Additionally, it can be used to model demolition waste, coal mine refuse and Chat.

1. Introduction

Mine tailings are the waste materials that remain after the extraction of the ore mineral. They are traditionally stored in tailings ponds located in the vicinity of the mine opening [1]. This method of storage has created several environmental catastrophes in the past and has prevented the use of a widely available resource.

In order to make use of mine tailings and reduce their bulky method of storage [2]proposed the thickened tailings disposal (TTD) procedure for the disposal of mine tailings. He showed that by the process of thickening the tailings to heavy slurry before disposal, it is possible to create a self-supporting deposit of tailings and to eliminate the traditionally used settling pond.

The thickened surface tailings disposal concept has been applied at various mining facilities for several decades, including the Kidd Creek Mine in the northern region of Ontario (Canada) and for managing the disposal of red muds produced by the alumina industry [3].

Success in the process and results of mine tailings hardening led researchers to further investigate the possibility of using these hardened tailings in the construction industry thereby utilizing a vastly available

Content from this work may be used under the terms of the Creative Commons Attribution 3.0 licence. Any further distribution of this work must maintain attribution to the author(s) and the title of the work, journal citation and DOI.
Published under licence by IOP Publishing Ltd

untapped resource and hindering, albeit to a certain extent, the harmful effects of unprocessed tailings on the living environment. This success supported by historical evidence that told of using tailings in the mine backfilling business further strengthened this new approach.

The earliest reported use of mine tailings as a construction material was its use as a mine backfill material. Use of mine tailings for backfilling started in the early years of the potash industry by using dry and wet residues [4]. Several technologies were developed in the last 80 years. One of these, the slurry backfill technology is considered a very effective backfill method, because of its combination of tailings and brine disposal. Since 1908, the slurry backfill technology has been successfully applied in the German potash industry [4]. Success in these early attempts led researchers to attempt a thorough investigation approach of mine backfilling using more advanced technologies and procedures.

Several research studies discussed the applicability of using mine tailings in several types of construction applications. Examples include the works of Fall et al. [5], Fall et al. [6], Benzaazoua et al. [7], Yang et al. [8], Roy at al. [9], Ercikdi et al. [10], Fall and Pokharel [11], Helinski et al. [12], Ercikdi et al. [13], Ugama et al. [14] and Liu et al. [15]. However, none had comprehensively addressed the issue of using mine tailings as construction material for temporary roads in cold climates. Simulation of the strength and weathering characteristics of these tailings using a powerful numerical technique was lacking in these studies, as well.

The main objective of this study is the development of a new sustainable approach to road construction using mining waste in cold climate through the development of a Discrete Element Method (DEM) program as a tool for simulating the strength characteristics of the newly developed Musselwhite tailings-binder matrices.

Investigations consist of two phases: experimental and computational. The Experimental Phase 1 is divided to 3 stages: characteristics of materials, formulation of tailing matrices and assessment of matrices usefulness as a read construction material. The Experimental Phase is thoroughly discussed in Mahmood [16]. Phase 2 consists of 3 stages: development of an adequate program predicting matrices usefulness for road material and verification of the program for strength and weathering tests. This paper discusses the verification of the strength tests for Musselwhite tailings matrices as verification for weathering tests was discussed elsewhere [16].

2. Binding materials

The binding materials for the mine tailings, which have been tested in this study, consist of Type I ordinary Portland cement (OPC), fly ash, slag and Calsifrit.

Unconfined compressive strength of fly ash lies in the range 1.38 to 6.895 MPa with permeability between 10^{-5} to 10^{-8} cm/s [17]. The fly ash was obtained from St. Laurence Cement Company in Longueil, Quebec.

Ground granulated blast furnace slag used in this study was obtained from Lafarge North America Cement Company in Montreal, Quebec under the trade name NewCem®. It is a light grey odorless powder produced in accordance with ASTM C 989 [18]. It is composed of (30-50) % Calcium Oxide, (0-20) % Magnesium Oxide with less than 1% Crystalline Silica. Its specific gravity is between 2-3, water pH 8-11. Its boiling point is above 1000 °C [19].

Calsifrit™ is a totally amorphous siliceous material, a matrix of calcium and sodium fluoro-aluminosilicate. This homogeneous solid substance has a blackish grey color, possesses a high reactivity potential and shows cementitious properties when finely ground. This product was obtained from the manufacturer NovaFrit International. Table 4 shows its chemical composition. Calsifrit has a pH (1% solution in water) of 6.5-7.5, a melting point of 800 °C, a specific gravity of 1.3 with a moisture content of less than 10% [20].

3. The tailings

IOP Conf. Series: Journal of Physics: Conf. Series **1049** (2018) 012104 doi:10.1088/1742-6596/1049/1/012104

Musselwhite tailings have a specific gravity of 3.26 and their physical properties are shown in Table 1. Atomic absorption analysis performed on the tailings revealed that the tailings contained heavy metals in the following proportions (mg/kg): Cu (144), Fe (87200), Cr (67), Zn (36), Pb (180).

Table 1. Physical properties of Musselwhite tailings

Type of Tailings	Musselwhite
D_{10} (mm)	0.0045
D_{50} (mm)	0.016
D_{60} (mm)	0.023
D_{30} (mm)	0.012
C_u	5
C_z	1.39
$P_{4.75mm}$ (%)	100
$P_{0.075mm}$ (%)	74.9
Initial moisture content (%)	30.15
USCS	SM

Where:
D_{10} = diameter corresponding to 10% finer,
D_{50} = diameter corresponding to 50% finer,
D_{60} = diameter corresponding to 60% finer,
D_{30} = diameter corresponding to 30% finer,
C_u = uniformity coefficient = D_{60} / D_{10},
C_z = coefficient of gradation = $D_{30}^2 / (D_{10} D_{60})$,
$P_{4.75mm}$ (%) = percentage passing sieve no.4,
$P_{0.075mm}$ (%) = percentage passing ASTM sieve # 200,
SW-SM = well graded sand with silt,
SM = silty sand,
SP = poorly graded sand,
USCS [21] = unified soil classification system.

4. Preparation of cylindrical specimens

Cylindrical specimens of tailings matrices with dimensions (44 diameter x 74 height) mm were molded for this purpose in accordance with ASTM D 4842 [22]. These cylindrical specimens were, then, left to cure in the moisture chamber for 28 days in the case of the wetting and drying samples, and for 43 days for the freezing and thawing. Unconfined compression tests were conducted on these tailings matrices after subjecting them to 12 cycles of drying at 60 °C and wetting by being submerged in distilled water at room temperature. The remaining samples were compressed after 12 cycles of freezing at -20 °C in a freezing cabinet and thawing in distilled water at room temperature (22 °C). Table 2 shows the codes used for the cylindrical specimens of the tailings matrices. Speed of descent of the platen was kept constant at 70 mm/min. Details of the wetting/drying and freezing/thawing weathering resistance tests are further explained in Mahmood [16].

Table 2. Cylindrical tailings matrices specimen codes.

Tailings Type	Weathering Type	Code	Specimen	Binder/ Tailings	OPC/ Binder (%)	Calsifrit/Binder (%)	Fly ash/Binder (%)	Slag/Binder (%)
Musselwhite	Wetting/ Drying	MW1	1, 2, 3, 4, 5, 6	0.375	100	0	0	0
		MW3	1, 2, 3, 4, 5, 6	0.375	90	10	0	0
		MW6	1, 2, 3, 4, 5, 6	0.375	75	25	0	0
		MWF1	2, 3, 4, 5, 6	0.375	75	0	25	0
		MWF3	1, 2, 3, 4, 5, 6	0.375	75	10	15	0
		MWF5	1, 2, 3, 4, 5, 6	0.375	75	20	5	0
		MWS1	1, 2, 3, 4, 5, 6	0.375	75	0	0	25
		MWS3	1, 2, 3, 4, 5, 6	0.375	75	10	0	15
		MWS5	1, 2, 3, 4, 5, 6	0.375	75	20	0	5
Musselwhite	Freezing/ Thawing	MW'1	1, 2, 3, 4, 5, 6	0.375	100	0	0	0
		MW'3	1, 2, 3, 4, 5, 6	0.375	90	10	0	0
		MW'6	1, 2, 3, 4, 5, 6	0.375	75	25	0	0
		MWF'1	2, 3, 4, 5, 6	0.375	75	0	25	0
		MWF'3	1, 2, 3, 4, 5, 6	0.375	75	10	15	0
		MWF'5	1, 2, 3, 4, 5, 6	0.375	75	20	5	0
		MWS'1	1, 2, 3, 4, 5, 6	0.375	75	0	0	25
		MWS'3	1, 2, 3, 4, 5, 6	0.375	75	10	0	15
		MWS'5	1, 2, 3, 4, 5, 6	0.375	75	20	0	5

IOP Conf. Series: Journal of Physics: Conf. Series **1049** (2018) 012104 doi:10.1088/1742-6596/1049/1/012104

5. Execution and verification of Tailings-DEM™ modeling of UCS

In order to operate, the program Tailings-DEM™ uses a parameter called the matrix mass coefficient. It is computed through the following steps:

1) The unconfined compressive stress is found from the statistical predictive equation derived previously in Mahmood [16]. The UCS was found by entering the following matrix components (in percentage): Calsifrit/binder (A), slag/binder (C) and binder/tailings (E) into Equation (1) as follows:

$$UCS = -6.856 + 0.14\ A + 0.092\ C + 0.6198\ E \qquad (1)$$

From the UCS above, it is possible to find the maximum compressive force (in N) for each of the matrix mixtures used in this equation.

2) Subsequently, the maximum load (in N) found experimentally [16] for the same mixtures used in step (1) above, is used to find a correlation with a factor defined as the matrix mass coefficient. This correlation is found by initially taking three values of the experimental maximum load and finding (by trial and error) their representative matrix mass coefficients (in kg) that gives the closest match with the computational values. The three values represent the maximum and minimum data boundary shown in Figure 1 and an arbitrary middle point. Subsequently other values for the matrix mass coefficient are introduced into the program to find the closest match with the remaining experimental values. This procedure generated the following linear Equation (2) shown in Figure 1 that relates the experimental maximum load and the matrix mass coefficient:

$$y = 2\,(10)^9 x - 276.9 \qquad (2)$$

Where: y is the experimental maximum load (in N) and x is the matrix mass coefficient (in kg).

The matrix mass coefficient is the major input used in the computer program to find the computational values including the computational maximum force for each tailings matrix mixture.

IOP Conf. Series: Journal of Physics: Conf. Series **1049** (2018) 012104 doi:10.1088/1742-6596/1049/1/012104

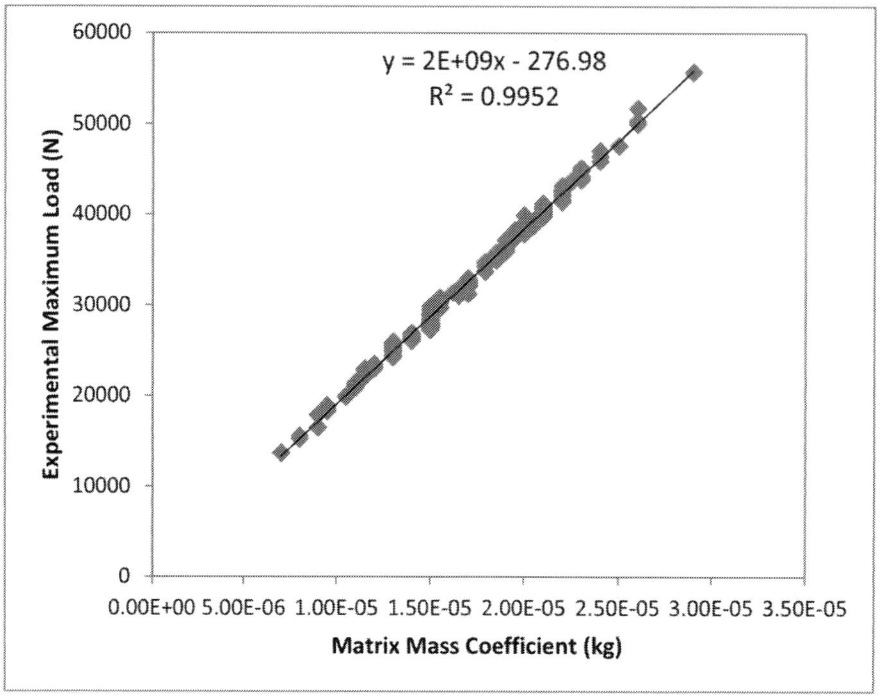

Figure 1. Correlation between the experimental maximum load and the matrix mass coefficient

3) The computational maximum force values were produced in step (2) as a byproduct of using the matrix mass coefficient. Hence, the same matrix mass coefficient now represents these new computational maximum force values and their corresponding experimental counterparts. Using regression techniques produced Equation (3) shown in Figure 2, that relates the newly found computational maximum forces and their respective matrix mass coefficients:

$$y = 2(10^9)\,x + 39.27 \qquad (3)$$

Where: y is the computational maximum load (in N) and x is the matrix mass coefficient (in kg).

4) Equation (3) is used to find the matrix mass coefficient when the statistical UCS value is found from Equation (1). This matrix mass coefficient is the major input used in the computer program to get the corresponding computational maximum force value.

5) Computational maximum force values found in step (4) are then compared with their experimental counterparts.

Table 3 shows the values generated from the procedure explained in steps 1 to 5.

Figure 2 shows the correlation between values of the experimental and computational maximum loads [16] found using the procedure explained above.

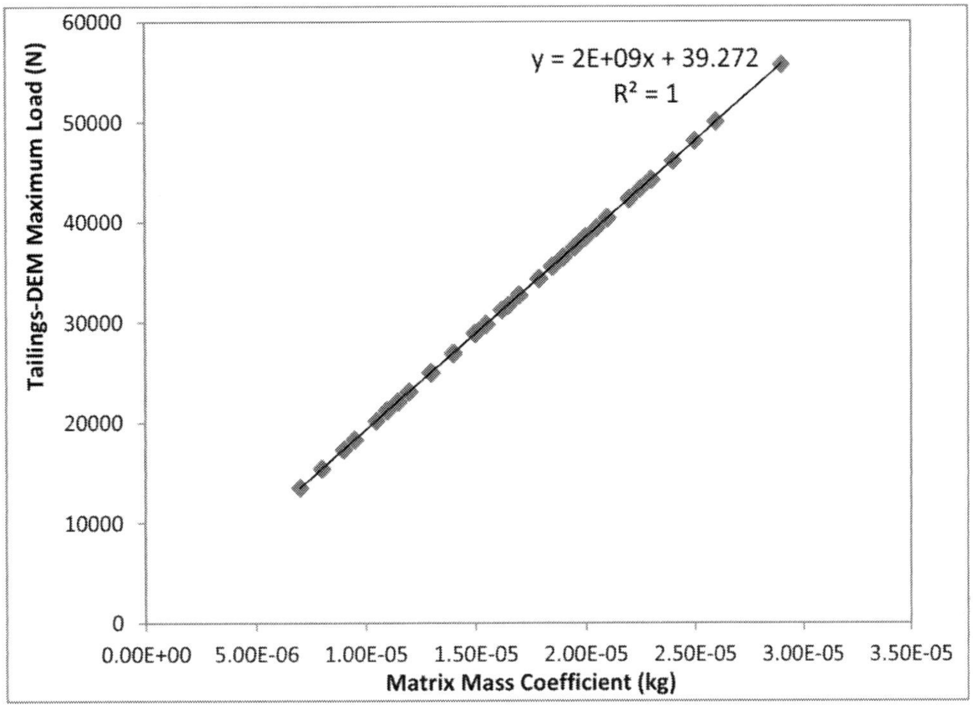

Figure 2. Correlation between the computational maximum load and the matrix mass coefficient

Subsequently, the user is prompted to enter the time increment in seconds. This time increment is part of the original stress and strain equations [16]. Equation (4) checks for its compatibility with the value $\sqrt{m/k}$. If the time increment is more than or equals the latter value, the user is asked to re-enter the time increment until compatibility is assured. Figure 3 shows the correlation between the experimental and numerical maximum loads.

$$\Delta t < 2\,x\,\frac{\sqrt{m}}{k} \qquad (4)$$

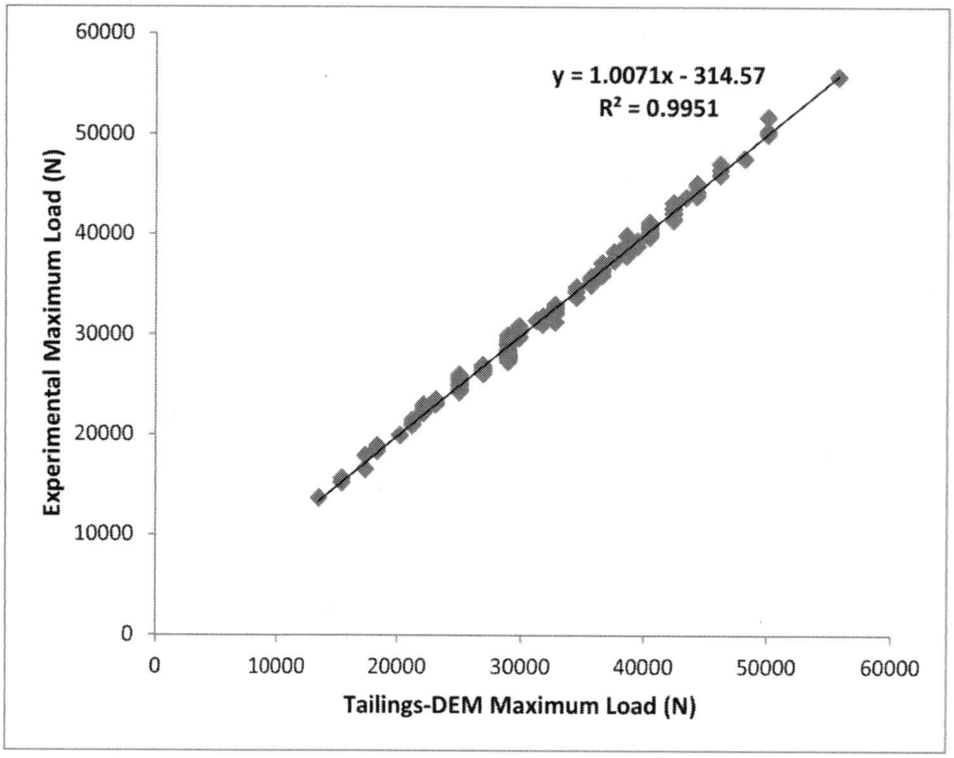

Figure 3. Correlation between the experimental maximum load vs. the Tailings-DEM maximum load

Subsequent to the input operations, triangular particles, that represent the divisions, are arranged in a grid like fashion and dispersed uniformly throughout a rectangular area that represents the cylinder under compression.

Table 3. Numerical procedure used in calculating the matrix mass coefficient

Stat. Max. Eq. load (N)	Weight (kg) from Eq. (5.15)	Max. Load from Program Run (N)	Difference (%)	Exp. Max Load (N)	Difference (%)
36696.49	1.84E-05	35200	4.08	37146.57	5.24
38825.24	1.94E-05	37300	3.93	39001.62	4.36
42018.35	2.1E-05	40400	3.85	38378.2	-5.27
36696.49	1.83E-05	35200	4.08	34926.59	-0.78
38825.24	1.94E-05	37300	3.93	39868.32	6.44
40953.98	2.05E-05	40400	1.35	39275.31	-2.86
40193.71	2.01E-05	38500	4.21	43517.59	11.53
40923.57	2.04E-05	38500	5.92	42422.81	9.25
41653.42	2.08E-05	40400	3.01	44810.04	9.84
24916.18	1.24E-05	23900	4.08	20527.17	-16.43
27044.92	1.35E-05	26000	3.86	24951.91	-4.20

30238.04	1.51E-05	29000	4.09	29528.71	1.79
24916.18	1.24E-05	23900	4.08	25712.18	7.05
27044.92	1.35E-05	26000	3.86	29346.25	11.40
29173.67	1.46E-05	28100	3.68	33132.37	15.19
28413.4	1.42E-05	27300	3.92	30091.31	9.28
29143.25	1.46E-05	28100	3.58	30547.47	8.012
29873.11	1.49E-05	28900	3.26	26609.29	-8.61
36696.49	1.83E-05	35200	4.08	35747.68	1.53
38825.24	1.94E-05	37300	3.93	40005.17	6.76
42018.35	2.1E-05	40400	3.85	41206.39	1.96
40953.98	2.05E-05	40400	1.35	42574.86	5.11
40193.71	2.01E-05	38500	4.21	39077.64	1.48
40923.57	2.04E-05	38500	5.92	42316.37	9.02
41653.42	2.08E-05	40400	3.01	37830.81	-6.79
24916.18	1.24E-05	23900	4.08	25195.2	5.14
27044.92	1.35E-05	26000	3.86	31839.92	18.34
30238.04	1.51E-05	29000	4.09	35473.99	18.25
24916.18	1.24E-05	23900	4.08	27035.04	11.6
27044.92	1.35E-05	26000	3.86	25408.07	-2.33
29173.67	1.46E-05	28100	3.68	25316.84	-10.99
28413.4	1.42E-05	27300	3.92	24860.68	-9.81
29143.25	1.46E-05	28100	3.58	25453.69	-10.4
29873.11	1.49E-05	28900	3.26	30577.88	5.49

After the particles arrangement, the program loading loop starts with increments of 100 N. It is thought that this increment furnishes sufficient data points for graph plotting and data manipulation.

At the start of loading, the rectangular area gets divided into 16 sectors equal in area to facilitate contact detection among the particles. A counter for each sector stores the number of particles dispersed in it. Then execution moves to the first sector to check for contact detection among particles. The first sector particle is taken and checked for contact with other particles. Once contact is found, its contact meter is increased by one up to a maximum of 4 contact points with 4 other particles. It is thought that the limiting geometry of these triangles does not permit interaction with more than 4 particles without encountering serious computational errors. Equation (5) shows the numerical condition that checks for contact among neighboring particles. Initial trials have shown that this is the minimum distance possible to achieve reasonable results.

$$(x_1 - x_i)^2 + (y_1 - y_i)^2 \leq 50 \text{ mm} \qquad (5)$$

Where:

x_1, y_1 are the Cartesian coordinates of the particle under consideration, and

x_i, y_i are the coordinates of the particle to be checked for contact proximity

This contact is the mechanism that activates stresses and strains [16]. After checking the first particle for contact, the sector particle counter is increased and the procedure repeated for the remaining sector particles.

Then execution moves from the first sector to the second sector and the whole contact procedure is repeated again until all sectors are checked. After that, execution moves to the loading stage again and another increment of 100 N is applied. This continues until a displacement of 72 mm in the vertical direction is reached. This displacement is scaled by 10 to account for the differences in the experimental and computational modeling and to account for the 2-dimensional approximation. When this displacement is reached, the specimen failure is assumed to take place and the program stops. As such, the program is capable of finding x, y stresses and strains for its member particles.

6. Computational and experimental results

It is necessary to verify the accuracy and reliability of the proposed computational model using laboratory-approved experimental results. Hence the UCS results of Mahmood [16] were used in the verification and analysis of the computational data found using this computational program.

It is known that solidified waste in the field undergoes continuous weathering cycles. Therefore, strength testing will not be complete without investigating the effect of weathering cycles on the compression UCS strength.

After undergoing the weathering tests, the matrices specimens of tailings were tested under uniform conditions to find their strength in the UCS apparatus. Figures 4 to 8 show the UCS experimental and computational results comparison for Musselwhite tailings matrices for the matrix binder combinations shown above. The experimental specimens had been subjected to freezing/thawing weathering cycles before undergoing the UCS compression testing. The following shows these matrix binder combinations in groups of figures:

1) Figure 4 shows the 100% OPC,

2) Figure 5 shows the 90% OPC, 10% Calsifrit,

3) Figure 6 shows the 75% OPC with 25% Calsifrit,

4) Figure 7 shows the 75% OPC with 25% fly ash,

IOP Conf. Series: Journal of Physics: Conf. Series **1049** (2018) 012104 doi:10.1088/1742-6596/1049/1/012104

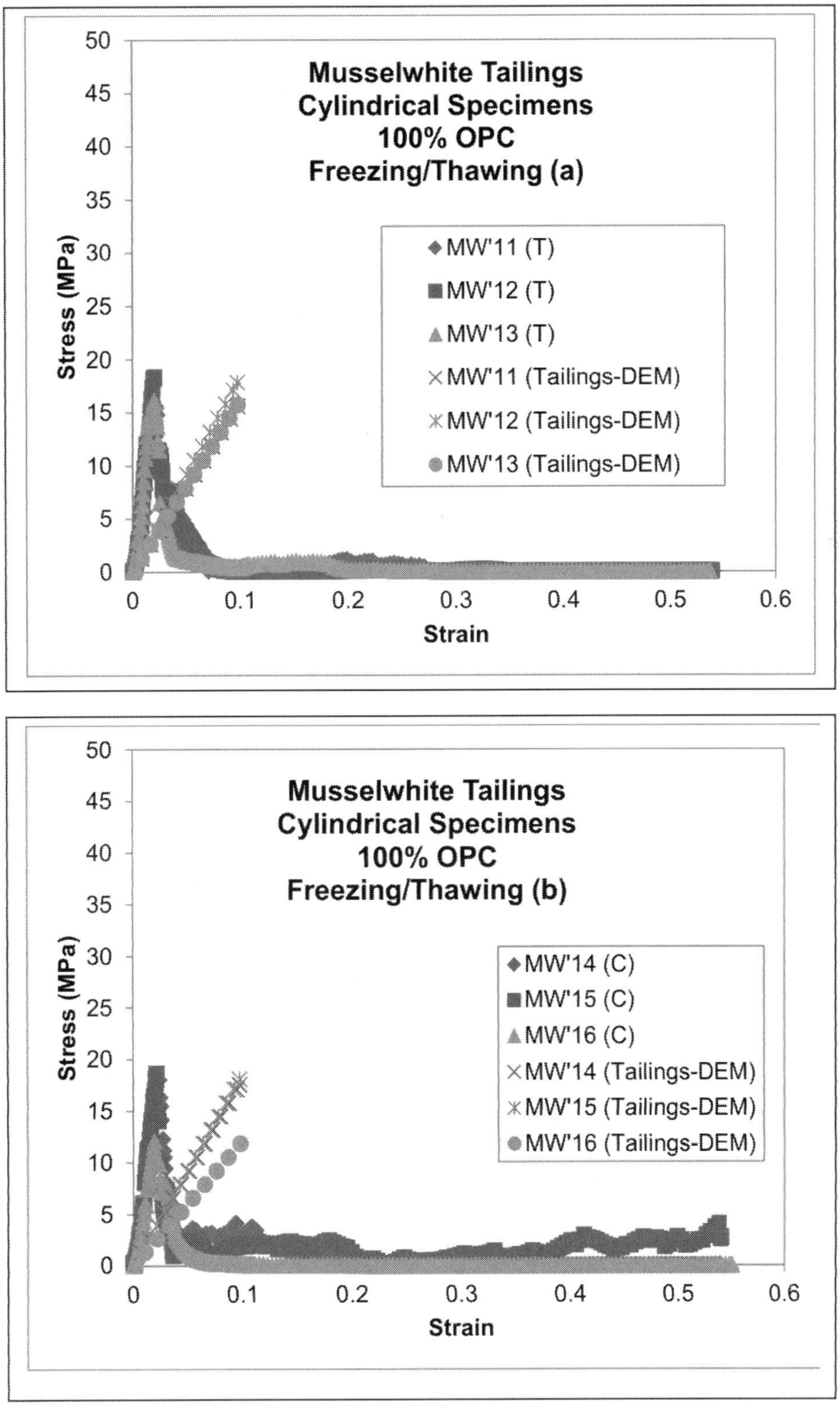

Figure 4. Computational and experimental UCS values for Musselwhite samples MW'11(a)-MW'16(b) after freezing/thawing

IOP Conf. Series: Journal of Physics: Conf. Series **1049** (2018) 012104 doi:10.1088/1742-6596/1049/1/012104

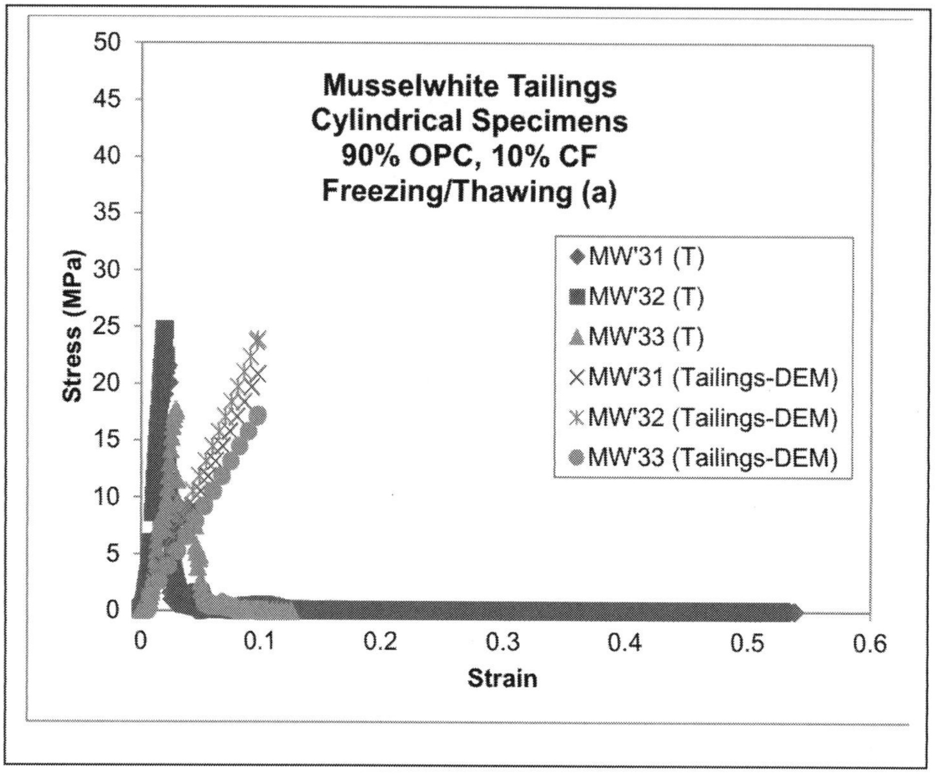

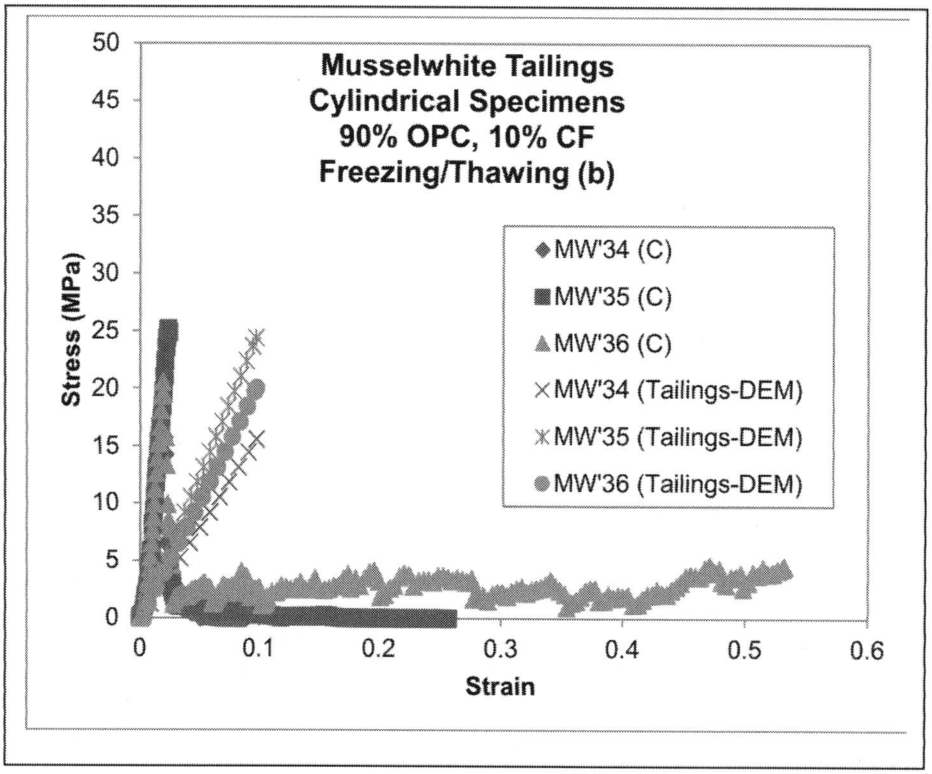

Figure 5. Computational and experimental UCS values for Musselwhite samples MW'31(a)-MW'36(b) after freezing/thawing

IOP Conf. Series: Journal of Physics: Conf. Series **1049** (2018) 012104 doi:10.1088/1742-6596/1049/1/012104

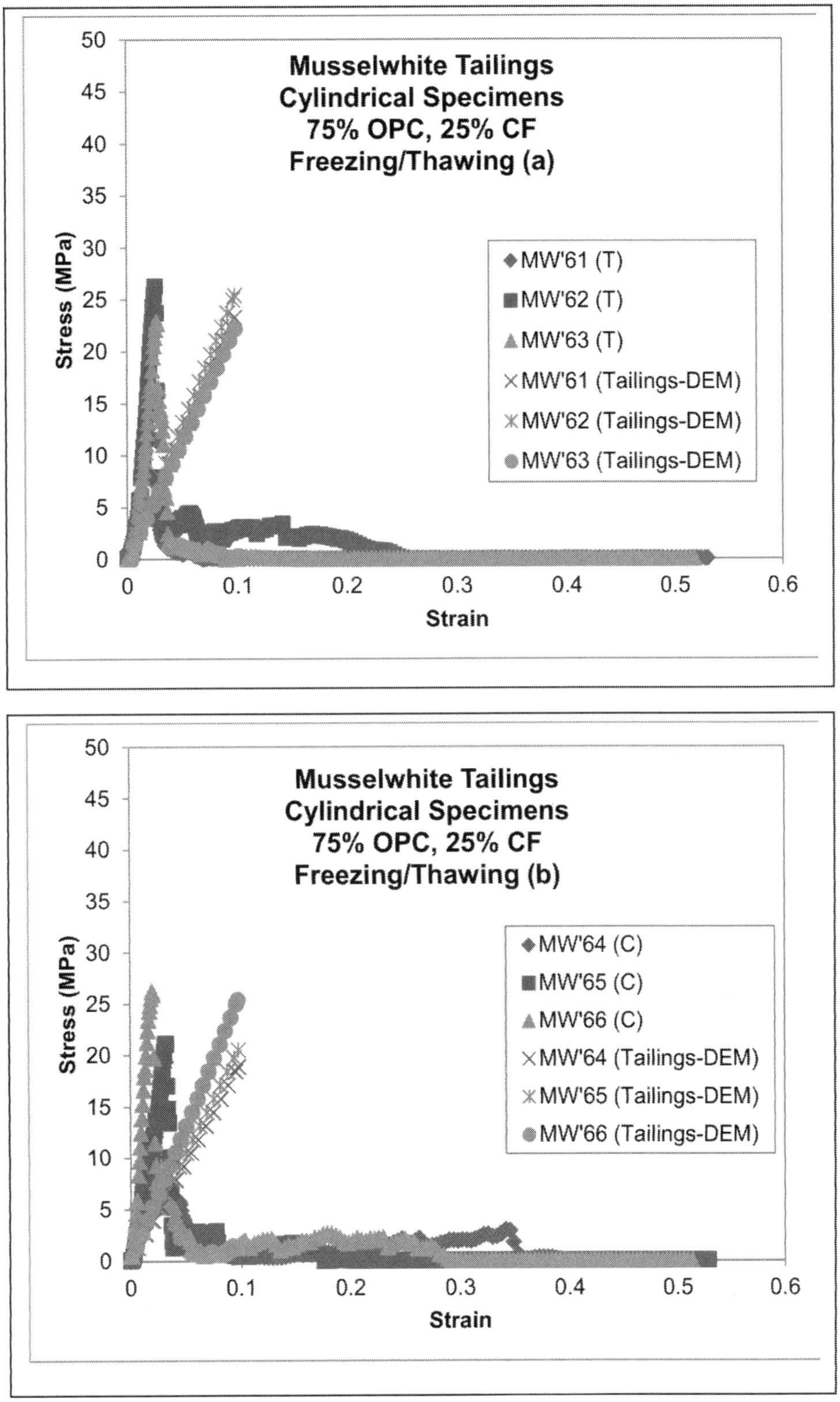

Figure 6. Computational and experimental UCS values for Musselwhite samples MW'61(a)-MW'66(b) after freezing/thawing

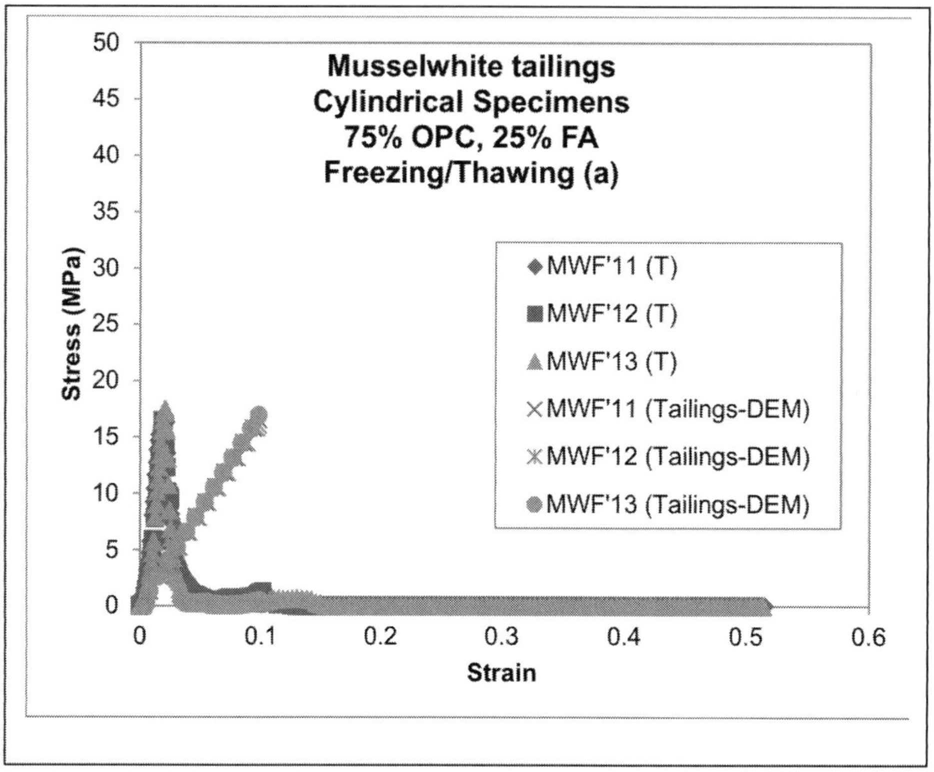

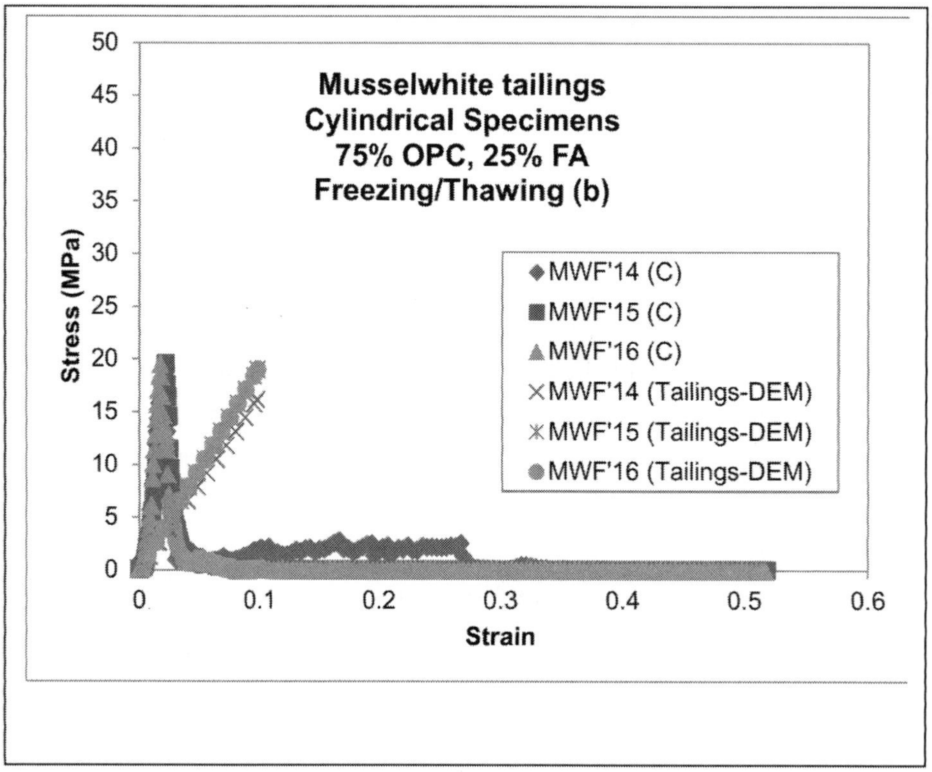

Figure 7. Computational and experimental UCS values for Musselwhite samples MWF'11(a)-MWF'16(b) after freezing/thawing

IOP Conf. Series: Journal of Physics: Conf. Series **1049** (2018) 012104 doi:10.1088/1742-6596/1049/1/012104

7. Discussion of Tailings-DEM™ strength modeling

Experimental and computational UCS values for the tailings matrices have similar types of curves indicating that computational modeling of this test was to a large extent accurate. Marginal differences between experimental and computational values could be attributed to a few factors.

The main difference between the computational and the experimental programs is the approximation to 2D modeling when designing this program. Also, binder constituents were largely modeled through the UCS in Equation (1), which was derived statistically. As explained in Mahmood [16], this equation found, out of the four binders used, the ratio of Calsifrit and slag to be most effective when calculating the UCS. The effect of cement could not be associated directly with the UCS. Its effect, however, was evident in conjunction with the other combinations, specifically, the binder/tailings ratio. Also, the experimental tests used a constant platen speed that was not modeled in the computational program.

The computer program is capable of modeling other types of wastes and binders. The matrix mass coefficient shown above can be used with other types of wastes. By first finding the experimental UCS for the waste matrix in question, the matrix mass coefficient can be incorporated into the program and used as detailed above.

Generally, the computational data followed experimental results closely for these tailings matrices for the cases shown. Maximum values and slopes for the experimental and computational results followed closely. However, the freezing /thawing specimens have higher slopes for the same tailings matrices indicating the move from ductile to more brittle behavior.

The marginal differences between the computational and experimental UCS results could be attributed to the experimental errors in preparing the specimens tested. It is known that soil cement matrices suffer from result variability as explained in Mahmood [16]. Furthermore, the presence of the metals; Fe, Zn and Pb in Musselwhite matrices have further complicated the outcome of the UCS experiments.

As water and cement are added to Musselwhite tailings, zinc hydroxide anions form and compete effectively with other anions for available adsorption sites to form a low permeability membrane coating the cement particles [23]. Several researchers reported that Zn retarded setting through formation of this low permeability membrane of calcium hydroxyzincate ($CaZn_2(OH)_6.H_2O$) around the cement particles [23], [24], [25],[26], and Pb acted as a retardant agent by inhibiting the cement hydration [27]. The formation of a membrane around cement particles with the precipitation of calcium hydroxyzincate ($CaZn_2(OH)_6.H_2O$) can prevent water and ion transport needed for cement hydration. The retardation or suppression of cement hydration adversely affects strength development. In the presence of lead, hydration occurs at a much slower rate than the reference [27], and the UCS values tend to decrease. Coating of tri-calcium silicate $3CaO.SiO_2$ (C_3S) of cement with Pb or complexation of Pb [25], [28], [29] might have retarded the hydration of cement.

The initial water content of the samples might have also been a contributing factor to the decrease in the UCS values. The initial water content of 30.15% for Musselwhite plus the water/cement ratio of 3/10 might have been excessive for cement hydration especially as the Calsifrit ratio of the samples increased. Visual observation during the experiments revealed that increasing the amount of Calsifrit increased the fluidity of these mixes. Conner [17]reported that as the water/cement ratio increased, the percentage of larger pores and therefore the permeability of the product increased, as well. These experimental results were used in the multiple regression analysis that led to the creation of predictive Equation (1).

Another reason could be the assumptions used in defining the program Tailings-DEM™. The assumptions were based on assuming the particles being right-isosceles triangles. This assumption had affected the other assumption of contact mechanism as was above. In reality, however, soil particles have varying shapes and smaller sizes and their contact mechanism is much more complicated and varied. In addition, most soils contain different amounts of humidity, an affect that was not modeled in Tailings-DEM™.

8. Tailings-DEM™ applications

Since Musselwhite tailings were predominantly sandy, Tailings-DEM™ can be incorporated to model other types of aggregates, such as sands and gravely sands. This program can be utilized to model sands with a particle size distribution close to Musselwhite tailings.

Tailings-DEM™ can be applied to the following types of waste materials:
1) Demolition waste,
2) Coal mine refuse,
3) Chat.

8.1 Demolition waste

The Ministry of Industry in Canada defines the construction and demolition waste as "waste materials from the construction and demolition of roads, bridges and buildings such as wood gypsum and metal", [30]. Hence most tailings components are common with demolition waste.

Construction and demolition debris are one of the most abundant types of man-made wastes found in Canada. For example in Alberta, construction and demolition waste accounts for about 25% of the total amount of municipal solid waste sent to landfills [31].

Compressive strengths for construction and demolition debris cubes at the age of 28 days range from 11.05 to 28.25 MPa [32]. This makes them structurally close, in terms of strength, to the Musselwhite tailings modeled in this study.

The matrix mass coefficient for the construction and demolition debris samples can be obtained from these studies and used in this computer program along with the stiffness coefficient. The same study produced experimental values of the modulus of elasticity [32]. These values can be substituted for the stiffness coefficient as the first input to the program.

8.2 Coal mine refuse

Coal refuse is a by-product of the coal mining industry. Coal refuse is, generally, defined by a minimum ash content combined with a maximum heating value measured on a dry basis [33]. This material consists primarily of non-combustible rock with some attached carbon material that cannot be effectively separated [33].

Coal continues to be one of the primary sources of energy in several countries including the United States where mining of coal results in the production of large quantities of coal refuse. It is estimated that over 500 million tons of coal mining and preparations refuse are produced annually in the United States [34]. Cementation and stabilization of coal mine refuse for proper re-use in construction applications removes most of the harmful environmental effects resulting from land filling.

Kumar et al. [34] report that several studies have been conducted with the purpose of investigating the use of stabilized and un-stabilized mixtures of coal mine refuse and fly ash, considered as the two most important by-products of the coal industry, for construction of highway embankments and base courses [35], [36], [37], [38], [39], [40].

The matrix mass coefficient can be produced from the compressive strength of coal refuse mixtures. Crandall [41] reports that Knissel and Helms [42] provided values of uniaxial compressive strength of coal mine washery refuse samples of up to 4 MPa. These values were the result of 28 days curing with a ratio of 10% cement [41].

The matrix mass coefficient defined in this study can be successfully used with cement stabilized coal mine refuse, after experimentally finding the proper UCS values and stiffness coefficients as detailed above.

8.3 Chat

Mine tailing piles (chat) are comprised mainly of angular chert fragments and contain residual amounts of lead sulfide (galena), zinc sulfide (sphalerite) minerals and their weathering products [43]. The difference between chat and tailings material is that chat is a fine to coarse dolomite produced during the early milling process, which used density separation [44]. Chat is transported mechanically by conveyor and deposited in large piles. Tailings are produced in later years using a wet chemical process and have typically smaller fragments than chat [44]. The tailings are hydraulically deposited into impoundments known as tailings ponds [45].

Chat has been used for the stabilization of roadway bases with additives such as fly ash and cement kiln dust for stabilization and solidification purposes [46].

An experimental study by Wasiuddin et al. [46] to optimize the use of chat in hot mix asphalt for pavement application showed great potential in the use of chat for both surface and base course layers.

Teredesai et al. [47] reported UCS values of up to 5.2 MPa for cylindrical specimens of chat stabilized with class C fly ash and cement kiln dust and cured for 28 days.

The matrix mass coefficient can be used in conjunction with chat mechanical properties to optimize the use of chat with other binders and determine its suitability for construction purposes. Higher binder content may need to be used with angular chat fragments.

9. Conclusions

- A novel numerical tool was developed to model the strength characteristics of tailings matrices. The tool was designed using the compiler Code Blocks and the computer programming language C++,

- The DEM tool developed (Tailings-DEM™) is capable of simulating, with great accuracy, the UCS strength values of the Musselwhite tailings matrices tested,

- Since Musselwhite tailings were predominantly sandy, the program can be incorporated to model other types of aggregates, such as sands and gravely sands. Tailings-DEM™ can be utilized to model sands with a particle size distribution close to Musselwhite tailings. Additionally, it can be used to model demolition waste, coal mine refuse and Chat.

Acknowledgment
The authors would like to acknowledge with gratitude the support that Dr. Michelle Nokken and Dr. Catherine Mulligan of Concordia University (Montreal, Canada) have provided for this study.

10. References

[1] Simieritsch T, Obad J and Dyer S 2009 Tailings plan review- an assessment of oil sands company submissions for compliance with ERCB Directive 074: tailings performance criteria and requirements for oil sands mining schemes Pembina Institute Box 7558 Drayton Valley Alberta T7A 1S7.

[2] Robinsky E 1978 Tailings disposal by the thickened discharge method for improved economy and environmental control Tailings Disposal Today **2** pp. 75-92 1979 by Miller Freeman Publications Inc. San Francisco California *In* Proceedings of the 2nd International Tailings Symposium May 1978 Denver Colorado.

[3] Robinsky E I 1999 Tailing dam failures need not be disasters-The Thickened Tailings Disposal (TTD) system CIM Magazine **92**(1028) 140-142.

[4] Wasayo S 2003 Suitability of mine tailings for shotcrete as a ground support. M.A.Sc. thesis Mining Engineering Dalhousie University Halifax Nova Scotia.

[5] Fall M, Benzaazoua M and Ouellet, S 2005 Experimental characterization of the influence of tailings fineness and density on the quality of cemented paste backfill Minerals Engineering **18**(1) 41–44.
[6] Fall M, Benzaazoua M and Saa E G 2008 Mix proportioning of underground cemented tailings backfill Tunnelling and Underground Space Technology **23**(1) 80–90.
[7] Benzaazoua M, Bussiere B, Demers I, Aubertin M, Fried E and Blier A 2008 Integrated mine tailings management by combining environmental desulphurization and cemented paste backfill: Application to mine Doyon Quebec Canada Minerals Engineering **21**(4) 330–340.
[8] Yang H, Chen C, Pan L, Lu H, Sun H and Hu X 2009 Preparation of double-layer glass-ceramic/ceramic tile from bauxite tailings and red mud Journal of the European Ceramic Society **29**(10) 1887–1894.
[9] Roy S, Adhikari G R and Gupta R N 2007 Use of gold mill tailings in making bricks: a feasibility study Waste Management and Research **25**(5) 475–482.
[10] Ercikdi B, Kesimal A, Cihangir F, Deveci H and Alp I 2009 Cemented paste backfill of sulphide-rich tailings: importance of binder type and dosage Cement and Concrete Composites **31**(4) 268–274.
[11] Fall M and Pokharel M 2010 Coupled effects of sulphate and temperature on the strength development of cemented tailings backfills: Portland cement-paste backfill. Cement and Concrete Composites **32**(10) 819–828.
[12] Helinski M, Fahey M and Fourie A 2010 Coupled two-dimensional finite element modeling of mine backfilling with cemented tailings Canadian Geotechnical Journal **47**(11) 1187-1200.
[13] Ercikdi B, Cihangir F, Kesimal A, Deveci H and Alp I 2010 Effect of natural pozzolans as mineral admixture on the performance of cemented-paste backfill of sulphide-rich tailings Waste Management and Research **28**(5) 430–435.
[14] Ugama T I, Ejeh S P and Amartey D Y 2014 Effect of tailings on the properties of concrete **6** (10) IISTE.
[15] Liu T, Li X, Guan L, Liu P, Wu T, Li Z and Lu A 2016 Low-cost and environment-friendly ceramic foams made from lead–zinc mine tailings and red mud: Foaming mechanism, physical, mechanical and chemical properties Ceramics International **42**(1B).
[16] Mahmood Ali A 2012 Experimental and computational assessment of tailings binder matrices for construction purposes in cold regions. Ph. D. thesis Concordia University Department of Building Civil and Environmental Engineering Montreal Quebec.
[17] Conner J 1990 Chemical fixation and solidification of hazardous wastes Van Nostrand Reinhold New York.
[18] ASTM C 989 2006 Standard specification for ground granulated blast-furnace slag for use in concrete and mortars Annual Book of ASTM Standards American Society for Testing and Materials Philadelphia Pennsylvania Vol. 04.02.
[19] Lafarge North America 2007 12950 Worldgate Dr. Ste. 500 Herndon Virginia 20170.
[20] NovaFrit International 2006 1200 Garnier Street Ville Ste-Catherine Quebec J0L 1E0.
[21] Das Braja M 2000 Fundamentals of geotechnical engineering. Brooks/Cole Publishing Company California.
[22] ASTM D 4842-90 1996 Standard test method for determining the resistance of solid wastes to freezing and thawing Annual Book of ASTM Standards, American Society for Testing and Materials Philadelphia Pennsylvania Vol. 11.04.
[23] Asavapisit S, Fowler G and Cheeseman C R 1997 Solution chemistry during cement hydration in the presence of metal hydroxide wastes Cement and Concrete Research **27**(8) 1249–1260.
[24] Li X D, Poon C S, Sun H, Lo I M C and Kirk D W 2001. Heavy metal speciation and leaching behaviors in cement based solidified/stabilized waste materials Journal of Hazardous Materials **82**(3) 215–230.

[25] Gervais C and Ouki SK 2002 Performance study of cementitious systems containing zeolite and silica fume: effects of 4 metal nitrates on the setting time, strength and leaching characteristics Journal of Hazardous Materials **93**(2) 187-200.
[26] Fernández Olmo I, Somarriba P and Irabien A 2003 Mechanical and environmental properties of cement/foundry waste systems *In* Proceedings of WASCON International Conference on the Environmental and Technical Implication of Construction with Alternative Materials 4-6 June 2003 San Sebastian Spain.
[27] Salihoglu G, Pinarli V, Salihoglu N K and Karaca G 2007 Properties of steel foundry electric arc furnace dust solidified/stabilized with Portland cement Journal of Environmental Management **85**(1) 190-197.
[28] Thevenin G and Pera J 1999 Interactions between lead and different binders Cement Concrete Research **29**(10) 1605–1610.
[29] Palomo A and Palacios M 2003 Alkali-activated cementititous materials: alternative matrices for the immobilization of hazardous wastes part II. Stabilization of chromium and lead Cement Concrete Research **33**(2) 289–295.
[30] Saotome T 2007 Development of construction and demolition waste recycling in Ontario School of Engineering Practice SEP 704 McMaster University Ontario.
[31] Alberta Environment 2011 Construction and demolition waste reduction program Technical Info Alberta Environment Pollution Prevention and Conservation www.environment.alberta.ca/info
[32] Kumutha R and Vijai K 2010 Strength of concrete incorporating aggregates recycled from demolition ARPN Journal of Engineering and Applied Sciences **5**(5) 64-71.
[33] EPA 2008 Coal refuse Materials Characterization Paper in Support of the Advanced Notice of Proposed Rulemaking –Identification of Nonhazardous Materials That Are Solid Waste. US Environmental Protection Agency Ariel Rios Building 1200 Pennsylvania Ave. NW Washington DC 20004.
[34] Kumar S, Puri V K, Das Braja M and Devkota B C 2001 Geotechnical properties of fly ash and lime- fly ash stabilized coal mine refuse Electronic Journal of Geotechnical Engineering **6**.
[35] Butler P E 1974 Utilization of coal mine refuse in the construction of highway embankments *In* Proceedings of the First Symposium on Mine and Preparation Plant Refuse Disposal Louisville Kentucky pp. 237 –255.
[36] Moulton L K, Anderson D A, Seals R K and Hussian S M 1974 Coal refuse: an engineering material *In* Proceedings of the First Symposium on Mine and Preparation Plant Refuse Disposal National Coal Association Louisville Kentucky pp. 1-25.
[37] Wilmoth R C and Scott R B 1974 Use of coal mine refuse and fly ash as a road base material *In* Proceedings of the First Symposium on Mine and Preparation Plant Refuse Disposal October 1974 Louisville Kentucky pp. 263 – 275.
[38] Drake W B 1976. Coal refuse in highway embankment and aggregates *In* Proceedings of the 2nd Kentucky Coal Refuse Disposal and Utilization Seminar University of Kentucky Lexington Kentucky pp. 17–19.
[39] Head W J, McQuade P V and Anderson R B 1982 Coal refuse and fly ash compositions: potential highway base course materials Transportation Research Board Transportation Research Record 839: 11 – 19.
[40] EPRI 1989 Ash utilization in highway: Pennsylvania demonstration project Electric Power Research Institute EPRI Report No. GS – 6431 June 1989 Palo Alto California.
[41] Crandall Wallace E 1992 SME Mining Engineering handbook Volume 2 Chapter 19.3 The Society for Mining Metallurgy and Exploration Englewood Colorado.
[42] Knissel W and Helms W 1983 Strength of cemented rockfill from washery refuse results from laboratory investigations *In* Proceedings of the International Symposium on Mining with Backfill 7-9 June 1983 Lulea Sweden pp. 31-37.

[43] ODEQ 2003 Summary report of washed and unwashed mine tailings (chat) from two piles at the Tar Creek superfund site Ottawa County Oklahoma Technical Report Oklahoma Department of Environmental Quality 707 N. Robinson Oklahoma City Oklahoma 73102.

[44] US Department of Health and Human Services 1998 Health consultation: A note of explanation. Technical Report US Department of Health and Human Services 200 Independence Avenue S.W. - Washington D.C. 20201.

[45] EPA 1994. Design and evaluation of tailings dams Technical Report EPA 530-R-94-038 US Environmental Protection Agency Ariel Rios Building 1200 Pennsylvania Ave. NW Washington DC 20004.

[46] Wasiuddin N M, Hamid W B, Kolothody N, Gause C F, Zaman M M and Nairn R R 2005 A laboratory study to optimize the use of raw chat in hot mix asphalt for pavement application: final report Technical Report Oklahoma Department of Environmental Quality 707 N Robinson Oklahoma City Oklahoma 73102.

[47] Teredesai R, Musharraf Z, Miller G A and Nairn R 2005 Cementitious stabilization of raw chat for roadway base application Technical Report Oklahoma Department of Environmental Quality Oklahoma City Oklahoma.

A Benchmarking of the Integrated Train Driver Performance Model

Jalil Azlis-Sani[1], Siti Zawiah Md Dawal[2], Norhayati Mohmad Zakwan[3]

[1]Faculty of Mechanical and Manufacturing Engineering, Universiti Tun Hussein Onn Malaysia, 86400 Batu Pahat, Johor, Malaysia
[2]Department of Mechanical Engineering, Faculty of Engineering, University of Malaya. 50603 Kuala Lumpur, Malaysia
[3]Department of Business Administration, Faculty of Management, Universiti Teknologi Malaysia, UTM Skudai, 81310, Johor, Malaysia

E-mail: azlis@uthm.edu.my

Abstract. Human factors / ergonomics (HFE) focuses on the interaction of human with the environment. The traditionally concept of HFE has always focused on the interrelationship between three main elements of human-machine-environment. Therefore, significant to this research was that an integrated Malaysian train driver performance model has been successfully developed for identifying the significant factors that influence train driver performance, highlighting the interaction between human factors, human activities and its context. There are a number of models and frameworks in existence, designed to fulfil the various contexts of particular research. This newly developed model was compared with past models from the previous researches. Through this benchmarking approach, this study has looked at the significant factors in holistic and comprehensive perspective without ignoring other potential domains of factors. In conclusion, the previous referred models mostly from the UK and Australia addressed similar arising issues of their train drivers. Fatigue, driving activity and sleepiness were among factors mostly investigated. Yet, the characteristics of the performance were similar between Malaysia, UK and Australia. However, this integrated model provides better and more complete model evaluating human performance.

Human factors / ergonomics (HFE) focuses on the interaction of human with the environment. The traditionally concept of HFE has always focused on the interrelationship between three main elements of human-machine-environment. The human element is a major component in any relationship with machine and the environment [1]. This relationship can be best described using the human performance model [2]. A development in HFE has seen the introductions of new terms to redefine machine and environments, in the form of the physical environment ('things'), the organisational environment and the social environment [3-5]. This concept of HFE is applicable in the evaluation of individuals and daily work activities, and has been incorporated in assessment of employees in the industries. Numerous studies have been conducted to evaluate the relationship between human and these major environmental elements. Although the focus of these studies may vary, with some looking on the physical environment while others may address the issue on organisation aspects, their

Content from this work may be used under the terms of the Creative Commons Attribution 3.0 licence. Any further distribution of this work must maintain attribution to the author(s) and the title of the work, journal citation and DOI.
Published under licence by IOP Publishing Ltd

objectives remained the same, which is to improve the performance and the well-being of employees by integrating human into a better system [6].

Performance is an important aspect for the industry, and is a priority to be achieved either by an individual or an organisation especially in handling complex tasks or working in demanding situations [7]. Job performance is a valuable element and an important dependent variable for achieving high quality work output and services [8]. To remain competitive and maintain a high level of job performance, the employee itself is the main focus to be taken care of [9].

Performance of the employee and the system is important in the transportation industry, especially in public transportations, where high performance of the organisation would result in increased safety conditions [10]. Ignorance of the performance of employees may lead to undesirable results. Accidents in the transportation industry frequently occur especially on the road. However, for the railway, ships and aviation industries the frequency of accidents is very low [11] although the occurrence of accidents usually results in a large number of injuries, casualties and devastations. Thus, it is in the interest of these industries to seriously maintain high work performance, improve the safety levels and awareness to avoid catastrophes [12]. HFE has been increasingly accepted an important tool to improve human performance and safety at every level, to ensure a safe transportation system [13, 14].

Past studies on evaluation of train driver performance has mostly focused on only one or very few relationships between influencing factors. There is yet to be any effort in integrating these influencing factors, with no studies conducted for train drivers in Malaysia. This paper will attempt to address the integrated developed model on Malaysian train driver performance and the comparison between existing models available around the globe.

2. Human performance in the railway system

The train driver is amongst the most important group in the railway system. In Malaysia, even with the introduction of electric trains in 1995, the responsibilities and functions of the drivers are still important. Among the challenges faced by train drivers today are the increasing number of train services on single or double tracks with a variety of freight trains, passenger trains, electric commuter trains and fast electric train (Electric Train Service – ETS). These require increased attention of the driver and the crew to avoid any difficulties or accidents throughout the train journey. However, studies on human performance in the railway industry are still in its infancy in Malaysia, and literatures on the subject are thus unavailable. For this reason, most of the literature review for this paper has relied on literatures from other countries, especially from western nations.

Performances of workers are important for the designers or engineers during the design or improvement of equipment's, interfaces, jobs or systems [15]. Understanding of the human performance is not only done when the system is in place and working, but can be as early on in design stage, which may reduce the potential of error and improve the design, safety and effectiveness. Human performance in ergonomics is not limited to safety, comfort and satisfaction. It can support the reputation of the organisation, and its business strategy to stay competitive and sustainable [16].

In the railway industry, error-free operation is very critical to ensure the safety of the system. Circumstances from any train-related accidents would incur high costs and damage [11]. Railway safety does not focus on the train locomotives only, but also expands on aspects of the passengers, track staff, control room staff and other relevant staff. Train drivers are one of the important stakeholders for ensuring safety of passengers, the train locomotives and the system [17, 18]. A landmark train accident in Ladbroke Grove in 5th October 1999, with 400 injuries and 31 deaths, has triggered an extensive evaluation of railway safety in the UK. It was reported that human error was the

IOP Conf. Series: Journal of Physics: Conf. Series **1049** (2018) 012105 doi:10.1088/1742-6596/1049/1/012105

main cause of the accident, due to signal passed at danger (SPAD) and the subsequent head-on collision events [19]. Although rail ergonomics is evolving slower than other branches of transport ergonomics [17], there is an increase in the level of awareness on the importance of human factors in the railway industry, and is very crucial for improving system reliability, safety and human performance [13].

Human capabilities and limitations are the main factors which would contribute to the safety of the individual and the system. These factors should be considered during early stages of the design; corresponding to the needs, knowledge and characteristics of the workers [20]. Improvements may be delayed if there is lack of understanding on human performance of the organization [21]. Human performance measures are required to be used as a predictor in addressing the increasing demand for improvements in system and safety, or as an objective tool to evaluate the need to improve the design of work, equipment and the environments.

Table 1 shows a list of thirty two previous studies over the past twenty five years on human performance in railway since 1986. At least seventeen factors affecting the performance of train drivers were discussed in these studies.

Table 1: Previous studies on factors relating to human performance in rail.

	Variables	Cognitive	Mental workload	Physical workload	Alertness	Awareness	Stress	Working task	Vigilance	Fatigue	Environment	Working conditions	Sleep	Shift work	Job rotation	Safety culture	Human performance
1	Austin and Drummond [27]		•				•		•								
2	Edkins and Pollock [28]				•				•								
3	Clarke [22]															•	
4	Ugajin [23]															•	
5	Felici, Sujan [29]	•														•	
6	Joshi, Kaufman [25]													•		•	
7	Kecklund, Ingre [30]		•		•		•						•	•			
8	Härmä, Sallinen [31]									•			•			•	
9	Hockey and Carrigan [32]									•						•	
10	Cothereau, de Beaurepaire [33]										•						
11	McLeod, Walker [34]	•				•											
12	Jansson, Olsson [35]	•				•											
13	Hamilton and Clarke [26]				•				•	•						•	
14	Lamond, Darwent [36]				•								•				
15	Pickup, Wilson [37]		•														
16	Farrington-Darby, Wilson [38]															•	
17	Murali [39]													•		•	
18	Jones, Dorrian [40]									•							
19	Shepherd and Marshall [41]																•
20	Dorrian, Roach [42]									•							
21	Luke [43]													•		•	
22	Philip and Åkerstedt [44]									•			•				
23	Dorrian, Roach [45]	•				•				•							
24	Chau, Gauchard [46]													•			
25	Jay, Dawson [47]									•				•			
26	Darwent, Lamond [48]												•				
27	Kumar and P.K. [49]						•									•	•
28	Baysari, McIntosh [50]																•
29	Birlik [51]															•	
30	Koohi [52]														•		
31	Ku and Smith [53]									•				•			
32	Dorrian, Baulk [54]		•	•						•			•				

IOP Conf. Series: Journal of Physics: Conf. Series **1049** (2018) 012105 doi:10.1088/1742-6596/1049/1/012105

Each article have at least discussed two or more factors, except some studies which have deliberated on the general safety and human performance as the main discussion. Interrelations between factors on the performance of the driver were also discussed in order to further understand the influence of factors on one another.

It shows that safety / safety culture was the predominating factor discussed by the researchers (12 out of 32). As highlighted previously, the level of safety is crucial and will always be the highest priority in the railway industry. Three studies have exclusively discussed on safety [22-24]. They highlighted the importance of the safety culture to reduce the risk of accidents and to improve the level of safety in the organization. Other researchers have examined the relationship between the safety and shift work of the train driver. Joshi, Kaufman [25] have studied the interactions of the train driver's reaction towards the train control system in shift work; and its potential for safety risk. Researchers were also interested to evaluate human performance in relation to the interactions of safety with other factors.

Hamilton and Clarke [26] proposed a model to describe performance of the driver in relation to infrastructure features and operational conditions. Through the proposed model, driver behaviour could be predicted by using performance times for discrete actions. It had also utilised cognitive task analysis (CTA) to predict workload and operator performance time. This study was focused on cognitive and mental workload of the driver. It was hoped to identify human performance problems for the train driver empirically.

3. The Performance Model

In this study, fourteen factors were initially evaluated by hypothesis testing, from which seven factors were identified to be significant [55]. Table 2 lists the seven factors which influence the performance of train drivers in Malaysia, which are job-related tension (internal conflict), fatigue, occupational stress, driving task, work environment, working condition and safety. The integrated performance model of Malaysian train drivers is presented graphically in Figure 1.

Table 2: Factors affecting performance of the Malaysian train drivers

Domain	Hypothesis	
	H1	Occupational stress in Malaysia
Human	**H2a**	Job related tension (internal conflict)
	H4	Fatigue
Activity	**H7**	Driving task
	H8a	Work environment
Context	**H9a**	Safety culture
	H10	Working condition

*(Factor) has a significant effect on the performance of train drivers

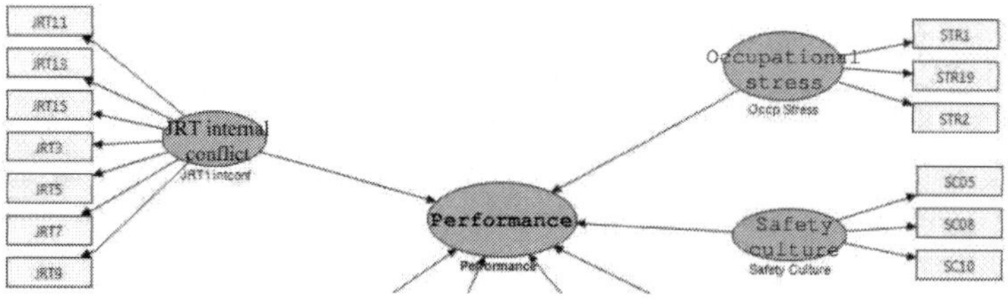

Figure 1: Malaysian train driver integrated performance model

The integrated train driver performance model was developed based on SEM-PLS algorithm, utilising shapes (oval and rectangles) and arrows to show the relationship between constructs and indicator variables [56-58]. A simplified representation is shown in Figure 2, consisting of 3 concentric circles.

The innermost circle is the main objective of the study on the performance of the train driver. The outermost circle is divided into three to indicate the three domains influencing the performance of the train drivers, which are human, context and activity. The middle circle represents the factors which affects the performance of the train driver. For clarity, the colour schemes for the factors are chosen to be similar to their respective associated domain. This middle circle is dynamic, meaning that for future studies; factors can be inserted or removed to suit the circumstances of that particular study. However, the factors added would still be grouped under the three major domains. The simplified version of the integrated Malaysian train driver performance model is deemed to be easily understood by the general public and could be used as guidance for the management and stakeholders in managing human capital in the TOC.

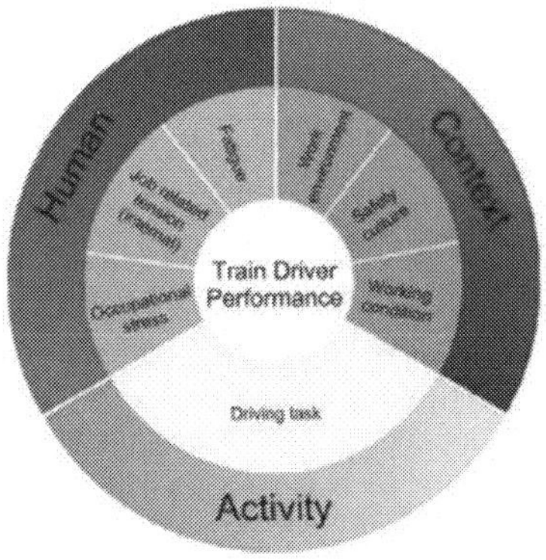

Figure 2: An integrated Malaysian train driver performance model

4. Benchmarking of the model with other related studies

There are a number of models and frameworks in existence, designed to fulfil the various contexts of particular research. This newly developed model can be compared with past models from the literatures. Table 3 shows an overview of the comparison between the proposed model of this study with other existing models and literature on train drivers from around the world.

The comparison is made based on several details. The first column of table shows the authors and the year the article is published, to indicate the timeline of the study. The table also shows brief descriptions of the model or framework, the measurement methods and its sampling factor. The most important comparison is the key indicators selected for each particular model or framework. The comparison can be made on the choice of factors for the domains of human, activity or context.

In most models, the 'human' domain was the main focus and interest, indicating the preferential importance of 'human' in performance study of employees. However, 'human' is a domain containing a large number of factors, inclusive of cognitive, job satisfaction, and gender. Fletcher and Dawson [59] has developed a work-related fatigue model of 193 train drivers in the UK, focusing on fatigue by evaluating sleepiness and alertness using sleep / work diaries and fatigue model due to hours of work. The emphasis was on physiological factor, i.e. sleepiness and fatigue with relation to hours of work. Performance of the train driver was measured through alertness level and objective performance test (OSPAT). However, this model did not consider the environmental factor (context domain).

Two other models related to railway have focused on the cognitive performance of the train driver; and its relationship with work environment and job-task. In 2005, Hamilton & Clarke [26], through their Cognitive Task Analysis (CTA) model, investigated cognitive ability of the train drivers in the UK and their reaction with signals and signs along the track to check on signal pass at danger (SPAD) scenario. Similarly, McLeod, Walker [34] have also studied cognitive aspect of train drivers in the UK, focusing on Automatic Warning System (AWS) which was also related to SPAD scenario. These two-research work had focused on the cognitive response of the train driver while driving to reduce SPAD incidents.

Six articles from Australia had focused particularly on fatigue of the train drivers. Lamond, Darwent [36] studied alertness of fifteen train drivers of Adelaide – Melbourne relay trip by assessing their sleepiness through activity monitoring and sleep diary. Dorrian, Roach [42] have used simulation studies to evaluate the alertness of the train drivers and their fatigue levels. The study was further extended in 2007 to relate fatigue and alertness with incorrect response of cognitive disengagement towards safety of the train driving. Jap, Lal [60] also conducted a simulation study of fifty male train drivers to investigate fatigue using EEG (electroencephalography) activity monitoring. Darwent, Lamond [48] have studied on the sleepiness and performance of the long relay train drivers between Adelaide and Perth, focusing on the relationship between sleep and performance of the train drivers. In addition, the work environment and their working condition were also considered. Although three basic influential domains were addressed, it had only considered certain criteria, i.e. sleepiness (human), driving, work environment and working condition. Fatigue during extended rail operations was also investigated by Jay, Dawson [47], which utilised a 5-min response task (RT) and 7-point Samn-Perelli Fatigue Checklist on nine male drivers. From the six articles on the evaluation of Australian train drivers, five studies have centred solely on fatigue of the train drivers, using either simulation study or actual driving, focusing on long-haul train operations. Another study was examined sleepiness of the train drivers and its relationship with their driving activities, work environment and working condition. The limitations of these past studies were the sole focus on only fatigue or sleepiness, without consideration to other factors which may also influence the performance of the train drivers. The absence of consideration for other factors can misdirect the research direction and may result in inaccurate conclusions in investigations involving human performance as a whole.

Table 3: Benchmarking of the integrated model with previous literatures

Authors	Brief descriptions	Measurement method	Target group, sample	Domain		
				Human	Activity	Context
Fletcher and Dawson [59]	• Field-based validations of a work-related fatigue model	• Sleep/work diaries, wore actigraphs • Fatigue model • Subjective alertness test (VAS) • Objective performance test (OSPAT) •	• 193 train drivers • UK	• Sleepiness • Fatigue • Alertness	• Hours of work	•
Hamilton & Clarke [26] ***CTA Model***	• Train driver performance in interaction with infrastructure features • Development of a human factors SPAD hazard checklist •	• Line speed on driver interaction with signals and signs •	• 250 train drivers • UK	• Cognitive	• Driving	• Infrastructure
McLeod, Walker [34] ***Situational Model***	• Train driver performance in interaction with Automatic Warning System (AWS) • AWS provides audible alert for SPAD	• Structured interviews with train drivers	• 20 scenarios of AWS usage • Train drivers • UK	• Cognitive	• Driving	•
Lamond, Darwent [36]	• Train drivers' sleep and alertness during short relay operations	• Activity monitor • Sleep diary • Actiware-sleep software	• 15 drivers • Adelaide – Melbourne relay trip • Australia	• Alertness • Sleepiness	• Driving	•
Dorrian, Roach [42]	• Effects of fatigue on train handling during speed restrictions	• Simulation driving • 10-min PVT • 100 mm VAS	• 20 male train drivers from Queensland depots • Australia •	• Fatigue • Alertness	• Simulated driving	•
Dorrian, Roach [45]	• Effects of fatigue increasing inefficiency and accident risk	• Simulation driving • 10-min PVT • 100 mm VAS	• 20 male train drivers from Queensland depots • Australia •	• Fatigue • Alertness	• Simulated driving	• Incorrect responses • Cognitive disengagement • Safety
Darwent, Lamond [48]	• sleep and performance of train drivers during an extended freight-haul operation	• Mini-Mitter Actigraph-L activity monitoring devices – to assess sleep/wake states • Sleep diary	• 10 male train drivers • Long relay Adelaide – Perth • Australia	• Sleep • Performance	• Extended freight-haul driving	• Work environment • Working condition
Jay, Dawson [47]	• Driver fatigue during extended rail operations	• 5-min response task (RT) – using PVT • 7-point Samn–Perelli Fatigue Checklist	• 9 male drivers • Australia	• Fatigue	•	•
Baysari, Caponecchia [61]	• Review of 40 rail safety investigation to understand contribution of HFE to railway accidents	• Report review • Adoptation of HFACS (Human Factors Analysis and Classification System) •	• Rail safety • Australia			• Accidents and safety
Jap, Lal [60]	• Comparison of EEG activity among train drivers during monotonous driving	• Simulation driving • EEG activity monitoring	• 50 male train drivers • Australia	• Fatigue • Mental fatigue		
Jalil Azlis-Sani [55]	• Integrated train driver performance model	• 27-items questionnaire	• 229 respondents • Train drivers • Malaysia	• Occupational stress • Job related tension • Fatigue •	• Work (driving) task	• Work environment • Working conditions • Safety culture

Only one literature did a review of forty rail safety investigations; to understand the contribution of human factors and ergonomics (HFE) to railway accidents. This literature was absolutely discussed on accidents and safety of the rail industry in Australia.

Subjective measurements are often used in investigation involving performance of workers. The technique is widely used in ergonomics study to understand certain issues in human factors, although whether the approach is acceptable as scientific measurement has been a topic of debate [62]. Nevertheless, there are several objective measurements methods that have been used in ergonomics study [63], although their applications have been very limited especially on the number of respondents to be measured. An example of the use of objective measurements is the application of Electroencephalogram (EEG) signals to investigate mental fatigue [60, 64]. However, in this study; subjective measurements were mainly used, utilising the paper-pencil questionnaire survey to gather information from 229 train drivers.

In contrast to previously developed models, the proposed model in this study presents an integrative model. The three main domains of human, activity and context were identified from past literatures and have been evaluated thoroughly. Occupational stress, job related tension and fatigue were the results of improper task-design and poor work environment and working conditions. These factors were investigated simultaneously to study its relationship with the performance of the train driver. In addition, this model includes consideration of the safety culture, whereas other models tend to study safety aspects separately. Thus, this model integrates the various important aspects and factors in a complete package for the evaluation of human performance. The model is characterising to be dynamic, in which factors can be customised within the three main domains to create an improved human performance study in HFE.

In conclusion, the previous referred models mostly from the UK and Australia addressed similar arising issues of their train drivers. Fatigue, driving activity and sleepiness were among factors mostly investigated. Yet, the characteristics of the performance were similar between Malaysia, UK and Australia. However, this integrated model provides better and more complete model evaluating human performance.

5. Conclusion

To date, it was found that an integrated model of train driver performance has yet to be fully developed, especially in Malaysia. This paper attempts to address this limitation by developing an integrated model of train driver performance, specifically for train drivers in Malaysia. This model will provide a comprehensive understanding of human performance by incorporating three main domains of human performance; human – activity – context. Past studies have evaluated these factors either individually or with limited interactions between the factors. It can then be subsequently used by researchers as an important reference point to venture into evaluating other affecting factors of human performance, in different area of studies. In addition, train operating companies (TOC) can utilise this framework as a guideline to improve the design of workplace and tasks, train infrastructure as well as raising the level of awareness among employees. This will ensure that the company remains profitable, competitive and safe, and maintains a high level of performance of employees and the organisation.

Acknowledgment

The authors are supported by the University of Malaya Postgraduate Grants and Fundamental Research Grant Scheme (FRGS) Vot 1542, Office of Research, Innovation, Commercialization and Consultation (ORICC) Universiti Tun Hussein Onn Malaysia. The authors would also like to thank Keretapi Tanah Melayu Berhad on the contribution and participation for this study.

References

[1] Branton, P. 1987 In praise of ergonomics - a personal perspective *International Reviews of Ergonomics* 1-20
[2] Wilson, J.R. 1990 A framework and a context for ergonomics methodology in *Evaluation for Human Work* J.R. Wilson and E.N. Corlett, Editors, Taylor & Francis 1-39.
[3] Parsons, K.C. 2000 Environmental ergonomics: a review of principles, methods and models. *Applied Ergonomics* **31(6)** 581-594.

[4] Wilson, J.R. 2000 Fundamentals of ergonomics in theory and practice *Applied Ergonomics* **31(6)** 557-567.

[5] Carayon, P. and M.J. Smith 2000 Work organization and ergonomics *Applied Ergonomics* **31(6)** 649-662

[6] Dul, J., et al. 2012 A strategy for human factors/ergonomics: developing the discipline and profession *Ergonomics* **55(4)** 377-395

[7] Klein, M.C.A., R.M. van Lambalgen, and J. Treur 2010 An Agent Model for Analysis of Human Performance Quality. in *Web Intelligence and Intelligent Agent Technology (WI-IAT)* 2010 IEEE/WIC/ACM International Conference

[8] Kahya, E. 2009 The effects of job performance on effectiveness *International Journal of Industrial Ergonomics* **39** 96-104

[9] Layer, J.K., W. Karwowski, and A. Furr 2009 The effect of cognitive demands and perceived quality of work life on human performance in manufacturing environments *International Journal of Industrial Ergonomics* **39(2)** 413-421

[10] Haque, M.M., H.C. Chin, and A.K. Debnath 2013 Sustainable, safe, smart—three key elements of Singapore's evolving transport policies *Transport Policy* **27** 20-31

[11] Evans, A.W. 2011 Fatal train accidents on Europe's railways: 1980-2009 *Accident Analysis & Prevention* **43(1)** 391-401

[12] Silla, A. and V.-P. Kallberg 2012 The development of railway safety in Finland *Accident Analysis & Prevention* **45** 737-744

[13] Clarke, T. 2005 The Ergonomics Programme at Network Rail *Rail Human Factors: Supporting the Integrated Railway*, ed J.R. Wilson, et al. (Ashgate) p 19 - 21

[14] Wickens, C.D., et al. 2004 *An Introduction To Human Factors Engineering* 2nd Edition (Pearson - Prentice Hall)

[15] Wilson, J.R. 2014 Fundamentals of systems ergonomics/human factors *Applied Ergonomics* **45(1)** 5-13

[16] Dul, J. and W.P. Neumann 2009 Ergonomics contributions to company strategies *Applied Ergonomics* **40(4)** 745-752

[17] Wilson, J.R. and B.J. Norris 2005 Rail human factors: Past, present and future *Applied Ergonomics* **36(6)** 649-660

[18] Wilson, J.R. 2006 Interaction with virtual environments *International Journal of Human-Computer Studies* **64(3)** 157-157

[19] Stanton, N.A. and G.H. Walker 2011 Exploring the psychological factors involved in the Ladbroke Grove rail accident *Accident Analysis & Prevention* **43(3)** 1117-1127

[20] Shahrokhi, M. and A. Bernard 2009 A framework to develop an analysis agent for evaluating human performance in manufacturing systems *CIRP Journal of Manufacturing Science and Technology* **2(1)** 55-60

[21] Genaidy, A., et al. 2007 The work compatibility improvement framework: An integrated perspective of the human-at-work system *Ergonomics* **50(1)** 3-25

[22] Clarke, S. 1998 Safety culture on the UK railway network *Work & Stress* **12(3)** 285-292

[23] Ugajin, H. 1999 Human Factors Approach to Railway Safety *Quarterly Report of RTRI* **40(1)** 5-8.

[24] Farrington-Darby, T., L. Pickup, and J.R. Wilson 2005 Safety culture in railway maintenance. *Safety Science* **43(1)** 39-60

[25] Joshi, V.V., L.M. Kaufman, and T.C. Giras 2001 Human behavior modeling in train control systems *Reliability and Maintainability Symposium Proceedings* 183-188

[26] Hamilton, W.I. and T. Clarke 2005 Driver performance modelling and its practical application to
railway safety *Applied Ergonomics* **36(6)** 661-670
[27] Austin, A. and P.D. Drummond 1986 Work problems associated with suburban train driving *Applied Ergonomics* **17(2)** 111-116
[28] Edkins, G.D. and C.M. Pollock 1997 The influence of sustained attention on Railway accidents *Accident Analysis & Prevention* **29(4)** 533-539
[29] Felici, M., M.A. Sujan, and M. Wimmer 2000 Integration of functional, cognitive and quality requirements : A railways case study *Information and Software Technology* **42(14)** 993-1000
[30] Kecklund, L., et al. 2001 The TRAIN - project: Railway safety and the train driver information environment and work situation - A summary of the main results *Signalling Safety* (London)
[31] Härmä, M., et al. 2002 The effect of an irregular shift system on sleepiness at work in train drivers
and railway traffic controllers *Journal of Sleep Research* **11(2)** 141-151
[32] Hockey, B. and N. Carrigan 2003 Human Factors in Railway Systems : Implications for Safety. *Rail Research UK* (University of Leeds) 23
[33] Cothereau, C., et al. 2004 Professional and medical outcomes for French train drivers after "person under train" accidents: three year follow up study *Occupational and Environmental Medicine* **61** 488-494
[34] McLeod, R.W., G.H. Walker, and N. Moray 2005 Analysing and modelling train driver performance *Applied Ergonomics* **36(6)** 671-680
[35] Jansson, A., E. Olsson, and L. Kecklund 2005 Acting or reacting? A cognitive work analysis approach to the train driver task *Rail Human Factors: Supporting the Integrated Railway*, ed J.R. Wilson, et al. (Ashgate) 40 - 49
[36] Lamond, N., D. Darwent, and D. Dawson 2005 Train drivers' sleep and alertness during short relay operations *Applied Ergonomics* **36(3)** 313-318
[37] Pickup, L., et al. 2005 The Integrated Workload Scale (IWS): A new self-report tool to assess railway signaller workload *Applied Ergonomics* **36(6)** 681-693
[38] Farrington-Darby, T., J.R. Wilson, and B.J. Norris 2005 Investigating Train Driver Behaviour: The Use of Lineside Information When Regulating Speed *Rail Human Factors: Supporting the Integrated Railway*, ed J.R. Wilson, et al. (Ashgate) 60 - 69
[39] Murali, M.C. 2005 The Human Element in Train Safety on Indian Railways *International Railway*
Safety Conference (Cape Town)
[40] Jones, C.B., et al. 2005 Working hours regulations and fatigue in transportation: A comparative analysis *Safety Science* **43(4)** 225-252
[41] Shepherd, A. and E. Marshall 2005 Timeliness and task specification in designing for human factors in railway operations *Applied Ergonomics* **36(6)** 719-727
[42] Dorrian, J., et al. 2006 The effects of fatigue on train handling during speed restrictions *Transportation Research Part F: Traffic Psychology and Behaviour* **9(4)** 243-257
[43] Luke, T., Brook-Carter,N, Parkes, AM, Grimes, E and Mills 2006 A An investigation of train driver visual strategies *Cogn Tech Work* **8** 15–29
[44] Philip, P. and T. Åkerstedt 2006 Transport and industrial safety, how are they affected by sleepiness and sleep restriction? *Sleep Medicine Reviews* **10(5)** 347-356
[45] Dorrian, J., et al. 2007 Simulated train driving: Fatigue, self-awareness and cognitive disengagement *Applied Ergonomics* **38(2)** 155-166
[46] Chau, N., et al. 2007 Contributions of occupational hazards and human factors in occupational injuries and their associations with job, age and type of injuries in railway workers *International Archives of Occupational and Environmental Health* **80(6)** 517-525
[47] Jay, S.M., et al. 2008 Driver fatigue during extended rail operations *Applied Ergonomics* **39(5)** 623-629

[48] Darwent, D., N. Lamond, and D. Dawson 2008 The sleep and performance of train drivers during
an extended freight-haul operation Applied Ergonomics **39(5)** 614-622
[49] Kumar, A. and S. P.K. 2008 Human Error Control in Railways *Jordan Journal of Mechanical and Industrial Engineering* **2(4)** 183 - 190
[50] Baysari, M.T., A.S. McIntosh, and J.R. Wilson 2008 Understanding the human factors contribution to railway accidents and incidents in Australia *Accident Analysis & Prevention* **40(5)** 1750-1757
[51] Birlik, G. 2009 Occupational Exposure to Whole Body Vibration - Train Drivers *International Scientific Journal of Occupational Health and Safety* **47** 5-10
[52] Koohi, I. 2009 Accidents Analysis of Rail Transportation Industry in Iran *World Applied Sciences*
Journal **7(3)** 358-365
[53] Ku, C.-H. and M.J. Smith 2010 Organisational factors and scheduling in locomotive engineers and conductors: Effects on fatigue, health and social well-being *Applied Ergonomics* **41(1)** 62-71
[54] Dorrian, J., S.D. Baulk, and D. Dawson 2011 Work hours, workload, sleep and fatigue in Australian Rail Industry employees *Applied Ergonomics* **42(2)** 202-209
[55] Jalil Azlis-Sani 2015 Development of an integrated train driver performance model (Theses: University of Malaya)
[56] Jalil Azlis-Sani, S.Z.M.Dawal, Norhayati Mohmad Zakwan and Mohd Faizal Mohideen Batcha 2013 Measuring Human Performance of Malaysian Train Drivers : A Factor Analysis *17th International Conference on Industrial Engineering Theory, Applications and Practice Proceeding* (Pusan National University, Busan, Korea)
[57] Jalil Azlis-Sani, S.Z.M.Dawal, Norhayati Mohmad Zakwan and Mohd Faizal Mohideen Batcha 2013 Measuring Human Performance of Malaysian Train Drivers : Developing Model Using PLS Approach *17th International Conference on Industrial Engineering Theory, Applications and Practice* (Pusan National University, Busan, Korea)
[58] Jalil Azlis-Sani, S.Z.M.Dawal, Zakuan Norhayati Mohmad 2013 Validity and Reliability Testing
on Train Driver Performance Model Using a PLS Approach *Advanced Engineering Forum*
[59] Fletcher, A. and D. Dawson 2001 Field-based validations of a work-related fatigue model based on hours of work *Transportation Research Part F: Traffic Psychology and Behaviour* **4(1)** 75-88
[60] Jap, B.T., S. Lal, and P. Fischer 2011 Comparing combinations of EEG activity in train drivers during monotonous driving *Expert Systems with Applications* **38(1)** 996-1003
[61] Baysari, M.T., et al. 2009 Classification of errors contributing to rail incidents and accidents: A comparison of two human error identification techniques *Safety Science* **47(7)** 948-957
[62] Annett, J. 2002 Subjective rating scales: science or art? *Ergonomics* **45** 966-987
[63] Fagarasanu, M. and S. Kumar 2002 Measurement instruments and data collection: a consideration
of constructs and biases in ergonomics research *International Journal of Industrial Ergonomics* **30(6)** 355-369
[64] Kar, S., M. Bhagat, and A. Routray 2010 EEG signal analysis for the assessment and quantification of driver's fatigue *Transportation Research Part F: Traffic Psychology and Behaviour* **13(5)** 297-306

IOP Conf. Series: Journal of Physics: Conf. Series **1049** (2018) 012106 doi:10.1088/1742-6596/1049/1/012106

Modelling an Integrated Human Performance Model of the Train Driver

Jalil Azlis-Sani[1], Siti Zawiah Md Dawal[2], Norhayati Mohmad Zakwan[3]

[1]Faculty of Mechanical and Manufacturing Engineering, Universiti Tun Hussein Onn Malaysia, 86400 Batu Pahat, Johor, Malaysia
[2]Department of Mechanical Engineering, Faculty of Engineering, University of Malaya. 50603 Kuala Lumpur, Malaysia
[3]Department of Business Administration, Faculty of Management, Universiti Teknologi Malaysia, UTM Skudai, 81310, Johor, Malaysia

E-mail: azlis@uthm.edu.my

Abstract. The purpose of this paper is to address the development an integrated performance model of train drivers. A human performance measure was developed by integrating several significant factors of the train driver. The evaluation was conducted on train drivers of a major train operating company (TOC) in Malaysia, focusing only on drivers of intercity passenger trains and freight trains. 229 respondents had participated in the quantitative paper-and-pencil survey conducted, and the data obtained was subsequently analysed using SPSS software and was then tested using structural equation modelling (SEM)-PLS approach in the SmartPLS software to determine the relationship among the significant factors of train driver performance. Fourteen factors were hypothesized and tested under the three main domains, namely human domain; activity domain; and lastly, the context domain. The results indicated that fatigue, job related tension (internal conflict) and occupational stress; under human domain were found to be the significant factors which influence the performance of train drivers. For the activity domain, hypothesis testing proved that driving task, was significant factor. In the context domain; three factors were found to be significant. These include safety culture, working environment and working condition. To summarize, the study identified a total of seven significant factors which include occupational stress, job related tension – internal conflict, fatigue, driving task, work environment, safety culture and working condition. However, the results have failed to support the remaining seven factors of job related tension (external conflict), job satisfactions, sleep, job demand, work facilities, and safety issue.

1. Introduction

The performance of train drivers has been studied during the past decade, especially in the United Kingdom after the train accident at Ladbroke Grove on 5th October 1999 [1-3]. These studies have focused on factors such as cognitive, workload, fatigue, sleepiness and analysis of task. There were also studies on accident analysis, new design of cab, safety and operating system [4-9].

However, there are very limited literatures on the performance of train driver as an integrated model which relates the various influencing factors. Past research has focused on assessment of individual

Content from this work may be used under the terms of the Creative Commons Attribution 3.0 licence. Any further distribution of this work must maintain attribution to the author(s) and the title of the work, journal citation and DOI.
Published under licence by IOP Publishing Ltd

influencing factors, overlooking the relationship possibilities between them. Furthermore, these factors were not categorised systematically, whether based on its common characteristics or on the HFE basic interacting domains of human – machine – environment. Understanding the relationship between the performance of train drivers and factors influencing their performance would enable their integration into the development of a performance model [10]. The availability of the performance model would offer a holistic approach in evaluation of train driver performance to ensure the safety of the train journey as well as the overall system [11]. In addition, the performance evaluation would improve the quality of service, reduce degree of risk and avoid occurrence of accidents. Most studies on train driver performance have been conducted in European countries, especially in the United Kingdom. To date, there has yet to be any study conducted on performance measurement of train drivers in Malaysia.
This study consists of three major domains influencing the performance of the train drivers in Malaysia namely drivers' activities, context and human domains. These domains cover the generic factors of human performance as proposed by Bailey [12], Baines, Asch [10] and Chang and Yeh [13]. A total of sixty factors were extracted from their studies, although to consider all of the proposed factors in this study, given the constraint of time and resources, is quite impossible. However, the selections of domains and factors for this study by using factor analysis are considered sufficient within the scope of the research work. These three domains (activities, context and human) are considered to be sufficient in representing human performance within the time frame of the research.

The major reference of this study is based on the research by Ryan, Wilson [14] on the development of the Rail Ergonomics Questionnaire (REQUEST) instrument with several adaptations and modifications. This comprehensive instrument was designed especially for railway workers and has been refined and developed over a number of years. Additions to the REQUEST instrument were made based on the studies of Strahan, Watson [15], John [16-18], Johns and Hoaking [19], Gradisar, Lack [20], Austin and Drummond [21] and Dawal, Taha [22].

The present study was conducted on the train drivers of a major train operating company (TOC) in Malaysia. Since all the train drivers were males, there were no data for female respondents. The survey was conducted on intercity passenger and freight train drivers. These types of trains are characterised by their long-haul journeys with a minimum of four hours trips, which may have a significant impact on the safety of the rail network.

2. Methodology

In this study, subjective rating scales would be the main method used to gather responses from the subjects and the environment [23, 24]. Demand and effects of the employees' wellbeing would be investigated to evaluate their performance [25]. A questionnaire survey would be used to obtain individual responses, which is known to be the best method for collecting perceptions and opinions from the driver with regard to their performance [14].

2.1 Self-administered questionnaire

The survey questionnaire used was developed from the combination of several existing validated measurements of past research; which are worded in English, were carefully translated into Bahasa Malaysia (Malaysian language) and verified by experts to ensure linguistic and contextual accuracy. The translations were necessary since most of the respondents were well-versed only in Bahasa Malaysia. The instrument consisted of 148-items for measuring nine variables.

Structural equation modelling (SEM) technique is used to evaluate the model and to determine the relationships between variables. Since the research is in the exploratory stage in determining the relationship between variables in measuring the performance of the train driver, the PLS-SEM approach is considered to be the most suitable.

3. Result

3.1 Evaluation of the measurement model

A systematic analysis using a partial least square (PLS) approach is used, which presents the results in two steps [26, 27]. The first step examined the validity and reliability of the survey items in the measurement model, whereas the second step analyses the structural model. The measurement model consists of composite reliability to evaluate internal consistency, and convergent validity by using average variance extracted (AVE). Discriminant validity was assessed using the Fornell-Larcker criterion and cross loadings.

3.1.1 Convergent validity

Convergent validity was assessed through factor loadings, composite reliability and average variance extracted (AVE) [28]. The loading of the items should be greater than 0.5 with composite reliability (CR) values of 0.7 [28]. The average variance extracted (AVE) proposed by Fornell and Larcker (1981) was to measure the variance amount from the indicators relative to measurement error. For convergent validity, the AVE values should be more than 0.5 [27].

Table 1 summarizes the results of the measurement model, showing that the values for loadings and CR for all the domains of human, activity and context are generally greater than the cut-off value for loadings and CR, showing that all the constructs have met the recommended values and are thus reliable and valid. The values highlighted in grey boxes shows the improved values after deletion of unreliable items.

3.1.2 Discriminant Validity

Discriminant validity is another assessment of the constructs (factors) to test the validity of the measurement. Discriminant validity tests whether believed unrelated constructs are indeed unrelated. Thus, the cross-construct correlations should be very low, whereas it should measure strongly the construct it attempts to reflect [27].

Table 1: Results of the measurement model

Domain	Factors	Items	Loading	AVE	CR
Human	Fatigue	FF01	0.718	0.549	0.878
		FF02	0.839		
		FF03	0.840		
		FF04	0.533		
		FF06	0.748		
		FF07	0.725		
	Job related tension (external conflict)	JRT1	0.732	0.598	0.748
		JRT14	0.813		
	Job related tension (internal conflict)	JRT11	0.696	0.586	0.813
		JRT13	0.514		
		JRT15	0.570		
		JRT3	0.644		
		JRT5	0.677		
		JRT7	0.695		
		JRT9	0.526		
	Job Satisfaction 1	JS1	0.620	0.557	0.861
		JS2	0.806		
		JS3	0.666		
		JS4	0.847		
		JS5	0.768		
Human	Job Satisfaction 2	JS6	0.863	0.696	0.872
		JS7	0.907		
		JS8	0.722		
	Occupational Stress	STR1	0.801	0.570	0.798
		STR19	0.672		
		STR2	0.786		
	Sleep	FSL02	0.928	0.504	0.742
		FSL03	0.577		
		FSL07	0.564		
Activity	Driving task	DT1	0.700	0.579	0.788
		DT2	0.764		
		DT3	0.768		
	Job Demand	JCH1	0.750	0.507	0.748
		JCH2	0.835		
		JCH5	0.513		
Context	Working condition	SI06	0.610	0.503	0.702
		WC6	0.634		
		WC7	0.678		
	Safety Culture	SC05	0.810	0.589	0.811
		SC08	0.793		
		SC10	0.695		
	Safety Issue	SI01	0.819	0.585	0.848
		SI02	0.771		
		SI04	0.666		
		SI05	0.794		
	Working Environment 2	WE12	0.781	0.643	0.842
		WE14	0.908		
		WE15	0.704		
	Work Facilities	WC4	0.649	0.703	0.819
		WC5	0.992		
	Performance	PMS2008	0.754	0.561	0.836
		PMS2009	0.797		
		PMS2010	0.754		
		PMS2011	0.687		

The discriminant validity of this study is presented in Table 2, showing the average variance extracted (AVE) of the factor (construct), indicated in bold, is much greater than the squared correlations for each construct. This indicates adequate discriminant validity is achieved. Since the measurement model demonstrated adequate convergent validity and discriminant validity, this indicated that each factor (construct) was unique and captured phenomena not represented by other factors in the model [29].

Table 2: Discriminant validity of construct

	Work cond-ition	Driving task	Fatigue	JRT 2 extconf	JRT1 intconf	Job Demand	Job Sat 1	Job Sat 2	Occp Stress	Perfor-mance	Safety Culture	Safety Issue	Sleep	Work Env2	Work Facilities
Working condition	**0.503**														
Driving task	0.025	**0.579**													
Fatigue	0.066	0.160	**0.550**												
JRT 2 extconf	0.030	0.223	0.095	**0.598**											
JRT1 intconf	0.046	0.226	0.213	0.219	**0.586**										
Job Demand	0.014	0.002	0.000	0.000	0.006	**0.507**									
Job Sat 1	0.006	0.051	0.106	0.038	0.100	0.045	**0.557**								
Job Sat 2	0.002	0.000	0.004	0.002	0.009	0.000	0.007	**0.696**							
Occp Stress	0.016	0.107	0.115	0.026	0.111	0.008	0.047	0.006	**0.571**						
Performan ce	0.035	0.024	0.035	0.013	0.041	0.003	0.017	0.002	0.030	**0.560**					
Safety culture	0.035	0.039	0.077	0.034	0.127	0.000	0.180	0.000	0.030	0.041	**0.589**				
Safety Issue	0.041	0.001	0.031	0.001	0.000	0.040	0.002	0.000	0.007	0.005	0.001	**0.584**			
Sleep	0.003	0.015	0.088	0.031	0.014	0.001	0.025	0.001	0.003	0.005	0.002	0.003	**0.506**		
Work Env2	0.013	0.029	0.010	0.013	0.051	0.008	0.060	0.000	0.029	0.011	0.045	0.001	0.001	**0.643**	
Work Facilities	0.063	0.016	0.048	0.006	0.098	0.006	0.059	0.006	0.034	0.023	0.099	0.003	0.002	0.028	**0.701**

*Diagonals (in bold) represent the AVE while the off diagonals represent the squared correlations

3.2 Evaluation of the structural model

Once it has been determined that the factors are reliable and valid, the subsequent process is to analyse the structural model by testing the hypothesis [26, 27]. The assessment of the structural model is used to determine whether the concept which has been selected is empirically confirmed.

Table 3 : List of the hypotheses of the modified model

Domain	Hypothesis	
Human	**H1**	Occupational stress
	H2a	Job related tension (internal conflict)
	H2b	Job related tension (external conflict)
	H3a	Job satisfaction (1)
	H3b	Job satisfaction (2)
	H4	Fatigue
	H5	Sleepiness

Activity	**H6**	Job demand
	H7	Driving task
Context	**H8a**	Working environment
	H8b	Work facilities
	H9a	Safety culture
	H9b	Safety issue
	H10	Working condition

E.g.: (Factor) has a significant effect on the performance of train driver

IOP Conf. Series: Journal of Physics: Conf. Series **1049** (2018) 012106 doi:10.1088/1742-6596/1049/1/012106

The modified model consists of three domains with fourteen hypotheses. Table 3 lists the hypotheses of the modified model after factor analysis process. Hypothesis testing was used to examine the path loadings and -value between constructs of the inner model. Furthermore, bootstrapping technique was implemented with t-statistics to test for significance. Table 4 shows the path coefficients' and the results of hypothesis testing.

From the seven hypotheses drawn for human domain, three factors were found to be not significant. The non-significant factors were job related tension (external conflict) and two factors of job satisfactions. Fatigue was positively related ($t= 1.371$, $p<0.1$) with the performance of the train driver. Other factors tested for hypotheses for the human domain were job related tension (internal conflict) ($t=1.998$, $p<0.05$) and occupational stress ($t=4.448$, $p<0.01$) towards performance of the train driver.

For the activity domain, the driving task was found to have positive relationship to performance as its hypothesis is significant ($t = 2.170$, $p<0.05$). Conversely, the job demand was found as a non-significant factor on train driver performance ($t = 0.428$).

The context domain has five factors, from which only three factors had significant hypothesis. The significant factors having direct positive effects on the performance of train drivers were safety culture ($t=2.065$, $p<0.05$), working environment ($t=2.417$, $p<0.01$) and working condition ($t=1.935$, $p<0.05$). On the other hand, safety issue and working condition were found to be non-significant to the performance of the driver.

Table 4: Path coefficients' and hypothesis testing

Cluster	Hypothesis		Beta	Standard Error	t-value	Decision
Human	**H1**	Occupational Stress → Performance	0.328	0.074	4.448***	Significant
	H2a	Job related tension (internal conflict) → Performance	0.148	0.074	1.998**	Significant
	H2b	Job related tension (external conflict)→ Performance	-0.043	0.073	0.587	Not significant
	H3a	Job Satisfaction 1 → Performance	-0.042	0.071	0.584	Not significant
	H3b	Job Satisfaction 2 → Performance	-0.045	0.069	0.648	Not significant
	H4	Fatigue → Performance	-0.085	0.073	1.371*	Significant
	H5	Sleep → Performance	-0.026	0.080	0.327	Not significant
Activity	**H6**	Job Demand → Performance	-0.037	0.086	0.428	Not significant
	H7	Driving task → Performance	0.141	0.065	2.170**	Significant
Context	**H8a**	Working environment→ Performance	-0.176	0.073	2.417***	Significant
	H8b	Work facilities → Performance	0.076	0.070	1.089	Not significant
	H9a	Safety culture→ Performance	-0.133	0.065	2.065**	Significant
	H9b	Safety Issue → Performance	0.060	0.084	0.712	Not significant
	H10	Working condition→ Performance	0.120	0.062	1.935**	Significant

Note: *(10%) $p<0.1 = 1.28$; **(5%) $p<0.05 = 1.645$; *** (1%) $p<0.01=2.33$ | One-tailed test

4. The performance Model

Fourteen factors were initially evaluated by hypothesis testing, from which seven factors were identified to be significant. Figure 1 shows the overall graphical representation of the model. From the fourteen hypotheses tested, only seven were accepted which were found to have a strong influence on the performance of the train driver. These hypotheses are: H1 (occupational stress), H2a (job related tension – internal conflict), H4 (fatigue), H7 (driving task), H8a (work environment), H9a (safety culture) and H10 (working condition).

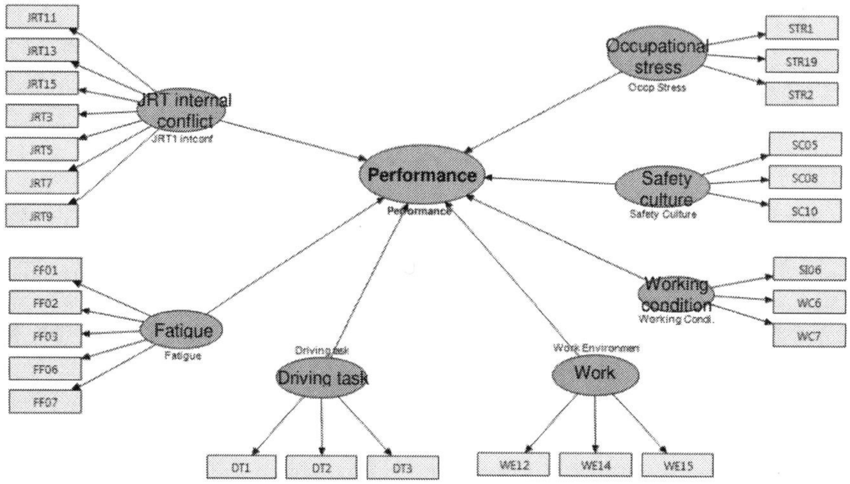

Figure 1: Malaysian train driver integrated performance model

5. Conclusion

Three domains, namely human, activity and context, were proposed to categorise the fourteen factors. The relationships between factors were evaluated iteratively using SEM-PLS. The measurement model was also evaluated through composite reliability, prioritizing the indicators according to their individual reliability. This evaluation technique validated that all items with high loading were measuring a particular factor but were loaded lower on other factors. The convergent validity was also measured and confirmed that the items used were valid, reliable, and correlated positively to measure the same construct (factor). Discriminant validity was performed to assess the validity of the measurements, to ensure that constructs were not measuring other constructs or overlapping constructs. It is expected that the cross-construct correlations should be very low but would measure strongly the construct it attempts to reflect. It was found that all factors have average variance extracted (AVE) values greater than 0.5, indicating that each factor was unique and captured phenomena not represented by other factors in the model.

From the originally proposed structural model having fourteen hypotheses, only seven hypotheses were found to be significant. These are job-related tension (internal conflict), fatigue, occupational stress, driving task, work environment, working condition and safety culture. The results highlighted that the three domains i.e. human, context and activity with their seven factors are important in developing an integrated train driver performance model for Malaysia.

Acknowledgment

The authors are supported by the University of Malaya Postgraduate Grants and Fundamental Research Grant Scheme (FRGS) Vot 1542, Office of Research, Innovation, Commercialization and Consultation (ORICC) Universiti Tun Hussein Onn Malaysia. The authors would also like to thank Keretapi Tanah Melayu Berhad on the contribution and participation for this study.

References

[1] Wilson, J.R. and B.J. Norris 2005 Special issue on rail human factors *Applied Ergonomics* **36(6)** 647-648
[2] Stanton, N.A. and G.H. Walker 2011 Exploring the psychological factors involved in the

Ladbroke Grove rail accident *Accident Analysis & Prevention* **43(3)** 1117-1127
[3] Sutherland, C. and L. Groombridge 2001 The Paddington rail crash: identification of the deceased
following mass disaster *Science & Justice* **41(3)** 179-184
[4] Darwent, D., N. Lamond, and D. Dawson 2008 The sleep and performance of train drivers during
an extended freight-haul operation *Applied Ergonomics* **39(5)** 614-622
[5] Dorrian, J., et al. 2006 The effects of fatigue on train handling during speed restrictions *Transportation Research Part F: Traffic Psychology and Behaviour* **9(4)** 243-257
[6] Dorrian, J., et al. 2007 Simulated train driving: Fatigue, self-awareness and cognitive disengagement *Applied Ergonomics* **38(2)** 155-166
[7] Edkins, G.D. and C.M. Pollock 1997 The influence of sustained attention on Railway accidents *Accident Analysis & Prevention* **29(4)** 533-539
[8] Farrington-Darby, T., J.R. Wilson, and B.J. Norris 2005 Investigating Train Driver Behaviour: The Use of Lineside Information When Regulating Speed *Rail Human Factors: Supporting the Integrated Railway*, ed. J.R. Wilson, et al, (Ashgate) 60 - 69
[9] Jay, S.M., et al. 2008 Driver fatigue during extended rail operations *Applied Ergonomics* **39(5)** 623-629
[10] Baines, T.S., et al. 2005 Towards a theoretical framework for human performance modelling within manufacturing systems design *Simulation Modelling Practice and Theory* **13(6)** 486-504
[11] Williamson, A., et al. 2011 The link between fatigue and safety *Accident Analysis & Prevention* **43(2)** 498-515
[12] Bailey, R.W. 1996 Human performance engineering : designing high quality, professional user interfaces for computer products, applications, and systems. 3rd ed. (New Jersey: Prentice Hall)
[13] Chang, Y.H. and C.H. Yeh 2010 Human performance interfaces in air traffic control *Applied Ergonomics* **41(1)** 123-129
[14] Ryan, B., et al. 2009 Developing a Rail Ergonomics Questionnaire (REQUEST) *Applied Ergonomics* **40(2)** 216-229
[15] Strahan, C., B. Watson, and A. Lennon 2008 Can organisational safety climate and occupational stress predict work-related driver fatigue? *Transportation Research Part F: Traffic Psychology and Behaviour* **11(6)** 418-426
[16] Johns, M.W. 1991 A New Method for Measuring Daytime Sleepiness: The Epworth Sleepiness Scale *Sleep* **14(6)** 540-545
[17] Johns, M.W. 1992 Reliability and Factor Analysis of the Epworth Sleepiness Scale *Sleep* **15(4)** 376-381
[18] Johns, M.W. 1993 Daytime Sleepiness, Snoring and Obstructive Sleep Apnea: The Epworth Sleepiness Scale *Chest* **103** 30-36
[19] Johns, M.W. and B. Hoaking 1997 Excessive Daytime Sleepiness: Daytime Sleepiness and Sleep
Habits of Australian Workers *Sleep* **20(10)** 844-849
[20] Gradisar, M., et al. 2007 The Flinders Fatigue Scale: Preliminary Psychometric Properties and Clinical Sensitivity of a New Scale for Measuring Daytime Fatigue associated with Insomnia *Journal of clinical sleep medicine: official publication of the American Academy of Sleep Medicine* **3(7)** 722-728
[21] Austin, A. and P.D. Drummond 1986 Work problems associated with suburban train driving *Applied Ergonomics* **17(2)** 111-116
[22] Dawal, S.Z., Z. Taha, and Z. Ismail 2009 Effect of job organization on job satisfaction among shop floor employees in automotive industries in Malaysia *International Journal of Industrial Ergonomics* **39(1)** 1-6

IOP Conf. Series: Journal of Physics: Conf. Series **1049** (2018) 012106 doi:10.1088/1742-6596/1049/1/012106

[23] Annett, J. 2002 Subjective rating scales: science or art? *Ergonomics* **45** 966-987
[24] Wilson, J.R. and C.E. Nigel, eds. 1995 Evaluation of Human Work. 2nd ed. Taylor & Francis
[25] Wilson, J.R. 1995 Chapter 1 : A Framework and a Context for Ergonomics Methodology, *in Evaluation of Human Work*, W. John R and E.N. Corlett, Editors (Taylor and Francis)
[26] Anderson, J.C. and D.W. Gerbing 1991 Predicting the Performance of Measures in a Confirmatory Factor Analysis With a Pretest Assessment of Their Substantive Validities *Journal of Applied Psychology* **Vol. 76(No. 5)** 732 - 740
[27] Chin, W.W. 2010 How to Write Up and Report PLS Analyses: Handbook of Partial Least Squares, V. Esposito Vinzi, et al., Editors. (Springer Berlin Heidelberg) 655-690.
[28] Hair, J.F., et al. 2010 Multivariate Data Analysis. 7th ed. (Prentice Hall)
[29] Hair, J.F., et al. 2014 A Primer on Partial Least Squares Structural Equation Modeling (PLS-SEM) (SAGE Publications, Inc.) 328.